PROPERTIES OF ORDINARY
WATER-SUBSTANCE

In all its Phases:
Water-vapor, Water, and all the Ices

Compiled by
N. ERNEST DORSEY, PHYSICIST
National Bureau of Standards
Washington, D. C.

American Chemical Society
Monograph Series

BOOK DEPARTMENT
REINHOLD PUBLISHING CORPORATION
430 PARK AVENUE   NEW YORK 22, N.Y.
1940

COPYRIGHT, 1940, BY
REINHOLD PUBLISHING CORPORATION

All rights reserved

Second Printing, November 1953

Third Printing, February 1957

Printed in the United States of America by
INTERNATIONAL TEXTBOOK PRESS, SCRANTON, PA.

GENERAL INTRODUCTION

American Chemical Society Series of
Scientific and Technologic Monographs

By arrangement with the Interallied Conference of Pure and Applied Chemistry, which met in London and Brussels in July, 1919, the American Chemical Society was to undertake the production and publication of Scientific and Technologic monographs on chemical subjects. At the same time it was agreed that the National Research Council, in coöperation with the American Chemical Society and American Physical Society, should undertake the production and publication of Critical Tables of Chemical and Physical Constants. The American Chemical Society and the National Research Council mutually agreed to care for these two fields of chemical development. The American Chemical Society named as Trustees, to make the necessary arrangements for the publication of the monographs, Charles L. Parsons, secretary of the society, Washington, D. C.; the late John E. Teeple, then treasurer of the society, New York; and Professor Gellert Alleman of Swarthmore College. The Trustees arranged for the publication of the A. C. S. series of (a) Scientific and (b) Technologic Monographs by the Chemical Catalog Company, Inc. (Reinhold Publishing Corporation, successors) of New York.

The Council, acting through the Committee on National Policy of the American Chemical Society, appointed editors (the present list of whom appears at the close of this introduction) to have charge of securing authors, and of considering critically the manuscripts submitted. The editors endeavor to select topics of current interest and authors recognized as authorities in their respective fields.

The development of knowledge in all branches of science, especially in chemistry, has been so rapid during the last fifty years, and the fields covered by this development so varied that it is difficult for any individual to keep in touch with progress in branches of science outside his own specialty. In spite of the facilities for the examination of the literature given by Chemical Abstracts and by such compendia as Beilstein's Handbuch der Organischen Chemie, Richter's Lexikon, Ostwald's Lehrbuch der Allgemeinen Chemie, Abegg's and Gmelin-Kraut's Handbuch der Anorganischen Chemie, Moissan's Traité de Chimie Minérale Générale, Friend's and Mellor's Textbooks of Inorganic Chemistry and Heilbron's Dictionary of Organic Compounds, it often takes a great deal of time to coördinate the knowledge on a given topic. Consequently when men who have spent years in the study of important subjects are willing to coördinate their knowledge and present it in concise, readable form, they perform a service of the highest value. It was with a clear recognition of the usefulness of such work that the American Chemical Society undertook to sponsor the publication of the two series of monographs.

Two distinct purposes are served by these monographs: the first, whose fulfillment probably renders to chemists in general the most important service, is to present the knowledge available upon the chosen topic in a form intelligible to those whose activities may be along a wholly different line. Many chemists fail to realize how closely their investigations may be connected with other work which on the surface appears far afield from their own. These monographs enable such men to form closer contact with work in other lines of research. The second purpose is to promote research in the branch of science covered by the monograph, by furnishing a well-digested survey of the progress already made, and by pointing out directions in which investigation needs to be extended. To facilitate the attainment of this purpose, extended references to the literature enable anyone interested to follow up the subject in more detail. If the literature is so voluminous that a complete bibliography is impracticable, a critical selection is made of those papers which are most important.

AMERICAN CHEMICAL SOCIETY
BOARD OF EDITORS

Scientific Series:—
  WILLIAM A. NOYES, *Editor*,
  S. C. LIND,
  W. MANSFIELD CLARK,
  LINUS C. PAULING,
  L. F. FIESER.

Technologic Series:—
  HARRISON E. HOWE, *Editor*,
  WALTER A. SCHMIDT,
  E. R. WEIDLEIN,
  F. W. WILLARD,
  W. G. WHITMAN,
  C. H. MATHEWSON,
  THOMAS H. CHILTON,
  BRUCE K. BROWN,
  W. T. READ,
  CHARLES ALLEN THOMAS.

Preface

Origin.

This compilation was begun under the auspices of a committee of the National Bureau of Standards, the late E. W. Washburn, Chief Chemist, being chairman. Its purpose is to present either specifically or by reference all the material likely to be of interest to anyone studying the properties of ordinary water-substance, *i.e.*, that of the usual isotopic composition.*

After the general plan had been decided upon, the compiler was left entirely free to determine the details. He is greatly indebted to many for advice; but he alone is responsible for all errors of judgment in the selection of information to be included, in the form of presentation,† in the explanations and discussions, and in all the other details involved in the making of such a compilation.

Plan.

The plans of the committee called for (a) assembling from the *International Critical Tables* all data pertaining to the properties of the ordinary water-substance in all its phases, (b) revision and extension of those data in the light of more recent work, (c) inclusion of types of data that had been omitted from the *Critical Tables,* either through oversight or because of the nature of the plan adopted for those Tables, and (d) the arrangement of the whole so as to facilitate its use. The committee desired that the data be grouped in accordance with the several phases of the substance, and their combinations.

This general plan has been adhered to. But the term "data" has been interpreted broadly, for there is much non-numerical information that should be available to one studying the water-substance.

The units in which the numerical data are expressed are always plainly indicated, and the significance of the data is explicitly stated wherever there seems to be any danger of their being misunderstood by one not well informed in the field concerned. In some cases pertinent formulas are given or derived, and a computed quantity (*e.g.,* disposable energy of formation) is accompanied by the basic data and the formula employed in deriving it. Most of this will seem to the expert to be very elementary and needless, but having more suitable sources of his own, he will seldom seriously consult this one for information in his own field. It is not the expert, but he who is not especially well acquainted with the field concerned, who must be considered.

---

* A review of our knowledge of the properties of the isotope deuterium oxide (D_2O) has been published by H. C. Urey and G. K. Teal.[1]

† Certain changes in the tabular presentations have been made by the Publisher in accordance with the style adopted for the A.C.S. monographs.

[1] Urey, H. C., and Teal, G. K., *Rev. Mod. Phys.*, **7**, 34-94 (1935).

Groups of interpolation formulas (*e.g.*, for the thermal expansion of water, Table 100) have been compared by means of skeleton tables. This has revealed some persisting errors in recognized compilations, and some oft-quoted formulas that are totally worthless. Such formulas should not appear in future compilations.

In some cases (*e.g.*, the Verdet constant for water) an arbitrary expression suitable for computation has been set up as a norm against which to compare several independent sets of data, and by which interpolation can be readily performed. In no case is it claimed that such a norm represents the data. It is merely an expression that can be easily evaluated and that runs along near the data, so that deviations from it can be readily compared and studied. It is especially valuable when a wide range of the independent variable is covered by several discrete and not satisfactorily overlapping groups of observations. This procedure also has revealed some persisting errors.

Descriptive information is often given in the form of direct quotation.

Scope.

Information is given regarding the properties of pure, ordinary water-substance in all its phases—water-vapor, water, and the several ices—and regarding the phenomena and data pertaining to its synthesis and dissociation and to its transition from phase to phase; but, except as presently noted, no information is given regarding its behavior in the presence of another substance. Similar information about the water-substance as it occurs in nature has been given when readily available.

The effect of the presence of air is considered, as are also the solubilities and diffusivities in water of the atmospheric and noble gases, of hydrogen, ozone, carbon monoxide, and ammonia, and the diffusion of water-vapor in air, hydrogen, and carbon dioxide, and through a few solids. All other information given concerning water and another substance is merely incidental to an understanding of the behavior of the water-substance itself.

Some types of information that might not be expected in such a compilation as this have been included. For example: The preparation of dust-free water and of monocrystals of ice, the color of water and of the sea, and the volumes of water menisci.

Period Covered.

It is hoped that no important article bearing upon the subject and appearing before 1938 has been overlooked; but only the most important of those appearing after June 30, 1937, and coming to the compiler's attention, have been considered. In accordance with the initial plan, the *International Critical Tables* has as far as possible been depended upon for information antedating January 1, 1923, and the compiler himself has searched the journals from 1922 to 1938. In many cases, data from the *International Critical Tables* have been supplemented by other early data; and in fields not covered by them, the compiler has tried to consult all the significant reports, early as well as recent. He fully realizes that he has only partially succeeded in this attempt, and will be glad to have important omissions brought to his attention.

Acknowledgments.

It is a pleasure to the compiler to acknowledge his great indebtedness to the many who have assisted him in divers ways: to the *International Critical Tables* and its several experts, and to the National Academy of Sciences for their gracious permission to use the data and other information published in those tables; to Messrs. Friedr. Vieweg & Sohn, Braunschwieg, Germany, for their kind permission to use Tables 41 and 42 of the Wärmetabellen prepared by Holborn, Scheel, and Henning, and published in 1919; to the various investigators who have assisted him by correspondence regarding their own work; to his fellow associates, members of every interested division of the National Bureau of Standards, for information, advice, and criticism; and especially to Dr. Lyman J. Briggs, Director of that Bureau, for his unfailing patience and encouragement, without which the work could not have been done.

N. ERNEST DORSEY.

National Bureau of Standards,
Washington, D. C.
September, 1938.

Contents

	PAGE
GENERAL INTRODUCTION	iii
PREFACE	v

Introduction

1. ARRANGEMENT AND DOCUMENTATION. 1
   Arrangement 1
   Documentation 1

2. SYMBOLS, UNITS, AND EQUIVALENTS 3
   Italics 3
   Symbols of Units 3
   Prefixes 4
   Systems of Units 4
   Conversion Factors 6
   Symbols, Units, and Equivalents . . . . (Table 1) 6

I. Synthesis and Dissociation

IA. Synthesis

3. UNION OF HYDROGEN AND OXYGEN (BIBLIOGRAPHY) 11

4. COMPOSITION OF WATER 12

5. HEAT OF FORMATION OF H_2O 12
   Water-vapor: Heat of Formation 14
   Heat of Formation of Water-vapor. . . . . (Table 2) 16
   Water: Heat of Formation 17
   Heat of Formation of Water. . . . . (Table 3) 18
   Ice: Heat of Formation. 18

6. DISPOSABLE ENERGY OF FORMATION OF H_2O AT CONSTANT PRESSURE 19
   Water-vapor: Disposable Energy of Formation 22
   Disposable Energy in the Formation of Water-vapor . (Table 4) 22
   Water: Disposable Energy of Formation 24
   Disposable Energy in the Formation of Water. . . . (Table 5) 24
   Ice: Disposable Energy of Formation 25

IB. Dissociation

7. DISSOCIATION OF WATER-VAPOR 25
   Thermal Dissociation of Water-vapor. 25
   Thermal Dissociation of Water-vapor. . . . . (Table 6) 28
   Variation of α with KRT/P. . . . . (Table 7) 30
   Thermal Dissociation of Water-vapor: Effect of Pressure and Temperature . . . . (Table 8) 32
   Photochemical Dissociation of Water-vapor 33
   Ionic Dissociation of Water-vapor 33
   Dissociation of Water-vapor in the Glow Discharge . (Table 9) 34

x PROPERTIES OF ORDINARY WATER-SUBSTANCE

	PAGE
8. DISSOCIATION OF WATER	35
Photochemical Dissociation of Water	35
Ionic Dissociation of Water	35
Heat of Ionization of Water .... (Table 10)	36
Disposable Energy in the Ionization of Water .... (Table 11)	36
Alpha-ray Dissociation of Water	36

II. Single-phase Systems

IIA. Water-vapor

9. MOLECULAR DATA FOR WATER-VAPOR	37
Mean Free Path of Molecules of Water-vapor	37
Miscellaneous Molecular Data for Water-vapor ..... (Table 12)	38
Kinetic Data for Molecules of Water-Vapor .... (Table 13)	40
Molecular size	41
Various Values Assumed for $K \equiv \pi d^2 n L$ .... (Table 14)	42
Estimates of the Effective Size of the Molecule of Water-vapor (Table 15)	43
Moments of Inertia of the Molecule of Water-vapor	44
Dipole Moment of the Molecule of Water-vapor	46
Polarizability of the Molecule of Water-vapor	49
Anisotropy of the Molecule of Water-vapor	49
Models of the Molecule of Water-vapor	51
Constants of the Triangular Model of H_2O ..... (Table 16)	52
Association of the Molecules of Water-vapor	54
Estimates of the Extent of the Molecular Association of Saturated Water-vapor .... (Table 17)	56

10. INTERACTION OF WATER-VAPOR AND CORPUSCULAR RADIATION	56
Alpha Particles	56
Electrons	57
Ionization Potential and Energy	57
Ionization Potential and Energy: Water-vapor .... (Table 18)	58
Ionization by Accelerated Electrons	59
Ionization of Water-vapor by Accelerated Electrons: Strong Fields .... (Table 19)	59
Miscellaneous Data for the Interaction of Accelerated Electrons with Water-vapor: Weak Fields .... (Table 20)	59
Excited Atoms and Molecules	60
Mobility of Ions in Water-vapor	60

11. VISCOSITY OF WATER-VAPOR	61
Formulas (a) Pressure=1 kg*/cm²	62
(b) Any Pressure	64
(c) Saturation Pressure	66
Sutherland Constant for Water-vapor .... (Table 21)	63
Viscosity of Water-vapor .... (Table 22)	64
Viscosity at the Critical Point	66
Viscosity of Fog	66
Viscosity of Moist Air	67
Comparison of Various Values for the Viscosity of Water-vapor at or below 1 kg*/cm² .... (Table 23)	67

12. ACOUSTIC DATA FOR WATER-VAPOR	68
Velocity of Sound in Water-vapor	68
Velocity of Sound in Water-vapor .... (Table 24)	69
Absorption of Sound by Water-vapor	70
Moist Air	70
Velocity of Sound in Moist Air .... (Table 25)	71

CONTENTS xi

PAGE

13. DIFFUSION OF WATER-VAPOR INTO GASES AND THROUGH SOLIDS 72
    Diffusion into Gases 72
        Diffusion of Water-vapor into Gases . . . . (Table 26) 73
    Diffusion of Water-vapor through Solids 73
        Copper 74
        Rubber. 76
            Permeability of Rubber to Water-vapor . . . . (Table 27) 75
        Miscellaneous Materials 77
        Diffusion of Water-vapor through Various Solids . (Table 28) 76

14. PRESSURE-VOLUME-TEMPERATURE ASSOCIATIONS FOR DILATED
WATER-VAPOR 78
    Equations of State. 78
    Isopiestics (kg*/cm²) of the Specific Volume of Dilated Water-vapor
        (Table 29) 79
    Isopiestics (lb.*/in.²) of the Specific Volume of Dilated Water-vapor
        (Table 30) 81
    Specific Volume of Dilated Water-vapor, and its Defect (Table 31) 81
    Defect of Specific Volume of Dilated Water-vapor: Isopiestics
        (kg*/cm²) . . . . (Table 32) 82
    Defect of Specific Volume of Dilated Water-vapor: Isopiestics
        (atm.) . . . . (Table 33) 84
    Defect of Specific Volume of Dilated Water-vapor: Isopiestics
        (lb.*/in.²) . . . . (Table 34) 85
    Isometrics of the Pressure (atm.) of Dilated Water-vapor (Table 35) 85
    Isometrics of the Pressure (kg*/cm²) of Dilated Water-vapor
        (Table 36) 87
    Defect of Pressure (atm.) of Dilated Water-vapor: Isometrics
        (Table 37) 88
    Defect of Pressure (kg.*/cm.²) of Dilated Water-vapor: Isometrics
        (Table 38) 91

15. THERMAL ENERGY OF DILATED WATER-VAPOR 91
    Formulas and their Coefficients: Specific Heat, Enthalpy, and Entropy
        of Dilated Water-vapor. . . . . (Table 39) 92
    Specific Heat of Dilated Water-vapor. 91
        Values of c, $p \to 0$, and its Integrals. — Dilated Water-vapor
        (Table 40) 95
    Specific Heat of Dilated Water-vapor at Constant Pressure: Pre-
        ferred Values . . . . (Table 41) 96
    Specific Heat of Dilated Water-vapor at Constant Pressure: Vari-
        ous Sets of Values . . . . (Table 42) 96
    Specific Heats of Water-vapor at 1 atm: Comparison of Interpola-
        tion Formulas . . . . (Table 43) 100
    Mean Internal Specific Heat of Dilated Water-vapor at Constant
        Pressure . . . . (Table 44) 102
    Specific Heat of Dilated Water-vapor at Constant Volume
        (Table 45) 103
    Mean Specific Heat of Dilated Water-vapor at Constant Volume
        (Table 46) 105
    Molecular Specific heat of Dilated Water-vapor at Constant
        Volume. (Table 47) 105
    Mean Molecular Specific Heat of Dilated Water-vapor at Constant
        Volume. (Table 48) 107
    Ratio of the Principal Specific Heats of Dilated Water-vapor
        (Table 49) 110
    Ratio of the Principal Specific Heats of Air Saturated with Water-
        vapor. . . . . (Table 50) 111
    Enthalpy of Dilated Water-vapor . . . . (Table 51) 111
        Enthalpy of Dilated Water-vapor: Comparison of Several Sets of
        Values . . . . (Table 52) 114

PROPERTIES OF ORDINARY WATER-SUBSTANCE

PAGE

    Isopiestic Variation in the Enthalpy of the Water Substance through the Critical Temperature . . . . (Table 53)   115
    Entropy of Dilated Water-vapor. . . . . (Table 54)   116
    Various Thermal Data for Dilated Water-vapor: Computed from Spectroscopic Data. . . . . (Table 55)   117
    Joule-Thomson Coefficient for Dilated Water-vapor . . . (Table 56)   119

16. THERMAL CONDUCTIVITY OF DILATED WATER-VAPOR   121
    Thermal Conductivity of Dilated Water-vapor. . . . . . (Table 57)   122

17. REFRACTIVITY OF DILATED WATER-VAPOR.   123
    Refractivity of Dilated Water-vapor . . . . (Table 58)   124

18. ABSORPTION OF RADIATION BY WATER-VAPOR   125
    Absorption of Infrared Radiation by Water-vapor . . . . . (Fig. 1)   127
    Absorption of Radiation by Water-vapor . . . . (Table 59)   128

19. EMISSIVITY OF WATER-VAPOR   131
    Emissivity of Water-vapor . . . . (Table 60)   132

20. LUMINESCENCE OF WATER-VAPOR   133
    Fluorescence   134
    Rayleigh Scattering   134
    Raman Scattering   135
        Raman Spectrum of Water-vapor . . . . (Table 61)   135
    Scattering of X-rays   136

21. SPECTRA OF WATER-VAPOR   136
    Absorption Spectrum.   136
        Lines and Bands in the Absorption Spectrum of Water-vapor (Table 62)   137
        Some Regions of Notable Transparency in the Absorption Spectrum of Water-vapor . . . . (Table 63)   141
        Analyses of the Absorption Vibration Spectrum of Water-vapor (Table 64)   142
        Molecular Constants Involved in the Vibration Spectrum of Water-vapor. (Table 65)   144
        Fine-structure of Absorption Bands of Water-vapor: Sources of Data . . . . (Table 66)   145
        Some Rotation Terms in the Infrared Spectrum of Water-vapor (Table 67)   146
    Emission Spectrum.   149
        Bands in the Emission Spectrum of Water-vapor. . . (Table 68)   150

22. DIELECTRIC CONSTANT OF WATER-VAPOR   151
    Dielectric Constant of Water-vapor . . . . (Table 69)   152
    Variation of the Dielectric Constant of Water-vapor with the Temperature and the Density . . . . (Fig. 2)   153

23. DIELECTRIC STRENGTH OF WATER-VAPOR.   154
    Voltage Difference at which Current Begins to flow through Water-vapor. . . . . (Table 70)   155

24. GLOW DISCHARGE IN WATER-VAPOR   156
    Terms and Symbols   156
    Qualitative Relations.   158
    Numerical Data   161
        Vanishing of the Positive Column in Water-vapor . . (Table 71)   157
        Effect of a Magnetic Field upon the Crookes Dark Space in Water-vapor. . . . . (Table 72)   158

CONTENTS xiii

PAGE

Rate of Dissociation of Water-vapor in the Glow Discharge
(Table 73) 158
Distribution of Electrons in the Crookes Dark Space in Water-vapor
(Table 74) 159
Dissociation of Water-vapor in the Glow Discharge . . (Table 75) 160
Current-voltage Relation for Water-vapor between a Point and a
Plane. . . . . (Table 76) 160

IIB. Water

25. MOLECULAR DATA FOR WATER 161

Structure of Water 161
Linkage of Molecules 161
Evidence for Association . . . . (Table 77) 166
Properties of Dihydrol and of Trihydrol . . . . (Table 78) 168
Estimates of the Ice-content of Water . . . . . (Table 79) 169
Establishment of Equilibrium 170
Architecture of Water   172
Miscellanea 175
Dipole Moment of the Molecule of Water. 175
Apparent Dipole Moment of Water . . . . (Table 80) 176
Polarizability of the Molecule of Water. 177
Anisotropy of the Water Molecule 177

26. INTERACTION OF WATER AND CORPUSCULAR RADIATION . . . 178

Alpha Particles 178
Beta Rays 178
Neutrons 178

27. TENSILE RUPTURE OF WATER 179

28. INTERNAL PRESSURE OF WATER 181

Internal Pressure of Water . . . . (Table 81) 181

29. VISCOSITY OF WATER 182

Viscosity of Water 183
0 °C to 109 °C. (Table 82) 183
0 °C to 100 °C. (Table 83) 184
Comparison of Values . . . . (Table 84) 184
Viscosity of Saturated Water Below 0 °C and above 100 °C . (Table 85) 185
Viscosity of Compressed Water . . . . (Table 86) 186
Viscosity of Sea-water . . . . (Table 87) 188
Effect of Various Factors 189
Relation between Viscosity and Other Properties. 190

30. RIGIDITY OF WATER 190

31. ACOUSTIC DATA FOR WATER 191

Velocity of Sound in Water 191
Velocity of Sound in Water . . . . (Table 88) 192
Velocity of Sound in Natural Waters. . . . . (Table 89) 193
Velocity of Sound in Sea-water . . . . (Table 90) 194
Acoustic Resistivity of Water 195
Absorption of Sound by Water 196
Absorption of Ultrasonic Radiation by Water . . . . (Table 91) 197

xiv PROPERTIES OF ORDINARY WATER-SUBSTANCE

		PAGE
32.	**PRESSURE-VOLUME-TEMPERATURE ASSOCIATIONS FOR WATER.**	198
	Uniformity of Water	198
	Variability in Water	206
	Density of Water	225
	Dissolved Air: Effect on Density	251
	Effect of Dissolved Air on the Density of Water . (Table 92)	198
	Density of Compressed Water at a Pressure of 1 Atmosphere (Table 93)	199
	Specific Volume of Compressed Water at a Pressure of 1 Atmosphere (Table 94)	203
	Specific Volume of Compressed Water: Pressure Exceeding 1 Atmosphere . (Table 95)	207
	Isopiestic Variation of the Specific Volume of Compressed Water with the Temperature (Fig. 3)	230
	Isometric Association of the Pressure and Temperature of Compressed Water (Table 96)	225
	Isopiestic Thermal Expansion of Water (Table 97)	230
	Mean Isopiestic Coefficient of Thermal Expansion of Water (Table 98)	232
	Interpolation Formulas for the Thermal Expansion of Water (Table 99)	234
	Comparison of Interpolation Formulas for the Thermal Expansion of Water (Table 100)	235
	Thermal Slopes of the Isometrics of Water (Table 101)	238
	Mean Isometric Thermal Coefficient of Pressure of Water (Table 102)	238
	Isothermal Compression of Water (Table 103)	239
	Mean Isothermal Compressibility of Dilated Water (Table 104)	241
	Mean Isothermal Compressibility of Water between Pressures 1 and P (Table 105)	242
	Mean Isothermal Compressibility of Water between Pressures P_1 and P_2 (Table 106)	245
	Adiabatic Compressibility of Water (Table 107)	246
	Density of Sea-water: Pressure = 1 Atmosphere (Table 108)	248
	Specific Volume of Sea-water: Pressure Greater than 1 Atmosphere (Table 109)	248
	Isothermal Compressibility of Natural Water	252
	Isothermal Compressibility of Natural Waters. (Table 110)	253
	Adiabatic Compressibility of Natural Waters	254
33.	**MECHANICAL EQUIVALENT OF THE CALORIE**	254
	Mechanical Equivalent of the Calorie (Table 111)	255
34.	**THERMAL ENERGY OF WATER**	256
	Specific Heat of Water	264
	Effect of Dissolved Air on the Specific Heat of Water (Table 112)	257
	Specific Heat of Compressed Water at 1 Atm or at Constant Volume (Table 113)	257
	Mean Specific Heat of Water at 1 Atm (Table 114)	259
	Specific Heat of Compressed Water at Constant Pressure (Table 115)	259
	Specific Heat of Compressed Water: Limit as Temperature Approaches that of Saturation (Table 116)	260
	Specific Heat of Compressed Water at Constant Volume (Isopiestics) (Table 117)	261
	Specific Heat of Compressed Water at Constant Volume (Isometrics) (Table 118)	262
	Ratio and Difference of the Principal Specific Heats of Water: 1 Atm or Saturation (Table 119)	262
	Ratio and Difference of the Principal Specific Heats of Water under High Pressure (Table 120)	264
	Various Isopiestic Thermal Data for Water (Table 121)	264

CONTENTS

		PAGE
Enthalpy of Compressed Water	(Table 122)	265
Entropy of Compressed Water.	(Table 123)	267
Heat of Isothermal Compression of Water.	(Table 124)	268
Decrease in Internal Energy of Water on Isothermal Compression	(Table 125)	269
Joule-Thomson Coefficient for Water		269
Joule-Thomson Coefficient for Water	(Table 126)	270
Heat Liberated by Adiabatic Compression of Water	(Table 127)	271
Isentropic Increase in the Temperature of Water from Saturation to and above the Critical Pressure	(Table 128)	272
Specific Heat of Sea-Water	(Table 129)	272

35. THERMAL CONDUCTIVITY OF WATER 273

Thermal Conductivity of Water	(Table 130)	273
Thermal Conductivity of Compressed Water	(Table 131)	274
Thermal Conductivity of Sea-water	(Table 132)	275

36. TEMPERATURE OF MAXIMUM DENSITY OF WATER 275

| Temperature of Maximum Density of Water. | (Table 133) | 276 |
| Temperature of Maximum Density of Sea-water | (Table 134) | 277 |

37. REFRACTIVITY OF WATER 279

Variation with the Temperature		280
Effect of Electric Field.		283
Other References.		287
Index of Refraction of Water in the Visible Spectrum: Preferred Values	(Table 135)	281
Reduction of the Index of Refraction of Water from Air to Vacuum	(Table 136)	282
Various Values of the Refraction of Water at 20 °C	(Table 137)	283
Refraction of Water at 20 °C: Comparison of Data	(Table 138)	286
Dispersion Formulas for Water	(Table 139)	287
Refraction of Water at Various Temperatures	(Table 140)	288
Mean Temperature Coefficient of Index of Refraction of Water	(Table 141)	290
Variation of Refraction of Water with Temperature: Comparison of Formulas	(Table 142)	291
Temperature Gradient of the Index of Refraction	(Table 143)	293
Variation of the Refraction of Water with the Pressure	(Table 144)	294
Refraction of Natural Waters	(Table 145)	295

38. REFLECTION OF LIGHT BY WATER 296

Reflection of X-rays by Water	(Fig. 4)	297
Reflectivity of Water	(Table 146)	298
Albedo of Water	(Table 147)	299

39. LUMINESCENCE OF WATER 300

Types of Luminescence. Definitions and General Characteristics		300
Fluorescence and Phosphorescence		300
Tyndall Scattering		301
Rayleigh Scattering.		301
Raman Scattering		302
Electron and β-ray Luminescence		304
Mechanical Luminescence		304
Fluorescence of Water		304
Rayleigh Scattering by Water		305
Polarization and Intensity of Light Laterally Scattered by Water	(Table 148)	306
Raman Scattering by Water		307
Polarization and Intensity of the Raman Bands of Water		308

	PAGE
Polarization and Angular Scattering of the Raman Bands of Water (Table 149)	309
Effect of Temperature on the Raman Scattering by Water	309
Shift of Raman Lines of Water with Change in Temperature (Table 150)	310
Effect of Solutes on the Raman Scattering by Water	311
Analysis of the Raman Spectrum of Water (Table 151)	312
Interpretation of the Raman Spectrum of Water	313
Analysis of the Raman Band near $\delta\nu = 3400$ cm^{-1} (Table 152)	313
Raman Spectrum of Water (Table 153)	314
Abridged Raman Spectrum of Water (Table 154)	316
Electron and β-ray Luminescence of Water	317
Mechanical Luminescence of Water	317

40. **Preparation of Dust-free Water** 318

41. **Diffraction of X-rays by Water** 319
    - Periodicities in the Diffraction of X-rays by Water ... (Table 155) 320
    - Diffraction of X-rays by Water: Intensity and Effect of Temperature (Table 156) 321

42. **Absorption and Scattering of X-rays and of γ-rays by Water** 321
    - Absorption of X-rays and of γ-rays by Water .... (Table 157) 322
    - Angular Distribution of the Radiation (x and γ) Scattered by Water (Table 158) 324
    - Coefficients of Scattering of X-rays and of γ-rays by Water (Table 159) 324

43. **Absorption and Transmission of Radiation by Water** .. 325
    - Pure Water 326
      - Monochromatic Absorptivity of Water .... (Table 160) 326
      - Monochromatic Absorptivity of Water in the Range $\lambda = 310$ to 800 $m\mu$ .... (Table 161) 330
      - Absorptivity of Water: Effects of Pressure and Temperature (Table 162) 332
      - Total Transmissivity of Water .... (Table 163) 332
      - Penetration of Solar Radiation into Water .... (Table 164) 333
    - Radiation Filters Containing a Layer of Water 334
    - Natural Waters 334
      - Monochromatic Absorptivity of Sea-water .... (Table 165) 335
      - Effective Absorptivity of Some Coastal and Inland Waters (Table 166) 336
      - Penetration of Daylight into the Sea .... (Table 167) 337

44. **Emissivity of Water** 339

45. **Photoelectric Effects for Water** 339
    - Photovoltaic Effect for Water 339
    - Photoelectric Emission by Water 339
      - Photoelectric Emission by Water .... (Table 168) 340

46. **The Spectrum of Water** 341
    - Under-water Sparks 341
    - Absorption Spectrum 341
      - Absorption Spectrum of Water .... (Table 169) 341
      - Analyses of the Absorption Spectrum of Water .. (Table 170) 343
      - Effect of Temperature and Pressure upon the Absorption Spectrum of Water .... (Table 171) 346

CONTENTS

		PAGE
47.	THE COLOR OF WATER AND OF THE SEA	346
	Pure Water	346
	Transmitted Light	346
	Scattered Light	347
	The Sea	348
48.	OPTICAL ROTATORY POWER OF WATER	350
49.	DIELECTRIC PROPERTIES OF WATER	350
	Symbols and Definitions	350
	Types of Dielectrics	351
	Dipole Theory	352
	Free Reorientability	353
	Drude-Debye Relations	355
	Restricted Reorientability	357
	Dielectric Constant of Water	357
	Miscellanea	357
	Saturation	358
	Variation with Frequency	359
	Dielectric Constant of Water at 17 °C: Variation with the Frequency (Fig. 5)	359
	Dielectric Constant of Water at 17 °C (Table 172)	360
	Absorption Bands in the Electrical Spectrum of Water (Table 173)	363
	Variation with Temperature	363
	Variation of the Dielectric Constant of Water with the Temperature (Table 174)	364
	Variation of the Polarizability of Water with the Temperature (Table 175)	366
	Variation with Pressure	367
	Variation of the Dielectric Constant of Water with the Pressure (Table 176)	368
	Dielectric Constant of Sea-water	368
	Dielectric Absorption	368
	Variation of the Dielectric Constant of Water with $2n^2\kappa/\lambda_0$ and with $2n^2\kappa\lambda_0$ (Fig. 6)	370
	The Absorption Index of Water (Table 177)	371
	Dielectric Absorption of Water at 22 °C (Table 178)	372
	Transition Wave-lengths for Water (Table 179)	372
50.	CONDUCTION OF ELECTRICITY BY WATER	373
	Conductivity of water	374
	Electrical Conductivity of Pure Water (Computed) (Table 180)	374
	Equivalent Conductivity	375
	Equivalent Conductivities of the Ions of Water (Table 181)	376
	Electrolytic Ionization of Water	377
	Ionization Exponent and Product for Water (Table 182)	378
	Ionization Exponent for Water: Comparison of Values (Table 183)	379
	Conductivity of Rain-water	380
	Conductivity of Sea-water	380
	Electrical Conductivity of Sea-water (Table 184)	381
51.	KERR ELECTRO-OPTIC EFFECT FOR WATER	381
52.	ELECTRICAL DISCHARGE IN WATER	383
	Arc	383
	Brush	383
	Corona	383
	Spark	384

xviii PROPERTIES OF ORDINARY WATER-SUBSTANCE

	PAGE
53. MAGNETIC SUSCEPTIBILITY OF WATER	384
Specific Susceptibility of Water at 20 °C . . . . (Table 185)	385
Variation of the Specific Susceptibility of Water with the Temperature (Table 186)	386
54. VERDET CONSTANT OF WATER	388
Effect of Temperature	388
Dispersion of the Verdet Constant	389
The Verdet Constant of Water . . . . (Table 187)	390
Deviations of the Observed and Computed Values of the Verdet Constant for Water from Those Defined by Formula (4) . . . (Fig. 7)	393
Deviations of Other Observed and Computed Values of the Verdet Constant of Water from Those Defined by Formula (4) . . . . (Fig. 8)	394
55. MAGNETIC BIREFRINGENCE OF WATER	395

IIC. Ice

56. FOREWORD	395
57. TYPES OF ICE	395
58. APPEARANCE OF ICE–I	398
59. FORMS AND FORMATION OF ICE	398
Crystallographic Structure	398
Structure of Ice in Bulk	401
Internal Melting	404
Flowers of Ice	405
Formation of Frazil or Needle Ice	407
Formation of an Ice Sheet	407
Rate of Thickening of Ice Sheet . . . . (Table 188)	409
Growth and Orientation of Crystals	409
Recrystallization	412
Regelation	412
Purity of Ice	414
Production of Homogeneous Ice	414
Monocrystals of Ice	415
Freezing of Supercooled Water	416
Icicles	417
Hail	418
Snow and Frost	419
Glaciers	420
Sea-ice	423
60. MOLECULAR DATA FOR ICE	424
Association of Molecules in Ice	424
Structure of the Molecule of Ice	425
61. INTERACTION OF ICE AND CORPUSCULAR RADIATION	428
62. ADHESIVENESS OF ICE	428
63. SLIDING FRICTION OF ICE	428
64. DEFORMABILITY OF ICE	429
Descriptive Treatment	431
Linear Compression	431
Extension	432
Flexure	434

CONTENTS

	PAGE
Punching	435
Penetration	436
Flowing	437
Flow of Ice through an Annular Gap . . . . . (Table 189)	442
Recovery from Stress	443
Brittleness	444
Quantitative Treatment	444
Young's Modulus	444
Young's Modulus of Ice . . . . (Table 190)	445
Poisson's Ratio	446
Rigidity	446
Rigidity of Ice . . . . (Table 191)	447
Tensile Strength	447
Strength of Linear Compression	448
Strength of Ice in Linear Compression . . . (Table 192)	449
Shearing Strength	449
Hardness	450
Plasticity and Viscosity	451
Viscosity of Ice . . . . (Table 193)	454
Viscosity of River Ice . . . . (Table 194)	456
Viscosity of Glacier Ice . . . . (Table 195)	456
Sustaining Power of an Ice Sheet	457

65. DEFORMABILITY OF SNOW ... 458

Hardness of Snow: Variation with Depth in Snow-blanket (Table 196)	459
Hardness of Snow: Effect of Tamping . . . . (Table 197)	459
Strength and Hardness of Snow . . . . (Table 198)	460

66. ACOUSTIC AND OTHER VIBRATIONAL DATA FOR ICE ... 460

Velocity of Transmission	460
Velocity of Waves in Ice . . . . (Table 199)	461
Reflectivity	462

67. PRESSURE-VOLUME-TEMPERATURE ASSOCIATIONS FOR ICE ... 462

Density of Snow	462
Density of Ice–I at 1 Atmosphere	462
Density of Ice–I at Atmospheric Pressure . . . . . (Table 200)	466
Densities and Specific Volumes of the Ices at their Melting-points (Table 201)	467
Densities of the Ices not at their Melting-points	468
Thermal Expansion of Ice (Cubical)	468
Isopiestic Coefficient of Cubical Expansion of Ice . (Table 202)	469
Specific Volume of Ice from Dilute Solutions . . . (Table 203)	470
Compressibility of Ice	471
Ice–I	471
Ice–VI	472
Ice–VII	472

68. COEFFICIENT OF LINEAR EXPANSION OF ICE 472

Thermal Coefficient of Linear Expansion of Ice . . . . (Table 204)	473

69. THERMAL ENERGY OF ICE–I ... 474

Specific Heat of Ice	474
Apparent Isopiestic Specific Heat of Ice. . . . . . (Table 205)	477
True Isopiestic Specific Heat of Ice. . . . . (Table 206)	479
Entropy of Ice	480
Various Isopiestic Thermal Data for Ice . . . . . . (Table 207)	480

PROPERTIES OF ORDINARY WATER-SUBSTANCE

		PAGE
70.	THERMAL CONDUCTIVITY OF ICE AND OF SNOW	481
	Single Crystals	481
	Ice in Bulk	482
	Thermal Conductivity and Diffusivity of Ice . . . (Table 208)	482
	Snow	483
	Thermal Conductivity and Diffusivity of Snow . . (Table 209)	483
71.	REFRACTIVITY OF ICE	484
	Indices of Refraction of Ice . . . . (Table 210)	485
72.	REFLECTIVITY OF ICE AND OF SNOW	485
	Ice	485
	Snow	486
73.	LUMINESCENCE OF ICE	487
	Fluorescence of Ice	487
	Rayleigh Scattering by Ice	487
	Raman Scattering by Ice	487
	Effect of Temperature	487
	The Raman Spectrum of Ice . . . . (Table 211)	488
74.	DIFFRACTION OF X-RAYS BY ICE	489
	Diffraction of X-rays by Ice . . . . (Table 212)	489
75.	ABSORPTION AND TRANSMISSION OF RADIATION BY ICE AND BY SNOW	489
	Ice	491
	Monochromatic Absorptivity of Ice . . . . . . (Table 213)	490
	Transmissivity of Ice for Black-body Radiation . . (Table 214)	491
	Snow	491
	Transmission of Radiation by Snow . . . . . . (Table 215)	492
	Glaciers and *Névés*	493
76.	EMISSIVITY OF ICE AND OF SNOW	493
77.	PHOTOELECTRIC EMISSION BY ICE	494
	Relative Photoelectric Sensitivity of Ice . . . . (Table 216)	494
78.	ABSORPTION SPECTRUM OF ICE	494
79.	OPTICAL ROTATION BY ICE	495
80.	DIELECTRIC PROPERTIES OF ICE	495
	Dielectric Constant of Ice	496
	Drude - Debye Constants for the Dielectric Constant of Ice (Table 217)	497
	Dielectric Constant of Ice: Observed and Computed (Table 218)	498
	Dielectric Constant of Ice . . . . (Table 219)	499
	Dielectric Constant of Ice: Variation of ε' with $(1+a^2\nu^2)^{-1}$ (Fig. 9)	501
	Isothermal Variation of the Dielectric Constant of Ice with the Frequency of the Field . . . . (Fig. 10)	502
	Thermal Variation of the Dielectric Constant of Ice for 120 Charges and Discharges per Second . . . . (Fig. 11)	503
	Dielectric Absorption of Ice	504
	Phase Defect for Ice: Observed and Computed . . (Table 220)	504
	Phase Defect for Ice . . . . (Table 221)	505
	Dielectric Strength of Ice	505

CONTENTS xxi

PAGE
81. ELECTRICAL CONDUCTIVITY OF ICE 505
    Apparent Electric Conductivity of Ice: Observed and Computed
        (Table 222) 506
    Electrical Conductivity of Ice . . . . (Table 223) 508
    Thermal Variation of the Electrical Resistance of Ice . . (Table 224) 509

82. MISCELLANEOUS ELECTRICAL DATA FOR ICE 510
    Pyroelectric Effect 510

83. MAGNETIC SUSCEPTIBILITY OF ICE 510

III. Multiple-phase Systems

84. SURFACE-TENSION OF WATER 511
    Factors Possibly Affecting the Surface-tension 512
    Surface-tension of Water . . . . (Table 225) 514
        Thermal Variation of the Surface-tension of Water . (Table 226) 516
    Molecular Surface Energy 519
        Molecular Surface Energy of Water . . . . (Table 227) 521
    Angle of Contact. 522
        Contact Angle between Air-water Surfaces and Dry Solids
        (Table 228) 523
    Effect of Overlying Gas upon the Surface-tension 524
        Effect of Overlying Gas upon the Surface-Tension of Water
        (Table 229) 525
    Miscellanea 526
        Floating Bubbles and Drops. 526
        Depression under Reduced Pressure 526
        Transition Layers 527
        Surface Films 527
        Relations between the Surface-tension and Other Properties . . 527
        Movements of Bubbles 528
        Voltaic Effects 528
        Stability of Doubly Gas-faced Liquid Films 528

85. SOLUBILITY OF SELECTED GASES IN WATER 528
    Definitions and Symbols 528
        Mean Coefficient of Absorption (0 to P) of Selected Gases in
        Water . . . . (Table 230) 529
        Solubility of Air in Water . . . . (Table 231) 534
        Mean Molecular Coefficient of Absorption (0 to p) of Selected
        Gases by Water . . . . (Table 232) 536
        Effect of Pressure on the Solubility of Gases in Water . (Table 233) 540
    Miscellanea 545
        Effect of Pressure on the Solubility of A, H_2, He, and N_2 in Water
        (Table 234) 546
        Solubility of Atmospheric Gases in Sea-water . . . (Table 235) 548
    References 551

86. RATE OF SOLUTION OF GASES IN WATER 552
    Aeration of Quiescent Water . . . . (Table 236) 553
    Entrance Coefficient of Gases into Water . . . . . . (Table 237) 553
    Exit Coefficient of Gases from Water . . . . (Table 238) 554
    Absorption of Oxygen by a Thin Film of Water . . . (Table 239) 554

87. DIFFUSION OF GASES IN WATER 555
    Diffusivities of Selected Gases in Water . . . . (Table 240) 556

xxii   PROPERTIES OF ORDINARY WATER-SUBSTANCE

PAGE

88. PRESSURE-VOLUME-TEMPERATURE ASSOCIATIONS FOR SATURATED
    WATER AND STEAM 556
    Critical Data 557
        Critical Constants of Water . . . . (Table 241)   558
    Saturated Vapor 559
        Vapor-pressure—Definition 559
            Past History, Effect of 560
            Catalysts, Effect of 560
            Gas, Effect of 560
            Curvature of Surface, Effect of 568
            Tension, Effect of 570
            Solute, Effect of 574
            Adsorbed Water, Effect of 574
            Formulas 574
        Vapor-pressure of Water (atm): -5 to $+374$ °C . . (Table 242) 560
        Vapor-pressure of Water (mm-Hg): 0 to -16 °C . . (Table 243) 562
        Vapor-pressure of Water (millibars): 0 to -50 °C . (Table 244) 563
        Vapor-pressure of Water (mm-Hg): 0 to 102 °C . . (Table 245)   563
        Comparison of Smoothed Values for the Vapor-pressure of Water
            (mm-Hg, atm) . . . . (Table 246)   566
        Thermal Rate of Variation in the Vapor-pressure of Water (atm)
                                                          (Table 247) 568
        Comparison of Thermal Rates of Variation in the Vapor-pressure of
            Water (mm-Hg) . . . . (Table 248)   570
        Temperature of Saturated Water-vapor: 0.0075 to 225 kg*/cm²
                                                          (Table 249) 570
        Density and Specific Volume of Saturated Water-vapor 575
        Specific Volume of Saturated Water-vapor . . . . . (Table 250) 575
        Density of Saturated Water-vapor . . . . (Table 251) 577
        Density of Water-vapor Saturated in the Presence of a Foreign
            Inert Gas . . . . (Table 252) 577
    Saturated Liquid 579
        Boiling Point 579
        Boiling Points of Water . . . . (Table 253) 580
        Comparison of Values for the Boiling Points of Water
                                                          (Table 254) 580
            Effect of a Solute 582
        Density and Specific Volume of Saturated Water 583
        Specific Volume of Saturated Water . . . . (Table 255) 583

89. THERMAL ENERGIES OF SATURATED WATER AND SATURATED
    STEAM 584
        Enthalpy of Saturated Water and of Saturated Steam . (Table 256) 585
        Specific Heat of Saturated Water-vapor . . . . . . (Table 257) 586
        Entropy of Saturated Water and of Saturated Steam . . (Table 258) 586

90. STEAM-TABLES AND DIAGRAMS 588
        A List of Some Recent Steam-tables and Diagrams and of Reports of
            Extended Work Pertaining Thereto . . . . (Table 259) 589
        International Skeleton Steam-Tables, 1934 . . . . . (Table 260) 591

91. FUGACITY AND ACTIVITY OF WATER 594
        A Comparison of Some Bases for an Estimate of the Fugacity of Satu-
            rated Water . . . . (Table 261) 596
        Activity and Fugacity of Water at 50 °C . . . . . . (Table 262) 597

92. PRESSURE-TEMPERATURE ASSOCIATIONS FOR EQUILIBRIUM
    BETWEEN ICE AND ANOTHER PHASE 597
        Triple Points 597
            Triple Points of the Water Substance . . . . . . (Table 263) 598

CONTENTS

	PAGE
Ice and Water-vapor	598
Vapor-pressure of Ice–I	598
Vapor-pressure of Ice–I . . . . (Table 264)	600
Vapor-pressure of Ice–I, Deviations of Observations from the Values Defined by Formula (1) . . . . . . (Fig. 12)	599
Density and Specific Volume of Vapor Saturated with Respect to Ice	601
Ideal Specific Volume and Density of Vapor Saturated with Respect to Ice . . . . (Table 265)	601
Ice and Water	602
Melting-point of Ice	602
Ice–I: Normal Melting-point and Triple Point	603
Absolute Temperature of the Ice-point . . (Table 266)	602
Melting-points of the Ices under Hydrostatic Pressure (Table 267)	603
Effect of a Solute	604
Melting-points in Aqueous Solutions of Certain Gases (Table 268)	605
Melting-point in Sea-water . . . . (Table 269)	606
Ice and Ice	608
Pressure-temperature Associations for Equilibrium between the Several Pairs of Types of Ice . . . . (Table 270)	607
93. PHASE DIAGRAM FOR WATER AND THE ICES	608
Phase Diagram for Water and the Ices (Bridgman) . . . . (Fig. 13)	608
94. SURFACE CHARGES ON WATER AND ON ICE	609

IV. Phase Transition

95. ENERGY CHANGES ACCOMPANYING PHASE TRANSITION.	612
External Work and Change in Volume during Phase Transition (Table 271)	612
Latent heat of Phase Transition	613
Latent Heat of Vaporization	613
Latent Heat of Sublimation	614
Latent Heat of Fusion	615
Latent Heat of Change in Phase . . . . (Table 272)	616
Latent Heat of Ice to Ice	618
Miscellanea	618
Disposable Energy from Isopiestic Change in Phase	619
Disposable Energy from Isopiestic Change in Phase (Table 273)	619
96. VAPORIZATION AND CONDENSATION	620
Escape and Capture of Molecules	620
General Relations	620
Coefficient of Capture	621
Temperature Adjustment	622
Change in Association	622
Evaporation	622
Superheating	622
Some Factors Affecting Evaporation	623
Curvature of Surface	623
Blanketing Layers and Surface Films	624
Convection	625
Wind	625
Aspect of Surface	625
Electric Charge	625
Cooling by Evaporation	625
Some Data Pertaining to the Evaporation of Water (Table 274)	626

xxiv   PROPERTIES OF ORDINARY WATER-SUBSTANCE

	PAGE
Rate of Evaporation	626
Formulas for the Rate of Evaporation (Table 275)	627
Effective Partial Pressure of Blanketing Vapor . (Table 276)	630
Various Observed Rates of Evaporation of Water (Table 277)	630
Evaporation From Large Outdoor Areas of Water (Table 278)	631
Small Drops	631
Condensation	632
Supersaturation	633
Nuclear condensation	633
Condensation of Water-vapor on Nuclei (Table 279)	634
State of Water-vapor at the Cloud-limit (Table 280)	635
Condensation on Snow in the Open (Table 281)	637
Condensation on Metals	637

97. FREEZING AND MELTING 637
   Ice Needles 637
   Supercooling of Water 638
   Superheating of Ice 643
   Rate of Freezing and of Melting 644
      Velocity of Crystallization of Supercooled Water . . (Table 282) 645
      Rate of Melting: Effect of Tension 646
   Crystalloluminescence 646

98. TRANSITION OF ICE TO ICE 647

99. MISCELLANEOUS CHANGES ACCOMPANYING PHASE TRANSITION 648
   Change in Refraction with Change in Phase ...... (Table 283) 648
   Change in Absorption Spectrum with Change in Phase . (Table 284) 648
   Raman Spectra 649
      Change in Raman Spectrum with Change in Phase . (Table 285) 650
   Magnetic Susceptibility 650

V. Miscellanea

100. MISCELLANEOUS PHENOMENA AND DATA 651
   Penetration of Solids by Water............... 651
   Thermal Anomalies of Water 651
   Impact of Solids upon Water 651
   Volume of the Water Meniscus: Special .... (Table 286) 651
   Volume of the Water Meniscus: General .... (Table 287) 653
   Radiation from an Ideal (Black-body) Radiator .... (Table 288) 653
   Vision under Water 654
   Sea-water 654
      Composition of the Salt of Sea-water .... (Table 289) 655
   Surprises 655
   Interpolation 656

INDEX 657

Introduction

General information that will facilitate the use of this compilation is given in this Introduction. It is classified under two heads: (1) arrangement and documentation, and (2) symbols, units, and equivalents. Information regarding the origin, plan, and scope of this volume, and the period covered by it, will be found in the Preface.

1. ARRANGEMENT AND DOCUMENTATION

Arrangement.

As may be seen in the table of contents, information regarding the water-substance has been assembled in five broad groups: (I) Synthesis and dissociation; (II) Single-phase systems, subdivided into (IIa) Water-vapor, (IIb) Water, (IIc) Ice; (III) Multiple-phase systems; (IV) Phase transition; and (V) Miscellanea. Each of these is subdivided into smaller units, devoted to a closely related group of data or phenomena.

As far as possible, the subgroups are arranged according to the nature of the phenomena involved, and in the following order: atomic or molecular, mechanical, acoustic, thermal, optical, electrical and magnetic.

Everything that either involves the presence of a second phase (*e.g.*, surface tension) or requires the presence of a second phase in order to insure the existence of the assumed condition (*e.g.*, pressure of saturated vapor) has been placed in Group III (Multiple-phase systems), excepting such (*e.g.*, latent heat) as have to do with the act of transition from one phase to another. The last form the contents of Group IV (Phase transition). In Group V (Miscellanea) have been placed certain odd bits of information that do not fit in well elsewhere, and a brief note on interpolation.

Documentation.

All data and descriptive information, and most of the formulas and statements regarding theories, are accompanied by references to the sources from which they have been obtained or on which they rest. In many cases references to secondary sources that have come to the compiler's attention and seem to be of importance or interest to one seriously studying the subject are given also, although not otherwise used in the compilation.

With the exception of references taken from the *International Critical Tables,* and a few others plainly indicated as secondary, each book and article cited has been examined by the compiler with reference to the information accredited to it; and so far as it was practical, the limiting pages have been determined for the references from the *Critical Tables* also; but of the papers covered by those references, only those pertaining to fields that had to be reworked have been studied by the compiler.

If the results of an investigation have been published more than once, whether in whole or in part, in abstract, as advance notice, or as a correction, a reference to each such publication that has come to the compiler's attention is given, except that no reference is given to the *Bulletin of the American Physical Society,* the abstracts in that being contained in the *Physical Review* also, to which reference is made.

The relations between the several articles as regards extent and content, but not time of publication, are indicated by means of the following symbols used for separating the references, each referring primarily to the two between which it appears: = means that the articles differ in no essential feature other, perhaps, than language; — means that one is a slightly revised copy of the other, the changes may or may not be of importance; → and ← mean that the one at the head of the arrow is a shorter report (abstract, review, etc.) than that at the other end. · The shorter is always at the head of the arrow, whatever its date.

In all cases, the information to which such a string of references relates has been derived from the first of the string, unless the contrary is explicitly stated.

Occasionally it has seemed desirable to refer to a particular page or illustration, or to give the characteristic designation of the article as a separate unit in some recognized series, or to indicate whether the article is an abstract (A), a letter to the editor (L), or a review (R). In such cases, this parenthetical information is enclosed in parentheses and placed after the final page number of the article, the letters A, L, and R being used as here indicated.

The year given in a reference is generally that given on the title page of the volume containing the paper referred to. If the title page indicates that the volume covers portions of two or more years, then the last of those years is the one generally given. Some such rigid rule is necessary if simplicity is to be secured, for the covers of some reprints have a date that is incompatible with the dates on the title page of the volume containing the article in question. For example, the title page of volume 70 of the *Proceedings of the American Academy of Arts and Sciences* states that it covers from May 1934 to May 1935, and it carries 1936 as the date of publication. To the nine papers that it contains are attached the following dates:

Paper	Received	Presented	Cover date
1	October 1, 1934	October 10, 1934	March, 1935
2	October 8, 1934	February 13, 1935	March, 1935
3	December 14, 1934	December 12, 1934	April, 1935
4	February 4, 1935	February 13, 1935	May, 1935
5	May 9, 1935	March 13, 1935	August, 1935
6	August 6, 1935	March 13, 1935	December, 1935
7	October 15, 1935	October 9, 1935	December, 1935
8	December 3, 1935	December 11, 1935	February, 1936
9	December 12, 1935	December 11, 1935	February, 1936

A reference to any paper in that volume will in the present compilation be dated 1935.

2. Symbols, Units, and Equivalents

To facilitate the use of this volume, the symbols used in any major section have usually been defined therein, the exceptions being mainly the well-standardized symbols, including those of the units. These and a few others that may be unfamiliar to some have been assembled in Table 1, where certain of them are defined, and some numerical values and conversion factors are given. Of the symbols for units, however, only the simpler are included, from which the more complicated may be constructed in the well-known manner described in a following paragraph. Many of the symbols, other than those for units, are occasionally used with other meanings, but the context, and especially the definitions in the accompanying text, will enable the user to determine the proper interpretation in each case.

Italics.

In accordance with the established custom of the English language, English letters used out of character, that is, in senses different from those ascribed to them in usual English script, are printed in distinctive type, in italics if the context is roman, and conversely. Those used in character may also be printed in distinctive type for very special purposes. Exceptions occur, which are usually in the direction that leads to conformity with other members of the group of related symbols.

Of the abbreviating symbols used in this compilation, true abbreviations and the symbols for the units of measure, for the chemical elements, and for the names of mathematical functions are printed in roman; most of the others are printed in italics, the letters composing them being obviously used out of character.

Symbols of Units.

Symbols of units are printed in Greek or Roman letters, never in italics, and generally do not end with a period, the principal exceptions to the last being the symbols for the British units (in., ft., lb., etc.), and the period is not always used with them. In writing the synthetic symbol for a derived unit, the symbols of the units that form a product are separated by a period without additional space (g·cm); those that form a ratio or a quotient are either combined as a fraction $\left(\dfrac{\text{cm}}{\text{sec}}\right)$ or separated by the shilling mark (cm/sec) or the symbols in the divisor are written with negative exponents and separated from those of the dividend by a period without additional space (cm·sec^{-1}, g·cm·sec^{-2}). In all such cases any period that might otherwise form part of the symbol of an individual unit is omitted, thus avoiding a duplication of the period in the interior; the final period is unnecessary, as the resulting combination is of a type that is never used except as a symbol for a unit. The periods indicating multiplication are placed above the line.

The elements of the symbol of a unit designated by a compound name in which the elements of the compound are not to be understood as being combined by processes analogous to multiplication and division, are connected by a hyphen (ft-c = foot-candle, mm-Hg = millimeter of mercury).

In general, the same form of symbol (the singular) is used whatever the magnitude of the numeral to which it is attached. When convenient, the symbol is preceded, without spacing, by an integral power of ten (positive or negative), to indicate that the unit used is so related to that designated by the unmodified symbol (2.9986 10^{10}cm/sec = 2.9986 × 10^{10} cm/sec).

Prefixes.

The following metric prefix symbols and prefixes are used, each indicating that the ratio of the unit to that corresponding to the symbol or name to which the prefix is attached is that shown by the corresponding number, *e.g.*, 1 microgram = 0.000 001 g, 1 kilogram = 1000 g.

μ- = micro- = 0.000 001, m- = milli- = 0.001, c- = centi- = 0.01, k- = kilo- = 1000, mega- = 1 000 000. There is no generally accepted single-character symbol for mega. (A capital M is sometimes used, but that may be misread for the commonly used Roman numeral for 1000; the Roman symbol ($\overline{\text{M}}$) for a million seems to be more appropriate.) Each of these symbols has always the significance here given; m- never stands for either micro- or mega-, but solely for milli-. The final vowel of the prefix is dropped from mega- and from micro- when the next letter is a vowel.

These prefixes are combined, one to another, as need be, the one nearer to unity preceding the other (mμ- = millimicro- = 10^{-9}; kmega- = kilomega- = 10^9) never the other way round. Either singly or combined they may be directly attached to any unit of, or based upon, the metric system.

Systems of Units.

The normal system of units based upon the centimeter, the gram (unit of mass), and the second, and commonly designated as the cgs system, is generally used; but the practical absolute electrical units, *i.e.*, ohm = 10^9 cgsm, ampere = 0.1 cgsm, volt = 10^8 cgsm, joule = 10^7 cgsm = 10^7 ergs, watt = 10^7 erg/sec, and their international counterparts are also used; and occasionally the British units (ft., lb., etc.).

The cgs electrostatic system of units is denoted by the symbol cgse; the electromagnetic, by cgsm. The same three symbols may be used to denote the appropriate unit in the corresponding system, whatever the nature of that unit, as cgsm is used in the preceding paragraph.

2. SYMBOLS, UNITS, AND EQUIVALENTS

As the names of the several international electrical units are the same as those of their counterparts in the practical absolute system, it is occasionally necessary to distinguish between the two. In such cases, the qualifier "(Int.)" accompanies the symbol for the unit when that must be interpreted as the international unit. If this qualifier does not appear, the unit is the absolute one, or the number to which the symbol is attached is not known with sufficient accuracy to justify making a distinction between the two. It is only with respect to the joule that it is ever necessary to make the distinction in this compilation.

It must be remembered that the international joule used in experimental work is defined in terms of the international ohm and the international volt, and that the certification of resistances and of standard cells is always in terms of the concrete standards of the certifying laboratory; those standards define the international units for that laboratory. The amount of energy corresponding to the international joule as defined by such laboratory units of resistance and voltage has varied from laboratory to laboratory, and from time to time, the variation steadily decreasing as the concrete standards have been improved in permanence and in reproducibility. But even as late as 1931 it was very significant. At that time the international joule of Great Britain, as defined by the ohm and the volt, was greater than those of Germany, France, and this Bureau, the differences being, respectively, 1.91, 1.98, and 0.62 parts in 10 000 (hundredths of a per cent).[1] Since that comparison, some of the national laboratories have revised the values assigned to their concrete standards; and now (1938) the discrepancies between the values of the international joules of the several countries amount to no more than 3 or 4 parts in 100 000 (3 or 4 thousandths of a per cent). It is confidently expected that with the present arrangements for systematic intercomparisons and the use of the better standards now available, the discrepancies will be kept continuously well below those existing in 1931. At present (1938), the international joule as defined by the standards of the National Bureau of Standards lies between 1.0002 and 1.0003 absolute joules, and is probably nearer the lower value.

From this discussion it is obvious that when one has to do with an accuracy of a hundredth of a per cent, or higher, it is impossible to translate observations expressed in international joules into absolute joules without loss of accuracy, unless one knows what particular international joule was used, and how it is related to those for which comparisons with the absolute joule have been made. This is true irrespective of the accuracy of the absolute measurements.

For this reason the compiler has not attempted to convert reported data from one of these joules to the other, unless the author himself has given the conversion factor that he believed to apply.

[1] Vinal, G. W., *Bur. Stand. J. Res.*, **8**, 729-749 (RP448) (1932).

Conversion Factors.

Such conversion factors as seem appropriate are given immediately above the numerical data in each table, in the following form: Unit of $P = 1$ kg*/cm² $= 0.967841$ atm $= 735.559$ mm-Hg $= 0.980665$ bar. This indicates that a number, say a, standing in the P-column, represents a pressure of a kg*/cm², of $0.967841a$ atm, of $735.559a$ mm-Hg, of $0.980665a$ bars, all of which pressures are equal, one to another.

Table 1.—Symbols, Units, and Equivalents

With a few exceptions, this table contains only the simple units of measure appearing frequently in this compilation, certain well-standardized symbols that are not always defined in the accompanying text, and some symbols and names that may be unfamiliar to the user, and none of the well-known mathematical symbols. Some of these symbols, other than those for the units, are sometimes used in ways not indicated here; those uses are sufficiently explained where they occur. The synthetic symbols for the derived units of measure are formed from the simple ones in the well-known way already described.

In the first column are given alphabetically the abbreviating symbols and the names of those units for which there are no such symbols. In the second are given the names of the units or other quantities corresponding to the symbols, and the definitions and equivalents of the units, or as much of this as seems desirable.

Greek symbols are given at the end of the table.

A	Angstrom, a unit of length; $1A = 10^{-8}$ cm $= 0.1$ mμ.
A	A pressure equal to 1 atm; its numerical value depends upon the unit of pressure.
atm	Normal atmosphere, a unit of pressure; 1 atm $= 1.01325$ megadynes/cm² $= 1.03323$ kg*/cm² $= 1.01325$ bars. Note: In Germany, "at" is frequently used to denote a pressure of 1 kg*/cm² $= 0.96784$ atm.
bar	Bar, a unit of pressure; 1 bar $= 1$ megadyne/cm².
°C	Degree centigrade. The degree interval on the scale of the centigrade thermometer, on which the normal melting point of ice is called 0, and the normal boiling point of water is called 100. Unless something else is clearly indicated, the intervals are to be counted from that zero.
c-	Centi-, a prefix meaning 1/100.
c_p	Specific heat at constant pressure.
c_v	Specific heat at constant volume.
cal	Gram calorie, a unit of heat. Unless another value is specified, it is assumed that 1 cal $= 4.185$ joules.
cal$_{15}$	15°-calorie. Similarly for cal$_{20}$, cal$_m$. See Section 33.

2. SYMBOLS, UNITS, AND EQUIVALENTS

Table 1—*(Continued)*

cgs	A symbol used to designate either the system of normal units based up the centimeter, the gram (unit of mass), and the second, or a unit of that system.
cgse	A symbol for any unit of the cgs electrostatic system, and for the system itself.
cgsm	A symbol analogous to cgse, referring to the cgs electromagnetic units and system.
cm	Centimeter, a unit of length; 1 cm = 0.01 m = 0.032808 ft.
cm-Hg	Centimeter of mercury, a unit of pressure; 1 cm-Hg = 13.3322 kilodynes/cm^2 = 13.3322 millibars. By definition, 1 cm-Hg is the pressure exerted by a vertical column of mercury 1 cm long, at a place where the acceleration of gravity is 980.665 cm/sec^2, and when the density of the mercury is 13.5951 g/cm^3 and either the free surface of the column is flat or a proper correction has been made for the effect of its curvature.
d	Density; seldom used. Also, derivative.
dyne	The cgs unit of force; the force that will give to a mass of one gram an acceleration of 1 cm/sec^2.
E	Internal or intrinsic energy of a substance or system.
e	The number 2.71828..., the base of the natural system of logarithms.
erg	The cgs unit of work and energy; the work done by a constant force of one dyne while the point of application of the force moves one centimeter in the direction in which the force acts; 1 erg = 1 cm·dyne.
°F	Degree Fahrenheit. The degree interval on the scale of the Fahrenheit thermometer, on which the normal melting point of ice is marked 32, and the normal boiling point of water is marked 212. Unless something else is clearly indicated, the intervals are to be counted from the origin defined by these numbers.
ft.	Foot, unit of length; 1 ft. = 30.480 cm.
g	The acceleration of gravity. By international agreement all data involving the value of g are to be reduced to the basis of g = 980.665 cm/sec^2 = 32.1740 ft/sec^2, which is called the "normal" value of g.
g	Gram, a unit of mass; 1 g = 0.0022046 lb (avdp.)
g*	Gram weight, a unit of force; the weight of a mass of 1 g at a place where g = 980.665 cm/sec^2; 1 g* = 980.665 dynes.
gfw	Gram formula weight, a unit of mass; as many grams as there

Table 1—*(Continued)*

<table>
<tr><td></td><td>are units in the formula weight. In any specific case the pertinent formula should be clearly indicated.</td></tr>
<tr><td>gfw-H$_2$O</td><td>Gram formula weight of H$_2$O; 1 gfw-H$_2$O = 18.0154 grams of the water-substance.</td></tr>
<tr><td>g-mole</td><td>Gram mole, a unit of mass; as many grams as there are units in the molecular weight.</td></tr>
<tr><td>H</td><td>Enthalpy; heat content; total heat. $\Delta H = \Delta(E + pv)$, where Δ means "the increase of," p = pressure, v = volume, E = intrinsic energy. In the case of the water-substance, H is commonly used to denote the value of ΔH in going from saturated water (liquid) at 0 °C to the indicated state.</td></tr>
<tr><td>h</td><td>Planck's constant of action; $10^{27} h$ = 6.56 erg·sec.</td></tr>
<tr><td>in.</td><td>Inch, a unit of length; 1 in. = 2.5400 cm.</td></tr>
<tr><td>(Int.)</td><td>International. (Int.) accompanies the symbols for units belonging to the international electrical system.</td></tr>
<tr><td>j</td><td>Joule, a unit of energy; 1 j = 10^7 ergs = 10 megergs.</td></tr>
<tr><td>°K</td><td>Degree Kelvin. The degree interval on that thermodynamic scale of temperature which has 100 degrees between 0 °C and 100 °C. Unless something else is clearly indicated, the intervals are to be counted from the absolute zero, which in this compilation is assumed to be 273.1 °K below 0 °C, unless another value is definitely specified, cf. Table 266.</td></tr>
<tr><td>k</td><td>Boltzmann's constant; molecular gas-constant. $k \equiv R/N = 1.372 \; 10^{-16}$ erg/°K per molecule.</td></tr>
<tr><td>k-</td><td>Kilo-, a prefix meaning 1000.</td></tr>
<tr><td>kc</td><td>Kilocycle = 1000 cycles.</td></tr>
<tr><td>kg</td><td>Kilogram; 1 kg = 1000 g.</td></tr>
<tr><td>kg*</td><td>Kilogram weight; 1 kg* = 1000 g* = 980665 dynes.</td></tr>
<tr><td>kj</td><td>Kilojoule; 1 kj = 1000 j.</td></tr>
<tr><td>km</td><td>Kilometer; 1 km = 1000 m.</td></tr>
<tr><td>l</td><td>Liter, a unit of volume; 1 l = 1000.027 cm^3.</td></tr>
<tr><td>lb</td><td>Pound; a unit of mass; 1 lb = 453.59243 g.</td></tr>
<tr><td>lb*</td><td>Pound weight, a unit of force. The weight of a mass of 1 lb at a place where g = 980.665 cm/sec^2; 1 lb* = 0.44482 megadyne.</td></tr>
<tr><td>(lb* in^2)$_L$</td><td>A pressure of one pound per square inch at a place (London) where the acceleration of gravity is 981.16 cm/sec^2; 1 (lb*/in^2)$_L$ = 1.000505 lb*/in^2 = 68.982 millibars.</td></tr>
<tr><td>log$_e$</td><td>Logarithm to the base e; natural logarithm. Log$_e x$ = 2.302585 log$_{10}x$.</td></tr>
<tr><td>log$_{10}$</td><td>Logarithm to the base 10; common logarithm.</td></tr>
</table>

2. SYMBOLS, UNITS, AND EQUIVALENTS

Table 1—*(Continued)*

M	Molecular weight. For H_2O, $M = 18.0154$.
m	Mass.
m-	Milli-, a prefix meaning 1/1000.
m	Meter, a unit of length; 1 m = 100 cm = 3.2808 ft.
mega-	A prefix meaning 1 000 000.
mg	Milligram; 1 mg = 0.001 g.
mg*	Milligram weight, a unit of force; 1 mg* = 0.001 g*.
micro-	A prefix meaning 1/1 000 000.
micron	A unit of length (see μ).
ml	Milliliter; 1 ml = 0.001 1 = 1.000 027 cm^3.
mm	Millimeter; 1 mm = 0.001 m = 0.1 cm.
mm-Hg	Millimeter of mercury, a unit of pressure (see cm-Hg); 1 mm-Hg = 0.1 cm-Hg = 1.33322 millibars.
ms	Millisecond; 1 ms = 0.001 sec.
mμ-	Millimicro-, a prefix meaning 1/1 000 000 000.
mμ	Millimicron; 1 mμ = 0.001 μ = 10^{-9} m = 10^{-7} cm = 10A.
N	Avogadro's number; the number of molecules per gram-mole. $N = 6.061 \times 10^{23}$.
P, p	Pressure.
p	Poise; see below.
poise	The cgs unit of viscosity. It is the viscosity of a liquid which, when streaming lamellarly, exerts upon one side of an internal layer parallel to the lamellas a drag of 1 dyne/cm^2 in the direction of the velocity v when the value of dv/dx at that side of the layer is 1 cm/sec per cm, dx being an element of the normal drawn outward from that side of the layer; 1 poise = 1 g/cm·sec = 1.0197 mg*·sec/cm^2 = 1 dyne·sec/cm^2.
R	The gas-constant. $R = 8.315$ j/g-mole·°K.
radian	A unit of angle. The angle of which the arc is equal to the radius; 1 radian = 57° 17′ 44.8″.
r.m.s.	Root-mean-square. The square root of the mean of the squares of the individual values of the quantity indicated.
sat	Saturated; at saturation. Used chiefly as a subscript.
steradian	A unit of solid angle. The solid angle subtended at the center by a spherical surface equal in area to the square of the radius of the sphere. The solid angle subtended at its center by a hemisphere is 2π steradians.
T	Temperature, on the thermodynamic scale.
t	Temperature, on the centigrade scale. Occasionally, time.

Table 1—*(Continued)*

ton	A unit of mass; 1 (short) ton = 2000 lb, 1 (long) ton = 2240 lb.
ton*	Ton weight, a unit of force; 1 (short) ton* = 2000 lb*, 1 (long) ton = 2240 lb*.
v	Volume. Velocity.
v^*	Specific volume, the volume of a unit of mass of the substance.
W	Disposable energy. See Section 6.
γ	Ratio of the principal specific heats of a substance; $\gamma = c_p/c_v$.
Δ	A deviation or difference. Δx = an increase in x.
δ	A deviation or difference.
ϵ	Dielectric constant.
η	Viscosity.
κ	Magnetic susceptibility. Electrical conductivity.
Λ	Equivalent conductivity (electrical).
λ	Wave-length.
μ	Magnetic permeability. Joule-Thomson coefficient. Moment of a dipole. Coefficient of absorption.
μ	Micron, a unit of length; 1 $\mu = 10^{-6}$ m = 0.001 mm = 10 000A.
μ-	Micro-, a prefix meaning 1/1 000 000.
μ-Hg	Micron of mercury, a unit of pressure; 1 μ-Hg = 0.001 mm-Hg.
ν	Wave-number; $\nu = 1/\lambda$. Number. Frequency.
π	Pi, the ratio of the circumference of a circle to its diameter; $\pi = 3.14159\ldots$
ρ	Density. Depolarization factor.
σ	Density (seldom used).
τ	Time. Transmissivity.
χ	Specific susceptibility (magnetic).
$\sim$	Cycle.
$\sim$/sec	Cycle per second.

I. Synthesis and Dissociation

IA. SYNTHESIS

3. UNION OF HYDROGEN AND OXYGEN

The chemical reactions that occur in mixtures of hydrogen and oxygen, with or without the admixture of another gas; the way they vary with the temperature, pressure, illumination, and composition of the mixture; the ignition temperatures and explosion limits of such mixtures, and the way these vary with the size, form, and material of the containing vessel; and all the other various phenomena associated with the reactions that occur in such mixtures:—all these lie beyond the scope of this compilation, which in the field of the synthesis of water is limited to the stoichiometric composition of water, the heat of formation, and the maximum work that can be obtained from the reaction when it is carried out at constant pressure.

Those desiring information regarding the chemical reactions themselves, the attendant phenomena, and the way they vary with the conditions are referred to the compilations by W. A. Bone and D. T. A. Townend,[1] A. Skrabal,[2] and C. Winther.[3] Those desiring more recent data, conclusions, and inferences can obtain from the papers listed in the accompanying notes [4-28] a general idea of the present status of the subject, and references

---

[1] Bone, W. A., and Townend, D. T. A., *Int. Crit. Tables*, **2**, 172-195 (1927).
[2] Skrabal, A., *Idem*, **7**, 113-152 (1930).
[3] Winther, C., *Idem*, **7**, 159-173 (1930).
[4] Alyea, H. N., and Haber, F., *Z. physik. Chem.*, (B), **10**, 193-204 (1930).
[5] Andreew, K. K., and Chariton, J. B., *Trans. Faraday Soc.*, **31**, 797-804 (1935).
[6] Bestchastny, A. L., et. al., *Physik. Z. Sowj.*, **5**, 562-579 (1934).
[7] Breton, J., and Laffitte, P., *Compt. rend.*, **202**, 316-318 (1936).
[8] Chapman, D. L., and Reynolds, P. W., *Proc. Roy. Soc. (London)* (A), **156**, 284-306 (1936).
[9] David, W. T., *Phil. Mag.* (7), **20**, 65-68 (1935). *Nature*, **138**, 930 (L) (1936); **139**, 67-68 (L) (1937).
[10] Drop, J., *Rec. Trav. Chim. Pays-Bas*, **54**, = *(4)*, **16**, 671-679 (1935).
[11] van Heiningen, J., *Rec. Trav. Chim. Pays-Bas*, **55**, 65-75, 85-100 (1936).
[12] Hinshelwood, C. N., et. al., *Proc. Roy. Soc. London*, (A), **118**, 170-183 (1929); **122**, 610-621 (1929); **124**, 219-227 (1929); **130**, 640-654 (1931); **134**, 1-7 (1931); **138**, 311-317 (1932); **141**, 29-40 (1933). *Trans. Faraday Soc.*, **28**, 184-191 (1932). *Nature*, **131**, 361-362 (1933). *Z. Elektroch.*, **42**, 445-449 (1936).
[13] Hinshelwood, C. N., and Williamson, A. T., "The reaction between hydrogen and oxygen," Oxford University Press, 1934.
[14] Jost, W., *Z. Elektroch.*, **41**, 183-194, 232-250 (1935); **42**, 461-467 (1936).
[15] Lewis, B., and von Elbe, G., *J. Chem'l Phys.*, **2**, 537-546 (1934); *J. Am. Chem. Soc.*, **59**, 656-662 (1937); **59**, 970-975 (1937).
[16] Lindeijer, E. W., *Rec. Trav. Chim. Pays-Bas*, **56**, 97-104, 105-118 (1937).
[17] Maas, J. H., and Ewing, C., *J. Phys'l Chem.*, **37**, 13-15 (1933).
[18] Malinowski, A. E., and Skrynnikow, K. A., *Physik. Z. Sowj.*, **7**, 43-48 (1935).
[19] Miyanishi, M., *Sci. Papers Inst. Phys. Chem. Res. (Tokyo)*, **26**, 70-76 (1935); **27**, 47-51, 52-58 (1935).
[20] Mole, G., *Proc. Phys. Soc. (London)*, **48**, 857-864 (1936).
[21] Poljakow, M. W., et. al., *Acta Physicochim. URSS*, **1**, 551-553, 817-820, 821-832 (1935); **2**, 211-214, 397-400 (1935).

to other pertinent publications. The list does not pretend to include all the important papers, nor even all the most important ones, but only those which happen to be readily at hand. In certain cases of multiple authorship the references have been listed under the name of a single author, even though that name appears in some of the cases as junior author.

4. The Composition of Water

A review of the best work that has been done on the stoichiometric composition of water has been written by J. R. Partington.[29] He concluded that the ratio of the combining volumes at 0 °C and a pressure of 760 mm-Hg is $O_2/H_2 = 1/2.00288$.

If the atomic weight of H is 1.0077,[30] and that of O is 16.0000, the molecular weight of H_2O is $M = 18.0154$.

Natural water contains 1 D_2 to about 6500 H_2.[31]

5. Heat of Formation of H_2O

By the heat of formation $(Q_{T,P})_\phi$ of a certain phase (ϕ) of a substance at pressure P and absolute temperature T is meant the amount of heat evolved when one gram-formula-weight (gfw) of ϕ at P and T is formed from its elements at T and the total pressure P. It is the decrease in the enthalpy ($H = E + pv$) at fixed T and P per gfw of the substance formed.

Its value for any P and T within the domain in which the phase ϕ can exist can be determined from its value for one such pair of values by means of relation (1) in which δ may be computed by means of any one of the relations (2a), (2b), etc.

$$(Q_{T,P})_\phi = (Q_{T'P'})_\phi - \delta \tag{1}$$

$$\delta = \int_{T'}^{T} (\Delta C)_P \cdot dT + [P(\Delta v)_P - P'(\Delta v)_{P'}]_T + [(\Delta E)_P - (\Delta E)_{P'}]_T \tag{2a}$$

$$= \int_{T'}^{T} (\Delta C)_P dT + [P(\Delta v)_P - P'(\Delta v)_{P'}]_{T'} + [(\Delta E)_P - (\Delta E)_{P'}]_{T'} \tag{2b}$$

$$= \int_{T'}^{T} (\Delta C)_P \cdot dT - \int_{P'}^{P} [\Delta(\mu C)]_T dp \tag{2c}$$

---

[22] Prettre, M., and Laffitte, P., *Compt. rend.*, **187**, 763-765 (1928); **188**, 397-399 (1929).
[23] Prettre, M., *Compt. rend.*, **196**, 1891-1893 (1933); **201**, 962-964 (1935); **204**, 1734-1736 (1937). *J. de. Chim. Phys.*, **33**, 189-218 (1936).
[24] Rodebush, W. H., *et al.*, *J. Am. Chem. Soc.*, **59**, 1924-1931 (1937); *J. Phys'l Chem.*, **41**, 283-291 (1937).
[25] Semenoff, N., *Z. Physik*, **48**, 571-582 (1928); *Z. physik. Chem.* (*B*), **2**, 161-168, 169-180 (1929). (*B*), **28**, 43-53, 54-64 (1935). *Chem. Rev.*, **6**, 347-379 (1929).
[26] Semenova, N., *Acta Physicochim. URSS*, **6**, 25-42 (1937).
[27] Sokolik, A., and Shtsholkin, K., *Acta Physicochim. URSS*, **1**, 311-317 (1934).
[28] Tauzin, P., *Compt. rend.*, **196**, 1605-1607 (1933); **197**, 1046-1049 (1933).
[29] Partington, J. R., "The Composition of Water," 1928.
[30] *Int. Crit. Tables*, **7**, 44 (1926).
[31] Hall, N. F., and Jones, T. O., *J. Am. Chem. Soc.*, **58**, 1915-1919 (1936), and Gabbard, J. L., and Dole, M., *Idem*, **59**, 181-185 (1937).

5. HEAT OF FORMATION

$$= \int_{T'}^{T} (\Delta C)_P dT - \int_{P'}^{P} [\Delta(\mu C)]_T dp \tag{2d}$$

Here, ΔC, Δv, and ΔE denote, respectively, the increase in the specific heat at constant pressure, in the specific volume, and in the internal energy per unit mass, when the elements taken in the right proportions combine to form the substance, the pressure and temperature being those indicated by the subscripts, and the unit of mass being in all cases that of one gfw of the substance formed; the Joule-Thomson coefficient μ is $(\delta T/\delta p)_a$; p is a variable pressure; the a indicates that the change is adiabatic; and $\Delta(\mu C)$ is the excess of μC for ϕ over that for the equivalent mixture of the uncombined elements.

If ϕ and all the elements involved may be regarded as ideal gases in the domain considered, then $(\Delta C)_P = (\Delta C)_{P'}$, $\mu = 0$, and all except the first term in each expression for δ is zero, giving $\delta = \int_{T'}^{T} (\Delta C)_P dT$. Also, if ϕ is either a liquid or a solid, and is formed from gaseous elements, its volume will usually be negligible as compared with that of the gases from which it is formed, and then δ again reduces to its first term if the gases are ideal and the changes in the volume and in the internal energy with the pressure are negligible for ϕ.

This simplification is frequently assumed, but it is valid only under the condition just stated. In many cases, especially when the pressure approaches or equals that of saturation of one of the substances, data are not available for determining the error so introduced. For water-vapor, there are no data for μ in that region, and those in the region of superheat indicate that μ increases rapidly as saturation is approached.

If P and T are such associated values that the phases ϕ and ϕ' are in equilibrium, and if L_{TP} is the latent (absorbed) heat per gfw on passing from ϕ' to ϕ at P and T, then $(Q_{TP})_\phi = (Q_{TP})_{\phi'} - L_{TP}$.

From these relations it is possible to compute the heat of formation of any phase of a substance for any allowable P and T if the value of (Q_{TP}) for any one phase and one set of values of P and T is known, together with the values of the required auxiliary quantities.[32]

Whenever the data in the following paragraphs are expressed in international joules and are of such a precision as to justify a distinction between that unit and the absolute joule, the symbol "(Int.)" will precede the symbol for the unit. This symbol does not appear when the precision of the data is too low to justify any distinction between the units, or when the data are expressed in terms of the absolute unit; in either case, the unqualified symbol for the unit may properly be interpreted as indicating the absolute unit.

---

[32] For a review of the available data, see Bichowsky, F. R., and Rossini, F. D., "The Thermochemistry of the Chemical Substances," New York, Reinhold Publishing Corp., 1936.

Water-vapor: Heat of Formation.

If the heat of formation of water may be regarded as independent of the pressure when that does not exceed one atmosphere, as is essentially true, then the heat of formation of water-vapor at 25 °C and the pressure of water-vapor saturated at that temperature may be derived from Rossini's value (p. 17) for the heat of formation of water (285.775 (Int.) kj/gfw = 285.890 kj/gfw at 25 °C and one atmosphere) and the latent heat of vaporization at 25 °C, 43.939 (Int.) kj/gfw = 43.956 kj/gfw.[33] It is

$$(Q_{298.1, 23.76 \text{ mm}})_g = 241.84 \text{ (Int.) kj/gfw} = 241.93 \text{ kj/gfw-}H_2O$$
$$= 13.424 \text{ (Int.) kj/g} = 13.429 \text{ kj/g}$$

With this value those given in *International Critical Tables* (242.0 at 18 °C [34] and 241.8 at 25 °C [35]) agree remarkably well.

Bichowsky based his value on the sources from which he derived his value for the heat of formation of water (p. 17), together with the sources given below,[36-41] which treat of the heat of vaporization. (From the way in which these references are given in the *Critical Tables* one might infer that the data in them were used alone, instead of in conjunction with those contained in the immediately preceding references for the formation of water.)

Randall's value was not derived from the data contained in the references heading the table in which it appears, but seems to have been arrived at in the following manner: From the work of A. Schuller and V. Wartha,[42] of C. von Than,[43] and of J. Thomsen,[44] G. N. Lewis[45] deduced the value 68470 cal/gfw for the heat of formation of water at 0 °C and 1 atm, and from that the value 68270 cal/gfw at 25 °C. From the work of T. W. Richards and J. H. Mathews,[40] and of A. W. Smith,[41] G. N. Lewis and M. Randall [46, p. 477] chose 540.0 cal/g or 9730 cal/gfw as the value of the latent heat of vaporization of water at 0 °C, and from that obtained 10450 cal/gfw at 25 °C. Whence the heat of formation of water-vapor is 57820 cal/gfw at 25 °C and 1 atm.

---

[33] Osborne, N. S., Stimson, H. F., and Fiock, E. F., *Bur. Stand. J. Res.*, **5**, 411-480 (RP209) (1930).
[34] Bichowsky, F. R., *Int. Crit. Tables*, **5**, 176 (1928).
[35] Randall, M., *Idem*, **7**, 231 (1930).
[36] Carlton-Sutton, T., *Proc. Roy. Soc. (London) (A)*, **93**, 155-176 (1917).
[37] Griffiths, E. H., *Phil. Trans. (A)*, **186**, 261-341 (1895).
[38] Henning, F., *Ann. d. Physik* (4), **21**, 849-878 (1906); (4), **29**, 441-465 (1909); (4), **58**, 759-760 (1919).
[39] Holborn, L., Scheel, K., and Henning, F., "Wärmetabellen," Vieweg, Braunschweig, 1919.
[40] Richards, T. W., and Mathews, J. H., *Proc. Am. Acad. Arts Sci.*, **46**, 511-538 (1911) = *J. Am. Chem. Soc.*, **33**, 863-888 (1911).
[41] Smith, A. W., *Phys. Rev.*, **33**, 173-183 (1911).
[42] Schuller, A., and Wartha, V., *Ann. d. Physik (Wied.)*, **2**, 359-383 (1877).
[43] von Than, C., *Idem*, **13**, 84-105 (1881).
[44] Thomsen, J., *Idem, (Pogg)*, **148**, 368-404 (1873).
[45] Lewis, G. N., *J. Am. Chem. Soc.*, **28**, 1380-1395 (1906).
[46] Lewis, G. N., and Randall, M., "Thermodynamics and the Free Energy of Chemical Substances," New York, McGraw-Hill Book Co., 1923.

5. HEAT OF FORMATION

Lewis and Randall [46, p. 59] state that this calorie is equivalent to 4.182×10^7 ergs, and Randall [47] states that it is equivalent to 4.182 joules. Each statement implies that this calorie is to be regarded as equivalent to 4.182 absolute joules; and that interpretation is borne out by A. W. Smith's article [41] to which they refer. They have essentially followed Smith, who averaged the result obtained from the absolute (mechanical) determination by O. Reynolds and W. H. Moorby [48] with that from the electrical determination by H. T. Barnes [49] as converted to the basis of the 1911 international joule. This procedure tacitly assumes that the international joule of 1911 is essentially identical with the absolute joule, and that the average so obtained may be regarded, as it is by Smith, as the value of the calorie in terms of the absolute joule.

If the value 4.182 given by Lewis and Randall is used as the conversion factor, 57280 cal/gfw becomes 241.8 kj/gfw, the value given by Randall (see p. 14). But Lewis and Randall [45, p. 59] state that their unit is the 15°-calorie, which is actually close to 4.185 joules, making 57280 cal/gfw = 242.0 kj/gfw. Which of these two is to be preferred, and what importance is to be attached to the digit following the decimal point can be determined only by a detailed study of every constituent that enters into the value 57280 cal/gfw. For those constituents that rest upon a direct determination in terms of the 15°-calorie, the factor 4.185 is to be preferred; whereas a determination of the proper factor, or factors, to be used for those constituents that rest upon electrical measurements requires a knowledge both of the exact value of each of the units actually employed in the measurements, and of the numerical factor by means of which the data were converted to calories. Each of these may differ from observer to observer. Such detailed study has not been attempted by the present compiler.

M. Randall [50] accepts formulas (3) as satisfactory representations of the molecular specific heats, the unit being 4.182 joule/gfw H_2O.

$$(C_p)_{H_2} = 6.5 + 0.0009T \tag{3a}$$
$$(C_p)_{O_2} = 6.5 + 0.0010T \tag{3b}$$
$$(C_p)_{H_2O} = 8.81 - 0.0019T + 0.00000222T^2 \tag{3c}$$

Whence,

$$\Delta C \equiv (C_p)_{H_2O} - [(C_p)_{H_2} + 0.5(C_p)_{O_2}] = -0.94 - 0.0033T + 0.00000222T^2$$
$$= -3.931 - 0.0138T + 0.00000928_4T^2 \text{ joule/gfw-}H_2O$$

Accepting this as a satisfactory expression of ΔC when P is the pressure of water-vapor saturated at 25 °C, formula (4) is obtained:

$$(Q_T)_g = (Q_{298.1})_g - 1.703 + 0.393108(T/100) + 0.069003(T/100)^2 - 0.00309468(T/100)^3 \text{ kj/gfw} \tag{4}$$

---

[47] Randall, M., *Int. Crit. Tables*, **7**, 226 (1930).
[48] Reynolds, O., and Moorby, W. H., *Phil. Trans.* (A), **190**, 301-422 (1897).
[49] Barnes, H. T., *Idem*, **199**, 149-263 (1902); *Proc. Roy. Soc. (London)* (A), **82**, 390-395 (1909).
[50] Randall, M., *Int. Crit. Tables*, **7**, 231 (1930).

Taking $(Q_{298.1})_g = 241.930$ kj/gfw, this becomes

$$(Q_T)_g = 240.227 + 0.393108(T/100) + 0.069003(T/100)^2 - 0.00309468(T/100)^3 \text{ kj/gfw} \quad (5)$$

$$= 241.75 + 0.700758(t/100) + 0.0436485(t/100)^2 - 0.00309468(t/100)^3 \text{ kj/gfw} \quad (6)$$

the temperature being t °C $= (273.1 + t)$ °K $= T$ °K. These apply directly to a pressure of 23.76 mm-Hg, but insofar as all three gases may be regarded as ideal they apply to any pressure at which water-vapor can exist at t °C and the formulas for the specific heats remain valid.

Likewise, accepting Rossini's value for the heat of formation of water at 25 °C, but using formulas (7) for the specific heats and a lower value

Table 2.—Heat of Formation of Water-vapor

$(Q_T)_g = (Q_{TP})_g =$ heat evolved in the formation of 1 gfw-H_2O of vapor at temperature T °K and pressure P from H_2 and O_2 at the same total pressure (P) and temperature; $(q_T)_g = (Q_T)_g/18.0154 =$ heat of formation of 1 g H_2O; $(Q_T)_g = (Q_{273.1})_g + \Delta Q$.

With the exception of the last column, the tabular values are based on $(Q_{298.1})_g = 241.93$ kj/gfw and the values of the specific heat accepted by Randall (see text); the values in the last column have been computed by equation (8), (Chipman).

The values apply only to those pressures at which water-vapor can exist at the temperature considered, and at which the formula for ΔC is valid.

Unit of Q and $\Delta Q = 1$ kj/gfw-H_2O; of $q = 1$ kj/g. Temp. $= t$ °C.

t	T	$(Q_T)_g$ Randall	$(q_T)_g$	ΔQ	Chipman $(Q_T)_g$
0	273.1	241.75	13.419	0.00	241.96
15	288.1	241.86[a]	13.425	0.11	242.11
18	291.1	241.88	13.426	0.13	242.13
20	293.1	241.89	13.427	0.14	242.16
25	298.1	241.93[b]	13.429	0.18	242.20
50	323.1	242.11	13.439	0.36	242.44
100	373.1	242.49	13.460	0.74	242.91
200	473.1	243.30	13.505	1.55	243.80
300	573.1	244.16	13.553	2.41	244.62
500	773.1	245.96	13.653	4.21	246.07
1000	1273.1	250.03	13.879	8.28	248.61
1500	1773.1	251.64	13.968	9.89	249.59
2000	2273.1	248.47	13.792	6.72	248.99
2500	2773.1	238.20	13.222	−3.55	(246.83)
3000	3273.1	218.50	12.129	−23.25	(243.09)

[a] A. D. Crow and W. E. Grimshaw[52] state that the value 58 kcal/gfw ($= 242.7$ kj/gfw) at 15 °C is accepted by the Research Department, Woolwich, England.

[b] W. F. Giauque and M. F. Ashley[53] derive the value 57.823 kcal/gfw ($= 241.99$ kj/gfw) at 25° C.

---

[51] Chipman, J., *Ind. Eng. Chem.*, **24**, 1013-1017 (1932).
[52] Crow, A. D., and Grimshaw, W. E., *Phil. Trans.* (A), **230**, 39-73 (1931).
[53] Giauque, W. F., and Ashley, M. F., *Phys. Rev.* (2), **43**, 81-82 (L) (1933).

5. HEAT OF FORMATION

(43.70 kj/gfw) for the latent heat of vaporization, J. Chipman [51] obtained an expression equivalent to (8).

$$(C_p)_{H_2} = 6.70 + 0.0007T \text{ cal/gfw} \qquad (7a)$$
$$(C_p)_{O_2} = 6.50 + 0.0010T \text{ cal/gfw} \qquad (7b)$$
$$(C_p)_{H_2O} = 7.20 + 0.0027T \text{ cal/gfw} \qquad (7c)$$

$$\Delta C = -2.75 + 0.0015T \text{ cal/gfw} = -11.509 + 0.00628T \text{ j/gfw}$$
$$(Q_T)_g = 239.05 + 1.151(T/100) - 0.03139(T/100)^2 \text{ kj/gfw}$$
$$= 241.959 + 0.9795(t/100) - 0.03139(t/100)^2 \text{ kj/gfw} \qquad (8)$$

valid for the range $300 \lesssim T \lesssim 2300$ °K.

Values computed by each of these formulas (6, 8) are given in Table 2.

Water: Heat of Formation.

The heat evolved in the formation of one gfw (18.0156 g)* of (liquid) water at 25 °C and 1 atm (A) from H_2 and O_2 at the same temperature and total pressure has been found by F. D. Rossini [54] to be

$$(Q_{298.1,A})_w = 285.775 \pm 0.040 \text{ (Int.) kj} = 285.890 \text{ kj †}$$
$$= 15.8626 \text{ (Int.) kj/g} = 15.8690 \text{ kj/g}$$

This is believed to be the best value at present available.

To the same heat of formation, F. R. Bichowsky [55] has assigned the value 286.2 kj/gfw at 18 °C, based upon the work of various experimenters [56-68]; and M. Randall the values 285.5 kj/gfw at 25 °C, 285.8 at 18 °C, and 286.3 at 0 °C. The sources upon which Randall's values rest have already been given (p. 14).

So far as H_2 and O_2 may be considered ideal and both the compressibility and the variation of the internal energy of water may be ignored, this same value will hold good for any pressure at which water can exist as a

---

* This is the value used by Rossini; it is 0.0002 g greater than that given in the Critical Tables and generally used in this compilation.

† This is negligibly affected by an error in reduction, the corrected value being 285.782 (Int.) kj = 285.897 kj [Rossini, F. D., Bur. Stand. J. Res., **7**, 329-330 (RP343) (1931)].

[54] Rossini, F. D., Bur. Stand. J. Res., **6**, 1-35 (RP259) (1931) → Proc. Nat. Acad. Sci., **16**, 694-699 (1930) → Science, (N.S.) **72**, 378 (1930).
[55] Bichowsky, F. R., Int. Crit. Tables, **5**, 176 (1929).
[56] Abria, Compt. rend., **22**, 372-373 (1846).
[57] Andrews, T., Phil. Mag. (3), **32**, 321-339 (1848).
[58] Berthelot, M., Ann. de Chim. et phys. (5), **23**, 176-187 (1881).
[59] Berthelot, M., and Matignon, C., Idem (6), **30**, 547-565 (1893).
[60] Despretz, C., Idem (2), **37**, 180-181 (1828).
[61] Dulong, Compt. rend., **7**, 871-877 (1838).
[62] Favre, P. A., and Silbermann, J. T., Ann. de chim. et phys. (3), **34**, 357-450 (1852).
[63] Grassi, C., J. pharm. et chim. (3), **8**, 170-181 (1845).
[64] Mixter, W. 'G., Amer. J. Sci. (4), **16**, 214-228 (1903).
[65] Rümelin, G., Z. physik. Chem., **58**, 449-466 (1907).
[66] Schuller, A., and Wartha, V., Ann. d. Physik (Wied.), **2**, 359-383 (1877).
[67] von Than, C., Ber. deut. chem. Ges., **10**, 947-952 (1877); Ann. d. Physik (Wied.), **13**, 84-105 (1881).
[68] Thomsen, J., "Systematisk gennemförte termokemiske undersögelsers numeriske og teortiske resultater." 1882-1886.

liquid at 25 °C, and the value at any other temperature and pressure may be determined by means of the relation $(Q_{TP})_w = (Q_{298.1,A})_w - \int_{298.1}^{T}(\Delta C)_w dT$, the pressure being not less than that of water-vapor saturated at T °K.

No entirely satisfactory formulation for the specific heat of water at constant pressure is available, but for temperatures between 0 and 100 °C its variation is so slight (maximum 18.03, minimum 18.01 cal/gfw·°C) that the error introduced by regarding it as constant and equal to 18.02 may, for our present purposes, be ignored. By combining this expression with those for the specific heat of the gases (3) one obtains the expression $(\Delta C)_w = 8.27 - 0.0014T$ cal/gfw·°K $= 34.58 - 0.00585T$ j/gfw·°K, the calorie being here taken as equivalent to 4.182 j (p. 15). In his compilation M. Randall[69] has taken $(\Delta C)_w = 8.0$ cal/gfw·°K $= 33.45$ j/gfw·°K, independent of T. The first is to be preferred; it leads to the relation $(Q_{TP})_w = Q_{298.1,A})_w + 0.823 - 3.298(t/100) + 0.0292(t/100)^2$ kj/gfw, which becomes $(Q_{TP})_w = 286.713 - 3.298(t/100) + 0.0292(t/100)^2$ kj/gfw-H_2O when $(Q_{298.1,A})_w = 285.890$. Values computed by means of that formula are given in Table 3. Under the same conditions, Randall's expression for $(\Delta C)_w$ leads to $(Q_{TP})_w = 286.726 - 3.346(t/100)$ kj/gfw.

Table 3.—Heat of Formation of Water

$(Q_{TP})_w$ = heat evolved in the formation of 1 gfw-H_2O of liquid at the temperature t and pressure P from H_2 and O_2 at the same temperature t and total pressure P. The pressure P may have any value at which water can exist as a liquid at temperature t.

These values have been computed by means of the more exact formula involving the square of t; by adding to each the associated δ, the corresponding value defined by the formula based on Randall's approximate value for $(\Delta C)_w$ will be obtained.

Unit of $(Q_{TP})_w$ and of δ = 1 kj/gfw-H_2O. Temp. = t °C.

t	$(Q_{TP})_w$	δ	t	$(Q_{TP})_w$	δ
0	286.713	0.013	30	285.726	−0.003
10	286.383	0.009	40	285.398	−0.010
15	286.219[a]	0.006	50	285.071	−0.018
18	286.120	0.004	60	284.744	−0.026
20	286.054	0.003	80	284.093	−0.043
25	285.890	0	100	283.444	−0.064

[a] A. D. Crow and W. E. Grimshaw[52] have stated that the value at 15 °C accepted by the Research Department, Woolwich, England, is 68.4 kcal/gfw (= 286.25 kj/gfw).

Ice: Heat of Formation.

From Rossini's value for the heat of formation of water (p. 17), the excess (0.82 kj/gfw, Table 3) of the heat of formation of water at 0 °C

[69] Randall, M., *Int. Crit. Tables*, **7**, 232 (1930), the coefficient of $T \log T$ in the expression in the second line under $H_2O(l)$ at top of the page.

6. DISPOSABLE ENERGY

over that at 25 °C, and the latent heat of fusion of ice (0.3336 × 18.0154 = 6.01 kj/gfw, Table 272), it follows that the heat evolved in the formation of one gfw-H_2O of ice at 0 °C and one atmosphere, from gaseous H_2 and O_2 at the same temperature and pressure, is

$$(Q_{273.1,A})_i = 292.72 \text{ kj/gfw-}H_2O$$
$$= 16.248 \text{ kj/g}$$

All of this refers to ice-I, the usual type that melts at 0 °C under a pressure of one atmosphere.

Insofar as H_2 and O_2 may be considered ideal, and both the compressibility and the variation of the internal energy of ice-I may be ignored, this same value will hold for any pressure under which ice-I can exist at 0 °C; and the value at any other temperature and pressure may be determined from the relation $(Q_{TP})_i = (Q_{273.1,A})_i - \int_{273.1}^{T} (\Delta C)_i dT$, but only within the domain in which ice-I can exist and within which the limitations just imposed are fulfilled.

Combining the expression $(0.5057 + 0.001863t \text{ cal}_{20}/\text{g·°C})$ found by H. C. Dickinson and N. S. Osborne [70] for the specific heat of ice with those (p. 15) for the specific heats of H_2 and O_2, one obtains $(\Delta C)_i = -4.282 + 0.1320t$ j/gfw·°C $= -0.040331 + 0.01320(T/100)$ kj/gfw·°K. Whence, $(Q_{TP})_i = (Q_{273.1,A})_i + 0.4282(t/100) - 0.660(t/100)^2$ kj-gfw, which yields the following tabulated values when $(Q_{273.1,A})_i = 292.72$ kj/gfw-H_2O. In the range 0 to -40 °C those values are only 0.01 j/gfw smaller than the corresponding ones based on the value of $(\Delta C)_i$ given in M. Randall's compilation;[71] viz.,

$$(\Delta C)_i = (\Delta C)_w + (\Delta C)_{w \to i} = (+8.0 - 9.11 + 0.0336t)4.182$$
$$= -4.642 + 0.1405t \text{ j/gfw·°C}.$$

t	0	-10	-20	-30	-40	°C
$(Q_{TP})_i$	292.72	292.67	292.61	292.53	292.44	kj/gfw-H_2O

6. DISPOSABLE ENERGY OF FORMATION OF H_2O AT CONSTANT PRESSURE

Of the greatest amount $(W_{TP} + P\Delta v)$ of external work that can be obtained from a process in which the final total pressure and temperature are the same as the initial ones, and the final volume exceeds the initial by the amount Δv, the amount W_{TP} is related to the corresponding heat evolution (Q_{TP}) of the process as indicated by the relation $(d/dT)(W_{TP}/T) = -Q_{TP}/T^2$. But of the total energy available under the best conditions within the stated limitations the amount $P\Delta v$ has to be expended in over-

---

[70] Dickinson, H. C., and Osborne, N. S., *Bull. Bur. Stand.*, 12, 49-81 (SP248) (1915).
[71] Randall, M., *Int. Crit. Tables*, 7, 232, 1930, $H_2O(l)$ and $H_2O(l) = H_2O(s)$.

coming the pressure, thus leaving only W_{TP} disposable for other work. For this reason, W_{TP} is sometimes called the technical work, or the technical energy, freed by the process; here it will be called the disposable energy. It is the decrement of the quantity $(E + Pv - TS)$ that was denoted by ζ in Gibbs' papers, and that has been called the thermodynamic potential at constant pressure, the free energy at constant pressure, and the Gibbs function, but is called by G. N. Lewis and M. Randall [46] simply the free energy, which term had already been used by Helmholtz and others to denote the quantity $(E - TS)$. Here, E denotes the internal energy, S the entropy, and P, v, and T denote the pressure, volume, and absolute temperature, respectively.

It may readily be shown that $W_{TP} = W_{TP_1} - \int_{P_1}^{P} (\Delta v)_T dp$. If each of the substances involved in the process is either an ideal gas or one having a volume that is negligible as compared with Δv, then $\Delta v = (\Delta n)RT/P$, where Δn is the increase in the number of molecules of the gaseous substances per gfw of the compound formed, and $W_{TP} = W_{TP_1} - (\Delta n)RT \log_e(P/P_1)$. It is convenient, and a common practice, to split the logarithm into two parts, $\log(P/P_1) = \log(A/P_1) + \log(P/A)$, A denoting the pressure of one normal atmosphere, and to tabulate $w \equiv W_{TP_1} - (\Delta n)RT \log_e(A/P_1)$ instead of W_{TP}, the latter being given by the relation $W_{TP} = w - (\Delta n)RT \log_e(P/A)$. If the assumed phases with which we are concerned can exist under the pressure A when at the temperature T, then w is the value of W_{TP} at the pressure of one normal atmosphere; otherwise, as in the case of water-vapor at temperatures below 100 °C, w is a purely fictitious quantity, *i.e.*, it is merely the numerical value obtained by adding to W_{TP} the numerical value of the quantity $(\Delta n)RT \log_e(P/A)$. The somewhat common practice of stating without reservation that w is the value of W_{TP} at one atmosphere is undesirable.

If $Q_{T_1P_1}$ and $W_{T_1P_1}$ are, respectively, the heat of formation and the disposable energy corresponding to T_1 and P_1, and if $\Delta C = a + b_1 T + b_2 T^2 + \ldots b_n T^n$, then the value of W at the same pressure (P_1) and any temperature (T) within the allowable range is given by expression (9).

$$W_{TP_1} = Q_{T_1P_1} - (Q_{T_1P_1} - W_{T_1P_1})(T/T_1) + K - BT + f(T) + a\{T \log_e(T/T_1) - (T - T_1)\} \tag{9}$$

where

$$K = \tfrac{1}{2}b_1 T_1^2 + \tfrac{1}{3}b_2 T_1^3 + \tfrac{1}{4}b_3 T_1^4 + \ldots + \frac{1}{n+1} b_n T_1^{n+1}$$

$$B = b_1 T_1 + \tfrac{1}{2}b_2 T_1^2 + \tfrac{1}{3}b_3 T_1^3 + \ldots + \frac{1}{n} b_n T_1^n$$

$$f(T) = \tfrac{1}{2}b_1 T^2 + \tfrac{1}{6}b_2 T^3 + \tfrac{1}{12}b_3 T^4 + \ldots + \frac{1}{n(n+1)} b_n T^{n+1}$$

6. DISPOSABLE ENERGY

If $(T - T_1)$ is small, then $\{T \log_e(T/T_1) - (T - T_1)\}$ is small also, being equal to

$$T_1\left\{\left(1 + \frac{T - T_1}{T_1}\right)\log_e\left(1 + \frac{T - T_1}{T_1}\right) - \frac{T - T_1}{T_1}\right\}.$$

Whenever $\Delta v = (\Delta n)RT/P$, the value of W for any associated temperature (T) and pressure (P) within the allowable range is given by the expression $W_{TP} = W_{TP_1} - (\Delta n)RT \log_e(A/P_1) - (\Delta n)RT \log_e(P/A)$, which is equivalent to (10), a form more convenient for computation.

$$W_{TP} = W_{T_1P_1} - (\Delta n)RT_0\log_e(A/P_1) + aT_0\log_e(T_0/T_1) + \tau\{a + (Q_{T_1P_1} - W_{T_1P_1})/T_1\} + \tau^2(\tfrac{1}{2}D_1 + \tfrac{1}{3}D_2\tau + \ldots + \frac{1}{n+1}D_n\tau^{n-1}) -$$

$$t\left\{(Q_{T_1P_1} - W_{T_1P_1})/T_1 + (\Delta n)R\log_e(A/P_1) - a\log_e(T_0/T_1) + \tau(D_1 + \tfrac{1}{2}D_2\tau + \ldots + \frac{1}{n}D_n\tau^{n-1})\right\} + aT_0\left(\frac{T}{T_0}\log_e\frac{T}{T_0} - \frac{t}{T_0}\right) +$$

$$\tfrac{1}{2}D_1t^2 + \tfrac{1}{6}D_2t^3 + \ldots + \frac{1}{n(n+1)}D_nt^{n+1} - (\Delta n)RT\log_e(P/A) \qquad (10)$$

where

$$T = T_0 + t, \; T_1 = T_0 + \tau$$
$$D_1 = b_1 + b_2T_0 + b_3T_0^2 + \ldots + b_nT_0^{n-1}$$
$$D_2 = b_2 + 2b_3T_0 + 3b_4T_0^2 + \ldots + (n-1)b_nT_0^{n-2}$$
$$D_3 = b_3 + 3b_4T_0 + 6b_5T_0^2 + \ldots + \frac{(n-2)(n-1)}{2}b_nT_0^{n-3}$$
$$\cdots \cdots \cdots \cdots \cdots \cdots \cdots$$
$$D_m = b_m + mb_{m+1}T_0 + \frac{m(m+1)}{2}b_{m+2}T_0^2 + \ldots +$$
$$\frac{m(m+1)\ldots(n-1)}{1\cdot 2\cdot 3 \ldots \cdot (n-m)}b_nT_0^{n-m}$$
$$\cdots \cdots \cdots \cdots \cdots \cdots \cdots$$
$$D_n = b_n$$

If two phases, ϕ_1 and ϕ_2, of a substance are in equilibrium at T' and P', then $(W_{T'P'})_{\phi_1} = (W_{T'P'})_{\phi_2}$ and $(Q_{T'P'})_{\phi_1} = (Q_{T'P'})_{\phi_2} + L_{12}$, where L_{12} is the latent heat absorbed by the substance in passing from phase ϕ_1 to phase ϕ_2.

Quite recently, H. Zeise [72] has reviewed the several methods for obtaining w, and has derived the essential formulas.

Wherever it is necessary in the following paragraphs to know that the given data are expressed in terms of the international joule, as distinguished from the absolute, the symbol "Int." will accompany the symbol for the unit (see p. 13).

[72] Zeise, H., *Z. Elektroch.*, **39**, 758-773, 895-909 (1933).

Water-vapor: Disposable Energy of Formation.

In order to obtain from expressions (9) and (10) the value of W_{TP}, it is necessary to know the value of W_{TP} for some one set of associated values of P and T. Assuming the value, $(W_{298.1,A})_w = 56.560 \times 4.182 = 236.53$ kj/gfw-H_2O, given by M. Randall [71] for the disposable energy in the formation of water at 25 °C and 1 atm, and remembering that at the pressure (23.76 mm-Hg) of water-vapor saturated at 25 °C the disposable energy of formation at 25 °C is the same for the vapor as for the liquid, one obtains the value $(W_{298.1,23.76mm})_g = 223.65$ kj/gfw. Whence, taking

Table 4.—Disposable Energy in the Formation of Water-vapor

The greatest amount of external work that can be obtained from the combination of H_2 and O_2 at temperature T and total pressure P, to form water-vapor at the same temperature and pressure is $(W_{TP})_g + P\Delta v$, where Δv is the accompanying increase in volume, and P does not exceed either that at which water-vapor can exist at temperature T or that at which the formula used for ΔC is valid; $(W_{TP})_g \equiv w + 0.5RT \log_e(P/A)$, where w takes the values tabulated below, and A = pressure of one normal atmosphere; $w_g \equiv w_{12}/18.0154$; $P\Delta v = -0.5RT$.

Computations by means of formulas (12), (13), and (14) are indicated by subscripts; w_{12} is essentially the negative of M. Randall's [71] $\Delta F°$. For values computed from molecular and spectroscopic data, see A. R. Gordon.[83]

Unit of w and $0.5RT = 1$ kj/gfw-H_2O; of $w_g = 1$ kj/g. Temp. = t °C.

t	T	w_{12}	$0.5RT$	w_g	w_{13}	w_{14}
0	273.1	229.11	1.135	12.718	229.25	229.12
15	288.1	228.41	1.198	12.679	228.55	228.41
18	291.1	228.27	1.210	12.671	228.40	228.27
20	293.1	228.18	1.218	12.666	228.30	228.18
25	298.1	227.95[a]	1.239	12.653	228.07	227.94
50	323.1	226.76	1.343	12.587	226.87	226.75
100	373.1	224.36	1.551	12.454	224.44	224.32
200	473.1	219.40	1.967	12.178	219.38	219.29
300	573.1	214.27	2.383	11.894	214.14	214.19
500	773.1	203.6	3.214	11.30	203.3	203.3
1000	1273.1	175.0	5.293	9.71	174.8	174.9
1500	1773.1	145.1	7.372	8.05	145.6	145.8
2000	2273.1	115.4	9.450	6.41	116.3	117.0
2500	2773.1	87.0	11.529	4.83	87.4	89.3
3000	3273.1	61.3	13.608	3.40	58.9	63.4

[a] Randall gives $54.507 \times 4.182 = 227.95$ kj/gfw; W. F. Giauque and M. F. Ashley [53] give $54.670 \times 4.185 = 228.79$ kj/gfw at 25 °C; F. D. Rossini [84] derives from his own value for the heat of formation of water (p. 17), together with various published spectroscopic values for the entropies of the gases, 228.57 (Int.) kj/gfw at 25 °C (=228.67 kj/gfw).

---

[73] Bjerrum, N., *Z. physik. Chem.*, 79, 513-536 (1912).
[74] Langmuir, I., *J. Am. Chem. Soc.*, 28, 1357-1379 (1906).
[75] Löwenstein, L., *Z. physik. Chem.*, 54, 715-726 (1905).
[76] Nernst, W., and v. Wartenberg, H., *Nachr. k. Ges. Wiss., Göttingen*, 1905, 35-45 (1905).
[77] v. Wartenberg, H., *Ber. deut. physik. Ges.*, 8, 97-103 (1906); *Z. physik. Chem.*, 56, 513-533 (1906).

$(Q_{298.1, 23.76mm})_g = 241.93$ kj/gfw (p. 14), accepting Randall's expression for ΔC (p. 15), and using expression (9), one obtains

$$(W_{TP})_g = 240.227 - 0.003931 T \log_e T - 0.0690 (T/100)^2 + 0.0015473 (T/100)^3 - 0.016885 T + 0.0041575 T \log_e(P/A) \text{ kj/gfw-H}_2\text{O} \qquad (11)$$

which differs from the one given by Randall solely by the presence of the last term and by small changes in the first and in the next to the last terms. The last two result from the use of a slightly different value for the heat of formation of water, and the first is necessary for the completion of the expression. Obviously, expression (11) ceases to have a physical significance when P exceeds the pressure of water-vapor saturated at T °K. The alternative form corresponding to (10) is

$$(W_{TP})_g = 229.110 - 0.046291 t - 0.05622 \left(\frac{t}{100}\right)^2 + 0.0015473 \left(\frac{t}{100}\right)^3 - 1.0736 \left\{ \frac{T}{T_0} \log_e \left(\frac{T}{T_0}\right) - \frac{t}{T_0} \right\} + 0.0041575 T \log_e(P/A) \text{ kj/gfw-H}_2\text{O} \quad (12)$$

It seems that the expression given by Randall rests upon the value he gives for the heat of formation of water (p. 14), together with his conclusions regarding the thermal dissociation of water-vapor, as derived from the work of N. Bjerrum,[73] I. Langmuir,[74] L. Löwenstein,[75] W. Nernst and H. v. Wartenberg,[76] H. v. Wartenberg,[77] with a consideration of that of F. Haber and F. Fleischmann,[78] F. Faber and G. W. A. Foster,[79] A. Holt,[80] W. Nernst,[81] and Schmidt,[82] these being the references given in that section of his table.

Using other data (see p. 17), Chipman obtained an expression equivalent to (13)

$$(W_{TP})_g = 239.05 - 1.151 (T/100) \log_e T + 0.03139 (T/100)^2 + 2.783 (T/100) + 0.0041575 T \log_e(P/A) \text{ kj/gfw-H}_2\text{O}. \qquad (13)$$

$$= 229.25 - 4.6529 (t/100) - 3.1434 \{(1 + t/273.1) \log_e(1 + t/273.1) - (t/273.1)\} + 0.03139 (t/100)^2 + 0.0041575 T \log_e(P/A) \text{ kj/gfw-H}_2\text{O},$$

and E. D. Eastman [82a] one equivalent to (14)

$$(W_{TP})_g = 239.50 - 0.8412 (T/100) \log_e T - 0.0090605 (T/100)^2 + 0.0007366 (T/100)^3 + 0.9374 (T/100) + 0.0041575 T \log_e(P/A) =$$

[78] Haber, F., and Fleischmann, F., *Z. anorg. allgem. Chem.*, 51, 245-288 (1906).
[79] Haber, F., and Foster, G. W. A., *Idem*, 51, 289-314 (1906).
[80] Holt, A., *Phil. Mag.* (6), 13, 630-635 (1907).
[81] Nernst, W., *Z. anorg. allgem. Chem.*, 45, 126-131 (1905).
[82] Schmidt, *Diss.*, Berlin, 1921.
[82a] Eastman, E. D., *Bur. Mines, Circ. 6125:* p. 15 (1929).
[83] Gordon, A. R., *J. Chem'l Phys.*, 1, 308-312 (1933).
[84] Rossini, F. D., *private communication*, 1935.

$$229.12 - 4.6558(t/100) - 2.2973\{(1 + t/273.1)\log_e(1 + t/273.1) - t/273.1\} - 0.003026(t/100)^2 + 0.0007366(t/100)^3 + 0.0041575 T \log_e(P/A) \text{ kj/gfw-H}_2\text{O}. \tag{14}$$

Values computed by each of these formulas are given in Table 4.

Water: Disposable Energy of Formation.

Accepting for the disposable energy in the formation of water at 25° C and 1 atm the value given by M. Randall,[71] viz., $(W_{298.1,A})_w = 56.560 \times 4.182 = 236.53_4$ kj/gfw-H$_2$O, for the heat of formation $(Q_{298.1,A})_w = 285.890$ kj/gfw-H$_2$O (p. 17), and for $(\Delta C)_w$ the expression $(\Delta C)_w = 34.58 - 0.00585T$ j/gfw-H$_2$O·°C (p. 18), one obtains expression (15), in which A = pressure of one normal atmosphere. It has no physical significance if P is less than the vapor-pressure of water at t °C.

$$(W_{TP})_w = 240.70_9 - 0.16845t - 0.00000292t^2 + 9.444\{(1 + t/273.1) \times \log_e(1 + t/273.1) - t/273.1\} + 0.01247T \log_e (P/A) \text{ kj/gfw-H}_2\text{O}. \tag{15}$$

Values computed by means of (15) are given in Table 5.

Table 5.—Disposable Energy in the Formation of Water

The greatest amount of external work that can be obtained from the combination of H$_2$ and O$_2$ at temperature T and pressure P, not less than the vapor-pressure of water at T °K, to form water at the same temperature and pressure is $(W_{TP})_w + P(\Delta v) \equiv w + 1.5RT \log_e(P/A) + P(\Delta v)$, where A = pressure of 1 atm, and (Δv) is the increase in volume on passing from the mixed gases to the resultant water; $P(\Delta v) = -1.5RT = -0.01247T$ kj/gfw-H$_2$O. The following values of w have been computed by formula (15); $w_g = w/18.0154$.

Unit of w and $1.5RT = 1$ kj/gfw-H$_2$O; of $w_g = 1$ kj/g; temp. $= t$ °C $= T$ °K.

t	T	w	$1.5RT$	w_g
0	273.1	240.709	3.406	13.3613
10	283.1	239.031	3.531	13.2682
15	288.1	238.196	3.593	13.2218
20	293.1	237.363	3.656	13.1756
25	298.1	236.535[a]	3.718	13.1296
30	303.1	235.708	3.780	13.0837
40	313.1	234.059	3.905	12.9922
60	333.1	230.804	4.154	12.8115
80	353.1	227.585	4.404	12.6328
100	373.1	224.402	4.653	12.4561

[a] W. F. Giauque and M. F. Ashley[53] have concluded that $W = 56.720 \times 4.185 = 237.37$ kj/gfw at 25 °C. Had this value been used instead of Randall's, the computed values of W would have exceeded those in the table by $0.77 + 0.0028t$ kj/gfw.

If for $(\Delta C)_w$ one uses Randall's approximation (33.45 j/gfw·°C, p. 18), keeping all else as before, one obtains a formula differing slightly from (15), but yielding essentially the same values except at the higher temperatures. But even at 100 °C the value so obtained is only 0.006 kj/gfw smaller than that given in Table 5. It will be noticed that this difference at 100° C is only a tenth of that similarly produced in the computed value of the heat of formation (Table 3).

Ice: Disposable Energy of Formation.

We have found for water that $(W_{273.1,A}) = 240.71$ kj/gfw (Table 5); hence $(W_{273.1,A})_i = 240.71$ kj/gfw also, water and ice-I (the common type of ice) being in equilibrium at that temperature and pressure. The heat of formation of ice is $(Q_{273.1,A})_i = 292.72$ kj/gfw (p. 19), and $(\Delta C)_i = -0.040331 + 0.01320(T/100)$ kj/gfw °K (p 19). Hence the disposable energy in the formation of ice from H_2 and O_2 at the same T and P may be computed by means of (16).

$$(W_{TP})_i = 240.71 - 0.1904t - 11.014\{(T/T_0)\log_e(T/T_0) - t/T_0\} + 0.660(t/100)^2 + 0.01247T\log_e(P/A)$$
$$\equiv w + 0.01247T\log_e(P/A) \text{ kj/gfw-}H_2O \qquad (16)$$

$T_0 = 273.1$, $T = 273.1 + t$, temperature $= t$ °C, $A =$ pressure of one normal atmosphere. This expression has no physical significance at temperatures above the melting point of ice under pressure P, but may be extrapolated to $w = 235.94$ for $T = 298.1$, which is the same as the value $(56.418 \times 4.182 = 235.94)$ given in M. Randall's compilation.[71]

t	0	−10	−20	−30	−40	−50	°C
w	240.71	242.62	244.51	246.40	248.30	250.20	kj/gfw-H_2O
$0.01247T$	3.41	3.28	3.16	3.03	2.91	2.78	kj/gfw-H_2O

IB. DISSOCIATION

7. Dissociation of Water-vapor

Thermal Dissociation of Water-vapor.

When a mixture of H_2, O_2, and H_2O-vapor is in thermal equilibrium, $k_1(n_w/v)^2 = k_2(n_H/v)^2(n_O/v)$, where v is the volume of the mixture; n_w, n_H, and n_O are, respectively, the number of moles of H_2O, H_2, and O_2; k_1 is half the number of moles of H_2O-vapor that dissociate in unit time when the concentration (n_w/v) is unity; and k_2 is half the number of moles of H_2O formed per unit of time in a mixture of H_2 and O_2 in which the concentration $(n_H/v, n_O/v)$ of each gas is unity. The factor $\frac{1}{2}$ arises from the fact that two molecules of H_2O are necessarily involved in each case, the reaction being $2H_2O \rightleftarrows 2H_2 + O_2$. The values of the k's vary with the temperature and with the unit of concentration, but, to at least a first approximation, are independent of the actual concentrations. The numerical value of k_1 varies as the square of the size of the unit of concentration, and that of k_2 as the cube.

Their ratio ($K = k_1/k_2$) is the ratio of the number of moles of H_2O-vapor that dissociate to the number formed in the same time when the concentration of each of the three gases is unity. It is called the *constant of dissociation*, and in this case it has the dimensions of a concentration. The reciprocal of K might logically be called the constant of combination, but unfortunately, it also is occasionally called the constant of dissociation. Obviously, any function of K and the temperature is an isothermal constant, but that does not justify the confusion introduced by calling such arbitrary functions constants of dissociation. Here we shall restrict that term to the ratio k_1/k_2 and shall denote it by K; $K = (n_H)^2(n_O)/(n_w)^2 v$.

Insofar as the three gases may be considered ideal, this expression for K is equivalent to (17) in which the p's indicate the partial pressures.

$$KRT = (p_H)^2(p_O)/(p_w)^2 \tag{17}$$

The product KRT, which in this case is of the dimensions of a pressure, is frequently denoted by K_p and called the dissociation constant; and so is its square root. We shall not use that term to denote either KRT or $(KRT)^{\frac{1}{2}}$.

If the mixed gases have been derived from the dissociation of W moles of H_2O, and if α is the fraction of W dissociated when the mixture is in equilibrium, then $n_w = (1 - \alpha)W$, $n_H = \alpha W$, $n_O = \alpha W/2$, and the relations already found become

$$K = \frac{\alpha^3}{2(1 - \alpha)^2}\left(\frac{W}{v}\right) \tag{18}$$

$$KRT = \frac{\alpha^3}{(1 - \alpha)^2(2 + \alpha)}(P) \tag{19}$$

Here, P is the total pressure, and $18.0154 W/v$ is the actual density of the mixture. The first of these expressions (18) is always true if the reaction is $2H_2O \rightleftarrows 2H_2 + O_2$ and the relation stated at the beginning of this section applies; but the second (19) can apply only in the domain in which the gases may be regarded as ideal. The computation of α when the value of KRT/P is known may be facilitated by the use of Table 7.

If both sides of (18) are divided by C_1 (= the concentration of one gfw per liter) they become dimensionless; similarly if (19) is divided by A, the pressure of one normal atmosphere. It may be shown that $(d/dT)(\log_e KRT/A) = 2(Q_{TP})_g/RT^2 = -2(d/dT)(W_{TP}/RT)_g$, where $(Q_{TP})_g$ and $(W_{TP})_g$ have each the same significance as on pp. 12, 19. Whence from (11) one obtains (20) for $P = A$,

$$-\frac{RT}{2}\log_e\left(\frac{KRT}{A}\right) = 240.247 - 0.003931 T \log_e T - 0.0690\left(\frac{T}{100}\right)^2 +$$

$$0.001548_3\left(\frac{T}{100}\right)^3 - I_d T \text{ kj/gfw-}H_2O \tag{20}$$

7. DISSOCIATION OF VAPOR

where I_d, a constant of integration, is the same quantity as that designated by I in M. Randall's compilation,[1] but is expressed in a different unit.

By means of this formula and the values of KRT/A derived from the observed values of α, the corresponding values of I_d can be computed; $R = 8.315$ j/gfw·°K $= 0.8206$ l·atm/gfw·°K. These values of I_d are given in Table 6; their mean is 15.54 j/gfw·°K. By means of it and relation (20) values of K/C_1 and of α for $P = A$ have been computed; they also are given in the table.

B. Lewis and J. B. Friauf[2] have derived the following expressions:

$$0.5 \log_{10}(A/KRT) = \frac{57295}{4.573T} - 0.848 \log_{10}T - 1.474 \frac{T}{10000} +$$
$$7.78 \left(\frac{T}{10000}\right)^2 - 8.72 \left(\frac{T}{10000}\right)^3 + 0.0616; \; T \lesssim 2800 \, °K \quad (21)$$

$$0.5 \log_{10}(A/KRT) = \frac{14412}{T} + 2.0975 \log_{10}T - 0.7610 \frac{T}{10000} -$$
$$2.047\left(\frac{T}{10000}\right)^2 + 2.473\left(\frac{T}{10000}\right)^3 - 10.4417; \; 2800\,°K \lesssim T \lesssim 3000\,°K \quad (22)$$

$$0.5 \log_{10}(A/KRT) = \frac{13317.3}{T} - 0.34492 \log_{10}T - 4.3285;$$
$$T \gtrsim 3000 \, °K \quad (23)$$

Values computed by these equations are given in the second section of Table 6; for each of the temperatures 2800 and 3000, values have been computed by each of the equations applicable thereto.

The preceding discussion relates only to the thermal dissociation of water-vapor into uncharged molecules of H_2 and O_2. The questions arise: Does an increase in temperature give rise to other types of dissociation? Are any of the products of dissociation electrically charged?

K. F. Bonhoeffer and H. Reichardt[3] have concluded, from their study of the thermal variation in the optical absorptivity of water-vapor, that at temperatures above 1200 °C the amount of vapor dissociating into $H_2 + 2OH$ is a little greater than that into $2H_2 + O_2$. The OH was electrically neutral.

G. M. Woods and T. C. Poulter[4] found no detectable electrical conduction through water-vapor heated to 500 °C. The sensitivity of the method employed was not great.

---

[1] Randall, M., *Int. Crit. Tables*, **7**, 231 (1930).
[2] Lewis, B., and Friauf, J. B., *J. Am. Chem. Soc.*, **52**, 3905-3920 (1930).
[3] Bonhoeffer, K. F., and Reichardt, H., *Z. physik. Chem.* (A), **139**, 75-97 (1928).
[4] Woods, G. M., and Poulter, T. C., *Proc. Iowa Acad. Sci.*, **33**, 172-173 (1926).

Table 6.—Thermal Dissociation of Water-vapor

Values given in the first section of the main table are based upon the observational data used in M. Randall's compilation.[1] Those data have been checked against the original publications for errors in transcription, and all computations have been made independently for this table. The quantity $(A/KRT)^{0.5}$ is denoted by K in Randall's compilation, where it is called the constant of dissociation.

α = ratio of the number of gfw-H_2O dissociated into $H_2 + 0.5O_2$ to the total number of gfw of H_2O contained in a mixture of the three gases when the mixture is in equilibrium at the pressure and temperature indicated, and has the analytical composition H_2O. $K = (n_H)^2(n_O)/(n_w)^2 v$, where v = total volume; n_w, n_H, and n_O = the number of gfw of H_2O, H_2, and O_2, respectively, contained in that volume; A = the pressure of one normal atmosphere; C_1 = the concentration of one gfw per liter, the gases being assumed to be ideal; and I_d is a constant of integration (see text). The computed values (comp) have been derived from equation (20) with $I_d = 15.54$, its mean value. K/C_1 and KRT/A are each dimensionless; obviously, they are, respectively, the numerical values of K and of KRT when the unit of K is 1 gfw/l and that of KRT is 1 atm, but in writing such equations as (20) the A and the C_1 must be retained if mathematical nonsense is to be avoided.

D. M. Newitt [5] has given the following values for the high temperatures and pressures encountered in explosions of mixtures of H_2 and O_2. P_i, P_m = initial and maximum pressures, respectively; T_m = maximum temperature; α_m = value of α corresponding to P_m and T_m. The values of K have been computed by the compiler from the tabulated values of P_m, T_m, and α_m.

P_i	3	10	25	50	75	100	125	150	175	atm
P_m	23	78	198	395	602	808	1010	1200	1400	atm
T_m	2585	2630	2660	2655	2700	2715	2720	2680	2690	°K
100 α_m	2.2	1.8	1.7	1.4	1.4	1.5	1.3	1.2	1.0	
$10^6 K/C_1$	0.6	1.1	2.3	2.6	5.2	6.3	5.1	4.8	3.4	

Newitt regards these values as superior to all earlier ones. In the computation of T_m and α_m, he introduced a correction for the variation of the co-volume with the temperature.

K. F. Bonhoeffer and H. Reichardt [3] have given the following values for the dissociations $H_2O \to 0.5H_2 + OH$ (subscript OH) and $H_2O \to H_2 + 0.5O_2$ (subscript O). The latter essentially agree with those of the main table.

T	1000	1300	1500	1705	1900	2155	2505	°K
$0.5 \log_{10}(A/KRT)_{OH}$	10.5$_5$	7.1$_5$	5.7$_0$	4.5$_0$	3.6$_0$	2.7$_5$	1.7$_5$	
$0.5 \log_{10}(A/KRT)_{O}$	10.0$_5$	7.0$_0$	5.7$_1$	4.6$_4$	3.8$_0$	3.0$_4$	2.15$_5$	

7. DISSOCIATION OF VAPOR

Table 6—*(Continued)*

For values computed from molecular and spectroscopic data, see A. R. Gordon.[6]

Unit of $I = 1$ j/gfw·°K. Temp. $= T$ °K. A/KRT and K/C_1 are dimensionless.

T	1 atm $100\alpha_{obs}$	$(A/KRT)^{0.5} = 10^n B$ Value B	n	$\log_{10}$	I	n	$K/C_1 = 10^{-n}B'$ B'_{obs}	B'_{comp}	$-\log_{10}(K/C_1)$ Obs	Comp	1 atm $100\alpha_{comp}$	Ref.[a]
1325	0.00325	7.63	6	6.883	14.85	16	1.58	1.86	15.801	15.731	0.00343	La
1354	0.0049	4.12	6	6.615	15.93	16	5.30	4.82	15.276	15.317	0.00475	La
1393	0.0069	2.47	6	6.393	15.01	15	1.43	1.63	14.843	14.788	0.00719	La
1397	0.0078	2.05	6	6.312	16.04	15	2.08	1.84	14.683	14.736	0.00750	NW
1433	0.0103	1.35	6	6.131	14.99	15	4.65	5.30	14.333	14.275	0.0108	La
1455	0.0142	8.36	5	5.922	16.35	15	12.0	9.88	13.921	14.005	0.0133	La
1474	0.0141	8.45	5	5.927	14.04	14	1.16	1.66	13.936	13.779	0.0159	La
1480	0.0189[b]	5.44	5	5.736	17.00	14	2.78	1.96	13.555	13.708	0.0168	NW
1531	0.0255	3.47	5	5.540	15.08	14	6.61	7.38	13.180	13.132	0.0265	La
1550	0.0287	2.91	5	5.464	14.53	14	9.29	11.82	13.032	12.927	0.0311	La
1561	0.034	2.26	5	5.354	15.50	13	1.54	1.54	12.814	12.811	0.0341	NW
1705	0.102	4.34	4	4.638	15.60	12	3.80	3.74	11.421	11.428	0.1014	Lo
1783	0.182	1.82	4	4.260	16.36	11	2.07	1.69	10.684	10.772	0.170	Lo
1863	0.354	6.71	3	3.827	18.60[c]	10	1.46	0.69	9.838	10.159	0.277	Lo
1968	0.518	3.78	3	3.578	16.18	10	4.33	3.71	9.363	9.430	0.492	Lo
2155	1.18	1.09	3	3.039	15.31	9	4.73	4.99	8.325	8.302	1.20	W
2257	1.77	5.93	2	2.773	15.31	8	1.54	1.62	7.814	7.790	1.80	W
2337	2.8	2.95	2	2.470	17.35	8	5.98	3.84	7.222	7.412	2.43	S
2505	4.5	1.43	2	2.155	16.31	7	2.38	1.98	6.623	6.704	4.23	S
2642	4.3	1.53	2	2.185	10.70[c]	7	1.97	6.30	6.705	6.201	6.28	B
2684	6.2	8.70	1	1.940	13.95	7	6.00	8.74	6.222	6.058	7.01	S
2698	7.5	6.46	1	1.810	15.98	7	10.8	9.72	5.966	6.012	7.27	B
2731	8.2	5.65	1	1.752	16.02	6	1.40	1.04	5.854	5.905	7.89	S
2761	6.6	7.96	1	1.901	12.21[c]	6	0.70	1.55	6.157	5.811	8.49	B
2834	9.8	4.26	1	1.629	15.21	6	2.37	2.57	5.625	5.590	10.05	B
2929	11.1	3.50	1	1.544	14.15	6	3.40	4.75	5.469	5.323	12.3	B
3092	13.0	2.71	1	1.433	12.14[c]	6	5.36	12.16	5.270	4.915	16.7	S
				Mean	15.54							

B. Lewis' and J. B. Friauf's[2] formulas (see p. 27) yield the following values.

T	$(A/KRT)^{0.5} = 10^n B$ Value B	n	$\log_{10}$	$K/C_1 = 10^{-n}B'$ Value B'	n	$-\log_{10}$	1 atm 100α
1000	9.30	9	9.968	1.41	21	20.851	2.85×10^{-5}
1200	6.48	7	7.811	2.42	18	17.616	7.81×10^{-4}
1400	1.842	6	6.2652	2.57	15	14.591	8.38×10^{-3}
1600	1.267	5	5.1028	4.74	13	12.324	0.0499
1800	1.576	4	4.1976	2.73	11	10.565	0.201
2000	2.974	3	3.4734	6.89	10	9.162	0.609
2200	7.62	2	2.8817	9.55	9	8.020	1.51
2400	2.451	2	2.3894	8.45	8	7.073	3.20
2500	1.490	2	2.1733	2.194	7	6.6587	4.39
2600	94.2	0	1.9739	5.29	7	6.2769	5.91
2800	41.6	0	1.6189	2.517	6	5.5991	9.95
2800	41.4	0	1.6165	2.545	6	5.5943	9.98
2900	28.7	0	1.4577	5.11	6	5.2919	12.55
3000	20.4	0	1.3098	9.75	6	5.0109	15.4
3000	20.4	0	1.3099	9.75	6	5.0111	15.4
3200	11.0	0	1.0421	3.14	5	4.5035	22.3
3500	5.00	0	0.6988	1.39	4	3.8558	34.4
4000	1.75	0	0.2432	9.94	4	3.0026	55.0

[5] Newitt, D. M., *Proc. Roy. Soc. (London) (A)*, **119**, 464-480 (1928).
[6] Gordon, A. R., *J. Chem'l Phys.*, **1**, 308-312 (1933).

Table 6—*(Continued)*

B. Lewis and G. von Elbe,[7] have given the following values, with which those given by E. Justi and H. Lüder [8] essentially agree. Each denotes $(KRT/A)^{0.5}$ by the symbol K_p. Two types of dissociation are considered: $H_2O \rightleftarrows H_2 + \tfrac{1}{2}O_2$ and $H_2O \rightleftarrows \tfrac{1}{2}H_2 + OH$. The first is denoted in the table by the symbol H, O; the second by H, OH.

T	$(A/KRT)^{0.5} = 10^n B$ H, O B	n	H, OH B	n	$\log_{10}(A/KRT)^{0.5}$ H, O	H, OH
300	5.9	39	2	43	39.77	43.3
400	1.82	29	5	31	29.26	31.7
600	4.36	18	1	20	18.64	20.0
800	1.90	13	1.17	14	13.28	14.07
1000	1.12	10	3.39	10	10.05	10.53
1200	7.94	7	1.48	8	7.90	8.17
1400	2.19	6	2.95	6	6.34	6.47
1600	1.58	5	1.58	5	5.20	5.20
1800	1.86	4	1.55	4	4.27	4.19
2000	3.31	3	2.51	3	3.52	3.40
2200	8.13	2	5.50	2	2.91	2.74
2400	2.57	2	1.55	2	2.41	2.19
2600	1.00	2	5.50	1	2.00	1.74
2800	4.27	1	2.19	1	1.63	1.34
3000	2.04	1	9.98	0	1.31	0.999

[a] *References:*
B = Bjerrum, N., *Z. physik. Chem.*, **79**, 513-536 (1912).
La = Langmuir, I., *J. Am. Chem. Soc.*, **28**, 1357-1379 (1906).
Lo = Löwenstein, L., *Z. physik. Chem.*, **54**, 715-726 (1905).
NW = Nernst, W., and Wartenberg, H. v., *Nachr. v. d. k. Ges. Wiss. Göttingen*, **1905**, 35-45 (1905).
S = Siegel, W., *Z. physik. Chem.*, **87**, 641-668 (1914).
W = von Wartenberg, H., *Ber. deut. physik. Ges.*, **8**, 97-103 (1906).
[b] Frequently quoted incorrectly as 0.0184.
[c] Omitted in taking the mean.

Table 7.—Variation of α with KRT/P

From (19), $\alpha^3 = (1 - \alpha)^2(2 + \alpha)(KRT/P)$. If α is small, α^3 varies as $2KRT/P$. If α exceeds 4 per cent, its approximate value can be read from this table, and the computation of a more exact value is facilitated by using in the calculation the approximate value here given for $(1 - \alpha)^2(2 + \alpha)$. For example, if KRT/P is 0.645, one sees at once that α lies between 62 and 63 per cent, and that $(1 - \alpha)^2(2 + \alpha)$ **is** about 0.374. Whence $\alpha^3 = 0.374(0.645)$ or α = 62.2 per cent. **Should** greater precision be desired, use this value of α to compute $(1 - \alpha)^2(2 + \alpha)$, finding 0.3746; substitute this value in the cubic, and solve, finding α = 62.28. Take the mean of this and the value (62.2) previously found, which gives α = 62.24 per cent.

[7] Lewis, B., and von Elbe, G., *J. Am. Chem. Soc.*, **57**, 612-614 (1935).
[8] Justi, E., and Lüder, H., *Forsch. Gebiete Ingenieurw.*, **6**, 209-216 (1935).

7. DISSOCIATION OF VAPOR

Table 7—*(Continued)*

KRT/P and α are each dimensionless.

KRT/P	P/KRT	$(P/KRT)^{0.5}$	$(1-\alpha)^2(2+\alpha)$	100α
34.05×10^{-6}	29370	171.4	1.880	4
1.186×10^{-4}	8430	91.8	1.820	6
2.909×10^{-4}	3440	58.6	1.760	8
4.210×10^{-4}	2375	48.7	1.731	9
5.879×10^{-4}	1701	41.2	1.701	10
7.96×10^{-4}	1256	35.44	1.671	11
0.001051	951	30.84	1.642	12
0.001363	734	27.09	1.612	13
0.001735	576	24.00	1.582	14
0.002173	460	21.45	1.553	15
0.002688	372	19.29	1.524	16
0.003317	301	17.35	1.481	17
0.003975	251.6	15.86	1.466	18
0.004770	209.6	14.48	1.437	19
0.00568	176.0	13.27	1.408	20
0.00671	149.0	12.21	1.379	21
0.00788	126.9	11.26	1.351	22
0.00921	108.6	10.42	1.322	23
0.01076	92.9	9.64	1.294	24
0.01234	81.0	9.00	1.266	25
0.01420	70.4	8.39	1.238	26
0.01628	61.4	7.84	1.210	27
0.01857	53.8	7.33	1.182	28
0.02112	47.3	6.88	1.154	29
0.02396	41.7	6.46	1.127	30
0.02708	36.93	6.077	1.100	31
0.03054	32.74	5.722	1.073	32
0.03435	29.11	5.395	1.046	33
0.03859	25.91	5.090	1.019	34
0.04319	23.15	4.811	0.9929	35
0.04828	20.71	4.551	0.9666	36
0.0539	18.55	4.307	0.9406	37
0.0600	16.67	4.083	0.9149	38
0.0668	14.98	3.870	0.8893	39
0.0741	13.50	3.674	0.8640	40
0.0812	12.32	3.508	0.8389	41
0.0910	10.99	3.315	0.8141	42
0.1007	9.93	3.151	0.7895	43
0.1113	8.98	2.997	0.7652	44
0.1230	8.13	2.851	0.7411	45
0.1356	7.37	2.715	0.7173	46
0.1496	6.68	2.584	0.6938	47
0.1650	6.06	2.462	0.6706	48
0.1818	5.50	2.345	0.6476	49
0.2000	5.00	2.236	0.6250	50
0.2202	4.54	2.131	0.6026	51
0.2422	4.13	2.032	0.5806	52
0.2663	3.76	1.939	0.5589	53
0.2930	3.41	1.847	0.5575	54
0.3220	3.105	1.762	0.5164	55
0.3545	2.821	1.680	0.4956	56
0.3896	2.567	1.602	0.4752	57
0.4287	2.333	1.527	0.4551	58
0.4720	2.119	1.456	0.4354	59
0.5192	1.926	1.388	0.4160	60
0.5720	1.748	1.322	0.3970	61
0.6300	1.587	1.260	0.3783	62
0.6948	1.439	1.200	0.3600	63
0.7662	1.305	1.142	0.3421	64

Table 7—(Continued)

KRT/P	P/KRT	(P/KRT)^0.5	$(1-\alpha)^2(2+\alpha)$	100α
0.8460	1.182	1.087	0.3246	65
0.9352	1.069	1.034	0.3075	66
1.035	0.966	0.983	0.2908	67
1.146	0.873	0.934	0.2744	68
1.271	0.787	0.887	0.2585	69
1.411	0.709	0.842	0.2430	70
1.571	0.636	0.798	0.2279	71
1.750	0.571	0.756	0.2132	72
1.955	0.512	0.715	0.1990	73
2.189	0.457	0.676	0.1852	74
2.453	0.408	0.638	0.1719	75
2.760	0.362	0.602	0.1590	76
3.115	0.321	0.567	0.1465	77
3.522	0.284	0.533	0.1346	78
4.008	0.250	0.500	0.1230	79
4.602	0.217	0.466	0.1120	80
9.575	0.104	0.323	0.0641	85

Table 8.—Thermal Dissociation of Water-vapor: Effect of Pressure and Temperature

The data in this table have been computed by means of equations (19) and (20) with $I_d = 15.54$, the mean of Table 6; they are valid only to the extent to which the constants in equation (20) are valid. The compiler expresses no opinion regarding their validity at the higher temperatures, but the values of α here given essentially agree with those given by N. Bjerrum [9] in a similar table covering the same domain. See also Newitt's values at the head of Table 6. G. Chaudron [10] has expressed the opinion that the specific heats of the gases at high temperature are not known with sufficient precision to enable one to infer with certainty the dissociation at such temperatures.[11]

P = total pressure; α = fraction of H_2O-vapor dissociated when the analytical composition of the mixed gases is that of H_2O; K = constant of dissociation (see Table 6); C_1 = concentration of 1 gfw/l.

Unit of $P = 1$ atm. K/C_1 is dimensionless. Temperature = T °K.

T	$P \to$ $-\log_{10}(K/C_1)$	0.1	1	10	100
			100α		
1000	21.964	0.000056	0.000026	0.0000121	0.0000056
1500	13.478	0.0434	0.0202	0.00936	0.00434
2000	9.225	1.24	0.579	0.269	0.125
2500	6.724	8.77	4.17	1.96	0.914
3000	5.139	27.7	14.1	6.85	3.24
3500	4.113	51.1	29.4	15.1	7.34
4000	3.465	67.8	44.1	24.4	12.2
4500	3.098	76.6	54.7	31.7	16.5
5000	2.950	80.	58.7	35.6	18.8

[9] Bjerrum, N., *Z. physik. Chem.*, **79**, 513-536 (1912).
[10] Chaudron, G., *Bull. soc. chim. de France* (4), **37**, 657-679 (1925).
[11] See also Gordon, A. R., and Barnes, C., *J. Phys'l Chem.*, **36**, 1143-1151 (1932).

Photochemical Dissociation of Water-vapor.

A. Coehn and G. Grote [12] have reported that under the radiation from a Hg-arc in a quartz tube the equilibrium amount of dissociation of water-vapor at a pressure of 0.825 atm is the same at 240 °C as at 150 °C, *i.e.*, 0.124 per cent, which is the same as for thermal dissociation at about 1730 °C. They give the following values for 150 °C, P being the initial pressure of the water-vapor, and α being the fraction dissociated when equilibrium is attained:

P	0.825	0.585	0.448	0.370	atm
α	0.1237	0.1992	0.2960	0.3618	per cent
$P\alpha$	0.102	0.116	0.132	0.135	

From these data, which may be represented by the formula $P(\alpha + 0.0681) = 0.1600$, the observers infer that the reaction is of the first order. Two hours were required for establishing equilibrium.

R. S. Mulliken [13] has suggested that the strong continuous absorption band beginning suddenly near $\lambda = 1800A$ and initiating the ultraviolet absorption of water-vapor represents a dissociation or a predissociation process.

H. Senftleben and I. Rehren [14] and J. R. Bates and H. S. Taylor [15] observed a dissociation into H and OH when a mixture of the vapors of water and of mercury was illuminated by the light from a suitable line of the mercury spectrum (*cf.* p. 60).

A. Tian [16] is occasionally quoted as having studied the photochemical dissociation of water-vapor, but the object of his investigations was liquid water, and he was of the opinion that the vapor was not involved (*cf.* p. 35).

A. Terenin and H. Neujmin [17] and H. Neujmin and A. Terenin [18] have reported that under the action of radiation in the range $\lambda = 1300$ to $1500A$, the excited OH radical is split off from the water-vapor molecule, carrying an abnormally great amount of rotational energy; and that the threshold energy for that dissociation is 207 kcal/gfw-H_2O (= 866 kj/gfw-H_2O).

Ionic Dissociation of Water-vapor. See also Sections 10 and 24.

In 1925, F. Hund [19] calculated the energy that would be required to remove one or both H-ions from H_2O constructed like his atomic model (p. 45), finding 370 ± 30 and 920 ± 40 kcal/gfw, respectively. These values are equivalent to 16.0 and 39.8 electron-volts per molecule or 1550 and 3850 kj/gfw. Recently, J. C. Slater [20] has concluded that the energy of such a triangular molecule is -9.3 electron-volts.

---

[12] Coehn, A., and Grote, G., Nernst, "Festschrift," p. 136-167, 1912.
[13] Mulliken, R. S., *J. Chem'l Phys.*, **1**, 492-503 (1933).
[14] Senftleben, H., and Rehren, I., *Z. Physik.*, **37**, 529-538 (1926).
[15] Bates, J. R., and Taylor, H. S., *J. Am. Chem. Soc.*, **49**, 2438-2456 (1927).
[16] Tian, A., *Compt. rend.*, **152**, 1012-1014 (1911); *Ann. de Phys.* (9), **5**, 248-365 (1916).
[17] Terenin, A., and Neujmin, H., *J. Chem'l Phys.*, **3**, 436-437 (L) (1935).
[18] Neujmin, H., and Terenin, A., *Acta Physicochim. URSS*, **5**, 465-490 (1936).

K. F. Bonhoeffer [21] and K. F. Bonhoeffer and H. Reichardt [3] have obtained the following values: $H_2 + 2OH = 2H_2O + 128$ kcal; $H + H = H_2 + 101$ kcal. They concluded that the dissociation $H_2O \rightarrow H + OH$ requires 111 or 115 kcal/gfw = 464 or 481 kj/gfw = 4.8 or 5.0 electron-volts per formula molecule. But E. Gaviola and R. W. Wood [22] found that a greater value, probably about 5.2 electron-volts per molecule, is required for

Table 9.—Dissociation of Water-vapor in the Glow Discharge
(See also Table 75)

A. Güntherschulze and H. Schnitger [29] give the following data for the dissociation $2H_2O \rightleftarrows 2H_2 + O_2$ in the glow discharge, primarily in the cathode portion. Partial pressure of H_2O is p_1, of the $(2H_2 + O_2)$ is p_2; T °K is the temperature at which the thermal dissociation at those pressures would equal that actually observed in the glow discharge at the temperature t °C; α is the dissociation coefficient as already defined (p. 26).

Unit of p_1 and of $p_2 = 1$ mm-Hg; α is dimensionless. Temp. = t °C; for T see heading.

t	p_1	p_2	100α	T
−47.0	42	6.1	8.8	1915
−42.0	74	18.1	14.0	2095
−37.0	134	46.6	18.9	2170
−32.0	227	70.0	17.1	2190
−27.0	387	78.0	11.8	2120
−22.0	636	117.4	11.0	2140
−17.0	1027	145.	8.6	2115
−12.0	1627	329.	11.9	2245
0.0	4579	1078.	13.6	2380
+17.0	14530	5720.	24.9	2750
18.2	15670	4850.	17.1	2600
22.6	20560	10070.	24.8	2810

$H_2O \rightarrow H + OH$, and that such dissociation occurs only once in about ten thousand collisions between the molecules of H_2O and of excited Hg. And more recently, H. Senftleben and O. Reichemeier [23] have reported 5.05 ± 0.04 electron-volts for $H_2O \rightarrow H + OH$, and that the value increases with the temperature.

R. S. Mulliken [24] has concluded from thermal data that the work per H-atom required to ionize H_2O completely is $110 \times 4.185 = 460$ kj, or 920 kj/gfw-H_2O. M. Magat [25] from spectroscopic data derived for $H_2O \rightarrow 2H^+ + O^-$ the energy $(269 \pm 3) \times 4.185 = 1126$ kj/gfw, and from thermal data $218 \times 4.185 = 912$ kj/gfw, whence he concluded that the primary products of the thermal dissociation are not normal atoms of H and O, but

---

[19] Hund, F., *Z. Physik*, **32**, 1-19 (1925).
[20] Slater, J. C., *Phys. Rev.* (2), **38**, 1109-1144 (1931).
[21] Bonhoeffer, K. F., *Z. Elektrochem.*, **34**, 652-654 (1928).
[22] Gaviola, E., and Wood, R. W., *Phil. Mag.* (7), **6**, 1191-1210 (1928).
[23] Senftleben, H., and Reichemeier, O., *Physik. Z.*, **34**, 228-230 (1933).
[24] Mulliken, R. S., *Phys. Rev.* (2), **40**, 55-62 (1932).
[25] Magat, M., *Compt. rend.*, **197**, 1216-1220 (1933).

normal H and activated (O_{1_D}) oxygen. He stated that this had been previously suggested by Haber and Bonhoeffer.

D. W. Mueller and H. D. Smyth [26] have reported that an electronic bombardment of H_2O-vapor yields negative O-ions and OH-ions, as well as negative H-ions. The number of the H-ions formed per electron is a maximum when the energy of the electrons is about 8 electron-volts; whereas the O-ions and the OH-ions both continue to increase in number with the energy as it exceeds 20 electron-volts.

The energy expended in producing the dissociation accompanying a glow discharge in water-vapor is 11 electron-volts per molecule of H_2O dissociated.[27]

The chemical reactions of water-vapor dissociated by an electrical discharge have been studied by H. C. Urey and G. I. Lavin.[28]

8. Dissociation of Water

Photochemical Dissociation of Water.

When water (liquid) is exposed to the radiation from a quartz-enclosed mercury arc, only H_2 is initially freed, the H_2O being converted into H_2O_2; later, from the dissociation of the accumulated H_2O_2, O_2 appears; and finally equilibrium is established, the liberated gas having the composition $2H_2 + O_2$. There is no decomposition of pure water unless the radiation contains waves shorter than 1900A. Between 8 and 20 °C the velocity of decomposition increases as the temperature rises, averaging 1.37 per cent per degree C.[16]

Ionic Dissociation of Water (See also Electrolytic Ionization, Section 50).

The heat Q_{ion} absorbed in the ionization of water ($H_2O \rightarrow H^+ + OH^-$) has been expressed by formulas equivalent to the following:

$(Q_{ion})_{Ro} = 57.370 - 0.242(t-18) + 0.00063(t-18)^2$ kj/gfw,
$\quad 10 \lesssim t \lesssim 35$ °C [30];

$(Q_{ion})_{Rl} = 122.16 - 0.2216T$ kj/gfw [31];

$(Q_{ion})_{LG} = 57.07 - 0.217(t-20)$ kj/gfw, temperatures near 20 °C [32],

$(Q_{ion})_{HH} = 91.760 - 5.9352(T/100) - 1.9862(T/100)^2$ kj/gfw.[33]

These formulas lead to the values in Table 10.

[26] Mueller, D. W., and Smyth, H. D., *Phys. Rev.* (2), **38**, 1920 (A) (1931).
[27] Linder, E. G., *Phys. Rev.* (2), **38**, 679-692 (1931).
[28] Urey, H. C., and Lavin, G. I., *J. Am. Chem. Soc.*, **51**, 3290-3293 (1929).
[29] Güntherschulze, A., and Schnitger, H., *Z. Physik*, **103**, 627-632 (1936).
[30] Rossini, F. D., *Bur. Stand. J. Res.*, **6**, 847-856 (RP309) (1931).
[31] Randall, M., *Int. Crit. Tables*, **7**, 232 (1930).
[32] Lambert, R. H., and Gillespie, L. J., *J. Am. Chem. Soc.*, **53**, 2632-9 (1931).
[33] Harned, H. S., and Hamer, W. J., *J. Am. Chem. Soc.*, **55**, 2194-2206 (1933).

Table 10.—Heat of Ionization of Water

Unit of $Q_{\text{Ion}} = 1$ kj/gfw-H_2O; temp. $= t$ °C $= T$ °K.

t	0	10	15	18	20	25	30	40	50	60
T	273.1	283.1	288.1	291.1	293.1	298.1	303.1	313.1	323.1	333.1
$(Q_{\text{Ion}})_{\text{Ro}}$	....	59.35	58.10	57.37	56.89	55.71	54.56	(52.35)		
$(Q_{\text{Ion}})_{\text{RI}}$	61.64	59.43	58.32	57.65	57.21	56.10	54.99	52.78	50.56	48.35
$(Q_{\text{Ion}})_{\text{LG}}$	....		(58.16)	57.50	57.07	(55.99)				
$(Q_{\text{Ion}})_{\text{HH}}$	60.74	59.05	58.18	57.65	57.30	56.42	51.85	53.71	51.85	49.95

Table 11.—Disposable Energy in the Ionization of Water

The greatest amount of external work that can be obtained from an isothermal ionization of liquid water ($H_2O \rightarrow H^+ + OH^-$) at atmospheric pressure A is $W_{\text{ion}} + A\Delta v$; $A\Delta v$ is negligible, $W_{\text{ion}} = -122.16 - 0.2216 T \log_e T + 1.40457 T$ kj/gfw-H_2O [34]; $w_{\text{ion}} = W_{\text{ion}}/18.0154$. As W_{ion} is negative, work is absorbed during the process. The temperature is t °C, $T = 273.1 + t$.

Unit of $W_{\text{Ion}} = 1$ kj/gfw; of $w_{\text{Ion}} = 1$ kj/g.

t	W_{Ion}	w_{Ion}
0	− 78.07	− 4.334
10	− 78.72	− 4.370
15	− 79.07	− 4.389
20	− 79.44	− 4.410
25	− 79.82[a]	− 4.431
30	− 80.23	− 4.453
40	− 81.10	− 4.502
60	− 83.05	− 4.610
80	− 85.26	− 4.733
100	− 86.90	− 4.824

[a] R. F. Newton and M. G. Bolinger [35] have reported $W_{\text{Ion}} = -79.92$ kj/gfw at 25 °C.

Alpha-ray Dissociation of Water.

The decomposition of water by alpha particles, and the nature of the products formed, have been studied by C. E. Nurnberger.[36]

[34] Randall, M., *Int. Crit. Tables*, **7**, 232 (1930).
[35] Newton, R. F., and Bolinger, M. G., *J. Am. Chem. Soc.*, **52**, 921-925 (1930).
[36] Nurnberger, C. E., *J. Phys'l Chem.*, **38**, 47-69 (1934); *J. Chem'l Phys.*, **4**, 697-702 (1936).

II. Single-phase Systems

IIA. WATER-VAPOR

9. MOLECULAR DATA FOR WATER-VAPOR

Some miscellaneous data for water-vapor and the numerical values accepted for certain constants appearing throughout this section are given in Table 12, and certain kinetic data for each of a series of temperatures extending from 1 °K to 3000 °C are given in Table 13. More detailed information is given in the following sections and tables. For theory and data consult Boltzmann,[1] Meyer,[2] Jeans,[3] and Dushman.[4]

Mean Free Path of Molecules of Water-vapor.

The effective mean free path (L_e) of a molecule of a gas is derived from the observed viscosity (η) by means of the relation $L_e = 3\eta/\rho\bar{v} = 3(\pi R/8M)^{1/2} \cdot (\eta T^{1/2}/p) = 3N(\pi/8RM)^{1/2} \cdot (\eta/nT^{1/2})$. When all quantities are expressed in cgs units, this becomes, for water-vapor, $L_e = 4039(\eta T^{1/2}/p) = 2.944(\eta/nT^{1/2}) \cdot (10^{19})$. Here ρ = density, $\bar{v}$ = mean translational speed of thermal agitation of the molecules, R = gas constant per g-mole, M = molecular weight, p = pressure, T °K = absolute temperature, N = number of molecules per g-mole, n = number of molecules per unit of volume.

These relations are derived from the simple kinetic theory of gases devoid of intermolecular attraction. In that case the free path of a molecule between two consecutive collisions is straight, and L_e is a certain average length of all such free paths. In actual gases there is intermolecular attraction, which causes the paths to be curved, increasing the frequency of collision and reducing the lengths of the individual paths. In such cases the value of L_e found by means of the relations just given is the mean free path in a gas having the existing values of M, η, ρ, and $\bar{v}$, but devoid of intermolecular attraction. It may be called the effective mean free path of the actual gas.

The curvature produced in the path by attraction between the molecules will, obviously, decrease as the translational kinetic energy of the molecules increases. Wherefore W. Sutherland[5] suggested that $L_e = L_s/(1 + CT^{-1})$, where L_s is what would be the corresponding mean free path in the actual gas if there were no intermolecular attraction. It may be called the Suther-

---

[1] Boltzmann, L., "Vorlesungen über Gastheorie," Leipzig, J. A. Barth, 1896, 1898.
[2] Meyer, O. E., "Die Kinetische Theorie der Gase," Breslau, Maruschke & Berendt, 1899.
[3] Jeans, J. H., "The Dynamical Theory of Gases," Cambridge Univ. Press, 1921.
[4] Dushman, S., *Gen. Elec. Rev.*, **18**, 952-958, 1042-1049, 1159-1168 (1915).
[5] Sutherland, W., *Phil. Mag.* (5), **36**, 507-531 (1893).

Table 12.—Miscellaneous Molecular Data for Water-vapor

From the values, R = gas-constant per g-mole = 8.315 10^7erg/(g-mole·°K, N = Avogadro's number = number of molecules per g-mole = 6.061 × 10^{23}, M = molecular weight of H_2O = 18.0154, and η = viscosity of water-vapor = 96 micropoise (= 96 × 10^{-6} g/cm·sec) at 20 °C, and on the assumption that the molecule of water-vapor is H_2O, the following values are obtained either directly or from the kinetic theory of gases. The first three, designated as A, B, and D, respectively, are products that occur frequently in expressions derived from the kinetic theory.

Unit of pressure (p) = 1 barye = 1 dyne/cm²; of viscosity = 1 poise = 1 g/cm·sec. Temperature is T °K.

Symbol	Quantity		At T °K.	At 20 °C.	Unit
A	$A \equiv (2\pi RM)^{½}$	($\log_{10}A$ = 4.98684; cgs units)	970.2	970.2	1 m·sec⁻¹·°K⁻½
B	$B \equiv (\pi/8RM)^{½} = \pi/2A$	($\log_{10}B$ = $\bar{5}$.20928; cgs units)	1.619	1.619	10^{-5} sec·°K^½·cm⁻¹
D	$D \equiv (R/M)^{½}$	($\log_{10}D$ = 3.33211; cgs units)	2148	2148	1(erg/g·°K)^½
M	Molecular weight		18.0154	18.0154	1 g/g-mole
N	Avogadro's number = number of molecules per g-mole		6061	6061	10^{20}
R	Gas-constant per g-mole		8.315	8.315	10^6 erg/g-mole·°K
m	Mass of one molecule = M/N		29.72	29.72	10^{-24} g
m_s	Mass striking 1 cm² per sec = mn_s = $p(M/2\pi RT)^{½}$		$185.7pT^{½}$	$10.85p$	10^{-6} g
n_g	Number of g-moles per cm³ = p/RT		$1202.6pT^{-1}$	$4.10p$	10^{-11}
n	Number of molecules per cm³ = Nn_g		$728.9pT^{-1}$	$2.49p$	10^{13}
n_s	Number of molecules that strike 1 cm² of the wall of the container in 1 sec = $n\bar{v}/4 = nP/AT^{½}$		$62.4pT^{-½}$	$3.65p$	10^{17}
$\bar{v}$	Mean velocity of the molecules = $D(8T/\pi)^{½}$ (See Table 13)		$34.28T^{½}$	587	1 m/sec
v_2	Square root of the mean squared velocity of the molecules = $D(3T)^{½} = \bar{v}(3\pi/8)^{½}$ (See also Table 13)		$37.21T^{½}$	637	1 m/sec

9. VAPOR: MOLECULAR DATA

E_m	Average translational kinetic energy of thermal agitation per molecule = $0.5\,mv_1^2 = 1.5RT/N$ (See Table 13)	$2.057T$	602	10^{-16} erg
E_g	As for E_m except that E_g is referred to the g-mole = $NE_m = 1.5RT$	$1.247T$	366	10^8 ergs
d_e	Effective diameter of a molecule = $(2KAT^{1/2}/3\pi^2N\eta)^{1/2}$ (See *Molecular size*, p. 41).	$104.0(KT^{1/2}/\eta)^{1/2}$	$4.4K^{1/2}$	10^{-8} cm
d_s	Sutherland diameter = $d_e/(1 + CT^{-1})^{1/2}$ (See *Molecular size*, p. 41). If $C = 650\,°K$, then d_s is	$104.0\{KT^{1.5}/(T + 650)\eta\}^{1/2}$ 1.058	$2.4K^{1/2}$ 1.058	10^{-8} cm
K	See Table 14 for various values. Preferred value			
L_e	Effective mean free path of a molecule $= 3\eta/\rho\bar{v} = 3NB\eta/nT^{1/2}$ (See also p. 37) Also, $3\eta/\rho\bar{v} = 3D\eta(\pi T/8)^{1/2}/p$	$2.944(10^6\eta/nT^{1/2})$ $4039\eta T^{1/2}/p$	$16.51/n$ $6.6/p$	10^{13} cm 1 cm
L_s	Sutherland mean free path = $L_e(1 + CT^{-1})$ (See p. 37) If $C = 650\,°C$	$L_e(1 + 650T^{-1})$ $29.44(10^6\eta/T^{1/2})$	$21.7/p$ 165	1 cm 10^{12} cm
nL_e	Total length of initial free paths in 1 cm³ = $3NB\eta/T^{1/2}$			10^{12} cm
nL_s	Total length of initial Sutherland paths in 1 cm³ = $nL_e(1 + CT^{-1})$. If $C = 650\,°K$, nL_s is		531	10^{12} cm
ν	Number of collisions per cm³ per sec = $n/\tau = n\bar{v}/L$: $\nu_e = n\bar{v}/L_e = 8Np^2/3\pi R\eta T$	$6.19p^2/\eta T$	$219p^2$	10^{15}
τ	Mean free time = $L/\bar{v}$; $\tau_e = L_e/\bar{v} = 3\pi\eta/8p$	$1178\eta/p$	$0.113/p$	10^{-3} sec
η	Viscosity of water-vapor. Value at 20 °C is taken as		96	10^{-6} g/cm·sec

40 PROPERTIES OF ORDINARY WATER-SUBSTANCE

land mean free path. The "Sutherland constant" (C) is characteristic of the gas, and measures the intensity of the intermolecular attraction. This suggestion of Sutherland's leads to the relation $\eta = AT^{1.5}/(T + C)$, where $A \equiv nL_s(8RM)^{1/2}/3N\pi^{1/2}$ is a constant. This relation agrees quite closely with the observations on many gases over wide ranges in T, the pressure not exceeding a few atmospheres. But as the pressure increases, the isothermal value of η for water-vapor increases (see Section 11), and the Sutherland relation ceases to hold. This means that nL_s is not a constant, as the simple theory demands, but varies with the pressure, the variation depending upon T.

Unless qualified in some way, the term "mean free path" is ambiguous. In common practice, it usually means either L_e or L_s. For water-vapor

Table 13.—Kinetic Data for Molecules of Water-vapor

Information regarding the symbols not defined here and the numerical values of the basic constants will be found in Table 12.

If $\rho =$ density, $p = \frac{1}{3}\rho(v_2)^2 = RTn_g$; whence $(v_2)^2 = 3RT/M$. If the distribution of velocities is Maxwellian, $\bar{v} = v_2\sqrt{8/3\pi} = 0.9213\,v_2$. Insofar as the principle of equipartition of energy applies, each g-mole of a resting gas, containing only one type of molecule and in thermal equilibrium, contains an amount of kinetic energy equal to $0.5RT$ for each of the degrees of freedom characteristic of the molecule.

$m =$ mass of one molecule of $H_2O = 29.72 \cdot 10^{-24}$g; $(v_2)^2 = 1.3846T \cdot 10^7$(cm/sec)^2; $v_2 = 3721.1\sqrt{T}$ cm/sec; $\bar{v} = 3428.2\sqrt{T}$ cm/sec; $E_m =$ average translational kinetic energy of thermal agitation of a molecule $= 20.57T \cdot 10^{-17}$erg; $E_g =$ the same kinetic energy of a g-mole $= NE_m = 12.472T \cdot 10^7$ ergs.

Unit of $\bar{v}$ and $v_2 = 1$ m/sec; of $E_m = 10^{-16}$ erg; of E_g and of $0.5RT = 10^8$ ergs.

t	T	$\bar{v}$	v_2	E_m	E_g	$0.5RT$
−272.1	1	34.3	37.2	2.1	1.2	0.42
−271.1	2	48.5	52.6	4.1	2.5	0.83
−268.1	5	76.6	83.2	10.3	6.2	2.08
−263.1	10	108.4	117.7	20.6	12.5	4.16
−250	23.1	164.8	178.8	47.5	28.8	9.60
−200	73.1	293.1	318.1	150.4	91.2	30.39
−150	123.1	349.5	379.4	253.2	153.5	51.18
−100	173.1	451.0	489.6	356.1	215.9	71.97
−50	223.1	512.0	555.8	458.9	278.2	92.7
0	273.1	566.5	614.9	562.	340.6	113.5
25	298.1	591.9	642.5	613.	371.8	123.9
100	373.1	662.2	718.8	767.	465.3	155.1
200	473.1	745.7	809.4	973.	590.0	196.7
300	573.1	820.7	890.8	1179.	715.	238.3
500	773.1	953.2	1034.6	1590.	964.	321.4
1000	1273.1	1223.2	1327.7	2619.	1588.	529.3
1500	1773.1	1443.5	1566.9	3647.	2211.	737.2
2000	2273.1	1634.5	1774.1	4676.	2835.	945.0
3000	3273.1	1961.3	2128.9	6733.	4082.	1360.8

9. VAPOR: MOLECULAR DATA

nL_e is about 1.6×10^{14} cm at room temperature and saturation pressure, and increases with both pressure and temperature to about twice that value at the limits of the experimental values for the viscosity; the corresponding values of nL_s are about 5 and 8×10^{14} cm, varying not greatly with T but subject to variations in the chosen value of C (see Table 21). At a pressure of 1 barye (1 dyne/cm²) L_e varies from about 6 cm at 0 °C to about 20 cm at 350 °C, the corresponding values of L_s being 20 and 45 cm respectively.

Molecular Size.

A general discussion of the various methods available for estimating the sizes of molecules, atoms and ions, together with numerical values and a bibliography of 153 titles has been published by K. F. Herzfeld.[6]

In the simple kinetic theory of gases, the term *molecular diameter* is used to indicate the diameter of the equivalent elastic spheres by which the molecules may be regarded as replaced when one is concerned solely with the mechanical effects arising from the thermal agitation of the molecules of the gas when in a state closely approximating the ideal. This diameter is an average minimum value of the distance between the centers of colliding molecules under the stated conditions. It is an average, because the minimum distance may vary from one collision to another, because the molecules have various velocities, are neither necessarily spherical nor surrounded by fields having spherical symmetry, and possibly interpenetrate, more or less, at each collision.

Insofar as the molecules may be replaced by elastic spheres between which there are no forces except when the spheres are in actual collision, the path between two consecutive collisions will be straight, and the average value (L) of all such paths will be related to d, the diameter of the spheres, as indicated by the equation $nL = K/\pi d^2$, n being the number of molecules per unit volume and K being a number determined by the distribution of the velocities of the molecules and by the kind of average free path represented by L. Some of the values that have been computed for K are given in Table 14.

It seems to be agreed that Chapman's highest value ($K = 1.058$) is the most appropriate for use in formulas involving the viscosity. But all the other values tabulated have been used, and molecular diameters so computed are not infrequently tabulated without any indication of the value of K to which they refer, even though different entries refer to different values, and some refer to d_e (computed from L_e) and some to d_s (computed from L_s).

The area πd^2 is often called the cross-sectional area of the sphere of action. But when that term is used, d having the significance here considered, it should be so qualified as to indicate that it refers solely to those intermolecular collisions that occur as the result of thermal agitation; for

[6] Herzfeld, K. F., *Jahrb. d. Radioak.*, 19, 259-334 (1922).

the radius of the sphere of action may have a different value in relation to other events. From the observed viscosity of water-vapor it may be inferred that d_e is of the order of 3 to 5A (1A = 10^{-8} cm), and d_s of 2A; the corresponding values of πd^2 are of the orders of 28, 78, and 13A^2, respectively.

Table 14.—Various Values Assumed for $K = \pi d^2 nL$

K is a pure number.

Condition	K
I. Gas at rest; L = mean value of the free paths:	
(a) All molecules, except one, continuously at rest	1.
(b) All molecules have the same speed, chaotically distributed	0.750
(c) Maxwellian distribution of velocities $K = 1/\sqrt{2}$	0.707
II. Gas streaming lamellarly; various types of averaging, each intended to be that appropriate to the equation $\eta = \frac{1}{3}\rho \bar{v} L$:	
(a) Distribution of velocities strictly Maxwellian.[1] $K = 3(0.3502)/\sqrt{2} = 1.0506/\sqrt{2}$	0.743
(b) Maxwellian distribution as modified by the streaming[2] $K = 3(0.3097)/\sqrt{2} = 0.9290/\sqrt{2}$	0.657
(c) Taking into account the persistence of the streaming velocity (it was ignored by Boltzmann and by Meyer), and assuming a Maxwellian distribution as modified by streaming, J. H. Jeans[3] finds by one approximation $K = 1.317/\sqrt{2}$	0.931
and by another $K = 1.382/\sqrt{2}$	0.977
(d) By a somewhat different procedure, S. Chapman[7] finds for condition (c) $K = 3(0.491)(1+\epsilon)/\sqrt{2} = 1.473(1+\epsilon)/\sqrt{2} = 1.042(1+\epsilon)$, where $(1+\epsilon)$ depends upon the intermolecular attraction, and lies between 1.000 and 1.016. If $\epsilon = 0$, K is	1.042
If $\epsilon = 0.016$, K is	1.058

Unfortunately, the term *molecular diameter* is not infrequently used in other senses. This has led to great confusion, and not infrequently to comparison of values that are not comparable. Some of these other uses are illustrated in Table 15. Each of the values there tabulated has been called molecular radius. Some have been called radius of molecular action, and some half the radius of molecular action.

[7] Chapman, C., *Phil. Trans., (A)*, **216**, 279-348 (1916).

Table 15.—Estimates of the Effective Size of the Molecule of Water-vapor

r = "radius" of the molecule as derived in the manner indicated. If the procedure defines a diameter (d), then $r = d/2$. For additional explanations, see text.

Unit of $r = 1A = 10^{-8}$ cm.

Basis	r
Viscosity at 100 °C and 1 atm; $r = d_e/2$	2.08
Viscosity at 100 °C and 1 atm; $r = d_s/2$	1.25
Van der Waals' equation, assuming that the co-volume b is 4 times the volume of all the molecules in a g-mole, that the critical pressure and temperature are $P_c = 217.7$ atm. and $T_c = 273 + 374$ °C $= 647$ °K, and that $b = RT_c/8P_c$	1.44
As before, except that $b = RT_c/15P_c$	1.17
Kamerlingh-Onnes' equation of state and Jakob's data for the density. G. Holst [8] finds	3.2
The relation $\pi d^3 n/6 = (\epsilon - 1)/(\epsilon + 2)$, ϵ = dielectric constant	2.9
If in the liquid state the molecules were cubes and were closely packed at 4 °C, then the length of the edge of each cube would be $e = (M/N\rho)^{1/3} = 2r$	1.55
If in the liquid state the molecules were spheres and were closely packed at 4 °C (tetrahedral packing), then $d = e \cdot 2^{1/6} = 2r$	1.74
If δ is mean distance between the centers of adjacent molecules of a liquid in which the velocity of sound is V and the thermal conductivity is λ, then $\delta = (2kV/\lambda)^{1/2} = 2r$, where k is the Boltzmann gas constant $= R/N = 1.372 \times 10^{-16}$ (erg/°K) per molecule. For water at 4 °C, $\lambda = 5.61 \times 10^4$ erg/cm·sec·°C, $V = 1425$ m/sec.	1.32
Assuming that the effective radius of a molecule of mercury vapor is 1.80A (the value computed from the viscosity, but whether in terms of L_e or of L_s, and if the latter, for what temperature, is not stated, M. W. Zemansky [9] determined the greatest value of $r = \Delta - 1.80$A that is consistent with the production of a stated effect upon the radiation emitted by the mercury molecule, Δ being the distance between the center of a water-vapor molecule and that of a colliding molecule of mercury vapor. His data yield the following values:	
For quenching the radiation	−0.80
For depolarizing the radiation	+5.37
For broadening the spectral lines	6.48

[8] Holst, G., *Proc. Akad. Wet. Amsterdam*, **19**, 932-937 (1917).
[9] Zemansky, M. W., *Phys. Rev.* (2), **36**, 919-934 (1930).

Table 15—*(Continued)*
Basis

If Δ is the greatest mean distance between the center of a molecule of water-vapor and that of an electron, consistent with the molecule's essential blocking of the advance of the electron, then the value of Δ for each of several values of the kinetic energy of the colliding electron can be determined from a curve published by E. Brüche.[10] They are as follows, Δ being taken as r:

Energy = 4	electron·volts	2.24
Energy = 7.8	electron·volts	2.32
Energy = 9	electron·volts	2.39
Energy = 16	electron·volts	2.30
Energy = 25	electron·volts	2.03
Energy = 36	electron·volts	1.83

Formulas connecting the diameter of a molecule of a vapor with the surface tension and other properties of the corresponding liquid have been derived from assumptions that have not been generally accepted. Thus:

S. Mokroushin [11] obtains a value, which after correction to the basis of 6.06×10^{23} molecules per g-mole, yields.... 2.72

S. P. Owen [12] obtains 1.62

In the triangular molecule (Table 16) the distance from the center of the oxygen to that of either hydrogen nucleus is approximately 1.00

Moments of Inertia of the Molecule of Water-vapor.

Band spectra are attributed to the rotation of the molecule. The fundamental frequencies, from which those of the maxima of the several bands are obtained by summations and differences, are determined by the frequency of rotation of the molecule, as are also the constant frequency-differences between the consecutive lines of a band and between the components of certain doublets. The quantum theory establishes a relation between these frequencies, or differences in frequency, and the changes in the rotational energy of the molecule. From that, the pertinent moment of inertia of the molecule can be obtained.

For example, ignoring all complications, the quantum theory requires that the integral of the angular momentum over a complete cycle shall be an integral (n) multiple of the Planck constant of action ($h = 6.56 \ 10^{-27}$ erg·sec); that is, $4\pi^2 I \bar{\nu} = nh$, I being the effective moment of inertia, and $\bar{\nu}$ the frequency of the rotation of the molecule. Hence, the rotational energy is $E = \frac{1}{2}I(2\pi\bar{\nu})^2 = \dfrac{1}{8\pi^2 I}(h^2 n^2)$. As n changes from one integer

---

[10] Brüche, E., *Ann. d. Physik*, (5), **1**, 93-134 (1929).
[11] Mokroushin, S., *Phil. Mag.* (6), **48**, 765-768 (1914).
[12] Owen, S. P., *Proc. Univ. Durham Phil. Soc.*, **6**, 308-311 (1932).

9. VAPOR: MOLECULAR DATA

to the next, E changes by $\Delta E = \dfrac{h^2}{8\pi^2 I}(2n+1)$; and on the quantum theory, this must equal $h\nu$, where ν is the frequency of the associated radiation. Whence it is evident that each unit change in n causes ν to change by $\Delta\nu = h/4\pi^2 I$, which gives rise to a spectrum of lines spaced at equal intervals of frequency. From the observed values of $\Delta\nu$, I can be computed:

$$I = h/4\pi^2 \Delta\nu = 1.66_1/\Delta\nu = 55.4/(^1/\lambda_1 - {}^1/\lambda_2)\ 10^{-40}\ \text{g·cm}^2$$

the unit of λ being 1 cm, and of ν being 1 wave/sec. In other cases the procedure is somewhat similar. In every case I is the effective moment of inertia under the conditions characteristic of the vibrations by means of which it has been obtained; it differs from that pertaining to the static molecule.[13, 14] An early review of the subject was published by A. Eucken,[15] and a later one may be found in a monograph by C. Schaefer and F. Matossi.[16]

The absorption spectrum of water-vapor is so complex that until recently its interpretation has been very incomplete and subject to dispute, but the difficulties are now being rapidly overcome. Three different moments of inertia are involved, indicating that the atoms in a molecule lie at the vertices of a triangle.[13, 14, 15; 16, p. 235-245; 17, 18, 19, 20, 21] The values of the moments of inertia as derived by F. Hund [22] from the observations by Eucken [15] have been much quoted and used; they are $I_A = 0.98$, $I_B = 2.25$, $I_C = 3.20$ 10^{-40}g·cm^2, or $I_A = 0.59$, $I_B = 1.34$, $I_C = 1.91$ proton·angstrom.[2] But probably the best values are those given by K. Freudenberg and R. Mecke [14] and based on the structure of 17 water-vapor bands studied by Mecke and his associates. They are as follows, the unit being $10^{-40}\ \text{g·cm}^2 = 0.5969\ m_H A^2$, m_H indicating the mass of a hydrogen atom:

$$I_1 = 0.996 + 0.045\sigma + 0.026\pi - 0.098\delta$$
$$I_2 = 1.908 + 0.014\sigma + 0.033\pi - 0.034\delta$$
$$I_3 = 2.981 + 0.047\sigma + 0.062\pi + 0.062\delta$$
$$\Delta = 0.077 - 0.012\sigma + 0.003\pi + 0.194\delta$$

where $\Delta \equiv I_3 - (I_1 + I_2)$, and (σ, π, δ) are the quantum integers involved in the vibration considered. For the stationary (unvibrating) state they give $I_1 = 1.009$, $I_2 = 1.901$, $I_3 = 2.908$.

The values found by P. Lueg and K. Hedfeld [21] from a study of 3 bands are $I_A = 0.97$, $I_B = 2.13$, $I_C = 3.07$, but R. Mecke and W. Baumann [13] think that the data they used are not satisfactory.

[13] Mecke, R., and Baumann, W., *Physik. Z.*, **33**, 833-835 (1932).
[14] Freudenberg, K., and Mecke, R., *Z. Physik*, **81**, 465-481 (1933).
[15] Eucken, A., *Jahrb. d. Radioak.*, **16**, 361-411 (1920).
[16] Schaefer, C., and Matossi, F., "Das Ultrarote Spektrum," Berlin, Julius Springer, 1930.
[17] Witt, H., *Z. Physik*, **28**, 249-255 (1924).
[18] Mecke, R., *Physik. Z.*, **30**, 907-910 (1929).
[19] Mahanti, P. C., *Physik. Z.*, **32**, 108-110 (1931).
[20] Mecke, R., *Z. Physik*, **81**, 313-331 (1933).
[21] Lueg, P., and Hedfeld, K., *Z. Physik*, **75**, 512-520 (1932).

From the data available in 1927, R. T. Birge [23] derived the following values for the OH-ion, h being taken as $6.557 \ 10^{-27}$erg·sec and c as $2.99796 \ 10^{10}$cm/sec; moment of inertia $= 1.634 \ 10^{-40}$g·cm^2 at the upper state of excitation, and 1.500 at the lower, the corresponding separation of the nuclei being 1.022A and 0.979A. On the same basis, but from more recent data, D. Jack [24] derived essentially the same values for the moments of inertia, i.e., 1.633 and 1.498 10^{-40}g·cm^2. While from their own measurements of a newly discovered band, H. L. Johnston, D. H. Dawson, and M. K. Walker [25] derive for the states 2_Σ and 2_π, respectively, the values $10^{40} I = 1.591$ and 1.454 g·cm^2, and the separations 1.009A and 0.964A.

A review of the methods available for estimating the moments of inertia and certain related data has been published by A. Eucken [15]; more detailed treatments may be found in various treatises, such as C. Schaefer and F. Matossi's "Das Ultrarote Spectrum." The derivation of formulas required for the interpretation of the spectrum has been published by F. Lütgemeier,[26] H. A. Kramers and G. P. Ittmann,[27] D. M. Dennison,[28] and others.

Dipole Moment of the Molecule of Water-vapor.

Several distinct phenomena exhibited by some substances, but not by others, can be most satisfactorily explained by assuming that the molecules involved in them contain rigid, or nearly rigid, electrical dipoles, the distance between the poles being significantly less than the diameter of the molecule. Water-vapor exhibits phenomena that can be explained in this way. Discussions of the general subject have been published by P. Debye [29]; and by the Faraday Society,[30] and of certain aspects of it by R. S. Mulliken.[31]

In 1903, M. Reinganum [32] attributed intermolecular forces to such dipoles, and on that assumption computed the moments that must be assigned to them for each of a number of liquids. But the magnitude of these moments can be obtained more directly from a study of the dielectric constants, as has been pointed out by P. Debye.[33]

He showed that if each molecule contains a rigid electrical dipole as

---

[22] Hund, F., *Z. Physik*, **31**, 81-106 (1925).
[23] Birge, R. T., *Int. Crit. Tables*, **5**, 415 (1929).
[24] Jack, D., *Proc. Roy. Soc. (London) (A)*, **118**, 647-654 (1928).
[25] Johnston, H. L., Dawson, D. H., and Walker, M. K., *Phys. Rev. (2)*, **43**, 473-480 (1933)
→ *(2)*, **43**, 374 (A) (1933).
[26] Lütgemeier, F., *Z. Physik*, **38**, 251-263 (1926).
[27] Kramers, H. A., and Ittmann, G. P., *Idem*, **53**, 553-565 (1929); **58**, 217-231 (1929); **60**, 663-681 (1930).
[28] Dennison, D. M., *Rev. Mod. Physics*, **3**, 280-345 (1931).
[29] Debye, P., "Handb. d. Radiol." (E. Mark, Ed.), **6**, 597-786 (1925); "Polar Molecules," New York Chemical Catalogue Co. (Reinhold Publishing Corp.), 1929.
[30] The Faraday Society, *Trans.*, **30**, 679-904 (1934).
[31] Mulliken, R. S., *J. Chem'l Phys.*, **3**, 573-585 (1935).
[32] Reinganum, M., *Ann. d. Physik (4)*, **10**, 334-353 (1903).
[33] Debye, P., *Physik. Z.*, **13**, 97-100, 295 (1912).
[34] Debye, P., "Polar Molecules," p. 8, New York, Chemical Catalog Co. (Reinhold Publishing Corp.), 1929.

9. VAPOR: MOLECULAR DATA

well as electrons that are elastically bound, relation (1) should be satisfied,

$$\frac{\epsilon - 1}{\epsilon + 2} \cdot \frac{1}{\rho} = a + \frac{b}{T} \tag{1}$$

the intermolecular action of the fields due to the molecules themselves being considered to a first approximation only. In this equation, ϵ = dielectric constant, ρ = density, a and b are essentially positive constants characteristic of the substance, and T is the absolute temperature. The first term (a) depends upon the elastically bound electrons and determines the optical dispersion of the substance; the second constant (b) depends upon the strength, or moment, of the dipole. If the electrical quantities are expressed in electrostatic units, then $a = \dfrac{4\pi}{3} \cdot \dfrac{N}{M} \alpha$ and $b = \dfrac{4\pi N}{9kM} \mu^2$, where N = number of molecules per mole, M = molecular weight, k = molecular gas-constant = R/N, R being the gas-constant per mole, μ = moment of the dipole, and α the moment induced by unit field. The quantity α is called the "polarizability"[34] or the "deformability"[35] of the molecule. $\alpha \equiv e^2 \Sigma_p (n/f)_p$ where e = electronic charge, n_p = number, per molecule, of the elastically bound electrons for which the force of restitution is f_p per unit displacement, and Σ_p = summation for all values of p. When in an electrostatic field of strength E, such a molecule will have an induced electric moment of amount $\mu' = \alpha E$. The value of a is related to the index of refraction (n) and to the density (ρ) as follows: $a\rho = (n^2 - 1)/(n^2 + 2)$.

Taking the unit of $\rho = 1$ g/cm^3, $N = 6.06_1 \times 10^{23}$ molecules per g-mole, $M = 18.0154$, $k = 1.372 \, 10^{-16}$ erg per molecule and per °K (I.C.T. values), we find for water-vapor:

$$10^{24} \alpha = 7.10 \sqrt{a} \text{ cgse units}$$
$$10^{20} \mu = 5.40 \sqrt{b} \text{ cgse units}$$

R. Gans[36] has considered in greater detail the effect of the intermolecular fields upon ϵ; and convenient methods for using the more complicated equations so obtained have been given by H. Isnardi,[37] P. Lertes,[38] and C. P. Smyth.[39] But the simpler formulas given by Debye are amply accurate for our present purposes.

The available data for water-vapor are shown in Figure 2, Section 22. $(\epsilon - 1)T/\rho$ being plotted against T. As $(\epsilon + 2)$ is practically equal to 3, equation (1) requires that the points shall lie on a right line, sloping upward as T increases, and cutting the axis of ordinates at $(\epsilon - 1)T/\rho = 3b$. The data published by J. D. Stranathan[40] seem to be by far the most satis-

---

[35] Rao, I. R., *Indian J. Phys.* **2**, 435-465 (1928).
[36] Gans, R., *Ann. d. Physik (4)*, **64**, 481-512 (1921).
[37] Isnardi, H., *Physik. Z.*, **22**, 230-233 (1921).
[38] Lertes, P., *Z. Physik*, **6**, 257-268 (1921).
[39] Smyth, C. P., *Phil. Mag. (6)*, **45**, 849-864 (1923); *J. Am. Chem. Soc.*, **46**, 2151-2166 (1924).
[40] Stranathan, J. D., *Phys. Rev. (2)*, **48**, 538-549 (1935) → **47**, 794(A) (1935), extending **45**, 741 (1934).

factory. They lead to $b = 1149.6 \pm 7.8$, giving

$$10^{18}\,\mu = 1.83_1 \text{ cgse}$$
$$\mu = 0.383 \text{ electron·angstrom}$$

Another recent set of data is that by L. G. Groves and S. Sugden,[41] leading to $10^{18}\,\mu = 1.84 \pm 0.01$ cgse.

Of earlier data, the most satisfactory set is that published by R. Sänger and O. Steiger [42]; but the published details are not sufficient to enable one to form an independent estimate of the accuracy of the final results. They lead to a value of $10^{18}\,\mu$ lying between 1.84 and 1.88 cgse, a little greater than Stranathan's value.

The observations of C. T. Zahn [43] are not inconsistent with these values for μ, but are too scattered to justify an attempt to derive an independent value from them. The same is true of M. Jona's,[44] which are admittedly affected by an unknown percentile error.

The data of K. Bädeker,[45] though frequently quoted, and used by G. Holst [46] in his computation of μ, are plainly in error and are worthless for that purpose, as they impose a negative value upon the essentially positive a.

Less directly obtained estimates have been made, based upon more or less questionable assumptions and in some cases involving rough empirical relations. Thus, G. Holst [46] has computed from Kamerlingh-Onnes' equation of state, as fitted to Jakob's data for the density, the value $10^{18}\,\mu = 2.6$ cgse; whereas J. K. Syrkin [47] has concluded that $10^{20}\,\mu = 1.66 T_{\text{crit}}/(P_{\text{crit}})^{1/2}$, the unit of P being 1 atm, thus finding for water-vapor $10^{18}\,\mu = 0.73$ cgse. (Syrkin does not state what units he uses, but they seem to be as here given; however, certain of the relations that he gives appear to be mutually inconsistent, no matter what probable guess is made regarding the units.)

A. Kirrmann [48] has reviewed the various methods that have been proposed for estimating the value of μ, giving numerical values for various substances and a bibliography of 42 entries. Numerous errors occur in both tabulation and bibliography, and in at least some instances data derived from observations on the liquid phase by procedures now known to be incorrect are not distinguished from those derived satisfactorily from observations on the corresponding gas phase. A more recent table of dipole

---

[41] Groves, L. G., and Sugden, S., *J. Chem. Soc. (London)*, **1935**, 971-974 (1935).
[42] Sänger, R., and Steiger, O., *Helv. Phys. Acta*, **1**, 369-384 (1928); republished by Sänger, R., *Physik. Z.*, **31**, 306-315 (1930), and by Sänger, R., Steiger, O., and Gächter, K., *Helv. Phys. Acta*, **5**, 200-210 (1932).
[43] Zahn, C. T., *Phys. Rev.* (2), **27**, 329-340 (1926).
[44] Jona, M., *Physik. Z.*, **20**, 14-21 (1919).
[45] Bädeker, K., *Z. physik. Chem.*, **36**, 305-335 (1901).
[46] Holst, G., *Proc. Akad. Wet. Amsterdam*, **19**, 932-937 (1917).
[47] Syrkin, J. K., *Z. anorg. allgem. Chem.*, **174**, 47-56 (1928).
[48] Kirrmann, A., *Rev. gén. des Sci.*, **39**, 598-603 (1928).

moments, collected by N. V. Sidgwick, has been published by G. C. Hampson and R. J. G. Marsden.[49]

Polarizability of the Molecule of Water-vapor.

The term "polarizability" of the molecule is used by Debye to denote that portion of the molecular electric moment induced per unit electrical field as a result of the displacement of elastically bound electrons. It is the quantity already denoted by α, and is related to the a of equation (1) as follows: $a = (4\pi N/3M)\alpha$, which in the case of water-vapor becomes $10^{24}\alpha = 7.10a$ cgse units (see preceding topic: *Dipole Moment*). Both Raman and Rao call it the (mean) "deformability" of the molecule. The quantity a is related to the index of refraction (n) and the density (ρ) in this manner: $a\rho = (n^2 - 1)/(n^2 + 2)$.

From the last stated relation, I. R. Rao[35] has computed the value $10^{24}\alpha = 1.50$ cgse. From the a (0.224) derived from Stranathan's observations (1935) (Section 22), we find

$$10^{24}\alpha = 1.59 \text{ cgse}$$
$$\alpha = 1.59 \text{ electron·angstrom per } (e/A^2) \text{ field}$$
$$= 1.59 A^3$$

That is, if a molecule of water-vapor is placed in an electro-static field having n times the strength of the field that exists at the distance of 1 angstrom (10^{-8} cm) from an isolated electron (charge $= e = 4.774 \times 10^{-10}$ cgse), then that field will induce in the molecule an electric moment equivalent to that of two charges of opposite signs, each equal to e, separated by the distance of $1.59n$ angstroms, n being assumed to be so small that the induced moment is sensibly proportional to the strength of the inducing field.

This polarizability is generally ascribed almost exclusively to the oxygen. The polarization of the oxygen by its attendant hydrogens gives rise to a very significant negative component in the permanent dipole moment of the water-vapor molecule. See P. Debye,[34, p. 68+] and I. R. Rao.[35]

Anisotropy of the Molecule of Water-vapor.

Light scattered at an angle of 90° by water-vapor, and by certain other gases and vapors, is not completely polarized. This suggests that the polarization induced in the molecule by the incident light depends upon the orientation of the molecule, the latter being optically anisotropic. Similarly, any existence of electric or of magnetic double refraction would indicate that the molecule is anisotropic with reference to those forces also. For theoretical treatment of the subject, see T. H. Havelock[50]; C. V. Raman and K. S. Krishnan,[51] F. Hund,[52] and P. Debye.[53]

---

[49] Hampson, G. C., and Marsden, R. J. G., *Trans. Faraday Soc.*, **30**, appendix, 86 pp. (1934).
[50] Havelock, T. H., *Phil. Mag. (7)*, **3**, 158-176 (1927).
[51] Raman, C. V., and Krishnan, K. S., *Idem*, 713-723, 724-735 (1927).

If the electrical moments induced along the principal axes of the molecule by unit external fields parallel to those axes are A, B, and C, then, in the notation of Raman and of Rao

δ = factor measuring the anisotropy of the molecule =

$$\frac{A^2 + B^2 + C^2 - (AB + BC + CA)}{(A + B + C)^2}$$

r = depolarization factor = ratio of the weaker plane-polarized component of the transversely scattered light to the stronger one.

A, B, C = deformabilities of the molecule.

a, b, c = deformabilities of the atom or ion.

α = (mean) deformability of the molecule = $\frac{3}{4\pi N_1}\left(\frac{n^2 - 1}{n^2 + 1}\right)$, n = index of refraction, N_1 = number of molecules per unit volume. Debye calls α the polarizability.

When magnetic anisotropy is to be considered, the components of the susceptibility (or magnetic deformability) are referred to the same axes as A, B, and C; and are commonly denoted by A', B', and C'. Similarly in other cases.

For water-vapor, $B = C$, approximately; whence, writing $\Delta \equiv A/C$, $\delta =$

$$\left(\frac{\Delta - 1}{\Delta + 2}\right)^2; \; r = \frac{2(\Delta - 1)^2}{4(\Delta - 1)^2 + 5(2\Delta + 1)}; \; \alpha = \frac{A + 2C}{3} = C\left(\frac{\Delta + 2}{3}\right).$$

I. R. Rao [54] found $r = 0.0191$, $r_0 = 0.0199$; $\delta = 0.0166$ at 120 °C, r_0 being the value corresponding to a very low density. In a later paper [55] Rao suggested that the atoms and ions may themselves be anisotropic, and concluded that the anisotropy of O^{--} is greater in H_2O than in CO_2, and "seems to be anomalous." He attributed the entire anisotropy of H_2O to the O^{--}, and gave for it $\Delta = A/C = a/c = 1.45$ to 1.53.

All of this refers to the optical anisotropy of the molecule. There appears to be no available data from which the magnetic anisotropy of a molecule of water-vapor can be computed.

It should be remarked that the notation in this field is confused. For example, S. W. Chinchalkar [56] defines the optical anisotropy as

$$\delta = \frac{(A - B)^2 + (B - C)^2 + (C - A)^2}{(A + B + C)^2}$$

which is twice the value used by Raman and Rao.

[52] Hund, F., *Z. Physik*, **43**, 805-826 (1927).
[53] Debye, P., "Handb. d. Radiologie" (E. Marx, Ed.), **6**, 754-776, Leipzig, 1925.
[54] Rao, I. R., *Indian J. Phys.*, **2**, 61-96 (1928).
[55] Rao, I. R., *Idem*, **2**, 435-465 (1928).
[56] Chinchalkar, S. W., *Indian J. Phys.*, **6**, 165-179 (1931).

Models of the Molecule of Water-vapor.

Descriptive models of the molecule have been derived from chemical data. These need not detain us beyond mentioning that C. Friedel,[57] J. W. Brühl,[58] H. E. Armstrong,[59] T. M. Lowry and H. Burgess,[60] J. Piccard,[61] C. P. Smyth,[62] M. Smith,[63] M. L. Huggins,[64] and others have presented reasons for supposing that oxygen is tetravalent, that the electrons of the oxygen atom have a tetrahedral distribution about the nucleus, and that the H_2O-molecule is a tetrahedron, two of its vertices being occupied by H and two by electrons serving as bonds for secondary valencies.

On this basis, S. W. Pennycuick[65] has constructed models for ice and for water.

Another mode of approach is by way of physical data. This leads to numerical results and to an idea of the probable stability of the model proposed. Only two types of model are possible. Either the three molecules lie in a straight line, or they do not; the first gives a linear, and the second a triangular model. The latter is in qualitative accord with that mentioned in a preceding paragraph.

It may be shown that the linear model is either non-polar or unstable.[34, pp. 63-68] Hence this model, which was considered by T. H. Havelock[50] and proposed by F. J. v. Wisniewski,[66] must be condemned, as we know that the H_2O-molecule is polar (see *Dipole Moment*, p. 46).

The triangular model is stable [34, pp. 68-76] if the triangle is isosceles, the oxygen being at the unique vertex, and if the oxygen is so polarizable that $\alpha/r^3 > 1/8$, where r is the distance from the nucleus of the oxygen atom to that of either hydrogen atom, and α is the polarizability (p. 47). Furthermore, if the unique internal angle of the triangle is 2θ, $\alpha = (1/8)(r/\sin\theta)^3 = (r^2/b)^3$ and the total electric moment of the molecule is $\mu = 2er\cos\theta(1 - 1/8\sin^3\theta) = 2ea(1 - r^3/b^3)$, where a is the altitude of the triangle, and b the base.

Such a triangular arrangement has been discussed by W. Heisenberg,[67] M. Born and W. Heisenberg,[68] F. Hund,[22] T. H. Havelock,[50] P. Debye,[34, pp. 63-68] A. S. Coolidge,[69] and others. Regarded as a rigid body, it has three principal moments of inertia: I_a about the axis along a, I_b about that parallel to b and passing through the center of mass of the system, and I_c about the axis perpendicular to the plane of the three atoms

---

[57] Friedel, C., *Bull. Soc. Chim. de France (N.S.)*, **24**, 160-169, 241-250 (1875).
[58] Brühl, J. W., *Ber. deut. chem. Ges.*, **28**, 2866-2868 (1895) ← *Z. physik. Chem.*, **18**, 514-518 (1895); *Ber. deut. chem. Ges.*, **30**, 162-172 (1897).
[59] Armstrong, H. E., *Compt. rend.*, **176**, 1892-1894 (1923).
[60] Lowry, T. M., and Burgess, H., *J. Chem. Soc. (London)*, **123**, 2111-2124 (1923).
[61] Piccard, J., *Helv. Chim. Acta*, **7**, 800-802 (1924).
[62] Smyth, C. P., *Phil. Mag. (6)*, **47**, 530-544 (1924).
[63] Smith, M., "Chemistry and Atomic Structure," 1924.
[64] Huggins, M. L., *Phys. Rev. (2)*, **27**, 286-297 (1926).
[65] Pennycuick, S. W., *J. Phys'l Chem.*, **32**, 1681-1696 (1928).
[66] v. Wisniewski, F. J., *Z. Physik*, **47**, 567-568 (1928).
[67] Heisenberg, W., *Z. Physik*, **26**, 196-204 (1924).
[68] Born, M., and Heisenberg, W., *Idem*, **23**, 388-410 (1924).
[69] Coolidge, A. S., *Phys. Rev. (2)*, **42**, 187-209 (1932).

and passing through the center of mass. If M = mass of an oxygen atom = 26.40 10^{-24}g, m = that of a hydrogen atom = 1.676 10^{-24}g, and a and b are expressed in centimeters, then $I_a \equiv 0.5mb^2 = 0.838b^2$ 10^{-24}g·cm^2; $I_b \equiv 2Mm\, a^2/(M+2m) = 2.974a^2$ 10^{-24}g·cm^2; and $I_c = I_a + I_b$. Whence, $10^{-12}a = 0.580(I_b)^{0.5}$, $10^{-12}b = 1.092(I_a)^{0.5}$, $\tan\theta = b/2a = 0.942\,(I_a/I_b)^{0.5}$.

Spectral data show that the molecule H_2O has indeed three principal moments of inertia, and furnish the means for determining their values (Tables 64 and 65). The largest is evidently I_c, but it is not obvious which of the others should be assigned to I_a. Of the two possibilities, that which assigns the smaller value to I_a gives θ the smaller value, but neither assignment leads to values of α and μ, computed by means of the Debye formulas just given, that accord with the experimental values (see Table 16). Indeed, W. G. Penney and G. B. B. M. Sutherland [70] have concluded that attempts to obtain the exact form of the molecule from spectroscopic data are unprofitable. But numerous attempts of that kind have been made. From them it was at first concluded that I_a, and therefore θ, should have the smaller of the two possible values, this choice yielding values of

Table 16.—Constants of the Triangular Model of H_2O

The atoms are at the vertices of an isosceles triangle, the O being at the unique vertex of angle 2θ; a = altitude, b = base joining the two H's, r = slant height of the triangle. The principal moments of inertia are I_a about the axis parallel to a, I_b about that parallel to b, and I_c about that perpendicular to the plane of the triangle, all three axes passing through the center of mass of the system. For a rigid system $I_c = I_a + I_b$; for a non-rigid system, $I_c = I_a + I_b$ for the unvibrating molecule, and $I_c = I_a + I_b + \Delta$ for the vibrating molecule. The I's, being obtained from spectroscopic data, refer to the vibrating molecule and Δ is a positive quantity, generally differing from zero. In that case, Mecke and his associates derived θ from the values of I_a and I_b, but r from $I_1 + \Delta$ and I_2, regarding $I_1 + \Delta$ as playing the part played by I_1 in the static molecule; here I_1 is the smaller and I_2 is the greater of the moments I_a and I_b. For the static molecule H_2O and cgs-units, $10^{-24}a^2 = 0.3364 I_b$, $10^{-24}b^2 = 1.194 I_a$, $10^{-24}r^2 = 0.3364 I_b + 0.2985 I_a$, $\tan\theta = 0.9416(I_a/I_b)^{0.5}$; for the associated triangle obtained by interchanging the values assigned to I_a and I_b, the elements are $a' = b/1.884$, $b' = 1.884a$, $r' = r\left[1.1269 - 0.2395\left(\dfrac{a}{r}\right)^2\right]^{0.5}$, $\tan\theta' = 0.8873 \cot\theta$. Until recently, it was not obvious from the spectral data which of the two smaller moments should be assigned to I_b, but Mecke has concluded that I_b must be the smaller, which makes $2\theta > 90°$. Previously, the other choice was preferred. The constants of both of the possible triangles are here tabulated; some of them have been computed by the compiler from the data given in the sources indicated.

---

[70] Penney, W. G., and Sutherland, G. B. B. M., *Proc. Roy. Soc. (London) (A)*, **156**, 654-678 (1936).

Table 16—(Continued)

The polarizability (α) and the dipole moment (μ) as computed from these values by means of Debye's formulas (see p. 51) $\alpha = (r^2/b)^3$, $\mu = 2ea[1 - (r/b)^3]$, do not agree with the observed values. For the obtuse-angled triangles $10^{24}\alpha$ varies from 0.2 to 0.3 cm^3 and $10^{18}\mu$ from 3.9 to 4.4 cgse; for the acute-angled ones, $10^{24}\alpha$ varies from 0.3 to 1 cm^3, and $10^{18}\mu$ from 0.7 to 3.2 cgse. The observed values are $10^{24}\alpha = 1.59$ cm^3, $10^{18}\mu = 1.83$ cgse.

Unit of $r, a, b = 10^{-8}$ cm = 1A

Source[a]	Obtuse angle				Acute angle			
	θ	r	a	b	θ[b]	r	a	b
FM	52.3°	0.952	0.583	1.507	34.5°	0.969	0.800	1.098
FM'	52.5[c]	0.965	0.601	1.509	34.2[c]	0.982	0.801	1.132
LH	54.5	0.98	0.57	1.6	32.3	1.00	0.849	1.074
VC	50	1.00	0.64	1.53	36.7	1.01	0.812	1.206
EH	55.2	1.00	0.57	1.64	31.7	1.02	0.870	1.074
ES	55	1.02	0.58	1.67	32	1.07	0.91	1.13
W	56	0.99	0.55	1.63	30.9	1.02	0.865	1.036
M	48	0.86	0.58	1.28	38.6	0.87	0.680	1.093
Pd	60							
Pl	57.5							

[a] *Sources:*

EH = A. Eucken [15] and F. Hund [74a]; $I = 0.98$, $I = 2.25$, $I = 3.20 \; 10^{-40}$ g·cm^2.
ES = E. Eucken,[15] R. Sänger, and O. Steiger.[42]
FM = K. Freundenberg and R. Mecke.[14] Stationary state, $I_1 = 1.009$, $I_2 = 1.901$, $I_3 = 2.908$ 10^{-40} g·cm^2.
FM' = Same as FM except that it is for the vibrating state $\sigma = \pi = \delta = 0$, $I_1 = 0.996$, $I_2 = 1.908$, $I_3 = 2.981$, $I_3 - (I_1 + I_2) = 0.077$; θ is derived from $I_1 = 0.996$ and $I_2 = 1.908$; r, a, and b from $I_1' = 0.996 + 0.077 = 1.073$ and $I_2 = 1.908$.
LH = P. Lueg and K. Hedfeld.[21]
M = R. Mecke.[75]
Pd = J. Piccard.[61]
Pl = E. K. Plyler.[76]
VC = J. H. van Vleck and P. C. Cross.[77]
W = Spectral data of H. Witt [78]; $I_1 = 0.91$, $I_2 = 2.23$, $I_3 = 3.14$; the values of the I's and of the data in the table were computed by the compiler.

[b] From observations on the spectrum of laterally scattered light, J. Cabannes and A. Rousset [79] infer that 2θ [$\theta(?)$] is about 23°, which does not agree with the other values.

[c] Here the θ was computed from the $I_1 = 0.996$ and $I_2 = 1.908$, while a, b, and r were computed from $I_1' = I_1 + 0.077 = 1.073$ and $I_2 = 1.908$, which corresponds to another value of θ.

α and μ that are the nearer to the experimental ones. But from an extended mathematical treatment J. C. Slater [71] has concluded that if the triangular configuration is to be in equilibrium, θ must slightly exceed 45°, which lies

---

[71] Slater, J. C., *Phys. Rev.* (2), **38**, 1109-1144 (1931).
[72] Wilson, E. B., Jr., *J. Chem'l Phys.*, **4**, 526-528 (1936); Randall, H. M., Dennison, D. M., Ginsburg, N., and Weber, L. R., *Phys. Rev.* (2), **52**, 160-174 (1937).
[73] Mulliken, R. S., *Phys. Rev.* (2), **40**, 55-62 (1932).
[74] Mulliken, R. S., *J. Chem'l Phys.*, **1**, 492-503 (1933); Idem, **3**, 506-514, 586-591 (1935).
[74a] Hund, F., *Z. Physik*, **32**, 1-19 (1925).
[75] Mecke, R., *Physik. Z.*, **30**, 907-910 (1929).
[76] Plyler, E. K., *Phys. Rev.* (2), **38**, 1784 (L) (1931); **39**, 77-82 (1932).
[77] van Vleck, J. H., and Cross, P. C., *J. Chem'l Phys.*, **1**, 357-361 (1933).
[78] Witt, H., *Z. Physik*, **28**, 249-255 (1924).
[79] Cabannes, J., and Rousset, A., *Compt. rend.*, **194**, 706-708 (1932).

between the two values fixed by the I's, and nearer the larger one. From spectral data, Mecke [13, 14] likewise concluded that only the larger value of the angle is satisfactory. Several sets of values that have been assigned to the elements of the triangular model are given in Table 16.

So far we have proceeded as if the molecule were rigid. It is not. Rotation causes distortion which is sufficient to produce marked spectroscopic effects under certain conditions.[72]

The view now coming into favor is that of R. S. Mulliken,[73] who writes: "In general no attempt is made to treat the molecule as *consisting of* atoms or ions. Attempts to regard a molecule as consisting of specific atomic or ionic units held together by discrete numbers of bonding electrons or electron-pairs are considered as more or less meaningless, except as an approximation in special cases, or as a method of calculation. ... A molecule is here regarded as a set of *nuclei,* around each of which is grouped an electron configuration closely similar to that of a free atom in an external field, except that the outer parts of the electron configurations surrounding each nucleus usually belong, in part, jointly to two or more nuclei..." The symmetry of the water molecule is that of an isosceles triangle, and the electron configuration, in spectroscopic notation, is given by him as $1s^2 2s^2 2pa^2 2pb^2 2pc^2$. "The order in which the symbols are written is that of decreasing firmness of binding." See also his article on electronic structure.[74]

Association of the Molecules of Water-vapor.

Far from saturation, water-vapor behaves like an ideal gas with molecules of the composition H_2O.

As saturation is approached, both the density and the specific heat of water-vapor increase with abnormal rapidity, which indicates something of the nature of an association that increases as saturation is approached. The amount of association cannot be great, however, for the total departure of the density from that of an ideal gas is only a few per cent unless the pressure is high (see Section 14). The correct interpretation of these observations is difficult and not without a considerable degree of arbitrariness, as similar effects arise from the intermolecular forces that are taken into account by van der Waals' equation of state, though molecular aggregations caused by them are generally thought to be too transitory to be considered as associated molecules. Whether this distinction is justified is not entirely clear. As a result, interpretations differ, and it is necessary to consider in each case the assumptions on which the interpretation rests. In reference to the general subject, papers by J. W. Ellis [80] and by E. J. M. Honigmann [81] are of interest.

It is generally assumed that if there is an association, the composition of the vapor is indicated by the expression $(1 - x)H_2O + x(H_2O)_2$,

---

[80] Ellis, J. W., *Phys. Rev. (2)*, **38**, 693-698 (1931).
[81] Honigmann, E. J. M., *Die Naturwiss.*, **20**, 635-638 (1932).

x being the fraction of the molecules which are double. The ratio of the mass of double molecules to the total mass of vapor is $2x/(1 + x)$.

From his observations at pressures not exceeding 80 per cent of that corresponding to saturation, and generally much lower, T. Shirai [82] concluded that $x = 0$; he also concluded that the high values of x that had been computed by E. Bose [83] are not acceptable. The latter conclusion is reached also by A. W. C. Menzies.[84] Much earlier, S. Weber [85] had concluded that the molecular weight of water-vapor is 20 at -80 °C, and M. Knudsen [86] that it was 21.1 at -75 °C.

Assuming that the entire departure of the density of water-vapor from that of the ideal gas of molecular weight 18.0154 is due to the presence of double molecules, changing in number with the temperature and the pressure, H. Levy [87] constructed an equation of state which satisfactorily represented the available data for the density and the specific heat; and W. Nernst [88] derived in the same manner the extent of the association in the saturated vapor.

On the same assumption, but by a different procedure, A. Battelli [89] had previously derived from his own observations corresponding, but markedly different, values, those for the higher temperatures being impossible.

H. L. Callendar [90] concluded that the density of superheated water-vapor can be expressed by an equation of state of the form $p(v - b) = RT - v_c p$, in which $v_c = 26.3(373/T)^{10/3}$ cm^3/g. He calls v_c the "coaggregation volume," but does not interpret it more particularly; he takes $b = 1$ cm^3/g. If v_c is interpreted as the amount by which the specific volume is reduced by the formation of $(H_2O)_2$ from H_2O, then $x = v_c/(v - b)$. M. Jakob has discussed Callendar's ideas and theory in some detail,[91] and so has J. H. Awbery.[92]

O. Maass and J. H. Mennie,[93] following a different procedure, found still other values.

Values of x corresponding to each of these several procedures are given in Table 17.

In a short note presented before the American Physical Society, H. T. Barnes and W. S. Vipond [94] announced observations indicating that the

---

[82] Shirai, T., *Bull. Chem. Soc., Japan*, **2**, 37-40 (1927).
[83] Bose, E., *Z. Elektrochem.*, **14**, 269-271 (1908).
[84] Menzies, A. W. C., *J. Am. Chem. Soc.*, **43**, 851-857 (1921).
[85] Weber, S., *Comm. Phys. Lab. Leiden*, **150**, 3-52 (1915).
[86] Knudsen, M., *Ann. d. Physik. (4)*, **44**, 525-536 (1914).
[87] Levy, H., *Verh. deut. physik. Ges.*, **11**, 328-335 (1909).
[88] Nernst, W., *Idem*, **11**, 313-327, 336-338 (1909).
[89] Battelli, A., *Ann. chim. phys. (7)*, **3**, 408-431 (1894).
[90] Callendar, H. L., "Properties of Steam," 1920.
[91] Jakob, M., *Engineering (London)*, **132**, 143-146, 651-653, 684-686, 707-709 (1931).
[92] Awbery, J. H., *Rep. Prog. Phys. (Phys. Soc. London)*, 161-197 (1934).
[93] Maass, O., and Mennie, J. H., *Proc. Roy. Soc. (London) (A)*, **110**, 198-232 (1926).
[94] Barnes, H. T., and Vipond, W. S., *Phys. Rev.*, **28**, 453 (A) (1909). See also, Barnes, H. T., "Ice Engineering," pp. 32-33, Montreal, Renouf Publ. Co., 1928.

vapor arising from dry ice is initially polymerized to the same extent as ice itself, but quickly breaks down to water-vapor of the usual type, with the absorption of about 80 calories of heat per gram (335 joules/g).[94] It seems that the details of this work have not been published. No other article suggesting anything of the kind has come to the compiler's attention.

From a study of the variation in the Raman spectrum with the density and temperature, S. A. Ukholin [95] has concluded that the molecules of water-vapor vibrate as if uninfluenced by their neighbors if the mean distance between them exceeds 10A (0.001 μ), corresponding to a density of 0.03 g/cm^3. But if the mean distance does not exceed 8A ($\rho \lessapprox 0.06$ g/cm^3) then there is an interaction; some of the molecules remain in the immediate vicinity of others for an interval that is long as compared with the period of the spectral vibration.

Table 17.—Estimates of the Extent of the Molecular Association of Saturated Water-vapor
(See text* for references and remarks)

The composition of the vapor is assumed to be $(1-x)H_2O + x(H_2O)_2$, $2x/(1+x)$ and $(1-x)/(1+x)$ being the fractions of the total mass that consist of double and of single molecules, respectively.

Source→ Temperature = t °C.	Battelli 1894	Nernst 1909	Callendar	Maass and Mennie, 1926
		100x		
0	0.6	0.05	0.04	
10	0.7	0.08	0.06	
20	1.0	0.14	0.10	
50	1.2	0.41	0.35	
98			1.6[a]	0.95[a]
100	1.8	1.75	1.6	
108			1.4[a]	0.71[a]
200	7.7		9.5	
350	> 200		60.	

[a] Pressure is 1 atm.
*For the very high estimates at −70 to −80 °C. by Knudsen and by Weber see p. 55.

---

10. Interaction of Water-vapor and Corpuscular Radiation

Alpha Particles.*

In water-vapor, the range of the alpha particles from polonium is 0.77 times their range in air at the same temperature and pressure.[96] As their range in air at 0 °C and 1 atm is 3.72 cm [97] and varies inversely as the density, their range in water-vapor at T °K and a pressure of p mm-Hg is

* Data from Kleeman, R. D., *Int. Crit. Tables*, 1, 370 (1926).
[95] Ukholin, S. A., *Compt. rend. Acad. Sci. URSS*, 16, 395-398 (1937).
[96] v. d. Marwe, C. W., *Phil. Mag.* (6), 45, 379-381 (1923).
[97] Geiger, H., *Z. Physik*, 8, 45-57 (1921).

10. VAPOR: CORPUSCULAR RADIATION

$R = 7.98T/p$ cm. Whence have been computed the following ranges (R) in water-vapor saturated at the temperature t:

t	0	15	20	25	50	100	°C
$R*$	476	180	133	100	27.9	3.92	cm

* More recent observations [98] indicate that the range of these particles in air at 0 °C and 1 atm is 3.690 ± 0.005 cm; hence R for water-vapor is in each case 0.81 per cent smaller than the value here tabulated.

Electrons.†

When an electron strikes a molecule it may become attached to it, if the velocity of the electron is low, forming a negative ion; but if that velocity is great, the molecule is more or less disrupted, or ionized, and radiation may be emitted.

Ionization potential and energy.—The energy required to ionize a molecule is generally expressed in terms of the potential difference (I) through which an electron must pass in order to be able to cause the ionization. That difference I is called the ionizing, or ionization, potential. If I is expressed in volts, the energy required for the ionization is I electron·volts = $1.59 I \times 10^{-19}$ joule per ionized molecule = $96.4 I$ kj (= $23.0 I$ kcal) per gfw-H_2O.

When water-vapor is bombarded by electrons, ions of numerous types are formed, depending upon the energy of the electrons. As this was not at first recognized, and the nature of the ions formed was not determined, it is not surprising that some marked differences exist between the values of the ionization potential reported by the various early workers.

Finding that the transfer $A^+ + H_2O \rightarrow H_2O^+ + A$ occurs so readily that it is impossible to eliminate the ion H_2O^+ however carefully the argon is dried, H. D. Smyth and E. C. G. Stueckelberg [104] concluded that either there is another ionization potential (well above 13 volts) corresponding to the removal of a different electron, or the collision in this case occasions an excitation to a higher level than that corresponding to ionization by electron impact. They were inclined to the second view, but at a later date Smyth [99] preferred the first.

The energies required for removing the several individual electrons from H_2O have been estimated by R. S. Mulliken [105] to be as follows, the electrons being designated in accordance with the notation currently used by spectroscopists: $2pc$, 13.2 electron-volts (observed); $2pb$, $2pa$, and $2s$, 16, 17 and 30 electron-volts, respectively; see also R. S. Mulliken.[106]

† See also *Ionic Dissociation*, Sections 7 and 8.
[98] Kurie, F N. D., *Phys. Rev.*, (2), **41**, 701-707 (1932).
[99] Smyth, H. D., *Rev. Modern Phys.*, **3**, 347-391 (384, 389) (1931).
[100] Mohler, F. L., *Int. Crit. Tables*, **6**, 72 (1929).
[101] Mackay, C. A., *Phys. Rev.* (2), **24**, 316-329 (1924); *Phil. Mag.* (6), **46**, 828-835 (1923).
[102] Barton, H. A., and Bartlett, J. H., Jr., *Phys. Rev.* (2), **31**, 822-826 → 154-155 (A) (1928).
[103] Smyth, H. D., *Rev. Modern Phys.*, **3**, 347-391 (385) (1931).
[104] Smyth, H. D., and Stueckelberg, E. C. G., *Phys. Rev.* (2), **32**, 779-783 (1928).
[105] Mulliken, R. S., *Phys. Rev.* (2), **40**, 55-62 (1932).

Table 18.—Ionization Potential and Energy: Water-vapor

I = ionization potential, E = ionization energy, each for the particular ion indicated; Int. = relative intensity of the indicated ionization when the pressure is about 0.5 μ-Hg (= 0.0005 mm-Hg) and the speed of the electrons is that generated by a potential difference of 50 volts.

The data credited to S, MS, B, and L are those given by H. D. Smyth [99] as a result of his study of all the pertinent information available in June, 1931. SM states that the intensity of H_3O^+ is approximately proportional to the square of the pressure; that of the other ions to the first power of the pressure.

Unit of I = 1 volt; of E = 10 kj/gfw-H_2O

Ion	Int.	I	E	Process	Ref.[a]
H_2O^+		13[b]	126	$H_2O \to H_2O^+$	S
H_2O^+	1000	12.7 (13.3, 14.2, 15.0) 16.0	122–154		SM
OH^+	200	18.9	182	$H_2O \to H + OH^+$	SM
OH^+		17.3[c]	167		MS
H_3O^+	200	Same as H_2O^+	122–154	$H_2O^+ + H_2O \to H_3O^+ + OH$	SM
H_3O^+		13.0	126		MS
H^+	200	13.5, 18.9	130, 182	$H \to H^+$; $H_2O \to H^+ + OH$	SM
H^+		19.2	185		B
O^+	20	18.5	178	$H_2O \to H_2 + O^+$	SM
O_2^+	6			$O_2 \to O_2^+$	SM
H_2^+	5	33.5	323	$H_2O \to H_2^+ + O^+$	SM
OH^-		[d]			SM
O^-		[e]			SM
H^-		6.6, 8.8[f]	64, 85		L
Heat of formation of H_2O			24		

		Recent determinations of I			
From spectroscopic data					
		12.4	16.5		H
6.92	9.2	12.9	16.7 17.8 24.5		R
		12.56 ± 0.02			P
By electron impact					
		12.59 ± 0.05			SB
7.60		10.15 12.35 13.55 16.75			TW

[a] References:
 B Bleakney (reported by Lozier, q.v.).
 H Henning, H. J., *Ann. d. Physik*, (5), **13**, 599-620 (1932).
 L Lozier, W. W., *Phys. Rev.* (2), **36**, 1417-1418 (1930).
 MS Mueller, D. W., and Smyth, H. D., *Idem*, **38**, 1920 (A) (1931).
 P Price, W. C., *J. Chem'l Phys.*, **4**, 147-153 (1936).
 R Rathenau, G., *Z. Physik*, **87**, 32-56 (1933).
 S Smyth, H. D., *Rev. Modern Phys.*, **3**, 347-391 (384, 389) (1931).
 SB Smith, L. G., and Bleakney, W., *Phys. Rev.* (2), **49**, 883 (A) (1936).
 SM Smyth, H. D., and Mueller, D. W., *Idem*, **43**, 116-120 (1933) → *Idem*, **42**, 902 (A) (1932).
 TW Thorley, N., and Whiddington, R., *Proc. Leeds Phil. Lit. Soc. (Sci.)*, **3**, 265-269 (1936).

[b] F. L. Mohler [100] gives 13.2 volts on basis of observations by C. A. Mackay.[101]

[c] H. A. Barton and J. H. Bartlett, Jr.,[102] reported 13 volts.

[d] SM states that OH^- first appears for $I > 15$ volts; as I increases, the intensity of OH^- increases to a maximum, then decreases nearly to zero at 25 volts, then increases indefinitely.

[e] SM states that as I increases, O^- first appears at 22 ± 3 volts, increases to a maximum near 31 ± 4 volts, then decreases, to rise again at 36 ± 4 volts.

[f] Lozier could find H^- only when I was very close to either 6.6 or 8.8 volts. "The significance of this remarkable result is not yet clear."[103] SM found 7.9 volts for H^-; they regard this as probably an unresolved combination of the two observed by L.

---

[106] Mulliken, R. S., *J. Chem'l Phys.*, **3**, 506-514 (1935).
[107] Gaviola, E., and Wood, R. W., *Phil. Mag.* (7), **6**, 1191-1210 (1928).

10. VAPOR: CORPUSCULAR RADIATION

E. Gaviola and R. W. Wood [107] have concluded that about 5.2, and *not* less than 4.9, electron-volts are required to dissociate H_2O into H and OH. See also von Bishop,[108] Grinfeld,[109] and Townsend.[110]

Ionization by accelerated electrons.—The number (α) of pairs of ions produced per cm of path by each electron of a stream driven through

Table 19.—Ionization of Water-vapor by Accelerated Electrons: Strong Fields

Adapted from the compilation by O. Stuhlman, Jr.,[111] from J. S. Townsend.[110]

α = number of pairs of ions produced (= number of electrons freed) by each electron per cm of path in an applied uniform field of X volts/cm, the pressure of the vapor being p mm-Hg. Room temperature. Townsend derives the formula $(\alpha/p)_c = 12.9\, e^{-289p/X}$

X/p	100	200	300	400	500	600	700	800	900	1000
αp	1.31	3.6	5.2	6.35	7.2	7.95	8.5	9.0	9.4	9.7
$(\alpha/p)_c$	0.71	3.0	4.9	6.25	7.2	7.96	8.54	8.9	9.35	9.7

Table 20.—Miscellaneous Data for the Interaction of Accelerated Electrons with Water-vapor: Weak Fields

N. E. Bradbury and H. E. Tatel [112] have reported that if p does not exceed 3 mm-Hg then no negative ions are formed if X/p is less than 10 (volt/cm) per mm-Hg, but that at higher values of p such ions are formed at lower values of X/p.

The following data are from V. A. Bailey and W. E. Duncanson.[113] The electrons move in a uniform field of strength X; h is the probability of an electron's becoming attached to a molecule with which it collides; k is ratio of mean energy of agitation of the electrons to that of the molecules; L is mean free path of an electron when the pressure (p) of the vapor is 1 mm-Hg; u is mean velocity of agitation of the electrons; W is their mean energy, and w their mean velocity of drift; β is the probability of attachment in the interval of time required for the electron to move 1 cm along the direction of X; λ is the electron's fractional loss of energy during a collision with a molecule.

Unit of X = 1 volt/cm; of p = 1 mm-Hg; of β = 1 in 100; of k = 1 to 1; of w and u = 1 km/sec; of W = 1 electron·volt; of L = 1 μ; of λ = 1%; of h = 1 in 10000.

X/p	β/p	k	w	W	u	L	λ	h
12	1.5	3.2	27	0.14	221	37	4.23	0.06
14	2.2	6.8	44	0.21	272	48	4.18	0.19
16	17	7.2	39	0.32	337	63	4.00	1.3
20	23	17	56	0.70	498	107	3.69	3.0
24	23	40	84	1.37	700	165	3.28	4.5
32	25	49	96	1.81	804	169	3.52	5.0

[108] von Bishop, E. S., *Physik. Z.*, **12**, 1148-1157 (1911).

[109] Grinfeld, R., *Univ. Nac. LaPlata, Estud. Cien. Fis. Mat.*, **4**, 283-293 (No. 82) (1928); **4**, 415-426 (No. 86) (1928); *Physik. Z.*, **31**, 247-252 (1930).

a gas by a uniform field (X) of sufficient strength to cause one electron to dislodge another from a molecule which it strikes varies with the density of the gas, but not otherwise with the temperature. If the pressure of the gas is p, then, at any given temperature, α/p depends solely upon X/p.[111] The factor by which the number of electrons in the stream is multiplied for each cm of its path is e^α, the number at the end of a path x cm long being $n = n_0 e^{\alpha x}$.

If the field is weak, the velocity of the impinging electron will be too low to dislodge another from the molecule, and it may itself be caught, forming a negative ion.

Excited Atoms and Molecules.

$A^+ + H_2O \rightarrow H_2O^+ + A$, and $H_2O^+ + O_2 \rightarrow O_2^+ + H_2O$.[104]

Collisions of excited Hg-atoms in the resonance level 2^3P_1 with normal water-vapor molecules may lead to several different processes. In most cases the Hg-atom is thrown down to the metastable 2^3P_0 level, in a few cases (about 1 in 10 000 collisions) the H_2O molecule is dissociated into H and OH, and in some cases (less than 1 in 1000) the complex quasimolecule Hg-H_2O is formed. The last dissociates, emitting a continuous band at 2800A.[107] See also [114].

Mobility of Ions in Water-vapor.

The mobility (K) of an ion is its velocity of migration per unit field intensity. Over a wide range of densities, the product of K multiplied by the density is a constant for a given gas. The quantity K_0 satisfying the formula $K_0 = K \dfrac{273p}{T}$, where K is the mobility observed at $T\ °K$ and a pressure of p atmospheres, has been called the "mobility constant" of the gas. The values found for K_0 are rather discordant, and opinions differ regarding the interpretation of the observed data. From recent data, it has been concluded that the mobility constant for normal ions in water-vapor is $K_0 = 0.62$ cm·sec^{-1} per volt·cm^{-1} for the positive ion and 0.56 for the negative, but it is probable that each value should be increased by 20 per cent.[115] Numerical values are from L. B. Loeb and A. M. Cravath.[116]

H. A. Erickson [117] finds that the molecule of H_2O in air gives up an electron to the final positive air ion, and thus forms an H_2O^+ ion of greater mobility; and that the reciprocal of the mobility of negative ions in moist air is linear in the relative humidity.

[110] Townsend, J. S., "Theory of Ionization of Gases by Collision," London, Constable, 1910.

[111] Stuhlman, O., Jr., *Int. Crit. Tables*, **6**, 121 (1929). From Bishop, E. S., *Physik. Z.*, **12**, 1148-1157 (1911).

[112] Bradbury, N. E., and Tatel, H. E., *J. Chem'l Phys.*, **2**, 835-839 (1934).

[113] Bailey, V. A., and Duncanson, W. E., *Phil. Mag.* (7) **10**, 145-160 (1930).

[114] Senftleben, H., and Rehren, I., *Z. Physik*, **37**, 529-538 (1926), and Bates, J. R., and Taylor, H. S., *J. Am. Chem. Soc.*, **49**, 2438-2456 (1927).

[115] Loeb, L. B., *Int. Crit. Tables*, **6**, 111 (1929).

[116] Loeb, L. B., and Cravath, A. M., *Phys. Rev.* (2), **27**, 811-812 (1926).

[117] Erickson, H. A., *Idem*, **32**, 792-794 (1928).

S. Chapman [118] has reported that when water is sprayed or when air is bubbled through water, ions of various mobilities are produced. When the numbers of ions are plotted against their mobilities one obtains as a background a broad flat curve with its maximum betwen the mobilities 0.05 and 0.10 cm/sec per (volt/cm), on which are superposed a number of peaks. In the sprayed liquid there are the same number of carriers of each sign, and the peaks occur at the mobilities 1.2, 0.3, and 0.2 for the negative carriers, and at 0.5 and 0.22 for the positive (first paper), or 1.5 for the negative and 0.9 for the positive (third paper). When air is bubbled through the liquid, the negative carriers are twice as numerous as the positive if the air tube is a small capillary, and 100 times as numerous if the tube is 12 mm in diameter (fourth paper); and the peaks occur at mobilities 1.2 and 0.25 for the negative and 0.7 and 0.3 for the positive (second paper), 1.5 and 0.3 for negative and 0.9 and 0.4 for positive (third paper), 1.9, 1.1, and 0.4 for negative and 1.1 and 0.4 for positive (fourth paper).

11. Viscosity of Water-vapor *

Four extended, but in part disagreeing, series of observations on the viscosity (η) of water-vapor have been reported: H. Speyerer (1925), range 1 to 10 kg*/cm², 107 to 347 °C; W. Schiller (1934), 1 to 30 kg*/cm², 100 to 300 °C; K. Sigwart (1936), 25 to 270 kg*/cm², 276 to 383 °C; and G. A. Hawkins, H. L. Solberg, and A. A. Potter (1935), 1 to 247 kg*/cm², 218 to 542 °C. Internal evidence indicates that the last is not satisfactory; for example, the effect of eddies in the wake of the falling body has been entirely ignored, although it must have been very perceptible in certain cases, and the instrumental temperature-coefficients used seem to be in error. It will not be considered further.

Of the first three, Speyerer and Schiller find that the values of η along an (η, p)-isotherm increase rapidly with the pressure (p), but the two sets differ markedly. Speyerer's values increase ever more rapidly as the pressure increases, whereas Schiller's exhibit marked irregularities. In contrast to them, Sigwart finds that η increases slowly and linearly with the pressure until the saturation pressure is rather closely approached, and then ever more rapidly. For the linear portion of the isotherms the slope is about 5 per cent per 100 kg*/cm², varying rather irregularly (2 to 7.6 per cent) from one isotherm to another. The total increase in η along the 275 °C isotherm to saturation does not exceed 1 per cent, and the increase to saturation is presumably less at lower temperatures. In striking contrast to this, Speyerer finds along the 270 °C isotherm an increase of 10 per cent on going from $p = 1$ to $p = 10$ kg*/cm², and Schiller finds 19 per cent for the same change.

[118] Chapman, S., *Phys. Rev. (2)*, **49**, 206 (A) (1936); **51**, 145 (A) (1937); **52**, 184-190 (1937); **53**, 211 (A) (1938).

* For complete references see p. 68.

Under such conditions it is difficult to speak with confidence regarding the relative merits of the several series, but the author is inclined to favor Sigwart's. All three are given in Table 22, Schiller's values having been obtained by scaling his graph, which is a small one.

Values given by Sigwart for temperatures and pressures outside the domain covered by his observations were derived by extrapolation on the assumption that at 1 kg*/cm² $\eta = 16.47 T^{1.5}/(T + 548)$ micropoise (μp), that formula giving values to which his can be satisfactorily extrapolated. But that formula is based on H. Vogel's (1914) assumption that the Sutherland constant (the constant in the denominator) is to be taken as 1.47 times the absolute temperature of the normal boiling point, and that J. Puluj's (1878) reported value (90.4 μp) for η at 0 °C is correct. The first may be accepted; but the second should be 88.4, as Puluj used incorrect constants in reducing his observations (see Table 22, note b). Probably this difference of 2 per cent lies within the range of experimental error, but it should not be forgotten.

Fortunately, all three sets of values corresponding to 1 kg*/cm², observed or extrapolated, agree within ± 2 per cent with one another and with the two short series by W. Schugajew (1934) and by C. H. Braune and R. Linke (1930), (see Table 23). So far as pressure is concerned, no distinction need be made between observations at 1 kg*/cm² and those at lower, but not excessively low, pressures.

No recent measurements at temperatures below 100 °C have been found; early ones are given in Table 22. Early observations were frequently made in terms of the viscosity of air, and were reduced to absolute units by assuming a value for air. In some cases the values were reduced to 0 °C on the basis of an assumed formula, generally requiring a knowledge of the absolute temperature (T_0 °K) of the ice point. Not infrequently an incorrect value for the viscosity of air or for T_0 was used, and sometimes an unsatisfactory formula. The resulting false values are commonly quoted, and appear in compilations of constants. An attempt has been made to correct such as are included in the following tables, the corrections that have been applied being explained in footnotes.

Formulas.

(*a*) *Pressure = 1 kg*/cm².*—In the simple kinetic theory of gases the viscosity is given by the formula $\eta = BT^{1/2}$, where B is characteristic of the gas and is proportional to $nL_e = K/\pi d^2$, n being the number of molecules per unit of volume, L_e the effective mean free path of a molecule, d the molecular diameter (the diameter of the spheres by which the molecules may be regarded as replaced), and K a numerical factor determined by the distribution of the velocities of the molecules and by the kind of average free path L_e is. But actually η does not vary as $T^{1/2}$. Hence nL_e must vary with T. The failure of the formula may be explained by an attraction between the molecules, causing their paths to be curved, and increasing the frequency with which collisions occur. Since this curvature will decrease as the translational kinetic energy of the molecules increases, Sutherland (1893) suggested that L_e should be replaced by $L_s/(1 + CT^{-1})$, where C is a constant (Sutherland's constant), characteristic of the gas, which measures the intensity of the intermolecular attraction, and L_s is what the free path would be if there were no such attraction. Then $\eta = AT^{1.5}/(T + C)$, A and C each being characteristic of the gas. This formula has been found to represent the observations on many gases over a wide range of temperatures. Similar expressions have been derived by S. Chapman (1916) and by J. H. Jeans (1916) with greater attention to mathematical rigor.

Unless observations are extended over a considerable range of temperature, C cannot be determined with precision. Consequently the experimental values that have been assigned to it vary greatly. Certain empirical formulas relating C to other quantities have been proposed. In 1910, A. O. Rankine observed that for the gases he investigated T_{crit}/C varied within a narrow range, the mean being 1.14; helium and hydrogen were marked exceptions. This relation suggested to Hans Vogel (1914) another between C and the normal boiling point (T_b), and he concluded that $C = 1.47 T_b$. These and certain experimental values are given in Table 21.

Table 21.—Sutherland Constant for Water-vapor

$$\eta = AT^{1.5}/(T + C)$$

Observations by André Fortier (1936) indicate that for air the value of C depends on T.

C	Source
548 °K	H. Vogel (1914); $C = 1.47 T_b$
568	A. O. Rankine's relation (1910); $C = T_{crit}/1.14$
650	L. L. Bircumshaw and V. H. Stott,[123] from C. J. Smith (1924).
673	H. Speyerer (1925); experimental.
961	H. Braune and R. Linke (1930); experimental.

Actually, the observational data for water-vapor at and below 1 kg*/cm² can, within experimental error, be represented essentially as well by means of a linear equation in the temperature (t °C) as by a Sutherland formula, and several observers give such equations. From a study of a great mass of data, Trautz (1931) concluded that $\eta = T^n \eta_{crit}/T_{crit}$, where n is, in general, a function of T, depends upon the type of substance, and approaches unity as T approaches T_{crit}. When n is unity, Trautz's formula is linear in t. An equation of this type with $n = 1$ and $\eta_{crit} = 266$ μp has been accepted by some (see Sigwart) as fairly satisfactory for water-vapor at 1 kg*/cm²; and this in spite of the fact that it yields too large a value at 0 °C. With Sigwart's value for η_{crit} (378 μp) a still higher zero value is obtained. The value $\eta_{crit} = 226$ μp (actually 228) seems to have been derived by Trautz in some unexplained way from the observations of others.

(b) *Any pressure.*—R. Plank (1933) has shown that Speyerer's data can be represented satisfactorily by the empirical formula (1) in which v^* liters per gram is the specific volume of the vapor.

$$\eta = (86.1 + 0.373t) \cdot (1 + 0.0175\, v^{*-1} + 0.0025\, v^{*-2}) \text{ micropoise} \quad (1)$$

Table 22.—Viscosity of Water-vapor

The values given in Section I are probably to be preferred in the domain $t = 100$ to $500\ °C$, $P = 1$ to $250\ kg^*/cm^2$. For temperatures below $100\ °C$, see Sections V and VI. Sets of values for $P = 1$ are compared in Table 23.

Unit of $\eta = 1$ micropoise $(=1\ \mu p) = 1\ \mu g/cm \cdot sec = 1.0197 \times 10^{-8}\ kg^* \cdot sec/m^2$; of $P = 1\ kg^*/cm^2 = 0.9678$ atm. Temperature $= t\ °C$.

I. From K. Sigwart.[124] Values for $P = 1$ have been computed by means of the formula $\eta = 16.47 T^{1.5}/(T + 548)$, which Sigwart regards as a satisfactory representation of the values obtained by extrapolating his observations. See text for bases of constants; $T \equiv 273 + t$. Dubiety at $P = 1$ is 1 or 2 per cent; at $400\ °C$, 3 per cent; at $500\ °C$, 10 per cent; along saturation line, 1 per cent from 50 to $300\ °C$, 2 per cent at $360\ °C$, and 3 per cent at $370\ °C$. At critical point, $\eta = 378 \pm 5$ per cent.

$P \rightarrow$ $t_{sat} \rightarrow$ t	1 99.1	10 179.0	20 211.4	50 262.7	100 309.5	150 340.6	200 364.1	220 372.0	250
					η				
t_{sat}	128	159	172	189	214	234	271	319	
100	128								
150	147								
200	166	166							
250	184	184	184						
300	201	202	203	206					
350	219	220	221	225	231	238			
400	235	236	237	238	244	248	256	272	286
450	252	238	239	241	243	245	249	257	263
500	268			241	243	245	248	251	255

II. From H. Speyerer.[125] He thinks that his observations at $P = 1$ can be satisfactorily represented by either of the formulas: $\eta = 88.33 + 0.3712t$ or $\eta = 86.8 T^{1.5}/(T + 673)$; he uses the first.

$P \rightarrow$ $t_{sat} \rightarrow$ t	1 99.1	2 119.6	4 142.9	6 158.1	8 169.6	10 179.0
			η			
t_{sat}	125.5	135.0	146.8	156.0	165.0	176.0
110	129.2					
150	144.1	146.0	149.3			
200	162.6	164.3	167.4	170.8	175.9	183.3
250	181.2	182.8	185.8	189.1	193.7	201.1
300	199.7	201.2	203.9	207.2	211.6	218.7
350	218.3	219.7	222.3	225.3	229.5	236.4

[119] Mokrzycki, G., *J. de phys.* (6), **7**, 188-192 (1926).
[120] Bircumshaw, L. L., and Stott, V. H., *Int. Crit. Tables*, **5**, 6 (1929).
[121] Millikan, R. A., *Phil. Mag.* (6), **19**, 209-228 (215) (1910).
[122] Stearns, J. C., *Phys. Rev.* (2), **27**, 116 (A) (1926).

11. VAPOR: VISCOSITY

Table 22—(Continued)

III. From W. Schiller.[126] Values obtained by scaling his graph; he estimates the dubiety to be not over 2 per cent. At critical point, $\eta = 692$.

t	P_{sat}	P_{sat}	1	5	10	15	20	25	30
100	1.03	126	126						
120	2.02	141	133						
140	3.68	159	140						
150	4.85	170	144						
160	6.30	182	147	171					
180	10.12	208	155	174					
200	15.8	236	162	177	212	234			
220	23.6	268	170	183	215	235	255		
240	34.1	298	178	190	219	237	257	271	288
250	40.5		181	193	221	238	258	272	290
260	47.9		185	196	222	240	259	273	291
280	65.4		192	204	227	244	262	276	293
300	87.6		200	211	232	244	265	280	295
374	225.2	692							

IV. From W. Schugajew.[127] His mean of several discordant sets of observations. Within his limits of error, η does not vary with P over the range investigated (0 to 93).

t	100	150	200	250	300	350	400
η	126	144	164	183	202	222	241

V. From compilation by L. L. Bircumshaw and V. H. Stott[123]; taken from a paper by C. J. Smith.[128] Various corrections have been made by the present compiler, and explained in the footnotes. Except as the contrary is indicated in the notes, the pressure was much less than 1 atm, and was far below saturation.

t	0	15	16.7	28.9	99.95	100	151.2	207.1
η	88.4[b]	92.0[c]	94.7[b]	99.7[d]	125[e]	127	145	168
Ref[a]	P	KW	P	V	MS	Sm	Sm	Sm

VI. From H. Braune and R. Linke.[129] These observations have not been published in detail, and the short table of only 11 lines of data referring to water-vapor contain at least two gross errors in computation, and one in transcription. These errors do not appear in the following values. Braune and Linke represent their observations by means of the formula $\eta = 22.36 T^{1.5}/(T + 961)$; the value (961) for the Sutherland constant is excessively high. At 20.2 °C the pressure was essentially P_{sat}; for others, 120 to 210 mm-Hg.

t	20.2	92.6	107.5	210.3	313.1	366.4	406.6
η	93.7	117.8	124.2	163.8	214.9	226.1	242.2

[123] Bircumshaw, L. L., and Stott, V. H., *Int. Crit. Tables,* **5**, 4 (1929).
[124] Sigwart, K., *Forsch. Gebiete Ingenieurw.,* **7**, 125-140 (1936).
[125] Speyerer, H., *Forsch. Gebiete Ingenieurw.,* **273**, 1-30 (1925) → *Z. Ver. Deuts. Ing.,* **69**, 747-752 (1925).
[126] Schiller, W., *Forsch. Gebiete Ingenieurw.,* **5**, 71-74 (1934).

Table 22—*(Continued)*

[a] References follow Table 23.

[b] After reduction to the basis of $\eta = 179.2$ for air at 16.7 °C.[130] The observations were made at 16.7 °C with an apparatus giving 183 for air at the same temperature, and yielded the value 96.7, which was published and is commonly quoted. The value 90.4 which he gives for water-vapor at 0 °C, and which is quoted in many tables, including the *Critical Tables,* was computed by him from his 96.7 value on the assumption that η is directly proportional to T and that 0 °C = 238 °K. On correcting the 96.7, using Sutherland's equation with $C = 548$, and placing 0 ° C = 273.1 °K, one finds $\eta = 88.4$ at 0 °C.

[c] After reduction to the basis of $\eta = 178.4$ for air at 15 °C.[130] It was published and quoted as 97.5 at 15 °C, but on the basis of $\eta = 189$ for air at 15 °C. The following entry in the *Critical Tables,* giving $\eta = 97.5$ at 20.6 °C, was taken from Smith, who took it from Landolt-Börnstein,[131] where it had been doubly entered: once as the value at 15° C, and again as the value at 20.6 °C, the mean of the temperatures at which observations were taken. In the 5th edition of those Tabellen the value is entered but once, and is assigned incorrectly to 18.6-21.6 °C.

[d] After reduction to the basis of $\eta = 170.9$ for air at 0 °C.[130] It was published and has been quoted in various tables, including the *Critical Tables,* as 100.6; but that was based on the assumption that $\eta = 172.4$ for air at 0 °C. Pressure was near that of saturation.

[e] After reduction to the basis of $\eta = 170.9$ for air at 0 °C, as in note *d*. It was published and is commonly quoted as 132, but that is on the basis of $\eta = 180$ for air at 0 °C. The pressure was near that of saturation.

(*c*) *Saturation pressure.*—R. Plank (1933) also found that Speyerer's values for the viscosity (η'') of saturated water-vapor is related to that (η') of saturated water at the same temperature as shown by formula (2):

$$1/\eta'' + 1/\eta' = 1.1003 - 0.00261t \text{ micropoise}^{-1} \qquad (2)$$

But Schiller and Sigwart each find that this law of rectilinear mean is valid only over a limited range of temperatures.

Viscosity at the Critical Point.

Three estimates of the value of η_{crit} have been found: Trautz (1931) by an unknown method inferred that $\eta_{\text{crit}} = 228.3$ μp. By plotting and extrapolating the mean of the fluidities (η^{-1}) of saturated vapor and of saturated water at common temperatures, Schiller (1934) inferred 692, and Sigwart (1936) 378 μp ± 5 per cent.

Viscosity of Fog.

G. Mokrzycki [119] has concluded that the viscosity of foggy air is given by the relation $\eta = \eta_a + 1.59\Delta$ poise, where η_a (taken as 0.000171) is the viscosity of dry air, and Δ g/cm^3 is the weight of the fog per cm^3 of foggy air. In his observations, $10^6\Delta$ lay between 1.5 and 15, and the diameters of the fog particles between 0.5 and 10 μ. The value found for η was independent of the size of the particles.

---

[127] Schugajew, W., *Physik. Z. Sowj.*, **5**, 659-665 (1934).
[128] Smith, C. J., *Proc. Roy. Soc. (London) (A)*, **106**, 83-96 (1924).

11. VAPOR: VISCOSITY

Viscosity of Moist Air.

In their compilation,[120] L. L. Bircumshaw and V. H. Stott quote R. A. Millikan [121] as finding that air saturated at 26 °C has a viscosity 1904/1863 = 1.022 times as great as dry air at the same temperature, and J. C. Stearns [122] as claiming "that the viscosity of air is *decreased* by saturating it with moisture, the decrease being $\frac{1}{3}$ per cent at 760 mm-Hg and 35 per cent at 14 mm-Hg pressures." It seems that this 6-line abstract is all that Stearns has ever published regarding his measurements. They are probably of little, if any, value.

Table 23.—Comparison of Various Values for the Viscosity of Water-vapor at or below 1 kg*/cm²

Several values from each of the sections of Table 22 are here compared with those defined by the formula $\eta_c = 16.12T^{1.5}/(T + 548)$, which gives Puluj's corrected value at 0 °C and uses Vogel's value for C.

The source of the data as well as the section of Table 22 in which they are given is indicated, and the individual observers of the several values given in *International Critical Tables* are designated. For additional information, see Table 22.

Unit of η = 1 micropoise. t = temperature °C.

Source→ Section of Table→ t	η_c	Sig. I	Spe. II	Schi. III	Schu. IV	ICT V	BL VI
0	88.4					88.4ª P	
15	94.1					92.0 KW	
16.7	94.7					94.7 P	
20.2	96.0						93.7
28.9	99.3					99.7 V	
92.6	123.0						117.8
99.95	125.8					125 MS	
100	125.8	128		126	126	127 Sm	
107.5	128.6						124.2
150	144.1	147	144.1	144	144		
151.2	144.5					145 Sm	
200	162.0	166	162.6	162	164		
207.1	164.6					168 Sm	
210.3	165.7						163.8
250	179.6	184	181.2	181	183		
300	196.8	201	199.7	200	202		
313.1	201.9						214.9
350	213.6	219	218.3		222		
366.4	219.0						226.1
400	230.0	235			241		
406.6	232.1						242.2
500	261.7	268					

ª In the review [132] of the thesis of F. Houdaille (Paris, 1896) it is stated that he found 88.5 for water-vapor at 0 ° C.

---

[129] Braune, H., and Linke, R., *Z. physik. Chem. (A)*, **148**, 195-215 (1930).
[130] *Int. Crit. Tables*, **5**, 2 (1929).
[131] Landolt-Börnstein, *Phys.-Chem. Tabellen*.
[132] Houdaille, F., *Fortschr. d. Physik*, **52**₁, 442-443 (1897).

References

ICT   Bircumshaw, L. L., and Stott, V. H., *Int. Crit. Tables*, **5**, 1-6 (1929).
BL   Braune, H., and Linke, R., *Z. phys. chem. (A)*, **148**, 195-215 (1930).
    Chapman, S., *Phil. Trans. (A)*, **216**, 279-348 (1916).
    Fortier, A., *Compt. rend.*, **203**, 711-712 (1936).
    Hawkins, G. A., Solberg, H. L., and Potter, A. A., *Trans. Am. Soc. Mech. Eng.*, **57**, 395-400 (FSP-57-11) (1935).
    Houdaille, F., *Fortschr. d. Phys.*, **52₁**, 442-443 (1897)   ← *Thesis*, Paris, 1896.
    Jeans, J. H., "The Dynamical Theory of Gases," 1916.
KW   Kundt, A., and Warburg, E., *Ann. d. Physik (Pogg.)*, **155**, 337-365, 525-550 (1875).
MS   Meyer, L., and Schumann, O., *Idem, (Wied.)*, **13**, 1-19 (1881).
    Millikan, R. A., *Phil. Mag. (6)*, **19**, 209-228 (1910).
    Mokrzycki, G., *J. de Phys. (6)*, **7**, 188-192 (1926).
    Plank, R., *Forsch. Gebiete Ingenieurw.*, **4**, 1-7 (1933).
P   Puluj, J., *Sitz. ber. Akad. Wiss, Wien, (Math-Phys.)*, **78₂**, 279-311 (1878).
    Rankine, A. O., *Proc. Roy. Soc. (London) (A)*, **84**, 181-192 (1910).
Schi   Schiller, W., *Forsch. Gebiete Ingenieurw.*, **5**, 71-74 (1934).
Schu   Schugajew, W., *Phys. Z. Sowj.*, **5**, 659-665 (1934).
Sig   Sigwart, K., *Forsch. Gebiete Ingenieurw.*, **7**, 125-140 (1936).
Sm   Smith, C. J., *Proc. Roy. Soc. London, (A)*, **106**, 83-96 (1924).
Spe   Speyerer, H., *Forsch. Gebiete Ingenieurw.*, **273**, 1-30 (1925) → *Z. d. Ver. deuts. Ing.*, **69**, 747-752 (1925).
    Stearns, J. C., *Phys. Rev. (2)*, **27**, 116 (A) (1926).
    Sutherland, W., *Phil. Mag. (5)*, **36**, 507-531 (1893).
    Trautz, M., *Ann. d. Physik (5)*, **11**, 190-226 (1931).
V   Vogel, H., *Idem, (4)*, **43**, 1235-1272 (1914).

12. Acoustic Data for Water-vapor

Velocity of Sound in Water-vapor.

In an ideal gas, the velocity of sound is independent of the pressure, and its square is directly proportional to the absolute temperature. For actual gases, especially near the region of liquefaction, departures from this relation should be expected. Nevertheless, many of the earlier data on the velocity of sound in gases and vapors were deduced on the assumption that this relation applies. Fortunately, it does apply fairly well for air at the temperatures that were used, and many of the observations consisted in determining merely the ratio of the velocity (v_t) in the gas or vapor under study at the temperature t to that ($u_{t'}$) in air at the temperature t'.

Then $\dfrac{v_t}{u_{t'}}\sqrt{\dfrac{T'}{T}}$ was called the ratio of the velocity at any temperature to that of air at the same temperature, and the product of this expression and the velocity in air at 0 °C was recorded as the velocity in the gas or vapor at 0 °C, although the observations may have been made at a quite different temperature. Actually, this ratio is not the ratio of the velocities at *any* temperature, but, so far as air satisfies the assumed relation, it is the ratio of the velocity in the gas or vapor at t to that in air at t. Hence the data can be reinterpreted when the value accepted by the observer for the absolute temperature of the ice-point is known; that can in certain cases be determined from the data he records. In some cases, $v_t\sqrt{T'}/u_{t'}\sqrt{T}$ has, incorrectly, been treated as if it were ratio of the velocity in the gas or vapor at t °C to that in air at 0 °C, thus leading to values that are entirely wrong.

12. VAPOR: ACOUSTICS

The last seems to be the explanation of the excessively low values ($V_{93} = 402.4$, $V_{96} = 410.0$ m/sec generally attributed to W. Jäger, and apparently originating in the Landolt-Börnstein *Tabellen*. They appear in the second and succeeding editions of the *Tabellen* (the first edition has not been examined) and in the *International Critical Tables*.

The first accounts for the value 401 m/sec given by A. Masson[133] for water vapor at 0 °C, and included in the tables mentioned. Actually, Masson's observations were made at 95 °C. Appropriate corrections have been applied to these in Table 24.

Table 24.—Velocity of Sound in Water-vapor

$\delta_c = V_c - V$ and $\delta_i = V_i - V$, where V is the observed velocity, and V_c and V_i are those defined by the empirical formulas given in the text; $T = 273.1 + t$ °C, the absolute temperature corresponding to t °C. Except as the opposite is indicated, the vapor was saturated at the temperature t °C.

Unit of V, δ_c, and $\delta_i = 1$ m/sec. Temp. $= t$ °C $= T$ °K

t	V	$V/\sqrt{T}$	Obs.	t	V	$V/\sqrt{T}$	δ_c	δ_i
27	432	24.94	Th[a]					
100	405[b]	21.0	T		—W. G. Shilling[f]—			
Int. Crit. Tables[c] (corrected)				100	471.5	24.41	0.0	− 1.5
93.1	458.7[d]	23.97	J	200	536.7	24.68	+ 1.2	+ 0.5
95	462[e]	24.08	M	300	593.2	24.78	+ 1.4	+ 0.7
96.6	467.7[d]	24.32	J	400	643.2	24.79	+ 1.0	0.0
100	471.5	24.41	S	500	688.2	24.75	0.0	− 1.2
110	413[b]	21.1	T	600	727.8	24.63	− 0.1	− 1.4
120	417.5[b]	21.1	T	700	762.5	24.44	+ 0.8	− 0.1
130	424.4[b]	21.1	T	800	795.3	24.28	+ 0.3	+ 0.3
1000	853.9	23.93	S	900	825.0	24.08	0.0	+ 1.4
				1000	853.9	23.93	− 2.2	+ 1.3

[a] G. E. Thompson.[136] Frequency = 108.6 kilocycles/sec.

[b] Velocity in tin tube, diameter = 1.4 cm. Pressure of vapor = 1 atm; hence the densities at 100, 110, 120, and 130 °C were, respectively, 1.00, 0.71, 0.51, and 0.38 of that corresponding to saturation at the indicated temperature.

[c] From compilation by A. L. Foley[137] and ascribed, as indicated to: A. Masson,[138] W. Jäger,[139] W. Treitz,[140] and W. G. Shilling.[141] For nature of corrections here applied, see preceding text.

[d] These data of Jäger's are given in the *Int. Crit. Tables* and elsewhere as $V_{93} = 402$, and $V_{96} = 410$; the densities of the vapor were, respectively, 0.55 and 0.70 of that corresponding to saturation at the temperature indicated.

[e] For this, most tables give Masson's $V_0 = 401$, though the observations were at 95 °C.

[f] Frequency = 3 kilocycles/sec.

---

[133] Masson, A., *Ann. Chim. et Phys. (3)*, 53, 257-293 (1858).
[134] Treitz, W., *Diss., Bonn.*, 1903.
[134a] Shilling, W. G., *Phil. Mag. (7)*, 3, 273-301 (1927).
[135] Irons, E. J., *Phil. Mag. (7)*, 3, 1274-1285 (1927).
[136] Thompson, G. E., *Phys. Rev. (2)*, 36, 77-79 (1930).
[137] Foley, A. L., *Int. Crit. Tables*, 6, 461-467 (463) (1929).
[138] Masson, A., *Compt. rend.*, 44, 464-467 (1857) → *Phil. Mag. (4)*, 13, 533-536 (1857).
[139] Jäger, W., *Ann. d. Physik (Wied.)*, 36, 165-213 (1889).
[140] Treitz, W., *Diss. Bonn.* (1903).
[141] Shilling, W. G., *Phil. Mag. (7)*, 3, 273-301 (1927).

The low values, $V_{110} = 413$, $V_{120} = 417.5$, and $V_{130} = 424.4$ m/sec, attributed to W. Treitz [134] in the tables mentioned are the velocities in tin tubes 14 mm. in diameter, and refer to superheated vapor at a pressure of 1 atm. His value $V_{100} = 404.8$ m/sec for the saturated vapor in the same tubes seems to have escaped the attention of the compilers. It seems that he did not measure the velocity in air in those tubes.

The data of W. G. Shilling [134a] were deduced by him from the velocity observed in a quartz tube 4 cm in diameter, the frequency being 3000 cycles/sec. They are approximately represented by the formula

$$V_c = 100 \sqrt{15.256 + 7.1135 \frac{t}{100} - 0.1384_7 \left(\frac{t}{100}\right)^2} \text{ m/sec.}$$

E. J. Irons [135] represents the same data by the formula $V_i = -883.4 + 404.4 T^{0.204}$ m/sec, where $T = 273 + t$ °C.

Absorption of Sound by Water-vapor.

No information concerning the absorption of sound by pure water-vapor has come to the author's attention.

Moist Air.

The addition of water-vapor to air increases both the velocity and the absorption of sound. For any given frequency the absorption varies with the humidity, passing through a well-marked maximum; for any given humidity the absorption varies with the frequency, passing through a well-marked maximum at a certain frequency that varies as a quadratic function of the humidity. These maxima may be more than 20 times as great as the absorption to be expected from the classical theory. This effect is associated with the presence of oxygen; it does not appear with pure nitrogen. It is thought to arise from an acceleration, produced by intercollisions of O_2 and H_2O molecules, of the otherwise very sluggish process of equipartition of added energy among the several degrees of freedom of the oxygen molecule, a process which in the absence of H_2O requires a time of the order of 0.01 sec for completion. Collision with two H_2O molecules is very much more effective than collision with but one. (See [142-147].)

The following illustrative data for moist air at 20 °C, frequency 3000 cycles/sec, have been taken from the first of Knudsen's papers [142] just mentioned. Here w = number of H_2O molecules per 100 molecules of the mixture, and the absorption coefficient m is that defined for a plane wave

---

[142] Knudsen, V. O., *J. Acoust. Soc. Am.*, **5**, 112-121 (1933) → *Phys. Rev. (2)*, **43**, 1051 (A) (1933).
[143] Kneser, H. O., *Idem*, **5**, 122-23 (1933).
[144] Knudsen, V. O., *Idem*, **6**, 199-204 (1935).
[145] Knudsen, V. O., and Obert, L., *Idem*, **47**, 256 (A) (1935).
[146] Kneser, H. O., and Knudsen, V. O., *Ann. d. Physik (5)*, **21**, 682-696 (1935).
[147] Knudsen, V. O., and Obert, L., *J. Acoust. Soc. Am.*, **7**, 249-253 (1936).

12. VAPOR: ACOUSTICS

by $I = I_0 e^{-mx}$ where x cm is the distance the wave travels while the intensity decreases from I_0 to I.

w	0.01	0.05	0.10	0.20	0.25	0.30	0.50	1.00	2.00
$10^4 m$	0.10	0.32	0.74	1.82	1.95	1.68	0.82	0.47	0.33

The effect of moisture upon the velocity of sound in air is "comparatively slight."[148] It was reported by C. D. Reid[150] to be represented at 20 °C by the formula $V_h = V_0 + 0.14h$, where h is the relative humidity; but that coefficient is *in error*—it is ten times as great as his observations justify. He finds for dry air $V_0 = 331.68$ m/sec, for air saturated at 20 °C, $V_h = 333.05$, giving $V_h - V_0 = 1.37$ m/sec for $h = 100$; hence, at 20 °C, $V_h = V_0 + 0.0137h$. Since for saturation at 20 °C the partial pressure of H_2O is $e = 17.51$ mm-Hg, these values lead to $V_h = V_0 (1 + Ae)$ where $10^4 A = 2.36$ per mm-Hg, the mean frequency being 130 kc/sec. This is not very different from the values found by others.

Table 25.—Velocity of Sound in Moist Air

The velocity (V_h) of sound in air in which the partial pressure of water-vapor is e mm-Hg may be conveniently represented by $V_h = V_d (1 + Ae)$, where V_d is the velocity in dry air, and A is an empirical coefficient that does not vary greatly with either the frequency, the temperature, or the humidity.

Unit of V and $\Delta = 1$ m/sec; of $\nu = 1$ kc/sec; of $e = 1$ mm-Hg; temp. $= t$ °C

I. C. D. Reid.[150] At 20 °C, $\nu = 130$, $10^4 A = 2.36$ after correction (see text).
II. C. Ishii.[151] Observations at 30 °C.

ν	288	730	1439	2000	2892
$10^4 A$	2.2_3	2.3	2.1_2	1.8_4	1.6_9

III. H. G. Muhammad.[152] $\nu = 0.994$; air either dry (V_d) or saturated with H_2O (V_{sat}); $\Delta = V_{sat} - V_d$; values of A computed by compiler.

t	V_{sat}	V_d	Δ	e	$10^4 A$
15	341.55	340.40	1.15	12.78	2.64
20	344.83	343.51	1.32	17.51	2.18
25	348.10	346.35	1.75	23.69	2.13
30	351.5	349.3	2.2	31.71	2.0
35	355.3	352.1	3.2	42.02	2.2
40	259.10	354.95	4.15	55.13	2.12
45	363.05	357.75	5.30	71.6	2.07
50	367.2	360.7	6.5	92.3	1.95
55	371.7	363.4	8.3	117.8	1.93
60	376.9	366.2	10.7	149.2	1.96
65	382.8	368.9	13.9	187.3	2.01
70	389.8	371.6	18.2	233.5	2.10
75	398.6	374.3	24.3	299.1	2.16
80	408	377	31	355.1	2.3

[148] Hubbard, B. R., *J. Acous. Soc. Am.*, **3**, 111-125 (1931).
[150] Reid, C. D., *Phys. Rev. (2)*, **35**, 814-831 (1930).
[151] Ishii, C., *Sci. Papers Inst. Phys. and Chem. Res. (Tokyo)*, **26**, 201-207 (No. 560) (1935).
[152] Muhammad, H. G., *Bull. Acad. Sci., Allahabad*, **3**, 269-294 (1934).

13. Diffusion of Water-vapor into Gases and Through Solids

By definition, the coefficient of diffusion of a substance is the quantity D in expression (1)

$$\frac{dm}{d\tau} = -D \frac{d\rho}{dx} dA \tag{1}$$

in which dm is the mass of the substance that flows in the direction of increasing x through the area dA, perpendicular to x, in the time $d\tau$, when the gradient of the concentration of the substance (mass per unit of volume) is $d\rho/dx$. Expression (1) is equivalent to (2)

$$dm/d\tau = -(D \cdot d\rho/dp) \cdot dA \cdot (dp/dx). \tag{2}$$

in which p is the partial pressure of the substance at the point x.

If the gas is ideal, and for the present purposes water-vapor may be considered ideal, expression (2) can be given the following forms, in which R is the universal gas constant expressed in gram-moles and in the same units of p and x as are used in expression (2); A_n is the pressure of 1 atm, expressed in the same units as p; dv_0 is the volume of dm at 0 °C and 1 atm; $(p_m)_0$ is the pressure exerted by dm when at 0 °C and confined in a given volume V; and T_0 °K and T °K are the absolute temperatures at 0 °C and at the temperature at which the diffusion occurs.

$$dm/d\tau = -(DM/RT_0) \cdot (T_0/T) \cdot dA \cdot (dp/dx) \text{ g/sec} \tag{3}$$
$$= -(D/RT_0) \cdot (T_0/T) \cdot dA \cdot (dp/dx) \text{ g-mole/sec} \tag{4}$$
$$dv_0/d\tau = -(D/A_n) \cdot (T_0/T) \cdot dA \cdot (dp/dx) \tag{5}$$
$$(dp_m)_0/d\tau = -(D/V) \cdot (T_0/T) \cdot dA \cdot (dp/dx) \tag{6}$$

The units of mass in (3) and (4) are fixed by the specification that the unit of mass occurring in R shall be the g-mole. In the other equations the only restriction is that corresponding quantities shall be expressed in the same units and that the units of volume and of area shall be equal, respectively, to the cube and the square upon the unit of length.

For water-vapor the constant factors occurring in (3), (4) and (5) when the units of length, area, and volume are, respectively, 1 cm, 1 cm², and 1 cm³, and $T_0 = 273.1$, take the following values for each of the two units of p commonly used in such work.

Unit of p	R	M/R	M/RT_0	$1/A_n$
1 atm	82.06	0.2195	$8.039(10^{-4})$	1
1 mm-Hg	62366	$2.889(10^{-4})$	$1.0577(10^{-6})$	$1.3158(10^{-3})$

Diffusion into Gases.

It has been found empirically that when one gas diffuses into another DP/T^n is constant, P being the total pressure and n being an empirically determined constant. Hence, if D_0 is the value of D at P_0 and T_0, then the

13. VAPOR: DIFFUSION

value at P and T will be $D = D_0(T/T_0)^n(P_0/P)$. The value of n depends upon both the gases.

The subject, diffusion, has recently been discussed in a series of papers by M. Trautz and W. Müller,[153] with special reference to the corrections that must be applied to experimental data on account of various disturbing effects inherent in the procedures followed. They concluded that Winkelmann's data are the only ones capable of yielding reliable values for water-vapor. See also H. Mache.[154]

Table 26.—Diffusion of Water-vapor into Gases

The coefficient of diffusion (D) satisfies the following formulas: $dm/d\tau = -D(d\rho/dx)\cdot dA$; $DP = D_0 P_0 (T/T_0)^n$. Here dm is the mass crossing dA in time $d\tau$, P is the total pressure, and n is an empirically determined constant depending on both gases. See text for further information.

Unit of $D = 1$ cm²/sec. Pressure $P = 1$ atm. Temp. $= t$ °C $= T$ °K

Gas→		CO_2	Air	H_2

I. M. Trautz and W. Müller.[155] Derived from observations by Winkelmann. See text.

	$n \to$	2.115	1.853	1.844
	$D_0 \to$	0.1384	0.219	0.747
		± 0.0015	± 0.001	± 0.003

II. Adapted from compilation [a] by W. P. Boynton and W. H. Brattain.[156]

| | $n \to$ | 2.00 | 1.75 | 1.75 |
t	T		D	
0	273	0.1387	0.220	0.7516
10	283	0.1490	0.234	0.8004
15	288	0.1544	0.242 [b]	0.8254
18	291	0.1576	0.246	0.8404
20	293	0.1598	0.249	0.8507
25	298	0.1653	0.257	0.8761
30	303	0.1708	0.264	0.9021
50	323	0.1939	0.295	1.009
70	343	0.2190	0.328	1.121
90	363	0.2452	0.362	1.238
100	373	0.2590	0.380	1.298

[a] The values given by Boynton and Brattain are based upon the work of G. Guglielmo,[157] F. Houdaille,[158] M. LeBlanc and G. Wuppermann,[159] and A. Winkelmann.[160]
[b] W. E. Summerhays [161] has reported for air $D = 0.282$ at 16.1 °C and $P = 1$ atm.

Diffusion of Water-vapor through Solids.

In the case of a gas diffusing through a septum of thickness Δx and area dA, the total pressure being the same on both sides of the septum,

[153] Trautz, M., and Müller, W., *Ann. d. Physik (5)*, **22**, 313-374 (1935).
[154] Mache, H., *Sitzber. Akad. Wiss. Wien, (Math.-Nat. Abt. IIa)*, **119**, 1399-1423 (1910).
[155] Trautz, M., and Müller, W., *Ann. d. Physik (5)*, **22**, 333-374 (1935).
[156] Boynton, W. P., and Brattain, W. H., *Int. Crit. Tables*, **5**, 62-65 (1929).
[157] Guglielmo, G., *Atti accad. sci. Torino*, **17**, 54-72 (1881); **18**, 93-107 (1882) = *Repert. d. Physik (Exner)*, **19**, 568-581 (1883).
[158] Houdaille, F., Thesis, Paris (1896) → *Fortschr. d. Physik*, **52**, 442-443 (1897).
[159] LeBlanc, M., and Wuppermann, G., *Z. physik. Chem.*, **91**, 143-154 (1916).
[160] Winkelmann, A., *Ann. d. Physik (Wied)*, **22**, 1-31, 152-161 (1884).
[161] Summerhays, W. E., *Proc. Phys. Soc. London*, **42**, 218-225 (1930).

partial pressures of the gas in question being continually constant on each side and differing by Δp, the diffusion equation is that obtained by replacing $d\rho/dx$ and dp/dx in formulas (1) to (5) by $\Delta\rho/\Delta x$ and $\Delta p/\Delta x$.

If the partial pressure of the gas in question is kept constant and equal to p_1 on one side of the septum, and increases on the other as a result of the accumulation of the transmitted gas in a vessel of fixed volume V, then the partial pressure (p_m) in that vessel will increase in accordance with expression (7)

$$dp_m/d\tau = (dp/d\rho)\cdot(d\rho/d\tau) = (dp/d\rho)\cdot(dm/d\tau)/V \qquad (7)$$

If $dp_m/d\tau$ is small, and if the distribution of gas throughout the thickness of the septum is essentially the same as if the partial pressures on the two sides of the septum had for a long time been continuously the same as they actually are at the instant in question, then—and only then—may the $dm/d\tau$ in (7) be validly replaced by its value as given by (1), (2), (3), or (4), as modified in the way just stated. Using (2) and remembering that $-\Delta p = p_1 - p_m$, (7) becomes (7a)

$$dp_m/d\tau = (D\cdot dA/V\cdot\Delta x)(p_1 - p_m) \qquad (7a)$$

of which the solution is (8)

$$\log_e[(p_1 - p_0)/(p_1 - p_m)] = D\cdot dA\cdot\tau/V\cdot\Delta x \qquad (8)$$

in which p_0 is the value of p_m at $\tau = 0$.

Copper.—The data pertaining to the diffusion of water-vapor through copper and given in the compilation by F. Porter [162] are based on the observations by N. B. Pilling [163] and by H. G. Deming and B. C. Hendricks.[164] They consist of a value for D/D_h for Cu at 700 °C from the former, and $D_h T_0/T$ for Cu at 500 and at 750 °C from the latter, D_h being the value of the coefficient for H_2. The values that Pilling actually gives are the relative values of $dp_m/d\tau$ of expression (7) for each of several gases. The ratio of any two of those values will be equal to the ratio of the D's for the two gases only if the conditions were such as to justify expression (7a), if p_m were always negligible in comparison with p_1, if p_1 were the same for each gas, and if the state of the copper were the same in both cases. The original article contains nothing regarding any of these items except the last. The same copper tube was used in all cases; but the purpose of the article was to prove that the structure of copper is profoundly modified by heating it in a reducing gas, and there is no indication of the order in which the observations on the diffusion of the several gases (H_2, H_2O, CO, CO_2) were made. Hence the value $D/D_h = 0.065$ at 700 °C inferred by Porter from those observations should be accepted only with great caution.

[162] Porter, F., *Int. Crit. Tables*, **5**, 76 (1929).
[163] Pilling, N. B., *J. Franklin Inst.*, **186**, 373-374 (1918).
[164] Deming, H. G., and Hendricks, B. C., *J. Am. Chem. Soc.*, **45**, 2857-2864 (1923).

13. VAPOR: DIFFUSION

The Deming and Hendricks paper contains a curve from which it is seen that H_2 passed through a certain septum of Cu at 700 °C at the rate of 0.020 mg/hr·cm² when $\Delta p/\Delta x = 1$ atm/mm; whence, by equation (3), $10^6 D_h = 22$ cm²/sec at 700 °C. If this Cu septum was equivalent in structure to that used by Pilling, and if Porter's inference from Pilling's observations is justified, then for water-vapor and Cu at 700 °C, $10^6 D = 22(0.065) = 1.43$ cm²/sec, and $dm/d\tau = 320(10^{-12})$ g/sec·cm·atm.

Recently, J. H. deBoer and J. D. Fast [165] have reported for water-vapor through Cu at 810 °C $dm/d\tau = 34(10^{-12})$ g/sec·cm·atm or $10^6 D = 0.150$ cm²/sec. This value is only slightly greater than 1/10 of that inferred in the preceding paragraph.

Table 27.—Permeability of Rubber to Water-vapor
(These sources are those quoted in *Int. Crit. Tables*, **2**, 272, and **5**, 76. See also Table 28.)

$dm/d\tau = D \cdot dA \cdot (\Delta \rho/\Delta x)$, $dv_0/d\tau = (DT_0/A_n T) \cdot dA \cdot (\Delta p/\Delta x)$, $\kappa \equiv DT_0/A_n T$, subscript h indicates that the quantity applies to hydrogen. For explanations of symbols, see text.

J. Dewar [169] made some determinations on Para rubber. It was initially 1 mm thick, but was stretched until Δx was about 0.01 mm. One side of the rubber was in contact with water, which was considered as equivalent to the surface being in contact with saturated vapor. Under these conditions, at 15 °C, he found $\kappa/\kappa_{air} = 163$. For another sample, also at 15 °C, he found $\kappa_{air}/\kappa_h = 0.178$. Whence for water-vapor at 15 °C $\kappa/\kappa_h = 29$.[a]

J. D. Edwards and S. F. Pickering [170] report data leading to the following values for a dental dam, for the two cases, assumed to be equivalent: (1) saturated vapor on one side; (2) water in contact with one side of the rubber.

Unit of $\Delta x = 1$ mm, of $D = 1$ cm²/sec, of $\kappa = 1$ cm²/sec·atm. Temp. = 25 °C

(1) Vapor				(2) Water			
Δx	$10^1 D$	$10^6 \kappa$[a]	κ/κ_h[b]	Δx	$10^6 D$	$10^6 \kappa$[a]	κ/κ_h
0.18	23.5	21.5	47	0.21	46.4	42.5	95
0.25	28.1	25.7	62	0.25	51.9	47.2	115

[a] In Porter's compilation [162] Dewar's value for κ/κ_h is given as 16, and the Edwards and Pickering values for $10^6 \kappa$ are given as 16.0 for (1) and 35 for (2). The values given in the table are those derived by the compiler from the data given in the original papers.

[b] In Whitby's compilation [168] the value of κ/κ_h for (1) is given as 55, essentially the mean of the two here tabulated.

---

[165] deBoer, J. H., and Fast, J. D., *Rec. Trav. Chim., Pays-Bas (4)*, **16**, 970-974 (1935).
[166] Barrer, R. M., *Nature*, **140**, 106-107 (L) (1937).
[167] Schumacher, E. E., and Ferguson, L., *Ind. Eng. Chem.*, **21**, 158-162 (1929).
[168] Whitby, G. S., *Int. Crit. Tables*, **2**, 272 (1927).
[169] Dewar, J., *Proc. Roy. Inst. Grt. Brit.*, **21**, 813-826 (1915).
[170] Edwards, J. D., and Pickering, S. F., *Sci. Papers Bur. Stand.*, **16**, 327-362 (S387) (1920) → *Chem. Met. Eng.*, **23**, 17-21, 71-75 (1920).

Rubber.*—Certain remarks on the diffusion process in rubber have been published by R. M. Barrer.[166] Since the publication of the *International Critical Tables*, E. E. Schumacher and L. Ferguson [167] have published their studies of the passage of water-vapor through rubber. The work was done under such conditions that expression (8) might be expected to apply. Their results are presented in the form of smooth curves of p_m versus τ. When values of p_m were read from those curves and log $(p_1 - p_m)$ was plotted against τ, the resulting graphs were not straight, but were convex toward the origin, showing that the observations do not satisfy expres-

Table 28.—Diffusion of Water-vapor through Various Solids

(For diffusion through copper see text; through rubber see also text and Table 27.)

Δx = thickness of film; D = coefficient of diffusion; p = vapor pressure on "wet" side of film, on other side $p = 0$.

Unit of $D = 1$ cm²/sec, of $p = 1$ mm-Hg, of $\Delta x = 1$ cm. Temp. = t °C

I. *Acetylcellulose and nitrocellulose.*[170a] Not more than one film was used in a series; D varies with p, its value being determined by the amount of water in the film. The films were prepared on a polished plate (glass or silver), and stripped. The side that had been against the plate had a high polish; if that is the "wet" side of the film, D is greater than if the diffusion took place in the opposite direction; in a case for which data are given the difference was 14 per cent.

	Acetylcellulose, unfilled $t = 25$ °C, $\Delta x = 0.010$ cm				Nitrocellulose, unfilled $t = 20$ °C, $\Delta x = 0.004$ cm	
p	23.76	20.90	13.78	4.99	17.59	14.90
$10^6 D$	109	102	81	83	164	142

Acetylcellulose filled with 1 part toluenesulfamide to 1.5 parts ethylphthalate.
$t = 25$ °C, $\Delta x = 0.0054$ cm, $p = 23.756$ mm-Hg

Filler (%)	0	10	25
$10^6 D$	99	74	45

II. *Miscellaneous materials.*[171]

Material	$10^3 \Delta x$	Δp	t	$10^6 D$
Asphalt sealing compound	82.0	22.8	25.0	3.30
Balata	38.4	22.8	25.0	5.16
Balata	38.4	10.5ª	25.0	4.82
Bakelite (molded)	55.6	22.8	25.0	13.8
Benzyl cellulose	68.6	22.8	23.9	30.3
Cellulose acetate	16.3	22.8	25.0	448
Cellulose acetate	15.6	22.8	25.0	465
Cellulose acetate (plasticized)	2.45	22.8	25.0	3.39
Cellulose film (waterproof)	4.40	22.8	25.0	234
Phenol fiber	78.0	22.8	25.0	14.0
Phenol fiber	78.8	22.8	25.0	15.0

* See also Tables 27 and 28.
[170a] Wosnessensky, S., and Dubnikow, L. M., *Koll. Z.*, 74, 183-194 (1936).
[171] Taylor, R. L., Herrmann, D. B., and Kemp, A. R., *Ind. Eng. Chem.*, 28, 1255-1263 (1936).

13. VAPOR: DIFFUSION

Table 28—(Continued)

Material	$10^3 \Delta x$	Δp	t	$10^6 D$
Polystyrene ($10^3 \Delta x$ from 54 to 209)	Various	18.0	21.1	11.3
Polystyrene ($10^3 \Delta x$ from 54 to 209)	Various	11.9[a]	21.1	11.2
Rubber hydrochloride (plasticized)	2.9	18.0	21.1	148
Vinyl chloride (plasticized)	48.1	18.0	21.1	10.7
Vinyl chloride (plasticized)	46.0	18.0	21.1	11.1
Wax, hydrocarbon	51	18.6	21.1	0.18
Rubber [b] (soft-vulcanized)	35.4	7.66	25.0	19.0
Rubber [b] (soft-vulcanized)	35.4	17.8	25.0	19.1
Rubber [b] (soft-vulcanized)	35.4	21.5	25.0	19.7
Rubber [b] (soft-vulcanized)	35.4	22.8	25.0	21.0
Rubber [b] (soft-vulcanized)	35.4	23.6	25.0	22.1
Rubber [b] (soft-vulcanized)	36.6	5.8[a]	25.0	22.7
Rubber (hard; i. e., 32% combined Sulfur)	48.5	22.8	25.0	4.33
Paragutta insulation[c]	46.9	22.8	25.0	5.51
Paragutta insulation[c]	44.9	22.8	25.0	5.34
Paragutta insulation[c]	46.7	10.5[a]	25.0	5.14
Paragutta insulation[c]	44.9	10.5	25.0	4.94
Guttapercha	32.4	22.8	25.0	4.25
Guttapercha	32.4	10.5[a]	25.0	4.15
Chloroprene polymer (vulcanized)	86.5	18.0	21.1	7.44
Polyethylene tetrasulfide	76.3	18.0	21.1	0.62

Temperature → Material	0.0	21.0	21.1	25	30	35
			$10^6 D$			
Polystyrene			11.3			13.3
Rubber (vulcanized)	13.2	19.5		21.0		25.2
Silk (varnished)					9.9	

[a] The partial pressure of H_2O on one side of the material was always zero except for the cases to which this note refers; for them the partial pressure on the low-pressure side was as follows: for $\Delta p = 10.5$ it was 12.3; for $\Delta p = 11.9$, 6.1; for $\Delta p = 5.8$, 17.8.

[b] Composition: Crepe 90, sulfur 1.5, zinc oxide 2.5, mineral rubber 3.0, paraffin 1.5, stearic acid 0.5, tetramethyl thiuram disulfide 0.5, phenyl-β-naphthylamine 1.0. Vulcanized in mold at 126 °C for 20 min.

[c] Submarine cable insulation.

---

sion (8). Furthermore, the values derived by the observers for what they call the permeability of rubber seem to have been based in each case on some kind of average of the slopes of the graphs over the first interval of 20 to 40 hours. Had an equal interval at a later time been used, much smaller values would have been obtained; their observations extended to 120 to 300 hours. For these reasons the reader is referred directly to their paper. They stated that D is inversely proportional to the thickness of the specimen.

For values reported in the *International Critical Tables*[162, 168] see Table 27.

Miscellaneous materials.—The articles cited in Table 28 contain much detailed data; they have not been critically examined by the compiler, who has merely converted the final values into the units here used.

14. Pressure-Volume-Temperature Associations for Dilated Water-Vapor

By dilated water-vapor is meant the vapor at a pressure that is less than its saturation pressure at the temperature concerned. The adjective "dilated" is less ambiguous than the more frequently used "unsaturated," which is equally appropriate to the supersaturated condition.

Equations of State.

Numerous equations of state for dilated water-vapor have been proposed, but none has been generally accepted (see [172-186]).

Only the following equations of state are further considered in this compilation: Linde's (1905), equation (1), which represents very exactly the observations of O. Knoblauch, R. Linde, and H. Klebe,[187] and in which $T_1 = 273.0 + t$.

$$\begin{aligned}Pv/m &= 47.1T_1 - P[1 + 2P(10^{-6})] \cdot [0.031(373/T_1)^3 - 0.0052] \\ & \qquad (\text{kg/m}^2) \cdot (\text{m}^3/\text{kg}) \\ &= 4.558_5 T_1 - P[1 + 0.02066P] \cdot [31(373/T_1)^3 - 5.2] \\ & \qquad \text{cm}^3 \cdot \text{atm/g} \end{aligned} \quad (1)$$

Callendar's (1920-1928) equation (2), based upon quite different considerations, in which $T = 273.1 + t$,

$$\begin{aligned} Pv/m &= 47T - P[0.0263(373.1/T)^{10/3} - 0.001] \ (\text{kg/m}^2) \cdot (\text{m}^3/\text{kg}) \\ &= 4.551_2 T - P[26.3(373.1/T)^{10/3} - 1] \ \text{cm}^3 \cdot \text{atm/g} \end{aligned} \quad (2)$$

and Keyes, Smith, and Gerry's equation (3),[188] representing their own observations at the Massachusetts Institute of Technology.

$$Pv/m = 4.55504 T_2 + PB \ \text{cm}^3 \cdot \text{atm/g} \quad (3)$$

in which $T_2 = 273.16 + t$, temperature being t °C, and

---

[172] Linde, R., *Mitt. Forsch. Geb. Ing.*, **21**, 57-92 (1905).
[173] Bose, E., *Z. Elektrochem.*, **14**, 269-271 (1908).
[174] Wohl, A., *Z. physik. Chem.*, **87**, 1-39 (1914).
[175] Holst, G., *Proc. Akad. Wet. Amsterdam*, **19**, 932-937 (1917).
[176] Callendar, H. L., "Properties of Steam," 1920; *World Power*, **1**, 274-280, 325-328 (1924); *Engineering (London)*, **126**, 594-595, 625-627, 671-673 (1928).
[177] Eichelberg, *Mitt. Forsch. Geb. Ing.*, **220**, 1-31 (1920).
[178] Tumlirz, O., *Sitzb. Akad. Wiss. Wien (2a)*, **130**, 93-133 (1921).
[179] Jazyna, W., *Z. techn. Physik*, **8**, 159-160 (1927).
[180] Newitt, D. M., *Proc. Roy. Soc. (London) (A)*, **119**, 464-480 (1928).
[181] Heck, R. C. H., *Mechan. Eng.*, **51**, 116-122 (1929).
[182] Hausen, H., *Forsch. Gebiete Ingenieurw.*, **2**, 319-326 (1931).
[183] Jakob, M., *Engineering (London)*, **132**, 143-146, 651-653, 684-686, 707-709 (1931).
[184] Naumann, F., *Z. physik. Chem. (A)*, **159**, 135-144 (1932).
[185] Sugawara, S., *Mem. Col. Eng. Kyoto Imp. Univ.*, **7**, 17-48 (1932).
[186] Keyes, F. G., Smith, L. B., and Gerry, H. T., *Mechan. Eng.*, **56**, 87-92 (1934); *Idem*, **57**, 164, 176 (1935); *Proc. Am. Acad. Arts & Sci.*, **70**, 319-364 (1935).
[187] Knoblauch, O., Linde, R., and Klebe, H., *Mitt. Forsch. Geb. Ing.*, **21**, 33-55 (1905).

Table 29.—Isopiestics (kg*/cm²) of the Specific Volume of Dilated Water-vapor

(See also Tables 30 and 31).

The values given in the first section are those of Keyes, Smith, and Gerry, computed by means of formula (3) for $v/m > 10$ cm³/g, and derived by graphical methods for smaller volumes. In the second are the values obtained by J. Havlíček and L. Miškovsky,[191] and in the third are those derived by M. Jakob [189] from the observed values of the specific heat at constant pressure. See also Table 32.

Unit of $p = 1$ kg*/cm² $= 980665$ dynes/cm² $= 0.96784$ atm; of $v/m = 1$ cm³/g. Temp. $= t$ °C

I. Keyes, Smith, and Gerry (1935)

$p \rightarrow$ t	1	5	10	25	50	100
100	1729.6					
150	1975.4					
200	2215.8	433.8	210.4			
250	2454.1	484.1	237.6	88.99		
300	2691.3	533.2	263.3	101.1	46.41	
350	2927.9	581.6	288.2	112.1	53.12	23.03
400	3164.1	629.6	312.7	122.6	59.05	27.05
450	3400.1	677.4	337.0	132.7	64.60	30.41
500	3636.0	725.0	361.1	142.7	69.92	33.45
550	3871.8	772.5	385.1	152.6	75.10	36.32

$p \rightarrow$ t	150	200	250	300	350	400
350	11.98					
400	16.10	10.31	6.366	3.022[a]	2.173[a]	
450	18.90	13.05	9.456	6.979	5.168	3.858
500	21.25	15.11	11.39	8.898	7.108	5.774
550	23.36	16.87	12.96	10.35	8.494	7.109

II. Havlíček and Miškovsky (1936)

$p \rightarrow$ t	1	25	50	100	150
100	1728.7				
150	1974.5				
200	2215.1				
250	2453.4	88.96			
300	2690.5	101.0	46.38		
350	2927.2	112.0	53.07	23.02	11.97
400	3163.3	122.5	59.02	27.04	16.11
450	3399.4	132.7	64.58	30.41	18.91
500	3635.1	142.7	69.92	33.47	21.27
550	3870.9	152.6	75.10	36.34	23.40

$p \rightarrow$ t	200	250	300	350	400
400	10.31	6.365			
450	13.07	9.473	6.970		
500	15.13	11.42	8.927	7.126	5.761
550	16.91	13.01	10.40		

[188] Keyes, F. G., Smith, L. B., and Gerry, H. T., *Mech. Eng.*, **57**, 164, 176 (1935); *Proc. Am. Acad. Arts and Sci.*, **70**, 319-364 (1935).

PROPERTIES OF ORDINARY WATER-SUBSTANCE

Table 29—(Continued)

III. Jakob (1912)

$p \rightarrow$	1	3	5	7	9
t			v/m		
110	1781.6				
120	1830.2				
130	1878.9				
140	1927.3	630.5			
150	1975.5	647.6			
160	2023.7	664.6	392.3		
170	2071.6	681.4	403.0	283.3	
180	2119.6	698.1	413.6	291.3	223.2
190	2167.4	714.6	423.9	299.2	229.6
200	2215.2	731.1	434.2	306.8	235.9
220	2310.7	763.9	454.4	321.7	247.9
240	2406.0	796.4	474.4	336.4	259.7
260	2501.1	828.8	494.2	350.9	271.2
280	2596.0	861.1	514.0	365.3	282.6
300	2690.9	893.2	533.7	379.5	293.9
350	2927.9	973.3	582.4	414.7	321.7
400	3164.3	1052.9	630.6	449.6	349.1
450	3400.6	1132.3	678.6	484.2	376.2
500	3636.4	1211.3	726.2	518.4	402.9
550	3872.2	1290.2	773.8	552.5[b]	429.5

$p \rightarrow$	11	13	15	17	19
t			v/m		
190	185.2				
200	190.6	159.1	135.9		
220	200.9	168.3	144.3	125.9	111.3
240	210.8	177.0	152.0	133.0	118.0
260	220.5	185.3[a]	159.5	139.7	124.2
280	230.0	193.5	166.8	146.3	130.2
300	239.3	201.6	173.9	152.7	135.9
350	262.4	221.4	191.3	168.3	150.1
400	285.0	240.7	208.2[b]	183.4	163.7
450	307.4	259.7	224.8	198.1	177.1
500	329.4	278.5	241.2	212.6	190.1
550	351.2	297.1	257.3	226.9	202.9

[a] It seems probable that one of these values given by Keyes, Smith, and Gerry for $t = 400$, $p = 300, 350$ is incorrect, see Table 32.

[b] Published values 556.8 for $p = 7$, $t = 550$, and 210.9 for $p = 15$, $t = 400$, are obviously affected by typographical errors.

$$B = B_0 + B_0^2 g_1(\tau P) + B_0^4 g_2(\tau P)^3 - B_0^{13} g_3(\tau P)^{12}, \tau = 1/T_2,$$
$$B_0 = 1.89 - 2641.62\tau(10)^{80870\tau^2},$$
$$g_1 = 82.546\tau - 1.6246\tau^2(10^5),$$
$$g_2 = 0.21828 - 1.2697\tau^2(10^5),$$
$$g_3 = 3.635(10^{-4}) - 6.768\tau^{24}(10^{64}).$$

This equation and the data given in the Academy paper supersede all others previously published by these observers.

(To p. 83)

[189] Jakob, M., *Z. Ver. deuts. Ing.*, **56**, 1980-1988 (1912).
[190] Knoblauch, O., and Jakob, M., *Mitt. Forsch. Geb. Ing.*, **35, 36**, 109-152 (1906); Knoblauch, O., and Mollier, H., *Idem*, **108-109**, 79-106 (1911).
[190a] Wohl, A., *Z. physik. Chem.*, **87**, 1-39 (1914).
[191] Havlíček, J., and Miškovsky, L., *Helv. Phys. Acta*, **9**, 161-207 (1936).

14. VAPOR: P-V-T DATA

Table 30.—Isopiestics (lb*/in²) of the Specific Volume of Dilated Water-vapor [192]

As published, the pressures refer to $g = 981.16$ cm/sec² (London); here, the unit so defined will be denoted by the subscript L; those without subscripts are based upon the international value, $g = 980.665$.

Unit of $p = 1(\text{lb*/in}^2)_L = 0.06898_2 \text{bar} = 0.06808_0 \text{atm}$; of $v/m = 1$ ft³/lb $= 62.428$ cm³/g.
Temp. $= t$ °C

p	t_{sat}	t_{sat}	250	300	350	400	450	500
					v/m			
400	228.8	1.1712	1.2517	1.4269	1.5872	1.7391	1.8867	2.0292
450	235.2	1.0419	1.0941	1.2562	1.4019	1.5392	1.6710	1.7991
500	241.2	0.9380	0.9662	1.1191	1.2532	1.3790	1.4988	1.6151
600	251.9	0.7805	0.7729	0.9124	1.0309	1.1387	1.2409	1.3392
700	261.2	0.6672		0.7643	0.8709	0.9667	1.0560	1.1422
800	269.7	0.5804		0.6515	0.7508	0.8379	0.9180	0.9943
900	277.4	0.5120		0.5627	0.6569	0.7372	0.8102	0.8795
1000	284.5	0.4570		0.4901	0.5816	0.6563	0.7239	0.7871
1200	297.2	0.3727		0.3767	0.4673	0.5354	0.5944	0.6490
1400	308.4	0.3111			0.3839	0.4482	0.5018	0.5504
1600	318.4	0.2635			0.3191	0.3823	0.4320	0.4762
1800	327.5	0.2255			0.2661	0.3302	0.3775	0.4182
2000	335.8	0.1936			0.2205	0.2879	0.3336	0.3719
2400	350.6	0.1433				0.2216	0.2670	0.3022
2800	361.3	0.1020				0.1703	0.2181	0.2517
3200	373.6	0.0629				0.1253	0.1800	0.2136
3600						0.0805	0.1491	0.1837
4000							0.1213	0.1589

Table 31.—Specific Volume of Dilated Water-vapor, and its Defect [193]
(Third International Steam-Table Conference, 1934.)

For the allowed tolerances and other data forming the skeleton tables then adopted for steam engineering, see Table 260.

$\Delta \equiv 4.70636 T/p - v/m$, $T = 273.1 + t$, units as stated below; $4.70636 T$ (cm³/g)·(kg*/cm²) $= 4.555 T$ cm³·atm/g. These values of Δ are comparable with those in Tables 32 and 33. Specific volume is v/m, pressure is p.

Unit of $p = 1$ kg*/cm² $= 0.96784$ atm; of $4.70636T = 1$ (cm³/g)·(kg*/cm²); of v/m and of $\Delta = 1$ cm³/g. Temp. $= t$ °C

$p \rightarrow$ t	1	5	10	25	50	4.70636T
			v/m			
100	1730					1755.9
150	1975					1991.3
200	2216	433.8	210.4			2226.6
250	2454	484.1	237.6	89.0		2461.9
300	2691	533.2	263.3	101.1	46.41	2697.2
350	2928	581.6	288.2	112.1	53.12	2932.5
400	3164	629.6	312.7	122.6	59.05	3167.8
450	3400	677.4	337.0	132.7	64.60	3403.2
500	3636	725.0	361.1	142.7	69.92	3638.5
550	3872	772.5	385.1	152.6	75.10	3873.8

[192] Callendar, H. L., *Proc. Inst. Mech. Eng.*, **1929**, 507-527 (1929).
[193] Third International Steam-Table Conference, 1934; *Mech. Eng.*, **57**, 710-713 (1935).

Table 31—(Continued)

$p \rightarrow$	75	100	125	150	200	250	300
t				v/m			
300	27.48						
350	33.22	23.03	16.66	11.98			
400	37.78	27.05	20.53	16.10	10.31	6.366	3.02
450	41.83	30.41	23.52	18.90	13.05	9.46	6.98
500	45.62	33.45	26.14	21.25	15.11	11.39	8.90
550	49.25	36.32	28.55	23.36	16.87	12.96	10.35

$p \rightarrow$	1	5	10	25	50	75	100	125	150	200	250	300
t							Δ					
100	26											
150	16											
200	11	11.7	12.3									
250	8	8.3	8.6	9.4								
300	6	6.2	6.4	6.8	7.53	8.48						
350	4	4.9	5.0	5.2	5.53	5.88	6.29	6.80	7.57			
400	4	4.0	4.1	4.1	4.31	4.46	4.63	4.81	5.02	5.53	6.305	7.54
450	3	3.2	3.3	3.4	3.46	3.55	3.62	3.71	3.79	3.97	4.15	4.36
500	2	2.7	2.7	2.8	2.85	2.89	2.93	2.97	3.01	3.08	3.16	3.23
550	1	2.3	2.3	2.4	2.38	2.40	2.42	2.44	2.46	2.50	2.54	2.56

Table 32.—Defect of Specific Volume of Dilated Water-vapor: Isopiestics (kg*/cm²)

Here are assembled the values of the amount (Δ) by which the several values of the specific volume of water-vapor, as given in Table 29, fall short of that of an ideal gas ($M = 18.0154$) under the same conditions of temperature and pressure. They may conveniently be used for computing the specific volume as defined by any of those sets of values and at any temperature and pressure within the range of the table.

$v/m = 4.70636T/p - \Delta$, where $T = 273.1 + t$, units being as stated below; $4.70636T$ (cm³/g)·(kg*/cm²) = $4.555T$ cm³·atm/g, the basis used in this compilation.

Example: at 300 °C and 25 kg*/cm², $\Delta = 6.8$ for the K.S.G. set, and 6.9 for the H.M. set; that is, this specific volume in the second set is 0.1 cm³/g smaller than in the first, its actual value being $(2697.21/25) - 6.9 = 101.0$ cm³/g, agreeing with the corresponding H.M. value in Table 29.

Unit of $p = 1$ kg*/cm² = 980665 dynes/cm² = 0.96784 atm; of $4.70636T = 1$ (cm³/g)·(kg*/cm²); of $\Delta = 1$ cm³/g. Temp. = t °C

Keyes, Smith, and Gerry (1935)

$p \rightarrow$	1	5	10	25	50	100	150	200	250	300	350	400
t							Δ					
100	26.3											
150	15.9											
200	10.8	11.5	12.3									
250	7.8	8.3	8.6	9.49								
300	5.9	6.2	6.4	6.8	7.53							
350	4.6	4.9	5.0	5.2	5.53	6.30	7.57					
400	3.8	4.0	4.1	4.1	4.31	4.63	5.02	5.53	6.305	7.538[a]	6.878[a]	
450	3.1	3.2	3.3	3.4	3.46	3.62	3.79	3.97	4.157	4.365	4.555	4.650
500	2.5	2.7	2.8	2.8	2.85	2.94	3.01	3.08	3.16	3.230	3.288	3.322
550	2.0	2.3	2.3	2.4	2.38	2.42	2.46	2.50	2.54	2.56	2.574	2.576

14. VAPOR: P-V-T DATA

Table 32—*(Continued)*

Havlíček and Miškovsky (1936)

p→ t	4.70636T	1	25	50	100	150	200	250	300	350	400
							Δ				
100	1755.94	27.2									
150	1991.26	16.8									
200	2226.58	11.4									
250	2461.90	8.5	9.52								
300	2697.21	6.7	6.9	7.56							
350	2932.53	5.3	5.3	5.58	6.30	7.58					
400	3167.85	4.5	4.2	4.34	4.64	5.01	5.53	6.306			
450	3403.17	3.8	3.4	3.48	3.62	3.78	3.95	4.140	4.374		
500	3638.49	3.4	2.8	2.85	2.91	2.99	3.06	3.13	3.201	3.268	3.335
550	3873.80	2.9	2.4	2.38	2.40	2.42	2.46	2.48	2.51		

Jakob (1912).

p→ t	4.70636T	1	3	5	7	9	11	13	15	17	19
							Δ				
110	1803.00	21.4									
120	1850.07	19.9									
130	1897.13	18.2									
140	1944.20	16.9	17.6								
150	1991.26	15.8	16.2								
160	2038.32	14.6	14.8	15.4							
170	2085.39	13.8	13.7	14.1	14.6						
180	2132.45	12.8	12.7	12.9	13.3	13.7					
190	2179.52	11.9	11.9	12.0	12.2	12.6	12.9				
200	2226.58	11.4	11.1	11.1	11.3	11.5	11.8	12.2	12.5		
220	2320.71	10.0	9.7	9.7	9.8	10.0	10.1	10.2	10.4	10.6	10.8
240	2414.83	8.8	8.5	8.6	8.6	8.6	8.7	8.8	9.0	9.0	9.1
260	2508.86	7.8	7.5	7.6	7.5	7.6	7.6	7.7	7.8	7.9	7.8
280	2603.09	7.1	6.6	6.6	6.6	6.6	6.6	6.7	6.7	6.8	6.8
300	2697.21	6.3	5.9	5.7	5.8	5.8	5.9	5.9	5.9	6.0	6.0
350	2932.53	4.6	4.2	4.1	4.2	4.1	4.2	4.2	4.2	4.2	4.2
400	3167.85	3.6	3.0	3.0	3.0	2.9	3.0	3.0	3.0[b]	2.9	3.0
450	3403.17	2.6	2.1	2.0	2.0	1.9	2.0	2.1	2.1	2.1	2.0
500	3638.49	2.1	1.5	1.5	1.4	1.4	1.4	1.4	1.4	1.5	1.4
550	3873.80	1.6	1.1	1.0	1.0[b]	0.9	1.0	0.9	1.0	1.0	1.0

[a] It seems probable that one of the values given for v/m at $t = 400$, $p = 300, 350$ (Table 29) is incorrect.

[b] From the corrected values as given in Table 29.

(Cont'd from p. 80)

Data computed by means of each of these equations will be found in the tables here given. In certain cases they are represented by the amounts by which they fall below the corresponding values defined by $Pv/m = 4.555T$ cm³·atm/g. These defects, being strictly comparable, afford a ready means for comparing the several sets of values; they also facilitate interpolation.

By graphical means, M. Jakob [189] has derived an extended table of specific volumes from the values found by Knoblauch and co-workers for the specific heat of water-vapor.[190] These values agree well with the direct observations of Knoblauch, Linde, and Klebe, represented also by Linde's equation (1), and extend far beyond the range covered by them. They are given in Table 29.

Several sets of directly observed values are also tabulated, either directly or in terms of deviations. Values that have been accepted more or less by engineers at various times may be found in the numerous steam tables, see Table 259.

(To p. 90)

Table 33.—Defect of Specific Volume of Dilated Water-vapor: Isopiestics (atm)

Here are assembled the several values of the amount (Δ) by which certain observed and computed values of the specific volume of dilated water-vapor fall short of that of an ideal gas ($M = 18.0154$) under the same conditions of temperature and pressure. They furnish a convenient means for comparing the several sets and for determining the specific volume, as defined by those sets, corresponding to any temperature and pressure within the range covered. At lower temperatures the relative departure from ideality, even at saturation, is small, $p\Delta/4.555T$ ($= \delta/v_c{}^*$ of Table 250) being < 0.5 per cent if $t \leqq 60$ °C, 0.9 per cent at 80 °C, 1.7 per cent at 100 °C, and 2.0 per cent at 110 °C.

Example: What is the value of pv/m at 160 °C and 3 atm on the basis of the "I" values (Int. Crit. Tables)? At 160 °C and 5 atm $\Delta = 15.2$, at 1 atm it is $15 - 1 = 14$; whence at 3 atm it is 14.6, giving $p\Delta = 43.8$, and $pv/m = 1972.8 - 43.8 = 1929.0$. The value given in I.C.T. is 1929.

$v/m = 4.555T/p - \Delta$ cm^3/g; units as stated below.

Unit of $p = 1$ atm $= 1.01325$ bars $= 1.03323$ kg*/cm^2, of $4.555T = 1$ cm^{3}atm/g, of $\Delta = 1$ cm^3/g; $T = 273.1 + t$. Temp. $= t$ °C

100$p \to$		17	21	30	37	45	57	61	67	68	72	73
t	4.555T						Δ (Shirai)[a]					
80	1608.4		1		16							
100	1699.5	20		14		12	7		8			
120	1790.6			7					7	7	3.4	
140	1881.7									10	4.3	5.0

$p \to$		____1____			____5____			____10____		
Ref[a]$\to$		I	L	C	I	L	C	I	L	C
t	4.555T					Δ				
110	1745.0	22	23	24						
120	1790.6	19	21	23						
130	1836.1	17	19	21						
140	1881.7	16	17	19						
150	1927.2	15	15	18						
160	1972.8		14	17	15.2	15.8	15.4			
170	2018.3		12	16	13.8	14.4	14.2			
180	2063.9		11	14		13.0	13.2			
185	2086.6		10	14		12.4	12.6	14.0	13.8	12.4
200	2155.0		9	13		10.8	11.2		11.9	11.0

[a] References:
  C   Callendar's equation of state, equation (2).
  I   Compilation by F. G. Keyes[194] based on observations by A. Battelli,[195] M. Jakob,[189] and O. Knoblauch, R. Linde, and H. Klebe.[187]
  L   Linde's equation of state, equation (1).
  S   T. Shirai.[196]

---

[194] Keyes, F. G., *Int. Crit. Tables*, **3**, 436 (1928).
[195] Battelli, A., *Ann. Chim. Phys.* (6), **26**, 394-425 (1892); (7), **3**, 408-431 (1894).
[196] Shirai, T., *Bull. Chem. Soc. Japan*, **2**, 37-40 (1927).

14. VAPOR: P-V-T DATA

Table 34.—Defect of Specific Volume of Dilated Water-vapor: Isopiestics (lb*/in²)
(From Callendar's data, see Table 30.)

$v/m = 66.906_6 T/p - \Delta$ cm³/g, the unit of p being 1 (lb*/in²)$_L$ where $g = 981.16$ cm/sec² (London); $66.906_6 T$ cm³·(lb*/in²)$_L$/g = $4.555T$ cm³·atm/g. Hence these values of Δ are directly comparable with those of Tables 31, 32, and 33. For values of t_{sat}, see Table 30.

Unit of p = 1 (lb*/in²)$_L$; of P = 1 atm = 1.01325 bars; of Δ = 1 cm³/g

$t \rightarrow$ $66.906_6 T \rightarrow$	250 34998.8	300 38344.2	350 41689.5	400 45034.8	450 48380.1	500 51725.5	t_{sat}
p	P			Δ			
400	27.232	9.36	6.78	5.14	4.02	3.17	2.63	10.84
450	30.636	9.47	6.79	5.12	3.99	3.19	2.63	10.53
500	34.040	9.68	6.82	5.14	3.98	3.19	2.62	10.26
600	40.848	10.08	6.95	5.13	3.97	3.17	2.60	9.82
700	47.656		7.06	5.20	3.99	3.19	2.59	9.42
800	54.464		7.26	5.24	3.99	3.17	2.58	9.16
900	61.272		7.48	5.31	4.02	3.18	2.57	8.96
1000	68.080		7.75	5.38	4.06	3.19	2.59	8.78
1200	81.696		8.43	5.57	4.10	3.21	2.59	8.53
1400	95.312			5.81	4.19	3.23	2.59	8.37
1600	108.928			6.14	4.28	3.27	2.60	8.28
1800	122.544			6.55	4.40	3.31	2.63	8.25
2000	136.160			7.08	4.54	3.64	2.65	8.28
2400	163.392				4.93	3.49	2.69	8.44
2800	190.624				4.45	3.66	2.77	8.83
3200	217.856				6.25	3.88	2.83	9.59
3600	245.088				7.48	4.13	2.90	
4000	272.320					4.52	3.01	

Table 35.—Isometrics of the Pressure (atm) of Dilated Water-vapor [197]

Four sets of experimental data, here indicated by superscripts *a, b, c,* and *d*, as finally corrected by the authors, are given. These include, with corrections, and supersede similar data previously reported from that laboratory,[198] and are the data from which formula (3) was derived.

Unit of v/m = 1 cm³/g, of P = 1 atm = 1.01325 bars = 1.03323 kg*/cm². Temp. = t °C

$v/m \rightarrow$ t	150.0^a	140.0^b	100.0^b	97.5^a	75.0^a	75.0^b
			P			
195		13.787				
196		13.898				
197		13.957				
198		14.003				
199		14.040				
200	13.220	14.074				
210	13.591	14.479				
212.5			19.612			
220	13.952	14.871	20.086	20.514		(26.238)^c
230	14.302	15.252	20.664	21.106	26.503	26.482

[197] Keyes, F. G., Smith, L. B., and Gerry, H. T., *Proc. Am. Acad. Arts Sci.*, **70**, 319-364 (1935).
[198] Smith, L. B., and Keyes, F. G., *Mech. Eng.*, **52**, 123-124 (1930); **53**, 135-137 (1931); **54**, 123-124 (1932); Keyes, F. G., and Smith, L. B., *Idem*, **53**, 132-135 (1931).

Table 35—*(Continued)*

$v/m \rightarrow$ t	150.0[a]	140.0[b]	100.0[b]	97.5[a]	75.0[a]	75.0[b]
				P		
240	14.647	15.627	21.217	21.688	27.309	27.303
250	14.986	16.014	21.769	22.258	28.099	28.082
260	15.323	16.366	22.297	22.816	28.869	28.854
270	15.658	16.745	22.825	23.365	29.618	29.592
280	15.990	17.087	23.356	23.905	30.355	30.345
290	16.319	17.448	23.863	24.437	31.081	31.055
300	16.646	17.802	24.378	24.964	31.794	31.767
310	16.971	18.146	24.881	25.485	32.496	32.476
320	17.297	18.502	25.391	26.001	33.188	33.154
330	17.620	18.839	25.892	26.514	33.870	33.857

$v/m \rightarrow$ t	57.5[a]	50.0[b]	50.0[c]	40.0[b]	40.0[c]	40.0[a]
				P		
250	34.983	39.052	39.388			
260	36.051	40.469	40.592			
264				48.689	48.827	
265					48.996	
270	37.090			49.753	49.855	49.856
280	38.104	42.988	43.088	51.417	51.511	51.533
290	39.101					53.123
300	40.078	45.366	45.483	54.613	54.708	54.665
310	41.035					56.169
320	41.980	47.688	47.751	57.643	57.711	57.642
330	42.915					59.096
340		49.914	49.985	60.586	60.646	
360		52.105	52.188	63.428	63.497	See end
380		54.257	54.308	66.225	66.279	of table
400		56.384	56.426	68.963	69.015	
420		58.465	58.503	71.658	71.682	
440		60.546	60.555	74.311	74.320	
460		62.605	62.608	76.966	76.947	

$v/m \rightarrow$ t	39.5[b]	30.0[b]	30.0[c]	20.0[d]	20.0[b]	20.0[c]
				P		
264	49.164					
270	50.322					
280	52.008					
281		63.622	63.730			
290	53.635	65.821	65.921			
300	55.229	68.145	68.220			
305					90.499	90.604
305.5				90.765		(91.143)[f]
310	56.745			92.582		
320	58.261	72.568	72.611	96.342	96.449	96.301
330	59.718			100.009		
340		76.832	76.841	103.586	103.627	103.654
350				107.103		
360		80.841	80.900	110.525	110.456	110.504
380		84.804	84.845	117.105	117.049	117.092
400		88.674	88.678	123.535	123.400	123.436
420		92.437	92.439	129.632	129.628	129.624
440		96.161	96.152	135.594	135.656	135.674
460		99.836	99.789	141.484	141.580	141.587

14. VAPOR: P-V-T DATA

Table 35—(Continued)

v/m→	17.5[d]	15.0[d]	12.5[d]	10[d]	7.5[d]	6.25[d]
t				P		
313	101.287					
320	104.286					
322		113.965				
330	108.690	118.413				
332.5			130.593			
340	112.988	123.725	135.657			
344.5				151.833		
350	117.154	128.838	142.215	156.607		
357.5					178.402	
360	121.235	133.857	148.566	165.137	181.389	
364.5						194.118
370			154.741	173.384	193.226	201.951
380	129.679	143.453	160.516	181.385	204.683	216.133
400	136.666	152.670	172.435	196.909	226.821	243.518
420	143.930	161.501	183.610	211.775	248.049	269.746
440	150.958	170.094	194.464	226.179	268.715	295.539
460	157.886	178.473	204.988	240.171	288.700	320.444

v/m→	5.0[d]	4.0[d]	3.0[d]	2.0[d]	v/m→	40[b]
t			P		t	P
370				220.025	310	56.192
371.5	210.784				320	57.638
375		219.842	220.644		330	59.102
380	225.819	230.931	234.370	267.878		
390		252.748	262.404	317.298		
400	260.715	274.367	290.947	367.654		
410		295.536[g]	319.801			
420	294.563	317.218	348.962			
430		338.537				
440	327.836	359.710				
460	360.173					

[a, b, c, d] These superscripts serve to identify the several sets of data.
[e] This is for $t = 227.5$ °C.
[f] This is for $t = 306.0$ °C.
[g] The authors state that this value is evidently in error; by interpolation they find 295.866.

Table 36.—Isometrics of the Pressure (kg*/cm²) of Dilated Water-vapor [199]

(See also Table 38.)

Experimental results as smoothed by the observers.

Unit of $v/m = 1$ cm³/g; of $p = 1$ kg*/cm² = 0.96784 atm. Temp. = t °C

v/m→	10	9	8	7	6	5.5	5	4.5
t				p				
350	162							
360	174	181						
370	184	193	202	209	213			
380	192	201	211	221	229	231	233	235
390	199	210	221	233	244	248	252	255
400	207	217	230	245	258	264	269	274

[199] Nieuwenburg, C. J., and Blumendal, H. B., *Rec. trav. chim. Pays-bas*, **51**, 707-714 (1932).

Table 36—(Continued)

$v/m \rightarrow$	10	9	8	7	6	5.5	5	4.5
t					p			
410	213	225	238	255	272	279	286	293
420	219	232	247	266	286	294	303	312
430	226	240	257	277	298	308	319	330
440	233	248	266	288	311	323	334	347
450	239	254	275	298	323	337	351	364
460	246	262	284	310	337	351	366	382
470	252	269	292	319	347	363	380	399
480	258	277	297	328	359	376	395	416

$v/m \rightarrow$	4	3.5	3.0	2.5	2.2	2.0	1.8	1.6
t					p			
380	236	238	241	246	256	280	344	511
390	259	262	268	280	300	329	403	600
400	281	287	296	313	339	379	462	
410	301	311	324	348	379	430	520	
420	322	334	351	381	423	477	582	
430	342	356	378	416	465	525		
440	361	379	405	450	504	571		
450	381	401	432	486	545			
460	400	424	459	520	591			
470	419	444	487	553				
480	439	467	515	589				

Table 37.—Defect of Pressure (atm) of Dilated Water-vapor: Isometrics

At the end of the table are certain values derived from the L (Linde) data of Table 33; the rest have been derived from the Keyes, Smith, and Gerry data of Table 35, the b data only being used when there are several sets referring to the same value of v/m.

The fractional defect is δ, defined by the relation $P = P_i(1 - \delta)$, where P_i ($=4.555 Tm/v$ atm) is the pressure of an ideal gas ($M = 18.0154$) at the same temperature and density, except for errors in the assumed value of the gas constant (R) and in T. Here $T = 273.1 + t$.

The values tabulated are $\delta(v^* + 1)$, it having been observed that this product varies less with v/m ($\equiv v^*$) than does either δv^* or $\delta(v^* + 2)$. These values enable one to compare readily this set of data with those from which the values in Table 38 were derived, and to derive the approximate value of P corresponding to any specific volume and temperature within the range of the table. For rough estimates of P, the proper value of $\delta(v^* + 1)$ may be derived from the table by inspection; for more precise values, use may be made of the fact that $\delta(v^* + 1)$ varies almost linearly with the density, t being constant; and almost linearly with $1/T$, v/m being constant.

As δ has been derived directly from experimental data, it is subject to experimental irregularities, and so are the values of P derived from it.

14. VAPOR: P-V-T DATA

Table 37—*(Continued)*

Unit of $v/m \equiv v^* = 1$ cm^3/g; of $4.555T = 1$ cm$^3 \cdot$atm/g; δ is dimensionless. Temp. = t °C

$v/m \equiv v^* \rightarrow$		150	140	100	97.5	75	57.5
t	$4.555T$			$\delta(v^* + 1)$			
195	2132.20		13.359				
196	2136.75		12.605				
197	2141.30		12.335				
198	2145.86		12.185				
199	2150.42		12.118				
200	2154.97	12.050	12.078				
210	2200.52	11.107	11.115				
220	2246.07	10.304	10.304	10.679	10.786		
230	2291.62	9.642	9.619	9.926	10.053	10.131	
240	2337.17	9.053	9.013	9.311	9.381	9.412	
250	2382.72	8.544	8.329	8.724	8.787	8.821	9.114
260	2428.27	8.073	7.957	8.259	8.263	8.270	8.560
270	2473.82	7.637	7.240	7.811	7.793	7.817	8.067
280	2519.37	7.245	7.118	7.366	7.375	7.345	7.625
290	2564.92	6.893	6.717	7.034	7.001	6.987	7.221
300	2610.47	6.569	6.384	6.670	6.659	6.636	6.857
310	2656.02	6.275	6.136	6.385	6.350	6.304	6.531
320	2701.57	5.982	5.809	6.074	6.070	6.049	6.230
330	2747.12	5.723	5.629	5.806	5.808	5.750	5.952
$t_{sat} \rightarrow$		191.90	195.26	211.98	213.33	227.30	
$P_i \rightarrow$		14.210	15.238	22.095	22.725	30.391	
$\delta(v^* + 1) \rightarrow$		12.913	12.542	11.466	11.304	10.300	

$v/m \equiv v^* \rightarrow$		50	40	39.5	30	20	17.5
t	$4.555T$			$\delta(v^* + 1)$			
250	2382.72	9.206					
260	2428.27	8.502					
270	2473.82		8.017	7.958			
280	2519.37	7.489	7.530	7.476			
290	2564.92			7.048	7.134		
300	2610.47	6.685	6.690	6.654	6.723		
310	2656.02			6.322			
320	2701.57	5.987	6.008	6.000	6.019	6.006	6.003
330	2747.12			5.724			5.681
340	2792.67	5.423	5.421		5.414	5.415	5.402
350	2838.22						5.136
360	2883.77	4.926	4.929		4.929	4.913	4.890
380	2974.87	4.492	4.491		4.489	4.475	4.387
400	3065.97	4.105	4.112		4.102	4.096	4.069
420	3157.07	3.777	3.776		3.770	3.755	3.740
440	3248.17	3.468	3.480		3.468	3.459	3.454
460	3339.27	3.193	3.200		3.195	3.193	3.193
$t_{sat} \rightarrow$		250.00	263.05		280.19	304.64	312.70
$P_i \rightarrow$		47.654	61.054		84.008	90.477	101.05
$\delta(v^* + 1) \rightarrow$		8.988	8.320		7.556	6.560	6.240

$v/m \equiv v^* \rightarrow$		15	12.5	10	7.5	6.25
t	$4.555T$			$\delta(v^* + 1)$		
330	2747.12	5.655				
340	2792.67	5.367	5.303			
350	2838.22	5.105	5.044	4.930		
360	2883.77	4.860	4.806	4.701	4.490	
370	2929.32		4.586	4.489	4.295	4.126

Table 37—(Continued)

$v/m \equiv v^*\to$		15	12.5	10	7.5	6.25
t	4.555T			$\delta(v^*+1)$		
380	2974.87	4.427	4.395	4.293	4.114	3.958
400	3065.97	4.049	4.009	3.935	3.784	3.651
420	3157.07	3.723	3.686	3.621	3.491	3.378
440	3248.17	3.432	3.397	3.340	3.226	3.127
460	3339.27	3.173	3.141	3.088	2.988	2.900
$t_{sat}\to$		321.56	332.03	343.89	357.11	364.05
$P_t\to$		113.74	130.32	151.34	177.98	193.49
$\delta(v^*+1)\to$		5.922	5.522	5.076	4.495	4.229

$v/m \equiv v^*\to$		5.0	4.0	3.0	2.0
t	4.555T		$\delta(v^*+1)$		
370	2929.32				2.549
375	2952.10		3.511	3.103	
380	2974.87	3.723	3.447	3.055	2.460
390	3020.42		3.327	2.957	2.370
400	3065.97	3.449	3.210	2.861	2.280
410	3111.52		3.100[a]	2.767	
420	3157.07	3.201	2.990	2.674	
430	3202.62		2.886		
440	3248.17	2.972	2.785		
460	3339.27	2.764			
$t_{sat}\to$		370.53	373.40		
$P_t\to$		209.09	216.41		
$\delta(v^*+1)\to$		3.860	3.530		

From column L of Table 33, $t = 200\,°C$, $4.555T = 2154.97$.

$P\to$	1	2	3	5	10
$v^* \equiv v/m \to$	2146	1068	708.0	420.2	203.6
$\delta(v^*+1)\to$	9.0	9.4	10.2	10.2	11.3

[a] If the observers' interpolated value is used (see Table 35, note) this becomes 3.098.

(Cont'd from p. 84)

In the reduction of his observations on the pressures developed by the explosion of mixtures of H_2 and air, D. M. Newitt [180] used for water-vapor the equation of state (4) proposed by A. Wohl,[190a] in which v^* is the specific volume (v/m).

$$P = RT/(v^* - b) - a/Tv^*(v^* - b) + c/T^2 v^{*3} \qquad (4)$$

He gives reasons for believing that b varies with the temperature. Using two different methods of estimation, he obtained the following two sets of values:

T	273	2500	2700	2900	3100	°K
$10^5 b$	160	69.5	68.8	68.1	67.4	liters/g-mole-H_2O
$10^5 b$	160	59.7	58.0	56.0	54.3	liters/g-mole-H_2O

All sets of observations of the density and of the specific heat of water-vapor indicate that something of the nature of an association sets in as the condition of saturation is approached (see p. 54).

Table 38.—Defect of Pressure (kg*/cm²) of Dilated Water-vapor: Isometrics

Derived from the data in Table 36 (Nieuwenburg and Blumendal).

$p = p_i(1 - \delta)$, where $p_i = 4.7064Tm/v$ kg*/cm², unit of $m/v = $ kg/cm³, of $4.7064T = 1$ (cm³/g)·(kg*/cm²), $4.7064T$ (cm³/g)·(kg*/cm²) $= 4.555T$ cm³·atm/g, $T = 273.1 + t$. Hence, these values of δ are strictly comparable with those of Table 37.

Unit of $v/m \equiv v^* = 1$ cm³/g, of $4.7064T = 1$(cm³/g)·(kg*/cm²); δ is dimensionless. Temp. $= t$ °C

$v/m \equiv v^* \rightarrow$		10	9	8	7	6	5.5	5	4.5
t	$4.7064T$				$\delta(v^* + 1)$				
350	2932.5	4.92							
360	2979.6	4.58	4.53						
370	3026.6	4.32	4.26	4.19	4.13	4.04			
380	3073.7	4.12	4.11	4.05	3.97	3.87	3.82	3.72	3.61
390	3120.8	3.98	3.95	3.90	3.81	3.72	3.66	3.58	3.48
400	3167.8	3.82	3.83	3.74	3.66	3.58	3.52	3.45	3.36
410	3214.9	3.72	3.70	3.67	3.55	3.45	3.40	3.33	3.24
420	3262.0	3.61	3.59	3.55	3.43	3.31	3.28	3.21	3.13
430	3309.1	3.49	3.48	3.41	3.31	3.22	3.18	3.11	3.03
440	3356.1	3.35	3.35	3.29	3.19	3.10	3.06	3.02	2.94
450	3403.2	3.27	3.28	3.19	3.10	3.01	2.96	2.91	2.85
460	3450.2	3.16	3.16	3.07	2.96	2.89	2.86	2.81	2.76
470	3497.3	3.08	3.07	2.98	2.88	2.82	2.79	2.74	2.67
480	3544.4	2.98	2.97	2.90	2.82	2.74	2.70	2.66	2.60

$v/m \equiv v^* \rightarrow$		4	3.5	3.0	2.5	2.2	2.0	1.8	1.6
t	$4.7064T$				$\delta(v^* + 1)$				
380	3073.7	3.46	3.28	3.06	2.80	2.61	2.45	2.24	1.91
390	3120.8	3.34	3.18	2.97	2.71	2.52	2.37	2.15	1.80
400	3167.8	3.23	3.07	2.88	2.63	2.45	2.28	2.06	
410	3214.9	3.13	2.98	2.79	2.55	2.37	2.20	1.98	
420	3262.0	3.02	2.89	2.71	2.48	2.29	2.12	1.90	
430	3309.1	2.93	2.80	2.63	2.40	2.21	2.05		
440	3356.1	2.85	2.72	2.55	2.32	2.14	1.98		
450	3403.2	2.76	2.64	2.48	2.25	2.07			
460	3450.2	2.68	2.56	2.40	2.18	1.99			
470	3497.3	2.60	2.50	2.33	2.12				
480	3544.4	2.52	2.42	2.25	2.05				

15. Thermal Energy of Dilated Water-vapor

In this section are considered the specific heat (c and C), the enthalpy or heat content, the entropy increase, the decrease in temperature on adiabatic free expansion (Joule-Thomson effect), and certain related quantities, all intimately related to the thermal energy.

Specific Heat of Dilated Water-vapor.

Data on the specific heat of dilated water-vapor are not entirely concordant, especially in the region adjacent to the state of saturation. The situation is complicated by the existence of several more or less contradictory steam tables involving extended extrapolation and much choice and

Table 39.—Formulas and their Coefficients: Specific Heat, Enthalpy, and Entropy of Dilated Water-vapor
(See also Table 43).

Two sets of formulas are given. I. Those by which F. G. Keyes, L. B. Smith, and H. T. Gerry [197] sum up the results of their program of research on the properties of steam; these supersede all others of the same type that they have published from time to time, and are to be used when the most precise values are desired. II. Those derived by J. Havliček and L. Miškovsky [191] from their own observations. The values defined by the two sets of formulas agree closely, as is shown by the values in Table 40.

c_p = specific heat at constant pressure, $c_{p \to 0}$ = limit approached by c_p as p approaches zero, p = pressure, H = enthalpy (heat content), S = entropy. Both H and S are measured from saturated water at 0 °C.

I. Keyes, Smith, and Gerry.

Unit of energy = 1 Int. joule, of mass = 1 g; of p = 1 kg*/cm²; temp. = t °C Int. scale; $T \equiv 273.16 + t$

1. Limiting value of the specific heat at constant pressure as the pressure is indefinitely reduced ($c_{p \to 0}$):

$$c_{p \to 0} = 1.47198 + 7.5566(10^{-4})T + 47.8365/T$$

2. Value of c_p for pressure p and temperature t. $c_p = c_{p \to 0} + A'p + B'p^2 + C'p^4 + D'p^{13}$, A', B', C', and D' having the following values:

t	$10^a A'$	a	$10^b B'$	b	$10^c C'$	c	$10^d D'$	d
100	1.5954	1	2.921	2				
120	1.1093	1	1.446	2	4.596	5		
140	7.979	2	7.612	3	1.492	5		
160	5.907	2	4.221	3	5.306	6		
180	4.483	2	2.447	3	2.035	6		
200	3.477	2	1.474	3	8.320	7		
220	2.748	2	9.183	4	3.595	7	2.812	21
240	2.208	2	5.890	4	1.628	7	1.330	22
260	1.801	2	3.877	4	7.683	8	7.416	24
280	1.488	2	2.611	4	3.756	8	4.744	25
300	1.244	2	1.794	4	1.892	8	3.431	26
320	1.051	2	1.255	4	9.787	9	2.765	27
340	8.967	3	8.928	5	5.175	9	2.449	28
360	7.717	3	6.444	5	2.787	9	2.348	29
380	6.693	3	4.713	5	1.524	9	2.387	30
400	5.848	3	3.489	5	8.423	10	2.485	31
420	5.143	3	2.612	5	4.692	10	2.415	32
440	4.550	3	1.974	5	2.622	10	1.637	33
460	4.049	3	1.506	5	1.463	10	−1.563	34
480	3.622	3	1.159	5	8.087	11	−1.226	34
500	3.255	3	8.794	6	4.390	11	−4.900	35

3. Value of the enthalpy (H) of dilated water-vapor as measured from saturated water at 0 °C; that is, H is the increase in the "heat content" on changing 1 g of saturated water at 0 °C into dilated water-vapor at temperature t and pressure p.

$$H = 2502.36 + \int_{273.16}^{T} c_{p \to 0}\, dT - (Ap + Bp^2 + Cp^4 + Dp^{13})$$

15. VAPOR: THERMAL ENERGY

Table 39—(Continued)

A, B, C, and D having the following values:

t	A	$10^b B$	b	$10^c C$	c	$10^d D$	d
100	12.1691	9.1692	1	2.6285	3		
150	6.8776	2.2182	1	1.8411	4		
200	4.4208	6.9814	2	2.0664	5		
250	3.0994	2.6285	2	3.1568	6	4.2871	19
300	2.3103	1.1242	2	5.9107	7	2.1742	22
350	1.8009	5.2816	3	1.2560	7	2.7131	25
400	1.4519	2.6635	3	2.8219	8	6.4847	28
450	1.2013	1.4184	3	6.0829	9	2.0549	30
500	1.0143	7.8789	4	9.6158	10	−8.7526	33
550	0.8705	4.5225	4	−1.1283	10	−9.9275	34

If unit of $p = 1$ atm, these coefficients must be increased numerically by the following amounts: A by 3.323, 3.322, 3.321, and 3.320 per cent, B 6.755 per cent, C 14.00 to 13.97 per cent, and D 52.95 per cent; the first three values for A refer, respectively, to 100, 150, and 200 °C, the fourth to all the others; the first for C refers to 100 °C, the other to all the others. All have been derived from the values of the coefficients as published.

4. Value S of the excess of the entropy of expanded water-vapor at temperature t and pressure p above that of saturated water at 0 °C is given by the formula

$$S = 6.8158 + \int_{273.16}^{T} (c_{p\to 0}) \frac{dT}{T} - 1.06242 \log_{10} p - (Kp + Lp^2 + Mp^4 + Np^{13})$$

K, L, M, and N having the following values:

t	$10^k K$	k	$10^l L$	l	$10^m M$	m	$10^n N$	n
100	2.603	2	2.555	3	6.730	6		
200	7.078	3	1.364	4	4.146	8	8.944	22
300	2.965	3	1.765	5	9.790	10	4.670	28
400	1.570	3	3.556	6	4.016	11	3.018	33
500	9.601	4	9.182	7	1.222	12	1.258	36

II. J. Havlíček and L. Miškovský.[191]

Each formula is given in duplicate, first with the Int. steam calorie (= 4.1860 Int. joules), and secondly with the Int. joule as the unit of energy.

Unit of mass = 1 g, of $p = 1$ kg*/cm²; temp. = t °C (Int. scale); $T \equiv 273.2 + t$.

$c_{p\to 0} = 0.4402 + 0.0095(t/100) + 0.00072(t/100)^2$ (Int. cal.)
$= 1.8427 + 0.03977(t/100) + 0.0030139(t/100)^2$ (Int. joule)

$H = 597.6 + \int_0^t c_{p\to 0} dt - \{d(p/10^6) + e(p/10^6)^2 + f(p/10^6)^5\}$ (Int. cal.)

$= 2501.6 + \int_0^t c_{p\to 0} dt - \{d(p/10^6) + e(p/10^6)^2 + f(p/10^6)^5\}$ (Int. joule)

$d = 716.64(100/T)^2 + 107.73(100/\tau)^2(3 + 440/\tau) - 1.026$ (Int. cal.)
$= 2999.86(100/T)^2 + 450.96(100/\tau)^2(3 + 440/\tau) - 4.295$ (Int. joule)

Table 39—*(Continued)*

$$e = 2.7981(10^7)(100/T)^8 - 0.0726 \quad \text{(Int. cal.)}$$
$$= 11.7128(10^7)(100/T)^8 - 0.3039 \quad \text{(Int. joule)}$$
$$f = 3.1242(10^{18})(100/T)^{22} - 3.8952(10^{17})(100/T)^{21} \quad \text{(Int. cal.)}$$
$$= 13.0779(10^{18})(100/T)^{22} - 16.3053(10^{17})(100/T)^{21} \quad \text{(Int. joule)}$$
$$\tau \equiv T - 220 = 53.2 + t$$

---

judgment in smoothing and reconciling the various sets of related data, and also by the repeated publication, with much emphasis on their supposed accuracy, of certain values computed by A. Leduc in 1913, which appear without a literature reference under the designation "best" in his contribution to the *International Critical Tables*.[200] These values of Leduc's assume that the general equation of state set up by him for normal gases is applicable to water-vapor, and that the values he used for certain auxiliary data for water-vapor are correct. It may without difficulty be shown that his equation does not satisfactorily represent the best experimental values for water-vapor, and that the auxiliary values he used do not accord with the best data now available. Therefore, whatever his computed data for the specific heat of water-vapor may have been worth in 1913, they can scarcely be accepted now. The more important steam tables will be found listed in Table 259; but little use is made of them in this section.

Data for the specific heat at high temperatures have either been computed from basic constants and spectroscopic data, or based upon the specific heat at constant volume, as inferred from the pressures generated when suitable mixtures of gases are exploded in a closed vessel, and upon the value of $c_p - c_v = T(\delta v/\delta T)_p (\delta p/\delta T)_v$ as computed from an assumed equation of state. Until recently they have been quite uncertain. The interpretation of the spectroscopic data has not been entirely clear.[201] In the explosion experiments the temperature is never uniform throughout the volume, the heat losses are difficult to determine, as are the effect of radiation and its absorption, and especially the dissociation; moreover there is not equilibrium between the molecules and the temperature of their immediate surroundings.[202]

In addition to the papers listed elsewhere in this section the following should be examined by those especially interested in this subject.

(1) Determinations by the explosion method:

Langen, A., *Mitt. Forsch. Geb. Ing.*, **8**, 1-54 (1903); Bjerrum, N., *Z. Elektroch.*, **17**, 731-735 (1911); Siegel, W., *Z. physik. Chem.*, **87**, 641-668 (1914); Gallina, V., *Ann. d. R. Scuola d'Ing. (Padova)*, **4**, 77-87 (1928); David, W. T., and Leah, A. S., *Phil. Mag. (7)*, **18**, 307-321 (1934); Lewis, B., and von Elbe, G., *J. Chem'l Phys.*, **2**, 659-664, 890 (L) (1934); **3**, 63-71 (1935); *J. Am. Chem. Soc.*, **57**, 612-614 (1935); *Phil. Mag. (7)*, **20**, 44-65 (1935); Schmidt, F. A. F., *Forsch. Gebiete Ingenieurw.*, **8**, 91-99 (1937).

[200] Leduc, A., *Int. Crit. Tables*, **5**, 82 (1929).

[201] Bonhoeffer, K. F., and Reichardt, H., *Z. physik. Chem. (A)*, **139**, 75-97 (1928); Justi, E., *Forsch. Gebiete Ingenieurw.*, **2**, 117-124 (1931); Gordon, A. R., and Barnes, C., *J. Phys'l Chem.*, **36**, 1143-1151 (1932).

[202] Chaudron, G., *Bull. Soc. Chim. France (4)*, **37**, 657-679 (1925); McCrea, W. H., *Proc. Cambridge Phil. Soc.*, **23**, 942-950 (1927); Bonhoeffer, K. F., and Reichardt, H., *loc. cit.*

15. VAPOR: THERMAL ENERGY

Table 40.—Values of $c_{p\to 0}$ and Its Integrals: Dilated Water-vapor.

By definition, $c_{p\to 0}$ is the limit approached by the specific heat at constant pressure as the pressure is reduced.

I. The following values have been computed by means of the formulas given in Table 39, and are designated as: I = Keyes, Smith, and Gerry's formula $c_{p\to 0} = 1.47198 + 7.5566(10^{-4})T + 47.8365/T$, $T = 273.16 + t$. II = Havlíček and Miškovský's $c_{p\to 0} = 1.8427 + 0.03977(t/100) + 0.0030139(t/100)^2$, $T = 273.2 + t$.

Unit of c_p = 1 Int. joule/g·°C. Temp. = t °C (Int. scale)

t	I $c_{p\to 0}$	II $c_{p\to 0}$	I $\int_0^t c_{p\to 0}dt$	II $\int_0^t c_{p\to 0}dt$	I $\int_0^t (c_{p\to 0})\frac{dt}{T}$	II $\int_0^t (c_{p\to 0})\frac{dt}{T}$
0	1.85352	1.8427	0	0	0	0
50	1.86421	1.8634	92.904	92.64	0.312300	0.31136
100	1.88216	1.8855	186.537	186.36	0.581600	0.58093
150	1.90479	1.9092	281.195	281.21	0.819589	0.81943
200	1.93063	1.9343	377.068	377.29	1.033684	1.03400
250	1.95875	1.9610	474.294	474.68	1.228970	1.23000
300	1.98855	1.9892	572.970	573.42	1.409065	1.40983
350	2.01964	2.0189	673.169	673.61	1.576637	1.57740
400	2.05172	2.0501	774.948	775.32	1.733709	1.73438
450	2.08459	2.0828	878.352	878.64	1.881851	1.88240
500	2.11810	2.1170	983.416	983.62	2.022305	2.02275
550	2.15212	2.1526	1090.168	1090.35	2.156070	2.15650

II. Adapted from tables by E. Justi and H. Lüder[203]; conversion to joules and to grams was made by the compiler.

Unit of mass and of energy are indicated. 1 cal = 4.186 joules. Temp. = t °C

Units→	g-mole cal	g-mole joule	g joule	g-mole cal	g-mole joule	g joule
t		$c_{p\to 0}$			$\int_0^t (c_{p\to 0})\frac{dt}{T}$	
20	7.98	33.41	1.855			
100	8.10	33.90	1.882	2.53	10.59	0.588
200	8.32	34.78	1.930	4.48	18.75	1.041
300	8.56	35.83	1.989	6.08	25.45	1.413
400	8.84	37.01	2.054	7.48	31.31	1.738
500	9.12	38.18	2.120	8.68	36.33	2.017
600	9.41	39.39	2.186	9.81	41.06	2.279
700	9.72	40.69	2.259	10.85	45.42	2.521
800	10.02	41.94	2.328	11.82	49.48	2.746
1000	10.58	44.28	2.458	13.58	56.84	3.155
1200	11.08	46.38	2.574	15.14	63.38	3.518
1400	11.52	48.22	2.678	16.58	69.40	3.852
1600	11.88	49.72	2.761	17.90	74.93	4.159
1800	12.19	51.03	2.832	19.13	80.08	4.445
2000	12.45	52.12	2.894	20.28	84.89	4.712
2500	12.95	54.21	3.009	22.80	95.44	5.298
3000	13.23	55.38	3.074	24.98	104.57	5.804

[203] Justi, E., and Lüder, H., *Forsch. Gebiete Ingenieurw.*, **6**, 209-216 (1935). Superseding Justi, E., *Idem*, **5**, 130-137 (1934).

(2) *Computation from fundamental constants and spectroscopic data:*

Henning, F., and Justi, E., *Wiss. Abh. d. Phys. Tech. Reichs.*, **14**, 171-174 (1930-31) = *Z. techn. Physik*, **11**, 191-194 (1930); Hausen, H., *Forsch. Gebiete Ingenieurw.*, **2**, 319-326 (1931); Zeise, H., *Z. Elektroch.*, **39**, 758-773, 895-909 (1933); Trautz, M., and Ader, H., *Z. Physik*, **89**, 12-14 (1934); Lewis, B., and von Elbe, G., *J. Chem'l Phys.*, **3**, 63-71 (1935); *J. Am. Chem. Soc.*, **37**, 612-614 (1935); Wilson, E. B., Jr., *J. Chem'l Phys.*, **4**, 526-528 (1936); Kassel, L. S., *Chem. Rev.*, **18**, 277-313 (1936); Murphy, G. M., *J. Chem'l Phys.*, **5**, 637-641 (1937).

(3) *Reviews, summaries, etc.:*

Callendar, H. L., *Phil. Trans. (A)*, **215**, 383-399 (1915); *World Power*, **1**, 274-280, 325-328 (1924); **3**, 302-312 (1925); Fischer, V., *Z. techn. Physik*, **5**, 17-21, 39-44, 83-88 (1924); Jazyna, W., *Idem*, **6**, 261-262 (1925); *Z. Physik*, **57**, 341-344 (1929); Plank, R., *Z. techn. Physik*, **5**, 397-404 (1924); Saunders, S. W., *J. Phys'l Chem.*, **28**, 1151-1166 (1924); Davis, H. N., and Keenan, J. H., *Proc. World Eng. Cong. (Tokyo)*, **4**, 239-264 (1931).

(4) *Steam tables, see Section 90.*

Table 41.—Specific Heat of Dilated Water-vapor at Constant Pressure: Preferred Values

Various other sets of values will be found in Tables 42 and 43.

The following values are those given by Keyes, Smith, and Gerry [197] and computed by means of the formula $c_p = c_{p \to 0} + A'p + B'p^2 + C'p^4 + D'p^{13}$ in which $c_{p \to 0}$ and the several coefficients take the values given in Tables 39 and 40, respectively; values for other values of t and P may be similarly computed.

Unit of $c_p = 1$ Int. joule/g $= 0.23907$ cal$_{15}$/g; of $P = 1$ kg^2/cm^2. Temp. $= t$ °C.

$t \to$ P	340	360	380	400 c_p	420	440	460
120	5.708	4.479	3.836	3.427	3.152	2.960	2.822
140	8.948	5.624	4.501	3.877	3.474	3.200	3.007
160		7.791	5.420	4.441	3.862	3.481	3.218
180			6.864	5.168	4.332	3.810	3.459
200				6.164	4.905	4.195	3.735

With the preceding values Keyes, Smith, and Gerry [197] compare those published by W. Koch,[204] finding the following differences: $\Delta =$ Koch $-$ KSG; unit of $\Delta = 1$ Int. joule/g.

$t \to$ P	340	360	380	400 1000Δ	420	440	460
120	$+ 80$	$- 13$	$- 23$	$+ 5$	$+ 37$	$+ 62$	$+ 78$
140	$+ 21$	$+ 34$	$- 52$	$- 43$	$- 5$	$+ 31$	$+ 56$
160		$- 2$	$- 34$	$- 68$	$- 41$	$+ 1$	$+ 38$
180			$- 46$	$- 58$	$- 59$	$- 22$	$+ 19$
200				$- 37$	$- 47$	$- 27$	$+ 6$

Table 42.—Specific Heat of Dilated Water-vapor at Constant Pressure: Various Sets of Values

(See also Tables 41, 43, and 55)

Preferred values are given in Table 41; those here given indicate values in use in 1932. They serve to show how the values run, and are useful in making rough approximations; many of them differ by several units in the second decimal place from the preferred values defined by the formula used in Table 41 (see bottom of that table).

15. VAPOR: THERMAL ENERGY

Table 42—(Continued)

1 joule/g = 0.9997 Int. joule/g = 0.2390 cal$_{15}$/g = 18.015 joule/g·mole = 4.305 cal$_{15}$/g-mole. 1 kg/cm^2 = 0.96784 atm = 0.98066 bar.
Unit of $P = 1$ kg*/cm^2, of $P_a = 1$ atm, of $P_b = 1$ bar = 10^6 dynes/cm^2, of $c_p = 1$ joule/g·°C, of $C_p = 1$ cal/g-mole·°C. Temp. = t °C.

I. O. Knoblauch and A. Winkhaus.[205]

P	0.5	1.0	2.0	6.0	10
P_a	0.4839	0.9678	1.9357	5.807	9.678
P_b	0.4903	0.9807	1.9613	5.884	9.807
t_{sat}	80.9	99.1	119.6	158.1	179.1
t			c_p		
t_{sat}	2.000	2.038^a	2.097^a	2.323	2.565
110	1.971	2.021			
120	1.963	2.009			
140	1.950	1.988	2.055		
160	1.946	1.971	2.026	2.310	
180	1.946	1.971	2.005	2.222	
200	1.954	1.971	1.996	2.151	2.381
220	1.958	1.971	1.996	2.105	2.260
240	1.967	1.975	1.996	2.072	2.185
260	1.975	1.980	1.996	2.059	2.143
280	1.984	1.988	2.005	2.055	2.122
300	1.992	2.000	2.013	2.055	2.109
320	2.000	2.009	2.021	2.059	2.105
340	2.013	2.017	2.030	2.063	2.105
360	2.026	2.030	2.038	2.072	2.109
380	2.034	2.038	2.051	2.080	2.118

P	12	14	16	18	20
P_a	11.614	13.550	15.485	17.421	19.357
P_b	11.768	13.729	15.691	17.652	19.613
t_{sat}	187.1	194.2	200.5	206.2	211.4
t			c_p		
t_{sat}	2.687^a	2.808	2.925^a	3.051	3.180^a
200	2.528	2.712			
220	2.360	2.482	2.620	2.787	2.984
240	2.264	2.348	2.440	2.544	2.678
260	2.201	2.260	2.331	2.402	2.490
280	2.164	2.214	2.264	2.318	2.377
300	2.147	2.184	2.226	2.268	2.310
320	2.138	2.168	2.201	2.235	2.268
340	2.134	2.159	2.184	2.214	2.239
360	2.130	2.151	2.176	2.197	2.222
380	2.134	2.151	2.168	2.189	2.210

II. O. Knoblauch and W. Koch.[206]

P	19.9		29.9		40.1		60.2	
P_a	19.260		28.938		38.810		58.264	
P_b	19.515		29.322		39.325		59.036	
	t	c_p	t	c_p	t	c_p	t	c_p
	332.9	2.277	241.6	3.353	252.8	3.881	280.3	4.625
	334.8	2.260	246.6	3.219	257.3	3.805	283.8	4.353

[204] Koch, W., *Forsch. Gebiete Ingenieurw.*, **3**, 1-10 (1932).
[205] Knoblauch, O., and Winkhaus, A., *Mitt. Forsch. Geb. Ing.*, **195**, 1-20 (1917) → *Z. Ver. deuts. Ing.*, **59**, 376-379, 400-405 (1915).
[206] Knoblauch, O., and Koch, W., *Z. Ver. Deut. Ing.*, **72**, 1733-1739 (1928) → *Mech. Eng.*, **51**, 147-150 (1929).

Table 42—(Continued)

$P \to$		19.9		29.9		40.1		60.2
		c_p	t	c_p	t	c_p	t	c_p
335.9		2.244	259.3	2.968	263.8	3.495	287.5	4.161
			283.8	2.671	274.0	3.261	301.1	3.679
			320.1	2.449	288.5	2.989	324.6	3.156
			353.9	2.378	322.7	2.650	334.6	3.051
			358.4	2.382	371.2	2.491	368.0	2.742
			386.3	2.315	403.7	2.411	369.2	2.683
			433.9	2.327	436.5	2.390	429.0	2.532
			437.0	2.340	436.8	2.390	437.7	2.537
			493.2	2.344	494.0	2.382		

P	80.0		100.2		120.1	
P_a	77.427		96.978		116.238	
P_b	78.453		98.263		117.778	
t		c_p	t	c_p		c_p
297.6		5.672	314.9	6.392	329.6	8.234
300.5		5.303	317.9	6.388	328.6	7.828
309.7		4.701	318.0	5.952	335.4	6.488
326.2		3.881	324.5	5.668	342.2	5.706
341.5		3.466	335.0	4.722	345.1	5.542
375.0		3.018	352.5	3.951	352.7	5.023
377.5		3.030	376.1	3.428	365.9	4.295
421.0		2.733	411.9	3.030	389.0	3.608
423.7		2.750	410.1	3.064	427.6	3.110

III. W. Koch.[204]

P	120	130	140	150	160
P_a	116.14	125.82	135.50	145.18	154.85
P_b	117.68	127.49	137.29	147.10	156.91
t_{sat}	323.11	329.25	335.04	340.51	345.71
t			c_p		
t_{sat}	8.625	9.743	11.020	12.502	14.307
340	5.791	7.009	8.973		
360	4.468	4.987	5.661	6.553	7.792
380	3.814	4.103	4.451	4.874	5.389
400	3.433	3.622	3.835	4.086	4.375
420	3.190	3.324	3.471	3.638	3.823
440	3.023	3.124	3.232	3.354	3.484
460	2.902	2.981	3.065	3.157	3.257

P	170	180	190	200
P_a	164.53	174.21	183.89	193.57
P_b	166.71	176.52	186.33	196.13
t_{sat}	350.66	355.38	359.88	364.18
t		c_p		
t_{sat}	16.560	19.428	23.121	27.885
360	9.730	13.511	22.828	
380	6.025	6.821	7.863	9.383
400	4.714	5.112	5.577	6.130
420	4.036	4.275	4.551	4.861
440	3.630	3.789	3.969	4.170
460	3.362	3.479	3.605	3.743

[207] Trautz, M., and Steyer, H., *Forsch. Gebiete Ingenieurw.*, **2**, 45-52 (1931).
[208] Jakob, M., *Z. Ver. Deuts. Ing.*, **56**, 1980-1988 (1912).
[209] Knoblauch, O., and Mollier, H., *Mitt. Forsch. Geb. Ing.*, **108** and **109**, 79-106 (1911) → *Z. Ver. Deuts. Ing.*, **55**, 665-673 (1911).

15. VAPOR: THERMAL ENERGY

Table 42—(Continued)

IV. M. Trautz and H. Steyer.[207]

P	150	200	250	300
P_a	145.176	193.568	241.960	290.352
P_b	147.100	196.133	245.166	294.200
t		c_p		
350	8.79			
400	4.18	7.32	9.67	14.35
450	3.10	4.14	4.39	5.19
500	2.51	3.35	4.18	4.98

V. From compilation by A. Leduc[200] based on observations by M. Jakob[208] and by O. Knoblauch and H. Mollier.[209]

P	1.033	2.066	4.133	6.199	10.333	14.465	20.665
P_a	1	2	4	6	10	14	20
P_b	1.013	2.026	4.053	6.080	10.132	14.186	20.265
t				p			
100	2.017						
120	1.996	2.093					
140	1.984	2.046					
160	1.975	2.021	2.155	2.356			
180	1.971	2.005	2.101	2.226	2.645		
200	1.971	1.996	2.067	2.151	2.381	2.808	
250	1.980	1.996	2.034	2.072	2.155	2.264	2.461
300	1.996	2.009	2.034	2.059	2.109	2.159	2.247
350	2.021	2.025	2.051	2.067	2.109	2.147	2.214
400	2.051	2.059	2.076	2.093	2.122	2.155	2.205
450	2.084	2.093	2.113	2.118	2.139	2.164	2.201
500	2.122	2.126	2.134	2.143	2.159	2.176	2.201
550	2.156	2.159	2.164	2.168	2.180	2.189	2.210

VI. Derived from c_v, as inferred from the pressures developed when confined mixtures of H_2 and O_2 are exploded, by means of an assumed equation of state: $c_p - c_v = T(\delta v/\delta T)_p^2 (\delta p/\delta T)_v$. Pressure = 1 atm.

Ref.[b]→	ICT	HH	E	Ref.[b]→	ICT	HH	E
t		c_p		t		c_p	
100	2.040		1.966	1000	2.297	2.335	2.462
200	2.010	1.946	1.996	1200	2.485	2.544	2.636
300	2.004		2.032	1400	2.708	2.796	2.837
400	2.010	1.980	2.074	1600	2.977		3.067
500	2.030		2.122	1800	3.289		3.320
600	2.062	2.055	2.178	2000	3.470		3.601
700	2.103		2.238	2200	3.586		3.902
800	2.158	2.172	2.305	2300	3.614		4.064
900	2.221		2.382				

VII. Computed by Gordon and Barnes[210] from basic constants, Steinwehr's equation of state,[211] and spectroscopic data. Pressure = 1 atm.

T	t	C_p	c_p	T	t	C_p	c_p
400	126.9	8.26	1.920	900	626.9	9.46	2.199
500	226.9	8.45	1.964	1000	726.9	9.74	2.264
600	326.9	8.68	2.017	1100	826.9	10.02	2.329
700	426.9	8.93	2.076	1200	926.9	10.29	2.392
800	526.9	9.19	2.136				

[210] Gordon, A. R., and Barnes, C., *J. Phys'l Chem.*, **36**, 1143-1151 (1932).
[211] Steinwehr, H. v., *Z. Physik*, **3**, 466-476 (1920).

Table 42—*(Continued)*

VIII. A. Leduc.[212] Computed by him from certain other data for water-vapor, assuming the validity of the general equation of state set up by him as applicable to all normal gases. In his compilation,[200] these values are called the "best," and are given without citation. They have been republished.[213] It will be noticed that they disagree with the experimental values in magnitude and in that the latter pass through a minimum as the temperature is decreased, the pressure remaining constant; whereas Leduc's computed values continually decrease.[c]

$t \rightarrow$	100	120	140	150	160
P_a			c_p		
1	1.81$_6$	1.83$_3$	1.87$_5$	1.93	2.00$_5$
2		1.87$_1$	1.91$_3$	1.95$_9$	2.02$_6$
3			1.95$_9$		
4				2.02$_6$	2.07$_6$
P_sat	1.81$_6$	1.87$_1$	1.97$_1$		

[a] For the limiting value of c_p at t_sat, the following values by M. Jakob[208] are given in the compilation by A. Leduc.[200]

P	1	2	4	8	12	16	20	kg*/cm²
t_sat	99.1	119.6	142.9	169.6	187.1	200.5	211.4	°C
c_p	2.01$_7$	2.08$_8$	2.23$_1$	2.51$_1$	2.81$_7$	3.13$_5$	3.49$_0$	joule/g·°C

[b] References:

E = E. D. Eastman.[214]
HH = L. Holborn and F. Henning.[215]
ICT = adapted from the compilation by A. Leduc,[200] apparently derived directly from Table C (p. 205) of Partington and Shilling's "The Specific Heat of Gases" (London, 1924). There, two sets of values are given. In each of them the same value of c_v is used, but in one set $c_p - c_v$ is said to have been derived from Berthelot's equation of state; and in the other, from Callendar's. The former is the smaller by 5.8, 1.6, 0.5, and 0.05 per cent at 100°, 200°, 300°, and 400 °C, respectively. The average of the two sets is given in ICT, but here those ascribed to Callendar's equation are used. The value of c_p is derived from c_v and $c_p - c_v$.

[c] Leduc[216] maintains that this experimentally observed minimum arises from experimental errors, and that his computed values are to be preferred. But it can be shown that the subsidiary data employed by him were not sufficiently accurate, that the test he employed to demonstrate the applicability of his equation of state to water-vapor was not sufficient, and that that equation does not satisfactorily represent the specific volume of water-vapor. Whatever may have been the worth of these computed values in 1913, one is scarcely justified in regarding them more highly than the corresponding experimental data available today. Additional information regarding his equation and his method of computation may be found in the *Comptes Rendus*.[217]

Table 43.—Specific Heats of Water-vapor at 1 atm: Comparison of Interpolation Formulas
(See also Table 39)

Callendar's formula is $c_p = 1.997_9 + 38.49(373.1/T)^{10/3} P_\text{atm}/T$ joules/g·°C; KSG's is that given in Table 39; the others are of the form $c = A + B(t/100) + C(t/100)^2 + D(t/100)^3$, the coefficients having the values

---

[212] Leduc, A., *Ann. de chim. et phys.* (8), **28**, 577-613 (1913).
[213] Leduc, A., *Jour. de Phys.* (6), **2**, 24-30 (1921).
[214] Eastman, E. D., *U. S. Bureau of Mines Tech. Paper* **445**, (1929).

15. VAPOR: THERMAL ENERGY

Table 43—*(Continued)*

given at the heads of the several columns. Eastman's formula (E) is not supposed to give the details of the actual variations, but merely values that are always near the true value. Each tabulated value of the specific heat has been computed by means of the indicated formula.

Temp. = t °C; unit of $c = 1$ joule/g·°C

Ref.[a]→	KSG	Ca	R	HSH	Sh	E	F
A			1.96_3	1.894_8	2.229_2	1.94_2	2.311
B			-0.015_9	$+0.0386_8$	-0.214_9	$+0.0205$	-0.745
C			$+0.0051_6$	0	$+0.0343_8$	$+0.00311$	$+0.46_0$
D			0	0	-0.00118_0	0	0
t				c_p			$(c_p)_{\text{sat}}$[b]
100	2.078	2.10_1	1.95_2	1.933_5	2.047_5	1.96_5	2.02_6
120		2.08_0	1.95_1	1.941_2	2.018_8	1.97_1	2.07_9
140		2.06_4	1.95_1	1.949_0	1.992_5	1.97_7	2.16_9
160		2.05_2	1.95_1	1.956_7	1.968_6	1.98_3	2.29_7
180		2.04_2	1.95_1	1.964_4	1.946_9	1.98_9	2.46_2
200	1.968	2.03_5	1.95_2	1.972_2	1.927_5	1.99_5	2.66_1
250		2.02_2	1.95_5	1.991_5	1.888_5	2.01_3	3.32_2
300	2.002	2.01_4	1.96_1	2.010_8	1.862_0	2.03_2	4.21_6
350		2.00_9	1.97_0	2.030_2	1.847_7	2.51_9	5.45_3[c]
400	2.058	2.00_6	1.98_2	2.049_5	1.844_2	2.07_4	
450		2.00_4	1.99_6	2.068_9	1.850_9	2.09_7	
500	2.122	2.00_2	2.01_3	2.088_2	1.866_7	2.12_2	
550		2.00_1	2.03_2	2.107_5	1.891_0	2.14_9	
600		2.00_0	2.05_3	2.126_8	1.922_6	2.17_7	
700		1.99_9	2.10_5	2.165_6	2.004_8	2.23_8	
800		1.99_9	2.16_6	2.204_2	2.106_1	2.30_5	
900		1.99_9	2.23_8	2.242_9	2.219_7	2.38_0	
1000		1.99_8	2.32_0	2.281_6	2.338_2	2.46	
1200		1.99_8	2.51_5	2.359_0	2.562_1	2.64	
1400		1.99_8	2.75_1	2.436_3	2.721_2	2.84	
1600		1.99_8	3.03_0	2.513_7	2.759_8	3.07	
1800		1.99_8	3.34_9	2.591_0	2.618_3	3.32	
2000		1.99_8	3.70_9	2.668_4	2.243_2	3.60	
2200		1.99_8	4.11_0	2.745_8	1.576_7	3.90	
2300		1.99_8	4.32_7	2.784_4	1.116_6	4.06	

Ref.[a]→	KSG	Ca	R	HSH	Sh	E	Sh
A			1.95_7	1.914	2.132_9	1.95_3	1.687_9
B			-0.00623	$+0.0193_4$	-0.0962_9	$+0.0112_9$	-0.177_2
C			$+0.0017_2$	0	$+0.0111_6$	$+0.00103_7$	$+0.0292_9$
D			0	0	-0.00029_5	0	-0.00093_1
t				Mean c_p between 100 and t °C			c_v
100	2.078	2.10_1	1.95_3	1.93_3	2.04_8	1.96_5	1.539
120		2.09_0	1.95_2	1.93_7	2.03_3	1.96_8	1.516
140		2.08_0	1.95_1	1.94_1	2.01_9	1.97_0	1.495

[215] Holborn, L., and Henning, F., *Ann. d. Physik (4)*, **23**, 809-845 (1907).
[216] Leduc, A., *Chem. Rev.*, **6**, 1-16 (1929).
[217] Leduc, A., *Compt. rend.*, **152**, 1752-1756 (1911); **153**, 51-54 (1911); **154**, 812-815 (1912).

Table 43—(Continued)

Ref[a]→ t	KSG	Ca	R	HSH	Sh	E	Sh c_v
			Mean c_p between 100 and t °C				
160		2.07_3	1.95_1	1.94_5	2.00_6	1.97_4	1.476
180		2.06_7	1.95_1	1.94_9	1.99_4	1.97_7	1.458
200	1.994	2.06_1	1.95_2	1.95_3	1.98_2	1.98_0	1.443
250	1.987	2.05_0	1.95_3	1.96_2	1.95_7	1.98_3	1.414
300	1.988	2.04_2	1.95_3	1.97_2	1.93_6	1.99_6	1.395
350	1.993	2.03_6	1.95_6	1.98_2	1.92_0	2.00_5	1.387
400	2.002	2.03_1	1.95_9	1.99_1	1.90_7	2.01_5	1.388
450	2.012	2.02_7	1.96_4	2.00_1	1.89_9	2.02_5	1.399
500	2.023	2.02_4	1.96_9	2.01_1	1.89_4	2.03_6	1.418
550	2.036	2.02_2	1.97_5	2.02_0	1.89_2	2.04_7	1.444
600		2.02_0	1.98_2	2.03_0	1.89_3	2.05_8	1.478
700		2.01_6	1.99_7	2.04_9	1.90_4	2.08_3	1.563
800		2.01_4	2.01_7	2.06_9	1.92_6	2.11_0	1.668
900		2.01_2	2.04_0	2.08_8	1.95_5	2.13_9	1.787
1000		2.01_0	2.06_7	2.10_7	1.99_1	2.17	1.914
1200		2.00_8	2.13_0	2.14_6	2.07_5	2.24	2.170
1400		2.00_7	2.20_7	2.18_5	2.16_3	2.31	2.393
1600		2.00_6	2.29_7	2.22_3	2.24_1	2.40	2.538
1800		2.00_5	2.40_2	2.26_2	2.29_5	2.49	2.559
2000		2.00_4	2.52_0	2.30_1	2.31_1	2.59	2.412
2200		2.00_3	2.65_2	2.33_9	2.27_5	2.70	2.053
2300		2.00_3	2.72_4	2.35_9	2.23_2	2.76	1.779

[a] References:

Ca = H. L. Callendar.[218]
E = E. D. Eastman.[214]
F = V. Fischer.[219]
HSH = L. Holborn, K. Scheel, and F. Henning.[220]
R = M. Randall.[221]
Sh = W. G. Shilling.[222]

[b] At the pressure of the saturated vapor, i.e., the value that would be found at that pressure by extrapolating the curve defined by observations taken at slightly lower pressures and at the temperature indicated.

[c] At critical temperature (374 °C) the equation gives $(c_p)_{sat} = 5.95_8$ joules/g·°C.

Table 44.—Mean Internal Specific Heat of Dilated Water-vapor at Constant Pressure [223]

By definition $\bar{c}_{ip} \equiv \bar{c}_p - p\Delta v/\Delta t$, where $\bar{c}_p$ is the mean specific heat at constant pressure over the range t to $(t + \Delta t)$, Δv is the accompanying increase in the specific volume, and p is the pressure expressed in appropriate units. Were the vapor ideal, $\bar{c}_{ip}$ would be $\bar{c}_v$, the specific heat at constant volume, and would be independent of p.

[218] Callendar, H. L., "Properties of Steam," pp. 98, 60, 61, 15 (1920).
[219] Fischer, V., Z. Physik, 43, 131-151 (1927).
[220] Holborn, L., Scheel, K., and Henning, F., "Wärmetabellen" (1919).
[221] Randall, M., Int. Crit. Tables, 7, 231 (1930).
[222] Shilling, W. G., Phil. Mag. (7), 3, 273-301 (1927).
[223] Derived from Callendar, H. L., Proc. Inst. Mech. Eng., 1929, 507-527 (1929).

15. VAPOR: THERMAL ENERGY

Table 44—*(Continued)*

Unit of $p = 1(\text{lb*/in}^2)_L{}^a$; of $P_a = 1$ atm $= 1.01325$ bars; of $\bar{c}_{tp} = 1$ joule/g.°C. Temp. $= t$ °C

p	$t \to$ P_a	t_{sat}	t_{sat}	250	300	350	400	450
					$\bar{c}_{tp}$ between t and 500 °C			
400	27.232	228.8	1.819	1.787	1.782	1.695	1.669	1.644
450	30.636	235.2	1.847	1.821	1.753	1.711	1.677	1.652
500	34.040	241.2	1.880	1.864	1.781	1.733	1.697	1.674
600	40.848	251.9	1.941	1.952	1.840	1.772	1.730	1.694
700	47.656	261.2	2.000		1.901	1.819	1.766	1.724
800	54.464	269.7	2.064		1.967	1.863	1.797	1.650
900	61.272	277.4	2.137		2.044	1.914	1.834	1.780
1000	68.080	284.5	2.200		2.132	1.967	1.877	1.816
1200	81.696	297.2	2.347		2.346	2.087	1.958	1.880
1400	95.312	308.4	2.511			2.231	2.051	1.941
1600	108.928	318.4	2.691			2.405	2.158	2.030
1800	122.544	327.5	2.889			2.622	2.278	2.306
2000	136.160	335.8	3.114			2.899	2.417	2.204
2400	163.392	350.6	3.665				2.776	2.424
2800	190.624	361.3	4.394				3.273	2.706
3200	217.856	373.6	5.476				4.055	3.084
3600	245.088						5.334	3.584
4000	272.320							4.314

[a]For $g = 981.16$ cm/sec^2, value at London.

Table 45.—Specific Heat of Dilated Water-vapor at Constant Volume

I. Derived from data in Table 47 where remarks and references are given. Pressure is very low.

Unit of $c_v = 1$ joule/g.°C; of $P = 1$ kg*/cm^2; of $P_a = 1$ atm; of $P_b = 1$ bar.
Temp. $= t$ °C $= T$ °K

Source[a] → t	T	PS	ChK	KJ	M	J	Misc.	T$_A$	T$_B$
				c_v					
0	273.1		1.391						
12.3	285.4							1.368	1.394
100.0	373.1	1.530	1.419	1.410	1.421	1.415	Br[a]		
104.56	377.66						1.570		
107.5	380.6							1.403	1.433
140.0	413.1			1.412					
160.0	433.1			1.417					
180.0	453.1			1.422					
190.0	463.1			1.426					
200.0	473.1	1.532	1.473	1.426	1.471	1.452			
202.6	475.7							1.442	1.475
210.0	483.1			1.431					
220.0	493.1			1.438					
230.0	503.1			1.442					
240.0	513.1			1.452					
250.0	523.1			1.456					
260.0	533.1			1.463					
270.0	543.1			1.473					
280.0	553.1			1.480					
290.0	563.1			1.517					
297.8	570.9							1.484	1.528
300.0	573.1	1.536	1.533	1.505		1.514			
310.0	583.1			1.494					
320.0	593.1			1.531					
330.0	603.1			1.542					
340.0	613.1			1.561					
350.0	623.1			1.575					

PROPERTIES OF ORDINARY WATER-SUBSTANCE

Table 45—(Continued)

t	Source→ T	PS	ChK	KJ	M	J	Misc.	T_A	T_B
					c_v				
360.0	633.1			1.594					
370.0	643.1			1.610					
380.0	653.1			1.626					
390.0	663.1			1.647					
392.9	666.0							1.528	1.594
400.0	673.1	1.545		1.668	1.591	1.606	K[a]		
410.0	683.1						1.621		
488.1	761.2							1.577	1.645
500.0	773.1	1.567					1.656		
526.9	800.0		1.680						
550.0	823.1						1.689		
583.2	856.3						H[a]	1.624	1.710
600.0	873.1	1.599			1.721		1.668		
650.0	923.1						1.686		
678.4	951.5							1.659	1.777
700.0	973.1	1.640					1.705		
750.0	1023.1						1.726		
773.9	1047.0						W[a]	1.684	1.842
800.0	1073.1	1.696			1.853		2.030		
850.0	1123.1						2.716		
868.9	1142							1.749	1.905
900.0	1173.1	1.759							
926.9	1200		1.944						
963.9	1237							1.824	1.963
1000.0	1273.1	1.836			1.978				
1058.9	1332							1.902	2.014
1153.9	1427							1.977	2.065
1200.0	1473.1	2.023			2.090				
1248.9	1522						P[a]	2.046	2.128
1250.0	1523.1						2.416		
1326.9	1600		2.163				B[a]		
1340.0	1613.1						2.335		
1343.9	1617							2.132	2.193
1400.0	1673.1	2.247			2.185				
1439.9	1713							2.230	2.267
1534.9	1808							2.337	2.353
1540.0	1813.1						2.530		
1600.0	1873.1	2.516							
1629.9	1903							2.432	2.432
1726.9	2000		2.323						
1800.0	2073.1	2.809							
1819.9	2093							2.644	
2000.0	2273.1	3.008							
2200.0	2473.1	3.127							
2300.0	2573.1	3.152							
Limit								2.769	

II. Computed from the c_p of Section I of Table 42 and the value of $c_p - c_v = T \times (\delta v/\delta T)_p (\delta p/\delta T)_v$ as given by Linde's equation of state (Section 14, eq. 1).

$P \rightarrow$	0.5	1	2	4	$P \rightarrow$	0.5	1	2	4
$P_a \rightarrow$	0.484	0.968	1.936	3.871	$P_a \rightarrow$	0.484	0.968	1.936	3.871
$P_b \rightarrow$	0.490	0.981	1.961	3.923	$P_b \rightarrow$	0.490	0.981	1.961	3.923
t		c_v			t		c_v		
110	1.487	1.513			160	1.470	1.481	1.506	1.571
120	1.481	1.506			170	1.472	1.484	1.499	1.557
130	1.477	1.497	1.532		180	1.473	1.486	1.495	1.542
140	1.471	1.492	1.523		190	1.477	1.488	1.494	1.530
150	1.469	1.487	1.512	1.588	200	1.482	1.490	1.494	1.520

[a] For reference see Table 47.

15. VAPOR: THERMAL ENERGY

Table 46.—Mean Specific Heat of Dilated Water-vapor at Constant Volume

(See also Table 51)

Derived from columns SS and M of Table 48, which see. The fourth digit of $\bar{c}_v$ is uncertain by several units.

Unit of $\bar{c}_v = 1$ joule/g·°C $= 0.2390$ cal$_{15}$/g·°C.   Temp. $= t$ °C $= T$ °K; range 0 to t °C

t	Ref.[a] T	SS $\bar{c}_v$	M	t	Ref.[a] T	SS $\bar{c}_v$	M
0	273.1	1.434	1.470	1600	1873	1.790	1.775
100	373	1.452	1.480	1700	1973	1.826	1.814
200	473	1.471	1.486	1800	2073	1.865	1.856
300	573	1.492	1.496	1900	2173	1.907	1.905
400	673	1.510	1.504	2000	2273	1.951	1.956
500	773	1.529	1.516	2100	2373	1.998	2.008
600	873	1.547	1.529	2200	2473	2.044	2.060
700	973	1.565	1.543	2300	2573	2.092	2.113
800	1073	1.586	1.564	2400	2673	2.141	2.170
900	1173	1.605	1.589	2500	2773	2.192	2.222
1000	1273	1.624	1.608	2600	2873	2.246	2.277
1100	1373	1.645	1.631	2700	2973	2.301	2.338
1200	1473	1.670	1.656	2800	3073		2.396
1300	1573	1.698	1.686	2900	3173		2.457
1400	1673	1.726	1.715	3000	3273		2.520
1500	1773	1.759	1.742				

[a] For references see Table 48.

Table 47.—Molecular Specific Heat of Dilated Water-vapor at Constant Volume [224]

(See also Table 45)

The several values are said to have been derived from the indicated sources. Those from PS are the same as those given as "best" by A. Leduc [200] and are defined by the formulas $C_v = 6.750 - 0.00119t + 2.34(t/1000)^2$ if $100 °C \leq t \leq 1700 °C$, and $C_v = -12.652 + 0.02214t - 4.67(t/1000)^2$ if $1700 °C \leq t \leq 2300 °C$. Trautz gives two sets of smoothed values: T_A, based on all the observations; T_B, based on only those from spectroscopic data. He thinks the B values are to be preferred to about 1200 °C. He states that the observed values have as far as possible been reduced to infinite volume.

Unit of $C_v = 1$ cal/gfw-H$_2$O·°C $= 0.05551$ cal/g·°C $= 0.2323$ joule/g·°C. Temp. $= t$ °C $= T$ °K. Pressure very low

t	Source[a] T	PS	ChK	KJ	M C_v	J	Misc.	T_A	T_B
0	273.1		5.99						
12.3	285.4							5.89	6.00
100.0	373.1	6.58$_5$	6.11	6.07	6.11$_6$	6.09	Br[a]		
104.56	377.66						6.76		
107.5	380.6							6.04	6.17
140.0	413.1		6.08						
160.0	433.1		6.10						
180.0	453.1		6.12						

[224] Trautz, M., *Ann. d. Physik* (5), **9**, 465-485 (1931).

Table 47.—*(Continued)*

t	T	PS	ChK	KJ	M	J	Misc.	T_A	T_B
					C_v				
190.0	463.1			6.14					
200.0	473.1	6.59$_3$	6.34	6.14	6.33$_1$	6.25			
202.6	475.7							6.21	6.35
210.0	483.1			6.16					
220.0	493.1			6.19					
230.0	503.1			6.21					
240.0	513.1			6.25					
250.0	523.1			6.27					
260.0	533.1			6.30					
270.0	543.1			6.34					
280.0	553.1			6.37					
290.0	563.1			6.53					
297.8	570.9							6.39	6.58
300.0	573.1	6.61$_2$	6.60	6.48		6.52			
310.0	583.1			6.43					
320.0	593.1			6.59					
330.0	603.1			6.64					
340.0	613.1			6.72					
350.0	623.1			6.78					
360.0	633.1			6.86					
370.0	643.1			6.93					
380.0	653.1			7.00					
390.0	663.1			7.09					
392.9	666.0							6.58	6.86
400.0	673.1	6.65$_2$		7.18	6.84$_9$	6.91$_6$	K^a		
410.0	683.1						6.98		
488.1	761.2							6.79	7.08
500.0	773.1	6.74$_4$					7.13		
526.9	800.0		7.23						
550.0	823.1						7.27		
583.2	856.3						H^a	6.99	7.36
600.0	873.1	6.88$_5$			7.40$_8$		7.18		
650.0	923.1						7.26		
678.4	951.5							7.14	7.65
700.0	973.1	7.06$_2$					7.34		
750.0	1023.1						7.43		
773.9	1047						W^a	7.25	7.93
800.0	1073.1	7.30			7.97$_6$		8.74		
850.0	1123.1						11.69		
868.9	1142							7.53	8.20
900.0	1173.1	7.57$_3$							
926.9	1200		8.37						
963.9	1237							7.85	8.45
1000.0	1273.1	7.90$_2$			8.51$_5$				
1058.9	1332							8.19	8.67
1153.9	1427							8.51	8.89
1200.0	1473.1	8.70$_8$			8.99$_8$				
1248.9	1522						P^a	8.81	9.16
1250.0	1523.1						10.4		
1326.9	1600		9.31				B^a		
1340.0	1613.1						10.05		
1343.9	1617							9.18	9.44
1400.0	1673.1	9.67$_3$			9.40$_8$				
1439.9	1713							9.60	9.76
1534.9	1808							10.06	10.13
1540.0	1813.1						10.89		

15. VAPOR: THERMAL ENERGY

Table 47.—(Continued)

t	T	PS	ChK	KJ	M	J	Misc.	T$_A$	T$_B$
					C_v				
1600.0	1873.1	10.83							
1629.9	1903							10.47	10.47
1726.9	2000		10.0						
1800.0	2073.1	12.09[b]							
1819.9	2093							11.38	
2000.0	2273.1	12.95							
2200.0	2473.1	13.46							
2300.0	2573.1	13.57							
Limit								11.92	

[a] Sources:
- B = Bjerrum, N., *Z. physik. Chem.*, **79**, 513-536 (1912).
- Br = Brinkworth, J. H., *Phil. Trans. (A)*, **215**, 383-438 (1915).
- ChK = *Chemikerkalender*, 1931; derived from spectroscopic data.
- H = Holborn, L., and Henning, F., *Ann. d. Physik (4)*, **18**, 739-756 (1905); **23**, 809-845 (1907).
- J = Jakob, Max, private communication to Trautz.
- K = Knoblauch and associates: Knoblauch, O., and Jakob, M., *Sitz. K. Bayer (Münch.) Akad. Wiss. (Math.-phys.)*, **35**, 441-446 (1905); *Mitt Forsch. Geb. Ing.*, **35**, **36**, 109-152 (1906) → *Z. Ver. Deuts. Ing.*, **51**, 81, 88, 124-131 (1907); Knoblauch, O., and Mollier, H., *Mitt. Forsch. Geb. Ing.*, **108**, **109**, 79-106 (1911) → *Z. Ver. Deuts. Ing.*, **55**, 665-673 (1911); Knoblauch, O., and Raisch, E., *Idem*, **66**, 418-423 (1922); Knoblauch, O., and Winkhaus, A., *Mitt. Forsch. Geb. Ing.*, **195**, 1-20 (1917) → *Z. Ver. Deuts. Ing.*, **59**, 376-379, 400-405 (1915).
- KJ = Knoblauch, O., and Jakob, M., *Mitt. Forsch. Geb. Ing.*, **35**, **36**, 109-152 (1906) → *Z. Ver. Deuts. Ing.*, **51**, 81-88, 124-131 (1907).
- M = Mecke, R., unpublished communication to Trautz; from band spectrum.
- Misc = Several sources: B, Br, H, K, P, and W.
- P = Pier, M., *Z. Elektroch.*, **15**, 536-540 (1909); **16**, 897-903 (1910).
- PS = Partington, J. R., and Shilling, W. G., "The Specific Heat of Gases," London, Benn, 1924.
- T = Trautz, M., *Ann. d. Physik (5)*, **9**, 465-485 (1931); see head of table for significance of A and B.
- W = Womersley, W. D., *Proc. Roy. Soc. (London) (A)*, **100**, 483-498 (1921); **103**, 183-184 (1923). Criticism by R. T. Glazebrook, *Idem*, **101**, 112-114 (1922).

[b] The published value (11.84) seems to have been computed by means of the formula valid below 1700 °C. This and several other values belonging in the PS column were assigned by Trautz to ChK.

Table 48.—Mean Molecular Specific Heat of Dilated Water-vapor at Constant Volume

The mean molecular specific heat at constant volume over the indicated range in temperature is $\bar{C}_v$. The values at high temperatures have been inferred from the pressures developed on exploding mixtures of H_2 and O_2, initial pressure being 1 atm; those at lower temperatures, from C_p and an assumed equation of state. Comments on the several series will be found with the appropriate references.

Unit of C_v = 1 cal/gfw-H_2O·°C = 0.05551 cal/g·°C = 0.2323 joule/g·°C.   Temp. = t °C = T °K

Range→ Ref.[a]→		0 to t °C					100 to t °C
t	T	CG	NW	SS	M	W	PS
				C_v			
0	273.1	6.00	5.99	6.17	6.33		
100	373.1	6.05	6.05	6.25	6.37	6.59	6.65
200	473.1		6.14	6.33	6.40	6.61	6.63
300	573.1		6.24	6.42	6.44	6.65	6.61
400	673.1			6.50	6.47	6.70	6.62

Table 48.—(Continued)

t	Range→ Ref.ᵃ→ T	CG	NW	0 to t °C SS C_v	M	W	100 to t °C PS
500	773.1	6.32		6.58	6.53	6.76	6.63
527	800.1		6.53				6.64
600	873.1			6.66	6.58	6.85	6.67
700	973.1			6.74	6.64	6.92	6.72
800	1073.1			6.83	6.73	7.02	6.78
900	1173.1			6.91	6.84	7.14	6.86
927	1200.1		7.09				6.89
1000	1273.1	6.82		6.99	6.92	7.29	6.96
1100	1373.1			7.08	7.02	7.50	7.07
1200	1473.1			7.19.	7.13	7.75	7.20
1300	1573.1			7.31	7.26	8.03	7.34
1327	1600.1		7.61				7.40
1400	1673.1			7.43	7.38	8.35	7.50
1500	1773.1	7.50		7.57	7.50	8.74	7.68
1600	1873.1			7.71	7.64	9.23	7.87
1700	1973.1			7.86	7.81	9.82	8.07
1727	2000.1		8.10				8.13
1800	2073.1			8.03	7.99	10.46	8.29
1900	2173.1			8.21	8.20	11.07	8.51
2000	2273.1	8.36		8.40	8.42	11.69	8.74
2100	2373.1	8.55		8.60	8.64		8.96
2127	2400.1		8.50				8.98
2200	2473.1	8.75		8.80	8.87		9.17
2300	2573.1	8.96		9.01	9.10		9.36
2400	2673.1	9.17		9.22	9.34		
2500	2773.1	9.39		9.44	9.57		
2527	2800.1		8.8				
2600	2873.1	9.62		9.67	9.80		
2700	2973.1	9.85		9.91	10.06		
2800	3073.1	10.09			10.31		
2900	3173.1	10.34			10.58		
3000	3273.1	10.60			10.85		
3100	3373.1	10.86					
3200	3473.1	11.13					
3300	3573.1	11.41					
3400	3673.1	11.70					
3500	3773.1	11.99					
3600	3873.1	12.29					
3700	3973.1	12.59					
3800	4073.1	12.91					
3900	4173.1	13.23					
4000	4273.1	13.56					

t	Range→ Ref.ᵃ→ T	18 to t °C WM C_v	t	Range→ Ref.ᵃ→ T	15 to t °C B	MW C_v	16 to t °C N
1758	2031	8.26	1400	1673		10.7	
1770	2043	8.20	1750	2023		10.6	
1781	2054	8.07ᵇ	1811	2084	7.92ᵈ		
1882	2155	8.32	1950	2223		10.8	
1973	2246	8.38ᵇ	2110	2383	8.54ᵈ		
2035	2308	8.54ᵇ	2120	2393		10.6	
2060	2333	8.54ᵇ	2377	2650	9.37ᵈ		
2092	2365	8.70	2663	2936	10.0		

15. VAPOR: THERMAL ENERGY

Table 48.—(Continued)

Range→ Ref[a]→ t	T	18 to t °C WM C_v	Range→ Ref[a]→ t	T	−15 to t °C B C_v	MW C_v	16 to t °C N
2104	2377	8.72	2908	3181	10.5		
2148	2421	8.65	3060	3333	10.9		
2182	2455	8.58[c]	2327	2600			10.25
2311	2584	8.78	2427	2700			10.32
2318	2591	8.73	2527	2800			10.40
2330	2603	8.76	2627	2900			10.44
2404	2677	8.66[c]	2727	3000			10.50

SS[a] t	Range→ T	C_v	273.1 to t °C = 0 to T °K t	T	C_v
−273.1	0	5.96	1327	1600	7.11
−173	100	5.96	1427	1700	7.23
−73	200	5.98	1527	1800	7.35
+27	300	6.03	1627	1900	7.49
127	400	6.09	1727	2000	7.64
227	500	6.16	1827	2100	7.80
327	600	6.23	1927	2200	7.98
427	700	6.31	2027	2300	8.16
527	800	6.39	2127	2400	8.35
627	900	6.47	2227	2500	8.55
727	1000	6.56	2327	2600	8.75
827	1100	6.64	2427	2700	8.96
927	1200	6.73	2527	2800	9.17
1027	1300	6.82	2627	2900	9.40
1127	1400	6.92	2727	3000	9.61
1227	1500	7.01			

[a] References and notes:

B = N. Bjerrum.[225] Those marked [d] are from observations by Pier. These values are quoted by B. Neumann [226] and from him by CG.
CG = A. D. Crow and W. E. Grimshaw,[227] values are defined by the formula: $C_v = 6 + (7/15) \cdot (t/1000) + (16/45) \cdot (t/1000)^2$, between 0 °C and t °C.
M = Muraour,[228] values are derived from the observations by Pier and Bjerrum. The values of mean C_v given in a compilation by A. Leduc [200] were taken from this list.
MW = G. B. Maxwell and R. V. Wheeler.[229]
N = D. M. Newitt.[230] He stated that there were no experimental values for the region between 1600 °K and 2000 °K.
NW = W. Nernst and K. Wohl.[231]
PS = Computed by the compiler from the formulas for C_v given by Partington and Shilling, see Table 47.
SS = F. Schmidt and H. Schnell.[232]
W = W. D. Womersley.[233] Values for $t < 1000$ were derived from earlier data obtained by others.
WM = K. Wohl and M. Magat.[234]

[b, c] See [234].

[d] See [a] B.

---

[225] Bjerrum, N., *Z. Elektroch.*, **18**, 101-104 (1912).
[226] Neumann, B., *Z. angew. Chem.*, **32**, 141-146 (1919).
[227] Crow, A. D., and Grimshaw, W. E., *Phil. Trans. (A)*, **230**, 39-73 (1931).
[228] Muraour, *Chim. et indus.*, **10**, 23-29 (1923).
[229] Maxwell, G. B., and Wheeler, R. V., *J. Chem. Soc.*, **1928**, 15-21 (1928).
[230] Newitt, D. M., *Proc. Roy. Soc. (London) (A)*, **125**, 119-134 (1929).
[231] Nernst, W., and Wohl, K., *Z. techn. Phys.*, **10**, 608-614 (1929).
[232] Schmidt, F., and Schnell, H., *Z. techn. Phys.*, **9**, 81-92 (1928); computed from available data.
[233] Womersley, W. D., *Proc. Roy. Soc. (London) (A)*, **100**, 483-498 (1921); **103**, 183-184 (1923); see criticism by Glazebrook, R. T., *Idem*, **101**, 112-114 (1922).

Table 49.—Ratio of the Principal Specific Heats of Dilated Water-vapor

$\gamma = c_p/c_v$

Unit of $P = 1$ kg*/cm²; of $P_a = 1$ atm; of $P_b = 1$ bar. Temp. $= t$ °C $= T$ °K

I. From PS of Table 45 and ICT of Section VI of Table 42. Pressure = 1 atm. Except for the change noted in Table 47, note b, these are the ones contained in the table of "best" values for the molecular specific heat in A. Leduc's compilation [200] and derived from Partington and Shilling, "The Specific Heat of Gases."

t	γ	t	γ	t	γ
100	1.334	700	1.283	1600	1.183
200	1.311	800	1.272	1800	1.171
300	1.305	900	1.263	2000	1.154
400	1.301	1000	1.251	2200	1.147
500	1.295	1200	1.228	2300	1.146
600	1.290	1400	1.205		

II. From the velocity of sound. W. G. Shilling.[222] Equation of state used was either Callendar's (Cal) or Berthelot's (Ber).

Eq.→ t	Cal. γ	Ber.	Eq.→ t	Cal. γ	Ber.
100	1.332	1.317	600	1.316	1.316
200	1.334	1.332	700	1.295	1.296
300	1.337	1.337	800	1.277	1.277
400	1.335	1.336	900	1.257	1.257
500	1.329	1.330	1000	1.241	1.241

III. From Section II of Table 45 and section I of Table 42.

P		0.5	1	2	4
P_a		0.484	0.968	1.936	3.871
P_b		0.490	0.981	1.961	3.923
t	T		γ		
110	383	1.325	1.336		
120	393	1.325	1.336		
130	403	1.325	1.333	1.352	
140	413	1.325	1.332	1.349	
150	423	1.325	1.331	1.348	1.376
160	433	1.324	1.331	1.345	1.372
170	443	1.322	1.328	1.343	1.368
180	453	1.321	1.326	1.341	1.365
190	463	1.320	1.325	1.339	1.362
200	473	1.318	1.323	1.336	1.360

IV. Values computed by A. Leduc [235] by his method, employing his equation of state for normal gases; and in his compilation [200] marked "best." See remarks in Section VIII of Table 42.

P_a \ $t→$	120	130	140	150	160
			γ		
1	1.365		1.364	1.333	1.314
2		1.37	1.36	1.344	1.326
3			1.37	1.356	1.34
4				1.37	1.35
P_{sat}	1.378[a]		1.380		

[234] Wohl, K., and Magat, M., *Z. physik. Chem.* (B), **19**, 117-138 (1932); derived from 3 sets (unmarked, b, and c) by Wohl, K., and von Elbe, G., *Idem*, **5**, 241-271 (1929).

Table 49.—(Continued)

V. Derived by H. G. Muhammad [236] from his determinations of γ for air saturated with water-vapor at the temperature indicated. Total pressure about 745 mm-Hg.

t	γ	t	γ	t	γ
15	1.3421	35	1.3723	55	1.3054
20	1.3568	40	1.3671	60	1.3015
25	1.3672	45	1.3406	65	1.2993
30	1.3736	50	1.3199	70	1.2976

[a]For 100 °C and P_{sat} he gives $\gamma = 1.373$.

Table 50.—Ratio of the Principal Specific Heats of Air Saturated with Water-vapor [236]

P = total pressure, $\gamma = c_p/c_v$ for the saturated air. Unit of P = 1 mm-Hg. Temp. = t °C.

t	P	γ	t	P	γ
15	742.1	1.4009	45	745.6	1.3949
20	743.4	1.4008	50	745.4	1.3893
25	744.3	1.4007	55	745.0	1.3828
30	741.8	1.4005	60	745.0	1.3765
35	743.3	1.4000	65	745.0	1.3697
40	745.1	1.3989	70	745.0	1.3618

Table 51.—Enthalpy of Dilated Water-vapor

Several sets of the better values are compared in Table 52. The enthalpy ("heat-content," "total heat") of dilated water-vapor as measured from saturated water at 0 °C is H; $c_p = (\delta H/\delta t)_p$; $\bar{c}_p = (H_t - H_{t_1})/(t - t_1)$ is the mean value of c_p over the indicated range; P_s kg*/cm² is the saturation pressure at the indicated temperature.

An earlier and more detailed table, in which the unit of pressure is 1 lb*/in², has been published by H. L. Callendar,[192] and very detailed tables and graphs may be found in the numerous steam tables (see Table 259).

Unit of P_a = 1 atm, of P = 1 kg*/cm², of H = 1 Int. joule/g.°C. Temp. = t °C (Int. scale)

I. Keyes, Smith, and Gerry.

Based on their formula given in Table 39; computed and published by themselves. Very detailed tables on the same basis, but in terms of °F and lb*/in², have been published by J. H. Keenan and F. G. Keyes, "Thermodynamic Properties of Steam," 1936. Only the last few digits of the value of H for P are given here, the others being the same as for the corresponding value of P_a. For example, the value of H for P = 25 and t = 300 is 3010.53.

[235] Leduc, A., *Ann. d. chim. et phys.* (8), **28**, 577-613 (1913); *J. de Phys.* (6), **2**, 24-30 (1921).
[236] Muhammad, H. G., *Bull. Acad. Sci., (Allahabad)*, **3**, 269-294 (1934).

PROPERTIES OF ORDINARY WATER-SUBSTANCE

Table 51.—(Continued)

P_a	P	100	150	200	250	300	200 300 $\bar{c}_p$
		—	—	H	—	—	
1		2675.35	2776.22	2874.79	2973.43	3072.94	1.9815
	1	5.80	6.46	4.94	3.54	3.02	1.9808
5				2854.71	2959.95	3063.11	2.0840
	5			5.58	60.50	3.53	2.0795
10				2826.07	2941.79	3050.26	2.2419
	10			8.06	4.01	1.11	2.2305
25					2877.66	3008.14	
	25				81.51	10.53	
50						2921.77	
	50					8.04	
P_{sat}	P_{sat}	2675.35	2744.84	2790.46	2799.00	2747.05	

P_a	P	350	400	450	500	550	450 550 $\bar{c}_p$
		—	—	H	—	—	
1		3173.67	3275.81	3379.48	3484.72	3591.58	2.1210
	1	3.73	5.86	9.52	4.76	1.61	
5		3166.10	3269.74	3374.48	3480.51	3587.97	2.1349
	5	6.40	70.14	4.68	0.68	8.12	
10		3156.37	3262.04	3368.16	3475.21	3583.44	2.1528
	10	7.00	2.52	8.57	5.55	3.74	
25		3125.45	3238.02	3348.74	3459.05	3569.70	2.2096
	25	7.17	9.34	9.80	9.91	70.43	
50		3067.51	3194.99	3314.84	3431.26	3546.30	2.3146
	50	71.51	7.87	7.08	3.03	7.83	
100		2918.67	3095.64	3240.77	3372.45	3497.70	2.5693
	100	30.01	102.67	5.80	6.36	500.80	
150		2677.85	2971.96	3156.96	3309.40	3446.65	2.8969
	150	710.39	85.28	65.64	15.21	51.67	
200			2809.41	3060.81	3240.77	3393.09	3.3228
	200		33.55	74.19	9.86	400.11	
250			2550.33[a]	2948.90	3166.89	3336.95	3.8805
	250		607.05[a]	68.32	79.21	46.15	
P_{sat}	P_{sat}	2562.58					

II. Havlíček and Miškovský. (See also Table 52.)

Based on their formula given in Table 39. Conversion into joules was made by the compiler.

P	100	150	200	250	300	200 300 $\bar{c}_p$
	—	—	H	—	—	
1	2674.0	2775.3	2874.1	2972.9	3072.5	1.984
25				2882.5	3008.5	
50					2927.3	

P	350	400	450	500	550	450 550 $\bar{c}_p$
	—	—	H	—	—	
1	3173.4	3275.5	3378.9	3484.0	3591.2	2.123
25	3125.3	3237.9	3348.8	3459.3	3569.8	2.210
50	3068.8	3195.6	3315.7	3432.1	3547.6	2.319
100	2928.1	3099.3	3242.9	3374.3	3500.3	2.574

15. VAPOR: THERMAL ENERGY

Table 51.—(Continued)

$t \rightarrow$ P	350	400	450	500	550	450 550 c_p
			H			
150	2707.5	2981.7	3162.1	3312.4	3450.9	3.888
200		2828.5	3070.4	3245.8	3398.6	3.282
250		2608.7	2964.5	3173.8	3343.8	3.793
300			2838.1	3096.8	3286.8	4.487
350				3013.1		
400				2922.2		

III. W. Koch.[237] Conversion into joules was made by the compiler.

$t \rightarrow$ P	200	225	250	275	300	325	200 300 c_p
				H			
1	2871.6	2920.6	2970.4	3020.6	3070.4	3121.1	1.988
5	2852.8	2903.8	2956.2	3008.9	3060.4	3411.4	2.076
10	2826.0	2882.5	2938.2	2993.0	3046.6	3099.3	2.206
25		2804.6	2878.3	2943.6	3004.3	3063.7	
50				2836.4	2922.7	2998.0	
75					2809.2	2920.1	
100						2818.8	

$t \rightarrow$ P	350	375	400	425	450	475	500
				H			
1	3171.7	3222.8	3273.9	3325.8	3378.1	3430.4	3483.6
5	3163.8	3216.1	3266.8	3320.3	3372.7	3426.6	3479.4
10	3153.3	3206.9	3258.8	3313.2	3366.8	3421.2	3474.8
25	3121.9	3179.7	3234.5	3292.7	3348.8	3405.3	3461.0
50	3066.2	3130.3	3191.4	3254.2	3315.3	3375.2	3434.2
75	3004.3	3078.8	3147.4	3214.8	3281.0	3345.0	3407.4
100	2930.2	3020.6	3100.2	3174.2	3245.0	3313.2	3378.9
125	2834.8	2952.0	3045.3	3129.4	3207.3	3278.5	3349.2
150	2701.2	2870.3	2983.8	3079.6	3164.2	3242.5	3316.6
175		2768.6	2913.4	3022.7	3116.5	3203.5	3284.3
200		2625.4	2830.2	2961.2	3065.4	3162.5	3250.0
225		2327.4	2729.3	2890.8	3011.4	3119.0	3214.8
250		1856.1	2602.8	2813.0	2956.2	3073.4	3176.8
275		1817.1	2429.1	2726.8	2896.7	3026.0	3137.8
300		1792.9	2194.3	2633.8	2833.9	2976.2	3096.0

IV. Third International Steam-Table Conference, 1934. For allowed tolerances, etc., see Table 260.

$t \rightarrow$ $P_{sat} \rightarrow$ P	100 1.03323	150 4.8535	200 15.857	250 40.560	300 87.611	350 168.63	400	450	500	550
					H					
1	2676	2777	2875	2973	3074	3174	3276	3380	3485	3592
5			2855	2959	3063	3166	3269	3374	3481	3588
10			2827	2940	3048	3155	3270	3369	3476	3584
25				2880	3006	3125	3238	3350	3461	3570
50					2924	3069	3195	3315	3433	3548
75					2816	3005	3149	3280	3405	3525

[237] Koch, W., Z. Ver. deuts. Ing., 78, 1160 (1934).

Table 51.—(Continued)

$t \rightarrow$	100	150	200	250	300	350	400	450	500	550
$P_{sat} \rightarrow$	1.03323	4.8535	15.857	40.560	87.611	168.63				
P						H				
100						2929	3099	3243	3375	3501
125						2834	3044	3204	3346	3477
150						2709	2982	3163	3316	3452
200							2833	3071	3249	3400
250							2607	2962	3177	3345
300								2196	2837	3097
P_{sat}	2676	2746	2792	2801	2747	2562				

[a] Values in Section I at 400 °C for P and $P_a = 250$ are not so good as the others, density > 0.1 g/cm³.

Table 52.—Enthalpy of Dilated Water-vapor: Comparison of Several Sets of Values

Adapted from a table published by J. Havliček and L. Miškovský.[191] Not checked by the compiler.

The values of H (enthalpy, counted from saturated water at 0 °C) are given for set A; for the other sets only the excess (ΔH) of each value over the corresponding one for A is given. For example, at $P = 1$ and $t = 150$ the value of H for A is 663.3; for P it is $663.3 - 0.2 = 663.1$.

Some of the values here given for the P set differ slightly from the corresponding ones in Table 51, which Havliček and Miškovský computed by means of their equation.

Unit of $P = 1$ kg*/cm², of H and $\Delta H = 1$ Int. steam cal/g = 4.1860 Int. joule/g. Temp. = t °C (Int. scale)

$t \rightarrow$ P	Ref.[a]	150	200	250	300	350 H and ΔH	400	450	500	550
1	A	663.3	686.9	710.4	734.2	758.2	782.6	807.3	832.4	858.0
	P	− 0.2	− 0.3	− 0.1	− 0.1	0.0	0.0	0.0	− 0.1	− 0.4
	G		− 0.9	− 0.8	− 0.7	− 0.5	− 0.5	− 0.3	− 0.2	
	I	− 0.1	− 0.4	− 0.3	− 0.2	− 0.2	− 0.2	− 0.1	− 0.1	− 0.2
25	A		688.4	719.2	747.1	773.8	800.2[b]	826.5	852.9	
	P		− 0.2	− 0.9	− 0.7	− 0.4	− 0.2	− 0.2	− 0.3	
	G		− 0.8	− 1.5	− 1.3	− 1.1	− 0.2	+ 0.3		
	C			− 1.0	− 1.5	− 1.3	− 0.6	− 0.4	0.0	
	I		− 0.6	− 1.2	− 0.8	− 0.5	− 0.2	0.0	− 0.3	
50	A			699.5	733.8	763.9	792.4	820.1	847.4	
	P			− 0.5	− 1.1	− 0.8	− 0.5	− 0.3	− 0.1	
	G			− 1.3	− 1.3	− 1.5	− 0.2	+ 0.3		
	C			− 1.4	− 1.2	− 1.0	− 1.1	− 0.4		
	I			− 1.1	− 0.9	− 0.8	− 0.8	− 0.2	− 0.1	
100	A				700.0	741.2	775.4	806.5	836.3	
	P				− 0.6	− 1.0	− 0.9	− 0.5	− 0.2	
	G				0.0	− 0.6	− 0.2	+ 0.7		
	C				− 0.9	− 1.6	− 1.8	− 0.8		
	I				− 0.5	− 1.2	− 0.9	− 0.5	− 0.2	
150	A					713.2	756.2	792.0	824.6	
	P					− 0.7	− 1.0	− 0.7	− 0.2	
	G					− 0.4	− 0.3	+ 0.3		
	C					− 1.5	− 1.4	− 0.2		
	I					− 1.1	− 0.9	− 0.2	− 0.2	

15. VAPOR: THERMAL ENERGY

Table 52.—(Continued)

$t \to$ P	Ref.[a]	150	200	250	300	350 H and ΔH	400	450	500	550
200	A						676.9	734.3	776.3	812.3
	P						− 0.8	− 0.8	− 0.9	− 0.3
	G						− 0.8	− 2.0	+ 0.1	
	I						− 0.4	− 0.9	− 0.3	− 0.3
250	A						623.3	708.7	759.3	799.5
	P						− 0.7	− 0.4	− 1.0	− 0.6
	G						− 1.5	− 2.5	− 0.4	
	I						− 0.8	− 1.2	− 0.5	− 0.6
300	A						525.1			
	P						+ 0.4			
	G						− 0.9			
	I						− 0.6			

[a] References: The authors do not give the references more specifically than here.
  A A.S.M.E.: Davis, Keenan, Keyes, Osborne, Smith. (These seem to be the equivalent of the values in Section I of Table 51.)
  C B.E.M.A.A.: Egerton, Callendar.
  G Germany: Hausen, Henning, Jakob, Koch. (This is identical with the set of values published by Koch and given in Table 51.)
  I Third International Steam Table Conference, 1934. See Section IV of Table 51.
  P M.A.P.: Havliček, Miškovský.

[b] The value for A at $P = 25$, $t = 450$ is printed 802.0, which is evidently too great; the corresponding value in Table 51 leads to 800.2.

Table 53.—Isopiestic Variation in the Enthalpy of Water Substance Through the Critical Temperature [191]

The following values of J. Havliček and L. Miškovský have been taken from their Table 1. The enthalpy (H) is measured from saturated water at 0 °C.

Since the saturation temperature for $P = 200$ is 364.1 °C, and for $P = 225$ is 373.9 °C, and the critical pressure and temperature are 225.5 kg*/cm² and 374.11 °C, respectively, it is obvious that all values lying above the heavy line refer to the liquid; and all below it, to the vapor.

Unit of $P = 1$ kg*/cm²; of $H = 1$ Int. steam cal/g = 4.1860 joule/g. Temp. = t °C (Int. scale)

$P \to$ t	200	225	250	275 H	300	350	400
20	24.55				26.79		28.92
100	103.8				105.54		107.22
200	205.5				206.0		206.9
300	318.2				316.7		316.2
350	392.6	389.45	387.8	385.4	383.9	381.45	379.3
360	415.6	410.0	406.1	402.9	400.5		
370	610.5	439.5	428.7	422.7	418.4		
375						421.3	416.4
380	640.4	596.5	470.0	448.0	440.7		
390	661.4	630.2	585.7	500.7	470.3		
400	646.1	652.6	622.6	581.6	525.5	478.5	463.8
410					582.2		

116   PROPERTIES OF ORDINARY WATER-SUBSTANCE

Table 53.—(Continued)

P→ t	200	225	250	275 H	300	350	400
450	733.5	720.2	708.3	693.1	677.9	644.5	607.3
500	775.4	767.5	758.3	749.4	639.8[a]	719.6	698.6
550	812.0	804.3	798.9				

[a] So printed; probably should be 739.8.

Table 54.—Entropy of Dilated Water-vapor
(See also Table 55)

The excess of the entropy of the vapor at the indicated temperature and pressure above that of saturated water at 0 °C is S; S_s is the value of S at the indicated pressure and the corresponding saturation temperature t_s; ΔS is the change in S when the temperature is changed over the indicated range while the pressure remains constant at the indicated value.

Unit of $P = 1$ kg*/cm² $= 0.96784$ atm; of $P_a = 1$ atm; of $p = 1$ (lb*/in²)L[a]; of S, S_s, and $\Delta S = 1$ Int. joule/g.°C. Temp. $= t$ °C (Int. scale)

I. Preferred values: F. G. Keyes, L. B. Smith, and H. T. Gerry.[197] Based on their formulas in Table 39; conversion from calories to joules (1 Int. steam calorie = 4.1860 Int. joules) was made by the compiler. Very detailed tables on the same basis but in terms of °F and lb*/in² have been published by J. H. Keenan and F. G. Keyes, "Thermodynamic Properties of Steam," 1936.

t→ P	100	200	300 S	400	500	400 500 ΔS
1	7.3686	7.8425	8.2221	8.5482	8.8375	0.2893
10		6.7022	7.1313	7.4699	7.7663	0.2964
50			6.2216	6.6570	6.9831	0.3261
100				6.2283	6.6093	0.3810
200				5.5820	6.1630	0.5810
250					5.9889	

II. Adapted from R. Mollier.[238] Not to be preferred to values in I.

t→ P	t_s	t_s	200	300 S	400	500	200 300	400 500 ΔS
1	99.1	7.373	7.860	8.241	8.560	8.833	0.382	0.273
2	119.6	7.145	7.531	7.918	8.238	8.512	0.386	0.274
3	132.9	7.012	7.336	7.727	8.049	8.323	0.391	0.275
4	142.9	6.917	7.195	7.591	7.914	8.190	0.395	0.276
6	158.1	6.781	6.992	7.397	7.724	8.001	0.405	0.277
8	169.6	6.684	6.841	7.257	7.587	7.866	0.416	0.279
10	179.0	6.607	6.719	7.147	7.481	7.761	0.426	0.280
20	211.4	6.353		6.790	7.144	7.432		0.288
30	232.8	6.189		6.562	6.939	7.235		0.296
40	249.2	6.061		6.382	6.788	7.093		0.305
50	262.7	5.952		6.222	6.666	6.980		0.314
100	309.5	5.546			6.232	6.610		0.378
150	340.5	5.240			5.882	6.367		0.484
200	364.2	4.905			5.512	6.168		0.656

15. VAPOR: THERMAL ENERGY

Table 54.—(Continued)

III. Adapted from H. L. Callendar.[192] Not to be preferred to values in I.

p	P_a	t_s	t_s	300	400	500	400 500 ΔS
			⎯⎯⎯⎯⎯⎯⎯⎯⎯⎯ S ⎯⎯⎯⎯⎯⎯⎯⎯⎯⎯				
400	27.232	228.8	6.252	6.611	6.986	7.296	0.310
500	34.040	241.2	6.168	6.480	6.870	7.176	0.306
600	40.848	251.9	6.096	6.366	6.773	7.085	0.312
800	54.464	269.7	5.974	6.168	6.614	6.938	0.324
1000	68.080	284.5	5.870	5.985	6.482	6.821	0.339
1200	81.696	297.2	5.777	5.802	6.368	6.722	0.354
1600	108.928	318.4	5.604		6.171	6.560	0.389
2000	136.160	335.8	5.437		5.989	6.426	0.437
2400	163.392	350.6	5.259		5.806	6.308	0.502
2800	190.624	363.1	5.057		5.606	6.200	0.594
3200	217.856	373.6	4.789		5.362	6.098	0.736
3600	245.088				5.031	6.000	0.969
4000	272.320					5.901	

^a Value at London, where $g = 981.16$ cm/sec².

Table 55.—Various Thermal Data for Dilated Water-vapor: Computed from Spectroscopic Data

G = Gibbs function (H − TS), often called the "free energy at constant pressure," H = enthalpy or heat content, S = entropy, T = absolute temperature, E_0 = internal energy at 0 °K, C_p = specific heat at constant pressure. G, H, and S are measured from 0 °K, *i.e.*, each is the increase in the corresponding property when the temperature is increased from 0 °K to that indicated.

I. A. R. Gordon.[239] It is assumed that the pressure is so low that the vibration of each H_2O molecule is essentially uninfluenced by the presence of neighboring molecules. Giauque has independently computed S for T = 298.1, 463.1, and 485.0 °K, using thermal data and obtaining values that are smaller than Gordon's by about 0.1 per cent.[240] In order that Gordon's value for R shall be the same as that adopted by the *International Critical Tables* and used elsewhere in this compilation it is necessary to use the relation 1 cal = 4.1873 joules in converting his values from calories to joules.

Temp. = T °K = t °C. 1 cal/g-mole = 0.23243 joule/gram

		1 cal per g-mole·deg			1 joule per gram·deg		
T	t	$-\dfrac{G-E_0}{T}$	S	C_p	$-\dfrac{G-E_0}{T}$	S	C_p
298.1	25.0	37.179	45.101	8.000	8.642	10.483	1.859
300	26.9	37.230	45.151	8.002	8.653	10.494	1.860
350	76.9	38.452	46.389	8.066	8.937	10.782	1.875

[238] Mollier, R., "Neue Tabellen und Diagramme," 5th ed., 1927.
[239] Gordon, A. R., *J. Chem'l Phys.*, **2**, 65-72, 549(L) (1934); supersedes *Idem*, **1**, 308-312 (1933).
[240] Giauque, W. F., and Stout, J. W., *J. Am. Chem. Soc.*, **58**, 1144-1150 (1936); Giauque, W. F., and Archibald, R. C., *Idem*, **59**, 561-569 (1937).

Table 55.—(Continued)

Unit→		1 cal per g-mole·deg			1 joule per gram·deg		
T	t	$-\dfrac{G-E_0}{T}$	S	C_p	$-\dfrac{G-E_0}{T}$	S	C_p
400	126.9	39.513	47.472	8.155	9.184	11.034	1.895
450	176.9	40.452	48.439	8.260	9.402	11.259	1.920
500	226.9	41.296	49.315	8.379	9.598	11.462	1.948
550	276.9	42.062	50.119	8.504	9.776	11.649	1.976
600	326.9	42.765	50.864	8.635	9.940	11.822	2.007
650	376.9	43.415	51.561	8.771	10.091	11.984	2.039
700	426.9	44.020	52.216	8.910	10.232	12.136	2.071
750	476.9	44.587	52.836	9.053	10.363	12.281	2.104
800	526.9	45.121	53.425	9.199	10.487	12.418	2.138
850	576.9	45.627	53.987	9.347	10.605	12.548	2.172
900	626.9	46.106	54.525	9.497	10.716	12.673	2.207
950	676.9	46.563	55.043	9.648	10.823	12.794	2.242
1000	726.9	46.999	55.542	9.799	10.924	12.910	2.278
1050	776.9	47.418	56.023	9.948	11.021	13.021	2.312
1100	826.9	47.820	56.489	10.095	11.115	13.130	2.346
1150	876.9	48.206	56.941	10.240	11.204	13.235	2.380
1200	926.9	48.579	57.380	10.382	11.291	13.337	2.413
1250	976.9	48.940	57.807	10.522	11.375	13.436	2.446
1300	1026.9	49.289	58.223	10.656	11.456	13.533	2.477
1400	1126.9	49.956	59.022	10.914	11.611	13.718	2.537
1500	1226.9	50.586	59.783	11.153	11.758	13.895	2.592
1750	1476.9	52.03	61.54	11.67	12.093	14.304	2.712
2000	1726.9	53.32	63.13	12.09	12.393	14.673	2.810
2250	1976.9	54.49	64.58	12.4	12.665	15.010	2.88
2500	2226.9	55.57	65.90	12.7	12.916	15.317	2.95
2750	2476.9	56.56	67.12	12.9	13.146	15.601	3.00
3000	2726.9	57.49	68.25	13.1	13.362	15.863	3.04

II. A. R. Gordon and C. Barnes.[241] The spectroscopic data upon which these values are based have been superseded by those of R. Mecke,[242] W. Baumann and R. Mecke,[243] and K. Freudenberg and R. Mecke.[244] This fact must be considered by the user. Irrespective of their accuracy, they are of interest in showing the distribution of S among the three types of motion. Subscripts: t = translational, r = rational, v = vibrational entropy. The value of S_t has been corrected for the departure of water-vapor from ideality, Steinwehr's equation of state [245] being used.

Unit of S_t, S_r, S_v, and $S = 1$ cal/g-mole·°C $= 0.23243$ joule/g·°C. Temp. $= t$ °C $= T$ °K; press. $= 1$ atm

t	T	S_t	S_r	S_v	S	$0.2324_3 S$
127	400	36.025	13.959	0.043	50.03	11.627
227	500	37.152	14.626	0.115	51.89	12.061
327	600	38.066	15.167	0.222	53.46	12.425
427	700	38.835	15.626	0.346	54.81	12.738
527	800	39.501	16.024	0.494	56.02	13.020
627	900	40.088	16.375	0.657	57.12	13.277
727	1000	40.611	16.688	0.831	58.13	13.511
827	1100	41.084	16.972	1.020	59.08	13.731
927	1200	41.516	17.231	1.215	59.96	13.957

Saturated vapor (press. = 26.739 mm-Hg)

| 27 | 300 | 41.278 | 13.104 | 0.008 | 54.39 | 12.642 |

15. VAPOR: THERMAL ENERGY

Table 55.—*(Continued)*

III. M. Trautz and H. Ader.[246] The following values (C_{vr}) for the rotational component of the specific heat at constant volume were computed from the spectroscopic data of Mecke and Baumann (1933). They concluded that this component attains its full value at a temperature only slightly above 50 °K. As usual, R is the gas constant.

T	5.0	6.66	10.0	12.5	20.0	25.0	40.0	50.0 °K
C_{vr}/R	0.1061	0.2252	0.4587	0.5998	0.9359	1.0813	1.3600	1.461

IV. E. B. Wilson, Jr.,[247] has concluded that the distortion caused by the rotation of the molecules affects the thermal properties of the vapor as indicated by the following formulas, ρ being a constant characteristic of the molecule, and a prime indicating what would be the value of the property if the molecule were rigid: $G = G' - \rho R T^2$, $S = S' + 2\rho R T^2$, $C_v = C'_v + 2\rho R T^2$. For water-vapor at very low pressure he gives $\rho = 2.04(10^{-5})$ (°K)$^{-2}$.

(H. M. Randall, D. M. Dennison, N. Ginsburg, and L. R. Weber [248] have concluded that no first order computation, such as that of Wilson's, is at all satisfactory when the rotation is so rapid as to cause a displacement of the spectral line in excess of about $\Delta \nu = 5$ cm^{-1}.)

Table 56.—Joule-Thomson Coefficient for Dilated Water-vapor

The Joule-Thomson coefficient (μ) is the decrease in temperature per unit drop in pressure when the expansion is adiabatic. It measures the internal latent heat of expansion. For the more permanent gases, it has been found that the ratio of the decrease in temperature to the associated decrease in pressure "is, for small pressures at all events, constant and independent of both the fall in pressure and the absolute value of the pressure." [249] That is, μ is independent of the pressure. This is nearly, but not quite true for water-vapor. For that, the variation of μ with the pressure, temperature being constant, is small, but not linear in the pressure; and similarly, when the pressure is constant μ is not linear in the temperature. A. Griessmann [250] erred in concluding that μc_p for water-vapor is independent of the temperature, the pressure being constant; the values he accepted for c_p are unsatisfactory. For a discussion of the experimental

---

[241] Gordon, A. R., and Barnes, C., *J. Phys'l Chem.*, **36**, 1143-1151 (1932).
[242] Mecke, R., *Z. Physik*, **81**, 313-331 (1933).
[243] Baumann, W., and Mecke, R., *Idem*, **81**, 445-464 (1933).
[244] Freudenberg, K., and Mecke, R., *Idem*, **81**, 465-481 (1933).
[245] von Steinwehr, H., *Z. Physik*, **3**, 466-476 (1920).
[246] Trautz, M., and Ader, H., *Z. Physik*, **89**, 12-14 (1934).
[247] Wilson, E. B., Jr., *J. Chem'l Phys.*, **4**, 526-528 (1936).
[248] Randall, H. M., Dennison, D. M., Ginsburg, N., and Weber, L. R., *Phys. Rev. (2)*, **52**, 160-174 (1937).
[249] Buckingham, E., *Bull. Bur. Std.*, **3**, 237-293(S57) (1907).

Table 56.—*(Continued)*

evidence for the validity of the assumption, often made, that μ obeys the law of corresponding states, see H. N. Davis.[251]

Unit of $P = 1$ kg*/cm² $= 0.9678$ atm; of $\mu = 1$ °C per (1 kg*/cm). Temp. $= t$ °C

I. DK = observations by Davis and Kleinschmidt[a]; KSG' = values computed by F. G. Keyes, L. B. Smith, and H. T. Gerry[252]; KSG'' = later values computed by them.[197] In each case the computation is based on an empirical formula set up by KSG to represent their determinations of the specific volume of dilated water-vapor. The two formulas differ: KSG'' supersedes KSG'.

P	DK	KSG'	KSG''	P	DK	KSG'	KSG''
		—— μ ——				—— μ ——	
		125 °C				260 °C	
1.125	4.801	4.564	4.908	1.60	1.485	1.553	1.495
		145 °C		3.16	1.536	1.557	1.506
1.405	3.730	3.805	3.928	10.55	1.549	1.564	1.539
2.81	3.998	3.816	4.009	14.77	1.548	1.563	1.548
		166 °C		15.11	1.577	1.563	1.549
1.60	2.973	3.159	3.180	20.00	1.551	1.558	1.550
1.76	3.092	3.161	3.182	25.30	1.545	1.550	1.543
2.85	3.209	3.167	3.228	32.40	1.533	1.536	1.526
5.62	3.264	3.173	3.270	39.60	1.511	1.518	1.499
		196 °C				300 °C	
1.60	2.368	2.472	2.405	1.70	1.163	1.202	1.168
3.52	2.409	2.481	2.451	3.16	1.193	1.204	1.174
7.04	2.557	2.485	2.500	8.44	1.192	1.209	1.188
7.04	2.474	2.485		14.00	1.207	1.211	1.199
7.60	2.522	2.484	2.505	15.00	1.197	1.211	1.199
10.55	2.570	2.480	2.500	20.25	1.201	1.210	1.202
10.55	2.490	2.480	2.500	32.70	1.187	1.201	1.199
		225 °C				347 °C	
1.60	1.882	1.985	1.910	1.60	0.932	0.914	0.905
3.16	1.995	1.990	1.930	7.38	0.929	0.918	0.918
7.04	1.978	1.997	1.970	15.00	0.928	0.920	0.922
10.55	1.948	1.995	1.985	35.00	0.919	0.917	0.926
14.77	1.987	1.990	1.990				
20.25	1.953	1.977	1.970				
20.25	2.001	1.977	1.970				

II. Some of preceding DK observations rearranged.

$P \rightarrow$	1.60	3.16	7.04	10.55	14.77	15.00	20.25	39.60
t				—— μ ——				
166	2.973							
196	2.368		2.516	2.530				
225	1.882	1.995	1.978	1.948	1.987		1.977	
260	1.485	1.536		1.549	1.548			1.511
300		1.193				1.197	1.201	
347	0.932					0.928		

[250] Griessmann, A., *Z. Ver. deut. Ing.*, **47**, 1852-1857, 1880-1884 (1903).
[251] Davis, H. N., *Proc. Am. Acad. Arts Sci.*, **45**, 241-264 (1910).
[252] Keyes, F. G., Smith, L. B., and Gerry, H. T., *Mech. Eng.*, **56**, 87-92 (1934).

16. VAPOR: THERMAL CONDUCTION

Table 56.—*(Continued)*

III. Near state of saturation. A. Griessmann.[250] Values of μ are uncertain by 2 in the first decimal place.

$P \to 1.5$ $t_{sat} \to 110.8$		2.0 119.6		2.5 126.8		3.0 132.9	
t	μ	t	μ	t	μ	t	μ
136.6	4.5	134.1	4.4	136.3	4.2	138.4	4.1
138.0	4.4	135.2	4.4	137.4	4.2	139.4	4.0
141.9	4.3	138.7	4.4	140.9	4.1	143.0	4.1
145.8	4.2	140.2	4.3	142.3	4.1	144.3	4.0
148.6	4.0	144.0	4.3	146.1	4.2	148.1	4.0
150.3	4.0	147.9	4.1	149.9	3.9	151.8	3.8
152.8	3.8	150.7	3.9	152.6	3.9	154.5	3.7
153.7	3.8	152.3	3.9	154.2	3.7	156.0	3.5
		154.7	3.7	156.5	3.6	158.3	3.6
		155.6	3.7	157.4	3.6	159.2	3.5

$P \to 1.5$ $t_{sat} \to 110.8$		4.0 142.9		4.5 147.2		5.0 151.1	
t	μ	t	μ	t	μ	t	μ
141.5	3.9	146.9	3.8	150.0	3.6	155.3	3.1
145.0	4.1	148.2	3.8	153.6	3.4	158.7	3.1
146.3	3.9	151.9	3.6	157.1	3.3		
150.1	3.7	155.4	3.4	159.7	3.3	$P \to 5.5$	
153.6	3.6	158.0	3.4			$t_{sat} \to 154.7$	
156.3	3.5	159.4	3.4			156.8	3.0
157.7	3.4					160.2	3.0
160.0	3.4						
160.9	3.4						

IV. J. R. Roebuck,[253] from H. N. Davis,[251] except as noted.

t	120	150	165	200	250	300	350	400
μ	5.33	3.63	3.182[b]	2.20	1.50	1.15	0.90	0.75

[a] Apparently no final account of this work has been published, but the DK values here given and included in the KSG paper were published by H. N. Davis and J. H. Keenan.[254] They stated (p. 926, see also second paper, p. 253): "The Harvard data have not yet been published, but the definitive experimental results are given in Table 14." Those are the DK values here given. Preliminary reports: H. N. Davis,[255] and R. V. Kleinschmidt.[256] In the last it is stated that the final report of the work has been presented to the Steam Research Committee, and that "we hope to be able to publish it in the near future." But it had not been published in 1929, and apparently it has not yet appeared. Although the values include the third place of decimals, duplicates differ by several units (5 to 8) in the second place.

[b] Determination by H. M. Trueblood,[257] $P = 3.86$ kg/cm².

---

16. THERMAL CONDUCTIVITY OF DILATED WATER-VAPOR

In 1931, M. Jakob [258] stated that the only data available for the thermal conductivity of water-vapor are those given by E. Moser [259] for 46 to

[253] Roebuck, J. R., *Int. Crit. Tables*, **5**, 146 (1929).
[254] Davis, H. N., and Keenan, J. H., *Mech. Eng.*, **51**, 921-931 (1929); *Proc. World Eng. Cong. (Tokyo)*, 1929, **4**, 239-264 (1931).
[255] Davis, H. N., *Mech. Eng.*, **46**, 85-87, 108 (1924).
[256] Kleinschmidt, R. V., Idem, **45**, 165-167 (1923); **46**, 84-85 (1924); **48**, 155-157 (1926).
[257] Trueblood, H. M., *Proc. Am. Acad. Arts Sci.*, **52**, 731-804 (1917).

100 °C. They are limited to low pressures, and are not very exact. Jakob quotes them as $k = 45.8$ microcal/cm·sec·°C at 46 °C and 56.6 at 100 °C, which are equivalent to 1.92 and 2.37 kiloerg/cm·sec·°C, respectively. From the same source, T. H. Laby and Edith A. Nelson [260] derive the values 1.80 and 2.17 kiloerg/cm·sec·°C. The difference seems to arise from the

Table 57.—Thermal Conductivity of Dilated Water-vapor

S. W. Milverton [262] has reported that his observations between 70 and 95 °C, at pressures between 100 mm-Hg and near-saturation, can be expressed within 1/3 per cent by a formula equivalent to the following, the unit of p being 1 mm-Hg:

$$10^4 k = 1.5058 + 0.009081t + 0.001266p - 0.00001130pt \text{ watt/cm·°C}$$

This formula is not to be assumed valid outside the ranges specified, nor for $p < 100$ mm-Hg; in particular, it leads to a negative temperature coefficient when p exceeds about 1.05 atm. Values defined by it are given in Section I, those lying beyond the limits set by Milverton are enclosed in parentheses.

Unit of $p = 1$ mm-Hg; of $P = 1$ kg*/cm² = 735.6 mm-Hg; of $k = 10^{-4}$ watt/cm·°C. Temp = t °C

I. Computed by means of Milverton's formula:

$p \rightarrow$ $t_{sat} \rightarrow$ t	100 51.6	200 66.4	300 75.9	400 83.0	500 88.7	750 99.6
55	(2.070)					
60	(2.110)					
70	2.189	2.236				
80	2.268	2.305	2.341			
90	2.348	2.373	2.398	2.423	2.448	
95	2.388	2.407	2.426	2.446	2.465	
100	(2.428)	(2.441)	(2.455)	(2.468)	(2.482)	(2.515)

II. Computed by means of the formula $k = 1.25 \eta c_v$ (see text).

$t \rightarrow$ P	110	120	130	140	150	160	170	180	190	200
1	2.44	2.50	2.56	2.62	2.68	2.74	2.81	2.88	2.95	3.03
2			2.66	2.71	2.76	2.82	2.87	2.93	3.00	3.07
4					2.96	3.00	3.05	3.09	3.13	3.18

III. D. L. Timrot and N. B. Varhaftik,[263] and N. B. Varhaftik [264] have reported that $k/\eta c_v = 1.361$, as mean value over the range $t = 70$ to 250 °C and $p = 5$ to 100 mm-Hg; $k/\eta c_v = 1.416$ at 288.8 °C, and 1.546 at 476.7 °C. Varhaftik states that the temperature coefficient of this ratio between 69 and 476 °C is 3.7 per cent of the value of the ratio at 100 °C.

---

fact that the values given in the thesis assume for the conductivity of air a value higher than that (2.23 kiloerg/cm·sec·°C) which Laby and Nelson

17. VAPOR: REFRACTION

regard as the best available. For very recent determinations below 100 °C, see Table 57.

Jakob advises that the conductivity at other temperatures and pressures be computed by means of the relation $k = 1.25 \eta c_v$, where η is the viscosity and c_v is the specific heat at constant volume. He estimates that the values so computed are uncertain by ± 5 per cent. The values in Section II of Table 57 have been so computed from the data in Tables 45 and 22.

The thermal conductivity of a mixture of air and water-vapor is not given by the simple additive relation. It is greatest for a mixture containing about 20 per cent of H_2O by volume. Grüss and Schmick [261] give the following values, k_a being the conductivity of dry air, $t = 80$ °C:

%H_2O	7.2	15.0	17.1	19.7	22.5	25.0	30.6	31.2	44.4	51.9
$1000(k-k_a)/k_a$	20	35	37	36	35	37	26	30	−1	−26

17. Refractivity of Dilated Water-vapor

The data in this section are restricted to the optical spectrum; for values of the dielectric constant, see Section 22.

The determinations of the refractivity of water-vapor by C. and M. Cuthbertson [265] were accepted by J. J. Fox and F. G. H. Tate, and given in their compilation [226]; here they will be denoted by the symbol CC. Earlier determinations by Mascart [267] and by Lorenz [268] are not discordant with the Cuthbertson values. The only more recent determinations that have come to the attention of the compiler are those by J. Wüst and H. Reindel [269] and by P. Hölemann and H. Goldschmidt [270]; they will be denoted by WR and HG, respectively.

All three (CC, HG, and WR) reduced their observations on the assumption that $r \equiv (n - 1)10^6$ is directly proportional to the density (d) of the gas, n being the index of refraction; and they expressed their results in terms of the value that r would have, on that assumption, if the density were such that the vapor contained as many formula weights of H_2O per liter as there are of H_2 in a liter of H_2 at 0 °C and 760 mm-Hg. This

---

[258] Jakob, M., *Engineering (London)*, **132**, 744-746, 800-804 (1931).
[259] Moser, E., *Thesis*, Berlin, 1913.
[260] Laby, T. H., and Nelson, Edith A., *Int. Crit. Tables*, **5**, 215 (1929).
[261] Grüss, H., and Schmick, H., *Wiss. Veröff. Siemens-Konz.*, **7**, 202-224 (1928).
[262] Milverton, S. W., *Proc. Roy. Soc. (London) (A)*, **150**, 287-308 (1935).
[263] Timrot, D. L., and Varhaftik, N. B., *Chem. Abstr.*, **31**, 6957 (1937) ← *Inst. Izvest. Vsesoyuz. Teplotekh.*, **1935**, No. 9, 1-12 (1935).
[264] Varhaftik, N. B., *Idem*, **31**, 6958 (1937) ← **1935**, No. 12, 20-23 (1935).
[265] Cuthbertson, C. and M., *Phil. Trans. (A)*, **213**, 1-26 (1913).
[266] Fox, J. J., and Tate, F. G. H., *Int. Crit. Tables*, **7**, 8, 11 (1930).
[267] Mascart, E. E. N., *Compt. rend.*, **86**, 321-323 (1878).
[268] Lorenz, L., *Ann. d. Physik (Wied.)*, **11**, 70-103 (1880).
[269] Wüst, J., and Reindel, H., *Z. physik. Chem. (B)*, **24**, 155-176 (1934).
[270] Hölemann, P., and Goldschmidt, H., *Idem*, **24**, 199-209 (1934).

value we shall denote by r_0. In making the reduction from r to r_0, each took for the formula weight of H_2 the value $M_H = 2.016$, and the corre-

Table 58.—Refractivity of Dilated Water-vapor

In each case the value for $\lambda = 5460.7$A was determined absolutely by a count of fringes, and the values for other λ's were determined relative to that. For $\lambda = 5460.7$A, the values found were $r_0 = 252.7 \pm 0.5$ (CC[a]), 256.9 ± 1.0 (WR[a]), and 252.1 ± 0.3 (HG[a]), where $r_0 = 0.8029\, r/d$ in the case of CC, and $0.8038\, r/d$ in the case of the other two (see text). Here d g/l is the density and $r = (n-1)10^6$, n being the index of refraction; λ is the wave-length of the light. The Lorentz-Lorenz expression for the molecular refraction is $R = 1000(n^2 - 1)M/(n^2 + 2)d$, $M =$ molecular weight (18.016).

The Cuthbertson dispersion formula is $r_0 = 29190/(118.86 - \lambda^{-2})$, and WR's, after reduction by 1.87 per cent in order to bring their values to the basis of the more accurate value of r_0 obtained by HG for $\lambda = 5460.7$A, is $r_0 = 34168/(138.89 - \lambda^{-2})$, the unit of λ in each case being $1\,\mu$. The computed values here tabulated have been obtained by means of these equations.

Unit of $\lambda = 1$ A $= 10^{-4}\mu = 10^{-8}$ cm; of $d = 1$ g/l; of $R = 1$ cm³/gfw. Temp. $= t$ °C

Ref.[a]→	CC	CC r_0	WR	CC-WR 100Δ	CC r/d	CC
λ	Obs.	Computed		Comp.	Observed	R
4799.9	254.95	254.89	253.94	95	317.52	3.814
5085.8	253.80	253.84	253.06	78	316.09	3.796
5209.1	253.45	253.44	252.70	74	315.65	3.791
5460.7	252.70	252.71	252.09	62	314.72	3.780
5769.5	251.95	251.95	251.44	51	313.79	3.769
5790.5	251.91	251.90	251.40	50	313.74	3.768
6538.5	250.69	250.67	250.22	45	312.22	3.750
6707.8	250.28	250.26	250.00	26	311.70	3.744

Hölemann and Goldschmidt[270]. $\lambda = 5460.7$

t	150	250	350	500
R	3.763	3.768	3.761	3.767

C. and M. Cuthbertson; 1936.[273] See remarks in text, N_0 and N_c = number of bands observed and calculated, respectively; direct count for $\lambda = 5462.23$ only.

λ	5462.23	4359.54	4078.97	4047.68	3342.42	3126.56	3022.37	2968.13
N_0	776.85	989.8	1064.7	1073.8	1334.5	1064.6[b]	1503.8	1536.8
N_c	776.85	989.6	1064.8	1073.8	1334.2	1064.0[b]	1502.7	1535.9

[a] References:
    CC C. and M. Cuthbertson.[265]
    WR J. Wüst and H. Reindel.[269]

[b] Obviously there is some gross error in these N's for $\lambda = 3126.56$A.

---

sponding value for water, $M_{H_2O} = 18.016$. But CC assumed that the density of H_2 at 0 °C and 760 mm-Hg is 0.089849 g/l, which is equivalent to a specific volume of 22.438 l/gfw; whereas the others assumed the specific volume under those conditions to be 22.415 l/gfw. Hence, for the CC values $r_0 = 0.8029 \, r/d$, the unit of d being 1 g/l; and for the other two, $r_0 = 0.8038 \, r/d$. None of them report the values of d that were used; it appears that the temperature was seldom much below 140 °C. Little physical significance should be attached to the particular values assigned by them to the constants in their interpolation formulas.

A. Bramley [271] has reported that the application of an electric field changes the value of n by an amount in excess of that caused by the attendant change in density arising from electrostriction.

The general subject of optical dispersion has been reviewed by S. A. Korff and G. Breit.[272]

Since the foregoing was written, C. and M. Cuthbertson [273] have reported two series of new observations. That of 1934 gave for the green mercury line $r_0 = 253.1$ and that of 1935 gave 252.5; the mean of these two absolute determinations "is almost exactly" 252.7, the value reported in 1913. In reducing their observations they used the same values of the molecular weights as before, but for the density of H_2 at 0 °C and 1 atm they used the value 0.08995 g/l, whereas before they used 0.089849 g/l. In their earlier paper they seem to have used wave-lengths in air; in this, wave-lengths *in vacuo*. They illustrate the agreement of their recent observations with their previously determined dispersion equation by means of the values given at the bottom of Table 58, N_e being the value defined by that equation when N for $\lambda = 5462.23\text{A}$ is 776.85.

P. Hölemann [274] has considered the change in the refraction when a substance passes from the vapor to the liquid state.

18. Absorption of Radiation by Water-Vapor

As the spectrum of a gas or vapor consists in large part of numerous narrow lines arranged in bands, and as the observed absorption is the average absorption over a spectral range that is usually greater than the width of a single line and generally great enough to embrace several lines, it is obvious that the simple relation $I = I_0 e^{-kl}$, applicable when the absorption is essentially constant over the spectral range covered by a single observation, will not apply in general to the observed absorption by gases and vapors. For them, the radiation corresponding to the regions between the lines will pass essentially unabsorbed, and in the simplest case the relation will be of the form $I = aI_0 + (1-a)I_0 e^{-kl}$ or $(I_0 - I)/I_0 = (1-a)(1-e^{-kl})$, where l is the length along the path of a parallel beam of radiation in the medium between the points where the intensities are I_0

---

[271] Bramley, A., *J. Franklin Inst.*, 203, 701-711 (1927).
[272] Korff, S. A., and Breit, G., *Rev. Mod. Phys.*, 4, 471-503 (1932).
[273] Cuthbertson, C. and M., *Proc. Roy. Soc. (London) (A)*, 155, 213-217 (1936).
[274] Hölemann, P., *Z. physik. Chem. (B)*, 32, 353-368 (1936).

and I, respectively, a is the fraction of the radiation that passes between the lines of absorption, and k is the coefficient of absorption corresponding to the lines in the region observed, and is assumed to be the same for all those lines. If this last assumption is not fulfilled, the expression for the absorption will be more complicated. As l increases, the absorption approaches $(1 - a)$, not unity; if l exceeds a certain value, the absorption is essentially independent of l. Furthermore, the apparent coefficient of absorption k', defined by $I = I_0 e^{-k'l}$, decreases as l is increased, unless $a = 0$, in which case $k' = k$. The decrease is very slow at first but ultimately k' varies as l^{-1}.

Such behavior has been reported. F. Paschen [275] observed that the absorption in the $\lambda = 4.3\,\mu$ band of CO_2 (pressure = 75 cm-Hg, temp = 17 °C) is essentially as great for a path of 7 cm as for one of 33 cm, that the absorption in the $2.7\,\mu$ band of CO_2 is about 28 per cent for a 7-cm path and 43 per cent for a 33-cm path, and that in the water-vapor band near $2.7\,\mu$ the absorption is about 60 per cent for the 7-cm path and about 80 per cent for the 33-cm one, the vapor being just under saturation at 100 °C. Whence one obtains for k' the following values, that for the shorter path being given first: CO_2, $k' = 0.045, 0.017$; H_2O, $k' = 0.13, 0.049$ cm^{-1}. In each case, the value for the longer path is markedly less than that for the shorter one.

Even if the absorption varied continuously throughout the spectrum, somewhat similar effects are to be expected in regions in which the variation with λ is great.

One is not justified in assuming that the observed absorption by a gas or vapor under a specified condition follows the exponential law. In general, observations for a single length of path do not suffice for the determination of the absorption for a path of a different length.

The variation of the absorption with the temperature and pressure of the gas or vapor has been studied by E. v. Bahr,[276] continuing work began by K. Ångström.[277] She found that many gases, including water-vapor, behave thus for infrared radiation: (1) When the density of the gas is decreased, and the length of the path is correspondingly increased so that the mass of gas traversed per unit cross-section of the path remains unchanged, then the percentage of the incident radiation absorbed is decreased. That is, the apparent coefficient k' decreases more rapidly than the density. (2) If to the expanded gas an inert and transparent gas is added until the total pressure is the same as before expansion, the mass of the expanded gas traversed per unit cross-section of the path remaining unchanged, then the amount of absorption is restored to its value before expansion. This is said to hold for total pressures up to at least 1 atm.

[275] Paschen, F., *Ann. d. Physik (Wied.)*, **51**, 1-39 (1894).
[276] v. Bahr, E., *Ann. d. Physik (4)*, **29**, 780-796 (1909) ← *Diss.*, Upsala (1909); *Idem*, **33**, 585-597 (1910); *Idem*, **38**, 206-222 (1912); *Verh. deut. physik. Ges.*, **15**, 673-677 (1913).
[277] Ångström, K., *Ark. Math., Astr., och Fys.*, **4**, no. 30 (1908).

18. VAPOR: ABSORPTION OF RADIATION

That is, for a given total pressure so produced, the absorption depends solely upon the mass of absorbing gas traversed per unit cross-section of the path, and not at all upon the actual specific volume of that gas. (3) The rate of increase in the absorption with the total pressure, produced as stated, is at first great, but finally becomes zero. For water-vapor she gives the following data for $\lambda = 2.7\ \mu$, the length of column and the mass of water-

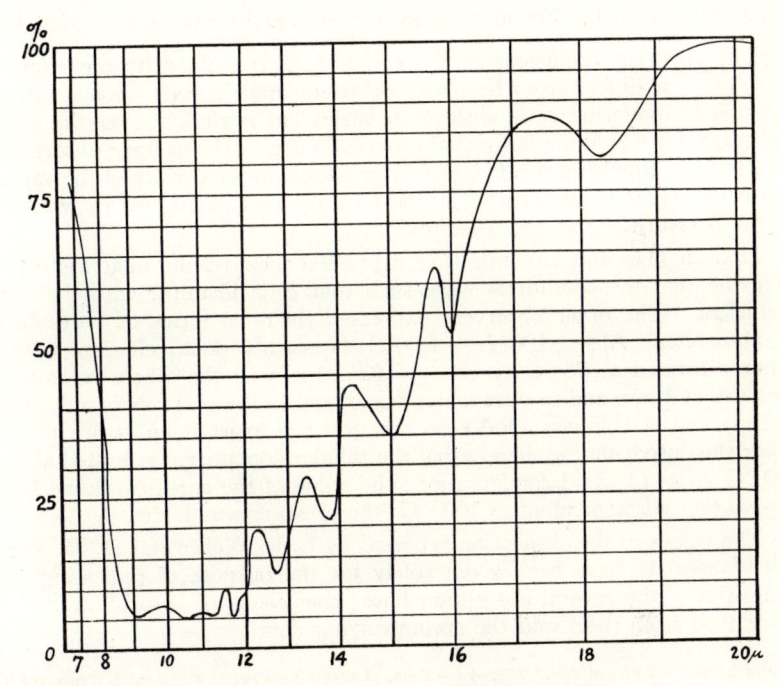

FIGURE 1. Absorption of Infrared Radiation by Water-vapor.

[Adaptation of a figure by H. Rubens and E. Aschkinass, *Ann. d. Physik (Wied.)*, **64**, 584-605 (1898).]
Abscissas = wave-length, unit = $1\ \mu = 10^{-4}$cm; ordinates = per cent of incident radiation that is absorbed by a certain column of water-vapor at a partial pressure of about 1 atm; length of column was about 75 cm. Temperature was somewhat over 100 °C.

vapor remaining unchanged, and the increase in total pressure being produced by adding dry air, which is essentially transparent for that radiation:

p	12	100	235	370	570	755	mm-Hg
$100(I_0-I)/I_0$	2.8	4.7	7.2	8.6	10.6	12.1	

(4) Such increase in total pressure does not increase the width of an absorption band, but does increase the intensity of the band at each point, including its maximum. (5) If the pressure is increased solely by heating, the width of an absorption band is increased, but the intensity at the maxi-

mum of the band is unchanged. As regards (4) and (5), water-vapor was not included in her investigations.

That is, the amount of absorption per molecule is influenced by the number of molecular impacts per second, and also by the mean kinetic

Table 59.—Absorption of Radiation by Water-vapor

(See also Figure 1. For absorption of x-rays and γ-rays, see Section 42)

The apparent coefficient of absorption (k') is defined by the relation $I = I_0 e^{-k'l}$; it differs from the true coefficient and depends upon l, unless the strictly monochromatic absorption varies but negligibly over the spectral range for which the individual observations give the average absorption (see text). Furthermore, k' is not proportional to the density of the vapor, and if the vapor is mixed with an inert transparent gas, k' varies with the partial pressure of that gas (see text).

Dreisch (D) and Granath (G) expressed their results in terms of k', believing that the conditions were such that k' is identical with the true coefficient. The other observers expressed theirs in terms of percentage of absorption, Abs $= 100(I_0 - I)/I_0$. Granath's data refer to water-vapor saturated at 25 °C (press. = 0.024 atm); all the others refer to a pressure of 1 atm and to pure water-vapor, unless the contrary is indicated. Rubens and Aschkinass (RA) do not give the exact temperature of the vapor they used, stating merely that the tubular container was heated above 100 °C, so as to avoid condensation; the value of the specific volume here used is that corresponding to 100 ° C; the pressure was 1 atm.

The values of the absorption reported by L. R. Weber and H. M. Randall [281] seem to have been given solely for the purpose of indicating the intensities of the several absorption lines; coefficients of absorption cannot be derived from them and the accompanying data.

Unit of $\lambda = 1\,\mu = 10^4$A; of Abs = 1%; of $l = 1$ cm; of $v/m = 1$ cm³/g; of $k' = 1$ cm⁻¹. Temp. $= t$ °C

λ	Abs	l	v/m	t	$1000 k'$	$k'v/m$	Ref.[a]
0.190	18	29	43310	25	7.0	300	G
0.195	51	241	43310	25	3.0	130	G
0.205	24	241	43310	25	1.2	50	G
0.95	8	109	1801	127	0.76	1.4	H
1.12	14	109	1801	127	1.4	2.5	H
1.35	11.0	25	1677	100	4.6	7.7	D
1.37	20.3	25	1677	100	8.7	14.6	D
1.37	75	109	1801	127	12.7	22.9	H
1.404	38.3	25	1677	100	19.3	32.4	D
1.45	24.8	25	1677	100	11.4	19.1	D

[278] Paschen, F., *Ann. d. Physik (Wied.)*, **50**, 409-443 (1893); **51**, 1-39, 40-46 (1894); **52**, 209-237 (1894); **55**, 287-300 (1894).
[279] Schmidt, H., *Ann. d. Physik (4)*, **29**, 971-1028 (1909).
[280] Fowle, F. E., *Astroph. J.*, **42**, 394-411 (1915); *Smithsonian Misc. Collect.*, **68**, No. 8 (publ. 2484), (1917); Smithsonian Physical Tables.
[281] Weber, L. R., and Randall, H. M., *Phys. Rev. (2)*, **40**, 835-847 (1932).

18. VAPOR: ABSORPTION OF RADIATION

Table 59.—*(Continued)*

λ	Abs	l	v/m	t	1000k'	k'v/m	Ref.[a]
1.50	10.1	25	1677	100	4.2	7.1	D
1.80	8.8	25	1677	100	3.7	6.2	D
1.83	84	109	1801	127	16.8	30.2	H
1.85	37.3	25	1677	100	18.7	31.4	D
1.885	47.6	25	1677	100	25.8	43.3	D
1.9	74	109	1801	127	12.4	22.3	H
1.935	37.5	25	1677	100	18.8	31.5	D
1.97	25.9	25	1677	100	12.0	20.1	D
2.0	8.6	25	1677	100	3.6	6.0	D
2.48	93	109	1801	127	24.4	43.9	H
2.55	47.5	25	1677	100	25.8	43.3	D
2.585	80.0	25	1677	100	64.4	108	D
2.618	91.8	25	1677	100	90.2	151	D
2.65	77.2	25	1677	100	59	99	D
2.82	91	109	1801	127	22.1	39.8	H
3.19	40	109	1801	127	4.7	8.5	H
3.26	27	109	1801	127	2.9	5.2	H
5.25	85	109	1801	127	17.4	31.3	H
7.0	75	75	1677	100+	18.5	31	RA
7.55	90	104	1801	127	22.1	39.8	H
7.90	72	104	1801	127	12.2	22.0	H
8.0	40	75	1677	100+	6.8	11.4	RA
8.2	32	104	1801	127	3.7	6.7	H
9.0	5	75	1677	100+	0.7	1.1	RA
10.0	7	75	1677	100+	1.0	1.7	RA
11.0	6	75	1677	100+	0.8	1.3	RA
11.5	10	75	1677	100+	1.4	2.3	RA
11.7	5	75	1677	100+	0.7	1.1	RA
12.0	9	75	1677	100+	1.3	2.2	RA
12.4	20	75	1677	100+	3.0	5.0	RA
12.4	34	104	1801	127	4.0	7.2	H
12.8	13	75	1677	100+	1.9	3.2	RA
13.0	18	75	1677	100+	2.7	4.5	RA
13.3	47	104	1801	127	6.1	11.0	H
13.4	28	75	1677	100+	4.4	7.4	RA
13.9	22	75	1677	100+	3.3	5.6	RA
14.0	28	75	1677	100+	4.4	7.4	RA
14.3	43	75	1677	100+	7.5	12.5	RA
14.4	61	104	1801	127	9.0	16.3	H
15.0	35	75	1677	100+	5.7	9.6	RA
15.5	52	75	1677	100+	9.8	16.4	RA
15.6	76	104	1801	127	13.7	24.7	H
15.7	63	75	1677	100+	13.2	22.2	RA
16.0	52	75	1677	100+	9.8	16.4	RA
16.5	74	75	1677	100+	18.0	30.1	RA
17.0	85	75	1677	100+	25.3	42.4	RA
17.0	88	104	1801	127	20.4	36.7	H
17.0	52	32.4	1801	127	22.6	40.7	H
17.3	61	32.4	1801	127	29.0	52.2	H
17.5	88	75	1677	100+	28.3	47.5	RA
18.0	82	75	1677	100+	22.9	38.4	RA
18.3	80	75	1677	100+	21.5	36.1	RA
18.3	61	32.4	1801	127	29.0	52.1	H
18.5	83	75	1677	100+	23.6	39.6	RA
19.0	93	75	1677	100+	35.5	59.5	RA

Table 59.—(Continued)

λ	Abs	l	v/m	t	1000k'	k'v/m	Ref.[a]
19.2	82	32.4	1801	127	52.9	95.2	H
19.5	98	75	1677	100+	52.1	87.4	RA
19.8	84	32.4	1801	127	56.5	101.8	H
20.0	99	75	1677	100+	61.3	103	RA
20.3	81	32.4	1801	127	51.2	92.2	H
32	60.4	40	1723	110	23.2	39.9	RWg
52	99.3	40	1723	110	124	213	RWg
108	80.4	40	1905	(?)[b]	40.8	78	RWd
110	80.4	40	1723	110	40.8	70.3	RWg
314	50.8	40	1723	110	17.7	30.5	RWg

Positions of other maxima (m) and of regions (w) of very little absorption.

λ	Ref.[a]	λ	Ref.[a]
47 w	We	79 m	We
50 m	We, Wi	79.3 m	Wi
52.5 m	Wi	90.9 m	Wi
54 w	We	91 w	We
58 m (?)	We	103 m (?)	We
56.6 m	Wi	108.9 m	Wi
62 w	We	115 w	We
63.7 m	Wi	116.8 m	Wi
66 m	We	125 w (?)	We
69.6 m	Wi	131.8 m	Wi
74.5 m	Wi	138 w (?)	We
75 w	We	167 m	Wi

Diathermacy of moist air for the complete radiation from an enclosure at the temperature t °C. Length of tube = 250 cm. Pressure = 1 atm; temp. = 70 °F = 21.1 °C, moisture content = 0.032 mm of precipitable water, giving $lm/v = 0.0032$ g/cm², $v/m = 78$ 1/g of vapor, and relative humidity = 70 per cent. (In the legend to the graphs, the temperature of the moist air is incorrectly given as 70 °C.)[282]

t	510	590	735	760	780[c]	820	850
100I/I₀	91.3	92.2	92.6	92.6	92.6	92.7	92.9
1000k'	0.36	0.33	0.31	0.31	0.31	0.30	0.29
k'v/m	2.6	2.4	2.2	2.2	2.2	2.2	2.1

[a] References and remarks:

D   T. Dreisch,[283] quoted by J. Becquerel and J. Rossignol.[284] The significance of the values tabulated by Dreisch is not entirely clear. The heading of the column indicates that the values are k' in our notation, and it is stated that $l = 1$ meter; but it seems that he means by the last merely that the unit of k' is 1 m⁻¹, which is the interpretation adopted by Becquerel and Rossignol, and used in this compilation. Neither the temperature, the pressure, nor the density of the vapor is explicitly stated, but the text indicates that the vapor was probably saturated at 100 °C, corresponding to a specific volume of 1677 cm³/g. Assuming this and the preceding interpretation of the tabular data to be correct, one finds that the ordinates of his Fig. 2 should each be 0.3 of the corresponding tabular value, which they are. The values here assigned to him have been derived from his data on the basis of

[282] Brown, S. L., *Phys. Rev. (2)*, **21**, 103–106 (1923).
[283] Dreisch, T., *Z. Physik*, **30**, 200–216 (1924).
[284] Becquerel, J., and Rossignol, J., *Int. Crit. Tables*, **5**, 269 (1929).

Table 59.—(Continued)

these assumptions. Becquerel and Rossignol appear to have erred in stating that the values of k' refer to vapor at 0 °C and 1 atm.

G   L. P. Granath[285] used vapor saturated at 25 °C.
H   G. Hettner[286] used vapor at 127 °C and 1 atm.
RA   H. Rubens and E. Aschkinass,[287] vapor saturated at 100 °C was passed into the middle of a metal tube open at each end and heated somewhat above 100 °C.
RWd   H. Rubens and R. W. Wood.[288]
RWg   H. Rubens and H. v. Wartenberg[289] used vapor at 110 °C and 1 atm, tube open at ends.
We   W. Weniger.[290]
Wi   H. Witte.[291]

ᵇ Neither the temperature nor the pressure is stated.

ᶜ Using a column of moist air 51 cm long, $v/m = 100500$ cm³/g of vapor, $t = 800$ °C, W. W. Coblentz[292] found absorption = 0.9 per cent, which corresponds to $1000k' = 0.18$, $k'v/m = 18$. For a column of dry air of the same length, the absorption was about 0.09 per cent, "which is the magnitude of the errors of observation." The pressure was 1 atm in each case. The measurement was incidental to another investigation.

---

energy of the impacting molecules; but the intimate effects of these two influences seem to differ.

In general, determinations made at a single partial pressure, a single temperature, and a single wave-length do not suffice for the determination of the absorption under other conditions.

Observations by F. Paschen [278] indicate that the infrared radiation from heated gases is of thermal origin, and satisfies Kirchhoff's law, cases involving obvious chemical and electrical effects being excluded; but the evidence is not entirely convincing.[279] See also, C. Schaefer and F. Matossi, "Das Ultrarote Spektrum," p. 104, 1930.

Tables for use in the determination of the effect of moisture upon the atmospheric transmission of solar and of terrestrial radiation have been published by F. E. Fowle,[280] together with his observations bearing thereon.

19. Emissivity of Water-vapor

The only direct measurements of the emissivity of water-vapor and of its variation with the thickness of the layer of vapor seem to be those mentioned in Table 60. Earlier estimates based upon the values of the spectral absorptivity in the regions effective—the bands at $\lambda = 2.67\ \mu$ and $6.6\ \mu$,

---

[285] Granath, L. P., *Phys. Rev. (2)*, **34**, 1045-1048 (1929) → **33**, 1073 (A) (1929).
[286] Hettner, G., *Ann. d. Physik (4)*, **55**, 476-496 (1918).
[287] Rubens, H., and Aschkinass, E., *Ann. d. Physik (Wied.)*, **64**, 584-605 (1898) = *Astrophys. J.*, **8**, 176-192 (1898).
[288] Rubens, H., and Wood, R. W., *Verh. deut. physik. Ges.*, **13**, 88-100 (1911).
[289] Rubens, H., and v. Wartenberg, H., *Physik. Z.*, **12**, 1080-1084 (1911) = *Verh. deut. physik. Ges.*, **13**, 796-804 (1911).
[290] Weniger, W., *J. Opt. Soc. Amer.*, **7**, 517-527 (1923).
[291] Witte, H., *Z. Physik*, **28**, 236-248 (1924).
[292] Coblentz, W. W., *Proc. Nat. Acad. Sci.*, **3**, 504-505 (1917); also *Sci. Papers Bur. Stand.*, **15**, 529-535 (S357) (1920).

Table 60.—Emissivity of Water-vapor

ϵ = ratio of the net radiation from the layer of water-vapor, or of a mixture of the vapor and CO_2-free air, to that from the ideal radiator (black body) at the same temperature. The intensity of radiation from the ideal radiator may be found in Table 288; τ = thickness of the layer of radiating gas; p = partial pressure of the vapor. Temp. = t °C = t_F °F.

Unit of $\tau = 1$ cm; of $p = 1$ atm; ϵ is dimensionless

I. Pure vapor. E. Schmidt [297]; values confirmed by E. Eckert.[295] The following values have been read from Schmidt's graph; $p = 1$ atm in all cases.

$\tau \rightarrow$	0.96	2.00	3.02	4.02	6.00	12.0	18.2
t				100ϵ			
100	9.0	14.4	17.9	20.8	23.6	30.2	34.3
200	8.1	13.2	16.6	19.4	22.3	29.0	33.2
300	7.2	11.8	15.2	17.8	20.8	27.6	31.8
400	6.1	10.2	13.7	16.0	19.2	26.0	30.2
500	5.2	8.8	12.0	14.4	17.4	24.3	28.4
600	4.4	7.8	10.5	12.6	15.6	22.6	26.6
700	3.8	6.7	9.2	11.2	13.8	20.7	24.8
800	3.2	6.0	8.2	10.0	12.4	19.0	23.0
900	2.8	5.2	7.3	9.0	11.3	17.4	21.4
1000	2.6	4.8	6.6	8.1	10.3	16.0	20.0

II. Vapor mixed with air. H. C. Hottel and H. G. Mangelsdorf.[298] The following values have been read from their graph. In all cases, $\tau = 51.2$ cm (1.68 ft.) and total pressure = 1 atm.

$p\tau \rightarrow$		0.256	0.51	1.02	2.04	4.09	8.5	25.6	51.2
$p \rightarrow$		0.0050	0.010	0.020	0.040	0.080	0.167	0.50	1.00
t_F	t				100ϵ				
200	93	1.36	2.6	5.0	8.5	14	23	39	50
400	204	1.15	2.2	4.1	7.6	12.5	21	36	48
600	316	0.98	1.9	3.6	6.7	11.0	18.5	34	45
800	427	0.82	1.6	3.2	5.9	10.0	16.8	31	43
1000	538	0.71	1.4	2.8	5.2	9.0	15.2	29	41
1200	649	0.63	1.3	2.5	4.6	8.0	14.0	27	39
1400	760	0.57	1.10	2.2	4.1	7.2	12.5	26	37
1600	871		0.96	1.98	3.7	6.5	11.5	24	35
1800	982		0.88	1.72	3.2	5.8	10.2	22	34
2000	1093		0.75	1.51	2.9	5.3	9.4	21	32
2200	1204		0.70	1.38	2.6	4.8	8.6	19	30
2400	1316		0.61	1.20	2.4	4.4	8.0	18	29
2600	1427		0.52	1.08	2.1	4.0	7.2	17	27
2800	1538			0.98	1.96	3.6	6.7	16	26
3000	1649			0.89	1.80	3.3	6.2	15	25

[293] Schack, A., *Z. Ver. deut. Ing.*, **68**, 1017-1020 (1924); *Z. techn. Physik*, **5**, 267-278 (1924); Idem, **7**, 556-563 (1926).

and the region $\lambda = 12$ to $25\,\mu$—had been made by A. Schack.[293] They involve a number of simplifying assumptions, including the following:

1. Where the absorption is the greatest in any band, the observed relative amount of radiation transmitted is exponentially related to the thickness; $I = I_0 e^{-k_m x}$, k_m being independent of x. In general, this is not true (see Section 18).

2. The integral effect of any given band can be represented by an expression of the form

$$I_0 \int_{\lambda_0}^{\lambda_0+\Delta\lambda} e^{-k_m x(\lambda-\lambda_0)/\Delta\lambda}\, d\lambda = \frac{I_0 \Delta\lambda (1 - e^{-k_m x})}{k_m x}$$

the band extending from λ_0 to $\lambda_0 + \Delta\lambda$.

3. The value of k_m observed when the temperature and pressure are relatively low will apply when they are high. Actually, the individual lines are broadened and the local variations in intensity throughout the band are, in part, wiped out as the temperature and the pressure are increased, either singly or together.

Owing to such assumptions and to the fact that the value found for k_m depends upon experimental details, these estimates are unsatisfactory, although they were of great value when made, being the only estimates then available.

M. Jakob [258] has discussed them, and has compared them with the results obtained by Schmidt; and H. C. Hottel [294] has published an English paraphrase of Schack's work, extended by graphs and some elaboration of detail.

The intensity of the radiation emitted at a given temperature by a given volume of water-vapor when mixed with a nonabsorbing and nonradiating gas is not determined solely by the amount of the vapor contained in that volume, *i.e.*, Beer's law is not valid for such a mixture.[295, 296]

20. Luminescence of Water-vapor

The types of luminescence here considered are these: Fluorescence, including phosphorescence; the Rayleigh scattering (also called the Tyndall effect); the Raman scattering; and the scattering of x-rays. The distinctive characteristics of these several effects, except the last, are considered in Section 39.

---

[294] Hottel, H. C., *Trans. Am. Inst. Chem. Eng.*, **19**, 173-205 (1927); *J. Ind. Eng. Chem.*, **19**, 888-894 (1927).
[295] Eckert, E., *Forschungsheft*, **387**, 1-20 (1937).
[296] Schmidt, E., and Eckert, E., *Forsch. Gebiete Ingenieurw.*, **8**, 87-90 (1937).
[297] Schmidt, E., *Forsch. Gebiete Ingenieurw.*, **3**, 57-70 (1932).
[298] Hottel, H. C., and Mangelsdorf, H. G., *Trans. Am. Inst. Chem. Eng.*, **31**, 517-549 (1935).

Fluorescence.

When water-vapor at atmospheric pressure is illuminated by light from the mercury line $\lambda = 2537$A, it emits an intense ultraviolet fluorescence having a continuous spectrum which has a maximum near 2537A, and extends several thousands of wave-numbers (cm^{-1}) toward the visible spectrum; its intensity is some thousands of times greater than that of the Raman-scattered light.[299]

When moist nitrogen is exposed to radiation of high frequency (ultra-Schumann), it luminesces, and the spectrum of the luminescent light contains the so-called water-band.[300] This was described as a fluorescence of water-vapor. But it is now agreed that the "water-band" actually arises from the OH molecule. Consequently, the observed luminescence cannot properly be described as a fluorescence of water-vapor. G. H. Dieke[301] so remarks, and explains the dissociation as arising from "impacts of the second kind" between H_2O molecules and excited N_2 molecules; in which case, the band should vanish if the N_2 were eliminated.

H. Neuïmin and A. Terenin[302] have stated that radiation from the Schumann region excites luminescence in water-vapor as a result of photo-dissociation combined with excitation of the resulting OH radical. The hydroxyl band at $\lambda = 3062$A is particularly strong. The luminescence is strongest when the pressure of the vapor is about 0.8 mm-Hg; it is reduced by the presence of CO or of H_2, but not by that of N_2 or of argon.

In speaking of the reaction that occurs in a mixture of H_2 and O_2, S. Horiba[303] has stated that in the transition region from nonexplosive to explosive reaction at the lower critical pressure there is luminescence, although the reaction velocity is measurably small, and heating seldom occurs.

Rayleigh Scattering.

The depolarization (see Section 39) of the light scattered by water-vapor is $\rho = 0.0199$, the density of the gas being low.[304] Such a defect in polarization has been explained by Lord Rayleigh[305] and by M. Born[206] as being due to an anisotropy of the molecule; but I. R. Rao[307] has concluded that it is the anisotropy of the oxygen ion that is involved. For a discussion of the anisotropy of the H_2O molecule and of its ions, see Section 9.

No direct measurement of the ratio of the intensity of the light laterally scattered by water-vapor to that of the incident light has been found, but from other ratios obtained by others, W. H. Martin and S. Lehrman[308] have computed the value $Ir^2/EV = 1.06 \times 10^{-8}$ per atmosphere, as applying to water-vapor at 27 °C. Here $I =$ intensity of the scattered light at the distance r from the scattering vapor of volume V, and E is the intensity of the incident (exciting) light.

[299] Rasetti, F., *Nuovo Cim.*, N. S., **8**, 191-193 (1931).

20. VAPOR: LUMINESCENCE

Raman Scattering.

As each frequency in the incident radiation gives rise to its own series of Raman lines, the interpretation of the observed spectrum is difficult, or even ambiguous, unless the incident radiation is unifrequent, which it usually is not. For a discussion of the general subject and the earlier data see K. W. F. Kohlrausch, "Der Smekal-Raman Effekt," 1931.

Table 61.—Raman Spectrum of Water-vapor

$\delta\nu$ = difference between the wave-number of the Raman line and that of the associated line in the incident radiation; $\lambda_R = \delta\nu$ is the wave-length corresponding to the fundamental frequency responsible for $\delta\nu$, and presumably has a representative of nearly the same value in the infrared spectrum of the vapor; No. = number of components observed.

Unit of $\delta\nu = 1$ cm^{-1}; of $\lambda_R = 1\,\mu = 10^4$A $= 10^{-4}$ cm

No.		$\delta\nu$			Ref.[a]
1		3655			DK
3		3654	1648	984	JW
2 (?)	3804 (?)	3650	none	none	Rk
1		3655			Ro
1		3654.5			Be
1		3646[b]			Uk
$\lambda_R \rightarrow$	2.63	2.74	6.06	10.2	

[a] References and remarks:
- Be   D. Bender.[309]
- DK   P. Daure and A. Kastler[310] used vapor saturated at 130 °C; found a single, fine Raman line with no satellite of comparable intensity.
- JW   H. L. Johnston and M. K. Walker.[311]
- Rk   D. H. Rank[312] used vapor at atmospheric pressure; sought for the JW lines at $\delta\nu$ 1648 and 984, but failed to find them. On long exposure found a line ($\nu = 21950$) that JW assigned to a 984 cm^{-1} shift from $\lambda = 4358$A, but which can just as well be regarded as a 3650 cm^{-1} shift from $\lambda = 3906$A. The $\delta\nu = 3804$ cm^{-1} is quite doubtful; the line was very weak, though the exposure was for 93 hours.
- Ro   I. R. Rao[313] concludes that the data available early in 1934 indicate that there is but a single line, that for which $\delta\nu = 3655$ cm^{-1}.
- Uk   S. A. Ukholin.[314]

[b] This is for density $\rho = 0.055$ g/cm^3; at low density the line becomes double, $\delta\nu = 3639$ and 3653 cm^{-1}.

---

[300] Wood, R. W., *Phil. Mag. (6)*, **20**, 707-712 (1910); Wood, R. W., and Hemsalech, G. A., *Idem*, **27**, 899-908 (1914); Meyer, C. F., and Wood, R. W., *Idem*, **30**, 449-459 (1915).
[301] Dieke, G. H., *Proc. Akad. Wet. Amsterdam*, **28**, 174-181 (1925).
[302] Neuimin, H., and Terenin, A., *Acta Physicochim. URSS*, **5**, 465-490 (1936).
[303] Horiba, S., *Sci. Rep. Tôhoku Imp. Univ. (Sendai)*, (1) Honda Anniv. Vol., 430-443 (1936).
[304] Rao, I. R., *Indian J. Phys.*, **2**, 61-96 (1927).
[305] Lord Rayleigh, *Phil. Mag. (6)*, **35**, 373-381 (1918).
[306] Born, M., *Verh. deut. physik. Ges.*, **19**, 243-264 (1917); **20**, 16-32 (1918).
[307] Rao, I. R., *Indian J. Phys.*, **2**, 435-465 (1928).
[308] Martin, W. H., and Lehrman, S., *J. Phys'l Chem.*, **26**, 75-88 (1922).
[309] Bender, D., *Phys. Rev. (2)*, **47**, 252 (L) (1935).
[310] Daure, P., and Kastler, A., *Compt. rend.*, **192**, 1721-1723 (1931).

Scattering of X-rays.

H. Gajewski [315] has found that his observed intensity (I) of the x-rays scattered by water-vapor at an angle θ to the direction of the incident beam agrees within experimental error with that calculated for the triangular molecule with the O-H distance = 0.86A and the H-H = 1.28A, which distances (see Table 16) are those published by R. Mecke.[316] Account was taken of the effect of the hydrogen as well as of the oxygen, of the interaction between the two, and of the Compton effect. His values are as follows, the unit of intensity being that at $\theta = 0$:

$2\sin(\theta/2)$	0	0.4	0.5	0.6	0.8	1.0	1.2	1.4	1.6	1.8
I	1.0	0.92	0.78	0.62	0.44	0.35	0.30	0.26	0.23	0.21

21. Spectra of Water-vapor

Absorption Spectrum.

The absorption spectrum of water-vapor extends from the far infrared to the far ultraviolet (see Table 62), and consists of a series of bands, each composed of numerous lines. In places, several bands overlap. Only recently has it been possible to arrange the bands in a satisfactory order, and to analyze their structure. Among the earlier attempts to analyze the bands of longer wave-length, may be mentioned those of C. R. Bailey,[317] C. R. Bailey, A. B. D. Cassie and W. R. Angus,[318] H. Deslandres,[319] J. W. Ellis,[320] A. Eucken,[321] G. Hettner,[322] F. Hund,[323] P. Lueg and K. Hedfeld,[324] R. Mecke,[316] E. K. Plyler,[325] W. W. Sleator and E. R. Phelps,[326] and H. Witt.[327]

The papers from Mecke's laboratory in 1932 [328] should be consulted for procedure and comments, but the numerical analysis then published is superseded by that of 1933.[329] The result of this analysis is given in Tables 64

---

[311] Johnston, H. L., and Walker, M. K., *Phys. Rev. (2)*, **39**, 535 (L) (1932).
[312] Rank, D. H., *J. Chem'l Phys.*, **1**, 504-506 (1933).
[313] Rao, I. R., *Phil. Mag. (7)*, **17**, 1113-1134 (1934).
[314] Ukholin, S. A., *Compt. rend. Acad. Sci. URSS*, **16**, 395-398 (1937).
[315] Gajewski, H., *Physik. Z.*, **33**, 122-131 (1932).
[316] Mecke, R., *Physik. Z.*, **30**, 907-910 (1929).
[317] Bailey, C. R., *Trans. Faraday Soc.*, **26**, 203-212, 213-215 (1930).
[318] Bailey, C. R., Cassie, A. B. D., and Angus, W. R., *Idem*, **26**, 197-202 (1930).
[319] Deslandres, H., *Compt. rend.*, **180**, 1454-1460, 1980-1986 (1925).
[320] Ellis, J. W., *Phil. Mag. (7)*, **3**, 618-621 (1927).
[321] Eucken, A., *Verh. deut. physik. Ges.*, **15**, 1159-1162 (1913); *Jahrb. d. Radio-akt.*, **16**, 361-411 (1920); *Z. Elektroch.*, **26**, 377-383 (1920).
[322] Hettner, G., *Z. Physik*, **1**, 345-354 (1920).
[323] Hund, F., *Idem*, **43**, 805-826 (1927).
[324] Lueg, P., and Hedfeld, K., *Idem*, **75**, 512-520 (1932).
[325] Plyler, E. K., *Phys. Rev. (2)*, **39**, 77-82 (1932).
[326] Sleator, W. W., and Phelps, E. R., *Astroph. J.*, **62**, 28-48 (1925).
[327] Witt, H., *Z. Physik*, **28**, 249-255 (1924).
[328] Mecke, R., *Die Naturwiss.*, **20**, 657 (1932); *Z. physik. Chem. (B)*, **16**, 409-420, 421-437 (1932); Mecke, R., and Baumann, W., *Physik. Z.*, **33**, 833-835 (1932).
[329] Mecke, R., *Z. Physik*, **81**, 313-331 (1933); Baumann, W., and Mecke, R., *Idem*, **81**, 445-464 (1933); Freudenberg, K., and Mecke, R., *Idem*, **81**, 465-481 (1933).

21. VAPOR: SPECTRA

Table 62.—Lines and Bands in the Absorption Spectrum of Water-vapor [*]

Herein are listed the recorded wave-lengths of the maxima or centers of the several bands, of the isolated lines that may not be involved in the fine-structure of the bands, and of certain series of lines (B, Ra, WbR) forming the fine-structure of bands. The last are included solely as illustrations of the nature of such structure; for additional data, reference should be made to the sources listed in Table 66. All wave-lengths recorded by B and by Ra are listed, but only the distinct maxima observed by D and by WbR are given.

Unit of $\lambda = 1\,\mu = 10^4 A = 10^{-4}$ cm; of $\nu = 1$ cm^{-1}

λ	ν	R^a	Ref.[b]	λ	ν	R^a	Ref.[b]
400^e	25.0	0	Ba	82.4	121.4		K
250^e	40.0		Ba	81.988	121.97		WrR
170^e	58.8		Ba	79.3	126.1		Wi
167.	59.9		Wi	79.	126.		Ru[1, 2]
150.	66.7		Wi	78.63	127.2		WrR
134.7	74.2		WrR	78.0	128.2		Ru[2]
132.3	75.6		WrR	77.66	128.77		WrR
132.2	75.6		Ru[2]	75.6	132.3		K
131.8	75.9		Wi	75.32	132.77		WrR
127.8	78.2		WrR	74.5	134.2		Wi
126.5	79.0		WrR	72.4	138.1		K
125.6	79.6		WrR	72.2	138.5		Ru[2]
121.7	82.2		WrR	71.79	139.30		WrR
116.8	85.6		Wi	70.98	140.88		WrR
113.1	88.4		WrR	69.6	143.7		Wi
111.7	89.5		WrR	66.62	150.10		WrR
108.9	91.8		Wi	66.0	151.5		K
108.1	92.5		WrR	66.	152.		Ru[1]
105.8	94.5		Ru[2]	65.8	152.0		Ru[2]
104.03	96.13		WrR	65.14	153.52		WrR
103. (?)	97.1		Ru[1]	63.7	157.0		Wi
100.96	99.05		WrR	63.34	157.88		WrR
99.5	100.5		Wi	60.01	166.64		WrR
99.415	100.59		WrR	58.2	171.8		K
98.559	101.46		WrR	58 (?)	172.		Ru[1]
95.613	104.59		WrR	57.7	173.3		Ru[2]
94.541	105.77		WrR	56.6	176.7		Wi
94.	106.		StW	56.3	177.6		K
93.199	107.30		WrR	52.5	190.5		Wi
92.662	107.92		WrR	52.0	192.3		K
90.9	110.0		Wi	52.	192.3		StW
89.919	111.21		WrR	50.	200.		Wi
88.520	112.97		WrR	50	200.		Ru[1, 2]
85.662	116.74		WrR	49.8	200.8		K
85.0	117.6		K	49.0	204.1		Ru[2]
84.690	118.08		WrR	48.0	208.3		K
83.196	120.20		WrR	44.2	226.2		K
83.	120.		Wi	44.1	226.8		Ru[2]

[*] Since this table was written the region 18 to 75 μ has been carefully mapped and analyzed by Randall and associates, 38 to 170 μ by Barnes and associates, and the ultraviolet spectrum beyond $\lambda = 1785\,\mu$ by Price (see text); and the 6324A band in the spectrum of the low sun has been studied by V. N. Kondratjev and D. I. Eropkin.[856]

Table 62.—(Continued)

λ	ν	R^a	Ref.[b]	λ	ν	R^a	Ref.[b]
40.6	246.3		K	12.82	780.0		RuHt
40.0	250.0		Ru[2]	12.65	790.5	15	WbR
38.6	259.1		K	12.42	805.2	s	RuHt
35.7	280.1		Ru[2]	12.4	806.		RuA
35.6	280.9		K	12.14	823.7	7	WbR
33.0	303.	s	Ht	11.89	841.0		RuHt
32.9	304.		Ru[2]	11.77	849.6	7	WbR
31.0	323.	s	Ht	11.66	857.7	s	RuHt
30.6	327.		Ru[2]	11.6	862.		RuA
28.9	346.		Ru[2]	11.47	871.8		RuHt
26.6	376.		Ru[2]	11.24	889.7		RuHt
25.0	400.		Ru[2]	11.06	904.2	5	WbR
24.72	404.5	38	WbR	10.94	914.1	s	RuHt
23.81	420.0	66	WbR	10.9	917.		RuA
23.8	420.		Ru[2]	10.80	925.9		RuHt
22.9	437.		Ru[2]	10.66	938.1		RuHt
22.61	442.1	31	WbR	10.42	959.7	9	WbR
22.38	446.8	42	WbR	10.30	970.9	s	RuHt
21.81	458.5	65	WbR	9.98	1002.		RuHt
21.6	463.		Ru[2]	9.74	1027.	s	RuHt
21.12	473.5	53	WbR	9.50	1053.		RuHt
20.62	485.0	28	WbR	9.30	1075.		RuHt
20.5	488.		Ru[2]	6.296	1588.3		M(T)
20.29	492.8	24	WbR	6.2673	1595.6	c	SPh, BC
19.87	503.3	30	WbR	6.26	1597.	c	Many
19.8	505.	s	Ht	3.168	3156.	c	PlS
19.8	505.		RuA	3.168	3156	c	M(T)
19.7	508.		Ru[2]	3.15	3175		Many
19.3	518.	32	WbR	3.109	3216		SPh
19.2	521.		Ru[2]	2.672	3742	c	SPh
19.02	525.8	21	WbR	2.663	3755		M(T), BC
18.37	544.4		RuHt	2.66	3759		Many
18.34	545.2	19	WbR	2.618	3820	9.0	D
18.21	549.1	14	WbR	2.05	4878		Many
17.57	569.2	16	WbR	2.00	5000		Many
17.5	571.		RuA	1.885	5305	2.6	D
17.33	577.0	s	RuHt	1.875	5333		M(T)
17.3	578.	14	WbR	1.87	5348		Many
17.0	588.		Ht	1.870	5348		SPh
16.89	592.1	14	WbR	1.46	6849		Many
16.80	595.2		RuHt	1.45	6896	c	McU
16.53	605.0	14	WbR	1.404	7122	1.9	D
16.00	625.0		RuHt	1.382	7236	c	SPh
15.99	625.4	40	WbR	1.38	7246	c	McU
15.73	635.7	32	WbR	1.379	7252		M(T)
15.7	637		RuA	1.37	7299		Many
15.62	640.2	s	RuHt	1.16	8621		Many
15.17	659.2	25	WbR	1.135	8810		M(T)
14.98	667.6		RuHt	1.13	8850	c	McU
14.5	690.	35	WbR	1.13	8850		Many
14.32	698.3	s	RuHt	0.964868	10364.1	2	B
14.3	699		RuA	0.964559	10367.4 ⎫	5	B
13.62	734.2		RuHt	0.964506	10368.0 ⎭	5	B
13.50	740.7	17	WbR	0.963758	10376.0	6	B
13.4	746.		RuA	0.963616	10377.6	6	B
13.34	749.6	s	RuHt	0.958184	10436.4 ⎫	3	B
13.06	765.7	17	WbR	0.958112	10437.2 ⎭	4	B

21. VAPOR: SPECTRA

Table 62.—*(Continued)*

λ	ν	R[a]	Ref.[b]	λ	ν	R[a]	Ref.[b]
0.958005	10438.4	4	B	0.932567	10723.1	4	B
0.957936	10439.1	1	B	0.932506	10723.8	1	B
0.957138	10447.8	4	B	0.932083	10728.7	5	B
0.956891	10450.5	4	B	0.931914	10730.6 ⎫	7	B
0.956616	10453.5	6	B	0.931611	10734.1 ⎭	6	B
0.955733	10463.2	5	B	0.930962	10741.6	7	B
0.955451	10466.3 ⎫	4	B	0.9060	11038.		M(T)
0.955348	10467.4 ⎭	6	B	0.9050	11050.		M
0.954393	10477.9	8	B	0.8230	12151.		M
0.953612	10486.4	5	B	0.8227	12155.		M(T)
0.952230	10501.7	10	B	0.7957	12568.		M(T)
0.951934	10505.0	4	B	0.77	12987		Many
0.951706	10507.4	9	B	0.7227	13837		M(T)
0.950171	10524.4 ⎫	8	B	0.7220	13850		M
0.950076	10525.5 ⎭	9	B	0.6994	14298		M(T)
0.949960	10526.8	4	B	0.6960	14368		M
0.949751	10529.1	6	B	0.69	14493		Many
0.949449	10532.4 ⎫	7	B	0.6530	15314		M
0.949341	10533.7 ⎭	5	B	0.6524	15328		M(T)
0.948197	10546.3	9	B	0.6324	15813		M(T)
0.948023	10548.4	6	B	0.5952	16801		M(T)
0.946116	10569.5	9	B	0.5924	16880		M(T)
0.945996	10570.9	8	B	0.5722	17476		M(T)
0.945698	10574.2	1	B	0.17844	56040	1	Ra
0.945622	10575.0	5	B	0.17705	56480 ⎫		Ra
0.945479	10576.6	4	B	0.17546	57000 ⎭		
0.945412	10577.4	4	B	0.17392	57500 ⎫		Ra
0.944339	10589.4	4	B	0.17153	58300 ⎭		
0.944089	10592.2	12	B	0.1700	58820		L
0.943790	10595.6	8	B	0.16948	58980 ⎫		Ra
0.943068	10603.7	6	B	0.16805	59510 ⎭		
0.942836	10606.3 ⎫	8	B	0.16585	60300 ⎫		Ra
0.942685	10608.0 ⎭	9	B	0.16421	60900 ⎭		
0.9420	10616.		M(T)	0.16194	61750 ⎫		Ra
0.941772	10618.3	6	B	0.15986	62560 ⎭		
0.941044	10626.5	6	B	0.15871	63010 ⎫		Ra
0.94	10638.		Many	0.15736	63560 ⎭		
0.938684	10653.2	9	B	0.15550	64310 ⎫		Ra
0.938122	10659.6	9	B	0.15422	64850 ⎭		
0.937974	10661.3	4	B	0.1392	71840	1	L
0.937774	10663.5	9	B	0.13405	74600	1	Ra
0.937158	10670.6	10	B	0.13722	72880 ⎫	W	Ra
0.936960	10672.8	7	B	0.13610	73470 ⎭		
0.936646	10676.4 ⎫	5	B	0.13567	73710 ⎫	w	Ra
0.936495	10678.1 ⎭	6	B	0.13462	74290 ⎭		
0.935893	10685.0 ⎫	7	B	0.13406	74590 ⎫	1	Ra
0.935755	10686.6 ⎭	8	B	0.13322	75060 ⎭		
0.935526	10689.2 ⎫	2	B	0.13242	75540 ⎫		Ra
0.935450	10690.0 ⎱	6	B	0.13177	75890 ⎭		
0.935365	10691.0 ⎰	4	B	0.13093	76380 ⎫		Ra
0.935306	10691.7 ⎭	4	B	0.13063	76560 ⎭		
0.934567	10700.1 ⎫	7	B	0.12973	77080 ⎫		Ra
0.934405	10702.0 ⎱	10	B	0.12910	77460 ⎭		
0.934252	10703.8 ⎭	7	B	0.12839	77890 ⎫		Ra
0.933947	10707.2	6	B	0.12790	78180 ⎭		
0.933464	10712.8 ⎫	6	B	0.12708	78690 ⎫		Ra
0.933355	10714.0 ⎭	7	B	0.12671	78920 ⎭		

Table 62.—(Continued)

λ	ν	R^a	Ref.[b]	λ	ν	R^a	Ref.[b]
0.12569	79560 ⎱		Ra	0.10592	94410 ⎱	S	Ra
0.12531	79810 ⎰			0.10500	95240 ⎰		
0.12441	80380 ⎱		Ra	0.10430	95880 ⎱	w	Ra
0.12428	80470 ⎰			0.10409	96080 ⎰		
0.12407	80600	sh	Ra	0.10277	97310	Ssh	Ra
0.12381	80770	sh	Ra	0.10133	98690	S	Ra
0.12240	81700 ⎱	S	Ra	0.10050	99500	s	Ra
0.12141	82370 ⎰			0.1000	100000	b	H
0.11934	83800		Ra	0.09996	100040 ⎱	w	Ra
0.11914	83940		Ra	0.09984	100170 ⎰		
0.11977	83490 ⎱		Ra	0.09960	100400		Ra
0.11902	84020 ⎰			0.09938	100620 ⎱		Ra
0.11769	84980 ⎱		Ra	0.09917	100830 ⎰		
0.11701	85470 ⎰			0.09811	101930	w	Ra
0.11520	86810 ⎱		Ra	0.0957	104500	b	Ra
0.11486	87070 ⎰			0.08568	116710		H
0.11334	88230		Ra	0.08506	117560		H
0.11293	88550 ⎱	s	Ra	0.08437	118520		H
0.11270	88730 ⎰			0.08359	119630		H
0.11236	89000 ⎱	w	Ra	0.08293	120580		H
0.11204	89250 ⎰			0.08218	121680		H
0.11164	89580 ⎱	s	Ra	0.08149	122710		H
0.11138	89790 ⎰			0.07948	125820		H
0.11127	89870 ⎱	w	Ra	0.07841	127530		H
0.11118	89940 ⎰			0.07756	128930		H
0.10984	91040	sh	Ra	0.07667	130430		H
0.10951	91320	w	Ra	0.07588	131790		H
0.10911	91630 ⎱	S	Ra	0.07508	133190		H
0.10883	91880 ⎰			0.0745	134200	b	H
0.10863	92050 ⎱	w	Ra	0.0740	135100	b	Ra
0.10839	92260 ⎰			0.0694	144100	b	Ra
0.10777	92800 ⎱	s	Ra	0.0590	169500	bs	Ra
0.10744	93080 ⎰			0.0504	198400	bc	Ra

[a] Remarks:

    b   region of continuous absorption begins here, and extends to shorter wave-lengths.

    bc   beginning of the "continuum," which from here on underlies the lines and bands.

    bs   strong absorption from here toward shorter wave-lengths, apparently made up of discrete bands (whether bs indicates anything essentially different from bc, is not clear).

    c   center of band.

    l   long wave-length limit of the band.

    W   very weak.

    w   weak.

    s   strong.

    S   very strong.

    sh   sharp line.

Numerals indicate the intensity on a scale peculiar to the observer.

[b] References:

    B   Brackett, F. S., *Astrophys. J.*, **53**, 121-132 (1921).
    Ba   v. Bahr, E., *Verh. deut. physik. Ges.*, **15**, 731-737 (1913).
    BC   Bartholomé, E., and Clusius, K., *Z. Electroch.*, **40**, 529-531 (1934).
    D   Dreisch, T., *Z. Physik*, **30**, 200-216 (1924).
    H   Henning, H. J., *Ann. d. Physik (5)*, **13**, 599-620 (1932).
    Ht   Hettner, G., *Idem (4)*, **55**, 476-496 (1918).

21. VAPOR: SPECTRA

Table 62.—*(Continued)*

K	Kühne, J., *Z. Physik*, **84**, 722-731 (1933).
L	Leifson, S. W., *Astrophys. J.*, **63**, 73-89 (1926).
M	Mecke, R., *Physik. Z.*, **30**, 907-910 (1929).
Many	Approximate value obtained by various observers.
M(T)	Mecke and collaborators (see Table 65).
McU	McAlister, E. D., and Unger, H. J., *Phys. Rev. (2)*, **37**, 1012 (A) (1931).
PlS	Plyler, E. K., and Sleator, W. W., *Idem*, **37**, 1493-1507 → 108 (A) (1931).
Ra	Rathenau, G., *Z. Physik*, **87**, 32-56 (1933).
Ru¹	Rubens, H., *Sitzb. k. preuss. Akad. Wiss. (Berlin)*, **1913**, 513-549 (1913).
Ru²	*Idem*, **1921**, 8-27 (1921).
RuA	Rubens, H., and Aschkinass, E., *Ann. d. Physik (Wied.)*, **64**, 584-605 (1898) = *Astrophys. J.*, **8**, 176-192 (1898).
RuHt	Rubens, H., and Hettner, G., *Verh. deut. physik. Ges.*, **18**, 154-167 (1916) ← *Sitzb. k. preuss. Akad. Wiss. (Berlin)*, **1916**, 167-183 (1916).
SPh	Sleator, W. W., and Phelps, E. R., *Astrophys. J.*, **62**, 28-48 (1925).
StW	Strong, J., and Woo, S. C., *Phys. Rev. (2)*, **42**, 267-278 (1932).
WbR	Weber, L. R., and Randall, H. M., *Idem*, **40**, 835-847 (1932).
Wi	Witt, H., *Z. Physik*, **28**, 236-248 (1924).
WrR	Wright, N., and Randall, H. M., *Phys. Rev. (2)*, **44**, 391-398 (1933).

e From Paschen's observations v. Bahr inferred that water-vapor exhibits no selective absorption for $\lambda = 400\,\mu$, approximately, and that there are bands at $\lambda = 250$ and $170\,\mu$.

Table 63.—Some Regions of Notable Transparency in the Absorption Spectrum of Water-vapor

E. K. Plyler and W. W. Sleator [357] have stated that there appears to be no absorption at the center ($3.168\,\mu$) of the band that is the harmonic of that near $6.26\,\mu$.

Unit of $\lambda = 1\,\mu = 10^4 A = 10^{-4}$ cm

Region of transparency λ	Reference
115	H. Rubens.[358]
91	[358]
75	[358]
62	[358]
47	[358]
3.168[a]	E. K. Plyler and W. W. Sleator.[357]
Between 1.38 and 1.87[b]	W. W. Sleator and E. R. Phelps.[359]
Between 1.87 and 2.67[b]	[359]
Between 0.137 and 0.178[c]	G. Rathenau.[337]

[a] See head of table.

[b] Between these two bands there is no absorption comparable with that in the band at $1.38\,\mu$.

[c] Even at the highest pressure used there was a transparent region between the two bands beginning, respectively, at $0.137\,\mu$ and $0.178\,\mu$.

---

and 65. M. Magat [330] has concluded that that analysis is certainly correct. R. Mecke [331] has remarked: "It is very interesting that the normal frequency vibrating with a momentum parallel to the axis of symmetry does not occur in the absorption spectrum, although one would expect this one

---

[330] Magat, M., *Ann. de Phys. (11)*, **6**, 108-193 (1936).
[331] Mecke, R., *Trans. Faraday Soc.*, **30**, 200-214 (1934).
[332] Cornell, S. D., *Phys. Rev. (2)*, **51**, 739-744 → 595 (A) (1937).
[333] Oldenberg, O., *Phys. Rev. (2)*, **46**, 210-215 (1934).

to have the strongest absorption. The reason why it does not absorb is not known."

The increase in the width of an absorption band when the pressure of the vapor itself is increased has been studied by S. D. Cornell,[332] who found for the half-width and its increase per atmosphere the values 0.57 ± 0.06 cm^{-1} and 0.13 cm^{-1}/atm for $\lambda = 0.94\ \mu$, and 0.40 ± 0.04 cm^{-1} and 0.29 cm^{-1}/atm for $\lambda = 11.35\ \mu$.

OH radicals are formed and excited when an electric discharge is passed through water-vapor; the excitation so produced gives the radical a relatively excessive amount of rotation, as is shown by its emission spectrum; its absorption spectrum shows only normal rotation.[333] These excited radi-

Table 64.—Analyses of the Absorption Vibration Spectrum of Water-vapor

The most recent formulas for the computation of the normal frequencies of the water-vapor molecule are those derived by L. G. Bonner[360] from the positions found by R. Mecke[361] for the band centers. He included terms in the fourth power of the distances and found $\omega_\sigma = 3796.0 - 39.5v_\sigma - 53.05v_\pi - 10.50v_\delta$, $\omega_\pi = 3674.8 - 53.05v_\sigma - 70.2v_\pi - 9.45v_\delta$, $\omega_\delta = 1615 - 10.50v_\sigma - 9.45v_\pi - 19.5v_\delta$.

The first close approximation to them was obtained by R. Mecke (Table 65); and a closer one by M. Magat,[362] who obtained $\omega_\sigma = 3795 - 40v_\sigma - 50v_\pi - 10v_\delta$, $\omega_\pi = 3670 - 50v_\sigma - 71v_\pi - 10v_\delta$, $\omega_\delta = 1615 - 10v_\sigma - 10v_\pi - 19.5v_\delta$, which formulas he still considers the best.[330]

Here the v's determine the harmonic; σ, π, δ indicate the types of vibration of the triangular molecule, σ is the "symmetrical" vibration, in which the O-atom oscillates along a line perpendicular to that joining the two H-atoms, π is the "antisymmetrical" vibration, in which the O-atom oscillates along a line almost parallel to that joining the two H-atoms, and δ is the "scissors" vibration, in which the H-atoms alternately approach and recede from one another. In every case, the other atoms oscillate in such

---

[334] Frost, A. A., and Oldenberg, O., *J. Chem'l Phys.*, **4**, 642-648 (1936).
[335] Kondratjew, V., and Ziskin, M., *Acta Physiochim. URSS*, **6**, 307-319 (1937).
[336] Schaefer, C., and Matossi, F., "Das ultrarote Spektrum," 1930.
[336a] Price, W. C., *J. Chem'l Phys.*, **4**, 147-153 (1936).
[337] Rathenau, G., *Z. Physik*, **87**, 32-56 (1933).
[338] Kühne, J., *Z. Physik*, **84**, 722-731 (1933).
[339] Randall, H. M., *J. Opt. Soc. Amer.*, **7**, 45-57 (1923).
[340] Lütgemeier, F., *Z. Physik*, **38**, 251-263 (1926).
[341] Kramers, H. A., and Ittmann, G. P., *Idem*, **53**, 553-565 (1929); **58**, 217-231 (1929); **60**, 663-681 (1930).
[342] Dennison, D. M., *Rev. Mod. Phys.*, **3**, 280-345 (1931).
[343] Mulliken, R. S., *Phys. Rev. (2)*, **40**, 55-62 (1932); *J. Chem'l Phys.*, **1**, 492-503 (1933).
[344] Van Vleck, J. H., and Cross, P. C., *J. Chem'l Phys.*, **1**, 357-361 (1933).
[345] Bonhoeffer, K. F., and Haber, F., *Z. physik. Chem. (A)*, **137**, 263-288 (1928).
[346] Justi, E., *Forsch. Gebiete Ingenieurw.*, **2**, 117-124 (1931).
[347] Basu, K., *Indian Physico-Math. J.*, **4**, 21-27 (1933); **6**, 55-64 (1935).
[348] Bosschieter, G., and Errera, J., *J. de Phys. (7)*, **8**, 229-232 (1937).

21. VAPOR: SPECTRA

Table 64.—*(Continued)*

a manner that the center of inertia of the molecule is unaffected by the oscillations.

The resulting wave number is given by the formula

$$\nu_0 = v_\sigma \omega_\sigma + v_\pi \omega_\pi + v_\delta \omega_\delta$$

The values of the v's assigned to each of the several band-centers, and the corresponding computed values for ν_0, are tabulated below.

Unit of $\nu_0 = 1$ cm^{-1}; of $\lambda = 1\ \mu = 10^4 \text{A} = 10^{-4}$ cm.

σ	v π	δ	ν_0 Obs.	Bonner	Magat ν_0 Computed	Mecke	σ	Magat[a] v π	δ	λ
			600[b]				0	0	1	16.7
0	0	1	1595.5	1595.5[c]	1595.5	1595	0	1	0	6.30
0	0	2	3152.0	3152.0[c]	3151	3150	0	2	0	3.17
			3500[b]				1	0	0	2.86
0	1	0	(3600)	3604.6						2.78
1	0	0	3756.5	3756.5[c]	3755	3756	0	2	1	2.66
1	0	1	5332.3	5331.0	5330.5	5331	1	0	3	1.88
1	1	0	7253.0	7255.0	7254	7247	1	2	1	1.38
1	1	1	8807.05	8810.6	8809.5	8802	2	0	3	1.14
1	2	0	10613.12	10613.1[c]	10612	10598				0.94
3	0	0	11032.36	11032.5[c]	11025	11034				0.91
1	2	1	12151.22	12149.8	12147	12133				0.82
3	0	1	12565.01	12565.0[c]	12560.5	12569				0.80
1	3	0	13830.92	13830.8[c]	13826	13809				0.72
3	1	0	14318.77	14318.8[c]	14324	14307				0.70
1	3	1	15347.91	15348.6	15340.5	15324				0.65
3	1	1	15832.47	15832.4[c]	15839	15822				0.63
1	3	2	16821.61	16827.4	16817	16799				0.60
1	4	0	16899.01	16908.1	16892	16880				0.59
3	2	0	17495.48	17464.7	17493	17445				0.57

[a] An alternative interpretation proposed by M. Magat[353] to include the Raman bands; he states that it accounts for the observed spectrum as well as does Mecke's interpretation given in the first column.

[b] Raman band.

[c] Used in computing the constants.

---

[349] Buskovitch, A. V., *Phys. Rev. (2)*, **45**, 545-549 (1934).

[350] Weizel, W., *Z. Physik*, **88**, 214-217 (1934).

[351] Bartholomé, E., *Z. Elektroch.*, **42**, 341-359 (1936).

[352] Hsu, J. H., *Chinese J. Phys.*, **1**, No. 3, 59-67 (1935).

[353] Randall, H. M., Dennison, D. M., Ginsburg, N., and Weber, L. R., *Phys. Rev. (2)*, **52**, 160-174 (1937) → *Idem*, **50**, 397 (A) (1936).

[354] Wilson, E. B., Jr., *J. Chem'l Phys.*, **4**, 526-528 (1936).

[355] Eliaševič, M., *Compt. rend. Acad. Sci. URSS, N. S.* **1934₂**, 540-541 (1934); **1934₃**, 250-252 (1934).

[355a] Barnes, R. B., Benedict, W. S., and Lewis, C. M., *Phys. Rev. (2)*, **47**, 918-921 (1935).

[356] Kondratjev, V. N., and Eropkin, D. I., *Compt. rend. Acad. Sci. URSS*, **1934₁**, 170-172-175 (1934).

Table 65.—Molecular Constants Involved in the Vibration Spectrum of Water-vapor [329]

$v_0 = v_\sigma \omega_\sigma + v_\pi \omega_\pi + v_\delta \omega_\delta$; $\omega_\sigma = 3795 - 39(v_\sigma + v_\pi)$, $\omega_\pi = 3670 - 70(v_\sigma + v_\pi)$, $\omega_\delta = 1615 - 20(v_\sigma + v_\pi + v_\delta)$. The moments of inertia are I_A, I_B, I_C, related to the tabulated quantities, A, B, C, respectively, by means of the equations $I_A = h/8\pi^2 cA = (27.658/A) \, 10^{-40}$ g·cm^2, etc., h being taken as 6.547×10^{-27} erg·sec, and c as $2.99796 \, 10^{10}$cm/sec; $10^{40}I_A = 0.996 + 0.045v_\sigma + 0.026v_\pi - 0.098v_\delta$, $10^{40}I_B = 1.908 + 0.014v_\sigma + 0.033v_\pi - 0.034v_\delta$, $10^{40}I_C = 2.981 + 0.047v_\sigma + 0.062(v_\pi + v_\delta)$ g·cm^2; $\Delta = I_C - (I_A + I_B)$; $\alpha = \angle$ HOH, $r = $ distance HH, H and O being the points occupied by the atoms H and O in the triangular water molecule; v_σ, v_π, v_δ are quantum numbers.

Unit of $\lambda = 1\mu = 10^4$A $= 10^{-4}$ cm; of $\nu_0 = 1$ wave/cm; of A, B, and $C = 1$cm^{-1}; of $r = 1$A $= 10^{-8}$ cm; of $\Delta = 10^{-40}$ g·cm^2.

λ	$v_\sigma \, v_\pi \, v_\delta$	ν_0(obs.)	A	B	C	α	r	Δ
	0 0 0	0	27.81[a]	14.50[a]	9.28[a]	105° 6'	0.970	0.077
6.296	0 0 1	1595.4	30.70	14.70	9.12	107° 30'	0.984	0.29
3.168	0 0 2	3152	35.8	15.0	9.0	111° 10'	0.984	0.46
2.663	1 0 0	3756.35	26.50	14.47	9.20	103° 45'	0.975	0.06
1.875	1 0 1	5332.3	29.0	14.6	8.8	106° 30'	0.995	0.30
1.379	1 1 0	7253	26.1	14.3	8.9	103° 40'	0.992	0.08
1.135	1 1 1	8807.0	28.60	14.7	8.71	105° 30'	1.002	0.33
0.9420	1 2 0	10613.25	25.25	13.89	8.75	103° 40'	0.998	0.075
0.9060	3 0 0	11032.33	24.45	14.18	8.87	102° 5'	0.991	0.032
0.8227	1 2 1	12151.23	27.84	14.17	8.66	105° 40'	1.001	0.248
0.7957	3 0 1	12565.01	26.75	14.41	8.74	104° 10'	0.997	0.212
0.7227	1 3 0	13830.91	24.70	13.71	8.61	103° 20'	1.005	0.075
0.6994	3 1 0	14318.73	24.04	13.95	8.69	102° 25'	1.004	0.050
0.6524	1 3 1	15347.90	27.15	13.97	8.52	105° 25'	1.012	0.247
0.6324	3 1 1	15832.47	(26.5)	(14.20)	(8.58)	104° 20'	1.013	(0.26)
0.5952	1 3 2	16821.62	29.28	14.03	8.34	107° 25'	1.023	0.400
0.5924	1 4 0	16898.81	25.02	13.60	8.50	103° 55'	1.014	0.115
0.5722	3 2 0	17495.44	23.72	13.84	8.32	102° 0'	1.026	0.160

v_δ		0				1			2	
v_σ	$v_\sigma \, v_\pi \, v_\delta$	α	100Δ	$v_\sigma \, v_\pi \, v_\delta$	α	100Δ	$v_\sigma \, v_\pi \, v_\delta$	α	100Δ
0	0 0 0	105° 6'	7.7	0 0 1	107° 30'	29	0 0 2	111° 10'	46
1	1 0 0	103° 45'	6	1 0 1	106° 30'	30			
1	1 1 0	103° 40'	8	1 1 1	105° 30'	33			
1	1 2 0	103° 40'	7.5	1 2 1	105° 40'	25			
1	1 3 0	103° 20'	7.5	1 3 1	105° 25'	25	1 3 2	107° 25'	40
1	1 4 0	103° 55'	11.5						
3	3 0 0	102° 5'	3.2	3 0 1	104° 10'	21			
3	3 1 0	102° 25'	5.0	3 1 1	104° 20'	26			
3	3 2 0	102° 0'	16.0						

[a] In their extended analysis (see p. 149), Randall and his associates used the following for the (0,0,0) values of A, B, and C: 27.8055, 14.499724, and 9.279276, respectively; these small changes from Mecke's values being made for the purpose of facilitating the computations.

21. VAPOR: SPECTRA

Table 66.—Fine-structure of Absorption Bands of Water-vapor: Sources of Data

This table indicates the range in wave-length covered by the observations, the wave-lengths (λ) of the centers of the bands falling within that range, and where a report of the observations may be found. H. M. Randall [339] has reviewed the earlier data.

Unit of $\lambda = 1\mu = 10000\text{A} = 10^{-4}$ cm.

λ_1	Range: to λ_2	Bands λ	Ref.[a]
170	38		B
135	60		WrR
75	18		R
24.9	10.13		WR
7.7	4.8	6.26	SP
7.6	4.8	6.26	vB
7.0	5.4	6.26	PlS, M
3.8	3.0	3.15	Ba
3.3	2.8	3.15	PlS, SP, M
2.8	2.5	2.67	PlS, S, M
2.65	1.35	1.37, 1.87, 2.67	Dr
1.9	1.8	1.87	PlS, SP, M, Pl
1.42	1.35	1.37	SP, M, Pl
1.15	1.11	1.13	LH, M
0.96	0.93	0.942	BM, Br, LH, H
0.92	0.89	0.906	LH
0.92	0.78	0.906, 0.823, 0.796	BM
0.73	0.71	0.723	BM
0.71	0.69	0.699	FM
0.66	0.64	0.652, 0.632	FM
0.63	0.56	0.595, 0.592, 0.572	FM
0.18	0.15		Ra, Pr
0.14	0.098		Ra, Pr

[a] References:
- B   Barnes, R. B., Benedict, W. S., and Lewis, C. M., *Phys. Rev. (2)*, **47**, 918-921 (1935).
- vB   v. Bahr, E., *Verh. d. physik Ges.*, **15**, 731-737 (1913).
- Ba   Barnes, R. B., *Phys. Rev. (2)*, **36**, 296-304 (1930).
- BM   Baumann, W., and Mecke, R., *Z. Physik*, **81**, 445-464 (1933).
- Br   Brackett, F. S., *Astrophys. J.*, **53**, 121-132 (1921).
- Dr   Dreisch, T., *Z. Physik*, **30**, 200-216 (1924).
- FM   Freudenberg, K., and Mecke, R., *Idem*, **81**, 465-481 (1933).
- H   Hsu, J. H., *Chinese J. Phys.*, **1**, No. 3, 59-67 (1935).
- LH   Lueg, P., and Hedfeld, K., *Z. Physik*, **75**, 512-520 (1932).
- M   Mecke, R., *Idem*, **81**, 313-331 (1933).
- Pl   Plyler, E. K., *Phys. Rev. (2)*, **39**, 77-82 (1932).
- PlS   Plyler, E. K., and Sleator, W. W., *Idem*, **37**, 1493-1507 (1931).
- Pr   Price, W. C., *J. Chem'l Phys.*, **4**, 147-153 (1936).
- R   Randall, H. M., Dennison, D. M., Ginsburg, N., and Weber, L. R., *Phys. Rev. (2)*, **52**, 160-174 (1937).
- Ra   Rathenau, G., *Z. Physik*, **87**, 32-56 (1933).
- S   Sleator, W. W., *Astrophys. J.*, **48**, 125-143 (1918).
- SP   Sleator, W. W., and Phelps, E. R., *Idem*, **62**, 28-48 (1925).
- WR   Weber, L. R., and Randall, H. M., *Phys. Rev. (2)*, **40**, 835-847 (1932).
- WrR   Wright, N., and Randall, H. M., *Idem*, **44**, 391-398 (1933).

Table 67.—Some Rotation Terms in the Infrared Spectrum of Water-vapor

Mecke and associates.[329]   (For other bands see references given in the text.)

λ_0 is the wave-length corresponding to $v_\sigma = v_\pi = v_\delta = 0$ (see Table 65)
Unit of $\lambda = 1\mu = 10A = 10^{-4}$ cm; of $\nu = 1$ cm^{-1}

$\lambda \rightarrow$ J_k	λ_0	6.262	3.168	2.663	1.875	1.379	1.135	0.9420	0.9060
0	0	1595.5		3756.5	5332.0	7253	8807.05	10613.12	11032.36
1_-1	23.78	1618.9		3780.1	5355.4	7276	8830.45	10636.5	11055.39
1_0	37.09	1635.4		3792.1	5370.3	7290	8844.46	10647.38	11065.62
1_1	42.31	1641.1		3797.3	5376.6	7295	8850.36	10652.44	11070.97
2_-2	70.09	1665.3		3826.0	5401.7	7321	8875.50	10680.43	11099.92
2_-1	79.43	1677.2		3834.1	5411.7	7327	8885.21	10687.33	11106.42
2_0	95.15	1694.1		3850.0	5428.6		8903.48	10702.95	11122.45
2_1	134.98	1742.5		3886.1	5468.		8944.85	10737.05	11153.25
2_2	136.24	1743.8		3887.7	5469.		8945.95	10738.36	11154.84
3_-3	136.76	1732.3		3892.0	5467.8	7385	8940.13	10743.60	11163.46
3_-2	142.23	1739.8		3896.1	5473.8	7390	8945.80	10747.07	11166.75
3_-1	173.31	1772.7		3928.0	5507.		8979.6	10777.43	11198.26
3_0	206.35	1813.0		3956.9	5540.0		9014.4	10804.89	11222.51
3_1	212.24	1820.2		3964.0	5547.8		9020.7	10811.45	11229.65
3_2	286.80	1908.0					(9103.5)	10876.06	11288.60
3_3	286.93	1908.2		(4030.0)	5630.		(9104.5)	10876.14	11288.60
4_-4	222.07			3975.5	5550.0	7468	9022.49	10823.87	11244.55
4_-3	224.81			3977.2	5553.5		9025.3	10825.52	11245.6
4_-2	275.50			4025.7	5608.5		9079.8	10883.15	11296.70
4_-1	300.44			4051.0	5635.5		9104.8	10903.47	11324.44
4_0	315.83			4064.2	5654		9121.9	10911.18	11331.58
4_1	384.03							10986.5	
4_2	385.44							10987.4	
5_-5	325.37			4075.9	5653	7567	9122.0	10921.67	11341.5
5_-4	326.56			4076.3			9123.2	10921.95	11342.8
5_-3	399.42							10999.0	11425.24
6_-6	446.95			} 4195	} 5772	} 7683	} 9239.1	11036.5	
6_-5	447.16							11036.4	
7_-7	586.28			4333	5909	7815	9372.5	11168.4	
7_-6	586.6							11168.7	

21. VAPOR: SPECTRA

$\lambda \rightarrow$ / J_K	0.8227	0.7957	0.7227	0.6994	0.6524 ν	0.5952	0.5924	0.5722
8₋₈	744.1							
8₋₇	774.2							
9₋₉,₈	920			4486		9523.7		
10₋₁₀,₉	1115			4663	6238	9693.0		
				4853	6429	9878		
					7969			
					8140			
					8328			
0	12151.22	12565.01	13830.92	14318.77	15347.91	16821.61	16899.01	17495.48
1₋₁	12173.77	12588.16	13853.25	14341.39	15370.50	16844.02	16920.91	17517.54
1₀	12187.75	12600.51	13864.20	14351.48	15383.61	16859.26	16932.20	17527.60
1₁	12193.31	12606.21	13869.33	14356.75	15389.06	16864.94	16937.32	17532.89
2₋₂	12218.99	12632.88	13896.46	14385.03	15413.94	16887.38	16964.05	17560.54
2₋₁	12227.82	12641.14	13903.80	14391.52	15423.09	16898.25	16971.39	17566.13
2₀	12244.54	12658.20	13919.05	14407.25	15439.45	16915.28	16986.62	17583.38
2₁	12285.38	12695.13	13951.93	14437.52	15479.12	16959.67	17021.83	17611.71
2₂	12286.79	12696.74	13953.36	14439.16	15480.16	16960.64	17023.14	17613.89
3₋₃	12282.17	12696.74	13958.17	14447.20	15476.22	16949.60	17024.64	17621.55
3₋₂	12287.20	12701.18	13962.18	14450.68	15481.57	16956.22	17029.14	17623.92
3₋₁	12319.80	12734.88	13992.61	14481.70	15514.01	16989.78	17059.56	17656.45
3₀	12353.80	12764.73	14018.98	14505.99	15546.62	17026.53	17088.47	17677.95
3₁	12360.07	12771.76	14025.58	14512.91	15552.93	17033.13	17094.61	17684.93
3₂	12436.76	12841.33	14088.12	14570.78	15628.00	17110.03	17166.84	17744.06
3₃	12437.11	12841.72	14088.50	14571.19	15628.29	17110.72	17167.29	17744.66
4₋₄	12362.86	12777.51	14038.24	14526.37	15556.87	17028.53	17103.05	17699.12
4₋₃	12365.22	12779.80	14038.26	14528.16	15560.67	17033.27	17105.32	17701.03
4₋₂	12418.03	12834.43	14088.26	14578.10	15610.99	17086.81	17158.13	17750.84
4₋₁	12443.81	12863.58	14116.1	14595.70	15635.36	17114.41	17176.18	17766.43
4₀	12460.23	12874.17	14123.58	14612.93	15651.79	17130.82	17202.09	17781.27
4₁	12530.29		14201.7	14669.74		17200.73	17269.44	17833.78
4₂	12531.68		14202.2	14669.82	15721.84	17202.15	17270.70	17835.67
5₋₅	12460.40	12875.42	14133.94	14622.07	15652.43	17128.50	17199.01	17794.88
5₋₄	12461.32	12876.70	14134.2					
5₋₃	12536.42	12950.89	14204.3					
6₋₆	12574.8	12990.5		14735.35	15764.59		17311.21	17903.00
6₋₅	12575.1	12991.0						
7₋₇	12706.6	13122.5		14865.74	15892.95			
7₋₆	12706.8	13122.7						
8₋₈,₇	12855.4	13272.1						

cals last for at least 1/8 second after the discharge ceases.[334] The absorption spectrum shows that these radicals are present also in the combustion zone of H_2 burning in O_2 (pressure 3 to 25 mm-Hg, furnace 470 to 550 °C), and in amount about 1000 times that corresponding to equilibrium.[335]

For a general discussion of the infrared spectra and for typical data obtained prior to 1930, see C. Schaefer and F. Matossi.[336]

In the far ultraviolet, water-vapor exhibits a continuous absorption from 1785 to 1550A; below 1550A there are several series of bands: one of doublets, $\Delta\nu = 170$ cm^{-1}, beginning at $\lambda = 1240$A and accompanied by another displaced from it toward shorter wave-lengths by about $\Delta\nu = 3170$ cm^{-1}; another at 1220A accompanied by a weaker one displaced toward shorter λ's by $\Delta\nu = 3240$ cm^{-1}; a third at 1091A; and a fourth at 1078A. These have been reported and analyzed by W. C. Price.[336a] The range 1700 to 957A had previously been studied by G. Rathenau,[337] but with a smaller dispersion.

The ultraviolet portion of the spectrum is regarded as arising from changes in the electron configuration of the molecule; the near infrared bands, from the oscillations of the constituent atoms about their positions of equilibrium in the molecule; and the fine-structure of those bands, from rotations of the molecule as a whole, the effect of the rotation being superposed upon that arising from the oscillations of the atoms. In the far infrared are lines arising from the rotation alone. They are narrow when the pure vapor is at a low pressure, but broaden as the pressure is increased by the admission of air, owing to the impact of the molecules of air upon those of the vapor.[338] As the H_2O molecule is triangular (see Section 9, Models) there are three principal axes of rotation, but only two of them seem to be spectroscopically effective.

An early report on the fine-structure of bands was published by H. M. Randall.[339] The detailed analysis of the water-vapor spectrum rests upon the theoretical work of F. Lütgemeier [340] and of H. A. Kramers and G. P. Ittmann.[341] D. M. Dennison [342] has reviewed the general subject; R. S. Mulliken [343] has studied and published the electronic configurations of the H_2O molecule; and in addition to those mentioned elsewhere, J. H. Van Vleck and P. C. Cross [344] have calculated from spectral data certain constants of the H_2O molecule. Certain relations connecting spectroscopic and thermal data have been considered by K. F. Bonhoeffer and F. Haber,[345] E. Justi,[346] and others. Their conclusions will be found in the sections devoted to the appropriate thermal data. In addition to those referred to elsewhere in this section, K. Basu,[347] G. Bosschieter and J. Errera,[348]

[357] Plyler, E. K., and Sleator, W. W., *Phys. Rev. (2)*, **37**, 1493-1507 (1931).
[358] Rubens, H., *Sitz. preuss. Akad. Wiss.*, 513-549 (1913).
[359] Sleator, W. W., and Phelps, E. R., *Astrophys. J.*, **62**, 28-48 (1925).
[360] Bonner, L. G., *Phys. Rev. (2)*, **46**, 458-464 (1934).
[361] Mecke, R., *Z. Physik*, **81**, 313-331 (1933).
[362] Magat, M., *Compt. rend.*, **197**, 1216-1220 (1933).
[363] Magat, M., *Compt. rend.*, **196**, 1981-1983 (1933).

A. V. Buskovitch,[349] and W. Weizel [350] have considered the interpretation of the spectrum of water-vapor.

The vibration of polyatomic molecules is the subject of a review by E. Bartholomé.[351]

1. Band at $\lambda = 0.94\,\mu$. The fine-structure of this band at $0.94\,\mu$ has been studied by F. S. Brackett (see Tables 62 and 66) and by J. H. Hsu.[352]

2. Region of $\lambda = 18$ to $75\,\mu$. The region 18 to $75\,\mu$ has been studied and analyzed with great care by H. M. Randall and his associates.[353] They find that the rotational stretching and deformation of the molecule may correspond to a change of over $200\ \text{cm}^{-1}$ in the position of a line. They have determined the amount of this correction and have obtained good agreement between observed and computed λ's up to quantum number $j = 12$. For $j = 11$ the vertex angle of the molecule is $98°\ 52'$, and the O-H distance is 0.9640A, whereas the static values are $104°\ 36'$ and 0.9558A. They have concluded that for components displaced much above $5\ \text{cm}^{-1}$ no first order correction for stretching, such as that proposed by E. B. Wilson, Jr.,[354] can be satisfactory. They give extended tables of energy levels. A partial analysis of earlier data has been attempted by M. Eliaševič.[355]

3. Region of 38 to $170\,\mu$. The absorption of atmospheric water-vapor has been mapped from 38 to $170\,\mu$ and analyzed by R. B. Barnes, W. S. Benedict, and C. M. Lewis.[355a] The analysis was based on Mecke's values of the constants (Tables 64 and 65) and was carried, but very incompletely, to $j = 11$. They concluded that the pure rotation spectrum of water-vapor is in complete agreement with the analysis.

Emission Spectrum.

The emission spectrum of water-vapor extends from the far infrared to the extreme ultraviolet, and consists of numerous bands, each containing many lines. Whether excited in the flame or by an electrical discharge, it is always accompanied by lines of hydrogen and of oxygen.

The ultraviolet bands have been studied in much detail. T. Heurlinger [364] found that most of the line of the 3064A band can be arranged in 12 branches forming what he called a band with doublet series; W. W. Watson [365] found that the 2811A band is of the same type, and concluded that both bands arise from the OH-molecule, formed in the flame and in the electric discharge. L. Grebe and O. Holtz [366] had previously suggested OH as the source of 3064. For details of the analyses of the bands, see E. C. Kemble,[367] D. Jack,[368] and especially R. S. Mulliken,[369] in addition to the authors cited in Table 68. R. T. Birge [370] has given certain con-

---

[364] Huerlinger, T., *Diss.*, Lund, 1918.
[365] Watson, W. W., *Astrophys. J.*, **60**, 145-158 (1924).
[366] Grebe, L., and Holtz, O., *Ann. d. Physik (4)*, **39**, 1243-1250 (1912).
[367] Kemble, E. C., *Phys. Rev. (2)*, **30**, 387-399 (1927).
[368] Jack, D., *Proc. Roy. Soc. (London) (A)*, **120**, 222-234 (1928).
[369] Mulliken, R. S., *Phys. Rev. (2)*, **32**, 388-416 (1928).
[370] Birge, R. T., *Int. Crit. Tables*, **5**, 415 (1929).

Table 68.—Bands in the Emission Spectrum of Water-vapor

The emission spectrum of water-vapor extends beyond the limits covered by this table. F. Paschen [378] observed three bands of $\lambda > 8.2\,\mu$, but was unable to determine their wave-lengths; and J. J. Hopfield [379] stated that the spectrum extends in the ultraviolet to about 900A = 0.09 μ. Each band contains many lines; each value of λ here given is that corresponding approximately to a maximum of emission, and serves to identify the band. In the infrared, the positions of the maxima are not exactly known, as the several determinations were made over 40 years ago, and differ among themselves.

λ = wave-length, $\nu = 1/\lambda$ = wave-number, Bd = band designation in the paper first mentioned in the last column. With the exception of P and PWl, each of the cited papers includes a study of the fine-structure of the indicated band.

Unit of $\lambda = 1\mu = 10^4$A $= 10^{-4}$ cm; of $\nu = 1$ cm^{-1}

λ	ν	Bd	Ref.[a]
8.2	1220		P
6.6	1510		P
5.6	1780		P
5.3	1880		P
2.8	3570		P
2.7	3700		N
1.8	5560		P, N
1.4	7140		P
0.3564[b]	28060		RWl
0.3484	28700	1, 2	JDW
0.3428	29170	0, 1	JDW, Ja
0.3328	30050		RWl
0.3185	31300	2, 2	DJ
0.3122	32030	1, 1	DJ, Ja
0.3064[c]	32640	0, 0	DJ, Di, GH, Ja
0.3021	33100		Wn
0.2875	34780	2, 1	DJ, Ja
0.2811	35570	1, 0	DJ, Wn, Ja
0.2676	37370	3, 1	CC
0.2608	38340		Ja
0.2447	40870	3, 0	CC

[a] References:
 CC Chamberlain K, and Cutter, H. B., *Phys. Rev. (2)*, **44**, 927-930 (1933).
 DJ Dawson, D. H., and Johnston, H. L., *Idem*, **43**, 980-991 (1933).
 Di Dieke, G. H., *Proc. K. Akad. Wet. Amst.*, **28**, 174-181 (1925); *Nature*, **115**, 194 (1925).
 GH Grebe, L., and Holtz, O., *Ann. d. Physik (4)*, **39**, 1243-1250 (1912).
 Ja Jack, D., *Proc. Roy. Soc. (London) (A)*, **115**, 373-390 (1927); *Idem*, **118**, 647-654 (1928).
 JDW Johnston, H. L., Dawson, D. H., and Walker, M. K., *Phys. Rev. (2)*, **43**, 473-480 (1933).
 N Neunhoeffer, M., *Ann. d. Physik (5)*, **2**, 334-349 (1929); *Idem*, **4**, 352-356 (1930).
 P Paschen, F., *Idem, (Wied.)*, **50**, 409-443 (1893); *Idem*, **51**, 1-39 (1894); *Idem*, **52**, 209-237 (1894); *Idem*, **53**, 334-336 (1894).
 RWl Rodebush, W. H., and Wahl, M. H., *J. Chem'l Phys.*, **1**, 696-702 (1933); *J. Am. Chem. Soc.*, **55**, 1742 (L) (1933).
 Wn Watson, W. W., *Astrophys. J.*, **60**, 145-158 (1924).

[b] Degraded toward the red.

[c] The band at 3064A seems to be associated with the formation of chemically "active gas" by the discharge.[379a]

stants of the OH-molecule, derived from spectral data. W. H. Rodebush and M. H. Wahl [371] think that the bands at 3564A and 3328A arise from ionized OH.

The infrared emission bands have been studied mainly by Paschen. He found that the position of the maximum of a band shifts, and the intensity increases, as the temperature of the vapor is increased. The direction of the shift depends upon the band. He concluded that the emission probably obeys Kirchhoff's law.[372] He gives the following examples of the shift with temperature [373]: $\lambda = 2.831$, 2.812, 2.717, and 2.661 μ, corresponding respectively to the temperatures 1460, 1000, 500, and 100 °C; $\lambda = 5.322$, 5.377, 5.416, 5.607, 5.900, and 5.948 μ, corresponding respectively to the temperature of the oxy-hydrogen flame, 1470, 1000, 600, 100, and 17 °C; $\lambda = 6.620$, 6.597, 6.563, 6.527, and 6.512 μ, for 1470, 1000, 600, 100, and 17 °C. For the 2.8 μ and the 6.6 μ bands the maximum shifts to longer wave-lengths as the temperature increases; for the 5.3 μ band the shift is in the opposite direction. E. v. Bahr [374] regarded the 6.6 μ and 5.3 μ bands as components of a double band, the separation varying with the absolute temperature (T) in such a manner that ($\nu_2 - \nu_1$) is proportional to $\sqrt{T}$; and showed that Paschen's observations satisfied that relation, ν_1 and ν_2 being $1/\lambda_1$ and $1/\lambda_2$, respectively.

Over the range 6.254 μ to 5.780 μ, and again at 5.188 μ, the emission is very weak, but it is strong at 5.636 μ, which lies between them.[375]

For a discussion of the occurrence of the OH radical in flames and in vapor through which a discharge has passed, see V. Kondratjew and M. Ziskin,[376] T. Kitagawa,[377] O. Oldenberg,[333] and A. A. Frost and O. Oldenberg.[334]

A review of our knowledge of the vibrations of polyatomic molecules has been published by E. Bartholomé.[351]

22. Dielectric Constant of Water-vapor

P. Debye [380] has shown that the dielectric constant of a gas composed of molecules that contain both permanent electrical dipoles and elastically bound electrons will satisfy relation (1)

$$\left(\frac{\epsilon - 1}{\epsilon + 2}\right)\frac{T}{\rho} = aT + b \tag{1}$$

[371] Rodebush, W. H., and Wahl, M. H., *J. Am. Chem. Soc.*, **55**, 1742 (L) (1933); *J. Chem'l Phys.*, **1**, 696-702 (1933).
[372] Paschen, F., *Ann. d. Physik (Wied.)*, **51**, 1-39, 40-46 (1894). See also Section 18.
[373] Paschen, F., *Idem*, **51**, 1-39 (1894); **53**, 334-336 (1894).
[374] v. Bahr, E., *Verh. deut. physik. Ges.*, **15**, 731-737 (1913).
[375] Paschen, F., *Ann. d. Physik (Wied.)*, **52**, 209-237 (1894).
[376] Kondratjew, V., and Ziskin, M., *Acta Physicochim. URSS*, **6**, 307-319 (1937); **7**, 65-74 (1937).
[377] Kitagawa, T., *Proc. Imp. Acad. Japan*, **12**, 281-284 (1936).
[378] Paschen, F., *Ann. d. Physik (Wied.)*, **50**, 409-443 (1893).
[379] Hopfield, J. J., *Nature*, **110**, 732-733 (1922).
[379a] Urey, H. C., and Lavin, G. I., *J. Am. Chem. Soc.*, **51**, 3290-3293 (1929).

where ϵ = dielectric constant, ρ = density, T = absolute temperature, and a and b are positive constants characteristic of the gas. There are various reasons for believing that the molecules of water-vapor are of that kind. (See Section 9, *Dipole moment*.)

Table 69.—Dielectric Constant of Water-vapor
(See also Figure 2)

For comments and references to original publications, see text. Debye's equation requires $(\epsilon - 1)T/\rho = 3(aT + b)$, approximately, a and b being positive constants characteristic of the gas or vapor. The best set of determinations is that of Stranathan, giving $3b = 3449 \pm 23$, $3a = 0.671 \pm 0.065$.

The investigator is indicated by his initial: B = Bädeker, J = Jona, SS = Sänger and Steiger, Z = Zahn.

Unit of $\epsilon = 1$ cgse; of $\rho = 1$ g/cm³; of $p = 1$ mm-Hg; temp. $= t$ °C

Stranathan. Computed from $3a = 0.671$, $3b = 3449$.

t	20	50	80	100	125	150	180	200
$(\epsilon - 1)T/1000\rho$	3.65	3.67	3.69	3.70	3.72	3.73	3.74	3.77

t	$10^6(\epsilon-1)$	$\frac{(\epsilon-1)T}{1000\rho}$	t	$10^6(\epsilon-1)$	$\frac{(\epsilon-1)T}{1000\rho}$
	SS, $\rho=0.0004181$			Z, $p=20$ (extrapolated)	
120	400.2	3.76₀	23.3	25	3.8₀
150	371.7	3.76₀	29.6	23	3.6₅
180	348.8	3.77₉	39.6	19	3.2₄
210	328.7	3.79₄	107.9	15	3.7₇
			165.0	12	3.9₄
	J, $p=760$			B, $p=760$	
117.3	612	4.20	140.0	765	5.90
124.4	584	4.16	142.2	767	5.98
128.9	575	4.19	143.2	736	5.76
142.0	530	4.12	145.8	694	5.50
150.7	511	4.15	148.6	648	5.21
163.1	489	4.21			
178.3	462	4.27			

As $(\epsilon + 2)$ is essentially equal to 3, and ρ is very nearly proportional to the pressure when T is invariable, relation (1) requires that $(\epsilon - 1)$ shall, for T constant, be essentially proportional to the pressure. Such proportionality for gases throughout the range 0 to 800 mm-Hg has been found by K. Wolf,[381] using frequencies of 1 to 10 megacycles per second; but it could be tested for water-vapor only over a much shorter range in pressure, owing to difficulty in maintaining the insulation. But very recently, J. D. Stranathan [382] has reported that he has found such proportionality to hold for water-vapor "to very near saturation," departures near saturation being due in many cases, perhaps in all, to the electrostatic polarization of the moisture adsorbed on the solid insulation, and not to leakage. From observations at 14 temperatures, ranging from 21.3 to 197.9 °C, he

[380] Debye, P., *Physik. Z.*, **13**, 97-100, 295 (1912).
[381] Wolf, K., *Ann. d. Physik (4)*, **83**, 884-902 (1927).

22. VAPOR: DIELECTRIC CONSTANT

found for the coefficients of equation (1) the values: $a = 0.2237 \pm 0.0217$, $b = 1149.6 \pm 7.8$. These are the best determinations now available.

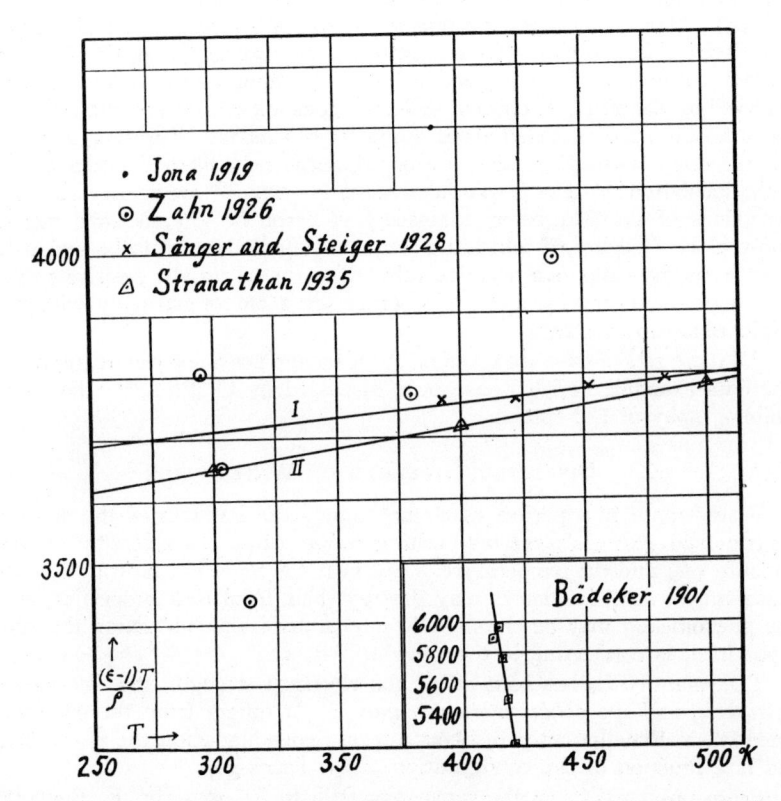

FIGURE 2. Variation of the Dielectric Constant of Water-vapor with the Temperature and Density.

Line II, determined by Stranathan's equation (see Table 69), is probably the best available representation of the variation of $(\epsilon - 1)T/\rho$ with T. Line I closely represents the data obtained by Sänger and Steiger; its equation is $(\epsilon - 1)T/\rho = 3576 + 0.448T$. The three points (triangles) on line II were computed by means of Stranathan's equation; all other specially marked points represent experimental values. Bädeker's data are totally at variance with the others. References are given in the text. ϵ = dielectric constant (cgse), ρ g/cm³ = density, T °K = absolute temperature.

The best previous data for the dielectric constant of water-vapor were those of R. Sänger and O. Steiger,[383] republished by Sänger,[384] and by R. Sänger, O. Steiger, and K. Gächter.[385]

[382] Stranathan, J. D., *Phys. Rev.* (2), **48**, 538-549 (1935) → **47**, 794 (A) (1935), extending **45**, 741 (A) (1934).
[383] Sänger, R., and Steiger, O., *Helv. Phys. Acta*, **1**, 369-384 (1928).
[384] Sänger, R., *Physik. Z.*, **31**, 306-315 (1930).
[385] Sänger, R., Steiger, O., and Gächter, K., *Helv. Phys. Acta*, **5**, 200-210 (1932).

Of the earlier determinations, those of C. T. Zahn,[386] were the best, and his results were used by H. L. Curtis in deriving the formula given in *International Critical Tables*[387] for the variation of ϵ with the temperature and the pressure. They are not inconsistent with those of Stranathan and of Sänger and Steiger, but are of such a low order of accuracy that the formula derived from them is quite erroneous. The abnormally rapid increase observed by Zahn in the apparent dielectric constant with the vapor-pressure when the pressure exceeded about a quarter of that for saturation, is probably not due to adsorption, as he supposed, but is more likely due to a failure in the insulation.[388] The determinations of M. Jona [389] are merely relative, the values of $(\epsilon - 1)$ being admittedly in error by an unknown factor. Those of K. Bädeker,[390] which are frequently quoted, seem to be seriously in error, as they impose a negative value upon the essentially positive coefficient a of equation (1), and are in other ways out of harmony with the results obtained by others.

All these sets of data are given in Table 69 and are displayed in Figure 2. The data available in 1926 have been discussed by O. Blüh,[391] who gives a bibliography of 172 entries.

23. Dielectric Strength of Water-vapor

Water-vapor is a perfect insulator* unless the strength of the applied electric field exceeds a certain critical value (E_c) characteristic of the pressure (p) and the temperature. But near the region of saturation, condensation on solid insulators may destroy their insulating properties, and this phenomenon may be erroneously interpreted as an indication that the vapor itself is conducting in that region.

The quantity E_c has been called the electrical strength, the *elektrische Festigkeit*, and the *cohésion diélectrique*.[392] It differs from the sparking potential, in that the latter involves a term depending upon the electrodes, and is a function of the configuration of the spark-gap.

From observations in the range $p = 0.05$ to 5.4 mm-Hg, E. Bouty[393] concluded that, at 22 °C, $E_c = 333 + 50.0p + 0.106/p^2$ volt/cm, the unit of p being 1 mm-Hg. This requires E_c to pass through a minimum (345 volt/cm) at $p = 0.16$ mm-Hg.

The over-all voltage (V_i) at which a current begins to pass between a given pair of electrodes, though obviously related to E_c, is not derivable

---
* The conductivity arising from the action of radiations of various kinds is ignored in this section; it is relatively low in most cases.

[386] Zahn, C. T., *Phys. Rev. (2)*, **27**, 329-340 (1926).
[387] Curtis, H. L., *Int. Crit. Tables*, **6**, 78 (1929).
[388] Wolf, K., *Physik. Z.*, **27**, 588-591, 830 (1926); *Ann. d. Physik (4)*, **83**, 884-902 (1927). See however, Stranathan.[382]
[389] Jona, M., *Physik. Z.*, **20**, 14-21 (1919).
[390] Bädeker, K., *Z. physik. Chem.*, **36**, 305-335 (1901).
[391] Blüh, O., *Physik. Z.*, **27**, 226-267 (1926).
[392] Bouty, E., *Compt. rend.*, **131**, 443-447, 469-471 (1900).
[393] Bouty, E., *Idem*, **131**, 503-505 (1900).

23. VAPOR: DIELECTRIC STRENGTH

from it, and may differ from the sparking potential. It has been studied recently by E. Weichelt [394] and by S. Franck.[395] Using as electrodes brass spheres 4 cm in diameter, placed with their supporting rods and the gap along the axis of a cylindrical metal cage 12 cm in diameter, Weichelt found the ratio of V_i for water-vapor to that for air, for the same gap and at the same temperature and pressure to be 1.16_6. The gaps ranged from 0.5 to 3.2 cm, and temperatures from 115 to 160 °C; the pressure was always 1 atm. The uncertainty of the individual measurements was about 2.5 per cent. For a given pressure, V_i for air, and hence for water-vapor, varies inversely as the absolute temperature.

Table 70.—Voltage Difference at which Current Begins to Flow Through Water-vapor [395]

See text for definitions and details.

Unit of V_i = 1 kilovolt; of gap = 1 cm

I. Comparison with air. Point and Plate.[a]

Gap →	1.0	2.5	5.0	7.5	10
Gas			$(V_+/V_-)_i$		
Vapor	1.100	1.057	1.057	1.014	1.015
Air	1.084	1.061	1.130	1.127	1.138

II. Vapor. Values read from graphs.

			Point and Plate			
Gap	1	2	4	6	8	10
$(V_+)_i$	4.6	4.9	5.2	5.4	5.6	5.7
$(V_-)_i$	4.2	4.6	5.0	5.3	5.5	5.6
		Plates			Spheres[b]	
Gap	0.66	1.33	2.0	0.5	1.0	1.5
V_i	18.8	33.9	48.7	14.1	27.3	37.4

[a] The subscript + means that the potential of the point is positive with reference to that of the plate.
[b] Diameters = 5 cm.

Franck, using a different type of apparatus and working at about 0.95 atm and about 100 °C, found V_i for water-vapor to be about 8 per cent greater than for air, the electrodes being either plates, or balls 5 cm in diameter; but if the gap is between a point and a plate, the ratio depends upon the length of the gap and upon whether the potential of the point is positive (V_+) or negative (V_-) with reference to that of the plate. If the gap exceeds about 3 cm, V_i for water-vapor is greater than that for air if the point is negative, and less if the point is positive. See Table 70.

The sparking potential (V_s) in water-vapor between clean spheres has been studied by I. Strohhäcker,[396] and that between a point and a plane by

---

[394] Weichelt, E., *Phys. Z.*, **32**, 182-183 (1931).
[395] Franck, S., *Z. Phys.*, **69**, 409-417 (1931).
[396] Strohhäcker, I., *Z. Physik*, **27**, 83-88 (1924).

S. Franck.[395] The first used spheres 2 cm in diameter and worked at room temperatures and pressures not exceeding 15 mm-Hg. When his observed values of V_s are plotted against px, p mm-Hg being the pressure and x cm being the length of the spark-gap, a curve continuously concave to the axis of px is obtained for air. The same is true for water-vapor if px exceeds about 2.5, corresponding to $V_s = 700$ volts approximately; but between $px = 2.5$ and $px = 1.5$, V_s changes very little, and as px is reduced below 1.5, V_s decreases rapidly. The curve for the vapor cuts that for air near the point $px = 5.8$, $V_s = 890$.

Franck used a point made by sharpening a brass rod, 10 mm in diameter, to a cone with a vertex angle of 12°, and worked at 99 °C and a pressure of 728 mm-Hg. The opposing plate was earthed. The electrodes were illuminated with ultraviolet light. When the point is negative, $V_{s-} = 23x$ kilovolts for water-vapor, and $18.4x$ kilovolts for air, the potential for the vapor exceeding that for air by 25 per cent of the latter. When the point is positive the curve connecting V_{s+} and x is concave toward the axis of x, and for a given x, V_{s+} for the vapor exceeds that for air by about 13 per cent of the latter. As read from the curve he gives, the values for water-vapor are as follows:

x	1	2	3	4 cm
V_{s+}	14	25	34	40 kilovolts

An alternating voltage of 50 cycles/sec gave for the vapor almost exactly the same curve as was obtained with the point positive.

24. Glow Discharge in Water-vapor

(For the voltage and the field-strength at which conduction begins, see Section 23.)

Terms and Symbols.

Six distinct regions can be readily distinguished in the typical electrical discharge through a gas at low pressure and contained in a long tube with an electrode at each end. They will be designated as follows, beginning at the cathode:

1. Cathode glow. A thin layer of velvety glow closely adhering to the cathode.
2. Crookes dark space. A nonluminous region which increases in length as the pressure is decreased.
3. Negative column. A luminous region which is often called the negative glow. In this region and near the end that adjoins the Crookes dark space the electric field along the discharge becomes very weak, essentially zero as compared with the over-all voltage on the tube.
4. Faraday dark space.
5. Positive column. A luminous region, in many cases striated, extending from the Faraday dark space, and by the shortest path, to the anode.

6. **Anode glow.** A thin glow closely adhering to the anode.

(a) If the distance between the electrodes is progressively decreased, the pressure of the gas and the over-all voltage (V) on the tube remaining unchanged, the positive column suddenly vanishes when that distance has been reduced to a certain value (D), and at the same time V suddenly decreases by an amount ΔV which is generally equal to the ionization potential of the gas.

(b) If d_c is the distance from the cathode to the boundary between the Crookes dark space and the negative column, d_c depends upon the pressure and V. If it exceeds a certain value $(d_c)'$ (about 0.7 mm in all cases), the application of a transverse magnetic field reduces it, but never below the value $(d_c)'$.

(c) The distance (d_0) from the cathode to the point in the negative column at which the strength of the field becomes essentially zero is somewhat greater than d_c, though the two distances are often confounded.

(d) The excess (V_-) of the potential at d_0 above that at the cathode is known as the cathode drop in potential. It does not differ greatly from the over-all voltage (V).

Table 71.—Vanishing of the Positive Column in Water-vapor

Adapted from A. Günther-Schulze [407] with the correction of certain arithmetical errors.

The increase in pD with p, when $V = 415$ volts and p exceeds 1.47 mm-Hg, is attributed to the heating that occurs at those higher pressures.

D mm = Distance between the electrodes when the column just vanishes (§ a); p mm-Hg = pressure of the vapor; V volts = over-all difference in the potentials of the electrodes.

\multicolumn{4}{c}{$V = 415$ volts}	\multicolumn{4}{c}{$p = 1.30$ mm-Hg}
p	D	pD	$10^5 D/V^2$	V	D	pD	$10^5 D/V^2$
6.75	4.10	27.7	2.38	427	11.4	14.8	6.25
6.16	4.27	26.3	2.48	496	14.1	18.3	5.74
4.75	4.83	23.0	2.80	597	21.7	28.2	6.09
3.55	5.00	17.8	2.90	681	31.2	40.6	6.74
2.37	7.00	16.6	4.06	800	40.8	53.0	6.37
1.47	9.40	13.8	5.46			Mean	6.24
1.04	14.6	15.2	8.98				
0.828	15.2	12.6	8.83				
0.560	25.2	14.1	14.63				

Mean of last 4 = 13.9

Wherever in the remainder of this section it seems desirable to refer to one of the preceding paragraphs it will be done by writing in parentheses the number or letter designating the paragraph, preceded by the sign §; thus (§ c) will refer to paragraph (c) above.

[397] Bonhoeffer, K. F., and Pearson, T. G., *Z. physik. Chem. (B)*, **14**, 1-8 (1931).
[397a] Brewer, A. K., and Kueck, P. D., *Phys'l Chem.*, **38**, 889-900, 1051-1059 (1934).
[398] Emeléus, K. G., and Lunt, R. W., *Trans. Faraday Soc.*, **32**, 1504-1512 (1936) (for Part II see Lunt).

Qualitative Relations.

In the glow discharge, water-vapor is dissociated and gives reactions characteristic of, but more active than, atomic hydrogen, reducing metallic salts not reduced by the latter.[379a] The belief of Urey and Lavin [379a] that the OH molecule is present has been confirmed. It is present and excited, and

Table 72.—Effect of a Magnetic Field upon the Crookes Dark Space in Water-vapor [408]

d_c = distance from the cathode to the farther end of the Crookes dark space (see §§ b and c); p = pressure of the vapor; H = strength of the magnetic field transverse to the discharge.

Unit of p = 1 mm-Hg; of d_c = 1 mm; of H = 1 gauss.

$p \rightarrow$ / H	1.00	3.00 (d_c)	19.23
0	7.41	2.30	0.81
120	3.77	—	—
240	2.51	—	—
360	1.55	1.54	0.61
720	0.80	0.95	0.62
1200	0.65	0.60	0.60

Table 73.—Rate of Dissociation of Water-vapor in the Glow Discharge [405]

I = current; v = volume of uncondensable gases produced per second (neither the pressure nor the temperature is stated); x = distance between the electrodes, which were plates 3 cm in diameter, the tube being 6 cm in diameter and 17 cm long. Observations were made at room temperature and a pressure of 0.75 mm-Hg.

v is not linear in I. Linder assumed that each H_2 in the gas corresponds to the dissociation of one H_2O. There was a slight deficiency of O_2, only 30.0 to 30.4 vol. per cent being found.

Unit of I = 1 ma = 0.001 ampere; of v = 1 mm³/sec; of x = 1 cm.

$x \rightarrow$ / I	1.0	1.9	3.0	4.3	7.75	9.75
			(v)			
1.0	0.239	0.159	0.166	0.150	0.173	—
1.5	—	—	—	—	—	0.239
2.0	—	—	—	—	—	0.343
3.0	0.465	0.531	0.520	0.510	0.531	0.571
5.0	0.717	0.822	0.916	0.862	0.929	0.909
10.0	1.430	1.725	1.900	1.860	{2.020 / 1.925}	2.230
15.0	2.58₅	2.88	3.03	2.98	—	3.34
17.0	—	—	—	—	3.61	—
20.0	3.42	3.93₅	4.14	4.15	—	{4.64 / 4.36}
25.0	4.42	4.77	4.87	5.00	—	{5.97 / 5.71}
26.0	—	—	—	—	5.53	—
28.0	—	—	—	5.62	—	—

24. VAPOR: GLOW DISCHARGE

possesses an excessive amount of rotation.[333] It can be detected spectroscopically for at least 1/8 second after the discharge stops [334]; but the unexcited OH molecule has a very short life, of the order of 0.001 sec.[397] The chemical reactions that occur in the glow discharge are numerous,

Table 74.—Distribution of Electrons in the Crookes Dark Space in Water-vapor [405]

I = current, n = total number of electrons per second that cross a transverse section of the Crookes dark space at a distance x from the cathode, n_0 = total number of electrons per second that leave the cathode, d_0 = distance from cathode to place of zero field (see § c). Observations at room temperature and a pressure of 0.75 mm-Hg; electrodes, plates 3 cm in diameter in tube 6 cm in diameter; distance between electrodes = 3.5 cm.

Unit of I = 1 ma = 0.001 ampere; of d_0 and x = 1 mm.

$I \rightarrow$ $d_0 \rightarrow$ x	1 9.6	3 7.5	5 6.9	10 6.1 — n/n_0 —	15 5.9₅	20 5.8₅	25 5.6
1	1.96	2.07	2.26	2.22	2.43	2.25	2.34
2	3.77	4.36	4.39	4.86	5.42	5.21	5.50
3	6.82	8.42	9.35	10.50	11.80	11.30	12.60
4	12.3	16.0	18.6	20.0	23.7	24.0	27.1
5	20.7	28.0	31.0	35.5	43.3	45.0	51.5
5.3							57.0
6.0	32.8	45.0	49.0	52.0			
6.5			56.0				
7.0	51.0	59.0					
8.0	68.8						
9.0	79.0						
d_0	72.0[a]	59.0	60.0	53.0	63.3	65.2	63.5

[a] This is obviously too small.

complex, and very sensitive to slight changes in the attendant conditions. A consideration of them lies beyond the scope of this compilation. Some of them are considered in the papers already mentioned; others in the following recent ones that have happened to come to the author's attention. In them references will be found to earlier work: A. K. Brewer and P. D. Kueck,[397a] K. G. Emeléus and R. W. Lunt,[398] K. H. Geib and P. Harteck,[399] K. H. Geib,[400] V. Kondratjew and M. Ziskin,[401] R. W. Lunt,[402] E. J. B. Willey.[403]

Although the spectrum of the positive column in stationary water-vapor is the same throughout its length, H. O. Kneser [404] has observed that if the

[399] Geib, K. H., and Harteck, P., *Idem*, **30**, 131-134, 140-141 (1934).
[400] Geib, K. H., *J. Chem'l Phys.*, **4**, 391 (L) (1936).
[401] Kondratjew, V., and Ziskin, M., *Acta Physicochim. URSS*, **5**, 301-324 (1936).
[402] Lunt, R. W., *Trans. Faraday Soc.*, **32**, 1691-1700 (1936) (for Part I see Emeléus).
[403] Willey, E. J. B., *Idem*, **30**, 230-245-246 (1934).
[404] Kneser, H. O., *Ann. d. Physik (4)*, **79**, 585-596 (1926).
[405] Linder, E. G., *Phys. Rev. (2)*, **38**, 679-692 (1931).
[406] Günther-Schulze, A., *Z. Physik*, **23**, 334-336 (1924).
[407] Günther-Schulze, A., *Z. Physik*, **30**, 175-186 (1924).

vapor is streaming along the line of discharge, then, under suitable conditions, the spectrum of the positive column varies progressively from one end to the other, the intensity of the water bands, as compared with those of the Balmer lines and the continuous spectrum of hydrogen, decreasing

Table 75.—Dissociation of Water-vapor in the Glow Discharge
See also Table 73. Adapted from E. G. Linder.[405]

I = current, d_0 = distance from cathode to zero field (see § c), V_- = cathode drop in potential (see § d), n_1 = number of electrons per second crossing the transverse section of the discharge at the distance d_0 from the cathode, $n_1 W_1$ = total energy of those n_1 electrons, $n_1 N$ = number of H_2 molecules formed per second, $k = W_1/N$. As k is found to be essentially constant, it follows that the number of molecules dissociated per second per electron entering the region of zero field is essentially proportional to the mean energy of those electrons.

Room temperature, $p = 0.75$ mm-Hg, electrodes were plates 3 cm in diameter and 3.5 cm apart in tube 6 cm in diameter.

Unit of $I = 1$ ma $= 0.001$ ampere; of $d_0 = 1$ mm; of $V_- = 1$ volt; of $W_1 = 1$ volt-electron/electron; of $N = 1 H_2$ per electron; of $k = 1$ electron-volt per H_2; of $(V_- - W_1)/V_- = 1$ per cent of V_-.

I	d_0	V_-	W_1	N	k	$\frac{V_- - W_1}{V_-}$
1	9.6	302	53.8	4.78	11.2	82.2
3	7.5	325	55.5	5.18	10.7	82.8
5	6.9	343	56.8	5.30	10.7	83.1
10	6.1	385	63.3	5.67	11.2	83.6
15	5.9₅	435	64.7	6.03	10.7	85.2
20	5.8₅	500	70.5	6.22	11.3	85.9
25	5.6	554	79.2	5.90	13.4	85.7

Table 76.—Current-voltage Relation for Water-vapor between a Point and a Plane
Read from curves given by S. Franck.[395]

Data for plate electrodes and low pressure are in Table 75.

Point was conical, vertex angle 12°, base 10 mm in diameter; plate was earthed; temperature 100 °C, pressure 725 mm-Hg.

I = current; x = distance from point to plane; V_+, V_- = potential of the point above, below, that of the plane.

Unit of $I = 1$ microampere; of $x = 1$ cm; of $V = 1$ kilovolt.

$x \rightarrow$	1		2.5		5.0		10	
I	V_+	V_-	V_+	V_-	V_+	V_-	V_+	V_-
0ᵃ	5.1₅	4.6	5.9	5.3	5.9			
10	6.9	6.8	10.2	10.6	14.5	13.6	17.9	17.0
20	8.2	8.8	13.1	14.0	19.2	18.5	23.2	24.0
30	9.4₅	10.3	15.6	16.4₅	22.1	22.1		
40	10.5₅	11.2		18.2		25.0		

ᵃ These values for $I = 0$ do not, but should, agree with those given in Table 70. The two sets of curves from which the values were read seem to be definitely discordant.

25. WATER: MOLECULAR DATA

rapidly as the vapor advances along the tube. He regarded this as an indication that the emitters of these bands (probably OH ions) dissociate under the continued action of the discharge.

E. G. Linder [405] finds that almost all the dissociation of water-vapor occurs very near the cathode, probably within the distance d_0 (see § c). Hence, the actual amount of dissociation of OH ions in the positive column must be a very small part of the total dissociation of the vapor in the tube.

Numerical Data.

A. Günther-Schulze [406] has concluded that 85 per cent of the energy delivered to the cathode comes from the cathode drop in potential. This is confirmed by the more recent work of E. G. Linder.[405]

Günther-Schulze [407] has concluded also that the thickness (d_+) of the anode glow is independent of the material used as anode and of the strength of the current. At such low pressures that d_+ is not greater than a small multiple of the mean free path of an electron, d_+ decreases as the pressure is increased; at higher pressures, d_+ is almost independent of the pressure. He gives the following values for water-vapor at room temperature:

p	1.11	2.66	3.55	14.45	mm-Hg
d_+	1.69	1.43	0.64	0.30	mm

The energy expended in producing the dissociation accompanying a glow discharge in water-vapor is 11 electron-volts per molecule of H_2O ionized (= 254 kcal/gfw = 1060 kj/gfw).[405]

IIB. WATER

25. MOLECULAR DATA FOR WATER

In this section are considered the structure of the liquid and the properties, or apparent properties, of the individual molecules, each regarded as identical with the molecule of the vapor.

Structure of Water.

The structure of the liquid will be considered under two heads: (1) *Linkage,* or association in its broadest sense, covering all types of bonding, temporary or permanent, between adjacent molecules of the vapor type; and (2) *architecture,* or the relative arrangement of the individual (vapor) molecules in a typical volume of the liquid or in a typical associated group, as the case may be.

Linkage of molecules.—On the basis of the kinetic theory of matter it is to be expected that collisions between dipole molecules will, in at least some cases, result in the colliding molecules remaining for a longer or shorter time close together and mutually oriented in a preferred manner, as a result of attraction between the dipoles; as has been suggested by

---
[408] Günther-Schulze, A., *Z. Physik,* **24**, 140-147 (1924).

J. Malsch [1] and others. While so bonded they act as a single unit. These associated molecules may consist of two or more simple (vapor) molecules, and may be themselves bonded more or less strongly with other molecules, associated or simple, if all the molecules are crowded rather closely together, as in the case of a liquid. This bonding together of associated molecules may result in the formation throughout the liquid of numerous mutually independent groups of many molecules (hundreds or thousands), all in any one group having at any instant a similar orientation, but each group undergoing a relatively slow but continuous change in its personnel, size, and orientation. This has been called the cybotactic state.[2] Or, in the extreme case, each simple (vapor) molecule may be bonded to its nearest neighbors, and they to theirs, so that the entire volume of liquid forms, in a certain sense, a single loosely bonded molecule, the several individual bonds being continually broken and replaced by others. The distance between adjacent molecules will vary from point to point in the liquid. Where they are very close together the attraction between them will tend to keep them together with a preferred mutual orientation, thus welding them, at least temporarily, into an associated molecule; and such molecules may in turn be built temporarily into a definite architectural form characteristic of the substance. Thus the entire volume of the liquid may at any instant be quasicrystalline, the direction of the crystal axes varying from point to point, even over minute distances, and the entire picture changing from instant to instant.

All these views have been held. The last, the quasicrystalline concept, seems to be the most favored at present, especially by those engaged in the x-ray study of the structure of substances. It should be remembered that the duration of the structures so revealed need not be many times greater than the period of the x-rays used, provided that the structures are being continually renewed in form, but not necessarily either in the same place or with the same orientation. That period, of the order of 10^{-18} sec, is exceedingly short as compared with the mean time between molecular encounters in the liquid (even if the free path between encounters were as short as 0.001 of the mean distance between the centers of adjacent molecules, that time would be of the order of 10^{-15} sec). Consequently, the "structure" so revealed may be nothing more than an indication of the most frequent type of collision or of association, whether temporary or permanent.

Our knowledge and inferences regarding the molecular groupings and bondings to be found in water have been reviewed and discussed in a Symposium on the Constitution of Water [3] and by H. M. Caldwell,[4] G. G. Longinescu,[5] O. Redlich,[6] and T. C. Barnes and T. L. Jahn [7]; and more

---

[1] Malsch, J., *Ann. d. Physik (5)*, **29**, 48-60 (1937).
[2] Stewart, G. W., *Phys. Rev. (2)*, **35**, 726-732 (1930).
[3] Symposium, *Trans. Faraday Soc.*, **6**, 71-123 (1910).
[4] Caldwell, H. M., *Chem. Rev.*, **4**, 375-398 (1927), bibliography of over 125 titles.
[5] Longinescu, G. G., *Idem*, **6**, 381-418 (1929).
[6] Redlich, O., *Monatsh. Chem.*, **53-54**, 874-887 (1929).

briefly in a section of R. Kremann's article *Wasser*.[8] The structure of liquids in general is the subject of a paper by J. Frenkel,[9] of a symposium before the Faraday Society,[10] and of a paper by K. F. Herzfeld.[11] Individual views have been expressed by many, some of whom are mentioned in the following paragraphs. Both Caldwell and Kremann, especially the latter, have considered more particularly the evidence derived from various criteria for normality, such as, depression of the freezing point, elevation of the boiling point, ratio of the absolute temperature of the normal boiling point to that of the critical point, ratio of the molecular heat of vaporization to the absolute temperature of the normal boiling point (the increase in entropy on vaporization at the pressure of one normal atmosphere), the temperature coefficient of the molecular surface energy, etc.

Many types of observation (see Table 77) and in particular those showing that the behavior of water differs in very many respects from that of the large number of liquids commonly described as normal, suggest that water is not a simple substance, but is an equilibrium mixture of two or more interconvertible types, or groupings, of molecules, the composition of the mixture varying with the temperature and the pressure. Such variation is usually regarded as gradual, but E. J. M. Honigmann[12] is inclined to regard it as proceeding discontinuously.

The constituents of the mixture may conceivably be either isomeric or polymeric or both. Or they may be merely portions of the liquid in which the molecules are temporarily more closely packed or more uniformly oriented or both, than in adjacent portions. Or the abnormality may result from a uniform packing different from that in normal liquids. All these views have been advanced.[13] At present, explanations in terms of packing and orientation seem to be the most favored,[14] but the evidence is conflicting. For example, J. W. Ellis reports spectroscopic data which indicate a much closer union than any that can fairly be regarded as arising from a mere packing of the molecules; whereas Stewart draws the opposite conclusion from x-ray data.

Most of those who regard the abnormalities of water as arising from an association of molecules—from the presence of groups containing only a few molecules of H_2O each—seem to accept the opinion set forth in detail

[7] Barnes, T. C., and Jahn, T. L., *Quart. Rev. Biol.*, **9**, 292-341 (1934).
[8] Doelter, C., *Handb. d. Mineralchem.*, **3**, 855-915 (1918).
[9] Frenkel, J., *Acta Physicochim. URSS*, **3**, 633-648, 913-938 (1935).
[10] Symposium, *Trans. Faraday Soc.*, **33**, 1-282 (1937) → *Nature*, **139**, 272-274 (rev.) (1937).
[11] Herzfeld, K. F., *J. Appl. Phys.*, **8**, 319-327 (1937), bibliography of 42 titles.
[12] Honigmann, E. J. M., *Naturwissenschaften*, **20**, 635-639 (1932).
[13] Longinescu, G. G., *Chem'l Rev.*, **6**, 381-418 (1929); Smyth, C. P., *Idem*, **6**, 549-587 (1929); Stewart, G. W., *Phys. Rev. (2)*, **43**, 1426 (A) (1930); **37**, 9-16 (1931); *Indian J. Phys.*, **7**, 603-615 (1932); Honigmann, E. J. M., *Die Naturwiss.*, **20**, 635-638 (1932); Bernal, J. D., and Fowler, R. H., *J. Chem'l Phys.*, **1**, 515-548 (1933); Fowler, R. H., and Bernal, J. D., *Trans. Faraday Soc.*, **29**, 1049-1056 (1933).
[14] Ellis, J. W., *Phys. Rev. (2)*, **38**, 693-698 (1931) → **38**, 582 (A) (1931); Herzfeld, K. F., *J. Appl. Phys.*, **8**, 319-327 (1937).
[15] Röntgen, W. C., *Ann. d. Physik (Wied.)*, **45**, 91-97 (1892).

by W. C. Röntgen,[15] but even then already old,* that water contains ice, that it is a saturated solution of ice in a liquid composed of simpler molecules. The most widely accepted assumption of this type seems to be that water is a mixture of $(H_2O)_2$ and $(H_2O)_3$, water-vapor not too near saturation being (H_2O), and the common type of ice (ice-I), not too near its melting point, being $(H_2O)_3$. In 1900, W. Sutherland[17] showed that many of the exceptional properties of water can be quantitatively accounted for on that assumption, and he derived the values of those properties that must be individually assigned to $(H_2O)_2$ and $(H_2O)_3$ in order to account for the observed values for water. These are given in Table 78. He proposed the names hydrol, dihydrol, and trihydrol, respectively, for the polymers (H_2O), $(H_2O)_2$, and $(H_2O)_3$; these names are now quite generally used.

More recently, the idea that hydrol is present in water, especially at temperatures approaching the boiling point, has been advanced. This idea, that water contains hydrol (steam) in solution, is an essential part of H. L. Callendar's theory.[18] And it seems to be involved in the hypothesis by which C. Barus[19] sought to explain certain observations on the evaporation of very small drops near room temperature; *viz.*, that water contains two constituents, one more volatile than the other.

But not a few who accept Röntgen's general suggestion take exception to Sutherland's interpretation; there is no general agreement as to the sizes of the individual groups of associated molecules. For example, G. B. B. M. Sutherland[20] accepts $(H_2O)_2$ but no higher polymer; C. Gillet[21] and H. E. Armstrong[22] have suggested that water is a mixture of (H_2O) and $(H_2O)_2$; H. H. Vernon[23] proposed $(H_2O)_2$ and $(H_2O)_4$; L. Schames[24] and H. Auer[25] preferred $(H_2O)_3$ and $(H_2O)_6$, the first regarding ice as a mixture of $(H_2O)_6$ and $(H_2O)_{12}$. The more highly associated molecule of water was regarded as $(H_2O)_6$ by S. W. Pennycuick[26]; as $(H_2O)_6$ or $(H_2O)_9$ by G. Tammann[27]; as $(H_2O)_6$, $(H_2O)_9$, $(H_2O)_{12}$, or $(H_2O)_{23}$ by J. Duclaux.[28] Some are noncommittal, *e.g.*, C. S. Hudson,[29] A. Kling and A. Lassieur.[30]

* Röntgen's suggestion had been foreshadowed for at least 12 years. H. A. Rowland, in commenting upon the fact that the specific heat of water is a minimum at about 30 °C, had written: "However remarkable this fact may be, ... it is no more remarkable than the contraction of water to 4°. Indeed, in both cases the water hardly seems to have recovered from freezing. The specific heat of melting ice is infinite. Why is it necessary that the specific heat should instantly fall, and then recover as the temperature rises? Is it not more natural to suppose that it continues to fall even after the ice is melted, and then to rise again as the specific heat approaches infinity at the boiling point?"[16]

[16] Rowland, H. A., *Proc. Am. Acad. Arts Sci.*, **15**, 75-200 (131) (1880); "Physical Papers," 343-468 (398-399) (1902).
[17] Sutherland, W., *Phil. Mag. (5)*, **50**, 460-489 (1900).
[18] Callendar, H. L., *Engineering (London)*, **126**, 594-595, 625-627, 671-673 (1928); see also Jakob, M., *Idem*, **132**, 651-653, 684-686, 707-709 (1931); Awbery, J. H., *Rep. Prog. Phys. (Phys. Soc. London)*, 161-197 (1934).
[19] Barus, C., *Am. J. Sci. (4)*, **25**, 409-412 (1908).
[20] Sutherland, G. B. B. M., *Proc. Roy. Soc. (London) (A)*, **141**, 535-549 (1933).

Neither is there agreement as to whether these more associated molecules exist as free individuals or form the blocks of a larger structural unit. Many regard them as free, but many others think that they are combined to form microcrystals of ice. H. Schade,[31] and H. Schade and H. Lofert [32] have advanced the idea that the polymers in water approach the size of colloid particles. But the light scattering units of which they are speaking may be merely regions where the molecules, as a result of their thermal agitation, are more closely packed than normal (see Section 39, Rayleigh scattering). H. T. Barnes [33] seems to think that as 0 °C is approached the $(H_2O)_8$ molecules in the water clump together to form invisible colloidal particles of ice; others think that they are temporarily associated into ill-defined groups of hundreds or thousands, all in any one group having essentially the same orientation,—the cybotactic state proposed by G. W. Stewart [34]; and still others think that all the molecules are more or less closely bonded together into one quasicrystalline mass.

Of the quasicrystalline theories, three types have been proposed. One requires a fairly rigid structure; this was proposed by J. D. Bernal and R. H. Fowler [35] and in a modified form by M. L. Huggins at about the same time.[36] The other two types of theory are based on structures that are loose and very transient,[37] and will be considered in the paragraphs devoted to the architecture of water. Here it suffices to state that the Bernal-Fowler theory seems to be preferred at present. It is accepted, either directly or in a slightly modified form, by the workers here noted,[38-46] many of whom advance arguments in its favor.

[21] Gillet, C., *Bull. Soc. Chim. Belg.*, **26**, 415-418 (1912).
[22] Armstrong, H. E., *Compt. rend.*, **176**, 1892-1894 (1923).
[23] Vernon, H. H., *Phil. Mag. (5)*, **31**, 387-392 (1891).
[24] Schames, L., *Ann. d. Physik (4)*, **38**, 830-848 (1912).
[25] Auer, H., *Idem (5)*, **18**, 593-612 (1933).
[26] Pennycuick, S. W., *J. Phys'l Chem.*, **32**, 1681-1696 (1928).
[27] Tammann, G., *Z. physik. Chem.*, **84**, 293-312 (1913); *Z. anorg. allgem. Chem.*, **158**, 1-16 (1926).
[28] Duclaux, J., *Compt. rend.*, **152**, 1387-1390 (1911); *J. chim. phys.*, **10**, 73-100 (1912).
[29] Hudson, C. S., *Phys. Rev.*, **21**, 16-26 (1905).
[30] Kling, A., and Lassieur, A., *Compt. rend.*, **177**, 109-111 (1923).
[31] Schade, H., *Z. Chem. Ind. Koll. (Koll. Z.)*, **7**, 26-29 (1910).
[32] Schade, H., and Lofert, H., *Koll. Z.*, **51**, 65-71 (1930).
[33] Barnes, H. T., in Alexander, J., "Colloid Chemistry," Vol. 1, pp. 435-443, Chemical Catalog Co., Inc. (Reinhold Publishing Corp.), 1926.
[34] Stewart, G. W., *Phys. Rev. (2)*, **35**, 1426 (A) (1930); **37**, 9-16 (1931); *Indian J. Phys.*, **7**, 603-615 (1932).
[35] Bernal, J. D., and Fowler, R. H., *J. Chem'l Phys.*, **1**, 515-548 (1933); see also Fowler, R. H., and Bernal, J. D., *Trans. Faraday Soc.*, **29**, 1049-1056 (1933); Bernal, J. D., *Idem*, **33**, 27-40-45 (1937): note E. Bauer's remarks (p. 43) about the last.
[36] Huggins, M. L., *J. Phys'l Chem.*, **40**, 723-731 (1936).
[37] Katzoff, S., *J. Chem'l Phys.*, **2**, 841-851 (1934); Warren, B. E., *J. Appl. Phys.*, **8**, 645-654 (1937).
[38] Ananthakrishnan, R., *Proc. Indian Acad. Sci.*, **2**, 291-302 (1935).
[39] Cartwright, C. H., *Nature*, **135**, 872 (L) (1935); **136**, 181 (L) (1935); *Phys. Rev. (2)*, **49**, 470-471 → 421 (A) (1936).
[40] Cross, P. C., Burnham, J., and Leighton, P. A., *J. Am. Chem. Soc.*, **59**, 1134-1147 (1937).
[41] Debye, P., *Chem'l Rev.*, **19**, 171-182 (1936).

V. Danilow [47] departs from the Bernal-Fowler theory in some particulars; E. N. daC. Andrade [48] is more inclined to favor the cybotactic theory advanced by Stewart; and P. Girard and P. Abadie [49] think that the semi-

Table 77.—Evidence for Association

Some properties of water from which conclusions regarding an association of molecules have been drawn.

1. *Magnetic susceptibility.* The variation in the susceptibility with the temperature indicates the presence of at least two types of molecule, the more polymerized decreasing in number as the temperature rises.[51-57]

2. *Spectrum.* The infrared spectrum of water indicates the presence of at least two types of molecule, their relative numbers varying with the temperature.[58-60] At one time, J. W. Ellis and B. W. Sorge [61] seemed inclined to accept the idea of polymerization, but later E. L. Kinsey and J. W. Ellis [62] were undecided, rather inclining toward the quasicrystalline theory as propounded by Bernal and Fowler.

3. *Raman effect: Scattering.* It has been held that the Raman spectrum of water also indicates the presence of at least two types of molecule, their relative numbers varying with the temperature.[63-69]

I. R. Rao concluded that there are three types of polymer involved (see Table 79). At one time, M. Magat [70] seemed to favor the polymer theory, but he later decided [71] that such a view is untenable, and that a quasicrystal-

---

[42] Errera, J., *J. de Chim. Phys.*, **34**, 618-626 (1937).
[43] Herzfeld, K. F., *J. Appl. Phys.*, **8**, 319-327 (1937).
[44] Kinsey, E. L., and Ellis, J. W., *Phys. Rev. (2)*, **49**, 105 (L) → 209 (A) (1936).
[45] Magat, M., *Ann. de Phys. (11)*, **6**, 108-193 (1936); *Trans. Faraday Soc.*, **33**, 114-120 (1937).
[46] Rehner, J., Jr., *Rev. Sci. Inst.*, **5**, 2-3 (1937).
[47] Danilow, V., *Acta Physicochim. URSS*, **3**, 725-740 (1935).
[48] Andrade, E. N. daC., *Phil. Mag. (7)*, **17**, 698-732 (1934).
[49] Girard, P., and Abadie, P., *Compt. rend.*, **202**, 398-400 (1936).
[50] Ukholin, S. A., *Compt. rend. Acad. Sci. URSS*, **16**, 395-398 (1937).
[51] Piccard, A., *Compt. rend.*, **155**, 1497-1499 (1912); *Arch. sci. phys. et nat. (4)*, **35**, 209-231, 340-359, 458-482 (1913).
[52] Cabrera, B., and Duperier, A., *J. de Phys. (6)*, **6**, 121-138 (1925).
[53] Mathur, R. N., *Indian J. Phys.*, **6**, 207-224 (1931).
[54] Johner, W., *Helv. Phys. Acta*, **4**, 238-280 (1931).
[55] Auer, H., *Ann. d. Physik (5)*, **18**, 593-612 (1933).
[56] Azim, M. A., Bhatnagar, S. S., and Mathur, R. N., *Phil. Mag. (7)*, **16**, 580-593 (1933).
[57] Cabrera, B., and Fahlenbrach, H., *An. Soc. Esp. Fis. y Quim.*, **32**, 525-537 (1934).
[58] Collins, J. R., *Phys. Rev. (2)*, **20**, 486-498 (1922); **26**, 771-779 (1925).
[59] Plyler, E. K., *J. Opt. Soc. Amer.*, **9**, 545-555 (1924).
[60] Ganz, E., *Ann. d. Physik (5)*, **28**, 445-457 (1937).
[61] Ellis, J. W., and Sorge, B. W., *J. Chem'l Phys.*, **2**, 559-564 (1934).
[62] Kinsey, E. L., and Ellis, J. W., *Phys. Rev. (2)*, **49**, 105 (L), 209 (A) (1936).
[63] Gerlach, W., *Naturwissenschaften*, **18**, 68 (L) (1930).
[64] Meyer, E. H. L., *Physik. Z.*, **31**, 510-511 (1930).
[65] Specchia, O., *Nuovo Cim. (N. S.)*, **7**, 388-391 (1930).
[66] Segrè, E., *Atti Accad. Linc. (6)*, **13**, 929-931 (1931).
[67] Sutherland, G. B. B. M., *Proc. Roy. Soc. (London) (A)*, **141**, 535-549 (1933).
[68] Rao, I. R., *Nature*, **132**, 480 (1933); *Proc. Roy. Soc. (London) (A)*, **145**, 489-508 (1934); *Phil. Mag. (7)*, **17**, 1113-1134 (1934).

Table 77.—(Continued)

line structure is required. R. S. Krishnan [72] could find in pure liquids no evidence of such large molecular clusters—comparable in size to a wavelength of light—as had been postulated by Plotnikow,[73] but he did find such clusters in certain binary liquid mixtures.

4. *Index of refraction.*[74]

5. *Color.* J. Duclaux [75] has suggested that the proportion of ice-molecules contained in water might be determined from a study of the thermal variation in the color. And H. T. Barnes [76] has stated that "just at the freezing point the color of the St. Lawrence River changes rapidly and old river men can tell the approach of the ice-forming period by the color."

6. *Diffraction of x-rays.* The diffraction pattern obtained when x-rays are passed through water indicates that the water has a quasicrystalline character, changing with the temperature; but opinions differ as to whether the crystalline character refers to small groups of molecules or to large.[77-84] For numerical data, see Tables 155 and 156.

7. *Dielectric constant.*[85, 86]

8. *Surface tension.*[87]

9. *Pressure-volume-temperature correlations.* G. Tammann [88] has pointed out that water should contain at least as many types of molecule as there are types of ice that can exist in equilibrium with it. There were four such types then known that are stable (see Figure 13). Of these he concluded that III, V, and VI are isomers, all arising from molecules of the type $(H_2O)_3$. He gives reasons for believing that ice-I is either $(H_2O)_9$, breaking up into $9(H_2O)$, or is $(H_2O)_6$, breaking up into $2(H_2O)_3$. From Bridgman's observations of the behavior of water under pressure, Tammann derived the values credited to him in Table 79. In the same table are the values derived by Jazyna and by Yoshida from the thermal expansion of water; they are much smaller than any of the others.

[69] Rao, C. S. S., *Proc. Roy. Soc. (London) (A)*, **151**, 167-178 (1934).
[70] Magat, M., *J. de Phys. (7)*, **5**, 347-356 (1934).
[71] Magat, M., *Ann. de Phys. (11)*, **6**, 108-193 (1936); *Trans. Faraday Soc.*, **33**, 114-120 (1937).
[72] Krishnan, R. S., *Proc. Indian Acad. Sci.*, **1**, 211-216, 915-927 (1935); **2**, 221-231 (1935).
[73] Plotnikow, J., and Splait, L., *Physik. Z.*, **31**, 369-372 (1930); Plotnikow, J., and Nishigishi, S., Idem, **32**, 434-444 (1931).
[74] Chéneveau, C., *Compt. rend.*, **156**, 1972-1974 (1913).
[75] Duclaux, J., *Rev. gén. des Sci.*, **23**, 881-887 (1912).
[76] Barnes, H. T., "Ice Engineering," p. 10, Montreal, Renouf Publishing Co. (1928).
[77] Keesom, W. H., and de Smedt, J., *J. de Phys. (6)*, **4**, 144-151 (1923); **5**, 126-128 (1924).
[78] Sogani, C. M., *Indian J. Phys.*, **1**, 357-392 (1927).
[79] Prins, J. A., *Z. Physik*, **56**, 617-648 (1929).
[80] Meyer, H. H., *Ann. d. Physik (5)*, **5**, 701-734 (1930).
[81] Debye, P., and Menke, H., *Physik. Z.*, **31**, 797-798 (1930).
[82] Good, W., *Helv. Phys. Acta*, **3**, 205-248, 436 (1930).
[83] Stewart, G. W., *Phys. Rev. (2)*, **35**, 726-732, 1426 (A) (1930); **37**, 9-16 (1931).
[84] Kinsey, E. L., and Sponsler, O. L., *Phys. Rev. (2)*, **40**, 1035-1036 (A) (1932); *Proc. Phys. Soc. (London)*, **45**, 768-779 (1933).
[85] Errera, J., *J. de Phys. (6)*, **5**, 304-311 (1924).
[86] Malsch, J., *Physik. Z.*, **33**, 383-390 (1932).
[87] Ramsay, W., and Shields, J., *J. Chem. Soc. (London)*, **63**, 1089-1109 (1893).

Table 77.—(Continued)

E. E. Walker[89] and M. F. Carroll[90] have discussed the association of liquids in the light of van der Waals' equation of state and the concept of corresponding states, and D. B. Macleod[91] has considered certain phases of it from the concept of "free space."

10. *Viscosity.* From his measurements of the viscosity of water at two temperatures and under various pressures, P. W. Bridgman[92] concluded that there is an association that decreases as the pressure is increased, vanishing at very high pressures.

11. *Rate of cooling.*[93]

Table 78.—Properties of Dihydrol and Trihydrol[17]

$$2(H_2O) = (H_2O)_2 + 6.8 \text{ kcal} = (H_2O)_2 + 28.5 \text{ kilojoules.}$$
$$3(H_2O)_2 = 2(H_2O)_3 + 19.1 \text{ kcal} = 2(H_2O)_3 + 80.0 \text{ kilojoules.}[a]$$

Property	$(H_2O)_2$	$(H_2O)_3$	Unit
Density (ρ)	1.0894	0.88	1 g/cm³
Coefficient of expansion $\left(-\dfrac{1}{\rho_0}\cdot\dfrac{d\rho}{dt}\right)$	9	2	10^{-4} (°C)$^{-1}$
Specific refractivity $\left(\dfrac{n^2-1}{n^2+2}\cdot\dfrac{1}{\rho}\right)$	0.20434	0.20968	1 cm³/g
Compressibility $\left(\dfrac{1}{\rho_0}\cdot\dfrac{d\rho}{dP}\right)$	16[a]	10 (?)	10^{-6} (atm)$^{-1}$
Surface tension	78.3	73.3	1 dyne/cm
Critical temperature	368	538	°C
Specific heat	0.8[a]	0.6[a]	1 cal/g·°C
Latent heat of fusion		16	1 cal/g
Latent heat of vaporization	257	250 ca.	1 cal/g
Viscosity (η)	3.0	38.1	10^{-3} poise
Pressure coefficient of viscosity		−34	10^{-5} (atm)$^{-1}$
Virial constant[b]	16.0	15.2	1 kiloatm·cm⁶/g²
Magnetic specific susceptibility[c]	−722.2	−701.3	10^{-9} cgsm

[a] J. Duclaux[94] concluded that the ice molecule is either $(H_2O)_9$ or $(H_2O)_{12}$, and was noncommittal regarding the formula for the "unpolymerized" water. He computed the following values, $t = 0$ °C:
  Heat of depolymerization of 1 g-molecule of dissolved ice = 4 kilocal.
  Specific heat of dissolved ice = 0.62 cal/g·°C.
  Specific heat of "unpolymerized" water = 0.99 cal/g·°C.
  Compressibility of "unpolymerized" water = 36×10^{-6} (atm)$^{-1}$.

[b] By the virial constant he means the quantity l in the equation of state $\left(p + \dfrac{l}{2v^2}\right)v$ $= RTf(v,T)$. Sutherland[95] concluded that $f(v,T) = \dfrac{25}{13}\{1 + \sqrt{T}\phi(v)\}$, whence $\dfrac{3}{4}l = v^3T\left(\dfrac{dp}{dT}\right)_v + \dfrac{25}{26}vRT - \dfrac{3}{2}pv^2$, the last term being negligible. He gives five methods for estimating the value of l for a liquid.

[c] Derived by L. Sibaiya[96] for water at 20 °C; for H₂O he finds −775.5 in the same units.

25. WATER: MOLECULAR DATA

Table 79.—Estimates of the Ice-content of Water

If the average molecule of the mixture can be represented by $(H_2O)_\beta$, β may be called the average degree of association of the mixture. O. Fruwirth [97] has tabulated 10 ways for getting β, the values ranging from 1.3 to 4.7. It is probable that β is the quantity called "degree of association" by M. A. Azim, S. S. Bhatnagar, and R. N. Mathur [98] for which they give the values 4.18 at 0 °C and 2.70 at 100 °C. It is not stated how the values were obtained. By the method of Ramsay and Shields (see p. 520), J. Timmermans and H. Bodson [99] derive $\beta = 3.0$ from their own observations, probably near room temperature.

c = total mass of all the ice molecules contained in a unit mass of water at t °C; n = number of molecules of the indicated type per 100 molecules of the mixture.

Unit of $p = 1$ kg*/cm² $= 0.968$ atm; of P $= 1$ atm.

I. Pressure = 1 atmosphere.

Ref.[a] → t	Su	Du	Ta	Re	Rao	Yo[b]	t	Miscellaneous 100c	Ref.[a]
			100c						
0	37.5	18.3	14.6	39.0	37	0.6	0	55	Sch
4					32	0.5	0	39	RC
10		14.3	10.4			0.3	0	1.7[c]	Ja
20	32.1	11.3	7.0	26.2		0.2	0	34	Jo
30		9.1	4.4			0.1	0	24	Pi
38					32		10	1.0[c]	Ja
40	28.4	7.4	2.5	15.4		0.05	20	28	RC
60	25.5	5.1		7.8		0.01	20	0.2[c]	Ja
80	23.4	3.7		3.4		0	20	30	Au
98					21		20	15	Go
100	21.7	2.8		0			20	28	Jo
							30	0[c]	Ja

Type → t	H_2O	$(H_2O)_2$ 100c (Rao)	$(H_2O)_3$	H_2O	$(H_2O)_2$ n (Rao)	$(H_2O)_3$
0	9	57	37	19	58	23
4	10	58	32	19.5	58.5	22
38	16	52	32	29	50	21
98	21	58	21	36	51	13
Ice, 0 °C	0	32	68	0	41	59

II. Variation with pressure.

p → t	1	500	1000	1500	2000	2500
				100c(Ta)		
−10	20.2	14.5	13.1	9.6	6.4	5.3
0	14.6	10.0	7.8	5.6	4.2	3.9
10	10.4	7.1	5.7	4.1	3.1	3.0
20	7.0	4.7	3.9	2.8	2.1	2.1
30	4.4	2.8	2.4	1.7	1.4	
40	2.5	1.4	1.3	0.7	0.7	0.8

[88] Tammann, G., *Z. physik. Chem.*, 84, 293-312 (1913); *Z. anorg. Chem.*, 158, 1-16 (1926).
[89] Walker, E. E., *Phil. Mag.* (6), 47, 111-126, 513-525 (1924).

Table 79.—(Continued)

$t \rightarrow$ P	0	20	40	60	80	100
			100c(Su)			
1	37.5	32.1	28.4	25.5	23.4	21.7
150	35.1	30.0	26.4	23.7	21.7	20.3
$-10^6 \Delta c/\Delta P$	160	147	133	120	107	93

Extreme estimates at 0 °C (Du).

P	0	100	500	1000	1500	2000	2500
100 c_{max}	18.3	16.9	12.2	8.1	5.4	3.6	1.6
100 c_{min}	18.3	16.8	11.8	7.3	4.0	2.2	0.5

[a] References:

Au Auer, H., *Ann. d. Physik (5)*, **18**, 593-612 (1933).
Du Duclaux, J., *J. de chim. phys.*, **10**, 73-109 (1912).
Go Good, W., *Helv. Phys. Acta*, **3**, 205-248, 436 (1930).
Ja Jazyna, W., *Z. Physik*, **58**, 429-435 (1929).
Jo Johner, W., *Helv. Phys. Acta*, **4**, 238-280 (1931).
Pi Piccard, A., *Arch. Sci. phys. et nat. (4)*, **35**, 209-231, 340-359, 458-482 (1913).
Rao Rao, I. R., *Nature*, **132**, 480 (1933); *Proc. Roy. Soc. (London) (A)*, **145**, 489-508 (1934).
RC Richards, T. W., and Chadwell, H. M., *J. Am. Chem. Soc.*, **47**, 2283-2302 (1925).
Re Redlich, O., *Monatsh. f. Chem.*, **53-54**, 874-887 (1929).
Sch Schames, L., *Ann. d. Physik (4)*, **38**, 830-848 (1912).
Su Sutherland, W., *Phil. Mag. (5)*, **50**, 460-489 (1900).
Ta Tammann, G., *Z. anorg. allgem. chem.*, **158**, 1-16 (1926).
Yo Yoshida, U., *Mem. Coll. Sci. Kyoto*, **19**, 271-277 (1936).

[b] Computed by means of his formula: $c = 0.006e^{-0.0644t}$ based on a study of the thermal variation of the density and of the specific heat.

[c] These values were derived from the thermal expansion of water.

---

crystalline state is well developed in water at room temperature, but do not commit themselves to the Bernal-Fowler theory. On the other hand, S. A. Ukholin [50] thinks that certain Bernal-Fowler interpretations of the thermal variations in the Raman spectrum of water are of quite doubtful validity.

Establishment of equilibrium.—If water is a mixture of two or more types of molecular structure, the relative proportions of the several types varying with the temperature, then the question arises: Does an appreciable time elapse after a change in temperature or in phase before the types are again in mutual equilibrium?

H. T. Barnes [100] has advanced the idea that equilibrium takes place slowly, at least at temperatures near 0 ° C, and has described experiments

---

[90] Carroll, M. F., *Idem (7)*, **2**, 385-402 (1926).
[91] Macleod, D. B., *Trans. Faraday Soc.*, **21**, 145-150, 151-159 (1925).
[92] Bridgman, P. W., *Proc. Amer. Acad. Arts Sci.*, **61**, 57-99 (1926).
[93] Vernon, H. M., *Phil. Mag. (5)*, **31**, 387-392 (1891).
[94] Duclaux, J., *J. de chim. phys.*, **10**, 73-109 (1912).
[95] Sutherland, W., *Phil. Mag. (5)*, **35**, 211-295 (241) (1893).
[96] Sibaiya, L., *Current Sci.*, **3**, 421-422 (1935).
[97] Fruwirth, O., *Sitz.-ber. Akad. Wiss. Wien, Abt. (IIb)*, **146**, 157-167 (1937) = *Monatsh. f. Chem.*, **70**, 157-167 (1937).
[98] Azim, M. A., Bhatnagar, S. S., and Mathur, R. N., *Phil. Mag. (7)*, **16**, 580-593 (1933).
[99] Timmermans, J., and Bodson, H., *Compt. rend.*, **204**, 1804-1807 (1937).
[100] Barnes, H. T., *Scientific Monthly*, **29**, 289-297 (1929).
[101] Barnes, T. C., and Jahn, T. L., *Proc. Nat. Acad. Sci.*, **19**, 638-640 (1933); *Quart. Rev. Biol.*, **9**, 292-341 (1934).

interpreted as showing that water may be temporarily deprived of the nuclei —assumed to be the ice-molecules in it—required for the formation of ice; and T. C. Barnes and T. L. Jahn [101] have reported experiments interpreted as showing that water freshly formed from steam freezes less quickly than that freshly formed by melting ice, the initial temperature and other conditions being the same for each specimen of water. The conclusion is drawn that an appreciable time is required for equilibrium to be established. In view of the uncertainty of our knowledge of the behavior of water as regards freezing, all such interpretations must be accepted with reservations. T. C. Barnes and associates [101, 102] have described experiments indicating that the growth of certain organisms in water freshly formed from steam differs from that in water freshly formed by melting ice.[102] The explanation offered attributes this difference to a difference in the ice-content of the two kinds of water; that is, these workers believe that a very appreciable time is required for the establishment of equilibrium.* Inspired by this work, A. P. Wills and G. F. Boeker [103] interpreted in the same way certain peculiar observations they obtained in their study of the magnetic susceptibility of water, but they found later that the peculiar behavior was the result of errors.[104] J. Zeleny [105] invoked such delay as an explanation of certain observations on the electrification of water-drops broken by an air-blast.

Other attempts to answer the question have led to the conclusion that equilibrium is attained very quickly, practically at once, some of the properties investigated being these:

(1) Vapor-pressure.[106-110]

(2) Magnetic susceptibility.[111-113]

(3) Infrared absorption.[114]

(4) Index of refraction.[115]

(5) Density. Using a method by which a difference of one in 10^6 in the density could be detected, M. Dole and B. Z. Wiener [116] compared the

* Such a conclusion is quite at variance with certain observations made by the compiler after this section had been written (see Section 97).

[102] Barnes, H. T. and T. C., *Nature*, **129**, 691 (1932); Barnes, T. C., *Proc. Nat. Acad. Sci.*, **18**, 136-137 (1932); Lloyd, F. E., and Barnes, T. C., *Idem*, **18**, 422-427 (1932); Barnes, T. C., and Larson, E. J., *J. Am. Chem. Soc.*, **55**, 5059 (1933).
[103] Wills, A. P., and Boeker, G. F., *Phys. Rev. (2)*, **42**, 687-696 (1932).
[104] Wills, A. P., and Boeker, G. F., *Idem*, **46**, 907-909 (1934).
[105] Zeleny, J., *Idem*, **44**, 837-842 (1933).
[106] Bonhoeffer, K. F., and Harteck, P., *Z. physik. Chem. (B)*, **5**, 293-296 (1929).
[107] West, W. A., and Menzies, A. W. C., *J. Phys'l Chem.*, **33**, 1893-1896 (1929).
[108] Wright, S. L., and Menzies, A. W. C., *J. Am. Chem. Soc.*, **52**, 4699-4708 (1930).
[109] Menzies, A. W. C., *Proc. Nat. Acad. Sci.*, **18**, 567-568 (1932).
[110] Egerton, A., and Callendar, G. S., *Phil. Trans. (A)*, **231**, 147-205 (A698) (1932).
[111] Auer, H., *Ann. d. Physik (5)*, **18**, 593-612 (1933).
[112] Cabrera, B., and Fahlenbrach, H., *Z. Physik*, **82**, 759-764 (1933).
[113] Wills, A. P., and Boeker, G. F., *Phys. Rev. (2)*, **46**, 907-909 (1934).
[114] Ellis, J. W., and Sorge, B. W., *Science (N. S.)*, **79**, 370-371 (1934).
[115] LaMer, V. K., and Miller, M. L., *Phys. Rev. (2)*, **43**, 207-208 (1933).
[116] Dole, M., and Wiener, B. Z., *Science (N. S.)*, **81**, 45 (1935).

density of freshly condensed steam with that of freshly melted ice, and found no difference.

In most cases, the conditions were not identical with those in the experiments of T. C. Barnes and his associates, water heated for a time to 100 °C being used instead of condensed steam, and water that had remained long at a much lower temperature, instead of freshly melted ice. Although this difference might be expected to reduce the magnitude of the effect sought, it would scarcely make the effect entirely negligible as compared with that corresponding to the other conditions.

G. Tammann and A. Elbrächter[117] attempted to answer the question by comparing the observed temperature change accompanying adiabatic expansion with that computed from the specific heat and the equation of state. Differences were observed, but were of such a kind that they could not be interpreted.

Other information regarding the controversy may be found in articles by T. C. Barnes,[118] W. D. Bancroft and L. P. Gould,[119] and by A. W. C. Menzies.[120] See also pp. 638, 644.

At one time H. B. Baker[121] thought that he had evidence that equilibrium between the molecular species in a pure liquid might be established slowly. However, it seems that his conclusions are not accepted.[122]

F. Vlès[123] has explained the appearance of a Tyndall cone of light scattered by "cold water from freshly melted very pure ice" as due to colloidal aggregates of trihydrol. Before this explanation can be accepted it must be shown that it did not arise from foreign particles suspended in the liquid, and dissolving slowly as the temperature rose.

Certain observations briefly reported by F. W. Gray and J. F. Cruickshank[124] as indicating that the thermal change in the magnetic susceptibility of water lags behind the temperature should be carefully studied and checked by other observers.

The Earl of Berkeley[125] has suggested that information regarding association might be obtainable from a study of changes produced in the index of refraction as the liquid is centrifuged.

Architecture of water.—It has long been known that oxygen behaves as if it had 4 valencies, two weaker than the others, and hydrogen as if it had two, one being weak (see, *e.g.*, T. M. Lowry and H. Burgess).[126] Making use of the first, H. H. Vernon[23] suggested that, between 4 and 100 °C, water consists of two H_2O units joined by a double bond between the O's; and below 4 °C of 4 H_2O units arranged in a square and united

---

[117] Tammann, G., and Elbrächter, A., *Z. anorg. allgem. Chem.*, 200, 153-167 (1931).
[118] Barnes, T. C., *Science (N. S.)*, 79, 455-457 (1934); 81, 200-201 (1935).
[119] Bancroft, W. D., and Gould, L. P., *J. Phys'l Chem.*, 38, 197-211 (1934).
[120] Menzies, A. W. C., *Science (N. S.)*, 80, 72-73 (1934).
[121] Baker, H. B., *J. Chem. Soc. (London)*, 121, 568-574 (1922); 1927, 949-958 (1927).
[122] Menzies, A. W. C., *J. Phys'l Chem.*, 33, 1893-1896 (1929); *Science (N. S.)*, 80, 72-73 (1934).
[123] Vlès, F., *Chem. Abst.*, 31, 592 (1937) ← *Arch. phys. biol.*, 13, 199-201 (1936).
[124] Gray, F. W., and Cruickshank, J. F., *Nature*, 135, 268-269 (L) (1935).
[125] Earl of Berkeley, *Nature*, 120, 840-841 (1927).

by single bonds between adjacent O's. H. E. Armstrong [127] regarded water as a mixture of H_2O ("hydrone") and of $(H_2O)_2$ formed by uniting an H and an OH, each individually, to the O of an H_2O. This type of $(H_2O)_2$ he called "hydronole." A. Kling and A. Lassieur [30] pictured water as a mixture of H_2O and a series of its polymers, regarding H_2O as existing in two isomeric forms: H—O—H, and H—OH. The suggestions of Armstrong and of Kling and Lassieur have been criticized by V. Auger.[128]

If the 4 pairs of electrons of the oxygen atom are arranged about the nucleus at the vertices of a regular tetrahedron,[129] then the simple molecule H_2O will be formed by adding an H to each of two vertices of that tetrahedron; the other two vertices may for convenience be called the electron vertices of the H_2O molecule. A pair of such tetrahedral molecules may unite, as a result of the attraction between the H vertices of one and the electron vertices of the other, to form a double molecule in any one of three ways: (a) by union at a single vertex (type V); (b) by union at the two vertices of an electron-H edge (type E); and (c) by union at the three vertices of a single face (type F), the one electron and two H vertices of a face of one uniting, respectively, with the one H and two electron vertices of a face of the other. Such tetrahedral molecules can be built up into structures that correspond to the observed x-ray pattern of ice, and serve as the basis of much of the recent discussion regarding the structure of water and of ice.

F. O. Anderegg,[130] and E. L. Kinsey and O. L. Sponsler [131] suggested a face union (F). The former regarded such double molecules as forming the chief molecular species in water, and the latter regarded them as the polymer that occurs in both ice and water, dissociation being into H^+ and $(H_3O_2)^-$.

S. W. Pennycuick[132] discussed the structure of water, especially in relation to such tetrahedral molecules, and concluded that the single vertex union (V) is the most probable. This not only yields a satisfactory crystal form, but also gives rise to rings $(H_2O)_6$ and to open chains. See also M. L. Huggins.[133]

The x-ray pattern for both ice and water indicates that, on the average, each oxygen atom has four or very nearly four others as near neighbors. Taking account of this in connection with the tetrahedral molecule, J. D.

---

[126] Lowry, T. M., and Burgess, H., *J. Chem. Soc. (London)*, **123**, 2111-2124 (1923).

[127] Armstrong, H. E., *Compt. rend.*, **176**, 1892-1894 (1923).

[128] Auger, V., *Idem*, **178**, 330-332 (1924).

[129] Smyth, C. P., *Phil. Mag.* (6), **47**, 530-544 (1924); Pennycuick, S. W., *J. Phys'l Chem.*, **32**, 1681-1696 (1928).

[130] Anderegg, F. O., *Chem. Abstr.*, **18**, 2282 (1924) ← *Proc. Indiana Acad. Sci.*, **1923**, 93-101 (1923).

[131] Kinsey, E. L., and Sponsler, O. L., *Proc. Phys. Soc. (London)*, **45**, 768-779 (1933) → *Phys. Rev.* (2), **40**, 1035-1036 (A) (1932).

[132] Pennycuick, S. W., *J. Phys'l Chem.*, **32**, 1681-1696 (1928).

[133] Huggins, M. L., *J. Phys'l Chem.*, **40**, 723-731 (1936) → *J. Am. Chem. Soc.*, **58**, 694 (L) (1936).

Bernal and R. H. Fowler [134] developed their theory of the quasicrystalline structure of water, which seems to be the favored one today. The exact crystal structures of ice and water are not determined unambiguously by the available data: there is an element of choice. They decided that three types of structures, passing continuously one into another, have to be considered: Type I is like tridymite; Type II is quartz-like; Type III is close-packed, like ammonia. Type I is ice and occurs in water at low temperatures; Type II predominates in water between 0 and 100 °C; Type III characterizes water at high temperatures, say between 150 °C and the critical point. There is no question of different kinds of molecules, but only of different arrangements of the same kind. Each small region of water has instantaneously a crystalline character, but in different regions the crystals are differently oriented, and each region is continuously changing in personnel and in crystal orientation. See also J. D. Bernal.[135]

B. E. Warren [136] does not admit as close or permanent a binding of the atoms into crystal forms as is postulated by Bernal and Fowler. He regards the crystal form as merely a kind of ideal that is more or less closely approached in water at any instant, but that is seldom, if ever, fully realized.

S. Katzoff [137] goes still further. He thinks that in water there is little if any periodicity in arrangement, and that what little there may be is entirely incidental. In his view, the important thing is the relative positions of adjacent molecules. They are probably held together in nearly the same manner as in the crystal, but except for that, the arrangement of the molecules is a random one. He found no evidence either for the definite "quartz-like" arrangement or for the extensive degree of close packing postulated by Bernal and Fowler; his observations were, in fact, incompatible with the assumption of a "quartz-like" arrangement. His proposed picture is that of a broken-down ice structure.

The earliest models placed the bonding H midway between the two O's, but it is probable that it is nearer to one than to the other, and that, of the 4 H's bonding to an O, two are nearer than the other two.[133, 134] And there are reasons for thinking that in H_2O one H is bound more firmly than the other (see L. Henry,[138] and G. Jacoby [139]).

A. Piekara [140] has pointed out that the rotating of a dipole molecule is affected by fields arising from two sources: the field due to its immediate neighbors and called by him the association field; and the resulting field arising from all the other, more distant, molecules, which he calls the Debye molecular field.

---

[134] Bernal, J. D., and Fowler, R. H., *J. Chem'l Phys.*, **1**, 515-548 (1933); Fowler, R. H., and Bernal, J. D., *Trans. Faraday Soc.*, **29**, 1049-1056 (1933).
[135] Bernal, J. D., *Trans. Faraday Soc.*, **33**, 27-40-45 (1937); and E. Bauer's remarks (p. 43).
[136] Warren, B. E., *J. Appl. Phys.*, **8**, 645-654 (1937).
[137] Katzoff, S., *J. Chem'l Phys.*, **2**, 841-851 (1934).
[138] Henry, L., *Bull. Classe Sci. Acad. Roy. Belg.*, **1905**, 377-393 (1905).
[139] Jacoby, G., *Ann. d. Physik (4)*, **72**, 153-160 (1923).
[140] Piekara, A., *Acta Phys. Polon.*, **6**, 130-143 (1937).
[141] Callendar, H. L., *Proc. Roy. Soc. London (A)*, **120**, 460-472 (1928); *Proc. Inst. Mechan. Eng.*, **1929**, 507-527 (1929).

25. WATER: MOLECULAR DATA

Miscellanea.—Both H. L. Callendar [141] and O. Maass and A. l.. Geddes [142] find that the liquid structure may persist at temperatures appreciably above the critical one—above that at which the meniscus vanishes. (*Cf.* Section 88, *Critical data.*)

Evidence thought to show that the structure of water in capillary spaces differs from that of water in bulk has been published by P. Gaubert,[143] who found birefringence in the film between two adjacent bubbles, and by É. Torporescu,[144] who observed certain voltaic effects. The observations reported by B. Derjaguin [145] as indicating that thin films of water possess a rigidity that increases as the film becomes thinner, conflict with the observations of R. Bulkley [146] and are probably to be otherwise explained, as it is not certain that Derjaguin had satisfactorily eliminated the effect of small suspended particles.

A. P. Wills and G. F. Boeker [147] infer from their magnetic measurements that "significant changes in molecular arrangement or association of the water molecules" may occur near 35 °C and near 55 °C. And M. Magat [148] concludes that many properties of water have anomalies near 40 °C, and regards these anomalies as arising from a change in the structure of water.

Dipole Moment of the Molecule of Water.

The value of the dipole moment (μ) of a single free molecule of H_2O as it occurs in water-vapor is such that $10^{19} \mu = 18.3_1$ cgse, see p. 48. A discussion of Debye's dipole theory, his formulas, and their limitations in the case of liquids may be found in Section 49.

All observations combine to show that Debye's formulas for freely reorientable dipoles, though applying to gases, are not applicable to dipole liquids, and are only approximately applicable to dilute solutions of dipole substances in nonpolar solvents. Nevertheless, those formulas have been quite generally used for determining what has been called the dipole moment of liquids. The procedure followed has been this: To the experimentally determined values of $(\epsilon - 1)/(\epsilon + 2)\rho$ is fitted an expression of the general form $a + b/T + cT$, sometimes with $c = 0$, and from this the apparent dipole moment denoted by μ_a. is determined by means of the relation $b = 4\pi N(\mu_a)^2/9Mk$. In these equations ϵ is the dielectric constant, ρ the density, T °K the absolute temperature, N the number of molecules per gram-mole (6.061 × 10^{23}), k the molecular gas constant (1.372 × 10^{-16} erg/°K·molecule), and M the formula weight, which is assumed to be the molecular weight of the molecule to which μ_a

---

[142] Maass, O., and Geddes, A. L., *Phil. Trans. (A)*, **236**, 303-332 (1937).
[143] Gaubert, P., *Compt. rend.*, **200**, 304-306, 679-680 (1935).
[144] Torporescu, É., *Idem*, **202**, 1672-1674 (1936) = *Bull. de Math. et Phys.*, Bucarest, **6**, 40-41 (1936).
[145] Derjaguin, B., *Z. Physik*, **84**, 657-670 (1933); *Phys. Z. Sowj.*, **4**, 431-432 (1933).
[146] Bulkley, R., *Bur. Stand. J. Res.*, **6**, 89-112 (RP264) (1931).
[147] Wills, A. P., and Boeker, G. F., *Phys. Rev. (2)*, **46**, 907-909 (1937).
[148] Magat, M., *J. de Phys. (7)*, **6**, 179-181 (1935); *Trans. Faraday Soc.*, **33**, 114-120 (1937).

refers. This quantity (μ_a) is what has been generally called the dipole moment of the liquid, and has been denoted by μ. Here it will be called the apparent dipole moment, and will be denoted by μ_a, μ being used to denote the dipole moment of a molecule of the gas phase. The ratio $(\mu_a/\mu)^2$ is the factor by which the theoretical value of b for freely reorientable molecules must be multiplied in order to obtain the value appropriate to the liquid. Its value is commonly indicated by that of μ/μ_a, the square root of its reciprocal.

The value to be preferred at present for the coefficient b for water is 107.13 °K·cm³/g (see Table 175),

whence $10^{19} \mu_a = 5.59$ cgse units per gfw-H$_2$O
or $\mu_a = 0.117_1$ electron·angstroms per gfw-H$_2$O
which gives $\mu/\mu_a = 3.27$.

This and other values that have been published for μ_a are given in Table 80. The value 1.7 given by L. Kockel was obtained by fitting to her

Table 80.—Apparent Dipole Moment of Water.

The apparent dipole moment (μ_a) is here used to denote the value derived from the coefficient (b) of T^{-1} in the expression for $(\epsilon - 1)/(\epsilon + 2)\rho$ in powers of T, by means of the relation $b = 4\pi N(\mu_a)^2/9kM$, M being the formula weight of H$_2$O (see text). Introducing the numerical values of the several quantities, one finds for water $1.850(10^{19}) \mu_a = \sqrt{b}$. For water-vapor $c = 0$ and μ_a in the expression for b is the actual dipole moment (μ) of the molecule H$_2$O ($10^{19} \mu = 18.3_1$ cgse, p. 00).

Unit of $\mu_a = 10^{-19}$ cgse units (= 0.02094 electron·angstrom) per gfw-H$_2$O

μ_a	μ/μ_a	Reference (see text for comments.)
5.59	3.27$_6$	Preferred values.
5.7	3.21	P. Debye [151]
5.689	3.21$_8$	M. Forró [152]
1.7	10.8	L. Kockel [153]
7.4	2.4$_7$	P. Lertes [154]

own observations Debye's expression for gases (that containing only the a and b terms), although her observations demand the c term also; and P. Lertes' value, 7.4, was derived from the torque exerted upon water by a rotating electric field. Although both these values are included in the table it is probable that little, if any, weight should be attached to them.

Values of μ_a as derived from solutions of water in nonpolar solvents may be found, with references, in the table compiled largely by N. V. Sidg-

---
[149] Hampson, G. C., and Marsden, R. J. B., *Trans. Faraday Soc.*, 30, Appendix (1934).
[150] Frank, F. C., *Proc. Roy. Soc. (London) (A)*, 152, 171-196 (1935).
[151] Debye, P., *Physik. Z.*, 13, 97-100, 295 (1912).
[152] Forró, M., *Z. Phys.*, 47, 430-445 (1928).
[153] Kockel, L., *Ann. d. Physik (4)*, 77, 417-448 (1925).
[154] Lertes, P., *Z. Phys.*, 4, 315-336 (1921); 6, 56-68 (1921).
[155] Cennamo, F., *Nuovo cim., (N. S.)*, 13, 304-309 (1936).

25. WATER: MOLECULAR DATA

wick and published by G. C. Hampson and R. J. B. Marsden,[149] and such results have been discussed by F. C. Frank [150] and others.

Polarizability of the Molecule of Water.

By the polarizability of a molecule of a substance is meant that portion of the molecular electric moment that is induced per unit electrical field as a result of the relative displacement of elastically bound electrons. It is the quantity α occurring in Debye's formula $(\epsilon - 1)/(\epsilon + 2)\rho = (4\pi N/3M)\cdot(\alpha + \mu^2/3kT) = a + b/T$, [see Section 49, eq. (2)], and is related to the optical index of refraction n in accordance with the formula $a = 4\pi N\alpha/3M = (n^2 - 1)/(n^2 + 2)\rho$. For water ($M = 18.0154$), this gives $10^{24}\alpha = 7.096a = 7.096(n^2 - 1)/(n^2 + 2)\rho$.

The value for a that is to be preferred at present is 0.2262 cm^3/g (see Table 175), which gives $10^{24}\alpha = 1.605$ cgse, or $\alpha = 1.605$ (angstrom)^3. On the other hand, using the values of n for the D-line, as given in Table 135 and of ρ as given in Table 93, one finds for 10, 20, and 30 °C the following values: $\alpha = 1.462_8$, 1.462_1, and 1.461_7A^3, respectively. F. Cennamo [155] obtained, from his own values of n, values essentially agreeing with these; it should be noticed that the values he gives are of a, not of α. The dielectric constant of water vapor leads to $\alpha = 1.59$A^3 for the vapor molecule (p. 49); and the optical index, to 1.50 (p. 49).

Anisotropy of the Molecule of Water.

For a discussion of the anisotropy of molecules, an explanation of terms and symbols, and references to the general subject, see Section 9, *Anisotropy*, and J. W. Beams.[156]

If the electrical moments induced along the principal electrical axes of the molecule by a unit electrical field parallel to those axes are A, B, and C, respectively, and if the magnetic moments similarly induced along the same axes are A', B', and C', then the factor (δ) measuring the optical anisotropy of the molecule is $\delta = \dfrac{A^2 + B^2 + C^2 - (AB + BC + CA)}{(A + B + C)^2}$ and the mean susceptibility per molecule of the unmagnetized substance is $\theta' \equiv (A' + B' + C')/3$. It is generally assumed that $B = C$, and that $B' = C'$, A being the greatest of the three induced electrical moments. Then $\delta = \left(\dfrac{A - C}{A + 2C}\right)^2$, $\theta' = (A' + 2C')/3$, and C'/A' serves to define the magnetic anisotropy; C'/A' is commonly called the magnetic anisotropy, although its value is unity for an isotropic molecule and zero for extreme anisotropy.

The data given by M. Ramanadham [157] lead to the following values, that of δ being taken from I. R. Rao [158] and based on the observations of

---

[156] Beams, J. W., *Rev. Mod. Physics*, **4**, 133-172 (1932).
[157] Ramanadham, M., *Indian J. Phys.*, **4**, 15-38 (1929).
[158] Rao, I. R., *Idem*, **2**, 61-96 (1928).
[159] Krishnan, K. S., *Phil. Mag. (6)*, **50**, 697-715 (1925).
[160] Chinchalkar, S. W., *Indian J. Phys.*, **6**, 165-179 (1931).

K. S. Krishnan[159]: $\delta = 0.00553$, $A'/\theta' = 1.14$, $C'/A' = 0.81$ if $B' = C'$. These values rest upon his value (-1.1×10^{-14}) for the coefficient of magnetic birefringence, and that is numerically greater than the more recent values ($-0.3_8 \times 10^{-14}$). From the latter, S. W. Chinchalkar[160] computes $A'/\theta' = 1.05$, whence $C'/A = 0.93$ if $B' = C'$.

The value of δ for water is only 1/3 as great as that (0.0166) for water-vapor (cf. p. 50).

26. Interaction of Water and Corpuscular Radiation

Alpha Particles.

The range of alpha particles in water (liquid) at 15 °C is given by W. Michl[161] as 32 microns for rays from Po, and by H. R. v. Traubenberg and K. Philipp[162] and K. Philipp[163] as 60 microns for rays from RaC'. (See S. Meyer.[164])

The decomposition of water by alpha particles, and the nature of the products formed, have been studied by C. E. Nurnberger.[165]

Beta Rays.

The coefficient of mass absorption of water (liquid) for the β rays from Ra-E is given by G. Fournier[166] as $\mu/\rho = 17.4$ cm²/g; that calculated by him from the absorption by H_2 and O_2 on the assumption that the coefficients of atomic absorption are not affected by the union of the atoms to form molecules is 16.0. He regards the difference as an evidence that water is abnormal.

For luminescence excited by β rays, see Section 39, *Electron Luminescence*.

Neutrons.

The coefficient of absorption of neutrons in water is $\mu = 0.027$ cm⁻¹.[167] A table based on Fermi's expression for the slowing down of neutrons by water has been published by G. Horvay.[168]

The scattering of neutrons by water has been studied by J. R. Dunning and G. B. Pegram[169] and by M. Deisenroth-Myssowsky, I. Kurtschatow, G. Latyschew, and L. Myssowsky.[170]

---

[161] Michl, W., *Sitz. Akad. Wiss. Wien (Abt. IIa)*, **123**, 1955-1963, 1965-1999 (1914).
[162] v. Traubenberg, H. R., and Philipp, K., *Z. Physik*, **5**, 404-409 (1921).
[163] Philipp, K., *Idem*, **17**, 23-41 (1923).
[164] Meyer, S., *Int. Crit. Tables*, **1**, 367-369 (1926).
[165] Nurnberger, C. E., *J. Phys'l Chem.*, **38**, 47-69 (1934); *J. Chem'l Phys.*, **4**, 697-702 (1936).
[166] Fournier, G., *Compt. rend.*, **183**, 200-203 (1926).
[167] Arzimowitsch, L., Kurtschatow, I., Latyschew, G., and Chramow, W., *Physik. Z. Sowj.*, **8**, 472-486 (1935). In this article the name of the last author is incorrectly spelled "Chromow"; it is corrected later. They do not state the unit used for μ, but it is probably the one (cm⁻¹) here given.
[168] Horvay, G., *Phys. Rev. (2)*, **50**, 897-898 (1937).
[169] Dunning, J. R., and Pegram, G. B., *Phys. Rev. (2)*, **45**, 768-769 (A) (1934).
[170] Deisenroth-Myssowsky, M., Kurtschatow, I., Latyschew, G., and Myssowsky, L., *Physik. Z. Sowj.*, **7**, 656-669 (1935).

27. Tensile Rupture of Water

Although it has long been known that liquids can, under suitable conditions, sustain relatively great tensile loads, it is doubtful if a true tensile rupture of a liquid is observable, unless it is when a liquid is forced to flow at high speed through a constricted section of a tube, as described by O. Reynolds[171] before the British Association for the Advancement of Science, in 1894. In all ordinary cases, the breaking of a column of liquid proceeds by a process of constriction arising from the action of surface tension; and when a column of liquid gives way under the action of a direct tension, the failure appears to occur at the liquid-solid boundary, not in the liquid itself unless that is known to contain a dissolved gas, in which case the failure seems to be associated with the presence at that point of a bubble of gas. Unless the principal radii of curvature (r_1, r_2) of the bounding surface are so small that the concept of an invariable surface-tension is not validly applicable to it, no element of liquid abutting upon a gas can sustain a tension exceeding $T\left(\dfrac{1}{r_1} + \dfrac{1}{r_2}\right)$, T being the surface-tension. For example, if it is valid to apply the idea of an invariable surface-tension to bubbles 0.00001 cm (= 1000A) in diameter, then the presence of a bubble of air of that size when under the tension will, of itself, cause a column of water to break at a tension of about 30 atm, however much greater the true tensile strength of water might be. On the other hand, the effect of completely dissolved gas upon the tension to which water can be practically subjected seems to be negligible, as shown particularly by the observations of M. Berthelot,[172] H. H. Dixon and J. Joly,[173] and H. H. Dixon.[174] Gas may also be the main cause of the observed failures at the liquid-solid boundary. All the recorded work was done before the advent of modern methods for outgassing solid surfaces. (See F. Donny,[175] O. Reynolds,[176] and J. Meyer.[177])

Two procedures, mainly, have been used for determining the tension to which water may be experimentally subjected. One is based upon the well known sticking of the mercury column to the top of a barometer when the tube and mercury are clean; it involves the determination of the length of the mercury column that can be similarly supported from a thin layer of water adhering to the top of the tube. This has been used by O. Rey-

---

[171] Reynolds, O., "Papers on Mechanical and Physical Subjects," Vol. **2**, pp. 578-587, Cambridge Univ. Press, 1901.
[172] Berthelot, M., *Ann. de chim. et phys. (3),* **30**, 232-237 (1850).
[173] Dixon, H. H., and Joly, J., *Phil. Trans. (B),* **186**, 563-576 (1895).
[174] Dixon, H. H., *Proc. Roy. Dublin Soc.,* **12**, 60-65 (1909); *Ann. Report Smithsonian Inst. for 1910,* 407-425 (1911).
[175] Donny, F., *Ann. de chim. et phys. (3),* **16**, 167-190 (1846) = *Ann. d. Physik (Pogg.),* **67**, 562-584 (1846).
[176] Reynolds, O., "Papers on Mechanical and Physical Subjects," Vol. **1**, pp. 394-398, Cambridge Univ. Press, 1900.
[177] Meyer, J., *Abh. deuts. Bunsen-Ges.,* **3**, No. 1, whole No. 6, 1911 = "Zur Kenntniss des negativen Druckes in Flüssigkeiten," W. Knapp, Halle, 1911.
[178] Reynolds, O., *Ibid.,* pp. 230-243, 394-8.
[179] Moser, J., *Ann. d. Physik (Pogg.),* **160**, 138-143 (1877).

nolds,[178] J. Moser,[179] and H. v. Helmholtz.[180] Tensions up to 3 atm were observed (Reynolds).

The other principal procedure was introduced by M. Berthelot[172] and is more suitable for detailed investigations. In this, the liquid is enclosed in a sealed tube which it nearly fills; by careful heating, the liquid is expanded until it completely fills the tube, exerting upon it a moderate pressure; then the temperature is slowly reduced. For a time, the liquid continues to fill the tube completely, but presently gives way with a snap, returning to the unstressed volume appropriate to the existing temperature. From the change in volume, Berthelot inferred that he had subjected water to a tension of 50 atm. The water was known to contain a small amount of dissolved air.

Improved apparatus employing Berthelot's procedure has been described and used by A. M. Worthington[181] and by J. Meyer[177] but the tensions did not exceed 34 atm. Meyer reported that the cooling curve is discontinuous at the instant of rupture.

H. H. Dixon[174] has subjected water containing dissolved air and fibers of wood to a tension of nearly 160 atm, using Berthelot's method. That is the highest recorded value that has been found.

From the adherence observed when flat, polished steel surfaces, very slightly wet with water, are wrung together, H. M. Budgett[181a] concluded that the tension on the water at the time of rupture approached 60 atm. Before being placed together, the surfaces were wiped until they appeared dry to a casual observer. The water remaining on them formed isolated, microscopic drops. After wringing them together and then sliding them apart, it was found that the drops had been drawn out into thin parallel lines. It was estimated that the area of the ruptured surface of water did not exceed 1/10 of the complete area of the steel surface. By actual test, it was found that the surfaces would not adhere unless there was a minute quantity of liquid between them, and that their adherence was essentially the same *in vacuo* as in the free atmosphere.

O. Reynolds[171] observed that, when water is forced through a tube having at one point a short length of greatly constricted cross-section, a characteristic hissing is heard if the velocity exceeds a certain value. He described such experiments at the meeting of the British Association, in 1894, and attributed the hissing to the boiling of the water under the reduced pressure existing at the constriction. Were that the correct explanation, the hissing would begin at a very low velocity if the temperature of the water were nearly 100 °C. This has been tested by S. Skinner and F. Ent-

[180] v. Helmholtz, H., *Verh. deut. physik. Ges. (Berlin)*, **6**, 16-18 (1887) = "Wiss. Abhand.," 3, 264-266, Leipzig, J. A. Barth, 1895.
[181] Worthington, A. M., *Phil. Trans. (A)*, **183**, 355-370 (1892).
[181a] Budgett, H. M., *Proc. Roy. Soc. London (A)*, **86**, 25-35 (1912).

wistle.[182] They found that the velocity at which hissing begins is not zero at 100°, but over the range studied (12 to 99 °C) is essentially proportional to $(t_c - t)$, indicating that it vanishes at, or near, the critical temperature (t_c). From this they concluded that the hissing arises from collapse following a true rupture of the water, and that the tensile strength of water vanishes at a temperature near the critical.

J. Larmor [183] has shown that if the van der Waals equation continuously applies, the tensile strength of the liquid will vanish if the temperature equals or exceeds $(27/32)T_c$; which for water is 273 °C.

See also E. Askenasy,[184] G. A. Hulett [184a]; and the compilation by T. F. Young and W. D. Harkins.[185]

28. Internal Pressure of Water

By the internal pressure of a liquid is meant the mean force of molecular attraction per cm² required to maintain the molecules at their existing aver-

Table 81.—Internal Pressure of Water

P_i = internal pressure; P = external pressure; v/v_{20} = ratio of the specific volume at t and P to that at 20 °C and 1 kg*/cm²; T °K = absolute temperature.

Unit of P and P_i = 1 megagram*/cm² = 1000 kg*/cm² = 967.8 atm. Temp. = t °C

Method[a]→ Ref.[b]→ t	State S	P = 1 kg*/cm² LtHt H P_i	Visc L	t	Isometrics $P+P_i = T(\partial P/\partial t)_v$ TR P	P_i
0	11.66	12.69	72.6		$v/v_{20} = 2.11$	
10	11.60		74.6	525	1.692	9.688
20	11.50	12.49	71.3	575	2.348	7.632
30	11.44	12.32	68.0	625	2.819	3.701
40	11.35	12.17	57.2		$v/v_{20} = 2.13$	
50	11.24		49.7	525	1.640	9.190
60	11.13	11.84	43.8	575	2.278	7.852
70	11.02		39.0	625	2.762	3.858
80	10.92	11.45	35.2		$v/v_{20} = 2.18$	
90	10.81		31.9	525	1.508	8.312
100	10.71	11.04	25.9	575	2.082	6.948
120		10.60		625	2.565	5.235
140		10.11			$v/v_{20} = 2.25$	
160		9.63		525	1.368	7.202
180		9.14		575	1.876	6.264
		Other values[c]		625	2.323	5.127

[a] Methods: LtHt = from the latent heat of vaporization. State = from an equation of state and the critical temperature. Visc = from the viscosity.

[b] References:
  H   Herz, W., *Z. Elektroch.*, **32**, 210-213 (1926).
  L   Lederer, E. L., *Koll. Beih.*, **34**, 270-338 (1932).
  S   Schuster, F., *Z. anorg. allgem. chem.*, **146**, 299-304 (1925).
  TR Tammann, G., and Rühenbeck, A., *Ann. d. Physik* (5), **13**, 63-79 (1932).

[c] From the effect of solutes upon the compressibility, P. G. Tait,[186] concluded that P_i = 5.7 at room temperature. P. Walden [187] computed for P_i at 100 °C the values: 11.0 from the latent heat, 8.4 from van der Waals' equation, and 4.4 from the surface tension.

age distances in opposition to the pressure arising from the thermal agitation of the molecules. Estimates of its value have been inferred in several ways from other types of data; it cannot be directly measured. See Table 81.

29. Viscosity of Water

From a consideration of all pertinent data available in 1924, N. E. Dorsey [188] concluded that the values for the viscosity of water given in Tables 82, 85, and 86 are the best that can be derived from those data. They are the result of an entirely independent study of the recorded data, and involve many complete recomputations. In a forthcoming paper, the procedure followed will be described in some detail, and replies to certain criticisms of the conclusions reached will be given.

Those values for the range 0 to 100 °C at 1 atm differ by a few tenths of a per cent, usually in excess, from the corresponding ones published earlier by E. C. Bingham and R. F. Jackson,[189] and have been criticised by Bingham.[190] The greatest difference is nearly 0.5 per cent, and occurs near 17 °C (see Table 84). As the Bingham-Jackson values have been much used in the standardization of viscosimeters, they are here reproduced in Table 83, and compared with others in Table 84. In very many cases the precision of the measurements relative to water is such that the difference between these two sets of values is of no consequence; but in every case it is very desirable that both the temperature and the assumed viscosity of the water be explicitly stated so that future investigators may know how the results should be revised in order to correct them for any error that may have been discovered in the value of that assumed viscosity.

P. Leroux [191] ascribes an uncertainty of not over 1 in 200 to his elaborate determinations in the range 1.5 °C to 44.5 °C; but their variation with the temperature is quite different from that of the values obtained by others. It is believed that this discrepancy is due to errors in the temperatures, as the method by which he inferred the temperature of the water is not satisfactory, and the discrepancy is such as would exist if the recorded temperatures were, in each case, intermediate between the actual temperature of the water and that of the room, its difference from the actual temperature of the water increasing as that departs more and more from the temperature of the room, whether above or below. His values are omitted from this

---

[182] Skinner, S., and Entwistle, F., *Proc. Roy. Soc. (London) (A)*, **91**, 481-485 (1915).
[183] Larmor, J., *Proc. Lond. Math. Soc. (2)*, **15**, 182-191 (1916).
[184] Askenasy, E., *Verh. Naturhist. mediz. Vereins (Heidelberg) (N. F.)*, **5**, 429-448 (1896).
[184a] Hulett, G. A., *Z. physik. Chem.*, **42**, 353-368 (1903).
[185] Young, T. F., and Harkins, W. D., *Int. Crit. Tables*, **4**, 434 (1928).
[186] Tait, P. G., *Beibl. zu Ann. d. Phys.*, **13**, 442-445 (1889) ← Results Voyage *Challenger*, "Phys. and Chem.", Vol. 2, part 4; *Proc. Roy. Soc. Edinburgh*, **15**, 426-427 (1888).
[187] Walden, P., *Z. physik. Chem.*, **66**, 385-444 (1909).
[188] Dorsey, N. E., *Int. Crit. Tables*, **5**, 10 (1929).
[189] Bingham, E. C., and Jackson, R. F., *Bull. Bur. Stand.*, **14**, 59-86 (SP298, Aug., 1916) (1919).
[190] Bingham, E. C., *J. Rheology*, **2**, 403-423 (1931).

compilation. The only other sets of careful determinations of the viscosity of water that have come to the author's attention since 1924 are those given in sections II and III of Table 85. The determinations by G. A. Hawkins, H. L. Solberg, and A. A. Potter [192] are not satisfactory. The effect of eddies in the wake of the falling body has been ignored, and the temperature coefficients used for their instruments seem to be in error.

From observations by himself and by Beilby, L. Hawkes [193] inferred that the viscosity of water increases very greatly as the temperature is

Table 82.—Viscosity of Water: 0 °C to 109 °C

From compilation [a] by N. E. Dorsey.[188] See also text and Table 83.

The uncertainty in the tabulated values is probably of the order of 0.1 or 0.2 per cent between 0° and 40 °C, and of 0.5 to 1 per cent between 40 °C and 100 °C. Pressure = 1 atm. Temp. = $(t_1 + t_2)$ °C.

Unit of viscosity $(\eta) = 1$ millipoise $= 0.001$ cgs unit.

$t_2 \rightarrow$ t_1	0°	1°	2°	3°	4°	5°	6°	7°	8°	9°
0°	17.94	17.32	16.74	16.19	15.68	15.19	14.73	14.29	13.87	13.48
10	13.10	12.74	12.39	12.06	11.75	11.45	11.16	10.88	10.60	10.34
20	10.09	9.84$_3$	9.60$_8$	9.38$_0$	9.16$_1$	8.94$_9$	8.74$_6$	8.55$_1$	8.36$_3$	8.18$_1$
30	8.00$_4$	7.83$_4$	7.67$_0$	7.51$_1$	7.35$_7$	7.20$_8$	7.06$_4$	6.92$_5$	6.79$_1$	6.66$_1$
40	6.53$_6$	6.41$_5$	6.29$_8$	6.18$_4$	6.07$_5$	5.97$_0$	5.86$_8$	5.77$_0$	5.67$_5$	5.58$_2$
50	5.49	5.40	5.32	5.24	5.15	5.07	4.99	4.92	4.84	4.77
60	4.70	4.63	4.56	4.50	4.43	4.37	4.31	4.24	4.19	4.13
70	4.07	4.02	3.96	3.91	3.86	3.81	3.76	3.71	3.66	3.62
80	3.57	3.53	3.48	3.44	3.40	3.36	3.32	3.28	3.24	3.20
90	3.17	3.13	3.10	3.06	3.03	2.99	2.96	2.93	2.90	2.87
100	2.84	2.82	2.79	2.76	2.73	2.70	2.67	2.64	2.62	2.59

[a] Based on the observations of:

Bingham, E. C., and White, G. F., *Z. physik. Chem.*, **80**, 670-686 (1912); Grotrian, O., *Ann. d. Physik (Wied.)*, **8**, 529-554 (1879); Heydweiller, A., *Idem*, **59**, 193-212 (1896); Hosking, R., *Phil. Mag.* (5), **49**, 274-286 (1900); *Idem* (6), **7**, 469-484 (1904); *Idem*, **17**, 502-520 (1909); *Idem*, **18**, 260-263 (1909); *J. and Proc. Roy. Soc., N. S. Wales*, **42**, 34-56 (1908); *Idem*, **43**, 34-38 (1909); Lyle, T. R., and Hosking, R., *Phil. Mag.* (6), **3**, 487-498 (1902); Dr. Poiseuille, *Mém. Savans Etrang. Inst. Paris*, **9**, 433-544 (1846); *Compt. rend.*, **11**, 961-967, 1041-1048 (1840); **12**, 112-115 (1841); *Idem*, **15**, 1167-1187 (1842); Slotte, K. F., *Ann. d. Physik. (Wied.)*, **20**, 257-267 (1883); Sprung, A., *Idem (Pogg.)*, **159**, 1-35 (1876); Thorpe, T. E., and Rodger, J. W., *Phil. Trans. (A)*, **185**, 397-710 (1894); and Washburn, E. W., and Williams, G. Y., *J. Am. Chem. Soc.*, **35**, 737-750 (1913).

[191] Leroux, P., *Compt. rend.*, **180**, 914-916 (1925); *Ann. de Phys.* (10), **4**, 163-248 (1926).
[192] Hawkins, G. A., Solberg, H. L., and Potter, A. A., *Trans. Am. Soc. Mech. Eng.*, **57**, 395-400 (FSP-57-11) (1935).
[193] Hawkes, L., *Nature*, **123**, 244 (1929).
[194] Dufour, L., *Ann. d. Physik (Pogg.)*, **114**, 530-554 (1861).
[195] Sorby, H. C., *Phil. Mag.* (4), **18**, 105-108 (1859).
[196] Davy, Sir Humphry, *Phil. Trans.*, **143** (1822) = *Annals Phil., (N. S.)*, **5**, 43-49 (1823).
[197] König, W., *Ann. d. Physik (Wied.)*, **25**, 618-625 (1885).
[198] Dufour, H., "Séances Soc. Fr. de Phys.," pp. 6-7, 1887; reviewed in *Lum. Electr.*, **23**, 337 (1887).
[199] Raha, P. K., and Chatterjee, S. D., *Indian J. Phys.*, **9**, 445-454 (1935).
[200] Trautz, M., and Fröschel, E., *Ann. d. Physik* (5), **22**, 223-246 (1935).
[201] Pacher, G., and Finazzi, L., *Atti. R. Ist. Veneto di Sci., Let., Arti*, **59**$_2$ = (8), **2**$_2$, 389-402 (1900).
[202] Quincke, G., *Ann. d. Physik. (Wied.)*, **62**, 1-13 (1897).

Table 83.—Viscosity of Water: 0 °C to 100 °C [189]

Bingham and Jackson [189]

Their values for the viscosity (η) are those defined by the formula $\eta^{-1} + 120 = 2.1482 \{(t - 8.435) + \sqrt{8078.4 + (t - 8.435)^2}\}$, in which the values of the constants were so determined as to fit a formula of this form to a certain set of mean values derived by them from the data available.

Unit of $\eta = 1$ millipoise $= 0.001$ cgs unit. Temp. $= (t_1+t_2)$°C

t_1 \ $t_2 \rightarrow$	0	1	2	3	4	5	6	7	8	9
0	17.921	17.313	16.728	16.191	15.674	15.188	14.728	14.284	13.860	13.462
10	13.077	12.713	12.363	12.028	11.709	11.404	11.111	10.828	10.559	10.299
20	10.050	9.810	9.579	9.358	9.142	8.937	8.737	8.545	8.360	8.180
30	8.007	7.840	7.679	7.523	7.371	7.225	7.085	6.947	6.814	6.685
40	6.560	6.439	6.321	6.207	6.097	5.988	5.883	5.782	5.683	5.588
50	5.494	5.404	5.315	5.229	5.146	5.064	4.985	4.907	4.832	4.759
60	4.688	4.618	4.550	4.483	4.418	4.355	4.293	4.233	4.174	4.117
70	4.061	4.006	3.952	3.900	3.849	3.799	3.750	3.702	3.655	3.610
80	3.565	3.521	3.478	3.436	3.395	3.355	3.315	3.276	3.239	3.202
90	3.165	3.130	3.095	3.060	3.027	2.994	2.962	2.930	2.899	2.868
100	2.838									

Table 84.—Viscosity of Water: Comparison of Values

A comparison of (B) the Bingham and Jackson values of Table 83 with (D) those of Table 82 and with (M) Bingham and Jackson's "mean" of certain determinations. Those means seem to have served as the bases for the determination of the values of the 4 adjustable constants in their formula. The D values are here given as in *International Critical Tables;* the last figure has no significance.

Unit of $\eta = 1$ micropoise $= 10^{-6}$ cgs unit.

t	M−B	M	B	D	D−B	t	B	D	D−B
0	− 34	17887	17921	17934	+ 13	10	13077	13097	+ 20
5	− 33	15155	15188	15188	0	11	12713	12733	+ 22
10	− 16	13061	13077	13097	+ 20	12	12363	12390	+ 27
15	+ 2	11406	11404	11447	+ 43	13	12028	12061	+ 33
20	− 4	10046	10050	10087	+ 37	14	11709	11748	+ 38
25	+ 4	8941	8937	8949	+ 12	15	11404	11447	+ 43
30	+ 12	8019	8007	8004	− 3	16	11111	11156	+ 45
35	− 20	7205	7225	7208	− 17	17	10828	10875	+ 47
40	− 27	6533	6560	6536	− 24	18	10559	10603	+ 44
45	− 30	5958	5988	5970	− 18	19	10299	10340	+ 41
50	+ 3	5497	5494	5492	− 2	20	10050	10087	+ 37
55	+ 8	5072	5064	5072	+ 8	21	9810	9843	+ 33
60	+ 13	4701	4688	4699	+ 11	22	9579	9608	+ 29
65	+ 4	4359	4355	4368	+ 13	23	9358	9380	+ 22
70	+ 1	4062	4061	4071	+ 10	24	9142	9161	+ 19
75	− 5	3794	3799	3806	+ 7	25	8937	8949	+ 12
80	− 9	3556	3565	3570	+ 5	26	8737	8746	+ 9
85	− 14	3341	3355	3357	+ 2	27	8545	8551	+ 6
90	− 19	3146	3165	3166	+ 1	28	8360	8363	+ 3
95	− 13	2981	2994	2994	0	29	8180	8181	+ 1
100	− 17	2821	2838	2839	+ 1	30	8007	8004	− 3

29. WATER: VISCOSITY

Table 85.—Viscosity of Saturated Water Below 0 °C and above 100 °C
Unit of $\eta = 1$ mp $= 0.001$ cgs unit; of $P_{sat} = 1$ kg*/cm² $= 0.9678$ atm. Temp. $= t$ °C

I. Adapted from a compilation by N. E. Dorsey [188]; see text. For $t < 0$ °C, values are by G. F. White and R. H. Twining,[218] corrected and adjusted to accord with the values in Table 82. For $t > 100$ °C, values are from a table given by Heydweiller [219]; they are based on observations by M. de Haas [220] and have been so adjusted as to fit smoothly with the values tabulated by Thorpe and Rodgers (see Table 82, references) for temperatures below 100 °C. The three observations by de Haas (2.232 at 124.0 °C, 1.925 at 142.2 °C, and 1.805 at 153.0 °C) seem to be the only direct determinations of η at $t > 100$ °C that had been made before 1931 (cf. M. Jakob[221]).

t	P_{sat}	η	t	P_{sat}	η
− 2	0.0054	19.1	110	1.46	2.56
− 4	0.0046	20.5	120	2.02	2.32
− 5	0.0043	21.4	130	2.75	2.12
− 6	0.0040	22.0	140	3.68	1.96
− 8	0.0034	24.0	150	4.85	1.84
−10	0.0033	26.0	160	6.30	1.74[a]

II. Adapted from K. Sigwart.[222]

t	P_{sat}	η	t	P_{sat}	η
100	1.03	2.83	275	60.7	1.04
125	2.37	2.28	300	87.6	0.95[a]
150	4.85	1.86[a]	325	123.0	0.84
175	9.10	1.58	350	168.6	0.71
200	15.9	1.36[a]	360	190.4	0.63
225	26.0	1.23	370	214.7	0.53
250	40.6	1.13	374	225.2	0.378

III. V. Shugayev [223] has reported the following values:

t	115.5	143.5	156.5	160.0	173.0	196.5	210.0	°C
η	2.86	2.01	1.87	1.82	1.55	1.42	1.30	mp

[a] From the observed mobility of electrolytic ions, G. V. Hevesy [224] inferred the values: $\eta = 1.79$ at 156 °C, 1.21 at 218 °C, and 0.92 at 306 °C.

---

[203] Duff, A. W., *Phys. Rev.*, **4**, 23-38 (1896).
[204] Kimura, O., *Bull. Chem. Soc. Japan*, **12**, 147-149 (1937).
[205] Trautz, M., and Fröschel, E., *Ann. d. Physik (5)*, **22**, 223-246 (1935).
[206] Bulkley, R., *Bur. Stand. J. Res.*, **6**, 89-112 (1931).
[207] Bowden, F. P., *Physik. Z. d. Sowj.*, **4**, 185-196 (1933).
[208] Bastow, S. H., and Bowden, F. P., *Proc. Roy. Soc. (London) (A)*, **151**, 220-233 (1935) → Bowden and Bastow, *Nature*, **135**, 828 (L) (1935).
[209] Macaulay, J. M., *Nature*, **138**, 587 (L) (1936).
[210] Ostwald, W., and Genthe, A., *Zool. Jahrb. Abth. f. System., Geog., Biol. d. Thiere*, **18**, 3-15 (1903).
[211] Krümmel, O., and Ruppin, E., *Wissensch. Meeresunters. (N. F.)*, **9**, 27-36 (1906).
[212] Macleod, D. B., *Trans. Faraday Soc.*, **21**, 145-150, 151-159, 160-167 (1925).
[213] Sharma, R. K., *Chem. Abs.*, **20**, 2267 (1926) ← *Quart. J. Indian Chem. Soc.*, **2**, 310-311 (1925).
[214] Herz, W., *Z. anorg. allgem. Chem.*, **168**, 89-92 (1927).
[215] Silverman, D., and Roseveare, W. E., *J. Am. Chem. Soc.*, **54**, 4460 (1932).
[216] Lederer, E. L., *Koll.-Beih.*, **34**, 270-338 (1932).
[217] Frenkel, J., *Acta Physicochim. URSS*, **3**, 633-648 (1935).
[218] White, G. F., and Twining, R. H., *Am. Chem. J.*, **50**, 380-389 (1913).

Table 86.—Viscosity of Compressed Water

(For viscosity of saturated water ($P = P_\text{sat}$) see Table 85)

In Section I are given directly the values of the mean pressure coefficient of viscosity from 0 to P; from that and the value of η at 1 atm (essentially zero pressure) as given in Table 82, the viscosity under the pressure P can be computed. In Section II is given the only available set of data for compressed water above 100 °C.

Unit of $\eta = 1$ mp = 0.001 g/cm·sec = 0.001 cgs unit; of $P = 1$ kg*/cm² = 0.9678 atm; of $k = 10^{-6}$ per kg*/cm². Temp. = t °C.

I. Adapted from a compilation by N. E. Dorsey [188] with the addition of computed values.

From their study of the viscosity of aqueous solutions, G. Tammann and H. Rabe [225] inferred that Bridgman's (1926)[a] values for water at 10 °C are in error, and set up expressions for the variation in the viscosity of water with the pressure, valid for $P \gtreqless 2000$ kg*/cm². Those expressions are equivalent to the following: $10^6(\eta - \eta_0)/\eta_0 P \equiv 10^6 k = -134.9 + 0.05778P$ at 0 °C, $-37.63 + 0.02430P$ at 10 °C, $+41.91 + 0.01054P$ at 30 °C, and $+82.7$ at 75 °C. Later, E. L. Lederer [226] set up the following equation for water:

$$\log_{10}(\eta/\eta_0) \equiv \log_{10}(1 + kP) = \frac{1.650P}{1000T} + \frac{1369P \cdot \log_{10}T}{10^8} - 0.1300e^{-f}$$

where $f \equiv 10^{-6}(1350 t^2) + \frac{691}{P}$; temperature = t° C, $T \equiv 273 + t$ °C, and unit of $P = 1$ kg*/cm². They also tabulated the values of $\log_{10}(\eta/\eta_0)$ so computed for the values of P and T appearing in Bridgman's table. From these tabulated tables and from Tammann and Rabe's expressions were computed the several values of k here appearing under the heading "computed."

The observations of J. Sachs [227] and of E. Warburg and J. Sachs,[228] indicate that $10^{-6}k = -170$ per kg*/cm² at 20 °C for $P \gtreqless 150$ kg*/cm². This does not agree with the other observations.

$\eta = \eta_0(1 + kP)$, where η and η_0 refer to the same temperature, but the first refers to the pressure P and the second to zero pressure, which may for our present purposes be taken as 1 atm. From the k here given and the value of η_0 as given in Table 82, η may be computed. For example, here we find for 30 °C and $P = 10000$ kg*/cm², $k = +117 \times 10^{-6}$ per kg*/cm²; from Table 82 we find $\eta_0 = 8.00_4$ mp at 30 °C. Whence at 30 °C and 10 000 kg*/cm², $\eta = 8.00_4(1 + 117 \times 10^{-6} \times 10\,000) = 8.00_4(2.17) = 17.4$ mp.

[219] Heydweiller, A., *Ann. d. Physik (Wied.)*, **59**, 193-212 (1896).
[220] de Haas, M., *Comm. Phys. Lab. Leiden*, **12**, 1-8 (1894).
[221] Jakob, M., *Engineering (London)*, **132**, 744-746, 800-804 (1931).
[222] Sigwart, K., *Forsch. Gebiete Ingenieurw.*, **7**, 125-140 (1936).

Table 86.—(Continued)

Unit of $P = 1$ kg*/cm² = 0.9678 atm; of $k = 10^{-6}$ per kg*/cm². Temp. = t °C.

	Experimental					Ref.[a]	Computed							
							Tammann and Rabe				Lederer			
$t \rightarrow$	0	10.3	15	30	75		0	10	30	75	0	10.3	30	75
$\eta_0 \rightarrow$	17.94	12.99	11.45	8.004	3.81									
P	k						k							
23.8														
100		−200[b]	−55	−47[d]		R	−134	−37.0	+42.2	+82.7				
200	−214[c]		−63			C	−129	−35.2	+43.0	+82.7				
300			−51	−25[d]		C	−123	−32.8	+44.0	+82.7				
400	−128[e]		−54		+72	C	−118	−30.4	+45.1	+82.7			+94
500		−62	−46	+49[d]		C	−112	−27.9	+46.1	+82.7				
600	−124		−39	−17[d]		B, C	−106	−25.5	+47.2	+82.7	−64	−72	+46	
700	−105[e]		−33			C	−100	−23.0	+48.2	+82.7				
900			−30			C	−94	−20.6	+49.3	+82.7				
1000	−79	−46		+53	+76	B	−83	−15.8	+51.4	+82.7	−56	−69	+47	+94
1500	−45	−29		+57	+75	B	−77.1	−13.3	+52.4	+82.7	−40	−50	+53	+97
2000	−21.5	−16.0		+64	+81	B	−48.2	−1.2	+57.7	+82.7	−15	−26	+62	+100
3000	+8.0	+5.1		+76	+84	B	−19.3	+11.0	+63.0	+82.7	+13	+1	+74	+104
4000	+27.8	+20.2		+87	+90	B	(+38.4)	(+35.3)	(+73.5)	(+82.7)	+29	+19	+83	+110
5000	+44	+33.2		+95	+100	B					+43	+33	+91	+115
6000	+58	+43		+102	+109	B	(+154)	(+83.9)	(+94.6)	(+82.7)	+59	+43	+99	+120
7000		+52		+107	+117	B						+51	+106	+127
8000		+60		+112	+125	B			(+147)			+59	+113	+133
9000				+116	+136	B							+120	+141
10000				+117		B							+125	
11000				+119		B							+134	

Experimental (L. Hauser [229])

P	310	362		20	30	40	413	50	60	70	75	80	90	516
t	60[e]	55		−33	−7	+17		+34	+48	+62		+73	+82	40
k	+40	+60												+40

Table 86.—(Continued)

II. From K. Sigwart.[222]

$t \rightarrow$	100	125	150	175	200	225	250	275	300
$P_{sat} \rightarrow$	1.03	2.37	4.85	9.10	15.9	26.0	40.6	60.7	87.6
P					η				
P_{sat}	2.83	2.28	1.86	1.58	1.36	1.23	1.13	1.04	0.95
50	2.94	2.30	1.88	1.60	1.38	1.24	1.14		
100	3.01	2.44	1.94	1.64	1.41	1.26	1.15	1.05	0.96
200	3.24	2.53	2.03	1.70	1.46	1.28	1.19	1.08	0.99
300	3.43	2.67	2.14	1.76	1.51	1.32	1.21	1.11	1.02

$t \rightarrow$	325	350	360	370	374	400	410	430	450
$P_{sat} \rightarrow$	123.0	168.6	190.4	214.7	225.2				
P					η				
P_{sat}	0.84	0.71	0.63	0.53	0.378				
200	0.89	0.75	0.66						
300	0.93	0.82	0.76	0.71	0.69	0.40	0.33	0.29	0.28

[a] References:
- B Bridgman, P. W., *Proc. Am. Acad. Arts Sci.*, **61**, 57-99 (1926); *Proc. Nat. Acad. Sci.*, **11**, 603-606 (1926).
- C Cohen, R., *Ann. d. Physik (Wied.)*, **45**, 666-684 (1892).
- R Röntgen, W. C., *Idem*, **22**, 510-518 (1884).
- Others are given at head of table.

[b] At 9 °C.
[c] At 1 °C.
[d] At 23 °C.
[e] For the range 50 °C to 80 °C.

Table 87.—Viscosity of Sea-water [230]

Salt content is s grams of salt per kg of sea-water, η = viscosity of the sea-water, η_0 = viscosity of pure water at 0 °C, η_t = viscosity of pure water at t °C as given in Table 82. The second half of the table has been computed from the first, by the present compiler.

$s \rightarrow$	5	10	20	30	40	5	10	20	30	40
t			$1000\eta/\eta_0$					$1000\eta/\eta_t$		
0	1009	1017	1032	1045	1059	1009	1017	1032	1045	1059
5	855	863	877	891	905	1010	1019	1036	1052	1069
10	738	745	758	772	785	1011	1020	1038	1057	1075
15	643	649	662	675	688	1008	1017	1037	1058	1078
20	568	574	586	599	611	1010	1021	1042	1065	1086
25	504	510	521	533	545	1010	1022	1044	1069	1093
30	454	460	470	481	491	1017	1031	1053	1078	1100

---

[223] Shugayev, V., *Chem. Abs.*, **29**, 2804 (1935) ← *J. Exp. Theo. Phys. (U.S.S.R.)*, **4**, 760-765 (1934).
[224] Hevesy, G. V., *Z. Elektroch.*, **27**, 21-24 (1921).
[225] Tammann, G., and Rabe, H., *Z. anorg. allgem. chem.*, **168**, 73-85 (1927).
[226] Lederer, E. L., *Koll. Beih.*, **34**, 270-338 (1932).
[227] Sachs, J., *Diss.*, Freiburg, 1883.
[228] Warburg, E., Sachs, J., *Ann. d. Physik (Wied.)*, **22**, 518-522 (1884).
[229] Hauser, L., *Ann. d. Phys.* (5), **5**, 597-632 (1901).
[230] Krümmel, O., and Ruppin, E., *Wissensch. Meeresunters. (N. F.)*, **9**, (*Abt. Kiel*): 27-36 (1906) → Krümmel, O., "Handb. d. Ozeanog.," Vol. 1, 1907; Vol. 2, 1911.

reduced from $-9\,°C$ to $-12\,°C$, and that water becomes a vitreous solid at $-17\,°C$. But that is incorrect. It is incompatible with the data in Table 85, and is completely at variance with the compiler's observation that, to all appearances, water is essentially as mobile at $-20\,°C$ as it is at $0\,°C$. Furthermore L. Dufour [194] stated that he had observed suspended drops of water to be fluid (flüssig) at $-20\,°C$, and the observation of H. C. Sorby [195] that the mobility of the clear liquid enclosed in small cavities in natural quartz is essentially the same at $-20\,°C$ as at room temperature has been quoted as evidence that water is quite mobile at $-20°\,C$, and was so interpreted by him. The opinion, sometimes expressed, that such liquid inclusions are probably CO_2, not water, is contrary to the observations of Sir Humphry Davy [196] (see p. 642).

Effect of Various Factors.

Magnetic Field.—W. König [197] found that a magnetic field of 6300 to 7300 gauss transverse to the direction of flow of a solution of a paramagnetic salt produced no observable change in the viscosity of the solution. H. Dufour [198] thought that he had observed that a transverse magnetic field decreased the viscosity of mercury, but as pointed out in the review, his observations are probably to be explained by the force exerted by the field on the flowing mercury. P. K. Raha and S. D. Chatterjee,[199] using transverse fields up to 35000 gauss found no change for water, but definite changes for certain organic substances—for some an increase, for others a decrease. For a review of the subject, see M. Trautz and E. Fröschel.[200]

Electric Field.—G. Pacher and L. Finazzi [201] have found that an electrostatic field of nearly 27 kilovolt/cm transverse to the direction of flow of water produces no observable change in the viscosity. With other liquids the observed rate of flow in the field was very slightly different (not over 3 in 10,000) from that with no field, sometimes greater, sometimes less. They concluded that there is no true effect. This agrees with the observations of W. König [197] but not with the conclusions reached by G. Quincke [202] and by A. W. Duff,[203] which seem to be incorrect. (See criticism by Pacher and Finazzi [201]). For a solution of stearic acid in benzene O. Kimura [204] observed a marked increase in the viscosity. M. Trautz and E. Fröschel [205] have reviewed the subject.

Adjacent Solid.—(See also p. 512+). Some have suggested that when a viscous liquid is flowing over a solid there is a layer of the liquid of appreciable thickness, next to the solid, that remains at rest. But R. Bulkley [206] has found that in the case of oils, in which the effect should be especially pronounced, the thickness of such a stationary film does not exceed $0.02\,\mu$ to $0.03\,\mu$ ($1\,\mu = 0.0001$ cm). Similarly, F. P. Bowden [207] has been unable to find any evidence of surface forces acting at measurable distances; he could not work at distances smaller than 0.1 to $0.2\,\mu$, but was certain that there is no such long-range effect (up to $50\,\mu$) as some have reported. These results were confirmed by S. H. Bastow and F. P. Bowden.[208] They state: "No sign of induced rigidity was detected in liquids at a distance of 1000A [0.1 μ] from the surface even at temperatures near the freezing point. All the liquids investigated, except liquid crystals, were unable to withstand the slightest pressure without normal flow." Within experimental error, the viscosity of the thin film was the same as that of the liquid in bulk.

Although J. M. Macaulay [209] was undecided whether the high value (0.11 poise) that he computed from the rate at which water at 16 °C entered the gap between two parallel plates separated by 0.25 μ should be accepted as the actual value of the viscosity under those conditions, it seems most probable that it should not.

Dissolved Gas.—W. Ostwald and A. Genthe [210] have studied the effect of dissolved gas on the viscosity of water at 20 °C, and found as follows, η_0 being the viscosity of gas-free water:

Gas	N_2	O_2	CO_2	CH_4
η/η_0	1.017	0.990	1.007	0.998

Their report is lacking in detail. Ruppin found that the viscosity at 20 °C of air-saturated water is the same as that of gas-free water (see O. Krümmel and E. Ruppin.[211]

Relation between Viscosity and Other Properties.

Certain empirical and semitheoretical relations between the viscosity and other properties of the liquid have been proposed and discussed by D. B. Macleod,[212] R. K. Sharma,[213] W. Herz,[214] D. Silverman and W. E. Roseveare,[215] and E. L. Lederer.[216]

The theoretical expression $\eta = ATe^{-B/T}$ derived by J. Frenkel [217] for the temperature variation of the viscosity of a liquid does not represent the observations on water.

30. Rigidity of Water

(See also p. 189)

F. Michaud [231] has found that as the concentration of a jelly is progressively decreased, the rigidity becomes zero before the concentration does. This shows that the rigidity of water is zero. Like J. Colin,[232] he quite disagrees with the conclusion of T. Schwedoff [233] regarding the rigidity of liquids.

B. Derjaguin [234] has published the following values for the rigidity of very thin films of water:

Thickness of film	0.089	0.093	0.137	0.150	μ
Rigidity	1.9	1.7	0.04	0	10^8 g*/cm²

He suggests that the rigidity arises from chains of hundreds of oriented molecules extending from the solid boundaries into the liquid. His obser-

---

[231] Michaud, F., *Ann. de Phys. (9)*, **19**, 63-80 (1923).

vations are at variance with those of R. Bulkley[206] on the viscous flow of oils, in which such chains might be expected and had been thought to exist. Bulkley found that adjacent to a solid wall there is no stationary film as much as 0.03 μ in thickness. Behavior similar to that observed by Derjaguin would have resulted if the water had contained minute solid particles in suspension.

More recent work by F. P. Bowden and S. H. Bastow,[208] B. Derjaguin,[235] and J. M. Macauley [209] does not necessitate any change in the preceding statement.

31. Acoustic Data for Water

The greatest amount of vibratory energy that water at atmospheric pressure can transmit without cavitation is about 0.3 watt per cm^2.[236]

Velocity of Sound in Water.

The velocity of sound in water increases with the temperature to a maximum near 75 °C, and then decreases. In general, the velocity in other liquids decreases continuously.

Until 1927, the values obtained for the velocity of sound in water far from its boundaries, and for the variation of that velocity with the temperature, were quite discordant (cf. A. L. Foley [237]). This was due in large part to the measurements having been made in vessels that were not large as compared with the wave-length of the sound employed, so that a large, complicated, and unsatsifactorily determined correction had to be applied to the observed velocity to eliminate the effect of the walls of the vessel.

In 1927, J. C. Hubbard and A. L. Loomis [238] published a very concordant set of preliminary data for frequencies (ν) of 200 to 400 kilocycles per second, and for temperatures of 5 to 35 °C. At such high frequencies, the waves are so short that containing vessels of moderate size produce no effect upon the observed velocity. This was followed the next year by a final report, in which the range was 0 to 40 °C.[239] And more recently, at this Bureau, very careful determinations have been made by C. R. Randall,[240] at ν = 750 kilocycles/sec. In that work, the water was boiled in Pyrex glass immediately before being introduced into the apparatus, and extreme care was taken to ensure that the apparatus and contents had attained the temperature of the bath before measurements were made. The bath was thermostatically controlled to within ± 0.02 °C. The uncertainty

---

[232] Colin, J., *Compt. rend.*, **116**, 1251-1253 (1893).
[233] Schwedoff, T., *J. de Phys. (2)*, **8**, 341-359 (1889); **9**, 34-46 (1890); *(3)*, **1**, 49-53 (1892).
[234] Derjaguin, B., *Z. Physik*, **84**, 657-670 (1933); *Physik. Z. d. Sowj.*, **4**, 431-432 (1933).
[235] Derjaguin, B., *Nature*, **138**, 330-331 (L) (1936).
[236] Florisson, C., *Bull. Soc. Belge Élect.*, **52**, 165-170, 263-278, 339-348 (1936).
[237] Foley, A. L., *Int. Crit. Tables*, **6**, 464 (1929).
[238] Hubbard, J. C., and Loomis, A. L., *Nature*, **120**, 189 (1927).
[239] Hubbard, J. C., and Loomis, A. L., *Phil. Mag. (7)*, **5**, 1177-1190 (1928).
[240] Randall, C. R., *Bur. Stand. J. Res.*, **8**, 79-99 (RP402) (1932).
[241] Pooler, L. G., *Phys. Rev. (2)*, **35**, 832-847 (1930), superseding **31**, 157 (A) (1928).

Table 88.—Velocity of Sound in Water

V_c is the value defined by the empirical equation
$$V_c = 1404.4 + 4.8215t - 0.047562t^2 + 0.00013541t^3 = 1404.4 \times$$
$$\left\{ 1 + 3.4331_4 \left(\frac{t}{1000}\right) - 33.866_5 \left(\frac{t}{1000}\right)^2 + 96.419 \left(\frac{t}{1000}\right)^3 \right\} \text{ meters/sec};$$
it has a maximum at $t = 74.2$ °C. V_{obs} = observed velocity after application of a nominal correction for the finite size of the containing vessel (this correction is zero for the HL and the R values); $\Delta = V_{\text{obs}} - V_c$, $\delta = \Delta/V_c$. In all cases the mean pressure is 1 atm, the boundaries are nominally at infinity, and the temperature is t °C. The R values are to be preferred (see text).

Unit of V_c and of $\Delta = 1$ m/sec $= 3.2808$ ft/sec. Temp. $= t$ °C

t	V_c	R[a]	Δ HL[a]	P[a]	R[a]	1000δ HL[a]	P[a]	t	Earlier data V_c	Δ	Ref.[a]
0	1404.4	−0.9	+2.6		−0.6	+1.9		3.9	1422	−23	M
5	1427.3		+0.4			+0.3		7.6	1438	−29	M
10	1448.0	0	+0.8		0	+0.6		13.7	1462	−25	M
15	1466.5		+1.0			+0.7		25.2	1498	−41	M
20	1482.9	+0.2	+1.3		+0.1	+0.7		4.0	1423	− 5[b]	B
25	1496.3[c]		+1.8	−10.3		+1.2	−6.9	21.5	1487	− 6[b]	B
30	1509.9	0	0.0	−11.2	0	0.0	−7.4	8.1	1440	− 5	CS
35	1520.7		−0.1			−0.1		13	1459	−18[d]	D
40	1529.8	−0.3	+0.5	−11.5	−0.2	+0.3	−7.5	14	1463	−11	D
50	1543.5	0		−11.9	0		−7.7	18	1477	+12	D
60	1551.7	−0.2		−11.9	−0.1		−7.7	19	1480	−19[d]	D
70	1555.3	0		−12.2	0		−7.8	19	1480	+38	D
75	1555.6			−11.4			−7.3	19	1480	+ 9	D
80	1555.0	−0.4			−0.3			31	1512	− 7[d]	D
86	1553.4	−1.0			−0.6			19.5	1481	−19	C
								20	1483	−13	J

[a] References:
- B   Bungetzianu, D., *Bull. Soc. Roumaine des Sci., (Bucarest)*, **19**, 1224-1246 (1910); **21**, 208-267, 405-486 (1912); **22**, 182-214 (1913). Especially the last.
- C   Cisman, A., *J. de Phys.* (6), **7**, 345-352 (1926).
- CS   Colladon, D., and Sturm, C., *Mém. Sav. Etrang. Inst. Paris*, **5**, 267-347 (1838); *Ann. de Chim. et Phys.* (2), **36**, 113-159, 225-257 (1827); *Ann. d. Physik (Pogg.)*, **12**, 39-76, 161-197 (1828); "Mémoire sur la compression des liquides et la vitesse de son dans l'eau," 1827, C. Schuchardt, Geneva, 1887. (All seem to refer to the same work.)
- D   Dörsing, K., *Ann. d. Physik* (4), **25**, 227-251 (1908) ← *Diss.*, Bonn, 1907.
- HL   Hubbard, J. C., and Loomis, A. L., *Phil. Mag.* (7), **5**, 1177-1190 (1928); supersedes *Nature*, **120**, 189 (1927).
- J   Jonesca, V., *J. de Phys.* (6), **5**, 377-383 (1924).
- M   Martini, T., *Atti R. Ist. Veneto* (6), **4**, Appendice (1886) (not included in index) → *Beibl. Ann. d. Physik*, **12**, 566-568 (1888).
- P   Ref. 241.
- R   Ref. 240.

[b] These two values differ slightly from the corresponding ones given in Foley's compilation [237] and assigned to the same source. The one here given for 21.5 °C corresponds exactly to the value given by Bungetzianu; that for 4 °C was computed by the compiler from the data given by the observer.

[c] L. Bergmann,[247] using frequencies of 4.5 to 13 megacycles/sec and inferring the length of the waves from their observed diffraction of light, obtained for the velocity at 25 °C the very low value, 1465 m/sec. But S. Parthasarathy,[248] using a similar method and a frequency of 7.32 megacycles/sec, found 1494 m/sec at 24 °C.

[d] These three were obtained in the same glass tube and at the same frequency; the others by D were obtained in other tubes and, in part, at other frequencies.

---

[242] Boyle, R. W., Lehmann, J. F., and Morgan, S. C., *Trans. Roy. Soc. Canada, III (3)*, **22**, 371-378 (1928).
[243] Boyle, R. W., and Taylor, G. B., *Idem*, **21**, 79-83 (1927), superseding **19**, 197-203 (1925).
[244] Biquard, P., *Compt. rend.*, **188**, 1230-1232 (1929).

31. WATER: ACOUSTICS

Table 89.—Velocity of Sound in Natural Waters
(See also Table 90)

Excepting the references to Lübcke and to Dorsey, the following has been taken with slight changes in form from the compilation of A. L. Foley.[237]

In all oceans the average vertical velocity for depths of 3.5 to 8.0 km (2.2 to 5 miles) is 1528 to 1529 meters/second; for lesser depths it is less. In general, the horizontal velocity increases by 0.2 per cent per 1 °C increase in temperature, 0.11 per cent per 100 meters increase in depth, and 0.1 per cent per 1 per cent increase in salinity (total salts).[249] A. B. Wood and H. E. Browne[250] represent the data (6 to 17 °C, salinities near 3.5 per cent) obtained by A. B. Wood, H. E. Browne, and C. Cochrane[251] by means of the equation $V = 4626 + 13.8t - 0.12t^2 + 3.73s$ ft/sec $= 1410.0 + 5.21t - 0.036t^2 + 1.137s$ m/sec, where the salinity is s parts per 1000 and the temperature is t °C. The practical application of acoustics to coastal and oceanic surveying has been discussed by H. G. Dorsey,[252] who gives data indicating that the velocity is independent of the intensity of the source.

Unit of $V = 1$ meter/sec; of depth $(D) = 1$ meter; of salinity $(s) = 1$ per cent by weight. Temp. $= t$ °C.

Ocean: Horizontal Velocity.

Place	D	s	t	V	Ref.[a]
Open ocean	13	3[b]	14.5	1503.5	M
Block Island Sound, N. Y.	30	3.35	3.0	1453.3	S
Long Island Sound, N. Y.	30		13	1492.3	E
Isle of Wight		3.51	6	1474	WBC
Isle of Wight		3.52	7	1478	WBC
Isle of Wight		3.5	16.95	1511	WBC

Ocean: Vertical Velocity

Place	D	V	Ref.[a]
N. Atlantic	1288	1520	HS
Carribean Sea	338	1478	HS
Carribean Sea	1771	1486	HS
Pacific	1185	1505	HS
Pacific	2962	1493	HS
All oceans	>3500	1528	HS

Fresh Water

Water	t	V	Ref.[a]
Lake Geneva	8.1	1435	CS
Seine River	15	1437	W
Seine River	30	1528	W
Seine River	50	1652	W
Seine River	60	1725	W

[a] References:
    CS   See Table 88, references.
    E   Eckhardt, E. A., *Phys. Rev. (2)*, **24**, 452-455 (1924).
    HS   Heck, N. H., and Service, J. C., *U. S. Coast and Geod. Sur.*, *Spec. Publ. 108* (1924).
    M   Marti, *Compt. rend.*, **169**, 281-282 (1919).
    S   Stephenson, E. B., *Phys. Rev. (2)*, **21**, 181-185 (1923).
    W   Wertheim, G., *Ann. de Chim. et Phys. (3)*, **23**, 434-475 (1848) = *Ann. d. Physik (Pogg.)*, **77**, 427-445, 544-571 (1849).
    WBC Ref. 251.

[b] Density at 14.9 °C was 1.0245 g/cm³.

Table 90.—Velocity of Sound in Sea-water
(See also Table 89)

Adapted from the detailed practical tables by N. H. Heck and J. H. Service [253] based upon the very extensive tables of V. Bjerknes *et al.*[254]

The following values of the velocity (V) at 0 °C and various depths, salinity 35 g/kg, have been taken directly from the tables of Heck and Service. The values they give for other salinities (s = 31 to 37 g/kg) and temperatures (t = 0 to t_m °C) may be reproduced very closely by means of the formula

$$V_{s,t,d} = V_{35,0,d} + 2.39t - 0.028t^2 + \{0.83 - 18.0(10^{-6})d + 0.0075t\}(s - 35)$$

the units being those named below.

Heck and Service compute the velocity for each successive layer of 200 fathoms, inferring the distribution of temperature and salinity from observations at three depths, *i.e.*, surface, 200 fathoms, and bottom, and average these velocities to obtain the mean velocity (V_m). They find in actual practice that this mean velocity differs from that computed from the measured depth and the observed time required for sound to pass to the bottom and back, by an average of about 0.2 per cent (about 3 m/sec), the probable error for a single determination being 6 to 8 times as great, and single determinations differing from V_m by 3.5 per cent (52.5 m/sec). (In his compilation,[237] A. L. Foley seems to have had in mind the probable error of a single determination when he stated that values computed by the method of Heck and Service may differ by 20 m/sec from the actual value). Such differences include the errors inherent in the method of echo-sounding as well as those arising from an attempt to infer the distribution of temperature and salinity from observations at only three depths.

Unit of $d = d_1 + d_2 = $ 1 fathom = 6 ft. = 182.88 cm; of V = 1 fathom/sec = 1.829 m/sec.

$d_1 \rightarrow$ d_2	100	300	500 $V_{35,0,d}$	700	900	100	300	500 t_m	700	900
0	793	796	799	804	806	22	22	22	20	12
1000	809	813	816	820	825	8	6	4	4	3
2000	826	831	833	836	839	3	3	3	3	2
3000	844	848	850	855	857	2	2	2	2	2
4000	861	863	866	870		2	2	2	2	

$d_1 \rightarrow$ d_2	100	300	500	700	900 $V_{35,0,d}$ (Unit of	1100 $V = 1$ m/sec)	1300	1500	1700	1900
0	1450	1456	1461	1470	1474	1479	1487	1492	1500	1509
2000	1510	1520	1523	1529	1534	1544	1551	1554	1564	1567
4000	1574	1578	1584	1591						

[245] Špakovskij, B., *Compt. rend. Acad. Sci. URSS (N. S.)*, 1934₃, 591-594 (1934).
[246] Schaffs, W., *Z. Physik*, 105, 658-675 (1937).
[247] Bergmann, L., *Physik. Z.*, 34, 761-764 (1933).
[248] Parthasarathy, S., *Proc. Indian Acad. Sci. (A)*, 2, 497-511 (1935).
[249] Lübcke, E., *Z. techn. Physik*, 10, 386-388 (1929).
[250] Wood, A. B., and Browne, H. E., *Proc. Phys. Soc. (London)*, 35, 183-193 (1923).
[251] Wood, A. B., Browne, H. E., and Cochrane, C., *Proc. Roy. Soc. (London) (A)*, 103, 284-303 (1923).

in the values obtained is believed to be distinctly less than 0.1 per cent. They are the values to be preferred.

L. G. Pooler [241] has measured the velocity at 25 to 75 °C and $\nu = 1269$ to 2715 cycles per second, using a recently developed formula for correcting for the effect of the walls of the vessel. His values lie 0.7 to 0.8 per cent below those of Randall and of Hubbard and Loomis (see Table 88).

To facilitate comparison of the several sets of data, an empirical formula (1) of arbitrary form, but reproducing Randall's values at 10, 30, 50, and 70 °C, was set up. The values so determined are designated as V_c.

$$V_c = 1404.4 + 4.8215t - 0.047562t^2 + 0.00013541t^3 \text{ m/sec} \quad (1)$$

The excess of the reported velocity (V_{obs}) above V_c is given in Table 88 for each of a number of determinations.

The observations of R. W. Boyle, J. F. Lehmann, and S. C. Morgan [242] at 80 kc/sec, of Hubbard and Loomis (200 to 400 kc/sec), of R. W. Boyle and G. B. Taylor [243] (29 to 570 kc/sec), of Randall (750 kc/sec), and of P. Biquard [244] (1360 kc/sec) indicate that the velocity varies little, if at all, with the frequency. In fact, there is no certain reported evidence that the velocity in an unbounded volume of water changes at all as the frequency is varied from 100 cycles/sec to 1.4 megacycles/sec. The smaller values reported for the lower frequencies (see Table 88, columns headed "P" and "Earlier data") probably arise from the unsatisfactory nature of the correction which is necessitated in such cases by the presence of the walls of the vessel.

More recently, B. Špakovskij,[245] using values of ν up to 1000 kc/sec, has concluded that up to that frequency, at least, the velocity is constant within his experimental error (about 1 per cent), and W. Schaffs [246] using $\nu = 16381$ kc/sec, found the velocity to be 1467 m/sec at 17 °C. This is only 0.3 per cent smaller than the value defined by formula (1). Whence it seems probable that the velocity is independent of the frequency, at least up to 16 megacycles/sec.

Acoustic Resistivity of Water.

The acoustic resistivity of a material is defined as the amount by which the r.m.s. pressure in a plane sound wave must exceed the static pressure in order to confer upon the medium a unit r.m.s. velocity. (r.m.s. = square root of the mean square.) It is equal to $\sqrt{E\rho} = V\rho$, where E = bulk modulus of the material, ρ is its density, and V = velocity of sound = $\sqrt{E/\rho}$. For water at usual temperatures, ρ is essentially unity, and the acoustic resistivity is numerically equal to the velocity of sound. If the velocity is V meters per second, the resistivity is V gram/mm²·sec. In each, the modulus is that corresponding to the conditions existing during

---

[252] Dorsey, H. G., *J. Acoust. Soc. Amer.*, **3**, 428-442 (1932).
[253] Heck, N. H., and Service, J. H., *U. S. Coast and Geod. Sur., Spec. Publ. 108* (1924).
[254] Bjerknes, V., et al., "Dynamic Meteorology and Hydrography," *Carnegie Inst. of Washington Publ.* **88** (1910).

the passage of the wave, which approximate those characteristic of adiabatic compression and expansion.[255]

H. G. Dorsey [252] remarks that the increase in the acoustic resistance of sea-water with increase in temperature may not arise solely from the change in temperature, but may be due in part to an increase in the amount of suspended matter. He states that water churned up by the propeller of a ship absorbs sound completely, and that the reflectivity of ocean bottoms decreases in the order: soft mud (best), hard sand, broken coral, sea-grass (poorest).

In a more recent paper [256] he writes (p. 299): "The assumption that the acoustical impedance of warm water is greater than cold has not been disproved and work in the warm water of the Gulf of Mexico tends to confirm the assumption."

Absorption of Sound by Water.

At audio frequencies the absorption of sound by water is small, and is determined by the viscosity and thermal conductivity. But at high—ultrasonic—frequencies it is much greater and seems to involve something of the nature of molecular resonance. Furthermore, at those high frequencies the radiation has a degassing effect; and this absorbs additional energy if the water is not gas-free. (See C. Sörensen.[257])

Sörensen [257] has found that, although the absorption is greater if the water contains gas, the heating of the water by the radiation is the same as if the water were gas-free, the additional absorption arising from the work required to drive out the gas. He found as follows: for $\nu = 194, 380$, and 530 kc/sec, the work expended in removing the gas was, respectively, 51.2, 72.6, and 87.4 kilowatts per cm^3 of gas removed. The rate of removal was not constant, but steadily decreased as the water became more and more nearly gas-free.

He stated that the absorption passes through a maximum at some frequency below 194 kc/sec. H. Oyama [258] has stated that the heating of the water is a maximum at about 700 to 800 kc/sec, which seems to be entirely incompatible with the values in Table 91.

Claeys found the absorption to be greater in narrow tubes than in wider ones, and has suggested that the difference is associated with the presence of convection currents [259]; but Sörensen [260] found the absorption to be independent of the diameter of the tube.

Sörensen [260] has found that the coefficient of absorption (k, see Table 91) decreases as the temperature rises, but that this decrease is less rapid

---

[255] Sabine, P. E., *Int. Crit. Tables*, 6, 459 (1929).
[256] Dorsey, H. G., *J. Acoust. Soc. Amer.*, 7, 286-299 (1936).
[257] Sörensen, C., *Ann. d. Physik (5)*, 26, 121-137 (1936); 27, 70-74 (1936).
[258] Oyama, H., *Sci. Abs. (A)*, 39, 292 (1936) ← *J. Inst. Elec. Eng. (Japan)*, 55, 985-989 (1935).
[259] Claeys, J., and Sack, H., *Acad. Roy. de Belg., Bull. Cl. Sci. (5)*, 23, 659-671 (1937).
[260] Sörensen, C., *Ann. d. Physik (5)*, 27, 70-74 (1936).

31. WATER: ACOUSTICS

than that of $\eta/\rho v^3$, where η is the viscosity, ρ the density, and v the velocity of sound. The variation is not linear in the temperature, but near room temperatures dk/dt is quite close to -0.00024 cm^{-1} per 1 °C, the frequency being between 200 and 1000 kc/sec. E. Baumgardt,[261] on the other hand, concluded that k is proportional to $\eta/\rho v^3$, and reported the following values, all for $\nu = 7958$ kc/sec:

t	18.6	22.2	22.5	31.2	39.5 °C
$100k$	3.46	3.06	2.98	2.42	2.08 cm^{-1}

That k is not always proportional to ν^2 was pointed out by P. Biquard,[262] and is obvious from the data in Table 91. With toluene, Biquard[263] observed a lateral scattering of the radiation.

Claeys, Errera, and Sack[264] have suggested that the increased absorption at high frequencies may arise in part from a kind of hysteresis in the adiabatic compressibility of the water.

Table 91.—Absorption of Ultrasonic Radiation by Water

The coefficient k is that defined by the relation $I = I_0 e^{-kx}$ where $I_0 - I$ is the reduction in the intensity of a plane wave while traveling a distance x. Data have been published both in terms of k and of the corresponding exponent for the reduction in amplitude, which is only half as great as k, and it is not always easy to determine to which they refer.

Unit of $\nu = 10^6$ cycles/sec; of $k = 1$ cm^{-1}; of $k/\nu^2 = 10^{-14}$ sec^2/cm. Room temp.

ν	k	k/ν^2	Ref.[a]
0.194	0.017	45	S
0.380	0.015	10	S
0.530	0.014	5.0	S
0.950	0.011	1.22	S
1.44	0.00135	0.065	CES
2.03	0.00330	0.080	CES
2.79	0.0030	0.038	F
4.77	0.0121	0.053	CES
7.55	0.035	0.062	Bi
7.96	0.0346	0.055	Ba
7.97	0.040	0.064	Bi
8.37	0.027	0.038	F
11.14	0.066	0.054	CES
54	1.39	0.048	Bär
69	2.28	0.048	Bär
83	3.52	0.051	Bär

[a] References:
- Ba   Baumgardt, E.[261]
- Bär   Bär, R., *Helv. Phys. Acta*, **10**, 332-337 (1937).
- Bi   Biquard, P.[262,263]
- CES   Claeys, J., Errera, J., and Sack, H.[264]
- F   Fox, F. E., *Phys. Rev. (2)*, **52**, 973-981 (1937).
- S   Sörensen, C., *Ann. d. Physik (5)*, **26**, 121-137 (1936) = *Diss.*, Greifswalder, 1935.

---

[261] Baumgardt, E., *Compt. rend.*, **202**, 203-204 (1936).
[262] Biquard, P., *Ann. d. Phys. (11)*, **6**, 195-304 (1936).
[263] Biquard, P., *Compt. rend.*, **202**, 117-119 (1936).
[264] Claeys, J., Errera, J., and Sack, H., *Idem*, **202**, 1493-1494 (1936).
[265] Biquard, P., *Compt. rend.*, **193**, 226-229 (1931).

32. Pressure-Volume-Temperature Associations for Water
(For saturated water, see Section 88)

Water under the pressure of its pure saturated vapor is called saturated water; that under a higher pressure has been called compressed water; that under a lower pressure may be called dilated water. Above the critical temperature the substance will be classed as compressed water if the specific volume is less than that (3.1 cm³/g) at the critical point; as dilated vapor if the specific volume is greater than at the critical point. These terms are to be so understood wherever they occur in this compilation.

Uniformity of Water.

Until the discovery of the isotopes of hydrogen and oxygen—that is, until very recently—water as commonly purified by careful distillation was

(To p. 202)

Table 92.—Effect of Dissolved Air on the Density of Water
(See p. 251)

ρ = density of air-free water under a pressure of 1 atm; ρ_a = density of water saturated with air at a pressure of 1 atm.

Unit of ρ and of $\rho_a = 1$ g/cm³. Temp. = t °C.

t	$10^6(\rho-\rho_a)$	Observer[a]	t	$10^6(\rho-\rho_a)$	t	$10^6(\rho-\rho_a)$
5° to 8 °C	3.0	Chappuis		Marek		Marek
15.6	1.89	Frivold	6	3.3	13	2.7
0	2.5	Marek	7	3.4	14	2.5
1	2.7	Marek	8	3.4	15	2.2
2	2.9	Marek	9	3.3	16	1.9
3	3.1	Marek	10	3.2	17	1.6
4	3.2	Marek	11	3.1	18	1.2
5	3.3	Marek	12	2.9	20	0.4

[a] See text for comments on Marek's work and references to his and to Chappuis' papers. Frivold.[289]

---

[266] Lamb, A. B., and Lee, R. E., *J. Am. Chem. Soc.*, **35**, 1667-1693 (1681) (1913).
[267] Hall, N. F., and Jones, T. O., *J. Am. Chem. Soc.*, **58**, 1915-1919 (1936); Gabbard, J. L., and Dole, M., *Idem*, **59**, 181-185 (1937).
[268] Christiansen, W. H., Crabtree, R. W., and Laby, T. H., *Nature*, **135**, 870 (L) (1935).
[269] Mendelejev, J., *Compt. rend. Acad. Sci. URSS*, **8**, 105-108 (1935₃).
[270] Dole, M., and Wiener, B. Z., *Science (N. S.)*, **81**, 45 (1935).
[271] Peel, J. B., Robinson, P. L., and Smith, H. C., *Nature*, **120**, 514-515 (1927).
[272] Stott, V., and Bigg, P. H., *Int. Crit. Tables*, **3**, 24-26 (1928).
[273] Chappuis, P., *Trav. et Mém. Bur. Int. Poids et Mes.*, **13**, D1-D40 (1907).
[274] Thiesen, M., Scheel, K., and Diesselhorst, H., *Wiss. Abh. Phys.-Techn. Reichs.*, **3**, 1-70 (1900).
[275] Tilton, L. W., and Taylor, J. K., *J. Res. Nat. Bur. Stand.*, **18**, 205-214 (RP971) (1937).
[276] Bridgman, P. W., *Int. Crit. Tables*, **3**, 40 (1928); as corrected in accordance with the errata published in Vol. 7.
[277] Bridgman, P. W., *Proc. Am. Acad. Arts Sci.*, **47**, 439-558 (1912); **48**, 307-362 (1913).
[278] Bridgman, P. W., *J. Chem'l Phys.*, **3**, 597-605 (1935).

32. WATER: P-V-T DATA

Table 93.—Density of Compressed Water at a Pressure of 1 Atmosphere

(For sea-water see Table 108.)

Density = ρ, temperature = $(t_1 + t_2) = t$ °C. In the second column the complete value of ρ is given or indicated; in the following columns, only the last three or four digits, the preceding digits being understood to be those in the left-hand section of the second column, either on or above the line, unless there is a line over the first of the tabulated digits, in which case the immediately preceding digit will differ by one unit from that just specified. For example, Mohler finds $\rho = 0.997292$ at -12 °C and 0.996931 at -13 °C.

Unit of $\rho = 1$ gram per milliliter $= 0.999973$ g/cm³. Temp. $= (t_1+t_2) = t$ °C.

I. J. F. Mohler.[290]

$t_1 \rightarrow$ t_2	0	−1	−2	−3	−4	−5	−6	−7	−8	−9
−0	0.999 868	773	673	553	380	176	$\overline{9}$50	$\overline{7}$20	$\overline{5}$01	$\overline{2}$49
−10	0.997 935	636	292	$\overline{9}$31						

II. C. Despretz.[280] Not given in *International Critical Tables;* he studied the expansion of water between −9 and +100 °C.

t	0	−1	−2	−3	−4	−5	−6	−7	−8	−9
ρ	0.999 873	786	692	578	438	302	082	$\overline{8}$48	$\overline{6}$28	$\overline{3}$72
t	0	1	2	3	4	5	6	7	8	9
ρ	0.999 873	925	967	992	$\overline{0}$00	992	969	929	878	812

III. Revised Chappuis table.[275] See text. In the subsidiary columns C and T to the right of the ρ values in columns $t_1 = 0.0$ and $t_1 = 0.5$ are given the amounts, in units of the last place tabulated, by which each of the corresponding values for $t_1 = 0.0$ and $t_1 = 0.5$ in the tables published by Chappuis and by Thiesen, Scheel, and Diesselhorst, respectively, exceeds that here tabulated. Over the intermediate 0.5 °C range these differences may be linearly interpolated. Thus both of those tables may be recovered from this; and so may be the one in the *International Critical Tables*, that being merely the average of the other two. For example, at 17.3 °C this table gives $\rho = 0.9987515$, the C value is smaller than this by 10, the T value by 27, and the I.C.T. by 18 units in the last place, making those values 0.9987505, 0.9987488, and 0.9987497 respectively; from the tables themselves one finds exactly these same values. Under Δ is given the average increase in ρ per 0.1 °C increase in t for the one degree range covered by the line in which the value stands, the unit of Δ being that of the last tabulated digit of ρ.

Table 93.—(Continued)

$t_1 \rightarrow$ $t_2 \downarrow$	0.0	C	T	0.1	0.2	0.3	0.4	0.5	C	T	0.6	0.7	0.8	0.9	$+0.1°$ Δ
0	0.999 8676	+5	0	8743	8808	8871	8933	8993	+3	+1	9051	9107	9161	9214	+58.9
1	9265	+2	+1	9314	9362	9407	9451	9493	+1	+2	9534	9573	9610	9645	+41.3
2	9678	+1	+2	9710	9740	9769	9796	9821	0	+1	9844	9866	9886	9905	+24.4
3	9922	0	0	9937	9950	9962	9972	9981	0	0	9988	9993	9997	9999	+7.8
4	1.000 0000	0	0	9999	9996	9992	9986	9979	0	−1	9970	9960	9948	9934	−8.1
5	0.999 9919	0	−1	9902	9884	9864	9843	9820	−1	−2	9796	9770	9742	9713	−23.6
6	9683	−1	−3	9651	9618	9583	9546	9508	−1	−3	9469	9428	9386	9342	−38.6
7	9297	−1	−4	9250	9202	9153	9102	9049	0	−4	8995	8940	8883	8825	−53.2
8	8765	−1	−6	8704	8642	8578	8513	8446	−1	−7	8378	8309	8238	8166	−67.3
9	8092	−1	−8	8017	7941	7863	7784	7704	0	−10	7622	7539	7454	7368	−81.1
10	7281	+1	−10	7193	7103	7012	6919	6825	+1	−11	6730	6634	6536	6437	−94.5
11	6336	−5	−12	6234	6131	6027	5922	5815	−10	−14	5706	5597	5486	5374	−107.5
12	5261	−13	−15	5146	5030	4913	4795	4675	−15	−16	4554	4432	4309	4184	−120.2
13	4059	−19	−18	3932	3803	3674	3543	3411	−20	−19	3278	3143	3007	2870	−132.7
14	2732	−20	−20	2593	2453	2311	2168	2024	−21	−21	1879	1732	1584	1436	−144.6
15	1286	−20	−22	1134	0982	0828	0674	0518	−19	−23	0360	0202	0043	9882	−156.5
16	0.998 9721	−16	−24	9558	9394	9229	9062	8895	−14	−25	8726	8557	8386	8214	−168.0
17	8041	−12	−27	7867	7691	7515	7337	7158	−8	−27	6979	6798	6616	6433	−179.3
18	6248	−4	−28	6063	5877	5689	5501	5311	−2	−30	5120	4928	4735	4541	−190.2
19	4346	+1	−31	4150	3953	3754	3555	3355	+3	−32	3153	2950	2747	2542	−201.0
20	2336	+7	−33	2130	1922	1713	1503	1292	+9	−34	1080	0867	0653	0438	−211.5

[279] Despretz, C., see *Compt. rend.*, **4**, 124-130 (1837) → *Ann. d. Physik (Pogg.)*, **41**, 58-71 (1837).
[280] Despretz, C., *Ann. de Chim. et phys.* (2), **70**, 5-81 (1839).
[281] Salm-Horstmar, *Ann. d. Physik (Pogg.)*, **62**, 283-284 (1844).
[282] Keenan, J. H., *Mech. Eng.*, **53**, 127-131 (1931).
[283] Marek, W. J., *Ann. d. Physik (Wied.)*, **44**, 171-172 (1891).

32. WATER: P-V-T DATA

21	0.998 0221	+12 −35	0004	9̄786	9̄567	9̄346	9̄125	+14 −36	8̄903	8̄679	8̄455	8̄230	−221.8
22	0.997 8003	+16 −37	7776	7547	7318	7088	6856	+17 −38	6624	6390	6156	5921	−231.9
23	5684	+18 −39	5447	5208	4969	4729	4487	+19 −40	4245	4002	3758	3512	−241.8
24	3266	+20 −41	3019	2771	2522	2272	2021	+19 −42	1769	1516	1262	1007	−251.5
25	0751	+19 −43	0494	0237	9̄978	9̄718	9̄458	+18 −44	9̄196	8̄934	8̄671	8̄406	−261.0
26	0.996 8141	+17 −44	7875	7608	7340	7071	6801	+16 −46	6530	6258	5986	5712	−270.4
27	5437	+14 −46	5162	4886	4608	4330	4051	+11 −47	3771	3490	3208	2926	−279.5
28	2642	+10 −48	2358	2072	1786	1499	1211	+6 −49	0922	0632	0341	0049	−288.5
29	0.995 9757	+4 −49	9463	9169	8874	8578	8281	+1 −50	7983	7684	7384	7084	−297.4
30	6783	−3 −51	6480	6177	5874	5569	5263	−5 −51	4956	4649	4341	4032	−306.1
31	3722	−8 −52	3411	3099	2787	2473	2159	−12 −53	1844	1528	1211	0894	−314.7
32	0575	−14 −53	0256	9̄936	9̄615	9̄293	8̄970	−16 −54	8̄647	8̄322	7̄997	7̄671	−323.1
33	0.994 7344	−19 −54	7016	6688	6359	6028	5698	−22 −55	5366	5033	4700	4365	−331.4
34	4030	−23 −55	3694	3358	3020	2682	2343	−25 −56	2003	1662	1320	0978	−339.5
35	0635	−25 −57	0291	9̄946	9̄600	9̄254	8̄907	−24 −57	8̄559	8̄210	7̄860	7̄510	−347.6
36	0.993 7159	−23 −58	6807	6454	6100	5746	5391	−22 −58	5035	4678	4321	3962	−355.5
37	3604	−19 −59	3244	2883	2522	2160	1797	−15 −59	1433	1068	0703	0337	−363.4
38	0.992 9970	−10 −59	9603	9234	8865	8495	8125	−5 −60	7753	7381	7008	6634	−371.0
39	6260	+3 −60	5884	5508	5132	4754	4376	+13 −61	3997	3617	3236	2855	−378.7
40	2473	+24 −61	2090	1707	1323	0938	0552	+35 −61	0165	9̄778	9̄390	9̄001	−386.1
41	0.991 8612	+49 −62	8221	7830	7439	7046	6653	−62	6259	5864	5469	5073	−393.6
42	4675	−62											

[284] Chappuis, P., *Trav. et Mém. Bur. Int. Poids et Mes.*, **14**, D1-D63 (D4) (1910); Marek, W. J., *Idem*, **3**, (D81-D90) (1884).
[285] Adeney, W. E., Leonard, A. G. G., and Richardson, A., *Phil. Mag.*, (6), **45**, 835-845 (1923).
[287] Emeléus, H. J., *et al.*, *J. Chem. Soc. (London)* 1934, 1207-1219 (1934).
[288] Richards, T. W., and Harris, G. W., *J. Am. Chem. Soc.*, **38**, 1000-1011 (1916).
[289] Frivold, O. E., *Physik. Z.*, **21**, 529-534 (1920); **24**, 86-87 (1923).
[290] Mohler, J. F., *Phys. Rev.*, **35**, 236-238 (1912). In *Int. Crit. Tables*, **3**, 26 (1928), these have been rounded off to 4 significant figures.

Table 93.—*(Continued)*

IV. M. Thiesen.[290a] Included in I.C.T.[272]

$t_1 \rightarrow$ t_2	0	1	2	3	4	5	6	7	8	9
					ρ					
40°	0.99 22₄	18₆	14₇	10₇	06₆	02₄	$\overline{9}$8₂	$\overline{9}$4₀	$\overline{8}$9₆	$\overline{8}$5₂
50	0.98 80₇	76₂	71₅	66₉	62₁	57₃	52₅	47₅	42₅	37₅
60	0.98 32₄	27₂	22₀	16₇	11₃	05₉	00₅	$\overline{9}$5₀	$\overline{8}$9₄	$\overline{8}$3₈
70	0.97 78₁	72₃	66₆	60₇	54₈	48₉	42₉	36₈	30₇	24₅
80	0.97 18₃	12₁	05₇	$\overline{9}$9₄	$\overline{9}$3₀	$\overline{8}$6₅	$\overline{8}$0₀	7₃₄	$\overline{6}$6₈	$\overline{6}$0₁
90 ·	0.96 53₄	46₇	39₉	33₀	26₁	19₂	12₂	05₁	$\overline{9}$8₁	$\overline{9}$0₉
100	0.95 83₈									

(Cont'd from p. 198)

regarded as a perfectly definite, homogeneous substance, the same the world over. And this, in spite of certain observations indicating the contrary, some of which will be mentioned presently.

Since the discovery of the isotopes all this has changed. We now know that water is not a simple substance, the same everywhere, but is a mixture in which the relative amounts of the several constituents vary with the source and with the manner of purification. Fortunately, this variation is so small in the purified waters commonly used in the study of the properties of water that its effect upon the observed values of those properties is entirely negligible in most cases, thus justifying one in speaking, as in this compilation, of the properties of the ordinary water substance. But in those few cases in which extreme precision of measurement has been attained— in which errors from other sources do not exceed a part in a million or thereabouts—it is necessary to consider whether differences in the composition of different specimens of "pure water" may cause significant differences in the property being studied.

Such a property is the density of water. Over the range 0 to 40 °C values are published to one part in 10 million. But there are as yet no data that enable one to say with certainty whether the density of the "pure water" commonly used in such work is definite to that precision. It probably is not.

About 25 years ago, A. B. Lamb and R. E. Lee [266] reported that the densities of various samples of distilled water, expected to be identical, might differ by as much as 8 parts in 10^7, even when great care was taken. This was long before the discovery of the isotopes.

If a sample of water contained 1 D_2 to 6500 H_2—the average ratio [267]— then removing the D_2 would decrease its density by about 17 parts in a million. W. H. Christiansen, R. W. Crabtree, and T. H. Laby [268] have reported that the density of rain-water is reduced by 12.7 parts in 10^6 by

(To p. 206)

32. WATER: P-V-T DATA

Table 94.—Specific Volume of Compressed Water at a Pressure of 1 Atmosphere

Derived from the densities as given in the corresponding sections of Table 93, where references and comments will be found.

Specific volume = v^*, temperature is $(t_1 + t_2) = t$ °C.

In the second column the complete value of v^* is given or indicated; in the following columns, only the last three or four digits, the preceding digits being understood to be those in the left-hand section of the second column, either on or above the line, unless there is a line over the first of the tabulated digits, in which case the immediately preceding digit will differ by one unit from that just specified. For example, Mohler finds $v^* = 1.002\overline{7}15$ at $-12°$ C and 1.003078 at -13 °C.

It is interesting to notice that, whereas the specific volume of water at 4 °C is 1.000000 ml/g (= 1.000027 cm³/g) when under a pressure of one normal atmosphere, it is 0.999973 ml/g (= 1.000000 cm³/g) when the pressure is 1.53 atm (see Table 105).

Unit of $v^* = 1$ ml/g $= 1.000027$ cm³/g.   Temp. $= (t_1 + t_2) = t$ °C.

I. J. F. Mohler.[290]

$t_1 \rightarrow$ t_2	0	−1	−2	−3	−4 v^*	−5	−6	−7	−8	−9
0	1.000 132	227	327	447	620	825	$\overline{0}$51	$\overline{2}$82	$\overline{5}$01	$\overline{7}$54
−10	1.002 069	369	715	$\overline{0}$78						

II. C. Despretz.[280]

t	0	−1	−2	−3	−4	−5	−6	−7	−8	−9
v^*	1.000 127	214	308	422	562	699	918	$\overline{1}$53	$\overline{3}$73	$\overline{6}$31
t	0	1	2	3	4	5	6	7	8	9
v^*	1.000 127	075	033	008	000	008	031	071	122	188

III. Revised Chappuis table, 1937. In the subsidiary columns C and T to the right of the v^* values in columns $t_1 = 0.0$ and $t_1 = 0.5$ are given the amounts, in units of the last place tabulated, by which each of the corresponding values for $t_1 = 0.0$ and $t_1 = 0.5$ in the tables published by Chappuis and by Thiesen, Scheel, and Diesselhorst, respectively, exceeds that here tabulated. Over the intermediate 0.5 °C range these differences may be linearly interpolated. Thus both of those tables may be recovered from this, and so may be the one in the *International Critical Tables*, that being merely the average of the other two. Example: At 17.3 °C this table gives $v^* = 1.0012501$; the C value is 9, the T value is 27, and the I.C.T. value is 18 units in the last place greater than that. Whence the C, T, and I.C.T. values are, respectively, 1.0012510, 1.0012528, and 1.0012519, agreeing exactly with those in the tables.

Table 94.—(Continued)

$t_1 \rightarrow$ t_2	0.0	C	T	0.1	0.2	0.3	0.4 v^*	0.5	C	T	0.6	0.7	0.8	0.9
0	1.000 1324	−5	0	1257	1192	1129	1067	1007	−3	−1	0949	0893	0839	0786
1	0735	−2	−1	0686	0638	0593	0549	0507	−2	−2	0466	0428	0390	0355
2	0322	−1	−2	0290	0260	0231	0204	0179	0	−1	0156	0134	0114	0095
3	0078	0	0	0063	0050	0038	0028	0019	−1	0	0012	0007	0003	0001
4	0000	0	0	0001	0004	0008	0014	0021	−1	+1	0030	0040	0052	0066
5	0081	0	+1	0098	0116	0136	0157	0180	0	+2	0204	0230	0258	0287
6	0317	0	+3	0349	0382	0417	0454	0492	0	+3	0531	0572	0614	0658
7	0703	0	+4	0750	0798	0847	0898	0951	+1	+5	1005	1060	1117	1175
8	1235	+1	+6	1296	1358	1422	1487	1554	+1	+7	1622	1692	1763	1835
9	1908	+1	+9	1983	2060	2138	2217	2297	−1	+9	2379	2462	2547	2632
10	2720	−2	+10	2808	2898	2989	3082	3176	−2	+11	3271	3368	3466	3565
11	3665	+5	+13	3767	3869	3974	4080	4187	+10	+14	4296	4405	4516	4628
12	4741	+14	+15	4856	4972	5089	5208	5328	+15	+15	5448	5571	5694	5819
13	5945	+18	+17	6072	6201	6330	6461	6594	+19	+18	6727	6862	6997	7134
14	7273	+20	+19	7412	7553	7695	7838	7982	+21	+21	8128	8275	8423	8572
15	8722	+19	+22	8874	9026	9180	9335	9491	+18	+23	9649	9807	9967	$\bar{0}$128
16	1.001 0290	+15	+24	0453	0617	0783	0950	1117	+13	+26	1286	1456	1628	1800
17	1974	+11	+26	2148	2324	2501	2679	2858	+8	+28	3038	3220	3402	3586
18	3770	+5	+29	3956	4143	4331	4520	4711	+1	+30	4902	5094	5288	5483
19	5678	−1	+31	5875	6073	6272	6472	6673	−4	+32	6875	7079	7283	7488
20	7695	−8	+33	7902	8111	8321	8532	8743	−10	+34	8956	9170	9385	9601

32. WATER: P-V-T DATA

21	1.001	9818		$\bar{0}$036	$\bar{0}$255	$\bar{0}$475	$\bar{0}$697	$\bar{0}$919		$\bar{1}$142	$\bar{1}$366	$\bar{1}$592	$\bar{1}$818
22	1.002	2045	-12 $+35$	2274	2503	2734	2965	3198	-15 $+36$	3431	3666	3901	4138
23		4375	-16 $+38$	4614	4853	5094	5335	5578	-18 $+38$	5822	6066	6312	6558
24		6806	-19 $+39$	7054	7304	7554	7805	8058	-20 $+40$	8311	8566	8821	9077
25		9335	-20 $+41$	9593	9852	$\bar{0}$112	$\bar{0}$374	$\bar{0}$636	-20 $+42$	$\bar{0}$899	$\bar{1}$163	$\bar{1}$428	$\bar{1}$694
			-20 $+43$						-19 $+44$				
26	1.003	1961	-18 $+45$	2229	2498	2767	3038	3310	-17 $+45$	3582	3856	4131	4406
27		4682	-13 $+47$	4960	5238	5517	5798	6079	-12 $+47$	6361	6644	6928	7212
28		7498	-10 $+52$	7785	8072	8361	8650	8940	-7 $+49$	9232	9524	9817	$\bar{0}$111
29	1.004	0406	-5 $+51$	0702	0999	1296	1595	1894	-2 $+50$	2194	2496	2798	3101
30		3405	$+2$ $+51$	3710	4016	4322	4630	4938	$+5$ $+52$	5247	5558	5869	6180
31	1.005	6493	$+8$ $+53$	6807	7122	7437	7754	8071	$+11$ $+53$	8389	8708	9028	9349
32		9670	$+14$ $+54$	9993	$\bar{0}$316	$\bar{0}$640	$\bar{0}$966	$\bar{1}$292	$+15$ $+54$	$\bar{1}$618	$\bar{1}$946	$\bar{2}$275	$\bar{2}$604
33		2934	$+17$ $+55$	3266	3598	3930	4264	4599	$+19$ $+55$	4934	5271	5608	5946
34		6285	$+22$ $+56$	6624	6965	7306	7649	7992	$+23$ $+56$	8336	8680	9026	9372
35		9720	$+24$ $+57$	$\bar{0}$068	$\bar{0}$417	$\bar{0}$767	$\bar{1}$117	$\bar{1}$469	$+23$ $+57$	$\bar{1}$821	$\bar{2}$174	$\bar{2}$528	$\bar{2}$883
36	1.006	3239	$+22$ $+58$	3595	3952	4310	4669	5029	$+21$ $+59$	5390	5751	6113	6476
37		6840	$+19$ $+59$	7205	7570	7937	8304	8672	$+14$ $+59$	9040	9410	9780	$\bar{0}$152
38	1.007	0524	$+9$ $+60$	0896	1270	1644	2020	2396	$+3$ $+60$	2772	3150	3528	3908
39		4288	-5 $+61$	4669	5051	5433	5816	6200	-14 $+62$	6585	6971	7357	7744
40		8132	-24 $+62$	8521	8911	9301	9692	$\bar{0}$084	-36 $+63$	$\bar{0}$477	$\bar{0}$871	$\bar{1}$265	$\bar{1}$660
41	1.008	2056	-50 $+63$	2453	2850	3248	3647	4047	$+63$	4448	4849	5251	5654
42		6058											

Table 94.—(Continued)

IV. M. Thiesen,[290a] I.C.T.[272]

$t_1 \rightarrow$ t_2	0	1	2	3	4 v^*	5	6	7	8	9
40	1.00 78$_2$	82$_1$	86$_1$	90$_1$	94$_3$	98$_5$	$\overline{0}$2$_8$	$\overline{0}$7$_2$	$\overline{1}$1$_6$	$\overline{1}$6$_1$
50	1.01 20$_7$	25$_4$	30$_1$	34$_9$	39$_8$	44$_8$	49$_8$	54$_8$	60$_1$	65$_2$
60	70$_5$	75$_8$	81$_3$	86$_7$	92$_3$	97$_9$	$\overline{0}$3$_6$	$\overline{0}$9$_3$	$\overline{1}$5$_1$	$\overline{2}$1$_0$
70	1.02 27$_0$	33$_0$	39$_0$	45$_2$	51$_3$	57$_6$	63$_9$	70$_3$	76$_8$	83$_3$
80	89$_9$	96$_5$	$\overline{0}$3$_2$	$\overline{0}$9$_9$	$\overline{1}$6$_8$	$\overline{2}$3$_7$	$\overline{3}$0$_6$	$\overline{3}$7$_6$	$\overline{4}$4$_7$	$\overline{5}$1$_8$
90	1.03 59$_0$	66$_3$	73$_6$	71$_0$	88$_4$	95$_9$	$\overline{0}$3$_5$	$\overline{1}$1$_1$	$\overline{1}$8$_8$	$\overline{2}$6$_5$
100	1.04 34$_3$									

(Continued from p. 202)

removing its D_2, and that in the fractional distillation of tap water the first and the last fractions differed in density by 20.0 parts in 10^6. Whence they concluded "that, if precise relative determinations of the density of water which had been repeatedly distilled had been made at any time since accurate thermometry has been available, they would have disclosed the fact that natural water is not a simple substance." It seems probable that they intend the reader to understand that the distillation was to be fractional.

J. Mendelejev [269] has reported that the density of the purified water from Lake Baikal increases with the depth from which the sample was drawn, water from 1650 meters being 56 in 10^7 greater in density than that from the surface. This indicates a gravitational separation of the constituents. (Before distillation, the difference in the densities was about 120 in 10^7, over twice that after distillation.)

Variability in Water.

In the preceding paragraphs we have considered possible differences between different samples; here we consider possible changes in the same sample, changes arising from other factors than the existing temperature and pressure. The volume of a solution is not, in general, equal to the sum of the volumes of the solute and the solvent. This well-known fact does not concern us now. Some, accepting the idea that water is a mixture of polymers, have advanced the idea that the relative numbers of the several polymers can be disturbed, at least temporarily, but for relatively long periods, by various means, such as antecedent heating, chilling or freezing (see Section 25, *Establishment of equilibrium*). And it is conceivable that very minute amounts of a soluble impurity may markedly change the polymerization. But no evidence that any of these hypothetical effects are actually of practical significance has come to my attention.

(Go to p. 225)

[290a] Thiesen, M., *Wiss. Abh. Phys.-Techn. Reichs.*, **4**, 1-32 (1904).

32. WATER: P-V-T DATA

Table 95.—Specific Volume of Compressed Water: Pressure Exceeding 1 Atmosphere

(For sea-water see Table 108)

The table is divided into the following sections:

I. Amagat, 0 to 198 °C, 1 to 1000 atm, 13 temperatures, steps of 25 or 50 atm.

II. Amagat, 0 to 49 °C, 1 to 3000 atm, 10 temperatures, steps of 100 atm.

III. Bridgman, -20 to $+80$ °C, 1 to 12 000 atm, 13 temperatures, steps of 500 atm.

IV. Bridgman, -20 to $+100$ °C, 1 to 12 000 kg*/cm², 11 temperatures, steps of 500 or 1000 kg*/cm².

V. Tammann and Jellinghaus, -14 to $+15$ °C, 1 to 1500 kg*/cm², 26 temperatures, steps of 100 kg*/cm².

VI. Smith and Keyes, 0 to 360 °C, 1 to 350 atm, every 10°, steps of 25 or 50 atm.

VII. Tammann and Rühenbeck, 20 to 650 °C, 1 to 2500 kg*/cm², 9 temperatures, steps of 100 kg*/cm².

VIII. Adams, 25 °C, 1 to 12 000 bars, steps of 500 or 1000 bars.

IX. Trautz and Steyer; reference only.

In every case, values at 1 atm have been retained or inserted so as to facilitate comparisons with the preceding tables.

In the first 5 sections the specific volume has been indicated by the amount $(10^{-4}\Delta)$ by which it falls short of 1 ml/g; in the other three sections the specific volume is given directly. Except in the last section, successive differences have been printed, in distinctive type, between the values from which they have been derived. These differences serve several purposes. They show directly irregularities in the "run" of the values, some of which are disturbingly great; they show at once how the mean temperature coefficient of expansion at constant pressure, and the mean compressibility at constant temperature, each for a tabular step, vary throughout the range covered by the section, and facilitate their evaluation at any point in the table; and they furnish one more means for comparing the results obtained by different observers.

Unit of $v^* = 1$ ml/g $= 1.000027$ cm³/g (for this table the distinction between the ml and the cm³ is entirely negligible); of $P = 1$ atm $= 1.03323$ kg*/cm² $= 1.01325$ bars; of $p = 1$ kg*/cm²; of $p_b = 1$ bar. Temp. $= t$ °C

I. E. H. Amagat.[291] His values are expressed in terms of the specific volume at 0 °C and 1 atm; those here given were obtained by multiplying each of his by 1.0001319 so as to reduce them to the same basis as that of Tables 93 and 94; $v^* = 1 - 10^{-4}\Delta$.

[291] Amagat, E. H., *Ann. de Chim. et phys.* (6), **29**, 68-136, 505-574 (1893).

Table 95.—*(Continued)*

(Lines continued on p. 209)

$t \to$ P	0		5		10		15 Δ		20		30		40	
1	−1.3	−1.2	−0.1	+2.6	−2.7	6.0	−8.7	9.0	−17.7	25.7	−43.4	33.6	−77.0	42.5
	12.6		12.3		12.0		11.9		11.8					
25	+11.3	−0.9	+12.2	2.9	+9.3	6.1	+3.2	9.1	−5.9					
	12.9		12.4		12.3		12.0		11.9		23.6		22.5	
50	24.2	−0.4	24.6	3.0	21.6	6.4	15.2	9.2	+6.0	25.8	−19.8	34.7	−54.5	42.2
	12.7		12.1		11.8		11.6		11.4					
75	36.9	+0.2	36.7	3.3	33.4	6.6	26.8	9.4	17.4					
	12.5		12.0		11.7		11.4		11.3		22.1		22.3	
100	49.4	+0.7	48.7	3.6	45.1	6.9	38.2	9.5	28.7	26.4	+2.3	34.5	−32.2	42.3
	12.3		11.9		11.6		11.3		11.2					
125	61.7	1.1	60.6	3.9	56.7	7.2	49.5	9.6	39.9					
	12.2		11.9		11.5		11.3		11.1		21.7		21.9	
150	73.9	1.4	72.5	4.3	68.2	7.4	60.8	9.8	51.0	27.0	24.0	34.3	−10.3	42.7
	12.2		11.8		11.5		11.2		11.0					
175	86.1	1.8	84.3	4.6	79.7	7.7	72.0	10.0	62.0					
	12.1		11.7		11.4		11.1		10.9		21.5		21.1	
200	98.2	2.2	96.0	4.9	91.1	8.0	83.1	10.2	72.9	27.2	45.5	34.7	+10.8	42.5
	23.8		22.8		22.6		22.0		21.5		21.2		20.9	
250	122.0	3.2	118.8	5.1	113.7	8.6	105.1	10.7	94.4	27.7	66.7	35.0	31.7	42.4
	23.7		22.5		22.3		21.9		21.4		20.9		20.5	
300	145.7	4.4	141.3	5.3	136.0	9.0	127.0	11.2	115.8	28.2	87.6	35.4	52.2	42.2
	23.2		22.3		22.0		21.6		21.2		20.6		20.3	
350	168.9	5.3	163.6	5.6	158.0	9.4	148.6	11.6	137.0	28.8	108.2	35.7	72.5	42.3
	22.7		22.2		21.5		21.1		20.8		20.3		20.2	
400	191.6	5.8	185.8	6.3	179.5	9.8	169.7	11.9	157.8	29.3	128.5	35.8	92.7	42.5
	22.5		22.0		21.2		21.0		20.7		20.2		19.6	
450	214.1	6.3	207.8	7.1	200.7	10.0	190.7	12.2	178.5	29.8	148.7	36.4	112.3	42.2
	22.1		21.9		21.0		20.5		20.1		19.9		19.9	
500	236.2	6.5	229.7	8.0	221.7	10.5	211.2	12.6	198.6	30.0	168.6	36.4	132.2	42.3
	21.5		21.0		20.8		20.4		19.9		19.3		19.3	
550	257.7	7.0	250.7	8.2	242.5	10.9	231.6	13.1	218.5	30.6	187.9	36.4	151.5	42.0
	21.3		21.0		20.1		19.8		19.7		19.2		19.2	
600	279.0	7.3	271.7	9.1	262.6	11.2	251.4	13.2	238.2	31.1	207.1	36.4	170.7	42.2
	21.2		20.5		19.9		19.6		19.4		19.2		18.9	
650	300.2	8.0	292.2	9.7	282.5	11.5	271.0	13.4	257.6	31.3	226.3	36.7	189.6	42.4
	20.5		20.3		19.6		19.2		19.1		18.7		18.6	
700	320.7	8.2	312.5	10.4	302.1	11.9	290.2	13.5	276.7	31.7	245.0	36.8	208.2	42.5
	20.4		19.7		19.4		19.0		19.0		18.3		18.4	
750	341.1	8.9	332.2	10.7	321.5	12.3	309.2	13.5	295.7	32.4	263.3	36.7	226.6	42.6
	20.1		19.7		19.2		18.9		18.7		18.3		18.3	
800	361.2	9.3	351.9	11.2	340.7	12.6	328.1	13.7	314.4	32.8	281.6	36.7	244.9	42.7
	19.8		19.2		19.0		18.5		18.3		18.2		17.8	
850	381.0	9.9	371.1	11.4	359.7	13.1	346.6	13.9	332.7	32.9	299.8	37.1	262.7	42.5
	19.3		18.7		18.6		18.4		17.8		18.1		17.5	
900	400.3	10.5	389.8	11.5	378.3	13.3	365.0	14.5	350.5	32.6	317.9	37.7	280.2	42.5
					18.4		17.8		17.7		17.6		17.2	
950					396.7	13.9	382.8	14.6	368.2	32.7	335.5	38.1	297.4	42.7
							17.7		17.5		17.2		17.1	
1000							400.5	14.8	385.7	33.0	352.7	38.2	314.5	42.3

32. WATER: P-V-T DATA

Table 95.—*(Continued)*
(Lines continued from p. 208)

50		60		70		80		90		100		198
						—Δ—						
−119.5	49.6	−169.1	56.5	−225.6	63.2	−288.8	68.0	−356.8	74.4	−431.2		
22.8		23.1		23.4						24.8		
−96.7	49.3	−146.0	56.2	−202.2						−406.4	1130	−1536
22.2		22.7		23.4						25.0		46
−74.5	48.8	−123.3	55.5	−178.8						−381.4	1109	−1490
21.5		22.0		22.4						24.5		46
−53.0	48.3	−101.3	55.1	−156.4	61.4	−217.8				−356.9	1087	−1444
21.3		21.5		22.3		22.9				24.2		46
−31.7	48.1	−79.8	54.3	−134.1	60.8	−194.9				−332.7	1065	−1398
21.0		21.0		21.8		22.4				23.7		44
−10.7	48.1	−58.8	53.5	−112.3	60.2	−172.5				−309.0	1045	−1354
20.7		20.8		21.3		22.0				23.6		44
+10.0	48.0	−38.0	53.0	−91.0	59.5	−150.5	65.1	−215.6	69.8	−285.4	1025	−1310
20.2		20.4		20.6		21.7		22.4		23.2		42
30.2	47.8	−17.6	52.8	−70.4	58.4	−128.8	64.4	−193.2	69.0	−262.2	1006	−1268
20.0		20.3		20.4		21.0		22.2		22.8		40
50.2	47.5	+2.7	52.7	−50.0	57.8	−107.8	63.2	−171.0	68.4	−239.4	989	−1228
19.9		20.0		20.2		20.9		21.7		22.3		39
70.1	47.4	22.7	52.5	−29.8	57.1	−86.9	62.4	−149.3	67.8	−217.1	972	−1189
19.8		19.4		19.8		20.3		21.2		22.1		37
89.9	47.8	42.1	52.1	−10.0	56.6	−66.6	61.5	−128.1	66.9	−195.0	957	−1152
19.6		19.3		19.7		20.1		20.8		21.5		37
109.5	48.1	61.4	51.7	+9.7	56.2	−46.5	60.8	−107.3	66.2	−173.5	942	−1145
19.0		19.3		19.4		20.1		20.3		20.9		36
128.5	47.8	80.7	51.6	29.1	55.5	−26.4	60.6	−87.0	65.6	−152.6	926	−1079
18.7		19.2		19.1		19.6		20.0		20.6		35
147.2	47.3	99.9	51.7	48.2	55.0	−6.8	60.2	−67.0	65.0	−132.0	912	−1044
18.5		18.8		18.8		19.2		19.7		20.7		34
165.7	47.0	118.7	51.7	67.0	54.6	+12.4	59.7	−47.3	64.0	−111.3	899	−1010
18.3		18.4		18.7		19.0		19.3		19.8		34
184.0	46.9	137.1	51.4	85.7	54.3	31.4	59.4	−28.0	63.5	−91.5[a]	884	976
18.2		18.1		18.5		18.8		19.0		19.5		33
202.2	47.0	155.2	51.0	104.2	54.0	50.2	59.2	−9.0	63.0	−72.0	871	943
18.0		18.0		18.3		18.2		18.8		19.5		32
220.2	47.0	173.2	50.7	122.5	54.1	68.4[b]	58.6	+9.8	62.3	−52.5	859	−911
17.5		17.7		18.0		17.9		18.6		19.0		32
237.7	46.8	190.9	50.4	140.5	54.2	86.3	57.9	28.4	61.9	−33.5	846	−879
17.0		17.6		17.8		17.9		18.2		18.7		31
254.7	46.2	208.5	50.2	158.3	54.1	104.2	57.6	46.6	61.4	−14.8	833	−848
17.5		17.6		17.8		17.5		17.7		18.5		30
272.2	46.1	226.1	50.0	176.1	54.4	121.7	57.4	64.3	60.6	+3.7	822	−818

PROPERTIES OF ORDINARY WATER-SUBSTANCE

Table 95.—(Continued)

II. E. H. Amagat. From the same source as the preceding, and treated in the same manner. $v^* = 1 - 10^{-4}\Delta$.

P ↓ / t→	0.00	2.10	4.35	6.85	10.10	14.25	20.40	29.45	40.45	48.85
1	−1.3 / 50.5	−0.1 / 50.3	0.0 / 50.2	−0.6 / 48.8	−2.8 / 47.5	−7.6 / 47.3	−18.5 / 47.2	−41.7 / 45.4	−78.7 / 46.4	−114.1 / 43.8
	−1.2	−0.1	+0.6	+2.2	+4.8	+10.9	+23.2	+37.0	+35.4	
100	+49.2 / 49.0	+50.2 / 48.0	+50.2 / 47.0	+48.2 / 47.0	+44.7 / 46.5	+39.7 / 45.5	+28.7 / 45.0	+3.7 / 43.5	−32.3 / 42.0	−70.3 / 43.0
	−1.0	0.0	2.0	3.5	5.0	11.0	25.0	36.0	38.0	
200	98.2	98.2	97.2	95.2	91.2	85.2	73.7	47.2	+9.7	−27.3
	0.0	+1.0	2.0	4.0	6.0	11.5	26.5	37.5	37.0	
300	47.5 / 145.7	46.0 / 144.2	45.5 / 142.7	45.0 / 140.2	44.0 / 135.2	43.0 / 128.2	42.5 / 116.2	42.5 / 89.7	41.0 / 50.7	42.5 / +15.2
	+1.5	1.5	2.5	5.0	7.0	12.0	26.5	39.0	35.5	
400	45.5 / 191.2	45.0 / 189.2	44.5 / 187.2	43.5 / 183.7	43.0 / 178.2	42.5 / 170.7	41.5 / 157.7	41.5 / 131.2	40.5 / 91.2	41.0 / 56.2
	2.0	2.0	3.5	5.5	7.5	13.0	26.5	40.0	35.0	
	44.5	43.5	42.5	42.5	42.5	40.5	40.5	39.5	40.0	39.5
500	235.7 / 43.0	232.7 / 43.0	229.7 / 42.0	226.2 / 41.5	220.7 / 40.5	211.2 / 40.0	198.2 / 39.0	170.7 / 38.0	131.2 / 38.0	95.7 / 38.5
	3.0	3.0	3.0	3.5	5.5	9.5	13.0	27.5	39.5	35.5
600	278.7 / 41.5	275.7 / 41.5	271.7 / 41.5	267.7 / 40.5	261.2 / 40.0	251.2 / 39.5	237.2 / 38.5	208.7 / 37.5	169.2 / 37.0	134.2 / 37.5
	3.0	3.0	4.0	6.5	10.0	14.0	28.5	39.5	35.0	
700	320.2	317.2	313.2	308.2	301.2	290.7	275.7	246.2	206.2	171.7
	3.0	4.0	5.0	7.0	10.5	15.0	29.5	40.0	34.5	
800	41.0 / 361.2	41.0 / 358.2	40.5 / 353.7	39.0 / 347.2	38.5 / 339.7	38.5 / 329.2	38.0 / 313.7	36.5 / 282.7	35.5 / 241.7	36.0 / 207.7
	3.0	4.5	6.5	7.5	10.5	15.5	31.0	41.0	34.0	
900	39.5 / 400.7	39.0 / 397.2	38.0 / 391.7	37.0 / 384.2	37.5 / 377.2	37.5 / 366.7	36.0 / 349.7	35.5 / 318.2	34.5 / 276.2	35.0 / 242.7
	3.5	5.5	7.5	7.0	10.5	17.0	31.5	42.0	33.5	
	38.0	36.5	37.0	36.5	36.5	36.0	35.0	35.0	33.5	34.0
1000	438.7 / 35.5	433.7 / 36.0	428.7 / 35.0	420.7 / 36.0	413.7 / 35.0	402.7 / 35.0	384.7 / 34.0	353.2 / 34.0	309.7 / 33.0	276.7 / 33.0
	5.0	5.0	8.0	7.0	11.0	18.0	31.5	43.5	33.0	
1100	474.2 / 35.0	469.7 / 35.0	463.7 / 35.0	456.7 / 35.0	448.7 / 34.5	437.7 / 34.5	418.7 / 34.0	387.2 / 34.0	342.7 / 32.5	309.7 / 32.0
	4.5	6.0	7.0	8.0	11.0	19.0	31.5	44.5	33.0	
1200	509.2	504.7	498.7	491.7	483.2	472.2	452.7	421.2	375.2	341.7
	4.5	6.0	7.0	8.5	11.0	19.5	31.5	46.0	33.5	
1300	34.6 / 543.8	34.1 / 538.8	34.1 / 532.8	34.1 / 525.8	33.5 / 516.7	32.5 / 504.7	32.0 / 484.7	32.5 / 453.7	32.0 / 407.2	31.5 / 373.2
	5.0	6.0	7.0	8.9	12.0	20.0	31.0	46.5	34.0	
1400	33.5 / 577.3	33.0 / 571.8	33.0 / 565.8	33.0 / 558.8	32.1 / 548.8	32.1 / 536.8	31.5 / 516.2	31.5 / 485.2	31.5 / 438.5	31.0 / 404.2
	5.5	6.0	7.0	10.0	12.0	20.4	31.0	46.5	34.5	
	32.5	33.0	32.5	31.5	31.5	31.0	31.6	31.0	31.0	30.5

32. WATER: P-V-T DATA

1500	609.8 32.0	5.0	604.8 31.5	6.5	598.3 31.5	8.0	590.3 31.5	10.0	580.3 31.0	12.5	567.8 30.5	20.0	547.8 31.0	31.4	516.2 30.6	46.5	469.7 31.1	35.0	434.7 30.0
1600	641.8 31.5	5.5	636.3 30.5	6.5	629.8 30.5	8.0	621.8 31.0	10.5	611.3 30.5	13.0	598.3 30.0	19.5	578.8 29.5	32.0	546.8 29.0	46.0	500.8 30.0	35.9	464.7 29.5
1700	673.3	6.5	666.8	6.5	660.3	7.5	652.8	11.0	641.8	13.5	628.3	20.0	608.3	32.5	575.8	45.0	350.8	36.4	494.2
1800	703.3 30.0 29.5	6.5	696.8 30.0 29.5	7.0	689.8 29.5 29.0	7.5	682.3 29.5 28.5	11.5	670.8 29.0 28.5	13.0	657.8 29.5 29.0	20.5	637.3 29.0 28.5	33.0	604.3 28.5 27.5	44.5	559.8 29.0 28.0	37.0	522.8 28.6 28.0
1900	732.8 29.0	6.5	726.3 29.0	7.5	718.8 29.0	8.0	710.8 28.0	11.5	699.3 28.5	12.5	686.8 28.0	21.0	665.8 28.0	34.0	631.8 27.0	44.0	587.8 27.5	37.0	550.8 27.5
2000	761.8 29.0	6.5	755.3 28.5	7.5	747.8 28.0	9.0	738.8 28.0	11.0	727.8 28.0 27.5	13.0	714.8 27.5	21.0	693.8 27.0	35.0	658.8 26.0	43.5	615.3 26.5	37.0	578.3 26.5
2100	790.8 27.5	7.0	783.8 27.5	8.0	775.8 27.0	9.0	766.8 27.5	11.0	755.8 27.5	13.5	742.3 26.5	21.5	720.8 27.0	36.0	684.8 26.0	43.0	641.8 26.0	37.0	604.8 26.5
2200	818.3	7.0	811.3	8.5	802.8	8.5	794.3	11.0	783.3	14.5	768.8	21.0	747.8	37.0	710.8	43.0	667.8	36.5	631.3
2300	844.8 26.5 26.5	7.0	837.8 26.5	8.5	829.3 26.5	7.5	821.8 27.5	11.0	810.8 27.5	16.0	794.8 26.0	21.0	773.8 26.0	37.5	736.3 25.5	42.5	693.8 26.0	36.5	657.3 26.0
2400	871.3 26.5 25.5	7.5	863.5 26.0	8.5	855.3 26.0	7.5	847.8 26.0	11.0	836.8 26.0	16.5	820.3 25.5	21.5	798.8 25.0	37.0	761.8 25.5	42.5	719.3 25.5	36.5	682.8 25.5
2500	896.8	7.0	889.8	8.5	881.3	8.5	872.8	11.0	861.8	16.5	845.3	22.0	823.3	36.5	786.8	42.5	744.3	36.5	707.8 25.0
2600	921.8 25.0	7.5	914.3 24.5	7.5	906.8 25.5	9.5	897.3 24.5	16.0	886.3 24.5	16.0	870.3 25.0	22.5	847.8 24.5	36.5	811.3 24.5	42.5	768.8 24.5	36.5	732.3 24.5
2700	946.8 25.0	8.0	938.8 24.5	8.0	930.8 24.0	9.5	921.3 24.0	11.0	910.3 24.0	15.5	894.8 24.5	23.0	871.8 24.0	36.0	835.8 24.5	43.0	792.8 24.0	36.5	756.3 24.0
2800	970.8 24.0	8.0	962.8 23.0	8.0	954.8 24.0	9.5	945.3 23.5	11.5	933.8 23.5	15.5	918.3 23.5	23.0	895.3 23.5	35.5	859.8 23.5	43.5	816.3 23.5	36.5	779.8 23.5
2900	993.8 23.0 22.0	8.0	985.8 23.0	8.0	977.8 23.0 24.0	9.0	968.8 23.5	11.5	957.3 23.0	15.5	941.8 23.0	23.5	918.3 23.0	35.5	982.8 23.0	43.5	839.3 22.5	36.0	803.3 23.5 22.5
3000	1015.8	7.0	1008.8	7.0	1001.8	9.5	992.3	12.0	980.3	15.5	964.8	23.5	941.3	35.5	905.8	44.0	861.8	36.0	825.8

PROPERTIES OF ORDINARY WATER-SUBSTANCE

Table 95.—(Continued)

III. Bridgman.[276] See p. 214.

P ↓	−20	−15	−10	−5	0 Δ	+2	+10	+15	+20	+25	+30	+35	+40
1			−21 229 +208 207 415	−8 230 +222 204 426	−1 233 +232 201 433		−3 223 +220 187 407		−18 212 +194 185 379		−43 210 +167 179 346		−78 209 +131 178 309
500						+12		+26		+27		+36	
1000				−13 −14 −11		+26		+28		+33		+37	
1500		+613 171 784 142	182 597 176 773 140	181 507 158 765 137	172 605 153 758 129	+21	167 574 147 721 130	+32	163 542 144 686 120	+36	160 506 145 651 127	+37	160 469 143 612 129
2000	+791 144	+16 11	−10 +8	+2 7	+37		+35		+35		+39	
2500	935 123	+7 9	13	11	887 119 1006 109 1115	36	851 119 970 109 1079	36	815 118 933 109 1042	37	778 117 895 109 1004	37	741 118 859 109 968
3000	1058	8	913 121 1034 110 1144	902 121 1023 110 1133		36		37		38		36	
3500			16 15	11 11 1		36		37		38		36	
4000		98 1257	101 1245 93 1338 85	102 1235 91 1328 85	101 1216 91 1307 86	36	101 1180 93 1273 86	37	101 1143 93 1236 87	37	102 1106 93 1199 86	39	99 1067 93 1160 86
4500		12	10	10	34		37		37		39	
5000			10	10	1393 79 1472 75 1547	34	1359 81 1440 77 1517	36	1323 82 1405 76 1481	37	1286 81 1367 76 1443	40	1246 81 1327 76 1403
5500				1413 76 1489 71 1560	17		32		35		38		40
6000			1423	13	30		36		36		38		40
6500					72 1619	30	72 1589 69 1658 66	35	73 1554 70 1624 67	40	71 1514 69 1583 64	39	72 1475 67 1542 65
7000				65 1625	6		34		41		41	
7500							1724	33	1691 63 1754 59 1813	44	1647 61 1708 58 1766	40	1607 61 1668 57 1725
8000									46		40	
8500									47		41	
9000									57 1870 52	49	55 1821 52	42	54 1779 51

32. WATER: P-V-T DATA

P	50	Δ	60	Δ	70	Δ	80
1	+43	+49	−170 201	+57	−227 210	+63	−290 216
500	−121 203 +82 183 265	49 51 47	+31 187 218	48 52	−17 183 +166	57 53	−74 187 +113
1000	44						
1500	161 426 143 569 128	43 45 43	163 381 145 526 129	51 50	164 330 146 426 131	50 47	167 280 149 429 131
2000	43						
2500	697 117 814 109 923	44 45 45	655 118 773 110 883	48 46 45	607 120 727 111 838	47 45 44	560 122 682 112 794
3000							
3500							
4000	100 1023 94 1117 88	44 43 41	103 986 93 1079 87	44 41 38	104 942 96 1038 90	42 42 44	106 900 96 996 88
4500							
5000	1205 82 1287 76 1363	40 40	1166 82 1248 76 1324	39 39	1128 81 1209 75 1284	42 39	1084 83 1167 78 1245
5500							
6000							

P	50	Δ	60	Δ	70	Δ	80
6500	+72 +1435 68 1503	+40 39	+72 +1396 68 1464	+39 39	+72 +1356 69 1425	+40 39	+73 +1318 69 1387
7000							
7500	64 1567 61 1628 57	40 40	65 1529 61 1590 57	38 38	65 1490 61 1551 57	38 35	65 1452 62 1514 58
8000							
8500	1685 54 1739 51 1790	40 40 40	1647 54 1701 52 1753	38 38 37	1608 54 1662 53 1715	36 36 36	1572 54 1626 53 1679
9000							
9500	49 1839 48 1887 48	40 40	49 1802 48 1850 47	37 37	49 1764 48 1812 48	35 35	50 1729 48 1777 48
9500					1 73 48 1921 47 1968	43 42 41	1830 49 1879 48 1927
10000			1 22 49 1971	49 50			
10500							
11000	1935 46 1981 46 2027	40 40	1897 46 1943 46 1989	38 38	1860 46 1906 46 1952	37 37 37	1825 46 1871 46 1917
11500							
12000							2067

Table 95—(*Continued*)

III. P. W. Bridgman, (ICT)[276] with corrections indicated by the errata published with Vol. 7. Based on Bridgman *Proc. Amer. Acad. Arts Sci.* **48,** 307-362 (1913). As the values he accepted for the specific volumes at 1 atm are somewhat smaller than those in Table 94, all those for a given temperature have been increased by the same amount, so chosen as to make the value at 1 atm the same as that in Table 94. For the temperatures -15 and -20 °C, no value is recorded for 1 atm; values for these temperatures have been increased by 0.0001. His data have been criticized by Tammann,[292] partly on the ground that both the direct observations of Amagat and others and the data derived from solutions indicate that, for pressures exceeding 300 kg*/cm², the temperature at which the density is a maximum lies below that at which water and ice are in equilibrium, whereas Bridgman's observations indicate that it lies above that equilibrium temperature even when the pressure is as great as 1500 kg*/cm²; and partly on the shape of the isopiestics at temperatures below 0 °C. (See also Fig. 3 and Section IV.) Bridgman's table in the *Proceedings* (*loc. cit.*) contains values for every 5 °C, the pressure being expressed in kg*/cm². In the *International Critical Tables* and in this work, the pressure is expressed in atm. $v^* = 1 - 10^{-4}\Delta$.

Data on pages 212, 213; Fig. 3 on p. 230.

IV. P. W. Bridgman.[278] The following values of v^* have been obtained from his table of molecular volumes by dividing by 18.0154. He states: "The results now found do not check in fine detail with those found before; in particular the minimum and maximum of volume as a function of temperature on this isobar at 1500 kg found in 1912 [292a] and shown in Fig. 40 of the 1912 paper, has not been found this time." In fact, the isobars defined by these values more nearly resemble those found by Tammann than did the earlier ones (see Fig. 3). On comparing these values with those of 1912 (from which those in Section III were derived) systematic differences, often amounting to 2 or 3 units in the next to the last place in Δ (*i.e.*, to 2 or 3 ml/kg), are found. For example, at 50 °C and 4, 5, 6, 7, 8, 9, 10, 11 and 12 kg*/cm² these values exceed the 1912 ones by 1.0, 1.8, 2.2, 2.7, 3.2, 3.3, 3.1, 2.9, and 3.3 ml/kg, respectively.

Unit of $p = 1$ kg*/cm²; of $P = 1$ atm; $v^* = 1 - 10^{-4}\Delta$ ml/g. Temp. $= t$ °C

$t \rightarrow$ p	P	-20	-15	-10 Δ	-5	0	$+20$
1	0.9678					-1 17	-18
						231	*199*
500	483.9					$+230$ 49	$+181$
						194	*187*
1000	967.8					434 10 424 56	368
						172	*167* *156*
1500	1451.8		630 *10*	620 14	606 15	591 67	524

[292] Tammann, G., and Schwarzkopf, E., *Z. anorg. allgem. Chem.*, **174**, 216-224 (1928); Tammann, G., and Jellinghaus, W., *Idem*, **174**, 225-230 (1928).

32. WATER: P-V-T DATA

Table 95—(Continued)

$t \rightarrow$ p	P	−20		−15		−10	Δ	−5		0		+20
						156		152		148		148		148
2000	1935.7	797	11	786	14	772	18	754	15	739	67	672
		142		134		131		130		129		129
2500	2419.6	939	19	920	17	903	19	884	16	868	67	801
				121		120		116		117		115
3000	2903.5			1041	18	1023	23	1000	15	985	69	916
				108		106		108		106		100
3500	3387.4			1149	20	1129	21	1108	17	1091	75	1016
						100		98		97		96
4000	3871.4					1229	23	1206	18	1188	76	1112
						175		172		173		179
5000	4839.2					1404	26	1378	17	1361	70	1291
										150		144
6000	5807.0									1511	76	1435

$t \rightarrow$ p	P	20		40		50	Δ	60		80		100
1	0.9768	−18	61	−79	42	−121	50	−171	113	−284	151	−435
		199		199		215		212		221		252
500	483.9	+181	61	+120	26	+94	53	+41	104	−63	120	−183
		187		174		162		173		180		190
1000	967.8	368	74	294	38	256	42	214	97	+117	110	+7
		156		156		154		154		159		167
1500	1451.8	524	74	450	39	411	43	368	92	276	102	174
		148		142		139		140		142		147
2000	1935.7	672	80	592	42	550	42	508	90	418	97	321
		129		126		127		127		129		134
2500	2419.6	801	83	718	41	677	42	635	88	547	92	455
		115		115		116		117		119		121
3000	2903.5	916	83	833	40	793	41	752	86	666	90	576
		100		105		105		106		109		112
3500	3387.4	1016	78	938	40	898	40	858	83	775	87	688
		96		96		96		98		99		104
4000	3871.4	1112	78	1034	40	994	38	956	82	874	82	792
		179		170		170		170		177		180
5000	4839.2	1291	87	1204	40	1164	38	1126	75	1051	79	972
		144		151		152		153		155		157
6000	5807.0	1435	80	1355	39	1316	37	1279	73	1206	77	1129
				130		133		135		135		140
7000	6774.9			1485	36	1449	35	1414	73	1341	72	1269
				119		120		122		125		127
8000	7742.7			1604	35	1569	33	1536	70	1466	70	1396
				109		110		110		112		114
9000	8710.6			1713	34	1679	33	1646	69	1578	68	1510

[292a] Bridgman, P. W., *Proc. Amer. Acad. Arts Sci.*, **48**, 307-362 (1913).

216   PROPERTIES OF ORDINARY WATER-SUBSTANCE

Table 95—(Continued)

t → p	P	20	40		50 Δ		60		80		100
			101		102		102		104		105
10000	9678.4		1814	33	1781	33	1748	66	1682	67	1615
			96		95		95		96		98
11000	10646.2		1910	34	1876	33	1843	65	1778	65	1713
			84		86		87		88		88
12000	11614.1		1994	32	1962	32	1930	64	1866	65	1801

V. G. Tammann and W. Jellinghaus.[293] The published values have been multiplied by 1.00013 so as to convert them into ml/g. As these authors assign no value to the specific volume at 1 atm when t is below 0 °C, values taken from Table 94 have been inserted, enclosed in parentheses. The values in this section do not suffice to determine more closely than ±4 °C the temperature that corresponds to the maximum density along any isopiestic, but they do show the general progressive change in the isopiestics (see Fig. 3). $v^* = 1 - 10^{-4}\Delta$ ml/g.

t → p	−8		−7		−6		−5 Δ		−4		−3	
1	(−15)	−2	(−13)	−3	(−10)	−2	(−8)	−2	(−6)	−2	(−4)	−1
100												
200												
300											145	+1
											48	
400									193	0	193	+2
									43		42	
500							237	1	236	1	235	1
							43		40		44	
600					281	1	280	4	276	−3	279	+2
					43		43		46		42	
700			327	3	324	1	323	1	322	1	321	2
			41		42		33		42		41	
800	371	3	368	2	366	10	356	−8	364	2	362	2
	42		43		42		50		40		40	
900	413	2	411	3	408	2	406	2	404	2	402	9
	38		39		40		40		40		40	
1000	451	1	450	2	448	2	446	2	444	2	442	9

[293] Tammann, G., and Jellinghaus, W., *Z. anorg. allgem. Chem.*, **174**, 225-230 (1928).

32. WATER: P-V-T DATA

Table 95—(Continued)

t→ p	−8		−7		−6		−5	Δ	−4		−3	
	39						*33*					
1100	490		*11*				479				*10*	
	23						*28*					
1200	513		*6*				507				*8*	
	42						*39*					
1300	555ᵉ		*9*				546				*19*	
	34						*30*					
1400	589ᵉ		*13*				576				*21*	
	36						*32*					
1500	625ᵉ		*17*				608				*9*	

t→ p	−2		−1		0		+1 Δ		2		3	
1	(−3)	−1	(−2)	−1	−1	0	−1	−1	0	0	0	−1
			51		*51*		*52*		*50*		*51*	
100			49	−1	50	−1	51	+1	50	−1	51	0
			49		*47*		*46*		*47*		*46*	
200	98	0	98	+1	97	0	97	0	97	0	97	+1
	46		*45*		*46*		*46*		*46*		*45*	
300	144	+1	143	0	143	0	143	0	143	+1	142	*1*
	47		*47*		*46*		*45*		*44*		*44*	
400	191	+1	190	+1	189	+1	188	+1	187	+1	186	*1*
	43		*42*		*42*		*43*		*42*		*42*	
500	234	2	232	1	231	0	231	2	229	1	228	*1*
	43		*43*		*43*		*42*		*43*		*42*	
600	277	2	275	1	274	1	273	1	272	2	270	*2*
	42		*42*		*41*		*41*		*40*		*39*	
700	319	2	317	2	315	1	314	2	312	3	309	*1*
	41		*40*		*39*		*38*		*38*		*39*	
800	360	3	357	3	354	2	352	2	350	2	348	*3*
	33		*38*		*40*		*40*		*41*		*40*	
900	393	−2	395	1	394	2	392	1	391	3	388	*2*
	40		*42*		*40*		*39*		*37*		*38*	
1000	433	−4	437	3	434	3	431	3	428	2	426	*2*
					35							
1100					469				*12*			
					30							
1200					499				*9*			
					28							
1300					527				*1*			
					28							
1400					555				*1*			
					44							
1500					599				*32*			

PROPERTIES OF ORDINARY WATER-SUBSTANCE

Table 95—(Continued)

$t \rightarrow$ p	4		5		6		7 Δ		8		9	
1	+1	0	+1	−1	+2	−1	+3	+2	+1	+1	0	+1
	50		50		47		46		47		48	
100	51	0	51	2	49	0	49	1	48	0	48	1
	45		45		46		45		45		44	
200	96	0	96	1	95	1	94	1	93	1	92	2
	45		45		46		46		45		44	
300	141	0	141	0	141	1	140	2	138	2	136	2
	44		42		43		43		38		43	
400	185	2	183	−1	184	1	183	6	176	−3	179	2
	42		43		41		42		43		41	
500	227	1	226	1	225	0	225	6	219	−1	220	3
	41		40		39		37		41		38	
600	268	2	266	2	264	2	262	2	260	2	258	2
	40		40		41		39		39		38	
700	308	2	306	1	305	4	301	2	299	3	296	2
	37		37		40		40		34		40	
800	345	2	343	−2	345	4	341	8	333	−3	336	3
	41		41		38		38		42		36	
900	386	2	384	1	383	4	379	4	375	3	372	2
	38		37		36		38		38		37	
1000	424	3	421	2	419	2	417	4	413	4	409	2
					38							
1100					457							
					33							
1200					490							
					36							
1300					526							
					28							
1400					554							
					13							
1500					567							

$t \rightarrow$ p	10		11		12		13 Δ		14		15
1	−1	0	−1	0	−1	+1	−2	0	−2	+3	−5
	48		48		46		46		45		45
100	47	0	47	2	45	1	44	1	43	3	40
	43		41		42		42		41		43
200	90	2	88	1	87	1	86	2	84	1	83
	44		45		44		44		44		43

32. WATER: P-V-T DATA

Table 95—(Continued)

$t\to$ p	10		11		12	Δ	13		14		15
300	134	1	133	2	131	1	130	2	128	2	126
	43		42		42		41		41		41
400	177	2	175	2	173	2	171	2	169	2	167
	40		41		39		40		41		40
500	217	1	216	4	212	1	211	1	210	3	207
	39		37		38		37		35		36
600	256	3	253	3	250	2	248	3	245	2	243
	38		39		40		39		40		40
700	294	2	292	2	290	3	287	2	285	2	283
	39		38		38		37		37		37
800	333	3	330	2	328	4	324	2	322	2	320
	37		37		36		36		35		35
900	370	3	367	3	364	4	360	3	357	2	355
	37		37		36		38		38		37
1000	407	3	404	4	400	2	398	3	395	3	392
									34		
1100			28						429		
									37		
1200			24						466		
									29		
1300			31						495		
									37		
1400			22						532		
									22		
1500			13						554		

VI. L. B. Smith and F. G. Keyes.[294] These values have been taken from their table, which was computed by means of an empirical equation set up by themselves as a satisfactory representation of their observations. In that table the values of v^* are given to a unit in the sixth place of decimals. But they state that their equation "may be trusted to represent the behavior of liquid water to at least one part in 2000" (p. 294) and that the computed specific volume for 4 °C and one atmosphere differs by "one part in 6900 from the accepted value" (p. 295). In view of these statements it seemed justifiable to give here only 4 places of decimals, corresponding to an accuracy of at least one in 10,000. It will be noticed that the values given for $P = 1$ atm do not all agree with those in Table 94.

[294] Smith, L. B., and Keyes, F. G., *Proc. Amer. Acad. Arts Sci.*, **69**, 285-312 (1934) → *Mech. Eng.*, **56**, 92-94 (1934). Supersedes Keyes, F. G., and Smith, L. B., *Mech. Eng.*, **53**, 132-135 (1931).

Table 95—(Continued)

VI. Smith and Keyes. See p. 219.

Unit of $P = 1$ atm; of $v^* = 1$ ml/g. Temp. $= t$ °C

$P \rightarrow$ t	1	25	50	75	100	125	150	175	200	250	300	350
0	1.0002 _3_ 11	0.9991 _15_ 11	0.9980 _15_ 12	0.9968 _15_ 11	0.9957 _15_ 11	0.9946 _15_ 11	0.9935 _5_ 11	0.9925 _4_ 10	0.9914 _16_ 11	0.9893 _15_ 21	0.9873 _15_ 20	0.9853 _5_ 5
10	1.0005 _15_ 11	0.9994 _15_ 11	0.9983 _15_ 11	0.9972 _15_ 11	0.9961 _15_ 11	0.9950 _15_ 10	0.9940 _15_ 11	0.9929 _15_ 11	0.9918 _16_ 20	0.9898 _15_ 20	0.9878 _15_ 20	0.9858 _15_ 15
20	1.0020 _25_ 11	1.0009 _25_ 11	0.9998 _25_ 11	0.9987 _25_ 11	0.9976 _25_ 11	0.9965 _25_ 10	0.9955 _25_ 11	0.9944 _25_ 10	0.9934 _24_ 21	0.9913 _25_ 20	0.9893 _25_ 20	0.9873 _25_ 25
30	1.0045 _34_ 11	1.0034 _34_ 11	1.0023 _34_ 11	1.0012 _34_ 11	1.0001 _33_ 11	0.9990 _34_ 10	0.9980 _33_ 11	0.9969 _33_ 11	0.9958 _33_ 20	0.9938 _33_ 20	0.9918 _32_ 20	0.9898 _32_ 32
40	1.0079 _42_ 11	1.0068 _42_ 11	1.0057 _41_ 11	1.0046 _41_ 12	1.0034 _42_ 10	1.0024 _41_ 11	1.0013 _41_ 11	1.0002 _41_ 10	0.9992 _41_ 21	0.9971 _41_ 21	0.9950 _41_ 20	0.9930 _41_ 41
50	1.0121 _49_ 11	1.0110 _49_ 12	1.0098 _49_ 11	1.0087 _49_ 11	1.0076 _48_ 11	1.0065 _48_ 11	1.0054 _48_ 11	1.0043 _48_ 10	1.0033 _47_ 21	1.0012 _47_ 21	0.9991 _47_ 20	0.9971 _46_
60	1.0170 _56_ 11	1.0159 _56_ 12	1.0147 _56_ 11	1.0136 _55_ 11	1.0124 _56_ 12	1.0113 _55_ 11	1.0102 _55_ 11	1.0091 _55_ 11	1.0080 _55_ 21	1.0059 _54_ 21	1.0038 _54_ 21	.0017 _53_
70	1.0226 _63_ 11	1.0215 _62_ 12	1.0203 _62_ 12	1.0191 _62_ 11	1.0180 _61_ 11	1.0168 _61_ 12	1.0157 _61_ 11	1.0146 _61_ 11	1.0135 _60_ 22	1.0113 _60_ 21	1.0092 _59_ 22	1.0070 _59_
80	1.0289 _69_ 12	1.0277 _69_ 12	1.0265 _69_ 12	1.0253 _68_ 12	1.0241 _68_ 11	1.0230 _67_ 12	1.0218 _67_ 11	1.0207 _66_ 12	1.0195 _67_ 22	1.0173 _66_ 22	1.0151 _65_ 22	1.0129 _65_
90	1.0358 _76_ 12	1.0346 _75_ 13	1.0334 _74_ 13	1.0321 _75_ 13	1.0309 _74_ 12	1.0297 _73_ 12	1.0285 _73_ 12	1.0273 _73_ 11	1.0262 _72_ 23	1.0239 _71_ 23	1.0216 _70_ 22	1.0194 _70_
100	1.0434 _82_ 13	1.0421 _82_ 13	1.0408 _81_ 12	1.0396 _80_ 13	1.0383 _80_ 13	1.0370 _80_ 12	1.0358 _79_ 12	1.0346 _78_ 12	1.0334 _78_ 24	1.0310 _77_ 24	1.0286 _76_ 22	1.0264 _75_
110		1.0503 _87_ 14	1.0489 _87_ 13	1.0476 _86_ 13	1.0463 _85_ 13	1.0450 _85_ 13	1.0437 _84_ 13	1.0424 _84_ 12	1.0412 _83_ 25	1.0387 _82_ 25	1.0362 _82_ 23	1.0339 _80_
120		1.0590 _94_ 14	1.0576 _93_ 14	1.0562 _93_ 14	1.0548 _92_ 13	1.0535 _91_ 14	1.0521 _91_ 13	1.0508 _90_ 13	1.0495 _89_ 26	1.0469 _88_ 25	1.0444 _87_ 25	1.0419 _86_
130		1.0684 _101_ 15	1.0669 _100_ 14	1.0655 _99_ 15	1.0640 _98_ 14	1.0626 _97_ 14	1.0612 _96_ 14	1.0598 _96_ 14	1.0584 _95_ 27	1.0557 _93_ 26	1.0531 _91_ 26	1.0505 _90_
140		1.0785 _107_ 16	1.0769 _107_ 15	1.0754 _105_ 16	1.0738 _105_ 15	1.0723 _104_ 15	1.0708 _103_ 14	1.0694 _102_ 15	1.0679 _101_ 29	1.0650 _100_ 28	1.0622 _98_ 27	1.0595 _96_
150		1.0892 _115_ 16	1.0876 _113_ 17	1.0859 _113_ 16	1.0843 _111_ 16	1.0827 _110_ 16	1.0811 _109_ 15	1.0796 _108_ 16	1.0780 _107_ 30	1.0750 _105_ 30	1.0720 _104_ 29	1.0691 _102_
160		1.1007 _123_ 18	1.0989 _122_ 17	1.0972 _120_ 18	1.0954 _120_ 17	1.0937 _118_ 17	1.0920 _117_ 16	1.0904 _115_ 17	1.0887 _115_ 32	1.0855 _112_ 31	1.0824 _109_ 31	1.0793 _108_
170		1.1130 _132_ 19	1.1111 _130_ 19	1.1092 _129_ 18	1.1074 _126_ 19	1.1055 _126_ 18	1.1037 _124_ 18	1.1019 _123_ 17	1.1002 _121_ 35	1.0967 _119_ 34	1.0933 _117_ 32	1.0901 _114_

32. WATER: P-V-T DATA

Table 95—(Continued)

VII. G. Tammann and A. Rühenbeck.[295] Their values, expressed in terms of the specific volume at 20 °C and 1 atm, have been multiplied by 1.00177 in order to bring them to the same basis as Tables 93 and 94. Both temperature and pressure are carried far beyond the values at the critical point, 374.15 °C, 225.65 kg*/cm², but all the values of the specific volume except one are smaller than the critical volume.

Unit of p = 1 kg*/cm²; of P = 1 atm; of v^* = 1 ml/g. Temp. = t °C

$t \rightarrow$ p	P	20		100		200 v^*		300		400	
1		0.968		1.0018 57							
100	96.8	0.9961 46	382	1.0343 51	1092	1.1435 85					
200	193.6	0.9915 45	377	1.0292 51	1058	1.1350 71					
300	290.3	0.9870 43	371	1.0241 39	1038	1.1279 86					
400	387.1	0.9827 44	375	1.0202 40	991	1.1192 70	2442	1.3635 158			
500	483.9	0.9783 43	379	1.0162 27	961	1.1123 58	2354	1.3477 111	6277	1.9754 339	
600	580.7	0.9740 24	395	1.0135 28	930	1.1065 46	2301	1.3366 75	5049	1.8415 447	
700	677.5	0.9716 43	391	1.0107 27	912	1.1019 31	2272	1.3291 77	4677	1.7968 364	
800	774.2	0.9673 44	407	1.0080 28	908	1.0988 58	2226	1.3214 54	4390	1.7604 371	
900	871.0	0.9629 30	423	1.0052 27	878	1.0930 57	2230	1.3160 76	4073	1.7233 164	
1000	967.8	0.9599 40	426	1.0025 8	848	1.0873 58	2211	1.3084 75	3985	1.7069 152	
1100	1064.6	0.9559 30	458	1.0017 27	798	1.0815 45	2194	1.3009 76	3908	1.6917 204	
1200	1161.4	0.9529 30	461	0.9990 16	780	1.0770 43	2163	1.2933 65	3780	1.6713 218	
1300	1258.1	0.9499 31	475	0.9974 28	753	1.0727 44	2141	1.2868 56	3627	1.6495 217	
1400	1354.9	0.9468 30	478	0.9946 16	737	1.0683 45	2129	1.2812 64	3466	1.6278 168	
1500	1451.7	0.9438 30	492	0.9930 17	708	1.0638 31	2110	1.2748 48	3362	1.6110 164	
1600	1548.5	0.9408 30	505	0.9913 15	694	1.0607 18	2093	1.2700 52	3246	1.5946 151	

Table 95—(Continued)

$t \to$ p	P	20		100		200 v^*		300		400	
1700	1645.3	0.9378	520	0.9898	691	1.0589	2059	1.2648	3147	1.5795	
		44		16		31		41		138	
1800	1742.0	0.9334	548	0.9882	676	1.0558	2049	1.2607	3050	1.5657	
		44		16		31		52		86	
1900	1838.8	0.9290	576	0.9866	661	1.0527	2028	1.2555	3016	1.5571	
		17		16		32		40		125	
2000	1935.6	0.9273	577	0.9850	645	1.0495	2020	1.2515	2931	1.5446	
		30		27		31		41		72	
2100	2032.4	0.9243	580	0.9823	641	1.0464	2010	1.2474	2900	1.5374	
		40		6		32		52		99	
2200	2129.2	0.9203	614	0.9817	615	1.0432	1990	1.2422	2853	1.5275	
		20		16		31		40		59	
2300	2225.9	0.9183	618	0.9801	600	1.0401	1981	1.2382	2834	1.5216	
		17		16		31		41		111	
2400	2322.7	0.9166	619	0.9785	585	1.0370	1971	1.2341	2764	1.5105	
		16		16		18		40		72	
2500	2419.5	0.9150	619	0.9769	583	1.0352	1949	1.2301	2732	1.5033	

$t \to$ p	P	400		500		550 v^*		600		650	
900	871.0	1.7233	7476	2.4709							
		164		1304							
1000	967.8	1.7069	6336	2.3405	4016	2.7421					
		152		916		1385					
1100	1064.6	1.6917	5572	2.2489	3547	2.6036					
		204		676		1090					
1200	1161.4	1.6713	5100	2.1813	3133	2.4946	2915	2.7861	3546	3.1407	
		218		498		771		1065		1991	
1300	1258.1	1.6495	4820	2.1315	2860	2.4175	2621	2.6796	2620	2.9416	
		217		452		590		828		1317	
1400	1354.9	1.6278	4585	2.0863	2722	2.3538	2383	2.5968	2131	2.8099	
		168		348		572		736		860	
1500	1451.7	1.6110	4405	2.0515	2498	2.3013	2219	2.5232	2007	2.7239	
		164		334		373		655		773	
1600	1548.5	1.5946	4235	2.0181	2459	2.2640	1937	2.4577	1889	2.6466	
		151		276		423		510		601	
1700	1645.3	1.5795	4110	1.9905	2312	2.2217	1850	2.4067	1798	2.5865	
		138		241		356		416		628	
1800	1742.0	1.5657	4007	1.9664	2197	2.1861	1790	2.3651	1586	2.5237	
		86		218		306		415		539	
1900	1838.8	1.5571	3875	1.9446	2109	2.1555	1681	2.3236	1462	2.4698	
		125		182		304		447		524	

[295] Tammann, G., and Rühenbeck, A., *Ann. d. Physik* (5), **13**, 63-79 (1932).

Table 95—(Continued)

$t \rightarrow$ p	P	400		500		550 v^*		600		650	
2000	1935.6	1.5446	3818	1.9264	1987	2.1251	1538	2.2789	1385	2.4174	
		72		194		209		252		274	
2100	2032.4	1.5374	3696	1.9070	1972	2.1042	1495	2.2537	1363	2.3900	
		99		160		206		338		357	
2200	2129.2	1.5275	3635	1.8910	1926	2.0836	1373	2.2209	1334	2.3543	
		59		159		207		285		257	
2300	2225.9	1.5216	3535	1.8751	1878	2.0629	1295	2.1924	1362	2.3286	
		111		146		189		232		407	
2400	2322.7	1.5105	3500	1.8605	1835	2.0440	1252	2.1692	1187	2.2879	
		72		136		206		219		306	
2500	2419.5	1.5033	3436	1.8469	1765	2.0234	1239	2.1473	1100	2.2573	

VIII. L. H. Adams.[296] In the conversion of his ratios to specific volumes it has been assumed that the specific volume at 1 atm and 25 °C is 1.0029 ml/g (Table 94).

Unit of $p_b = 1$ bar, of $P = 1$ atm, of $v^* = 1$ ml/g. Temp. $= 25$ °C.

p_b	P	v^*	p_b	P	v^*	p_b	P	v^*
500	493.5	0.9817	4000	3947.7	0.8874	9000	8882.3	0.8166
1000	986.9	0.9635	5000	4934.6	0.8695	9630[d]	9504.1	0.8098
1500	1480.4	0.9472	6000	5921.5	0.8540	10000	9869.2	0.8059
2000	1973.8	0.9328	7000	6908.5	0.8402	11000	10856.2	0.7964
3000	2960.8	0.9071	8000	7895.4	0.8278	12000	11843.1	0.7876

IX. M. Trautz and H. Steyer.[297] These authors have published values for the specific volume of water in the range 50 to 300 atm and 0 °C to near saturation or to 370 °C. M. Jakob[298] regards these values as inferior to those reported by Keyes and Smith in 1931 (see this table, Section VI).

[a] Specific volume at 750 atm and 100 °C was published as 1.00912, giving $\Delta = 92.5$; here it has been taken as 1.00902.

[b] Specific volume at 850 atm and 80 °C was published as 0.99308, giving $\Delta = 67.9$; here it has been taken as 0.99303.

[c] Additional values:

$t \rightarrow$ p	−14		−13	
1300			561	6
			42	
1400	607	4	603	14
	40			
1500	647			

[d] Equilibrium of water and ice-VI at 25 °C; this pressure "is about 25 bars higher" than the one found by Bridgman.

[296] Adams, L. H., *J. Am. Chem. Soc.*, **53**, 3769-3813 (1931).
[297] Trautz, M., and Steyer, H., *Forsch. Gebiete Ingenieurw.*, **2**, 45-52 (1931) → Steyer, H., *Z. d. Ver. d. Ing.*, **75**, 601 (1930).
[298] Jakob, M., *Engineering (London)*, **132**, 143-146 (1931).

32. WATER: P-V-T DATA

(Continued from p. 206)

M. Dole and B. Z. Wiener [270] have reported that within their experimental error, which did not exceed one part in 10^6, the density of a given sample of water is independent of the thermal history of the sample.

But J. B. Peel, P. L. Robinson, and H. C. Smith [271] have reported very queer changes in the density of water that has remained for a day or more in contact with carbon or with thoria.

Density of Water.

For the temperature of maximum density, see Section 36. With the exceptions of Tables 108, 109, and 110, referring to natural waters, all the data given below refer to air-free water unless the contrary is stated. The values for the density and the specific volume at a pressure of one atmosphere are exceedingly accurate, but at higher pressures the data obtained by different observers do not always agree satisfactorily.

The values given by V. Stott and P. H. Bigg [272] for the density and the specific volume at one atmosphere and for the range 0 to 40 °C are the means of the corresponding values published by P. Chappuis [273] and by

(Go to p. 250.)

Table 96.—Isometric Association of the Pressure and Temperature of Compressed Water

Adapted from E. H. Amagat.[299] Derived from the same observations as his data in Table 95.

Successive differences have been printed, in distinctive type, between the values from which they have been derived; v^* = specific volume at t and P, v_0^* = that at 0 °C and 1 atm. Certain obviously erroneous values in Amagat's tables, arising apparently from errors in transcription, have been changed on the basis of that assumption, so as to smooth the run of the differences. They are marked, and the original values are given in a footnote.

The following isometrics of water at temperatures and pressures close to the critical have been published by C. J. v. Nieuwenburg and Miss H. B. Blumendal [300]; unit of v^* = 1 cm³/g of P = 1 kg*/cm²:

$t \to$ v^*	350	360 P	370
1.6	250	334	412
1.8	168	223	286
2.0		189	234

Their values for temperatures and pressures exceeding the critical will be found in Table 36.

[299] Amagat, E. H., *Ann. de chim. et phys.* (6), **29**, 68-136, 505-574 (1893).

[300] v. Nieuwenburg, C. J., and Blumendal, (Miss) H. B., *Rec. trav. chim. Pays-Bas*, **51**, 707-714 (1932).

Table 96—(Continued)

(Lines continued on p. 227)

Unit of $P = 1$ atmosphere $= 1.01325$ megadyne/cm². Temp. $= t$ °C; $v_0^* = 1.00013$ ml/g.

$t \rightarrow$ v^*/v^*_0	0		1		2 P		3		4	
0.99778	43.45 50.35	−0.90	42.55 50.90	−0.65	41.90 51.40	−0.28	41.62 51.78	+0.06	41.68 52.11	+0.32
525	93.80 51.3	−0.35	93.45 51.50	−0.15	93.30 51.95	+0.10	93.40 52.32	0.39	93.79 52.6	0.71
273	145.1 51.9	−0.15	144.95 52.3	+0.30	145.25 52.6	0.47	145.72 53.1	0.68	146.4 53.6	1.0
0.99020	197.0 53.1	+0.3	197.3 53.5	0.6	197.9 53.8	0.9	198.8 54.0	1.2	200.0 54.2	1.5
0.98766	250.1 53.5	0.7	250.8 54.0	0.9	251.7 54.4	1.1	252.8 54.9	1.4	254.2 55.3	1.8
513	303.6 54.8	1.2	304.8 55.1	1.3	306.1 55.3	1.6	307.7 55.7	1.8	309.5 56.1	2.1
260	358.4 55.6	1.5	359.9 56.0	1.5	361.4 56.6	2.0	363.4 56.8	2.2	365.6 57.2	2.4
0.98007	414.0 56.8	1.9	415.9 57.3	2.1	418.0 57.7	2.2	420.2 58.3	2.6	422.8 58.6	2.7
0.97753	470.8 58.1	2.4	473.2 58.4	2.5	475.7 58.7	2.8	478.5 59.0	2.9	481.4 59.5	3.1
499	528.9 59.6	2.7	531.6 59.9	2.8	534.4 60.3	3.1	537.5 60.7	3.4	540.9 60.9	3.6
0.97245	588.5 60.0	3.0	591.5 60.5	3.2	594.7 60.9	3.5	598.2 61.3	3.6	601.8 61.8	3.9
0.96991	648.5 62.2	3.5	652.0 62.5	3.6	655.6 62.9	3.9	659.5 63.1	4.1	663.6 63.5	4.4
736	710.7 62.6	3.8	714.5 62.9	4.0	718.5 63.0	4.1	722.6 63.2	4.5	727.1 63.6	4.7
482	773.3 64.2	4.1	777.4 64.7	4.1	781.5 65.4	4.3	785.8 66.1	4.9	790.7 66.6	5.2
0.96227	837.5 65.7	4.6	842.1 66.1	4.8	846.9 66.5	5.0	851.9 67.2	5.4	857.3 67.7	5.6
0.95972	930.2	5.0	908.2	5.2	913..4	5.7	919.1	5.9	925.0	6.2

$t \rightarrow$ v^*/v^*_0	0		5		10 P		15		20		30	
1.0200												
150												
100												
050												
025												36.5 21.9	77.0
015									3.2 20.8	55.2	58.4 22.4	77.6
005							5.2 10.2	18.8	24.0 10.5	56.8	80.4 11.6	78.7
1.0000	1.0				3.7	11.7	15.4	19.1	34.5	57.9	92.4	78.7

32. WATER: P-V-T DATA

Table 96—(Continued)
(Lines continued from p. 226.)

5		6		7		8		9		10
					P					
42.00	+0.60	42.60	+0.90	43.50	+1.25	44.75	+1.75	46.50	+2.15	48.65
52.50		52.85		53.20		53.45		53.6		53.6
94.50	0.95	95.45	1.25	96.70	1.50	98.20	1.9	100.1	2.1	102.2
52.9		53.2		53.6		54.0		54.4		54.8
147.4	1.3	148.7	1.6	150.3	1.9	152.2	2.3	154.5	2.5	157.0
54.1		54.4		54.7		55.1		55.3		55.5
201.5	1.6	203.1	1.9	205.0	2.3	207.3	2.5	209.8	2.7	212.5
54.5		54.9		55.3		55.7		56.1		56.5
256.0	2.0	258.0	2.3	260.3	2.7	263.0	2.9	265.9	3.1	269.0
55.6		56.0		56.4		56.6		56.8		57.0
311.6	2.4	314.0	2.7	316.7	2.9	319.6	3.1	322.7	3.3	326.0
56.4		56.7		57.0		57.2		57.7		57.9
368.0	2.7	370.7	3.0	373.7	3.1	376.8	3.6	380.4	3.5	383.9
57.5		58.0		58.3		58.9		59.2		59.9
425.5	3.2	428.7	3.3	432.0	3.7	435.7	3.9	439.6	4.2	443.8
59.0		59.2		59.5		59.7		60.1		60.5
484.5	3.4	487.9	3.6	491.5	3.9	495.4	4.3	499.7	4.6	504.3
60.0		60.5		61.0		61.5		61.8		62.0
544.5	3.9	548.4	4.1	552.5	4.4	556.9	4.6	561.5	4.8	566.3
61.2		61.6		62.0		62.2		62.5		62.8
605.7	4.3	610.0	4.5	614.5	4.6	619.1	4.9	624.0	5.1	629.1
62.3		62.6		62.9		63.6		64.0		64.3
668.0	4.6	672.6	4.8	677.4	5.3	682.7	5.3	688.0	5.4	693.4
63.8		64.1		64.6		64.7		65.1		65.6
731.8	4.9	736.7	5.3	742.0	5.4	747.4	5.7	753.1	5.9	759.0
64.1		64.5		65.0		65.6		66.1		66.6
795.9	5.3	801.2	5.8	807.0	6.0	813.0	6.2	819.2	6.4	825.6
67.0		67.5		67.7		67.8		68.0		68.4
862.9	5.8	868.7	6.0	874.7	6.1	880.8	6.4	887.2	6.8	894.0
68.3		68.8		69.2		69.7		70.1		70.3
931.2	6.3	937.5	6.4	943.9	6.6	950.5	6.8	957.3	7.0	964.3

40		50		60		70		80		90		100
							P					
						51.5	134.5	186.0	146.0	332.0	153.0	485.0
						109.0		112.0		113.0		118.0
				38.5	122.0	160.5	137.5	298.0	147.0	445.0	158.0	603.0
				111.5		114.5		117.5		120.0		122.5
		39.5	110.5	150.0	125.0	275.0	140.5	415.5	149.5	565.0	160.5	725.5
		114.0		118.0		122.0		113.0		114.5		118.0
57.0	96.5	153.5	114.5	268.0	129.0	397.0	131.5	528.5	151.0	679.5	164.0	843.5[a]
56.5		58.5		61.5		61.5		72.0		75.0		75.5
113.5	98.5	212.5	117.5	329.5	129.0	458.5	142.0	600.5	154.0	754.5	164.5	919.0
22.5		24.0		24.0		25.5		26.0		26.5		27.0
136.0	100.0	236.0	117.5	353.5	130.5	484.0	142.5	626.5	154.5	781.0	165.0	946.0
23.5		24.5		24.5		25.5		26.0		26.0		27.0
159.5	101.0	260.5	117.5	378.0	131.5	509.5	143.0	652.5	154.5	807.0	166.0	973.0
11.6		11.5		12.5		12.5		12.5		13.5		13.5
171.1[a]	100.9	272.0	118.5	390.5[a]	131.5	522.0	143.0	665.0	155.5	820.5	166.0	986.5

Table 96—(Continued)
(Lines continued on p. 229)

v^*/v^*_0 \ $t\to$	0		5		10		15 P		20		30	
	9.5				*10.1*		*10.6*		*10.8*		*11.3*	
0.9995	10.5	−2.1	8.4	5.4	13.8	12.2	26.0	19.3	45.3	58.4	103.7	79.3
	19.2		*19.6*		*20.4*		*20.5*		*21.6*		*22.8*	
85	29.7	−1.7	28.0	6.2	34.2	12.3	46.5	20.4	66.9	59.6	126.5	80.5
	19.3		*20.0*		*20.3*		*21.6*		*22.1*		*23.0*	
75	49.0	−1.0	48.0	6.5	54.5	13.6	68.1	20.9	89.0	60.5	149.5	81.5
	50.0		*52.2*		*53.0*		*54.9*		*56.0*		*58.0*	
50	99.0	+1.2	100.2	7.5	107.5	15.5	123.0	22.0	145.0	62.5	207.5	84.0
	51.0		*52.5*		*54.5*		*56.1*		*57.5*		*60.0*	
25	150.0	2.7	152.7	9.3	162.0	17.1	179.1	23.4	202.5	65.0	267.5	85.5
	51.0		*53.0*		*54.8*		*56.4*		*57.5*		*59.5*	
0.9900	201.0	4.7	205.7	11.1	216.8	18.7	235.5	24.5	260.0	67.0	327.0	88.0
	105.4		*108.5*		*112.2*		*115.0*		*118.0*		*123.0*	
0.9850	306.4	7.8	314.2	14.8	329.0	21.5	350.5	27.5	378.0	72.0	450.0	92.0
	109.6		*113.6*		*116.0*		*119.0*		*122.0*		*128.0*	
800	416.0	11.8	427.8	17.2	445.0	24.5	469.5	30.5	500.0	78.0	578.0	96.0
	113.0		*117.2*		*121.0*		*125.0*		*127.0*		*132.0*	
750	529.0	16.0	545.0	21.0	566.0	28.5	594.5[a]	32.5	627.0	83.0	710.0	100.0
	118.0		*121.0*		*125.5*		*127.5*		*131.0*		*137.0*	
700	647.0	19.0	666.0[a]	25.5	691.5	30.5	722.0	36.0	758.0	89.0	847.0	107.0
	121.5		*125.5*		*130.0*		*134.5*		*137.0*			
650	768.5	23.0	791.5	30.0	821.5	35.0	856.5[a]	38.5	895.0			
	127.0		*132.0*		*133.0*		*138.5*					
0.9600	895.5	28.0	923.5	31.0	954.5	40.5	995.0					

v^*/v_0^* \ $t\to$	0.00		10.10		20.40		29.45 P		40.45		48.85	
1.000	1.0				34	54	88	86	174	86	260	
	200				*225*		*232*		*243*		*248*	
0.990	201	15	216	43	259	61	320	97	417	91	508	
	217		*233*		*241*		*252*		*263*		*266*	
0.980	418	31	449	51	500	72	572	108	680	94	774	
	230		*245*		*260*		*273*		*283*		*293*	
0.970	648	46	694	66	760	85	845	118	963	104	1067	
	247		*264*		*280*		*289*		*306*		*315*	
0.960	895	63	958	82	1040	94	1134	135	1269	113	1382	
	275		*289*		*304*		*316*		*327*		*334*	
0.950	1170	77	1247	97	1344	106	1450	146	1596	*120*	1716	
	295		*312*		*324*		*330*		*342*		*360*	

[a] Published values of the pressure were as follows: 5°, 0.9700, $P = 266.0$; 15°, 0.9750, $P = 598.5$; 15°, 0.9650, $P = 850.5$; 40°, 1.0000, $P = 171.5$; 60°, 1.0000, $P = 395.5$; 60°, 0.9995, $P = 405.0$; 100°, 1.0050, $P = 853.5$; 0.00°, 0.915, $P = 2335$.

32. WATER: P-V-T DATA

Table 96—*(Continued)*
(Lines continued from p. 228.)

40		50		60		70		80		90		100
						P						
11.9		*12.5*		*12.5*		*13.0*		*13.5*		*13.0*		*13.5*
183.0	*101.5*	284.5	*118.5*	403.0[a]	*132.0*	535.0	*143.5*	678.5	*155.0*	833.5	*166.5*	1000.0
24.0		*25.0*		*24.5*		*25.0*		*26.5*		*26.5*		
207.0	*102.5*	309.5	*118.0*	427.5	*132.5*	560.0	*145.0*	705.0	*155.0*	860.0		
24.0		*24.0*		*25.5*		*26.0*		*26.0*		*27.0*		
231.0	*102.5*	333.5	*119.5*	453.0	*133.0*	586.0	*145.0*	731.0	*156.0*	887.0		
60.5		*63.0*		*64.0*		*65.5*		*67.0*		*69.0*		
291.5	*105.0*	396.5	*120.5*	517.0	*134.5*	651.5	*146.5*	798.0	*158.0*	956.0		
61.5		*63.5*		*65.0*		*66.0*		*68.0*				
353.0	*107.0*	460.0	*122.0*	582.0	*135.5*	717.5	*148.5*	866.0				
62.0		*62.5*		*65.0*		*67.0*		*68.5*				
415.0	*107.5*	522.5	*124.5*	647.0	*137.5*	784.5	*150.0*	934.5				
127.0		*131.5*		*135.0*		*139.0*						
542.0	*112.0*	654.0	*128.0*	782.0	*141.5*	923.5						
132.0		*136.0*		*140.0*								
674.0	*116.0*	790.0	*132.0*	922.0								
136.0		*148.0*										
810.0	*128.0*	938.0										
144.0												
954.0												

$t \rightarrow$	0.00		10.10		20.40		29.45		40.45		48.85
v^*/v_0^*						*P*					
0.940	1465	*94*	1599	*109*	1668	*112*	1780	*158*	1938	*138*	2076
	158		*165*		*171*		*184*		*189*		*191*
0.935	1623	*101*	1724	*115*	1839	*125*	1964	*163*	2127	*140*	2267
	162		*172*		*179*		*188*		*193*		*197*
0.930	1785	*111*	1896	*122*	2018	*134*	2152	*168*	2320	*144*	2464
	169		*178*		*189*		*197*		*206*		*205*
0.925	1954	*120*	2074	*133*	2207	*142*	2349	*177*	2526	*143*	2669
	176		*183*		*193*		*191*		*199*		*213*
0.920	2130	*127*	2257	*143*	2400	*140*	2540	*185*	2725	*157*	2882
	195		*204*		*205*		*216*		*228*		
0.915	2325[a]	*136*	2461	*144*	2605	*151*	2756	*197*	2953		
	183		*192*		*209*		*213*				
0.910	2508	*145*	2653	*161*	2814	*155*	2969				

230   PROPERTIES OF ORDINARY WATER-SUBSTANCE

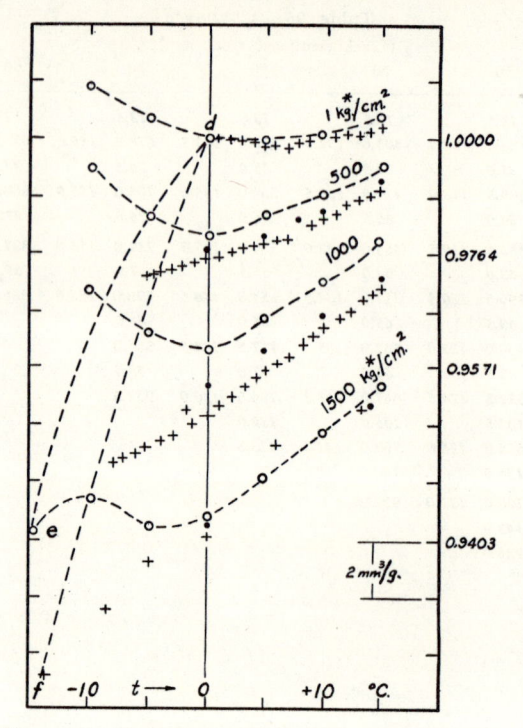

FIGURE 3. Isopiestic Variation of the Specific Volume of Water with the Temperature.

(See p. 214.)

[Adapted from G. Tammann and W. Jellinghaus, Z. anorg. allgem. Chem., 174, 225-230 (1928).] Ordinates represent changes in the specific volume, 1 division = 2 mm³/g; abscissas represent the temperatures, in °C.

For convenience, the several isopiestics have been relatively displaced, crowded together; to the right is given for each isopiestic the specific volume corresponding to the ordinate division next below the lowest point of the Bridgman isopiestic. The dots indicate the observations of E. H. Amagat [*Ann. chim. phys.* (6), **29**, 68-136, 505-574 (1893)]; the circles, those of P. W. Bridgman [*Proc. Am. Acad. Arts Sci.*, **47**, 439-558 (1912); **48**, 307-362 (1913)]; and the crosses, those of Tammann and Jellinghaus (*loc. cit.*). The lines *de* and *df* connect the several melting points as determined by Bridgman and by Tammann and Jellinghaus, respectively. The pressure corresponding to each isopiestic is indicated.

Table 97.—Isopiestic Thermal Expansion of Water

Adapted from a table computed by L. B. Smith and F. G. Keyes [294] by means of an empirical equation representing their observations. They give the values to five significant figures, but in view of what they say about the limitations of their equation (see Table 95, Section VI), three significant figures seem to be sufficient for this compilation. Similarly, it has not seemed necessary to give here their values for $P = 1, 25, 75, 125$, and 175 atm; those for $P = 1$, extending only to $t = 100$ °C, are, to the precision

32. WATER: P-V-T DATA

Table 97—(Continued)

of this table, identical with those for P_{sat}, and the others can be obtained without significant error by interpolation between those here given. Adjacent to each t, is given the corresponding value of the pressure (P_{sat}) of the saturated vapor. Each value of $(dv^*/dt)_p$ given under P_{sat} in the body of the table is the limiting value approached along the isopiestic corresponding to P_{sat} as t approaches that corresponding to saturation, *i.e.*, the associated t. The exponent n is given in the third column, and each value is to be used until another is given.

Examples: At 140 °C the saturation pressure is 3.57 atm; along that isopiestic $(dv^*/dt)_p$ at 140 °C is $10.48 \times 10^{-4} = 0.001048$. At 150 °C and $P = 50$ atm, $(dv^*/dt)_p = 1.10 \times 10^{-3} = 0.00110$.

Unit of P and of $P_{sat} = 1$ atm; of $(dv^*/dt)_p = 1$ ml/g·°C. Temp. $= t$ °C.

$P \rightarrow$ t	P_{sat}	n	P_{sat}	50	100	150	200	250	300	350
						$10^n (dv^*/dt)_p$				
0	0.006	5	−3.09							
10	0.012		+9.39							
20	0.023	4	2.00							
30	0.042		2.95	2.94	2.94	2.93	2.92	2.91	2.90	2.89
40	0.073		3.80	3.78	3.76	3.74	3.72	3.70	3.68	3.66
50	0.122		4.57	4.54	4.51	4.48	4.45	4.42	4.39	4.36
60	0.196		5.28	5.24	5.20	5.16	5.12	5.08	5.04	5.00
70	0.308		5.96	5.91	5.85	5.80	5.75	5.70	5.65	5.60
80	0.467		6.61	6.54	6.48	6.41	6.35	6.29	6.23	6.17
90	0.692		7.24	7.16	7.08	7.00	6.92	6.85	6.78	6.71
100	1		7.86	7.77	7.67	7.58	7.49	7.40	7.32	7.24
110	1.41		8.49	8.38	8.26	8.15	8.05	7.95	7.85	7.75
120	1.96		9.13	9.01	8.88	8.76	8.65	8.54	8.43	8.33
130	2.67		9.79	9.64	9.49	9.34	9.20	9.06	8.93	8.80
140	3.57		10.48	10.31	10.12	9.95	9.78	9.62	9.46	9.32
150	4.70	3	1.12	1.10	1.08	1.06	1.04	1.02	1.01	0.99
160	6.10		1.20	1.18	1.15	1.13	1.11	1.08	1.06	1.05
170	7.82		1.28	1.26	1.23	1.20	1.18	1.15	1.13	1.11
180	9.90		1.37	1.35	1.31	1.28	1.25	1.22	1.20	1.17
190	12.4		1.48	1.44	1.40	1.37	1.33	1.30	1.27	1.24
200	15.3		1.59	1.55	1.51	1.46	1.42	1.38	1.35	1.32
210	18.8		1.71	1.68	1.62	1.57	1.52	1.48	1.44	1.40
220	22.9		1.86	1.82	1.75	1.69	1.63	1.58	1.53	1.49
230	27.6		2.02	1.98	1.90	1.82	1.76	1.70	1.64	1.59
240	33.0		2.21	2.18	2.07	1.98	1.90	1.83	1.76	1.71
250	39.2		2.44	2.41	2.28	2.17	2.07	1.98	1.90	1.84
260	46.3		2.71	2.69	2.53	2.38	2.26	2.16	2.06	1.98
270	54.3				2.83	2.65	2.50	2.36	2.25	2.16
280	63.3				3.22	2.98	2.78	2.61	2.47	2.35
290	73.5				3.72	3.39	3.13	2.91	2.73	2.58
300	84.8				4.43	3.93	3.57	3.28	3.05	2.85
310	97.4				5.46	4.68	4.15	3.74	3.43	3.17
320	111					5.76	4.94	4.36	3.91	3.55
330	127					7.63[a]	6.08[a]	5.18	4.53	4.03
340	144					10.99[a]	7.89[a]	6.47	5.53	4.84
350	163						11.68[a]	8.28	6.72	5.76

[a] For $P = 175$, $10^3(dv^*/dt)_p = 6.68$ at 330 °C, 9.30 at 340 °C, and 14.39 at 350 °C.

Table 98.—Mean Isopiestic Coefficient of Thermal Expansion of Water

For slopes of isopiestics see Table 97. Values for other pressures and ranges in temperature may be readily obtained from the values given in Table 95.

Unit of $P = 1$ atm $= 1.0332$ kg*/cm²; of $p = 1$ kg*/cm². Temp. $= t$ °C

I. Pressure $= 1$ atm. Coefficients have been computed from the data in Table 94. They essentially agree with the earlier observations by G. T. Gerlach,[301] as quoted in the fifth edition of Landolt-Börnstein's *Tabellen*.

For α_1 the temperature is expressed in terms of the numbers x; the first line contains the values for each degree from 0 to 9; the second, for each 10° from 0 to 90; the third, for each 10° from 5 to 95. Examples: The mean coefficient between 40 and 41 °C is 38.92×10^{-5}; that between 45 and 46 °C is 43×10^{-5}.

$$\alpha_1 = 10^5(v_{t+1} - v_t)/v_t; \quad \alpha_{10} = 10^5(v_{t+10} - v_t)/10v_t.$$

$x \rightarrow$	0	1	2	3	4	5	6	7	8	9
t					α_1					
x	−5.89	−4.13	−2.44	−0.78	+0.81	+2.36	+3.86	+5.32	6.73	8.12
$10x$	−5.89	+9.45	21.20	30.75	38.92	46	52	59	64	70
$10x+5$	+2.36	15.67	26.18	34.97	43	49	56	61	67	73
t	0	10	20	30	40	50	60	70	80	90
α_{10}	+1.4ᵃ	15.0	25.7	34.6	42.2	49.2	55.6	61.5	67.2	72.7

II. Derived from the same observations as the specific volumes given in Table 95.

A = Amagat; taken from his more detailed table [291] which differs slightly from and supersedes his earlier table.[302] The latter is still frequently quoted in certain important compilations.

B = Bridgman: computed by the compiler from B's data in Table 95(III).

SK = Smith and Keyes: computed by the compiler from their table of specific volumes, with retention of more significant figures than are given in Table 95.

Unit of $P = 1$ atm; $\alpha_m = 10^5(v_2 - v_1)/(t_2 - t_1)v_1$

Interval→ Observer→ P	0° to 10° A	B	SK	10° to 20° A	B	SK	30° to 40° A	B	SK
					α_m				
1	1.4ᵃ	2	3.4	14.9	15	14.8	33.4	35	32.4
100	4.3ᵃ		3.9	16.5		15.2	34.5		33.5
200	7.2		4.3	18.3		15.3	35.0		33.5
300	9.8		4.8	20.5		15.5	35.7		33.2
500	14.9	12		23.6	27		37.0	37	
700	19.2			26.2			37.7		
1000	25.9	27		29.4	29		39.6	38	
2000	36.4	40		35.6	38		42.3	42	
4000		41			42			44	
6000		36			42			47	

[301] Gerlach, G. T., "Salzlösungen," Freiberg, 1859.
[302] Amagat, E. H., *Compt. rend.*, 105, 1120-1122 (1887).

32. WATER: P-V-T DATA

Table 98—(Continued)

Interval→	40° to 50°			60° to 70°			70° to 80°		
Observer→	A	B	SK	A	B	SK	A	B	SK
P					α_m				
1	42.2	43	41.6	55.6	56	55.3	61.8	62	61.5
100	42.2		41.3	54.8		54.6			60.6
200	42.6		40.9	53.9		53.9	60.0		59.8
300	42.3		40.6	52.8		53.4	59.0		58.9
500	42.9	50		52.3	48		56.6	57	
700	43.4			52.3			55.4		
1000	43.7	45		51.2	53			54	
2000		46			53			49	
4000		49			49			46	
6000		46			46			45	

III. Comparison of Bridgman's values of 1912 (B′) and of 1935 (B″). The B′ values have been computed from the specific volumes given in his table of 1912, and the B″ ones from those given in Table 95, Section IV. The 1912 table is the basis of the *International Critical Table* values, which are given in Table 95, Section III.

$$\alpha_m = 10^5(v_2 - v_1)/(t_2 - t_1)v_1$$

Unit of $p = 1$ kg*/cm²; of $P = 1$ atm

Interval→	0°–20°		20°–40°		40°–50°		50°–60°		60°–80°		80°–100°
Set→	B′	B″	B′	B″	B′	B″	B′	B″	B′	B″	B″
p	P					α_m					
0		8	8	30	30	42	42	49	49	59	56	74
500	483.9	19	24	33	31	44	26	49	53	55	52	60
1000	967.8	27	29	36	38	44	39	49	43	53	50	56
2000	1935.6	36	36	41	43	45	45	47	44	50	47	51
4000	3871.4	41	43	43	44	45	45	46	42	48	46	45
6000	5807	38	45	46	47	45	45	46	43	45	42	44
8000	7743			51		47	42	46	39	44	41	41
10000	9678					48	40	46	40	45	40	40
12000	11614					49	40	48	40	45	40	40

IV. W. Watson.[303] L. Bouchet[304] thinks that these values of α are vitiated by a systematic error. The pressures are all higher than the critical ($p_{crit} = 225.65$ kg*/cm²), and most of the temperatures exceed t_{crit} (374.15 °C). At the higher temperatures the thermal expansion is several times as great as that of the ideal gas. These coefficients are with reference to the volume (v_0) at room temperature and the indicated pressure. $\alpha' = 10^5(v_2 - v_1)/(t_2 - t_1)v_0$. For the ideal gas $\alpha' = 341$ if room temperature = 20 °C.

Unit of $p = 1$ kg*/cm²; of $P = 1$ atm.

t_1→		100	200	300	400	500	600	700	800
t_2→		200	300	400	500	600	700	800	900
p	P					α'				
400	387.1	99	600	3280	4300				
700	677.5	91	380	880	1760	1880	1880		
1000	967.8	84	280	520	1000	1060	1060	1060	1060

ᵃ The specific volume passes through a minimum in the interval here covered.

---

[303] Watson, W., *Proc. Roy. Soc. Edinburgh*, **31**, 456-477 (1911). Briefer report in *Ber. Sachs. Ges. Wiss, Leipzig (Math.-Phys.)*, **63**, 264-268 (1911).
[304] Bouchet, L., *Compt. rend.*, **178**, 554-556 (1924).

Table 99.—Interpolation Formulas for the Thermal Expansion of Water

Of the numerous interpolation formulas that have been proposed for the thermal expansion of water, those more frequently quoted are here assembled. In Table 100 the corresponding values defined by them are compared with one another and with the data of Tables 94 and 260. It will be seen that some of the formulas are entirely unsatisfactory; they should not appear in future hand-books or similar compilations.

For temperatures not exceeding 100 °C, the pressure is assumed to be 1 atm and invariable; for higher temperatures, it is the pressure of the saturated vapor, the water being continuously saturated; v_0 = actual volume at 0 °C, v_t = volume of the same mass at t °C. To each formula is assigned a key symbol serving to suggest its proposer and to identify the associated data in Table 100.

The development of a formula for liquids has been discussed by E. Salzwedel,[305] W. Jazyna,[306] W. Herz,[307] V. Fischer,[308] O. Tumlirz,[309] and others.

I. Formulas of the type $10^6(v_t - v_0)/v_0 = a + bt + ct^2 + dt^3 + et^4$; temp. = t °C.

Key[a]	a	b	c	d	$10^6 e$	Range, °C
Ch'	0	−67.464645	+8.934223	−0.07891946	0	0 to 10.3
Ch''	−54.7835	−55.24276	+7.945055	−0.04800150	0	10.3 to 13.0
Ch'''	−114.5565	−42.940141	+7.106115	−0.02905759	0	13 to 41
Hi	0	+108.679	+3.0074	+0.002873	−6.6457	100 to 200
La	0	−53.255	+7.61532	−0.0437217	+164.322	30 to 80
Pa[b]	0	−64.807	+8.6697	−0.26211	0	−10 to 4
Pi	0	−94.17	+1.449	−0.5985	0	−13 to 0
Sch	0	−64.268	+8.50526	−0.0678977	+401.209	0 to 33
We[c]	0	−64.5	+8.64	−0.267	0	−10 to +4
ZT[c]	+1100	+75	+3.5	0	0	110 to 140

II. Other formulas. Temp. = t °C.

Key[a]	Formula	Range, °C
T-C	$1 - \rho = \dfrac{v_t - 1}{v_t} = \dfrac{(t - 3.9863)^2}{508929.2} \cdot \dfrac{t + 288.9414}{t + 68.12963}$	0 to 40
Th'	$1 - \rho = \dfrac{v_t - 1}{v_t} = \dfrac{(t - 3.98)^2}{503570} \cdot \dfrac{t + 283}{t + 67.26}$	0 to 40
Th''	$1 - \rho = \dfrac{v_t - 1}{v_t} = \dfrac{(t - 3.98)^2}{466700} \cdot \dfrac{t + 273}{t + 67} \cdot \dfrac{350 - t}{365 - t}$	17 to 100
We[b]	$10^6 \left(\dfrac{v_t - v_4}{v_4} \right) = 8.2004(4 - t) + 5.44402(4 - t)^2 + 0.26698(4 - t)^3$	−10 to +4
ZT	$10^6 \left(\dfrac{v_t - v_0}{v_0} \right) = 51700 + 845(t - 110) + 3.5(t - 110)^2$	110 to 140

32. WATER: P-V-T DATA

Table 99—(Continued)

[a] Key and references:

Ch', Ch'', Ch'''	Chappuis, P., *Trav. et Mém. Bur. Int. Poids et Mes.*, **13**, D.1-D.40 (1907).
Hi	Hirn, G. A., *Ann. de chim. et phys.* (4), **10**, 32-92 (1867).
La	Landesen, G., *Schr. Naturf. Ges. Univ. Jurjeff (Dorpat)*, No. **14** (1904); based on his observations published in No. **11** (1902).
Pa	Panebianco, H., *Riv. di. min. e crist. ital.*, **38**, 3-11 (1909); reviewed in *Z. Kryst.*, **50**, 496-497 (1912).
Pi	Pierre, I., *Ann. de chim. et phys.* (3), **15**, 325-408 (1845); see Frankenheim, M. L., *Ann. d. Physik (Pogg.)*, **86**, 451-464 (1852).
Sch	Scheel, K., *Idem (Wied.)*, **47**, 440-465 (1892).
T-C	Tilton's improved formulation of the observations of Chappuis; see Tilton, L. W., and Taylor, J. K., *J. Res. Nat. Bur. Stand.*, **18**, 205-214 (RP971) (1937).
Th'	Thiesen, M., Scheel, K., and Diesselhorst, H., *Wiss. Abh. Physik. Techn. Reichsanstalt*, **3**, 1-70 (1900); *Ann. d. Physik (Wied.)*, **60**, 340-349 (1897).
Th''	Thiesen, M., *Wiss. Abh. Physik. Techn. Reichsanstalt*, **4**, 1-32 (1903).
We	Weidner, *Ann. d. Physik (Pogg.)*, **129**, 300-308 (1866).
ZT	Zepernick, K., and Tammann, G., *Z. physik. Chem.*, **16**, 659-670 (1895).

[b] Both the formulas, Pa and We, are based solely upon the data reported by Weidner; their difference, actually insignificant, results from certain arithmetical errors in Weidner's computation. An examination of Weidner's data shows that they do not justify the retention of more than three significant figures in each of the coefficients of the equation, and probably two would be enough.

In the Landolt-Börnstein *Tabellen*, the We formula is given incorrectly, the coefficients of the formula in Section II of this table being given as applicable to the formula of Section I. The same error in the 1905 edition of those *Tabellen* was pointed out by H. Panebianco in 1909, in the paper in which the Pa formula was derived. That formula is now included in the *Tabellen* as an independent one, but the error in the We formula is perpetuated. Furthermore, the reference given in the *Tabellen* for the Pa formula is *Riv. di min. e crist. ital.*, **38**, 3 (1912); the year should have been given as 1909, 1912 corresponding to vol. 41, in which there is nothing relating to the Pa formula.

[c] Actually published in the equivalent form shown in the second section of the table.

Table 100.—Comparison of Interpolation Formulas for the Thermal Expansion of Water

The several formulas, with their key symbols, are given in Table 99. For $t \leqq 100\,°C$ the pressure is 1 atm; for $t > 100\,°C$ the pressure is that of the saturated vapor. In the second column are given the values of $10^6(v_t - v_0)/v_0$ as determined from the data in Tables 94 and 260, the doubtful digits being overscored; v_0 is the volume at $0\,°C$. By adding to any of the values in column two the appropriate value of δ, the corresponding value of $10^6(v_t - v_0)/v_0$ as defined by the interpolation formula is obtained. Example: The value of $10^6(v_t - v_0)/v_0$ as defined by the Pi formula is $2583 - 210 = +2373$ at $12\,°C$, and $+195 + 3.9 = +198.9$ at $-2\,°C$.

Since the values in column Obs for the range 0 to $40\,°C$ have been derived by the T-C formula, the δ for that formula is zero throughout this range. Parentheses enclose values of δ that lie beyond the range of the formula. It is obvious that formulas Pi, We = Pa, La, Hi, and ZT are quite unsatisfactory.

[305] Salzwedel, E., *Ann. d. Physik (5)*, **5**, 853-886 (1930).
[306] Jazyna, W., *Z. Physik*, **58**, 429-435, 436-439 (1929).
[307] Herz, W., *Z. Elektroch.*, **32**, 460-462 (1926).
[308] Fischer, V., *Ann. d. Physik (4)*, **71**, 591-602 (1923).
[309] Tumlirz, O., *Sitz.-ber. Akad. Wiss. Wien (Abt. IIa)*, **130**, 93-133 (1921).

236 PROPERTIES OF ORDINARY WATER-SUBSTANCE

Value of $10^6(v_t - v_0)/v_0$ as defined by formula is Obs+δ. Temp.=t°C

Key[a]	Obs[b] $10^6\frac{(v_t-v_0)}{v_0}$ ‰	Pi	We=Pa	T-C	Ch'	Ch''	Ch'''	Th'	Th''	Sch	La	Hi	ZT
−12	+2583	−210	−160										
−10	1937	−252	−163										
−8	1369	−216	−163										
−6	919	−172	−74										
−4	488	−50											
−2	+195	+3.9	−29	(−21.9)									
+2	−100.2		+3.2	0	+0.4			−0.1		+5.2			
4	−132.4		−5.1	0	+0.4			+0.0		+7.2			
6	−100.7			0	+0.5			+0.3		+7.1			
8	−8.9			0	+0.6			+0.6		+6.0			
10	+139.6			0	+0.3	(−0.3)		+1.0		+4.4			
10.5	185.2			0	+0.1	+0.3		+1.1		+4.0			
11	234.1			0		+1.1		+1.3		+3.7			
11.5	286.3			0		+1.4		+1.4		+3.2			
12	341.7			0		+1.8		+1.5		+2.8			
12.5	400.3			0		+2.0		+1.6		+2.5			
13	462.0			0		+2.3	+2.3	+1.8		+2.2			
15	739.7			0			+2.4	+2.2	+3.3	+1.1			
20	1636.9			0			−0.3	+3.3	+5.1	+1.9			

δ

32. WATER: P-V-T DATA

t						
25	2800.7					
30	4207.5					
35	5838.8					
40	7679.9					
50	11940̄	0	+4.3	+6.5	+4.2	(+ 8.5)
60	16920̄	0	+5.2	+7.0	+10.9	+ 1.3
70	22570̄	0	+5.7	+6.4	(+21.7)	– 1.9
80	28860̄	0	+6.1	+4.2		– 3.1
90	35770̄	(+5.9)		+1.3		– 2.6
100	43300̄		–1.5	–4.4		–14.4
110	51400̄		+0.7	–8.3		–34.0
120	60200̄		+2.9	–9.2		–37.2
130	69600̄		–2.0	–4.4		
140	79700̄			–3.8		
150	90500̄					
160	102000̄					
180	127400̄					
200	156400̄					

–150			
–205	+240		
–266	+161		
–233	+160		
–209	+279		
–200	(+456)		
–209			
–619			
–2017			

[a] For the significance of the key symbols, except Obs (= observed), see Table 99.
[b] These so-called observed values have been derived from the data in Tables 94 and 260; values for $t<0$ are from Mohler's data; for t between 0 and 40 °C from those defined by Tilton's formula (T-C), for t between 40 and 100 °C from Thiesen's data, all given in Table 94; and for $t>100$ °C from the data in the International Skeleton Steam Table of 1934 (Table 260).

Table 101.—Thermal Slopes of the Isometrics of Water

(For mean isometric thermal coefficient of pressure, see Table 102)

Bridgman's data (B) have been read from his published graphs [P. W. Bridgman [292a, Figs. 9, 10]]; the others have been adapted from G. Tammann and A. Rühenbeck.[295]

$v_{1,20}$ and $v_{1,0}$ = specific volumes at 1 atm and at 20 °C and 0 °C, respectively; v^* = actual specific volume; pressure = 1 atm + p. $v_{1,20}$ = 1.00177 and $v_{1,0}$ = 1.00013 ml/g.

Unit of p = 1 kg*/cm² = 0.9678 atm.   Temp. = t °C

$t \rightarrow$	0	20	40	60	80	$t \rightarrow$	0	20	40	60	80
p			$(dp/dt)_v$(B)			p			$(dp/dt)_v$(B)		
0	−0.5	+4.5	9.1	12.5	15.0	10000			41.0	39.8	36.0
1000	+6.0	9.2		15.2	15.8	11000			43.0	41.6	37.5
2000	11.4	13.5	15.2	17.4	17.4	12000			44.1	42.3	37.6
3000	15.7	16.8	18.0	19.5	19.2	$t \rightarrow$	0	20	40	60	80
4000	18.8	19.7	20.6	21.8	21.5	$v^*/v_{1,0}$			$(dp/dt)_v$(B)		
5000	20.3	22.0	23.4	24.3	23.6	1.00	−1	+4.1	9.4	13.5	15.2
6000	17.8	25.4	27.0	27.0	26.0	0.95	+7.2	·11.0	14.0	17.1	18.0
7000		32.8	30.8	30.2	28.4	0.90	16.1	18.0	19.6	22.0	23.0
8000		38.5		33.2	30.8	0.87₅	19.8	21.2	23.8	25.6	26.5
9000			38.0	36.7	33.5	0.85	18.5	27.0	30.0	31.0	31.7

Tammann and Rühenbeck

$v^*/v_{1,20} \rightarrow$	2.11		2.13		2.18		2.25	
t	$(dp/dt)_v$	p	$(dp/dt)_v$	p	$(dp/dt)_v$	p	$(dp/dt)_v$	p
525	14.2	1692	13.6	1640	12.3	1508	10.7	1368
575	11.8	2348	12.0	2278	10.7	2028	9.6	1876
625	7.3	2812	7.4	2762	8.7	2565	8.3	2323

Table 102.—Mean Isometric Thermal Coefficient of Pressure of Water

(For slopes of isometrics at 0 to 80 and 500 to 600 °C, see Table 101)

Adapted from E. H. Amagat,[299] with correction of certain computational errors and with adjustment to accord with Table 96.

$\gamma = \dfrac{1}{P_1}\left(\dfrac{P_2 - P_1}{t_2 - t_1}\right)_v$; v = volume of a certain mass of water at P and t, v_0 = that of the same mass at 1 atm and 0 °C.

Unit of $\gamma = 1$ (°C)⁻¹.   Temperatures are t_1 °C and t_2 °C

$t_1 \rightarrow$	0	10	20	30	40	50	60	70	80	90
$t_2 \rightarrow$	10	20	30	40	50	60	70	80	90	100
v/v_0						$10^4 \gamma$				
1.0200								261₀	785	461
1.0150							317₀	857	493	355
1.0100						280₀	834	511	360	284
1.0050					169₀	747	481	331	286	241
1.0025				211₀	868	554	392	310	256	218
1.0015			1725₀	133₀	735	498	369	294	247	211
1.0005			237₀	974	633	451	348	281	237	203
1.000	27₀₀	833₀	168₀	852	590	435	337	274	234	202

32. WATER: P-V-T DATA

Table 102—(Continued)

$t_1 \rightarrow$ $t_2 \rightarrow$ v/v_0	0 10	10 20	20 30	30 40	40 50	50 60	60 70	70 80	80 90	90 100
				$10^4\gamma$						
0.9995	314	228₀	129₀	765	555	417	328	268	229	200
0.9985	152	956	891	637	493	381	310	259	220	
0.9975	112	633	680	545	443	358	294	248	213	
0.9950	86	349	431	406	360	304	260	225	198	
0.9925	80	250	321	320	303	265	233	207		
0.9900	79	199	258	269	259	238	213	191		
0.9850	74	149	190	204	207	196	181			
0.9800	70	124	156	166	172	167				
0.9750	70	108	132	141	158					
0.9700	69	98	117	126						
0.9650	69	89								
0.9600	66									

$t_1 \rightarrow$ $t_2 \rightarrow$ v/v_0	0.00 10.10	10.10 20.40	20.40 29.45	29.45 40.45	40.45 48.85	$t_1 \rightarrow$ $t_2 \rightarrow$ v/v_0	0.00 10.10	10.10 20.40	20.40 29.45	29.45 40.45	40.45 48.85
			$10^4\gamma$						$10^4\gamma$		
1.000			1760	889	588	0.935	62	65	75	75	79
0.990	75	194	258	276	259	0.930	62	62	73	71	74
0.980	74	110	160	172	165	0.925	61	62	71	69	67
0.970	70	92	124	127	129	0.920	59	62	65	66	69
0.960	69	83	100	108	106	0.915	58	57	64	65	
0.950	65	76	87	92	90	0.910	57	59	61		
0.940	64	68	74	81	85						

Table 103.—Isothermal Compression of Water

[See also Tables 104 (dilated water), 105 (mean between 1 and P atm.), and 106 (mean between P_1 and P_2 atm)]

E. Brander [310] has found that Bridgman's 1912 values (see Table 95) up to 6000 kg*/cm² can be represented by the empirical expression: $\log_{10}(1 + Ap) = B(v_1 - v_p)/v_p$ where v_1 and v_2 are the volumes of the same mass of water at the temperature considered and at the pressure of 1 and p kg*/cm², respectively, and A and B depend on the temperature only, taking the following values:

t	0	20	40	60	80	°C
$500A$	0.136	0.111	0.105	0.102	0.106	cm²/kg*
B	2.368	2.1668	2.128	2.084	2.101	dimensionless

C. Grassi,[311] working within the range 1 to 10 atm, found the isothermal compressibility $[\beta = -(1/v)(dv/dp)_t]$ for a given temperature to be independent of the pressure; it varied with the temperature, the following values being reported:

t	0	1.5	4.1	10.8	13.4	18.0	18.0	25.0	34.5	43.0	53.0	°C
$10^6\beta$	50.3	51.5	49.9	48.0	47.7	46.3	46.0	45.6	45.3	44.2	44.1	per atm

The following values have been selected from a table computed by

---

[310] Brander, E., *Soc. Scien. Fennica, Comm. Phys. Math.*, **7**, No. 7 (1934).

[311] Grassi, C., *Ann. de chim. et phys.* (3), **31**, 437-478 (1851) → *J. de Pharm. et Chim.* (3), **19**, 442-444 (1851).

Table 103—(Continued)

L. B. Smith and F. G. Keyes [294] by means of an empirical equation representing their observations, and have been rounded off to three significant figures (see remarks at head of Section VI of Table 95) theirs being given to five. They also give values for $P = 1, 25, 75, 125$, and 175 atm. Those for $P = 1$ atm, $t \gg 100$ °C are, within the precision of this table, identical with those for P_{sat}; those for the other pressures can be satisfactorily obtained by interpolation, except as noted.

Adjacent to each value of t is given the corresponding value of the pressure (P_{sat}) of the saturated vapor. Each value of $(dv^*/dp)_t$ given under P_{sat} in the body of the table is the limit approached along the isotherm as P approaches P_{sat}; v^* is the specific volume.

Unit of P and of $P_{sat}=1$ atm; of $(dv^*/dp)_t = 10^{-6}$ ml/g·atm. Temp.$=t$ °C

$P \rightarrow$ t	P_{sat}	P_{sat}	50	100	150	200	250	300	350
					$(dv^*/dp)_t$				
0	0.006	46.1	45.0	44.1	43.1	42.2	41.3	40.5	39.6
10	0.012	45.5	44.5	43.6	42.7	41.8	40.9	40.1	39.3
20	0.023	45.2	44.3	43.4	42.5	41.6	40.8	40.0	39.2
30	0.042	45.2	44.3	43.4	42.6	41.7	40.9	40.1	39.3
40	0.073	45.5	44.6	43.7	42.8	42.0	41.2	40.4	39.6
50	0.122	46.0	45.1	44.2	43.3	42.5	41.7	40.9	40.1
60	0.196	46.8	45.8	44.9	44.0	43.2	42.4	41.6	40.8
70	0.308	47.8	46.8	45.9	45.0	44.1	43.3	42.4	41.7
80	0.467	49.0	48.0	47.1	46.1	45.2	44.4	43.5	42.7
90	0.692	50.6	49.5	48.5	47.6	46.6	45.7	44.8	44.0
100	1	52.4	51.3	50.2	49.2	48.2	47.3	46.3	45.4
110	1.41	54.6	53.4	52.3	51.2	50.1	49.1	48.1	47.2
120	1.96	57.2	55.9	54.7	53.5	52.4	51.3	50.2	49.2
130	2.67	60.1	58.8	57.5	56.2	54.9	53.7	52.6	51.4
140	3.57	63.8	62.2	60.7	59.3	57.9	56.6	55.3	54.1
150	4.70	67.9	66.1	64.4	62.8	61.3	59.8	58.4	57.0
160	6.10	72.8	70.8	68.9	67.1	65.4	63.7	62.1	60.6
170	7.82	78.5	76.2	74.0	72.0	70.0	68.1	66.2	64.5
180	9.90	87.3	82.6	80.0	77.6	75.3	73.1	71.0	69.0
190	12.4	93.2	90.0	87.0	84.2	81.5	78.9	76.5	74.1
200	15.3	102.8	99.0	95.3	91.9	88.7	85.7	82.8	80.0
210	18.8	114.2	109.6	105.2	101.0	97.1	93.4	90.0	86.7
220	22.9	128.0	122.5	116.9	111.8	107.0	102.5	98.3	94.4
230	27.6	145	138	131	125	119	113	108	103
240	33.0	166	158	148	140	133	126	120	114
250	39.2	192	183	170	159	150	141	133	126
260	46.3	226	217	197	183	170	159	149	139
270	54.3	270		232	212	195	180	168	156
280	63.3			278 [a]	250	227	208	191	176
290	73.5			342 [a]	300	268	242	220	201
300	84.8			437	371 [a]	324	287	257	232
310	97.4			623	473 [a]	401 [a]	348	306	274
320	111				634 [a]	514 [a]	433	374	329
330	127				919	693 [a]	556	468	406
340	144				1653	992 [a]	746	603	512
350	163					1618 [a]	1054	795	651
360	184					3816	1693	1094	831

[a] Value of $(dv^*/dp)_t$ for $P = 75$ is 296 at 280°, 390 at 290°; for $P = 125$, 400 at 300°, 522 at 310°, 730 at 320°; for $P = 175$, 434 at 310°, 567 at 320°, 784 at 330°, 1125 at 340°, and 2318 at 350°.

Table 104.—Mean Isothermal Compressibility of Dilated Water
(Adapted from J. Meyer [312])

There is no discontinuity in the value of the compressibility (β_m) as P changes from positive (= pressure) to negative (= tension).

Values of P for which no value of β_m has been determined have been omitted from the table.

$\beta_m \equiv (v_1 - v_2)/v_1(P_2 - P_1)$; here, $P_1 = 0$, $v_1 =$ volume at zero pressure and the indicated temperature, $v_2 =$ volume of the same mass at the same temperature and P_2; the water always completely fills the vessel at P_2 and the indicated temperature, v_2 being the volume of the vessel under those conditions.

Unit of $P = 1$ atm $= 1.01325$ megadynes/cm²; of $\beta_m = 10^{-6}$ per atm. Temp. $= t$ °C

$t \rightarrow$ P	1	2	3	4	5	6	7	8
				β_m				
− 9								46.2
−10								46.7
−11	48.3							
−12		48.4	48.2		47.7	47.1	46.7	
−13			48.2	47.7	47.8			
−19							47.0	46.9
−20			47.6		47.2	47.0		
−21					47.4			
−24								47.1
−26			47.8		47.6	47.1		

$t \rightarrow$ P	9	10	11	12	13	14	15	16
				β_m				
− 2					46.3			
− 4						46.6		
− 5				46.7				46.5
− 6			46.4					
− 7					46.6			
− 8		46.3					46.8	
−10				46.5				
−11							46.5	
−12			46.5					
−15		46.6						
−18	46.9		46.7					
−20		46.8						
−23	46.9							

$t \rightarrow$ P	17	18	19	20	21	22	23	24
				β_m				
− 2						45.7		45.3
− 3						46.5		
− 4					46.8			
− 5								46.1
− 6					47.3			
− 7					47.1		46.2	
− 8					46.8			
− 9				47.2				
−10				47.3			46.2	46.4
−12				47.2		46.5		46.5

[312] Meyer, J., *Abh. d. Deutsch. Bunsen Ges.*, 3 No. 1, whole number 6 (1911).

Table 104—(Continued)

P \ t	17	18	19	20 β_m	21	22	23	24
−13							46.7	
−14			47.1			46.5	46.8	
−15						46.7		
−16			47.2		46.8		46.8	46.2
−17		47.4			47.0		47.1	
−18		47.4			46.8			46.3
−19				47.0	46.9			
−20		47.5				46.9		46.3
−21	47.4							46.4
−22	47.6					47.2	46.3	
−23						47.3		
−24							46.4	
−26					47.4			
−27					47.5			
−29				47.2				

P \ t	25	26	27	28 β_m	29	30	31	
+ 7				46.0				
+ 6	46.1			45.6				
+ 4	45.8			45.8				
+ 3	46.4							
+ 2				46.6				
− 2		45.4	46.0				45.4	
− 4		46.5						
− 5	46.1		45.8				45.1	
− 6						45.7		
− 7	46.2	46.3				45.1		
− 9	46.1					45.2		
−10		46.2						
−11						45.3		
−12	46.0					45.7		
−14	46.1				45.8			
−16					45.9			
−18					46.0			
−20				45.8				
−21				46.0				
−23				45.9				
−26			46.1					
−29			46.2					
−30		46.1						

Table 105.—Mean Isothermal Compressibility of Water Between Pressures 1 and P

(Between P_1 and P_2, Table 106; adiabatic compressibility, Table 107; natural waters, Table 110).

Values for other temperatures and pressures may be readily obtained from the specific volumes as given in Table 95.

Table 105—(Continued)

These mean compressibilities $[\beta_m = (v_1 - v_p)/(P - 1)v_1]$ cannot be satisfactorily represented by an expression of the form $\beta_m = a - bP$, which has been proposed by K. Drucker[313] and by A. L. T. Moesveld.[314] The former gives for $10^6 a$ and $10^6 b$, respectively, the values 47.0 and 0.0115 at 25 °C, and 46.3 and 0.015 at 35 °C; the latter gives 44.5 and 0.00492 at 25 °C. The unit of P is 1 atm in each case. W. Jazyna[315] has considered the calculation of the compressibility from other data.

$\beta_m = (v_1 - v_p)/(P - 1)v_1$ or $(v_0 - v_p)/Pv_0$, where v_0, v_1, and v_p are the volumes of the same mass of water at the same temperature but under the pressures 0, 1, and P, respectively. The values of v_1 have been taken from Table 94, unless others have been specified in the article cited. As t varies, β_m passes through a minimum at about 40 or 50 °C for Amagat's and most of Bridgman's observations, but near 30 °C for the values given by Smith and Keyes and for some of Bridgman's more recent data.

Unit of $P=1$ atm, of $P_b=1$ bar, of $p=1$ kg*/cm²; of $\beta_m=10^{-6}$ per unit of pressure. Temp.$=t$ °C.

$t \rightarrow$ P	0	10	20	30	40 β_m	50	60	80	100	Ref.[a]
1	51.5	48.4	46.4	45.3	44.8	44.7	45.0	46.7		R
	50.3	47.8	45.9	44.8	44.3	44.4	44.8	46.1	48.2	T
<10	50.3	48.3	46.0	45.3	44.5	44.1				G
25	52.5	50.0	49.1							A
	45.7	45.3	44.9	44.8	44.9	45.2	45.7	47.4	49.9	SK
50	52.0	49.6	48.3	48.0	45.5	45.9	46.3		48.7	A
	45.6	45.0	44.7	44.6	44.7	45.0	45.5	47.1	49.7	SK
	50.0				43.2		43.5	44.8		J
100	51.2	48.3	46.9	46.0	44.9	44.9	45.0			A
	51.0	48.0	47.5	45.7	46.5	43.7				A'
	45.0	44.5	44.2	44.1	44.2	44.6	45.0	46.6	49.1	SK
200	50.0	47.1	45.4	44.5	43.8	43.6	44.1	45.8	47.4	A
	50.0	47.2	46.2	44.5	40.1	43.1				A'
	44.1	43.6	43.3	43.2	43.4	43.7	44.2	45.8	48.2	SK
300	49.2	46.4	44.6	43.6	42.9	42.8	43.1	44.9	46.8	A
	49.2	46.1	45.0	43.8	43.0	42.8				A'
	43.2	42.7	42.4	42.4	42.5	42.9	43.3	44.9	47.2	SK
500	47.6	45.0	43.3	42.3	41.6	41.4	41.6	43.3	45.4	A
	47.5	44.8	43.4	42.4	41.7	41.6				A'
	46.6	44.6	42.3	41.8	41.5	40.1	39.5	42.0		B
1000			40.3	39.5	38.9	38.7	38.9	39.9	41.7	A
	44.0	41.7	40.3	39.4	38.6	38.7				A'
	43.4	41.0	39.6	38.7	38.4	38.2	38.2	39.2		B
2000	38.2	36.5	35.6	34.9	34.4	34.2				A'
	38.0	36.2	35.1	34.6	34.2	34.1	34.2	34.4	35.0	B
3000	33.9	32.7	31.9	31.5	31.1	31.0				A'
	33.6	32.4	31.7	31.1	31.0	30.8	30.9	31.5		B

[313] Drucker, K., Z. physik. Chem., 52, 640-704 (661) (1905).
[314] Moesveld, A. L. T., Idem, 105, 450-454 (1923).
[315] Jazyna, W., Z. Physik, 58, 858-860 (1929).

Table 105—(Continued)

$t \rightarrow$ p	−10	0	10	20	30	40	50	60	80	Ref.[a]
					β_m					
500	44.5	45.8	43.0	41.6	40.6	40.4	40.0	40.0	41.2	B
		46.4		39.7		39.4	40.4	41.8	43.2	B′
1000	41.8	42.2	39.9	38.5	37.6	37.3	37.1	37.1	38.0	B
		42.6		38.6		37.0	37.2	37.8	39.0	B′
1500	40.0	39.3	37.5	36.3	35.5	35.2	35.1	35.1	36.0	B
	42.5	39.5		36.0		35.0	35.0	35.7	36.3	B′
2000	38.6	37.0	35.4	34.4	33.7	33.4	33.2	33.4	34.1	B
	39.4	37.0		34.4		33.3	33.1	33.4	34.2	B′
3000	34.4	32.8	31.7	30.9	31.1	30.2	30.1	30.2	31.1	B
	34.6	32.9		31.1		30.2	30.1	30.2	30.8	B′
5000	28.3	27.4	26.7	26.2	26.1	25.8	25.7	25.8	26.2	B
	28.4	27.2		26.1		25.4	25.4	25.5	26.0	B′
7000			23.3	23.0	23.0	22.5	22.5	22.5	22.9	B
						22.2	22.2	22.2	22.6	B′
10000					19.3	19.1	19.1	19.1	19.3	B
						18.8	18.8	18.8	19.1	B′
12000						17.4	17.4	17.4	17.6	B
						17.1	17.2	17.2	17.4	B′

$t = 25$ °C. Unit of $P_b = 1$ bar $= 10^6$ dynes/cm^2.[296]

P_b	500	1000	1500	2000	3000	5000	7000	10000	12000
β_m	42.5	39.3	37.0	35.0	31.5	26.6	23.2	19.6	17.9

[a] References and remarks:

- A   E. H. Amagat. Values derived from Table 95, Section I.
- A′   E. H. Amagat. Values derived from Table 95, Section II.
- B   P. W. Bridgman. Values in the first section of the table (pressure in atm) derived from Table 95, Section III; those in the second section (pressure in kg*/cm^2) derived from his table,[292a] from which the values in Table 95, Section III, were derived.
- B′   P. W. Bridgman.[273] Values derived directly from the table there given.
- G   C. Grassi.[311] Between 1 and 10 atm he could detect no variation of β_m with the pressure. Values here given were obtained by interpolating between his values of β, which may be found in Table 103.
- J   R. S. Jessup.[316] Merely incidental determinations made for the purpose of testing his apparatus.
- R   C. R. Randall.[317] Values he derived from his determinations of the adiabatic compressibility (β_a) by means of the formula $\beta = \beta_a + T(dv/dt)^2/vc_p$; even at 86 °C the last term is less than 10 per cent of β_a.
- SK   L. B. Smith and F. G. Keyes.[294] Values derived directly from their table of specific volumes.
- T   D. Tyrer.[318] Values he deduced from his determinations of the adiabatic compressibility. (See R, above.)

---

[316] Jessup, R. S., *Bur. Stand. J. Res.,* **5,** 985-1039 (RP244) (1930).
[317] Randall, C. R., *Bur. Stand. J. Res.,* **8,** 79-99 (RP402) (1932).
[318] Tyrer, D., *J. Chem. Soc. (London),* 103, 1675-1688 (1913).

Table 106.—Mean Isothermal Compressibility of Water Between Pressures P_1 and P_2

(For $P_1 = 1$ or 0, see Table 105; for adiabatic compressibility, see Table 107; for natural waters, see Table 110.)

Values for other temperatures and pressures than those here given may be readily obtained from the specific volumes as given in Table 95 and from the more extended original tables from which these have been taken or computed.

A few values for 0 and 30 °C and for irregularly distributed pressures up to 95 atm have been published by Earl of Berkeley, E. G. J. Hartley, and C. V. Burton.[319] They do not differ significantly from those found by others, and are not given in this table.

$\beta_m = (v_1 - v_2)/(P_2 - P_1)v_1$, where v_1 and v_2 are the volumes of the same mass of water at the common temperature indicated and under the pressures P_1 and P_2, respectively.

Unit of $P = 1$ atm; of $\beta_m = 10^{-6}$ per atm.

P_1 \ t →	0	10	20	30	40	50	60	80	100	Ref.[a]
					β_m					
					$P_2 = P_1 + 25$					
1	52.5	50.0	49.1							A
	45.7	45.3	44.9	44.8	44.9	45.2	45.7	47.4	49.9	SK
25	51.6	49.2	47.6							A
	45.3	44.8	44.5	44.4	44.5	44.9	45.4	47.0	48.0	SK
75	50.2	47.0	45.3							A
	44.5	44.0	43.6	43.6	43.7	44.0	44.6	46.1	48.6	SK
125	49.1	46.3	44.6							A
	43.6	43.1	42.9	42.8	42.9	43.3	43.8	45.3	47.7	SK
175	48.8	46.0	43.8							A
	42.8	42.3	42.1	42.0	42.2	42.5	43.0	44.6	46.9	SK
					$P_2 = P_1 + 100$					
0	51.1	48.3	46.8	46.0	44.9	44.9	45.5	47.8		A
	44.6	44.1	43.8	43.7	43.8	44.1	44.6	48.7		SK
			45.8							RS
100	49.2	46.1	44.2	43.6	42.9	42.5	42.7	46.8		A
	43.3	42.8	42.6	42.6	42.7	43.0	43.5	45.1	47.4	SK
			44.8							RS
200	48.0	45.3	43.4	42.4	41.4	41.6	41.5	43.6	45.9	A
	41.7	41.3	41.1	41.1	41.2	41.6	42.0	43.5	45.7	SK
			42.4							RS
400	45.5	43.0	41.5	40.6	40.4	39.9	39.4	40.8	43.4	A
			39.9							RS
600	42.9	40.5	39.4	38.7	38.2	37.7	38.3	38.7	40.7	A
800	40.6	38.9	37.3	37.4	36.2	36.2	36.3	36.3	38.2	A
900			36.5	36.0	35.3	35.3	36.0	35.7	37.1	A
					$P_2 = P_1 + 500$					
1	47.5	44.7	43.4	42.4	41.7	41.6				A'
	46.6	44.6	42.3	41.8	41.5	40.1	39.5	42.0		B
	46.5	43.6								TJ

[319] Earl of Berkeley, Hartley, E. G. J., and Burton, C. V., *Phil. Trans. (A)*, **218**, 295-349 (1919).

Table 106—(Continued)

P_1 \ $t \rightarrow$	0	10	20	30	40	50	60	80	100	Ref.[a]
					β_m					
					$P_2 = P_1 + 500$					
500	41.6	39.5	38.0	35.5	36.2	36.6				A'
	41.0	38.1	37.7	36.5	36.2	36.9	37.5	37.1		B
	41.6	38.8								TJ
1000	35.8	34.8	33.8	33.7	33.0	32.5				A
	36.0	34.8	33.9	33.1	33.0	33.1	33.3	33.8		B
	34.5									TJ
1500	32.4	31.3	30.9	30.1	30.5	30.0				A
	32.6	31.2	30.4	30.5	30.0	29.9	30.2	30.6		B
2500	26.1	25.9	25.7	25.8	25.4	25.4				A
	26.1	26.0	25.7	25.4	25.5	25.1	25.2	25.8		B
4000	20.7	21.1	21.0	20.9	20.8	20.9	20.6	21.1		B
6000	17.0	17.0	17.1	16.6	16.8	16.7	16.6	16.7		B
8000			13.9	13.4	13.1	13.6	13.6	13.7		B
10000				11.6	11.8	11.8	11.7	11.6		B
11500					11.5	11.5	11.4	11.3		B

Temperatures above 100 °C

$t \rightarrow$		120	140	160	180	200	220	Ref.[a]
P_1	P_2			β_m				
25	50	53.1	58.0	64.8	74.0	86.4	104.4	SK
300	350	47.6	51.5	56.6	63.4	72.0	83.3	SK

$t \rightarrow$		240	260	280	300	320	340	360	Ref.[a]
P_1	P_2				β_m				
300	350	97.8	117.	144.	184.	251.	380	584	SK

t	190	200	210	220	230	240	250	260	Ref.[a]
P_1	13.0	19.8	19.0	23.2	27.7	33.2	29.6	46.1	
P_2	52.5	57.8	57.1	51.9	55.6	54.8	57.8	52.9	
β_m	104	114	101	91.	127	139	114	174	RY

[a] References:
- A E. H. Amagat.[320] Values have been selected from his more extended table.
- A' E. H. Amagat.[320] As for A, but from another of his tables.
- B P. W. Bridgman. Values computed from his specific volumes as given in Table 95, Section III.
- RS T. W. Richards and W. N. Stull.[321]
- RY W. Ramsay and S. Young.[322]
- SK L B. Smith and F. G. Keyes.[294] Values derived directly from their table of specific volumes.
- TJ G. Tammann and W. Jellinghaus. Values computed from their specific volumes as given in Table 95, Section V; the unit of pressure is in this case 1 kg*/cm², but the difference between that and 1 atm is of little importance here.

Table 107.—Adiabatic Compressibility of Water
(For natural waters, see Table 110.)

The most accurate available values for the adiabatic compressibility, β_a, of water are probably (R), those derived by C. R. Randall [317] from his

[320] Amagat, E. H.; *Ann. de chim. et phys.*, (6) **29**, 68-136, 504-574 (548, 549) (1893).
[321] Richards, T. W., and Stull, W. N., *Carnegie Inst. of Washington, Publ.* No. 7 (1903) → *J. Am. Chem. Soc.*, **26**, 399-412 (1904), and *Z. physik. Chem.*, **49**, 1-14 (1904).
[322] Ramsay, W., and Young, S., *Phil. Trans. (A)*, **183**, 107-130 (1892).

32. WATER: P-V-T DATA

Table 107—(Continued)

measurements of the velocity of supersonic vibrations ($\beta_a = 1/V^2\rho$, $V =$ velocity, $\rho =$ density). Earlier determinations (HL) by the same method were made by J. C. Hubbard and A. L. Loomis,[322a] and similar determinations (P), based on the velocity of waves of audio-frequency, have been made by L. G. Pooler.[323] D. Tyrer [318] determined β_a directly (T) from the observed expansion that accompanied a sudden reduction of pressure from 2 atm to 1 atm. These 4 sets of data (R, HL, P, and T) are given below.

J. Claeys, J. Errera, and H. Sack [324] have suggested that the adiabatic compressibility of water may exhibit a type of hysteresis.

Unit of $\beta_a = 1$ cm² per megadyne $= 1.01325$ per atm. Temp. $= t$ °C.

Method→ Source[a]→ t	Velocity R	HL	$10^6 \beta_a$ Direct P	T
0	50.77	50.53		49.59
5		49.06		
10	47.71	47.66[b]		47.13[c]
15		46.47		
20	45.54	45.48[d]		45.00
25		44.69[e]	45.42	
30	44.05	44.05	44.72	43.52
35		43.51		
40	43.08	43.02	43.72	42.68
50	42.48		43.09	42.24
60	42.25		42.90	42.07
70	42.28		42.95	41.90
75			43.01	
80	42.58			41.99
86	42.87			
90				42.14
100				42.34

[a] Source: See head-matter.
[b] D. Colladon and C. Sturm [325] found $10^6\beta_a = 49.5$ atm⁻¹ at 10 °C, which must be reduced by 1.65,[326] giving $10^6\beta_a = 47.8$ atm⁻¹ $= 47.2$ cm² per megadyne.
[c] R. W. Boyle, J. F. Lehmann, and S. C. Morgan [327] found from the velocity of supersonic waves that $10^6\beta_a = 46.1$ cm²/megadyne at 12 °C.
[d] A. Pasuinskii [328] found $10^6\beta_a = 45.5$ cm²/megadyne at 20 °C.
[e] S. Parthasarathy [329] found $10^6\beta_a = 44.9$ cm²/megadyne at 24 °C.

[322a] Hubbard, J. C., and Loomis, A. L., *Phil. Mag. (7)*, **5**, 1177-1190 (1928).
[323] Pooler, L. G., *Phys. Rev. (2)*, **35**, 832-847 (1930).
[324] Claeys, J., Errera, J., and Sack, H., *Compt. rend.*, **202**, 1493-1494 (1936).
[325] Colladon, D., and Sturm, C., *Mém. Sav. Etrang. Inst. Paris*, **5**, 267-347 (1838); *Ann. de chim. et phys.*, **36**, 113-159, 225-257 (1827); *Ann. d. Physik (Pogg.)*, **12**, 39-76, 161-197 (1828); "Mém. sur la compression des liquides et la vitesse de son dans l'eau," 1827; Ch. Schuchert, Genf, 1887.
[326] Bungetzianu, D., *Bull. Soc. Roumaine Sci. (Bucarest)*, **19**, 1224-1246 (1229) (1910).
[327] Boyle, R. W., Lehmann, J. F., and Morgan, S. C., *Trans. Roy. Soc. Canada, III (3)*, **22**, 371-378 (1928).
[328] Pasuinskii, A., *Acta Physiochim, URSS*, **3**, 779-782 (1935).
[329] Parthasarathy, S., *Proc. Indian Acad. Sci. (A)*, **2**, 497-511 (1935).
[330] Beattie, J. A., *Int. Crit. Tables*, **3**, 100 (1928).
[331] Knudsen, M. H. C., "Hydrographische Tabellen," Copenhagen, 1901.
[332] Buchanan, J. Y., *Proc. Roy. Soc. (London) (A)*, **23**, 301-308 (1875).
[333] Hill, E. G., *Proc. Roy. Soc. Edinburgh*, **27**, 233-243 (1907).

Table 108.—Density of Sea-water: Pressure = 1 Atmosphere
(For density at temperature of maximum density, see Table 134.)

Sea-water contains about 35 g of salts per kg of sea-water. For the composition of the salts, and variations in the salinity and the temperature of sea-water, see Section 100.

The following data are derived from a table by J. A. Beattie,[330] based primarily on the data given by M. H. C. Knudsen[331] and J. Y. Buchanan,[332] but with a consideration of those of E. G. Hill,[333] J. J. Manley,[334] Dittman,[335] F. L. Ekman,[336] R. Lenz,[337] C. O. Makaroff,[338] C. von Neumann,[339] and T. E. Thorpe and A. W. Rücker.[340]

Unit of $\rho = 1$ gram/ml; ($\rho = 1 + 10^{-3}\Delta$); % = per cent by weight; salts = total salts. Temp. = t °C. Pressure = 1 atm.

%Cl	% salts	0	5	10	15	20	25	30	35
					Δ				
0.1	0.184	140	149	120	58	− 34	−151	−292	−455
0.2	0.364	287	293	261	197	+104	− 15	−158	−322
0.3	0.545	433	436	402	335	241	+120	− 24	−189
0.4	0.725	579	579	542	474	377	256	+110	− 56
0.5	0.906	725	722	683	612	514	391	245	+ 77
0.6	1.086	871	865	823	751	651	526	379	210
0.7	1.267	1016	1007	963	889	787	661	513	343
0.8	1.447	1162	1150	1103	1027	924	796	647	476
0.9	1.628	1307	1292	1243	1165	1060	931	780	608
1.0	1.808	1452	1434	1383	1303	1196	1066	914	741
1.1	1.989	1597	1577	1523	1441	1333	1201	1048	874
1.2	2.169	1742	1719	1663	1579	1469	1336	1182	1007
1.3	2.350	1887	1861	1803	1717	1605	1472	1316	1140
1.4	2.530	2032	2003	1943	1855	1742	1607	1450	1274
1.5	2.711	2177	2146	2083	1993	1879	1742	1585	1407
1.6	2.891	2322	2288	2223	2131	2016	1878	1720	1541
1.7	3.072	2468	2431	2364	2270	2153	2014	1855	1675
1.8	3.252	2613	2574	2504	2408	2290	2150	1989	1809
1.9	3.433	2758	2716	2644	2547	2427	2286	2124	1944
2.0	3.613	2904	2859	2785	2686	2564	2422	2260	2079
2.1	3.794	3049	3002	2926	2825	2701	2558	2395	2214
2.2	3.974	3195	3145	3067	2964	2839	2695	2531	2349
2.3	4.155	3341	3289	3208	3104	2978	2831	2667	2484

Table 109.—Specific Volume of Sea-water: Pressure Greater than 1 Atm
(For the specific volume at 1 atm, take the reciprocal of the density as given in Table 108.)

The data in Section A of this table, taken from the compilation of L. H. Adams,[341] are from V. Bjerknes and J. W. Sandström[342] and are

[334] Manley, J. J., *Idem*, **27**, 210-232 (1907).
[335] Dittman, "Rep. Sci. Results, Physics and Chemistry, Voyage H. M. S. *Challenger*," Vol. 1, 1889.
[336] Ekman, F. L., *Kongl. Svenska Vet-akad. Handl.*, **9**, No. 4, 1870.
[337] Lenz, R., *Mém. de l'acad Sci. Russie* (7), **29**, No. 4, (1881) → *Fortschr. d. Physik*, **38₃**, 661-662 (1888).
[338] Makaroff, C. O., *J. Russ. Phys. Chem. Soc. (Chem.)*, **23**, II, 30-88 (1891).
[339] von Neumann, C., *Ann. d. Physik (Pogg.)*, **113**, 382 (1861).
[340] Thorpe, T. E., and Rücker, A. W., *Phil. Trans. (A)*, **166**, 405-420 (1876).

32. WATER: P-V-T DATA

Table 109—*(Continued)*

based upon the observations of V. W. Ekman.[343] Differences of successive values of Δ are printed in distinctive type; in the first two subsections they are between the values from which they have been derived, in the third subsection they are to the right of the greater Δ.

Those in Section B are from the detailed practical tables of N. H. Heck and J. H. Service [344] based upon the very extensive tables appended to the publication by Bjerknes and Sandström already mentioned. The values here tabulated were taken directly from the tables of Heck and Service; those they give for other salinities and temperatures may be reproduced, usually within 2 units in the 5th significant digit, by means of the formula

$$\Delta_{s,t,d} = \Delta_{35,0,d} - (6.48 + 0.00375d)t - 0.46t^2 + (\delta - 0.283t - 0.005t^2) \cdot (s - 35)$$

Within the limits they consider ($s = 31$ to 37, $t = 0\,°C$ to t_m, t_m being here tabulated), δ varies with d as shown in the final portion of the table. The units are as indicated below.

Salinity = s = total salts per kg of sea-water; specific volume = v_s = $1 - 10^{-5}\Delta$; pressure = P; depth below surface = $d = (d_1 + d_2)$; temperature = $t\,°C$; $\Delta_{s,t,d}$ = value of Δ for salinity s, temperature t, and depth d; $\Delta_{35,0,d}$ = value of Δ for $s = 35$ g/kg, $t = 0\,°C$, depth = d; P_d = pressure at depth d.

Unit of $P=1$ bar $=0.9869$ atm; of $s=1$ g/kg; of $v_s=1$ ml/g; of $d=1$ fathom $=6$ ft $=182.88$ cm.

Section A

$t \rightarrow$ P	0		4.97		9.97 Δ		14.96		19.96	
($s=31.130$)										
0	2440	35	2405	65	2340	89	2251	113	2138	
	888		*864*		*845*		*830*		*820*	
200	3328	59	3269	84	3185	104	3081	123	2958	
	832		*812*		*796*		*783*		*772*	
400	4160	79	4081	100	3981	117	3864	134	3730	
	784		*766*		*751*		*739*		*731*	
600	4944	97	4847	115	4732	129	4603	142	4461	
($s=38.525$)										
0	3004	45	2959	72	2887	96	2791	117	2674	
	866		*845*		*830*		*813*		*803*	
200	3870	66	3804	87	3717	113	3604	127	3477	
	813		*793*		*775*		*767*		*757*	
400	4683	86	4597	105	4492	121	4371	137	4234	
	766		*749*		*736*		*725*		*717*	
600	5449	103	5346	118	5228	132	5096	145	4951	

[341] Adams, L. H., *Int. Crit. Tables*, 3, 439-440 (1928).
[342] Bjerknes, V., and Sandström, J. W., *Carnegie Inst. Washington, Publ.* 88, Part I (1910).
[343] Ekman, V. W., *Conseil. perm. int. l'explor. de la Mer, Publ. de Circon.* No. 43, Copenhagen, 1. 08.
[344] Heck, N. H., and Service, J. H., *U. S. Coast and Geod. Survey, Spec. Publ.* No. 108, (1924).

Table 109—(Continued)

$t = 0\ °C;\ s = 35.00$

P		Δ	P		Δ	P		Δ
0	2736		400	4434	404	800	5940	361
100	3181	445	500	4827	393	900	6291	351
200	3612	431	600	5209	382	1000	6633	342
300	4030	418	700	5579	370			

Section B

$d_1 \rightarrow$	100	300	500	700	900	100	300	500	700	900
d_2			$\Delta_{35,0,d}$					t_m		
0	2736[a]	2985	3147	3309	3469	22	22	22	20	12
1000	3627	3783	3939	4093	4245	8	6	4	4	3
2000	4394	4544	4692	4837	4983	3	3	3	3	2
3000	5126	5271	5513	5553	5691	2	2	2	2	2
4000	5828	5965	6099	6234		2	2	2	2	

$d_1 \rightarrow$	100	300	500	700	900	100	300	500	700	900
d_2			δ					P_d		
0	76.3	75.3	75.0	74.3	73.8		55.4	92.3	129.3	166.5
1000	73.3	72.8	72.3	71.8	71.3	203.6	240.6	278.0	315.4	352.7
2000	70.8	70.3	69.8	69.5	69.0	390.2	427.5	465.2	502.6	540.4
3000	68.5	68.2	67.8	67.2	66.8	578.0	616.7	654.7	692.7	730.8
4000	66.2	66.0	65.8	65.2		768.8	807.0	845.1	883.6	

[a] This value appears to be too great, the difference between the value for $d = 100$ and $d = 300$ being 249, while the other differences for 200 fathoms are about 160. In fact, the value 2736 is that for the surface (See $P = 0$, end of Section A).

(Continued from p. 225.)

M. Thiesen, K. Scheel, and H. Diesselhorst.[274] Believing it better to keep the two sets distinct, I have not included that table in this compilation. The greatest difference between the two sets is 113 parts in 10^7, and occurs at 41 °C.

Thiesen, Scheel, and Diesselhorst used the method of balanced columns of liquids, and their observations indicate that metal was dissolved from the tubes by the water. The amount so dissolved during the course of their second series of determinations was inferred, from the change in the electrical conductivity of the water, to have been such as to affect the density by 10 parts in a million. They endeavored to eliminate the effect of such solution by suitably combining related sets of data. Their observations were not very closely spaced with reference to the temperature. They combined their observations so as to obtain the density (ρ) at exactly 5° intervals from 0 to 40 °C. From these 9 values they determined the constants in the formula

$$1 - \rho = \frac{(t - 3.98)^2}{503570} \cdot \frac{t + 283}{t + 67.26}$$

and by means of that formula computed the values given in their table.

Chappuis used the weight-thermometer method, using both glass and platinum-iridium bulbs; observations were made at many temperatures, closely spaced and well distributed. He represented them by a triad of formulas in powers of the temperature ($t\,°C$), the constants being determined with high precision (see Table 99). His table was computed by means of these formulas. Quite recently it has been found [275] that there are systematic differences between the values in that table and the actual observations of Chappuis, and that the actual observations can be more closely represented by the single formula

$$1 - \rho = \frac{(t - 3.9863)^2}{508929.2} \cdot \frac{t + 288.9414}{t + 68.12963} \tag{1}$$

than by Chappuis's triad; and a table has been computed by means of that formula. That table is given in full in this compilation, together with the amounts by which each of the other two tables differ therefrom.

The data given by P. W. Bridgman in his compilation [276] are based upon those he had previously published,[277] and, with his recent paper,[278] are the source of most of the data here attributed to him.

An early study of the expansion of water in the range -13 to $+100\,°C$ was made by C. Despretz,[279] but the compiler has not found those data; in the range -9 to $+100\,°C$ by the same investigator [280]; and in the range -4 to $-10\,°R$ by Salm-Horstmar.[281]

Tables and charts for compressed water, based, in the main, on data obtained at the National Bureau of Standards and at the Massachusetts Institute of Technology, have been published by J. H. Keenan.[282]

Dissolved Air: Effect on Density.

Dissolved air decreases the density of water. The frequently quoted differences reported by W. J. Marek [283] are the differences between the densities of water that has been freed from air by exhaustion just before measurement and those of water, at the same temperatures, that has merely been exposed to air for intervals of 1 to 3 days.[284] They refer to the rather ill-defined conditions generally encountered in practice, rather than to the extreme conditions of complete air-freedom and air-saturation of Chappuis' work. Furthermore, when they are applied to those values for the density of air-saturated water which are published in the same paper, they lead to values of the density quite different from those generally accepted for air-free water (Table 93). The paper does not contain sufficient details to enable one to determine either the cause of the discrepancy or the accuracy of the differences, which are, in fact, about 43 per cent smaller than those published in his earlier (1884) paper. The several reported results are shown in Table 92, p. 198. P. Chappuis [284] and W. A. Adeney, A. G. G. Leonard, and A. Richardson [286] have studied the aeration of quiescent water. The former found that at $13.5\,°C$ a layer 12 cm below the surface became half saturated in about a day, and 3/4 saturated in something over 4 days.

Recently, H. J. Emeléus et al.[287] have reported that saturating water at 20 °C with air reduces its density by an amount equal to that caused by increasing the temperature of the water by 0.01 °C; that is, by 2 in 10^6. That is much greater than the value given in Table 92. On the other hand, T. W. Richards and G. W. Harris[288] found by essentially the same method that the change in density under those conditions is less than 2 in 10^7, which is about one-half the tabulated value.

Isothermal Compressibility of Natural Water.

Three sets of data are given in Table 110. The first two have ultimately been either taken or derived from the tables given by V. Bjerknes and J. W. Sandström[342] which in turn are based upon the observations of V. W. Ekman.[343] The first was selected by L. H. Adams[341] for his compilation and gives β for each of several pressures. The second has been adapted from the detailed practical tables compiled by N. H. Heck and J. H. Service[344] from the tables of Bjerknes and Sandström. It gives the values of the compression ($\beta' \equiv -(dv^*/dp)_t$, v^* = specific volume) at 0 °C, salinity = 35 g/kg, for each of several depths below the surface. If we write $\beta'_{s,t,d}$ for the value of β' for sea-water of salinity s, temperature t, and depth d, then the values that Heck and Service give for the ranges $s = 31$ to 37, $t = 0$ to t_m may be reproduced very closely by means of the formula

$$10^5\beta'_{s,t,d} = 10^5\beta'_{35,0,d} - (0.0238 - 2(10^{-6})d)t + 0.000312t^2 - (0.015 - 1.1(10^{-6})d)\cdot(s - 35)$$

the units being those named below. The pressures corresponding to the several values of d may be found in Table 109.

The third set covers earlier data frequently quoted but less reliable than those covered by the other two. They were obtained by P. G. Tait[345] and supersede those published in his earlier papers.[346] He summarizes these data in the following formulas:

Spring water:

$$10^7\beta_m = 520 - 17p + p^2 - (355 + 5p)\frac{t}{100} + (3 + p)\frac{t^2}{100} \text{ per atm, and}$$

$$10^5\beta_m = \frac{186}{36 + p}\left(1 - \frac{3t}{400} + \frac{t^2}{10000}\right) \text{ per atm.}$$

Sea-water:

$$10^5\beta_m = \frac{179}{38 + p}\left(1 - \frac{t}{150} + \frac{t^2}{10000}\right) \text{ per atm, and at 0 °C}$$

$$10^5\beta_m = 481 - 21.25p + 2.25p^2 \text{ per atm.}$$

For solutions of NaCl at 0 °C, he gives:

$$10^5\beta_m = 186/(36 + p + s') \text{ per atm, when the solution contains } s'$$

grams of NaCl per 100 grams of water; s' was varied from 3.88 to 17.63. In neither pair are the two equations identical, but each was supposed to represent the data satisfactorily. In each of these equations, β_m is the mean compressibility between 1 atm and the pressure of p (long) tons* per sq. in.

The sea-water was not more particularly described; neither the composition nor the density is stated. Both it and the spring water were, presumably, nearly saturated with air, though nothing seems to have been said about this in the original articles. In the last section of the table certain values of β_m for pure, air-free water (from Table 105) are given together with the corresponding ones as derived from Tait's equations.

Symbols: $\beta = -\dfrac{1}{v_1}\left(\dfrac{\delta v}{\delta p}\right)_t$; $\beta_m = +\dfrac{1}{v_1}\left(\dfrac{v_1 - v}{p - A}\right)_t$; $\beta' = -\left(\dfrac{\delta v}{\delta p}\right)_t$; $v =$ specific volume; $p =$ pressure; A is value of p corresponding to 1 atm; v_1 is value of v at t °C and a pressure of 1 atm; the value of the pressure may be represented by P, P_b, or p, or indicated by b; $b =$ depth below the surface of the ocean; $s =$ salinity; $t_m =$ highest temperature for which Heck and Service give data against which the formula given above for $\beta'_{s,t,d}$ may be checked.

Table 110.—Isothermal Compressibility of Natural Waters

(For source of data, explanation of symbols, etc. see text.)

Unit of $P_b = 1$ bar $= 0.9869$ atm; of $P = 1$ atm; of $p = 1$ (long) ton*/in²; of $\beta = 10^{-6}$ per bar; of $\beta_m = 10^{-6}$ per atm; of $\beta' = 1$ (cm³/g) per bar; of $d = (d_1 + d_2) = 1$ fathom $= 6$ ft $= 182.88$ cm; of $s = 1$ g/kg. Temp. $= t$ °C.

A. Sea-water, 0 °C, 35 g salts per kg of sea-water.

P_b	β	P_b	β	P_b	β	P_b	β
0	46.4	300	42.3	600	38.7	400	35.6
100	45.0	400	41.0	700	37.6	1000	34.7
200	43.6	500	39.6	800	36.6		

B. Sea-water, 0 °C, 35 g salts per kg of sea-water. $\beta' \equiv \beta'_{35,0,d}$, $\beta' = (\delta v/\delta p)_t$

$d_1 \to$ d_2	100	300	500 $10^6\beta'$	700	900	100	300	500 t_m	700	900
0	4.50	4.44	4.39	4.33	4.28	22	22	22	20	12
1000	4.24	4.19	4.14	4.09	4.03	8	6	4	4	3
2000	4.00	3.94	3.91	3.87	3.83	3	3	3	3	2
3000	3.78	3.73	3.70	3.65	3.62	2	2	2	2	2
4000	3.58	3.55	3.52	3.47		2	2	2	2	2

C. $\beta_m \equiv (v_1 - v_p)/v_1(p-1)$; $10^6\beta_m = a + bt + ct^2$.

Water→ p	P	Spring a	$-b$	c	Sea a	$-b$	c
0	1	52.0	0.355	0.003	48.1	0.340	0.003
1	152	50.4	0.360	0.004	46.2	0.320	0.004
2	305	49.0	0.365	0.005	44.8	0.305	0.005
3	457	47.8	0.370	0.006	43.8	0.295	0.005

[345] Tait, P. G., *Beibl. zu. Ann. d. Physik*, **13**, 442-445 (1889) ← "Rep. Sci. Results Voy. H. M. S. *Challenger*, Phys. and Chem.," Vol. 2, Part 4, London, Edinburgh, and Dublin, 1888.
[346] Tait, P. G., *Proc. Roy. Soc. Edinburgh*, **12**, 223-224, 757-758 (1884); **15**, 84 (1887).

Table 110—(Continued)

$t \rightarrow$ Water$^a \rightarrow$ P	Spr.	0 Pure	Sea	Spr.	10 Pure β_m	Sea	Spr.	20 Pure	Sea
1	52.0	51.5	48.1	48.7	48.4	45.0	46.1	46.4	42.5
150	50.4	50.5	46.2	47.2	47.6	43.4	44.8	46.0	41.4
300	49.0	49.2	44.8	45.8	46.4	42.3	43.7	44.6	40.7
450	47.8	48.0	43.8	44.7	45.3	41.4	42.8	43.6	39.9

^{a} The 3 samples of water are spring (= Spr.), pure air-free (= Pure), and sea-water (= Sea). The data for the first and third have been computed by means of Tait's equations; those for the pure water have been taken from Table 105.

Adiabatic Compressibility (β_a) of Natural Waters.

From the observed velocity of sound generated by explosions in the sea, A. B. Wood, H. E. Browne, and C. Cochrane [347] have concluded that, for sea-water at 16.95 °C, under a mean pressure of 2 bars, and containing 35 g of salts per kg of sea-water, $10^6 \beta_a = 42.7$ per bar = 43.3 per atm. As usual, $\beta_a \equiv -\dfrac{1}{v}\left(\dfrac{\delta v}{\delta P}\right)_a$.

33. Mechanical Equivalent of the Calorie

By the mechanical equivalent of the calorie is meant the work required to produce the amount of heat designated as one calorie.

Several different calories have been used and must be distinguished if a higher accuracy than 1 or 2 in 1000 is desired. For this reason, among others, it is desirable to express quantities of heat in terms of a less ambiguous unit, such as the joule. For uncertainties in the value of the international joule, see Section 2.

Of the various calories that have been used, four are of particular importance, having received widespread recognition. They are designated and defined thus: 1 cal$_{15}$ = amount of heat required to raise 1 gm of water from 14.5 to 15.5 °C; 1 cal$_{20}$ = amount of heat required to raise 1 gram of water from 19.5 to 20.5 °C; 1 cal$_m$ = 1 mean calorie = 1/100 of the amount of heat required to raise 1 gram of water from 0 to 100 °C, and 1 cal (ST) = 1 cal (steam) = I Int. cal. = 1 steam-table calorie = 1/1000 of the heat that is equivalent to 1/860 international kilowatt-hour = 4.18605 Int. joules. For the first three, the water is to be under an air pressure of 1 atmosphere. The fourth, independent of the properties of any particular substance, was defined by the International Steam-Table Conference, meeting in London in 1929.[348]

A fifth calorie (cal$_{ms}$), a mean calorie based upon air-free saturated water, has been proposed by N. S. Osborne, H. F. Stimson, and E. F. Fiock [349] and defined as 1/100 of the change in the enthalpy ("heat con-

---

[347] Wood, A. B., Browne, H. E., and Cochrane, C., *Proc. Roy. Soc. (London) (A)*, **103**, 284-303 (1923).
[348] *Engineering (London)*, **128**, 751-752 (1929); *Z. Ver. deuts. Ing.*, **73**, 1856-1858 (1929); *Mech. Eng.*, **52**, 120-122 (1930).
[349] Osborne, N. S., Stimson, H. F., and Fiock, E. F., *Mech. Eng.*, **50**, 152-153 (1928).

tent") of 1 gram of saturated water on passing from 0 to 100 °C. This exceeds 1 cal_m by only about 0.001 joule.

R. Jessel[350] has held that the heat capacity of water is significantly affected by the presence of dissolved air, and that air-free water must be used if highly reproducible results are to be obtained. With that view T. H. Laby and E. O. Hercus[351] disagree. See also Table 112.

The various results obtained for the mechanical equivalent of the calorie have been reviewed and discussed by J. S. Ames,[352] E. H. Griffiths,[353]

Table 111.—Mechanical Equivalent of the Calorie

The first value for cal_{20} and for cal_m has been taken directly from the compilation by T. H. Laby and E. O. Hercus.[359]

Unit of work = 1 joule = 10^7 ergs, unless value is followed by I (= Int. joule)

Ref.[a]	cal_{15}	cal_{20}	cal_m	cal_{ms}
ICT accepted	4.185[b]	4.181	4.186	
ICT mean		4.1818	4.1853	
LH (1927)		4.1809		
OSF (1928)			4.188	4.1876(I)
OSG (1937)				4.1876(I)
RTB (1929)	4.1852			
HJ (1926)	4.1863			
L (1933)	4.186	4.182		

[a] References:
"ICT accepted" are values accepted by the *International Critical Tables* [1, 18 (1926)].
"ICT mean" are the means given by Laby and Hercus,[359] and are based upon the work of:
Barnes, H. T., *Proc. Roy. Soc. (London) (A)*, **82**, 390-395 (1909).
Bousfield, W. R., and W. E., *Phil. Trans. (A)*, **211**, 199-251 (1911).
Callendar, H. L., *Idem*, **212**, 1-32 (1912).
Day, W. S., *Phil. Mag. (5)*, **46**, 1-29 (1898).
Griffiths, E. H., *Proc. Roy. Soc. (London)*, **55**, 23-26 (1894); *Phil. Trans. (A)*, **184**, 361-504 (1894); *Idem*, **186**, 261-341 (1895).
Henning, F., *Ann. d. Physik (4)*, **58**, 759-760 (1919).
Jaeger, W., and v. Steinwehr, H., *Sitz. Preuss. Akad. Wiss. (Berlin)*, **1915**, 424-432 (1915); *Ann. d. Physik (4)*, **64**, 305-366 (1921).
Reynolds, O., and Moorby, W. H., *Phil. Trans. (A)*, **190**, 301-422 (1897).
Rispail, L., *Ann. de chim. et phys. (8)*, **20**, 417-432 (1910).
Rowland, H. A., *Proc. Amer. Acad. Arts Sci.*, **15**, 75-200 (1880).
Schuster, A., and Gannon, W., *Phil. Trans. (A)*, **186**, 415-467 (1895).
Sutton, T. C., *Phil. Mag. (6)*, **35**, 27-29 (1918).

LH   Laby, T. H., and Hercus, E. O., *Phil. Trans. (A)*, **227**, 63-92 (1927).
OSF   Osborne, N. S., Stimson, H. F., and Fiock, E. F.[349]
OSG   Osborne, N. S., Stimson, H. F., and Ginnings, D. C., *J. Res. Nat. Bur. Stand.*, **18**, 389-448 (RP983) (1937).
RTB   Value derived by R. T. Birge[356] from the observations of others.
HJ   The Reichsanstalt value as given by F. Henning and W. Jaeger, "Handb. d. Physik" (Scheel), Vol. 2, pp. 487-518 (497), 1926.
L   Conclusion of V. S. Lipine.[355]

[b] The value accepted for this compilation: 1 cal_{15} = 4.185 joules.

---

[350] Jessel, R., *Proc. Phys. Soc. (London)*, **46**, 747-763 (1934).
[351] Laby, T. H., and Hercus, E. O., *Idem*, **47**, 1003-1008-1011 (1935). See also Hercus, E. O., *Idem*, **48**, 282-284 (1936).
[352] Ames, J. S., *Rapports Cong. Int. Phys. (Paris)*, **1**, 178-213 (1900).
[353] Griffiths, E. H., "Dictionary of Applied Physics" (Glazebrook), Vol. 1, pp. 477-494 (1922).
[354] Laby, T. H., *Proc. Phys. Soc. (London)*, **38**, 169-172-175 (1926).
[355] Lipine, V. S., *Mém. prés à la VIII Conf. Gén. des Poids et Mes.*, 1933.
[356] Birge, R. T., *Rev. Mod. Phys.*, **1**, 1-73 (1929).
[357] Henning, F., and Jaeger, W., "Handb. d. Physik" (Scheel), Vol. 2, pp. 487-518, 1926.
[358] Fiock, E. F., *Bur. Stand. J. Res.*, **5**, 481-505 (RP210) (1930) = *Mech. Eng.*, **52**, 231-242 (FSP-52-30) (1930).
[359] Laby, T. H., and Hercus, E. O., *Int. Crit. Tables*, **5**, 78 (1929).

T. H. Laby,[354] and V. S. Lipine.[355] R. T. Birge [356] has discussed them with special reference to the actual values of the standards used; and F. Henning and W. Jaeger,[357] concluding that all determinations, except those made at the Physikalisch-Technischen Reichsanstalt, are vitiated by uncertainties regarding the actual values of the standards employed, rejected all except the Reichsanstalt's.

The several determinations of various thermal properties of saturated water have been reviewed by E. F. Fiock.[358]

34. Thermal Energy of Water

In this section are considered the specific heat (c and C), the enthalpy or heat content (H, $H = E + pv$, $(\delta H/\delta t)_p = c_p$), the "free energy at constant pressure" (the Gibbs function, G, $G = H - ST$, $G_T - G_{T_0} = -T \int_{T_0}^{T} (H/T^2) dT$), the entropy ($S$), the heat of isothermal compression (Q), the decrease in the internal energy on isothermal compression (D), the increase in temperature on adiabatic compression (Joule-Thomson effect), and certain related quantities, all intimately related to the thermal energy of water.

Data referring to the thermal energy of water have been reviewed and discussed by E. F. Fiock [358] and by M. Jakob.[360]

There are no direct determinations of the values of the specific heat of water at constant volume, of the ratios of the specific heats, or of their differences, but all of these can be computed from the observed compressibility, thermal expansion, and specific heat at constant pressure. Values so determined may be called *static* values. They can also be determined from the velocity of sound, in which case they may be described as *dynamic*. Likewise, the increase in temperature on adiabatic compression may be determined either statically, from the thermal expansion, or dynamically, from the observed drop in temperature that accompanies a sudden release of pressure.

If water consisted of a single species of molecule and if the internal state of a molecule were unaffected by gross dynamic changes in the substance, then no difference between the static and the dynamic values of those various thermal quantities would be expected. But there are reasons for believing that water may contain associated molecules of more than one type, and there is evidence indicating that the internal state of a molecule may be affected by gross dynamic changes in the substance. In which cases the static and the dynamic values of those thermal properties would be expected to differ, unless the times required to reëstablish equilibrium between the several types of molecules and between each type of molecule and the gross dynamic state of the substance are each negligibly short as

[360] Jakob, M., *Engineering (London)*, 132, 518-521, 550-551 (1931).

Table 112.—Effect of Dissolved Air on the Specific Heat of Water

R. Jessel [350] has stated that the presence of dissolved air increases the specific heat of water and lowers the temperature at which the minimum occurs; he presents the two sets of values of c here tabulated. He suggests that the calorie should be defined in terms of air-free water.

Laby and Hercus do not accept Jessel's conclusions; they conclude from thermodynamic considerations that the presence of dissolved air produces a negligibly small effect, and suggest that Jessel's observations may be explained by irregularities caused by escaping air-bubbles, as remarked by them in 1927.[369] In the discussion following the Laby-Hercus paper, Jessel maintains his position, and develops his view of the subject.

The compiler has determined, and tabulated below, the excess of each of Jessel's values of c above that of the corresponding number (n) defined by the formula $n = 4.185\left[1 - 0.233\left(\dfrac{t-15}{1000}\right) + 6.32\left(\dfrac{t-15}{1000}\right)^2\right]$; if $n \lesseqgtr 20\,°C$, n lies between the two sets of c.

Unit of $c = 1$ joule/g·°C $= 10^7$ ergs/g·°C. Temp. $= t$ °C

Ordinary			De-aerated		
t	c	$c-n$	t	c	$c-n$
12.3	4.1944	0.0066	16.8	4.1858	+0.0025
20.1	4.1857	0.0050	22.1	4.1780	−0.0014
25.2	4.1861	0.0083	32.9	4.1740	−0.0020
32.3	4.1890	0.0126	38.6	4.1732	−0.0035
37.1	4.1893	0.0129	43.7	4.1724	−0.0064
37.7	4.1892	0.0127	43.9	4.1736	−0.0053
42.5	4.1924	0.0142	50.2	4.1776	−0.0058
49.0	4.1916	0.0092	52.5	4.1773	−0.0083
			59.5	4.1772	−0.0168
			69.8	4.1845	−0.0265

Table 113.—Specific Heat of Compressed Water at 1 Atm or at Constant Volume

(For more highly compressed water see Table 115; for the limiting value as saturation is approached see Table 116; for sea-water see Table 129.)

The O values are the most accurate at present available. They are given directly; the others, by the amount (Δ) that each exceeds the corresponding O one.

Example: At 5 °C the A value is $4.20137 + 0.004 = 4.205$, the JS value is $4.20137 - 0.004 = 4.197$.

[361] Tammann, G., and Elbrächter, A., *Z. anorg. allgem. Chem.*, **200**, 153-167 (1931).
[362] Awbery, J. H., *Int. Crit. Tables*, **5**, 113 (1929).
[363] Randall, M., *Idem*, **7**, 232 (1930).
[364] Jaeger, W., and v. Steinwehr, H., *Ann. d. Physik (4)*, **64**, 305-366 (1921) → *Sitz. d. K. preuss. Akad. Wiss. (Berlin)*, **1915**, 424-432 (1915).
[365] Osborne, N. S., Stimson, H. F., and Fiock, E. F., *Mech. Eng.*, **51**, 125-127 (1929).
[366] Hofbauer, P. H., *Atti pont. acc. sci. nouvi Lincei*, **84**, 353-363, 581-586 (1931).
[367] Koch, W., *Forsch. Gebiete, Ingenieurw.*, **5**, 138-145 (1934) → *Z. Ver. deut. Ing.*, **78**, 1110 (1934).
[368] Havlíček, J., and Miškovský, L., *Helv. Phys. Acta*, **9**, 161-207 (1936).
[369] Laby, T. H., and Hercus, E. O., *Proc. Phys. Soc. (London)*, **47**, 1003-1008-1011 (1935); Hercus, E. O., *Idem*, **48**, 282-284 (1936).

Table 113—(Continued)

The data in Awbery's compilation are based upon $c_p = 4.190$ joules/g.°C at 15 °C; those in Randall's on 4.182; those of Jaeger and v. Steinwehr on 4.1842. Here they have all been reduced to the same basis by multiplying the respective values of the ratio $(c/c_{15})_p$ by 4.185, the value, in absolute joules, accepted by the *International Critical Tables*. As 4.185 Int. joules is essentially the value at 15 °C found by O, all the values in the table may be regarded as expressed in Int. joules.

The values given for c_v have been derived by the compiler from the O values of c_p by the R and HL values given in Table 119.

Unit of c_p, c_v, and $\Delta = 1$ Int. joule/g.°C. Temp. $= t$ °C (Int. scale)

Ref.[a]→ t	O $c_p = 1$	A	R ——— 1000 Δ. ———	JS	c_v
0	4.21753	−3	+5		4.2151
5	4.20137	+4	+5	−4	4.2012
10	4.19107	+3	+2	−1	4.1865
15	4.18463	0	0	0	4.1706
20	4.18073	−2	−1	0	4.1535
25	4.17856		−2	0	4.1348
30	4.17751		−4	−2	4.1147
35	4.17734		−4	−1	4.0939
40	4.17772	−3	−5	−1	4.0729
45	4.17860			+1	
50	4.17990		−5	+3	4.015
55	4.18153				
60	4.18354	−3	−4		3.976
65	4.18592				
70[b]	4.18873		−4		3.923
75	4.19191				
80	4.19551	−9	−3		3.852
85	4.19957				
90	4.20418		−2		3.790
95	4.20932				
100	4.21510	−22	−5		3.757

[a] References:

A Awbery, J. H.,[362] based upon the consideration of the work of Barnes, H. T., *Proc. Roy. Soc. (London) (A)*, **67**, 238-244 (1900); *Phil. Trans. (A)*, **199**, 149-263 (1902); *Trans. Roy. Soc. Canada III (3)*, **3**, 3-27 (1909); Reports B. A. A. S., **1909**, 403-404 (1909); Barnes, H. T., and Cooke, H. L., *Phys. Rev.*, **15**, 65-72 (1902); Bousfield, W. R., *Proc. Roy. Soc. (London) (A)*, **93**, 587-591 (1917); Bousfield, W. R. and W. E., *Phil. Trans. (A)*, **211**, 199-251 (1911); Callendar, H. L., *Idem*, **212**, 1-32 (1912); *Proc. Roy. Soc. (London) (A)*, **86**, 254-257 (1912); Cotty, M. A., *Ann. de chim. et phys. (8)*, **24**, 282-288 (1911); Dieterici, C., *Ann. d. Physik (4)*, **16**, 593-620 (1905); Griffiths, E. H., *Phil. Trans. (A)*, **184**, 361-504 (1894); "Thermal Measurement of Energy," Cambridge Univ. Press, 1901; "Dictionary of Applied Physics" (Glazebrook's), vol. 1, pp. 477-494, London, Macmillan, 1922; Guillaume, C. E., *Compt. rend.*, **154**, 1483-1488 (1912); Jaeger, W., and v. Steinwehr, H., *Ann. d. Physik (4)*, **64**, 305-366 (1921); Janke, *Diss.*, Rostock (1910); Lüdin, E., *Mitt. Naturw. Ges. Winterthur, Heft 2* (1900); Pagliani, S., *Nuovo Cim. (6)*, **8**, 157-188 (1914); Regnault, *Mém. Acad. Roy. Sci. Inst. France*, **21**, 729-748 (1847); Rowland, H. A., *Proc. Amer. Acad. Arts Sci.*, **15**, 75-200 (1880).

JS Jaeger, W., and v. Steinwehr, H.;[364] included in *Wärmetabellen* (1919) compiled by L. Holborn, K. Scheel, and F Henning.

O Osborne, N. S., *private communication*, 1938.

R Randall, M., *Int. Crit. Tables*, **7**, 232 (1930), based on work of Barnes, H. T., *Phil. Trans. (A)*, **199**, 159-263 (1902).

[b] A. Romberg [309a] has reported $c_{73} = (1.0040 \pm 0.0005)c_{20}$, whence $c_{73} = 4.197$, a very high value.

[309a] Romberg, A., *Proc. Am. Acad. Arts Sci.*, **57**, 375-387 (1922).

34. WATER: THERMAL ENERGY

Table 114.—Mean Specific Heat of Water at 1 Atm [370]

($\bar{c}_p$ = mean specific heat between t_1 and t_2)
Unit of $\bar{c}_p$ = 1 Int. joule/g.°C. Temp. t_1 and t_2 on Int. Centigrade scale.

$t_1 \rightarrow$ t_2	5	10	15 $\bar{c}_p$	20	25
50	4.18174	4.17998	4.17889	4.17826	4.17801
55	161	4.18006	910	860	847
60	170	027	948	910	906
65	195	069	4.18002	973	977
70	237	124	065	4.18048	4.18061
75	291	195	149	136	157

Table 115.—Specific Heat of Compressed Water at Constant Pressure

(For values at a pressure of 1 atm see Table 113.)

If c_p and c_{p_1} are the specific heats at the constant pressures p and p_1, respectively, and for the same temperature, then Δ_{p_1} is defined by the relation $c_p = c_{p_1}(1 + \Delta_{p_1})$. Values of c_{p_1} for each temperature and of $1000\Delta_{p_1}$ for each temperature and pressure are tabulated. Example: From Section I (Koch) we find for 260 °C and $p = 300$, $c_p = 4.944 (1 - 0.057) = 4.944 - 0.282 = 4.662$; likewise for 260 °C and $p = 50$, $c_p = 4.663 (1 + 0.060) = 4.663 + 0.280 = 4.943$.

Unit of p = 1 kg*/cm² = 0.9678 atm; of c_p = 1 joule/g.°C. Temp. = t °C

I. W. Koch.[367] Smoothed values based on thermal determinations; precision does not exceed 0.004 joules/g.°C; conversion from Int. steam cal. to joules by the compiler; 1 cal = 4.186 joules.

$p \rightarrow$ t	50 c_{50}	100	150	200 $-1000\Delta_{50}$	250	300
0	4.203	2	4	6	8	10
20	4.169	2	4	7	9	12
40	4.161	2	5	8	10	13
60	4.165	3	6	9	12	15
80	4.182	4	7	10	14	17
100	4.203	4	7	11	15	18
120	4.232	4	8	12	16	20
140	4.266	4	10	13	17	22
160	4.324	5	10	15	19	24
180	4.395	6	11	17	22	28
200	4.483	7	13	20	26	32
220	4.592	8	16	23	30	37
240	4.738	10	19	28	37	45
260	4.944	13	25	36	46	57

$p \rightarrow$ t	50	100	150 $+1000\Delta_{300}$	200	250	300 c_{300}
260	60	47	34	22	12	4.663
280		60	44	37	14	4.860
300		106	63	35	16	5.119
310			91	48	20	5.262
320			140	72	44	5.433
330			223	110	44	5.659
340			361	175	73	5.965
350				278	102	6.430

[370] Osborne, N. S., *Private communication*, 1938.

Table 115—*(Continued)*

II. M. Trautz and H. Steyer.[371] Computed from volumetric data and presented by small graphs; values read from graphs with a precision not exceeding 1 or 2 parts per 1000. Conversion to joules by the compiler.

$p \rightarrow$ t	50 c_{50}	100	150	200 $-1000\Delta_{50}$	250	300
0	4.169	1	5	6	9	10
50	4.169	4	6	8	11	14
100	4.203	4	8	10	15	17
150	4.282	6	13	18	22	28
200	4.479	10	19	24	34	42
250	4.998	20	31	45	59	74

t			$+1000\Delta_{300}$			c_{300}
250	80	58	41	31	16	4.630
300			74	43	20	5.295
350				310	162	8.79

III. P. W. Bridgman.[372] Computed from volumetric data. Compiler scaled his small graph and converted the values from calories to joules. Precision not greater than 1 or 2 parts in 1000.

$t \rightarrow$ $c_1 \rightarrow$ p	0 4.21_8	20 4.18_5	40 4.18_5 $-1000\Delta_1$	60 4.18_5	80 4.20_6
1000	+40	+52	+55	+46	+30
2000	62	74	84	75	+18
3000	76	84	99	94	+2
4000	87	88	113	102	−12
5000	98	95	124	107	−25
5500	109	100	128	108	−31
6000	135	106	130	109	−38
7000		108	128	106	−53
8000		94	122	102	−70
9000			118	97	−88
10000			114	90	−109
11000			108	83	−136
12000			100	75	−169

Table 116.—Specific Heat of Compressed Water: Limit as Temperature Approaches that of Saturation

The subscript "$t \rightarrow$ sat" is used to denote the limiting value approached as t approaches the temperature corresponding to saturation under the specified conditions. The value of $(c_v)_{t \rightarrow \text{sat}}$ may be obtained from that of $(c_p)_{t \rightarrow \text{sat}}$ by subtracting the corresponding value of $(c_p - c_v)$, given in Table 119.

Unit of $c_p = 1$ Int. joule/g.°C. Temp. $= t$ °C (Int. scale)

I. Preferred value at 100 °C is $(c_p)_{t \rightarrow \text{sat}} = 4.2151$ (see Table 113).

II. SK[a] Values computed by Smith and Keyes by means of an equation set up by them to represent several sets of data, including their own on the specific volume.

[371] Trautz, M., and Steyer, H., *Forsch. Gebiete Ingenieurw.*, **2**, 45-52 (1931).
[372] Bridgman, P. W., *Proc. Amer. Acad. Arts Sci.*, **48**, 307-362 (Fig. 11) (1912).

34. WATER: THERMAL ENERGY

Table 116—(Continued)

t	$(c_p)_{t \to sat}$	t	$(c_p)_{t \to sat}$	t	$(c_p)_{t \to sat}$
0	(4.2208)	100	4.2127	190	4.4514
10	(4.1877)	100	4.2127	200	4.4958
20	4.1772	110	4.2267	210	4.5465
30	4.1747	120	4.2435	220	4.6066
40	4.1763	130	4.2611	230	4.6755
50	4.1765	140	4.2839	240	4.7560
60	4.1808	150	4.3099	250	4.8423
70	4.1869	160	4.3340	260	4.9651
80	4.1930	170	4.3719		
90	4.2022	180	4.4137		

III. Miscellaneous values.

Ref.[a] →	O	SK	A	R
t		$(c_p)_{t \to sat}$		
100	4.2151	4.2127	4.193	4.210
125				4.235
150		4.3099	4.218	4.265
200		4.4958	4.250	4.294
250		4.8423	4.29	
300			4.34	

[a] References:
A   Awbery, J. H.[362] For work considered by him, see Table 113, reference note and head matter.
O   Osborne, N. S.[370]
R   Randall, M., *Int. Crit. Tables*, **7**, 232 (1930), based on work of Barnes, H. T., *Phil. Trans. (A)*, **199**, 148-263 (1902). See head of Table 113.
SK   Smith, L. B., and Keyes, F. G., *Proc. Amer. Acad. Arts Sci.*, **69**, 285-314 (1934) → *Mech. Eng.*, **56**, 92-94 (1934).

Table 117.—Specific Heat of Compressed Water at Constant Volume (Isopiestics)

Derived from a graph published by P. W. Bridgman [372, Fig. 12] and based upon his measurements of the compressibility. The values of c_v cannot be read more accurately than 1 or 2 parts in 1000.

In each case the volume is that corresponding to the indicated temperature and pressure. If c_1 and c_v = specific heat at constant volume for a pressure of 1 atm and of p kg*/cm², respectively, the temperature being the same in each case, then $\Delta \equiv (c_1 - c_v)/c_1$ or $c_v = c_1(1 - \Delta)$. The values tabulated for c_1 correspond to those read from the graph. Example: At 40 °C and 3000 kg*/cm², $c_v = 4.07_2(1 - 0.124) = 3.56_7$ j/g.°C.

Unit of p = 1 kg*/cm² = 0.9678 atm; of c_1 = 1 joule/g.°C. Temp. = t °C

$t \to$	0	20	40	60	80
$c_1 \to$	4.22_3	4.14_7	4.07_2	3.96_7	3.85_0
p			1000Δ		
1000	49	62	63	56	16
2000	83	98	102	91	0
3000	108	119	124	113	−13
4000	128	131	147	129	−23
5000	143	139	167	140	−33
6000	162	162	183	148	−43
7000		195	193	154	−54
8000		211	200	159	−65
9000			205	161	−80
10000			209	163	−100

Table 117—(Continued)

$t \rightarrow$	0	20	40	60	80
$c_1 \rightarrow$	4.22₃	4.14₇	4.07₂	3.96₇	3.85₀
p			1000Δ		
11000			209	161	−126
12000			205	155	−165

Table 118.—Specific Heat of Compressed Water at Constant Volume (Isometrics)

Derived from a graph published by P. W. Bridgman [372, Fig. 13] and based upon his measurements of the compressibility. The values of c_v cannot be read more accurately than 1 or 2 parts in 1000.

If $c_{1,0} = c_v$ for $v^* = 1$ and $t = 0$ °C, then $\Delta \equiv (c_{1,0} - c_v)/c_{1,0}$; or, the specific heat at constant volume (c_v) for the volume and temperature indicated is $c_v = c_{1,0}(1 - \Delta)$; $v^* =$ specific volume. Bridgman gives $c_{1,0} = 1.000$ cal/g·°C = 4.18_5 joules/g·°C.

Unit of $v^* = 1$ cm³/g. Temp. = t °C

$t \rightarrow$	0	20	40	60	80
v^*			1000Δ		
1.025					84
1.000	0	12	43	80	97
0.975	22	50	83	112	91
0.950	47	85	116	140	75
0.925	76		140	160	62
0.900	105	133	166	176	50
0.875	126	145	193	190	39
0.850	150	188	216	202	17
0.825		219	230	206	−24
0.800			228		

compared with that for the change in pressure involved in the dynamic method.

In the case of the decrease of temperature on adiabatic expansion, G. Tammann and A. Elbrächter [361] have sought evidence for such a difference. They have found differences, but have been unable to explain them in terms of the expected type of changes in the association (see Table 126).

Table 119.—Ratio and Difference of the Principal Specific Heats of Water: 1 Atm or Saturation

(For values at higher pressures, see Table 120; for sea-water, see note [b].)

If c_p and $c_v =$ specific heat of water at constant pressure and constant volume, respectively, $\gamma = c_p/c_v$; if $\alpha = (1/v)(dv/dt)_p$, $v =$ volume of any fixed mass of water, $T =$ absolute temperature, and $V =$ velocity of sound, then $\gamma - 1 = T\alpha^2 V^2/c_p$; $c_p - c_v = c_p(\gamma - 1)/\gamma$. Bridgman's values (B) for ($c_p - c_v$) have been derived from his graphs,[372, Figs. 11, 12] which are based upon his measurements of the compressibility.

The values at saturation are those derived from the limiting values

Table 119—(Continued)

approached by c_p and c_v as t approaches the temperature at which water is saturated at the coexisting values of p and v, respectively.

Unit of $(c_p - c_v) = 1$ Int. joule/g.°C = 0.23895 cal$_{15}$/g.°C. Temp. = t °C

I. Pressure = 1 atm = 1.03323 kg*/cm².

Basis[a]→ t	R	S $10^4(\gamma - 1) \equiv \delta_\gamma$	HL	R	HL $10^3(c_p - c_v) \equiv \delta_c$	B
0	5.81	5	5.85	2.45	2.46	
5			0.34		0.14	
10	10.84	2₀	10.86	4.53	4.54	
15			33.6[b]		14.00	
20	65.5	6₀	65.6	27.20	27.23	3₀
25			105.8		43.8	
30	152.7	14₀	152.5	62.8	62.7	
35			203.7		83.4	
40	257.5	27₀	257.5	104.9	104.8	11₀
50	385	38₀		155		
60	524	62₀		208		22₀
70	676	80₀		266		
80	840	100₀		324		36₀
86	942			361		
90	110[c]	110₀		414[c]		
100	123[c]	114₀		458[c]		

II. Values at saturation.
Derived from the values of c_p and c_v as computed by Smith and Keyes (see Table 116). $\delta_\gamma \equiv 10^4(\gamma - 1)$, $\delta_c \equiv 10^3(c_p - c_v)$.

t	δ_γ	δ_c	t	δ_γ	δ_c	t	δ_γ	δ_c
0	1.5	0.6	90	999	381.4	180	2903	993.9
10	13.4	5.6	100	1184	446.1	190	3268	1095.3
20	63.3	26.3	110	1381	512.7	200	3540	1175.5
30	143.5	59.1	120	1586	580.8	210	3831	1259.7
40	246.7	100.5	130	1808	652.1	220	4130	1346.2
50	369	148.5	140	2023	721.3	230	4440	1437.4
60	506	201.5	150	2257	793.9	240	4750	1531.9
70	658	258.7	160	2499	866.1	250	5120	1640.5
80	822	318.8	170	2743	941.7	260	5460	1752.4

[a] Except as otherwise noted, the bases on which these values rest are:
- R — C. R. Randall's determination of V^{373}; c_p from Table 113, Column A (or R if no value in A); $(1/v)(dv/dt)_p$ from equations of P. Chappuis, 0° to 40°, or of Thiesen 40° to 86° (Table 99); computation by the compiler.
- HL — J. C. Hubbard and A. L. Loomis [374] from their own determinations of the velocity of sound; c_p from Table 113 (Column JS), Chappuis' equations.
- B — P. W. Bridgman, from his determinations of the compressibility (cf. Tables 115 and 117).
- S — F. A. Schulze.[375] Computed by him from isothermal compressibility and c_p; the same values are given in each paper.

[b] For sea-water at 16.95 °C and pressure = 2 atm, salinity = 35 g/kg, $10^4(\gamma - 1) = 94 \pm 5$, from velocity of sound and isothermal compressibility.[347]

[c] Computed by D. Tryer [376] from the isothermal compressibility and c_p.

---

[373] Randall, C. R., *Bur. Stand. J. Res.*, **8**, 79-99 (RP402) (1932).
[374] Hubbard, J. C., and Loomis, A. L., *Phil. Mag. (7)*, **5**, 1177-1190 (1928).
[375] Schulze, F. A., *Z. physik. Chem.*, **88**, 490-505 (1914); *Physik. Z.*, **26**, 153-155 (1925).
[376] Tyrer, D., *J. Chem. Soc. (London)*, **103**, 1675-1688 (1913); *Z. physik. Chem.*, **87**, 169-181 (1914).

Specific Heat of Water.

In the *International Critical Tables,* two sets of values for the specific heat of water at constant pressure (c_p) are given: those compiled by J. H. Awbery [362] and those by M. Randall.[363] The first are considered the more accurate. Neither agrees with the set published by W. Jaeger and H. v. Steinwehr [364] and included in the "Wärmetabellen" (Vieweg, Braunschweig, 1919) compiled by L. Holborn, K. Scheel, and F. Henning (see Table 113). A graphical comparison of the more important published values has been given by N. S. Osborne, H. F. Stimson, and E. F. Fiock.[365] Among the various interpolation equations that have been proposed may be mentioned those by P. H. Hofbauer,[366] L. B. Smith and F. G. Keyes,[294] W. Koch,[367] and J. Havliček and L. Miškovský.[368] None of the earlier ones is satisfactory if more than moderate accuracy is desired.

Table 120.—Ratio and Difference of the Principal Specific Heats of Water under High Pressure

Derived from graphs constructed by P. W. Bridgman [372, Figs. 11, 12] * from his measurements of the compressibility. The specific heats cannot be read from the graphs to a higher accuracy than 1 or 2 parts in 1000. The values of γ have been computed from those of $(c_p - c_v)$ as determined from the graphs.

Unit of $(c_p - c_v) = 1$ joule/g. °C $= 0.23895$ cal$_{15}$/g.°C; of $p = 1$ kg*/cm² $= 0.9678$ atm

$t \rightarrow$	0	20	40	60	80	0	20	40	60	80
p			$1000(\gamma - 1)$					$100(c_p - c_v)$		
0	0	7	27	55	93	0	3	11	22	36
1000	7	21	37	67	77	3	8	14	25	29
2000	23	37	46	72	73	9	14	17	26	28
3000	34	49	56	77	77	13	18	20	27	30
4000	46	61	69	87	81	17	22	24	30	32
5000	50	62	83	97	83	18	22	28	33	33
6000	31	78	93	104	85	11	27	31	35	34
7000		117	113	113	91		39	37	38	37
8000		159	127	126	98		52	41	42	40
9000			139	139	99			45	45	41
10000			152	148	102			49	49	43
11000			158	153	102			51	51	44
12000			163	155	96			53	52	43

* These supersede Fig. 41 of his earlier paper [377] which, contrary to these, indicates that at the lower temperatures $(c_p - c_v)$ has a pronounced maximum at a pressure near 5000 kg*/cm².

Table 121.—Various Isopiestic Thermal Data for Water

C_p = specific heat at constant pressure = limit approached by the ratio $(\Delta q/\Delta T)_p$ as ΔT approaches zero, Δq being the heat that must be added to the substance in order to increase its temperature by the amount ΔT;

[377] Bridgman, P. W., *Proc. Amer. Acad. Arts Sci.,* **47**, 439-558 (550) (1912).

Table 121—(Continued)

$H =$ heat content (enthalpy), $H_0 = \int_0^T C_p dT$; $G = H - ST$ is the function that Gibbs denoted by ζ and that has been called the "free energy at constant pressure," $G_0 = -T \int_0^T (H_0/T^2) dT$; $S_0 =$ entropy. All these quantities refer to the gfw, and those with subscript ₀ are measured from 0 °K; the pressure is 1 atm; $i =$ ice, $w =$ water. For method employed in extrapolating C_p to 0 °K, see articles cited. The compiler has changed the units, and derived S_0 from H_0 and G_0. 1 gfw = 18.0154 g; 1 joule/gfw = 0.0551 j/g; 1 cal = 4.185 j.

Unit of C_p and $S_0 = 1$ j/(gfw.°K); of H_0 and $G_0 = 1$ kj/gfw. Temp. = T °K

Ref.[a]→ T	C_p	Simon H_0	$-G_0$	S_0	C_p	Miething H_0	$-G_0$	S_0
273i	41.0	5.35	5.00	37.9	50.2	5.49	5.02	38.4
273w	75.7	11.36	5.00	59.9	75.4	11.50	5.02	60.5
280	75.4	11.89	5.42	61.8	75.4	12.03	5.45	62.4
290					75.4	12.78	6.09	65.0
300	75.3	13.40	6.55	66.5	75.3	13.53	6.74	67.5
320					75.3	15.05	8.15	72.6
340	75.4	16.41	9.40	76.0	75.3	16.56	9.65	77.1
360					75.3	18.07	11.24	81.4
373	75.7	18.91	12.02	92.9	75.3	19.04	12.30	84.0

[a] References:
Miething, H., *Abh. d. Deuts. Bunsen Ges.*, No. **9** (1920), based upon the data of Nernst, W., *Ann. d. Physik (4)*, **36**, 395-439 (1911) and of Pollitzer, F., *Z. Elektroch.*, **19**, 513-518 (1913). Simon, F., "Handb. d. Physik" (Geiger and Scheel), vol. 10, p. 363, 1926, based on his own previously unpublished observations.

Table 122.—Enthalpy of Compressed Water

For observations through the critical region, see Table 53.

The enthalpy (H) of a substance is defined by the relation $H = (E + pv) - (E + pv)_0$, where $(E + pv)_0$ is the value of $(E + pv)$ for some state of the substance arbitrarily selected as the basis of reference. For water, the reference state is that of saturation at 0 °C. E is the intrinsic energy, v the volume, and p the pressure; H, E, and v each refers to a unit mass of the substance. The specific heat at constant pressure is $c_p = (dH/dt)_p$.

In some cases the value tabulated is the excess of the corresponding value of H, expressed in Int. steam calories per gram, above the numerical value of the corresponding Centigrade temperature. For example, in Section II, at 75 °C and $p = 50$ kg*/cm² observer S found H to be $75 + 1.0 = 76.0$ Int. steam cal/g = 318.1 Int. joule/g.

Table 122—(Continued)

I. Pressure = 1 atm. N. S. Osborne.[370]

Unit of H either = 1 Int. joule/g or 1 Int. steam cal/g = 4.18605 Int. joules/g, as indicated. Temp. = t °C (Int. scale)

Unit→ t	Joule H	Cal $10^4(H-t)$	Unit→ t	Joule H	Cal $10^4(H-t)$
0	0.1026	245	55	230.228	−12
5	21.147	517	60	251.140	−54
10	42.126	634	65	272.064	−70
15	63.064	652	70	293.000	−55
20	83.976	610	75	313.952	−5
25	104.874	532	80	334.920	+86
30	125.763	435	85	355.907	+223
35	146.650	331	90	376.917	+412
40	167.538	229	95	397.950	+659
45	188.428	134	100	419.011	+971
50	209.324	53			

II. Four sets of values, each indicated by the initial of the experimenter.

Unit of $p = 1$ kg*/cm²; of $H = 1$ Int. steam cal/g = 4.1860 Int. joule/g. Temp. t = °C

$t→$ p		0	20	25	40	50	60	75	80	100 $H-t$	120	125	140	150	160	175	180
50	HM[a]																
	K	1.18	1.16		1.06		0.96		0.89	0.9	1.1		1.4		1.9		2.6
	S	1.2		1.0		0.9		1.0		0.8		1.1		1.5		2.2	
	TS		1.18		1.02		1.18		0.88	0.95	1.05		1.40		2.00		2.82
100	HM		2.9							1.9							
	K	2.36	2.30		2.15		1.98		1.86	1.8	1.9		2.1		2.5		3.2
	S	2.2		2.2		2.0		1.9		1.7		2.0		2.3		3.0	
	TS		2.25		2.02		1.88		1.78	1.78	1.90		2.18		2.62		3.45
150	HM																
	K	3.53	3.43		3.23		3.01		2.82	2.2[b]	2.7		2.8		3.1		3.7
	S	3.6		3.3		3.0		2.9		2.6		2.9		3.1		3.9	
	TS		3.38		3.12		2.92		2.78	2.75	2.75		3.00		3.38		4.00
200	HM		4.55							3.8							
	K	4.71	4.57		4.31		4.03		3.78	3.6	3.5		3.6		3.8		4.2
	S	4.9		4.3		4.0		3.9		3.5		3.9		3.7		4.5	
	TS		4.48		4.18		3.95		3.72	3.65	3.62		3.78		4.08		4.60
250	HM																
	K	5.89	5.70		5.40		5.06		4.74	4.5	4.3		4.3		4.4		4.7
	S	6.0		5.7		5.0		4.9		4.3		4.6		4.5		5.3	
	TS		5.58		5.20		4.90		4.68	4.50	4.50		4.62		4.85		5.25
300	HM		6.79							5.54							
	K	7.07	6.83		6.48		6.09		5.70	5.4	5.1		5.0		5.0		5.3
	S	7.0		6.8		6.0		5.8		5.2		5.0		5.3		5.7	
	TS		6.65		6.25		5.92		5.60	5.45	5.3		5.40		5.58		5.90
350	HM																
	K																
	S	8.2		7.6		6.9		6.5		6.1			5.7		6.1		6.1
	TS																
400	HM	8.92								7.22							
	K																
	S	9.2		8.5		8.0		7.3		7.0			6.8		6.8		7.0
	TS																

34. WATER: THERMAL ENERGY

Table 122—(Continued)

$H - t$

t→ p		200	220	225	240	250	260	275	280	300	310	320	325	330	340	350	360	370
50	HM																	
	K	3.9	5.6		7.8		10.9											
	S	3.8		6.0		9.2												
	TS	4.00	5.52															
100	HM	4.1								20.7								
	K	4.3	5.8		7.9		10.7		14.6	20.4								
	S	4.3		6.4		9.2		13.8		21.2								
	TS	4.42	5.78		7.72				15.48	21.85								
150	HM																	
	K	4.7	6.0		7.9		10.5		14.1	19.2	22.5	26.7			32.3	40.1		
	S	4.8		6.9		9.0		13.3		19.6			31.0					
	TS	4.85	6.08		7.88				14.85	20.22								
200	HM	5.05								18.2						42.6	55.6	
	K	5.1	6.3		8.0		10.4		13.7	18.2	21.1	24.6		29.0	34.9	42.9		
	S	5.3		7.0		9.4		13.0		18.7			28.5			48.	64.	
	TS	5.32	6.50		8.10				14.32	19.00								
250	HM															37.8	46.1	58.7
	K	5.6	6.6		8.0		10.2		13.2	17.4	20.0	23.1		26.8	31.4	37.4		
	S	5.9		7.5		9.6		12.9		18.0			26.7			41	52	65
	TS	5.98	6.95		8.32				13.75	17.95								
300	HM	6.0								16.7						33.9	40.5	48.4
	K	5.9	6.8		8.32		10.1		12.8	16.7	19.0	21.8		25.0	28.9	33.7		
	S	6.5		7.8		9.8		12.8		17.4			25.2			37.5	44	52
	TS	6.45	7.42		8.72		10.45		13.50	17.05								
350	HM															31.45		
	K																	
	S	7.0		8.0		10.1		12.5		16.9			24.2			35	41	46
	TS																	
400	HM	6.9								16.2						29.3		
	K																	
	S	7.6		8.5		10.3		12.0		16.6			23.1			33	37	42
	TS																	

[a] References:
  HM Havlíček, J., and Miškovský, L., *Helv. Phys. Acta*, **9**, 161-207 (Tabelle 1) (1936).
  K   W. Koch.[307]
  S   Schlegel, E., *Z. techn. Phys.*, **14**, 105-107 (1933).
  TS M. Trautz and H. Steyer.[371]

[b] Koch's 2.2 at $p = 150$, $t = 100$ is surely wrong; probably it should be either 2.8 or 2.7.

Table 123.—Entropy of Compressed Water

For the excess of the entropy of water at 25 °C above that of ice at 0 °K, see Table 207.

The excess of the entropy of water at the indicated temperature (t) and pressure (p) above that of saturated water at 0 °C is S.

Two sets of data are given, distinguished by the initials of the experimenters.

Unit of $p = 1$ kg*/cm²; of $S = 1$ millical/g.°K (Int. steam cal.) = 4.1860 millijoule/g.°K. Temp. = t °C

p→ t	50	100	150	200	250	300	350	400	Ref.[a]
0	0.1	0.1	0.2	0.3	0.3	0.3			K
	0.1	0.2	0.2	0.3	0.3	0.4	0.2	0.2	S
20	70.7	70.6	70.5	70.4	70.3	70.2			K
25	87.5	87.2	86.8	86.6	86.2	85.8	85.4	85.2	S

Table 123—(Continued)

$p \rightarrow$ t	50	100	150	200	250	300	350	400	Ref.[a]
40	136.4	136.1	135.9	135.6	135.3	135.0			K
50	167.5	166.8	166.3	165.2	165.4	164.7	164.3	163.8	S
60	197.9	197.5	197.1	196.6	196.2	195.7			K
75	241.7	240.9	240.1	239.4	238.7	238.0	237.3	236.6	S
80	256.1	255.5	254.9	254.3	253.6	253.0			K
100	310.6	309.8	308.9	308.1	307.3	306.4			K
	310.9	309.9	309.1	308.2	307.4	306.4	305.8	304.9	S
120	363.9	362.8	361.8	360.8	359.7	358.7			K
125	376.0	375.0	373.8	372.8	371.8	370.8	369.8	369.7	S
140	414.2	412.9	414.7	410.4	409.2	407.9			K
150	438.1	436.8	435.6	434.3	433.3	431.9	431.0	430.0	S
160	462.6	461.2	459.7	458.2	456.8	455.3			K
175	497.3	495.8	494.2	492.9	491.4	490.0	488.7	487.3	S
180	509.6	507.9	506.2	504.5	502.8	501.1			K
200	555.4	553.4	551.4	549.4	547.5	545.5			K
	555.7	552.9	551.0	549.4	547.9	546.2	544.7	543.2	S
220	600.5	597.9	595.6	593.3	591.0	588.8			K
225	610.4	608.1	606.0	604.0	602.0	600.0	598.2	596.7	S
240	644.4	641.6	639.0	636.3	633.7	631.1			K
250	665.5	662.8	660.2	657.7	655.4	652.9	651.0	648.9	S
260	688.5	685.4	682.2	679.2	676.1	673.1			K
275		717.6	714.2	710.9	707.8	704.9	702.2	699.8	S
280		729.5	725.8	722.1	718.5	714.9			K
300		775.1	770.2	765.7	761.4	757.3			K
		775.1	770.2	765.7	761.9	757.9	754.6	751.5	S
310			793.3	788.0	783.2	778.7			K
320			817.5	811.0	805.5	800.4			K
325			833.9	826.8	821.0	815.5	811.2	806.3	S
330			843.5	835.1	828.4	822.6			K
340			872.9	861.1	852.4	845.4			K
350				890.3	878.2	869.2			K
				897.5	885.8	876.1	869.5	863.0	S
360				941.	919	903	894	886	S
370					954	930	917	907	S

[a] References:
K W. Koch.[367]
S Schlegel, E., Z. techn. Phys., 14, 105-107 (1933).

Table 124.—Heat of Isothermal Compression of Water

Adapted from a compilation by J. R. Roebuck,[378] based on a graph by P. W. Bridgman.[372, Fig. 7]

Q is the amount of heat that must be removed to keep the temperature unchanged when the pressure is increased from 1 atm to 1 atm + p.

[378] Roebuck, J. R., Int. Crit. Tables, 5, 147 (1929).

34. WATER: THERMAL ENERGY

Table 124—(Continued)

Unit of $p = 1$ kg*/cm² $= 0.9678$ atm; of $Q_c = 1$ cal$_{15}$/g; of $Q_j = 1$ joule/g

$t \rightarrow$ p	0	20	40 Q_c	60	80	0	20	40 Q_j	60	80
500	0.2	0.7	1.5	2.1	2.6	0.8	2.9	6.3	8.8	10.9
1000	0.6	1.6	2.9	4.1	5.0	2.5	6.7	12.1	17.2	21.0
2000	1.9	3.8	5.8	7.9	9.2	8.0	15.9	24.3	33.1	38.5
3000	4.0	6.4	8.7	11.4	13.1	16.8	26.8	36.4	47.8	54.9
4000	6.4	8.9	11.6	14.6	16.5	26.8	37.3	48.6	61.2	69.1
6000	10.6	14.0	17.3	20.9	23.2	44.4	58.7	72.5	87.6	97.2
8000		19.6	23.1	27.0	29.3		82.1	96.8	113.0	122.8
10000			28.7	32.9	35.3			120.2	137.8	147.9
12000			34.5	38.8	40.8			144.6	162.6	171.0

Table 125.—Decrease in Internal Energy of Water on Isothermal Compression

Adapted from P. W. Bridgman.[372, Fig. 8]

D = resultant decrease in the internal energy of water when the pressure on it is isothermally increased from 1 atm to p kg*/cm². The work (W) done on the water during such compression is the excess of the heat (Q) given out (Table 124) above D; $W = Q_j - D$.

Unit of $p = 1$ kg*/cm² $= 0.9678$ atm; of $D = 1$ joule/g

$t \rightarrow$ p	0	20	40 D	60	80
500	0.2	2.9	5.2	8.4	10.3
1000	0.9	5.7	10.0	15.5	18.8
2000	3.3	10.5	18.6	27.2	32.2
3000	6.5	15.3	25.1	35.6	42.3
4000	9.2	19.0	30.1	42.5	49.8
5000	10.7	22.0	34.3	48.1	56.1
6000	10.3	24.3	38.1	53.0	61.1
7000		26.6	41.5	57.1	65.5
8000		29.7	44.6	60.9	69.1
9000			47.5	64.4	72.4
10000			50.2	67.4	75.1
11000			52.8	70.1	77.4
12000			55.0	72.4	79.5

Joule-Thomson Coefficient for Water.

The Joule-Thomson coefficient (μ) is the decrease in temperature per unit drop in pressure, the expansion being adiabatic. It measures the internal latent heat of expansion, and is the increase in temperature on adiabatic compression.

The several sets of observations given in Table 126 are discordant, and it is to be noticed that at the lower temperatures the observed (dynamic) values (μ_o) differ significantly from the corresponding (static) ones (μ_c) computed from the specific heat and the coefficient of thermal expansion; see discussion in text, p. 256.

The experimental determination of μ has been discussed by K. J. Umpfenbach[379] and by G. Tammann.[380]

As the temperature is varied, the pressure limits remaining unchanged, a temperature (τ) may be found at which μ passes through zero, changing its sign. This is called the inversion temperature. The following values, τ_1 and τ_2, were computed by W. Koch [367] and by M. Trautz and H. Steyer,[371] respectively, from their determinations of the enthalpy:

p	50	100	150	200	250	300	kg*/cm²
τ_1	242.2	244.3		247.3		248.9	°C
τ_2	245.0	249.3	253.8	258.1	263.1	267.3	°C

Table 126.—Joule-Thomson Coefficient for Water

(See text also.)

$\mu = (dt/dp)_a$, or $(\Delta t/\Delta p)_a$, where Δp is of the order of 100 kg*/cm²; $\mu_c = \mu_o + \delta$, where μ_o is the observed (dynamic) value, and μ_c is the corresponding (static) value computed from the specific heat and the coefficient of thermal expansion; p is the mean of the initial and the final pressure; ( )$_a$ indicates that heat is neither added nor removed from the water.

Unit of $p = 1$ kg*/cm²; of μ and $\delta = 0.01$ °C per 100 kg*/cm²

I. Tammann and Elbrächter.[a]

p	μ_o 0 °C	δ	p	μ_o 30 °C	δ	p	μ_o 70 °C	δ
2850	20.8	+6.5	2918	22.2	+11.5	2745	25.0	+8.6
2712	17.4	+9.3	2734	21.5	+11.7	2500	33.0	+1.1
2572	20.2	+5.9	2554	22.2	+10.7	2342	32.2	+2.4
2461	19.0	+6.4	2388	21.8	+10.7	2192	35.1	−0.1
2351	18.5	+6.6	2244	24.0	+8.2	2072	33.6	−1.7
2216	19.2	+5.3	2118	24.4	+7.4	1951	32.3	+4.4
2071	17.9	+5.9	1984	20.9	+10.5	1816	35.1	+1.6
1942	19.2	+4.0	1806	25.5	+5.4	1676	34.8	+3.4
1824	19.5	+2.5	1641	20.6	+9.7	1544	37.0	+2.4
1704	18.2	+2.6	1501	20.3	+9.5	1418	42.5	−2.2
1580	19.0	+0.7	1350	21.8	+7.4	1290	38.2	−0.2
1452	18.0	0.0	1202	23.0	+5.6	1148	40.8	−0.6
1325	16.5	+0.1	1022	24.0	+3.8	1020	42.0	−1.3
1188	16.6	−0.1	864	22.1	+4.3	886	41.1	−0.7
1038	15.2	+0.2	722	23.9	+1.5	752	43.7	−2.0
838	13.8	−0.8	560	22.2	+3.3	611	44.4	−2.1
762	12.8	−0.9	488	24.8	+0.1	506	45.2	−1.2
630	10.9	−0.3	415	19.6	+5.2	431	46.4	−1.7
462	10.2	−2.2	300	23.9	+0.4	294	48.3	−2.6
285	8.2	−4.4	210	23.1	+0.6	158	49.6	−2.9
98	−0.7	+0.2	125	22.7	+0.4	56	50.3	−2.4
560	12.8	−2.9	46	22.6	−0.1	886	41.1	−2.5A
440	11.9	−4.5				752	43.7	−2.4A
330	9.8	−5.0				611	44.4	−0.2A
255	8.6	−5.2				506	45.2	+3.0A
198	2.9	−2.7				431	46.4	+2.3A
88	0.3	−0.9				294	48.3	+1.7A
						158	49.6	−1.9A
						56	50.3	−0.3A

[379] Umpfenbach, K. J., *Z. techn. Physik,* 12, 25-29 (1931).
[380] Tammann, G., "Über die Beziehungen zwischen den inneren Kräften und Eigenschaften der Lösungen," Leipzig, Voss, 1907; see Roebuck, J. R., *Int. Crit. Tables,* 5, 146, 147 (1929).

34. WATER: THERMAL ENERGY

Table 126—(Continued)

II. Pushin and Grebenshchikov.[b]

$t \rightarrow$	0	25	37	54	80
p			μ_o		
1	−13.0	+6.6	+26.0	+39.0	+49.2
500	−2.0	+13.0	27.3	37.1	46.8
1000	+6.4	16.7	27.9	35.7	44.5
1500	11.6	18.8	27.9	34.4	42.3
2000	15.0	20.3	27.9	33.5	40.6
2500	17.3	21.3	27.9	32.9	39.2
3000	18.9	22.3	28.4	32.5	38.2
3500		24.2	29.3	32.2	
4000		24.0			

III. Bridgman.[c] **(ICT)**

t	0	20	40	60	80
p			μ_o		
1	−1.6	+13.7	28.7	41.7	54.8
500	+6.8	17.5	30.0	41.7	50.0
1000	13.2	22.0	30.9	41.3	46.2
1500	18.3	24.8	31.6	40.6	42.7
2000	21.5	26.3	32.2	39.7	40.3
3000	25.1	28.0	32.5	38.1	36.7
4000	26.0	28.3	32.3	36.6	34.4
6000	19.4	28.9	33.6	34.9	30.8
8000		35.5	33.3	33.7	27.9
10000			33.3	33.0	25.7
12000			32.0	32.0	23.8

IV. W. Koch.[367]

$t \rightarrow$	0	100	200	240	250	260	300	350
p				μ_o				
50	−235	−178	−72	−2	+20	+45		
100	−235	−178	−74	−8	+12	+35	+200	
200	−236	−179	−76	−13	+6	+27	+138	+702
300	−237	−181	−79	−16	+2	+23	+115	+399

V. By an optical method Mascart [381] found for water at 16 °C and p about 2, $10^4 \mu = 11$ °C/atm.

P. G. Tait,[382] using a Cu-Fe thermocouple, found the following values (if his "ton" = 2240 lbs) for water at 15.5 °C; he stated that they are to be accepted with caution:

p	79	158	236	315	kg*/cm²
$10^4 \mu$	9	10	11	12	°C per (kg*/cm²)

[a] G. Tammann and A. Elbrächter.[361] The pressure was raised to about 3000 kg*/cm², and then reduced by a series of sudden releases, the decrease in temperature being observed for each step. Each step was about 150 kg*/cm². They believe that the error in μ_o in no case exceeds ±0.015 °C per 100 kg*/cm² at 0 °C and at 30 °C, nor ±0.030 at 70 °C. In computing μ_o, they used P. W. Bridgman's values for the thermal expansion, and for the lower pressures at 70° Amagat's also; the values from this second computation are here indicated by A.
[b] N. A. Pushin and E. V. Grebenshchikov,[383] included in part in the compilation by J. R. Roebuck (ICT).[384] It will be noticed that these values differ in the same general way as do those of Tammann and Elbrächter from the corresponding ones computed from Bridgman's data.
[c] P. W. Bridgman [372, Fig. 14] as given by J. R. Roebuck.[384]

Table 127.—Heat Liberated by Adiabatic Compression of Water
Adapted from M. Trautz and H. Steyer.[371]

The heat liberated by adiabatic compression is μc_p, where $\mu = (\delta t/\delta p)_a$ is the Joule-Thomson coefficient, and c_p is the specific heat at constant pressure.

[381] Mascart, *Compt. rend.*, **78**, 801-805 (1874).
[382] Tait, P. G., *Proc. Roy. Soc. Edinburgh*, **11**, 217-219 (1882).
[383] Pushin, N. A., and Grebenshchikov, E. V., *J. Chem. Soc. (London)*, **123**, 2717-2725 (1923).
[384] Roebuck, J. R., *Int. Crit. Tables*, **5**, 146 (1929).

Table 127—(Continued)

Unit of $p = 1$ kg*/cm²; of $\mu c_p = 10^{-4}$ joule/g(kg*/cm²)

$p \rightarrow$ t	p_{sat}	100 μc_p	200	300	$p \rightarrow$ t	p_{sat}	100 μc_p	200	300
10	959	950	946	942	120	691	707	720	733
20	923	921	917	913	140	628	649	670	678
30	898	895	893	890	160	548	586	603	628
40	874	871	872	871	180	456	502	532	578
50	851	850	850	850	200	352	398	440	490
60	827	828	830	830	220	222	272	335	398
70	804	808	810	813	240	54	100	184	272
80	781	787	793	797	260	−172	−134	−33	−84
90	758	768	774	782	280		−377	−272	−159
100	736	747	758	768					

Table 128.—Isentropic Increase in the Temperature of Water from Saturation to and above the Critical Pressure

Adapted from a table computed by J. H. Keenan.[385] S = excess of entropy above that at 0 °C and 1 atm; $p_{crit} = 218.39$ atm, $t_{crit} = 374.15$ °C (see Table 241).

Example: For saturated water, $S = 0.1$ when $t = 28.73$ °C, the corresponding pressure being 0.037 atm; if the pressure is increased from 0.037 to 218.39 atm, the temperature must at the same time be increased by 0.45 °C (i.e., to 29.18 °C) if S is to remain unchanged.

Unit of $p = 1$ atm; of $S = 1$ Int. cal/g.°C = 4.186 joule/g.°C; of t_{sat} and $\Delta t = 1$ °C

$p \rightarrow$ S	218.39 Δt	387.18	p_{sat}	t_{sat}
0.1	0.45	0.86	0.037	28.73
0.3	1.39	2.48	0.851	95.59
0.5	2.76	4.91	9.284	175.28
0.7	4.94	9.31	50.32	265.19
0.8	5.68	12.45	96.19	309.08
0.9	5.64		158.18	347.35

Table 129.—Specific Heat of Sea-Water

The values in the last pair of columns have been derived by O. Krümmel[386] from those published by J. Thoulet and A. Chevallier[387] and given in the preceding columns.

r = ratio of the specific heat of sea-water to that of pure water at the same temperature; s = salt content; ρ = ratio of the density of sea-water to that of pure water at the same temperature. In all cases the temperature was 17.5 °C.

Unit of $s = 1$ g salt per kg sea-water; ρ and r are ratios. Temp. = 17.5 °C

ρ	r	ρ	r	s	r
1.0025	0.986	1.0275	0.931	0	1.000
1.0050	0.977	1.0300	0.927	5	0.982
1.0075	0.968	1.0325	0.924	10	0.968
1.0100	0.963	1.0350	0.921	15	0.958
1.0125	0.957	1.0375	0.917	20	0.951

[385] Keenan, J. H., *Mech. Eng.*, **53**, 127-131 (1931).
[386] Krümmel, O., "Handb. d. Ozeanog.," Vol. **1**, 1907.
[387] Thoulet, J., and Chevallier, A., *Compt. rend.*, **108**, 794-796 (1889).

35. WATER: THERMAL CONDUCTION

Table 129—(Continued)

p	r	p	r	s	r
1.0150	0.952	1.0400	0.913	25	0.945
1.0175	0.948	1.0425	0.910	30	0.939
1.0200	0.944	1.0450	0.907	35	0.932
1.0225	0.940	1.0475	0.903	40	0.926
1.0250	0.935	1.0500	0.900		

35. THERMAL CONDUCTIVITY OF WATER

At the time that T. Barratt and H. R. Nettleton prepared their compilation,[388] there was no available determination of the thermal conductivity of water at temperatures above 100 °C, and they concluded that in the

Table 130.—Thermal Conductivity of Water
(For sea-water, see Table 132.)

The values attributed to B, BN, KH, and ML have been computed by means of their linear formulas as given below; those attributed to SS have been read from their curve. At temperatures below 100 °C the pressure was 1 atm; at higher temperatures it was a few atmospheres greater than the vapor pressure, but the increase produced in the conductivity by the highest pressure used scarcely equals the uncertainty in the observations (cf. Table 131).

Formulas, $\tau = (t - 20)$:

BN^a $k = 0.00587 (1 + 0.00281\ \tau)$ watt per cm·°C; 0 to 80 °C (ICT)
KH^a $k = 0.00623 (1 + 0.0012\ \tau)$ watt per cm·°C; 0 to 80 °C
ML^a $k = 0.00610 (1 + 0.0023\ \tau)$ watt per cm·°C; 0 to 60 °C
B^a $k = 0.00590 (1 + 0.00260\ \tau)$ watt per cm·°C; 0 to 80 °C

Unit of $k = 10^{-5}$ watt/cm·°C = 2.389 (10⁻⁶) cal/cm·sec·°C. Temp. = t °C.

Ref [a]→	SS	B	BN	KH	ML	t	SS k	t	k
t			k						
0	554	559	554	608	583	100	680	200	666
10	576	575	570	615	596	110	684	210	659
20	598	590	587	623	610	120	686	220	652
30	615	605	604	630	623	130	687	230	644
40	630	621	620	638	637	140	686	240	635
50	643	636	636	645	650	150	685	250	624
60	654	651	653	653	663	160	682	260	614
70	665	667	670	660	(677)	170	680	270	602
80	671	682	686	668	(690)	180	676	280	590
90	676					190	672	290	576
100	680					200	666	300	564

[388] Barratt, T., and Nettleton, H. R., *Int. Crit. Tables,* 5, 218-233 (218, 227) (1929).
[389] Jakob, M., *Ann. d. Physik (4),* 63, 537-570 (1920).
[390] Schmidt, E., and Sellschopp, W., *Forsch. Gebiete Ingenieurw.,* 3, 277-286 (1932).

Table 130—(Continued)

[a] References:

- **B**   Bates, O. K., *Ind. Eng. Chem.*, **28**, 494-498 (1936). Supersedes *Idem*, **25**, 431-437 (1933).
- **BN**   Compilation by Barratt, T., and Nettleton, H. R.,[388] based upon Bridgman, P. W., *Proc. Amer. Acad. Arts Sci.*, **59**, 141-169 (1923); Jakob, M., *Sitz. Preus. Akad. Wiss.*, 1920, 406-413 (1920); *Ann. d. Physik (4)*, **63**, 537-570 (1920); Lees, C. H., *Phil. Trans. (A)*, **191**, 399-440 (1898); Milner, S. R., and Chattock, A. P., *Phil. Mag. (5)*, **48**, 46-64 (1899); Weber, H. F., *Sitz. Preus. Akad. Wiss.*, 1885, 809-815 (1885); *Repert. d. Physik (Exner)*, **22**, 116-122 (1886), considering Chree, C., *Proc. Roy. Soc. (London) (A)*, **42**, 300-302 (1887); **43**, 30-48 (1887); Christiansen, C., *Ann. d. Physik (Wied.)*, **14**, 23-33 (1881); Graetz, L., *Idem*, **18**, 79-94 (1883); **25**, 337-357 (1885); Kohlrausch, F., *Diss.*, Rostock, 1904; Lorberg, H., *Ann. d. Physik (Wied.)*, **14**, 291-308 (1881); Mache, H., and Tagger, J., *Sitz. Akad. Wiss. Wien (Abt. IIa)*, **116**, 1105-1110 (1907); Wachsmuth, R., *Diss.*, Leipzig, 1892; *Ann. d. Physik (Wied.)*, **48**, 158-179 (1893); Weber, H. F., *Idem*, **10**, 103-129, 304-320, 472-500 (1880); Weber, R., *Idem (4)*, **11**, 1047-1070 (1903).
- **KH**   Kaye, G. W. C., and Higgins, W. F., *Proc. Roy. Soc. (London) (A)*, **117**, 459-470 (1928).
- **ML**   Martin, L. H., and Lang, K. C., *Proc. Phys. Soc. (London)*, **45**, 523-529 (1933).
- **SS**   Schmidt, E., and Sellschopp, W.[390]

---

range 0 °C to 80 °C the conductivity can be represented by the formula BN given in Table 130, the accuracy being of the highest and amply sufficient to justify the use of water as a standardizing substance. This formula was based largely on the work of M. Jakob [389] at the Physikalisch-Technischen Reichsanstalt.

Since then, quite different results have been obtained by G. W. C. Kaye and W. F. Higgins at the National Physical Laboratory, and by L. H. Martin and K. C. Lang (see KH and ML, Table 130); and a series extending to 270 °C has been published by E. Schmidt and W. Sellschopp.[390] The last indicates that the variation is not linear in t; that the conductivity reaches a maximum near 130 °C, and has nearly the same value at 300 °C as at 0 °C. These sets of observations were believed to be in error by not more than one or two per cent. Other isolated and less accurate determinations have been published by J. F. D. Smith [391] and by T. W. Classen and J. Nelidow.[392]

The theory of the conduction of heat by liquids has been discussed recently by A. Kardos,[393] and a series of interesting papers treating of certain thermomechanical properties of liquids and their relations to thermal conductivity has been published by R. Lucas [394] and by F. Perrin and R. Lucas.[395]

Table 131.—Thermal Conductivity of Compressed Water

Adapted from T. Barratt and H. R. Nettleton [396] and based on P. W. Bridgman.[397]

[391] Smith, J. F. D., *Ind. Eng. Chem.*, **22**, 1246-1251 (1930).
[392] Classen, T. W., and Nelidow, J., *Physik. Z. Sowj.*, **5**, 191-199 (1934).
[393] Kardos, A., *Forsch. Gebiete Ingenieurw.*, **5**, 14-24 (1934).
[394] Lucas, R., *J. de Phys. (7)*, **8**, 98S-99S, 410-428 (1937); *Compt. rend.*, **204**, 418-420, 1631-1632 (1937).
[395] Perrin, F., and Lucas, R., *Compt. rend.*, **204**, 960-961 (1937).
[396] Barratt, T., and Nettleton, H. R., *Int. Crit. Tables*, **5**, 218-233 (227) (1929).
[397] Bridgman, P. W., *Proc. Amer. Acad. Arts Sci.*, **59**, 141-169 (1923).

36. WATER: MAXIMUM DENSITY

Table 131—(Continued)

k_p and k_1 = thermal conductivity of water at the indicated temperature and under the pressures p and 1 atm, respectively; $p = 1$ atm $+ P$ kg*/cm². Temp. = t °C.

Unit of $P = 1$ kg*/cm² = 0.9678 atm = 0.9807 megadyne/cm²

P	$t\rightarrow$ 30 $1000(k_p - k_1)/k_1$	75	P	$t\rightarrow$ 30 $1000(k_p - k_1)/k_1$	75
1000	58	65	7000	332	345
2000	113	123	8000	366	379
3000	163	176	9000	398	412
4000	210	225	10000	428	445
5000	253	268	11000	456	476
6000	293	308	12000	F[a]	506

[a] Frozen.

Table 132.—Thermal Conductivity of Sea-water

In 1907, O. Krümmel [398] stated that direct determinations of the thermal conductivity of sea-water were then lacking. He computed a series of values for 17.5 °C and various salinities, based upon the conductivity of pure water and upon the assumption that the heat diffusivity (conductivity divided by product of density times specific heat) is the same for sea-water as for pure water. These values, so corrected as to accord with the value of the conductivity of water given in the *International Critical Tables*, were given in the compilation by T. Barratt and H. R. Nettleton,[399] and are reproduced below. See also J. E. Fjeldstad.[400] Salinity = s grams total salts per kg of sea-water.

Unit of $k = 10^{-5}$ watt/cm.°C; of $s = 1$ g/kg. Temp. = 17.5 °C

s	0	10	20	30	35	40
k	583	569	563	560	558	557

36. Temperature of Maximum Density of Water

That the density of water under a pressure of 1 atm is a maximum at a temperature (t_m) near 4 °C has long been known, but the exact value of t_m, 3.98 °C on the international hydrogen scale (3.98 to 4.01 for mercury-in-glass thermometers), was not established until around the beginning of this century.

A list of 36 early and widely varying estimates of t_m has been published by F. Rossetti.[401] The only ones in that period that need be considered are those of Despretz, included in the following table.

As the pressure is increased, t_m decreases, its rate of decrease, at least for the first few hundred atmospheres, being essentially constant and greater than that of the depression of the freezing point. The first attempt to

---

[398] Krümmel, O., "Handbuch der Ozeanographie," Vol. 1, p. 280, Stuttgart, J. Engelhorn, 1907.
[399] Barratt, T., and Nettleton, H. R., *Int. Crit. Tables*, 5, 218-233 (229) (1929).
[400] Fjeldstad, J. E., *Geofysiske Publ.*, 10 No. 7, 1933.
[401] Rossetti, F., *Ann. de chim. et phys.* (4), 10, 461-473 (1867) ← *Atti. Ist. Veneto*, 12, (1866).

estimate the variation of t_m with the pressure seems to have been that of J. D. van der Waals.[402] His estimates, based upon an equation which he fitted to 3 values of the compressibility as determined by C. Grassi[311] and upon the thermal expansion as given by the formula of H. Kopp[403] [$10^6(v-v_0)/v_0 = -61.045t + 7.7183t^2 - 0.03734t^3$, $0° \lesssim t \lesssim 25$ °C] are quoted in a much used handbook, and ascribed to Grassi, computation by van der Waals. Actually, Grassi is in no way responsible for those values, which, indeed, are discordant with all direct determinations. The data from which they were derived are unsuited to that purpose, and the equation which van der Waals used for the compressibility reproduces only those three of Grassi's determinations which were used in deriving it, giving at other temperatures values which are entirely impossible. For these reasons, those estimates by van der Waals are not included in this compilation, it being sufficient to remark that for the range 1 to 10.5 atm they lead to a mean depression of t_m of 0.072 °C per atmosphere, nearly three times that found experimentally for a wide range of higher pressures (see Table 133).

The presence of a solute likewise depresses the value of t_m; again, by more than it depresses the freezing point. To a first approximation, each depression is proportional to the concentration of the solution.

A. J. Bijl[404] has reported values of t_m for mixtures of water and finely divided sugar-charcoal. They fall below the value for water, the depression (Dt_m) of t_m depending upon the amount of charcoal; e.g., for 7.87 g charcoal and 15.03 g water, $Dt_m = 4.8$ °C = 0.61 °C/g-charcoal; for 4.55 g charcoal and 16.82 g water, $Dt_m = 2.7$ °C = 0.59 °C/g-charcoal. He suggests that the effect arises from the composition of the water (relative amounts of the several polymers) in the absorbed layer differing from that of water in bulk.

Table 133.—Temperature of Maximum Density of Water

(For sea-water, see Table 134.)

If P does not exceed a few hundred atmospheres, the temperature of maximum density (t_m) is approximately given by $t_m = 3.98_2 - a(P-1)$ °C. The earlier observations gave 3.98 to 4.01 when $P = 1$, depending upon the nature of the glass of the thermometer. Lussana[a] derives from his observations $a = 2.25$ °C per 100 atm, which probably is as good as we can do; that value is higher than the mean of those given below because they are based on $t_m = 4.00$ °C for $P = 1$, while his equation calls for 4.10 °C. If P is not too great, the freezing point is $t_f = -0.0075(P-1)$; hence $t_m = t_f$ when P is about 270 atm.

[402] van der Waals, J. D., *Beibl. zu Ann. d. Physik*, **1**, 511-513 (1877) ← *Med. Kon. Acad. Wet. Amsterdam, Afd. Nat. (2)*, **11**, 1-13 (1877).
[403] Kopp, H., *Ann. d. Physik (Pogg.)*, **72**, 1-62, 223-293 (44) (1847).
[404] Bijl, A. J., *Rec. trav. chim. Pays-Bas*, **46**, 763-769 (1927).

36. WATER: MAXIMUM DENSITY

Table 133—(Continued)

Unit of $P = 1$ atm; of $a = 1$ °C per atm. Temp. $= t_m$ °C

t_m	Ref.[a]	P	t_m	$100a$	Ref.[a]	P	t_m	$100a$	Ref.[a]
3.98_2	Best	−26.3	4.6	2.2	M11	166	0.40	2.18	L95
3.98_2	de C94	−20.5	4.5	2.3	M11	200	−0.44	2.23	L95
3.98_6	Ch97[b]	−12.9	4.3	2.2	M11	222	−0.91	2.22	L95
3.98_0	TSD00	+41.6	3.3	1.7	A93	251	−1.54	2.22	L95
3.98_3	de C03	93.3	2.0	2.2	A93	268	−1.82	2.18	L95
3.99	D37	145	0.6	2.4	A93	300	−2.57	2.20	L95
4.00	D39	197	ca.0	2.0	A93	322	−3.05	2.20	L95
4.07	R66	47	3.06	2.0	L95	150–600		2.4	T82
4.04	R68	58	2.75	2.19	L95	?		2.0	T88
4.10	W78	100	1.90	2.12	L95	600	0	0.6_7	PG23
3.96	S92	112	1.68	2.09	L95				
4.05	L95	148	0.77	2.20	L95				
3.97_2	Ma91	163	0.40	2.22	L95				

[a] References:

- A93   Amagat, E. H., *Compt. rend.*, **116**, 946-952 (1893).
- de C94   de Coppet, L. C., *Ann. de chim. et phys. (7)*, **3**, 240-269 (1894).
- de C03   de Coppet, L. C., *Idem*, **28**, 145-213 (1903).
- Ch97   Chappuis, P., *Ann. d. Physik (Wied.)*, **63**, 202-208 (1897).
- D37   Despretz, C., *Compt. rend.*, **4**, 124-130 (1837) → *Ann. d. Physik (Pogg.)*, **41**, 58-71 (1837).
- D39   Despretz, C., *Ann. de chim. et phys. (2)*, **70**, 5-81 (1839).
- L95   Lussana, S., *Nuovo Cim. (4)*, **2**, 233-252 (1895).
- M11   Meyer, J., *Abh. d. Deutsch. Bunsen-Ges.*, **3**, No. 1, whole No. 6, 1911.
- Ma91   Makaroff, C. O., *J. Russ. Phys. Chem. Soc. (Chem.)*, **23 II**, 30-88 (1891).
- PG23   Pushin, N. A., and Grebenshchikov, E. V.[383]
- R66   Rossetti, F., *Ann. de chim. et phys. (4)*, **10**, 461-473 (1867) ← *Atti Reg. Ist. Veneto (3)*, **12**, (1866).
- R68   Rosetti, F., *Atti Reg. Ist. Veneto (3)*, **13**, 1047-1093, 1419-1457 (1868) → *Ann. de chim. et phys. (4)*, **17**, 370-384 (1869). Each (R66, R68) abstracted in *Ann d. Physik (Pogg.) Erg. Bd.*, **5**, 258-275 (1871).
- S92   Scheel, K., *Ann. d. Physik (Wied.)*, **47**, 440-465 (1892).
- T82   Tait, P. G., *Proc. Roy. Soc. Edinburgh*, **11**, 813-815 (1882), from observations of Marshall, D. H., Smith, C. M., and Omond, R. T., *Idem*, **11**, 809-813 (1882).
- T88   Tait, P. G., *Beibl. zu Ann. d. Physik*, **13**, 442-445 (1889) ← "Report Sci. Res. Voy. H. M. S. Challenger, Phys. and Chem.," **2**, Part 4, London, 1888.
- TSD00   Thiesen, M., Scheel, K., and Diesselhorst, H., *Wiss. Abh. Physik.-Techn. Reichsanstalt*, **3**, 1-70 (1900) → *Z. Instk.*, **20**, 345-357 (1900).
- W78   Weber, L., *Beibl. zu Ann. d. Physik*, **2**, 696-699 (1878) ← *Jahresber. Comm. Wiss. Untersch. Deuts. Meere in Kiel*, **4-6** (1874-76), 1-22 (1878).

[b] The equations by means of which Chappuis represents his two ultimate series of observations between 0° and 10 °C, lead to $t_m = 3.978$ and 3.994, respectively; and their mean, which defines his definitive values (Table 99), gives $t_m = 3.986$ °C.[405]

Table 134.—Temperature of Maximum Density of Sea-water

D. H. Marshall, C. M. Smith, and R. T. Omond [406] have reported that when sea-water (not more particularly specified) is adiabatically expanded from P to 1 atm there is no resultant change in temperature if the associated temperature and pressure (P) have the following values: −5 °C, 153 atm; −8.5 °C, 306 atm; −11 °C, 458 atm; and −13 °C, 610 atm.

The following data all refer to a pressure of 1 atm. It seems probable that all values of t_m in Section I should be increased by about 0.035 °C (cf. value for $s = 0$ with Table 133); it is believed that the data in that section are in other respects to be preferred to those in Section II.

[405] Chappuis, P., *Trav. et Mém. Bur. Int. Poids et Mes.*, **13**, D1-D40 (1907).

[406] Marshall, D. H., Smith, C. M., and Ormond, R. T., *Proc. Roy. Soc. Edinburgh*, **11**, 809-813 (1882).

Table 134—(Continued)

In Section II, the values given for the salinity (s) have been estimated from the density by means of Table 109 except as the contrary is indicated, and are, together with Dt_m/s, only approximately correct. It will be noticed that the values (B) derived from the compilation by J. A. Beattie [407] do not entirely accord with the others; although they are said to have been based upon the observations by L, Ma, N, R, and W. It seems likely that the relation he used to connect ρ_0 with s differs from that used by the present compiler.

$\rho_0 \equiv 1 + 10^{-4}\Delta_0$ is the density at 0 °C; $\rho_{max} \equiv (1 + 10^{-4}\Delta_m)$ is the density at t_m; t_m °C = temperature of maximum density; $Dt_m \equiv (3.947 - t_m)$ in Section I, and $(3.98 - t_m)$ in Section II.

Unit of s = 1 g salt per kg sea-water; of ρ_0 and ρ_m = 1 g/cm³; of Dt_m/s = 0.1 °C per (g/kg)

I. O. Krümmel.[408]

s	Δ_m	t_m	Dt_m/s	s	Δ_m	t_m	Dt_m/s
0	0	3.947		20	160.7	−0.310	2.13
1	8.5	3.743	2.04	21	168.7	−0.529	2.13
2	16.9	3.546	2.00	22	176.7	−0.744	2.13
3	25.1	3.347	2.00	23	184.8	−0.964	2.14
4	33.3	3.133	2.04	24	192.9	−1.180	2.14
5	41.5	2.926	2.04	25	201.0	−1.398	2.14
6	49.6	2.713	2.06	26	209.1	−1.613	2.14
7	57.7	2.501	2.07	27	217.2	−1.831	2.14
8	65.8	2.292	2.07	28	225.3	−2.048	2.14
9	73.8	2.075	2.07	29	233.4	−2.262	2.14
10	81.8	1.860	2.09	30	241.5	−2.473	2.14
11	89.7	1.645	2.09	31	249.7	−2.687	2.14
12	97.6	1.426	2.10	32	257.8	−2.900	2.14
13	105.6	1.210	2.11	33	265.9	−3.109	2.14
14	113.5	0.994	2.11	34	274.0	−3.318	2.14
15	121.3	0.772	2.12	35	282.2	−3.524	2.14
16	129.2	0.562	2.12	36	290.4	−3.733	2.13
17	136.9	0.342	2.12	37	298.6	−3.936	2.13
18	144.8	0.124	2.12	38	306.8	−4.138	2.13
19	152.7	−0.090	2.13	39	315.0	−4.340	2.13
20	160.7	−0.310	2.13	40	323.2	−4.541	2.12
21	168.7	−0.529	2.13	41	331.4	−4.738	2.12

II. Various observers.

s	Δ_0	t_m	Dt_m/s	Ref.[a]	s	Δ_0	t_m	Dt_m/s	Ref.[a]
9.1[b]		2.4	1.7	B	42	335	−4.6	2.0	L
18.1[b]		0.5	1.9	B	47	381	−5.3	2.0	L
27.1[b]		−1.3	1.9	B	35	281	−4.74	2.5	N
36.1[b]		−3.2	2.0	B	35	281	−3.90	2.3	R
8	71	+2.2	2.2	L	33	267	−3.21	2.2	R
17	139	−0.4	2.6	L	34	273	−3.67	2.2	D
20	158	−0.8	2.4	L	7.9[b]		+2.43	2.0	W
25	204	−1.2	2.1	L	17.7[b]		+0.45	2.0	W
33	262	−3.7	2.3	L	26	208	−1.57	2.1	M
36	293	−4.2	2.3	L	35	281	−3.88	2.3	M

[407] Beattie, J. A., *Int. Crit. Tables*, **3**, 108 (1928).
[408] Krümmel, O., "Handb. d. Ozeanog.," Vol. 1, 1907.

Table 134—*(Continued)*

a References:

B   Beattie, J. A.[407] (based on L, Ma, N, R, and W).
D   Despretz, C., *Compt. rend.*, **4**, 435-440 (1837); *Ann. d. Physik (Pogg.)*, **41**, 58-71 (1837); *Ann. de chim et phys. (2)*, **70**, 5-81 (1839).
L   Lenz, R., *Mém. Acad. Sci. Russie (7)*, **29**, No. 4 (1881) (observations by Reszow).
M   Makaroff, C. O., *J. Russ. Phys. Chem. Soc. (Chem.)*, **23 II**, 30-88 (1891).
N   v. Neumann, C., *Ann. d. Physik (Pogg.)*, **113**, 382 (1861) ← *Diss.*, München, 1861.
R   Rossetti, F., *Atti Reg. Ist. Veneto Sci., Let.*, ed *Arti (3)*, **13**, 1047-1093, 1419-1457 (1868) → *Ann. de. chim. et phys. (4)*, **17**, 370-384 (1869) → *Ann. d. Physik (Pogg.) Erg. Bd.*, **5**, 258-275 (1871).
W   Weber, L., *Jahresber. Comm. Wiss. Unters. Deuts. Meere in Kiel*, **4-6**, (1874-1876), 1-22 (1878) = *Diss.*, Kiel, 1877 → *Beibl. Ann. d. Phys. (Wied.)*, **2**, 696-699 (1878).

b Given in the citation.

---

37. Refractivity of Water

The data for the refraction of water given in Table 135 are believed to be the best of the kind now available. The reduction of the observations on which they are based was not completed until after the rest of this section had been written; and since they have been received, the remainder of the section has been only slightly revised, mainly by increasing the number of entries for the visible spectrum in Table 137, which initially included only the better values.

Relative to the discussion of a possible dependence of the properties of water upon its recent thermal history (see p. 170+), V. K. LaMer and M. L. Miller [409] have sought for a difference between the index of refraction of water newly boiled, rapidly chilled, and measured at once, and that of water treated in a similar manner but kept at room temperature for three days before measuring the index. No difference was found; the precision was ±3 in 10⁶. It should, however, be remembered that the index of refraction of water exhibits no anomaly at 4 °C,[410, 411] and that both B. C. Damien [411] and V. S. M. v.d. Willigen [412] have reported that the indices of different samples of water, nominally identical, may differ appreciably. The former stated that, like the latter, he had observed that apparently identical specimens of water in which no impurity could be found chemically may have different indices, although any one given sample always had the same index.

Intercomparison of data obtained by various observers is aided by comparing those of each with the same interpolation formula. C. Chéneveau [413] accepted formula (1), which is due to F. F. Martens,[414] as valid at 18 °C and for wave-lengths (λ) in the range $\lambda = 0.224$ to $1.256\,\mu$; the unit of λ in the formula is $1\,\mu$, and the index is with reference to air at atmospheric pressure and at the same temperature as the water.

$$n^2 = 1.76148 - 0.013414\lambda^2 + 0.0065438/(\lambda^2 - 0.0132526) \qquad (1)$$

[409] LaMer, V. K., and Miller, M. L., *Phys. Rev. (2)*, **43**, 207-208 (1933).
[410] Jamin, J., *Compt. rend.*, **43**, 1191-1194 (1856).
[411] Damien, B. C., *Ann. Sci. École Norm. Sup. (2)*, **10**, 233-304 (272-278) (1881) → *J. de Phys. (1)*, **10**, 198-202 (1881).
[412] v. d. Willigen, V. S. M., *Arch. Mus. Teyler*, **1**, 74-116, 161-200, 232-238 (1868).

For the same unit of λ and under the same conditions except that the temperature is 20 °C, J. Duclaux and P. Jeantet[415] have given formula (2).*

$$n^2 = 1.76253 - 0.0133998\lambda^2 + 0.00630957/(\lambda^2 - 0.0158800) \quad (2)$$

Neither of these formulas fits the observed values satisfactorily if $\lambda < 0.25\,\mu$; but the second fits the more closely, and its fit can be improved by adding to it the term 10^c where $c = 107.73(0.064156 - \lambda^2) - 5$. That formula (2), as so modified, is used as the norm with which to compare the observed values (Tables 137 and 138). The three formulas are compared in Table 139, where the values of $dn_c/d\lambda$ for each of a number of values of λ are also given. It will be noticed that the dispersion of water in the ultraviolet exceeds that of quartz.

Variation with the Temperature (Tables 140, 141, 142, and 143).

Each of the several measurements of the variation of the index of refraction with the temperature has usually been summarized by a formula. A number of these formulas for water are given and compared in Table 142. Except that of Ketteler, they are all algebraic expressions involving only integral powers of t. Ketteler's is this: $(n^2 - 1)\cdot(v^* - \beta) = C(1 + \alpha e^{-kt})$, where v^* is the specific volume of the water, C is the value of $(n^2 - 1)v^*$ when the substance is in the gas phase and greatly expanded, and α, β, and k are constants fixed by the substance alone; C varies with λ, determining the dispersion.

It will be noticed that Ketteler's formula can be put in the form $(n^2 - 1) = f(\lambda)\cdot F(t)$, where f and F are each a function of a single variable, λ or t. Whence, for a given λ, $(n_1^2 - 1)/(n_2^2 - 1) = F(t_1)/F(t_2)$ and $(n_1^2 - n_2^2)/(n_1^2 - 1) = 1 - F(t_2)/F(t_1)$ are each independent of λ. Likewise, for a given t, similar ratios are independent of t. Flatow's extended series of observations on water ($\lambda = 2145$ to 5893A, $t = 0$ to 80 °C) does not satisfy these conditions.

Although the temperature coefficient of n is a function of λ, its variation with λ is not very rapid (see Table 141).

There is no evidence of an anomaly at 4 °C; the index continues to increase as the temperature is reduced below that temperature, and B. C. Damien,[411] whose observations extended to -8 °C, found no maximum. Others have, however, found that the index does pass through a maximum at a temperature (t_{mr}) not far from 0 °C, but the several observers do not agree regarding that temperature. J. Jamin[416] concluded that t_{mr} was

---
* A typographic error occurs in the paper cited, the value of the numerator of the third term being given as antilog $\overline{3}.00800$, whereas the computed values for n show that it should have been antilog $\overline{3}.80000$.

[413] Chéneveau, C., *Int. Crit. Tables*, **7**, 13 (1930); *Recueil de const. phys. (Soc. fr. de Phys.)*, Paris, 1913.
[414] Martens, F. F., *Ann. d. Physik (4)*, **6**, 603-640 (1901).
[415] Duclaux, J., and Jeantet, P., *J. de Phys. (6)*, **5**, 92-94 (1924).

near 0 °C; L. Lorenz [417] that it was +0.01 °C for the D-line ($\lambda = 5893$A) and +0.17° for Li ($\lambda = 6708$A); C. Pulfrich [418] placed it between -1 and -2 °C, his observations extending to -10 °C; E. Ketteler [419] placed it at -1.5 °C; and N. Gregg-Wilson and R. Wright [420] at -0.5 °C. L. W.

Table 135.—Index of Refraction of Water in the Visible Spectrum: Preferred Values

(L. W. Tilton and J. K. Taylor. Numerical data privately communicated by Tilton prior to the publication of the detailed account of their work,[429] which contains extensive tables covering the ranges 0 to 60 °C and 4000 to 7250A.)

The index is with respect to dry air at a pressure of 760 mm-Hg and at the same temperature as the water. Under δ are given the values of $10^5(n_c - n)$, n_c being the value defined by the expression given in the head-matter of Table 137, and n being the index here given for 20 °C.

Unit of $\lambda = 1\ \mu = 10^{-4}$ cm $= 10^4$ A. Temp. $= t$ °C

I. Index with respect to dry air at the same temperature (t °C) and a pressure of 760 mm-Hg.

$t \rightarrow$ λ	10	20	30	40	20 δ	15	25 $-(10^6)dn/dt$	35
0.70652	1.330704	1.330019	1.328993	1.327685	-2.0	68.8	102.9	130.9
0.66781	1567	0876	9843	8528	-0.2	69.4	103.6	131.7
0.65628	1843	1151	1.330116	8798	$+0.2$	69.6	103.8	131.9
0.58926	3690	2988	1940	1.330610	$+2.2$	70.7	105.0	133.2
0.58756	3744	3041	1993	0662	$+2.4$	70.7	105.0	133.2
0.57696	4085	3380	2331	0998	$+2.6$	70.9	105.2	133.5
0.54607	5176	4466	3411	2071	$+3.3$	71.4	105.8	134.1
0.50157	7070	6353	5289	3939		72.1	106.7	135.2
0.48613	7842	7123	6055	4702	$+3.3$	72.3	107.0	135.4
0.47131	8653	7931	6860	5504		72.6	107.3	135.8
0.44715	1.340149	9423	8347	6984	$+2.9$	73.0	107.8	136.5
0.43583	0938	1.340210	9131	7765	$+2.6$	73.2	108.1	136.8
0.40466	3476	2742	1.341656	1.340280	$+1.9$	73.8	108.9	137.8

II. Index for the D-lines of Na (Hartmann's mean $\lambda = 0.58926\,\mu$), with respect to dry air at the same temperature (t °C) and a pressure of 760 mm-Hg.

t	n	t	n	t	n	t	n	t	n	t	n
10	1.333690	15	1.333387	20	1.332988	25	1.332503	30	1.331940	35	1.331308
11	638	16	315	21	897	26	396	31	819	36	173
12	582	17	238	22	803	27	287	32	695	37	036
13	521	18	158	23	706	28	174	33	569	38	1.330896
14	456	19	075	24	606	29	059	34	440	39	754

Tilton and J. K. Taylor [421] find for the index with respect to air at the same temperature $t_{mr} = +0.19$ °C for the D-line and $+0.33$ °C for

---

[416] Jamin, J., *Compt. rend.*, **43**, 1191-1194 (1856).
[417] Lorenz, L., *Ann. d. Physik (Wied.)*, **11**, 70-103 (1880).

He($\lambda = 6678$A), and they compute for the absolute index at these wavelengths the values $t_{mr} = -0.05$ and $+0.09$ °C, respectively.

Both Jamin and Damien have reported that the act of freezing is preceded by an anticipatory decrease in the index. Damien [411] has described

Table 136.—Reduction of the Index of Refraction of Water from Air to Vacuum

If n_{air} is the index of refraction of water with reference to air at the same temperature and a pressure of 760 mm-Hg, then the index with reference to a vacuum is $n_{vac} = n_{air} + \Delta$, Δ depending upon the wave-length (λ) and the temperature. The following values of Δ are based upon the values found by W. F. Meggers and C. G. Peters [430] for the index of refraction of air; for air at a fixed pressure, $(n-1)$ is inversely proportional to the absolute temperature.

Unit of $\lambda = 1\mu = 10^4$ A $= 10^{-4}$ cm. Temp. $= t$ °C

I. Temperature = 20 °C.

λ	$10^5 \Delta$	λ	$10^5 \Delta$	λ	$10^5 \Delta$	λ	$10^5 \Delta$
0.20	45.5	0.24	41.4	0.35	37.8	0.55	36.3
0.21	44.1	0.25	40.8	0.40	37.2	0.60	36.2
0.22	43.0	0.27	39.8	0.45	36.8	0.70	36.0
0.23	42.1	0.30	38.8	0.50	36.5	0.90	35.7

II. The D-lines; $\lambda = 0.5893$.

t	$10^5 \Delta$	t	$10^5 \Delta$	t	$10^5 \Delta$	t	$10^5 \Delta$
−10	40.4	10	37.5	30	35.0	70	30.8
− 5	39.7	15	36.9	40	33.9	80	29.8
0	38.9	20	36.2	50	32.8	90	29.0
+ 5	38.2	25	35.6	60	31.7	100	28.1

the phenomenon thus: "J'ai eu bien souvent l'occasion ... d'observer une brusque diminution de l'indice sans cause apparente. L'image d'une raie étant superposée au réticule, on voyait tout à coup cette image se déplacer lentement et graduellement. Un instant après seulement, des aiguilles de glace se formaient dans le prisme. Comme le fait remarquer M. Jamin: La congélation se prépare pour ainsi dire à l'avance au moment où elle va s'opérer."

[418] Pulfrich, C., *Idem*, **34**, 326-340 (1888).
[419] Ketteler, E., *Idem*, **33**, 353-381, 506-534 (1888).
[420] Gregg-Wilson, N., and Wright, R., *J. Phys'l Chem.*, **35**, 3011-3014 (1931).
[421] Tilton, L. W., and Taylor, J. K., *private communication*, 1935.
[422] Bramley, A., *Phys. Rev. (2)*, **33**, 279, 640 (1929); *J. Opt. Soc. Amer.*, **21**, 148 (1931).
[423] Chéneveau, C., *Ann. de chim. et phys. (8)*, **12**, 145-228, 289-293 (1907).
[424] Tilton, L. W., *Bur. Stand. J. Res.*, **2**, 909-930 (RP64) (1929); **6**, 59-76 (RP262) (1931); **11**, 25-58 (RP575) (1933); **13**, 111-124 (RP695) (1934); **14**, 393-418 (RP776) (1935).
[425] Dufet, H., "Recueil de données numériques Optique." Publ. by Soc. Fr. de Physique, Gauthier-Villars, Paris, 1898.
[426] Korff, S. A., and Breit, G., *Rev. Mod. Phys.*, **4**, 471-503 (1932).
[427] Tilton, L. W., *J. Res. Nat. Bur. Stand.*, **17**, 639-650 (RP934) (1936).
[428] Tilton, L. W., and Taylor, J. K., *Idem*, **18**, 205-214 (RP971) (1936).
[429] Tilton, L. W., and Taylor, J. K., *J. Res. Nat. Bur. Stand.*, **20**, 419-477 (RP1085) (1938).
[430] Meggers, W. F., and Peters, C. G., *Bull. Bur. Stand.*, **14**, 697-740 (S327) (1918).

37. WATER: REFRACTIVITY

Effect of Electric Field. (For Kerr effect, see Section 51.)

When an electric field of high frequency is applied to water in a direction perpendicular to that of the propagation of light through it, certain effects are observed which were initially interpreted as indicating that water has a set of absorption bands (and consequently, exhibits anomalous dispersion) for waves 3 to 10 meters long. Later observations showed that such an explanation is incorrect.[422]

Table 137.—Various Values of the Refraction of Water at 20 °C

(For the preferred values in the Visible Spectrum, see Table 135; for reduction to vacuum, see Table 136; for a comparison of certain sets of values, see Table 138; for $\lambda > 5000 \mu$, see Table 172.)

These indices (n) are with reference to air at a pressure of approximately one atmosphere and at the same temperature as the water, usually 20 °C, but those quoted from Rubens (Rub) are for 12 °C, and those from Rubens and Ladenburg (RL) are for 18 °C. In these exceptional cases the precision of measurement is not great enough to justify a correction for so small an interval as 8 °C. If several sources are cited for the same value, that value is the mean of those from the several sources, and only rarely does any one of the individual values differ from that mean by so much as 2 in the fifth decimal place. If the group of references includes ICT, then the tabulated mean is exactly that given by Chéneveau in the *International Critical Tables*. For certain frequently studied wave-lengths, several values of n are given, the first being regarded as superior to the others. No distinction has been made between the values of n that have been determined absolutely and those that have been derived from such absolute values by means of relative measurements, either by the same or by another observer.

The somewhat arbitrarily chosen norm with which the several values are compared is

$$(n_c)^2 = 1.762530 - 0.0133998\, \lambda^2 + \frac{0.00630957}{\lambda^2 - 0.0158800} + 10^{[107.731(0.064156 - \lambda^2) - 5]}$$

the unit of λ is $1\,\mu$ (see text). The values of λ used in computing n_c are those here given, those given to 5 significant figures having been taken from the list of wave-lengths given by H. Kayser.[431]

It will be noticed that the ICT values agree with the DJ ones until $\lambda = 0.214\,\mu$ is reached, where a sudden break of 40 in the fifth decimal place occurs; with increasing values of λ, the discrepancy decreases until at $\lambda = 0.397\,\mu$ the two series again coincide, and thereafter continue to coincide to the end of the DJ series.

---

[431] Kayser, H., *Int. Crit. Tables*, **5**, 276-322 (1929).

Table 137—(Continued)

Unit of $\lambda = 1\,\mu = 10^{-4}$ cm $= 10^4$ A

λ	n	$10^6(n_c-n)$	Ref.[a]	λ	n	$10^6(n_c-n)$	Ref.[a]
Cu[b]	0.000154	0.99999652		Th	Fe	836	1.33989	+16.5	Du
Cu	0.000154	0.9999964		St	Cd	0.44157	981	+ 2.0	ICT, Fl
Cu	0.000154	0.9999963		Sm	He	715	945	+ 0.2	Ro
	0.1151	Met[c]		ICT, DJ	Cd	0.46782	815	+ 1.4	ICT, Fl
Ag	0.1829	1.46379	−41.9	ICT, DJ	Cd	782	817	− 0.6	Si
Ag	0.1832	264	−18.2	DJ	Cd	0.47999	750	− 1.9	ICT, Fl
Ag	0.1834	199	−19.8	DJ	Cd	999	753	− 4.9	Si
Ag	0.1835	141	+ 7.5	DJ	Hβ	0.48613	714	+ 1.6	ICT, OL, Sch, Si
Ag	0.1838	1.46060	− 3.6	DJ					
Ag	0.1839	013	+12.5	DJ	Hβ	613	715	+ 0.6	Wil, Wü
Ag	0.1849	1.45715	+11.8	DJ	Hβ	0.48613	1.33719	− 3.4	Br
Ag	0.1853	595	+16.3	DJ	Hβ	613	738	−22.4	Ka
Al	0.18547	528	+34.3	DJ	Hβ	613	710	+ 5.6	La
Al	0.18582	477	−13.9	DJ	Hβ	613	704	+11.6	Be, Da, Du
Al	627	343	− 5.0	ICT, DJ					
Al	0.19352	1.43595	+ 4.4	DJ	N	0.50032	645	− 0.3	DJ
Al	898	1.42572	− 5.9	ICT, DJ	Cd	858	609	− 6.8	Si
Zn	0.20255	1.41993	− 7.0	DJ	N	0.51795	567	− 3.0	DJ
Zn	619	1459	− 2.1	DJ	Mg	0.51836	549	+13.2	Du
Zn	0.21385	1.40500	+ 7.9	DJ	Cd	0.53380	499	− 1.3	ICT, Fl
Cd	444	437	+ 5.6	DJ	Tl	505	490	+ 2.6	ICT
Cd	444	397	+45.9	ICT, Fl	Tl	505	492	+ 0.6	Br, Sch
Al	740	128	+ 0.8	DJ	Tl	505	498	− 5.4	Si
Cd	946	1.39883	+40.4	ICT, Fl	Tl	505	485	+ 7.6	Rü
Al	0.22100	775	+ 1.5	DJ	Tl	505	481	+11.6	Du, Ket, Wm
Al	636	305	+ 0.4	DJ					
Cd	650	257	+34.2	ICT, Fl	Hg	0.54607	447	+ 2.9	ICT
Al	0.22691	1.39258	− 0.5	DJ	Hg	607	448₅	+ 1.4	Ja
Cd	0.23129	1.38878	+35.6	ICT, Fl	Hg	607	440	+ 9.9	Ro
Al	671	533	− 5.2	DJ	N	0.56795	370	+ 1.0	DJ
Hg	783	434	+19.4	Ro	Hg	0.57696	342	− 1.4	ICT
Au	0.24280	103	+37.1	ICT, Fl	Hg	696	340₂	+ 0.4	Ja
Hg	827	1.37809	+17.5	Ro	Hg	907	333	+ 0.8	ICT
Zn	0.25020	1.37734	−11.2	DJ	Hg	907	333₅	+ 0.3	Ja
Al	680	406	−13.2	DJ	He	0.58756	305	+ 1.5	Ro
Cd	730	349	+20.0	ICT, Fl	Na	0.58929	300	+ 1.0	ICT, Fl, Sch
Al	753	372	−12.6	DJ					
Hg	0.2576	338	+17.3	Ro	Na	0.58929	1.33299	+ 2.0	BBD, HP, Ve, Wa
Al	0.26317	119	−12.8	DJ	Na	929	301	0.0	DJ, Lo
Al	525	031	−12.4	DJ	Na	929	303	− 2.0	Br, Gi, OL, Ruo, RZ, Wil
Al	604	1.36998	−12.0	DJ					
Au	760	904	+19.2	ICT, Fl					
Cd	0.27486	637	+11.0	ICT, Fl	Na	929	310	− 9.0	Ka
Hg	0.28035	442	+16.1	Ro	Na	929	308	− 7.0	Si
Al	0.28163	428	−12.0	DJ	Na	929	293	+ 8.0	Du, Ket, Rü, Wm
Hg	0.28936	168	+ 8.5	Ro					
Al	0.30822	1.35671	+12.7	ICT, Fl	Na	929	286	+15.0	Be
Al	822	694	−10.4	DJ	Na	929	280	+21.0	La
Al	927	668	− 8.8	DJ	Hα	0.65628	115	+ 0.3	ICT, OL, Ro, Sch
Hg	0.31317	567	+ 5.1	Ro					
Hg	0.33415	165	− 1.6	Ro	Hα	628	130	−14.7	Ka
Cd	0.34036	044	+16.0	ICT, Fl	Hα	628	119	− 3.7	Br, Wil, Wü
Al	0.35871	1.34795	− 8.3	DJ					
Al	0.36016	774	− 6.9	DJ	Hα	628	109	+ 6.3	Da, Du, La, Si
Cd	117	738	+15.8	ICT, Fl					
Al	124	760	− 7.1	DJ	Hα	628	100	+15.3	Be
He	0.38886	432	− 2.6	Ro	He	0.66782	087	+ 0.4	Ro
Al	0.39440	366	+ 7.3	ICT, Fl	Li	0.67079	079	+ 1.3	ICT
Al	440	378	− 4.8	DJ	Li	079	082	− 1.7	Sch, Wm
Al	615	360	− 3.8	DJ	Li	079	087	− 6.7	Br
Ca	0.39685	1.34325	− 2.6	ICT, Wil	Li	079	076	+ 4.3	Ket, Lo, Rü
Ca	685	352	− 3.6	DJ					
N	950	328	− 4.1	DJ	Li	079	073	+ 7.3	Du
Hg	0.40466	284	− 7.9	Ro	He	0.70652	003	− 3.1	Ro
Hδ	0.41017	228	− 1.0	Wil	K	0.76820	1.32888	−11.7	ICT
Hδ	0.41017	208	+19.0	Du	K	820	884	− 7.7	Br, Sch
Fe	0.43258	029	+18.3	Du	K	820	897	−20.7	Si
Hγ	405	035	+ 1.6	ICT, Da		0.808	815	−11.1	Rub, Se
Hγ	405	038	− 1.4	La, Sch		0.871	1.3270	− 2.6	ICT, Rub
Hγ	405	045	− 8.4	Br, Si		0.871	68	+17.3	Se
Hγ	405	024	+12.6	Be		0.943	58	+ 3.5	ICT, Rub, Se
Hγ	405	015	+21.6	Du					
Hg	583	030	− 6.4	ICT, OL		1.000	1.323	+197.	RL
Hg	583	027	− 3.4	Ro		1.028	1.3245	+ 4.9	ICT, Rub, Se
Hg	583	022₈	+ 0.8	Ja					

Table 137—(Continued)

λ	n	$10^5(n_c-n)$	Ref.[a]	λ	n	$10^3(n-1.328)$	Ref.[a]
1.130	1.3230	+ 3.6	ICT, Rub, Se	12.0	1.187	−141	RL
				13.0	1.269	−59	RL
1.256	1.3210	+14.8	ICT, Rub, Se	15.0	1.332	+ 4	RL
				18.0	1.505	+177	RL
1.5	1.316	+214.	RL	25.5 to 26.	1.41[d]	+ 8$_2$	ICT, RH
1.617	1.3149	+36.	Se	46.9 to 53.6	1.36[d]	+ 3$_2$	ICT, RH
1.968	1.3078	+93.	Se	75.6 to 86.5	1.41[d]	+ 8$_2$	ICT, RH
2.0	1.300	+787.	RL	52	1.68[d]	+35$_2$	CE
2.327	1.2997	+74.	Se	63	1.77[d]	+44$_2$	CE
2.4	1.275	+2363.	RL	83	1.89[d]	+56$_2$	CE
2.6	1.253	+4040.	RL	100	2.01[d]	+68$_2$	CE
2.8	1.282	+574.	RL	117	2.04[d]	+71$_2$	CE
3.0	1.365	−8335.	RL	152	2.09[d]	+76$_2$	CE

λ	n	$10^3(n-1.328)$	Ref.[a]	λ	n	(n−1.33)	Ref.[a]
3.0	1.365	+37	RL	4000.	9.50[e]	8.17	Lam
3.2	1.456	+128	RL	4200.	5.33	4.00	ICT, T
3.4	1.437	+109	RL	6000.	9.40[e]	8.07	Lam
3.6	1.384	+56	RL	8000.	8.97[e]	7.64	Lam
3.8	53	+25	RL	8400.	5.68	4.35	T
4.0	38	+10	RL	11000	6.27	4.94	T
4.5	43	+15	RL	15000	6.62	5.29	T
5.0	30	+ 2	RL	18000	6.65	5.32	T
5.5	1.300	−28	RL	27000	8.45[f]	7.12	ICT, T
5.8	1.271	−57	RL				
6.0	1.324	− 4	RL	$10-6\lambda$	n	(n−1.33)	Ref.[a]
6.2	60	+32	RL	0.027	8.45[f]	7.12	ICT, T
6.5	34	+ 6	RL	0.12 to 0.19	9.0	7.67	Sr
7.0	27	− 1	RL	0.375	9.08	7.75	Dr
8.0	1.293	−35	RL	0.5 to 0.6	Normal[g]		
9.0	64	−64	RL	0.75	8.98	7.65	Dr
10.0	1.196	−32	RL	2.0	8.92	7.59	Dr
11.0	50	−178	RL	3.3 to 7.0	9.0	7.7	McCJ

[a] References:

BBD   Baxter. G. P., Burgess, L. L., and Daudt, H. W., *J. Am. Chem. Soc.*, **33**, 893-901 (1911).
Be   Bender, C., *Ann. d. Physik (Wied.)*, **39**, 89-96 (1890).
Br   Brühl, J. W., *Ber. d. D. Chem. Ges.*, **24**, 644-649 (1891).
CE   Cartwright, C. H., and Errera, J., *Proc. Roy. Soc. (London) (A)*, **154**, 138-157 (1936); *Acta Physicochim. URSS*, **3**, 649-684 (1935) and Cartwright, C. H., *Nature*, **135**, 872 (L) (1935).
Da   Damien, B. C.[411]
Dr   Drude, P., *Ann. d. Physik (Wied.)*, **59**, 17-62 (1896).
DJ   Duclaux, J., and Jeantet, P., *Jour. de Phys.* (6), **2**, 346-350 (1921); **5**, 92-94 (1924).
Du   Dufet, H., *Idem.* (2), **4**, 389-419 (1885).
Fl   Flatow, E., *Ann. d. Physik* (4), **12**, 85-106 (1903) ← *Diss.*, Berlin.
Gi   Gifford, J. W., *Proc. Roy. Soc. (London) (A)*, **78**, 406-409 (1907).
HP   Hall, E. E., and Payne, A. R., *Phys. Rev.* (2), **20**, 249-258 (1922); **18**, 326-327 (1922).
ICT   Compilation by Chéneveau, C., *Int. Crit. Tables*, **7**, 12-16 (J3) (1930), based on observations of DJ, λ = 0.115 to 0.199; of Fl, Ma, and Si, λ = 0.214 to 0.361; of many observers, λ = 0.394 to 0.768; of Rub, λ = 0.871 to 1.256; of RH, λ = 25.5 to 86.5; and of T, λ = 4000 and 27000.
Ja   Jasse, O., *Compt. rend.*, **198**, 163-164 (1934).
Ka   Kanonnikoff, J., *J. prakt. Chem. (N.F.)*, **31**, 321-363 (1885).
Ket   Ketteler, E., *Ann. d. Physik (Wied.)*, **33**, 353-381, 506-534 (1888).
La   Landolt, H., *Idem (Pogg.)*, **117**, 353-385 (1862).
Lam   Lampa, A., *Sitz. K. Akad. Wiss. Wien, (Abt. IIa)*, **105**, 587-600, 1049-1058 (1896).
Lo   Lorenz, L., *Ann. d. Physik (Wied.)*, **11**, 70-103 (1880).
Ma   Martens, F. F., *Idem* (4), **6**, 603-640 (1901).
McCJ McCarty, L. E., and Jones, L. T., *Phys. Rev.* (2), **29**, 880-886 (1927).
OL   Osborn. F. A., and Lester, H. H., *Idem*, **35**, 210-216 (1912).
RH   Rubens, H., and Hollnagel, H., *Verh. physik. Ges.*, **12**, 83-98 (1910).
RL   Rubens, H., and Ladenburg, E., *Idem*, **11**, 16-27 (1909).
Ro   Roberts, R. W., *Phil. Mag.* (7), **9**, 361-390 (1930).
Rub   Rubens, H., *Ann. d. Physik (Wied.)*, **45**, 238-261 (1892).
Rü   Rühlman, R., *Idem (Pogg.)*, **132**, 1-29, 177-203 (1867).
Ruo   Ruoss, H., *Idem (Wied.)*, **48**, 531-535 (1893).
RZ   Röntgen, W. C., and Zehnder, L., *Idem*, **44**, 24-51 (1891).
Sch   Schütt, F., *Z. physik. Chem.*, **5**, 348-373 (1890).

Table 137—*(Continued)*

Se Seegert, *Diss.*, Berlin, 1908.
Si Simon, H. T., *Ann. d. Physik (Wied.)*, **53**, 542-558 (1894) ← *Diss.*, Berlin, 1894.
Sm Smith, S. W., *Phys. Rev.* (2), **40**, 156-164 (1932).
Sr Seeberger, M., *Ann. d. Physik* (5), **16**, 77-99 (1933).
St Steps, H., *Idem*, **16**, 949-972 (1933) ← *Diss.*, Jena, 1932.
T Tear, J. D., *Phys. Rev.* (2), **21**, 611-622 (1923) → *Abstr. Bull. Nela Res. Lab.*, **1**, 623-629 (1925).
Th Thovert, J. F., *Jour. de Phys.* (7), **2**, 55S (Bull. 305) (1931).
Ve Verschaffelt, J., *Bull. Acad. Roy. Sci., Let, Beaux-Arts Belg.* (3), **27**, 69-84 (1894).
Wa Walter, B., *Ann. d. Physik (Wied.)*, **46**, 423-425 (1892).
Wil v. d. Willigen, V. S. M., *Arch. de Musée Teyler (Haarlem)*, **1**, 74-116, 161-200, 232-238 (1868); **2**, 199-217, 308-316 (1869).
Wm Wiedemann, E., *Ann. d. Physik (Pogg.)*, **158**, 375-386 (1876).
Wü Wüllner, A., *Idem*, **133**, 1-53 (1868). •

[b] The $K\alpha$ line of characteristic x-radiation from Cu.

[c] At $\lambda = 0.1151\,\mu$ the reflection is "metallic."

[d] For $\lambda = 25.5$ to 152, the indicated values of n refer to the "residual" rays left after multiple reflection from the following solids: CaF_2, $\lambda = 25.5$ to 26; NaCl $\lambda=46.9$ to 53.6; KBr, $\lambda=75.6$ to 86.5; KCl, $\lambda=63$; TlCl, $\lambda=100$; TlBr, $\lambda=117$, and TlI, $\lambda=152\mu$. W. Weniger [432] also, has found that in the interval $\lambda = 50\,\mu$ to $300\,\mu$ the index is of the same order of magnitude as in the visible spectrum.

[e] It will be noticed that Lampa's values are much greater than those of T and ICT at $\lambda = 4200\,\mu$ to $27000\,\mu$. Lampa used the deviation by a prism, whereas Tear (T) derived n from the reflectivity and the extinction coefficient. Lampa concluded that the dispersion is normal in the range $\lambda = 8$ mm $(=8000\,\mu)$ to 1.2 m $(=12000000\,\mu)$, but there is much absorption near $\lambda = 8000\,\mu$, and hence, presumably, anomalous dispersion.

[f] For $\lambda = 27000\,\mu$ $(=0.0270 \times 10^6\,\mu)$, ICT gives $n = 9.0$.

[g] In the range $\lambda = 50$ to 60 cm $(= 0.5$ to $0.6\ 10^6\mu)$ there is no anomalous dispersion if the water is pure, that reported by R. Weichmann [433] being due to impurities, perhaps to dissolved glass.[434]

Table 138.—Refraction of Water at 20 °C: Comparison of Data

For the most accurate data, see Table 135; for other values, see Table 137 in which is given the formula by which the arbitrary norm (n_c) was computed. The preferred values are those in column TT.

Unit of $\lambda = 1\,\mu = 10^4$ A $= 10^{-4}$ cm. Index is with reference to air at 20 °C and 1 atm

Ref [a]→ λ	n_c	TT	Be	Br	Da	Du	Fl	Ka	Ket	La	Lo	OL	Rü	Sch	Si	Wa	Wil	Wm	Wü
								$10^5(n-n_c)$											
0.39440	1.34373						−7												
0.39685	1.34349																+3		
0.43405	1.34037		−13	+7	−2	−22			+1			+1	+10						−1
0.44157	1.33983						−2												
0.48613	1.33716	−4	−11	+3	−11	−15		+22		−4		−2		−1	−2		0		+1
0.53380	1.33498						・+1												
0.53505	1.33493			−1		−11			−12			−8	−2	+5			−14		
0.58929	1.33301	−2	−15	+3		− 9	−1	+9	−10		0	+1	−7	−1	+7	−2	+1	−9	
0.65628	1.33115	0	−15	+4	−7	− 6		+15		−4		−1		+1	−5		+2		−6
0.67079	1.33080			+7		− 7			−6		−2		−4	+2			+2		
0.76820	1.32876			+11										+6	+21				

[a] References:
TT Tilton, L. W., and Taylor, J. K., see Table 135; other symbols as in Table 137.

---

[432] Weniger, W., *J. Opt. Soc. Amer.*, **7**, 517-527 (1923).
[433] Weichmann, R., *Ann. d. Physik (4)*, **66**, 501-545 (1921).
[434] Mie, G., *Physik. Z.*, **27**, 792-795 (1926).

37. WATER: REFRACTIVITY

Other References.

For a discussion of procedures, instruments, and sources of error, see C. Chéneveau,[423] and L. W. Tilton [424]; for a compilation of data prior to 1898, see H. Dufet [425]; for a recent review of optical dispersion, see S. A. Korff and G. Breit [426]; and for a discussion of the accurate representation of the index of refraction as a function of the wave-length and temperature, see L. W. Tilton [427] and L. W. Tilton and J. K. Taylor.[428]

Table 139.—Dispersion Formulas for Water

$$(n_c)^2 = 1.762530 - 0.0133998\,\lambda^2 + \frac{0.00630957}{\lambda^2 - 0.0158800} + 10^{[107.731(0.064156 - \lambda^2) - 5]}$$

$$(n_c')^2 = 1.762530 - 0.0133998\,\lambda^2 + \frac{0.00630957}{\lambda^2 - 0.0158800}$$

$$(n_c'')^2 = 1.761480 - 0.0134140\,\lambda^2 + \frac{0.00654380}{\lambda^2 - 0.0132526}$$

n_c is used in Table 137; n_c' was proposed by Duclaux and Jeantet (see ftn., p. 280); and n_c'' by Martens and accepted by Chéneveau. n_c' is for 20 °C and n_c'' is for 18 °C. The index is with reference to air at the same temperature, at or near 20 °C; unit of $\lambda = 1\,\mu$.

E. Flatow [435] represented his observations (2145A to 5893A) by means of 5 formulas of the type $n^2 = m - k\,\lambda^2 + m'\,\lambda^2/(\lambda^2 - \lambda_1^2)$, one for each temperature used. They may be put in the form $n^2 = a - b\,\lambda^2 + c/(\lambda^2 - \lambda_1^2)$, the values of the constants, referred to a vacuum, being these:

t	a	b	c	λ_1^2
0	1.76565	0.013414	0.00633201	0.0159088
20	1.76362	0.013414	0.00626020	0.0161138
40	1.75758	0.013414	0.00619982	0.0162410
60	1.74840	0.013414	0.00613429	0.0163328
80	1.73755	0.013414	0.00603399	0.0165482

The unit of λ and of λ_1 is $1\,\mu$. The value for λ_1 at 20° as published in the original article and reproduced in compilations is $\lambda_1 = 0.12604\,\mu$. That is obviously out of line with the others, and fails to reproduce the observations; it leads to $c = 0.00617174$ and $\lambda_1^2 = 0.158861$. It seems probable that the zero is a typographical error, and that the value should have been printed $\lambda_1 = 0.12694$. That leads to the values of c and λ_1^2 tabulated here, and fits the observations. Flatow's formulas are not considered further in this table. His values at 20 °C, referred to air, are in Table 137, where they are compared with n_c; the mean temperature coefficients derived from them are in Table 141.

Very exact formulas based on their own observations have been given by L. W. Tilton and J. K. Taylor.[429]

[435] Flatow, E., *Ann. d. Physik (4)*, **12**, 85-106 (1903).

Table 139—(Continued)

n_0 = observed value; values of $(n_0 - n_c)$ have been obtained, by interpolation, from Table 137; for $\lambda > 1.25$, $(n_0 - n_c)$ is great.

Unit of $\lambda = 1\,\mu$, $= 10^4$ A; of $dn_c/d\lambda = 10^{-6}$ per A

λ	n_c	$10^6(n'_c - n_c)$	$10^6(n''_c - n_c)$	$10^6(n_0 - n_c)$	$-dn_c/d\lambda$
(0.18)	(1.473230)	(−8979)	(−23121)		(362.6)
0.19	1.443823	−3652	−12943	+3	237.6
0.20	1.423934	−1406	−7740	+64	166.5
0.22	1.398714	−174	−3357	−15	96.0
0.25	1.377333	−5	−1278	+112	53.0
0.30	1.358842	0	−333	−4	25.7
0.40	1.343192	0	−85	+41	9.4₄
0.50	1.336462	0	−132	+3	4.8₁
0.60	1.332683	0	−192	−6	3.0₀
0.70	1.330140	0	−236	+90	2.1₀
0.80	1.328181	0	−271	+111	1.7₈
1.00	1.324969	0	−315	−42	1.5₀
1.25	1.321239	0	−348	−148	1.5₂
1.50	1.317272	0	−369		1.6₇
2.00	1.307866	0	−397		2.1₀
2.50	1.296069	0	−421		2.6₂
3.00	1.281653	0	−446		3.1₅

Table 140.—Refraction of Water at Various Temperatures
(See also Table 135.)

The values under HP and J have been derived from the corresponding formulas (see Table 142), the values beyond the range for which a formula is claimed to be valid being inclosed in parentheses. The others have been derived from the published experimental data. All the indices are with reference to air at the same temperature as the water.

Damien (Da) was positive that n continues to decrease as t decreases below zero; Jamin (J) stated that n is a maximum near zero, and with this the HP formula, resting on observations above 15 °C, agrees; see also text.

E. v. Aubel[436] has computed, on the basis of certain assumptions, that the index of refraction of water at the critical point is $n = 1.102$, probably for $\lambda = 6708$A.

It will be noticed that the values of δ for the Ja observations exhibit surprising jumps at many of the places where the value of t was changed abruptly by several degrees.

n_0 = value of n at $t = 0$; $n_0 - n = (n_0 - n)_{HP} + \delta$. For example, at −10 °C P's value for $10^5(n_0 - n)$ is 27, J's is 7; that is, P finds $n_{-10} = 1.33411 - 0.00027 = 1.33384$, and J finds $n_{-10} = n_0 - 0.00007$; J does not assign a value to n_0.

[436] v. Aubel, E., *Physik. Z.*, 14, 302-303 (1913).

37. WATER: REFRACTIVITY

Table 140—*(Continued)*

Unit of $\lambda = 1\text{A} = 10^{-8}$ cm. Temp. $= t\,°C$

Ref [a]→ λ→ n_0→ t	HP 5893 D 1.33401 $10^5(n_0-n)_{HP}$	P 5893 D 1.33411	Da 6563 Hα 1.33225	Da 4862 Hβ 1.33825 $10^6\delta$	Da 4341 Hγ 1.34155	J 5893 D
−10	(22)	+5				−15
− 8	(12)	+4	−20	−19	−19	−10
− 6	(6)	+1	−13	−12	−12	− 7
− 5	(4)	0				− 6
− 4	(2)	0	− 7	− 6	− 7	− 4
− 2	(0)	−1	− 3	− 2	− 3	− 2
+ 2	(0)	+1	+ 2	+ 3	+ 2	+ 3
4	(7)	0	0	0	0	+ 1
5	(10)	+1				+ 1
6	(13)	+2	0	+ 1	+ 1	+ 1
8	(21)	+1	+ 1	+11	+ 2	+ 2
10	(31)	0	+ 2	+ 5	+ 4	+ 1
15	63		+ 2	+11	+ 4	− 1
20	105		+12	+15	+15	− 3

Ref [a]→ λ→ n_0→ t	HP 5863 D 1.33401 $10^5(n_0-n)_{HP}$	Ja 5791 Hg 1.334277	Ja 5770 Hg 1.334348 $10^6\delta$	Ja 5461 Hg 1.335443	Ja 4358 Hg 1.341218
0.03	0.2			+ 1.8	
3.85	63.	− 23	− 23	− 19	− 16
5.71	120			− 29	− 22
5.76	122	− 32	− 30	− 24	− 20
6.55	151	− 37	− 35	− 32	− 23
6.63	154			− 29	− 21
7.88	206			− 32	− 23
8.09	216				− 29
8.52	235	− 37	− 35	− 33	− 23
8.85	252	− 42	− 44	− 39	− 27
9.15	266	− 37	− 36	− 34	− 24
9.44	281			− 36	− 28
9.65	292			− 34	− 25
14.06	563	− 66	− 63	− 53	− 34
15.00	631	− 65	− 62	− 56	− 34
15.24	649			− 62	− 42
15.96	703	− 72	− 70	− 60	− 50
21.44	1183	−109	−105	− 93	− 61
22.19	1256	− 97	− 97	− 85	− 48
23.20	1358	−100	− 99	− 88	− 50
23.31	1370	− 99	− 97	− 85	− 48
24.42	1486	−104	−103	− 88	− 51
24.87	1534	−100	− 99	− 84	− 45
27.67	1848	−115	−109	− 97	− 51
28.16	1905	−126	−114	−104	− 71
28.60	1958	−119	−114	− 99	− 51
28.65	1963	−125	−113	−103	− 65
29.25	2035	−123	−116	−105	− 49
39.51	3414	− 13	+ 2	+ 8	+ 61
41.34	3688	− 21	− 7	+ 4	+ 64
47.45	4654	− 15	− 1	+ 13	+ 89
52.04	5433	− 22	− 12	+ 14	+107
62.42	7348	− 61	− 57	− 27	+ 99
75.95	10165	+ 45	+ 38	+ 57	+287
89.63	13434				+288
92.25	14109	− 89	− 85	− 28	+278
93.53	14451	− 36	− 58		+271

Table 140—(Continued)

[a] References:
- Da Damien, B. C.[411]
- HP Hall, E. E., and Payne, A. R., *Phys. Rev. (2)*, **20**, 249-258 (1922).
- J   Jamin, J.[416]
- Ja Jasse, O., *Compt. rend.*, **198**, 163-164 (1934).
- P   Pulfrich, C.[418]

Table 141.—Mean Temperature Coefficient of Index of Refraction of Water

(See also Table 135.)

In the lower right-hand corner of the table are given a number of values for the mean coefficient between 15 and 25 °C. Most of them have been derived from the formulas by which the several observers represent their observations.

The values in the rest of the table have been derived from observations, except those for C, HP, Ket, and R, which were computed.

$\Delta t = (t_2 - t_1)$ °C; $\Delta n = n_2 - n_1$, n_1 and n_2 being the values of n at the temperatures t and t_2 °C respectively.

Unit of $\lambda = 1\text{A} = 10^{-8}$ cm. Temp. $= t$ °C

$t_1 \rightarrow$ λ $t_2 \rightarrow$	0 20	20 40	40 60	60 80	80 100	Ref.[a]
			$10^5 \Delta n/\Delta t$			
2145	5.3	12.8	19.2	22.8$_5$		Fl
2195	5.3$_5$	12.7$_5$	19.1	22.7$_5$		Fl
2268	5.3	12.9$_5$	18.9$_5$	22.4		Fl
2314	5.3$_5$	12.8	19.0	22.7		Fl
2429	5.5	12.8$_5$	18.8	22.6$_5$		Fl
2574	5.3	12.7	18.8	22.6		Fl
2677	5.3	12.5$_5$	18.6$_5$	22.5		Fl
2749	5.2$_5$	12.6	18.5$_5$	22.3$_5$		Fl
3082	5.0	12.4$_5$	18.4	22.1$_5$		Fl
3404	4.9	12.3	18.3	21.9$_5$		Fl
3613	4.9$_5$	12.1$_5$	18.1$_5$	21.8		Fl
3945	4.7$_5$	12.2	18.0$_5$	21.6$_5$		Fl
4341	6.0					Da
4417	4.6$_5$	12.1	17.9	21.5$_5$		Fl
4680	4.5$_5$	12.0	17.7$_5$	21.4$_5$		Fl
4801	4.4	11.8	17.7$_5$	21.5		Fl
4862	6.0					Da
5340	4.3	11.7$_5$	17.6	21.2$_5$		Fl
5350	4.2$_4$	11.7$_5$	17.0$_5$	21.2$_5$	24.5$_5$	Ket
5350	4.1$_3$	11.8$_5$	17.7$_5$	20.8	19.7$_5$	R
5893	4.2	11.6$_5$	17.6$_5$	21.0		Fl
5893	4.2	11.7	17.0	21.1	24.4	Ket
5893	5.2$_5$	12.2	16.9$_5$	21.0$_5$	25.7$_5$	HP
5893	4.1	12.0	17.4	21.0$_5$		C(ICT)
6563	5.8$_5$					Da
6708	4.2	11.6	16.9	21.0	24.2	Ket
6708	3.9	11.2$_5$	17.2$_5$	21.1	21.8	R

Table 141—(Continued)

Ref.[a]→ λ→ t	Ket 5350 Tl	R $10^5(n_{t\_10}-n_t)$	Ket 6708 Li	R	15 to 25 °C 5893 $-10^5 \Delta n/\Delta t$	Ref.[a]
10	19.1	20.8	18.9	19.6	9.17	HP
15	43.8	41.5	43.3	39.1	8.78	Wa
20	65.7	61.8	65.0	58.3	9.12	D
25	85.2	81.6	83.8	77.1	8.97	J
30	102	101	100	95	9.0	Wm
40	133	136	132	130		
50	160	166	158	160	7.93	Fl
60	181	189	180	185	7.89	R
70	204	205	201	205	8.0	G
80	221	211	219	217		
90	236	206	234	221	8.4	Ket
100	255	189	251	215	8.4	Sch

[a] References:
- C   Chéneveau, C., *Int. Crit. Tables*, **7**, 12-16 (1930).
- D   Dufet, H., *Jour. de Phys. (2)*, **4**, 389-419 (1885) ← *Bull. soc. min. France*, **8**, 171-304 (1885).
- Da Damien, B. C.[411]
- Fl   Flatow, E.[435]
- G   Gifford, J. W., *Proc. Roy. Soc. (London) (A)*, **78**, 406-409 (1907).
- HP Hall, E. E., and Payne, A. R., *Phys. Rev. (2)*, **20**, 249-258 (1922).
- J   Jamin, J.[416]
- Ket Ketteler, E., *Ann. d. Physik (Wied.)*, **33**, 353-381, 506-534 (1888).
- R   Rühlmann, R., *Idem (Pogg.)*, **132**, 1-29, 177-203 (1867).
- Sch Schütt, F., *Z. physik. Chem.*, **5**, 348-373 (1890).
- Wa Walter, B., *Ann. d. Physik (Wied.)*, **46**, 423-425 (1892).
- Wm Wiedemann, E., *Idem (Pogg.)*, **158**, 375-386 (1876).

Table 142.—Variation of the Refraction of Water with the Temperature: Comparison of Formulas

In the first section of the table are collected formulas that are quoted in one or more widely used compilations or that have been proposed recently; in the second is a skeleton table comparing the values defined by those formulas for the D-lines (5893A). Key letters indicate the several formulas and their sources. Values beyond the range in t assigned to a formula are enclosed in parentheses.

Flatow did not give a formula [c] connecting n and t, but gave five dispersion formulas, one for each of the temperatures used (see Table 139.)

Unit of v^*, β, and $C = 1$ cm^3/g; of $k = 1$ per 1 °C; of $\lambda = 1A = 10^{-8}$ cm. Temp. $= t$ °C

I. Formulas for the variation of n with t.

Ketteler's formula $(n^2 - 1)(v^* - \beta) = C(1 + \alpha e^{-kt})$, where α, β, and k are constants determined solely by the material, v^* is the specific volume, and C depends upon the substance and the frequency of the radiation, is in disagreement with Flatow's observations (see text), and the present compiler has been unable to check Ketteler's computations satisfactorily. Possibly the values of v^* used by Ketteler in deriving the constants, given below, were unsatisfactory at the higher temperatures.

Very precise formulas based on their own observations have been published by L. W. Tilton and J. K. Taylor.[437]

[437] Tilton, L. W., and Taylor, J. K., *J. Res. Nat. Bur. Stand.*, **20**, 419-477 (RP1085) (1938).

Table 142—(Continued)

Other formulas are of the type $10^5(n - n_0) = at + bt^2 + ct^3 + dt^4$.

Key[a]	$10^2 a$	$10^3 b$	$10^5 c$	$10^6 d$	Range in t	λ	Basis[b]
C[c]	−12.4	−199.3	0	+5.	0 to 80	(D)5893	Air, t
D	−125.5	−206.4	+4.35	+11.5	1 to 50	(D)5893	Air, R
HP[d]	−66.	−262.	+181.7	−7.55	15 to 100	(D)5893	Air, 20
J	−125.73	−192.9	0	0	0 to 30	(D)5893	Vac
Lo	+0.76	−280.3	+213.4	0		(D)5893	Vac
P[e]	−83.	−295.	+640.	0	−10 to 10	(D)5893	Air, t
M[f]	−20.0	−290.5	0	+5.00		(D)5893	
R	0	−201.4	0	+4.936	0 to 92	(D)5893	Air, 9
Wa	−120.	−205.	+50.	0	0 to 30	(D)5893	Air, $R(?)$
Lo	+9.52	−279.3	+213.4	0	0 to 30	(Li)6708	Air, $R(?)$
R	0	−196.6	0	+4.600	0 to 92	(Li)6708	Air, 9
R	0	−209.0	0	+6.046	0 to 92	(Tl)5350	Air, 9
O	−118.73	−207.09	+7.92	+10.939	2 to 38	(Hg)5461	Vac

Key[a]	$10^5 \alpha$	$10^5 \beta$	$10^5 C$	$10^5 k$	Range in t	λ	Basis[b]
Ket	246	20271	61574	2290	0 to 100	(Li)6708	Vac
Ket	246	20271	62035	2290	0 to 100	(D)5893	Vac
Ket	246	20271	62439	2290	0 to 100	(Tl)5350	Vac

II. (D) $\lambda = 5893$A: Comparison of formulas.

For simplicity, all formulas have been compared with the most recent one (HP); $10^5(n_0 - n_t)$ is given for the HP formula, and that plus δ is the corresponding quantity for the other indicated formula; $(n_0 - n_t)_{\text{vac}} = (n_0 - n_t)_{\text{air}} + 10^{-5}\Delta$; the tabulated values of Δ have been derived from Table 136.

Ref[a]	HP	Wa	D	J	P	Lo	C	R	Ket	
Date	1922	1892	1885	1856	1888	1880	1930	1867	1880	Δ
Basis[b]	Air, 20	Air, R	Air, R	Vac	Air, t	Vac	Air, t	Air, 9	Vac	
t	$10^5(n_0-n_t)$				δ					
−10	(22)				6					−1.5
−5	(4)				0					−0.8
+5	(10)	1	1	1	1	−3	−4	−5	−7	+0.7
10	(31)	1	2	1	0	−5	−10	−10	−12	1.4
15	63	−1	1	−1	(−6)	−7	−17	−18	−17	2.0
20	105	−3	1	−2		−10	−23	−25	−20	2.7
25	155	−4	0	−3		−13	−29	−31	−25	3.3
30	213	−6	0	−1		−18	−34	−35	−28	3.9
40	349	(−5)	0	(+10)		(−37)	−37	−38	−31	5.0
50	509		−7				−35	−36	−32	6.1
60	688		(−28)				−28	−27	−30	7.2
70	887						−22	−19	−27	8.1
80	1109						−28	−22	−28	9.1
90	1352						(−55)	−45	−37	9.9
100	1624						(−104)		−56	10.8

[a] References:
C   Same as Table 141.
D   Same as Table 141.
HP Same as Table 141.
J   Jamin, J.[416]
Ket Ketteler, E., *Ann. d. Physik (Wied.)*, **33**, 353-381, 506-534 (1888).
Lo Lorenz, L.[417]
O   Osborn, F. A., *Phys. Rev. (2)*, **1**, 198-210 (1913).
P   Pulfrich, C.[418]
R   Rühlmann, R., *Ann. d. Physik (Pogg.)*, **132**, 1-29, 177-203 (1867).
Wa Walter, B., *Idem (Wied.)*, **46**, 423-425 (1892).

37. WATER: REFRACTIVITY

Table 142—(Continued)

[b] Basis: (Air, t), (Air, 9), (Air, R), etc., and (Vac) indicate that the associated n is that with reference to air at the temperature (t) of the water, at 9 °C, at room temperature, etc., and with reference to vacuum, respectively. Quite frequently observers and compilers state that a value of n refers to "air at the same temperature (Air, t) when it actually refers to air at room temperature (Air, R) (see citations R and HP). Here the basis is given as (Air, R) unless there is definite evidence that it is some other. In all cases, the pressure of the air of reference is 1 atm.

The difference ($n_0 - n_t$) when referred to vacuum (Vac) exceeds its value when referred to air at a fixed temperature (Air, R; Air, 9; Air, 20) by only 3 parts in 10 000, an amount that is negligible when, as here, the index is not carried beyond the fifth place of decimals. But when the index is with reference to air at the same (varied) temperature as the water (Air, t), then the value of ($n_0 - n_t$) for (Vac) does, in general, differ significantly from that for (Air, t).

[c] The original source of this formula (C), given in Chéneveau's compilation, has not been ascertained. The Landolt-Börnstein *Tabellen* attributes it to F. F. Martens, who prepared that section of the *Tabellen*, and states that it is based on Flatow's observations, but it is not given in any of Martens' papers that have come to the attention of the compiler.

[d] In their synopsis, the authors state that the values of n when referred to vacuum are represented by the formula: $n = 1.33401 - 10^{-7}(66t + 26.2t^2 - 0.1817t^3 + 0.000755t^4)$, but in reality that formula, which agrees with the one here given, reproduces their values as referred to air, and given in their Table II; it does not reproduce their vacuum values, given in Table IV.

[e] This formula seems to have been derived by H. Dufet[425] from the observations of Pulfrich.

[f] The Landolt-Börnstein *Tabellen* attributes this formula to F. F. Martens, who prepared that section of the *Tabellen*, and states that it is based on the data of Pulfrich, but it does not satisfactorily represent those data and has not been found in Martens' papers. It does not appear elsewhere in this compilation, and is not used in the following sections of this table.

Table 143.—Temperature Gradient of the Index of Refraction of Water

(See also Table 135.)

In the first section, the values of dn/dt for $\lambda = 5893$A are given as derived from the several formulas listed in Table 142; the key symbols are the same in both tables. In the second, experimental values for 20 °C and various λ's are given.

If n is with reference to air at a fixed temperature (Air, R; Air, 9; Air, 20), then dn/dt has essentially the same value (within 3 parts in 10 000) as if n had been referred to a vacuum; such is not the case if n is with reference to air at the same (varied) temperature as the water (Air, t).

For $\lambda = 12.6$ to 24 cm, M. Seeberger[437a] has concluded that $-10^5 dn/dt = 16_{60}$, essentially independent of λ and of the temperature throughout the range 16 to 70 °C.

[437a] Seeberger, M., *Ann. d. Physik* (5), **16**, 77-99 (1933).

Table 143—(Continued)

I. Temperature gradient as derived from formulas. $\lambda = 5893 A(D)$.
$-10^5 dn/dt = -10^5(dn/dt)_{HP} + \delta$. Temp. $= t$ °C.

Ref.[a]→ Basis[b]→ t	HP Air, 20 $-10^5 dn/dt$	Wa Air, R	D Air, R	J Vac	P Air, t 100δ	Lo Vac	C Air, t	R Air, 9
-10	(-5.12)	$(+207)$	$(+99)$	$(+252)$	-186	(-113)	$(+134)$	$(+111)$
-5	(-2.10)	$(+121)$	$(+97)$	$(+143)$	-50	(-87)	$(+23)$	$(+9)$
0	$(+0.66)$	$+54$	$+60$	$+60$	$+17$	-67	-54	-66
$+5$	(3.15)	$+6$	$+16$	$+4$	$+15$	-52	-103	-114
10	(5.38)	-23	-6	-26	-57	-42	-129	-137
15	7.40	-39	-14	-36	(-204)	-44	-136	-142
20	9.20	-40	-11	-23		-56	-126	-130
25	10.82	-31	-4	$+8$		-81	-104	-106
30	12.29	-14	-1	$+54$		-124	-75	-74
40	14.83	$(+37)$	-21	$(+186)$		(-266)	-4	$+2$
50	17.01		-119				$+54$	$+66$
60	19.00		(-338)				$+72$	$+90$
70	20.99						$+18$	$+43$
80	23.16						-139	-104
90	25.68						(-426)	-382
100	28.75							-821

II. Experimental values at 20 °C.

λ	1862	1930	1988	1990	2144	2144	5893A
$-10^5 dn/dt$	8.2	9.3	6.9	10.5	8.9[c]	7.2	8.0
Basis[b]	Air, 20	Air, 20	Air, 15	Air, 20	Air, 20	Air, 15	Air, 15
Ref.[d]	DJ	DJ	G	DJ	DJ	G	G

[a] As in Table 142.
[b] Basis of reference, see Table 142.
[c] This value is attributed by DJ to E. Flatow.[435] DJ conclude that $-dn/dt$ passes through a maximum near $\lambda = 2000$A.
[d] References:
   DJ Duclaux, J., and Jeantet, P.[415]
   G   Gifford, J. W., *Proc. Roy. Soc. (London) (A)*, **78**, 406-409 (1907).

Table 144.—Variation of the Refraction of Water with the Pressure

$\Delta \equiv (n_2 - n_1)/(p_2 - p_1)$; $dn/dp = a - bp$.

G. Quincke[438] did not accept Zehnder's values (see below), believing that his own observations[439] indicated that when the temperature is constant, then $(n - 1)/\rho$ is a constant, ρ being the density.

For effect of pressure on dielectric constant see Table 176 and accompanying text.

Unit of Δ and of $a = 10^{-6}$ per atm; of $b = 10^{-6}$ per atm²; of $\lambda = 1$ A $= 10^{-8}$ cm. Temp. $= t$ °C

I. D-lines, $\lambda = 5893$. Pressure not exceeding 4 atm.

Mascart[440] found $\Delta = 15.2$ at 15 °C, and 16.1 at 5.5 °C.

The following observations (Δ_{obs}) from L. Zehnder[441] are closely reproduced by $\Delta_c = 16.84 - 0.129t + 0.0022t^2$.

[438] Quincke, G., *Ann. d. Physik (Wied.)*, **44**, 774-777 (1891).
[439] Quincke, G., *Idem*, **19**, 401-435 (1883).
[440] Mascart, *Compt. rend.*, **78**, 801-805 (1874).
[441] Zehnder, L., *Ann. d. Physik (Wied.)*, **34**, 91-121 (1888).

37. WATER: REFRACTIVITY

Table 144—(Continued)

t	Δ_{obs}	Δ_c	t	Δ_{obs}	Δ_c
−0.78	16.91	16.94	4.95	16.26	16.26
0.00	16.82	16.84	8.95	15.87	15.86
+0.06	16.85	16.83	9.00	15.91	15.86
0.42	16.78	16.79	13.05	15.55	15.53
1.05	16.69	16.71	13.28	15.56	15.52
2.62	16.51	16.52	17.83	15.25	15.24
2.67	16.53	16.51	18.01	15.25	15.23
2.92	16.48	16.48	18.03	15.25	15.23
3.10	16.44	16.46	23.27	14.98	15.03

II. Various wave-lengths.

Low pressures; 18 °C RZ[a]

λ	Δ
4861	15.40
6807	15.16

Pressures up to 1800 kg/cm^2; 25 °C PR[a]

λ	a	b
4060	15.02	0.003182
4360	14.65	0.002700[b]
5460	14.75	0.003132
5790	14.56	0.002990

[a] References:
PR Poindexter, F. E., and Rosen, J. S., *Phys. Rev. (2)*, **45**, 760 (A) (1934).
RZ Röntgen, W. C., and Zehnder, L., *Ann. d. Physik (Wied.)*, **44**, 24-51 (1891).
[b] This value accords with the published coefficient, but seems strangely out of line with the others.

Table 145.—Refraction of Natural Waters

$n = n_w + \Delta$, n_w being the index for pure water at the same temperature and for the same wave-length; salt content = s g per kg of sea-water.

C. Chéneveau [a] gives the following values for 20 °C and the D-lines:

	$10^5\Delta$
City water, Paris	4
River Seine	5
Water saturated with CO_2 at 1 atm	−3
Mediterranean Sea	400

The last seems to be entirely too small; see below.

Unit of $dn/dt = 10^{-6}$ per 1 °C; of $s = 1$ g salt per kg sea-water; of $\lambda = 1 \text{A} = 10^{-8}$ cm

Sea-water

J. W. Gifford (1907)[a] 15° C

λ	$10^5\Delta$	$-dn/dt$
7682.4 K	647	
7065.6 He	645	
6563.0 H	650	
5893.2 Na	653	78.5
5607.1 Pb	664	
5270.1 Fe	665	
4861.5 H	676	
4678.4 Cd	680	
4340.7 H	691	
3961.7 Al	698	
2748.7 Cd	819	74.7
2265.1 Cd	979	75.8
2194.4 Cd	Abs[b]	

SS (1889)[a] Mediterranean Sea 20° C

λ	$10^5\Delta$
(A) 7608	697
(B) 6870	691
(C) 6563	696
(D) 5893	706
(F) 4862	719
(h) 4102	739
(H) 3969	756

$n_{10} - n_{20} = 0.00085$

O. Krümmel[a] D-line; 18 °C

s	$10^5(n_s - n_w)$
5	97
10	194
15	290
20	386
25	482
30	577
35	673
40	769

Table 145—*(Continued)*

[a] References:
  Chéneveau, C., *Int. Crit. Tables*, **7**, 12-16 (13) (1930) from observations by Dufet, H., *Bull. Soc. Min. France*, **8**, 171-304 (1885); Soret, J. L., and Sarasin, E., *Compt. rend.*, **108**, 1248-1249 (1889).
  Gifford, J. W., *Proc. Roy. Soc. (London) (A)*, **78**, 406-409 (1907).
  Krümmel, O.[408]
  SS Soret, J. L., and Sarasin, E., *Compt. rend.*, **108**, 1248-1249 (1889) quoted by Krümmel.
[b] Strong absorption for $\lambda \gtrless 2194.4$A.

38. Reflection of Light by Water

When light strikes a boundary separating two media of different refractivities, some of it is specularly reflected, some is scattered (non-specularly reflected), and the remainder enters the second medium. The amount that is specularly reflected depends upon both the angle of incidence (i) and the polarization of the incident light. Unless $i = 0$, the reflected light is partially plane-polarized even when the incident light is not, for the component that has its electric vector perpendicular to the plane of incidence (*i.e.*, that is polarized in the plane of incidence) is more strongly reflected than the other. As i increases from 0 to 90°, the ratio of the reflectivities of the two components passes through a maximum. The angle at which this occurs is known as the Brewsterian angle and is given by the relation $\tan i = n$. At that angle the reflected light is almost completely plane-polarized, the reflectivity of the weaker component, in the case of an air-water surface, being only a few ten-thousandths of that of the stronger; C. V. Raman and L. A. Ramdas [442] found for that ratio 75×10^{-5}, while Rayleigh [443] found 42×10^{-5}. From their own measurements, Raman and Ramdas computed that the thickness of the transition layer in which the index changes from that of air in bulk to that of water in bulk is of the order of 5×10^{-8} cm; the diameter of a water molecule, as calculated from viscosity data, is about 2.6×10^{-8} cm.

The reflectivity (R) is defined by the ratio $R = I_r/I_i$ where I_i and I_r are the intensities of the incident and of the specularly reflected light, respectively. If the medium is transparent (both absorption and scattering negligible), the reflectivity is $R_p = \dfrac{\sin^2(i-r)}{\sin^2(i+r)}$, if the incident light is plane-polarized in the plane of incidence (electric vector parallel to the reflecting surface); and $R_n = \dfrac{\tan^2(i-r)}{\tan^2(i+r)}$ if it is plane-polarized normal to the plane of incidence; i is the angle of incidence and r that of refraction. If the incident light is unpolarized, the reflectivity is half the sum of these two expressions. As the index of refraction of the second medium with

[442] Raman, C. V., and Ramdas, L. A., *Phil. Mag. (7)*, **3**, 220-223 (1927).
[443] Lord Rayleigh, *Phil. Mag. (5)*, **33**, 1-19 (1892)="Collected Works," Vol. 3, pp. 496-512, 1902.

38. WATER: REFLECTION

reference to the first is $n = \sin i/\sin r$, these expressions may be put in the form:

$$R_p = \frac{[\sqrt{n^2 - \sin^2 i} - \cos i]^2}{[\sqrt{n^2 - \sin^2 i} + \cos i]^2} \text{ and } R_n = \frac{[n^2 \cos i - \sqrt{n^2 - \sin^2 i}]^2}{[n^2 \cos i + \sqrt{n^2 - \sin^2 i}]^2}$$

If $i = 0$, $R_p = R_n = (n-1)^2/(n+1)^2$. Throughout the visible spectrum the transparency of water is such that R for it can be satisfactorily computed by means of these formulas. For experimentally determined values of R see Table 146.

The scattering (non-specular reflection) of light by a free liquid surface probably arises from the roughening of the surface by the thermal agitation of the molecules. The higher the surface tension and the more nearly

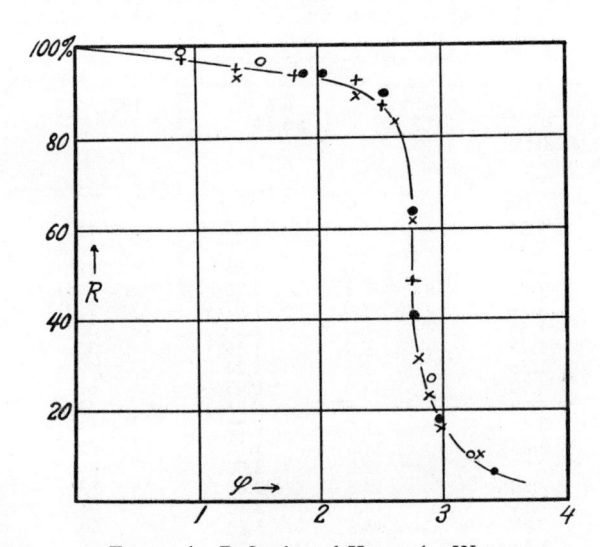

FIGURE 4. Reflection of X-rays by Water.

[Adapted from H. Steps, *Ann. d. Physik* (5), **16**, 949-972 (1933).]
Radiation used was the $K\alpha$ of Cu ($\lambda = 1.539$A); R = reflectivity; $90° - \phi$ = angle of incidence, unit = 0.001 radian = 0.057° = 3.44′.

equal are the indices of refraction of the adjacent fluids, the less the scattering. The observed ratio of the intensity of the light scattered by an air-water surface at room temperature to that scattered by plaster of Paris varies from 3×10^{-6} to 8×10^{-6}.[444]

V. E. Shelford and F. W. Gail [445] have reported that in calm, clear weather between 10 a.m. and 2 p.m. about 25 per cent of the light from the

[444] Raman, C. V., and Ramdas, L. A., *Proc. Roy. Soc. (London) (A)*, **109**, 150-157, 272-279 (1925). See also Ramdas, L. A., *Indian J. Phys.*, **1**, 199-234 (1927).
[445] Shelford, V. E., and Gail, F. W., *Publ. Puget Sound Biol. Sta.*, **3**, 141-176 (1922).

Table 146.—Reflectivity of Water

Observed values for unpolarized light. For the reflectivity of polarized light, see remarks in text; of x-rays, see Fig. 4.

Except as noted, the data have been taken from E. P. T. Tyndall's compilation [450] based on data obtained by A. K. Ångström,[451] K. Brieger,[452] F. Gehrts,[453] H. Rubens,[454] H. Rubens and E. Ladenburg.[455]

More recent data obtained by M. Weingeroff [456] and published as a small scale curve covering the range $\lambda = 11\,\mu$ to $\lambda = 17\,\mu$ essentially agree with these except at the longer wave-lengths, where he finds a somewhat smaller reflectivity.

$R = I_r/I_i$, where I_i and I_r are the intensities of the incident and the reflected light, respectively; i is angle of incidence; λ_{max} = wave-length at which R passes through a maximum.

Unit of $\lambda = 1\,\mu = 10^{-4}$ cm. Temp. = 20 °C

λ_{max}					Ref.[a]
3.2		6.3		19.5	RL
3.23	4.7	6.22			Re
3.045, 3.15, 3.28(?)					BM
3.18		6.40			MB1

Normal Incidence

λ	100R	λ	100R	λ	100R
0.3	2.33[b]	6.3	2.34	18.0	6.7
0.7	2.00[b]	6.5	2.10	18.5	7.5
1.0	1.98	7.0	1.95	19.0	8.4
1.5	1.95	7.5	1.75	20.0	8.9
2.0	1.74	8.0	1.67	21.0	8.2
2.4	1.45	8.5	1.60	23.	6.5[c]
2.6	1.25	9.0	1.44	33.	7.2[c]
2.8	1.35	9.5	1.24	52.	9.3[c]
3.0	3.40	10.0	0.95	52.	9.30[d]
3.2	4.10	10.5	0.85	63.	10.6[c]
3.23	3.4	11.0	0.75	63.	10.74[d]
3.4	3.25	11.5	1.10	82.	9.6[c]
3.5	2.95	12.0	2.00	83.	10.9[c]
4.0	2.20	12.5	2.00	83.	11.75[d]
4.5	2.14	13.0	3.10	94.	11.1[c]
5.0	2.00	13.5	3.40	100.	12.28[d]
5.5	1.68	14.0	4.10	108.	11.6[c]
5.6	1.67	14.5	4.80	117.	12.80[d]
5.8	1.40	15.0	5.30	117.	12.7[c]
5.9	1.50	15.5	5.4	152.	13.40[d]
6.0	2.00	16.0	5.3	310.	15.1[c]
6.1	2.28	16.5	6.0	4200.	48.4[f]
6.2	2.46	17.0	6.6		
6.22	2.4	17.5	6.9		

Normal Incidence		Unit of $\lambda = 1\,\mu$ $i = 50°$			
Unit of $\lambda = 1$ cm					
λ	100 R	λ	100 R	λ	100 R
		6.0	4.00	13.0	4.70
0.42	48.4[f]	7.0	3.25	14.0	6.65
0.84	51.6[f]	8.0	3.20	15.0	8.20
1.1	54.[f]	9.0	2.50	16.0	8.7
1.5	56.5[f]	10.0	2.10	17.0	9.8
1.8	58.[f]	11.0	1.80	18.0	11.3
2.7	64.[f]	12.0	2.80	19.0	13.7

38. WATER: REFLECTION

Table 146—(Continued)

Effect of temperature; normal incidence Unit of $\lambda = 1\ \mu = 10^{-4}$ cm. Temp. $= t\,°C$

$t \rightarrow$ λ	0	20 100R	30
117	11.8	12.7	13.2
310	14.9	15.1	17.1

a References:
- BM   Barnes, R. B., and Matossi, F., *Z. Physik*, **76**, 24-37 (1932).
- MBl Matossi, F., and Bluschke, *Idem*, **104**, 580-583 (1937). See also Matossi, F., and Fesser, H., *Idem*, **96**, 12-28 (1935).
- Re   Reinkober, O., *Idem*, **35**, 179-192 (1926).
- RL   Rubens, H., and Ladenburg, E., *Verh. physik. Ges.*, **10**, 226-227 (1908).

b Calculated from *n*.
c R. Rubens*; previously, he and H. Hollnagel† had found for the residual rays from KBr ($\lambda = 74$ to $88\ \mu$) $100R = 9.6$ at 19 °C, R for silver being taken as unity.
d C. H. Cartwright and J. Errara.‡
e H. Rubens and R. W. Wood.§
f J. D. Tear,** radiation was plane-polarized with the electric vector in the plane of incidence; $i = 8°\ 20'$.

Table 147.—Albedo of Water [457]

By definition, the albedo of a plane surface is $A = F_r/F_i$ where F_r is the total luminous flux reflected by the surface when uniformly illuminated by white light, the total luminous flux incident on the surface being F_i.

The following values were taken from an airplane at altitude H, and in some cases the reflected light passed through one or other of two color filters described simply as "green" and "red," respectively.

Unit of $H = 1$ ft $= 0.3048$ m; of $A = 10^{-3}$

Filter→ H	None	Green A	Red	Remarks
		Chesapeake Bay		
2000	97			Smooth
2000	38*a*	47	35	Well out
3000	36	40	45	Well out
		Potomac River		
10 to 20	69	55	104	
		Patuxent River		
2000	55			
3000	64			

a At Buzzards Bay, W. M. Powell and G. L. Clarke [458] found $A = 3$ to 4 per cent.

---

* Rubens, R., *Sitzb. preuss. Akad. Wiss. (Berlin) (Phys.-Math.)*, **1915**, 4-20 (1915).
† Rubens, R., and Hollnagel, H., *Ber. physik. Ges.*, **12**, 83-98 (1910).
‡ Cartwright, C. H., and Errera, J., *Proc. Roy. Soc. (London), (A)*, **154**, 138-157 (1936); *Acta Physicochim. URSS*, **3**, 649-684 (1935)→Cartwright, C. H., *Nature*, **135**, 872 (L) (1935).
§ Rubens, H., and Wood, R. W., *Verh. physik. Ges.*, **13**, 88-100 (1911).
** Tear, J. D., *Phys. Rev. (2)*, **21**, 611-622 (1923).

[446] Hulburt, E. O., *J. Opt. Soc. Amer.*, **24**, 35-42 (1934).
[447] Weniger, W., *J. Opt. Soc. Amer.*, **7**, 517-527 (1923).
[448] Schaefer, C., and Matossi, F., "Das Ultrarote Spektrum," Berlin, J. Springer, 1930.
[449] Korff, S. A., and Breit, G., *Rev. Mod. Phys.*, **4**, 471-503 (1932).
[450] Tyndall, E. P. T., *Int. Crit. Tables*, **5**, 258-259 (1929).
[451] Angström, A. K., *Phys. Rev. (2)*, **3**, 47-55 (1914).
[452] Brieger, K., *Ann. d. Physik (4)*, **57**, 287-320 (1918).
[453] Gehrts, F., *Idem*, **47**, 1059-1088 (1915).

sky is reflected by the surface of the sea; and E. O. Hulburt [446] has found that, when the sky is clear, the brightness of the rim of the sea is 25 per cent of that of the sky near the horizon, the surface of the sea being ruffled by a breeze of 5 to 25 knots, under which condition, "The reflecting facets of the sea which are visible to the observer are tilted up on an average of 15° from the horizontal." He defines the rim of the sea as the surface lying between the horizon and a line of sight making an angle of 3° with the surface of the sea.

For reviews and summaries see W. Weniger,[447] C. Schaefer and F. Matossi,[448] S. R. Korff and G. Breit.[449]

39. LUMINESCENCE OF WATER

Luminescence may be excited in water by various means: by light, x-rays, and gamma rays, by bombardment with electrons and beta rays, and by mechanical shock. These will be considered in the order named, after certain terms have been defined and certain general characteristics of the several types of luminescence excited by light have been briefly considered.

Types of Luminescence. Definitions and General Characteristics.

When a beam of light is passed through a medium, the medium becomes luminous, emitting light even at right angles to the incident beam. This luminescence may be very faint, and is observed most satisfactorily in a direction at right angles to the incident beam. It is usually partially polarized, even when the exciting light is unpolarized. The intensity of the component polarized in a given plane varies with the orientation of that plane about the line of sight, passing through a maximum (I_s) and a minimum (I_w). The ratio $\rho = I_w/I_s$ is called the depolarization factor, and $\Delta = 2I_w/(I_w + I_s) \equiv 2\rho/(1 + \rho)$ is called the depolarization. These two quantities should not be confused.

Four distinct types of such luminescence are recognized: Fluorescence and phosphorescence, Tyndall scattering, Rayleigh scattering, and Raman scattering.

Fluorescence and Phosphorescence. — Phosphorescence being merely long-lived fluorescence, the latter, and shorter, term will be used for both. Fluorescence differs from other types of luminescence in that the spectrum of its light depends solely upon the medium, and in that the light is not polarized except as polarization may be imposed upon it by the incident radiation. But any specified portion of the spectrum of the fluorescent light may appear only under certain conditions; *e.g.*, only when the exciter contains frequencies lying within a certain range, or exceeding a certain

---

[454] Rubens, H., *Verh. physik. Ges.*, **17**, 315-335 (1915).
[455] Rubens, H., and Ladenburg, E., *Idem*, **11**, 16-27 (1909); *Sitz. preus. Akad. Wiss*, **1908**, 274-284 (1908).
[456] Weingeroff, M., *Z. Physik*, **70**, 104-108 (1931).
[457] Kimball, H. H., and Hand, I. F., *Monthly Weather Rev.*, **58**, 280-281 (1930).
[458] Powell, W. M., and Clarke, G. L., *J. Opt. Soc. Amer.*, **26**, 111-120 (1936).

value, or contains corpuscles having a kinetic energy exceeding a certain value.

Tyndall Scattering.—If small foreign particles are distributed throughout the medium, they will scatter the light by reflection and diffraction. The spectrum of the light so scattered (Tyndall scattering) is the same as that of the incident light as modified by the color of the scattering particles. If they are colorless, the ratio of the intensity of the scattered to that of the incident light varies continuously throughout the spectrum, the rate of variation at any place depending upon the size of the particles, as well as upon the wave-length. If the particles are very small, the ratio varies inversely as the fourth power of the wave-length of the light, making the scattered light much bluer than the incident. If the incident light is unpolarized, the light so scattered at right angles to the incident beam will be completely plane-polarized in the plane of scattering (the electric vector being normal to that plane) if the particles are spherical and isotropic; otherwise, the polarization will not be complete.[459]

Rayleigh Scattering.—An exactly analogous scattering by pure, dust-free gases was predicted by Lord Rayleigh [460] and has been observed. It arises from the scattering by the molecules themselves, which here play exactly the same part as is played by the foreign particles in the Tyndall scattering. It is but a step to extend the same idea to liquids and solids, but in them the molecules are so closely packed that they cannot satisfactorily play the part of foreign particles. Nevertheless, liquids and solids do exhibit exactly this same type of scattering, the scattering "particles" in them being the slight variations from point to point, and from instant to instant, in the number of molecules per unit of volume, these variations arising from the thermal agitation of the molecules.[461] Somewhat similar variations in the concentration of the primary molecules (H_2O) may arise from the temporary association of these molecules into rather large groups, as suggested by H. Schade and H. Lohfert [462] and by G. W. Stewart,[463] who uses the adjective cybotactic to describe such a condition of association. It would seem that such groups also might act as scattering particles.

Scattering of this type by pure, dust-free media, whether liquid, solid, or gas, will be called Rayleigh scattering. To the light so scattered applies everything that has been said about Tyndall scattered light, except that the scattered light will never be completely polarized if there is interaction between the scattering units, no matter how nearly isotropic the units may be.[464] Each type of scattering—Tyndall and Rayleigh—is frequently called by either name.

[459] See Strutt, J. W. (later, Lord Rayleigh), *Phil. Mag. (4)*, **41**, 107-120, 274-279 (1871); **41**, 447-457 (1871). Lord Rayleigh, *Idem (5)*, **12**, 18-101 (1881); *(6)*, **35**, 373-381 (1918).
[460] Lord Rayleigh, *Phil. Mag. (5)*, **47**, 375-384 (1899).
[461] See v. Smoluchowski, M., *Ann. d. Physik (4)*, **25**, 205-226 (1908); Einstein, A., *Ann. d. Physik (4)*, **33**, 1275-1298 (1910); Raman, C. V., *Proc. Roy. Soc. (London) (A)*, **101**, 64-80 (1922).
[462] Schade, H., and Lohfert, H., *Kolloid Z.*, **51**, 65-71 (1930).
[463] Stewart, G. W., *Phys. Rev. (2)*, **35**, 726-732 → 1426 (A) (1930).
[464] Cabannes, J., *Jour. de phys. (6)*, **3**, 429-442 (1922).
[465] Cabannes, J., and Daure, P., *Compt. rend.*, **186**, 1533-1534 (1928).

In 1928 J. Cabannes and P. Daure [465] announced that the radiation in the Rayleigh scattered light appeared to have a slightly lower frequency (displacement of about 0.01A toward the red), and that the line appeared broader than in the incident light and was superposed on a rather sharply limited continuous background, the whole having the appearance of a winged line. This last is sometimes referred to as the Cabannes-Daure effect. The work was continued by J. Cabannes and P. Salvaire,[466] J. Cabannes,[467] and others; and more recently by W. Ramm,[468] who found that when the incident radiation is truly monochromatic the scattered line is a symmetrical triplet, of which the central line has, within experimental error, the same frequency as the incident radiation. All three components of the triplet are of about the same intensity, and the spacing agrees well with that called for by L. Brillouin's theory of scattering.[469] The triplet rests on a continuous background, as Cabannes observed. Ramm found only the triplet, no indication of any series of lines such as had been reported by E. Gross [470] and thought by him to be required by P. Debye's theory.[471] Ramm's conclusions rest on his study of the radiation that is scattered backward (turned 180° with reference to the incident light). Cabannes [467] has reported that the continuous background is almost completely depolarized.

In addition to those already mentioned, the theory and interpretation of such molecular scattering have been discussed by Y. Rocard,[472] A. Bogros and Y. Rocard,[473] J. Cabannes and Y. Rocard,[474] W. Ramm,[468] and A. Rousset.[475]

Raman Scattering.—In the preceding cases, the spectrum of the scattered light was determined either solely by the medium or solely by the incident light as modified by the color of the scattering particles. But C. V. Raman [476] observed that the spectrum of the scattered light contains lines that are foreign to the spectrum both of the incident light and of the medium. The scattering that gives rise to these lines is known as the Raman scattering—sometimes in Germany as the Smekal-Raman scattering, A. Smekal [477] having shown theoretically in a letter on another subject, that a scattering of this type is demanded by the quantum theory. See also Y. Rocard.[478]

On the quantum theory, which accounts fairly well for the observed phenomena, when a quantum of radiation of frequency $c\nu_i$ strikes an atom

---

[466] Cabannes, J., and Salvaire, P., *Compt. rend.*, **188**, 907-908 (1929).
[467] Cabannes, J., *Idem*, **191**, 1123-1125 (1930).
[468] Ramm, W., *Physik. Z.*, **35**, 756-773 (1934).
[469] Brillouin, L., *Ann. de Phys. (9)*, **17**, 88-122 (1922).
[470] Gross, E., *Naturwissenschaften*, **18**, 718 (L) (1930).
[471] Debye, P., *Ann. d. Physik (4)*, **39**, 789-839 (1912).
[472] Rocard, Y., *Ann. de Phys. (10)*, **10**, 116-179, 181-231, 472-488 (1928).
[473] Bogros, A., and Rocard, Y., *Jour. de Phys. (6)*, **10**, 72-77 (1929).
[474] Cabannes, J., and Rocard, Y., *Idem*, **10**, 52-71 (1929).
[475] Rousset, A., *Jour. de Phys. (7)*, **6**, 507-515 (1935); *Ann. de Phys. (11)*, **5**, 5-135 (1936).
[476] Raman, C. V., *Indian J. Phys.*, **2**, 387-398 (1928).
[477] Smekal, A., *Naturwissenschaften*, **11**, 873-875 (L) (1923).
[478] Rocard, Y., *Compt. rend.*, **186**, 1107-1109 (1928).

or an aggregation of atoms, there may be a transfer of a quantum of radiation of frequency cv_c characteristic of the atom or aggregation. The transfer may occur in either direction, and cv_c may be any of the characteristic frequencies of the material. Hence the scattered radiation will contain the additional frequencies $c(v_i + v_c)$ and $c(v_i - v_c)$, there being as many v_c's as there are frequencies characteristic of the material, and as many v_i's as there are frequencies in the incident radiation. The complete set of additional frequencies attendant upon any one incident frequency (cv_i) is characterized by a definite set of frequency differences ($c\delta v$), the same for every cv_i. This set of frequency differences is called the Raman spectrum of the medium. It consists of two parts. In one, called the antistokes Raman spectrum, the actual frequencies exceed cv_i; in the other, the stokes Raman spectrum, they are smaller than cv_i. Only those atoms and aggregations that are suitably excited can contribute to the first; whereas any that are unexcited may contribute to the second. Hence the stokes frequencies will be much the stronger, unless the medium is subjected to some definitely exciting action. Throughout the preceding, c = velocity of light, $v(\equiv 1/\lambda)$ = wave-number, λ = wave-length of the radiation corresponding to the frequency cv.

The relative intensities of the lines in either part—stokes or antistokes—of the Raman spectrum vary widely. Those corresponding to the fundamental frequencies of the medium are, in general, much stronger than those corresponding to the combination frequencies. On account of their low intensities, the Raman lines can be satisfactorily observed only in the laterally scattered radiation; they may be only partially plane-polarized, the amount of polarization varying from line to line. The appearance of a line, and its position, may vary with the temperature, and be changed by the addition of a solute.

Although the quantum theory accounts fairly well for the observed phenomena, it does not enable one to predict the Raman spectrum with certainty. The observed frequency differences seldom coincide exactly with the frequencies in the absorption spectrum; some are not represented in the absorption spectrum, and some of the frequencies found in that are not represented in the Raman spectrum. Nevertheless, the Raman effect enables one to obtain from studies in the visible and in the near ultraviolet spectrum much information regarding those characteristic vibrations that would otherwise have to be studied in the infrared, where observations are much more difficult. Therein lies one reason for the importance that is attached to the study of Raman spectra.[479] Bibliographies are published from time to time in the *Indian Journal of Physics,* and the status of the entire subject in 1931 has been set forth and discussed by K. W. F. Kohlrausch.[480]

---

[479] See also, Raman, C. V., *Indian J. Phys.*, **6**, 263-273 (1931); Raman, C. V., and Krishnan, K. S., *Idem*, **2**, 399-419 (1928); Ganesan, A. S., and Venkateswaran, S., *Idem*, **4**, 195-280 (1929); Bhagavantam, S., *Idem*, **5**, 237-307 (1930).
[480] Kohlrausch, K. W. F., "Der Smekal-Raman Effekt," Springer, Berlin, 1931.

Electron and β-ray Luminescence.—See page 317.
Mechanical Luminescence.—See page 317.

Fluorescence of Water.

When water is subjected to gamma-rays from radium, it emits a white luminescence that is visible to the dark-adapted eye, and that is more strongly absorbed by 1 mm of glass than by 5 mm of either quartz or rock salt. The spectrum of this luminescence is continuous throughout the range covered by observations (visible spectrum and ultraviolet to $\lambda = 2500A$), and probably extends to the shortest wave-length unabsorbed by water. It is richer in the short waves than is the radiation from a 0.5-watt incandescent electric lamp.[481]

P. A. Čerenkov (also spelled Tscherenkow)[482] has reported (1934) that the luminescence excited in water by γ-rays is partially polarized, the electric vector lying parallel to the line of propagation of the incident radiation, and is not reduced by the common quenchers of fluorescence (KI, $AgNO_3$, nitrobenzene), nor by heating to 100 °C. He has reported (1936) effects produced by a strong magnetic field, and has concluded (1937) that the luminescence arises from the action of the Compton electrons freed by the γ-rays, as was suggested by S. Wawilow,[483] and not by the γ-rays themselves. See also p. 317.

Irradiating water by x-rays gave rise to no luminescence (Čerenkov; 1934).

In 1925, K. S. Krishnan[484] concluded that when a beam of light is passed through water the laterally scattered light contains fluorescent radiation; and Y. Rocard[485] came to the same conclusion. The latter decided that this fluorescence is not due to $(H_2O)_n$ molecules, but probably to the presence of glass dissolved from the container.

S. J. Wawilow and L. A. Tummermann[486] have found that the fluorescence of water, which they describe as blue, is reduced very little if at all by boiling, but repeated distillation in quartz completely destroys it. Bubbling of either air or CO_2 through doubly distilled water increases the intensity of the fluorescence, but the bubbling of oxygen does not. This "fluorescence" may include the scattered light also.

A. Carrelli, P. Pringsheim, and B. Rosen[487] have stated that Berlin city water excited by $\lambda = 3650A$ exhibits a rather strong blue-violet fluorescence consisting of a very broad, ill-defined, continuous spectral band. The same fluorescence of essentially the same intensity was obtained with Kahlbaum's distilled water, but was almost absent from his conductivity water.

---

[481] Mallet, L., *Compt. rend.*, **183**, 274-275 (1926); **187**, 222-223 (1928); **188**, 445-447 (1929).
[482] Čerenkov, P. A. (also spelled Tscherenkow), *Compt. rend. Acad. Sci. URSS (N. S.)*, 1934₂, 455-457 (1934); **12**, 413-416 (1936); **14**, 101-105 (1937).
[483] Wawilow, S., *Idem*, 1934₂, 459-461 (1934).
[484] Krishnan, K. S., *Phil. Mag. (6)*, **50**, 697-715 (1925).
[485] Rocard, Y., *Compt. rend.*, **180**, 52-53 (1925).
[486] Wawilow, S. J., and Tummermann, L. A., *Z. Physik*, **54**, 270-276 (1929).
[487] Carrelli, A., Pringsheim, P., and Rosen, B., *Z. Physik*, **51**, 511-519 (1928).

The interpretation of observations purporting to measure the amount of fluorescence excited in water by the optical spectrum is difficult. The radiation scattered by the Tyndall, Rayleigh, and Raman effects being incompletely polarized, the presence of an unpolarized component in the scattered light is not a certain criterion for even the presence of fluorescence. But if the depolarization factor ρ is measured once for the total laterally emitted light, and again for the same light deprived solely of the fluorescent light, then from these two factors the ratio of the intensity $(2f)$ of the fluorescent light to that $(a + b)$ of the scattered can be determined, a and b being the I_s and the I_w of the scattered light (p. 300). In the first case, $\rho' = (a + f)/(b + f)$; and in the second, $\rho = a/b$; whence $2f/(a + b) = 2(\rho' - \rho)/(1 - \rho')\cdot(1 + \rho)$. In practice, the fluorescent light is removed by means of a filter cutting out the ultraviolet. When the filter is in the incident beam, ρ is measured; when in the scattered, ρ'. This assumes that the intensities of the additional Raman bands that are present in the second case contribute negligibly to the intensity of the scattered light. By this procedure the following results were obtained:

$2f/(a+b)$	Filter	Reference
0.069	Orange	Krishnan, K. S.[484]
0.033	Green	Ibid.
0.03	Quinine	Sweitzer, C. W., J. Phys'l Chem., **31**, 1150–1191 (1927).
0.13	Quinine(?)	Canals, E., and Peyrot, F., Compt. rend., **198**, 1992–1994 (1934).

Rayleigh Scattering by Water.

In obtaining the data given in the following tables, the investigators made no attempt to eliminate the effects of the Raman scattering or of fluorescence, except as is indicated; and the data for polarization likewise refer to the total laterally emitted light.

The intensity of the scattered light varies reversibly with the temperature, decreasing as the temperature rises,[488] but the published data cannot be interpreted quantitatively.

E. O. Hulburt[489] has shown that the observations by W. Beebe and G. Hollister[490] of the intensity of the light scattered horizontally by the sea at various depths can be satisfactorily accounted for by the Rayleigh scattering and the values he himself obtained for the coefficient of absorption, except in the first 250 feet, where the absorption has to be increased by an amount equivalent to the presence in each cubic centimeter of one mote one-tenth of a square millimeter in sectional area.

More recently, L. H. Dawson and E. O. Hulburt[491] have found that within their experimental error ($< 15\%$) the total light scattered by water in the range $\lambda = 2536A$ to $5790A$ varies with λ as demanded by the Einstein-Smoluchowski expression (see Table 148).

---

[488] Schade, H., and Lohfert, H., Kolloid Z., **51**, 65-71 (1930).
[489] Hulburt, E. O., J. Opt. Soc. Amer., **22**, 408-417 (1932).
[490] Beebe, W., and Hollister, G., Bull. N. Y. Zool. Soc., **33**, 249-263 (1930).
[491] Dawson, L. H., and Hulburt, E. O., J. Opt. Soc. Amer., **27**, 199-201 (1937) → Phys. Rev. (2), **51**, 1017 (A) (1937).

In 1930, J. Plotnikow and L. Šplait [492] described and studied what they called a longitudinal scattering. The work was continued by J. Plotnikow and S. Nishigishi [493] and others, with varying results, leading to a discussion, sometimes spirited. The original contention of Plotnikow and his associates has been upheld by B. Čoban [494] and by L. Šplait,[494a] but it seems almost certain that everything that is not spurious in the phenomena described can be accounted for by the well-known Tyndall and Rayleigh scattering.[495]

Table 148.—Polarization and Intensity of Light Laterally Scattered by Water

See remarks in text. The values credited to K and to RR (see references[a]) are essentially those given in the compilation by J. W. T. Walsh and H. Buckley.[496]

I = intensity of the laterally scattered light at a distance r from the volume V of water from which the light comes; E = intensity of the incident (exciting) light; ρ = depolarization factor.

I_w/I_f = ratio of intensity of light laterally scattered by water to that of the light similarly scattered by the indicated fluid under the same conditions; l = liquid, g = gas at 0 °C and 1 atm.

ρ_i, ρ_s = value of ρ when the indicated filter is in the incident, scattered, beam, respectively. The filter is specified by the color of the light transmitted.

λ = wave-length of the incident (exciting) light.

Unit of $\lambda = 1$ A $= 10^{-8}$ cm; ρ is a pure number.

I. Polarization.

Unfiltered		Filtered				
ρ	Ref.[a]	Ref.[a]→ Filter	G ρ_i	K ρ_i	K ρ_s	
0.067	M¹	Red	0.119			
0.05	C	Orange	0.108	0.085	0.118	
0.125	RR	Green	0.105	0.079	0.095	
0.106	G	Blue-gr.	0.105			
0.096	K	Blue	0.144	0.145	0.099	
0.11[b]	Ro					
0.109	S					
0.096	Ra					

II. Intensity.

Fluid		I_w/I_f	Ref.[a]	λ	$10^6 Ir^2/VE$	Ref.[a]
Ether	(l)	0.192	ML		1.77	ML
Ether	(l)	0.192	RR	4358	1.77	M²
Benzene	(l)	0.069	S	5461	0.72	M²
Toluene	(l)	0.060	M¹	5780	0.57	M²
Air	(g)	165.	RR			

[492] Plotnikow, J., and Šplait, L., *Physik. Z.*, **31**, 369-372 (1930).
[493] Plotnikow, J., and Nishigishi, S., *Idem*, **32**, 434-444 (1931).
[494] Čoban, B., *Acta Phys. Polon.*, **4**, 1-16 (1935).
[494a] Šplait, L., *Idem*, **4**, 329-330 (1935).
[495] Krishnan, R. S., *Proc. Indian Acad. Sci.*, **1**, 44-47, 211-216 (1935); Mitra, S. M., *Z. Physik*, **96**, 34-36 (1935); Vrkljan, V. S., *Acta Phys. Polon.*, **4**, 325-327 (1935); and Katalinič, M., *Koll. Z.*, **74**, 288-296 (1936); *Z. Physik*, **106**, 439-452 (1937).
[496] Walsh, J. W. T., and Buckley, H., *Int. Crit. Tables*, **5**, 266 (1929).

Table 148—*(Continued)*

III. L. H. Dawson and E. O. Hulburt.[491] If the molecule of water were isotropic, then, on the density-fluctuation theory of Einstein and Smoluchowski, the total radiation of wave-length λ that a unit volume of water at 22 °C would scatter per unit of solid angle (*i.e.*, per steradian) in directions perpendicular to the direction of propagation of the incident light would be $I_\lambda \alpha$, where α has the values here tabulated, and I_λ is the intensity of the incident radiation of wave-length λ. Since the molecules are anisotropic, the scattering will exceed $I_\lambda \alpha$. The amount of this excess in the ultraviolet is unknown; in the visible spectrum it is less than 30 per cent. The observations of Dawson and Hulburt in the range $\lambda = 253.6$ to 546.1 mμ agree relatively with the values here tabulated.

Unit of $\lambda = 1$ m$\mu = 10^{-7}$ cm; of $\alpha = 1$ cm^{-1} steradian^{-1}. Temp. $= 22$ °C $= 295$ °K

λ	$10^7 \alpha$	λ	$10^7 \alpha$	λ	$10^7 \alpha$	λ	$10^7 \alpha$
600	0.79	500	1.66	400	4.06	300	14.8
550	1.03	450	2.53	350	7.15	250	34.4

[a] References:
- C   Cabannes, J., *Jour. de Phys.* (6), **3**, 429-442 (1922).
- G   Gans, R., *Z. Physik*, **17**, 353-397 (1923) ← *Contrib. fac. de cienc. Univ. La Plata (Mat. Fis.)*, **3**, 251-315 (1923); *Z. Physik*, **30**, 231-239 (1924).
- K   Krishnan, K. S., *Phil. Mag.* (6), **50**, 697-715 (1925).
- M   Martin, W. H., M¹=*J. Phys'l Chem.*, **24**, 478-492 (1920); M²=*Idem*, **26**, 471-476 (1922).
- ML Martin, W. H., and Lehrman, S., *Idem*, **26**, 75-88 (1922).
- Ra Ramanadham, M., *Indian J. Phys.*, **4**, 15-38 (1929).
- Ro Rocard, Y., *Compt. rend.*, **180**, 52-53 (1925).
- RR Raman, C. V., and Rao, K. S., *Phil. Mag.* (6), **45**, 625-640 (1923).
- S   Sweitzer, C. W., *J. Phys'l Chem.*, **31**, 1150-1191 (1927).

[b] Rocard reports this value for green light, and says that ρ varies very little with the wave-length. If fluorescence had not been eliminated [?] then $\rho = 0.16$ to 0.18.

Raman Scattering by Water.

Water contrasts sharply with most other liquids in that its Raman spectrum consists of broad diffuse bands, some of which overlap. This, together with the fact that the spectrum of the mercury arc, which is the most satisfactory illuminant, contains a number of bright lines, makes interpretation of the observations difficult, unless care is taken to remove from the light of the arc all except one line, or a small group of closely spaced lines. That has been done by H. Hulubei,[497] H. Hulubei and Y. Cauchois,[498] M. Magat,[499] and J. H. Hibben [500]; but most of the recorded observations have been made with the unfiltered radiation.

The most prominent features of the Raman spectrum of water are two bands, one broad and centered near the wave-length corresponding to $\delta\nu = 3400$ cm^{-1}, the other narrow and centered near $\delta\nu = 1650$ cm^{-1}. There has been much discussion about the structure of the first (see Table 152). It probably has three components, the strongest having its maximum near $\delta\nu = 3400$ cm^{-1}, the one of intermediate strength near $\delta\nu = 3200$ cm^{-1}, and

[497] Hulubei, H., *Compt. rend.*, **194**, 1474-1477 (1932).
[498] Hulubei, H., and Cauchois, Y., *Idem*, **192**, 1640-1643 (1931).
[499] Magat, M., *Idem*, **196**, 1981-1983 (1933); *Jour. de Phys.* (7), **5**, 347-356 (1934).
[500] Hibben, J. H., *J. Chem'l Phys.*, **5**, 166-172, 994 (1937).

a very weak one near $\delta\nu = 3600$ cm^{-1}, but H. Hulubei [497] and M. Magat [501] failed to find the 3600 cm^{-1} component, although they sought for it; and E. H. L. Meyer [502] suggested that the apparent structure of this band is an optical illusion. For the early discussion of the subject see W. Gerlach [503] and E. H. L. Meyer.[504] More recently, Magat [505] has reported that he has found this 3600 cm^{-1} component, but only at temperatures above 37 °C.

I. R. Rao [506] has sought to interpret the observations in terms of the polymerization of water; M. Magat,[501, 505, 507] accepting the quasicrystalline theory of liquid structure, has sought to interpret them in terms of the several modes of vibration of the molecule when subjected to the action of its neighbors. Magat's view is the one more favored at present.

Most of the early work was limited to a study of the bands near $\delta\nu = 3400$ and 1650 cm^{-1}, but many other lines and bands have been mapped (see Table 153). These have been regarded by I. R. Rao and P. Koteswaram [508] as spurious, as arising from excitation by another spectral line than that supposed by the observer; but J. H. Hibben [509] seems to have shown conclusively that such is not the case, that in his work, at least, the lines in dispute cannot have arisen from excitation by any other line than that he supposed, and that at least the lines near $\delta\nu = 175, 500, 1659$, and 2150 cm^{-1} are true Raman lines. He had not observed the lines reported at $\delta\nu = 4023$ and 5100 cm^{-1}, but he gave reasons for believing that Rao and Koteswaram's criticism is inapplicable to them also.

G. Bolla [510] has reported a spurious multiplication of the Raman bands under certain instrumental conditions.

General reviews of certain phases of the work in this field have been recently published by M. Magat,[511] A. Kastler,[512] P. C. Cross, J. Burnham, and P. A. Leighton,[513] and J. H. Hibben.[514]

Polarization and Intensity of the Raman Bands of Water.—J. Cabannes [515] has found that for a given substance the polarization of any given Raman line or band is independent of the frequency of the exciting radiation, and that the amount of the polarization of the several Raman lines or bands, each corresponding to a different value of $\delta\nu$, may differ, the depolarization factor (ρ) lying between 0 and 1 (actually, 6/7 is the limiting

---

[501] Magat, M., *Jour. de Phys. (7)*, 5, 347-356 (1934).
[502] Meyer, E. H. L., *Physik. Z.*, 31, 510-511 (1930).
[503] Gerlach, W., *Physik. Z.*, 31, 695-698 (1930).
[504] Meyer, E. H. L., *Idem*, 31, 699-700 (1930).
[505] Magat, M., *Jour. de Phys. (7)*, 6, 64S-65S (1935).
[506] Rao, I. R., *Proc. Roy. Soc. (London) (A)*, 130, 489-499 (1931); *Nature*, 132, 480 (1933); *Proc. Roy. Soc. (London) (A)*, 145, 489-508 (1934); *Phil. Mag. (7)*, 17, 1113-1134 (1934).
[507] Magat, M., *Ann. de Phys. (11)*, 6, 108-193 (1936); *Trans. Faraday Soc.*, 33, 114-120 (1937).
[508] Rao, I. R., and Koteswaram, P., *J. Chem'l Phys.*, 5, 667 (L) (1937).
[509] Hibben, J. H., *Idem*, 5, 994 (L) (1937).
[510] Bolla, G., *Nature*, 128, 546-547 (L) (1931); 129, 60 (L) (1932).
[511] Magat, M., *Ann. de Phys. (11)*, 6, 108-193 (1936) (Bibliog. of 148 entries).
[512] Kastler, A., *Rev. gén. des Sci. (Paris)*, 47, 522-536, 559-566 (1936).
[513] Cross, P. C., Burnham, J., and Leighton, P. A., *J. Am. Chem. Soc.*, 59, 1134-1147 (1937).
[514] Hibben, J. H., *J. Wash. Acad. Sci.*, 27, 269-299 (1937).
[515] Cabannes, J., *Compt. rend.*, 187, 654-656 (1918).

value). Values of ρ are given in Table 149. See also F. Heidenreich.[516] Using $\lambda_{Hg} = 3650A$ as exciter, G. I. Pokrowski and E. A. Gordon [517] measured both the polarization $(1 - \rho)$ and the relative intensity of the band $\delta\nu = 3400$ cm^{-1} when scattered at an angle θ with the direction of the incident beam (Table 149).

Of the three components ($\delta\nu = 3200, 3450, 3600$ cm^{-1}) of the $\delta\nu = 3400$ cm^{-1} band, the second is the strongest at room temperatures; and the third the weakest.[518]

Table 149.—Polarization and Angular Scattering of the Raman Bands of Water

$\delta\nu$ = approximate shift defining the maximum of the band or component studied; ρ = depolarizing factor; I = relative intensity of the band as scattered at an angle θ with the direction of the incident beam.

Unit of $\delta\nu = 1$ cm^{-1}; ρ and I are ratios

$\delta\nu \rightarrow$ Ref.[a]	175	500	750	1650	3200	3450	3600
				ρ			
An	0.85	g	g	–	0.10–0.15	0.40–0.50	0.85
CdeR	–	–	–	0.4	<0.30[b]	0.30	–
Ra	–	–	–	–	0.60	0.48	0.75
Sp	–	–	–	–	0.62	0.52	0.54
CR	–	–	–	–	Yes[c]	No[c]	–

PG[a] Exciter, $\lambda_{Hg} = 3650A$; $\delta\nu = 3400$.

θ	20°	40°	60°	90°	120°	140°	160°
I	2.0	1.9	1.3	1.0	0.7	0.7	0.7
$(1-\rho)^d$	0.07	–	–	0.70 ± 0.02	–	–	0.7

[a] References:
An Ananthakrishnan, R., *Proc. Indian Acad. Sci.*, **3**, 201-205 (1936).
CDeR Cabannes, J., and de Riols, J., *Compt. rend.*, **198**, 30-32 (1934).
CR Cabannes, J., and Rousset, A., *Idem*, **194**, 706-708 (1932).
PG Pokrowski, G. I., and Gordon, E. A., *Ann. d. Physik (5)*, **4**, 488-492 (1930).
Ra Ramaswamy, C.[518]
Sp Specchia, O., *Nuovo Cim. (N. S.)*, **9**, 133-137 (1932).

[b] This line is said to be more polarized than the 3450 cm^{-1} one, hence the < 0.30.
[c] No numerical value given; the 3200 line is said to be depolarized; the 3450 one to be polarized.
[d] They call these values the "polarization," presumably meaning $(1 - \rho)$.

Effect of Temperature on the Raman Scattering by Water.—At 11.5 °C the intensity of the band $\delta\nu = 3400$ cm^{-1} is the same whether the band is excited by $\lambda_{Hg} = 3020, 2968$, or $2654A$; whereas at 55 °C the one excited by $\lambda_{Hg} = 2968A$ is about 20 per cent more intense than either of the others.[519] But P. Pringsheim and S. Slivitch [520] have reported that the relative intensities of the several repetitions of a given band ($\delta\nu = 3400$ cm^{-1}), each corresponding to one of the stronger lines of the mercury spectrum, are independent of the temperature of the water.

Both of the two broad bands observed at 4690A and 4250A (presumably

[516] Heidenreich, F., *Z. Physik*, **97**, 277-299 (1935).
[517] Pokrowski, G. I., and Gordon, E. A., *Ann. d. Physik (5)*, **4**, 488-492 (1930).
[518] Ramaswamy, C., *Nature*, **127**, 558 (L) (1931); Specchia, O., *Nuovo Cim. (N. S.)*, **9**, 133-137 (1932). See also Hibben, J. H., *J. Chem'l Phys.*, **5**, 166-172 (1937).
[519] Meyer, E. H. L., and Port, I., *Physik. Z.*, **31**, 509-510 (1930).
[520] Pringsheim, P., and Slivitch, S., *Z. Physik*, **60**, 581-585 (1930).

corresponding, respectively, to $\delta\nu$ about 3400 cm^{-1} and 1650 cm^{-1}) become narrower and sharper as the temperature is increased.[521]

The $\delta\nu$ = 140 cm^{-1} band, observed and studied between 4 °C and 97 °C by E. Segrè,[522] decreases in intensity as the temperature is increased.

As the temperature is increased, the maximum of the broad Raman band centered near $\delta\nu$ = 3400 shifts in the direction of increasing $\delta\nu$, and

Table 150.—Shift of Raman Lines of Water with Change in Temperature

(See text for comments and references to other work.)

$c\delta\nu$ is the frequency difference corresponding to the maximum intensity of the Raman line or band; $\delta\nu = \pm (1/\lambda_m - 1/\lambda_i)$, where λ_i and λ_m are the wave-lengths of the incident and the scattered radiation, respectively; c = velocity of light; ρ = density of the water.

Unit of $\delta\nu$ = 1 cm^{-1}; of ρ = 1 g/ml. Temp. = t °C

Ref.[a]→	Uk[b]		Ra[a]	
t	ρ	$\delta\nu$	t	$\delta\nu$
60	0.98	3448	0	3502
130	0.93	3497	4	3412
200	0.86	3524	38	3493
260	0.78	3520	98	3466
300	0.70	3530		
320	0.66	3528	Sp[a]	
350	—	3530	17	3406
380	0.33	3530	41	3417
360	0.133	3530	60	3429
350	0.096	{3530[c] / 3646}	80	3452
			91	3474
330	0.055	3646[d]		
310	0.025	3645	Me[a]	
250	0.0135	{3639[e] / 3653}	11.5	3414
			55	3430
200	0.007	{3639[e] / 3653}	92	3551

Hi[a]			
28 °C	88 °C	28 °C	88 °C
$\delta\nu$		$\delta\nu$	
144	149	2170	2118
440	450	3219	3222
1627	1629	3445	3460

[a] References:
- Hi Hibben, J. H.[500]
- Me Meyer, E. H. L., *Physik. Z.*, **31**, 510-511 (1930).
- Ra Rao, I. R., *Proc. Roy. Soc. (London) (A)*, **145**, 489-508 (1934).
- Sp Specchia, O., *Nuovo Cim. (N. S.)*, **7**, 388-391 (1930).
- Uk Ukholin, S. A., *Compt. rend. Acad. Sci. URSS*, **16**, 395-398 (1937).

[b] Ukholin worked with water sealed in quartz tubes and heated to various temperatures; the state of the water was specified by means of the temperature and the density, which in many cases was less than at the critical point ($\rho_{crit} = 0.33$).

[c] Here the band persists and a new line appears.

[d] Here the band has vanished and only the new line remains.

[e] The line has now split into two.

---

[521] Meyer, E. H. L., *Physik. Z.*, **30**, 170 (1929).
[522] Segrè, E., *Atti Accad. Linc. (6)*, **13**, 929-931 (1931).

the component of smallest $\delta\nu$ ($\delta\nu$ about 3200 cm⁻¹, corresponding to $\lambda_R =$ 3.13 μ) decreases in intensity, nearly vanishing as 100 °C is approached.[523] The intensities of the other two components of that band ($\delta\nu$ about 3450 cm⁻¹ and 3600 cm⁻¹, corresponding to $\lambda_R = 2.90\ \mu$ and 2.77 μ, respectively) change but little with the temperature,[524] the change of the last ($\delta\nu = 3600$ cm⁻¹) being an increase (Rao 1930).[523]

The original papers should be consulted. More recent work, covering the effect of temperature on each of several bands, may be found in the papers here noted.[525]

Certain data given by Hibben and Ukholin are included in Table 150. Magat [526] is of the opinion that the variation of the Raman lines of water with the temperature is peculiar near 40 °C; but G. Bolla [525] disagrees with him.

Effect of Solutes on the Raman Scattering by Water.—The relative intensities of the several repetitions of a given band ($\delta\nu = 3400$ cm⁻¹), each corresponding to one of the stronger lines of the mercury spectrum, are unchanged by the addition of a solute to the water.[520]

The solution in water of HNO_3 or of certain salts forming electrolytic solutions causes the components of the 3400 cm⁻¹ band to become sharper, the 3200 cm⁻¹ component to decrease in intensity, and the 3600 cm⁻¹ component to increase; the band is shifted in the direction of increasing $\delta\nu$. In concentrated solutions of HNO_3, the 3200 cm⁻¹ component is vanishingly weak; whereas the 3600 cm⁻¹ one is the strongest of the three.[527] In solutions of $NaNO_3$, this band is shifted as just stated, but the intensity of the 3600 cm⁻¹ component is decreased, vanishing in a 66 per cent solution.[528]

On the other hand, the solution of HCl decreases the intensity of both the components 3200 cm⁻¹ and 3600 cm⁻¹, and somewhat increases that of 3400 cm⁻¹ (Brunetti and Ollano [527]; Rafalowski [527]).

In contrast to the observers mentioned in the two preceding paragraphs, W. Gerlach,[530] who has reported that he finds only two components (3200 cm⁻¹, 3400 cm⁻¹) in the 3400 cm⁻¹ band, has stated that only the 3400 cm⁻¹ component was visible in solutions of LiCl and of $CaCl_2$, but both were visible in solutions of $ZnCl_2$ and of $CdCl_2$. In solutions of the alkali nitrates

[523] Bhagavantam, S., *Ind. J. Phys.*, **5**, 49-57 (1930); Meyer, E. H. L., *Physik. Z.*, **31**, 510-511 (1930); Nisi, H., *Jap. J. Phys.*, **7**, 1-32 (1931); Rao, I. R., *Nature*, **125**, 600 (1930); *Proc. Roy. Soc. (London) (A)*, **130**, 489-499 (1931); Specchia, O., *Nuovo Cim. (N. S.)*, **7**, 388-391 (1930); Ganesan, A. S., and Venkateswaran, S., *Indian J. Phys.*, **4**, 195-280 (1929).

[524] Bhagavantam, S., *Indian J. Phys.* **5**, 49-57 (1930); Ganesan, A. S., and Venkateswaran, S., *idem*, **4**, 195-280 (1929).

[525] Bolla, G., *Nuovo Cim. (N. S.)*, **12**, 243-246 (1935); Rao, C. S. S., *Proc. Roy. Soc. (London) (A)*, **151**, 167-178 (1935); Magat, M., *Jour. de Phys.* (7), **6**, 64S-65S (1935); *Ann. de Phys. (11)*, **6**, 108-193 (1936); Ananthakrishnan, R., *Proc. Indian Acad. Sci.*, **3**, 201-205 (1936); Cross, P. C., Burnham, J., and Leighton, P. A., *J. Am. Chem. Soc.*, **59**, 1134-1147 (1937); Hibben, J. H., *J. Chem'l Phys.*, **5**, 166-172 (1937); Magat, M., *Trans. Faraday Soc.*, **33**, 114-120 (1937); Ukholin, S. A., *Compt. rend. Acad. Sci. URSS*, **16**, 395-398 (1937).

[526] Magat, M., *Jour. de Phys.* (7), **6**, 64S-65S (1935); *Trans. Faraday Soc.*, **33**, 114-120 (1937).

[527] Brunetti, R., and Ollano, Z., *Atti Accad. Lincei* (6), **12**, 522-529 (1930); Rafalowski, S., *Bull. Int. de l'Acad. Polonaise (A)*, **1931**, 623-628 (1931) → *Nature*, **128**, 546 (1931); Rao, I. R., *Nature*, **124**, 762 (1929) ← *Proc. Roy. Soc. (London) (A)*, **127**, 279-289 (1930); *Nature*, **125**, 600 (1930); *Proc. Roy. Soc. (London) (A)*, **130**, 489-499 (1931).

[528] Cabannes, J., and de Riols, J., *Compt. rend.*, **198**, 30-32 (1934).

[530] Gerlach, W., *Naturwissenschaften*, **18**, 68 (L), 182-183 (L) (1930); *Physik. Z.*, **31**, 695-698 (1930).

the wave-length separation of the two components increased almost linearly with the concentration, the band shifted in the direction of decreasing $\delta\nu$, and the 3400 cm^{-1} component split into two, as the concentration increased. He stated that a broad unresolved band at λ = 4160A (Is it the $\delta\nu$ = 1650 cm^{-1} band?) is weak in solutions of LiCl, but is sharp and displaced in the direction of increasing $\delta\nu$ in solutions of CaCl$_2$. N. Embirikos [531] generally confirms Gerlach.

In ammonium solutions, the $\delta\nu$ = 1650 cm^{-1} band is shifted in the direction of increasing $\delta\nu$, the shift being small for the nitrate and the chloride, but exceptionally great (about 30 cm^{-1}) for the sulfate.[532] This band is but little affected by adding NaNO$_3$ to the water.[528]

Table 151.—Analysis of the Raman Spectrum of Water [542]
(For the band near $\delta\nu$ = 3400 cm^{-1} see Table 152.)

v_σ, v_π, and v_δ are the quantum numbers corresponding to the three fundamental modes of vibration of the free molecule (Table 64); v_1, v_2, v_3, and v_0 are four others corresponding to fundamental vibrations determined by the interaction of the molecule with its neighbors. The frequency of a given vibration is $c\nu$ where $\nu = v_\sigma \nu_\sigma + v_\pi \nu_\pi + \ldots + v_0 \nu_0$, and $c\nu_\sigma$, $c\nu_\pi$, ... $c\nu_0$ are the frequencies of the 7 fundamental vibrations.

v_σ	v_π	v_δ	v_1	v_2	v_3	v_0	Infrared ν_{obs}	Raman ν_{obs}	ν_{calc}
0	0	0	0	1	0	0	—	60	(60)
0	0	0	0	0	0	1	—	152–225	166^a
0	0	0	0	0	1	0	510	500	570^a
0	0	0	1	0	0	0	670	740	(700)
0	0	1	0	1	0	0	1710	—	1720
0	0	1	0	0	0	1	1850	—	1820
0	0	1	0	0	1	0	2135	2135	2160
1	0	0	0	0	1	0	4023	4024	3950
1	0	1	0	0	1	0	5590	—	5620

^{a} Frequencies near these may be derived from Bernal and Fowler's proposed structure of water (see p. 174).

See also C. C. Hatley and D. Callihan,[533] H. Hulubei,[497] C. Ramaswamy,[534] A. da Silveira and E. Bauer,[535] A. Hollaender and J. W. Williams,[536] E. Bauer,[537] M. Magat,[538] F. Cennamo,[539] J. H. Hibben,[500] P. A. Leighton and J. Burnham,[540] and for the earlier work, especially K. W. F. Kohlrausch.[480]

[531] Embirikos, N., *Physik. Z.*, **33**, 946-947 (1932).
[532] da Silveira, A., *Compt. rend.*, **195**, 521-523 (1932).
[533] Hatley, C. C., and Callihan, D., *Phys. Rev. (2)*, **38**, 909-913 (1931).
[534] Ramaswamy, C., *Indian J. Phys.*, **5**, 193-206 (1930).
[535] da Silveira, A., and Bauer, E., *Compt. rend.*, **195**, 416-418 (1932).
[536] Hollaender, A., and Williams, J. W., *Phys. Rev. (2)*, **34**, 994-996 (1929).
[537] Bauer, E., *Jour. de Phys. (7)*, **6**, 63S-64S (1935).
[538] Magat, M., *Jour. de Phys. (7)*, **6**, 64S-65S (1935).
[539] Cennamo, F., *Nuovo Cim. (N. S.)*, **13**, 304-309 (1936).
[540] Leighton, P. A., and Burnham, J., *J. Am. Chem. Soc.*, **59**, 424-425 (L) (1937).

39. WATER: LUMINESCENCE

Interpretation of the Raman Spectrum of Water.—Numerous attempts have been made to interpret the Raman spectrum in terms of the fundamental vibrations of the water molecule (see Tables 64 and 65) and in terms of the observed infrared spectrum of water. At first they were limited to the band near $\delta \nu = 3400$ cm^{-1} (see Table 152), but recently they have been extended by M. Magat [541] to other lines. He has concluded that

Table 152.—Analysis of the Raman Band near $\delta \nu = 3400$ cm^{-1}

The following values of $\delta \nu$ and of $\lambda_R \equiv 1/\delta \nu$ are those assigned by the several observers to the maxima of the components of the band. Some give $\delta \nu$; some, λ_R; and some, both; if only one of these quantities has been published, the compiler has computed the other from it. Such computed values are enclosed in parentheses. Hu and Mag sought for the 3600 component, but could not find it; the others cited in the second section of the table say nothing about it. E. H. L. Meyer [543] suggested that the apparent structure of this band is an optical illusion; for the resulting controversy, see W. Gerlach [503] and E. H. L. Meyer.[504]

Unit of $\delta \nu = 1$ cm^{-1}; of $\lambda_R = 1$ $\mu = 10^{-4}$ cm

$\delta \nu$			λ_R			Ref.[a]
(3195)	(3448)	(3610)	3.13	2.90	2.77	Bh[1]
3200	3435	3630	3.12	2.91	2.75	Bo[3,4]
3225	3469	3589	3.10	2.88	2.79	BO
3224	3436	3625	(3.10)	(2.91)	(2.76)	CRi
3230	3450	3560	(3.10)	(2.90)	(2.81)	CC
3199	3453	3609	3.13	2.90	2.77	GaV
3228	3435	3624	3.10	2.91	2.76	HaC
3206	3456	3578	3.12	2.89	2.79	N[1]
3180	3440	3630	(3.14)	(2.91)	(2.75)	Ry[1]
3208	3419	3582	(3.12)	(2.92)	(2.79)	Ro[3]
(3205)	(3413)	(3584)	3.12	2.93	2.79	Ro[4]
3084	3423	3628	3.24	2.92	2.75	Ro[5]
3217	3433	3582	(3.11)	(2.91)	(2.79)	Ro[6,7]
3278	3406	3569	3.04	2.92	2.80	Sp[1]
3246	3405	3554	3.08	2.93	2.81	Sp[1]
3208	3435	3595	3.11$_7$	2.90$_9$	2.77$_9$	Mean
3232	3422	—	(3.09)	(2.92)	—	Ge
3324	3513	—	(3.01)	(2.85)	—	DuKo[1,2]
3233	3443	none	(3.09)	(2.90)	—	Hu
3195	3394	—	(3.13)	(2.95)	—	Ki
3221	3435	none	(3.10)	(2.91)	—	Mag[2]
3195[b]	3437	—	(3.13)	(2.91)	—	N[2]
3217	3441	—	3.09$_1$	2.90$_7$	—	Mean

[a] References: See Table 153.
[b] Nisi states that when the exciter is $\lambda_{Hg} = 4047$A or 3650–3663A the band looks like a triplet, but that when the exciter is $\lambda_{Hg} = 2967$A it consists of only two (diffuse) bands, as here given, with no indication of a third. Magat (Mag[2]) used both $\lambda_{Hg} = 4358$A and $\lambda_{Hg} = 2537$A, and in both cases failed to find the 3600 cm^{-1} component.

[541] Magat, M., *Ann. de Phys. (11),* **6**, 108-193 (1936); *Trans. Faraday Soc.,* **33**, 114-120 (1937).
[542] Magat, M., *Ann. de Phys. (11),* **6**, 108-193 (1936).
[543] Meyer, E. H. L., *Physik. Z.,* **31**, 510-511 (1930).

in addition to the three independent frequencies of the free molecule, four others, representing the effects of other molecules upon the one that is scattering the radiation, must be considered. His scheme is given in Table 151. S. A. Ukholin [525] has concluded that it is unsatisfactory to ascribe the maxima near $\delta\nu = 3400$ and 3600 cm^{-1}, respectively, to Bernal and Fowler's types II and III of water (p. 174). See also R. Ananthakrishnan.[525]

Table 153.—Raman Spectrum of Water

Here are given all the more important reported values of $\delta\nu$ and of $\lambda_R (\equiv 1/\delta\nu)$. Some refer to the maxima of unresolved bands, some to the maxima of the components of bands.

Unit of $\delta\nu = 1$ cm^{-1}; of $\lambda_R = 1$ $\mu = 0.0001$ cm

Orig.[a]	$\delta\nu$	λ_R	Remarks[b]	Ref.[c]	Orig.[a]	$\delta\nu$	λ_R	Remarks[b]	Ref.[c]
rot.	60	160.7		Bo[3,4]	f	3221	3.105		Mag[2]
	134	74.6		An		3224	3.102		CRi
	236	42.4		An		3225	3.101		BO
	140	71.4		Se		3228	3.10		HaC
	144	69.4		Hi[1]		3230	3.096		CC
	172	58.1		Bo[3,4]		3231	3.095		An
a	175	57.1		Mag[2], Hi[2]		3233	3.093		Hu
	200	50.0		Mag[3], CBL		3246	3.081		Sp[1]
	340	29.4		De		3260	3.067		Sp[2]
	440	22.7		Hi[1]		3270	3.058		Ro[1]
	464	21.6		An		3278	3.051		Sp[1]
a	500	20.0		Mag[2,3], Hi[2]		3290	3.04		Po
	510	19.6		Bo[4]		3324	3.008		DuKo[1,2]
	550	18.2		CRi		3360	2.976		Ro[1,2]
	600	16.7		Mag[1,2]		3390	2.950		Ro[1]
	700	14.3		CRi, Mag[3]		3394	2.946		Ki
a	740	13.5		Mag[2]		3400	2.941		DeKg, Ry[2]
	754	13.3		An		3405	2.937		Sp[1,2]
	780	12.8		Bo[4]		3406	2.936		Sp[1]
	1627	6.15		Hi[1]		3410	2.93	$(H_2O)_2$	Ro[4]
	1645	6.08		Bo[4]		3419	2.925		DuKo[1], Ro[3]
	1650	6.06		CRi		3420	2.924		HuC
	1656	6.04		CBL		3423	2.921		Ro[5]
f	1659	6.03		Mag[2], Hi[2]		3428	2.917		An
	1665	6.00		An		3431	2.915		MeP
	1705	5.86		KU		3433	2.913		Ro[6,7]
	2130	4.69		Mag[3]	f	3435	2.911	Max	Bo[3,4], HaC, Mag[2]
	2135	4.68		Mag[2]		3436	2.910		CRi, An
a	2150	4.65		Bo[4], Hi[2]		3437	2.909		N[2]
	2170	4.61		Hi[1], CBL		3440	2.907		Ry[1], CBL
	2355(?)	4.25(?)		GaV		3443	2.904		Hu
	3084	3.24		Ro[5]		3444	2.904		GhK
	3100	3.226		Ry[2]		3445	2.903		Hi[1]
	3180	3.145		Ry[1]		3448	2.900		Bh[1]
	3190	3.135		CBL		3450	2.898		CC, CPR
	3195	3.130	H_2O	Bh[1], N[2], Ki		3453	2.896		GaV
	3199	3.126		GaV		3456	2.894		N[1]
	3200	3.125		Bo[3,4]		3469	2.883		Bo
	3205	3.12		Ro[4]		3474	2.878		KU
	3206	3.119		N[1]		3513	2.846		DKUo[1,2]
	3208	3.117		Ro[3]		3554	2.814		Sp[1]
	3214	3.111		An		3560	2.809		CC, Sp[2]
	3217	3.108	Max	Ro[6,7]		3569	2.802		Sp[1]
	3219	3.106		Hi[1]					

39. WATER: LUMINESCENCE

Table 153—(Continued)

Orig.[a]	$\delta\nu$	λ_R	Remarks[b]	Ref.[c]	Orig.[a]	$\delta\nu$	λ_R	Remarks[b]	Ref.[c]
	3578	2.795		N[1]		6042[d]	1.655[d]		BO
	3582	2.792	Max	Ro[3,4,6,7]	c	6747	1.482	x	Hu
	3589	2.786		BO	c	7246	1.380	x	Hu
	3600	2.778		Ry[2], An		7729	1.294		HuC
	3605	2.774		An		7757	1.289	x	Hu
	3609	2.771		GaV	c	8200	1.220	x	Hu
	3610	2.770	($H_2O)_3$	Bh[1]		8243	1.213		HuC
	3624	2.759		HaC	c	8660	1.155	x	Hu
	3625	2.758		CRi		8703	1.149		HuC
	3628	2.756		Ro[5]	c	9175	1.090	x	Hu
	3630	2.755		Bo[3,4], Ry[1]		9223	1.084		HuC
	3650	2.740		CBL	c	9569	1.045		Hu
	3990	2.506		Bo[4]		10039	0.996		Hu
	4000	2.500		CBL		10151	0.985	x	HuC
a	4023	2.486		Mag[2,3]	c	10635	0.940		Hu
c	5090	1.965		Mag[2]		10944	0.910		Hu
	5100	1.961		Mag[2]		11264	0.888	x	HuC
	5502	1.818		GaV					

[a] Origin of the line: a = associated molecules (Mag[2]); c = combination tones (Mag[2], Hu); f = fundamental frequency of the molecule (Mag[2]); rot = from rotation of the molecule (Mag[2]).

[b] Remarks: Square bracket indicates extent of band; Max = position of maximum, but DuKo, W. Gerlach [544] and Hu found in the band centered near 3400 cm⁻¹ only two maxima, which the first two place near 3324 and 3513 cm⁻¹, and Hu near 3233 and 3443. H_2O, $(H_2O)_2$, and $(H_2O)_3$ are the molecules to which I. R. Rao [545] assigns the associated values of $\delta\nu$. In that paper he replies to Su, who disagrees with him and has endeavored to interpret this band in terms of H_2O and $(H_2O)_2$ only; x indicates that Mag[2] reports that he did not find the line.

[c] References:

An	Ananthakrishnan, R., *Proc. Indian Acad. Sci.*, **2**, 291-302 (1935).
Bh[1]	Bhagavantam, S., *Ind. J. Phys.*, **5**, 49-57 (1930); Bh[2], *Idem*, **5**, 237-307 (1930).
Bo[1]	Bolla, G., *Nature*, **128**, 546-547 (1931); Bo[2], *Idem*, **129**, 60 (1932); Bo[3], *Nuovo Cim. (N. S.)*, **9**, 290-298 (1932); Bo[4], *Idem*, **10**, 101-107 (1933).
BO	Brunetti, R., and Ollano, Z., *Atti Accad. Lincei (6)*, **12**, 522-529 (1930).
CBL	Cross, P. C., Burnham, J., and Leighton, P. A., *J. Am. Chem. Soc.*, **59**, 1134-1147 (1937).
CC	Carelli, A., and Cennamo, F., *Nuovo. Cim. (N. S.)*, **10**, 329-332 (1935).
CPR	Carrelli, A., Pringsheim, P., and Rosen, B., *Z. Physik*, **51**, 511-519 (1928).
CRi	Cabannes, J., and de Riols, J., *Compt. rend.*, **198**, 30-32 (1934).
De	Daure, P., *Idem*, **186**, 1833-1835 (1928).
DeKa	Daure, P., and Kastler, A., *Idem*, **192**, 1721-1723 (1931).
DuKo[1]	Dadieu, A., and Kohlrausch, K. W. F., *Naturwissenschaften*, **17**, 625-626 (1929); DuKo[2], *J. Opt. Soc. Amer.*, **21**, 286-322 (1931).
GaV	Ganesan, A. S., and Venkateswaran, S., *Indian J. Phys.*, **4**, 195-280 (1929).
GhK	Ghosh, J. C., and Kar, B. C., *J. Phys'l Chem.*, **35**, 1735-1744 (1931).
HaC	Hatley, C. C., and Callihan, D., *Phys. Rev. (2)*, **38**, 909-913 (1931).
Hi[1]	Hibben, J. H., *J. Chem'l Phys.*, **5**, 166-172 (1937); (28 °C.) Hi[2], *Idem*, **5**, 994 (L) (1937).
Hu	Hulubei, H., *Compt. rend.*, **194**, 1474-1477 (1932).
HuC	Hulubei, H., and Cauchois, Y., *Idem*, **192**, 1640-1643 (1931).
Ki	Kinsey, E. L., *Phys. Rev. (2)*, **34**, 541 (1929).
KU	Kimura, M., and Uchida, Y., *Jap. J. Phys.*, **5**, 97-101 (1928).
Mag[1]	Magat, M., *Compt. rend.*, **196**, 1981-1983 (1933); Mag[2], *Jour. de Phys. (7)*, **5**, 347-356 (1934); Mag[3], *Idem*, **6**, 64S-65S (1935).
Me[1]	Meyer, E. H. L., *Physik. Z.*, **30**, 170 (1929); Me[2], *Idem*, **31**, 510-511 (1930); Me[3], *Idem*, **31**, 699-700 (1930).
MeP	Meyer, E. H. L., and Port, I., *Idem*, **31**, 509-510 (1930).
N[1]	Nisi, H., *Jap. J. Phys.*, **5**, 119-137 (1929); N[2], *Idem*, **7**, 1-32 (1931).
Po	Pokrowski, G. I., *Z. Physik*, **52**, 448-450 (1928).
Ro[1]	Rao, I. R., *Indian J. Phys.*, **3**, 123-129 (1928); Ro[2], *Nature*, **123**, 87 (1929); Ro[3], *Idem*, **124**, 762 (1929) ← *Proc. Roy. Soc. (London) (A)*, **127**, 279-289 (1930); Ro[4], *Nature*, **125**, 600 (1930); Ro[5], *Proc. Roy. Soc. (London) (A)*, **130**, 489-499 (1931); Ro[6], *Phil. Mag. (7)*, **17**, 1113-1134 (1934); Ro[7], *Proc. Roy. Soc. (London) (A)*, **145**, 489-508 (1934).

[544] Gerlach, W., *Naturwissenschaften*, **18**, 68 (L) (1930).

[545] Rao, I. R., *Proc. Roy. Soc. (London) (A)*, **145**, 489-508 (1934).

Table 153—(Continued)

Ry[1]   Ramaswamy, C., *Indian J. Phys.*, **5**, 193-206 (1930); Ry[2], *Nature*, **127**, 558 (1931).
Se   Segrè, E., *Atti Accad. Lincei* (6), **13**, 929-931 (1931).
Sp[1]   Specchia, O., *Nuovo Cim.* (N. S.) **7**, 388-391 (1930); Sp[2], *Idem*, **9**, 133-137 (1932).

[d] These values are printed as 6642, 4.67, which are not self-consistent. It seems likely that there are two typographical errors; a 6 for a zero, and a 4 for a 1.

Table 154.—Abridged Raman Spectrum of Water

In this table the values that in the compiler's opinion may refer to the same band are connected by braces; in the case of the two prominent bands, centered near $\delta\nu = 1650$ cm^{-1} and 3400 cm^{-1}, only the extreme values and certain others of special interest are given, and the references in the "observer" columns apply to the band and not especially to the individual value on the line with them, except that the two "no's" on the line with $\delta\nu = 3610$ indicate that the maximum placed by several near 3610 was sought but not found by those observers. In the "observer" columns, r = recorded, no = sought but not found.

Orig.[b]	$\delta\nu$	λ_R	Bo[3,4]	CRi	Hu	HuC	Mag[2,3]	Misc
rot.	60	160.7	r					
	140 }	71.4 }						
	144 }	69.4 }						Se
	172 }	58.1 }	r					Hi[1]
a	175 }	57.1 }					r	
	200	50.0					r	Hi[2]
	340	29.4						De
	440	22.7						Hi[1]
	500 }	20.0 }					r	Hi[2]
a	510 }	19.6 }	r					
	550 }	18.2 }		r				
	550 }	18.2 }		r				
	600 }	16.7 }					r	
	700 }	14.3 }		r			r	
a	740 }	13.5 }					r	
	780 }	12.8 }	r					
	1627	6.15						Hi[1]
	1645 }							
f	1659 }	6.03	r	r			r	KU, Hi[2]
	1705 }							
	2130 }	4.69 }					r	
	2135 }	4.68 }					r	
a	2150 }	4.65 }	r					Hi[2]
	2170 }	4.61 }						Hi[1]
	2355(?)	4.25 (?)						GaV
	3084 }	3.24 }						
	3195	3.13						
f	3221 }	3.10 }	r	r	r	r	r	Many
	3400 }	2.94 }						
	3582	2.79						
	3610 }	2.77 }			no		no	
	3630 }	2.75 }						
	3990 }	2.50 }	r					
a	4023 }	2.49 }					r	
c	5090 }	1.96 }					r	
	5100 }	1.96 }					r	
	5502	1.81						GaV

39. WATER: LUMINESCENCE

Table 154—(Continued)

Orig.[b]	$\delta\nu$	λ_R	Bo[3,4]	CRi	Hu	HuC	Mag[2,3]	Misc
	6042[c]	1.66[c]						BO
c	6747	1.48			r		no	
c	7246	1.38			r		no	
	7729⎱	1.29⎱				r		
	7757⎰	1.29⎰			r		no	
c	8200⎱	1.22⎱			r		no	
	8243⎰	1.21⎰				r		
c	8660	1.12			r		no	
	8703	1.15				r		
c	9175⎱	1.09⎱			r		no	
	9223⎰	1.08⎰				r		
c	9569	1.04			r			
	10039	1.00			r			
	10151	0.98					r	no
c	10635	0.94			r			
	10944	0.91			r			
	11264	0.89					r	no

[a] Observer: Misc = miscellaneous observers. For significance of the symbols designating the observers, see references in Table 153.
[b] Origin of the line: See Table 153, note a.
[c] See note d in Table 153.

Electron and β-ray Luminescence of Water.

High-speed electrons, such as β-rays, excite in water the same kind of luminescence as do γ-rays (p. 304). Its intensity is not reduced by the common quenchers of fluorescence, nor by heating; it is partially polarized, the electric vector being parallel to the path of the electrons; its angular distribution is unsymmetrical, being much more intense in the direction of motion of the electron than in the reverse direction.[546] I. Frank and I. Tamm [547] seek to explain this asymmetry in terms of electrons moving with velocities exceeding that of light in the medium (here water). Čerenkov [546, p. 105-108] reported the following relative values of the intensity (I) of the light emitted at an angle ϕ with reference to the direction of motion of the exciting electrons:

ϕ	0°	15°	30°	37.5°	45°	60°	75°	90°
I	63	68	73	53	31	12.5	6.0	3.4

Mechanical Luminescence of Water.

In 1934 H. Frenzel and H. Schultes [548] announced that redistilled water luminesces under the action of ultrasonic vibrations, but that degassed water does not. L. A. Chambers,[549] using a frequency of 8.9 kc/sec observed such luminescence by 14 pure substances (including water) and some solutions, but none by 22 other pure substances. He has reported that the intensity of the luminescence varies "inversely with the temperature" and directly as $\mu\eta$, μ being the dipole moment and η the coefficient of

[546] Čerenkov, P. A. (Tscherenkov), *Compt. rend. Acad. Sci. URSS*, **14**, 101-105, 105-108 (1937).
[547] Frank, I., and Tamm, I., *Idem.*, **14**, 109-114 (1937).
[548] Frenzel, H., and Schultes, H., *Z. physik. Chem. (B)*, **27**, 421-424 (1934).
[549] Chambers, L. A., *J. Chem'l Phys.*, **5**, 290-292 (1937) → *Phys. Rev. (2)*, **49**, 881 (A) (1937).

viscosity of the substance. There is no visible luminescence if $10^{18}\mu\eta < 1.94$ cgse. The light "originates in cavitated areas or at the surface of the cavities." Similar observations have been made by V. L. Levšin and S. N. Rževkin.[550] They regard the light as an effect of the electrical potential differences that arise when the cavity is formed. They state that the light first appears at the liquid boundary, usually the lower, increases gradually in intensity and extent, until the entire volume of water is luminous, lasts for a time, and then abruptly vanishes.

40. Preparation of Dust-free Water

It is difficult to obtain a liquid free of suspended particles—exceedingly difficult in the case of water, much less difficult in the case of more mobile liquids. Various methods have been used and described in some detail by the workers here noted.[551]

By taking extreme precautions, Lallemand attained partial success with distillation. Spring did not succeed with distillation, obtained some success with filtration through animal black, and success with both electrical separation (cataphoresis) and gelatinous precipitation (envelopment). Biltz obtained his best results by precipitation with $Zn(OH)_2$, but found filtering through unglazed porcelain (Pukall filter) to be fairly satisfactory. Martin used repeated distillation *in vacuo* and without ebullition, fractional distillation in the same manner, envelopment, and cataphoresis. In his first paper, he reported that the remanent luminescence of water "is constant in intensity irrespective of the method of purification employed." In his second paper he stated that he believes this to be the first conclusive evidence for the scattering of light by pure substances. In that paper he also stated that the use of quartz vessels led to no improvement. Garrard has said that if there is no ebullition—no bubbling or bumping—during the distillation *in vacuo,* neither the actual temperature at which it is done nor the difference in the temperatures of the boiler and receiver affects the efficiency of the process for the removal of motes; that it is impossible to get dust-free water if the receiver contains either a piece of copper, or of vulcanized rubber, or a few cm³ of mercury; that if the receiver containing dust-free water is shaken, the water "is invariably contaminated again with motes," and that such recontamination is not prevented by shaking, rinsing back, and redistilling even as many as 20 times, and the same is true if the receiver is of quartz.

Sweitzer did not obtain satisfactory results with either ultrafiltration or centrifuging at 30 000 r.p.m., but did with envelopment. The least time for clearing when aluminum hydroxide was used as enveloper was two

[550] Levšin, V. L., and Rževkin, S. N. (Lewschin and Rschevkin), *Compt. rend. Acad. Sci. URSS,* **16,** 399-404 (1937).

[551] Lallemand, A., *Ann. de chim. et phys. (4),* **22,** 200-234 (1871); Spring, W., *Rec. trav. chim. Pays-Bas,* **18,** 153-168 (1899); Biltz, W., *Nachr. Ges. Wiss. Gottingen (Math.-Phys.),* 1904, 300-310 (1904); Marain, W. H., *Trans. Roy. Soc. Canada III (3),* **7,** 219-220 (1913); *J. Phys'l Chem.,* **24,** 478-492 (1920); Garrard, J. D., *Trans. Roy. Soc. Canada III (3),* **18,** 126-127 (1924); Sweitzer, C. W., *J. Phys'l Chem.,* **31,** 1150-1191 (1927); Schade, H., and Lohfert, H., *Koll. Z.,* **51,** 65-71 (1930); and Magat, M., *Jour. de Phys. (7),* **5,** 347-356 (1934).

weeks. No impairment was observed to result from prolonged standing in Pyrex vessels. Schade and Lohfert used the methods employed with success by Spring and by Biltz, and also that employed by W. Gerlach[544] *i.e.*, repeated distillation from a copper vessel. They state that the last is not inferior to the others if proper precautions are taken. They found that quartz vessels are not suitable, that water standing in such vessels very soon gives evidence of containing particles in suspension, presumably on account of solution of the quartz. On the other hand, carefully cleaned vessels of hard Jena glass were entirely satisfactory. Magat used doubly distilled water filtered through collodion.

W. H. Martin and S. Lehrman [552] have stated that water distilled in lead glass scatters 50 per cent more light than that distilled in sodium glass, and that the results for sodium glass, for Pyrex, and for fused quartz are all alike.

See also the papers here noted.[553]

41. Diffraction of X-rays by Water

When a slender pencil of x-rays is passed through water and impinges normally upon a photographic plate, the diffracted rays form upon the plate a principal dark ring, outside of which is a fairly uniform darkening which is rather sharply bounded along a ring concentric with the first. A more careful study has revealed 4 concentric rings at each of which the darkening of the plate passes through a maximum (see Table 155); but only the first is prominent. The intensities of these maxima, and to a less extent their positions, are affected by the absorption of the radiation by the water; W. Good [554] seems to have been the first to attempt to correct his data for water for this effect.

J. Thibaud and J. J. Trillat [555] have pointed out that, if the incident radiation contains both the general spectrum and the characteristic radiation, there will in general be two systems of rings, one arising from the characteristic radiation, and the other from the maximum of the general spectrum. Their relative intensities will depend upon the thickness of the layer of water. For a copper target and 40 kv, both systems will show if the water is only 1 mm thick, but only the second if it is 8 mm.

Explanation of the rings is far from simple.[556] Early observers attempted to explain them either as arising from diffraction by neighboring, more or less polymerized, molecules [557] or as originating within the molecules. At that time, R. W. G. Wyckoff [558] inclined to the latter, but stated that the data "do not exclude the possibility of their arising from characteristic

[552] Martin, W. H., and Lehrman, S., *J. Phys'l Chem.*, **26**, 75-88 (1922).

[553] Ananthakrishnan, R., *Proc. Indian Acad. Sci.*, **2**, 29-302 (1935); Magat, M., *Ann. de Phys. (11)*, **6**, 108-193 (1936); Mayer, J., and Pfaff, W., *Z. anorg. allgem. Chem.*, **242**, 305-314 (1935); and Malfitano, G., *J. de Chim. Phys.*, **19**, 32-33 (1921). The last gives some instructions for the preparation of collodion filters.

[554] Good, W., *Helv. Phys. Acta*, **3**, 205-248, 436 (1930).

[555] Thibaud, J., and Trillat, J. J., *Jour. de Phys. (7)*, **1**, 249-260 (1930).

[556] Amaldi, E., *Physik. Z.*, **32**, 914-919 (1931).

associations of molecules." C. V. Raman and K. R. Ramanathan [559] have developed a theory accounting for certain of the characteristics of the diffraction of x-rays by liquids on the basis of the fluctuations in density arising from thermal agitation. H. H. Meyer [560] was of the opinion that

Table 155.—Periodicities in the Diffraction of X-rays by Water

$d = \lambda / \left(2 \sin \dfrac{\phi}{2}\right)$ = equivalent grating space, ϕ = angular deviation corresponding to a maximum of the intensity of the diffracted rays, λ = wave-length of the x-rays incident upon the water; $t\,°C$ = temperature.

When an author reports λ and d, but no ϕ, the values of ϕ, as computed from λ and d, are enclosed in parentheses.

J. A. Prins [564] has reported that inside the main ring ($d = 3A$) the darkening of the plate remains nearly constant until the place corresponding to $d = 17A$ is reached, beyond which it decreases rapidly to a low limit.

More recently, a maximum between $d = 4A$ and $d = 5A$ has been observed, but it is not very pronounced.[565, 566]

Unit of λ and of $d = 1A = 10^{-8}$ cm $= 10^{-4}\mu$

λ		ϕ			d			t	Ref.[a]
0.712	13.44°	24°		3.04	1.71				ICT
1.54	29	46		3.07	1.97				ICT
1.54	27.3			3.25					So
1.539	(27.9)	(43.0)	(71.3°)	3.193	2.10	1.32		3	M
1.539	(28.5)	(42.8)	(70.1)	3.130	2.11	1.34		20	M
0.7090	(13.0)	(19.2)	(30.4)	(46.9°) 3.135	2.13	1.35	0.89	20	M
1.539	(28.8)	(42.8)	(66.7)	3.095	2.11	1.40		40	M
				3.27	2.11				St[1]
				3.24	2.11	1.13		21	St[2]
1.54	30.5	41		2.93[b]	2.20[b]				G

[a] References:
  G   Good, W.[554]
  ICT From compilation by Wyckoff, R. W. G., *Int. Crit. Tables*, 1, 338-353 (351) (1926). Values are based upon observations of Keesom, W. H., and DeSmedt, J., *Proc. Akad. Wet. Amsterdam*, 25, 118-124 (1922) [$\lambda = 0.712$]; 26, 112-115 (1923) [$\lambda = 1.54$].
  M   Meyer, H. H., *Ann. d. Physik (5)*, 5, 701-734 (1930).
  So Sogani, C. M., *Indian J. Phys.*, 1, 357-392 (1927).
  St[1] Stewart, G. W., *Phys. Rev. (2)*, 35, 1426 (A) (1930); St[2], Stewart, G. W.[561]

[b] After correcting for absorption. The other values in the table are not so corrected.

---

the main ring ($d = 3A$, see Table 155) arose from radiations scattered by adjacent molecules, and the next ($d = 2.1$) from those scattered by the

[564] Prins, J. A., *Z. Physik*, 56, 617-648 (1929).
[565] Katzoff, S., *J. Chem'l Phys.*, 2, 841-851 (1934).
[566] Warren, B. E., *J. Appl. Phys.*, 8, 645-654 (1937).
[557] Keesom, W. H., and DeSmedt, J., *Jour. de Phys. (6)*, 4, 144-151 (1923); 5, 126-128 (1924).
[558] Wyckoff, R. W. G., *Amer. J. Sci. (5)*, 5, 455-464 (1923).
[559] Raman, C. V., and Ramanathan, K. R., *Proc. Indian Ass. Cultiv. Sci.*, 8, 127-162 (1923).
[560] Meyer, H. H., *Ann. d. Physik (5)*, 5, 701-734 (1930).
[561] Stewart, G. W., *Phys. Rev. (2)*, 37, 9-16 (1931).
[562] Debye, P., and Menke, H., *Physik. Z.*, 31, 797-798 (1930).
[563] Stewart, G. W., *Phys. Rev. (2)*, 35, 726-732 (1930).

constituents of large complex groups of molecules; while G. W. Stewart [561] thought both of them arose from a single kind of molecular group. He advocated abandonment of the idea of an association into groups of a few molecules each, and its replacement by that of what he termed the cybotactic condition. In that condition, groups of hundreds or thousands of molecules have temporary existence, with ill-defined boundaries, and have a certain internal regularity. This accords with the conclusion of P. Debye and H. Menke,[562] that liquid mercury has a kind of quasi-crystalline structure. Evidence for the existence of the cybotactic condition in liquids has been summarized by G. W. Stewart.[563]

Table 156.—Diffraction of X-rays by Water: Intensity and Effect of Temperature [560]

I is the intensity relative to that (taken as 100) corresponding to d_1 at the same temperature, d = equivalent grating space.

Unit of $d = 1A = 10^{-8}$ cm. Temp. $= t\ °C$

t	$d_1{}^a$	I_1	$d_2{}^a$	I_2	d_3	I_3
3	3.193	100	2.10	21	1.32	5
20	3.130	100	2.11	18	1.34	5
40	3.095	100	2.11	14	1.40	5
Uncertainty	±0.003		±0.01		±0.03	

[a] G. W. Stewart [561] has reported that, as t increases, d_1 decreases at the rate of 0.0014A per 1 °C, d_2 increases, and at higher temperatures (above 40 °C) the maximum corresponding to d_2 vanishes. His observations extended from 2 °C to 98 °C.
S. Katzoff [565] has reported that changing the temperature from 3 °C to 90 °C does essentially nothing to the diffraction pattern beyond reducing the prominence of its features. B. E. Warren [566] has stated that the distance between adjacent O's increases from 2.9A at 1.5 °C to 3.0A at 83° C. This corresponds to an increase of about 1 in 10 in the specific volume, whereas the actual change (Table 94) is only 3.1 in 100.

42. Absorption and Scattering of X-rays and of γ-rays by Water

Until quite recently, it was thought that in their passage through matter such high-frequency electromagnetic radiation as x-rays and γ-rays disturb the massive nuclei of the atoms but little, their direct effects being restricted to an interaction with the extra-nuclear electrons. But now it is known that they excite the nucleus, which subsequently emits certain characteristic radiations. L. H. Gray and G. T. P. Tarrant [567] have reported that this characteristic radiation from water consists of two components; for one, the coefficient of absorption in lead is $\mu_{Pb} = 0.85$ cm^{-1}; for the other, $\mu_{Pb} = 1.96$ cm^{-1}.

The gross amount of energy expended in exciting the nuclei is, however, small as compared with that involved in the interactions between the radiation and the extra-nuclear electrons. Consequently, only the latter will be considered in the rest of this section, which accords with what has been the usual treatment of the subject.

The interaction between the radiation and an extra-nuclear electron may

[567] Gray, L. H., and Tarrant, G. T. P., *Proc. Roy. Soc. (London) (A)*, **136**, 662-691 (1932).

Table 157.—Absorption of X-rays and of γ-rays by Water

The absorption by water of x-rays generated by 100 kilovolts is equivalent to that by lead that is only 0.004 as thick.[571] In the region $\lambda = 0.1$ to 0.5A, the relation between λ^3 and μ is not linear.[572] The apparent absorption (μ_a) of water for the radiation from Ra-C, perhaps affected by the presence of secondary radiation arising from the cosmic radiation, varies with the thickness (x) of the water as follows [573]:

x	2	5	10	25 cm
1000 μ_a	17	25	28	38 cm^{-1}

For a study of the variation in the quality and the intensity of x-rays as they pass through water, and of the way these vary with the size of the beam, see F. Vierheller.[574]

Except as the contrary is indicated, the following data have been taken from a compilation by J. A. Gray,[575] which is based upon the work of J. Chadwick,[576] J. Chadwick and A. S. Russell,[577] C. W. Hewlett,[578] A. R. Olson, E. Dershem, and H. H. Storch,[579] F. K. Richtmyer,[580] E. G. Taylor,[581] and K. A. Wingårdh.[582]

$I = I_0 e^{-\mu x}$; $\mu_m = \mu M / \rho N_0$; ρ = density; M = formula-weight (H$_2$O) = 18.015; N_0 = number of molecules per g-mole = 6.061 × 10^{23}; for the significance of other symbols, see text.

Unit of $\lambda = 1A = 10^{-8}$ cm; of $\mu = 1$ cm^{-1}; of $\mu_m = 10^{-23}$ cm^2 per molecule

λ	μ	μ_m	λ	μ	μ_m
Cosmic[a]	0.00020	0.0006	0.340	0.290	0.861
Cosmic[a]	0.00075	0.0022	0.360	0.309	0.918
Cosmic[a]	0.00157	0.0047	0.380	0.330	0.980
Cosmic[a]	0.00518	0.0154	0.400	0.352	1.05
Cosmic[a]$_1$	0.0183	0.054	0.420	0.376	1.12
0.0047[b]	0.0437	0.130	0.440	0.400	1.19
Ra$_{10}$[c]	0.0472	0.140	0.500	0.500	1.48
Ra$_3$[d]	0.0558	0.166	0.550	0.600	1.78
0.059[d]$_1$	0.133	0.395	0.586	0.686	2.04
0.100	0.167	0.496	0.631	0.812	2.41
0.110	0.171	0.508	0.709	1.08	3.21
0.120	0.175	0.520	0.783	1.38	4.10
0.130	0.178	0.529	0.881	1.95	5.79
0.140	0.180[e]	0.536	0.929	2.18	6.47
0.150	0.183	0.545	0.977	2.52	7.48
0.160	0.187	0.555	1.539	9.00[f]	26.7
0.170	0.190	0.564			
0.180	0.194	0.576	Stumpen, H, *Physik. Z.*, 50, 215-227 (1928).		
0.190	0.198	0.588	0.158	0.186	0.55$_3$
0.200	0.201	0.597	0.211	0.204	0.60$_6$
0.220	0.212	0.630	0.264	0.238	0.70$_7$
0.240	0.223	0.662	0.317	0.280	0.83$_2$
0.260	0.234	0.695	0.370	0.335	0.99$_6$
0.280	0.246	0.731	0.423	0.384	1.14$_1$
0.300	0.259	0.769	0.475	0.482	1.43$_3$
0.320	0.273	0.811	0.56	0.649	1.92$_9$

[a] The reported apparent coefficients of absorption (μ_a) of the ultrapenetrating "cosmic" radiations vary from 0.0002 to 0.0052.[583]

[a]$_1$ Found by J. H. Sawyer [584] for the shower-producing cosmic rays.

[b] γ-rays from Th-C″, filtered through 6.8 cm of Pb.[585]

Table 157—(Continued)

[e] γ-rays from Ra-B and Ra-C, filtered through 10 mm of Pb; value given is μ_a, the apparent coefficient.
[d] Like the preceding, except that the filter is 3 mm of Pb.
[d₁] Reported by W. V. Mayneord and J. E. Roberts.[586]
[e] Using a reflection method and nearly homogeneous x-rays of effective $\lambda = 0.14A$, H. Fricke, O. Glasser, and K. Rothstein[587] found $\mu = 0.0256$.
[f] Reported by H. Steps,[588] who used the Cu K_a-radiation.

---

result in (1) a transfer to the electron of the entire energy of the impinging quantum, thus destroying the radiation and removing the electron from its energy level; or (2) a transfer of a portion of the quantum energy to the electron, thus removing the electron and scattering the remaining energy of the quantum as a quantum of reduced frequency (Compton effect); or (3) an elastic impact in which the electron is not removed, but the direction of the path of the quantum is changed. All of these reduce the intensity of the transmitted beam.

If a unifrequent beam of parallel rays of such high-frequency radiation of intensity I_0 impinges normally upon a slab of material (the absorber) of thickness x, the intensity (I) of the transmitted beam, at a point far beyond the absorber and in the prolongation of the incident beam, is given by the formula $I = I_0 e^{-\mu x}$, where μ is called the coefficient of absorption. The coefficient μ is made up of three parts, each related to one or two of the three results already enumerated. One (τ), called the coefficient of fluorescent or of photoelectric absorption, arises from the complete transfer of energy mentioned in result (1); another (σ_a), called the coefficient of true absorption due to scattering, arises from the partial transfer mentioned in result (2); and the third (σ_s), called the coefficient of true scattering, arises from the scattering or deviation of radiation mentioned in results (2) and (3). Thus $\mu = \tau + \sigma_a + \sigma_s$, which is often written $\mu = \tau + \sigma$, σ denoting $\sigma_a + \sigma_s$.

The vacancies left by the removal of electrons mentioned in results (1) and (2) are quickly filled, and that is accompanied by an emission of radia-

---

[568] Bruzau, M., *Ann. de Phys. (10)*, **11**, 5-140 (1929); Rees, W. J., and Clark, L. H., *Phil. Mag. (7)*, **16**, 691-703 (1933).
[569] Rees, W. J., and Clark, L. H., *Brit. J. Radiol.*, **5**, 432-444 (1932).
[570] Schindler, H., *Z. Physik*, **72**, 625-657 (1931).
[571] Kaye, G. W. C., and Owen, E. A., *Proc. Phys. Soc. (London)*, **35**, 33D-39D (1923).
[572] Richtmyer, F. K., *Phys. Rev. (2)*, **21**, 478 (1923).
[573] Hoffmann, G., *Physik. Z.*, **27**, 291-297 (1926).
[574] Vierheller, F., *Physik. Z.*, **28**, 745-757 (1927).
[575] Gray, J. A., *Int. Crit. Tables*, **6**, 8-22 (16, 21) (1929).
[576] Chadwick, J., *Proc. Phys. Soc. (London)*, **24**, 152-156 (1911-12).
[577] Chadwick, J., and Russell, A. S., *Proc. Roy. Soc. (London) (A)*, **88**, 217-229 (1913).
[578] Hewlett, C. W., *Phys. Rev. (2)*, **17**, 284-301 (1921).
[579] Olson, A. R., Dershem, E., and Storch, H. H., *Idem*, **21**, 30-37 (1923).
[580] Richtmyer, F. K., *Idem*, **18**, 13-30 (1921).
[581] Taylor, E. G., *Idem*, **20**, 709-714 (1922).
[582] Wingårdh, K. A., *Diss.*, Lund, 1923.
[583] Kramer, W., *Z. Physik*, **85**, 411-434 (1933). See also Myssowsky, L., and Tuwim, L., *Idem*, **35**, 299-303 (1925); **44**, 369-372 (1927); Millikan, R. A., and Cameron, G. H., *Phys. Rev. (2)*, **28**, 851-868 (1926); Millikan, R. A., and Otis, R. M., *Idem*, **27**, 645-658 (1926); Millikan, R. A., *Nature*, **116**, 823-825 (1925); *Proc. Nat. Acad. Sci.*, **12**, 48-54 (1926).

Table 158.—Angular Distribution of the Radiation (x and γ) Scattered by Water

The various theoretical formulas for the distribution of the scattered radiation are all of the form $I_s = IAF^2(1 + \cos^2\theta)$, where I is the intensity of the incident radiation, A is the universal constant $e^4/2m_0^2c^4 = 3.96 \times 10^{-26}$ cgs electrostatic units, F is a function of θ and of the structure of the atoms involved, and θ is the angle between the direction of propagation of the incident beam and that of the scattered beam of intensity I_s. For water, the values in the first section of this table, from a compilation by J. A. Gray,[589] were obtained by W. Friedrich and M. Bender,[590] the radiation being the platinum Kα doublet, $\lambda = 0.19$A; they find for μ the very high value 0.236 cm^{-1}.

θ	$(1+\cos^2\theta)$	$\left[\dfrac{F_\theta}{F_{90}}\right]^2$	$\left(\dfrac{I_{s,\theta}}{I_{s,90}}\right)$	θ	$(1+\cos^2\theta)$	$\left[\dfrac{F_\theta}{F_{90}}\right]^2$	$\left(\dfrac{I_{s,\theta}}{I_{s,90}}\right)$
10°	1.97	1.50	2.95	90°	1.00	1.00	1.00
20	1.88	0.90	1.67	100	1.03	1.02	1.05
30	1.75	0.94	1.65	110	1.12	0.98	1.11
40	1.59	0.92	1.50	120	1.25	0.96	1.21
50	1.41	0.96	1.35	130	1.41	0.97	1.38
60	1.25	0.98	1.23	140	1.59	0.98	1.55
70	1.12	1.04	1.11	150	1.75	1.01	1.78
80	1.03	1.04	1.07	160	1.88	1.02	1.94

X-rays: $\lambda = 0.31$A. Backhurst, I. *Phil. Mag.* (7), 321–351 (1934).

θ	30	40	50	60	70	80°
$I_{s,\theta}/I_{s,90}$	1.453	2.044	1.729	1.429	1.220	1.099
θ	100	110	120	130	140	150°
$I_{s,\theta}/I_{s,90}$	1.002	1.033	1.107	1.225	1.356	1.484

Table 159.—Coefficients of Scattering of X-rays and of γ-rays by Water

Adapted from a compilation by J. A. Gray.[591] It is not always possible to determine with certainty whether the coefficient found is that of the true scattering (σ_s) or that of the total scattering (σ).

Unit of $\lambda = 1$A $= 10^{-8}$ cm; of σ and $\sigma_s = 1$ cm^{-1}

λ^a	σ_s	σ	Source[b]
Ra[c]	0.0383		Neukirchen
0.161		0.185	Statz
0.240		0.206	Statz
0.285		0.170	Statz
0.32	0.198		Mertz
0.43	0.206		Mertz
0.501		0.201	Statz
0.54	0.210		Mertz
0.66	0.216		Mertz
0.79	0.228		Mertz

[a] Effective wave-length of filtered x-rays, as defined by the equation: $(\mu/\rho)_{Al} = 14\lambda^{2.92}$, $\rho =$ density; unit of $\mu/\rho = 1$ cm^2/g, of $\lambda = 1$A $= 10^{-8}$ cm.

[b] Sources:
Mertz, P., *Phys. Rev.* (2), **28**, 891-897 (1926); Neukirchen, J., *Z. Phys.*, **6**, 106-117 (1921); Statz, W., *Idem*, **11**, 304-325 (1922).

[c] Gamma rays from Ra-B and Ra-C, filtered through 2.6 cm. of lead.

tion characteristic of the atom. In the long run, that emission is uniformly distributed in all directions and so might logically be called scattered radiation, but it is not so called.

The scattered radiation mentioned in (2) and (3), which is the only radiation technically described as scattered, is most intense in a direction that coincides, or nearly coincides, with the direction of propagation of the incident beam.

The intensity of the transmitted beam at points near the absorber is abnormally great on account of the presence of scattered radiations. At distances that are great as compared with the transverse dimensions of the absorber, the intensity of the scattered radiations decreases approximately as the inverse square of the distance from the absorber, while that of the transmitted beam, which by hypothesis is composed of parallel rays, is independent of that distance. Values of μ computed from measurements made very near the absorber will be called "apparent" coefficients of absorption and will be denoted by μ_a.

When γ-rays pass through water, the scattered rays showing the Compton effect have a wave-length about three times as great as that of the incident rays.[568] Additional information on the quality of the radiation scattered by water when traversed by x-rays and by γ-rays is given by W. J. Rees and L. H. Clark [569]; and H. Schindler [570] has studied the secondary radiation excited in water by the "cosmic" radiation.

43. Absorption and Transmission of Radiation by Water

(For x-rays, γ-rays, and cosmic radiation, see Section 42; for corpuscular radiation, Section 26; for scattering, Sections 39 and 47.)

The fraction of the incident radiation transmitted by a given layer of water depends upon the amount laterally scattered by the water, as well as upon the amount truly absorbed, *i.e.*, converted into another form of energy. But the distinction has seldom been observed in reporting experimental data, the entire reduction in intensity being generally described as absorption. In some cases there is no necessity for maintaining the distinction, the scattering being experimentally negligible; but in other cases such is not the case.

Information regarding the scattering of radiation will be found in Sections 39 and 47, on luminescence and on the color of water, respectively;

[584] Sawyer, J. H., *Phys. Rev. (2)*, **50**, 25-26 (1936).
[585] Chao, C. Y., *Proc. Nat. Acad. Sci.*, **16**, 431-433 (1930).
[586] Mayneord, W. V., and Roberts, J. E., *Nature*, **136**, 793 (L) 1935).
[587] Fricke, H., Glasser, O., and Rothstein, K., *Phys. Rev. (2)*, **25**, 581 (1925).
[588] Steps, H., *Ann. d. Physik (5)*, **16**, 949-972 (1933).
[589] Gray, J. A., *Int. Crit. Tables*, **6**, 8-22 (19) (1929).
[590] Friedrich, W., and Bender, M., *Ann. d. Physik (4)*, **73**, 505-553 (1924).
[591] Gray, J. A., *Int. Crit. Tables*, **6**, 8-22 (17) (1929).

that for absorption and transmission, with or without scattering, will be given here.

Pure Water.[592]

The transparency of water is practically limited to the range $\lambda = 0.17$ to $1.0\,\mu$. Within that range the transmissivity is great except near the limits, but outside it the absorption is very great until in one direction

Table 160.—Monochromatic Absorptivity of Water

(For x-rays and γ-rays see Table 157.)

The As, Dr, Kr, Ow, RA, and RL data have been taken directly from a compilation by J. Becquerel and J. Rossignol,[595] who attribute the RL data to D. A. Goldhammer.[596] The As values beyond $\lambda = 0.75\,\mu$ correspond to the successive maxima and minima as reported by him; values in parentheses do not appear in the I.C.T.

A. Esau and G. Bäz [596a] have studied the absorption of water in the range $\lambda = 2.8$ to 10 cm, presenting the data graphically.

$I = I_0 e^{-kx}$ where x is the length of path, in water, that corresponds to a reduction in the intensity of a beam of parallel rays from I_0 to I.

Unit of $\lambda = 1\mu$ (in last section, 1 cm); of $k = 1$ cm^{-1}. Room temp.

Ref.[a]→ Year→ $10^3\lambda$	Kr 1901	Kr$_c$ 1901	Ts 1928	Ho 1933	DH 1934	Ha[b] 1935
			100k			
179[b]						
182.9			242[c]			
185.4			111			
186	58.4	68.8				93.0
186.2			86			
187.8			48			
190						30.0
191.6			20			
193	14.0	16.6				
193.5			16[d]			
198.0			9.4			
199.0			8.3			
200.0	7.9	9.0	7.3		8.0	2.95
203.0			5.1			
206.6			3.6			
208.4			6.0[d]			
210	5.2	6.1				1.45
210.4			3.3			
214.6			4.6[d]			
218.8			4.1[d]			
220	4.8	5.7	3.6[d]	6.4	3.3	0.86
230	2.9	3.4				0.47

[592] For general reviews and discussions, see Schaefer, C., and Matossi, F., "Das Ultrarote Spektrum," 1930, Lecomte, J., *Trans. Faraday Soc.*, **25**, 864-876 (1929); Fowle, F. E., *Smithsonian Misc. Collect.*, **68**, No. 8 (Publ. 2484) (1917); Dawson, L. H., and Hulburt, E. O., *Opt. Soc. Amer.*, **24**, 175-177 (1934).
[593] Leifson, S. W., *Astroph. J.*, **63**, 73-89 (1926).
[594] Schaeffer, E. J., Paulus, M. G., and Jones, H. C., *Physik. Z.*, **15**, 447-453 (1914).
[595] Becquerel, J., and Rossignol, J., *Int. Crit. Tables*, **5**, 268-271 (1929).
[596] Goldhammer, D. A., "Dispersion und Absorption des Lichtes," Leipzig, 1913.
[596a] Esau, A., and Bäz, G., *Physik. Z.*, **38**, 774-775 (1937).

43. WATER: ABSORPTION AND TRANSMISSION

Table 160—(Continued)

Ref.[a]→ Year→ $10^8\lambda$	Kr 1901	Kr_e 1901	Ts 1928	Ho 1933	DH 1934	Ha[b'] 1935
				100k		
240	2.7	3.2		4.8	1.35	0.29
250						0.25
260	2.2	2.5		3.7	0.92	0.21
280				2.5	0.77	0.12
300	1.3	1.5		2.1	0.64	0.09
320				1.6	0.43	
340				0.9	0.28	
360				0.9	0.19	
380				0.9	0.13	
400				0.5	0.08	

For range $\lambda = 310$ to $800\ m\mu$ see Table 161.

Ref.[a]→ Year→ λ	As 1895	Co 1922	Dr 1924	Pl 1924	Ref.[a]→ Year→ λ	As 1895	Co 1922	Dr 1924	Pl 1924
		k					k		
0.745				0.044	1.45	(36.0)		20.1	
0.845				0.044	1.475			29.9	
0.850			0.069		1.50	38.4		26.4	
0.900[e]			0.0161		1.56			15.0	
0.950			0.0311		1.60	(21.0)		9.0	
0.970		0.448			1.677	(13.6)		5.2	
0.980				0.142[f]	1.708	11.4			
0.995	0.416		0.472		1.75	16.0		7.5	
1.05			0.368		1.85			12.7	
1.085			0.333		1.90			31.5	
1.095	0.188				1.95			86	
1.13	0.29		0.60		1.956	125			
1.17			1.12		1.97	111		104	
1.20		1.22			2.00		103	70	
1.21			1.30	1.28[f]	2.08	42		35.6	
1.243	1.22				2.10			31.6	
1.25	(1.21)		1.24		2.147	27.8			
1.281	(1.17)				2.15			24.7	
1.30	(1.20)		1.48		2.237	(32)		19.6	
1.35	(1.61)		2.14		2.30			25.9	
1.40	(23.4)		3.05		2.35	(61)		33.0	
1.44		2.94			2.40			40.3	

Ref.[a]→ Year→ λ	As 1895	RL 1909	Ref.[a]→ Year→ λ	As 1895	RL 1909
	k			k	
2.6	(190)	530	6.09	2530	
2.8		2240	6.2	2060	2000
3.0		7330	6.5	1040	1030
3.02	2730		6.73	870	
3.2	(2590)	(6640)	6.765	880	
3.4		1440	6.92	820	
3.6		490	6.955	830	
3.93	204		7.0	(820)	810[g]
4.5	(447)	450	7.11	820	
4.70	545		7.275	845	
5.27	308		7.41	790	
5.42	342		7.44	810	
5.47	335		7.49	800	
5.8	928	910	7.545	810	
6.0	2120	2140	7.65	765	

Table 160—(Continued)

Ref.→ Year→ λ	As 1895 k	RL 1905	Ref.→ Year→ λ	As 1895 k	RL 1909
7.70	785		8.38	695	
7.83	765		8.43	755	
7.88	775		8.49	725	
7.94	690		9.0		700
8.0	(766)	755	10.0		705
8.065	785		11.0		1200
8.13	765		12		2590
8.16	785		13		2890
8.22	715		15		3570
8.28	765		18		2990

Ref.[a]→ Year→ λ	RA 1898	RH 1910	RW 1911 k	Ow 1912	CE 1936
24	>46				
52	>46				
61	>46				1160
63					
80		990			1020
83					
100					710
108			460	423	460
117					
152					360
314				242	320

Unit of $\lambda = 1$ cm; of $k = 1$ cm^{-1}

Ref.[a]→ Year→ λ	Ec 1913	Te 1923	Za 1927 k	Se 1933	Kn 1937	Rü 1918	Ref.[a]→ Year→ λ	Ec 1913	Te 1923	Za 1927 k	Se 1933	Kn 1937	Rü 1918
0.42		38.3					13.6				0.38		
0.84		22.3					14.0				0.40		
1.1		16.5					14.48					0.47	
1.5		15.3					15.29					0.45	
1.75	19.4						18.41					0.31	
1.8		16.1					19.				0.47		
2.7		10.5					20.44					0.26	
3.7	5.84						23.			0.170			
4.80					3.53		57.38						0.0261[h]
5.34					2.92		63.76						0.0228
5.7	2.38						67.98						0.0156
6.48					2.11		74.80						0.0172
							84.68						0.0050
8.05					1.53		98.72						0.0057
8.16					1.32		105.92						0.0099
8.80	0.86				1.17		128.81						0.0028
9.55					0.99		144.12						0.0015
10.10					0.90		144.50						0.0033
10.87					0.76		159.66						0.0014
11.12					0.74		183.80						0.00018
11.80					0.67		221.17						0.00076
12.6				0.52			225.51						0.00060
13.41					0.56		242.44						0.00149

[a] References:

- As Aschkinass, E., *Ann. d. Physik (Wied.)*, **55**, 401-431 (1895).
- CE Cartwright, C. H., and Errera, J., *Proc. Roy. Soc. (London) (A)*, **154**, 138-157 (1936) ← *Acta Physicochim. URSS*, **3**, 649-684 (1935) → Cartwright, C. H., *Nature*, **135**, 872 (L) (1935); *Idem*, **136**, 181 (L) (1935).
- Co Collins, J. R., *Phys. Rev. (2)*, **20**, 486-498 (1922).
- DH Dawson, L. H., and Hulburt, E. O., *J. Opt. Soc. Amer.*, **24**, 175-177 (1924); Supersedes Hulburt, E. O., *Idem*, **17**, 15-22 (1928).
- Dr Dreisch, T., *Z. Physik*, **30**, 200-216 (1924).
- Ec Eckert, E., *Verh. physik. Ges.*, **15**, 307-329 (1913).
- Ha Haas, E., *Biochem. Z.*, **282**, 224-229 (1935) (Temp. = 24 °C).
- Ho Hodgman, C. D., *J. Opt. Soc. Amer.*, **23**, 426-429 (1933).

43. WATER: ABSORPTION AND TRANSMISSION

Table 160—*(Continued)*

Kn Knerr, H. W., *Phys. Rev. (2)*, **52**, 1054-1067 (1937) → *Idem*, **51**, 1007 (A) (1937).
Kr Kreusler, H., *Ann. d. Physik (4)*, **6**, 412-423 (1901). Values as published in *Int. Crit. Tables*; see Kr$_c$.
Kr$_c$ Kr as corrected by the compiler.

That Kreusler's published coefficients were seriously in error (by a factor of the order of 10) was pointed out by E. O. Hulburt,[597] and Becquerel and Rossignol multiplied them by 10, getting the Kr values. By working backward from Kreusler's coefficients, through his recorded absorptions, the present compiler found that he had taken the length of the absorbing column as about 16.8 cm. For certain substances, he used a trough 16.85 cm long, but he states that he used for water a special quartz trough that was 20 mm long. Assuming that his reported absorptions are correct and that his trough was 20 cm long, we obtain the values given under Kr$_c$.

Ow Owen, D., *Electrician (London)*, **68**, 504-507 (1912).
Pl Plyler, E. K., *J. Opt. Soc. Amer.*, **9**, 545-555 (1924).
RA Rubens, H., and Aschkinass, E., *Ann. d. Physik (Wied.)*, **65**, 241-256 (1898).
RH Rubens, H., and Hollnagel, H., *Verh. physik. Ges.*, **12**, 83-98 (1910).
RL Rubens, H., and Ladenburg, E., *Idem*, **11**, 16-27 (1909). The I.C.T. gives D. A. Goldhammer ["Dispersion und Absorption des Lichtes," Leipzig, 1913] as the source of these values.
Rü Rückert, E., *Ann. d. Physik (4)*, **55**, 151-176 (1918).
RW Rubens, H., and Wood, R. W., *Verh. physik. Ges.*, **13**, 88-100 (1911).
Se Seeberger, M., *Ann. d. Physik (5)*, **16**, 77-99 (1933).
Te Tear, J. D., *Phys. Rev. (2)*, **21**, 611-622 (1923).
Ts Tsukamoto, K., *Rev. d'Optique*, **7**, 89-108 (1928).
Za Zakrzewski, K., *Bull. Int. Acad. Polon. Sci. Let. (A)*, 1927, 489-503 (1927).

[b'] The Ha values refer to 24 °C; absorption varies rapidly with the temperature, see Table 162. CO_2 content is of slight effect.

[b] S. W. Leifson[593] has stated that a thin film of water condensed on a window will absorb all radiation having $\lambda < 0.179\,\mu$; see also H. Ley and B. Arends.[598]

[c] In the Ts paper the value for $\lambda = 0.1829\,\mu$ is given as $k = 4.72$, which corresponds to only 1/10 of the reported transmission; the value (2.42) here given accords with that transmission.

[d] For these values, Ts used commercial distilled water; for the others, redistilled, conductivity = $6(10^{-8})$ ohm^{-1}·cm^{-1}. The value (1.16) he gives for $\lambda = 0.1935\,\mu$ is obviously wrong; his recorded absorption leads to $k = 0.155$, essentially 0.16.

[e] J. Kaplan[599] has published a curve showing the following maxima and minima:

λ	0.88$_5$	0.89	0.89$_7$	0.90	0.91	0.91$_5$	0.91$_8$	0.93	0.93$_2$	0.94 μ
1000k	8	9	6	15	9	15	9	15	14	25

[f] Pl's values for $\lambda = 0.98$ and 1.21 are for water at 0 °C.

[g] Given in the Becquerel and Rossignol compilation[595] as $k = 890$, but it was obtained from the RL data, and they lead to $k = 808$.

[h] From $\lambda = 50$ cm to $\lambda = 60$ cm there is much absorption, but G. Mie[600] has found that there is no anomalous dispersion if the water is pure, and that traces of glass (?) in solution give rise to bands of the type reported by R. Weichmann.[601]

λ becomes of the order of a meter, and in the other it becomes of the order of 0.00005 μ, there being, however, a great gap (0.17 $\mu > \lambda > 0.00015\,\mu$) in which no data are available. A thin film of water condensed on a window absorbs all radiation for which $\lambda < 0.18\,\mu$ [593]; a layer 1 cm thick absorbs 38 per cent of the radiation of $\lambda = 0.995\,\mu$ ($k = 0.472$ cm^{-1}), and 95 per cent of that of $\lambda = 1.4\,\mu$ ($k = 3.05$ cm^{-1}), and a layer only 10 μ thick

[597] Hulburt, E. O., *J. Opt. Soc. Amer.*, **17**, 15-22 (1928).
[598] Ley, H., and Arends, B., *Z. physik. Chem. (B)*, **4**, 234-238 (1929).
[599] Kaplan, J., *J. Opt. Soc. Amer.*, **14**, 251-256 (1927).
[600] Mie, G., *Physik Z.*, **27**, 792-795 (1926).
[601] Weichmann, R., *Ann. d. Physik (4)*, **66**, 501-545 (1921) → *Physik. Z.*, **22**, 535-544 (1921).

absorbs 9.0 per cent of that of $\lambda = 2.8\,\mu$ ($k = 2240$ cm^{-1}). From $\lambda = 0.4$ to $0.52\,\mu$ the transparency is great.

It has been found that, whereas aqueous solutions of salts that do not form hydrates absorb nearly the same as does a layer of pure water of the same thickness as the water in the sample of solution under examination, those of salts that form hydrates absorb less than do such thicknesses of pure water.[594]

Table 161.—Monochromatic Absorptivity of Water in the Range $\lambda = 310$ to 800 mμ.

The As, Au, and Ma data have been taken directly from the compilation by J. Becquerel and J. Rossignol [595]; the Ew data come from the same source, but each value has been multiplied by 2.303, so as to reduce the values to the basis used in this table. The value in parentheses does not appear in the I.C.T.

$I = I_0 e^{-kx}$ where x is the length of the path, in water, that is needed to reduce the intensity of a beam of parallel rays from I_0 to I.

Unit of $\lambda = 1$ m$\mu = 10^{-7}$ cm; of $k = 1$ cm^{-1}. Room temp.

Ref.[a]→ Year→ λ	Ew 1895	As 1895	Au 1904	Au$_e$ 1904	Pi 1918	Ma 1922	Sa 1931	LS 1932	Ho 1933	DH 1934
						$10^5 k$				
310							840			
313							690			
320							580		1600	430
325							512			
300							461			
340							382		900	280
360							281		900	190
370							200			
380							148		900	130
400							72		500	80
415	81						46			
420	74						41			
430	53						30			
436						12				
440	37						23			
450	28	20					18			
460	25						15*			
470	28				34		15			
480	30	20					15			
490	32		2				15			
494					30					
500	35	20					16	200		
510	37	22					17			
520	41	18	2	20			19			
522					30					
530	44	8	3				21			
539					22					
540	48	9	11				24			40
546						34				
550	53	36	26				27	150		
557					56					
558					36					

43. WATER: ABSORPTION AND TRANSMISSION

Table 161—(Continued)

Ref.[a]→ Year→ λ	Ew 1895	As 1895	Au 1904	Au$_c$ 1904	Pi 1918	Ma 1922	Sa 1931	LS 1932	Ho 1933	DH 1934
					$10^5 k$					
560	62	30	40				30			
570	76	20	43				38			
578						64				
579				78	56					
580	97	26	50							
589					96					140
590	161	78	89				85			
600	246	160	165		165		125			
602				188						
607				224						
610	272	190	220				160			
616				264						
618					206					
620	285	212	240				178			
630	299	224	250				181			
636					225					
640	315	235	275				200			
643				309						
648					236					
650	340	250	305				210			
658				339						
660	373	280	325							
663					245					
670	421	300								340
680	485	340								
690	575	400								390
700	690	550						450		
710	890	790								
720	1080	1150								
730	1310	1750								
740		2300								
750		2410						3000		
775		(2410)								
800		2040								

G. Hüfner and E. Albrecht [602] reported the following values for bands of width Δλ centered on the indicated λ. Temperature was 17 to 18 °C.

λ	449	468	487	506	527	552	576	602	631	664	mμ
Δλ	6	6	8	8	8	11	11	18	18	13	mμ
$10^5 k$	28	27	49	43	53	76	114	250	282	394	cm^{-1}

[a] References:

- As   Aschkinass, E., *Ann. d. Physik (Wied.)*, **55**, 401-431 (1895).
- Au   v. Aufsess, O. F., *Ann. d. Physik (4)*, **13**, 678-711 (1904).
- Au$_c$ Au as corrected by DH.
- DH   Dawson, L. H., and Hulburt, E. O., *J. Opt. Soc. Amer.*, **24**, 175-177 (1934). Supersedes Hulburt, E. O., *Idem*, **17**, 15-22 (1928).
- Ew   Ewan, T., *Proc. Roy. Soc. (London) (A)*, **57**, 117-161 (1895). His published values of ϵ are defined by the relation $I = I_0 (10)^{-\epsilon x}$, cf. Ewan, T., *Phil. Mag. (5)*, **33**, 317-342 (1892). The compiler has multiplied each of them by 2.303 so as to reduce them to the basis of this table. The values given in *Int. Crit. Tables* are incorrect, being ϵ, not k.
- Ho   Hodgman, C. D., *J. Opt. Soc. Amer.*, **23**, 426-429 (1933).
- LS   Lange, B., and Schusterius, C., *Z. physik. Chem. (A)*, **159**, 303-305 (1932); **160**, 468 (1932).
- Ma   Martin, W. H., *J. Phys'l Chem.*, **26**, 471-476 (1922).
- Pi   Pietenpol, W. B., *Trans. Wisconsin Acad. Sci., Arts, Let.*, **19**$_1$, 562-593 (1918).
- Sa   Sawyer, W. R., *Contrib. Canadian Biol. (N. S.)*, **7**, 75-89 (No. 8) (1931).

[602] Hüfner, G., and Albrecht, E., *Ann. d. Physik (Wied.)*, **42**, 1-17 (1891).

Table 162.—Absorptivity of Water: Effects of Pressure and Temperature

J. R. Collins [603] has found that changing the pressure from 120 atm to 5000 atm produces no change in the absorption of water in the range $\lambda = 0.71$ to $1.05\ \mu$.

He has also found [604] that an increase in temperature shortens the wavelengths at which the absorption has maxima, and changes the coefficients of absorption corresponding to the maxima. E. Ganz [605] has confirmed Collins' observations on the band at $\lambda = 0.77\ \mu$. He seems to state that the coefficient of mass absorption (k/ρ) for any fixed λ varies linearly with the temperature for each of the bands $\lambda = 0.77\ \mu$ and $\lambda = 0.84\ \mu$, ρ being the density of the water; but he may mean that k so varies.

Using $\lambda = 12.6$ cm, M. Seeberger [606] found that the absorption decreases rapidly as the temperature rises.

$I = I_0 e^{-kx}$.

Unit of $\lambda_{max} = 1\mu$; of k_{max} and $k = 1$ cm^{-1}; temp. $= t$ °C.

I. J. R. Collins.[604]

\multicolumn{2}{c}{0 °C}	\multicolumn{2}{c}{95 °C}	\multicolumn{2}{c}{0 to 95 °C}
λ_{max}	k_{max}	λ_{max}	k_{max}	$\Delta\lambda_{max}$	Δk_{max}
0.775	0.0280	0.740	0.0380	−0.035	0.0100
		0.845	0.0472		
0.985	0.430	0.970	0.606	−0.015	0.176
1.21	1.28	1.17	1.38	−0.04	0.10
1.45	29.8	1.43	28.7	−0.02	−1.1
1.96	108	1.94	108	−0.02	0

II. M. Seeberger.[606] $\lambda = 12.6$ cm.

t	15.5	16	30	50
k	0.53	0.52	0.28	0.09

III. E. Haas.[607] Illustrative. He concludes that this increase in k is not due to the increase in the dissociation of the water.

$1000\lambda \rightarrow$	\multicolumn{2}{c}{186}	\multicolumn{2}{c}{190}
t	k		k	
17	0.670	0.690	0.280	0.295
37	0.930		0.420	0.420

Table 163.—Total Transmissivity of Water

As the absorptivity of a substance varies with the frequency of the radiation, numbers expressing the total transmissivity are of significance only with respect to a specified source of radiation and to a specified thickness of the substance.

In the following, τ is the transmissivity for radiant energy, and τ_l is that for light. If I and I_0 are the energies of the transmitted and of the incident radiation, respectively, and if L and L_0 are the corresponding luminous

---

[603] Collins, J. R., *Phys. Rev. (2)*, **36**, 305-311 (1930) → **35**, 1433 (A) (1930).
[604] Collins, J. R., *Idem*, **26**, 771-779 (1925).
[605] Ganz, E., *Ann. d. Physik (5)*, **26**, 331-348 (1936) ← *Diss.*, München, 1936.
[606] Seeberger, M., *Ann. d. Physik (5)*, **16**, 77-99 (1933).
[607] Haas, E., *Biochem. Z.*, **282**, 224-229 (1935).

43. WATER: ABSORPTION AND TRANSMISSION

Table 163—(Continued)

intensities, then $\tau = 100\ I/I_0$ and $\tau_l = 100\ L/L_0$. The temperature of the radiating source is $t\ °C = T\ °K$; the thickness of the transmitting layer of water is x.

Unit of $x = 1$ cm; of $\lambda = 1\mu = 10^{-4}$ cm; of τ and $\tau_l = 1\%$

I. Ideal radiator: "Black-body."

$x = 0.05$ (Br)[a]

t	τ	t	τ
300	2.9	720	15.2
370	4.0	730	15.3
475(?)	5.3	760	15.7
540	9.0	810	18.1
560	9.9	820	18.7
580	10.3	870	20.6
625	11.4	933	22.8
670	13.3	940	23.0
680	13.5	960	23.6
685	13.8		

$x \rightarrow$	1 (ICT)[a]		10 (ICT)[a]	
T	τ	τ_l	τ	τ_l
2400[b]	22.5	99.90	0	98.96
3600[c]	50	99.92	31.5	99.16
5000[d]	69	99.93	54	99.26

II. Various sources.

		$x \rightarrow$	0.1	0.2	0.5	1.0	2.0
Source	T				τ (FC)[a]		
Iron	1000		3.1	1.4	0.5	0.25	0.10
Carbon	2150		35	28	20	15	11
Tungsten	2970		66	59	51	43	36
Sun	—		85	81	76	71	65

	(We)[a]		$x \rightarrow$	0.0019		0.0038
Source		Filter	λ[e]		τ	
Welsbach mantle	None		107			20.0
Mercury arc	None			55.5		33.0
Mercury arc	2 mm quartz			60.3		38.4
Mercury arc	Cardboard		310	62.7		39.8

[a] References:
Br   Brown, S. L., *Phys. Rev. (2)*, **21**, 103-106 (1923).
FC Forsythe, W. E., and Christison, F. L., *J. Opt. Soc. Amer.*, **21**, 150 (1931).
ICT Compilation by Walsh, J. W. T., and Buckley, H., *Int. Crit. Tables*, **5**, 264-268 (1929), based upon Aschkinass, E., *Ann. d. Physik (Wied.)*, **55**, 401-431 (1895), v. Aufsess, O. F., *Diss.*, Munich, 1903 → *Ann. d. Physik. (4)*, **13**, 678-711 (1904), and Ewan, T., *Proc. Roy. Soc. (London) (A)*, **57**, 117-161 (1894).
We Weniger, W., *J. Opt. Soc. Amer.*, **7**, 517-527 (1923).
[b] Corresponds to a tungsten filament vacuum lamp burning at 9 lumens per watt.
[c] Corresponds to a plain carbon arc.
[d] Approximately noon sun-light.
[e] This is the wave-length of the principal radiation.

Table 164.—Penetration of Solar Radiation into Water

Computed by W. Schmidt [608] on the basis of the coefficients of absorption as determined by Aschkinass,[609] and the distribution of energy in the solar spectrum. All values for the radiation refer to energy; none to luminosity.

[608] Schmidt, W., *Sitz. Akad. Wiss., Wien (2A)*, **117**, 237-253 (1908).
[609] Aschkinass, *Ann. d. Physik (Wied.)*, **55**, 401-431 (1895).

Table 164.—(Continued)

A = fraction of total incident radiation absorbed by a layer of water of thickness x, and lying within the indicated range in wave-length. For example: The amount of radiation in the range $\lambda = 0.9$ to $1.2\,\mu$ that is absorbed from sunlight by a column of water 1 cm thick is 5.60 per cent of the total solar energy incident upon the surface of the column.

τ = the rate at which the temperature of an exceedingly thin layer of water at the depth x would rise if it retained in itself all the solar energy that it absorbs. (The incidence is presumably normal.)

Unit of $\lambda = 1\mu = 0.001$ mm; of $\tau = 1$ °C per min.

$x \rightarrow$ λ	0.01mm	0.1mm	1mm	1cm	10cm	1m	10m	100m	∞
					1000Å				
0.2 to 0.6	0.00	0.00	0.00	0.1	0.8	7.5	65.0	223.1	237.0
0.6 to 0.9	0.00	0.04	0.7	6.3	54.8	230.1	350.2	359.7	359.7
0.9 to 1.2	0.08	0.67	6.8	56.0	170.6	178.8	178.8	178.8	178.8
1.2 to 1.5	0.54	4.78	23.3	69.5	86.6	86.6	86.6	86.6	86.6
1.5 to 1.8	1.82	16.28	53.0	80.0	80.0	80.0	80.0	80.0	80.0
1.8 to 2.1	2.00	14.05	25.0	25.0	25.0	25.0	25.0	25.0	25.0
2.1 to 2.4	0.84	6.42	24.2	25.3	25.3	25.3	25.3	25.3	25.3
2.4 to 2.7	0.94	5.24	7.2	7.2	7.2	7.2	7.2	7.2	7.2
2.7 to 3.0	0.19	0.40	0.4	0.4	0.4	0.4	0.4	0.4	0.4
Sum	6.41	47.88	140.6	269.8	450.7	640.9	818.5	986.1	1000.0
τ	6.68[a]	3.69	0.71	0.071	0.0071	0.0008	0.00008	0.000003	

[a] Surface layer.

Radiation Filters Containing a Layer of Water.

A cell 0.5 mm thick containing distilled water and provided with quartz or fluorite windows transmits essentially no radiation beyond the approximate range $\lambda = 0.17\,\mu$ to $1.5\,\mu$, and throughout that range the transmission is high except near the limits. Increasing the thickness restricts the range but slightly.

In his compilation, K. S. Gibson[610] gives the spectral ranges that may be isolated by the use of such water filters, either alone or in combination with other filters. He cites the following users of water filters: W. W. Coblentz,[611] cell 10 mm thick with thin quartz windows; T. Lyman,[612] cell 0.5 mm thick with fluorite windows, and cell 20 mm thick with quartz windows; H. Kreusler,[613] cell 20 mm thick with quartz windows.

Natural Waters.

The transmission of light by many coastal and inland waters is subject to wide fluctuations caused by variations in the turbidity and in the plankton. The amount of plankton varies with the season and the weather; and the turbidity with the amount of detritus, sand, and soil, whether brought in by streams or surface drainage, or stirred up from the bottom. Measurements of the transmission and of the effective absorptivity under such conditions are of no general value, but are of significance with reference to the

[610] Gibson, K. S., *Int. Crit. Tables*, **5**, 271-274 (1929).
[611] Coblentz, W. W., *Sci. Papers Bur. Stand.*, **17**, 725-750 (S438) (1922).
[612] Lyman, T., "Spectroscopy of the extreme ultra-violet," 1914.
[613] Kreusler, H., *Ann. d. Physik (4)*, **6**, 412-423 (1901).

actual plankton growth at the place and time considered; many such measurements have been made for the purpose of obtaining data for correlating plankton growth with the illumination existing at various depths.[614]

A few illustrative sets of measurements on waters of this kind will be found in Table 166.

O. F. v. Aufsess [615] has published curves showing the spectral absorption of the waters from several lakes. He was of the opinion that every departure of the color from that of pure water is due to the presence of foreign bodies, and that variations in the turbidity of a given lake change its color but little.

Table 165.—Monochromatic Absorptivity of Sea-water

The following data refer to samples taken from the open sea far from land, and at the depths indicated; the effect of scattering is probably small. E. O. Hulburt [616] studied samples from the Pacific, the Gulf Stream, and the Caribbean Sea, and could detect no difference in their absorptivities. Tsukamoto concluded that bromides are the cause of the great absorptivity for $\lambda < 220$ mμ. The earliest study of the ultraviolet absorption by sea-water seems to have been by J. L. Soret.[617]

$I = I_0 e^{-kx}$; I is the intensity of a parallel beam of radiation at a distance x along the beam beyond the point where the intensity is I_0, both points being in the water.

Unit of $k = 1$ cm^{-1}; of depth = 1 meter; of $\lambda = 1$ m$\mu = 0.001\mu = 10$A. Room temp.

	K. Tsukamoto[a]				E. O. Hulburt[a]		
Depth→ λ	Surface	3m 1000k	15m	Depth→ λ	Surface	Tap[b] 1000k	Dist[b]
212.3	1079		1247	254	154	104	79
213.6	646			266	131	74	48
216.1	160		361	280	90	46	35
217.5	145		206	303	39	16	12
217.9		267		313	21	7	5
221.0		58		366	3.0	2	2
221.8	120		185	436	0.23		0.12
226.4	74		137	546	0.35		0.341
227.6		42		578	0.7		0.640
231.7		44		612	2.3		2.3
233.6		70	118				

[a] References:
Hulburt, E. O.[597]
Tsukamoto, K., *Compt. rend.*, **184**, 221-223 (1927).

[b] Tap = water supply of Washington, D. C.; Dist = doubly distilled water. Values for Dist and $\lambda = 436, 546, 578$ were taken from W. H. Martin,[618] and $\lambda = 612$ from O. F. v. Aufsess.[615]

---

[614] Regnard, P. M. L., *J. Chem. Soc. (London)*, **60**, 2 (1891) ← *Mém. Soc. Biol. (Paris)*, **42**, 288 + (1890) → *Compt. rend. Soc. Biol.* (9), **11**, 289 (1890); Shelford, V. E., and Gail, F. W., *Publ. Puget Sound Biol. Sta.*, **3**, 141-176 (1922) (Bibliog. of 29 entries); Poole, H. H., *Sci. Proc. Roy. Dublin Soc.* (N. S.), **18**, 99-115 (1925); Poole, H. H., and Atkins, W. R. G., *J. Marine Biol. Assoc. United Kingdom* (N. S.), **14**, 177-198 (1926); **15**, 455-483 (1928).

[615] v. Aufsess, O. F., *Ann. d. Physik* (4), **13**, 678-711 (1904) ← *Diss.*, München, 1903.

[616] Hulburt, E. O., *J. Opt. Soc. Amer.*, **13**, 553-556 (1926).

[617] Soret, J. L., *Jour. de Phys.* (1), **8**, 145-158 (1879) ← *Arch. des sci. phys. et nat.*, **61**, 322-359 (1878); **63**, 89-112 (1878) → *Compt. rend.*, **86**, 708-711 (1878).

[618] Martin, W. H., *J. Phys'l Chem.*, **26**, 471-476 (1922).

Table 166.—Effective Absorptivity of Some Coastal and Inland Waters

Illustrations of the variability of such data when obtained under natural conditions.
$I = I_0 e^{-kx}$. Layers are specified by the depths of their bounding planes below the surface of the water.

Unit of $\lambda = 1$ mμ; of $k = 1$ cm^{-1}

I. Various Waters. C. D. Hodgman.[a]

Water[b]→ λ	Dist	Erie	Tap	Shaker	Br	Br'	Marsh	Snow
				100k				
220	6.4	134	28	150	150	99	99	101
240	4.8	92	5.3	106	92	60	69	58
260	3.7	67	4.8	78	69	44	55	46
280	2.5	53	4.8	60	53	34	46	39
300	2.1	41	3.0	48	39	32	41	34
320	1.6	34	2.1	39	30	25	30	30
340	0.9	28	0.9	30	23	21.4	20.0	28
360	0.9	20.7	0.5	21.4	17.1	18.7	16.4	23
380	0.9	14.3	0.5	14.3	11.1	15.7	13.1	20.7
400	0.5	10.6	0.5	8.1	6.4	7.6	9.9	17.7

II. Gunflint Lake, Minn. H. A. Erikson.[a]

λ→	447	466	500	529	545	563	581	623
100k	1.046	0.740	0.501	0.357	0.357	0.322	0.309	0.407

III. Sea, total depth = 9 m; photographic. Martin Knudsen.[a]

λ→ Layer	400	450	500	550	600	650
			100k			
1 to 3 m	0.58	0.30	0.16	0.19	0.30	0.38
3 to 5 m	0.40	0.27	0.21	0.20	0.28	0.38
5 to 7 m	0.64	0.39	0.27	0.38	0.49	0.60
1 to 8 m		0.300	0.184	0.181	0.275	

IV. San Juan Archipelago.
Photronic cell with filters B, G, and R; B transmits from $\lambda = 410$ to $\lambda = 500$, max. at 465; G from $\lambda = 500$ to $\lambda = 590$, max. at 540; R uniformly from $\lambda = 600$ to $\lambda = 700$. C. L. Utterback and J. W. Boyle.[a]

Filter→ Layer	B	G	R	Filter→ Layer	B	G	R
		100k				100k	
	Station 1. k constant				Station 4.		
1 to 6 m	0.248	0.201	0.435	0 to 5 m	0.232	0.187	0.456
	Station 2. k constant			5 to 10	0.215	0.190	0.396
50 to 20 m	0.338	0.273	0.483	10 to 15	0.187	0.171	0.336
	Station 3.			15 to 20	0.213	0.168	0.362
0 to 10 m	0.265	0.207	0.383	20 to 25	0.233	0.202	0.262
10 to 15	0.275	0.208	0.378	25 to 30	0.216	0.211	0.209
15 to 20	0.280	0.223	0.356	30 to 35	0.163	0.183	
20 to 25	0.179	0.210	0.212	35 to 40	0.156	0.184	
25 to 30	0.172	0.206	0.169	40 to 45	0.095	0.138	
30 to 35	0.165	0.188		45 to 50		0.089	

43. WATER: ABSORPTION AND TRANSMISSION

Table 166—(Continued)

V. Alaskan coastal waters. Apparatus as for preceding. C. L. Utterback.[a]

Filter→ Layer	B	G 100k Station H	R	Filter→ Layer	B	G 100k Station R	R
0 to 10 m	0.165	0.126	0.407	0 to 5 m	0.322	0.231	0.427
10 to 15	0.068	0.120	0.414	5 to 10	0.579	0.231	0.457
15 to 20	0.087	0.124	0.287	10 to 15	0.391	0.351	0.575
20 to 25	0.248	0.116	0.104	15 to 20	0.091	0.410	0.271
25 to 30	0.400	0.113	0.051	20 to 25	0.061	0.234	0.095
30 to 35	0.322	0.118		25 to 40		0.110	
		Station SS				Station CS	
0 to 5 m	0.558	0.397	0.563	0 to 5 m	0.166	0.168	0.410
10 to 20	0.256	0.172	0.370	5 to 10	0.164	0.169	0.408
20 to 25		0.171		10 to 15	0.161	0.167	

[a] References:
Erikson, H. A., *J. Opt. Soc. Amer.*, **23**, 170-177 (1933).
Hodgman, C. D., *Idem*, **23**, 426-429 (1933).
Knudsen, M., Cons. perm. intern. l'explor. mer, Publ. de Circons No. 76, 1922.
Utterback, C. L., *J. Opt. Soc. Amer.*, **23**, 339-341 (1933).
Utterbach, C. L., and Boyle, J. W., *Idem*, **23**, 333-338 (1933).

[b] Waters studied:
Dist = distilled water; Erie = Lake Erie; Tap = water supply of Cleveland, Ohio, drawn from Lake Erie; Shaker = Shaker Lake, Ohio; Br and Br' = two brooks; Marsh = water from an open marsh; Snow = melted snow.

Table 167.—Penetration of Daylight into Sea-water
(Cf. Table 164.)

I. Absorptivity. Otto Krümmel.[a] Based on observations of Regnard (citation not given) using a selenium cell; expressed in form $I = I_0 e^{-kx}$, where x is depth beneath surface. Unit of $x = 1$ m; of $k = 1$ m^{-1}.

x	1	2	3	5	7	9	11
k	0.124	0.087	0.063	0.043	0.032	0.026	0.025

II. Visual brightness; B_h = looking horizontally, B_v = looking vertically upward, B_h observed by W. Beebe and G. Hollister.[a] $B_{h'}$ and $B_{v'}$ computed [b] by E. O. Hulburt,[a] assuming the total illumination of the surface to be 10,000 candle/ft^2 = 10.764 ca/cm.^{2} Unit of $B = 1$ ca/cm^2; of depth = 1 ft.

Depth	$10^3 B_h$	$10^3 B_{h'}$	$10^3 B_{v'}$	Depth	$10^3 B_h$	$10^3 B_{h'}$	$10^3 B_{v'}$
0	.		10764	300	0.17	0.17	17
50	11.4	11.3	1140	350	0.13	0.125	13
100	3.6	3.66	366	500	0.024	0.066	6
200	0.4	0.44	44	800	0.024	0.014	1
250	0.23	0.23	23				

III. Spectral distribution of visual brightness at a depth of 800 ft. in the sea, the brightness at the surface being i_0; brightness looking horizontally = b_h, looking vertically upward = b_v; unit is arbitrary. Values of b_v and b_h as computed by E. O. Hulburt [a] for sea-water containing no suspended particles. Unit of $\lambda = 1$ m$\mu = 0.001$ μ.

Table 167—(Continued)

λ	i_0	$10^4 b_\lambda$	$10^2 b_v$	λ	i_0	$10^4 b_\lambda$	$10^2 b_v$
400	2.10	0.00	0.00	500	4.27	7.40	7.20
420	2.87	0.00	0.00	520	4.25	2.54	3.68
440	3.55	0.158	0.115	540	4.19	0.09	0.27
460	4.08	2.30	1.64	560	4.10	0.00	0.00
480	4.24	6.54	5.12	580	3.97	0.00	0.00
500	4.27	7.40	7.20	600	3.83	0.00	0.00

IV. Spectral composition of light in the sea. Derived by W. R. G. Atkins [619] from photographic data obtained by Grein in the Mediterranean Sea in 1913 and 1914. Designation of colors: R = red, OY = orange-yellow, G = green, BG = blue-green, B = blue, BV = blue-violet. $I \equiv I_c/I_t$, I_c and I_t = intensity of the indicated color and total intensity, respectively, both at the same depth (d). Unit of $d = 1$ m.

Color→ d	R	OY	G	BG	B	BV
			1000*I*			
1	96.7	165.7	165.7	165.7	198.9	207.3
5	0.98	1.18	117.3	117.3	254.4	508.8
10	0.34	1.06	89.64	89.64	282.2	537.1
20	0.018	1.05	4.68	17.26	279.7	697.2
50	0.0025	0.069	4.53	5.04	486	504
75		0.054	4.73	14.2	193.6	787.5
100		0.0052	1.56	1.73	346.2	650.8
200			3.18	8.06	37.16	952
500			12.27	30.63	30.8	920.3
1000				74.6	37.31	881.1
1500						(1000)

V. Yearly means from photoelectric measurements. H. H. Poole and W. R. G. Atkins.[620]

(a) No regular seasonal changes in opacity.

(b) Level of maximum absorption changes during the day, presumably from migration of zoöplankton.

(c) Mean value of k in $I = I_0 e^{-kd}$ for the layers 0 to 20, 20 to 40, and 40 to 60 m, are 0.150, 0.120, and 0.111 m⁻¹, respectively.

(d) For most turbid, $k = 0.228$ m⁻¹; for least, $k = 0.06$ at depth of 25 m.

(e) The percentage of the incident light that reaches a depth of 20 m is 6.62, 40 m is 0.72, and 60 m is 0.085.

(f) Down to 25 m, the horizontal illumination is 0.54 of the vertical.

(g) A Secchi disk is just visible when its illumination is 16 per cent of that of the surface of the water.

[a] References:
Beebe, W., and Hollister, G., *Bull. N. Y. Zool. Soc.*, **33**, 249-263 (1930).
Hulburt, E. O., *J. Opt. Soc. Amer.*, **22**, 408-417 (1932).
Krümmel, O., "Handb. d. Ozeanog.," Vol. **1**, 1907.

[b] Computed on the assumption that the true coefficients of absorption as determined by Hulburt are correct, that scattering is caused by thermal fluctuations in the concentration of the molecules, and that in the first 250 ft there is turbidity equivalent to one mote 0.1 mm² in sectional area in each cm³ of water. The last is introduced in order to make the computed values fit Beebe's observations.

44. Emissivity of Water

For radiation of wave-length greater than 1 μ, the absorptivity of water is very great (Section 43) and the reflectivity is small (Table 146); whence one may conclude that water will radiate very nearly as an ideal (black body) radiator, and that its emissivity will not be much less than unity. An experimental determination is difficult, and until recently there was only that of K. Siegl,[621] which for some reason leads to a value that is surely too low. In contrast to that, E. Schmidt[622] found that $E = 0.985 \pm 0.001$ and does not vary with the temperature of the water, which was varied from 10 °C to 50 °C. Here E is the ratio of the radiation from water to that from an ideal radiator at the same temperature (Table 288). He found that a layer of water 0.1 mm thick radiates as strongly as does a thick layer.

45. Photoelectric Effects for Water

Two types of photoelectric effect are exhibited by water: one is photovoltaic, and the other has to do with the emission of electrons from an illuminated surface of water.

Photovoltaic Effect for Water.

R. Audubert[623] observed that if one of two metal plates immersed in an aqueous solution of an electrolyte is illuminated and the other not, the two plates acquire a difference in electrical potential. His investigation of this phenomenon forms the subject of a series of papers.[624] He concludes that water is essential to the phenomenon, and that it is photolyzed, *i.e.*, separated into O and OH, by the light.

Photoelectric Emission by Water.

Using a mercury arc shining through a fluorite window, W. Zimmermann[625] found that the illumination of a clean, fresh, water surface resulted in no emission of photo-electrons, but as the surface aged an emission analogous to that from solids appeared and grew. This he attributed to dust deposited from the air.

On the other hand, W. Obolensky,[626] using a spark between terminals of aluminum, observed an emission that varied with the filtration of the light, and that occurred only for wave-lengths shorter than about $\lambda = 200$ mμ. (See Table 168.) Likewise, L. Couson and A. Molle[627] found an

---

[619] Atkins, W. R. G., *J. du Cons. Int. Expl. Mer.*, **7**, 171-211 (1932).
[620] Poole, H. H., and Atkins, W. R. G., *J. Marine Biol. Assoc. United Kingdom (N. S.)*, **16**, 297-324 (1929).
[621] Siegl, K., *Sitz. Akad. Wiss Wien (Abt. IIa)*, **116**, 1203-1230 (1907).
[622] Schmidt, E., *Forsch. Gebiete Ingenieurw.*, **5**, 1-5 (1934).
[623] Audubert, R., *Compt. rend.*, **189**, 800-802 (1929).
[624] Audubert, R., *Compt. rend.*, **189**, 1265-1267 (1929); **193**, 165-166 (1931); **194**, 82-84 (1932); **196**, 475-478, 1588-1590 (1933); *Jour. de Phys. (7)*, **5**, 486-496 (1934); Audubert, R., and Lebrun, G., *Compt. rend.*, **198**, 729-731 (1934).
[625] Zimmermann, W., *Ann. d. Physik (4)*, **80**, 329-348 (1926).

emission when the exciting light lay in the range $\lambda = 200$ mμ to 130 mμ, and concluded that it was not due to casual contamination of the surface. They suggested that these electrons come from the oxygen in the water molecule.

More recently, P. Görlich,[628] using an evacuated monochromator with an optical train of fluorspar, has confirmed the observations of Obolensky and of Couson and Molle. He has shown that the limiting wave-length at which the emission vanishes as λ is increased lies between 203 and 204 mμ; that it is independent of the actual value of the low conductivity of the water, which in his tests was varied from 3.3×10^{-6} to 3.2×10^{-4}(ohm·cm)$^{-1}$; and that it is the same for concentrated aqueous solutions of $AgNO_3$, $NaCl$, Na_2SO_4, or $K_4Fe(CN)_6$ as it is for pure water. He also studied the variation of the emission with the wave-length of the incident light (see Table 168).

In contrast to the preceding results stand those of H. Greinacher.[629] He used an electron counter, a flowing jet of water, and radiation from a quartz mercury-lamp, the filtration being 3 mm of fused quartz. Radiation so filtered is supposed to contain no wave-length shorter than 220 mμ, yet he reported a marked emission of electrons. The water contained air and lime.

Table 168.—Photoelectric Emission by Water

The ICT data were obtained with light from an Al-spark, filtered as indicated and containing no wave-length shorter than λ_{min}; the intensity of the emission by CuO when illuminated by the fluorite-filtered radiation corresponded to $S = 40,000$.

For the Görlich data, I is the intensity of the emission by water when illuminated by monochromatic light of wave-length λ and of a certain arbitrarily fixed intensity.

Unit of $\lambda = 1$ m$\mu = 10$A $= 10^{-7}$ cm

Filter	ICT[a] λ_{min}	S	Görlich[a] λ	I
Glass	330	0	204.1[b]	0
Calcite (CaCO$_3$)	220	0	203.0	4.0
Quartz and air	177	11	198.9	17.5
Quartz (SiO$_2$)	145	15	189.9	50.0
Fluorite[c] (CaF$_2$)	125	100	185.4	52.0
			176.3	21.4
			171.9	11.1

[a] Sources:
ICT From the compilation by A. L. Hughes, *Int. Crit. Tables*, **6**, 67-69 (1929); based on Obolensky, W., *Ann. d. Physik (4)*, **39**, 961-975 (1912).
Görlich, P., *Idem, (5)*, **13**, 831-850 (1932).

[b] H. Greinacher [629] observed an emission when $\lambda > 220$ mμ (see text).

[c] Fluorite with not more than 3 mm of air at atmospheric pressure.

---

[626] Obolensky, W., *Ann. d. Physik (4)*, **39**, 961-975 (1912).
[627] Couson, L., and Molle, A., *Arch. Sci. Phys. et Nat. (5)*, **10**, 231-242 (1928).
[628] Görlich, P., *Ann. d. Physik (5)*, **13**, 831-850 (1932).
[629] Greinacher, H., *Helv. Phys. Acta*, **7**, 514-519 (1934).

46. THE SPECTRUM OF WATER
(For Raman spectrum, see Section 39.)

Under-water Sparks.

The spectrum of the light from an electric discharge (spark, arc, etc.) occurring in water consists of a continuous, and frequently intense, background extending far into the ultraviolet; of lines characteristic of the electrodes; in some cases, of the spectrum of water-vapor; usually of lines due to hydrogen, but those due to oxygen are absent except in the brush discharge, where they have been observed by H. Smith.[630] B. Setna[630a] has reported the presence of the following "water bands" in the spectra of under-water sparks: $\lambda = 7760, 7933, 8226, 8475$A. See also [631].

A brief report on the emission of light by spark discharges in liquids has been made by J. A. Anderson,[632] in which other additional references are given. The subject has been studied primarily for the purpose of ascertaining how the spectrum of the electrode material is changed by the high pressure that exists in such discharges.

Absorption Spectrum.

The absorption spectrum of water at a given temperature is independent of the recent thermal history of the water, being the same for water from recently melted ice as for that from recently condensed steam.[633]

Table 169.—Absorption Spectrum of Water

The approximate position of other, generally less pronounced, bands may be determined from the data in Tables 160 and 161, and from the curves published by RL[a] and Re.[a]

λ_{obs} and λ_m are, respectively, the observed value of the wave-length of greatest absorption and the mean of the better determinations of that wave-length; ν_{obs} and ν_m are, respectively, the reciprocals of λ_{obs} and λ_m. Only the bands of pronounced absorption are represented in the λ_m column; and the values there given have been taken from El².[a]

The width and intensity of a band are in some cases indicated by letters, b = broad, n = narrow, st = strong, wk = weak.

Unit of $\lambda = 1\mu = 10^{-4}$ cm; of $\nu = 1$ cm⁻¹.

$\lambda_m, (\nu_m)$	λ_{obs}	ν_{obs}	Ref.[a]	$\lambda_m, (\nu_m)$	λ_{obs}	ν_{obs}	Ref.[a]
0.55	0.55	18200	As	0.75	0.745c	13420	El', Pl
(18200)	0.599b }	16690	MRB	(13300)	0.762	13120	Ga
	0.635 }	15750	MRB		0.768	13020	Ma
	0.60	16700	RuLa		0.77	12980	As
0.63	0.655b	15270	MRB		0.842	11880	Ga
(15900)	0.656 }	15240	MRB	0.85	0.845c	11830	El', Ma, Pl
	0.670 }	14920	MRB	(11760)	0.89	11200	Ka
	0.700b	14280	MRB, RuLa		0.90	11100	Ka

[630] Smith, H., *Phil. Mag. (6)*, **27**, 801-823 (1914).
[630a] Setna, B., *Indian J. Phys.*, **6**, 29-34 (1931).
[631] Bloch, L. and E., *Compt. rend.*, **174**, 1456-1457 (1922); Finger, H., *Verh. physik. Ges.*, **11**, 369-376 (1909); *Z. wiss. Photog.*, **7**, 329-356, 369-392 (1909); Konen, H., *Ann. d. Physik (4)*, **9**, 742-780 (1902); Konen, H., and Finger, H., *Z. Elektrcch.*, **15**, 165-169 (1909); Liveing, G. D., and Dewar, J., *Phil. Mag. (5)*, **38**, 235-240 (1894); Toriyama, Y., and Shinohara, U., *Nature*, **132**, 240 (1933).

Table 169—(Continued)

$\lambda_m, (\nu_m)$	λ_{obs}	ν_{obs}	Ref.[a]	$\lambda_m, (\nu_m)$	λ_{obs}	ν_{obs}	Ref.[a]
	0.915	10930	Ka	2.97 b, st	2.97[e]	3370	Pa
	0.93	10800	Ka	(3370)	3.03	3300	ES
	0.973	10280	La		3.04	3290	Re
0.98	0.98[e]	10200	Ma, Pl		3.06	3270	As, RL
(10200)	0.995	10050	Dr		3.08	3250	Er
	1.00	10000	As		3.30	3030	Re
1.18	1.20	8330	Ma		4.66	2146	Re
(8470)	1.21	8260	Dr	4.70 wk	4.70	2128	As, Cb, RL, ES
	1.215[e]	9230	Pl	(2128)	4.72[e]	2119	Pa
	1.25	8000	As		5.56	1798	PlC
	1.44	6940	Pl		5.83	1715	PlC
	1.445	6920	ES		6.05	1653	Re
	1.45	6900	Do		6.06[e]	1650	Pa
1.46	1.47	6800	Ma		6.08	1645	RL
(6850)	1.475	6780	Dr, St	6.1 n, st	6.10	1639	As, AF, Cb
	1.48	6760	Gr	(1639)	6.20	1613	ES
	1.50	6670	As, Cb		6.8	1470	Wi
	1.51	6620	Pa		7.1	1410	AF, RL
	1.74	5750	El[3]		7.3	1370	AF
	1.79	5590	El[3], Co		8.2	1220	AF
	1.93	5180	Es		8.6	1160	Wi
	1.94	5150	As		8.8	1140	AF
	1.95	5130	Ma		9.5	1050	RL
	1.954	5120	St		10.4	960	AF
	1.96	5100	Do, ES		12.5	800	AF
	1.97	5080	Dr		15.8	633	AF, RL
1.98n	1.98	5050	Gr		19.5[d]	513	El[3]
(5050)	2.05	4880	Cb, Pa		20.	500	Ca
	2.79[d]	3580	El[3]		20.2	495	AF
	2.90[d]	3450	El[3]		60.	167	Ca
	2.94	3400	Er		83.	120	CaE
	2.95	3390	Cb, Ma, St		2000	5.0	Te
					7000	1.4	Te
					20000[f]	0.5	Te

Region 2 cm to 28 cm is still to be investigated.
Between 28 cm and 300 cm there are probably no bands (Kn).

[a] References:

AF	Abney, W. de W., and Festing, E. R., *Proc. Roy. Soc. (London) (A)*, **35**, 328-341 (1883).
As	Aschkinass, E., *Ann. d. Physik (Wied.)*, **55**, 401-431 (1895).
Ca	Cartwright, C. H., *Nature*, **136**, 181 (L) (1935); *Phys. Rev. (2)*, **49**, 470-471 (1936).
CaE	Cartwright, C. H., and Errera, J., *Proc. Roy. Soc. (London) (A)*, **154**, 138-157 (1936).
Cb	Coblentz, W. W., *Carnegie Inst. Washington*, Publ. No. **35**: pp. 331 (56-58, 185), 1905; *Bull. Bur. Stand.*, **7**, 619-663 (S168) (1911).
Co	Collins, J. R., *Phys. Rev. (2)*, **52**, 88-90 (1937).
Do	Donath, B., *Ann. d. Physik (Wied.)*, **58**, 609-611 (1896).
Dr	Dreisch, T., *Z. Physik*, **30**, 200-216 (1924).
El[1]	Ellis, J. W., *J. Opt. Soc. Amer.*, **8**, 1-9 (1924).
El[2]	Ellis, J. W., *Phil. Mag. (7)*, **3**, 618-621 (1927).
El[3]	Ellis, J. W., *Phys. Rev. (2)*, **38**, 693-698→582 (A) (1931).
Er	Errera, J., *J. de Chim. Phys.*, **34**, 618-626 (1937).
ES	Ellis, J. W., and Sorge, B. W., *J. Chem'l Phys.*, **2**, 559-564 (1934).
Ga	Ganz, E., *Ann. d. Physik (5)*, **26**, 331-348 (1936) = *Diss.*, München, 1936.
Gr	Grantham, G. E., *Phys. Rev. (2)*, **18**, 339-349 (1921).
Ka	Kaplan, J., *J. Opt. Soc. Amer.*, **14**, 251-256 (1927).
Kn	Knerr, H. W., *Phys. Rev. (2)*, **52**, 1054-1067 (1937).
La	Lambly, J. E., *Phys. Rev. (2)*, **31**, 706 (A) (1928).
Ma	Magat, M., *Jour. de Phys. (7)*, **5**, 347-356 (1934).
MRB	McLennan, J. C., Ruedy, R., and Burton, A. C., *Proc. Roy. Soc. (London) (A)*, **120**, 296-302 (1928).
Pa	Paschen, F., *Ann. d. Physik (Wied.)*, **53**, 334-336 (1894).
Pl	Plyler, E. K., *J. Opt. Soc. Amer.*, **9**, 545-555 (1924).
PlC	Plyler, E. K., and Craven, C. J., *J. Chem'l Phys.*, **2**, 303-305 (1934).
Re	Reinkober, O., *Z. Physik*, **35**, 179-192 (1926).
RL	Rubens, H., and Ladenburg, E., *Verh. physik. Ges.*, **11**, 16-27 (1909).
RuLa	Russell, W. J., and Lapraik, W., *J. Chem. Soc. (London)*, **39**, 168-173 (1881).
St	Stansfeld, B., *Z. Physik*, **74**, 460-465 (1932).
Te	Tear, J. D., *Phys. Rev. (2)*, **21**, 611-622 (1923).
Wi	Williams, D., *Phys. Rev. (2)*, **49**, 869 (A) (1936).

[632] Anderson, J. A., *Int. Crit. Tables*, **5**, 433 (1929).
[633] Ellis, J. W., and Sorge, B. W., *Science (N. S.)*, **79**, 370-371 (1934).

[b] Sharp edge of band.
[c] Pl's observations refer to 0 °C.
[d] These three values are used by El[3] in his interpretation of the spectrum; the 2.79 and 2.90 μ represent 2 of the 3 Raman-spectrum components corresponding to the 3 μ band; the 19.5 μ band "is somewhat hypothetical" and interpreted as arising from the mutual vibrations of two (H_2O)'s.
[e] These Pa values varied with the thickness of the water.
[f] For greater values of λ, see Tables 177 and 178.

Table 170.—Analyses of the Absorption Spectrum of Water

Uncertainties in the values of the wave-lengths corresponding to the maxima of the several bands, and the complexity resulting from the overlapping of bands (see first section of this table), make the analysis of the spectrum difficult. Furthermore, there are differences of opinion as to the manner in which the vibrations of a molecule are affected by its neighbors. Some hold that the fundamental vibrations of the isolated molecule (those characteristic of the vapor) play no part in the case of the liquid, being completely suppressed by the action of neighboring molecules (e.g., Errera, 1937[a]; Bosschieter and Errera, Compt. rend., 204, (1937[a])); some hold that the liquid is a mixture of polymers, each having fundamental vibrations of its own (Ellis, 1931[a]; Cabannes and de Diols, 1934[a]; Rao, 1934[a]); and others hold that the fundamental vibrations of the isolated molecule remain predominant but, perhaps, restricted and with frequencies more or less modified, and that to these must be added additional fundamental vibrations arising from the bonding of molecules into a more or less definite, though transient, structure (e.g., Magat, 1936[a]).

Until there has arisen some generally accepted opinion regarding these matters, the original articles should be consulted by those interested in the interpretation of the absorption spectrum of water. Some of the recent or more typical of those articles are listed in footnote [a] to this table.

As illustrative of the analyses that have been proposed, those by Ellis and by Magat are here given. They are the most detailed. Ellis first interpreted in terms of two fundamentals ($\lambda = 2.97$ and 6.1μ) all the values in column λ_m of Table 169, except $\lambda = 4.70 \mu$. But in order to obtain that and the two bands that he discovered at $\lambda = 1.79$ and 1.74μ, he replaced the $\lambda = 2.97 \mu$ by two others ($\lambda = 2.90$ and 2.79μ) and included a fourth fundamental ($\lambda = 19.5 \mu$). The last is somewhat doubtful, and the two that replace the $\lambda = 2.97 \mu$ are derived from the Raman spectrum (see notes to Table 169). Magat first interpreted the spectrum in terms of three fundamentals ($\lambda = 2.9, 3.1,$ and 6.0μ), and later added 4 other fundamentals arising from the interaction of adjacent molecules. All five of these analyses are given below. In Magat's analyses n_1, n_2, n_3 have the same values in both cases, and are regarded as corresponding respectively to the σ, π, and δ (see Table 64) vibrations of the free molecule. In Ellis' analyses the frequencies have been so numbered as to correspond as closely as may be with the similarly numbered ones in Magat's.

Table 170—(Continued)

In all cases, the calculated frequency ($c\nu$) of a band is given by the relation $\nu = n_1\nu_1 + n_2\nu_2 + \ldots$, where $\nu = 1/\lambda$, $\nu_1 = 1/\lambda_1, \ldots, \lambda_1, \lambda_2, \ldots$ being the wave-lengths of the assumed fundamental vibrations, the n's being small integers, and c being the velocity of light. The values of the n's corresponding to each of the calculated λ's are tabulated; the fundamental λ's are those for which one n is unity and all the others are zero. $I =$ infrared; $R =$ Raman spectrum; "single" = only one of the n's occurs in the calculation of λ_{calc}; "comb." = both n's occur. Except in the first of Ellis' analyses, all the n's listed occur in the calculation, each with its indicated value.

Unit of $\lambda = 1\ \mu = 10^{-4}$ cm; of $\nu = 1$ cm^{-1}

Ellis (1927)[a]

λ_{obs}	Single n_2	n_3	Comb. n_2	n_3	n_3	n_2 λ_{calc}	Comb.
6.1	0	1	0	0	6.1		
2.97	1	2	0	0	3.05	2.97	
1.98	0	3	1	1	2.03		1.96
1.46	2	4	1	2	1.52	1.45	1.48
1.18	0	5	2	1	1.22		1.17
0.98	3	6	2	2	1.015	0.97	0.98
0.85	0	7	2	3	0.875		0.84
0.75	4	8	3	2	0.765	0.725	0.73
0.63	{ 0 { 5	9 10	0 0	0 0	0.675 0.61	0.58	
0.55	0	11	0	0	0.555		

Ellis (1931)[a]

λ_{obs}	n_1	n_2	n_3	n_6	λ_{calc}
19.5	0	0	0	1	19.5
6.1	0	0	1	0	6.1
4.7	0	0	1	1	4.65
2.90	0	1	0	0	2.90
2.79	1	0	0	0	2.79
1.79	0	1	1	1	1.785
1.74	1	0	1	1	1.745

Ellis and Sorge (1934)[a]

λ_{obs}	n_1	n_2	n_3	n_0	λ_{calc}	ν_{calc}
6.20	0	0	1	0	6.21	1610
4.70	0	0	0	1	4.69	2130
3.11	0	0	2	0	3.11	3220
2.92	0	1	0	0	2.92	3430
2.80	1	0	0	0	2.80	3570
1.93	{ 0 { 1	1 0	1 1	0 0	1.98 1.93	5040 5180
1.78	{ 0 { 1	1 0	0 0	1 1	1.80 1.75	5560 5700
1.47	0	2	0	0	1.46	6860
1.42	2	0	0	0	1.39	7170
1.18	{ 0 { 2	2 0	1 1	0 0	1.18 1.14	8470 8750

46. WATER: SPECTRUM

Table 170—*(Continued)*

Magat (1934)[a]

I	λ_{obs}	R	n_1	n_2	n_3	λ_{calc}	ν_{calc}
		6.028	0	0	1	6.028	1659
		3.105	0	1	0	3.105	3221
2.95		2.911	1	0	0	2.911	3435
1.95		1.965	1	0	1	1.964	5092
1.48			1	1	0	1.502	6660
1.20			1	1	1	1.203	8310
0.98			3	0	0	0.970	10310
0.84			3	0	1	0.836	11960
0.77			3	1	0	0.736	13590

Magat (1936)[a]

I	λ_{obs}	R	n_1	n_2	n_3	n_4	n_5	n_6	n_7	λ_{calc}	ν_{calc}
		167	0	0	0	0	1	0	0	167	60
		66–44	0	0	0	0	0	0	1	60.2	166
19.6		20.0	0	0	0	0	0	1	0	19.6	510
14.9		13.5	0	0	0	1	0	0	0	14.3	700
6.02			0	0	1	0	0	0	0	6.02	1660
5.85			0	0	1	0	1	0	0	5.81	1720
5.40			0	0	1	0	0	0	1	5.49	1820
4.684		4.684	0	0	1	0	0	1	0	4.63	2160
3.14			0	1	0	0	0	0	0	3.05	3280
2.90			1	0	0	0	0	0	0	2.898	3450
2.486		2.485	1	0	0	0	0	1	0	2.53	3950
1.95			1	0	1	0	0	0	0	1.957	5110
1.789			1	0	1	0	0	1	0	1.779	5620
1.47			1	1	0	0	0	0	0	1.486	6730
1.18			1	1	1	0	0	0	0	1.192	8390
			1	2	0	0	0	0	0	1.011	9890
0.98			3	0	0	0	0	0	0	0.966	10350
			1	2	1	0	0	0	0	0.866	11550
0.84			3	0	1	0	0	0	0	0.833	12010
			1	3	0	0	0	0	0	0.763	13110
0.75			3	1	0	0	0	0	0	0.734	13630
			1	3	1	0	0	0	0	0.677	14770
0.66			3	1	1	0	0	0	0	0.654	15290
			1	3	2	0	0	0	0	0.609	16430

[a] References:

Bosschieter, G., and Errera, J., *Compt. rend.*, **204**, 1719-1721 (1937); **205**, 560-562 (1937); *Jour. de Phys. (7)*, **8**, 229-232 (1937).
Cabannes, J., and de Riols, J., *Compt. rend.*, **198**, 30-32 (1934).
Carrelli, A., *Nuovo Cim. (N. S.)*, **14**, 245-256 (1937).
Cartwright, C. H., *Nature*, **136**, 181 (L) (1935); *Phys. Rev. (2)*, **49**, 470-471, 421 (A) (1936).
Cartwright, C. H., and Errera, J., *Acta Physicochim. URSS*, **3**, 649-684 (1935); *Proc. Roy. Soc. (London) (A)*, **154**, 138-157 (1936).
Ellis, J. W., *Phil. Mag. (7)*, **3**, 618-621 (1927); *Phys. Rev. (2)*, **38**, 693-698 (1931).
Ellis, J. W., and Sorge, B. W., *J. Chem'l Phys.*, **2**, 559-564 (1934).
Errera, J., *J. de Chim. Phys.*, **34**, 618-626 (1937).
Kinsey, E. L., and Ellis, J. W., *Phys. Rev. (2)*, **49**, 105 (L), 209 (A) (1936).
Magat, M., *Jour. de Phys. (7)*, **5**, 347-356 (1934); *Ann. de Phys. (11)*, **6**, 108-193 (Bibliog. of 148) (1936).
Piekara, A., *Acta Phys. Polon.*, **6**, 130-143 (1937).
Rao, I. R., *Proc. Roy. Soc. (London) (A)*, **145**, 489-508 (1934).
Williams, D., *Phys. Rev. (2)*, **49**, 869 (A) (1936).

Table 171.—Effect of Temperature and Pressure on the Absorption Spectrum of Water

J. R. Collins [637] found no change in the absorption spectrum of water when the pressure was increased from 120 atm to 5000 atm, the range covered being $\lambda = 0.71\ \mu$ to $1.05\ \mu$.

Collins (1937)[a] found no marked change in the band at $1.79\ \mu$ when the temperature was increased from 4 to 137 °C. Ganz (1936)[a] has reported that the intensity of the $0.77\ \mu$ band increases with the temperature, that the band near $0.84\ \mu$ is scarcely detectable at room temperature, but is well-defined and sharp at 87 °C, and (1937)[a] that the band at $4.7\ \mu$ is still visible at 84 °C.

Of historical interest only are the early observations of W. J. Russell and W. Lapraik [638] who were unable to detect visually any change in the absorption spectrum when the water was heated from 20 °C to 60 °C.

λ = wave-length at which the absorption is a maximum for the band considered.

Unit of $\lambda = 1\ \mu = 10^{-4}$ cm

Collins (1925)[a]			Stansfeld (1925)[a]		
λ 0 °C	λ 95 °C	Decr. in λ	λ 20 °C	λ 97 °C	Decr. in λ
0.775	0.740	0.035	1.475	1.468	0.007
	0.845		1.954	1.945	0.009
0.985	0.970	0.015			
1.21	1.17	0.04			
1.45	1.43	0.02			
1.96	1.94	0.02			

[a] References:
Collins, J. R., *Phys. Rev. (2)*, **26**, 771-779 (1925); **52**, 88-90 (1937).
Ganz, E., *Ann. d. Physik (5)*, **26**, 331-348 (1936) = *Diss.*, München, 1936; *Idem*, **28**, 445-457 (1937).
Stansfeld, B., *Z. Physik*, **74**, 460-465 (1932).

47. THE COLOR OF WATER AND OF THE SEA *

Pure Water.

Transmitted light.—It is generally stated that long columns of pure water appear blue by transmitted light. Lord Rayleigh [639] has written that W. Spring somewhere stated that in columns 4 or 5 meters long the color of pure water is a fine blue, only to be compared with the purest sky-blue as seen from a great elevation; but that when the incident light was white he himself has never obtained "a blue answering to Spring's description."

W. Spring [640] is positive that the blue color pertains to the water itself,

* W. D. Bancroft [642] has published a review and summary of many of the earlier papers treating of this subject.

[637] Collins, J. R., *Phys. Rev. (2)*, **36**, 305-311 (1930) → *Idem*, **35**, 1433 (A) (1930).
[638] Russell, W. J., and Lapraik, W., *J. Chem. Soc. (London)*, **39**, 168-173 (1881).
[639] Lord Rayleigh, *Proc. Roy. Inst. Grt. Brit.*, **19**, 765-771 (1910) = *Nature*, **83**, 48-50 (1910).
[640] Spring, W., *Rec. trav. chim. Pays-Bas*, **18**, 1-8, 153-168 (1899).
[641] Spring, W., *Ibid.*, **17**, 359-367 (1898).

and that the presence of suspended matter modifies that color, introducing a greenish tint, and in some cases rendering the water colorless, as in some lakes. A very small amount of exceedingly small particles of hematite in suspension will suffice for the last.[641]

From a study of the waters from a number of lakes, O. F. v. Aufsess[615] concluded that every departure from the blue of pure water arises from the presence of foreign substances, and that the color of a given lake is only slightly affected by changes in the turbidity.

In contrast with the preceding we have: (a) the statement of J. W. Lovibond[643] that a 4-foot stratum of distilled water is equivalent in color to a combination of the Lovibond filters Yellow 1.0 and Blue 1.45, which color is an unsaturated green; and (b) the fact that the values of the absorptivities given in Table 161 indicate that the color of a stratum of water 20 meters or less in thickness is very definitely green-blue, quite similar in hue to that of the spectrum near $\lambda = 0.49\,\mu$.[643a] It should be remembered that none of the water used in obtaining the data in Table 161 was really optically empty, even the best being only so dust-free as may be obtained by repeated distillations (cf. Section 40), and in some cases no correction was made for the effect of the windows.

It has been suggested that the blueness of water arises from the relatively few molecules of ice that are dissolved in the water. Such was the opinion of J. Duclaux.[644] He regarded pale green as the color of hydrol itself, and suggested that a study of the variation in the color of water with the temperature might yield data from which the relative proportions of the two polymers (ice and hydrol) could be computed. This is in line with Barnes' statement that the color of the St. Lawrence River changes as freezing becomes imminent (see Section 58).

Scattered light.—A. Turpain[645] has called attention to a series of papers published some 60 years ago by Alexandre Lallemand, and apparently forgotten. Lallemand[646] observed that most, probably all, liquids scatter light laterally, even when devoid of suspended particles. He called this fluorescence, and suggested that it arose in part from a kind of molecular reflection or diffusion, and in part from free vibrations of the molecules. That arising in the first way will be partially or completely polarized and of the same frequency as the incident light exciting it; that arising in the second will be unpolarized and of a different (he says longer) wave-length. He attributed sky light also to molecular scattering.

C. V. Raman[647] has shown that the intensity of the light molecularly scattered by a liquid, and the attendant coefficient of absorption, can be calculated on the "theory of fluctuations"—the theory that the number of

[642] Bancroft, W. D., *J. Franklin Inst.*, **187**, 249-271, 459-485 (1919).
[643] Lovibond, J. W., "Light and Colour Theories," p. 21, 1915.
[643a] Gibson, K. S., and Keegan, H. J., *J. Opt. Soc. Amer.*, **27**, 58 (A) (1937).
[644] Duclaux, J., *Rev. gén. des Sciences*, **23**, 881-887 (1912).
[645] Turpain, A., *Compt. rend.*, **197**, 1107-1109 (1933).
[646] Lallemand, A., *Ann. de chim. et phys.* (4), **22**, 200-234 (1871); (5), **8**, 93-136 (1876).
[647] Raman, C. V., *Proc. Roy. Soc. (London) (A)*, **101**, 64-80 (1922).

molecules per unit of volume varies slightly from point to point, on account of the thermal agitation of the molecules. The intensity so calculated for dust-free (optically empty) water is 160 times that of dust-free air; an observed value was 175 times. Likewise the computed coefficients of absorption for $\lambda = 0.494$ and $0.522\,\mu$, where there is negligible selective absorption, are 0.000029 and 0.000022 cm^{-1}, respectively; while the experimental value found by Aufsess for each was 0.00002 cm^{-1} (see Table 161).

He states that a sufficiently thick layer of pure water exhibits by molecular scattering a deep blue color more saturated than sky light and of comparable intensity. "The colour is primarily due to diffraction, the absorption only making it of a fuller hue."

The light scattered laterally by thin columns of water comes mainly from suspended particles, and is not blue, partaking largely of the color of the incident light as modified by the selective reflection of the particles.[640] Spring held the erroneous opinion that there is no scattering by the water itself.

The deep blue color of Crater Lake, Oregon, probably arises from the scattering by the water itself, modified, more or less, by the color of the sky.[648]

For methods employed for obtaining optically empty water, see Section 40.

The Sea.

Although long columns of pure water are blue by transmission, and the light scattered by thick layers of water is blue, there has been much discussion regarding the blueness of the sea. Three effects may contribute to the color: (1) the reflection of the blue sky; (2) the proper color of the water, including the effect of molecular scattering; and (3) the scattering of small particles held in suspension.

Lord Rayleigh[639] attributed most of the color to the reflection of the sky, and thought that only a very unimportant fraction of it is to be accounted for by the scattering by small particles. Whether he intended the last to include molecular scattering is not entirely clear. To these conclusions, C. V. Raman[647] and J. Y. Buchanan[649] have taken exception, attributing the blueness to the water itself. Buchanan based his objection on his observation that when quiet water, as in the screw-well of the *Challenger,* is viewed vertically under such conditions as to exclude reflected sky, it appears to be of a beautiful dark-blue color. Truly, as Rayleigh remarked in reference to the water at Capri, this light, scattered or internally reflected by the water, came in large part from the sky. That would enhance the blueness if the sky were blue, but it can scarcely be accepted as the sole explanation of the observations recorded by Buchanan. In reality, the discussion seems to have arisen, at least in part, from differ-

---

[648] Pettit, E., *Proc. Nat. Acad. Sci.,* **22,** 139-146 (1936).
[649] Buchanan, J. Y., *Nature,* **84,** 87-89 (1910).

ences in the interpretation of the vague term "color of the sea." Does it mean the color seen when the surface of the sea is viewed from above and at such a distance that the line of sight makes a large angle with the vertical, there being nothing to obstruct the reflection of the sky from the portion observed? If so, the intensity of the reflected light will be so great, as compared with that coming from the interior of the water, that the observed color will depend almost entirely upon that of the sky. This is what Rayleigh seems to have had in mind. But if the term "color of the sea" means the color of the light coming directly from its interior, unmixed with light reflected from the surface, then the color of the sky will play a very subordinate part. Under intermediate conditions, the relative importance of each of those sources of color will vary with the condition. Again, one may mean by "color of the sea" the color of the light transmitted by a very long column of sea-water when the incident light is white. That will not be exactly the same as the color of the light coming directly from the interior of the sea, but the difference will not be great when there are many white clouds; it may be called the color by transmission.

E. O. Hulburt,[650] confining his observations to the "rim of the sea," the region bounded by lines of sight making angles of 0° and of 3°, respectively, with the surface of the sea—which region "comprises more than 99/100 of the total area of the sea within the view of the observer," if he is less than 1000 ft. above sea-level—has found that the rim of the sea when ruffled by breezes of 5 to 25 knots takes its color from the sky at an altitude of 30°, and that its brightness is about 0.25 of that of the sky near the horizon, the sky being clear. The light reflected from the sea is, of course, polarized. Hulburt has studied that polarization.

During his bathysphere descents into the open sea, W. Beebe [651] observed the light scattered horizontally by the sea. At a depth of 600 ft. he comments on the blueness of the illumination, which was still "brilliant" at 800 ft. When he looked out and down he "saw only the deepest, blackest blue imaginable, a color which in the spectrum [as seen in his small spectroscope] had vanished four hundred feet above, overlaid and superseded by violet." At 1400 ft. "the outside world .... was, however, a solid, blue-black world, one which seemed born of a single vibration—blue, blue, forever and forever blue." (See also Table 167). The persistence of the visual sensation of blue after the spectroscope shows that the light is violet is again referred to as puzzling.[652] E. O. Hulburt [653] has suggested that this sensation of blueness may arise from a fluorescence of the eye itself.

According to J. Y. Buchanan,[649] only three color-types are required for describing the color of the surface-water of the ocean. (1) Deep olive-green, because of chlorophyll; observed near the edge of polar ice, and in

---

[650] Hulburt, E. O., *J. Opt. Soc. Amcr.*, **24**, 35-42 (1934).
[651] Beebe, W., *Bull. N. Y. Zool. Soc.*, **33**, 201-232 (1930).
[652] Beebe, W., *Science (N. S.)*, **80**, 495-496 (1934).
[653] Hulburt, E. O., *Idem*, **81**, 293-294 (1935).

certain other places. (2) Indigo. As one goes south from the Arctic the surface water assumes a pronounced indigo color, which persists until latitude 40° is passed. (3) Ultramarine. As one goes north from the equator the color persists as a pure and brilliant ultramarine until latitude 30° is passed. The passage between ultramarine and indigo is usually very rapid; the area of mixing is restricted.

48. Optical Rotatory Power of Water

(For the Faraday effect and the Verdet constant, see Section 54.)

The natural optical rotatory power of water (the ability of water to rotate the plane of polarization of a beam of light passing through it) is believed to be precisely zero. That it is indeed exceedingly small is shown by such observations as those of F. Bates and R. F. Jackson [654] who observed for two 200-mm tubes an average rotation of only $-0.0011°$, the extreme single observations being $-0.0043°$ and $+0.0003°$, each occurring with the same tube. These values lie within the range of their experimental error.

A. A. Bless,[655] whose error of setting was less than 0.02°, was unable to confirm F. Allison's conclusion [656] that water possesses a slight power to rotate the plane of polarization of light that passes through it while it is subjected to the action of x-rays.

49. Dielectric Properties of Water

The dielectric properties of water are determined by the value of its dielectric constant (ϵ) and of its absorption index (κ), and by the way these values vary with the conditions. Electrical conductivity is not a dielectric property of the substance, but it must be considered in the interpretation of experimental observations.

Symbols and Definitions.

Only the most frequently used symbols are listed here; others are defined where they appear.

(ϵ, ϵ', ϵ'', ϵ_0, ϵ_1) Dielectric constant. The value of the dielectric constant (ϵ), expressed in electrostatic units, is defined as the ratio of the mutual electrical capacity of a given pair of equipotential surfaces, fixed with reference to each other, when immersed in the dielectric to their capacity when immersed in a vacuum. If the polarizing of the dielectric is accompanied by a dissipation of energy, the apparent dielectric constant is a complex quantity, $\epsilon = \epsilon' - i\epsilon''$; $i \equiv \sqrt{-1}$. In such cases the real part (ϵ') is commonly called the dielectric constant. In certain cases, notably water, the dielectric constant (ϵ_0) at optical frequencies is much smaller than that (ϵ_1) under static conditions.

---

[654] Bates, F., and Jackson, R. F., *Bull. Bur. Stand.*, 13, 67-128 (S268) (1915).
[655] Bless, A. A., *Phys. Rev. (2)*, 33, 121-122 (A) (1929).
[656] Allison, F., *Idem*, 31, 158-159 (A) (1928).

(κ, e) Absorption index (κ). If a plane, simple harmonic, electromagnetic wave is traveling through a dielectric in the direction of z, its amplitude at $z = 0$ being A_0 and at $z = z$ being A, then $A = A_0 e^{-2\pi\kappa n z/\lambda_0}$, e (2.7183) being the base of the natural system of logarithms, and n and λ_0 having the values defined below. As the intensity (I) of such a wave varies as A^2, $I = I_0 e^{-4\pi\kappa n z/\lambda_0}$. Either of these two equivalent formulas may be used to specify the significance of κ.

(n, c, λ, λ_0) The index of refraction (n) is c/V where c (2.9979 × 10^{10} cm/sec) is the phase velocity of the wave in a vacuum, and V is that in the dielectric. Also, $n = \lambda_0/\lambda$ where λ_0 is its wave-length in a vacuum and λ is that in the dielectric.

(k', s) The conductivity (k') is, by definition, equal to the longitudinal electrical conductance of a cylinder of the material of unit length and unit cross-sectional area. It is the reciprocal of the volume resistivity. By definiton, $s \equiv k'\lambda_0/\epsilon c$ when all quantities are expressed in the same system of units; say, the cgse. If, however, k' is expressed in the cgsm system, and ϵ in the cgse, then $s = ck'\lambda_0/\epsilon$. Both expressions have been used, frequently without any specific statement about the units. Here, only the first will be used, the one in which all quantities are expressed in the same system.

(ω, ν) A simple harmonic oscillation will be expressed either as $B_0 \sin \omega t$ or as the real part of $B_0 e^{i\omega t}$; $i \equiv \sqrt{-1}$. Then $\omega = 2\pi\nu$, where ν is the frequency of the oscillation.

(ϕ, θ) The phase defect (ϕ) of a dielectric may be defined as follows. If a given pair of equipotential surfaces, fixed with reference to each other, distant from all others, and immersed in the dielectric, are subjected to a difference in potential defined by $V = V_0 \sin \omega t$, then the current will be $J = J_0 \cos(\omega t - \phi)$, and the polarization of the dielectric will be $D = D_0 \sin(\omega t - \theta)$. Both ϕ and θ depend upon the properties of the dielectric. They differ from zero only when the process of polarizing the dielectric is accompanied by a dissipation of energy, and the value of ϕ is given by the relation $\tan \phi = \epsilon''/\epsilon'$. If the dissipation arises solely from the process of polarizing the dielectric, k' being zero, then $\theta = \phi$; but if the dissipation arises solely from the conductivity, then $\theta = 0$, and $\tan \phi = 2s = 2k'\lambda_0/\epsilon c = 2k'/\epsilon\nu$.

Types of Dielectrics.

Two extreme types of dielectrics may be distinguished: (1) The ideal leaky dielectric, which is equivalent to an ideal, non-conducting and unabsorbing dielectric in parallel with a conductor of low conductivity. The combination exhibits both dispersion (variation of ϵ with λ_0) and absorption;

$$\epsilon = n^2(1 - \kappa^2), \quad \kappa = \frac{2s}{1 + (1 + 4s)^{1/2}}, \quad \tan \phi = 2s = 2k'\lambda_0/\epsilon c = 2k'/\epsilon\nu, \quad \theta = 0.$$

Insofar as current-voltage relationships are concerned, the dielectric acts as if it were a non-conducting, absorbing dielectric with the dielectric constant $\epsilon_e = \epsilon - i4\pi k'/\omega = n^2(1 - i\kappa)^2 = n^2(1 - \kappa^2) - 2in^2\kappa$. This case will not concern us further. Hereafter it will be assumed that either k' is zero or, if not, that effects arising from it have been independently eliminated.

(2) The non-conducting dielectric, either absorbing or non-absorbing, the last being but a limiting case. For such a dielectric, $\epsilon = n^2(1 - i\kappa)^2 = n^2(1 - \kappa^2) - 2in^2\kappa = \epsilon' - i\epsilon''$, $\tan \phi = \epsilon''/\epsilon'$. As these expressions for ϵ are of identically the same form as those for the effective dielectric constant (ϵ_e) in the preceding case, $\epsilon''\omega/4\pi = \epsilon''\nu/2$ may be called the apparent conductivity of the dielectric (it is frequently called simply the conductivity, or more recently, the dipole conductivity, although k' is, by hypothesis, zero). For such dielectrics ($k' = 0$), the dissipation of energy, arising solely from the act of changing the polarization, is intimately bound up with the true dielectric properties. The resulting absorption of energy is, therefore, commonly called dielectric absorption. (It was first described as "anomalous," to indicate that it did not arise from the conductivity as commonly measured.) It is in this sense that the term "dielectric absorption" is used in this compilation. It should not be confused with what Maxwell called electric absorption, which is merely one of the phenomena that accompany dielectric absorption.

Dipole Theory.

In the modern dipole theory of dielectrics, developed by Debye, the molecule is pictured as containing, in addition to the elastically bound electrons and ions of the earlier theories, a rigid or semirigid permanent electrical dipole firmly attached to the molecule, so that both move as a single unit. For certain types of molecules the moment of the dipole may be zero; for those substances the dipole theory adds nothing to the earlier ones. Thirty-five years ago M. Reinganum [657] suggested that certain molecules contained dipoles of constant moment, but whether the dipole could rotate without rotating the molecule itself was left an open question.

The presence of dipoles confers upon the dielectric two new types of polarization: (1) that produced by the aligning of the axes of the dipoles with the direction of the field, as a result of a reorientation of the molecule as a whole; and (2) that produced by the mutual angular displacement of the axes of adjacent dipoles that are elastically coupled, as two magnets might be, by their mutual attraction. This last is exactly similar to that associated with oscillators of other types, and needs no further consideration here, being completely covered by the well-known treatment of optical dispersion, together with P. Drude's extension of that to the case of great damping.[658]

Of the first of these types of polarization, two subtypes need to be considered: (1) that characterized by a free reorientability (free rotation) of each molecule at every instant, and (2) that characterized by a restricted reorientability (restricted rotation), each molecule being elastically bound

---

[657] Reinganum, M., *Ann. d. Physik (4)*, **10**, 334-353 (1903).

[658] Drude, P., *Z. physik. Chem.*, **23**, 267-325 (1897) esp. *Ann. d. Physik (Wied.)*, **64**, 131-158 (1898).

49. WATER: DIELECTRIC PROPERTIES

to its neighbor or neighbors in such a way that the moment of the resultant dipole is not zero, or each molecule being sometimes bound and sometimes free. Until very recently only the first of these subtypes had been mathematically considered.

Free reorientability.—Debye has considered in detail the case of freely reorientable dipoles. His results have been published in numerous papers, and the basic treatment has been given in his book, "Polar Molecules," (1929),[658a] from which much of the following has been derived, and to which reference will be made by means of the symbol PM followed by the number of the page.

In Debye's treatment it is assumed (a) that adjacent dipoles are not elastically bound to one another; (b) that each dipole plays the same part as every other; (c) that such a rotation of the molecule as attends the aligning of the axis of the dipole with the field is resisted by a torque of a viscous nature; (d) that the alignment of the axes is being continually disturbed by the thermal agitation of the molecules; and (e) that the Clausius-Mossotti expression for the molar polarizability of the dielectric, $P = (M/\rho)(\epsilon - 1)/(\epsilon + 2)$, is applicable. The last assumption implies that in the computation of the electric field at any point in the interior of the dielectric, the dielectric may be treated as a continuous medium, and the effect of a vanishingly small volume of the dielectric immediately surrounding the point may be ignored. As the theory deals with a molecular medium, the validity of this assumption has been questioned; and a steadily increasing amount of experimental data has forced the conclusion that this simple theory is certainly not quantitatively applicable to strongly dipolar pure liquids. But for years it was applied to them, and this must be borne in mind when studying the work of that period.

On the basis of these assumptions, Debye (PM, 27) derived, by statistical methods, expression (1) for the case of a constant field

$$\frac{P}{M} \equiv \left(\frac{\epsilon - 1}{\epsilon + 2}\right)\frac{1}{\rho} = \frac{4\pi N}{3M}\left[\alpha + \frac{\mu}{F} L\left(\frac{\mu F}{kT}\right)\right] \quad (1)$$

where $L(x) = \coth x - \frac{1}{x} = \frac{x}{3}\left(1 - \frac{x^2}{15} + \frac{2x^4}{315} - \ldots\right)$. If only the first, or the first two, terms in this expansion are retained, (1) becomes (2) or (3)

$$\frac{P}{M} \equiv \left(\frac{\epsilon - 1}{\epsilon + 2}\right)\frac{1}{\rho} = \frac{4\pi N}{3M}\left[\alpha + \frac{\mu^2}{3kT}\right] \equiv a + \frac{b}{T} \quad (2)$$

$$\frac{P}{M} \equiv \left(\frac{\epsilon - 1}{\epsilon + 2}\right)\frac{1}{\rho} = \frac{4\pi N}{3M}\left[\alpha + \frac{\mu^2}{3kT}\left(1 - \frac{\mu^2 F^2}{15k^2T^2}\right)\right] \equiv a + \frac{b}{T} - \frac{f}{T^3} \quad (3)$$

Here P is the molar polarizability of the substance (often called molar polarization), M molecular weight, ρ density, N Avogadro's number (6.061×10^{23} molecules per g-mole, k Boltzmann's molecular gas constant

---

[658a] Reinhold Publishing Corp., 1929.

1.372 × 10⁻¹⁶ erg/°K per molecule), T °K the absolute temperature, F the strength of the internal field, α the polarizability of the molecule by elastic displacement of the electrons, and μ the moment of the dipole. The quantities α and μ refer to the individual molecule; a, b, and f are merely symbols to be used for brevity. It is obvious that in all cases the value of ϵ can be derived from P by means of the relation

$$\epsilon = (1 + 2P\rho/M)/(1 - P\rho/M) \quad (3a)$$

As the intensity of the field is increased, the dipole contribution to the polarization approaches a condition of saturation, and the corresponding term in the polarizability P begins to decrease. This is shown by the presence of the F^2 term in expression (3). By means of formula (3a) one can readily derive from (3) expression (4) in which ϵ_x is the dielectric constant in an external field of intensity X and ϵ_l is that in a weak field, it being remembered that on the Mossotti hypothesis the internal field is $F = X + 4\pi I/3$ and $\epsilon X = X + 4\pi I$, I being the polarization of the medium; this gives us $F = X(\epsilon + 2)/3$. This ϵ is strictly ϵ_x, but as ϵ_x differs but little from ϵ_l the latter may be used in translating F into X in expression (3).

$$\epsilon_x = \epsilon_l - \frac{4\pi}{45} \cdot \frac{N\rho\mu^2}{MkT}\left(\frac{\epsilon_l + 2}{3}\right)^4 \cdot \left(\frac{\mu X}{kT}\right)^2 \quad (4)$$

or

$$\frac{\epsilon_l - \epsilon_x}{\epsilon_l} = \frac{4\pi}{45} \cdot \frac{N\rho\mu^2}{MkT\epsilon_l} \cdot \left(\frac{\epsilon_l + 2}{3}\right)^4 \cdot \left(\frac{\mu X}{kT}\right)^2 \quad (5)$$

Debye has used n to denote $N\rho/M$, the number of molecules per unit of volume. In addition to this difference in notation, it will be noticed that the magnitude of the negative term in (4) is only a third as great as that of the one given by Debye (PM, 111); Debye's value for ϵ_x is less than that given by (4). The procedure that he followed (PM, 110) leads to an expression of the form $\epsilon_x = (1 + 2\delta)/(1 - \delta)$, which may be expanded into $\epsilon_x = 1 + 3\delta + 3\delta^2 + \ldots$, since $\delta < 1$. If all powers of δ higher than the first are neglected, one obtains Debye's expression for ϵ_x, an expression that errs on the side of being too small. But δ may be, and in the case of water is, not much smaller than unity, in which case its higher powers cannot validly be neglected. This may be the explanation of the loss of the factor 1/3 from Debye's expression, which seems to be still current.[659]

For water at 20 °C (ϵ_l = 80.4 cgse, approximately) formula (5) reduces to

$$(\epsilon_l - \epsilon_x)/\epsilon_l X^2 = 0.01024(10^{18}\,\mu)^4 \text{ \% per (10 cgse field strength)}^2 \quad (6)$$
$$= 0.00114\,(10^{18}\,\mu)^4 \text{ \% per (kilovolt/cm)}^2 \quad (7)$$

For the H_2O molecule, $(10^{18}\,\mu)^4$ is approximately $11\frac{1}{4}$ cgse units (see p. 48), and $(\epsilon_l - \epsilon_x)/\epsilon_l X^2 = 0.0128\%$ per (kilovolt/cm)².

Under the action of an electric field, molecules that are "elastically"

---

[659] See Debye, P., *Chem'l Rev.*, **19**, 171-182 (1936).

49. WATER: DIELECTRIC PROPERTIES

aeolotropic tend to orient themselves so that the direction of maximum polarizability lies along the direction of the field. Consequently the presence of such molecules will give rise to effects that are strictly analogous to those produced by molecules containing fixed dipoles (c.f. PM, 109).

For a sinusoidal field of frequency ν, Debye (PM, 90) finds

$$\frac{P}{M} \equiv \left(\frac{\epsilon-1}{\epsilon+2}\right)\frac{1}{\rho} = \frac{4\pi N}{M}\left(\alpha + \frac{\mu^2}{3kT} \cdot \frac{1}{1+i\omega\tau}\right) \quad (8)$$

where $\omega = 2\pi\nu$ and τ is the relaxation time, which he defines as the time required for the polarization (*not* the polarizability P) to decrease to $1/e$ of its value after the constant inducing field is withdrawn. If the frictional torque resisting the turning of the molecule is $\zeta d\theta/dt$, $d\theta/dt$ being the angular velocity of the molecule, then $\tau = \zeta/2kT$ (PM, 94).

Drude-Debye relations.—At extremely high frequencies the massive molecules cannot follow the field, and formula (8) reduces to its first term. The value of the dielectric constant under those conditions will be denoted by ϵ_0. At very low frequencies (8) reduces to (2); that static value of the constant will be denoted by ϵ_1. Then from (8) the following isothermal formulas may be derived, it being tacitly assumed, in (8) as well as in the following formulas, that after leaving the optical spectrum there is one, and only one, type of singularity in ϵ, and that that is of the type here covered.

There is nothing about these isothermal equations that is peculiar to the dipole theory [660]; they may all be obtained from Drude's paper [660a] as has been remarked by J. Malsch,[661] W. Ziegler,[661a] and others. For this reason they are here called the Drude-Debye relations.

But the forms of the expressions given by Drude are not always the same as those given by Debye (PM, 90-94), and the time factor (a) that he used differs from Debye's τ by a factor that is a constant for any one substance. Both forms, as well as varieties of each, are here given; as one form may, for certain purposes, be preferable to another.

For brevity, write $x \equiv \omega\tau(\epsilon_1+2)/(\epsilon_0+2) \equiv a\nu$ and $\eta^2 \equiv (\epsilon_1^2 + \epsilon_0^2 x^2)/(1+x^2)$; as usual, $\omega = 2\pi\nu$. Then the several quantities occurring in the expression $\epsilon = \epsilon' - i\epsilon'' = n^2(1-i\kappa)^2$ may be computed by means of the following formulas:

$$\epsilon' = n^2(1-\kappa^2) = (\epsilon_1 + \epsilon_0 x^2)/(1+x^2) = \epsilon_0 + (\epsilon_1 - \epsilon_0)/(1+x^2) =$$
$$\epsilon_1 - (\epsilon_1 - \epsilon_0)x^2/(1+x^2) \quad (9)$$

$$\epsilon'' = 2n^2\kappa = (\epsilon_1 - \epsilon_0)x/(1+x^2) = (\epsilon_1 - \epsilon_0)a\nu/(1+a^2\nu^2) \quad (10)$$

$$2n^2 = \eta + \epsilon'; \quad 2n^2\kappa^2 = \eta - \epsilon' \quad (11)$$

$$\tan \phi = \epsilon''/\epsilon' = 2\kappa/(1-\kappa^2) = (\epsilon_1 - \epsilon_0)x/(\epsilon_1 + \epsilon_0 x^2) =$$
$$[(\epsilon_1 - \epsilon')(\epsilon' - \epsilon_0)]^{1/2}/\epsilon' \quad (12)$$

---

[660] Cf. Oplatka, G., *Helv. Phys. Acta,* **6**, 198-209 (1933); Murphy, E. J., *Trans. Electroch. Soc. (Amer.),* **65**, 133-142 (1934).
[660a] Drude, P., *Ann. d. Physik (Wied.),* **64**, 131-158 (1898).
[661] Malsch, J., *Ann. d. Physik (5),* **19**, 707-720 (1934).
[661a] Ziegler, W., *Physik. Z.,* **35**, 476-503 (1934).

$$\kappa = [(\eta - \epsilon')/(\eta + \epsilon')]^{1/2} = (\epsilon_1 - \epsilon_0)x/(\eta+\epsilon')\ (1+x^2) =$$
$$\tan\ (\phi/2) \qquad (13)$$

From the expression defining x and a, we find

$$a = 2\pi\tau(\epsilon_1 + 2)/(\epsilon_0 + 2);\ \tau = a(\epsilon_0 + 2)/2\pi(\epsilon_1 + 2) \qquad (14)$$

For deriving the value of a from the observed values of ϵ' and $\epsilon'' \equiv 2n^2\kappa$, Drude advises the use of one or another of the following formulas:

$$a^2 = (\epsilon_1 - \epsilon')/(\epsilon' - \epsilon_0)v^2;\ a = (\epsilon_1 - \epsilon')/2n^2\kappa v;$$
$$a = 2n^2\kappa/(\epsilon' - \epsilon_0)v \qquad (15a, 15b, 15c)$$

From (10) one finds

$$2x = A - (A^2 - 4)^{1/2},\ \text{where}\ A \equiv (\epsilon_1 - \epsilon_0)/2n^2\kappa \qquad (16)$$

from which τ can be found from the relation:

$$\tau = x(\epsilon_0 + 2)/2\pi(\epsilon_1 + 2)v \qquad (17)$$

It will be noticed that when the frequency is such (v_s) that $x = 1$, then $a = 1/v_s$, $\tau = (\epsilon_0 + 2)/(\epsilon_1 + 2)2\pi v_s$, and $\epsilon' = (\epsilon_1 + \epsilon_0)/2$. This frequency, at which ϵ' is midway between ϵ_1 and ϵ_0, may be called the transition frequency. It is characteristic of the substance, fixing the values of the coefficients a and τ, which have throughout this treatment been regarded as independent of the value of v. The corresponding wave-length in a vacuum $(\lambda_s = c/v_s)$ may be called the transition wave-length; $\lambda_s = ca = \lambda_0 av = \lambda_0 x$, where $\lambda_0 = c/v$. The quantity $c\tau$ also defines a characteristic wave-length; one that is much smaller than the transition wave-length. Several other characteristic wave-lengths may be defined (see following table). A reader must be on the alert, for authors do not always state clearly which of the several characteristic wave-lengths is being considered. As they all serve to fix the region in which the dielectric constant changes rapidly, each may be called a transition wave-length, but here the term will be restricted to the one for which ϵ' is midway between ϵ_1 and ϵ_0. This is the one most frequently designated by λ_s and called by Germans "Sprungwellenlange."

In the following table, the second column contains the values of x_c^2 or $a^2 v_c^2$ ($x \equiv av$) corresponding to the criterion specified in the first column

Criterion	x_c^2 or $a^2 v_c^2$	Symbol	Characteristic $\lambda_c = \lambda_s/x_c$ For $\epsilon_0=1.8, \epsilon_1=80.4$
$\epsilon' = (\epsilon_0 + \epsilon_1)/2$	1	λ_s	$\lambda_s = \lambda_s = 136\lambda_\tau$
Max. of $n^2\kappa$	1	$\lambda_{n^2\kappa}$	$\lambda_{n^2\kappa} = \lambda_s = 136\lambda_\tau$
$n^2 = (\epsilon_0 + \epsilon_1)/2$	$\frac{1}{2}\left[\frac{\epsilon_1-\epsilon_0}{\epsilon_1+\epsilon_0}+\left\{4+\left(\frac{\epsilon_1-\epsilon_0}{\epsilon_1+\epsilon_0}\right)^2\right\}^{1/2}\right]$	λ_n	$\lambda_n = 0.794\lambda_s = 108\lambda_\tau$
Max. of $n\kappa$	$(3\epsilon_1+\epsilon_0)/(\epsilon_1+3\epsilon_0)$	$\lambda_{n\kappa}$	$\lambda_{n\kappa} = 0.594\lambda_s = 81\lambda_\tau$
Max. of κ	ϵ_1/ϵ_0	λ_κ	$\lambda_\kappa = 0.150\lambda_s = 20\lambda_\tau$
Relaxation	$4\pi^2(\epsilon_1+2)^2/(\epsilon_0+2)^2$	λ_τ	$\lambda_\tau = 0.00734\lambda_s = \lambda_\tau$

("relaxation" means the time (τ) required for the polarization to become reduced to $1/e$ of its value, the impressed field being zero), and in the last column are given the values for water at 20 °C of each of these characteristic wave-lengths in terms of transition wave-length (λ_s) and of λ_τ as defined by the relation $\lambda_\tau = c\tau$.

If in (15b) and (15c) a is replaced by its equivalent λ_s/c and ν by c/λ_0, then one readily obtains

$$\epsilon' = \epsilon_1 - (2n^2\kappa/\lambda_0)\lambda_s \text{ and } \epsilon' = \epsilon_0 + 2n^2\kappa\lambda_0/\lambda_s \quad (18a, 18b)$$

That is, if there is but one such jump as we are considering and there is no absorption band beyond the optical spectrum, then ϵ' will be linear in both $2n^2\kappa/\lambda_0$ and in $2n^2\kappa\lambda_0$, and the values of ϵ_1 and ϵ_0 will be given by the intercept of these lines on the axis of ϵ'.

Restricted reorientability.—If as the result of their mutual action the dipoles of adjacent molecules are elastically coupled, as two magnets may be, then their combined moment will be less than the sum of the single, separate moments, and may in the limit be zero. Such coupling may be either permanent or temporary. All of which will tend to reduce the magnitude of that portion of the dielectric constant that is contributed by the dipoles, and of the variation of that portion with the strength of the field.

Debye [662] has recently considered this problem, and has derived formulas for the case in which each molecule is so elastically bound to a direction fixed in space that the potential energy of the molecule contains a term of the form $-E \cos \theta$ where θ is the angle that the axis of the dipole makes with the fixed direction. That fixed direction varies from molecule to molecule, and for each molecule it is continually changing as a result of the thermal agitation of the molecules. The compiler has not yet been able to perceive clearly the physical significance of these assumptions. Debye concludes that E is very great, and that the effect of restraints of the kind just specified is to multiply the b and the f of formula (3) by $2C$ and $3C^4$ respectively, C standing for E/kT. As the f of formula (3) is proportional to the last term in (4), the restrictions here imposed require that also to be multpilied by $3C^4$.

For other suggested explanations of the failure of liquids to conform to Debye's formulas for freely reorientable dipole molecules, and for criticisms of Debye's treatment of restricted reorientability, see M. Forró,[663] J. Frenkel,[664] G. Hettner,[665] J. Malsch,[666] L. Onsager,[667] and A. Piekara.[668]

Dielectric Constant of Water.

Miscellanea.—The degree of accord between various theories and the experimentally determined values for various substances has been discussed by O. Blüh.[669] The data for water, ice, and steam are considered in detail. The article concludes with a bibliography of 172 entries. More recently,

---

[662] Debye, P., *Acad. roy. Belg. Bull. Cl. Sci.* (5), **21**, 166-174 (1935); *Physik. Z.*, **36**, 100-101, 193-194 (1935). *Chem'l Rev.*, **19**, 171-182 (1936).
[663] Forró, M., *Z. Physik*, **47**, 430-445 (1928).
[664] Frenkel, J., *Acta. Physiochim. URSS*, **4**, 341-356 (1936).
[665] Hettner, G., *Physik. Z.*, **38**, 771-774 (1937) → *Verh. Physik. Ges.* (3), **18**, 57 (1937).
[666] Malsch, J., *Ann. d. Physik* (5), **29**, 48-60 (1937).
[667] Onsager, L., *J. Am. Chem. Soc.*, **58**, 1486-1493 (1936).
[668] Piekara, A., *Acta Phys. Polon.*, **6**, 130-143 (1937).

W. Ziegler [670] has done the same, but with special reference to the dispersion and absorption of electric waves. He gives a bibliography of 159 entries. For discussions of the difficulties inherent in the use of waves along wires immersed in the dielectric, see E. Frankenberger.[671] Among other things he concludes that changes in the surface of the wire may introduce systematic errors of several tenths of a per cent in n, the index of refraction.

P. Drude [672] has concluded that dissolved air produces no observable effect upon the value found for ϵ. Both he and A. Deubner [673] comment on the fact that the successive individual determinations of ϵ for a given sample of water are more concordant than are the determinations for different samples. Although this suggests that nominally identical specimens of water have different values of ϵ Drude dismisses such a suggestion as being extremely improbable.

G. Jacoby [674] has derived an expression relating the dielectric constant to the forces binding the atoms in the molecule. From this he concludes that one of the hydrogens in H_2O is bound more loosely than the other.

The dielectric constant of a film of water 2 microns (0.0002 cm) thick is the same as that of water in bulk.[675] The following papers touch upon the dielectric constant of water, but are not mentioned elsewhere in this section: James Dewar and J. A. Fleming [676] discuss the data obtained for water prior to 1897, and [677] give a single very low value for 1 °C. R. Fürth,[678] M. Jezewski,[679] H. Joachim,[680] Y. Matsuike,[681] E. B. Rosa,[682] J. F. Smale,[683] S. Tereschin,[684] B. B. Turner,[685] and G. U. Yule,[686] each reports a single determination, generally incidental to another problem or for the purpose of testing a proposed procedure. W. Nernst [687] reports a few preliminary determinations.

Saturation.—On Debye's simple theory of freely reorientable dipoles, the dielectric constant of a dipole substance should exhibit "saturation" effects in intense fields, the dielectric constant (ϵ_x) in a strong field of intensity X being less than that (ϵ_l) in a weak field, by the relative amount

[669] Blüh, O., *Phy. Z.*, **27**, 226-267 (1926).
[670] Ziegler, W., *Phys. Z.*, **35**, 476-503 (1934).
[671] Frankenberger, E., *Ann. d. Physik (5)*, **1**, 948-962 (1929).
[672] Drude, P., *Ann. d. Physik (Wied.)*, **59**, 17-62 (1896).
[673] Deubner, A., *Ann. d. Physik (4)*, **84**, 429-456 (1927) = *Diss.*, Freiburg.
[674] Jacoby, G., *Ann. d. Physik (4)*, **72**, 153-160 (1923).
[675] Kallmann, H., and Dorsch, K. E., *Z. physik. Chem.*, **126**, 305-322 (1927).
[676] Dewar, James, and Fleming, J. A., *Proc. Roy. Soc. (London) (A)*, **61**, 2-18 (1897).
[677] Dewar, James, and Fleming, *Idem*, **62**, 250-266 (1898).
[678] Fürth, R., *Physik. Z.*, **25**, 676-679 (1924); **44**, 256-260 (1927).
[679] Jezewski, M., *Bull. intern. acad. polon. (A)*, **1920**, 88-102 (1920).
[680] Joachim, H., *Ann. d. Physik. (4)*, **60**, 570-596 (1919).
[681] Matsuike, Y., *Sci. Rep. Tohôku Imp. Univ. (Sendai)*, **14**, 445-452 (1925).
[682] Rosa, E. B., *Phil. Mag. (5)*, **31**, 188-207 (1891).
[683] Smale, J. F., *Ann. d. Physik (Wied.)*, **57**, 215-222 (1896).
[684] Tereschin, S., *Idem*, **36**, 792-804 (1889).
[685] Turner, B. B., *Z. physik. chem.*, **35**, 385-430 (1900).
[686] Yule, G. U., *Ann. d. Physik (Wied.)*, **50**, 742-751 (1893).
[687] Nernst, W., *Z. physik. chem.*, **14**, 622-663 (1894).

49. WATER: DIELECTRIC PROPERTIES

$(\epsilon_l - \epsilon_x)/\epsilon_l = AX^2$, A being a factor depending upon the substance and its temperature (see eq. 5). For water, his expression, when corrected (p. 354), leads to the computed value $A_c = 0.0128$ per cent per (kilovolt/cm)^2.

But J. Malsch [688] has found for water $A = 0.000011$ per cent per (kilovolt/cm)^2, which is only 1/1000 of A_c. Debye's theory for restricted reorientability (p. 357) can be made to account for such a factor.

Variation with Frequency.—From a study of all pertinent data available in 1924 (a few later data were studied and placed in the report as it was being prepared for the printer), H. L. Curtis [689] concluded that, for ν not exceeding 100 megacycles/sec, the dielectric constant (ϵ) of water at a pressure of one atmosphere is independent of the frequency, and is about 81.2 at 17 °C. At higher frequencies, beginning near $\nu = 600$ megacycles/sec, ϵ steadily decreases as shown in Figure 5.

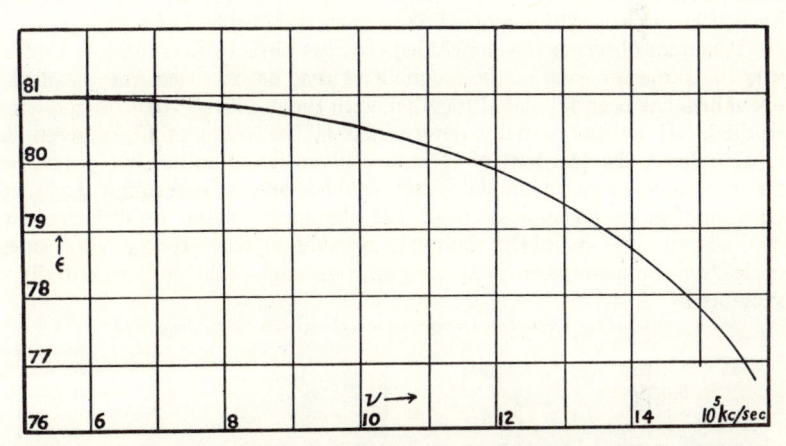

FIGURE 5. Dielectric Constant of Water at 17 °C; Variation with the Frequency.
[From compilation by H. L. Curtis and F. M Defandorf, *Int. Crit. Tables*, **6**, 77 (1929).]
Unit of ν (frequency) is 100 megacycles per second; of ϵ (dielectric constant) is 1 cgse.

The most exact determinations of the dielectric constant of water now available at the lower frequencies (under 100 megacycles/sec) are probably those of J. Wyman, Jr.[690] and of F. H. Drake, G. W. Pierce, and M. T. Dow.[691] They found, respectively, 81.47 and 81.54 at 17 °C. Their mean, 81.50, is used in Table 174 and is probably as close an estimate of the correct value as can be obtained from the data available. It differs by only 0.7 per cent from the mean (80.9) of the entries in the first section of Table 172, the two plainly abnormal values (88 and 73) being ignored.

[688] Malsch, J., *Physik. Z.*, **29**, 770-771 (1928); **30**, 837-839 (1929).
[689] Curtis, H. L., *Int. Crit. Tables*, **6**, 77, 78 (1929).
[690] Wyman, J., Jr., *Phys. Rev. (2)*, **35**, 623-634 (1930).
[691] Drake, F. H., Pierce, G. W., and Dow, M. T., *Idem*, **35**, 613-622 (1930).

The existence of bands of anomalous dispersion at frequencies below 1000 megacycles/sec seems to be generally discredited at present. Although such bands have been reported by A. R. Colley,[692] K. Iwanow,[693] and R. Weichmann,[694] others have been unable to confirm their existence, and at least some of them may have been caused by an impurity, probably glass, dissolved in the water (see G. Mie,[695] E. Frankenberger [696]). Likewise the report by A. Bramley [697] of a region of selective absorption, and hence, presumably, of anomalous dispersion, was later found by himself to be incorrect.[698] The regions of low frequency in which water has been especially investigated for the existence of absorption bands are listed in Table 173.

Table 172.—Dielectric Constant of Water at 17 °C

The best available value for frequencies below 600 megacycles per second is 81.50 cgse unit. It seems improbable that there are bands of absorption at frequencies below 1000 megacycles/sec, although such bands have been occasionally reported (see text and Table 173).

When an observer has concluded that his observations indicate that ϵ or n is the same for each of the frequencies that he has used, the mean of all his values has been tabulated together with the number (No.) of frequencies studied. If he has actually determined n, the value of n^2 is given, and ϵ (actually ϵ', the real part of ϵ) is also given if the author has either computed it himself or given data that enables one to determine it. Under Osc. or Os. is an indication of the damping of the oscillations used: d = damped, d-- = slightly damped, u = undamped. Below 1000 megacycles/sec the absorption of the water is so slight that n^2 is essentially the same as ϵ.

Unit of $\lambda_0 = 1$ cm, of $\nu = 1$ megacycle/sec, of $\epsilon = 1$ cgse unit. Temp. = 17 °C

λ_0	ν	ϵ	n^2	No.	Osc.	Ref.[a]
5.0(10⁸)	60(10⁻⁶)	81.0[b]			u	Ca
5.0(10⁴)	0.6	81.2			u	Ko
3.5(10⁴)	0.86	81.9			u	Ky
2.0(10⁴)	1.5	88.1			u	PH
1.1(10⁴)	2.7	82.5			u	As
8109–368	3.70–81	81.5[c]		5	u	Wy
2550–392	12–76	81.5	w[d]	2	u	DD
1110–309	27–97		79.4	3	d	CZ
700–332	43–90		81	7	u	MJ
600–310	50–97		80.5	2	d	Co
444–242	68–124	73	w[d]	4	u	Sa
345–230	87–130		81.07	7	u	Al
321–36.7	93–818		80.68	7	u	He
290–230	103–130		81.00	4	u	No
276–124	109–242		80.0	2	u	So

[692] Colley, A. R., *Phys. Z.*, **10**, 471-480 (1909) ← *J. Russ. Fis.-Chim. Obsc. (Phys.)*, **39**, 210-233 (1907).

[693] Iwanow, K., *Ann. d. Physik (4)*, **65**, 481-506 (1921).

[694] Weichmann, R., *Idem*, **66**, 501-545 (1921) → *Phys. Z.*, **22**, 535-544 (1921).

[695] Mie, G., *Phys. Z.*, **27**, 792-795 (1926).

[696] Frankenberger, E., *Ann. d. Physik (4)*, **82**, 394-412 (1917); (5), **1**, 948-962 (1929).

[697] Bramley, A., *J. Franklin Inst.*, **206**, 151-157 (1928); **207**, 315-321 (1929) → *Phys. Rev. (2)*, **33**, 279 (A), 640 (A) (1929).

[698] Bramley, A., *Phys. Rev. (2)*, **34**, 1061 (L) (1929); *J. Opt. Soc. Amer.*, **21**, 148 (A) (1931).

Table 172—(Continued)

λ_0	ν	ϵ	n^2	No.	Osc.	Ref.[a]
271–268	111–112		80.82	12	u	De
266	113		81.5		u	Ho
250–60	120–500		80.4	14	d –	Rt
200–25.8	150–1160		81.8	8	u	Ma
200	150		79.7		d	Dr'
75	400		80.2		(d	Dr'
75	400		81.7		d	Dr''
73.4–23.8	409–1260		81.1	73	d –	Fr'
68.4–55.2	438–542		80.3		3	Cy
63.7–59.5	471–504		81.3[e]		d –	Rp
63–49	476–612		79.8	20	u	GA
56.6–52.2	530–574		80.5	5	d –	Fr''
56.7–28.4	529–1056		80.1	34	u	AO
49.9–39.4	601–762		81.0[e]		d –	RP
45.6	658		81.1			Dev
35.0–31.0	857–968		81.1[e]		d –	Rp

λ_0	Ref.[a] Osc. $\nu/1000$	Rp d	Go u	Kn u n^2	Se u	Ec d	Te[f] d	La d	n^2 Misc. Os. Ref.[a]	Kn u	Ec d ϵ'	Te d
25.8	1.16								82.2 d Ma			
24.43	1.23	79.1										
24.06	1.25	78.7										
23.82	1.26		98.8									
23.06	1.30	78.0										
21.59	1.39	77.8										
20.55	1.46	77.8										
20.44	1.47			77.4						77.3		
19.0	1.57				80							
17.76	1.69		91.4									
17.42	1.72		87.4									
17.02	1.76		87.6									
16.83	1.78			77.4						77.2		
15.29	1.96			77.4						71.1		
14.48	2.06			77.4						77.1		
14.0	2.14				84							
13.6	2.21				82							
13.45	2.23								77.8 u AO			
12.65	2.37		83.2									
12.6	2.38				80							
11.80	2.54			77.4						77.0		
11.48	2.62	81.0										
11.11	2.69			77.4						77.0		
10.20	2.94			77.4						76.9		
9.85	3.05			77.4						76.9		
9.55	3.16			77.4						76.9		
8.80	3.43			77.4	78.3					76.8	77.9	
8.53	3.51		74.0									
8.05	3.73			77.4						76.7		
7.1	4.22								80.2 Mz			
6.48	4.63			77.4						76.3		
6.2	4.84								86.5 Mz			
5.7	5.26				78.8						77.6	
5	6.0								77.8[g] d Co			
4	7.5								76.1 d El			
3.7	8.1				64.4						61.4	
2.7	11.1					72.5						67.4
1.8	15.6					45.4						40.0

Table 172.—(Continued)

λ_0	Ref.[a]→ Osc.→ $\nu/1000$	Rp d	Go u	Kn u n^2	Se u	Ec d	Te[f] d	La d	Misc. n^2	Os. Ref.[a]	Kn u	Ec d ϵ'	Te d
1.75	17.2			63.8							56.5		
1.5	20.0					45.0							41.6
1.1	27.3					40.4							38.4
0.84	35.7					33.4							31.2
0.8	37.5							82.7					
0.6	50.0							89.7					
0.42	71.4					29.6							27.9
0.4	75.0							91.4					
0.005	6000			(Optical)					1.85				

[a] References:

Al	Alimova, M., *Ann. d. Physik (5)*, **9**, 176-178 (1931). Alimova, M. M., and Nowosilzew, N. S., *Idem*, **19**, 118-120 (1934).
AO	v. Ardenne, M., Gross, O., and Otterbein, G., *Phys. Z.*, **37**, 533-544 (1936).
As	Astin, A., *Phys. Rev. (2)*, **34**, 300-309 (1929).
Ca	Carman, A. P., *Phys. Rev. (2)*, **24**, 396-399 (1924).
Co	Cole, A. D., *Ann. d. Physik (Wied.)*, **57**, 290-310 (1896).
Cy	Colley, A. R., *Phys. Z.*, **10**, 329-340 (1909) ← *J. Russ. Fis.-Chim. Obs. (Phys.)*, **38**, 431-435 (1906).
CZ	Cohn, E., and Zeeman, P., *Vers. kon. Akad. Wet. Amsterdam*, **4**, 108-116 (1896).
DD	Drake, F. H., Pierce, G. W., and Dow, M. T., *Phys. Rev. (2)*, **35**, 613-622 (1930).
De	Deubner, A., *Ann. d. Physik (4)*, **84**, 429-456 (1927) = *Diss.*, Freiburg.
Dev	Devoto, G., *Gazz. chim. Ital.*, **60**, 208-212 (1930).
Dr'	Drude, P., *Ann. d. Physik (Wied.)*, **58**, 1-20 (1896).
Dr''	Drude, P., *Idem*, **59**, 17-62 (1896).
Ec	Eckert, E., *Verh. physik. Ges.*, **15**, 307-329 (1913).
El	Elle, D., *Ann. d. Physik (5)*, **30**, 354-370 (1937) ← *Diss.*, Jena.
Fr'	Frankenberger, E.[071]
Fr''	Frankenberger, E., *Idem (4)*, **82**, 397-412 (1927). See also Mi.
GA	Girard, P., and Abadie, P., *Compt. rend.*, **191**, 1300-1302 (1930).
Go	Goldsmith, T. T., Jr., *Phys. Rev. (2)*, **51**, 245-247 (1937).
He	Heim, W., *Z. Hochfreq.-techn.* (= *Jahrb. d. drahtl. Teleg.*), **30**, 160-167, 176-183 (1927).
Ho	Holborn, F., *Z. Phys.*, **6**, 328-338 (1921).
Kn	Knerr, H. W., *Phys. Rev. (2)*, **52**, 1054-1067 (1937) → *Idem*, **51**, 1007 (A) (1937).
Ko	Kockel, L., *Ann. de Phys. (4)*, **77**, 417-448 (1925).
Ky	Kyropoulos, S., *Z. Physik*, **40**, 507-520 (1926).
La	Lampa, A., *Sitz. Akad. Wiss. Wien (Abt. IIa)*, **105**, 587-600, 1049-1058 (1896).
MJ	McCarty, L. E., and Jones, L. T., *Phys. Rev. (2)*, **29**, 880-886 (1927).
Ma	Mazzotto, D., *Att. accad. naz. Lincei (5)*, **5₂**, 301-308 (1896).
Mi	Mie, G., *Phys. Z.*, **27**, 792-795 (1926).
Mz	Miesowicz, M., *Bull. intern. acad. polon. (A)*, **1934**, 95-102 (1934).
No	Novosilzew, N., *Ann. d. Physik (5)*, **2**, 515-536 (1929).
PH	Powers, W. F., and Hubbard, J. C., *Phys. Rev. (2)*, **15**, 535-536 (A) (1920).
Rp	Rukop, H., *Ann. d. Physik (4)*, **42**, 489-532 (1913).
Rt	Rückert, E., *Idem*, **55**, 151-176 (1918).
Sa	Sauzin, M., *Compt. rend.*, **171**, 164-167 (1920).
Se	Seeberger, M., *Ann. d. Physik (5)*, **16**, 77-99 (1933).
So	Southworth, G. C., *Phys. Rev. (2)*, **23**, 631-640 (1924).
Te	Tear, J. D., *Idem*, **21**, 611-622 (1923).
Wy	Wyman, J., Jr., *Phys. Rev. (2)*, **35**, 623-634 (1930).

[b] From the constant of an electrometer with its plates immersed in water.

[c] From the frequency of an oscillator immersed in water.

[d] Both DD and Sa measured n, but reported only the values of ϵ as derived from the n's by correcting for the electrical conductivity.

[e] Rukop determined n for 51 wave-lengths well distributed over the range $\lambda_0 = 63.7$ to 20.6 cm, and concluded that n is constant over the ranges 63.7 to 59.5, 49.9-39.4, and 35.0 to 31.0 cm. Between those regions and for $\lambda_0 < 31$ cm his observations were less uniform, and he thought there might be anomalous dispersion; but he could not find the sharp bands reported by Colley, and for which he was searching.

[f] Tear measured the reflectivity and the absorption, and computed n from them.

[g] Cole measured the reflectivity for $\lambda_0 = 5$ cm, and computed n from that on the assumption that there was no absorption.

49. WATER: DIELECTRIC PROPERTIES

Table 173.—Absorption Bands in the Electrical Spectrum of Water

It seems probable that water has no absorption bands at frequencies below 1000 megacycles/sec,[699] but bands have been reported in certain of the regions here tabulated, all of which have been carefully searched for bands; the number of bands reported in each case is stated in column 3. The range 250 to 450 megacycles/sec should be reëxamined. λ_0 = wavelength in air, ν = frequency.

Unit of $\nu = 100$ megacycles/sec; of $\lambda_0 = 1$ cm

ν	λ_0	Bands	Ref.[a]
0.8 to 1.0	375 to 300	2 sets	Br[b]
0.84 to 1.06	260 to 280	None	AN
1.0 to 1.4	300 to 220	None	No
1.0 to 1.4	300 to 220	None	Al
2.4 to 2.5	124 to 120	None	Iw[b]
2.5 to 4.5	120 to 67	Many	Iw[b]
4.5 to 5.0	67 to 60	None	Iw[b], Cy'
4.7 to 14.6	64 to 20	3	Rp
5.3 to 8.4	56 to 36	Many	Cy''
5.4 to 10.7	56 to 28	None	AO
5.6, 7.1, 7.4	54, 42, 32	3	We[b]
5.3 to 5.9	56 to 52	None	Mie[b], Fr'
4.1 to 8.1	73 to 37	None	Fr''
15 to 60	20 to 5	None	Kn
150, 430, 1500	2.0, 0.7, 0.2	3	Te[b]

[a] References:
- Al   See Table 172, references, first of Al.
- AN   See Table 172, references, second of Al.
- AO   See Table 172, references.
- Br   Bramley, A.[697] Reported selective absorption; hence anomalous dispersion may be inferred. He concluded later that the observed effects were spurious, at least in part.[698]
- Cy'   Colley, A. R., see Table 172, Cy.
- Cy''   Colley, A. R., *Physik. Z.*, **10**, 471-480 (1909) ← *J. Russ. Fis.-Chim. Obs. (Phys.)*, **39**, 210-233 (1907).
- Fr'   Frankenberger, E., *Ann. d. Physik (4)*, **82**, 394-412 (1927).
- Fr''   Idem (5), **1**, 948-962 (1929).
- Iw   Iwanow, K., *Ann. d. Physik (4)*, **65**, 481-506 (1921). He calls the value of $2\pi\nu/10$ frequencies; in this table the values of ν itself are given, as computed from his values of $\lambda_0/2$.
- Kn   Knerr, H. W., *Phys. Rev. (2)*, **52**, 1054-1067 (1937).
- Mie   Mie, G.[695] He concluded from the observations of Fr' that the bands reported by We are spurious.
- No   Novosilzew, N., see Table 172, references.
- Rp   Rukop, H., see Table 172, references.
- Te   Tear, J. D., *Phys. Rev. (2)*, **21**, 611-622 (1923). He reported selective absorption at these frequencies; hence anomalous dispersion may be inferred.
- We   Weichmann, R., *Ann. d. Physik (4)*, **66**, 501-545 (1921) → *Physik. Z.*, **22**, 535-544 (1921). Mie concluded that these bands are spurious.

[b] See note accompanying the corresponding reference.

Variation with Temperature.—(See also pp. 353-355). At the time of the compilation by H. L. Curtis and F. M. Defandorf [700] the available data on the variation of the dielectric constant of water with the temperature, especially at the higher temperatures, were far from satisfactory. The best that could be done was to give the linear equation $\epsilon = 80 - 0.4(t - 20)$, equivalent to $\epsilon = 81.2 [1 - 0.005(t - 17)]$, the coefficient of t being restricted to a single significant figure. It was known that the dielectric

[699] Malsch, J., *Ann. d. Physik (5)*, **19**, 707-720 (1934).
[700] Curtis, H. L., and Defandorf, F. M., *Int. Crit. Tables*, **6**, 74-81 (1929).

constant has no maximum at 4 °C, but the observations of F. Ratz [701] indicated that there might be a very indistinct maximum between 0 °C and 1 °C.

Since then, both L. Kockel [702] and J. Wyman, Jr.[703] have published consistent data for the entire range, 0 °C to 100 °C; F. H. Drake, G. W. Pierce and M. T. Dow [704] for the range 10 °C to 60 °C; and R. T. Lattey, O. Gatty, and W. G. Davies [705] two short series in the range, 14 °C to 18 °C. All of these are closely represented by formula (18c)

$$\epsilon = \epsilon_{17}\left[1 - 4.696\left(\frac{t-17}{1000}\right) + 10.2_8\left(\frac{t-17}{1000}\right)^2\right] \qquad (18c)$$

Values which appear to be of a much lower accuracy have been published by A. C. Cuthbertson and O. Maass [706] for 0, 15, 25, 50, and 75 °C; and E. P. Linton and O. Maass [707] have published a formula said to represent their own (unpublished) observations over the range 0 °C to 50 °C. This formula does not agree satisfactorily with the other data and, owing to the meagerness of the information published, cannot be critically appraised except by such comparison.

Formula (18c) also represents quite satisfactorily the data published by F. Ratz, those over an 8.5° range that are summarized by A. R. Colley [708] by means of an equation, those from 0° to 50 °C defined by a formula pro-
(Continued on Page 366)

Table 174.—Variation of the Dielectric Constant of Water with the Temperature

The several values are compared with those of (ϵ_f) defined by the preferred formula (18c), which is used in the following form:

$$\left(\frac{\epsilon}{\epsilon_{17}}\right)_f = 1 - 4.696\left(\frac{t-17}{1000}\right) + 10.2_8\left(\frac{t-17}{1000}\right)^2 \; ; \; \epsilon_f = 81.50(\epsilon/\epsilon_{17})_f$$

the value 81.50 being the mean of the ϵ_{17} values obtained by Wyman and by Drake, Pierce, and Dow (see p. 359).

Values of ϵ_f and of $(\epsilon/\epsilon_{17})_f$ are tabulated, and the values reported by the several observers are indicated by means of the quantities Δ and δ, which are defined by the relations $(\epsilon/\epsilon_{17}) = (\epsilon/\epsilon_{17})_f + \Delta$ and $\epsilon_{obs} = \epsilon_f(1 + \delta)$. At the bottom of each column of Δ is given the value of ϵ_{17} used by the observer in computing his smoothed values. For example, Wyman's $\epsilon_{17} = 81.47$, his value of ϵ/ϵ_{17} at 0 °C is $1.082_8 - 0.003 = 1.080$; consequently his ϵ_0 is $81.47(1.080) = 87.99$; by means of his formula (see text) we find $\epsilon_0 = 78.54(1.1205) = 88.00$.

[701] Ratz, F., *Z. physik. Chem.*, **19**, 94-112 (1896).
[702] Kockel, L., *Ann. d. Physik (4)*, **77**, 417-448 (1925).
[703] Wyman, J., Jr., *Phys. Rev. (2)*, **35**, 623-634 (1930).
[704] Drake, F. H., Pierce, G. W., and Dow, M. T., *Phys. Rev. (2)*, **35**, 613-622 (1930).
[705] Lattey, R. T., Gatty, O., and Davies, W. G., *Phil. Mag. (7)*, **12**, 1019-1025 (1931).
[706] Cuthbertson, A. C., and Maass, O., *J. Am. Chem. Soc.*, **52**, 483-489 (1930).
[707] Linton, E. P., and Maass, O., *Idem*, **54**, 1863-1865 (1932).

49. WATER: DIELECTRIC PROPERTIES

Table 174—(Continued)

It will be noticed that, except for fixed percentile errors, most of the short sets of older data given in the last sections of the table agree quite satisfactorily with the values (ϵ_f) defined by formula (18c).

Unit of $\epsilon = 1$ cgse unit; of $\Delta = 0.001$; of $\delta = 0.1\%$. Temp. $= t$ °C.

References[a]→			Ko Obs.	Ko	Wy	DD[b]	LD	Cy	LM	Dr	ICT[e]		Ratz[d] Observed	
							Computed Δ							
t	ϵ_f	$(\epsilon/\epsilon_{17})_f$										t	$(\epsilon/\epsilon_{17})_f$	Δ
0	88.25	1.082₈	+2	0	−3		−1			+5	+2	+1	0.1	1.082₃	−8
10	84.22	1.033₄	+4	0	−1	0	−1	+1	0	+1	+2	0.33	1.081₁	−3
20	80.35	0.985₉	−6	0	0	0	0	−1	+1	0	0	0.75	1.079₀	−2
25	78.49	0.963₁		0	+1	0	+1	−2	+2	−2	−2	0.80	1.078₈	+2
30	76.67	0.940₇	−3	0	+1	+1	+1	−4	+4	−3	−5	1.03	1.077₆	+1
40	73.14	0.897₄	−5	+1	+2	+3	+2		+9	−5	−10	1.57	1.074₉	0
50	69.78	0.856₂	−3	0	+2	+6	+2		+12	−8	−18	1.80	1.074₇	−1
60	66.60	0.817₁	−5	0	+2	+9	+3			−12	−28	2.20	1.071₈	+1
70	63.57	0.780₃	−2	0	+2		+2			−16	−40	3.04	1.067₆	0
80	60.71	0.745₀	+1	−1	0		+3				−55	3.43	1.065₆	0
90	58.02	0.712₂	+2	−2	−1		+2				−71	4.00	1.062₆	−2
100	55.58	0.681₀	+2	−4	−2		+1				−89	4.16	1.062₀	−2
ϵ_{17}	81.50	81.50		81.2	81.47	81.54	81.19	80.26	82.0	81.65	81.2		$\epsilon_{17} = 81.24$	

Lattey, Gatty, and Davies (Two series, 1 and 2) LD[a].

		(1)	(2)			(1)	(2)			(1)	(2)
t	ϵ_f	δ		t	ϵ_f	δ		t	ϵ_f	δ	
14.0	82.92	−4.2		33.9	75.26	+0.3		61.15	66.24		+9.1
15.6	82.04		+2.2	39.5	73.32	−3.4		63.0	65.67	−3.5	
20.0	80.36	−1.4		42.05	72.44		+1.4	72.4	62.87		−0.2
21.45	79.81		−1.4	46.0	71.11	+0.5		74.5	62.26	+1.3	
26.25	78.04		−4.2	53.0	68.97	−10.2		80.15	60.68		−1.8
31.7	76.06	−6.8		54.2	68.41		+1.2	81.3	60.35	+3.6	

Mean δ: Series 1, -2.7; Series 2, $+0.8$

	ϵ_f	δ		t	ϵ_f	δ		t	ϵ_f	δ
	Devoto[a]			28.5	77.21	−8.9		27.2	77.7	−86
3.8	86.70	−3.5		30.6	76.44	−7.8		31.7	76.1	−99
5.7	85.92	−4.3		33.8	75.30	−11.2		35.3	74.8	−103
7.6	85.25	−4.9		36.3	74.42	−11.6				
7.8	85.09	−1.1		39.2	73.42	−11.6			Mean	−96
9.8	84.30	−3.8								
11.4	83.67	−3.1			Jezewski[a]				Coolidge[a]	
12.8	83.12	−4.5						3.5	86.8	−1.1
15.0	82.27	−3.9		2.3	87.32	+39		13.6	82.8	+1.2
16.4	81.73	−4.3		16.5	81.69	+76		19.0	80.7	+2.5
16.5	81.69	−5.9		30.1	76.63	+76		24.7	78.6	0.0
17.1	81.46	−5.4		47.2	70.70	+68		39.0	73.5	+1.3
17.7	81.23	−3.7		66.9	64.49	+45				
18.5	80.93	−5.3		92.	57.50	+33			Mean	+1.2
18.7	80.85	−8.0		99.5	55.62	+32			Herrwagen[a]	
19.3	80.63	−6.7						4.70	86.34	−9.8
19.6	80.51	−5.6			Mean	+53		9.85	84.28	−8.9
20.9	80.04	−5.5						12.75	83.14	−8.4
21.7	79.71	−4.9			Cohn[a]			14.65	82.40	−8.6
25.0	78.48	−8.0						16.35	81.75	−8.7
25.2	78.42	−9.8		9.5	84.4	−97		20.75	80.07	−6.4
26.7	77.86	−8.5		10.5	84.0	−104				
28.3	77.28	−10.2		16.8	81.6	−99			Mean	−8.5
				19.8	80.4	−96				

[a] References:
Cohn, E.[709]
Coolidge, W. D.[710]
Cy Colley, A. R. See Table 172, references.
DD Drake, F. H., Pierce, G. W., and Dow, M. T.[704]
Devoto, G. See Table 172, references.

Table 174—*(Continued)*

Dr   Drude, P.[672]
Heerwagen, F., *Ann. d. Physik (Wied.)*, **49**, 272-280 (1893).
ICT   Curtis, H. L., and Defandorf, F. M., *Int. Crit. Tables*, **6**, 74-81 (1929).
Jezewski, M., *Jour. de Phys. (6)*, **3**, 293-308 (1922).
Ko   Kockel, L.[702]
LD   Lattey, R. T., Gatty, O., and Davies, W. G.[705]
LM   Linton, E. P., and Maass, O., *J. Am. Chem. Soc.*, **54**, 1863-1865 (1932).
Ratz, F., *Z. physik. Chem.*, **19**, 94-112 (1896).
Wy   Wyman, J., Jr.[703]

[b] The progressive increase in ϵ as t increases above 30 °C, as well as the abnormally small values (not here recorded) found below 10 °C, are probably the result of errors in determining the temperature.

[c] As the coefficient of t is given to only one significant figure, even the first figure of the number is uncertain.

[d] Ratz gives values for $t =$ 4.43, 5.37, 5.47, 10, 20, and 30 °C also; for which, $\Delta = -2, +1, 0, -2, -18,$ and -5, respectively.

Table 175.—Variation of the Polarizability of Water with the Temperature

The molar polarizability of a dielectric is $P = (M/\rho)(\epsilon - 1)/(\epsilon + 2)$, and Debye's theory for freely reorientable dipole molecules leads to the relation $P/M \equiv (\epsilon-1)/(\epsilon+2)\rho = a + b/T$ (eq. 2) where $b = 4\pi N\mu^2/9MkT$; hence $PT/M \equiv (\epsilon-1)T/(\epsilon+2)\rho = b + aT$. For water, the reorientability of the molecules is restricted, and PT/M is not linear in T; it is however closely given by the relation

$$PT/M \equiv (\epsilon - 1)T/(\epsilon + 2)\rho = 107.13 + 0.2262T + 0.00127550T^2$$

where T °K is the absolute temperature. The calculated values given below are those defined by this formula; the observed ones are determined from the indicated values of ϵ and ρ, the former being the values of ϵ_f given in Table 174. If $b = 107.13$, $10^{19}\mu = 5.59$ cgse units per gfw of H_2O (see p. 176).

Unit of $\epsilon = 1$ cgse; of $\rho = 1$ g/cm³. Temp. $= t°$ C.

t	T	ϵ	ρ	PT/M Obs	PT/M Calc	O−C
0	273.1	88.25	0.9999	264.03	263.87	+0.16
10	283.1	84.22	0.9997	273.33	273.34	−0.01
20	293.1	80.36	0.9982	282.92	282.94	−0.02
25	298.1	78.49	0.9971	287.81	287.84	−0.03
30	303.1	76.67	0.9957	292.83	292.80	+0.03
40	313.1	73.15	0.9922	302.98	302.92	+0.06
50	323.1	69.79	0.9881	313.30	313.30	0.00
60	333.1	66.60	0.9832	323.93	323.91	+0.02
70	343.1	63.57	0.9778	334.83	334.80	+0.03
80	353.1	60.71	0.9718	345.92	345.94	−0.02
90	363.1	58.02	0.9653	357.32	357.33	−0.01
100	373.1	55.58	0.9584	369.00	368.98	+0.02

(Continued from Page 364)

posed by P. Drude,[672] the values obtained by E. Cohn,[709] by W. D. Coolidge,[710] and by F. Heerwagen.[711] See Table 174; formulas are given below.

[708] Colley, A. R., *Physik. Z.*, **10**, 329-340 (1909).
[709] Cohn, E., *Ann. d. Physik (Wied.)*, **45**, 370-376 (1892).
[710] Coolidge, W. D., *Idem*, **69**, 125-166 (1899).
[711] Heerwagen, F., *Idem*, **49**, 272-280 (1893).

Formula (18c) leads to $d\epsilon/dt = -0.383$ for 1 °C at 17 °C and -0.370 at 25 °C, whereas v. Ardenne, Groos, and Otterbein [712] state that they found -0.36 at 18 °C, and R. King [713] states that his observations between 10 and 40 °C (displayed solely as a small-scale graph) satisfy a linear relation, the slope being -0.360 (printed 0.0360).

Values reported by C. B. Thwing [713a] and by C. Niven [714] appear to be untrustworthy. Preliminary measurements at several temperatures below 50 °C have been reported by W. C. Röntgen,[715] and M. Seeberger [716] has reported a few measurements, of low accuracy, in the range 15 to 50 °C, $\lambda_0 = 12.6$ to 19.0 cm (see Table 177, note b).

K. Iwanow [717] has reported that the variation of ϵ with t depends somewhat upon the frequency unless t exceeds 50 °C, and that it is abnormal within an absorption band. He used frequencies between 240 and 500 megacycles/sec, in which region he reported many absorption bands (see Table 173).

The formula here given (18c) and those proposed by the several observers may, for comparison, be thrown into the following forms, in which $\tau \equiv (t-25)/1000$:

$\epsilon = 78.49 [1 - 4.70_6\tau + 10.6_7\tau^2]$	Same as formula (18c)
$\epsilon = 78.54 [1 - 4.60\tau + 8.8\tau^2]$	Wyman
$\epsilon = 78.57 [1 - 4.61\tau + 15.5\tau^2]$	Drake, Pierce and Dow
$\epsilon = 78.7_7 [1 - 4.48\tau + 13.1_5\tau^2]$	Drude
$\epsilon = 78.2 e^{-4.7\tau}$	Kockel
$\epsilon = 78.2_5 e^{-4.61_6\tau}$	Lattey, Gatty and Davies
$\epsilon = 79.2 [1 - 4.28\tau + 21.2\tau^2 - 410\tau^3]$	Linton and Maass
$\epsilon = 78(1 - 5\tau)$	International Critical Tables
$n_t = n_{25} - 22.0\tau$; $n = \sqrt{\epsilon}$	Colley

Variation with Pressure.—The dielectric constant of water increases with the pressure. Working at 16.3 °C and $\nu = 60$ megacycles/sec, G. Falckenberg [718] found that ϵ increased by 0.72 when the pressure was increased from 7 to 200 atm., giving a mean value of $d\epsilon/dp = 0.0037$ cgse per atm.

More recently, S. Kyropoulos,[719] working at 20 °C and $\nu = 0.86$ megacycles/sec, has measured ϵ over the range $P = 1$ kg*/cm² to $P = 3000$ kg*/cm². His values are closely given by the empirical formula (19):

$$\epsilon = 80.79 [1.0273 + 0.0000372P - 10^{-0.000448(P+3490)}] \qquad (19)$$

[712] v. Ardenne, Groos, and Otterbein, *Physik. Z.*, **37**, 533-544 (1936).
[713] King, R., *Rev. Sci. Inst.*, **8**, 201-209 (1937).
[713a] Thwing, C. B., *Z. physik. Chem.*, **14**, 286-300 (1894).
[714] Niven, C., *Proc. Roy. Soc. (London) (A)*, **85**, 139-145 (1911).
[715] Röntgen, W. C., *Ann. d. Physik (Wied.)*, **52**, 593-603 (1894).
[716] Seeberger, M., *Idem (5)*, **16**, 77-99 (1933).
[717] Iwanow, K., *Ann. d. Physik (4)*, **65**, 481-506 (1921).
[718] Falckenberg, G., *Ann. d. Physik (4)*, **61**, 145-166 (1920).
[719] Kyropoulos, S., *Z. Physik*, **40**, 507-520 (1926).

the unit of P being 1 kg*/cm² = 0.9678 atm (see Table 176). It will be noticed that his value (80.79) for 20 °C and 1 atm is about 0.55 per cent greater than the value (80.35) given in Table 174. Using his values of ϵ, the densities (ρ) found by Bridgman (see Table 95), and the molecular weight $M = 18.00$, Kyropoulos computed the value of the molecular polarizability—$M(\epsilon - 1)/(\epsilon + 2)\rho$—and of $(\epsilon - 1)/\rho$, finding each to vary monotonously throughout the range in P. On going from 1 to 3000 kg*/cm², $M(\epsilon - 1)/(\epsilon + 2)\rho$ goes from 17.39 to 15.85; and $(\epsilon - 1)/\rho$, from 79.9 to 82.6.

An early determination by W. C. Röntgen [715] showed that the excess in the value of ϵ at 500 atm over that at 1 atm is of the order of one per cent.

Table 176.—Variation of the Dielectric Constant of Water with the Pressure

The values (ϵ_k) tabulated by Kyropoulos (see text) are very closely equal to those (ϵ_c) computed by means of the formula $\epsilon_c = 80.79\,(1.0273 + 0.0000372P - E)$, where $\log_{10} E = -0.000448\,(P + 3490)$, the unit of P being 1 kg*/cm². For convenience, the values of E and of δ, defined by $\epsilon_c = 80.79\,(1 + \delta)$, are tabulated. Temperature = 20 °C; $\nu = 0.86$ megacycles/sec.

Unit of $P = 1$ kg*/cm² = 0.968 atm; of $\epsilon = 1$ cgse unit

P	ϵ_k	ϵ_c	100δ	100E
1	80.79	80.79	0	2.73
7[a]		80.83	0.05	2.71
206[a]		81.83	1.29	2.21
500	83.07	83.18	2.96	1.63
1000	85.20	85.22	5.48	0.97
1500	87.03	87.03	7.73	0.58
2000	88.72	88.73	9.82	0.35
2500	90.34	90.34	11.82	0.21
3000	91.90	91.91	13.77	0.12

[a] G. Falckenberg [718] working at 16.3 °C, $\nu = 60$ megacycles/sec, and $p = 7$ and 200 atm (7.23 and 206.1 kg*/cm²) found $\epsilon_{206} - \epsilon_7 = 0.72$, as compared with the value 1.00 here given for ϵ_c.

Dielectric Constant of Sea-water.—R. L. Smith-Rose [720] is of the opinion that the dielectric constant of sea-water is about the same as that of distilled water. He attributes the very high apparent values, obtained when audiofrequencies and low radiofrequencies are used, to the effect of polarization films formed at the boundaries between the electrodes and the water.

Dielectric Absorption. (For notation and definitions, see first pages of this Section.)

P. Drude [721] observed that water and certain other substances absorb electric waves much more strongly than can be accounted for by their elec-

---

[720] Smith-Rose, R. L., *Proc. Roy. Soc. (London) (A)*, **143**, 135-146 (1933).

[721] Drude, P., *Ann. d. Physik (Wied.)*, **58**, 1-20 (1896); *Z. physik. Chem.*, **23**, 267-325 (1897); *Ann. d. Physik (Wied.)*, **64**, 131-158 (1898).

49. WATER: DIELECTRIC PROPERTIES

trical conductivity. This additional absorption, above that due to the conductivity, is what is here called dielectric absorption. He attributed it to the presence of the hydroxyl group (OH). W. D. Coolidge [710] referred to the same phenomenon. It arises from a dissipation of energy inherent in the process of polarizing the medium. (See p. 352). For a discussion of the subject on the dipole theory, see P. Debye,[722] and P. Debye and W. Ramm.[723]

A. B. Bryan [724] investigated the phase relations between the current and the emf applied to condensers containing various dielectrics. For water at 23.5 °C and frequencies (ν) of 0.2 to 1.4 megacycles/sec he found that the phase defect was $\phi = 0.8° + (2.09/\nu)°$, the unit of ν being 1 megacycle/sec. The first term is ascribed to the true dielectric absorption, and the second to the conductivity of the water. It may be easily shown that the part of ϕ arising from the conductivity is $(2/\nu\rho\epsilon)(10^{-6})$ radians $= (114.3/\nu\rho\epsilon)(10^{-6})°$, the unit of ν being as before, and the resistivity (ρ) and ϵ being expressed in the same system of units. Equating this to the second term of the observed ϕ, one finds $\rho\epsilon = 55$ microseconds; and if $\epsilon = 81$ cgse unit, $\rho = 0.68 \times 10^{-6}$ cgse unit $= 0.6$ megohm·cm. The resistivity of best "conductivity" water is nearly 240 megohm·cm, much higher than that used by Bryan.

If the first term of this ϕ arises from dielectric absorption, then its tangent is equal to $\epsilon''/\epsilon' = 2n^2\kappa/\epsilon'$. Hence $2n^2\kappa = \epsilon'$ tan $0.8° = 81(0.0140) = 1.13$, if $\epsilon' = 81$ cgse; whence $\kappa = 0.0070$.

Using better water, J. Granier [725] found tan $\phi = 0.02$, the portion contributed by the conductivity being 0.006. This gives for $2n^2\kappa$ the same value as was found by Bryan. For the Paris water supply ϕ was 6 times as great.

A. Esau and G. Bäz [726] have studied both the reflectivity and the absorption of water over the range $\lambda_0 = 2.8$ to 10 cm ($\nu = 10,700$ to 3000 times 10^6 cycles/sec), and have presented their results in the form of small graphs. From the heating produced by a field of 28.4×10^6 cycles/sec ($\lambda_0 = 1055$ cm) C. Schmelzer [727] concluded that the apparent conductivity ($\epsilon''\omega/4\pi$) of the water used ranged from 14 to 20 times 10^{-7} (ohm·cm)$^{-1}$. Hence ϵ'' ($\equiv 2n^2\kappa$) ranges from 0.09 to 0.13, and κ from 0.0006 to 0.0008, it being assumed that $n^2 = \epsilon' = 81$. For other determinations see Tables 177 and 178.

The most extended set of values of n and κ now available for water is that by H. W. Knerr.[728] They are given in Table 178 together with certain values derived from them by the compiler. The several values of λ_s have been computed by means of the formulas indicated, numbered as in the text.

[722] Debye, P., *Trans. Faraday Soc.*, **30**, 679-689 (1934); *Physik Z.*, **35**, 101-106 (1934).
[723] Debye, P., and Ramm, W., *Ann. d. Physik (5)*, **28**, 28-34 (1937).
[724] Bryan, A. B., *Phys. Rev. (2)*, **22**, 399-404 (1923).
[725] Granier, J., *Bull. Soc. Fr. Élec. (4)*, **3**, 333-482 (1923) = Thesis, Paris.
[726] Esau, A., and Bäz, G., *Physik. Z.*, **38**, 774-775 (1937).
[727] Schmelzer, C., *Ann. d. Physik (5)*, **28**, 35-53 (1937).
[728] Knerr, H. W., *Phys. Rev. (2)*, **52**, 1057-1067 (1937).

The discordance between the several sets of values, and their progressive increase with λ_0, except in the case of (g), indicate that in the case of water the conditions are not so simple as is assumed in deriving the much quoted formulas (8) to (17); that for water we are not justified in assuming that there is only one singularity, and that of the highly dissipative type, in the frequency range below the optical spectrum. This is borne out by the graphs in Figure 6. By formulas (18a) and (18b) each should be

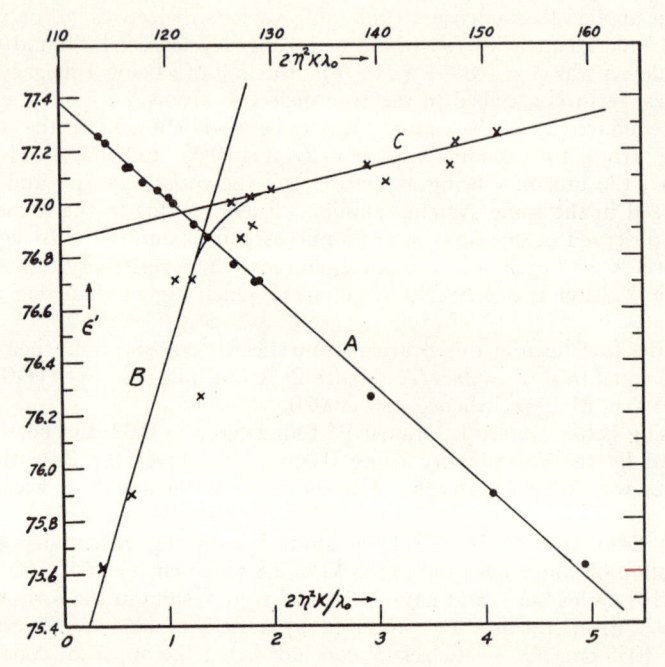

FIGURE 6. Variation of the Dielectric Constant of Water with $2n^2\kappa/\lambda_0$ and with $2n^2\kappa\lambda_0$.

Observations of H. W. Knerr [*Phys. Rev.* (2), **52**, 1054 to 1067 (1937)] at 22 °C. If formulas 18a and 18b were satisfied, the observations would lie along two right lines, one cutting the axis of ϵ' at n^2 and the other at ϵ_1. The observations are linear (dots) in $2n^2\kappa/\lambda_0$, line A, but the intercept is 77.38, whereas $\epsilon_1 = 79.61$. They are not linear (crosses) in $2n^2\kappa\lambda_0$. See text. The equations of the 3 lines are
(A) $\epsilon' = 77.38 - 0.365\ (2n^2\kappa/\lambda_0)$.
(B) $\epsilon' = 60.50 + 0.1327\ (2n^2\kappa\lambda_0)$.
(C) $\epsilon' = 75.83 + 0.009_4\ (2n^2\kappa\lambda_0)$.

a right line, and their intercepts on the axis of ϵ' should be ϵ_1 and ϵ_0. The graph of (18a) is a right line, but its intercept is 77.38, whereas $\epsilon_1 = 79.61$; that of (18b) consists of two right lines, their intercepts being 60.5 and 75.8, whereas $\epsilon_0 = 1.9$.

All of which indicates that a serious doubt attaches to the significance of the several values of λ_s that have been published, and that the values themselves cannot be satisfactorily compared unless the frequencies and

49. WATER: DIELECTRIC PROPERTIES

Table 177.—The Absorption Index of Water
(See also Table 178.)

By definition $I = I_0 e^{-4\pi\kappa n z/\lambda_0}$. The effect of a change in temperature is indicated in footnote b.

Unit of $\lambda_0 = 1$ cm; of $\nu = 10^6$ cycles/sec. Temp. $= t$ °C

λ_0	ν	t	κ	$n\kappa$	Ref.[a]
150,000	0.2		0.0070	0.063	Br
21,000	1.4		0.0070	0.063	Br
470	63.8		0.0070	0.063	Gr
242	124	17	0.0032	0.029	Ru
226	133	17	−0.0012	−0.011	Ru
221	136	17	0.0015	0.013	Ru
184	163	17	0.0003	0.003	Ru
160	188	17	0.0020	0.018	Ru
144.5	207.5	17	0.0042	0.038	Ru
144.1	208.0	17	0.0019	0.019	Ru
129	232	17	0.0032	0.029	Ru
106	283	17	0.0093	0.083	Ru
99	303	17	0.0050	0.045	Ru
85	353	17	0.0037	0.033	Ru
75	400	17	0.0114	0.102	Ru
68	441	17	0.0094	0.084	Ru
64	469	17	0.0129	0.116	Ru
57.4	522	17	0.0133	0.119	Ru
57.4	522	17	0.013	0.12	Fr
52	576	17	0.013	0.12	Fr
23	1300	20.2	0.069	0.62	Za
19.0	1580	15	0.079	0.71	Se
14.0	2140	17	0.049	0.45	Se
13.6	2210	16	0.045	0.41	Se
12.6	2380	16	0.058	0.52	Se[b]
12.6	2380	15.5	0.059	0.53	Se
8.8	3410	15	0.07	0.60	Ec
5.7	5260	21	0.12	1.08	Ec
4.0	7500	18	0.136	1.18	El
3.7	8100	14	0.21	1.72	Ec
2.7	11100	20(?)	0.268	2.26	Te
1.8	16700	20(?)	0.349	2.32	Te
1.75	17100	24	0.35	2.7	Ec
1.5	20000	20(?)	0.276	1.83	Te
1.1	27300	20(?)	0.230	1.44	Te
0.84	35700	20(?)	0.262	1.49	Te
0.42	71400	20(?)	0.240	1.29	Te

[a] References:
  Br Bryan, A. B.[724]
  Ec Eckert, E., *Verh. physik. Ges.*, **15**, 307-329 (1913).
  El Elle, D., *Ann. d. Physik (5)*, **30**, 354-370 (1937) ← *Diss.*, Jena.
  Fr Frankenberger, E., *Ann. d. Physik (4)*, **82**, 394-412 (1927).
  Gr Granier, J.[725]
  Ru Rückert, E., *Ann. d. Physik (4)*, **55**, 151-176 (1918).
  Se Seeberger, M.[716]
  Te Tear, J. D., *Phys. Rev. (2)*, **21**, 611-622 (1923).
  Za Zakrzewski, K., *Bull. intern. acad. polon. (A)*, **1927**, 489-503 (1927).

[b] Seeberger gives also the following for $\lambda_0 = 12.6$ cm:

t	15.5	16	30.0	50.0
100κ	5.9	5.8	3.2	1
$100 n\kappa$	53	52	28	9
n^2	80.6	80.3	76.6	81

the formulas actually used in computing them are known. Often the formula used is not stated. Some of the published values of λ_s are given in Table 179.

Table 178.—Dielectric Absorption of Water at 22 °C [728]
(For definitions and explanation of symbols, see first pages of this Section.)

$\epsilon = n^2(1 - \kappa^2) - 2in^2\kappa = \epsilon' - i\epsilon''$. In the following formulas (15a), (15b), (15c), and (16), from the text, put $\epsilon_0 = 1.90$ and $\epsilon_1 = 79.61$; in formula (g), similar to (15b), the 77.38 has been obtained from the linear graph in Figure 6. The mean n ($\bar{n}$) has been used in all computations.

(15a) $\lambda_s = \lambda_0 \{(\epsilon_1 - \epsilon')/(\epsilon' - \epsilon_0)\}^{1/2}$; (15b) $\lambda_s = \lambda_0 (\epsilon_1 - \epsilon')/2n^2\kappa$

(15c) $\lambda_s = \lambda_0 \{2n^2\kappa/(\epsilon' - \epsilon_0)\}$; (16) $\lambda_s = \lambda_0 \left\{ \dfrac{A - (A^2 - 4)^{1/2}}{2} \right\}$

(g) $\lambda_s = \lambda_0(77.38 - \epsilon')/2n^2\kappa$; $A \equiv (\epsilon_1 - \epsilon_0)/2n^2\kappa$

The quantities ν, ϵ', ϵ'', ϕ, and λ_s have been computed by the compiler. Conductivity of the water was 10^{-5} to 10^{-6} (ohm·cm)$^{-1}$; its contribution to ϵ'' at these frequencies is negligible.

Unit of ν = 1000 megacycles/sec; of λ_0 = 1 cm; of ϕ = 1°; of λ_s = 1 cm

ν	λ_0	n	κ	ϵ' $n^2(1-\kappa^2)$	ϵ'' $2n^2\kappa$	ϕ	16	15c	15a	15b	g
6.25	4.80	—	0.153	75.63	23.7	17.4	1.63	1.54	1.11	0.81	0.354
5.62	5.34	—	0.141	75.90	21.8	16.0	1.64	1.58	1.20	0.91	0.362
4.63	6.48	8.85	0.123	76.27	19.0	14.0	1.70	1.70	1.37	1.14	0.378
3.72	8.05	8.78	0.097	76.71	15.0	11.1	1.62	1.61	1.58	1.56	0.360
3.68	8.16	—	0.097	76.71	15.0	11.1	1.64	1.63	1.61	1.58	0.365
3.40	8.80	8.77	0.093	76.77	14.4	10.6	1.69	1.69	1.72	1.73	0.373
3.14	9.55	8.72	0.086	76.87	13.3	9.8	1.68	1.69	1.82	1.68	0.366
3.05	9.85	8.80	—	—	—	—	—	—	—	—	—
2.97	10.10	—	0.082	76.92	12.7	9.3	1.70	1.70	1.91	2.14	0.366
2.94	10.20	8.79	—	—	—	—	—	—	—	—	—
2.76	10.87	—	0.075	77.00	11.6	8.6	1.66	1.68	2.02	2.44	0.357
2.70	11.12	8.83	0.074	77.02	11.5	8.5	1.69	1.70	2.07	2.50	0.348
2.54	11.80	8.85	0.071	77.05	11.0	8.2	1.70	1.72	2.18	2.75	0.354
2.24	13.41	—	0.068	77.08	10.5	7.7	1.85	1.88	2.45	3.23	0.383
2.07	14.48	8.82	0.062	77.14	9.6	7.2	1.82	1.85	2.62	3.72	0.361
1.96	15.29	8.76	0.062	77.14	9.6	7.2	1.93	1.86	2.77	3.93	0.382
1.78	16.83	8.85	—	—	—	—	—	—	—	—	—
1.63	18.41	—	0.052	77.23	8.0	5.9	1.91	1.95	3.27	5.48	0.346
1.47	20.44	8.78	0.048	77.26	7.4	5.5	1.96	2.00	3.62	6.50	0.331
Mean		$8.80 = \bar{n}$									

Table 179.—Transition Wave-lengths for Water

The transition wave-length (λ_s) is defined as the wave-length in air corresponding to the frequency at which the real part (ϵ') of the dielectric constant is half the sum of the static value (ϵ_1) and the square of the optical index of refraction for infrared waves. There are reasons for doubting that water fulfills the conditions assumed in deriving the formulas employed

Table 179—(Continued)

in computing λ_s (see text). Knerr's values, computed by the compiler, have been taken from Table 178. The formulas are numbered as in Table 178 and in the text.

All these estimates of λ_s lie within the range 0.36 and 6.5 cm; hence λ_τ (= 0.00734λ_s, see p. 356) must lie within the range 0.0026 and 0.048 cm. But the value of τ computed by Debye by means of the oft-quoted formula $\tau = 4\pi\eta r^3/kT$, where η (= 0.01 poise) is the coefficient of viscosity, r (= 2 × 10^{-8} cm) is radius of the molecule, k (= 1.327 × 10^{-16} erg/°K per molecule) is the Boltzmann gas constant, and T (= 293 °K) is the absolute temperature (all referring to water at 20 °C), is τ = 0.27 × 10^{-10} sec corresponding to λ_τ = 0.81 cm—a value that is 17 to 310 times as great as that indicated by the tabulated values.

Unit of λ_0 and of λ_s = 1 cm. Temp. = t °C

λ_0	t	λ_s	Eq.	Ref.[a]
5 to 20	22	1.6 to 2.0	16	Knerr
5 to 20	22	1.5 to 2.0	15c	Knerr
5 to 20	22	1.1 to 3.6	15a	Knerr
5 to 20	22	0.8 to 6.5	15b	Knerr
5 to 20	22	0.362	g	Knerr
4	18	1.13	11	Elle
2.8 to 10	19	1.85	15a	Esau and Bäz
1050	20	1.2 to 1.7	10	Schmelzer

[a] References:
Elle, D. See Table 177, references.
Esau, A., and Bäz, G., *Physik. Z.*, **38**, 774-775 (1937).
Knerr, H. W., *Phys. Rev. (2)*, **52**, 1054-1067 (1937) → *Idem*, **51**, 1007 (A₄₆) (1937). (Computation by compiler, see Table 178.)
Schmelzer, C., *Ann. d. Physik (5)*, **28**, 35-53 (1937).

50. Conduction of Electricity by Water

The National Research Council has awarded to G. A. Hulett a grant for a study of the electrical conductivity of pure water.[729]

The observation of O. Risse,[730] that when water is exposed to x-rays its conductivity is increased and its pH is decreased, is probably to be explained by the heating of the water by thermal radiation from the anticathode.[731] H. Fricke and E. R. Brownscombe,[732] using methods more sensitive than those employed by Risse, were unable to detect any formation of either H_2O_2 or O_2 when water is exposed to x-rays, even though as much as 150 kiloroentgens were used.

All the following data regarding the electrical conduction of water, with the exception of those for natural waters (p. 380 and Table 184) and those published since 1929, have been obtained from the *International Critical Tables,* either directly or by means of formulas there given.

[729] Hulett, G. A., *Science*, **77**, 215 (1933).
[730] Risse, O., *Z. physik. Chem. (A)*, **140**, 133-157 (1929).
[731] Schnurmann, R., *Idem*, **150**, 110-114 (1930).
[732] Fricke, H., and Brownscombe, E. R., *Phys. Rev. (2)*, **44**, 240 (1933).

For the purposes of this compilation, the data have been grouped under three heads: (1) Conductivity of water, (2) Equivalent conductivity of the ions, (3) Electrolytic ionization.

Conductivity of Water.

The conductivity (κ) of a substance is the reciprocal of its volume resistivity, which is the longitudinal resistance per unit of length of a uniform cylinder of the substance of unit cross-sectional area. The dimensions of κ are given by the formula

$$\frac{\text{length}}{(\text{area} \times \text{resistance})} = (\text{resistance} \times \text{length})^{-1}$$

This quantity κ is also called both the specific conductance and the volume conductivity.

The best water obtained by Kohlrausch and Heydweiller is reported to have had at 10 °C the conductivity $\kappa = 4.2 \times 10^{-8}$ (ohm·cm)$^{-1}$.[733] This is about 10 per cent greater than that computed from the equivalent conductivities of the ions and the electrolytic ionization of water (see Table 180).

Table 180.—Electrical Conductivity of Pure Water (Computed)

The electrical conductivity (κ) of pure water is related to the equivalent conductivities of its ions (Λ_H and Λ_{OH}), its density (ρ), and its ionization product (K) as indicated by the equation $\kappa = (\Lambda_H + \Lambda_{OH})\rho\sqrt{K}/1000$. The following three sets of values of κ have each been computed from the values of Λ_H and Λ_{OH} as given in Table 181, the values of ρ as given in Table 93, and the values of K as defined (see N. Bjerrum [733]) in one of the following ways: (1) by the "best" values of p_w (see ICT of Table 182); (2) by the formula of A. Heydweiller,[738] $\log_{10} K^{-1} = 6099.6/(273 + t) + 24.25 \log_{10}(273 + t) - 66.4678$; or (3) by the formula of G. N. Lewis and M. Randall,[739] $\log_{10} K^{-1} = 6384.7/(273.1 + t) + 26.676 \log_{10}(273.1 + t) - 73.424$. (For definition of Λ and of K, see following paragraphs.)

Unit of $\kappa = 1$ (ohm·cm)$^{-1}$. Temp. $= t$ °C

$K \rightarrow$ t	(1) $\log_{10}\kappa$	$10^8 \kappa$	(2) $\log_{10}\kappa$	$10^8 \kappa$	(3) $\log_{10}\kappa$	$10^8 \kappa$
0°	8̄.07₅	1.19	8̄.064	1.16	8̄.067	1.17
10	8̄.36₃	2.3₁	8̄.357	2.28	8̄.360	2.29
15	8̄.49₇	3.1₄	8̄.491	3.10	8̄.493	3.11
18	8̄.57₄	3.7₅	8̄.568	3.70	8̄.569	3.71
25	8̄.74₁	5.5₁	8̄.736	5.44	8̄.736	5.44
35	8̄.96₁	9.1₄	8̄.954	9.00	8̄.952	8.96
40	7̄.06₀	11.5	7̄.051	11.25	7̄.048	11.17
50	7̄.23₄	17.1	7̄.232	17.06	7̄.226	16.83

---

[733] Bjerrum, N., *Int. Crit. Tables*, **6**, 152 (1929), from Kohlrausch, F., and Heydweiller, A., *Ann. d. Physik (Wied.)*, **53**, 209-235 (1894); *Z. physik. Chem.*, **14**, 317-330 (1894); also, Partington, J. R., *Int. Crit. Tables*, **6**, 142 (1929).

[734] Kling, A., and Lassieur, A., *Ann. de Chim. (10)*, **15**, 201-227 (1931); *Compt. rend.*, **201**, 203-204 (1935); *Jour. de Phys. (7)*, **7**, C.P. 21 (1936) ← *Doc. Sci.*, **4**, 225-229 (1935).

[735] Gostkowski, K., *Z. physik. Chem. (A)*, **170**, 149-152 (1934); *Acta Phys. Polon.*, **3**, 75-80 (1934).

[736] Thiessen, P. A., and Hermann, K., *Z. Elektroch.*, **43**, 66-69 (1937).

In recent years A. Kling and A. Lassieur [734] have persistently maintained that the exact conditions necessary for obtaining water of minimum conductivity are not known, and have presented experimental evidence which they think justify that conclusion. But the lowest conductivity that they reported (2×10^{-6}) was much greater than that found by Kohlrausch and Heydweiller.

K. Gostkowski [735] has described a paraffin-lined still from which, by low temperature distillation, he has obtained water for which $10^8\kappa = 7$(ohm·cm)$^{-1}$ at 0 °C. And P. A. Thiessen and K. Hermann [736] have described a procedure by which they obtained 400 cm³/hr of water having $10^8\kappa = 6.5$ to 8 (ohm·cm)$^{-1}$ at 25 °C, and have stated that by the use of an additional stage in the distillation they got $10^8\kappa = 5.95$ (ohm·cm)$^{-1}$ at 25 °C; they give the "theoretical" value as 5.52.

A. Deubner [737] has reported that the conductivity of water in contact with air may decrease from $10^6\kappa = 6$ or 8 (ohm·cm)$^{-1}$ as the water left the still, to 1.5 (ohm·cm)$^{-1}$ after it had been in contact with air for about a day. The explanation is not known, but he considered several possibilities.

Equivalent Conductivity.

The equivalent conductivity (Λ) of an electrolytic solution is defined by the relation $\Lambda = \kappa/c$, where c, called the equivalent concentration, is the number of equivalents (more explicitly, electrolytic equivalents) of the solute per unit volume of the solution. The equivalent conductivity is the axial conductance of a cylindrical volume of the solution, of unit length and of such a cross-sectional area that it contains one equivalent of the solute. Its dimensions are given by the formula:

$$\frac{\text{volume per equivalent}}{\text{resistance} \times \text{length}} = \frac{\text{area per equivalent}}{\text{resistance}}$$

The electrolytic, or electrochemical, equivalent of an ion is defined as the total mass of such ions needed to carry a combined resultant charge equal to a unit quantity of electricity. It may be expressed in terms of any one of various units, such as a gram per coulomb, or a gram per faraday. In the following, as in most physicochemical work, the second of these units will be employed, a faraday being understood to denote the resultant charge carried by one gram-formula weight of a univalent ion, which charge is not far from 96,500 coulombs.[740] When expressed in that unit, the electrolytic equivalent of an ion is equal to the formula-weight of the ion divided by its valence. Likewise, the electrolytic equivalent of a solute will be understood to mean the formula-weight of the solute divided by the combined

---

[737] Deubner, A., *Ann. d. Physik (4)*, **84**, 429-456 (1927).
[738] Heydweiller, A., *Ann. d. Physik (4)*, **28**, 503-512 (1909).
[739] Lewis, G. N., and Randall, M., "Thermodynamics and the free energy of chemical substances," New York, 1923.
[740] *Int. Crit. Tables*, **1**, 17 (1926).

valences of the similarly charged ions to which each formula-molecule may give rise.

If Λ is the equivalent conductivity of a solution in which the total equivalent concentration of the similarly charged ions is c_0, and if $c_1, c_2, c_3, \ldots$ are the equivalent concentrations of the several species of ions to which the solute actually gives rise, then it is possible to assign to each ion a quantity $(\Lambda_1, \Lambda_2, \Lambda_3, \ldots)$, called its equivalent conductivity, which depends only upon the temperature, upon the natures of the ion and of the solvent, and upon the value of c_0; which satisfies the formula $\Lambda c_0 = \Lambda_1 c_1 + \Lambda_2 c_2 + \Lambda_3 c_3 + \ldots$; and which approaches a definite limit as c_0 is reduced indefinitely, that limit being characteristic of the ion and independent of the natures and numbers of the other species of ions present. If the solute, which may be a mixture of substances, is completely dissociated, then $c_0 = c$, the equivalent concentration of the solute; in other cases, $c_0 < c$. If there is only a single solute and if it is completely dissociated, giving rise to only two species of ions, then $c_1 = c_2 = c$ and $\Lambda = \Lambda_1 + \Lambda_2$. This is the case for pure water, in which the molecules (taken as H_2O) that are ionized are regarded as those of a solute, the others as those of the solvent. In this case the ions are H^+ and OH^-, and the concentration is always very low (see next paragraph), so low that Λ_1 and Λ_2 have practically their limiting, constant values corresponding to zero concentration.

The relation of the structure of the molecule to the conductivity of water has been considered by many, some of the more recent being J. D. Bernal and R. H. Fowler,[741] J. D. Bernal,[742] John Rehner, Jr., [743] and G. Wannier.[744]

Table 181.—Equivalent Conductivities of the Ions of Water.[745]

Unit of $\Lambda = 1$ cm²·equivalent⁻¹ ohm⁻¹; temp. $= t$ °C.   Concentration indefinitely low

t	Λ_H	Λ_{OH}	$\Lambda_H + \Lambda_{OH}$
0	229.0	118	347.0
10	275.6	149	424.6
15	300.4	164.5	464.9
18	315.2	174[a]	489.2
25	350.0	196	546.0
35	399.6	228	627.6
40	421.4	244	665.4
50	464.3	276	740.3

[a] In the introduction to the section on Electrical Conductivity of Aqueous Solutions,[746] this value is given as 173.8, making $\Lambda_H + \Lambda_{OH} = 489.0$. At the same place and for 18 °C the following values are given also:

$$\frac{1}{\Lambda_H}\left(\frac{d\Lambda_H}{dt}\right) = 0.01573, \quad \frac{1}{\Lambda_{OH}}\left(\frac{d\Lambda_{OH}}{dt}\right) = 0.018 \text{ per } °C.$$

---

[741] Bernal, J. D., and Fowler, R. H., *J. Chem'l Phys.*, **1**, 515-548 (1933).
[742] Bernal, J. D., *Trans. Faraday Soc.*, **30**, 787 (1934).
[743] Rehner, John, Jr., *Rev. Sci. Inst. (N. S.)*, **5**, 2-3 (1934).
[744] Wannier, G., *Ann. d. Physik* (5), **24**, 545-568, 569-590 (1935) = *Diss.*, Basel.
[745] Kendall, J., *Int. Crit. Tables*, **6**, 259-304 (259) (1929).
[746] Anon, *Int. Crit. Tables*, **6**, 230, Table 3 (1929).

50. WATER: ELECTRICAL CONDUCTION

Electrolytic Ionization of Water.—(For photochemical dissociation and the energy involved in ionic dissociation, see Section 8.)

Let the symbol [H$^+$] denote the number of moles of H$^+$ ions per 1000 units of mass of water*; that is, per 1000/18.0154 moles of H$_2$O; and similarly for [OH$^-$]. If water dissociates solely in the manner H$_2$O $\rightleftarrows$ H$^+$ + OH$^-$, then for pure water, [H$^+$] = [OH$^-$]. If the addition of a solute adds H$^+$ ions to the solution, some of these will combine with OH$^-$ ions to form neutral molecules. This will continue until the product K = [H$^+$]·[OH$^-$] is the same as it would be for pure water, but now, [H$^+$] does not equal [OH$^-$]. Similarly if the solute adds OH$^-$ ions. (A. Kling and A. Lassieur [747] have suggested that the dissociation of water may not be restricted to the type just mentioned.)

The quantity K is called the ionization product, and $p_w \equiv \log_{10} K^{-1}$ is called the ionization exponent of water.

From these definitions and those in the earlier portions of this section, it follows at once that for pure water $1000c/\rho$ = [H$^+$] = [OH$^-$] = $\sqrt{K}$, and $\dfrac{1}{1000c} = \dfrac{1}{\rho\sqrt{K}}$ where ρ = density. If the unit of mass = 1 gram and of volume = 1 milliliter, then $1000c$ is the number of gram-equivalents of either ion per liter of pure water.

The symbol pH is commonly used to denote the common logarithm (base 10) of the reciprocal of the number of gram-equivalents of H$^+$ per liter of solution (*i.e.*, of the reciprocal of the "hydrogen-ion concentration"). Hence, for pure water pH = $\log_{10}(\rho\sqrt{K})^{-1} = 1/2 p_w - \log_{10}\rho$ if the units are those just stated.

E. Truog [748] has stated that, for water at 25 °C in equilibrium with air of average CO$_2$ content, pH = 5.7 to 5.8, and not 7, as frequently assumed. This value is affected but little by changes in temperature, the effect of changes in the solubility of CO$_2$ being largely offset by changes in the ionization. But if the air has been carefully freed of CO$_2$ and of NH$_3$, the pH of water at 25 °C will be about 7. The time required for exposed water to come into equilibrium with the CO$_2$ in the air is very brief (see Section 86).

Similarly S. B. Ellis and S. J. Kiehl [749] found for the purest water pH = 7.01 at 27.5 °C; and that value has been confirmed by J. A. Cranston and H. F. Brown.[750]

---

* Although it is customary and convenient to define the quantity [H$^+$] either in this way or in terms of a liter instead of a kilogram, such a definition introduces a purely arbitrary numerical factor. It would be more consistent with sound scientific custom to define [H$^+$] as the ratio of the number of moles of H$^+$ ions to the total number of moles (*i.e.*, of formula-weights) of H$_2$O, and similarly in other cases.

[747] Kling, A., and Lassieur, A., *Compt. rend.*, **181**, 1062-1064 (1925); *Ann. de Chim.* (10), **15**, 201-227 (1931).

[748] Truog, E., *Science (N. S.)*, **74**, 633-634 (1931).

[749] Ellis, S. B., and Kiehl, S. J., *J. Am. Chem. Soc.*, **57**, 2145-2149 (1935).

[750] Cranston, J. A., and Brown, H. F., *Trans. Faraday Soc.*, **33**, 1455-1458 (1937).

In striking contrast with other observers, A. Kling and A. Lassieur [751] have reported the very low value pH = 5.8, and maintain it in the face of criticism by R. Cliquet-Pleyel [752] although the conductivity of their water was very high.[753]

Table 182.—Ionization Exponent and Product for Water

(For comparison of experimental values, see Table 183.)

In computing the H and the LR values, the logarithms were carried one place farther than they are given in this table. The tabulated values of K correspond to the more exact logarithms, whence such apparent discrepancies as occur between K and p_w at 25 °C. For the pH value of water in contact with air, see text.

$K = [H^+] \cdot [OH^-]$. Unit of $[H^+]$ and of $[OH^-] = 1$ g-mole of ion per 1000 g water

Source[a] → t	ICT	H $p_w = \log_{10}K^{-1}$	LR	ICT	H $10^{14}K$	LR
0	14.93	14.952	14.946	0.117	0.112	0.113
5	14.72	14.741	14.736	0.190	0.182	0.184
10	14.53	14.541	14.537	0.295	0.287	0.291
15	14.34	14.352	14.348	0.457	0.445	0.448
18	14.23	14.242	14.240	0.589	0.572	0.576[b]
20	14.16	14.171	14.170	0.692	0.674	0.677
25	13.99	14.000	14.000	1.02	0.999	1.000[b]
30	13.83	13.838	13.840	1.48	1.45	1.45
35	13.67	13.684	13.687	2.14	2.07	2.06
37	13.61	13.624	13.628	2.46	2.38	2.35
40	13.52	13.537	13.543	3.02	2.91	2.87
45	13.39	13.397	13.406	4.07	4.01	3.93
50	13.26	13.265	12.276	5.50	5.44	5.30
60	13.03	13.019	13.036	9.33	9.58	9.21
70	12.82	12.796	12.820	15.1	16.0	15.2
80	12.63	12.595	12.626	23.4	25.4	23.7
90	12.45	12.413	12.451	35.5	38.6	35.4
100	12.29	12.249	12.295	51.3	56.4	50.8
150	11.63	11.641	11.729	234	229	186
200	11.26	11.293	11.428	550	509	373
250	11.17	11.119	11.302	676	761	499
300	11.40	11.062	11.295	398	866	507
306	11.46	11.062	11.300	347	867	501

Harned and Geary (HG)[a]

Source[a] → t	HH,HC	HM	HD $10^{14}K$	HG	Mean	Mean $\log_{10}K^{-1}$
0	(0.115)	0.1134	0.1132	0.1125	0.1133	14.9458
5	(0.186)	0.1850	0.1842	0.1834	0.1846	14.7333
10	(0.293)	0.2919	0.2921	0.2890	0.2920	14.5346
15	(0.452)	0.4505	0.4504	0.4500	0.4503	14.3465
20	0.681	0.6806	0.6806	0.6815	0.6809	14.1669
25	1.008	1.007	1.007	1.009	1.008	13.9965
30	1.471	1.470	1.467	1.466	1.468	13.8333
35	2.088	2.091	2.088	2.090	2.089	13.6801
40	2.916	2.914	(2.891)	2.920	2.917	13.5351
45	4.016	4.017		4.023	4.018	13.3960
50	5.476	5.482		5.465	5.474	13.2617

[751] Kling, A., and Lassieur, A., *Ann. de Chim.* (10), **15**, 201-227 (1931); *Compt. rend.*, **201**, 203-204 (1935).

50. WATER: ELECTRICAL CONDUCTION

Table 182—(Continued)

[a] Sources:

ICT	Bjerrum, N.,[733] those designated as "best values"; they are based upon the experimental values listed in Table 183. Values for temperatures below 100 °C rest upon no determination of potential or of hydrolysis based upon dissociation as inferred from the conductivity; at higher temperatures the greatest weight is given to the determination of potential.
H	A. Heydweiller's formula[738] (see also Bjerrum, N.[733]), $\log_{10} K^{-1} = \frac{6099.6}{273 + t} + 24.25 \log_{10}(273 + t) - 66.4678.$
HC	Harned, H. S., and Copson, H. R., *J. Am. Chem. Soc.*, **55**, 2206-2215 (1933).
HD	Harned, H. S., and Donelson, J. G., *Idem*, **59**, 1280-1284 (1937).
HG	Harned, H. S., and Geary, C. G., *Idem*, **59**, 2032-2035 (1937).
HH	Harned, H. S., and Hamer, W. J., *Idem*, **55**, 2194-2206, 4496-4507 (1933).
HM	Harned, H. S., and Mannweiler, G. E., *Idem*, **57**, 1873-1876 (1935).
LR	Formula given by G. N. Lewis and M. Randall.[739] (See Bjerrum, N.[733]), $\log_{10} K^{-1} = \frac{6384.7}{273.1 + t} + 26.676 \log_{10}(273.1 + t) - 73.424.$

[b] In his compilation,[754] M. Randall gives for the "equilibrium constant (activities)" $\times 10^{14}$ the values 0.114 at 0 °C, 0.58 at 18 °C, and 1.005 at 25 °C, based on the data of G. N. Lewis and M. Randall,[755] and of R. Lorenz and A. Böhi.[756]

Table 183.—Ionization Exponent for Water: Comparison of Values

For final value derived by Bjerrum, and for values computed by means of formulas, see Table 182.

All data in this table have been taken from the compilation of N. Bjerrum.[733] Recently, E. J. Roberts (Ro)[c] has reported for 25 °C, $10^{14} K = 0.988 \pm 0.004$; *i.e.*, $\log_{10} K^{-1} = 14.005 \pm 0.002$.

Method[a]	Cond	Hydrolysis				Potential				
		Ionization		Activity		Ion-concentration			Activity	
Property[b]	KH	Lun	Kan	Lun	Kan	LB	PT	Mic	Sor	LR[d]	Bu[e]
Ref.[c]						$p_w = \log_{10} K^{-1}$					
0	14.93		15.05		14.99	14.87				14.945	14.926
10	14.52	14.51		14.47							
15		14.34		14.30							
18	14.22		14.34		14.27	14.15		14.13	14.14	14.239	14.222
25	13.98	13.98	14.09	13.94	14.03	13.92	13.91	13.89		13.998	13.980
30						13.76					
37								13.72			
								13.50		13.626	13.590
40		13.53		13.49		13.41		13.42			
50	13.25	13.29		13.25		13.06				13.273	
60						12.90					
70						12.67					
80						12.46					
90		(NK)[c]				12.37					
100		12.28								12.29	
156		11.57	(Sos)[c]								
218		11.19									
306		11.46									

[a] Methods: Cond = computed from the conductance of the purest water. The values are said to have been computed by means of the Heydweiller formula (H of

---

[752] Cliquet-Pleyel, R., *Jour. de Phys. (7)*, **7**, C.P. 21 (1936) ← *Doc. Sci.*, **4**, 104-113 (1935); *Chem. Abst.*, **31**, 7727 (1937) ← *Doc. Sci.*, **5**, 65-70 (1936).

[753] Kling, A., and Lassieur, A., *Jour. de Phys. (7)*, **7**, C.P. 21 (1936) ← *Doc. Sci.*, **4**, 225-229 (1935); Lassieur, A., *Chem. Abs.*, **31**, 7727 (1937) ← *Doc. Sci.*, **5**, 11-15 (1936).

[754] Randall, M., *Int. Crit. Tables*, **7**, 224-313, 347-353 (232) (1930).

[755] Lewis, G. N., and Randall, M., *J. Am. Chem. Soc.*, **36**, 1969-1993 (1914).

[756] Lorenz, R., and Böhi, A., *Z. physik. Chem.*, **66**, 733-751 (1909); *cf.* G. N. Lewis and M. Randall.[739]

Table 183—(Continued)

Table 182), but they differ appreciably from those so computed by the present compiler (see Table 182).
Hydrolysis = computed from the hydrolysis of weak acids and of weak bases.
Potential = computed from the potentials of cells having a hydrogen electrode in either an acid or an alkaline solution.

[b] Property utilized: Both ionization and ion-concentration were determined from the observed conductance. Activity = activity-coefficient (f) of the ions; in the method of hydrolysis, it was computed by means of the formula: $\log_{10} f^{-1} = 0.3 \sqrt{c_i}$; c_i = concentration of either species of ion (unit = 1 g-mole per kg of water); in the method of potential, it was calculated by thermodynamic methods.

[c] References:
BU	Bjerrum, N., and Unmack, A., *Kgl. Danske Videnskab. Selskab, Math-fys. Medd.*, **9**, No. 1, pp. 208 (1929).
Kan	Kanolt, C. W., *Carnegie Inst. Wash., Publ.*, **63**, 283-298 (1907).
KH	Kohlrausch, F., and Heydweiller, A.[733]
LB	Lorenz, R., and Böhi, A.[756]
LR	Lewis, G. N., and Randall, M.[739]
Lun	Lundén, H., *J. de chim. phys.*, **5**, 574-608 (1907).
Mic	Michaelis, L., "Die Wasserstoffionenkonzentration," Berlin, J. Springer, 1914.
NK	Noyes, A. A., and Kato, Y., *Carnegie Inst. Wash., Publ. No.* **63**, 151-190 (1907).
PT	Poma, G., and Tanzi, B., *Z. physik. Chem.*, **79**, 55-62 (1912).
Ro	Roberts, E. J., *J. Am. Chem. Soc.*, **52**, 3877-3881 (1930).
Sor	Sörensen, S. P. L., *Biochem. Z.*, **21**, 131-304 (1909).
Sos	Sosman, R. B., *Carnegie Inst. Wash., Publ.* **63**, 191-235 (1907).

[d] Observations were made at 25 °C, and values of K at other temperatures were computed from that at 25 °C by means of the LR formula of Table 182, which assumes that the heat of neutralization is $(29210 - 53T)$ cal$_{20}$ per equivalent.

[e] The Debye-Hückel square-root formula[757] was assumed as the limiting law at infinite dilution. $\Lambda = f_\lambda \Lambda_0$, $f_\lambda = 1 - A\sqrt{c_i}$ where A depends upon the solvent, the solute, and the temperature, and c_i = concentration of either species of ion; the unit of c_i must accord with those used in A, as $A\sqrt{c_i}$ is dimensionless.

Conductivity of Rain-water.

For freshly fallen rain-water at 17.6 °C, H. Schmidt[758] found $10^6 \kappa = 128$ (ohm·cm)$^{-1}$. For purest water at 17.6 °C, $10^6 \kappa = 0.036$ (ohm·cm)$^{-1}$, Table 180.

Conductivity of Sea-water.

The composition of sea-water varies from place to place, and from time to time, depending upon the evaporation and the inflow of fresher water from streams, ice-bergs, and precipitation; its electrical conductivity likewise varies. Sea-water averages about 35 g of salts per kg. Data for variations in the salinity (S), composition of the salt, and temperatures of the oceans are given elsewhere (see p. 654). In the following, the unit of S is 1 g salt per kg of sea-water.

E. Ruppin[759] has found that the electrical conductivities at 0, 15, and 25 °C, respectively, may be computed by means of the expressions: $10^6 \kappa_0 = 978S - 5.96S^2 + 0.0547S^3$; $10^6 \kappa_{15} = 1465S - 9.78S^2 + 0.0876S^3$; $10^6 \kappa_{25} = 1823S - 12.76S^2 + 0.1177S^3$; and O. Krümmel[760] has stated that Knudsen

---
[757] Debye, P., and Hückel, E., *Physik. Z.*, **24**, 305-325 (1923).
[758] Schmidt, H., *Jahrb. d. drahtlos. Teleg.*, **4**, 636-638 (1911).
[759] Ruppin, E., *Wiss. Meeresunters. (N. F.)*, **9**, (Abt. Kiel), 178-183 (1906) → *Z. anorg. Chem.*, **49**, 190-194 (1908).
[760] Krümmel, O., "Handb. d. Ozeanog.," Vol. **1**, 1907.

has found that, whatever the value of S, $\log_{10}\kappa_t = \log_{10}\kappa_{15} + \alpha(t-15)$, where $\alpha = 0.01135$ when $t = 0$ °C, and $\alpha = 0.00928$ when $t = 25$ °C.

The conductivity is the same for all frequencies from zero to 100 kilocycles/sec,[761] but R. L. Smith-Rose [720] has reported that the conductivity of a sample taken from the English Channel increased, at 20 °C, from 0.043 (ohm·cm)$^{-1}$ at 0.5 kc/sec to 0.060 (ohm·cm)$^{-1}$ at 10,000 kc/sec. He found a mean temperature coefficient of κ to be 2.7 per cent per 1 °C, between 0 and 40 °C; Rivers-Moore [761] derived three per cent from observations in the range 12.0 to 18.3 °C; and H. Schmidt [758] derived 1.5 from observations in the range 18 to 21.4 °C. All of these are markedly greater than Knudsen's values, and rest on less extensive data.

Table 184.—Electrical Conductivity of Sea-water [760]

Accepting Ruppin's formulas (see text) connecting κ and the salinity (S) and Knudsen's values for the temperature coefficient (α) of κ, and assuming that the variation in α is linear in the temperature, Krümmel has computed the following values.

Unit of $S = 1$ g salt per kg sea-water; of $\kappa = 1$ (ohm·cm)$^{-1}$. Temp. $= t$ °C

$S \rightarrow$ t	5	10	15	20	25	30	35	40
				$10^4\kappa$				
0	48	92	135	176	216	254	293	331
5	55	107	156	203	248	292	335	378
10	63	122	178	231	283	332[a]	382	430
15	71	138	201	261	319	375	431	486
20	79	154	225	292	357	420[b]	482	543
25	88	171	249	323	394	464	532[c]	601
30	97	187	273	354	433	510	585	660

[a] For sea-water from off the coast near Hastings, England, S not stated, Balth.v.d.Pol [761] found $10^4\kappa = 377$ at 12.5 °C.
[b] For surface water from the North Sea, S not stated, H. Schmidt [758] found $10^4\kappa = 397$ at 20 °C.
[c] E. G. Hill [762] measured κ at 25 °C for sea-water of various concentrations lying between Cl = 20.439 ($S = 37$) and (Cl = 21.533 ($S = 39$), obtaining values about 2.5 per cent smaller than those given by Ruppin's formula. But T. Shedlovsky [703] found for a sample from Tortugas, 25 °C, Cl = 19.92 g/kg ($S = 36.10$), $\kappa = 0.05419$, only 0.8 per cent smaller than the tabular values indicate.

51. Kerr Electro-optic Effect for Water

When water is subjected to a uniform electric field it becomes slightly birefringent, behaving like a positive uniaxial crystal with its axis in the direction of the field. If n_0 and n_e denote, respectively, the ordinary and the extraordinary index of refraction for light of wave-length λ *in vacuo*, then $(n_e - n_0)/\lambda E^2 = C$ is independent of the field (E) and is known as the Kerr electro-optic constant, or the coefficient of electric birefringence. The value of C depends upon λ and the temperature as well as upon the

---

[761] v. d. Pol, Balth., Jr., *Phil. Mag.* (6), **36**, 88-94 (1918); Rivers-Moore, H. R., *Electrician* (London), **82**, 174-176 (1919).
[762] Hill, E. G., *Proc. Roy. Soc. Edinburgh*, **27**, 233-243 (1907).
[703] Shedlovsky, T., private communication from Dr. L. R. Blinks through Mr. R. S. Ould.

nature of the substance. T. H. Havelock [764] has developed a theory that seems to accord well with the observed facts. On this theory $C\lambda n/(n^2-1)^2$ is a constant for a given temperature, n being the ordinary index of refraction for light of wave-length λ. This theory, as well as that of Cotton and Mouton, requires that $(n_e - n)/(n_0 - n) = -2$. The experimental determination of this ratio of the absolute retardations is rendered extremely difficult by the presence of electrostriction and of thermal effects. In those cases in which the effect of these has been eliminated, the ratio has been found to be -2.[765] Apparently it has not been determined for water.

M. Pauthenier [766] reports that at 17 °C and for the D-line the value of C for water is 1.23 times its value for CS_2. This is the value accepted by H. Mouton [767] as the best available at the time his compilation was prepared. Accepting, with Pauthenier, L. Chaumont's data for CS_2,[768] Havelock's formula for the variation of C with λ and n, and F. F. Martens' formula for the refraction of CS_2,[769] increasing that n by 0.00090 in order to reduce it from 18° to 17 °C, we have as follows for CS_2 at 17 °C: For $\lambda = 0.54607\,\mu$, $10^7 C = 3.6315$, $n = 1.63811$, $10^{11} C\lambda n/(n^2-1)^2 = 1.1463$; for $\lambda = 0.57801\,\mu$, $10^7 C = 3.3580$, $n = 1.63171$, $10^{11} C\lambda n/(n^2-1)^2 = 1.1459$. Using the mean value $10^{11} C\lambda n/(n^2-1)^2 = 1.1461$, we find for $\lambda = 0.58931\,\mu$ and $n_{17} = 1.62974$, $10^7 C = 3.2728$ for CS_2, the D-lines, and 17 °C. This essentially agrees with the value that may be derived from the data given in Mouton's compilation.[767]

Whence for water at 17 °C and the D-lines ($\lambda = 5893\text{A}$)

$$C = 4.03\ 10^{-7}\text{cgse}$$
$$= 0.0363\ \text{cm/volt}^2$$

Taking $n = 1.33324$, this gives for the Havelock constant for water at 17 °C the value $C\lambda n/(n^2-1)^2 = 5.24\ 10^{-11}\text{cgse} = 4.72\ 10^{-6}\text{cm}^2/\text{volt}^2$. In his review of the molecular field problem, F. G. Keyes [770] accepts the value (reference not given) $C\lambda/n = 14.4\ 10^{-12}\text{cgse}$ for water at 20 °C. That leads to $C = 3.25\ 10^{-7}\text{cgse}$ for the D-lines and water at 20 °C, which can be reconciled with the preceding value for 17 °C only by assuming the very great temperature coefficient of 8 per cent per 1 °C. That for CS_2 is only 0.5 per cent per 1 °C; using that with the 17 °C value for C leads to 17.6 10^{-12}cgse for $C\lambda/n$ for water at 20 °C, which is the value given by P. Debye [771] on the basis of Pauthenier's observations. (For the quantity $C\lambda/n$, 1 cgse = 1 $\epsilon\cdot\text{cm}^3/\text{erg}$, ϵ = the unit of dielectric constant; for C, 1 cgse = 1 $\epsilon\cdot\text{cm}^2/\text{erg}$.)

If the electric field is oscillatory and its frequency is near that of a

---

[764] Havelock, T. H., *Proc. Roy. Soc. (London) (A)*, **77**, 170-182 (1905); **80**, 28-44 (1907); *Phys. Rev.*, **28**, 136-139 (1909).
[765] Pauthenier, M., *Ann. de Phys. (9)*, **14**, 239-306 (1920); *Jour. de Phys. (6)*, **2**, 183-196 (1921).
[766] Pauthenier, M., *Jour. de Phys. (6)*, **2**, 384-389 (1921).
[767] Mouton, H., *Int. Crit. Tables*, **7**, 109-113 (110) (1930).
[768] Chaumont, L., *Ann. de Phys. (9)*, **4**, 61-100, 101-206 (1915); **5**, 17-78 (1916).
[769] Martens, F. F., *Ann. d. Physik (4)*, **6**, 603-640 (632) (1901).
[770] Keyes, F. G., *Chem'l Rev.*, **6**, 175-216 (1929).
[771] Debye, P., *Handb. d. Radiol. (Marx)*, **6**, 597-786 (770) (1925).

characteristic system in the substance, the Kerr constant may differ markedly from its value for static fields. Effects, at first thought to be of this kind, were reported for water by A. Bramley,[772] but were later shown by him to have another origin.[773]

52. Electrical Discharge in Water

High-potential discharges between electrodes immersed in water are of various types—arc, brush, corona, spark—depending upon the nature of the circuit, and are accompanied by mechanical disturbances of the liquid. Such disturbances may exist—owing to electrostatic forces between the electrode and the electrically charged water—even when the discharge is feeble. M. Katalinic[774] has described the production of waves and of sprays when a high-voltage a.c. potential is applied to a wire electrode dipping in, or lying just below, the surface. He states that at 0.0004 cm from the electrode the potential gradient may be over 2 megavolts/cm when the applied voltage is 1200 volts.

Arc.

The spectrum of under-water arcs between carbon electrodes has been studied by H. Konen[775]; and H. D. Carter and A. N. Campbell[776] have studied the electrical products formed, using electrodes of various materials. The latter found that the rate of evolution of gas increases with the temperature of the water, and is independent of the pressure, at least to 23 atm, and that the temperature of the arc depends upon the nature of the electrodes and upon the temperature of the water, increasing about 1500 °C when the temperature of the water is increased from 5 °C to 100 °C. They give a bibliography of 46 entries. J. W. Shipley[777] has studied the arcing that occurs under certain conditions in the a.c. electrolysis of water, using solutions of NaOH.

Brush.

The color and spectrum of the brush discharge in water, and their variations with the conditions, have been studied by H. Smith.[778]

Corona.

Two types of impulse corona in water have been described by Y. Toriyama and U. Shinohara.[779] Using point-to-plate electrodes and impulses of 10 to 100 microseconds duration, they obtained a pink discharge having a line spectrum. With a higher crest voltage and a duration of only 0.1

[772] Bramley, A., *J. Franklin Inst.*, **206**, 151-157 (1928); *Phys. Rev. (2)*, **33**, 640 (1929).
[773] Bramley, A., *J. Opt. Soc. Amer.*, **21**, 148 (1931).
[774] Katalinic, M., *Z. Physik*, **77**, 257-270 (1932).
[775] Konen, H., *Ann. d. Physik (4)*, **9**, 742-780 (1902).
[776] Carter, H. D., and Campbell, A. N., *Trans. Faraday Soc.*, **28**, 479-496, 634-644 (1932) → *Trans. Electroch. Soc.*, **63**, 419-423 (1933).
[777] Shipley, J. W., *Trans. Am. Electrochem. Soc.*, **55**, 105-116 (1929).
[778] Smith, H., *Phil. Mag. (6)*, **27**, 801-823 (1914).
[779] Toriyama, Y., and Shinohara, U., *Nature*, **132**, 240 (1933).

microsecond, the corona was white, and the illustration they give indicates that the spectrum was continuous.

Spark.

The spectra of sparks in liquids and between electrodes of various kinds have been much studied. A report on the subject, including a bibliography of 23 entries, has been published by J. A. Anderson.[780] From such sparks, bubbles of gas are projected with considerable velocity, and in amounts greater than can be accounted for by electrolysis. For the water-spectrum of such sparks, see Section 46.

Y. Toriyama and U. Shinohara [781] have found that there is no direct relation between the conductivity of a liquid dielectric and its break-down voltage: the break-down is an electronic phenomenon, the conduction, an ionic one. Using needle points separated by the distance d, and "impulse voltage chopped at the tail of the impulse wave," they found the following values for the break-down difference of potential (V) for water of conductivity 1.43×10^{-4} (ohm·cm)$^{-1}$:

d	0.4	0.8	1.2	1.6	2.0	2.4	mm
V	17.5	20.0	22.5	25.2	27.7	30.6	kv

These values of V were read from their graph.

53. Magnetic Susceptibility of Water

The permeability (μ), the (volume) susceptibility (κ), the specific susceptibility (χ), and the density (ρ) of the medium are so related to the magnetic induction (B) and the resultant intensity (H) of the magnetic field that $B = \mu H = (1 + 4\pi\kappa)H$ and $\chi = \kappa/\rho$. The specific susceptibility (χ) is also called the coefficient of magnetization; if it is negative, the medium is said to be diamagnetic.

There is nothing to indicate that the value of χ for water depends upon the strength of the field. Fields of 1.2 to 40 kilogauss have been used.[782]

Detailed discussions of methods and of the several determinations of the susceptibility of water and its variation with the temperature have been published by P. Sève,[783] A. Piccard,[784] W. Johner,[785] and recently, again by Sève.[786]

The most precise determinations yet made are those by A. Piccard and A. Devaud [787] giving $10^9\chi = -719.92 \pm 0.11$ cgsm ($10^9\kappa = -718.64$) at 20 °C; and by H. Auer [788] giving $10^9\chi = -721.83 \pm 0.48$ cgsm at 20 °C.

---

[780] Anderson, J. A., *Int. Crit. Tables*, **5**, 433 (1929).
[781] Toriyama, Y., and Shinohara, U., *Phys. Rev. (2)*, **51**, 680 (L) (1937).
[782] Hayes, H. C., *Phys. Rev. (2)*, **3**, 295-305 (1914); Wills, A. P., *Idem*, **20**, 188-189 (1905); de Haas, W. J., and Drapier, P., *Ann. d. Physik (4)*, **42**, 673-684 (1913).
[783] Sève, P., *Ann. de chim. et phys. (8)*, **27**, 189-244, 425-493 (1912) → *Jour. de Phys. (5)*, **3**, 8-29 (1913).
[784] Piccard, A., *Arch. sci. phys. et nat. (4)*, **35**, 209-231, 340-359, 458-482 (1913).
[785] Johner, W., *Helv. Phys. Acta*, **4**, 238-280 (1931) = *Diss.*, Bern, 1930.
[786] Sève, P., *Congrès Internat. d'Élect.*, Sect. 2, Report 11, Paris, 1932.
[787] Piccard, A., and Devaud, A., *Arch. sci. phys. et nat. (5)*, **2**, 455-485 → 410 (1920).
[788] Auer, H., *Ann. d. Physik (5)*, **18**, 593-612 (1933).

53. WATER: MAGNETIC SUSCEPTIBILITY

The great difference between these two values (over 3 times the sum of their estimated uncertainties) is yet to be explained. The value accepted by the experts for the *International Critical Tables* is -720, essentially that obtained by Piccard and Devaud.

The surprising results reported by A. P. Wills and G. F. Boeker,[789] indicating a marked variability and a kind of hysteresis in the susceptibility of water, have not been confirmable,[788, 790] and seem to have been due to leaks in the apparatus.[791] But even after these have been eliminated the trend of the slope of the (χ, t) curve changes abruptly near 35 °C and near 55° C.[791, 792]

Cabrera and Fahlenbrach [790] found that the value of χ for water from freshly melted ice was the same as that for water that had not recently

Table 185.—Specific Susceptibility of Water at 20 °C

It is believed that the first and the third entry represent the same observations.

Unit of χ and $\kappa = 1$ cgsm

Year	Ref.[a]	$-10^9 \chi$	$-10^9 \kappa$
1912	Weiss and Piccard	719.3	720.6
1912	Sève	719.1[b]	720.4
1913	Piccard	719.3	720.6
1913	de Haas and Drapier	721	722
1914	Ishiwara	720	721
1920	Piccard and Devaud	719.92 ± 0.11	721.19
1929	*Int. Crit. Tables*	720	721
1933	Auer	721.83 ± 0.48	723.11

[a] References:
Auer, H., *Ann. d. Physik (5)*, **18**, 593-612 (1933).
de Haas, W. J., and Drapier, P., *Ann. d. Physik (4)*, **42**, 673-684 (1913).
Int. Crit. Tables. K. Honda, T. Ishiwara, T. Soné, and M. Yamada, *Int. Crit. Tables.* **6**, 354, 356 (1929). Based on work of W. J. de Haas and P. Drapier, T. Ishiwara, A. Piccard and A. Devaud.[787, p. 410] P. Sève,[788, vol. 27] P. Weiss and A. Piccard, and A. P. Wills. *Phys. Rev.*, **20**, 188-189 (1905); references to the first two papers and to the next to the last are given elsewhere in this list.
Ishiwara, T., *Sci. Rep. Tôhoku Imp. Univ., Sendai (1)*, **3**, 303-319 (1914).
Piccard, A., *Arch. sci. phys. et nat.* (4), **35**, 209-231, 340-359, 458-482 (1913).
Piccard, A., and Devaud, A., *Idem (5)*, **2**, 455-485 → 410 (1920).
Sève, P., *Ann. de chim. et phys.* (8), **27**, 189-244, 425-493 (1912).
Weiss, P., and Piccard, A., *Compt. rend.*, **155**, 1234-1237 (1912).

[b] This is the mean of the values for the two methods, as corrected by Piccard *loc. cit.*, and reduced to 20 °C by means of the coefficient 0.00012.

---

been frozen, the temperature being the same in both cases. But F. W. Gray and J. F. Cruikshank,[793] using a method by which the value of χ could be observed continuously, have reported that the numerical value of χ for water from freshly melted ice increased for a time, reaching a maximum about 20 minutes after the melting, and then falling abruptly to a constant value. They seek to explain this variation in terms of the molecular structure proposed by Bernal and Fowler (p. 174).

[789] Wills, A. P., and Boeker, G. F., *Phys. Rev. (2)*, **42**, 687-696 (1932).
[790] Cabrera, B., and Fahlenbrach, H., *Z. Physik*, **82**, 759-764 (1933).
[791] Wills, A. P., and Boeker, G. F., *Phys. Rev. (2)*, **46**, 907-909 (1934).
[792] Seely, S., *Phys. Rev. (2)*, **52**, 662 (L) (1937).
[793] Gray, F. W., and Cruikshank, . F., *Nature*, **135**, 268-269 (L) (1935).

O. Specchia [794] has suggested that an interference method might be used to advantage for measuring the change in level upon which rest many of the determinations of χ.

The effect of variations in temperature upon the value of χ for water is so slight that it is difficult to measure. In the earlier measurements it was not only masked, but actually reversed, by errors and parasitic effects. In 1932 it seemed that $(\chi_t - \chi_{20})/\chi_{20} = \alpha(t - 20)$ with $10^4\alpha = +1.31$, but later work indicates that the relation is not linear (see Table 186).

Explanations of the variation of χ with the temperature, generally in terms of changes in the polymerization, have been proposed by A. Piccard,[784, 795] W. Johner,[785] R. N. Mathur,[796] M. A. Azim, S. S. Bhatnagar, and R. N. Mathur,[797] B. Cabrera and H. Fahlenbrach,[798] G. Tammann,[799] and K. Honda and Y. Shimiza [800]; and Mathur [796] has remarked that any progressive change in the uniformity with which the molecules are aligned by the field should result in a corresponding change in χ. From the observa-

Table 186.—Variation of the Specific Susceptibility of Water with the Temperature

From the data available in 1932 it appeared that the specific susceptibility of water varied linearly with the temperature, the most probable value of $\alpha \equiv (1/\chi_{20}) \cdot (d\chi/dt)$ being 0.000131 per 1 °C.[786] But the observations by Piccard (P 1913)[a] and by Mathur (Mr 1931)[a] did not accord with that conclusion, and neither do the more recent observations by Auer (A 1933)[a], Cabrera and Fahlenbrach (CF 1934)[a], Wills and Boeker (WB 1934)[a], and Seely (See 1937)[a]. The values tabulated by O. Specchia and G. Dascola [803] for ordinary water seem to be seriously in error.

The formula published by Cabrera and Duperier (CD 1924, 1925)[a] and quoted in some compilations is now admitted to be wrong, actually defining a variation in the wrong direction; that given by Wills and Boeker (WB 1934)[a] for the range 20 to 66 °C is $\chi/\chi_{20} = 1 - 1.3(t - 20)/10^4 - 0.7(t - 20)^2/10^6$, and defines the values given under WB in the second section of this table.

In the first section of the table are given the several values of the linear coefficient (α) that have been proposed and used. In the second are the several sets of values that have been obtained for $-\chi_{20}$ and for $-(\chi - \chi_{20})$. From them the corresponding values of χ for each of the tabulated values of t may be obtained; e.g., the observations of P give the values $-10^9(\chi - \chi_{20}) = -0.9$ at 10° C and $-10^9\chi_{20} = 719.3$, whence $-10^9\chi = 719.3 - 0.9 = 718.4$ cgsm at 10 °C.

[794] Specchia, O., *Atti. Accad. Naz. d. Lincei* (6), **7**, 574-576 (1928).
[795] Piccard, A., *Compt. rend.*, **155**, 1497-1499 (1912).
[796] Mathur, R. N., *Indian J. Phys.*, **6**, 207-224 (1931).
[797] Azim, M. A., Bhatnagar, S. S., and Mathur, R. N., *Phil. Mag.* (7), **16**, 580-593 (1933).
[798] Cabrera, B., and Fahlenbrach, H., *Compt. rend.*, **197**, 379-381 (1933).
[799] Tammann, G., *Z. Physik*, **91**, 410-412 (1934).
[800] Honda, K., and Shimiza, Y., *Sci. Rep. Tôhoku Imp. Univ. (Sendai)*, **25**, 939-945 (1937).

53. WATER: MAGNETIC SUSCEPTIBILITY

Table 186—*(Continued)*
Unit of $\chi = 1$ cgsm; of $\alpha = 1$ per 1 °C. Temp. $= t$ °C

I. Linear coefficients. Not to be preferred.

Ref.[a]	WP	Me	J, CJP	Sè	ABM	CF
Year	1912	1916	1930	1932	1933	1933
$10^4 \alpha$	1.2	1.00	1.31	1.31	1.39	1.15
t	0 to 80	0 to 130	0 to 100		20 to 70	0 to 100

II. Various sets of values of χ_{20} and $\chi - \chi_{20}$. The Sè set has been computed by means of the linear coefficient $10^4 \alpha = 1.31$ that was accepted in 1932; the others, except ABM, are experimental values, and do not vary linearly with the temperature; $-10^9 \chi = -10^9 \chi_{20} - 10^9 (\chi - \chi_{20})$.

Ref[a]→ Year→ $-10^9 \chi_{20}$ t	Sè 1932 720[b]	P 1913 719.3	Me 1916 720[b]	Mr 1931 720[b]	ABM 1933 720[b]	A 1933 721.83	WB 1934 720[b]	CF 1934 720[b]	See 1937 720[b]
					$-10^9 (\chi - \chi_{20})$				
0	−1.9	−2.0	−2.0				−2.1	−1.66	−2.1
1	−1.8					−2.87			
5	−1.4	−1.5				−2.01	−1.4		
10	−0.9	−0.9	−0.0			−1.16	−1.0	−0.83	−1.1
15	−0.5	−0.5		−0.5		−0.52	−0.5		
20	0	0	0	0	0	0	0	0	0
25	+0.5	+0.3		+1.8		+0.41	+0.5		
30	0.9	0.8	+1.5		1	0.75	0.9	+0.83	+1.0
35	1.4	1.0		3.3		1.03	1.3		
40	1.9	1.4	2.0		2	1.29	1.7	1.66	1.7
45	2.4	1.7		5.1		1.54	2.0		
50	2.8	1.9	3.1		3	1.78	2.4	2.48	2.2
55	3.3	2.2		7.2			2.7		
60	3.8	2.4	3.1		4	2.23	2.9	3.31	3.0
65	4.2	2.6		8.4			3.2		
70	4.7	2.7	3.9		5	2.71	3.4	4.14	4.4
75	5.2	2.8		9.7					5.2
80	5.6	2.9	4.4		6			4.97	
85	6.1	3.0							
90	6.6	3.0	5.7					5.80	
95	7.1	3.1							
100	7.5	3.2	6.2						
110	8.5		7.2					7.06	
120	9.4		7.5					7.06	
130	10.4		7.7					7.06	

[a] References:
- A   Auer, H.[788]
- ABM   Azim, M. A., Bhatnagar, S. S., and Mathur, R. N., *Phil. Mag.* (7), **16**, 580-593 (1933).
- CD   Cabrera, B., and Duperier, A., *Ann. Soc. Esp. Fis. Quim.*, **22**, 160-167 (1924); *Jour. de Phys.* (6), **6**, 121-138 (1925).
- CF   Cabrera, B., and Fahlenbrach, H., *Z. Physik*, **82**, 759-764 (1933); *Ann. Soc. Esp. Fis. Quim.*, **31**, 401-411 (1933); *Z. Physik*, **89**, 166-178 (1934). See also, *Ann. Soc. Esp. Fis. Quim.*, **32**, 525-537, 538-542 (1934).
- CJP   Cabrera, B., Johner, W., and Piccard, A., *Compt. rend.*, **191**, 589-591 (1930).
- J   Johner, W.[785]
- Me   Marke, A. W., *Fortsch. Physik*, **73**₂, 74 (1918) ← *Overs. K. Danske Vid. Selsk. Forh. (Kopenhagen)*, **5** and **6**, 395-413 (1916).
- Mr   Mathur, R. N.[796]
- P   Piccard, A.[784]
- Sè   Sève, P.[786]
- See   Seely, S., *Phys. Rev.* (2), **52**, 662 (L) (1937).
- WB   Wills, A. P., and Boeker, G. F.[791]
- WP   Weiss, P., and Piccard, A., *Compt. rend.*, **155**, 1234-1237 (1912).

[b] Assumed by the compiler in deriving the values of $(\chi - \chi_{20})$.

[801] Rao, I. R., *Proc. Roy. Soc. London (A)*, **145**, 489-508 (1934).

tions of Cabrera and Fahlenbrach and Rao's estimate of the relative abundance of the several polymers,[801] L. Sibaiya [802] has computed the value of χ for each of the polymers assumed by Rao, finding: $10^9\chi_{20} = -775.5$ for (H_2O), -722.2 for (H_2O)$_2$, and -701.3 for (H_2O)$_3$.

54. Verdet Constant of Water

When plane-polarized light passes a distance l through a substance in a uniform magnetic field of strength H, the angle between the direction of H and the direction of advance of the light being θ, then the plane of polarization is rotated through an angle α, in the direction in which a right-handed screw lying along l must be turned in order that its advance shall be in the same direction as that of the component along l of the light, such that $\alpha = VlH \cos \theta$, V being a factor determined by the substance, its temperature, and the wave-length of the light. If V is positive and θ is zero, the rotation is in the direction of the amperian currents that are equivalent to the field H.

This phenomenon is often called the Faraday effect, and the factor V is commonly called the Verdet constant.

The effect appears very quickly after the field is applied. The exact amount of lag, if any, is not known. Not only does the early conclusion of J. W. Beams and F. Allison [804] that the lag for water exceeds that for CS_2 by 1.1 mμsec seem to be incorrect as to the numerical value (cf. F. Allison [805]), but doubt has even been cast upon their interpretation of their observations (see, e.g., J. W. Beams and E. O. Lawrence,[806] E. Gaviola,[807] F. G. Slack, R. L. Reeves, and J. A. Peoples, Jr.[808]). There is at present no generally accepted experimental evidence of any lag at all. Similar remarks apply to Allison's early conclusions that the difference between the lags for water and for CS_2 vanishes when the liquids are exposed to x-rays, and that such exposure increases slightly the value of the Verdet constant.[808a]

Effect of Temperature.

For the D-lines ($\lambda = 0.5893 \mu$) and within the range 4 °C to 97.7 °C, $10^4 V_t = 131.1 - 0.00400t - 0.000400t^2$ minute-of-arc per cm·gauss.[809] This formula is equivalent to that given by A. Cotton and R. Lucas,[810] and leads to the following values:

---

[802] Sibaiya, L., *Current Sci.*, **3**, 421-422 (1935).
[803] Specchia, O., and Dascola, G., *Nuovo Cim. (N. S.)*, **12**, 606-609 (1935).
[804] Beams, J. W., and Allison, F., *Phys. Rev. (2)*, **29**, 161-164 (1927).
[805] Allison, F., *Idem*, **30**, 66-70 (1927).
[806] Beams, J. W., and Lawrence, E. O., *J. Franklin Inst.*, **206**, 169-179 (1928).
[807] Gaviola, E., *Phys. Rev. (2)*, **33**, 1023-1034 (1929).
[808] Slack, F. G., Reeves, R. L., and Peoples, J. A., Jr., *Phys. Rev. (2)*, **46**, 724-727 (1934).
[808a] Allison, F., *Phys. Rev. (2)*, **31**, 158-159 (A) (1928).
[809] Rodger, J. W., and Watson, W., *Phil. Trans. (A)*, **186**, 621-655 (1895) → *Z. physik. Chem.*, **19**, 322-363 (1896).
[810] Cotton, A., and Lucas, R., *Int. Crit. Tables*, **6**, 425 (1929).

t	0	20	40	60	80	100 °C
$10^4 V_D$	131.1	130.8₆	130.3₀	129.4₂	128.2₂	126.7₀

For water, the ratio of V_D to the density is essentially constant if $t < 20$ °C, but above 20 °C the ratio increases almost linearly with t, about $1.4' \times 10^{-6}$cm^{-1} gauss^{-1} cm^3/g·deg.[809].

It is generally assumed, and the data available in 1932 indicated (*e.g.*, F. Schwers [811]), that V_λ/V_D is essentially independent of the common temperature, at least over a moderate range including 20 °C.

But the recent observations by F. G. Slack, R. L. Reeves, and J. A. Peoples, Jr. [812] suggest that V_λ/V_D may vary slowly with the temperature; they report as follows, V_{546} being the value of V for $\lambda = 0.546\,\mu$.

t (°C)	15	20	30	40	45
$10^4 V_D$	132.0	131.9	132.0	131.8	131.7
$10^4 V_{546}$	156.5	156.0	154.6	153.5	153.0
V_{546}/V_D	1.185	1.183	1.171	1.165	1.162

Their value of V_D for $t = 20$ °C being 0.77 per cent greater than the one generally accepted (see Table 187).

Dispersion of the Verdet Constant.

As just stated, $(V_\lambda/V_D)_t$ is essentially independent of t if t is not far from 20 °C, but the several sets of measurements of V for a fixed temperature and various wave-lengths exhibit annoying discrepancies, and a direct graphical comparison of them is not satisfactorily accurate. Furthermore, it is quite laborious to determine the deviation of each value from that demanded either by the equation (1) proposed by P. Joubin [813]

$$V\lambda = 0.002788_5 (n - 77.65\lambda dn/d\lambda) \tag{1}$$

or by that (2) given by S. S. Richardson [814] *

$$n\lambda^2 V = 0.003265_5 \left[\left(\frac{\lambda^2}{\lambda^2 - 0.01891} \right)^2 + 0.7381_0 \right] \tag{2}$$

or by that (3) used by U. Meyer [815]

* In *International Critical Tables*, **6**, 425, where the expression in parentheses is written in the form $\left(\dfrac{\lambda^2}{\lambda^2 - \lambda_1^2} \right)^2$, λ_1 is incorrectly given as 1260.4A. It should be 1375.1A, as Richardson gives $\lambda_1^2 = 0.01891\,\mu^2$.

Exactly similar formulas, but with different constants, have been proposed by G. Bruhat and A. Guinier,[816] $n\lambda^2 V = 0.0039470\,[0.43090 + \lambda^4/(\lambda^2 - 0.01680)^2]$, and by I. T. Pierce and R. W. Roberts,[817] $n\lambda^2 V = 0.0038723\,[0.45658 + \lambda^4/(\lambda^2 - 0.01719)^2]$, as representative of their own sets of observations. Each thinks that the formula given defines V to within 1 in 1000; the first, for the visible and ultraviolet spectrum; the second for the infrared.

[811] Schwers, F., *Bull. Acad. Roy. Belg.*, **1912**, 719-752 (1912).
[812] Slack, F. G., Reeves, R. L., and Peoples, J. A., Jr., *Phys. Rev.* (2), **46**, 724-727 (1934).
[813] Joubin, P., *Ann. de chim. et phys.* (6), **16**, 78-144 (1889).
[814] Richardson, S. S., *Phil. Mag.* (6), **31**, 232-256 (1916).
[815] Meyer, U., *Ann. d. Physik* (4), **30**, 607-630 (1909).
[816] Bruhat, G., and Guinier, A., *Jour. de Phys.* (7), **4**, 691-714 (1933).
[817] Pierce, I. T., and Roberts, R. W., *Phil. Mag.* (7), **21**, 164-176 (1936).

Table 187.—The Verdet Constant of Water
(See Figs. 7 and 8)

For the D-lines ($\lambda = 0.5893\,\mu$) the generally accepted value of V for water at 20 °C is $V_{20} = 0'.01309$ per cm·gauss = 3.808 microradians/cm·gauss, which essentially agrees with that (0.01308) given by A. Cotton and R. Lucas.[818]

V_n is the norm computed by means of the arbitrary formula (4,5) constructed so as to represent closely the observations of Mi and of I (see text), the required values of Δ being those here given. Except the Ri values, computed by means of the equation as given by Richardson, each set of values has been so adjusted by a flat percentile correction, as to give $V = 0'.01309$ for the D-lines.

Examples: For 20 °C and $\lambda = 0.2428\,\mu$, Ri's equation gives $10^4 V = 1159.8 + 5.6 = 1165.4$; for 20 °C and $\lambda = 0.2482\,\mu$, the value found by Mi was 1084, by Ro 1086, and by BG 1079, none of these giving the fifth digit.

Unit of $\lambda = 1\,\mu = 10^{-4}$ cm; of V and $\Delta = 1'$ per cm·gauss. Temp. = t °C

I. Best values for the D-lines ($\lambda = 0.5893$).

t	V_t	$10^4 V_{20}$	References[a]
20	0.01309	130.9	Rodger and Watson (1895)
18	0.01309	130.9	Agerer (1905)
20	0.01309	130.9	Richardson (1916)
17	0.01306	130.7	Stephens and Evans (1927)

II. The better values throughout the spectrum. Temp. = 20 °C.

λ	References[a]→ $10^4\Delta$	$10 V_n$	Mi	Ri[b]	Ro $10^4(V-V_n)$	BG	L	v.S
0.2428	+22.8	1159.8		+5.6				
82	15.2	1079.6	+4		+6	−1		
96	13.6	1060.2					−12	
0.2536	10.0	1008.5	0					
37	9.8	1007.2			+10			
40	9.6	1003.8				+1		
76	7.2	961.9	−4					
0.2652	3.1	881.3				+1.8		
55	2.9	878.3	+2		+12			
0.2700	+0.8	835.8	0					
50	−1.1	792.4					−12	
53	−1.1	790.0	0					
0.2804	−2.5	749.5	0			+3.4		
05	−2.6	748.7			+7			
94	−4.0	686.7	0					
0.2925	−4.3	667.0	−1					
68	−4.6	641.3	0					
0.3023	−4.7	610.9	0					
34	−4.8	605.0		+1.6				
0.3100	−4.8	571.7					−10	
26	−4.8	559.7	−1					
30	−4.8	558.0				+2.9		
31	−4.8	557.5			+8			
32	−4.8	557.1	−1					
0.3303	−4.9	486.1		+1.2				
41	−4.9	472.4			+5	+2.7		
42	−4.9	472.0	0					

[818] Cotton, A., and Lucas, R., *Int. Crit. Tables*, **6**, 425 (1929).

54. WATER: VERDET CONSTANT

Table 187—*(Continued)*

λ	References[a] $10^4\Delta$	$10^4 V_n$	Mi	Ri[b]	Ro $10^4(V-V_n)$	BG	L	v.S
0.3580	−4.4	398.5						−5.1
0.3609	−4.4	390.7					−5	
11	−4.4	390.1		+1.2				
12	−4.4	389.9		+1.3				
52	−4.2	379.8	−1					
55	−4.2	379.1				+2.4		
63	−4.2	377.0	−1					
65	−4.1	376.6			+5			
0.3729	−3.9	361.6						−4.0
0.3886	−3.4	328.0					−3	
0.3907	−3.4	323.8	−2					
62	−3.2	313.5		+1.0				
0.4046	−3.0	297.2					−2	
47	−3.0	298.4	0		+3			
54	−3.0	297.3				+1.6		
78	−2.9	293.2	−1					
0.4199	−2.6	274.3					−4	
0.4307	−2.3	258.9						−2.2
08	−2.3	258.8					−4	
40	−2.2	254.5		+0.7				
41	−2.2	254.4		+0.7				
58	−2.2	252.1			0	+1.3		
59[c]	−2.1	252.1	−1					
0.4400	−2.0	246.9	0					
05	−2.0	246.3					−3	
0.4505	−1.8	234.1	+0.8					
29	−1.8	232.4					−3	
0.4605	−1.6	223.0	+0.5					
78	−1.4	215.4		+0.5				
0.4705	−1.3	212.7	+0.6					
0.4805	−1.1	203.2	+0.7					
61	−1.0	198.2						−3.2
0.4905	−0.9	194.2	0.0					
16	−0.9	193.3	0		+0.1			
21[c]	−0.9	192.8					−4	
58	−0.8	189.8	+0.2					
0.5005	−0.8	185.8	0.0					
0.5105	−0.6	178.1	0.0					
0.5210	−0.5	170.4	−0.1					
70	−0.4	166.2						−1.5
0.5310	−0.4	163.6	+0.1					
0.5410	−0.3	157.1	−0.2					
61[c,d]	−0.3	154.0	+1	+0.1	−0.6			
0.5515	−0.2	150.8	+0.2					
0.5615	−0.2	145.0	+0.1					
0.5715	−0.1	139.7	−0.3			I[a]	I'[a]	
80[c]	−0.1	136.4			−0.9			
0.5815	0	134.7	−0.3					
0.5893[d]	0	130.9		0.1	0.0	0.0		0.0
0.5920	0	129.6	+0.2					
0.6000	0	125.9					0	
20	0	125.1	+0.3					
0.6104	+0.1	121.5			+0.3			
20	0.1	120.8	+0.1					
0.6220	0.1	116.7	+0.4					
0.6320	0.1	112.8	+0.5					

Table 187—(Continued)

λ	References[a]→ $10^4\Delta$	10^4V_n	Mi	Ri[b]	Ro $10^4(V-V_n)$	BG	L	v.S
0.6420	0.2	109.2	+0.3					
0.6530	0.2	105.4	+0.4					
63	0.2	104.3						−0.6
0.6708	0.22	99.6			+0.2			
0.7000	0.26	91.2				−0.2		
65	0.26	89.4		0.0				
0.8000	+0.3	69.0				−1.7	+1	
80	0.3	67.7		0.0				
71	0.4	58.0		0.0				
0.900	0.4	54.3				−3.4		
1.000	0.4	43.8				−3.0	0	
28	0.4	41.4		0.0				
1.100	0.4	33.6				−0.4		
1.200	0.3	30.2				−1.4		
50	0.3	27.8					+1	
56	0.3	27.6		0.0				
1.300	0.2	25.6				+0.7		

[a] References:
- A   Agerer, F., *Sitz. Akad. Wiss. Wien (Abt. IIa)*, **114**, 803-830 (1905).
- BG Bruhat, G., *Jour. de Phys. (7)*, **5**, 152 (1934) ← Bruhat, G., and Guinier, A., *Idem*, **4**, 691-714 (1933) → *Compt. rend.*, **197**, 1028-1030 (1933).
- I   Ingersoll, L. R., *Phys. Rev.*, **23**, 489-497 (1906).
- I' Ingersoll, L. R., *J. Opt. Soc. Amer.*, **6**, 663-681 (1922).
- L   Landau, St., *Physik. Z.*, **9**, 417-431 (1908).
- Mi Miescher, E., *Helv. Phys. Acta*, **3**, 93-133 (1930); **4**, 398-408 (1931).
- Ri Richardson, S. S.[814]
- Ro Roberts, R. W., *Phil. Mag. (7)*, **9**, 361-390 (1930).
- RW Rodger, J. W., and Watson, W.[809]
- SE Stephens, D. J., and Evans, E. J., *Phil. Mag. (7)*, **3**, 546-565 (1927).
- v. S von Schaik, W. C. L., *Arch. Néerl. des Sci. Exact et Nat.*, **17**, 372-390 (1882); **21**, 406-431 (1887).

[b] These values are those defined by Richardson's equation; his observed values are as follows:

λ	0.3034	0.3033	0.3611	0.3962	0.4341	0.4678	0.4958	0.5893
$10^4(V\lambda-V_D)$	+0.7	+2.7	+1.4	+1.3	+1.0	+0.4	+0.3	0.0

[c] R. de Mallemann, P. Gabiano, and F. Suhner [818a] have reported the following values, reduced from 11.5 °C to 20 °C by the compiler:

λ	0.436	0.492	0.546	0.578	μ
10^4V	254.1	194.8	154.2	136.4	

[d] F. G. Slack, R. L. Reeves, and J. A. Peoples, Jr.[812] have reported the following high values for 20 °C: $10^4V = 156$ for $\lambda = 0.5461$, and 131.9 for $\lambda = 0.5893\,\mu$.

$$nV = 0.005606\lambda^2/(\lambda^2 - 0.013253)^2 \qquad (3)$$

(n = index of refraction; λ = wave-length, unit = $1\,\mu$; the first constant in each equation has been so chosen as to make $V_D = 0'.01309$ per cm·gauss.)

Consequently, a simpler expression (4) has been adopted as a norm with which to compare both the observations and the proposed equations

$$V_n - \Delta = \frac{4.2347}{1000\lambda^2}\left(\frac{\lambda^2}{\lambda^2 - 0.012097}\right)^2 \qquad (4)$$

or

[818a] de Mallemann, R., Gabiano, P., and Suhner, F., *Compt. rend.*, **202**, 837-838 (1936)—Gabiano, P., *Jour. de Phys. (7)*, **7**, 84S (1936).

54. WATER: VERDET CONSTANT

$$(V_n - \Delta)^{-0.5} = 15.367\lambda - \frac{0.18589}{\lambda} \tag{5}$$

The values of Δ (see Table 187), varying slowly and continuously with λ, have been so chosen as to make V_n approximately represent the observations of Miescher (Mi of Table 187) and of Ingersoll (I and I' of Table 187). From the value of $(V_n - \Delta)^{-0.5}$ that of $V_n - \Delta$ can be directly obtained by the use of Barlow's Tables of Squares, etc., either by entering

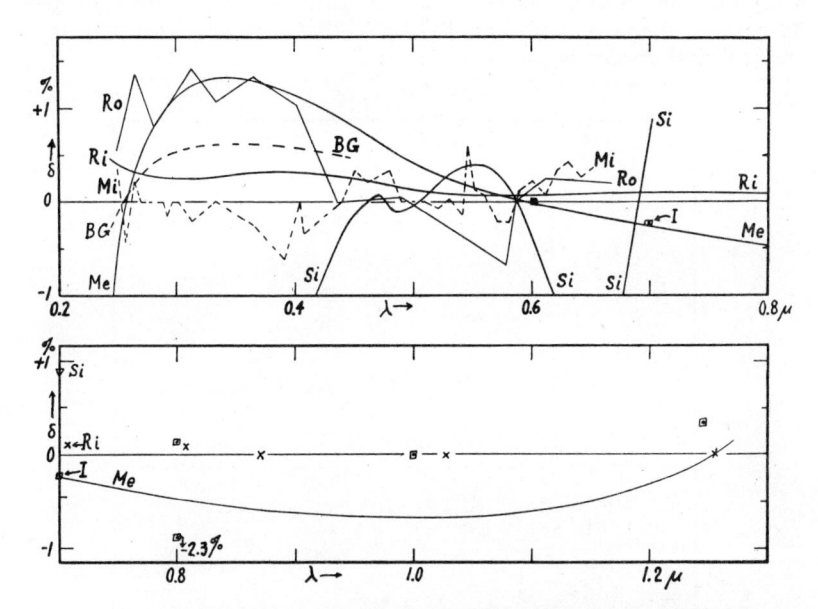

FIGURE 7. Deviations of the Observed and Computed Values of the Verdet Constant for Water from those Defined by Formula (4).

$\delta = (V - V_n)/V_n$, where V_n is the value defined by the arbitrarily chosen norm, formula (4); λ = wave-length. Unit of $\delta = 0.01$; of $\lambda = 1\,\mu = 0.0001$ cm.
 Each set of data, except Richardson's curve (Ri), has been multiplied by such a constant as to make $\delta = 0$ for the D-lines ($\lambda = 0.5893\,\mu$), but in that region the Me-curve has accidentally been drawn a little too high, not enough to be of real significance. In the upper section of the figure ($\lambda < 0.8\,\mu$), the Ri and Me curves represent, respectively, the formulas given by Ri and by Me; the values given by Si are said to have been read by him from a smoothing curve, and define the curve here given; all the BG-values lie within 0.1 per cent of the curve so marked. The successive individual determinations by Ro, also those by Mi, are connected by straight lines, and those by I are indicated by squares. In the lower section, the Me-curve is continued, the computed Ri-values are indicated by crosses, the I-values by squares, and the Si-value by an inverted triangle. One of the I-values near $\lambda = 0.8$ belongs far below the boundary of the figure, at -2.3 per cent, as indicated.

References:
BG   Bruhat, G., *Jour. de Phys.* (7), **5**, 152 (1934) ← Bruhat and Guiner, *Idem*, **4**, 691-714 (1933).
I   Ingersoll, L. R., *Phys. Rev.*, **23**, 489-497 (1906); *J. Opt. Soc. Amer.*, **6**, 663-681 (1922).
Me   Meyer, U., *Ann. d. Physik (4)*, **30**, 607-630 (1909).
Mi   Miescher, E., *Helv. Phys. Acta*, **3**, 93-133 (1930); **4**, 398-408 (1931).
Ri   Richardson, S. S., *Phil. Mag.* (6), **31**, 232-256 (1916).
Ro   Roberts, R. W., *Idem* (7), **9**, 361-390 (1930).
Si   Siertsema, L. H., *Arch. Néerl. des Sci.* (2), **6**, 825-833 (1901).

the column of $\sqrt{n}$ and taking the corresponding number in the column of $1/n$, or by entering the column of $1/n$ and taking the corresponding number in the column of n^2.

The various deviations from V_n are shown in Figs. 7 and 8, and those of the better series of observations are given in Table 187 in such a way that the individual observations in those series may be recovered if desired. It will be noticed from the graphs that Joubin's equation, though commonly included in compilations of data, is entirely unsatisfactory; whether the same is true of Meyer's is not clear, as that does approximately represent Roberts' observations.

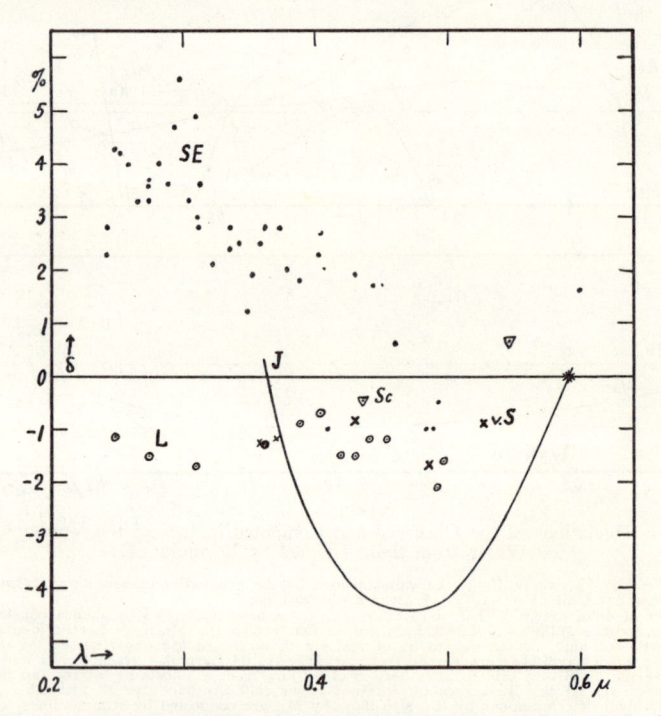

FIGURE 8. Deviations of Other Observed and Computed Values of the Verdet Constant of Water from those Defined by Formula (4).

$\delta = (V - V_n)/V_n$, where V_n is the value defined by the arbitrarily chosen norm, formula (4); λ = wave-length. Unit of $\delta = 0.01$; of $\lambda = 1\,\mu = 0.0001$ cm.

This differs from Fig. 7 both in scale and in data. The computed J-curve is obviously unsatisfactory. Its constant was so chosen as to make the curve pass through the star ($\delta = 0$ for the D-lines).

References:
    J   Joubin, P., *Ann. chim. phys.* (6), **16**, 78-144 (1889).
    L   (circles) St. Landau, *Physik. Z.*, **9**, 417-431 (1908).
    Sc   (inverted triangles) Schwers, F., *Bull. Acad. Roy. de Belg.*, **1912**, 719-752 (1912).
    SE   (dots) Stephens, D. J., and Evans, E. J., *Phil. Mag.* (7), **3**, 546-565 (1927).
    vS   (crosses) van Schaik, W. C. L., *Arch. Néerl. des Sci.*, **17**, 372-390 (1882); **21**, 406-431 (1887).

55. Magnetic Birefringence of Water

In a magnetic field, water is negatively birefringent, its index of refraction (n_p) for light in which the electric vector is parallel to the field is less than that (n_t) for light in which that vector is transverse to the field. If λ = wave-length of the light in a vacuum, and H is the strength of the magnetic field, then the coefficient (C_m) of magnetic birefringence (sometimes called the Cotton-Mouton constant) is defined by the equation $C_m = (n_p - n_t)/\lambda H^2$.

For water, C_m is very small, of the order of one-thousandth of the value for nitrobenzene. M. A. Haque [819] found $10^{14} C_m = -0.3_9$ cm^{-1} gauss^{-2}, and S. W. Chinchalkar [820] and H. A. Boorse [821] each found -0.3_7. An earlier, less accurate, measurement by M. Ramanadham [822] gave -1.1, nearly 3 times the values found by the others. A. Cotton and T. Belling [823] have reported -0.14, about $\frac{1}{3}$ of the value found by Haque.

IIC. ICE

56. Foreword

Of the several treatises dealing with ice, the one that seems to be by far the most comprehensive that has come to the compiler's attention is that by A. B. Dobrowlski: "Historja Naturalna Lodu" ("The Natural History of Ice"). The manuscript was completed in 1916, but the volume was not published until 1923. It is printed in Polish (a language not read by the compiler) and contains 940 pages, including a French translation of the introduction and table of contents, and an author index of over 1000 names. Its purpose and scope are thus defined: "*L'Histoire Naturelle de la Glace* est un essai de synthèse des recherches faites, dans la nature et dans le laboratoire, sur la glace de tout aspect et de toute origine. C'est une sorte d'index de tous les problèmes relatifs à ce corps si important, et si peu connu encore, avec un exposé de l'histoire de chacun de ces problèmes, des résultats acquis, des questions litigieuses et des lacunes."

57. Types of Ice

For crystallographic forms of ice, see Section 59; for x-ray studies of ice, see Sections 60 and 74.

Besides ordinary ice, more particularly designated as ice-I, and vitreous ice obtainable at low temperatures, six other distinct varieties of ice, each having a definite region of stability, are known [1]; two forms of ice-I have been reported [2]; and ice that is denser than water, though formed at a

---

[819] Haque, M. A., *Compt. rend.*, **190**, 789-790 (1930).
[820] Chinchalkar, S. W., *Indian J. Phys.*, **6**, 165-179 (1931).
[821] Boorse, H. A., *Phys. Rev.* (2), **46**, 187-195 (1934).
[822] Ramanadham, M., *Indian J. Phys.*, **4**, 15-38 (1929).
[823] Cotton, A., and Belling, T., *Compt. rend.*, **198**, 1889-1893 (1934).

[1] Bridgman, P. W., *Proc. Amer. Acad. Arts Sci.*, **47**, 439-558 (1912); *J. Chem'l Phys.*, **3**, 597-605 (1935); **5**, 964-966 **(1937)**.
[2] Seljakov, N. J., *Compt. rend. Acad. Sci. URSS*, **10**, 293-294 (1936); **14**, 181-186 (1937).

pressure of less than 1 atm, has been reported twice,[3] but in neither case could its formation be repeated. Shaw states that it is denser than water; and Cox says it "sank slowly to the bottom and remained there with, perhaps, one third of an inch of clear water above it." (A few rather casual observations by the compiler have suggested that these conclusions may rest upon an illusion. The capillary pull of the gas-liquid surface between the walls of the bulb and the ice will depress the ice, and if the surface of the ice is concave, the concavity may become filled with water, producing a striking impression that the ice is fully submerged. The requirements being rather exacting, their accidental fulfillment will not be frequent.) Evidence for the existence of several unstable types of ice was presented by G. Tammann, but not generally accepted; and nothing has been heard of those unstable types for many years.[4] In particular, his claims for the one he called ice-IV led to a discussion with Bridgman,[4] and resulted in Bridgman's leaving that designation open when he published his 1912 paper; but now he has adopted it as the designation of an ice recently discovered.[5] Bridgman's ice-IV should not be confused with Tammann's supposed ice of the same designation.

Ignoring for the present the vitreous and unstable types, and the dense ice reported from Canada, no type except the familiar variety, specifically denoted as ice-I, can exist under pressures much less than 2000 atm unless the temperature is very low, but at low temperatures the transformation of both ice-II and ice-III to ice-I proceeds so slowly, even at atmospheric pressure, that Tammann[6] has succeeded in removing them from the pressure chamber and examining them. He found ice-III to be a colorless, pellucid aggregate of coarse crystallites, which slowly swelled and broke up into a coarse white powder. In contact with warm objects it acquired a porcelain-like appearance.[7]

It has been said that vitreous ice is formed when small drops of water are quickly chilled to a low temperature, say to $-12\,°C$ or lower.[8] Beilby stated that this "ice" is perfectly transparent, shows under a microscope no evidence of crystalline structure, but at once crystallizes throughout when mechanically strained by a light pressure with a polished steel burnisher. Hawkes[8] regarded it as merely supercooled water, but that conflicts with the observations of H. C. Sorby,[9] L. Dufour,[10] and others, who have observed that water remains fluid even at temperatures ranging from $-12\,°C$ to $-20\,°C$. Sorby is especially definite in his statement of its fluidity. Using volumes of several cubic centimeters, the compiler has noticed that water at $-20\,°C$ appears to the eye to be as fluid as it is at

---

[3] Cox, J., *Trans. Roy. Soc. Canada, Sect. III (2)*, **10**, 3-4 (1904); Shaw, A. N., *Idem (3)*, **18**, 187-189 (1924).

[4] See Tammann, G., *Ann. d. Physik (4)*, **2**, 1-31 (1900); *Z. physik. Chem.*, **84**, 257-292 (1913); **88**, 57-62 (1914); Bridgman, P. W., *Proc. Amer. Acad. Arts Sci.*, **47**, 439-558 (1912); *Z. physik. Chem.*, **86**, 513-524 (1913); **89**, 252-253 (1915).

[5] Bridgman, P. W., *J. Chem'l Phys.*, **3**, 597-605 (1935).

[6] Tammann, G., "The States of Aggregation," New York, 1925; *Z. anorg. Chem.*, **63**, 285-305 (1909). Cf. Bridgman, P. W., *J. Franklin Inst.*, **177**, 315-332 (1914).

room temperature. Although a vitreous solid may be an undercooled liquid of great viscosity, there seems to be no valid reason for assuming, as is frequently done, that every vitreous solid is of that nature; unless, of course, the term liquid is so defined as to cover everything that is neither gaseous nor crystalline. F. Simon[11] has presented reasons and experimental data in support of the idea that there are in reality two types of vitreous "solid." One is a true glass and the other is a supercooled liquid, the latter being in internal thermodynamic equilibrium, and the former not. Nevertheless, in view of the work now to be mentioned, it seems probable that Beilby's observations are in some way erroneous. The work should be repeated.

Using larger volumes of pure water, the late E. W. Washburn of the National Bureau of Standards failed to obtain vitreous ice even when the water was quickly chilled with liquid air, but when the viscosity was increased by the addition of a little sugar (about 1 per cent) vitreous ice was obtained[12]. And E. F. Burton and W. F. Oliver[13] reported that when water-vapor condenses on copper at $-110\,°C$, or lower, the resulting ice is vitreous, and as the temperature is then raised, say to $-50\,°C$, the ice gradually crystallizes. This same phenomenon—vapor being condensed to an amorphous solid on a cold surface, and that solid becoming crystalline when the temperature is increased—had been previously reported by L. R. Ingersoll and S. S. DeVinney[14] for the case of nickel sputtered in hydrogen. The ice formed by condensation at $-80\,°C$ has the normal crystalline arrangement; as the temperature of formation is lowered, the crystal structure becomes less regular.

Of the known varieties of crystalline ice, all except the familiar one (ice-I) are denser than water under the same conditions of temperature and pressure. One consequence of this is that the pressure that is exerted by the freezing of water in a confined space can under no circumstance greatly exceed 2000 kg*/cm² (say 30,000 lb*/sq. in.), because the bulky ice-I cannot exist under such pressures. See phase diagram, Section 93, or H. T. Barnes.[15]

In what follows, we shall confine our attention, unless the contrary is clearly indicated, to the familiar variety of ice (ice-I), that which melts at $0\,°C$ when under a pressure of 1 atm.

[7] Cf. Dewar, J., *Chem. News,* **91**, 216-219 (1905).
[8] Beilby, G., "Aggregation and Flow of Solids," p. 195+, London, 1921; Hawkes, L., *Nature,* **123**, 244 (1929).
[9] Sorby, H. C., *Phil. Mag. (4),* **18**, 105-108 (1859).
[10] Dufour, L., *Ann. d. Physik (Pogg.),* **114**, 530-554 (1861).
[11] Simon, F., *Z. anorg. allgem. Chem.,* **203**, 219-227 (1932); *Trans. Faraday Soc.,* **33**, 65-73 (1937).
[12] Washburn, E. W., Oral communication, 1933. Work not published.
[13] Burton, E. F., and Oliver, W. F., *Proc. Roy. Soc. (London) (A),* **153**, 166-172 (1935); →*Nature,* **135**, 505-506 (L) (1935).
[14] Ingersoll, L. R., and DeVinney, S. S., *Phys. Rev. (2),* **26**, 86-91 (1925).
[15] Barnes, H. T., "Ice Engineering," p. 91, Montreal, Renouf Publishing Co., 1928.

58. Appearance of Ice-I

Ordinary ice, Ice-I, is crystalline, and has a conchoidal fracture and a vitreous luster. In large masses, it is vividly blue, owing to the scattering of light by its large molecules.[15, pp. 8, 9, 18] It has long been held that water contains ice in solution, the amount increasing as the temperature decreases (see p. 164). This causes the color of water to change as the temperature falls. Barnes says: "It is a remarkable sight in winter to watch the varying shades of the river water as the temperature changes. Just at the freezing point the color changes rapidly and old river men can tell the approach of the ice forming period by the color."[15, p. 10]

59. Forms and Formation of Ice

All interested in this subject should read the very interesting and beautifully illustrated publications by W. A. Bentley[16] and by W. A. Bentley and W. J. Humphreys.[17] They are much broader than the titles indicate, most of the topics considered in this section being discussed.

The information contained in this section is almost entirely descriptive, and has been arranged under the following heads:

Crystallographic structure
Structure of ice
Internal melting
Flowers of ice
Formation of frazil, or needle, ice
Formation of an ice-sheet
Growth and orientation of crystals
Recrystallization
Regelation
Purity

Production of homogeneous ice
Monocrystals
Freezing of supercooled water
Icicles
Hail
Snow and frost
Glaciers
Sea-ice
Icebergs (see Glaciers, Sea-ice)

Crystallographic Structure.

(For ratio of axes and for fine-structure, as revealed by x-rays, see Section 60.)

X-ray examination indicates that ice-II is characterized by a side-centered orthorhombic cell containing 8 molecules,[18] and ice-III by a body-centered orthorhombic cell containing 16 molecules and having $a:b:c = 1.73:1:1.22$.[19] Nothing is known of the crystallographic structure of any of the other ices except ice-I, to which we now turn.

The published data relative to the crystallographic structure of ice-I are confused by changing nomenclature, and are otherwise conflicting. The

---

[16] Bentley, W. A., *Monthly Weather Rev.*, **29**, 212-214 (1901); **35**, 348-352, 397-403, 439-444, 512-516, 584-585 (1907).

[17] Bentley, W. A., and Humphreys, W. J., "Snow Crystals," New York, McGraw-Hill Book Co., 1931.

[18] McFarlan, R. L., *J. Chem'l Phys.*, **4**, 60-64 (1936) → *Phys. Rev. (2)*, **49**, 199 (A) (1936).

[19] McFarlan, R. L., *Idem*, **4**, 253-259 (1936) → *Idem*, **49**, 644 (A) (1936).

59. ICE: FORMS AND FORMATION

subject is further complicated by recent observations indicating that ice-I can exist in either of two forms: α-ice, which is hexagonal and appears when water near 0 °C freezes, and β-ice, which is rhombohedral and appears when water that is supercooled by at least a few degrees freezes.[20]

From a consideration of all observations prior to 1906, P. H. Groth [21] thought it probable that ice-I belongs in the ditrigonal-pyramidal class of the trigonal system; A. E. H. Tutton [22] thought that more recent work, such as that of F. Rinne,[23] shows that it belongs in the hexagonal bipyramidal class of the hexagonal system. The subject has been reviewed still more recently by A. B. Dobrowolski,[24] who has concluded from his study of thousands of ice crystals that the symmetry of ice is that of the *ditrigonal-pyramidal* class of the trigonal system.

W. Altberg and W. Troschin[25] have described some unusual forms of ice found in the ice cave near Kungur in the Ural mountains; and G. Tammann and K. L. Dreyer [26] have given a popular account of the freezing of water and of some of the peculiarities of artificial ice.

E. S. Dana [27] states that ice crystallizes in the hexagonal system. This is accepted by H. T. Barnes,[28] who states that the crystals are probably hemimorphic, that the crystal faces are rarely distinct, and that the crystals are hard to measure.

L. J. Spencer [29] assigned ice to the holosymmetric class of the rhombohedral division of the hexagonal system. This probably harmonizes the observations of the early observers, who variously reported that the crystals were hexagonal and rhombohedral; in some cases, the latter term seems to have been used as an equivalent of the former. That the primary form was the rhombohedron in the cases reported by H. Abich,[30] by Sir David Brewster,[31] and by E. D. Clarke [32] seems beyond question; but the assertion of F. Leydolt [33] that in his extended study of ice from various sources, he had found only rhombohedric forms, possibly means no more than that they belonged to the hexagonal system, without distinction between the rhombohedral and the hexagonal classes of that system. Likewise the many

---

[20] Seljakov, N. J., *Compt. rend. Acad. Sci. URSS*, **10**, 293-294 (1936); **11**, 227 (1936); **14**, 181-186 (1937).
[21] Groth, P. H., "Chemische Krystallographie," Vol. **1**, p. 66 (1906).
[22] Tutton, A. E. H., "Crystallography and Practical Crystal Measurement," Vol. **1**, p. 543 (1922).
[23] Rinne, F., *Ber. Sächs. Ges. Wiss. (Math.-Phys.)*, **69**, 57-62 (1917).
[24] Dobrowolski, A. B., *Bull. Soc. Fr. Mineral.*, **56**, 335-346 (1933).
[25] Altberg, W., and Troschin, W., *Naturwissenschaften*, **19**, 162-164 (1931).
[26] Tammann, G., and Dreyer, K. L., *Idem*, **22**, 613-614 (1934).
[27] Dana, E. S., "A Textbook of Mineralogy," 3rd ed., p. 411, revised by W. E. Ford, New York, John Wiley & Sons, 1922.
[28] Barnes, H. T., "Ice Formation," p. 74, New York, John Wiley & Sons, 1906; "Ice Engineering," p. 18.
[29] Spencer, L. J., "Encyclopedia Britannica," 11th ed., vol. **7**, p. 581, 1910; 14th ed., vol. **6**, p. 819 (1929).
[30] Abich, H., *Ann. d. Physik (Pogg.)*, **146**, 475-482 (1872).
[31] Brewster, Sir David, *Phil. Mag. (3)*, **4**, 245-246 (1834).
[32] Clarke, E. D., *Trans. Cambr. Phil. Soc.*, **1**, 209-215 (1822).
[33] Leydolt, F., *Sitz. Akad. Wiss. Wien (Math.-nat.)*, **7**, 477-487 (1851).

reports of hexagonal forms, such as those by W. A. Bentley,[16] T. H. Holland,[34] P. A. Secchi,[35] J. Smithson,[36] and J. Tyndall,[37] are probably to be interpreted as indicating that the crystals belonged to the hexagonal system, without indicating to which of the two divisions of that system they should be assigned.

Although Leydolt found ice crystals of only a single system and doubted the existence of other types, there are reports indicating that ice-I may, under conditions not yet defined, crystallize in the cubic system. For example, A. E. Nordenskjöld [38] has observed in the frost coating a windowpane, the outer air being at -8 to $-12\,°C$, small rectangular forms which he decided could not possibly belong to the hexagonal system, but probably belonged to the rhombic, or possibly to the cubic. Similar forms have been observed by H. P. Barendrecht [39] to separate at low temperatures from solutions of water in acetaldehyde and in certain alcohols, including ethyl alcohol. By polariscopic observations, he found that these forms belonged to the cubic system. It was not practical to separate them from the viscous mother-liquor, but from the fact that the same type of crystal was obtained from all these solutions, he concluded that the crystals were ice and not a hydrate. Similar observations have been reported by F. Wallerant [40] who interprets them as indicating that the cubic crystals pertain to a type of ice that is stable only under high pressure. He remarked that in an alcoholic solution the cubic crystals are stable as long as the water content does not exceed 55% by weight; but if it does exceed that value, the cubic crystals, initially formed by supercooling the solution, transform into rhombohedric crystals of ordinary ice enclosing numerous (une infinité) small isotropic crystals. P. Tschirwinsky [41] has questioned the validity of the conclusions of Barendrecht and of Wallerant; he believed that the cubic crystals they obtained were those of the hydrates, such as $C_2H_5(OH) + 3H_2O$ of which the existence, he said, had been shown by Mendelejeff in 1865. It contains 54.1 per cent of water, which essentially coincides with the 55 per cent limit found by Wallerant for the stability of the crystals. R. Hartmann [42] also has reported a rectangular type of crystal skeleton obtained from aqueous solutions; but from the observation that the "freezing point" of supercooled water inoculated with such skeletons is the same as when it is inoculated with a crystal of the ordinary hexagonal type, he concluded that the skeletons actually belong to the hexagonal system.

[34] Holland, T. H., *Nature,* **39,** 295 (1889).
[35] Secchi, P. A., *Bull. meteor. Osserv. Coll. Romano,* **15,** 73-74 (1876).
[36] Smithson, J., *Ann. Philos. (N. S.),* **5,** 340 (1823).
[37] Tyndall, J., "The Forms of Water in Clouds and Rivers, Ice and Glaciers," New York, D. Appleton & Co., 1872.
[38] Nordenskjöld, A. E., *Ann. d. Physik (Pogg.),* **114,** 612-627 (1861).
[39] Barendrecht, H. P., *Z. physik. Chem.,* **20,** 234-241 (1896); *Z. anorg. Chem.,* **11,** 454-455 (1896).
[40] Wallerant, F., *Bull. Soc. Franç., Minéral.,* **31,** 217-218 (1908).
[41] Tschirwinsky, P., *Ann. Geol. et Min. Russie (French résumé),* **14,** 280-282 (1912) → *N. Jahrb. Min., Geol., Paläon.,* **1914₂,** 349 (1914).
[42] Hartmann, R., *Z. anorg. Chem.,* **88,** 128-132 (1914).

The subject has been reviewed by O. Mügge [43] and briefly summarized by W. H. Barnes [43a] who states that a very complete review, containing more than 100 citations, may be found in A. B. Dobrowolski's "Historja Naturalna Lodu."

J. M. Adams [44] has stated that the ice-crystal is asymmetric with respect to its basal (0001) plane, crystals twinned on that plane being separable into two kinds: (a) Those that may develop pits at the ends of the c-axis, and (b) those that may develop a cavity at the middle of that axis. And J. Smithson [36] has stated that when hail is sufficiently regular for satisfactory measurement, it always consists of two hexagonal pyramids joined base to base. *"One of the pyramids is truncated,"* and "The two pyramids appeared to form by their junction an angle of about 80 degrees."

From his study of the way ice yields to stresses of various kinds, J. C. McConnel [45] found that "a crystal behaves as if it were built up of an infinite number of indefinitely thin sheets of paper fastened together with some viscous substance which allows them to slide over each other with considerable difficulty; the sheets are perfectly inextensible and perfectly flexible. Initially they are plane and perpendicular to the optic axis; and when by the sliding motion they become bent, the optic axis at any point is still normal to the sheet at that point." It will be noticed that these sheets are parallel to the planes of Tyndall's flowers of ice (p. 405), and to those in which he observed melting to occur when ice is subjected to linear compression along the optic axis (p. 431). Quincke stated that "the planes of easiest cleavage in natural ice crystals (laminated structure, displacement without bending) are due to invisible layers of liquid salt solution which are embedded in the crystals, normal to the optic axis, or often in other positions." [46] Whether the sheets of McConnel's are to be identified with certain of these layers of Quincke's is not clear.

Structure of Ice in Bulk.

However uniform the ice may be, melting begins at the boundaries between the individual ice crystals, and as it proceeds, the crystals become more and more separated one from another, and the ice becomes "rotten." There is between the crystals a material with a lower melting point than the crystals themselves, and this material, when molten, dissolves the surface of the ice-crystals in contact with it, even when its temperature is below the normal melting point of the crystal. This material surrounds each crystal, enclosing it in a cell. These correspond to the foam-cells that are postulated in the theory of solidification that was developed by G. Quincke in a series of papers published some 30 years ago, and that was applied by him to explain the formation of ice and of glacier grains.[47]

[43] Mügge, O., *Centralbl. Min., Geol., Paläon.*, **1918**, 137-141 (1918).
[43a] Barnes, W. H., *Proc. Roy. Soc. (London) (A)*, **125**, 670-693 (1929).
[44] Adams, J. M., *Proc. Roy. Soc. (London) (A)*, **128**, 588-591 (1930) → *Phys. Rev. (2)*, **36**, 788 (A) (1930).
[45] McConnel, J. C., *Proc. Roy. Soc. (London)*, **48**, 259-260 (1890); **49**, 323-343 (1891).
[46] Quincke, G., Paragraph 35 in *Nature*, **72**, 543-545 (1905); *Proc. Roy. Soc. (London) (A)*, **76**, 431-439 (1905).

A brief outline of Quincke's hypothesis is given by H. T. Barnes.[48] On this hypothesis, the liquid contains at least two species of molecules, one vastly in excess of the other. The species present in smaller amount he regards as salt, but that is not at all necessary. Under the action of the intermolecular forces, there will, in general, be a segregation of each species; and those present in small amounts may form a net-work throughout the liquid, enclosing the more abundant species of molecules in numerous adjacent cells, resembling cells of foam. In the case of water, the contents of the cells (the purer water) freezes first; then the cell-walls freeze. Each individual crystal, each glacier grain, corresponds to a single cell. As the temperature rises, the cell-walls melt first. The resulting liquid (an aqueous solution) bathes the surfaces of the enclosed crystals, lowering their melting point, and so causes those surfaces to melt at a temperature below the normal melting point of the crystal itself. Thus, the crystals become separated more and more from one another.

Quincke has advanced the very extreme suggestion that "ice is a liquid jelly, with foam-walls of concentrated oily salt solution, which enclose foam-cells containing viscous, doubly refracting, pure or nearly pure water. The further the temperature falls below 0°, the greater is the viscosity of both liquids—in the walls and in the interior of the foam-cells—and the less the plasticity of the ice. Ice crystals at temperatures below 0° consist of a doubly refracting viscous liquid, and are intermediate between the soft crystals of serum albumen and ordinary crystals of quartz, felspar, etc."[49]

Whether these cell walls (Quincke) generally exist in water before freezing begins may be open to question. But as soon as freezing begins they will certainly begin to form. When each small volume of water crystallizes, it rejects the impurities it originally contained. Each minute crystal is surrounded by a layer of water more impure than that from which it was formed. As the crystal grows, this layer is continuously pushed out, becoming ever more impure, but always hugging the crystal, until it meets a similar layer surrounding a neighboring crystal. Further growth in that direction means a thinning of the combined layer and an increase in its concentration. Thus each crystal becomes enclosed in a material having a lower melting-point than itself. Diffusion will, of course, tend to equalize at each instant the concentration of the impurities throughout the volume that is still liquid; and adsorption may tend to retain at the crystal surface more of one type of impurity than of another.

The actual existence of an intercrystallic material differing in properties from the crystals is conclusively shown by the fact that melting always begins at the boundaries between the crystals. Furthermore, E. K. Plyler [50]

---

[47] Quincke, G., *Ann. d. Physik (4)*, **18**, 1-80 (1905); summary of conclusions is given in *Proc. Roy. Soc. London (A)*, **76**, 431-439 (1905) and in *Nature*, **72**, 543-545 (1905).
[48] Barnes, H. T., "Ice Formation," p. 79-82.
[49] Quincke, G., Paragraphs 3, 4, 36 of *Nature*, **72**, 543-545 (1905), and *Proc. Roy. Soc. (London) (A)*, **76**, 431-439 (1905).
[50] Plyler, E. K., *J. Elisha Mitchell Soc.*, **41**, 18 (1925).

observed that infrared radiation is much more strongly absorbed in the region between the crystals than in the crystals themselves; in some cases the boundary region was found to be less than $8\,\mu$ (0.0008 cm) thick. Ice cleaves most readily between crystals, that is, along the cell walls. The hypothesis of such cell walls accords with certain observation reported by G. Beilby,[51] and with the observation that "the more slowly artificial ice is frozen, and the less salt it contains, the more transparent it is, and rigid, and the more difficult to split with a knife." [15, p. 19; 47, p. 79]

The purer the water, the smaller is the amount of intercrystallic material, but that one is ever justified in considering that material as negligible seems improbable, especially when one realizes that the action of this material in separating the crystals is a surface phenomenon, and recalls how minute an amount of an impurity may suffice to produce a profound change in the properties of a surface.

Quincke [47, pp. 78, 79] states that by repeated fractional freezing, with a discarding of the unfrozen fraction, the ice becomes increasingly purer and composed of ever larger grains, but he has never succeeded in obtaining ice without grains; *i.e.*, with no intercrystallic material.

J. N. Finlayson [52] writes: "The crystals first formed continually enlarge and the interstices between them become filled with smaller crystals not so regularly oriented. The crystals, however, do not completely unite and there is a definite cleavage plane formed between them."

J. Y. Buchanan [53] has shown that the crystals of ice formed from a dilute salt solution are themselves free of salt, but that some of the solution is retained in the interstices between the crystals; at the same time he called attention to the pronounced effect this intercrystallic material may have upon the physical properties of the ice, even when the amount of salt present is excessively minute. The presence of such intercrystallic material should always be remembered when ice in bulk is being studied. In many cases it profoundly affects the results obtained.

The slow disintegration of ice when exposed to light and air near 0 °C has been described by E. Schmid.[54] It is frequently stated that such disintegration does not occur when the ice is completely submerged; but M. Faraday [55] observed that ice that was completely submerged in water contained in a vessel surrounded by an ice-jacket, and thus kept so near 0 °C that a cubic inch of ice was not dissolved in a week, became after several days "so dissected at the surfaces as to develop the mechanical composition of the masses, and to show that they were composed of parallel layers about a tenth of an inch thick, of greater and lesser fusibility, which layers appear, from other modes of examination, to have been horizontal in

---

[51] Beilby, G., "Aggregation and flow of solids," pp. 140-143, 1921.
[52] Finlayson, J. N., *Canadian Engineer*, **53**, 101-103 (1927).
[53] Buchanan, J. Y., *Proc. Roy. Soc. Edinburgh*, **14**, 129-149 (1887) → *Nature*, **35**, 608-611 (1887); **36**, 9-12 (1887); *Proc. Roy. Inst. Grt. Brit.*, **19**, 243-276 (1908).
[54] Schmid, E., *Ann. d. Physik (Pogg.)*, **55**, 472-476 (1842).
[55] Faraday, M., *Proc. Roy. Soc. (London)*, **10**, 440-450 (1860).

the ice whilst in the act of formation." However, L. Hawkes [56] has stated that F. J. Hugi [57] observed that "in cloudy weather a lump of ice melts as a whole, preserving a smooth outer surface, but in the sun's rays melting takes place at the intergranular boundaries." Such apparently incompatible observations are probably to be explained by the vast difference in the rates of melting.

A. Erman [58] has reported many interesting and unusual observations made on ice in Siberia.

Internal Melting.

When a block of ice that is above water is exposed to light, small cavities partly filled with water may be seen to form throughout the body of the ice, provided that the temperature of the ice is not too low. The most noted and beautiful of these are Tyndall's "flowers of ice," which will be considered in the next section. At the same time, water will be formed in cavities that previously had been dry. As the melting proceeds, the crystals become separated one from another, and each crystal becomes split into laminas perpendicular to the optic axis.[59] In speaking of the splitting of glacier grains into such laminas, Buchanan says (p. 263) : "It is only the grains that are exposed to the sky, and above water, that are so analysed; and prolonged exposure of this kind reduces a grain to the last stage of dilapidation. The grains beneath the surface, *whether of ice or water,* are almost completely unattacked." (Cf. Plyler's observations on the absorption of the infrared, p. 403.)

In speaking of melting in the interior of the blocks into which the ice becomes broken up, E. Schmid [54] said that small bubbles form first, and then thread-like cavities grow out from them. Could he have been speaking of what Tyndall later called flowers of ice?

If a cavity contains an inclusion, such as soot, which is a good absorber of radiant energy, the melting is easily understood. The inclusion abstracts energy from the radiation, becomes heated, and melts the ice. The same explanation was proposed for the melting when the only inclusion is air; but J. Tyndall [60] showed that the absorptivity of air is entirely too small for it to act in that way.

Nevertheless, internal melting occurs; and "proves that the interior portions of a mass of ice may be melted by radiant heat which has traversed other portions of the mass without melting them."

Tyndall suggested that the localization of internal melting at air-bubbles depends upon the existence of the free surface surrounding the bubble, that it is a surface phenomenon (see below) ; G. Quincke [47] attributed all inter-

[56] Hawkes, L., *Geol. Mag.,* **67**, 111-123 (1930).
[57] Hugi, F. J., *Edinb. New. Phil. J.,* **10**, 337-338 (1831).
[58] Erman, A., *Phil. Mag. (4),* **17**, 405-413 (1859).
[59] Buchanan, J. Y., *Proc. Roy. Inst. Grt. Brit.,* **19**, 243-276 (1908).
[60] Tyndall, J., *Proc. Roy. Soc. London,* **9**, 76-80 (1858) → *Ann. d. Physik (Pogg.),* **103**, 157-162 (1858).

nal melting to the presence of small amounts of impurities, which accords with Buchanan's conclusions [53]; and the Thomson brothers [61] were of the opinion that internal stresses were of prime importance. All may contribute, their relative importance varying with the conditions, but in very many cases the stresses are surely the most important. But whatever may be their practical value, Tyndall's views are interesting and suggestive. They should be compared with those of Faraday regarding regelation (p. 412).

From his carefully described observations on the gradual liquefaction of masses of ice by the formation and growth of drops of water within them, Tyndall inferred that the melting temperature differs from point to point, oscillating about its normal value. He wrote: "Through weakness of crystalline structure, or some other cause, some portions of a mass of ice melt at a temperature slightly under 32 °F., while others of stronger texture require a temperature slightly over 32° to liquefy them. The consequence is, that such a mass, raised to the temperature of 32°, will have some of its parts liquid and some solid." These variations in the melting point are attributed by W. Thomson (Lord Kelvin) [61, p. 141+] solely to variations in the stress. That some portions melt below 0 °C is certain, and accords with the conclusions of Quincke [47, p. 17] and of Buchanan; but whether any portions not subjected to such a tension as exists at the boundaries of a cavity completely filled with the melt derived from the ice that originally filled it can remain solid above 0 °C is another question. Tyndall thought they could, provided they have no free surface.

This he endeavors to explain as follows: "Regarding heat as a mode of motion, the author [Tyndall] shows that the liberty of liquidity is attained by the molecules at the surface of a mass of ice before the molecules at the centre of the mass can attain this liberty. Within the mass each molecule is controlled in its motion by the surrounding molecules. But if a cavity exists at the interior, the molecules surrounding that cavity are in a condition similar to those at the surface; and they are liberated by an amount of motion which has been transmitted through the ice without prejudice to its solidity. The author proves, by actual experiment, that the interior portions of a mass of ice may be liquefied by an amount of heat which has been *conducted* through the exterior portions without melting them."

Flowers of Ice.

J. Tyndall [60] observed that when a beam of light is passed through a block of ice, its path rapidly becomes dotted with numerous points resembling shining bubbles of air. Their appearance is accompanied by a clicking sound. When examined with a lens, these "flowers of ice" are found to consist of a bright central spot surrounded by six coplanar petals composed of water. Neighboring flowers lie in parallel planes which are independent

[61] Thomson, J., *Proc. Roy. Soc. (London)*, **8**, 455-458 (1857); **10**, 152-160 (1859); **11**, 198-204 (1861); **11**, 473-481 (1861); Thomson, W. (Lord Kelvin), *Idem*, **9**, 141-143 (1858); **9**, 209-213 (1858).

of the direction of the beam of light. It has been found that the plane of each flower is perpendicular to the optic axis of its associated crystal, and that the petals are parallel to the secondary axes of the crystal. Hence, they serve as a simple means for determining the orientations of the several individual crystals. Tyndall stated that the flowers are formed in planes parallel to those of freezing, but that some apparent exceptions had been noted. He goes on to say: "In some masses of ice, apparently homogeneous, the flowers were formed on the track of the beam, in planes which were in some cases a quarter of an inch apart."

E. Hagenbach-Bischoff, sometimes referred to as Hagenbach, stated that the flowers are sometimes circular, and that they ordinarily begin as circles, that the plane of a flower is perpendicular to the optic axis, and that its arms are parallel to the secondary axes of the crystal.[62]

G. Quincke [63] has stated that "at the edge of Tyndall's liquefaction figures, while they are in the process of enlarging, or on the bursting of the foam-walls of artificial ice as it melts, one often sees periodic vortex movements. These arise from a periodic capillary spreading out (Ausbreitung) of the salt solution of the foam-walls at the boundary between pure water and air or vacuum." It seems probable that those movements are very closely related to those observed by Tyndall when he compressed ice in the direction of the optic axis (see p. 431).

The clicking that accompanies the formation of the flowers was explained by Tyndall somewhat as follows: As the melting progresses the water formed adheres to the walls and so is subjected to tension, its volume being normally less than that of the ice from which it is formed. As soon as rupture occurs at any point, the tension is at once relieved, and the water immediately contracts, leaving a vacuous cavity. This sudden contraction gives rise to the click that is heard.

J. M. Adams and W. Lewis [66] have observed that when the water in such a cavity is refrozen it "never completely fills the space, but at the center of each pattern, there remains a group of minute cavities precisely bounded by natural faces." These they call "negative crystals." "The question as to the source of the space occupied by the negative crystals remains for the present unanswered." These remarks refer more particularly to cavities that have grown beyond the point at which the flowers of ice are well formed.

L. Hawkes [56] has suggested that the flowers of ice mark the positions of particles of dust included in the ice, the melting resulting from the radiation absorbed by the dust.

For additional information see note.[65]

[62] Hagenbach-Bischoff, E., *Arch. des Sci. Phys. et Nat., Genève (3)*, **23**, 373-390 (1890).
[63] Quincke, G., Paragraph 37 of *Nature*, **72**, 543-545 (1905); *Proc. Roy. Soc. (London) (A)*, **76**, 431-439 (1905).
[65] Hess, H., "Die Gletscher," p. 13, 1904; Barnes, H. T., "Ice Engineering," p. 18, 19; "Ice Formation," p. 77, 78; *Trans. Roy. Soc. Canada III (3)*, **3**, 3-27 (22) (1909); Hagenbach-Bischoff, E.[62]
[66] Adams, J. M., and Lewis, W., *Rev. Sci. Inst. (N. S.)*, **5**, 400-402 (1934).

Formation of Frazil, or Needle Ice.

The fine needle-like ice that is often distributed throughout the volume of rapidly moving water is frequently called "frazil." It causes much trouble in the operation of hydraulic power plants. Its formation has been described in some detail by H. T. Barnes [67] in terms of his colloidal theory.[68] He states that when the growing ice particles can be first seen microscopically they are disk-like, and devoid of crystal form; they flocculate and grow into true crystals. In rapidly moving water the crystals are broken up before they can become large, and are thoroughly mixed with the water, giving it a cloudy appearance. This is the beginning of frazil. It forms throughout the body of the river if the water is supercooled, even if the supercooling does not exceed a few thousandths of a degree. It is adhesive, forms agglomerations, and is carried under the surface ice, forming hanging dams, which may reach down to the river bed. When the temperature of the water is a few thousandths of a degree below 0 °C the frazil aggregates are very strong and tenacious, but when the water is at or above 0 °C they become soft and spongelike. He states that sunlight not only warms the water, but has a direct action on the colloidal ice particles, destroying their "agglomerating properties." It seems probable that the adhesion and agglomeration of the particles of frazil are related to the phenomena described by L. Dufour.[69] For additional information regarding colloidal and frazil ice, see the articles already cited and H. T. Barnes [70] and P. P. von Weimarn and W. Ostwald.[71]

Formation of an Ice Sheet.

(For the freezing of water cooled much below 0 °C, see p. 416.)

The formation of surface ice has been described in detail by H. T. Barnes,[15, pp. 60, 61] H. Hess,[65, p. 11] G. Tammann and K. L. Dreyer,[72] G. Seligman,[73] and others.

As the water of a pond is being cooled it is warmer than the air; consequently currents of warm air rise from the center of the surface, and cold air sweeps in from the sides. This chills the lateral waters, and is itself warmed thereby; thus the center of the surface remains warmer than the edge, and if the air temperature is not very low, the center may remain unfrozen long after the banks are bordered with ice. Until the temperature of the water has dropped to 4 °C, convection keeps it fairly uniform throughout the depth; with further chilling, the colder layers remain at the top. If surface freezing is to occur, either the temperature of the air

---

[67] Barnes, H. T., "Ice Engineering," p. 6+, 108, 109; *Scientific Monthly*, **29**, 289-297 (1929).
[68] See Barnes, H. T., "Colloid Chemistry," J. Alexander, ed., vol. **1**, pp. 435-443, New York, Reinhold Publishing Corp., 1926.
[69] Dufour, L., *Ann. d. Physik (Pogg.)*, **114**, 530-554 (1861).
[70] Barnes, H. T., "Colloid Symposium Monograph," Vol. **3**, 103-111, New York, Reinhold Publishing Corp., 1925; "Ice Formation," 1906.
[71] von Weimarn, P. P., and Ostwald, W., *Koll. Z.*, **6**, 181-192 (1910).
[72] Tammann, G., and Dreyer, K. L., *Naturwissenschaften*, **22**, 613-614 (1934).
[73] Seligman, G., *Proc. Roy. Inst. Grt. Brit.*, **29**, 463-483 (1937) → *Nature*, **139**, 1090-1094 (1937).

must be well below 0 °C, or the loss of heat by direct radiation must be great; otherwise the surface will receive heat from the lower layers more rapidly that it can lose it. When the temperature of the surface water has dropped sufficiently (Barnes says to 0 °C, but it probably drops appreciably below that [72]; crystallization begins at the banks, and needles of ice shoot out over the surface of the water. These branch and broaden until the entire surface of the water along the banks is covered with a thin layer of ice. How far this layer extends toward the center depends upon the existing conditions. If the entire surface is supercooled at the time the crystallization begins, the needles may extend rapidly throughout the entire surface, which will soon be completely covered with ice. Under other conditions the initial layer of ice will not reach the center of the surface, but will gradually grow toward it. When the surface has become iced over, the ice layer gradually thickens as heat passes out by conduction through it. H. T. Barnes [74] has stated that the growth in thickness is by an accumulation on the underside of the sheet of layers of disks of ice, "very much like stacked Chinese coins." ' Some have thought the initiation of freezing at the bank, instead of elsewhere, arises from an increased cooling caused by conduction of heat from the water and through the shore material, but in view of the low conductivity of rock, sand, and clay, it seems that this effect must be exceedingly small as compared with that due to the heat carried away convectively by the air.

These descriptions must be supplemented by important observations made many years ago by F. Klocke.[75] He observed that, if the chilling is severe, the lengths of the needles that shoot out over the surface as the water begins to freeze are parallel to their optic axes; for these needles the optic axis is parallel to the surface of the water. The needles broaden, on one side mainly; this is not an outgrowth of the needle. The optic axis of this newer ice is perpendicular to the surface. The main sheet of ice from the beginning consists of this newer ice, the needles seeming to be extraneous impurities. If the freezing occurs near 0 °C, the optic axes of even the needles are perpendicular to the surface.

A. E. Nordenskjöld [76] reports that, on August 31, 1878, when the sky was clear except near the western horizon, and the temperature of the water near the surface was between +1 and +1.6 °C, and that of the air on the vessel between +1.5 and +1.8 °C, "ice was seen to form on the calm, mirror-bright surface of the sea. This ice consisted partly of needles, partly of a thin sheet. The formation of (this) ice was clearly a sort of hoar-frost phenomenon, caused by radiation of heat."

After the surface of quiet water has become covered with ice, the sheet has been found to grow in thickness (x) in accordance with the empirical

[74] Barnes, H. T., *Scientific Monthly*, **29**, 289-297 (1929).
[75] Klocke, F., *Jahrb. Mineral., Geol.*, **1879**, 272-285 (1879).
[76] Nordenskjöld, A. E., "The Voyage of the *Vega* around Asia and Europe," New York, Macmillan & Co., pp. 317, 318, 1882.

formula $x + x^2/2 = -\dfrac{\tau K t}{LS}$, where τ is the time, t the temperature of the air, K the thermal conductivity of ice = 0.0057 cal/cm²·sec per °C/cm, L the latent heat = 80 cal/g, and S the density of ice = 0.9166 g/cm³.[15, p. 63] Whence we get Table 188. This formula presumably supersedes the one, $x^2 = -K\tau t$, given by P. Vedel,[77] who recommended an experimental determination of K in each particular case.

Table 188.—Rate of Thickening of Ice-Sheet

If the temperature of the air is t °C and that of the water is 0 °C, τ is the time required for the sheet to become x cm thick. The sheet is assumed to be free from snow. (See text.)

t \ x	−5	−10	−20	−30	−40
			τ		
1	64 min	32 min	16 min	11 min	8.0 min
2	2.9 hr	1.4 hr	43 min	29 min	21 min
10	1.79 da	21.4 hr	10.7 hr	7.1 hr	5.4 hr
15	3.80 da	1.90 da	22.8 hr	15.2 hr	11.4 hr
20	6.55 da	3.28 da	1.64 da	26.2 hr	19.7 hr
30	14.29 da	7.15 da	3.57 da	2.38 da	1.79 da
60	55.4 da	27.69 da	13.85 da	9.23 da	6.92 da
90	123.3 da	61.6 da	30.8 da	20.6 da	15.4 da

Growth and Orientation of Crystals.

U. Yoshida and S. Tsuboi [78] report (a) that all directions parallel to the basal plane of the hexagonal crystal of ice are equally suited for growth, which occurs more readily in these directions than in a direction that is perpendicular to that basal plane;* (b) that ice formed on the surface of calm water exposed to cold air during a fine night is usually composed of monocrystals of considerable size with their basal planes nearly parallel to the surface; (c) that ice columns formed in the ground usually take the form of long prisms, each consisting of several smaller prisms of monocrystals about 0.5 mm. in diameter (cf. p. 419); and (d) that the lower end of an icicle usually consists of a slender monocrystal. In all these cases the growth is parallel to the basal plane of the crystal. If a monocrystal of ice only slightly below 0 °C is placed in contact with a drop of water, the water freezes slowly, and the crystallographic axes of the new ice are "entirely the same" as that of the mother crystal (cf. p. 411).

Likewise, F. Leydolt [33] reported, as a conclusion from his extended polariscopic study of ice, that the optic axis of the ice is always perpendicular to the surface of the ice-sheet, whether that has been formed on a river, a pond, or in a small vessel. And still earlier, G. Tammann and

---

* H. D. Megaw [79] has stated that the direction of fastest growth is along the normal to the plane (11$\bar{2}$0).

[77] Vedel, P., *J. Franklin Inst.*, 140, 355-370, 437-455 (1895).
[78] Yoshida, U., and Tsuboi, S., *Mem. coll. sci., Kyoto (A)*, 12, 203-207 (1929).
[79] Megaw, H. D., *Nature*, 134, 900, 901 (L) (1934).

K. L. Dreyer,[72] and G. Seligman[73] had reported likewise; and R. Mallet[80] had announced that if any crystallizable material is "suddenly cooled from a state of fusion or solution, by a plane surface of low temperature, the crystals in forming arrange themselves perpendicularly to the refrigerating plane," in the direction of the flow of heat.

The orientation of the crystals in ice formed on the surface of calm water is generally reported to be as just described.[81] But J. C. McConnel and D. A. Kidd [81a] have observed surface ice in which the crystals were arranged with the optic axis nearly horizontal, and hence with the basal planes nearly perpendicular to the surface. They write (p. 334): "Some of the ice of the St. Moritz lake is built up of vertical columns, from a centimeter downward in diameter, and in length equal to the thickness of the clear ice, *i.e.,* a foot or more. A horizontal section, exposed to the sun for a few minutes, shows the irregular mosaic pattern of the divisions between the columns. The thickness of each column is not perfectly uniform. Sometimes indeed one thins out to a sharp point at the lower end. Each column is a single crystal, and the optic axes are generally nearly horizontal. Some experiments on freezing water in a bath lead us to attribute this curious structure to the first layer of ice having been formed rapidly, in air, for instance, below $-6\,°C$. We found that if the first layer had been formed slowly, and was therefore homogeneous with the axis vertical, a very cold night would only increase the thickness of the ice while maintaining its regularity."

In the usual case of an ice sheet formed on calm water, the optic axes of the crystals are both perpendicular to the refrigerating surface and parallel to the direction of gravity. Whence have arisen two explanations of this orientation. A. Bertin [82] announced that the direction of the optic axis is determined by that of the refrigerating surface, being always perpendicular to that; and in confirmation he reported certain observations in which the refrigerating surface was inclined to the horizontal. His conclusions have been confirmed by F. Klocke.[75] On the other hand, O. Mügge [81] maintained that the direction was determined gravitationally in the case of calm water; the most rapid growth taking place in the basal plane, plates normal to the optic axes are formed, and these will, obviously, float with the optic axes vertical. If there are currents in the water, as will be the case if the refrigerating surface is not horizontal, these currents will affect the orientation of the plates. In such a way he attempts to set aside Bertin's observations.

It seems most probable that Bertin's view is correct, that as the ice crystal is initially formed its optic axis is perpendicular to the refrigerating surface—or more specifically, it is parallel to the temperature gradient—

[80] Mallet, R., *Phil. Mag. (3),* **26,** 586-593 (1845).
[81] See also, Mügge, O., *Neues Jahrb. Mineral., Geol.,* **1895**_{II}, 211-228 (1895); Reusch, E., *Ann. d. Physik (Pogg.),* **121,** 573-578 (1864); Bertin, A., *Ann. de chim. et phys. (3),* **69,** 87-96 (1863); von Engeln, O. D., *Am. J. Sci. (4),* **40,** 449-473 (1915); Finlayson, J. N.[52]
[81a] McConnel, J. C., and Kidd, D. A., *Proc. Roy. Soc. (London),* **44,** 331-367 (1888).
[82] Bertin, A., *Ann. de chim. et de phys. (5),* **13,** 283-288 (1878).

except in so far as mechanical or other disturbances cause it to take some other direction. When the crystal is formed on the surface of calm water, the gravitational effect mentioned by Mügge will cooperate in keeping the axis vertical; currents in the water may force the axis out of line with the temperature gradient.[83] When crystallization begins on the surface of water that has been cooled significantly below 0 °C, it will give rise to temperature gradients in the plane of the surface, and these may account for the fact that the optic axes of the needles that first shoot out over the surface are horizontal, as noted by F. Klocke (p. 408), and that the crystalline structure of the first layer of ice is irregular (Bertin). Whether the same explanation is applicable to the ice, with optic axis horizontal, observed by McConnel and Kidd (p. 410) is not clear.

These unqualified conclusions apply only to the initial crystals formed on the surface of water previously free from ice. If there are ice crystals already present—blocks of ice, snow crystals—then a new effect comes into play: namely, the tendency for the new crystal to take the orientation of the old crystal from which it springs. The actual orientation will depend upon the relative magnitudes of the two tendencies and upon the difference in the orientations to which they would individually give rise. J. Y. Buchanan [59] has described the irregular crystalline structure of surface ice initiated by a floating block of glacier-ice.

In the preceding paragraphs we have considered the direction of the optic axis only, and have seen that under certain conditions, frequently realized, the individual crystals over a considerable expanse of ice are so oriented that their optic axes are parallel, or very nearly so. Turning to the secondary axes we find no such regularity (Mügge [81]). The orientation of neighboring flowers of ice, which indicate the orientation of the secondary axes, may be most varied.

K. R. Koch [84] has suggested that it is very probable that the orientation of the individual crystals may be uniform only in thin layers of ice, and that, in a thick block, it may differ from layer to layer. This seems to refer to the optic axis as well as to the secondary axes.

U. Yoshida and S. Tsuboi [78] state that when the freezing is very rapid, the ice crystals are oriented at random, even when the freezing is induced by a mother crystal.

When frost forms on ice, the optic axis of each little crystal added by the vapor is parallel to that of the ice crystal to which it is attached.[85] In general, the optic axes of adjacent crystals of ice are not parallel, in which case, the frost crystals point in various directions. The ice crystals formed by condensation on highly chilled metal plates are oriented at random (Yoshida and Tsuboi [78]).

C. R. Elford [86] has published photographs of triangular crystals extend-

---

[83] See also, Barnes, H. T., *Nature*, **83**, 276 (1910); von Engeln, O. D., *Am. J. Sci. (4)*, **40**, 449-473 (1915).
[84] Koch, K. R., *Ann. d. Physik (4)*, **45**, 237-258 (1914).
[85] Plyler, E. K., *J. Opt. Soc. Amer.*, **9**, 545-555 (1924).
[86] Elford, C. R., *Monthly Weather Rev.*, **64**, 83 (1936).

ing downward from heavy surface ribs in the ice sheet that formed on a mud-puddle that had dried up during the night.

Recrystallization.

If a vitreous solid is maintained above a certain temperature, generally well below its melting point, it will in time become crystalline. This limiting temperature, below which the change will not occur, is called the temperature of recrystallization. When a differential stress causes the substance to flow while at a temperature between that of recrystallization and the melting-point, the flowed substance recrystallizes as soon as the stress is relieved; but if the temperature is below that of recrystallization, the flowed material assumes the vitreous state. The temperature of recrystallization of ice is considerably below $-12\,°C$.[87] For an example of marked recrystallization caused by stress, see O. D. von Engeln.[81] See also p. 438.

R. Mallet [80] has written: "If a crystallizable body be heated near to but not up to its fusing point, by the application of heat in one plane, a crystalline structure perpendicular to the plane is immediately developed." And again: "In general, change of temperature beyond certain limits develops in crystallizable bodies a crystalline structure in the direction of the wave of heat, whether into or out of the mass of the body."

The term recrystallization is also used to denote the growth of one crystal at the expense of another. In order to distinguish this phenomenon from the preceding, it will be called migratory recrystallization.

G. Beilby [88] failed to observe any indication of migratory recrystallization during the slow warming and melting of a film of ice initially at $-11\,°C$, but G. Tammann and K. L. Dreyer [89] seem to have observed migratory recrystallization of ice, and the fact that glacier grains increase in size and decrease in number as the ice ages and moves under complex stresses shows that, however it may arise, there is in that case such an effect.[62, 90] And H. T. Barnes [74] has stated that surface ice becomes coarser with age, the large crystals consuming the smaller ones. Furthermore, J. Thomson [91] has shown that there must be migratory recrystallization whenever it is possible for molecules to pass by any means from a stressed crystal to one that is less stressed. An obvious means by which they may so pass is by fusion, or sublimation, and resolidification, the temperature being suitable.

Regelation.

M. Faraday,[92] observed that when two pieces of ice, each at $0\,°C$, are brought into contact they freeze together, and do this even in a vacuum or

---

[87] Beilby, G., "Aggregation and Flow of Solids," p. 196, 1921.
[88] Beilby, G., "Aggregation and Flow of Solids," 1921.
[89] Tammann, G., and Dreyer, K. L., *Z. anorg. allgem. Chem.*, **182**, 289-313 (1929).
[90] Vallot, J., *Compt. rend.*, **156**, 1575-1578 (1913).
[91] Thomson, J., *Proc. Roy. Soc. (London)*, **11**, 473-481 (1861).
[92] Faraday, M., *Proc. Roy. Inst. Grt. Brit.*, 1850; Exp. Res. in Chem. and Phys., pp. 372-374, 377-382 (1859); *Proc. Roy. Soc. (London)*, **10**, 440-450 (1860).

in water, and when brought together with the least possible pressure. Tyndall called this "regelation." Both he and Faraday explained it on the hypothesis that a thin layer of water bounded on each side by ice will freeze when it would not do so under other conditions. Having set forth his hypothesis that superficial portions of ice melt at a lower temperature than do others, Tyndall [60] writes: "The converse of this takes place when two pieces of ice at 32 °F., with moist surfaces, are brought into contact. Superficial portions are by this act virtually transferred to the center; and as equilibrium soon sets in between the motion of the tenuous film of moisture between the pieces of ice and the solid on each side of it, the consequence is shown to be that the film freezes, and cements the two pieces of ice together."

This explanation was vigorously upheld by Faraday [92] and as vigorously attacked by the Thomson brothers,[93] who ascribed regelation to a melting produced by stress, followed by solidification when the stress is relieved, such melting and refreezing having been shown by J. Thomson [91, 94] to be necessary consequences of varying stress. Even in a vacuum the surfaces are moist, and there is a pressure due to capillary action; when the ice is in water, there are pressures arising from the unavoidable currents set up when the two pieces are brought together; and in all cases there are pressures arising from mechanical disturbances transmitted from without. This explanation—melting produced by stress and followed by freezing when the stress is relieved—is the one now generally accepted.

In contrast to the preceding explanations, L. Pfaundler [95] thought that regelation was to be explained by the presence in melting ice of molecules of liquid as well as of solid, the two species continually changing one into the other, but just how he expected this to bring about the observed effect is not clear.

E. W. Brayley [96] thought that there was a close analogy between regelation and the union of polished plates of glass in the manufacture of plate glass. And that suggests a possible similarity between it and the adhesion between flat plates, whether of glass or of metal, that have been wrung together.

The completeness with which two blocks of ice will freeze together increases with the pressure and its duration, and depends upon the relative orientations of the crystals of the two blocks. If the orientations are exactly the same in the two blocks, then they freeze together completely, the plane of union differing in no respect from any other parallel plane in either block. If the principal axes are parallel, but the subordinate axes of the crystals in one block are not parallel to those of the crystals in the other, the blocks freeze together so completely that the plane of separation cannot

[93] Thomson, J., *Proc. Roy. Soc. (London),* **10**, 152-160 (1859); **11**, 198-204 (1861); Thomson, W. (Lord Kelvin), *Idem,* **9**, 141-143 (1858).
[94] Thomson, J., *Trans. Roy. Soc. Edinburgh,* **16**, 575-580 (1849).
[95] Pfaundler, L., Müller-Pouillet's "Lehrbuch der Physik," 9th ed., vol. 2$_2$, p. 595, 1898; Cf. *Sitz. Akad. Wiss. Wien,* **59**$_2$, 201-206 (1869).
[96] Brayley, E. W., *Proc. Roy. Soc. (London),* **10**, 450-460 (1860).

be detected by polariscopic observations, but can be by the production of Tyndall's flowers of ice, which show the difference in the orientations of the crystals in the two blocks. When tested by compression, the block yields first in that plane if the temperature of the room is over 0 °C, but the initial yield bears no relation to that plane if the temperature is below 0 °C. If the principal axes of the crystals of one block are perpendicular to those of the other, the union may again be invisible, but its strength is less than in the preceding case.[97]

Purity of Ice.

H. T. Barnes [98] has frequently emphasized the purity of ice, especially of that formed on the underside of a thick sheet growing over flowing water. That the great bulk of the impurities carried by the water is eliminated under these conditions seems certain, but the elimination of "every trace of foreign matter" would seem to require very special conditions.

G. Quincke [47] stated that the purity of the resulting ice is continuously increased by successive fractionings by freezing; but that he never succeeded in obtaining ice that was free of grains, that formed a single crystal.

F. Witt [99] has found that the proportion of radon included in the ice that first forms when water containing radon is frozen depends upon the rapidity of the freezing. When the freezing was as slow as 0.0005 cm³/sec, the concentration of the radon in the ice was only 3 or 4 per cent of that in the water; but if the rate of freezing exceeded 0.001 cm³/sec the amount increased rapidly with the rate.

Production of Homogeneous Ice.

In 1845, C. Brunner [100] described his attempts to produce homogeneous ice suitable for use in a determination of the density of ice. On cooling while exposed to the air, carefully boiled-out distilled water takes up so much air that when it is frozen it contains many bubbles, especially in the portion that was the last to freeze. Covering the water with turpentine immediately after the boiling kept out the air fairly well, but the ice was then full of cracks. Following a suggestion, which he attributes to F. C. Achard,[101] he exposed one side of the vessel, containing the water to be frozen, to a low temperature and the opposite to a temperature above 0 °C. The ice so formed contained some air-bubbles, but not nearly so many as that formed in the usual manner. He found that selected river-ice was much superior to any ice he succeeded in freezing in the laboratory.

L. Dufour [102] boiled-out and froze water in a Torricellian vacuum, air

---

[97] See Hess, H.[65], pp. 25, 26; Heim, A., *Ann. d. Physik (Pogg.) Erg. Bd.*, **5**, 30-63 (1871); Hagenbach-Bischoff, E.[62]
[98] Barnes, H. T., "Colloid Symposium Monographs," **3**, 103-111, Reinhold Publishing Corp., 1925; "Colloid Chemistry" (J. Alexander, ed.), Vol. **1**, pp. 435-443, 1926.
[99] Witt, F., *Sitz. Akad. Wiss. Wien (Abt. IIa)*, **139**, 195-202 (1930).
[100] Brunner, C., *Ann. d. Physik (Pogg.)*, **64**, 113-124 (1845).
[101] Achard, F. C., "Chem.-Phys. Schriften," Berlin, 1780.
[102] Dufour, L., *Compt. rend.*, **54**, 1079-1082 (1862).

pressure not over 0.5 mm. The ice contained a few very small bubbles. It was opalescent and very homogeneous. He states that the opalescence was not due to air, but to the crystalline structure of the ice, or to internal crevices.

R. Bunsen [103] introduced boiled-out distilled water into one arm of a U-tube initially filled with air-free mercury. The water was introduced while hot, was boiled in the tube, and the tube was then sealed so as to exclude all air. The water was frozen gradually from the top downward. Thus he obtained a cylinder of ice that was entirely (völlig) free of air-bubbles, and that was equal to the best crystal glass in clearness and transparency.

G. Forbes [104] used the same procedure, and stated: "The ice formed was quite uniform, very clear, and when cloven by planes perpendicular to the plane of freezing, split easily, showing the crystalline structure with great clearness."

G. Quincke [47] stated that by repeated fractionations, by means of freezing and thawing, the ice becomes ever purer and purer, and the crystals larger, but that he had never succeeded in obtaining ice that was not an aggregate of many crystals. The removal of air by alternate freezing and rapid thawing had been described, and the method seems to have been used by Duvernoy [105] nearly 40 years before.

A. Leduc [106] introduced hot, boiled-out distilled water into an exhausted vessel, and froze it progressively from the bottom to the top. The upper portions contained bubbles. Even after three such freezings *in vacuo* there were small bubbles in the portion last frozen. A fourth freezing appeared to produce no further improvement. Likewise, G. Bode [107] has stated that neither the boiling-out of distilled water nor its repeated freezing in a vacuum is sufficient to insure a clear sheet of ice.

H. Hess [108] stated that by slow, long-continued freezing very homogeneous ice-sheets may be formed on the water in large reservoirs. On thawing, these sheets break up into vertical, columnar pieces.

Monocrystals of Ice.

A portion of ice is defined by J. M. Adams and W. Lewis [109] as monocrystallic if the flowers of ice formed in it all lie in parallel planes and are similarly oriented. They actually use the faces of the negative crystals (p. 406), formed when the water in those cavities is refrozen, as indices of the orientation of the flowers.

They have reported that monocrystals "with dimensions of the order of 10 cm were readily produced" by the following method. A mono-

[103] Bunsen, R., *Ann. d. Physik (Pogg.)*, **141**, 1-31 (1870).
[104] Forbes, G., *Proc. Roy. Soc. Edinburgh*, **8**, 62-69 (1873).
[105] Duvernoy, *Ann. d. Physik (Pogg.)*, **117**, 454-463 (1862).
[106] Leduc, A., *Compt. rend.*, **142**, 149-151 (1906).
[107] Bode, G., *Ann. d. Physik (4)*, **30**, 326-336 (1909).
[108] Hess, H., "Die Gletscher," 1904.
[109] Adams, J. M., and Lewis, W., *Rev. Sci. Inst. (N. S.)*, **5**, 400-402 (1934) → *Phys. Rev. (2)*, **46**, 328 (A) (1934).

crystallic fragment of commercial ice is cut with two faces parallel and presenting any desired aspect to the principal axis. One of these faces is frozen to the outer surface of a metal vessel containing a freezing solution (−10 °C), and the opposite face is dipped just below the surface of distilled water at 0 °C in an ice-jacketed vessel. They boiled the water and cooled it rapidly to 0 °C. The main body of the growth is monocrystallic with the seed. As the length increases, growth becomes slow, and it may be desirable to make a new start, using a seed of greater cross-section, cut from the recently formed monocrystal. Parasitic crystals tend to start where the air-water surface meets the seed, and should be removed from time to time.

H. D. Megaw [79] has grown single crystals in a capillary tube of Lindemann glass in a copper-wire holder cooled by means of a mixture of acetone and solid carbon dioxide. And J. Meyer and W. Pfaff [110] have observed spicules of ice which they thought were monocrystals.

Freezing of Supercooled Water. (See also Section 97.)

That water can be cooled much below 0 °C without freezing, is well known. G. Oltramare [111] has stated that both R. Pictet and L. Dufour had carried the supercooling to −40 °C (no citation). The greater the supercooling the more readily does freezing occur, at least within certain limits, but the only certain way to initiate freezing intentionally is to "seed" the water with a suitable crystal. The smallest particle of ice will at once grow rapidly when placed in supercooled water, branching and spreading until either all the water has become frozen or the temperature has risen to 0 °C. All other methods commonly quoted as efficacious are so irregular in their actions, or have been shown to fail so signally under certain conditions, often ill-defined, that one must conclude that they are, at best, only secondarily involved in the initiation of the freezing that is sometimes, or often, observed when they are employed. On the other hand, certain conditions seem to oppose freezing. These are considered in Section 97. Initiation of freezing is not the mere negative of opposition to freezing, although the contrary might be inferred from many of the articles on this subject.

The freezing of water that has been greatly supercooled has been described by H. Hess,[108, p. 12] who states that the resulting ice generally takes the form of hexagonal prisms or plates, reminiscent of snow-crystals. H. T. Barnes and H. L. Cooke [112] had more difficulty in cooling water through 0 °C than in continuing the cooling to lower temperatures, and stated that when freezing occurred at low temperatures (approximately −10 °C) "the ice formed all through the mass of the water." They also reported that "care had to be taken that the thermometer bulb never touched the bottom of the flask when the water was supercooled, as it almost

[110] Meyer, J., and Pfaff, W., *Z. anorg. allgem. Chem.*, 224, 305-314 (1935).
[111] Oltramare, G., *Arch. des sci. phys. et nat.* (3), 1, 487-501 (1879).
[112] Barnes, H. T., and Cooke, H. L., *Phys. Rev.*, 15, 65-72 (1902).

invariably caused freezing to take place. The ice would start at the point of contact and immediately spread out all through the mass of the water."

Cooling water to -10 or $-13\,°C$ is not so difficult as one might infer from the published accounts (see p. 638).

When water contained in a wide tube, or other vessel, is frozen rapidly, it is not unusual to see spicules of ice, from a millimeter to several centimeters in length, growing from the surface, or from the walls of the vessel. These occur when the surface of the water, including the film on the walls, is frozen first. This traps the remaining water, so that its pressure rises as further freezing occurs, until presently it breaks through at some point of weakness, gushing out in a jet that freezes at once into a tube that continues to grow at its tip. The break usually occurs in the surface ice, but the pressure may be transmitted up the film on the surface of the vessel to a weak place in the covering ice.[72, 113] If heat is being abstracted from the water rapidly, but only a small portion of the water was supercooled when freezing began, the growth of the spicule may be slow. But if at the initiation of freezing the bulk of the water was considerably supercooled, the spicule shoots out with a surprising velocity. Such growths have been observed by the compiler, and always under conditions that seem to demand the explanation just offered. On the other hand, J. Meyer and W. Pfaff [110] have described similar spicules which they think arise from condensation of the vapor, and which they believe are monocrystals.

Icicles.

No record of any careful study of the crystalline structure of icicles has come to the compiler's attention except that of F. Leydolt.[33] As a result of his polariscopic study of icicles, he announced that the optic axis is always radial, normal to the geometrical axis of the icicle. Nevertheless, other quite positive though mutually contradictory statements occur. It will suffice to mention four.

If the icicle is produced by slow and continuous freeing, it consists of a single crystal with its optic axis horizontal; and if it hangs from the eaves of a roof, its optic axis is not only horizontal, but is also perpendicular to the eaves. If the growth is interrupted by a drop in the temperature, and is later renewed by the thawing of ice or snow on the roof, then the icicle acts as the chilling surface, and the axes of the new crystals are everywhere perpendicular to the surface of the original icicle. So says H. Hess,[108, p. 12] but it is possible that his statements are colored by his belief that the optic axis is always parallel to the heat-stream at the instant the crystal is formed. The truth of the second sentence of this paragraph is especially doubtful as it seems more reasonable to expect that the axis of the new crystal will be parallel to that of the crystal on which it is being formed.

In marked contrast to the preceding, J. C. McConnel and D. A.

[113] Bally, O., *Helv. Chim. Acta*, **18**, 475-476 (1935).

Kidd [81a, p. 334] have stated: "An icicle is an example of ice formed of very minute crystals irregularly arranged." A. Bertin [82] said that the crystallization of an icicle is confused, and U. Yoshida and S. Tsuboi [78] have stated that the lower end of an icicle usually consists of a slender single crystal.

Hail.

Treatises and journals devoted to meteorology, as well as those devoted to physics, should be consulted by one interested in the nature of hail and in the conditions under which it is formed. The experiments and discussions by L. Dufour,[69, 114] G. Oltramare,[111] and K. C. Berz,[115] are interesting and suggestive. Some of Dufour's experiments are considered in Section 97.

Hailstones may be very large. G. Oltramare [111] has stated that they may weigh as much as 500 grams (no citation). Captain Blakiston [116] has reported an ice storm in which large stones fell. He weighed blocks of 3.5 and 5 ounces (100 to 140 g), and pieces the size of a brick were said to have been seen. P. A. Secchi [117] has described a violent hailstorm in which many of the hailstones consisted of clusters of hexagonal prisms terminated at their outer ends by pyramids [illustrations], some of the crystals being a centimeter long and correspondingly wide. Many of the clusters were 5 to 6 cm in diameter, and some weighed as much as 300 grams.*

The velocity of the uprush of air that is required to support a spherical hailstone of density 0.7 g/cm³ at a height of about 5 km above sea-level has been computed by W. J. Humphreys [118] to be as follows:

Diameter	1	2	2.5	3	3.5	4	5	inches
Velocity	55	78	91	109	136	185	219	miles/hour

He gives corresponding values for other densities between 0.9 and 0.5 g/cm³. Similar data have been given by M. A. Giblett [119] and by G. Grimminger.[120]

J. Smithson [121] observed that hail which is sufficiently regular for measurement usually consists of two hexagonal pyramids joined base to base, one of the pyramids being truncated, and the angle formed by the junction of the pyramids being about 80 degrees; and F. Leydolt [33] has

---

* Accounts of severe hailstorms, in some of which hailstones were reported as large as 13 to 20 inches in circumference, some weighing from one to four pounds, may be found in *Nature*, 125, 32, 656, 728, 765, 800, 840, 877, 913, 956, 994 (1930); 126, 41, 81, 117, 153, 188, 224, 262, 385, 457, 663, 669, 976, 1012 (1930); 137, 219-220 (1936).

[114] Dufour, L., *Arch. des sci. phys. et nat. (N. S.)*, 10, 346-371 (1861).
[115] Berz, K. C., *Kolloid Z.*, 41, 196-220 (1927).
[116] Blakiston, Captain, *Proc. Roy. Soc. London*, 10, 468 (1860).
[117] Secchi, P. A., *Bull. Meteorol. Osserv. Coll. Romano*, 15, 73-74 (1876).
[118] Humphreys, W. J., *Monthly Weather Rev.*, 56, 314 (1928).
[119] Giblett, M. A., *J. Roy. Aeronaut. Soc. Grt. Brit.*, 31, 509-540-549 (1927).
[120] Grimminger, G., *Monthly Weather Rev.*, 61, 198-200 (1933).
[121] Smithson, J., *Ann. Phil. (N. S.)*, 5, 340 (1823).

reported that the optic axes of the constituent crystals in a hailstone are radial.

Descriptions of hailstones and of the microscopic appearance of sections of them have been published by J. H. L. Flögel.[121a]

Snow and Frost.

Snow crystals are formed by inverse sublimation, by the passage of the molecules directly from the gaseous to the solid state (Hess).[65, p. 9] G. Tammann [121b] states that snow is not formed at temperatures above −4 °C, that precipitation above −4 °C takes the form of rain. Obviously, the rain may freeze, producing hail. G. Stüve [121c] has concluded that gaseous nuclei of condensation give rise to drops of water only; that soluble salts give rise to drops if the temperature at which the condensation begins is above −20 °C, and to stars of snow if the temperature is lower; and that insoluble hygroscopic nuclei give rise to needles of ice at all temperatures below 0 °C.

The typical crystal of snow or frost is 6-rayed, but innumerable modifications are found. The finest collection of photographs of snow-crystals and frost figures is that of W. A. Bentley, containing over 4000 negatives. He states that in the 45 years of his study he has never seen two snow crystals that were exactly alike. He has published an extended study of such crystals,[122] and recently, in conjunction with W. J. Humphreys, has published a beautifully illustrated volume (230 pages) entitled "Snow Crystals" (1931). Microphotographs of snow and rime have been published by G. Stüve [121c]; detailed studies of crystals of snow and frost, both natural and artificial, have been carried out by U. Nakaya and associates,[123] and by G. Hellman,[124] I. B. Schukewitsch,[125] and A. Erman.[126] Snow and the structure and properties of snow fields and the changes they undergo have been studied by G. Seligman,[73, 127] and G. Seligman and C. K. M. Douglas.[128]

When a thin layer of mud freezes during a cold night, the ice often takes the form of loosely packed hexagonal columns, often hollow, each carrying on its top a grain of sand or a bit of earth; or the entire bundle of columns may be covered with a continuous roof of earth. It is commonly stated that the optic axis of the ice in these columns is always vertical,[108, p. 12] but F. Klocke [75] found this to be seldom true. He found that each column was generally an aggregate of small, approximately parallel needles meeting at sharp angles, and having their optic axes variously

---

[121a] Flögel, J. H. L., *Ann. d. Physik (Pogg.)*, **146**, 482-486 (1872).
[121b] Tammann, G., "Aggregatzustände," p. 219, 1922.
[121c] Stüve, G., *Gerlands Beitr. zu Geophys. (Köppen Bd. 1)*, **32**, 326-335 (1931).
[122] Bentley, W. A., *Monthly Weather Rev.*, **29**, 212-214 (1901); **35**, 348+, 397+, 439+, 512+, 584+ (1907).
[123] Nakaya, U., and associates, *J. Fac. Sci. Hokkaido*, 1934-1936; see Seligman, G., *Nature*, **140**, 345-348 (1937).
[124] Hellman, G., "Schneekristalle," 1893.
[125] Schukewitsch, I. B., *Bull. Acad. Imp. Sci. St. Petersburg (6)*, **4**, 291-302 (1910).
[126] Erman, A., *Phil. Mag. (4)*, **17**, 405-413 (1859).
[127] Seligman, G., *Nature*, **140**, 345-348 (1937); *J. Roy. Meteorol. Soc.*, **63**, 93-103 (1937).
[128] Seligman, G., and Douglas, C. K. M., "Snow Structure and Ski Fields," xii + 555, 1936.

directed, so that when placed in parallel light between crossed nicols it was impossible to orient a column in such a way as to obtain darkness. The composite character of the columns has been mentioned by Yoshida and Tsuboi also (see p. 409).

Glaciers.

Glaciers and the icebergs to which they give rise are merely compacted masses of snow and frost. They contain large amounts of entrapped air. H. T. Barnes [129] has found that the volume of the entrapped air, when under a pressure of 1 atmosphere and at the temperature of the iceberg, may be from 7 to 15 per cent of the total volume of the berg.

Icebergs derived from glaciers may explode with considerable violence, as a result of their internal strains. A. E. Nordenskjöld [130] likens them to immense Prince Rupert drops. He gives (pp. 319, 320) the following description of the breaking up of such bergs: "Glacier-ice shows a great disposition to fall asunder into small pieces without any perceptible cause. It is full of cavities, containing compressed air which, when the ice melts, bursts its attenuated envelope with a crackling sound like that of the electric spark. Barents relates that on the 20/10th August 1596 he anchored his vessel to a block of ice which was aground on the coast of Novaya Zemlya. Suddenly, and without any perceptible cause, the rock of ice burst asunder into hundreds of smaller pieces with a tremendous noise, and to the great terror of all the men on board. Similar occurrences on a smaller scale I have myself witnessed."

As the surface of the glacier melts, it becomes fissured and pitted, and acquires a granular structure. These "glacier grains" vary in size from that of snow crystals to several hundred cubic centimeters [108, pp. 166, 167]; E. Hagenbach-Bischoff [62] says they may be 10 or even 15 cm broad. Each grain is a single crystal, although its boundaries are very irregular, and generally curved, and its refraction may be complicated by the strains that exist in it; Tyndall's flowers of ice are circular.[131] "Glacier ice is a sort of conglomerate of these grains, differing, however, from a conglomerate proper in that there is no matrix, the grains fitting each other perfectly." [132] It would, perhaps, be better to say that the matrix consists solely of imperceptibly thin layers separating the grains and enclosing each in a separate cell. (See p. 401 + .) S. Skinner has studied the fine structure of the surface-ice of glaciers by the use of plaster casts.[133]

In general, the optic axes of the several grains seem to be arranged quite at random,[81a, 62, 131] but Drygalski has reported that the optic axes of the several grains in the old inland Antarctic ice-sheet and of those in the deeper layers of Antarctic icebergs are similarly directed (see G. Tammann [121b, pp. 214, 215]). C. Grad,[134] and A. Bertin [82] have reported the same

---

[129] Barnes, H. T., "Ice Engineering," p. 346, 1928.
[130] Nordenskjöld, A. E., "The Voyage of the *Vega* around Asia and Europe," 1882.
[131] Klocke, F., *Neues Jahrb. Mineral., Geol.*, 1881$_I$, 23-30 (1881).
[132] McConnell, J. C., and Kidd, D. A., *Proc. Roy. Soc. (London)*, 44, 331-367 (1888).
[133] Skinner, S., *Proc. Camb. Phil. Soc.*, 11, 33-36 (1901).

thing for the deeper portions of the termini of glaciers, the direction of the optic axis being vertical. Grad attributed this orientation to the prolonged action of pressure, and likened it to the temporary birefringence produced in glass by suitable stresses. Bertin seems to have thought that it is to be accounted for in somewhat the same way as the vertical direction of the optic axes of the crystals in a sheet of ice formed on calm water. O. Mügge [81] would explain it as a special case of his experiment in which a bar of ice supported near its ends and loaded in the middle, the optic axis being perpendicular to the length of the bar and inclined to the vertical, tends to rotate so as to bring the optic axis toward the vertical.

On the other hand, J. Müller [135] found no such general uniformity in the directions of the axes, though in small portions of the ice, scattered here and there through the lower end of the glacier, the axes were vertical.

The grains increase in size, and consequently decrease in number, as the ice ages. This occurrence in the body of a glacier does not result from the freezing of water seeping down from the melting surface, for it has been found that the body of the glacier is impervious to liquids, and that at depths exceeding a meter the temperature is rarely as high as 0 °C (Vallot [90] and Forel as quoted by Hagenbach-Bischoff [136]). It is probably due to migratory recrystallization (p. 412)—to a sublimation and recondensation similar to that by means of which large rain-drops devour smaller ones,[74, 62] assisted by the tendency of crystals in close contact and under pressure to freeze together solidly if their axes are similarly directed (p. 413 and [62, 136]), and by the melting and refreezing that accompany stresses and their variations (p. 437+ and E. Hagenbach-Bischoff [137]).

But through the upper layers of the glacier, especially through the *névé*, water can percolate. Hence if the surface temperature during the day is above 0° C there will be melting, percolation of water, and freezing of that water at greater depths. Thus the observed internal layers of ice may be formed, the grains may grow, and possibly an actual glacier might be formed in this manner,[138] but only in a relatively thin superficial layer can the growth of the grains be so affected.[139]

Those interested in the structure of a snow field and in changes it undergoes as it is gradually converted into a glacier, should read the papers by G. Seligman,[73, 127] and G. Seligman and C. K. M. Douglas.[128]

A glacier flowing down a mountain is squeezed together where its bed narrows, exhibiting definite lines of flow. Where the bed widens the ice splits, forming crevasses; these form suddenly, but widen slowly.[140]

---

[134] Grad, C., *Compt. rend.*, **64**, 44-47 (1867); see Heim, A., *Ann. d. Physik (Pogg.) Erg. Bd.*, **5**, 30-63 (1871).

[135] Müller, J., *Ann. d. Physik (Pogg.)*, **147**, 624-626 (1872).

[136] Hagenbach-Bischoff, E., *Z. Kryst.*, **20**, 309-310 (1892) ← *Verh. d. naturf. Ges. Basel*, **8**, 821-832 (1889).

[137] Hagenbach-Bischoff, E., *Z. Kryst.*, **11**, 110-111 (1885) ← *Verh. naturf. Ges. Basel*, **7**, 192-216 (1882).

[138] Devaux, J., *Compt. rend.*, **185**, 1602-1604 (1927).

[139] See also, Emden, R., *Neue Denkschr. allgem. Schweiz Ges. ges. Naturwiss.*, **33** (1892).

[140] Barnes, H. T., "Ice Formation," p. 92, 93, 1906.

The mechanics of glaciers has been discussed recently by M. Lagally.[141] Proposed explanations of the motion of glaciers have given rise to various disputes, such as the recent one between R. T. Chamberlin[142] and O. D. Engeln.[143]

Of the many phenomena that may be involved in the motion of glaciers, the one most commonly invoked to explain the general flow and the compression of the glacier where the bed narrows is that of the lowering of the melting-point by pressure, together with the regelation that follows a release of pressure.[144] G. Beilby,[88, pp. 194-200] however, has pointed out that this can be effective only when the temperature of the ice is near 0 °C, a pressure of 138 atm (corresponding to a free column of ice 5000 feet high) causing a depression of only 1 °C in the melting-point.

In contrast to the great pressure required to lower the melting-point by a small amount, he found that ice at -11 °C can be flowed by a steel burnisher exerting a pressure not exceeding 30 or 40 lbs. per sq. in. (say, 2 atm). In order to lower the melting-point by 11 °C, the pressure would have to be of the order of 1500 atm, corresponding to a free column of ice 55,000 ft. (over 10 miles) high.

He concluded that "true molecular flow, which," under differential stresses, "occurs alike at the external and internal surfaces of crystalline aggregates, has, therefore, a wider and more fundamental relation to the phenomena of ice flow than fusion and regelation." J. H. Poynting's (1881) conclusions (p. 431) regarding the melting of ice by pressure to which the resulting water is not subjected, and R. W. Wood's (1891) experiments (p. 439) on the compression of ice in which lead pellets were embedded, are of interest in this connection.

J. Vallot[90] has studied the variation in the temperature with the depth below the surface of a glacier. He found that the diurnal variation did not extend below one meter, nor the annual variation below 6 or 7 meters. At a depth of a meter the temperature rose above 0 °C only under exceptional conditions. The work was done on Mt. Blanc, at an elevation of about 4.3 km.

An explanation of the veined structure often observed in glaciers, based on the melting and refreezing of ice under the action of changing stresses, has been offered by W. Thomson (Lord Kelvin).[145]

H. Hess[108, p. 14] has stated that lightning had never been known to strike a glacier. But is it not probable that he has recorded merely a lacuna in our observations?

For a recent summary of our knowledge of glaciers, see H. Hess.[146]

---

[141] Lagally, M., *Gerlands Beitr. Geophys., Suppl. Bd.*, **2**, 1-94 (Bibliog. 112) (1933).
[142] Chamberlin, R. T., *Science (N. S.)*, **80**, 526-527 (1934).
[143] Engeln, O. D., *Idem*, **80**, 401-403 (1934); **81**, 459-461 (1935).
[144] Thomson, J., *Proc. Roy. Soc. (London)*, **8**, 455-458 (1857).
[145] Thomson, W. (Lord Kelvin), *Proc. Roy. Soc. London*, **9**, 209-213 (1858).
[146] Hess, H., Müller-Pouillets *Handbuch der Physik*, 11 ed., Vol. 5, pp. 355-397, 1928.

Sea-ice.

The ice formed from sea-water contains some salt. This salt differs in composition from that contained in the sea. For example, the ratio of the sulfates to the chlorides is greater in the ice than in the sea. In the process of freezing a selection is made, certain constituents of the salt are in a measure retained, while others are more completely eliminated.[147] It is to be expected that the retained salt is not contained in the ice-crystals themselves, but lies in the boundaries between the crystals. In newly formed sea-ice the salt is quite uniformly distributed in the proportion of 4 or 5 parts per 1000 of ice. As the ice ages, there is a migration of the salt from the interior to the surface. In a case reported by Drygalski, the salinity decreased in two months from 4 or 5 parts per 1000 to only 1 or 2 (see H. T. Barnes [148]), and A. E. Nordenskjöld [130] stated: "The water which is obtained by melting sea-ice is not completely free from salt, but the older it is the less salt does it contain" (p. 321).

J. Y. Buchanan [53] has shown that the salt contained in sea-ice is not contained in the ice crystals, nor as a solid inclusion, but in brine entrapped between the crystals. When the ice is first formed, the composition of that salt is the same as that in the water from which the ice was formed, but as the ice ages, the composition changes, owing to various secondary effects. Owing to its composite structure—ice-crystals surrounded by brine—sea-ice melts progressively, the amount melted at any given temperature being just enough to make the concentration of the resulting intercrystallic brine such that the melting point of ice bathed in the brine is the given temperature.

In this composite structure and in the progressive melting is to be found the explanation of the many differences between the behavior of sea-ice and of ice formed from fresh water. For example, sea-ice not only melts below 0 °C, but as the temperature is raised from a low value, the volume of the ice reaches a maximum and then decreases as the temperature is further increased. This decrease indicates that melting has already begun, although the ice appears to be as solid as ever; the melting is at the boundaries of the crystals, where they are bathed with the intercrystallic brine. Thus, ice containing 2.73 parts of chlorine per 1000 began to contract at -14 °C, and that containing 6.49 parts began at -18 °C. Buchanan has shown that such contraction can be explained in the manner indicated. Pettersson [147] gives several sets of data showing such contraction.

Buchanan stated: "At the winter quarters of the *Vega* brine was observed oozing out of sea-water ice and liquid at -30 °C. It was very rich in calcium and especially magnesium chlorides. In fact, *it is probably quite impossible by any cold occurring in nature to solidify sea-water."*

Thin layers of sea-ice are white, from mechanically suspended salt, and

---

[147] Pettersson, O., *"Vega*-Expeditiones Vetensk. Jakt.," Vol. 2, pp. 249-323, 1883; as reviewed in *Beibl. Ann. d. Physik,* **7,** 834-841 (1883).
[148] Barnes, H. T., "Ice Engineering," p. 231, 1928.

are so mobile that a small wave may travel through them without breaking them up.[148]

The following description of the formation of sea-ice is adapted from one that Barnes [148, p. 232] credits to J. B. Woodyatt. At first there appears a sort of thin slush on the surface of the sea. From a distance, its appearance resembles that of oil on water. It forms a cohering and flexible surface; the wash from a ship distorts the surface, but does not break it up. This slush forms into little discs about 4 inches in diameter, which gradually grow in diameter. They have no power of cohesion. They are pushed about by wind and water until several are piled partly on top of one another, making aggregates about 2 feet in diameter, the intervening spaces being filled with the slush. The lapping of the water deposits slush in ridges both on their tops and along their edges. Clumpets with such slushy edges tend to stick together when they meet, but even a small wave will pull them apart or slide one on top of another. Presently the intervening slush hardens, cementing the surface, then the ice grows rapidly.

At other times the growth proceeds quite differently, giving rise to much clearer and more brittle ice, with vertical cleavage planes. This type of ice forms on still water in very cold weather.

60. Molecular Data for Ice

Numerous suggestions regarding the nature and structure of the ice molecule have been advanced, but no generally accepted conclusion has yet been arrived at. Here the compiler will do no more than indicate a few of those suggestions, and cite certain publications in which the subject is discussed in some detail. The corresponding sections for the vapor (9) and the liquid (25) should be consulted.

Association of Molecules in Ice.

H. M. Chadwell [149] has reviewed the several suggestions regarding the molecular structure of water and of ice, and the evidence on which they rest. He gives a bibliography of over 100 titles.

The most widely held opinion seems to be that the molecule of ice-I is $(H_2O)_3$, called trihydrol, but several have regarded it as more complex, and G. B. B. M. Sutherland [150] concluded that nothing more complex than $(H_2O)_2$ is needed. To Sutherland's conclusions, I. R. Rao [151] seriously objects, favoring the trihydrol theory.

In contrast with the preceding, others, including W. H. Bragg [152] and W. H. Barnes,[153] are of the opinion that a mere space lattice of ions is preferable to any type of polymerization as a representation of the structure

---

[149] Chadwell, H. M., *Chem'l Rev.*, **4**, 375-398 (1927).
[150] Sutherland, G. B. B. M., *Proc. Roy. Soc. (London) (A)*, **141**, 535-549 (1933).
[151] Rao, I. R., *Idem*, **145**, 489-508 (1934).
[152] Bragg, W. H., *Proc. Phys. Soc. (London)*, **34**, 98-103 (1922).
[153] Barnes, W. H., *Proc. Roy. Soc. (London) (A)*, **125**, 670-693 (1929).

of ice-I. To that, T. M. Lowry and M. A. Vernon [154] do not agree, giving reasons for believing that there is a polymerization in which additional bonds come into play. They postulate a network of single bonds between quadrivalent oxygen and bivalent hydrogen.

From his determinations of the dielectric constant of ice-I and of the way it varies with the temperature and the frequency, J. Errera [155] thought it probable that at the freezing point there is no distinction between the molecules of water and of ice. That does not accord with the more common opinion that the proportion of trihydrol in water at 0 °C is less than 50 per cent of the whole (Table 79).

I. R. Rao [151] has expressed the opinion that it is not possible to derive from existing x-ray data any definite conclusion regarding the extent of the association in either ice or water.

The complexity of the ice molecule has been considered also by P. N. Chirvinskii,[156] J. Duclaux,[157] R. de Forcrand,[158] E. J. M. Honigmann,[159] and L. Schames.[160]

From the great values of the pressure-derivative of their mutual equilibrium temperatures (small values of dP/dt, Table 270), G. Tammann [161] has concluded that ices I, III, V, and VI have all the same molecular weight, their molecules being isomeric, differing one from another in the distances between the constituent atoms, but not in the grouping of them.

Structure of the Molecule of Ice.

The structure of the molecule of ice, as regards the arrangement of the atoms, the distances between them, and their bonding, is intimately related to the ultimate crystalline structure of ice, as revealed by means of x-rays. Both will be considered in this section, the second being considered first. For the actual values of the periodicities observed, see Table 212. H. M. Chadwell's review,[149] with bibliography, should be consulted.

Only ice-I, ice-II, and ice-III have as yet been studied by x-rays. R. L. McFarlan has concluded that the lattice pattern of ice-II is that of a side-centered orthorhombic cell having $a = 7.80$Å, $b = 4.50$Å, and $c = 5.56$Å, and containing 8 molecules [162]; and that that of ice-III is a body-centered orthorhombic cell having $a = 10.20$Å, $b = 5.87$Å, and $c = 7.17$Å, and containing 16 molecules.[163]

The x-ray studies of ice-I are numerous and conflicting (see A. B. Dobrowolski [164]). A key to the last may perhaps be found in the recent

---

[154] Lowry, T. M., and Vernon, M. A., *Trans. Faraday Soc.*, **25**, 286-291 (1929).
[155] Errera, J., *J. de Phys.* (6), **5**, 304-311 (1924).
[156] Chirvinskii, P. N., *Chem. Abst.*, **17**, 2525 (1923) ← *Bull. Soc. Russe amis l'étude l'univers (Petrograd)*, **7**, 6-10 (1918).
[157] Duclaux, J., *J. de chim. phys.*, **10**, 73-109 (1912).
[158] de Forcrand, R., *Compt. rend.*, **140**, 764-767 (1905).
[159] Honigmann, E. J. M., *Naturwissenschaften*, **20**, 635-638 (1932).
[160] Schames, L., *Ann. d. Physik* (4), **38**, 830-848 (1912).
[161] Tammann, G., "Aggregatzustände," pp. 143, 144; 1922.
[162] McFarlan, R. L., *J. Chem'l Phys.*, **4**, 60-64 (1936) → *Phys. Rev.* (2), **49**, 199 (A) (1936).
[163] McFarlan, R. L., *J. Chem'l Phys.* **4**, 253-259 (1936) → *Phys. Rev.*, **49**, 644 (A) (1936).

announcement by N. Seljakov [165] that ice-I occurs in two forms: α-ice, belonging to one of the following 4 classes, 3 of the hexagonal system—dihexagonal-bipyramidal (D_{6h}), trapezohedral (D_6), and dihexagonal-pyramidal (C_{6v})—and one—the ditrigonal-bypyramidal (or holohedral) class (D_{3h})—of the trigonal system; and β-ice, belonging either to the rhombohedral class (C_{3i}) or the pyramidal class (C_3) of the trigonal system. The lattice unit is essentially the same for each form: $a = 4.52 \pm 0.03$A, $c = 7.34 \pm 0.04$A, $c/a = 1.60 \pm 0.02$. He discusses the discrepancies in the reported values in the light of his observations.

A recent x-ray analysis of ice-I has been made by W. H. Barnes,[153] using both single crystals obtained from commercial, artificially frozen ice and thin plates of clear, flawless ice grown on the surface of a basin of water exposed to the air during cold weather. He obtained Laue photographs of the plates at -78.5 °C, and rotation and oscillation photographs of the single crystals at -20 °C. The former showed that the crystals can be referred to hexagonal axes, and the latter gave for the unit cell:

$$a = 4.53_5\text{A}, \quad c = 7.41\text{A}, \quad c/a = 1.634$$

and content = 4 hydrol (H_2O) molecules. The dimensions are believed to be correct within a few parts in 1000.

These results essentially agree with those ($a = 4.52$, $c = 7.32$, $c/a = 1.62$) of D. M. Dennison [166] for ice obtained by plunging into liquid air a capillary tube filled with water. Dennison's data are given in R. W. G. Wyckoff's compilation [167] together with the citations: W. H. Bragg,[152] R. Gross,[168] F. Rinne,[169] and A. St. John.[170]

On the other hand, A. St. John,[170] using single crystals obtained by freezing water in exposed open pans, found $a = 4.74$A, from which he computes $c = 6.65$A, accepting Dana's value $c/a = 1.4026$. Both H. T. Barnes [148, p. 20] and W. H. Barnes [153, p. 672] suggest that the difference between these and the other values may arise from a real difference in the structure of the specimens used.

Laue photographs of powdered ice at -9, -13, -78, and -183 °C were taken by W. H. Barnes,[153, p. 672] but no indication that the structure depends at all upon the temperature was found.

On the other hand, E. F. Burton and W. F. Oliver [171] have found that the structure of the ice formed by freezing water-vapor onto a cold metal surface below -80 °C varies with the temperature, becoming vitreous if the temperature is below about -110 °C. If the temperature of the vitreous ice is raised above -110 °C its structure becomes more organized, but

---

[164] Dobrowolski, A. B., *Bull. Soc. Fr. Mineral.*, **56**, 335-346 (1933).
[165] Seljakov, N., *Compt. rend. Acad. Sci. URSS*, **10**, 293-294 (1936); **11**, 227 (1936); **14**, 181-196 (1937).
[166] Dennison, D. M., *Phys. Rev. (2)*, **17**, 20-22 (1921).
[167] Wyckoff, R. W. G., *Int. Crit. Tables*, **1**, 338-353 (341) (1926).
[168] Gross, R., *Centralbl. Min., Geol., Palaon*, **1919**, 201-207 (1919).
[169] Rinne, F., *Ber. Sächs. Ges. Wiss. (Math.-Phys.)*, **69**, 57-62 (1917).
[170] St. John, A., *Proc. Nat. Acad. Sci.*, **4**, 193-197 (1918).
[171] Burton, E. F., and Oliver, W. F., *Proc. Roy. Soc. (London) (A)*, **153**, 166-172 (1935).

does not reach the normal structure before −80 °C is reached (see Table 212).

From his own observations, D. M. Dennison [166] inferred a close-packed hexagonal lattice consisting of two sets of interpenetrating prisms; whereas W. H. Barnes [153, p. 672] concluded from his that the structure is either ditrigonal bipyramidal ($D_{3h}{}^4$) or dihexagonal bipyramidal ($D_{6h}{}^4$), with the probabilities in favor of the latter, and that an ionic structure is to be preferred. He proposed the $D_{6h}{}^4$ structure with an H placed at the middle of each line joining a pair of adjacent O's. He essentially agrees with W. H. Bragg,[172] who regards the molecular structure of ice as hexagonal and differing from that of diamond simply by the replacement of each C by an O, and the insertion of an H between the members of each pair of O's. Such a structure imposes on O a covalence of 4 and on H that of 2. Some of the conclusions that flow from the assumption of such covalences have been discussed by S. W. Pennycuick.[173] See also T. M. Lowry and M. A. Vernon,[154] and R. de Forcrand.[158]

On the other hand, E. L. Kinsey and O. L. Sponsler [174] infer a different structure from the same observations by W. H. Barnes,[153, p. 672] one in which occur the units H^+ and $(H_3O_2)^-$, the latter having the form of a double tetrahedron with the O's at the extremities of the axis.

W. H. Bragg [172] has concluded that the distance between the centers of adjacent O-atoms is 2.76A, between neighboring atoms lying in the same plane is 4.52A, and between consecutive basal planes is 3.67A; and M. L. Huggins,[175] that the radius of the H-atom in ice, defined as "the distance from nucleus to valence electron-pair," is 0.73A.

Reasons have been presented for believing that each O-atom in ice is tetrahedrally surrounded by 4 others, and that each H-atom lies on the line connecting 2 adjacent O-atoms, but nearer to one of those atoms than to the other. L. Pauling [176] gives these distances as 0.95A and 1.81A; and P. C. Cross, J. Burnham, and P. A. Leighton [177] give them as 0.99A and 1.77A.

That the structure of ice is like that of tridymite was proposed by J. D. Bernal and R. H. Fowler [178] (see p. 174), and seems to be widely accepted. But W. H. Barnes [179] has cautioned against a too hasty or uncritical acceptance of that proposal, doubting if any modification of the structure proposed by himself was yet necessary. See also, M. L. Huggins,[180] L. Pauling,[176] W. F. Gaiuque and J. W. Stout.[181]

[172] Bragg, W. H., *Proc. Phys. Soc. (London),* **34,** 98-103 (1922).
[173] Pennycuick, S. W., *J. Phys'l Chem.,* **32,** 1681-1696 (1928).
[174] Kinsey, E. L., and Sponsler, O. L., *Phys. Rev. (2),* **40,** 1035-1036 (A) (1932); *Proc. Phys. Soc. (London),* **45,** 768-779 (1933).
[175] Huggins, M. L., *Phys. Rev. (2),* **21,** 205-206 (1923).
[176] Pauling, L., *J. Am. Chem. Soc.,* **57,** 2680-2684 (1935) → *Nature,* **137,** 327 (1936).
[177] Cross, P. C., Burnham, J., and Leighton, P. A., *J. Am. Chem. Soc.,* **59,** 1134-1147 (1937).
[178] Bernal, J. D., and Fowler, R. H., *J. Chem'l Phys.,* **1,** 515-548 (1933).
[179] Barnes, W. H., *Trans. Roy. Soc. Canada III (3),* **29,** 53-59 (1935).
[180] Huggins, M. L., *J. Am. Chem. Soc.,* **58,** 694 (L) (1936); *J. Phys'l Chem.,* **40,** 723-731 (1936).
[181] Giauque, W. F., and Stout, J. W., *J. Am. Chem. Soc.,* **58,** 1144-1150 (1936).

61. Interaction of Ice and Corpuscular Radiation

No data have been found regarding either the absorption of corpuscular radiation by ice (range of α-rays, absorption of β-rays, etc.) or the effect of such radiation upon ice (electron emission excited by +ions, by electrons, etc.); but T. H. Johnson [182] has shown that atomic beams (hydrogen) are reflected from ice in preferential directions.

62. Adhesiveness of Ice

J. W. McBain and D. G. Hopkins [183] have reported that the freezing of a thin film of water between two plates of fused silica produces a joint that is "very strong" in shear. The freezing was done with solid CO_2, and the test was presumably made at that temperature.

63. Sliding Friction of Ice

J. Joly [184] seems to have been the first to point out that the slipperiness of ice and the "biting" of a skate are to be explained by the melting of ice under pressure. As the ice is melted, the curved runner of the skate sinks into the ice until the bearing surface becomes of such a size that the load is just insufficient to lower the melting point below the temperature of the surroundings. The sinking of the runner gives rise to the "bite," and the layer of water, acting as a lubricant, causes the slipperiness. When the ice is very cold, the sinking will be slight, and, as is well known, the "bite" is poor; then, hollow-grinding is of advantage, as it reduces the area of the bearing surface, thus increasing the pressure. The same explanation of the slipperiness of ice was advanced later by O. Reynolds.[185] He stated that Nansen, "in his book on Greenland," says that at very low temperatures ice completely loses its slipperiness.

When one solid body slides over another, the coefficient of friction (f) and the angle of repose ϕ are related as follows to the normal force (L) with which one body is pressed against the other and the tangential force (F) required to slide one body over the other with a constant velocity: $F = fL$, $\tan \phi = f$, where f is independent of L, unless that is so great as to distort the surfaces; and ϕ is the angle at which the surface of contact must be inclined to the horizontal if the upper body, when once started, slides with constant velocity, F and L arising solely from the action of gravity.

Applying the preceding to the friction between ice and aluminum at a constant temperature near $-6.5\,°C$, H. Morphy [186] has found two distinct values for f, depending upon the load; the bearing surfaces were small, but of unknown area. If the load was under 14.3 g*, then $f = 0.36 \pm 0.01$,

---

[182] Johnson, T. H., *Nature*, **120**, 191 (1927).
[183] McBain, J. W., and Hopkins, D. G., *Dept. Sci. Ind. Res. (Gt. Brit.)*, "2nd Report Adhesives Res. Com.," pp. 34-89 (41) (1926).
[184] Joly, J., *Proc. Roy. Dublin Soc. (N. S.)*, **5**, 453-454 (1886).
[185] Reynolds, O., "Papers on Mechanical and Physical Subjects," Vol. **2**, pp. 734-738, Cambridge Univ. Press, 1901; ← "Mem. and Proc. Manchester Lit. Phil. Soc.," Vol. **43**, 1899.
[186] Morphy, H., *Phil. Mag. (6)*, **25**, 133-135 (1913).

ϕ about 20°; if the load exceeded 15 g*, $f = 0.17 \pm 0.01$, ϕ about 9.5°. Within each range, f was independent of the load.

The inverse case, *i.e.*, ice sliding down an inclined plane, has been studied by W. Hopkins [187]; and H. Moseley [187a] reported that he had repeated Hopkins's experiments, with many modifications, and had verified his conclusions.

Using a rough-hewn slab of sandstone so mounted that it could be rotated about an axis parallel to its plane and perpendicular to the grooves made by the tool, Hopkins found that the angle of repose for a block of polished marble resting on it was 20°. Ice loaded to a pressure of 150 lb*/ft² slid down the slab at the following unaccelerated rates:

Inclination (°)	3	6	9	12	15
Rate (in/hr)	0.31	0.52	0.96	2.0	2.5

At an inclination of 20° the motion was accelerated. At an inclination of 9° the removal of 2/3 of the load reduced the rate by nearly half. Even at 1° there was a perceptible motion. On a smooth, but unpolished, slab of the same kind of stone there was a perceptible motion when the inclination was only 40', and on polished marble there was motion at the "smallest possible inclination. The motion, in fact, afforded almost as sensitive a test of deviation from horizontality as the spirit-level itself."

In all these cases the ice "melted continuously but very slowly" at its surface in contact with the slab. He attributed the unaccelerated motion "to the circumstance of the lower surface of the ice being in a state of constant, though slow distintegration." The water acted as a lubricant.

When there was no melting, the angle of repose on sandstone was about the same for ice as for marble, and about equal to the inclination (20°) for the initiation of accelerated motion when there was melting.

64. Deformability of Ice

The ways in which the volume and form of a block of ice vary with its uniform temperature and with the uniform hydrostatic pressure to which it is subjected will be found in Section 67. Such changes in form are slight, depending solely upon the anisotropy of the block.

Changes produced by non-uniform stresses, and the way they vary with the temperature, will be considered here. Under such stresses, the form of a block may be changed enormously.

Although a large number of experiments of various kinds bearing upon the behavior of ice under non-uniform stresses have been reported, only a few of them are of such a kind as to yield numerical data of general applicability. Others, however, are of much interest, and contribute valuable descriptive information regarding the characteristics of ice. For these reasons they deserve a place in such a compilation as this. The experiments cannot be satisfactorily separated into mutually exclusive groups.

---

[187] Hopkins, W., *Phil. Mag. (3)*, **26**, 1-16 (1845).
[187a] Moseley, H., *Idem (4)*, **42**, 138-149 (1871).

They will be considered under the following heads, those yielding no numerical data of general applicability being termed descriptive, even though many of them contain quantitative data.

Descriptive treatment	Quantitative treatment
Linear compression	Young's modulus
Extension	Poisson's ratio
Flexure	Rigidity
Punching	Tensile strength
Penetration	Strength, linear compression
Flowing	Shearing strength
Recovery	Hardness
Brittleness	Plasticity and viscosity
	Sustaining power

Whatever other phenomena may be involved, most permanent deformations produced in ice by non-uniform stresses involve a melting where the stress is great, the flowing of the supercooled water thus produced, and the refreezing of that water when its pressure has been reduced—all in accordance with the conclusions reached by J. Thomson.[188]

It should be remembered that such melting can occur only when the melting point of ice under the stress is lower than the existing temperature of the ice, and also, as shown in the second of Thomson's articles, that when the stress is borne by the ice alone such positive melting will occur whatever the nature of the stress—pressure, tension, torsion, etc.—and the attendant lowering of the temperature will lead to the formation elsewhere of unstressed ice from the melt, except as such formation is impeded by the absence of nuclei suitable for its initiation. See also E. Riecke.[189]

On account of such effects, a mass of ice might flow and exhibit plasticity although its individual crystals remained unbroken and perfectly elastic.

The early papers treating of melting under pressure and of regelation should be read, *i.e.*, those already cited and J. Thomson,[190] W. Thomson (Lord Kelvin),[191] M. Faraday,[192] and the topic Regelation in this volume (p. 412) should be consulted.

J. Johnston and L. H. Adams [193] have given reasons for believing "that every *permanent* deformation of a crystalline aggregate is conditioned by, and consequent upon, a real melting," which melting is in general to be ascribed to an inequality in the pressures on the liquid and on the solid.

---

[188] Thomson, J., *Trans. Roy. Soc. Edinburgh*, **16**, 575-580 (1849); *Proc. Roy. Soc. (London)*, **11**, 473-481 (1861).

[189] Riecke, E., *Ann. d. Physik (Wied.)*, **54**, 731-738 (1895).

[190] Thomson, J., *Proc. Roy. Soc. (London)*, **8**, 455-457 (1857); **10**, 152-160 (1859); **11**, 198-204 (1861).

[191] Thomson, W. (Lord Kelvin), *Phil. Mag. (3)*, **37**, 123-127 (1850); *Proc. Roy. Soc. (London)*, **9**, 141-143, 209-213 (1858).

[192] Faraday, M., *Proc. Roy. Soc. (London)*, **10**, 440-450 (1860); "Exp. Res. in Chem. and Phys.," pp. 372-374, 377-382, 1859.

[193] Johnston, J., and Adams, L. H., *Am. J. Sci. (4)*, **35**, 205-253 (211) (1913).

64. ICE: DEFORMABILITY

They accept J. H. Poynting's conclusion [194] that, when ice is subjected to a pressure of P atm more than that to which the water in contact with it is subjected, the melting point is lowered by $0.0862P$ °C, 11.5 times as many degrees as it would have been if this additional pressure had been exerted upon the water as well as the ice. In that paper, Poynting gives experimental evidence that such is the case. Johnston and Adams state that such a lowering is accepted by Roozeboom, Ostwald, LeChatelier, and Nernst, but not by G. Tammann.[195] Whether, and in how far, such melting accounts for the ease with which ice at -11 °C was flowed by G. Beilby [196] remains to be determined (see p. 422).

Both J. Y. Buchanan [197] and G. Quincke [198] were of the opinion that the mechanical deformation of ice is greatly facilitated, if not conditioned, by the presence between the crystals of thin films of liquid (a solution) of low melting point. They believed that such films are always present.

Discussions of the deformability of ice may be found in such treatises as those of G. Tammann,[199] H. Hess,[108] G. Beilby,[88] and H. T. Barnes,[200] as well as in the scientific journals.

Descriptive Treatment.

Linear Compression.—In his abstract of the paper which he presented before the Royal Society of London, Dec. 17, 1857, J. Tyndall [60] described thus the behavior of a cylinder of ice subjected to longitudinal pressure. The cylinder "was placed between two slabs of boxwood and subjected to a gradually increasing pressure. Looked at perpendicular to the axis, cloudy lines were observed drawing themselves across the cylinder. Looked at obliquely, these lines were found to be the sections of dim surfaces which traversed the cylinder, and gave it the appearance of a crystal of gypsum whose planes of cleavage had been forced out of optical contact by some external force.

"The surfaces were not of plates of air, for they are formed when the compressed ice is kept under water. They also commence sometimes in the center of the mass, and spread gradually on all sides till they finally embrace the entire transverse section of the cylinder. A concave mirror was so disposed that the diffuse light of day was thrown upon the cylinder while under pressure. The hazy surfaces produced by the compression of the mass were observed to be in a state of intense commotion, which followed closely upon the edge of the surface as it advanced through the solid. It is finally shown that these surfaces are due to the liquefaction of the ice in planes perpendicular to the pressure.

---

[194] Poynting, J. H., *Phil. Mag. (5)*, **12**, 32-48 (1881).
[195] Tammann, G., *Ann. d. Physik (4)*, **7**, 198-224 (1902); "Krystallisieren und Schmelzen," pp. 173-181, 1903.
[196] Beilby, G., "Aggregation and Flow of Solids," pp. 194-200, 1921.
[197] Buchanan, J. Y., *Nature*, **35**, 608-611 (1887); **36**, 9-12 (1887).
[198] Quincke, G., *Proc. Roy. Soc. (London) (A)*, **76**, 431-439 (1905) → *Nature*, **72**, 543-545 (1905).
[199] Tammann, G., "Aggregatzustände," 1922.
[200] Barnes, H. T., "Ice Formation," 1906; "Ice Engineering," 1928.

"The surfaces were always formed with great facility parallel to those planes in which the liquid flowers [flowers of ice] already described are produced by radiant heat, while it is exceedingly difficult to obtain them perpendicular to those planes. Thus, whether we apply heat or pressure, the experiments show that ice melts with peculiar facility in certain directions." (Cf. W. Thomson (Lord Kelvin).[201]

Whence we may conclude that his description of the appearance and the formation of the surfaces applies primarily to cylinders so cut that the length is parallel to the optic axis of the ice; and that, under the action of linear compression along the axis, fusion occurs in discrete planes which are perpendicular to the axis.[202]

The intense commotion observed by Tyndall is probably related quite closely to that reported by Quincke[198] as occurring at the edge of an enlarging flower of ice (q.v.), and which he attributes to "a periodic capillary spreading out (Ausbreitung) of the salt solution of the foam-walls at the boundary between pure water and air or vacuum," except that in Tyndall's experiment a diffusion, rather than a capillary, phenomenon is probably involved.

In 1885, Koch reported that when subjected to a constant pressure (19 kg*/cm^2) along its axis a certain cylinder of ice shortened at the following rates: At -5.7 °C, 0.9×10^{-4} per hr; at -2.5 °C, 17×10^{-4} per hr; at -0.9 °C, 126×10^{-4} per hr. The direction of the optic axis is not stated, but it was probably parallel to the axis of the cylinder.[203]

Three years later, McConnel and Kidd reported that blocks cut from a uniform sheet of ice and subjected to a pressure of 3.7 kg*/cm^2 perpendicular to the optic axis appeared to yield at the rate of only 0.1×10^{-4} per hr, but that even this was probably entirely spurious.[204] The temperature seems to have been near 0 °C. The observations extended over 4 days. Even if this apparent yielding were true, and if the yielding were proportional to the pressure, these blocks under 19 kg*/cm^2 would yield only 0.5×10^{-4} per hr, as compared with the 126×10^{-4} per hr observed by Koch when the pressure was probably parallel to the axis.

For 3 pieces of glacier ice under a pressure of 3.2 kg*/cm^2 the yielding was 3.5×10^{-4}, 5.6×10^{-4}, and 0.7×10^{-4} per hr, respectively; the optic axes of the grains were randomly oriented. Observations extended over 5 days.

Similar observations have been reported by von Engeln.[205]

Extension.—The presence of numerous crevasses in every glacier, and the suddenness with which they are frequently formed, led students of glaciers to conclude that ice can yield to tension only elastically or by frac-

---

[201] Thomson, W. (Lord Kelvin), *Proc. Roy. Soc. (London)*, **9**, 141-143, 209-213 (1858).
[202] See also Tyndall's "The Forms of Water in Clouds and Rivers, Ice and Glaciers," New York, D. Appleton & Co., 1872.
[203] Koch, K. R., *Ann. d. Physik (Wied.)*, **25**, 438-450 (1885).
[204] McConnel, J. C., and Kidd, D. A., *Proc. Roy. Soc. (London)*, **44**, 331-367 (1888).
[205] von Engeln, O. D., *Am. J. Sci. (4)*, **40**, 449-473 (1915).

ture; and this, in turn, gave rise to many experimental investigations of the behavior of ice under tension.

Observations of Pfaff [206] and the more extended ones of Fabian [207] showed that ice yields progressively when subjected to a continuing, constant tension. Main [208] stated: "Ice subjected to tension stretches continuously by amounts which depend on the temperature and the tensile stress. When the stress is great and the temperature not very low, there are extensions amounting to 1 per cent of the length per day. So continuous and definite is the extension that it can even be measured from hour to hour. These extensions took place at temperatures which preclude the possibility of melting and regelation. The extension increases continuously with all stresses above 1 kilo per square cm, and at all temperatures between $-6\,°C$ and freezing." He, and also Fabian, used bars of ice which had been frozen in a mold, and which, therefore, were probably conglomerates of crystals very variously oriented.

McConnel and Kidd [204] observed that not only the rate, but even the existence, of a progressive elongation depends upon the structure of the ice. They carried out extensive experiments on the stretching of bars cut in specified orientations from uniform sheets of ice. (By a uniform sheet is meant one in which the optic axis has the same direction at every point. They speak of using single crystals, but their report contains nothing to indicate that their crystallographic tests sufficed to distinguish between a single crystal and a homogeneous sheet containing many crystals similarly oriented as regards their optic axes. Cf. Mügge.) The work was extended by experiments on the bending of bars, by McConnel [45] and by Mügge [81] (p. 434).

They found that when the optic axis is perpendicular to the line of tension there is no measurable progressive stretching even when the stress has half the breaking value and the temperature is near $0\,°C$. But if the optic axis is inclined $45°$ to the line of tension there is a marked progressive stretching. There seem to have been no measurements on bars in which the optic axis is parallel to the line of tension. Bars of glacier ice, grains varying in diameter from 2 mm to 30 or 100 mm, stretched rapidly, as did mechanically molded ice of which the structure was surely very irregular. "The change in the rate of extension produced by an alteration in the tension, was in every case altogether out of proportion to the magnitude of the latter."

They found that an icicle, which "is an example of ice formed of very minute crystals irregularly arranged," (cf. p. 417, *Icicles*) stretched very slowly indeed. They were loath to ascribe the slow stretching to the multicrystalline structure, but later experiments by McConnel [209] convinced him

---

[206] Pfaff, F., *Ann. d. Physik (Pogg.)*, **155**, 169-174 (1875); *Sitz-ber. phys.-med. Soc. Erlangen*, **7**, 72-77 (1875).
[207] Fabian, O., *Rep. Exp.-Physik (Carl's)*, **13**, 447-457 (1877) from *Sitz.-Ber. Krakauer Akad. Wiss. (Math.-Nat. Kl.)*, **4**, → Hess, H.[108], pp. [22], [23]
[208] Main, J. F., *Proc. Roy. Soc. London*, **42**, 329-330, 491-501 (1887).
[209] McConnel, J. C., *Proc. Roy. Soc. London*, **49**, 323-343, Exp. 8 (1891).

that the presence of intercrystalline faces does hinder plastic flow "by fettering the sliding of the layers in the separate crystals."

Flexure.—Reusch [209a] reported, in 1864, that he had given a thin strip of ice a permanent set by bending it carefully by hand; and that a bar of ice, 10 by 1.2 cm by 0.3 cm thick, suspended horizontally by two slings 8 cm apart, and loaded in the middle with 180 grams, became visibly bent in 20 to 30 minutes, and the depression of its middle section increased to 6 or 8 mm before the bar broke. The temperature of the room was a few degrees above 0 °C, and the length and the breadth of the bar were probably parallel to the plane of freezing; that is, the optic axis was probably vertical.

Since then, many experiments have been made upon the progressive bending of bars of ice supported horizontally on narrow blocks near their ends, and loaded midway between the blocks.

From such experiments, Pfaff [206] concluded that near its melting point ice behaves like wax, the continued application of a force, no matter how small, producing a permanent deformation. Nothing is said regarding the direction of the optic axis, but it was probably vertical.

McConnel and Kidd [204] seem to have been the first to study the behavior of bars cut from ice of known uniform structure and in a specified direction with reference to the crystalline axes.

From their observations, and from other observations on bending, made by himself, McConnel [45] concluded that "a crystal [of ice] behaves as if it were built up of an infinite number of indefinitely thin sheets of paper fastened together with some viscous substance which allows them to slide over each other with considerable difficulty; the sheets are perfectly inextensible and perfectly flexible. Initially they are plane and perpendicular to the optic axis; and when by the sliding motion they become bent, the optic axis at any point is still normal to the sheet at that point. Thus, when a bar with the optic axis transverse to its length is placed so that the axis is horizontal, and the sheets of paper consequently vertical and longitudinal, it refuses to take any plastic bend, however long the weight be applied. If the bar be now turned over, so that the sheets of paper are horizontal, quite a short interval suffices to produce a decided permanent depression of the middle of the bar."

Similar experiments by Mügge,[81] Tammann,[195, 1st] and Tammann and Salge [210] confirmed and complemented McConnell's conclusions. As a result it may be stated that when the optic axis is horizontal and perpendicular to the length of the bar no permanent bend is produced, however long the load is applied. But if the same rod is rotated about its length so that the optic axis is vertical, a very short interval suffices to produce a decided permanent depression (McConnel, Mügge). If the bar is placed midway between these two positions, so that the optic axis, perpendicular to the length, is inclined 45° to the vertical, and the ends are so clamped that they cannot rotate about the longitudinal axis, then the application

---

[209a] Reusch, E., *Ann. d. Physik (Pogg.)*, **121**, 573-578 (1864).
[210] Tammann, G., and Salge, W., *Neues Jahrb. Mineral., Geol.*, Beilage Bd., **57A**, 117-130 (1928).

of the load causes the portion between the supports to bend and to rotate about the longitudinal axis in such direction as to place the optic axis more nearly vertical (Mügge). In all these cases the optic axis at any point of the deformed bar occupies the same position in the transverse section of the bar as it did before the bar was deformed.

"When the optic axis was longitudinal, the bar bent indeed, but not very readily, and the general behaviour was more obscure" (McConnel). This case involves a punching effect (see below).

By experiments on a rod cut from ice in which the planes of freezing were inclined to the optic axis by 50°, McConnel [45, 2nd] (Experiment 14) showed that the phenomena just described are indeed related to the direction of the optic axis, and not to that of the planes of freezing. He concludes (Experiment 8) that the presence of interfaces between crystals hinders plastic flow "by fettering the sliding of the layers in the separate crystals."

Although somewhat differently pictured by him, Mügge's observations indicate that the sheets imagined by McConnel are permanently deformable without change in area, the permanence of the deformation being, perhaps, imposed by the "viscous substance" that binds the sheets together, rather than by the nature of the sheets themselves.

More recent, but apparently less extensive, work by M. Matsuyama [210a] confirms in the main those observations of McConnel and of Mügge upon which it touches, but Matsuyama concludes that the bending of such rods depends upon the relative displacements of the individual crystals rather than upon the distortions of the crystals themselves.

Such a picture of the structure of an ice crystal as was proposed by McConnel and extended by Mügge, supplemented by Tyndall's observation (1858) [60] that compression along the optic axis causes liquefaction in planes perpendicular to the axis, is of great value in any attempt to interpret observations having to do with the deformation of ice. (See also M. Faraday [55] and J. Thompson.[211])

Punching.—(See also Shearing strength, p. 449) O. Mügge [81] observed that when a bar of ice, cut with its length parallel to the optic axis, is supported horizontally on two blocks and most of the portion between the blocks is loaded uniformly by means of weights suspended from a stirrup resting on the top of the bar and nearly as broad as the distance between the blocks, then the portion directly under the stirrup is gradually forced downward without change in the direction of its optic axis. It is, in effect, punched from the bar. The entire periphery of the punched out portion is frequently marked by horizontal lines, but the ice remains perfectly clear, and exhibits no sign of cracks nor of optical anomalies due to strain. However the bar was rotated about the optic axis, no significant difference in the ease with which it can be punched was found; neither did a change in temperature from $-3°$ to $-16\,°C$ produce any marked effect.

[210a] Matsuyama, M., *J. Geol.*, 28, 607-631 (1920).
[211] Thomson, J., *Proc. Roy. Soc. London*, 11, 198-204 (1861).

There seems to be a minimum load below which punching does not occur. In one case a 5 kg* load produced no observable effect in 24 hrs, but when the load was increased to 7 kg*, the deformation was rapid.

When the length of the bar is perpendicular to the optic axis there is no such punching, however the axis may be oriented in the vertical plane.

When a portion is so punched from a bar whose length is parallel to the direction of the optic axis, the extent to which the ice has been punched from its initial position decreases progressively as one passes from either edge of the stirrup to the neighboring supporting block. Except possibly in extreme cases, the optic axis in these portions also remains horizontal. These lateral portions are marked by a series of faulting planes.

G. Tammann and W. Salge,[210] using a narrow stirrup to transmit the transverse punching force, have found that a longitudinal pressure of a few kilograms per cm^2 parallel to the optic axis may increase the number of these faulting planes some ten-fold if the temperature is -1 °C, but causes essentially no change in the number if the temperature is as low as -6 °C. If z is the punching force that must be applied in order to cause these planes to appear, to initiate a true punching, then at a fixed temperature, $\log_{10}(z_0/z) = bp$, p being the axially directed pressure; b is positive,* an increase in p producing a decrease in z. The value of z_0 depends upon instrumental details as well as upon the temperature. For certain conditions, they found the following values, the unit of z being 1 kg*, of p being 1 kg*/cm^2: $t = -1$ °C, $z_0 = 1.5$, $b = 0.52$; $t = -6$ °C, $z_0 = 2.2$, $b = 0.34$; $t = -12$ °C, $z_0 = 2.8$, $b = 0.11$. These values indicate that both z_0 and b are linear in t; $z_0 = 1.4(1 - 0.086t)$, $b = 0.56(1 + 0.066t)$. The lower the temperature the greater is the force required to initiate punching and the less is the effect of axially directed pressure.

Penetration.—If a solid object, such as a metal rod, tube, or ball, is pressed normally against a surface of ice, the pressure being maintained continuously, it gradually sinks into the ice, and the immediately surrounding ice rises in the form of a hillock. The rate at which a given object sinks under the action of a fixed pressure decreases as the temperature is reduced. Thus Pfaff [206] reports a case in which the object sank 3 mm in 2 hrs when the temperature was between -1 and 0 °C, 1.25 mm in 12 hrs when -4 to -3 °C, and, with 2.5 times that load, sank only 1 mm in 5 days when the temperature was between -6 and -12 °C. Somewhat similar observations have been reported by T. Andrews.[212]

If the force upon the object is inclined to the surface, the ice rises in a hillock in front of the object (Bianconi [213]). In all these cases, large blocks of ice were used, and the pressure was probably normal to the surface of freezing, and therefore, parallel to the optic axis, though there is in the

---

* The logarithmic expression given in their paper seems to be affected by a typographic error, as it requires this b to be negative, which conflicts with their computed values.

[212] Andrews, T., *Proc. Roy. Soc. London*, **40**, 544-549 (1886).
[213] Bianconi, J. J., *Compt. rend.*, **82**, 1193-1194 (1876).

papers nothing to indicate the direction. Andrews described his data for the rates of penetration at various temperatures as measures of the hardness at these temperatures, and as such they have been quoted, though the property they measure differs from all those commonly classed under that term.

J. T. Bottomley's experiments,[214] in which a loaded block of ice passed through a horizontal sheet of wire gauze, and a loaded wire cut through a block of ice, without in either case permanently damaging the block, are illustrations of both penetration and regelation (p. 412). Bottomley explained them correctly on the theoretical considerations of J. Thomson.[188] The pressure of the wire causes the ice to melt, chilling the wire, the water, and the contiguous ice; the water flows around the wire, thus becoming relieved of stress, and freezes; the heat liberated by the freezing warms the wire, and that the adjacent ice, replacing the heat abstracted by the previous melting, thus preparing the way for a repetition of the process. As shown by Bottomley, the process becomes exceedingly slow—evanescent —if the cold water is drained off before freezing, or if the wire be replaced by a cord. In the first case, the wire is deprived of the heat liberated by the refreezing of the water, and therefore it and the adjacent ice soon become chilled to the temperature corresponding to the melting point of ice under the existing stress; then the melting ceases except as heat is conducted along the wire from the surroundings. In the second case, the cord, a poor conductor of heat, is warmed by the freezing water only on its upper side, and its lower side and the adjacent ice soon become so cool that melting ceases. G. S. Turpin and A. W. Warrington [215] repeated Bottomley's experiments, arriving at the same conclusions. They explain Pfaff's observations at temperatures above $-1\,°C$ in a similar manner.

See also H. Hess.[108, pp. 14-18; 216]

Flowing.—An aggregation of irregular blocks of ice may be welded together by pressure into an apparently uniform mass, and ice may be made to flow through small openings and tubes.

The phenomena considered by J. Thomson (p. 430), *i.e.*, fusion under stress, flow of the melt, and regelation, are contributing factors in most, if not in all, and the controlling factors in many, of the laboratory experiments that demonstrate the flowing of ice (see also J. Thomson [144]). But there are other factors that need to be considered, which may in extreme cases become of prime importance, especially when the mass of ice is great and its temperature is low.

Using crystals of NaCl in their saturated aqueous solution, J. Thomson [217] demonstrated experimentally the welding and molding of crystals by stresses borne by them but not by the adjacent (saturated) liquid.

O. D. von Engeln [218] has reported the following observations on the

[214] Bottomley, J. T., *Nature*, **5**, 185 (1872).
[215] Turpin, G. S., and Warrington, A. W., *Phil. Mag. (5)*, **18**, 120-123 (1884).
[216] Hess, H., *Ann. d. Physik (4)*, **36**, 449-492 (1911).
[217] Thomson, J., *Proc. Roy. Soc. (London)*, **11**, 473-481 (1861).
[218] von Engeln, O. D., *Am. J. Sci. (4)*, **40**, 449-473 (1915).

welding of ice by pressure. Into a copper cylinder 4 in. in diameter and 12 in. high, with walls 1/16 in. thick, he placed a rough-hewn plug of pond-ice with the component crystals parallel to the length of the cylinder. The space between the plug and the cylinder was filled with water, and the water was frozen, thus filling the cylinder with solid ice. A longitudinal pressure of 500 lb*/in², carried in part by the copper cylinder, was applied for several hours; the pressure was then increased to 720 lb*/in², and was maintained over night; the next morning it was increased to 750 and at 4 P.M. to 1400 lb*/in², which was left on for 36 hrs. Then, at 20 °F (−6.7 °C), the highest temperature reached during the test, the lowest being −4 °F (−20 °C), the pressure was relieved, the cylinder was gently warmed, and the core of ice was slipped out. There had been no distortion of the metal cylinder. The core of ice was "of crystal clearness and homogeneous, showing no line of separation to mark the juncture of the rough-hewn prism of pond ice and the water frozen around it. The most striking result, however, was the fact that *the ice mass had been completely recrystallized*. The original pond-ice core was inserted with principal axes parallel to the pressure direction, the new crystals extended across the cylinder with their principal (and longer) axes at right angles to the pressure direction." They extended "straight across the cylinder instead of radially inward as might have been expected by analogy to the structure of cakes of can-frozen artificial ice."

The crystals were elongated and their terminals were wedge-shaped, thus contrasting sharply with glacier grains. He thought that the difference was due "to the fact that the conditions of our experiment permitted of no movement in the ice mass involved." In other experiments, differing from the preceding principally in the use of a softer metal cylinder, of cracked ice or snow with the crevices sometimes filled with water and frozen, sometimes not, and of a piston fitting into the cylinder so that the entire pressure was borne by the ice, the cylinders were distorted, allowing portions of the ice mass to move. In these experiments also there was a complete recrystallization, but the structure was granular, and the grains were "variously oriented crystallographically." (Such variously oriented grains had been previously reported by A. v. Obermayer.[228])

Whence he concluded "that a granular ice can be developed from snow by pressure with accompanying movement and at air temperatures eliminating the possibility of pressure melting and regelation." The pressures required are near those at which a cube of ice yields when not supported laterally.

The flowing of ice from a large mass through a small opening or tube is merely a special aspect of the penetration experiments of Pfaff,[206] Bianconi,[213] and Andrews,[212] and is illustrated on a grand scale by the flowing of a glacier through a valley of varying width.

The early tendency to attribute such flowing solely to the lowering of the melting-point by pressure, and the subsequent regelation when the

64. ICE: DEFORMABILITY

pressure was reduced, led to a vaguely held idea that the flow is actually that of a liquid, the portion of the ice under pressure being actually liquefied. This was thoroughly disproved by R. W. Wood [222] who showed by direct experiments that leaden balls embedded in the upper portion of a block of ice remain there even when the block is subjected to a continuously increasing pressure up to 7, 12, and finally 40 tons* per sq. in.;† at the highest pressure, water squirted through the pores of the iron cylinder containing the ice. He also showed that general liquefaction should not have been expected, for, when ice is initially in equilibrium, the application of pressure causes thawing throughout the entire volume of the ice. That abstracts heat, and proceeds only until the entire mass has been reduced to the melting point that corresponds to the applied pressure. This reduced temperature will facilitate the flow of heat from the surroundings into the ice, and will thus facilitate thermal melting, which will be added to that caused by the pressure. All the water so formed is subjected to the pressure, and will escape through all available channels, leaving the bulk of the ice almost as solid as before the pressure was applied. Not only is the resulting water distributed throughout the entire volume of the ice, but the fraction of the ice melted by the direct effect of the pressure is actually small. For example, if the initial temperature of the ice were 0 °C and no heat were allowed either to enter or to leave it, then under a pressure of 7 short-tons*/sq. in. the temperature would fall to -8.5 °C and less than 6 per cent of the original ice would melt; at 15 short-tons*/sq. in. the temperature would be -22 °C, less than 1/6 of the ice would be melted, and the system would be at the triple point (water, ice-I, ice-III); a further increase in pressure would cause freezing and a rise in temperature (see Sections 92 and 93). Some years after this work of Wood's, Sir James Dewar [219] performed the same experiment with lead shot embedded in ice, and obtained the same result. They did not sink at all, but were very irregularly distorted. He worked at -80 °C and used a pressure of 100 tons*/sq. in.‡ (15,000 atm.), believing that the melting-point would by that pressure be reduced to -80 °C. We now know (Sections 92 and 93) that at that temperature and pressure ice exists in the form of ice-VI, not of ice-I, nor of water. (See also Section 93, Fig. 13.)

Wood was interested in adiabatic melting, in distinguishing between "thermo-molten" and "pressure-molten" ice, and approximately obtained that condition, there being no way in which the apparatus could obtain heat except from the surrounding air. He comments on its intense cooling. It is probable that the temperature dropped approximately as has been indicated, and that the fraction of the ice that was actually molten was, as he stated and as the position of the balls indicated, very small.

† These seem to be the short tons (2000 lbs*).
‡ Long ton of 2240 lbs.*
[219] Dewar, Sir James, *Chem. News*, **91**, 216-219 (1905) ← *Proc. Roy. Inst. Gt. Brit.*, **17**, 418-426 (1903).

On the other hand, Dewar endeavored to keep the ice at $-80\,°C$ by embedding that part of his apparatus in solid CO_2; at that temperature, as we now know, the liquid phase does not exist at any pressure. There is no reason for expecting the shot to sink.

In contrast to Wood and Dewar, A. Mousson[220] had previously performed a similar experiment under approximately isothermal conditions and at such a temperature (-18 to $-21\,°C$) that, as we now know, the liquid phase does exist at certain pressures lower than the maximum (*ca.* 13,000 atm) reached by him. Under such isothermal conditions, the amount of melting is not limited in the manner considered by Wood, and it is to be expected that during a portion of the process the entire mass of ice will be molten, and that the metal object (in this case a rod of copper) will then fall to the bottom. That he observed. By far the greater portion of the ice was, of course, melted by heat drawn from the surrounding freezing mixture, the true pressure-melting being small; the pressure conditioned, rather than caused, the complete melting. Similar experiments have been reported by J. B. Boussingault.[221]

If an outlet is provided for the molten ice, it will flow toward it, and will progressively freeze as the pressure on it decreases, the fraction that freezes depending upon the conditions.

Thus there arises a spurious flowing of the ice, in which ice is melted at one point, and the resulting water flows to another, where it is refrozen. Such spurious flowing is always to be expected, but the line of advance of the ice so formed—the birthplace of that ice—is limited to those surfaces at which there is a steep gradient of pressure.

In addition to this, there is a true flowing in which a solid mass of ice undergoes changes in shape. This is shown by the manner in which embedded bodies and superficial landmarks (stakes, rocks, etc.) are carried along by glaciers, and by the flow of ice, under pressure, through contracting conical nozzles, and possibly by the corresponding flow through small apertures.

R. W. Wood [222] found that at a pressure of 3 tons*/sq. in. ice began to flow through a lateral hole (1/12 in. in diameter) in his compression chamber. It flowed slowly and steadily as a clear cylinder of ice, which broke off when 6 or 8 inches long. At 4.5 tons*/sq. in., "it seemed fairly to spurt from the orifice"; it flowed irregularly, sticking for a second or two and then yielding suddenly. The temperature was about $0\,°C$.

Sir James Dewar [219] has stated that near $0\,°C$ ice can easily be extruded in the form of a wire of clear, transparent ice; at $-80\,°C$ such flow still occurred, but "the ice wire was now made up of what looked like a set of disc-like scales"; at a temperature near that of liquid air, "no pressure the

---

[220] Mousson, A., *Ann. d. Physik (Pogg.)*, **105**, 161-174 (1858).
[221] Boussingault, J. B., *Compt. rend.*, **73**, 77-79 (1871) ← *Ann. de chim. et phys.* (4), **26**, 544-547 (1872) → *Ann. d. Physik (Pogg.)*, **144**, 326-329 (1871).
[222] Wood, R. W., *Am. J. Sci.* (3), **41**, 30-33 (1891).

apparatus would stand caused any flow, but only intermittent explosive ejections."

A similar experiment in which ice was forced to flow through a lateral tube 3/4 in. (1.9 cm) in diameter has been reported by O. D. von Engeln.[218] At a temperature of 22 °F (−5.5 °C), cracked ice was put in a pressure cylinder and by means of a closely fitting piston, a pressure of 3400 lb*/in² (239 kg*/cm²) was applied at 5 P.M. During the night the temperature fell to 4 °F (−15.5 °C), but rose to 14 °F (−10 °C) by 10:30 A.M., at which time the pressure had fallen to 3000 lb*/in² (211 kg*/cm²) and the extruded core of solid ice was 3 in. (7.6 cm) long; at 4:30 P.M. the core was $5\frac{1}{2}$ in. (14 cm) long, and the temperature was 29 °F (−1.7 °C). The temperature remained approximately that until 10 A.M., when the core was 14 in. (35.5 cm) long. The extruded rod was "perfectly clear, glassy and compact," even at the lowest temperature; the individual crystals were variously oriented; "shear lines and breccia bands could be identified but there were some apparently real crystal boundaries."

In 1902, H. Hess [223] forced ice from a cylinder through an attached conical nozzle, and measured the rate of advance of the piston acting upon the ice. In those experiments, essentially the entire flow was through the nozzle. He found there was no flow unless the pressure exceeded a certain value (p_{min}) depending upon the temperature and upon the ratio of the area of the emerging cylinder of ice to that of the initial cylinder. At 0 °C, he found the following values for (ratio, p_{min}): (1/9, 345), (1/6.3, 230), (1/3.1, 100), and (1/1.67, 30), the products of ratio times p_{min} being 38, 36, 32, and 18. The values of (t, p_{min}) for the fixed ratio 1/6.3 were (0, 230), (−3 to −5, 250), and (−10, 270). In both cases the unit of p is 1 kg*/cm².

When the pressure is kept constant, the rate of flow increases progressively; and after a rate of flow has been established, a reduced pressure will maintain it. This is explained by H. Hess [224] as due to the lubricating action of the water produced by the melting caused by the pressure.

Phenomena attending the flow of ice through apertures, and the flow of glaciers, have been studied and discussed by A. v. Obermayer [228] also.

A type of experiment that is intermediate between the simple experiments on penetration and those on the flow of ice through conical nozzles is that in which the ice is enclosed in a cylinder having a diameter greater than that of the piston to be forced into the ice. Under such conditions the piston can advance only as the ice is compressed or as an equivalent volume of ice (or water) is transferred through the gap between the piston and the walls of the cylinder. An interesting experiment in which the gap is wide is described by Hess.[224]

In such experiments, G. Tammann,[229] employing small gaps, found that, for any given temperature and pressure, the rate of advance of the

---

[223] Hess, H., *Ann. d. Physik (4)*, **8**, 405-431 (1902). See also, "Die Gletscher," pp. 28-31, and *Ann. d. Physik (4)*, **36**, 449-492 (1911).
[224] Hess, H., *Ann. d. Physik (4)* **36**, 449-492 (1911).

piston is normal unless the pressure is at least essentially as great as that (p_m) at which ice melts at the given temperature. An increase in temperature or in pressure increases the rate; and as the gradually increased pressure passes through p_m, the rate changes abruptly from a small value to a very much greater one. The rates observed for a given apparatus are shown in Table 189. In a later paper [210] he represents these data by the formula $\log_{10} r = k(p - p_0)$, where r is the rate of advance of the piston, p is the applied pressure, and k and p_0 are constants for a given temperature and apparatus. Actually, the choice of the values to be assigned to k and to p_0 involves quite an element of judgment, and those chosen satisfy only those observations for which the tabulated value of r exceeds unity. Values of k and p_0 are given at the bottom of Table 189. Similar experiments have been described by N. Slatowratsky and G. Tammann.[225]

Table 189.—Flow of Ice through an Annular Gap [229]

r = rate of advance of the piston into a cylinder of slightly greater diameter; pressure on the piston = p kg*/cm²; temperature = t °C. For a given p and t, r is normal if $p \lessapprox p_n$; at $p = p_f$ the rate was too great for measurement; p_m is the hydrostatic pressure at which ice-I melts at t °C. G. Tammann and W. Salge [210] represent these data by the formula $\log_{10} r = k(p - p_0)$, where k and p_0 have the values here given. (See text.)

Unit of $p = 1$ kg*/cm²; of $r = 0.0004538$ cm/min = 4.538 μ/min.

$t \rightarrow$ p	−5.7	−10.7	−15.7	−21.7
		r		
100	0.9	0.03	—	—
200	4.1	0.3	—	—
300	11.8	2.00	0.1	—
400	22.5	4.1	0.3	0.15
500	49.5	8.3	1.5	0.3
600	95.0	19.0	5.1	0.5
700	—	34	12.6	2.5
800	—	60	2.0	7.0
900	—	101	—	13.5
1000	—	170	—	20.5
1100	—	—	—	30
1200	—	—	—	53
1300	—	—	—	65
p_n	642	1116	1729	2000
p_f	665	1130	1787	2100
p_m	678	1225	1681	2170
$10^4 k$	35	29	34	20
p_0	10	190	390	360

In reference to the flowing of glaciers, R. M. Deeley and P. H. Parr[226] remark: "We have seen that glacier ice consists of crystal granules which not only shear freely along planes at right angles to the optic axis, but also undergo changes at their bounding surfaces which enable the mass to suffer

---

[225] Slatowratsky, N., and Tammann, G., *Z. physik. Chem.*, **53**, 341-348 (1905).
[226] Deeley, R. M., and Parr, P. H., *Phil. Mag. (6)*, **26**, 85-111 (1913).

continuous distortion under stress. The ability of glacier ice to spread out into piedmonts whose upper surfaces are very nearly level also shows that such shear may take place under very small stresses."

M. Matsuyama [210a] has expressed the opinion that in the distortion of ice composed of parallel crystals the surfaces between adjacent crystals play a more important role than the gliding planes perpendicular to the optic axis.

The hillock formed around an object forced into a block of ice in simple experiments on penetration, that formed ahead of a loaded object pushed along a surface of ice (Bianconi [213]), and the spreading that sometimes occurs at the loaded section when a horizontal bar of ice supported at its ends is loaded in the middle, are all special cases of flowing. The last, the spreading of the bar, occurs only when the optic axis is transverse to the length of the bar, and the load is at least of the order of 12 kg*/cm^2 (half the crushing load). See Hess.[227]

Recovery from stress.—When ice is relieved from stress, it partially returns to its unstressed form and size, provided that the stress has been neither excessive nor too long-continued. Part of the recovery is immediate, and part is progressive. It is the latter, the so-called elastic aftereffect, and especially as it relates to nonuniform stress, that forms the subject of this section. This progressive recovery has been noticed by many.

K. R. Koch [230] has given a few data showing the magnitude and the slowness of the recovery, which indicate that the lower the temperature, the greater is the amount of the progressive recovery, and the longer it takes. For an experiment at -12.5 to $-15\,°C$ he records, in arbitrary units, immediate recovery = 12.5, total recovery in 25 min 25 sec = 48.0, followed by an additional recovery of 10 in the next 10 hr 11 min. The total recovery was over 4.6 times the immediate recovery, and required hours.

In another experiment at $-1.5\,°C$ he records: Immediate recovery = 30, total in 5 min 15 sec = 41.8, no change in the next minute. Here, the total was only 1.4 times the immediate, and required only 5 min.

In his study of the bending of bars, J. C. McConnel [231] gave special attention to recovery from strain. He wrote: "In several cases .... after a heavy weight was removed, a slight gradual unbending of the bar took place. At first I thought this a mere consequence of the irregular elastic strains on the bar, the parts most severely strained gradually bending back the rest. But the magnitude of the recovery seems, on closer examination, to put this explanation out of the question, and I have now little doubt that it is a true molecular effect.... I conclude, then, that we have to deal with

[227] Hess, H., "Die Gletscher," p. 21; *Ann. d. Physik (4)*, **8**, 405-431 (1902).
[228] v. Obermayer, A., *Sitzb. Akad. Wiss. Wien* [2a], **113**, 511-566 (1904).
[229] Tammann, G., *Ann. d. Physik (4)*, **7**, 198-224 (1902).
[230] Koch, K. R., *Ann. d. Physik (Wied.)*, **25**, 438-450 (1885).
[231] McConnel, J. C., *Proc. Roy. Soc. (London)*, **49**, 323-343 (1891).

a real tendency of the forcibly displaced sliding layers to slide back. The rate of recovery, rapid at first, soon falls off."

M. Matsuyama [210a] has given certain data for the recovery of rods of ice from torsion about the axis of the rod, but information concerning the time allowed for that recovery seems to be lacking.

Brittleness.—E. Reusch [209a] observed that brilliant cracks, like those produced in glass by means of a diamond, can be produced in ice by pressing upon it with a convex knife-blade; and that these cracks can be formed even when the ice is in a warm room, and consequently, is covered with a layer of water. This indicates that ice is brittle even when near its melting point. G. Tammann and W. Müller [232] have stated that at 0 °C ice is as brittle as is rock salt when 700 °C below its own melting point.

Nevertheless, ice yields progressively to the action of differential stresses even when below its melting point. It is both plastic and brittle.

In speaking of the work of E. Brown,[233] H. T. Barnes says that ice splinters considerably when sawed at temperatures near 0 °F (-17.8 °C), but it can be sawed at 30 °F (-1.1 °C) with comparatively little difficulty.[234]

Quantitative Treatment.

Young's modulus.—The determination of the value of Young's modulus of elasticity (E) of ice by the usual static methods is rendered very difficult by the progressive yielding of ice to stress, and by its partial progressive recovery (elastic after-effect) when the stress is removed. Indeed, it is questionable whether significantly useful values can be obtained by such methods (see Boyle and Sproule, 1931). Nevertheless, most of the values commonly cited have been so derived, and are given in the last two sections of Table 190, as a matter of historical interest.

On the other hand, the dynamic methods based upon the velocity of propagation of high-frequency vibrations lead to values of E that are unaffected by the progressive yielding (see Boyle and Sproule, 1931). They alone deserve serious consideration.

Values obtained by the static method exhibit wide variations, and have been interpreted as indicating that the value of E depends upon the angle between the stress and the optic axis. This has not been borne out by the results obtained by the dynamic method, which indicate that values of E found for specimens that are nominally identical may differ by some 10 to 15 per cent, and that the observed variations with the orientation of the optic axis are of about the same magnitude. That is, there is no certain dependence of E upon the orientation of the crystal.

The observations of Trowbridge and McRae (1885), and some of those of Hess (1902, 1904) indicate that, when the apparent value of E is derived from the bending of a loaded horizontal bar, the planes of freezing being

[232] Tammann, G., and Müller, W., *Z. anorg. allgem. Chem.*, **224**, 194-212 (1935).
[233] Brown, E., Rep. Joint Board Eng. for St. Lawrence River.
[234] Barnes, H. T., "Ice Engineering," p. 224, 1928.

Table 190.—Young's Modulus of Ice

The most accurate values are those of Boyle and Sproule, of which the individual determinations in any given case lie within a range of 3.5 per cent after correction has been made for differences in the temperatures; different specimens, nominally identical, occasionally differed by 15 per cent. They used longitudinal vibrations of frequencies between 7 and 13 kilocycles/sec.

Values obtained by the static method are unreliable (see text) and mainly of historical interest.

θ = angle between the length of the specimen and the normal to the surface of freezing; E = Young's modulus; l = length of the specimen, w = width, τ = thickness in direction of application of the load; Op. Ax. = optical axis.

Unit of $E = 1$ kg*/mm² $= 1424$ lb*/in² $= 98.1$ megadyne/cm². Temp. $= t\,°C$

I. Dynamic method. Longitudinal vibrations except as noted.

	Boyle and Sproule[a]; $\theta=0°$					Boyle and Sproule[a]; $t=-26\,°C$		
	-9	-10	-30	-35	θ	$0°$	$45°$	$90°$
E	947	967	1040	1110	E	970	900	990[b]
E_c	954	960	1060	1090	E			945[c]
	$E_c \equiv 909(1 - 0.00558t)$				E_c	1040		

Miscellaneous values; $\theta = 90°$.

Vibration	——— Longitudinal ———				——— Trans. ———		?
t	-26	-26	-6	-4	-7	$0(?)$	?
E	990	945[c]	880	960	884	236	710[d]
Ref.[a]	BS	BS	TMcR	RS	K85	R	Ko

II. Static method. Bending of loaded horizontal bars.

Sheet ice.

Unless another direction is indicated, the load is applied perpendicular to the surface of freezing, if $\theta = 90°$. (From the data given on p. 611 of Matsuyama's article it may be seen that the unit he used and called the "c.g.s." unit is actually 1 gram-weight per cm², not 1 dyne/cm².)

Op. Ax. t	$\|l$	$\|\tau$	$\|w$	Ref.[a]	θ t	$0°$	$90°$ E	Ref.[a]
0 to -1	182		383	H	-5.4		642[e]	K85
-2 to -5	59			H	-5 to -7		860	TMcR
-1 to -5		254	418	H	-6.5 to -7.8	609	622	K13
(?)	67	194	336	H	-6.5 to -7.8		656[f]	K13
-3.5	185	60	92	Ma	-9		696	K85
					(?)		1120	K14
					(?)		950[e]	Mo
					(?)		500	B

Granular ice, natural and compressed.

t	E	Type of ice	Ref.[a]
-1 to -3	285	Large grains	H
0 to -3	226	Small grains	H
-0.5	300	Mixed sand and water, frozen	H
-2	150	Snow compacted by 33 atm	H
0 to -0.2	49	Snow compacted by 20 atm	H
0 to -3	280	Crystals randomly oriented	H
$-6(?)$	620[g]	Crystals disordered[h]	TMcR
?	190[i]	Crystals disordered[h]	F

Table 190—(Continued)

[a] References:

B	Bevan, B., *Phil. Trans.*, **116**₂, 304-306 (1826).
BS	Boyle, R. W., and Sproule, D. O., *Can. J. Res.*, **5**, 601-618 (1931).
F	Fabian, O., *Repert. Exp. Phys. (Carl)*, **13**, 447-457 (1877).
H	Hess, H., *Ann. d. Physik (4)*, **8**, 405-431 (1902); "Die Gletscher," (1904).
K13, K14, K85	Koch, K. R., *Ann. d. Phys. (4)*, **41**, 709-727 (1913); **45**, 237-258 (1914); *(Wied.)*, **25**, 438-450 (1885).
Ko	Köhler, R., *Z. Geophys.*, **5**, 314-316 (1929).
Ma	Matsuyama, M.[210a]
Mo	Moseley, (Canon) H., *Phil. Mag. (4)*, **42**, 138-149 (1871).
R	Reusch, E., *Ann. d. Phys. (Wied.)*, **9**, 329-334 (1880).
RS	Reich, M., and Stierstadt, O., *Phys.*, **32**, 124-130 (1931).
TMcR	Trowbridge, J., and McRae, A. L., *Am. J. Sci., (3)*, **29**, 349-355 (1885).

[b] By a similar method, M. Ewing, A. P. Crary, and A. M. Thorne, Jr.,[235] found $E = 935$ kg*/cm² at t between -5 and $-15\,°C$; and obtained the same value for ice artificially frozen in a vertical tube as for a rod cut with its length parallel to the surface of freezing.

[c] Length perpendicular to the preceding, but both parallel to the surface of freezing.

[d] Isotropic lake ice.

[e] Direction of application of the load is not stated.

[f] Direction of application of the load is parallel to the surface of freezing.

[g] By transverse vibrations.

[h] The ice was frozen in a metal tube; the orientation of the crystals is neither ordered nor perfectly at random.

[i] Computed from the observed elongation under tension within his estimate of the elastic limit (load ≲ 0.51 kg*/cm²).

---

horizontal (optic axis vertical), as the load increases the apparent value of E decreases to a minimum, and then increases as the breaking load is approached. This variation is, however, small as compared with that of the apparent E from specimen to specimen.

Data reported by O. Fabian (1877) indicate that ice is almost perfectly elastic for tensions not exceeding 0.5 kg*/cm², but it should be noticed (Table 190) that the value he obtains for E under such conditions is very low.

Poisson's ratio.—As computed from the observed velocity of longitudinal waves, Poisson's ratio for ice is 0.365 ± 0.007, and is the same for ice frozen in a vertical tube as for a rod cut with its length parallel to the surface of freezing,[235] and from the horizontal velocity of waves in an isotropic ice-sheet, R. Köhler [236] derived the value 0.30. From static observations, B. Weinberg [237] had derived the value 0.38 ± 0.49; the direction of extension is not clearly indicated.

Rigidity.—The rigidity of ice has been derived by B. Weinberg,[237, 238] K. R. Koch,[239] M. Matsuyama,[210a] and C. D. Hargis,[240] from observations on the torsion of bars, Weinberg deducing it from the apparent viscosity by means of an extension of Maxwell's theory (see *Plasticity*, p. 451),

---

[235] Ewing, M., Crary, A. P., and Thorne, A. M., Jr., *Physics*, **5**, 165-168 (1934).
[236] Köhler, R., *Z. Geophys.*, **5**, 314-316 (1929).
[237] Weinberg, B., *Z. Gletscherkunde*, **1**, 321-347 (1907).
[238] Weinberg, B., *Ann. d. Physik (4)*, **22**, 321-332 (1907).
[239] Koch, K. R., *Idem*, **45**, 237-258 (1914).
[240] Hargis, C. D., *Phys. Rev. (2)*, **19**, 526, 527 (1922).

64. ICE: DEFORMABILITY

and the others from the twist produced by a known torque, care being taken to eliminate the effect of plastic yielding and of the elastic after-effect. Weinberg calculated also the relaxation time (τ), and the strain (λ) corresponding to the elastic limit. There appears to be an error in the equation used by him in those computations (see p. 453), but the size of the error so introduced into the computed values is not known. From the numerical data on p. 611 of Matsuyama's paper it may be shown that the unit which he designates by "c.g.s." is actually 1 gram-weight per (cm^2·radian), although in the table on p. 615 he compares such values with others which are actually expressed in terms of 1 cgs unit = 1 dyne/(cm^2·radian).

The dynamic determination—from high-frequency torsional vibration of rods—by M. Ewing, A. P. Crary, and A. M. Thorne, Jr.[235] yields a value that is markedly higher than those obtained by the static methods, but which is probably to be preferred, though as yet unsupported.

Table 191.—Rigidity of Ice

N = modulus of rigidity; θ = angle between length of cylinder and optic axis; λ = shear corresponding to the elastic limit; τ = relaxation time (p. 452); E = Young's modulus; $t\,°C$ = temperature.

N is determined from the twisting of cylinders by torques about the axis of figure. Weinberg obtained $N = 1.0\,(1 - 0.13t)\,10^{10}$ dyne/(cm^2·radian) when $\theta = 0$, and $N = 0.8\,(1 - 0.65t)$ for glacier ice. Matsuyama reports for $\theta = 90°$, $N = 1.16\,(1 - 0.080t - 0.0017t^2)\,10^9$ dyne/(cm^2·radian). Gl = glacier ice.

Unit of N = 1 kmegadyne/(cm^2·radian) = 1019 kg*/(cm^2·radian); of τ = 1 sec; of λ = 1 microradian = 0.206"

I. Preferred value. Dynamic method. ECT.[b]

$N = 91.7 \pm 0.5$; $t = -5$ to -15 °C. Water frozen slowly in vertical brass tube.[a]

II. Static or slow oscillation method.

$\theta \rightarrow$	0	90	Gl	0	Gl	0	Gl	0	90	Ref.[b]
t	\multicolumn{3}{c}{N}	\multicolumn{3}{c}{λ}	\multicolumn{2}{c}{τ}	\multicolumn{2}{c}{E/N}	
—	27.2	29.4	—	—	—	—	—	4.03	3.18	K
0	10	—	8	56	10	960	480	—	—	W
−5	17	—	34	34[c]	3	1670	720	—	—	W
−5	—	1.6	—	—	—	—	—	—	—	M
−6	1.8	—	—	—	—	—	—	—	—	M
—	28.2[d]	—	—	—	—	—	—	—	—	H

[a] At an unstated temperature and for an isotropic ice-sheet, R. Köhler[236] found $N = 27$.

[b] References:
  ECT Ewing, M., Crary, A. P., and Thorne, A. M., Jr.[235]
  H   Hargis, C. D.[240]
  K   Koch, K. R.[239]
  M   Matsuyama, M.[210a]
  W   Weinberg, B.[238]

[c] In finding λ, N was taken as 16.5.
[d] Value of θ is unknown; the ice was frozen in a brass tube.

Tensile strength.—The tensile strength of ice may be expected to depend upon the structure of the ice and upon the direction of the line of stress with reference to the optic axis of the crystal, or crystals, of which the specimen is composed. No data on the tensile strength of single crystals or of ice of known uniform structure have come to the compiler's attention. The recorded values, at unstated temperatures, range from 2.4 to 16 kg*/cm², and Reusch computes 68 kg*/cm² as the maximum tension at the instant of rupture of a centrally loaded rod supported near its ends.†

Working at −8 °C and loading the specimen at the rate of 0.1 kg*/(cm²sec), H. Romanowicz and E. J. M. Honigman [241] found in three tests the mean values 16.1, 18.3, and 17.7 kg*/cm², the highest observed value being 24.8, and the lowest 14.8 kg*/cm².

Strength in linear compression.—The values reported for the linear compressive stress required to rupture a block of ice vary from 5 to 125 kg*/cm² (70 to 1800 lb*/in²), the values most frequently found lying near 25 kg*/cm² (360 lb*/in²). This wide variation is in large part due to variations in the structure of the ice, but in part to the technique employed. Barnes,[234, p. 228] quoting Prof. E. Brown,[233] says that the observed crushing stress depends upon the rate at which the stress is applied. That rate is seldom reported. The crushing stress may depend also upon the size of the specimen (G. van Diesen, 1871).

For 7-cm cubes loaded at the rate of 3 kg*/(cm²sec), H. Ramanowicz and E. J. M. Honigman [241] found in three tests the mean values 40.0, 43.0, and 44.1 kg*/cm², the extreme observed values were 54.4 and 34.0; the ice was formed by freezing water in cubic forms a little larger than the desired finished block.

No data for an isolated crystal have come to the compiler's attention, the data available referring to blocks of natural or of ordinary artificial ice, which consist of multitudes of crystals seldom arranged in more than approximate uniformity. In some cases the average direction of the optic axes was inferred from the direction of the planes of freezing, being assumed to be normal to those planes. For the values obtained in such cases, see Table 192.

Barnes [242] has reported that when blocks of ice are subjected to linear compression they may be heard to crack at approximately half the pressure required to crush them; they then stiffen perceptibly. He was unable to see these cracks. When the line of pressure is normal to the surfaces of freezing "the ice bursts sideways into innumerable long needles, resembling a cake of ice which has all but fallen to pieces in the sun." When the line of pressure was parallel to those surfaces "the block cracked lengthwise and transversely without shattering." The required pressure is somewhat greater in the first case than in the second.

---

† See: Barnes, H. T., Hayward, J. W., and McLeod, N. M., *Trans. Roy. Soc. Canada III, (3)*, **8**, 29-49 (1914); Fabian, O., *Rep. f. Exper.-Physik (Carl)*, **12**, 397-404 (1876) ← *Sitz.-ber. d. Krakauer Akad. Wiss. (Math.-Nat. Kl.)* Vol. 3; Finlayson, J. N.[243]; Hess, H., "Die Gletscher," p. 23; (Canon) Henry Moseley, *Phil. Mag. (4)*, **39** 1-8 (1870); Reusch, E., *Ann. d. Physik (Pogg.)*, **121**, 573-578 (1864).

[241] Romanowicz, H., and Honigman, E. J. M., *Forsch. Gebiete Ingenieurw.*, **3**, 99 (1932).

[242] Barnes, "Ice Engineering," 1928, p. 220. *Trans. Roy. Soc. Canada, III (3)*, **8**, 19-22 (1914).

64. ICE: DEFORMABILITY

Von Engeln reported that when a pressure approaching the crushing value was released, the ice frequently cracked, and if the release was rapid it actually broke apart, showing that it retained its elasticity. But if the pressure was maintained near the crushing value the ice yielded by flow without breaking.

Under given conditions, the strength increases as the temperature is reduced.

Taking 400 lb*/in² as the crushing stress, Barnes, Hayward, and McLeod computed the following values for the greatest possible thrusts per transverse linear foot (30.5 cm) when the ice has the thicknesses indicated.

Thickness	6	8	10	12 in.
	15.2	20.3	25.4	30.5 cm
Thrust	28800	38100	48000	57600 lb*/ft
	42900	57200	71500	85700 kg*/m

Other references:†

Table 192.—Strength of Ice in Linear Compression

The following data refer to ice of which the structure is believed to be uniform, the optic axes of the crystals being perpendicular to the surface of the water on which the ice was formed. Stress $\|$ ($\perp$) indicates that the compression is parallel (perpendicular) to the optic axis.

Unit of Strength $\doteq$ 1 kg*/cm² = 14.24 lb*/ft² = 0.981 megadyne/cm². Temp. = t °C

Stress→ t	$\|$	$\perp$	Source	Observer[a]
	Strength			
0	26	25	River	Barnes
−2.2		21	River	Brown
−10		49	River	Brown
−16.6		62	River	Brown
−11.7	124	72	River	Finlayson
−7	70	25	Pond	von Engeln

[a] Barnes, H. T.[242]; Brown, E.[233, 234], pp. 223–228; von Engeln, O. D., *Am. J. Sci.* (4), **40**, 449–473 (1915); Finlayson, J. N.[243]

Shearing strength.—J. N. Finlayson [243] reported that at temperatures above 20 °F (−6.7 °C) the shearing load had to be applied in his experiments "quite rapidly in order to secure satisfactory results."

"In cases where the specimens were sheared in the direction parallel to the optical axes of the crystals, beautiful conchoidal fractures were frequently obtained, indicating that the specimens had sheared along the walls of crystals."

The average of his values for the shearing strength was 114 lb*/in² (8.0 kg*/cm²) perpendicular to the optic axis, and 98 lb*/in² (6.9 kg*/cm²) parallel to that axis. Observations extending from −10 °F to +30 °F

† Barnes, H. T., Hayward, J. W., and McLeod, N. M., *Trans. Roy. Soc. Canada, III (3)*, **8**, 29-49 (1914); van Diesen, G., *Vers. en Med. K. Akad. Wet., Amsterdam* (2), **5**, 325-331 (1871); von Engeln, O. D., *Am. J. Sci.* (4), **40**, 449-473 (1915); Finlayson, J. N.[243]; Hess, H., "Die Gletscher"; Moseley, (Canon) Henry, *Phil. Mag.* (4), **39**, 1-8 (1870); Anon., *Z. ges. Kälte-Ind.*, **33**, 84-85 (1926), a few values derived from observations made at the government's testing bureau at Copenhagen are cited anonymously.

[243] Finlayson, J. N., *Canadian Engineer*, **53**, 101-103 (1927).

($-23.3\,°C$ to $-1.1\,°C$) gave no indication that the strength varies with the temperature. Individual determinations differ widely. A few tests of artificial ice indicated that its shearing strength is about 80 per cent of that of river ice.

"No marked elastic limit was noticeable before the specimens sheared off; but there was evidence of a slow realignment of crystals under pressure, as the load was found to fall off if the head of the testing machine were brought to rest during the test."

These conclusions of Finlayson are quoted by H. T. Barnes.[234, p.217]

Weinberg's data (Table 191) indicate that the elastic limit in shear is certainly less than 2 kg*/cm² and probably not over 1 kg*/cm², the limiting shear being a few seconds of arc (0.6″ to 11″).

Canon Henry Moseley [244] found no detectable shearing of a cylinder of ice 1.5 inches in diameter when the shearing force was 112.5 lbs*, but an appreciable shearing when the load was 121 lbs*; the temperature was stated to have been below freezing. The shearing apparatus consisted of two boards of hard wood held together by guides, and sliding one over the other; a 1.5-inch hole, to take the ice, was bored through each board. With this apparatus he found that a load of 208 lbs* caused shearing at the rate of 0.016 in/min; working in air at 74 to 75 °F he found that 200 lbs* gave a rate of 0.025 in/min for solid, natural ice, and 0.036 in/min for regelated ice formed by hammering cracked ice into the hole in the shearing apparatus.[245] His conclusion was that the shearing strength of such compacted ice is about 75 lb*/in² (= 5.3 kg*/cm²)[246].

Hardness.—Three kinds of hardness are commonly recognized. One indicates the resistance to abrasion; a second, the resistance to denting under the action of a dead load; and a third, the height of rebound of a specified object dropped in a specified manner.

The first is generally used in the description of minerals, and is indicated most frequently in terms of a scale defined arbitrarily by a specified set of minerals. On Moh's scale, the hardness of ice is generally given as 1.5; that is, its resistance to abrasion lies about midway between that of talc and that of gypsum.[234, p. 18; 247, 248] E. S. Dana [249] does not state the hardness of ice, but on the basis of several references, it is believed that either earlier editions of this work or J. D. Dana [250] gave the value of 1.5.

The second and the third kinds of hardness are commonly used in the description of metals, and are specified, respectively, by what are known as the Brinell hardness number and the Shore scleroscope hardness. No data for either of these have been found for ice, although there is no obvious

---

[244] Moseley, Canon Henry, *Phil. Mag. (4)*, **42**, 138-149 (1871).
[245] Moseley, Canon Henry, *Phil. Mag. (4)*, **39**, 1-8 (1870).
[246] Moseley, Canon Henry, *Proc. Roy. Soc. (London)*, **17**, 202-208 (1869).
[247] Van Horn, F. R., "General and Special Mineralogy," p. 458, published by the author, Cleveland, 1903.
[248] Bayley, W. S., "Descriptive Mineralogy," p. 147, New York, 1917.
[249] Dana, E. S., "A Text-Book of Mineralogy," 3rd ed., revised and enlarged by W. E. Ford, New York, 1922.
[250] Dana, J. D., "A Manual of Mineralogy."

reason for anticipating any serious difficulty in determining the Shore scleroscope hardness of ice at any temperature. It would, without doubt, vary with the structure of the ice, and, for homogeneous ice, it would depend upon the direction of the crystallographic axes of the individual crystals. The Brinell hardness number for ice would have no significance except at temperatures so low that the rate at which ice yields progressively under the action of a constant load applied to a small area of its surface is negligible; say, at temperatures below -30 °C.

Certain experiments on the rates at which loaded rods and tubes penetrate into ice at given temperatures have been made by several experimenters, including T. Andrews [251] who designated his data as measures of hardness. They have been so quoted, *e.g.,* by H. T. Barnes,[234, p. 47; 252] although it is evident that they refer to progressive deformability rather than to hardness in any of the senses in which that term is commonly used. In this compilation they have been assigned to the section *Penetration* (p. 436).

Plasticity and viscosity.—Ice is a plastic solid. That is, under the action of small differential stresses it seems to be perfectly elastic, suffering no permanent change in form, but if the stress exceeds a certain small "elastic limit," its deformation continuously increases. Other things being the same, the nearer the temperature is to 0 °C, the more rapidly does the deformation increase. When the stress is a shear of the type produced by an axial twist applied to one end of a cylinder while the other is held fixed, the value of the strain at the elastic limit, as computed by B. Weinberg [237, 238] is independent of the temperature. The phenomena are complicated by the effect of the stress upon the melting point of the ice, especially of the impure intercrystallic material. Many early observers concluded that under differential stresses ice yields progressively, however small the stress may be, especially when the temperature is near 0 °C: J. Thomson,[253] F. Pfaff,[206] J. J. Bianconi,[213] T. Andrews.[251]

Both the elastic limit and the rate of yielding vary with the structure of the ice, and, when the structure is uniform, with the directions of the stresses with reference to the optic axes of the constituent crystals. They would also be expected to vary with the amount and nature of the impurities contained in the intercrystallic material. It seems that this last has not yet been considered by those who have investigated the plasticity of ice.

Under the titles plasticity, viscosity, and hardness, many observations of the manner in which ice yields to differential stresses have been reported. Most of them, though very interesting, have been made under such conditions or reported in such deficiency of detail as to make quantitative interpretation impossible. They will be found in the earlier portions of this section. Those from which the observers have attempted to derive numerical values for what they call the viscosity will be considered here. But first

---

[251] Andrews, T., *Proc. Roy. Soc. (London),* **40,** 544-549 (1886).
[252] Barnes, H. T., "Ice Formation," p. 66, New York, 1906.
[253] Thomson, J., *Proc. Roy. Soc. (London),* **8,** 455-458 (1857).

it is necessary to define the terms we shall use. Consider a plastic material bounded on opposite sides by planes that are parallel and distant x one from the other. If one of these planes is kept at rest and the other in motion in its own plane with a velocity v, then each will experience a drag amounting to P units per unit area, such that $(P - p) = \mu v/x$, μ being a property of the solid but independent of the value of P, v, and x. Unless P exceeds p, v is zero. The force required to produce the strain corresponding to the elastic limit is p units per unit area. This is the form of equation demanded by J. Clerk Maxwell's theory of viscosity [254] as extended by T. Schwedoff [255] to include the case in which there is a definite fixed elastic limit different from zero. It may for convenience be regarded as a definition of a plastic solid, and is in effect the definition adopted by E. C. Bingham.[256] Whether any specified solid that is commonly described as plastic satisfies this definition is another question, and one that need not detain us. It is merely a matter of definition; if the solid does not behave in accordance with that definition, then it is not purely plastic in the sense in which we shall use the term.

As our defining equation follows from an extension of Maxwell's theory of viscosity, which provides for other phenomena observed in the study of the shearing of ice, it is desirable to recall the essentials of that theory. He regards a viscous substance as consisting of one or more types of molecular aggregate. When any aggregate is strained by a relative motion of adjacent parts of the substance, it gradually breaks up, relaxing the strain, and the parts then form new associations, not necessarily of the same type of aggregation as before. Thus the strain is gradually relieved unless continually renewed by a continuous relative motion imposed from without upon the adjacent parts of the substance. If the rate of relaxation is directly proportional to the strain, then, if left to itself, the strain decreases exponentially with the time, and the time required for it to decrease to e^{-1} ($=0.3679$) of its value was called by Maxwell the relaxation time, and denoted by T. The viscosity of the substance is equal to the product of the modulus of rigidity multiplied by the relaxation time.

In general, the value of T will differ from one type of aggregate to another. When a substance containing several types of aggregates is strained, those for which T is small will soon relax, throwing additional stress upon those for which T is great, and they in turn and in some measure protect from stress the newly formed aggregates for which T is small. Thus when the distorting stress is maintained constant the effective viscosity will increase with the duration of the stress; at the same time, the velocity with which the distortion increases will decrease.

When the distorting stress is removed, there will be a partial and progressive recovery of the original form. The aggregates that had been

---

[254] Maxwell, J. C., *Phil. Mag. (4)*, **35**, 129-145, 185-217 (1868); "Scientific Papers," **2**, 26-78 (1890) = *Phil. Trans.*, **157**, 49-88 (1866).
[255] Schwedoff, T., *Jour. de Phys. (2)*, **8**, 341-359 (1889); **9**, 34-46 (1890).
[256] Bingham, E. C., *Bull. Bur. Standards*, **13**, 309-353 (SP 278) (1916).

64. ICE: DEFORMABILITY

strained less than their elastic limit will recover at once, and in so doing will strain other aggregates. Those will yield elastically or viscously, depending upon the amount they are strained, thus introducing new strains; and so on. In the end, the substance will be subjected to permanent internal strains, unless the elastic limit of each type of aggregate is actually that of zero strain, in which case the substance is purely viscous.

If the aggregates do not relax unless the strain exceeds a certain value, the substance is plastic. In general, that limiting strain will differ from aggregate to aggregate, and may have any value from zero (purely viscous) to infinity (perfectly elastic for all stresses). The behavior of the substance under shearing stress will vary accordingly.

Those who have studied ice have called Px/v the viscosity. We shall call it the effective viscosity, and shall denote it by μ_e; and we shall call the quantity denoted by μ in the equation $(P - p) = \dfrac{\mu v}{x}$ the viscosity. This does not accord entirely with the somewhat confused nomenclature used by those interested in the study of plastic materials used in the arts, but it is logical and is justified by the manner in which the defining equation was derived from Maxwell's picture of the structure of viscous substances. In our notation $\mu_e = \mu + px/v$.

Although several experimenters have observed that μ_e for ice increases as v becomes small, B. Weinberg [257] appears to be the only one who has attempted to separate the two terms composing μ_e. From a consideration of the progressive yielding of cylinders of ice, each clamped at one end and subjected at the other to a constant torque about the axis of figure, he concluded that his observations can be quite satisfactorily expressed by formula (1). (In the original paper, the negative sign has been omitted from the exponent.)

$$\mu_e = \mu_0 \left(a - \frac{b}{t} \right)^{-t} + \frac{c}{\psi} \quad (1)$$

Here, the temperature is t °C, the rate of shear is ψ radians per sec, and μ_0 is the value of the viscosity at 0 °C and $\psi = \infty$. He does not state how long the stress had lasted when ψ was observed. For river ice, the geometrical axis of the cylinder being parallel to the optic axes of the constituent and parallel crystals, he gives $\mu_0 = 9.5$ megamegapoises, $a = 1.12$, $b = 0.54$ °C, and $c = 0.5$ megadyne·radian/cm². For glacier ice he gives $\mu_0 = 3.8$ megamegapoises, $a = 1.32$, $b = 0.65$ °C, and $c = 0.08$ megadyne·radian/cm². From these he computes the modulus of rigidity, the relaxation time, and the greatest shear for which there is no permanent deformation (see Table 191). This maximum shear at 0 °C is 56 microradians for the river ice, and 10 for the glacier ice. These computations are vitiated by an error analogous to that considered by E. Buckingham [258] in his dis-

---

[257] Weinberg, B., *Ann. d. Physik (4)*, **22**, 321-332 (1907). Superseding and extending *Idem*, **18**, 81-91 (1905).
[258] Buckingham, E., *Proc. Amer. Soc. Testing Materials*, **21**, 1154-1161 (1921).

cussion of an equation used by Bingham to represent the flow through a capillary tube. Values defined by means of formula (1) are given in Table 193.

Table 193.—Viscosity of Ice [257]

$$\mu_e = \mu_0\left(a - \frac{b}{t}\right)^{-t} + \frac{c}{\psi}$$

; temperature $= t\,°C$; rate of shear $= \psi$ radians/sec, corresponding to a difference of v meters/year in the velocities of two planes of slipping that are 100 meters apart. In his paper of 1905 Weinberg [257] gives $\mu_t = (12.44 - 4.02t + 0.277t^2) \times 10^{12}$ poises when the mean value of ψ is about 10^{-8} radian/sec; this formula is probably not so good as the other. Computation by the compiler.

Unit of $\mu = 10^{12}$ poises, of ψ and of v as already indicated; temp. $= t\,°C$

I. River ice. Planes of slipping are perpendicular to optic axis.
$\mu_0 = 9.5$, $a = 1.12$, $b = 0.54\,°C$, $c = 5 \times 10^5$ poise·radian/sec.

$\psi \rightarrow$ $v \rightarrow$ t	$10^{-8}(?)$ $31.6(?)$ μ_t	5×10^{-9} 15.8	10^{-8} 31.6	10^{-7} 316^a μ_e	5×10^{-7} 631	∞ ∞
0	12.4	110	60	14.5	10.5	9.5
−0.1	12.8	112	62	16.5	12.5	11.5
−0.5	14.5	114	64	19.1	15.1	14.1
−1.0	16.7	116	66	20.8	16.8	15.8
−2.0	21.6	118	68	23.3	19.3	18.3
−3.0	27.0	121	71	25.9	21.9	20.9
−4.0	33.0	124	74	28.6	24.6	23.6
−5.0	39.5	126	76	31.5	27.5	26.5
−7.5	58.2	135	85	40.3	36.3	35.3
−10.0	80.3	147	97	52.3	48.3	47.3
−12.5	93.0	163	113	67.3	63.9	62.9
−15.0	135	184	134	88.6	84.6	83.6

II. Glacier ice.
$\mu_0 = 3.8$, $a = 1.32$, $b = 0.65\,°C$, $c = 8 \times 10^4$ poise·radian/sec.

$\psi \rightarrow$ $v \rightarrow$ t	10^{-9} 3.16	5×10^{-9} 15.8	10^{-8} 31.6	10^{-7} 316^a μ_e	5×10^{-7} 631	∞ ∞
0	83.8	19.8	11.8	4.6	4.0	3.8
−0.1	84.7	20.7	12.7	5.5	4.8	4.7
−0.5	86.1	22.1	14.1	6.9	6.3	6.1
−1.0	87.5	23.5	15.5	8.3	7.6	7.5
−2.0	90.3	26.3	18.3	11.1	10.5	10.3
−3.0	93.8	29.8	21.9	14.6	14.0	13.8
−4.0	98.4	34.4	26.4	19.2	18.5	18.4
−5.0	104.4	40.4	32.4	25.2	24.5	24.4
−7.5	129	65.0	57.0	49.8	49.2	49.0
−10.0	178	114.4	106.4	99.2	98.6	98.4
−12.5	278	214	206	198	198	198
−15.0	475	411	403	396	395	395

[a] At this rate, two planes of slipping that are 10 cm apart will differ in velocity by 36 microns per hour.

In his earlier paper, Weinberg gives formula (2) for the shearing of river ice in a direction perpendicular to the optic axis, and at rates of approximately 0.01 microradian per second.

$$\mu = 12.44 - 5.02 t_R + 0.355 t_R{}^2 \text{ megamegapoises}$$
$$= 12.44 - 4.02 t + 0.227 t^2 \qquad (2)$$

where t_R is the temperature on the Réaumur scale (not Centigrade scale, as first published; see the 1907 paper). Values computed by means of this formula are given in the second column of the first part of Table 193.

From an extended study of the bending of horizontal rectangular bars of ice, supported near each end, and loaded in the middle, H. Hess [258a] had already deduced values for the apparent viscosity. These values were computed by means of the formula $\mu_e = \dfrac{lP}{4abv} = \dfrac{M}{v}$, where $M \equiv \dfrac{lP}{4ab}$, $P =$ load, $l =$ length between supports, $a =$ vertical thickness, $b =$ horizontal breadth, $v =$ velocity of depression of the mid-point of the bar $= l\psi/2$, where ψ radians/sec is the rate of shear; ψ does not exceed a few times 10^{-8}. These values of μ_e are less than a hundredth as great as those found by Weinberg, and vary with the length of time the load has been applied. He reported that there was no detectable change in μ_e in the range 0 to -6.8 °C. For three bars cut from the same sheet of uniform ice he recorded the values given in Table 194. For granular ice he obtained values of the same order of magnitude. Successive bendings in opposite directions produced no change in the value of μ_e, the load being moderate. From his observations, he concluded that, under moderate loads, μ_e increases with the duration of the load, and after about 5 min the rate of increase is essentially constant; but under loads near the breaking value, μ_e decreases as the duration increases. Such variations may be forecast from the extension of Maxwell's theory. For additional details, reference should be made to the original paper.

Similar observations on the bending of loaded bars of ice had been made 11 years earlier by J. C. McConnel [231] in his very important reconnaissance of the behavior of such loaded bars. He thought it unprofitable to attempt to compute the viscosity from his observations, but R. M. Deeley [259] has made such computations, finding for shears perpendicular to the optic axis values of the same order (10^{10} poises) as those obtained by Hess. The compiler has been unable to obtain from McConnel's data the actual values published by Deeley. McConnel's observations show that the apparent viscosity for shear parallel to the optic axis is many times (perhaps 100) as great as that for a shear perpendicular to that axis.

From the observations of J. C. McConnel and D. A. Kidd [260] on the progressive elongation of bars of ice subjected to longitudinal traction, R. M. Deeley and P. H. Parr [226] have computed the apparent viscosity,

[258a] Hess, H., *Ann. d. Physik (4)*, **8**, 405-431 (1902); "Die Gletscher" (1904).
[259] Deeley, R. M., *Proc. Roy. Soc. London (A)*, **81**, 250-259 (1908).
[260] McConnel, J. C., and Kidd, D. A., *Proc. Roy. Soc. (London)*, **44**, 331-367 (1888).

finding values varying from 9 to 900 megamegapoises, depending upon the temperature, the structure of the ice, and the direction of the shear.

Table 194.—Viscosity of River Ice [258a]

Values were derived from the bending of horizontal, rectangular bars supported at the ends and loaded at the middle. $P = $ load; $M = $ bending moment per unit of cross-sectional area $= Pl/4ab$; $\mu_\tau = $ value of the apparent viscosity as computed from the rate of shear τ sec after the load was applied; $l = $ length between supports; $a = $ vertical thickness; $b = $ horizontal breadth; vertical and horizontal refer to position of bar when loaded for test. All three bars were cut from the same sheet of ice.

Unit of $P = 1$ g*, of $M = 1$ g*·cm^{-1}, of $\mu = 10^{12}$ poises

Axis→	Parallel to l			Parallel to a		
$P\rightarrow$	2000	5000	6000	1000	1500	2000
$M\rightarrow$	1350	3400	4000	1600	2350	3100
τ		μ_τ			μ_τ	
15	0.065	0.105	0.0055	0.075	0.100	0.080
60	0.175	0.115	0.036	0.075	0.110	0.070
120	0.100	0.130	0.0365	0.075	0.090	0.110
300	0.110	0.160	0.035	0.080	0.120	0.120
1200		0.120				

	Optic axis parallel to b			
$P\rightarrow$	1000	1500	2000	3000
$M\rightarrow$	1500	2250	3000	4450
τ		μ_τ		
15	0.037	0.037	0.024	0.110
60	0.080	0.110	0.060	0.090
120	0.120	0.100	0.100	
300	0.210	0.190	0.170	

In the same paper, Deeley and Parr summarize the more important values reported for glacier ice, as given here in Table 195. They remark:

Table 195.—Viscosity of Glacier Ice
Adapted from R. M. Deeley and P. H. Parr.[226]

Unit of $\mu_e = 10^{12}$ poises

Observer		Computer		μ_e
Dr. Main	1888	R. M. Deeley	1912	6.0
McConnel and Kidd	1888	R. M. Deeley	1912	84.5
B. Weinberg	1907	B. Weinberg	1907	8.0
Blümcke and Hess	1907	B. Weinberg	1906	17.4
Tyndall and others	——	R. M. Deeley	1908	78.9
Blümcke and Hess	1910	B. Weinberg	1910	17.5
Blümcke and Hess	1910	Deeley and Parr	1913	147.7[a]
Blümcke and Hess	1910	Deeley and Parr	1913	125[a]

[a] From motion of glaciers in the winter.

"We have seen that glacier ice consists of crystal granules which not only shear freely along planes at right angles to the optic axis, but also undergo changes at their bounding surfaces, which enable the mass to suffer continuous distortion under stress. The ability of glacier ice to spread out into piedmonts whose upper surfaces are very nearly level also shows that such shear may take place under very small stresses." This accords with the small value of the shear (λ) that corresponds to the elastic limit. From Table 191, we find that λ does not exceed 60 microradians, nor does the coefficient of rigidity exceed about 3×10^{10} dynes/cm²·radian; hence ice will yield continuously if the shearing stress exceeds 1.8 megadynes/cm² = 1.8 kg*/cm², a very small value.

For other values of the viscosity of glacier ice, as derived from the observed flow of each of several glaciers, see R. M. Deeley.[261]

From the damping of torsional vibrations of a cylinder of ice about its axis, C. D. Hargis [240] obtained the values $\mu_e = 3.7$ megamegapoises when the period was 0.286 sec, and 6.21 when it was 0.448 sec. The cylinder was obtained by freezing water in a brass tube.

All the preceding may be summarized thus: (1) None of the available data for the plasticity or for the viscosity of ice is entirely satisfactory. (2) Values of μ_e derived from the bending of bars are of the order of 10^{10} poises, those from the axial torsion and those from the longitudinal stretching of bars are of the order of 10^{12} poises. (3) Although McConnel's and McConnel and Kidd's data indicate that μ_e for shear parallel to the optic axis is about 100 times as great as for shear perpendicular to that axis, Hess's data indicate that the difference is slight. (4) The value of μ_e increases as the rate of shear decreases (Table 193). (5) When the stress is kept constant, μ_e increases with the time the stress has been applied. Whether this involves other phenomena than those pertaining to the variation with the rate of shear cannot be determined from the data now available. (6) The value of μ increases very rapidly as the temperature decreases, a decrease of 10 °C being accompanied by a 5-fold increase in μ for river ice, and a 26-fold increase for glacier ice. This increase in μ causes a marked, but in general a smaller, increase in μ_e. (7) An attempt to fit the data of Hess, of McConnel, and of McConnel and Kidd to Weinberg's equation has been unsuccessful. (8) Owing to the absence of important data, to significant variations in the procedures followed, and to variations in the structure and the purity of the ice used, it is impossible to correlate satisfactorily the data obtained by different observers.

Sustaining power of an ice sheet.—A knowledge of the load that a sheet of ice of given thickness can sustain while resting upon water is of considerable importance, especially in military operations. Ordinary experience teaches, as pointed out by F. A. Forel,[262] that this load depends upon the state of the ice. Old ice that has been exposed to the sun and to air not much below 0 °C becomes split by a multitude of vertical cracks into irregular prismatic needles, *i.e.*, it becomes rotten. Such ice has little sustaining

[261] Deeley, R. M., *Geol. Mag. (5)*, 9, 265-269 (1912).
[262] Forel, F. A., *Rev. Sci.*, 51, 379 (1893).

power; no estimate of that power can be given, as it varies greatly with the existing condition of the ice. It is only of new ice still in the process of formation that numerical data can be given with any confidence. It is to such ice that the following figures refer.

An anonymous note [263] quotes from the "Echo de l'Armée" the following values as having been determined under the authorization of the French military establishment: When 4 cm thick, such ice will bear the weight of one man; when 9 cm thick, infantry marching in open formation; when 12 cm, artillery train of 8-cm guns; when 14 cm, train of 12-cm guns; when 16 cm, siege guns with loaded caissons; when 29 cm thick, it will carry almost any load that would be placed on it.

P. Vedel [264] has stated that the "army rules" were as follows: 2-in. (5-cm) ice will support a man or properly spaced infantry; 4-in. (10-cm) ice, a man on horseback, cavalry, light guns; 6-in. (15-cm) ice, such fieldpieces as 80-pounders; 8-in. (20-cm) ice, battery of artillery with carriages and horses, but not over 1000 lb* per sq. ft. on sledges; 10-in. (25-cm) ice, an army, an innumerable multitude; 15-in. (38-cm) ice, railroad tracks and trains. He stated that 24-in (61-cm) ice withstood the impact of a loaded railroad passenger car falling 60 ft (which he estimated at 1500 ft·tons*), but broke under the impact of a locomotive and tender (which he estimated at 3000 ft·tons*). Tables purporting to give the maximum safe load for the ice on a circular lake and for that on a canal are included in the article, but as the value he gives for the Young's modulus in an earlier portion of the paper is about 1000 times too great, it is feared that the data of those tables are untrustworthy. The present compiler has not attempted to check the computations.

65. Deformability of Snow

The deformability of snow and its variation with the depth of the overlying snow have recently been studied by M. Kuroda.[265]

For obtaining an estimate of the hardness, he used a brass-tipped wooden cone, vertex angle = 90°, dropped from a stated height, and measured (D) the surface diameter of the indentation produced.

For measuring the tensile strength he used telescoping sheet-metal forms of the general shape of the axial section of metal specimens intended for similar tests. One of these forms was pressed into a layer of undisturbed snow carefully taken up on a glass plate, and the force required to pull it apart was measured.

For obtaining the shearing strength he used a flat block sliding snugly in a slot cut in another block; through the center of the compound block and perpendicular to the plane of the sliding one was cut a rectangular hole. By means of a suitable sheet-metal form, prisms of snow that fitted

---
[263] Anon., *Idcm*, **51**, 318 (1893).
[264] Vedel, P., *J. Franklin Inst.*, **140**, 355-370, 437-455 (1895).
[265] Kuroda, M., *Sci. Papers Inst. Phys. and Chem. Res. (Tokyo)*, **12**, 69-81 (1929).

65. SNOW: DEFORMABILITY

Table 196.—Hardness of Snow: Variation with Depth in Snow-blanket
Adapted from M. Kuroda.[265]

The hardness is indicated by the surface diameter (D) of the conical indentation produced by dropping a brass-tipped wooden cone, vertex angle = 90°, from a stated height (h) above the surface of snow under study. By carefully removing the overlying snow, that surface was placed at any desired depth (d) below the undisturbed surface of the natural snow-blanket; $h = 0$ indicates that the cone was placed gently upon the surface and sank under its own weight. Values are given for two blankets; those in the first column $h = 0$ refer to one, and the others, to the other. In the original paper the values of h and d for the second blanket appear to have been interchanged, they are the reverse of those here given.

Unit of h, d, and D = 1 cm

$h \rightarrow$ d	0 D	0	5 D	10	20
0	16.8		16.0	19.2	
2	14.0				
10	10.0		4.7	9.1	11.7
15	4.4				
20	4.3		4.6	8.0	12.0
30	4.0				
40	4.5	3.4		7.6	12.0
50	4.0				
60	5.0				
70	3.3	3.0		5.4	8.0
80	3.6				

Table 197.—Hardness of Snow: Effect of Tamping [265]

By carefully removing the overlying snow, the surface under study could be brought to any desired depth (d) below the surface of the natural snow-blanket. The hardness of one portion of the surface so cleared was determined at once; another portion was tamped by a single dropping from a height H of a load of 2700 g* with a rectangular base 18 by 27 cm, and the hardness was then determined. D and h have the same significance as in Table 196.

Two snow-blankets were studied; temperature of snow, about −11 °C.

$h = 20$ cm. Unit of H, d, D = 1 cm

$H \rightarrow$ d	U^a	20 D	U^a	40 D	60
0	4.5		14.0		
10	8.5		8.0		
12	8.5	4.0			
13			5.5	2.0	1.5
16	6.0	3.0			
19	6.5	3.0			
20	5.2		4.5	2.5	2.0
22	5.2	4.8			
25	4.5		4.0	3.3	3.0
26	4.0	3.5			
30	3.0	3.0	3.5	3.5	3.5
45	4.5	4.5			

^{a} U = untamped.

the hole could be cut out and placed in it; the force then required to withdraw the sliding block was measured.

He gives curves showing the grain size, the density, D, and the temperature, throughout the thickness of a natural snow-blanket 9 meters thick. The size of the grains varied but little until the ground was approached, where the snow was several months old; the density varied from 0.35 g/cm³ at a depth of 50 cm to 0.65 at 700 cm; from a depth of 200 cm to that of 700 cm the hardness was essentially constant; the temperature was lowest (-0.9 °C) at mid-depth.

Table 198.—Strength and Hardness of Snow [265]

T = tensile strength, S = shearing strength, D = hardness as in Table 196, t_s °C = temperature of the snow. For details, see text.

Unit of $D = 1$ cm; of T and $S = 1$ g*/cm²

Snow[a]	A	B	C	D
t_s	-9.0	0	-2.0	
T	63	33	93	
S	3	2.5	20	43
D	20	20	155	

[a] Snow: A = fresh and powdery; B = wet and soft; C = surface crusted; D = surface more crusted than C.

---

66. Acoustic and Other Vibrational Data for Ice

(For the elastic constants of ice, see Sections 64 and 67; for density, Section 67.)

Acoustic data are those pertaining to longitudinal vibrations, to those in which the displacement is in the direction of propagation of the train of waves. Like other solids, ice can transmit transverse vibrations also, those in which the displacement is perpendicular to the direction of propagation; and a thin sheet of ice can transmit flexural vibrations. All these types of vibration are considered in this section.

Velocity of Transmission.

As ice is crystalline, the velocity of a given type of vibration might be expected to vary with the direction of propagation through the crystal, and R. Köhler [266] thought that his observations on an ice sheet 30 cm thick indicated such an effect for waves generated by the firing of explosives. But R. W. Boyle and D. O. Sproule [267] concluded, from their observations on longitudinal ultrasonic waves in rods of ice, that any such difference lies within the experimental error (see Table 199). It would seem that their observations are the more readily interpretable.

Flexural waves in an ice sheet 11 to 38 cm thick, resting on water 1.2 to 6 m deep, exhibit marked dispersion; the group velocity is $12.2\sqrt{\tau\nu}$

[266] Köhler, R., *Z. Geophys.*, 5, 314-316 (1929).
[267] Boyle, R. W., and Sproule, D. O., *Can. J. Res.*, 5, 601-618 (1931).

m/sec where τ cm is the thickness of the ice, and the frequency is ν cycles/sec, ν varying from about 13 to 600.[268]

Table 199.—Velocity of Waves in Ice
(For flexural waves, see text.)

θ = angle between the direction of advance of the wave and the normal to the planes of freezing (the optic axis); ν = frequency; v = velocity of propagation. The BS[a] observations for $\theta = 0$ are given within 0.9 per cent by $v = 3.12 (1 - 0.0025t)$ km/sec, which for $t = -26\,°C$ is 3.33; the differences between this and the several values tabulated for $-26\,°C$ represent inherent variations in the samples.

Unit of $v = 1$ km/sec; of $\nu = 1$ kilocycle/sec. Temp. $= t\,°C$

I. Longitudinal vibrations. Rods of ice.

BS[a] $\theta = 0°$ $\nu = 13$		BS[a] $t = -26\,°C$ $\nu = 7$ to 12		RS[a] $t = -4\,°C$ $\nu = 1.55$		ECT[a] $\nu = 1.31$ to 4.97	
t	v	θ	v	θ	v	θ	v
−9	3.18	0°	3.22	90°	3.23	90°	3.174
−10	3.21	90°	3.24			Vbl[b]	3.150
−30	3.33	90°	3.18[c]			Mean[b]	3.163 ± 0.009
−35	3.43	45°	3.11				

II. Longitudinal. Explosions. III. Transverse. Explosions.

Form	v	Ref.[a]	Form	v	Ref.[a]
Sheet	3.40[d]	ECT	Rod	1.914 ± 0.006[e]	ECT
Sheet	3.41	ECT	Sheet	1.846 ± 0.005	EC
30 cm Sht.	3.23	K	Sheet	1.70	K
Solid	3.15[d]	ECT	Glacier	1.60	M
Glacier	3.40	M	Glacier	1.69	M
Glacier	3.60	M	Glacier	1.67	M
Glacier	3.57	M	Glacier	1.69	S
Glacier	3.49	S	Glacier	1.60	S
Glacier	3.41	S	Glacier	1.82	S
Glacier	3.70	S	Névé	1.35	M
Névé	3.14	M			

[a] References and notes:
BS R. W. Boyle and D. O. Sproule [267]; rods cut from river ice.
EC M. Ewing and A. P. Crary.[268]
ECT M. Ewing, A. P. Crary, and A. M. Thorne, Jr.[269]; rods cut from river ice; explosions in the ice sheet; temperature −5 to −15 °C.
K R. Köhler [266]; lake ice 30 cm thick.
M H. Mothes.[270]
RS M. Reich and O. Stierstadt [271]; ice formed from distilled water.
S E. Sorge [272]; inland ice sheet, Greenland.

[b] Vbl indicates that the ice was formed by freezing water slowly in a vertical brass tube; under which conditions, the crystals are variously oriented. Mean = mean of all observations on rods of whichever kind.
[c] Direction of propagation was perpendicular to that in preceding case, but in both it was parallel to the planes of freezing.
[d] Computed from the observed velocity in thin rods. Sheet = thin plate, unsupported. Solid = infinite solid.
[e] Torsional vibrations of thin rod of Vbl ice (see note [b]), $\nu = 0.81$ kcycle/sec.

[268] Ewing, M., and Crary, A. P., *Physics*, **5**, 181-184 (1934) → Crary and Ewing, *Phys. Rev.* (2), **45**, 749 (A) (1934).

Reflectivity.

As compared with steel, or even with granite, ice is a very poor reflector of ultrasonic vibrations ($v = 84$ kcycle/sec). See R. W. Boyle and G. B. Taylor.[273]

67. PRESSURE-VOLUME-TEMPERATURE ASSOCIATIONS FOR ICE

All data pertaining to the specific volume and the density of ice and of snow, and to their variations with the temperature and the hydrostatic pressure, are assembled in this section. The density has not been directly determined for any type of ice except the usual one (ice-I), and no determination of the density of snow-crystals has come to the compiler's attention. (Deformability of ice and of snow, see Sections 64 and 65. Linear expansion of ice, see Section 68.)

Density of Snow.

The density of freshly fallen snow varies greatly, depending upon the aerodynamic conditions attending its deposition; the density at any point of a snow blanket increases with the age of the blanket, even in the absence of fusion.[274] The density in a natural blanket of snow increases nonlinearly with the depth.[234, p. 25] M. Kuroda [275] found 0.35 g/cm³ at a depth of 50 cm, and 0.65 at 700 cm. Values as low as 0.004 have been recorded for freshly fallen snow.[276]

The density of the persistent *névé* in the Pyrenees at altitudes of 2.5 to 3.4 km varies from 0.51 to 0.59 g/cm³ in August to September, and from 0.53 to 0.65 in October.[274] Devaux thought that this apparent increase was real. The samples were probably taken from near the surface. E. Sorge [277] has found that the density of the *névé* on the inland ice-sheet of Greenland is 0.51 g/cm³ at depths of 30 to 118 cm, varying inappreciably with the depth.

Density of Ice-I at 1 Atmosphere.

The density of the ice of glaciers is, as one would expect, lower than that of clear compact ice; values varying from 0.86 to 0.91 g/cm³ have been reported by J. Devaux.[278]

---

[269] Ewing, M., Crary, A. P., and Thorne, A. M., Jr., *Physics,* **5,** 165-168 (1934) → *Phys. Rev. (2),* **45,** 749 (A) (1934).
[270] Mothes, H., *Z. Geophys.,* **3,** 121-134 (1927); **5,** 120-144 (1929).
[271] Reich, M., and Stierstadt, O., *Physik. Z.,* **32,** 124-130 (1931).
[272] Sorge, E., *Z. Geophys.,* **6,** 22-31 (1930).
[273] Boyle, R. W., and Taylor, G. B., *Trans. Roy. Soc. Canada, III (3),* **20,** 245-257 (1926).
[274] Devaux, J., *Compt. rend.,* **185,** 1147-1149 (1927).
[275] Kuroda, M., *Sci. Papers Inst. Phys. & Chem. Res. (Tokyo),* **12,** 69-81 (1929).
[276] Keränen, J., *Annales Acad. Sci. Fennicae = Suomalaisen Tiedeak. Toimit. (A),* **13,** No. 8 (1920).
[277] Sorge, E., *Z. Geophys.,* **6,** 22-31 (1930).
[278] Devaux, J., *Compt. rend.,* **185,** 1602-1604 (1927).

The better of the recorded values for the density of ice-I at 0 °C and 1 atm vary from 0.918 to 0.916 g/ml. E. L. Nichols [279] has reported values indicating that the density is actually subject to such variations. His values indicate that the density of freshly formed natural ice is 0.91795 g/ml; of natural ice 1 year old, 0.91632; and of artificial ice-mantles frozen by means of solid CO_2 and ether, 0.91603. But H. T. Barnes,[280] using ice from the St. Lawrence River, failed to find such great variation. He found for new ice 0.91662, for ice 1 year old 0.91648, and for ice 2 years old 0.91637, the mean of all being 0.91649 g/ml. Although these values indicate a slight progressive decrease in the density as the ice ages, the change in 2 years is only about 1/6 of that reported by Nichols for 1 year. J. H. Vincent [281] also believes that he has shown "that the same specimen of water may assume different densities on freezing." The values he finds vary from 0.9155 to 0.9163.

At least a portion of the differences in the densities recorded is probably due to the fact that the water from which the ice was formed contained other substances in solution. J. Y. Buchanan [282] has shown that, when ice forms in an aqueous solution, some of the solution is entrapped between the crystals, and remains incompletely frozen so long as the temperature is above the cryohydric point. This makes the apparent density too great. The effect may be appreciable even when the solution is extremely dilute, as dilute as good distilled water. (See Table 203.) The pressure caused by the expansion attending the freezing of a portion of a volume of water entirely surrounded by ice will tend to keep the remainder of the volume in the liquid state, if the temperature is not far from zero. This also increases the apparent density.

The subject of the variability in the apparent density of ice should be given careful study, using ice of various ages, from various sources, and frozen under various controlled and recorded conditions. Special attention should be given to the possible effect of supercooling, of the presence of air, and of the size of the grains and its progressive increase, as reported by R. Emden,[283] not entirely forgetting the so-called "dense ice" reported by Cox (1904) and by Shaw (1924) (see Section 57). The effects of stresses, arising either from a difference in the expansions of ice and its container or from the expansion that occurs on freezing, should be considered more carefully than they have been in the past. It might be worth while trying to use monocrystals, either grown directly or obtained by disintegrating a carefully frozen sheet of ice by means of radiation. And one should not forget the suggestion [284] that variations in the density may

[279] Nichols, E. L., *Phys. Rev.*, **8**, 21-37 (1899).
[280] Barnes, H. T., *Idem*, **13**, 55-59 (1901) → *Physik. Z.*, **3**, 81-82 (1901) = Barnes, H. T., and Cooke, H. L., *Trans. Roy. Soc. Canada, III (2)*, **8**, 143-155 (1902).
[281] Vincent, J. H., *Phil. Trans. (A)*, **198**, 463-481 (1902) → *Proc. Roy. Soc. (London)*, **69**, 422-424 (1902).
[282] Buchanan, J. Y., *Proc. Roy. Soc. Edinburgh*, **14**, 129-149 (1887); *Nature*, **35**, 608-611 (1887); **36**, 9-12 (1887); *Proc. Roy. Inst. Grt. Britain*, **19**, 243-276 (1908).
[283] Emden, R., *Neue Denkschr. d. allgem. schweiz. Ges. Naturwiss.*, **33**, 43 pp. (1892).
[284] Eméleus, H. J., et al., *J. Chem. Soc. (London)*, **1934**, 1207-1219 (1934).

arise from actual changes in the composition of the water, such as a concentration of deuterium oxide during the process of freezing; even though the possibility of so concentrating deuterium oxide seems to have been disproved by the work of V. K. LaMer, W. C. Eichelberger, and H. C. Urey,[285] and of G. Bruni.[286] In the last two of these papers, Bruni disproves the contrary conclusion drawn in an earlier paper by G. Bruni and M. Strada.[287]

The more reliable determinations of the density of ice are listed in Table 200, where Kopp's distinctly abnormal value is also given. Those for natural and for artificial ice have been listed separately, simply because it has been suggested that they may differ. No such difference is at all obvious from the table.

The values may, however, be assorted into three distinct groups. One group includes the values obtained by C. Brunner [288] for river ice, by E. L. Nichols [279] for new ice, by L. Dufour [289] for ice from boiled-out water frozen in air at a very low pressure, and by A. Leduc [290] for ice from which the air had been removed with great care. The average of these values is 0.9178, and the average deviation from this is 0.0002.

At the other extreme is the group of values obtained from direct measurements of the difference in the specific volumes of ice and of water. This contains the unsatisfactory determination of H. Kopp,[291] the essentially identical values (0.9157) obtained by Plücker and Geissler [292] and by H. Endo,[293] and the slightly higher ones (0.9160) found by J. H. Vincent [281] and by Nichols [279] for artificial ice. The mean, omitting Kopp's value, is 0.9158.

The others fall into the third group, averaging 0.9165, and ranging from 0.9163 to 0.9166, the mean variation being about 0.0002.

The reason for such variations and for such a grouping of the values remains to be determined.

The values have usually been reported in terms of the density of water at 0 °C, and consequently differ slightly from the values here given. Still other changes occasionally seemed justified. They are described in the following remarks concerning the several determinations.

J. Plücker and Geissler [292] concluded that whenever the freezing occurs in the same way the density of the ice is always the same. They used a unique type of thermometer having in the bulb a distinct, but communicating compartment for the water under study. The value tabulated has

[285] LaMer, V. K., Eichelberger, W. C., and Urey, H. C., *J. Am. Chem. Soc.*, **56**, 248-249 (1934).
[286] Bruni, G., *Att. d. R. Acc. Naz. Lincei (6)*, **20**, 73-75 (1934)—*J. Am. Chem. Soc.*, **56**, 2013-2014 (1934).
[287] Bruni, G., and Strada, M., *Ibid*, **19**, 453-458 (1934).
[288] Brunner, C., *Ann. d. Phys. (Pogg.)*, **64**, 113-124 (1845).
[289] Dufour, L., *Compt. rend.*, **50**, 1039-1040 (1860); **54**, 1079-1082 (1862).
[290] Leduc, A., *Idem*, **142**, 149-151 (1906).
[291] Kopp, H., *Ann. d. Chem. u. Pharm. (Liebig)*, **93**, 129-232 (1855) → *Ann. de chim. et Phys. (3)*, **47**, 291-296 (1856).
[292] Plücker, J., and Geissler, *Ann. d. Phys. (Pogg.)*, **86**, 238-279 (265-279), (1852).
[293] Endo, H., *Sci. Rep. Tôhoku Imp. Univ. (Sendai) (1)*, **13**, 193-218 (1924-25).

been derived from their determinations of the difference between the specific volumes of ice and water. With one exception, omitted in deriving the mean, their observations are exceedingly concordant.

H. Kopp [291] measured the expansion that occurs when water freezes, using turpentine as the dilatometric liquid. Although his dilatometer was closed with a cork, he obtained two closely agreeing values giving for the density 0.908. Like other determinations by this method, the value is lower than that found by any other method, but this particular value is undoubtedly too low. (In the French abstract the values given for the contraction are wrong.)

L. Dufour [289] used a flotation method, and ice from boiled-out water frozen in a vacuum (air-pressure not over 0.5 mm-Hg). In the earlier work he used a mixture of alcohol and water, and found that a density of 0.9175 g/cm^3 just below 0 °C was required to support completely submerged ice. This mixture was found to dissolve the ice slightly. In the later work, he used a mixture of chloroform and petroleum, worked between −0.5 °C and −8 °C, and accepted 0.000158 (°C)$^{-1}$ as the coefficient of cubical expansion of ice. Sixteen determinations of the density of ice at 0 °C lay in the range 0.9168 to 0.9193 and averaged 0.9178.

The observations of R. Bunsen [294] vary linearly with the temperature (τ) at which the water was frozen. They may be represented by the formula $d = 0.91663 - 0.000047\tau$; the lower the temperature of freezing, the greater the density. In how far this arises from the difference in the expansion of ice and of glass remains to be determined. The value corresponding to $\tau = 0$ is the one here tabulated.

J. v. Zakrzewski [295] used Bunsen's method, but froze the water at a constant temperature not far from 0 °C. The mean of three very closely concordant determinations of the density at −0.701 °C, the temperature of freezing, gave $d_{-0.701} = 0.916710$ which reduces to $d_0 = 0.91661$ g/ml if $10^6\beta = 153$ (see Table 202). A single measurement at −4.720 °C gave $d_{-4.720} = 0.916995$ or $d_0 = 0.91633$; if the temperature recorded as −4.720 was actually −2.720, then $d_0 = 0.91661$, agreeing closely with the mean of the three at −0.701 °C. The −4.7 °C value has not been entered in the table.

E. L. Nichols [279] determined the density by weighing ice in air and in petroleum. The highest value for the specific gravity with reference to water at 0 °C was 0.91808 for a recently formed, natural icicle; the lowest was 0.91615 for an ice mantle frozen by means of ether and solid carbon dioxide, temperature about −70 °C. Other measurements of a tentative kind are described.

J. H. Vincent,[281] using a novel device, weighed in mercury the buoyancy of water and of the ice formed from it. The quantity directly determined was the difference in the specific volumes. The buoyancy of the ice was weighed at several temperatures in the range −0.4 °C to −10 °C, and the

[294] Bunsen, R., *Ann. d. Phys. (Pogg.)*, **141**, 1-31 (1870).
[295] v. Zakrzewski, J., *Ann. d. Phys. (Wied.)*, **47**, 155-162 (1892).

value at 0 °C was obtained by linear extrapolation. His values for the density at 0 °C vary from 0.915460 to 0.916335; he took as the weighted mean 0.9160. Two freezings of the same specimen of water gave densities differing by 57 parts in 100 000. He regarded this difference as real.

A. Leduc [290] took great pains to remove all air from the water from which the ice was formed. For determining the density, he froze the water in a specific-gravity flask of the Regnault type. The freezing proceeded gradually from the bottom up into the capillary; a mixture of ice and salt was used, the temperature being between $-5°$ and -10 °C.

Table 200.—Density of Ice-I at Atmospheric Pressure

(Comments on the several determinations may be found in the text.)

Dewar's (J. 1902) corrected (see text) value for -188.7 °C is $d_{-188.7} = 0.936$; this with the coefficient (Table 202) $10^6\beta = 153$ gives 0.948 for the density at absolute zero (-273.1 °C), which probably exceeds the true value. From $d_0 = 0.9166$ and $10^6\beta = 153$ one derives $d_{-188.7} = 0.943$.

The value accepted by J. R. Clarke [300] for the density at 0 °C is $d_0 = 0.9168 \pm 0.0005$ g/ml (ICT).

d_t = density at t °C; τ °C is the temperature at which the ice was frozen, value for $\tau = 0$ °C is obtained by extrapolation.

Unit of $d = 1$ g/ml $= 0.999973$ g/cm^3

Natural Ice			Artificial Ice		
Observer	d_0	Age	Observer	d_0	Notes
Brunner (1845)	0.91788		Plücker and Geissler (1852)	0.9156$_7$	Ch. in vol.
Nichols (1899)	0.91795	New[a]	Kopp (1855)	0.908	Ch. in vol.
Nichols (1899)	0.91632	1 yr.	Dufour (1860)	0.9175	Flotation
Barnes (1901)	0.91662	New	Dufour (1862)	0.9178	Flotation
Barnes (1901)	0.91648	1 yr.	Bunsen (1870)	0.91663	$\tau = 0$
Barnes (1901)	0.91637	2 yr.	v.Zakrzewski (1892)	0.91661	$\tau = -0.701$
			Nichols (1899)	0.91603	$\tau = -70$
			Vincent (1902)	0.9160	Ch. in vol.
			Leduc (1906)	0.9176	$\tau = -5$ to -10
			Endo (1925)	0.9157$_1$	Ch. in vol.

[a] Mean of three values: Icicles 0.91804, 0.91789, and new pond ice 0.91792 g/ml.

H. Endo [293] determined the change in specific volume from observations of the buoyancy, in lamp oil (kerosene?) at various temperatures, of a silica vessel containing the specimen. The ice was frozen by means of a mixture of ether and solid carbon dioxide.

J. Dewar [296] weighed in air and in liquid air "pieces of clear ice cut from large blocks." The density of the liquid air was determined in terms of that of liquid oxygen boiling under a pressure of 1 atm, assumed to be 1.137. Thus he obtained for ice at -188.7 °C the value 0.930$_0$, individual

[296] Dewar, J., *Proc. Roy. Soc. (London)*, **70**, 237-246 (1902) = *Chem. News*, **85**, 277-279, 289-290 (1902). Same data in *Proc. Roy. Inst. Grt. Brit.*, **17**, 418-426 (1903) → *Chem. News*, **91**, 216-219 (1905).

values ranging from 0.926_5 to 0.933_2. Reducing this value to the basis of 1.144_7 [297] for the density of oxygen under the stated conditions, we find 0.936_0. It is interesting to note that this is only 0.75 per cent less than

Table 201.—Densities and Specific Volumes of the Ices at their Melting-points.

(For change in volume on transition of ice to ice, see Table 271.)

The following data have been obtained by combining those (Bridgman's) in Tables 95 and 271, interpolating or extrapolating the water data where necessary.

d = density, v^* = specific volume, mp °C = melting point corresponding to P.

Unit of $P = 1$ atm = 1.01325 megadyne/cm²; of $d = 1$ g/ml; of $v^* = 1$ ml/g

Type of Ice	mp	P	Ice d_i	Ice v_i^*	Water d_w	Water v_w^*	$10^4(v_i^* - v_w^*)$
I	0.0	1	0.9168	1.0908	0.9921	1.0008	+900
	−5.0	590	0.9297	1.0756	1.0267	0.9740	+1016
	−10.0	1090	0.9397	1.0642	1.0504	0.9520	+1122
	−15.0	1540	0.9444	1.0589	1.0671	0.9371	+1218
	−20.0	1910	0.9481	1.0547	1.0830	0.9234	+1313
	−22.0[a]	2045	0.9483	1.0545	1.0878	0.9193	+1352
III	−17.0	3420	1.1595	0.8624	1.1293	0.8855	−231
	−17.0[a]	3420	1.1609	0.8614	1.1293	0.8855	−241
	−18.5	2820	1.1513	0.8686	1.1127	0.8987	−301
	−20.0	2430	1.1476	0.8714	1.1101	0.9085	−371
	−22.0[a]	2045	1.1459	0.8727	1.0878	0.9193	−466
V	0.16[a]	6175	1.2657	0.7901	1.1865	0.8428	−527
	0.0	6160	1.2653	0.7903	1.1862	0.8430	−527
	−5.0	5270	1.2596	0.7939	1.1707	0.8542	−603
	−10.0	4360	1.2488	0.8008	1.1511	0.8687	−679
	−15.0	3680	1.2421	0.8051	1.1357	0.8805	−754
	−17.0[a]	3420	1.2396	0.8067	1.1293	0.8855	−788
	−20.0	3040	1.2338	0.8105	1.1194	0.8933	−828
VI	+40.0	11 990	1.3616	0.7344	1.2604	0.7934	−590
	+30.0	10 250	1.3528	0.7392	1.2415	0.8055	−663
	+20.0[b]	8710	1.3492	0.7412	1.2250	0.8163	−751
	+15.0	8040	1.3464	0.7427	1.2158	0.8225	−798
	+10.0	7390	1.3430	0.7446	1.2063	0.8290	−844
	+5.0	6880	1.3407	0.7459	1.1986	0.8343	−884
	+0.16[a]	6175	1.3312	0.7512	1.1865	0.8428	−916
	0.0	6160	1.3308	0.7514	1.1862	0.8430	−916
	−5.0	5620	1.3236	0.7555	1.1774	0.8493	−938
	−10.0	5110	1.3158	0.7600	1.1682	0.8560	−960
	−15.0	4640	1.308	0.765	1.159	0.863	−980
VII[c]	20.0	48 400	1.67	0.60			

[a] Triple point, water and two ices.
[b] At 25.0 °C the melting pressure is 9630 bars (=9504 atm) and $v_i^* - v_w^* = -0.0714$ cm³/g, Bridgman's value being -0.0707.[303]
[c] The data for water needed for the computation of the densities are not available; values of $(v_i^* - v_w^*)$ are given in Table 271. This value for v_i is from P. W. Bridgman.[304]

[297] ICT, *Int. Crit. Tables*, **3**, 20 (1928).

that computed on the assumption that the mean coefficient of expansion between 0 °C and -190 °C is equal to the coefficient (153×10^{-6}) found near 0 °C.

No comments need be made regarding the other determinations; references for them have already been given. For discussions of the experimental values, see E. L. Nichols,[279] H. T. Barnes and H. L. Cooke,[298] and W. A. Roth.[299]

Densities of the Ices not at their Melting-points.

In addition to values that may be derived from the compressibilites, the coefficients of thermal expansion, and the specific volumes, the following values have been reported.

Ice-II. From x-ray data for ice-II at -155 °C and atmospheric pressure the density 1.21 g/cm^3 has been derived.[301]

Ice-III. From x-ray data for ice-III at -155 °C and atmospheric pressure McFarlan has concluded that the density is 1.103 g/cm^3.[302]

Ice-VI. L. H. Adams[303] has reported the following values for the specific volumes of ice-VI ($v_i{}^*$) and of water ($v_w{}^*$), each at 25.0 °C and under the indicated pressure:

P_b	7000	8000	9000	10,000	11,000	12,000	bars
P	6890	7895	8882	9869	10,856	11,843	atm
$10^4 v_w{}^*$	8402	8278	8166	8059	7964	7876	cm^3/g
$10^4 v_i{}^*$	7509	7463	7417	7371	7325	7279	cm^3/g
d_i	1.3317	1.3399	1.3482	1.3567	1.3652	1.3738	g/cm^3

(See also Table 201).

Thermal Expansion of Ice (Cubical).
(For linear expansion, see next Section.)

In the reports of the *Vega* expedition, O. Pettersson[305] gives data indicating that the thermal coefficient of cubical expansion of ice is about 170×10^{-6} per 1 °C if the temperature is not above -3 °C, and decreases as the temperature rises above that point, becoming negative (*i.e.*, there is a contraction) as the melting point is closely approached. The purer the water from which the ice is frozen, the smaller is the variation in the coefficient, and the nearer to the melting point does the contraction first appear.

This strange behavior, which may account in part for the conclusion of O. Fort[306] that the early observations of A. Petzholdt[307] indicated that ice

---

[298] Barnes, H. T., and Cooke, H. L., *Trans. Roy. Soc. Canada, III (2),* **8,** 143-155 (1902).
[299] Roth, W. A., *Z. physik. Chem.,* **63,** 441-446 (1908).
[300] Clarke, J. R., *Int. Crit. Tables,* **3,** 43 (1928).
[301] McFarlan, R. L., *J. Chem'l Phys.,* **4,** 60-64 (1936) → *Phys. Rev. (2),* **49,** 199 (A) (1936).
[302] McFarlan, R. L., *J. Chem'l Phys.,* **4,** 253-259 (1936) → *Phys. Rev. (2),* **49,** 644 (A) (1936).
[303] Adams, L. H., *J. Am. Chem. Soc.,* **53,** 3769-3813 (1931).
[304] Bridgman, P. W., *J. Chem'l Phys.,* **5,** 964-966 (1937).
[305] Pettersson, O., *Beibl. Ann. d. Physik,* **7,** 834-841 (1883).
[306] Fort, O., *Ann. d. Physik (Pogg.),* **66,** 300-302 (1845).
[307] Petzholdt, A., "Beiträge zur Geognosie von Tyrol," 1843.

expands as the temperature falls, is thus accounted for by J. Y. Buchanan [282]: The ice formed from a dilute solution contains no salt, but some of the solution is entrapped between the crystals. Ice separates from that solution until the concentration of the solution has risen to such a value that the existing temperature is that at which there is equilibrium between the solution and ice. Thus the volume of liquid enclosed by the ice varies with the temperature, but is never zero so long as the temperature is above the cryohydric point. The actual change in volume as the temperature rises is the resultant of two effects: (a) the expansion of the ice; (b) the contraction attending the melting that is required to adjust the concentration of the entrapped solution. Above some temperature, the latter overbalances the former. This explanation assumes that the overall volume of a block of ice follows the changes in the volume of the entrapped material.

Table 202.—Isopiestic Coefficient of Cubical Expansion of Ice

For ice-VI, P. W. Bridgman [312] has stated that within the range 0 to 20 °C $(\partial v^*/\partial t)_p = 120 \times 10^{-6}$ (cm³/g) per 1 °C.

Values for ice-I are tabulated below; $\beta \equiv \dfrac{1}{v}\left(\dfrac{\partial v}{\partial t}\right)_p$; with the exception of Pettersson's observations (see text), β seems to be essentially independent of t; the pressure is nominally 1 atm.

Unit of $\beta = 10^{-6}$ per 1 °C. Temp. = t °C

Observer	Year	β	Range of t
Brunner [310]	1845	122	−0.8 to −19.5
Plücker and Geissler [292]	1852	155[a]	0 to −24
Pettersson [305]	1883	170	t below[b] −3
v. Zakrzewski [205]	1892	77	−0.7 to −4.7
Vincent [281]	1902	152	−0.4 to −10
3 × linear coefficient [c]		155	0 to −20
Value of choice		153	

[a] In taking the mean, one value (170) has been omitted.
[b] See p. 468.
[c] See Table 204.

He showed, by calculation, that the contraction observed by Pettersson in the case of ice formed from ordinary distilled water is the same as that which arises in the manner just indicated when the ice is formed from a solution of sodium chloride that is so dilute as to contain only 7 grams of Cl per 10^6 grams of water (Table 203). He also calculated, for various concentrations, the temperature at which the apparent volume of the ice is a maximum, and the amount of liquid then contained in it (Table 203). If the solution contains 1 part of Cl to 10^6 parts of water, the maximum apparent volume occurs at nearly a quarter of a degree below 0 °C; this solution is "in the category of distilled waters."

Apparently, no other investigator of the density or of the thermal dilatation of ice has either recorded an apparent contraction of ice as its rising

Table 203.—Specific Volume of Ice from Dilute Solutions
Adapted from J. Y. Buchanan.[313]

The solute is NaCl. C = concentration of the solution at 0 °C; W_0 = volume at 0 °C of the solution that goes to the formation of ice of volume V at the indicated temperature (t or t_m); w_0 = volume at 0 °C of the unfrozen brine contained in V. Let β and β_1 denote the coefficients of cubical expansion of ice and of water, respectively; ρ_0 = ratio of the density of ice at 0 °C to that of water at the same temperature; and $k \equiv w_0/W_0$; k is independent of W_0, but depends upon C and t in such a way that $\lambda \equiv kt/C$ varies but slightly. Hence a close approximation to k for any pair of intermediate values of t and C can be obtained from the values of λ tabulated below. $\rho_0 V = W_0 [1 - k(1-\rho_0) + \{\beta + k(\rho_0\beta_1 - \beta)\}t]$; the relative contraction due to incomplete freezing is $(W_0 - \rho_0 V)/W_0 \equiv \Delta$. Taking $\rho_0 = 0.9169$, $10^6\beta = +153$, $10^6\beta_1 = -194$, we have

$$0.9169 V = W_0 [1 - 0.0831k + (10^{-6})(153 - 331k)t]$$

Replacing k by its equivalent $\lambda C/t$, regarding λ as constant, and solving the equation $(\partial V/\partial t)_{\lambda C} = 0$, one obtains $t_m = -23.3\sqrt{-\lambda C}$ for the temperature at which V is a maximum.

$(V/W_0)_P$ = value of V/W_0 observed by Pettersson for ice formed from ordinary distilled water.

(In deriving the following values, Buchanan used for ρ_0, β, and β_1, values that differ slightly from those just given; in particular, he took $10^6\beta = 160$, which gives $t_m = -22.75\sqrt{-\lambda C}$.)

Unit of $C = 1$ g-Cl per megagram of water = 0.0000282 gfw-NaCl per kg of water; of $\lambda = 1$ °C/(1 g-Cl per g water)

t	$100k$	$C = 7$ V/W_0	$(V/W_0)_P$	$-\lambda^a$
−0.07	1.000	1.08979	1.08980	100
−0.10	0.700	1.09006	1.09007	100
−0.15	0.467	1.09028	1.09038	100
−0.20	0.350	1.09037	1.09048	100
−0.40	0.175	1.09054	1.09057	100

C	t_m	$100k$	1000Δ	$-\lambda^a$
10000	−20.5	5.73	7.51	118
5000	−16.6	3.37	5.46	112
2500	−10.75	2.46	3.60	106
1250	−7.8	1.60₆	2.48	100.2
1000	−7.0	1.420	2.22	99.4
500	−4.9	1.000	1.56	98.0
250	−3.5	0.695	1.11	97.3
125	−2.55	0.469₆	0.77	95.8
100	−2.3	0.4183	0.70	96.2
10	−0.725	0.1363	0.22	98.8
1	−0.2275	0.0438	0.07	99.7
0.1	−0.0725	0.01377	0.02	99.8
0.01	−0.02275	0.004306	0.01	98.0

[a] These values of λ have been derived from the tabulated values of C, t or t_m, and k, which have been taken from Buchanan's paper.

temperature approaches 0 °C, or considered the effect discussed by Buchanan. That effect (the inclusion of unfrozen liquid) has however been considered, and seemingly observed, by A. W. Smith [308] and by H. C. Dickinson and N. S. Osborne [309] in their determinations of the specific heat of ice.

Other determinations of the cubical expansion of ice are those of C. Brunner [310] using natural ice and hydrostatic weighings; of J. Plücker and Geissler [292] using an ingenious double-bulb thermometer; of J. v. Zakrzewski,[295] who, using Bunsen's method for determining the density, obtained for the coefficient a value that is only half as great as that found by others; and of J. H. Vincent [281] using a novel hydrostatic method. (Table 202.)

Sir James Dewar [311] has inferred that the coefficient decreases as the temperature is greatly reduced, from the fact that ice at 0 °C cracks in all directions when dropped into liquid air (-188.7 °C), but ice that has been slowly cooled to -188.7 °C does not crack when it is dropped into liquid hydrogen (-252.7 °C). This conclusion accords with the recent determination of the linear expansion by M. Jakob and S. Erk (1928), see Table 204.

Compressibility of Ice.

Ice-I.—In his compilation, L. H. Adams[314] gives the value obtained by T. W. Richards and C. L. Speyers [315] for the isothermal compressibility of ice-I, namely $\gamma \equiv -\frac{1}{v_0}\left(\frac{\partial v}{\partial p}\right)_t = 12 \times 10^{-6}$ per bar at -7 °C and 300 bars. Those authors stated that the value of γ between 300 and 500 bars is probably not over 3 per cent less than it is between 100 and 300 bars. (1 bar = 1 megadyne/cm²; 'they called it a megabar, thus departing from international custom.)

In contrast to this, P. W. Bridgman [316] computed from his observations that at 0 °C and 1 atm $\gamma = 37 \times 10^{-6}$ per bar, a value 3 times as great as the former. Richards and Speyers [315] state that Bridgman has admitted that his calculated value is untrustworthy, and, on their suggestion that the difference may in part be due to γ having a large temperature coefficient, he recalculated γ obtaining the following values for $10^6\gamma$: 0 °C, 33; -5 °C, 23; -7 °C, 21; $-10°$ C, 19; -15 °C, 18. These are still much greater than theirs. They state that Bridgman agreed with them in thinking that the

---

[308] Smith, A. W., *Phys. Rev. (2)*, **17**, 193-232 (1903).
[309] Dickinson, H. C., and Osborne, N. S., *Bull. Bur. of Stand.*, **12**, 49-81 (S248) (1915).
[310] Brunner, C., *Ann. d. Physik (Pogg.)*, **64**, 113-124 (1845).
[311] Dewar, Sir James, *Proc. Roy. Inst. Grt. Britain*, **17**, 418-426 (1903) = *Chem. News*, **91**, 216-219 (1905).
[312] Bridgman, P. W., *Proc. Am. Acad. Arts. Sci.*, **48**, 307-362 (1912).
[313] Buchanan, J. Y., *Proc. Roy. Inst. Grt. Brit.*, **19**, 243-276 (251, 257) (1908).
[314] Adams, L. H., *Int. Crit. Tables*, **3**, 49-51 (50) (1928).
[315] Richards, T. W., and Speyers, C. L., *J. Am. Chem. Soc.*, **36**, 491-494 (1914).
[316] Bridgman, P. W., *Proc. Am. Acad. Arts Sci.*, **47**, 439-558 (1912).

discrepancy may be due to a softening of the ice just before melting. (Is this a reference to the phenomenon discussed by Buchanan, p. 469?)

Ice-VI.—P. W. Bridgman [312, p. 362] found for ice-VI the value: $-(dv^*/dp)_t = 4.6 \times 10^{-6}$ cm^3/g per bar, essentially constant throughout the ranges 0 to 20 °C and 6000 to 10,000 kg*/cm^2. (In the *International Critical Tables*,† this value is incorrectly recorded as what is here denoted by γ.) Bridgman now thinks that that value is probably too high, but he has not yet obtained consistent values for the compressibility at these high pressures for either ice-VI or water.[317]

Ice-VII.—Mean compressibility of ice-VII between 45,000 and 50,000 kg*/cm^2 is about 3/4 of that between 20,000 and 25,000 kg*/cm^2; on increasing the pressure from 20,000 to 45,000 kg*/cm^2 the specific volume of ice-VII decreases by 0.039 cm^3/g; at 50,000 kg*/cm^2 and room temperature the specific volume is about 0.60 cm^3/g.[317]

68. Coefficient of Linear Expansion of Ice

As ice is a crystalline substance, it is to be expected that its thermal coefficient of linear expansion ($\alpha \equiv (\delta l/l_0 \delta t)_p$ where l = length, t = temperature, p = pressure) will vary with the angle between l and the optic axis. No data on the expansion of single crystals of ice have been found, but curves representing such data for zinc, which also crystallizes in the hexagonal system, have been obtained by Grüneisen and Goens, and reproduced by M. Jakob and S. Erk.[318] They indicate that the expansion of zinc in the direction of the hexagonal axis is much greater than it is transverse to that axis, and that in the latter direction the expansion is negative when the temperature is very low.

The only satisfactory series of determinations of α for ice over a considerable range in temperature is that of Jakob and Erk.[318] They used rods of ice frozen slowly in paper tubes, the freezing proceeding radially from outside in. Polariscopic examination gave no indication of any regular orientation of the axes of the constituent crystals, but the variation of α with t was strikingly similar to that for zinc in which l is perpendicular to the optic axis; whence, the authors concluded that the axes of the constituent crystals had a pronounced radial component.

The only other determinations for specimens in which the crystals were thought to have been oriented fairly uniformly are those of C. A. v. Schumacher,[319] of Pohrt,[320] and of A. Moritz,[321] summed up by W. Struve,[322] quoted by H. Moseley,[323] and extending the work of W. Struve.[324] They indicated that α is essentially independent of the direction of the optic axis.

† Vol. 3, p. 50.
[317] Bridgman, P. W., *J. Chem'l Phys.*, **5**, 964-966 (1937).
[318] Jakob, M., and Erk, S., *Wiss. Abh. Physik.-Techn. Reichsanst.*, **12**, 301-316 (1928-29) = *Z. gesamt. Kälte-Ind.*, **35**, 125-130 (1928).
[319] v. Schumacher, C. A., *Mém. Acad. St. Pétersbourg, Math-Phys.* (6), **4**, 307-357 (1847).
[320] Pohrt, Published with those of Moritz (1847).
[321] Moritz, A., *Idem*, **4**, 358-384 (1847).
[322] Struve, W., *Idem*, **4**, 294-306 (1847).

68. ICE: LINEAR EXPANSION

Other frequently quoted determinations are those of E. L. Nichols[325]; of T. Andrews,[326] distributed from 0 to -34.5 °C and represented by $10^6\alpha = 88.79 + 3.800t + 0.06654t^2$; and the very erratic ones of W. H.

Table 204.—Thermal Coefficient of Linear Expansion of Ice
(For coefficient of cubical expansion, see Table 202.)

The only satisfactory determinations are those of Jakob and Erk (1928); the others here given are frequently quoted.

Values in parentheses have been computed by means of the appropriate equation given in the text. The specimen used by Andrews was frozen in a cylinder having a height equal to its diameter (2 ft); A and R indicate that the direction of the observed expansion was axial and radial, respectively. θ = assumed angle between the optic axis and the length (l) of the specimen; Vbl indicates that θ is variable, and that the axes are irregularly distributed.

$$\alpha = \frac{1}{l_0}\left(\frac{\partial l}{\partial t}\right)_p;\ \alpha_m = \frac{l_2 - l_1}{l_0(t_2 - t_1)};\ l_0 = \text{value of } l \text{ at } 0\ °C.$$

Unit of α and of $\alpha_m = 10^{-6}$ per 1 °C. Temp. = t °C

Jakob and Erk (1928).[318]

t	α	t	α	t	α	t	α	t	α	t	α
0	52.7	−50	45.6	−100	33.9	−150	16.8	−200	+0.8	−250	−6.1
−10	51.7	−60	43.7	−110	30.6	−160	13.0	−210	−1.3		
−20	50.5	−70	41.5	−120	27.3	−170	9.5	−220	−3.3		
−30	49.0	−80	39.2	−130	23.9	−180	6.3	−230	−4.5		
−40	47.4	−90	36.7	−140	20.4	−190	3.3	−240	−5.5		

Observer	θ	t_2	t_1	α_m	θ	t_2	t_1	α_m
v. Schumacher [319]	90°	−1.2°	−27.5°	51.4	Vbl R	0	−8.9	73.6
Pohrt [320]	90	−1.1	−26.8	51.1	Vbl R	−8.9	−17.8	50.3
Moritz [321]	0	−1.7	−28.5	51.8	Vbl R	−17.8	−29.5	36.9
Struve [324]	Vbl	−1.2	−27.5	53.0	Vbl R	−29.5	−34.5	35.5
Nichols [325]	Vbl	−8	−12	54.0	Vbl R	−1	−28	(51.7)
Sawyer [327]	Vbl	0	−18	(53.0)	Vbl R	0	−18	(61.8)
Sawyer [327]	Vbl	−1	−28	(42.8)	Vbl A	0	−17.8	37.0

Sawyer,[327] quoted by H. T. Barnes, J. W. Hayward, and N. M. McLeod,[328] covering the range −8 to −20 °C, and approximately represented by $10^6\alpha = 69.6 + 1.85t$. Neither of the last two sets accords well with the values obtained by others; each is represented by a graph in the compilation by J. R. Clarke.[329]

From x-ray studies at 0 °C and −66 °C, H. D. Megaw[330] has con-

[323] Moseley, H., *Phil. Mag. (4)*, **39**, 1-8 (1870).
[324] Struve, W., *Ann. d. Physik (Pogg.)*, **66**, 298-300 (1845).
[325] Nichols, E. L., *Phys. Rev.*, **8**, 184-186 (1899).
[326] Andrews, T., *Proc. Roy. Soc. (London)*, **40**, 544-549 (1886).
[327] Sawyer, W. H., *Proc. Maine Soc. Civ. Eng.*, **1**, 27 (1911).
[328] Barnes, H. T., Hayward, J. W., and McLeod, N. M., *Trans. Roy. Soc. Canada III (3)*, **8**, 29-49 (1914).
[329] Clarke, J. R., *Int. Crit. Tables*, **3**, 43-45 (43) (1928).
[330] Megaw, H. D., *Nature*, **134**, 900-901 (L) (1934).

cluded that the base (a) and the height (c) of the unit cell of ice have the following values: $a = 4.5135$A and $c = 7.3521$A at 0 °C, and $a = 4.5085$A and $c = 7.338$A at $-66°$ C. These give for the mean coefficient of linear expansion (α_m) between these temperatures the values $10^6 \alpha_m = 17$ for a and 29 for c, which are much smaller than would be inferred from any of the measurements on ice in bulk that are given in Table 204.

69. Thermal Energy of Ice-I

This section is devoted to the following types of data for the ordinary type of ice (ice-I): the isopiestic specific heat (c_p), the enthalpy or heat content ($H = E + pv$); the "free energy at constant pressure" ($G = H - ST$); and the entropy (S). No determination of either the heat of isothermal compression or the Joule-Thomson coefficient has been found for ice.

$$\int_{t_0}^{t} c_p dt = H_t - H_{t_0}; \quad -T \int_{T_0}^{T} (H/T^2)\, dT = G_T - G_{T_0};$$

$$\int_{t_0}^{t} (c_p/T)\, dt = S_t - S_{t_0}.$$

Specific Heat of Ice.

In order to obtain the true specific heat of ice, proper allowance must be made for the progressive melting caused by the presence of included water containing dissolved impurities.[331] Such inclusion is always present and gives rise to an apparent specific heat which exceeds the true, the excess increasing rapidly as 0 °C is approached. Even with the purest water used by Dickinson and Osborne, its effect was appreciable at -5 °C, and became very marked above -0.5 °C. Every nominal determination of the true specific heat of ice is to be regarded with suspicion unless the observer has clearly shown that this effect is negligible in his case, or has properly corrected for it. As early as 1904, A. D. Bogojawlensky[331a] had concluded that the specific heat of a pure crystal is linear in the temperature.

Here we shall distinguish between the apparent specific heat (c_a) of ice and the true specific heat (c) which would be found were the ice perfectly pure. Obviously, the former will vary with the specimen, and the latter must be derived from the former. Apparently the only observers who have attempted to derive c from their own determinations are Dickinson and Osborne,[309] who used four samples of very carefully purified water, and carried out the work with a precision unattained as yet by others. Over the range covered (-0.5 to -40 °C) they found that their results can be expressed by the formula $c_a = a + bt - d/t^2$ in which d varies with the specimen, but a and b do not. Hence, they regarded $a + bt$ as the value

---

[331] See Person, C. C., *Ann. de chim. et phys. (3)*, **30**, 73-81 (1850); Buchanan, J. Y.[282]; Smith, A. W., *Phys. Rev. (2)*, **17**, 193-232 (1903); Dickinson, H. C., and Osborne, N. S.[309]
[331a] Bogojawlensky, A. D., *Schrift. Natur. Ges. Univ. Jurjeff (Dorpat)*, 1904.

69. ICE: THERMAL ENERGY

of c, and interpreted d/L, which decreases rapidly as the purity is increased, as the initial freezing-point of the solution obtained when the specimen is completely melted, L being the latent heat of fusion of ice at 0 °C. For the specimens used by them, $d/L = -0.00125, -0.00120, -0.00095$, and -0.00005 °C, respectively.

The best extended series of observations covering the range -2.9 to -189.5 °C is probably that of Nernst and his associates.[332] They represent their data by the formula $c_a = a + bt - d/t$, in which the last term varies as $1/t$, whereas that in the Dickinson and Osborne formula varies as $1/t^2$. Their precision was not as high as that of Dickinson and Osborne.

Still lower temperatures (-189 to -250.6 °C) are covered by the observations of F. Pollitzer.[333] At such low temperatures $c_a = c$ except for experimental errors.

These and earlier observations were considered by J. H. Awbery in the derivation of the two values given for ice in his compilation.[334]

A second set of values is given by W. H. and E. K. Rodebush.[335] They have been computed by means of a formula, apparently unpublished, that was fitted to the observations of F. Pollitzer[333] and of Nernst (as quoted by Pollitzer) and made $c = 0$ at 0 °K.

More recent determinations of c_a have been published by O. Maass and L. J. Waldbauer [336] and by W. H. Barnes and O. Maass.[337] These values are exactly those quoted by H. T. Barnes [338] and credited by him to "Maass and Barnes, W. H., 1927." The first covered the range -3 to -182.7 °C; the second, involving refinements in the method, covered the range -2.6 to -78.6 °C. In each case, the quantity measured was the total heat required to convert ice at $-t_1$ °C to water at $+t_2$ °C; t_2 was 16.5 °C in the first and 25 °C in the second. The quantity actually measured much exceeded the amount of heat accounted for by the specific heat of ice; consequently the precision with which the specific heat and its variation can be determined from those data is much lower than the precision of the data themselves. Only the latter precision is stated, which in the second article is said to be ±0.05 per cent or better. Actually, that is merely the precision of reproducibility for the same specimen under nominally identical conditions. No data are given from which any other precision can be determined. In each case, only a single specimen of ice seems to have been used.

The total heat (H) required to change the specimen from ice at t °C to water at 16.5 °C (25 °C in the second) was represented by a formula of the type $H = a + bt + dt^2 + et^3$ and c_a was obtained by differentiating that equation, giving $c_a = -b - 2dt - 3et^2$. Although no great precau-

---

[332] Nernst, W., Koref, F., and Lindemann, F. A., *Sitz. Preus. Akad. Wiss.*, 1910, 247-261 (1910); Nernst, W., *Idem*, 1910, 262-282 (1910).
[333] Pollitzer, F., *Z. Elektroch.*, 19, 513-518 (1913).
[334] Awbery, J. H., *Int. Crit. Tables*, 5, 95-105 (95) (1929).
[335] Rodebush, W. H., and E. K., *Idem*, 5, 89 (1929).
[336] Maass, O., and Waldbauer, L. J., *J. Am. Chem. Soc.*, 47, 1-9 (1925).
[337] Barnes, W. H., and Maass, O., *Can. J. Res.*, 3, 205-213 (1930).
[338] Barnes, H. T., "Ice Engineering," p. 38, 1928.

tion was taken to ensure the purity of the water (ordinary distilled water was redistilled from an all-platinum still) there is nothing in the formulations to indicate any excessive increase in c_a as t approaches 0 °C, but those observations lying not below -110 °C in the first series can be satisfactorily represented by a formula of the type used by Dickinson and Osborne, d/L being taken as -0.0045 °C. The observers were of the opinion that at temperatures not exceeding -2.6 °C the ice was completely frozen, and consequently $c_a = c$. That opinion is not consistent with the observations either of Dickinson and Osborne or of Nernst, who used the best conductivity water. Actually, the precision with which they have determined the small quantities from which c_a and its variation have to be derived is much lower than they seem to have realized. It may be very conservatively taken as Δ/h where Δ is the excess of the observed value of H over that defined by their formula, and h is the excess of the observed value of H over that at 0 °C as defined by the formula; in accordance with their point of view, h is the heat required to raise the ice to 0 °C without melting it; the values of h, and they alone, are involved in the determination of c_a and its variation. All their values of h and of Δ are given below; it will be noticed that for no observation at a temperature exceeding -11 °C is Δ/h less than 1 per cent of h.

—————— Maass and Waldbauer ——————				—————— Barnes and Maass ——————			
t	H_{obs}	h	Δ	Δ/h	t	H_{obs}	h	Δ	Δ/h
-3.18	97.50	1.66	$+0.12$	$+7.2\%$	-2.60	105.69	1.24	-0.03	-2.4%
-28.6	109.37	13.53	$+0.08$	$+0.6$	-4.60	106.73	2.28	$+0.05$	$+2.2$
-58.6	122.06	26.22	-0.11	-0.4	-10.15	109.40	4.95	$+0.05$	$+1.0$
-78.6	129.86	34.02	-0.24	-0.7	-15.17	111.75	7.30	$+0.03$	$+0.4$
-110.0	141.25	45.41	$+0.23$	$+0.5$	-19.99	113.95	9.50	-0.03	-0.3
-138.7	149.45	53.62	$+0.34$	$+0.6$	-25.01	116.31	11.86	$+0.01$	$+0.1$
-182.7	158.16	62.32	$+0.08$	$+0.1$	-30.05	118.69	14.24	$+0.11$	$+0.8$
					-50.05	127.28	22.83	$+0.02$	$+0.1$
					-78.57	138.21	33.76	-0.04	-0.1

In Table 205 the values of c_a as defined by the two formulas are compared. One might expect that the somewhat lower accuracy of the earlier work would be in large part offset by the much greater range of temperature over which the observations are spread, but the two formulas lead to markedly divergent results. Even at -80 °C they differ by 7 per cent, the more recent giving the lower value. The observers' suggestion that a contributing factor to such divergence is the fact that the sublimation point of solid CO_2 was taken as -78.2 °C in the earlier paper instead of as -78.5 °C is not satisfactory, because according to their own formulation that change in temperature will change c_a by less than 0.2 per cent, which is negligible as compared with 7 per cent.

Until such marked discrepancies shall have been satisfactorily explained, confidence cannot be placed in either formula. This is especially unfortunate because the earlier work was in part intended to serve as a check upon the accuracy of similar measurements made upon other substances.

69. ICE: THERMAL ENERGY

Table 205.—Apparent Isopiestic Specific Heat of Ice

The apparent specific heat of ice (c_a) exceeds the true (c) on account of the presence of impurities (see text); it varies from specimen to specimen. The best determinations are those of Dickinson and Osborne (DO), who found that $c_a = 0.5057 + 0.001863t - 79.75d/t^2$ cal$_{20}$/g.°C, where d = initial freezing point of the completely fused ice. Their values of d varied from -0.00005 to -0.00125 °C; if the impurity were NaCl and d were -0.000062 the concentration would be 0.0001 per cent by weight. Their observations did not extend below -40 °C.

Nernst and his associates extended their observations (N) to -189 °C and represented them by the formula $c_a = 0.470_2 + 0.0153_2 t - 0.77_7/t$.

Pollitzer's observations (P) covered the range -189 to -250.6 °C.

The observations of Maass and Waldbauer (MW), and of Barnes and Maass (BM) were represented, respectively, by $c_a = 0.485 + 0.000914t - 5.46(10^{-6})t^2$ and $c_a = 0.48733 + 0.0009325t - 9.828(10^{-6})t^2$. For reasons stated in the text, implicit confidence cannot be placed in those values.

Excepting Pollitzer's, the values in Section I have been computed from these 4 equations, those lying beyond the range of the observations being enclosed in parentheses; $\delta \equiv (BM - MW)/MW$.

The values of C in Section II have been taken directly from the paper cited; conversion has been made by the compiler.

Unit of c_a = 1 cal$_{20}$ per g.°C = 4.181 joules per g.°C; of δ = 1%.

I. Various determinations.

Ref[a]→ d_1→ t	-0.006	DO[b] -0.0006	-0.00006	N, P	MW	BM	δ
			c_a				
0	...	...	...	N	0.485	0.487₃	+0.4
-5	0.5155	0.4983	0.4966	0.6180	0.480	0.482₄	+0.4
-10	0.4919	0.4876	0.4871	0.5326	0.475	0.477₀	+0.4
-20	0.4696	0.4685	0.4684	0.4784	0.464	0.464₈	+0.2
-30	0.4503	0.4498	0.4498	0.4501	0.453	0.450₅	-0.7
-40	0.4315	0.4312	0.4312	0.4283	0.440	0.434₃	-1.4
-60	(0.3940)	(0.3939)	(0.3939)	0.3913	0.410	0.396₀	-3.4
-80	(0.3568)	(0.3567)	(0.3567)	0.3571	0.377	0.349₈	-7.1
-100	(0.3194)	(0.3194)	(0.3194)	0.3248	0.339	(0.295₈)	-12.7
-120		(0.2821)		0.2929	0.297	(0.233₉)	-21.2
-140		(0.2449)		0.2612	0.250	(0.164₂)	
-160		(0.2076)		0.2299	0.199	(0.086₅)	
-180		(0.1704)		0.1987	0.144	(0.001₀)	
				P			
-189				0.186			
-200		(0.1331)					
-213.7				0.130			
-222.3				0.110			
-230.1				0.098			
-235.4				0.085			
-239.2				0.070₀			
-244.8				0.048₈			
-248.7				0.037₀			
-250.6				0.030₃			

Table 205—(Continued)

II. W. F. Giauque and J. W. Stout.[339] Mole fraction soluble impurity estimated to be 3×10^{-6}. Unit of C = 1 cal/g-mole·°K; molecular weight taken as 18.0156; 1 cal = 4.1832 Int. joules.

0 °C = 273.1 °K. Temp. = t °C = T °K. Unit of c_a = 1 cal/g.

T	t	C	c_a
10	−263.1	0.066[e]	0.0037
20	−253.1	0.490	0.0272
30	−243.1	0.984	0.0546
40	−233.1	1.466	0.0814
50	−223.1	1.896	0.1052
60	−213.1	2.304	0.1279
70	−203.1	2.701	0.1499
80	−193.1	3.075	0.1707
90	−183.1	3.448	0.1914
100	−173.1	3.796	0.2107
110	−163.1	4.130	0.2292
120	−153.1	4.434	0.2461
130	−143.1	4.728	0.2624
140	−133.1	4.993	0.2772
150	−123.1	5.265	0.2922
160	−113.1	5.550	0.3081
170	−103.1	5.845	0.3244
180	−93.1	6.142	0.3409
190	−83.1	6.438	0.3574
200	−73.1	6.744	0.3743
210	−63.1	7.073	0.3926
220	−53.1	7.391	0.4103
230	−43.1	7.701	0.4275
240	−33.1	8.103	0.4448
250	−23.1	8.326	0.4622
260	−13.1	8.642	0.4797
270	−3.1	8.960	0.4974

[a] References:
- BM Barnes, W. H., and Maass, O.[337]
- DO Dickinson, H. C., and Osborne, N. S.[309]
- MW Maass, O., and Waldbauer, L. J.[336]
- N   Nernst, W., Koref, F., and Lindemann, F. A.,[332] and Nernst, W.[332]
- P   Pollitzer, F.[333]

[b] For these values, 1 cal₂₀ = 4.183 joules, the value used by the authors in translating their electrical measurements into what they call cal₂₀.

[e] This value for 10 °K has been derived by Giauque and Stout from the following values obtained in 1923 by Simon and privately communicated to them:

T	9.47	9.88	10.46	11.35	11.55	12.10	12.85 °K
t	−263.63	−263.22	−262.64	−261.75	−261.55	−261.00	−260.25 °C
C	0.056	0.063	0.075	0.096	0.102	0.118	0.141
c_a	0.0031	0.0035	0.0042	0.0053	0.0057	0.0066	0.0078

From the behavior of gadolinium anthraquinone sulfonate to which much water had been added, D. P. MacDougall and W. F. Giauque[340] have inferred that between 0.2 and 4 °K the specific heat of ice does not exceed 0.01 cal/g-mole·°K.

[339] Giauque, W. F., and Stout, J. W., *J. Am. Chem. Soc.*, **58**, 1144-1150 (1936).
[340] MacDougall, D. P., and Giauque, W. F., *J. Am. Chem. Soc.*, **58**, 1032-1037 (1936).

69. ICE: THERMAL ENERGY

Table 206.—True Isopiestic Specific Heat of Ice

For temperatures not below $-40\ °C$ the best determinations are those of Dickinson and Osborne, which may be expressed by the formula $c = 0.5057 + 0.001863t$ cal$_{20}$ per (g.°C) $= 2.115_3 + 0.00779_3 t$ joule per (g.°C). (They used the relation 1 cal$_{20}$ = 4.183 j.) Values so computed are given in column DO, those beyond the range of the observations being in parentheses.

Values observed by Nernst and by Pollitzer are given in column N, P; those at the higher temperatures are actually c_a, not c.

The two sets of values from *International Critical Tables* are given under A and R.

Unit of $c = 1$ joule/(g.°C) = 0.2392 cal$_{20}$/(g.°C). Pressure not exceeding 1 atm.

Ref[a]→ t	DO	A	R	N,P
0	2.115	2.06 ± 0.01	2.1$_2$	N
−10	2.037			2.28
−20	1.959	1.94 ± 0.01		2.00
−23.1	1.935		1.9$_3$	
−30	1.882			
−40	1.804	1.82 ± 0.01		
−60	(1.648)	1.68 ± 0.02		
−70	(1.570)			1.56
−73.1	(1.546)		1.$_{32}$	
−80	(1.492)	1.54 ± 0.02		
−100	(1.336)	1.39 ± 0.01		
−120	(1.180)			1.23
−123.1	(1.152)		1.$_{21}$	
−140	(1.025)			
−150	(0.946)	1.030 ± 0.010		
−160	(0.868)			
−170	(0.790)			0.89$_6$
−173.1	(0.766)		0.8$_9$	
−180	(0.712)			P
−189	(0.642)			0.77$_8$
−200	(0.556)	0.653 ± 0.013		
−213.7	(0.450)			0.54$_6$
−222.3	(0.383)			0.46$_2$
−223.1	(0.377)		0.4$_8$	
−230.1	(0.321)			0.41$_1$
−235.4	(0.280)			0.35$_7$
−239.2	(0.251)			0.29$_2$
−244.8	(0.207)			0.20$_4$
−248.7	(0.177)			0.15$_5$
−250	(0.167)	0.151 ± 0.004		
−250.6	(0.162)			0.128
−273.1	(−0.013)		0	

[a] References:

A   Compilation by Awbery, J. H.,[334] based on the observations of Armstrong, H. E., *Proc. Roy. Inst. Grt. Brit.*, **19**, 354-412 (1908), Barnes, H. T., *Trans. Roy. Soc. Canada*, III (3), **3**, 3-27 (1909), Dickinson, H. C., and Osborne, N. S.,[309] Jackson, F. G., *J. Am. Chem. Soc.*, **34**, 1470-1480 (1912), Nernst, W.,[332] and *Ann. d. Physik* (4), **36**, 395-439 (1911), Nernst, W., Koref, F., and Lindemann, F. A.,[332] Person, C. C.,[331] Pollitzer, F., *Z. Electroch.*, **17**, 5-14 (1911); **19**, 513-518 (1913); Regnault, V., *Ann. d. Physik (Pogg.)*, **77**, 99-109 (1849).

DO   Dickinson, H. C., and Osborne, N. S.[309]
N   Nernst, W., as quoted by Pollitzer, see P.
P   Pollitzer, F.[333]
R   Rodebush, W. H. and E. K.,[335] see text.

Entropy of Ice.

If entropy is measured from 0 °K, then the entropy of water-vapor at 25 °C as computed from spectroscopic data is 45.10 cal/g-mole·°K [341]; whereas the value found from the specific heat of ice and of water, together with the latent heats of transition, is only 44.28, a difference of 0.82 cal/g-mole·°K.[342] Suggestions for explaining this discrepancy have been offered by W. F. Giauque and M. F. Ashley,[343] who definitely established the existence of the discrepancy, and by L. Pauling.[344] Each suggestion attributes it to a failure of ice to attain the ideal state at the lowest temperature reached experimentally; the first assumes the persistence of the *ortho* and *para* molecular states at extremely low temperature, and the second assumes a certain amount of disorder. See also Table 207.

Table 207.—Various Isopiestic Thermal Data for Ice

C_p = specific heat at constant pressure; $H_0 = \int_0^T C_p dT$ is the enthalpy (heat content); $G_0 = H_0 - ST = -\int_0^T (H_0/T^2) dt$ is the "free energy at constant pressure"; S_0 = entropy. All are for a pressure of 1 atm, and those with subscript $_0$ are measured from ice at 0 °K; i = for ice, w = for water. For more complete definitions, see Table 1 and Section 6. The values accepted by K. K. Kelley [354] essentially agree with the corresponding ones here given under R and S. 1 cal = 4.185 joules; 1 j/gfw = 0.05551 j/g.

Unit of C_p and S_0 = 1 j/(gfw.°K); of H_0 and G_0 = 1 kj/gfw. Temp. = T °K; 1 gfw = 18.0154g.

Ref*→ T	R	S C_p	M	R	S H_0	M
0	0	0	0	0	0	0
10		0.28	0.1		0.0006	0.0002
20		1.77	1.8		0.010	0.007
40		6.57	6.6		0.093	0.092
50	8.6		8.5	0.183		0.168
60		10.01	10.2		0.260	0.262
80		12.9	13.3		0.491	0.498
100	16.1	15.7	16.1	0.823	0.778	0.795
120			18.4			1.141
150	21.7	21.6	22.0	1.775	1.712	1.74
170			24.2			2.21
200	23.8[b]	27.5	27.9	3.005	2.94	2.99
220			30.5			3.58
250	34.7	34.9	35.4	4.54	4.48	4.56
273.1 i	38.2	41.0	50.2	5.38	5.35	5.49
273.1 w		76.0	75.4		11.36	11.50

[341] Gordon, A. R., *J. Chem'l Phys.*, **2**, 65-72 (1934).
[342] Giauque, W. F., and Stout, J. W., *J. Am. Chem. Soc.*, **58**, 1144-1150 (1936).
[343] Giauque, W. F., and Ashley, M. F., *Phys. Rev. (2)*, **43**, 81-82 (L) (1933).
[344] Pauling, L., *J. Am. Chem. Soc.*, **57**, 2680-2684 (1935).
[345] Kelley, K. K., *Bur. Mines (U. S.), Bulletin* 350 (1932).

70. ICE: THERMAL CONDUCTION

Table 207—(Continued)

Ref.[a]→ T	R	S $-G_0$	M	R	S S_0	M
0	0	0	0	0	0	0
10		0.00022	0.00006		0.085	0.025
20		0.0036	0.00226		0.69	0.46
40		0.0416	0.0348		3.36	3.18
50	0.044		0.0750	4.78		4.87
60		0.1424	0.132		6.69	6.57
80		0.3094	0.294		9.99	9.90
100	0.508	0.541	0.526	13.31	13.20	13.21
120			0.821			16.35
150	1.345	1.391	1.39	20.8	20.69	20.8
170			1.82			23.7
200	2.545	2.60	2.60	27.7	27.7	27.9
220			3.19			30.8
250	4.120	4.16	4.17	34.6	34.4	34.9
273.1 i	4.940	5.00	5.02	37.8	37.9	38.4
273.1 w	4.940	5.00	5.02	65.5[c]	59.9	60.5

[a] References:
  M Miething, H., *Abh. deuts. Bunsen-Ges.*, No. 9 (1920), based upon the data of Pollitzer (See R) and of Nernst, W., *Ann. d. Physik (4)*, **36**, 395-439 (1911), steps of 10 °K.
  R Rodebush, W. H. and E. K., *Int. Crit. Tables*, **5**, 84-91 (89) (1929), based upon the data of Pollitzer, F.[333]; they measure both H and G from the uncombined gases at 0 °K and 1 atm. That is, their values for H and G fall below those given for H_0 and G_0 by the heat of formation of ice at 0 °K and 1 atm, which they have taken as 282.6 kj/gjw.
  S Simon, F., "Hand. d. Physik," (Geiger and Scheel), vol. **10**, 363 (1926) based on his own previously unpublished observations.

[b] Apparently erroneous.
[c] For 25 °C (298.1 °K).

---

70. THERMAL CONDUCTIVITY OF ICE AND OF SNOW

Single Crystals.

The thermal conductivity of a single crystal of ice has not been studied, but indirect evidence indicates that the conductivity along the optic axis exceeds that in a direction perpendicular thereto.[346] The difference is probably small. More recently, J. M. Adams [347] has reported observations which he thinks suggest "that the polar character of the crystal extends to the mechanism of thermal conduction in it."

The only recorded numerical data bearing upon the subject seem to be the following, k_v and k_h denoting the conductivities perpendicular and parallel, respectively, to the planes of freezing.

	k_v	k_h	
Forbes (1873)	9.32	8.90	milliwatt/cm·°C
Straneo (1897)	21.9	21.0	milliwatt/cm·°C

Those by Forbes are surprisingly small.

[346] Barratt, T., and Nettleton, H. R., *Int. Crit. Tables*, **5**, 231 (1929). Based upon: Barnes, H. T., *Nature*, **83**, 276 (1910), Forbes, G., *Proc. Roy. Soc. Edinburgh*, **8**, 62-69 (1873), Straneo, P., *Atti. Acc. Lincei (5)*, **6₂**: 299-306 (1897).

[347] Adams, J. M., *Proc. Roy. Soc. London (A)*, **128**, 588-591 (1930) → *Phys. Rev. (2)*, **36**, 788 (A) (1930).

Ice in Bulk.

When data were being prepared for the *International Critical Tables*, the most extended series available on the thermal conductivity of ice was that by C. H. Lees.[348] More recent work by M. Jakob and S. Erk,[349] extending to $-130\,°C$, is probably to be preferred to all others now available. They found that when there is a flow of heat between a block of ice and a metal plate frozen to it, there is always a discontinuity in the temperature at the junction, the discontinuity increasing as the temperature is lowered.

Apparently, the earliest recorded attempt to measure the thermal conductivity of ice is that of F. Pfaff[350] leading to the surprising conclusion that the conductivity of ice is 0.82 that of iron (*i.e.*, to $k = 508$ milliwatt/cm·°C).

Table 208.—Thermal Conductivity and Diffusivity of Ice

(For single crystals, see text.)

The preferred values for the conductivity (k) are those (JE[a]) by Jakob and Erk. The (VD[a]) values have been computed by Van Dusen's equation ($k = 20.9\,(1 - 0.0017t)$ milliwatt/cm·°C), which was set up prior to the work by JE, and approximately represents the values found by Lees.

Thermal diffusivity ($k/\rho c$, ρ = density, c = specific heat) is 0.011 cm²/sec if $t > -30\,°C$ (VD[a]); is 0.0114 at $0\,°C$ (SH[a]), based on F. Neumann.[351]

Unit of $k = 1$ milliwatt/(cm·°C)

Ref.[a]→ t	JE k	VD	Ref.[a]→ t	JE k	VD	Ref.[a]→ t	JE k	VD	Ref.[a]→ t	VD k
0	22.4[b]	20.9	−50	27.8	22.7	−100	34.7	24.4	−150	26.2
−10	23.2	21.3	−60	29.1	23.0	−110	36.4	24.8	−160	26.6
−20	24.3	21.6	−70	30.5	23.4	−120	38.1	25.2	−170	26.9
−30	25.5	22.0	−80	31.8	23.7	−130	40.2	25.5		
−40	26.6	22.2	−90	33.1	24.1	−140		25.9		

[a] References:
- JE Jakob, M., and Erk, S.[349]
- SH Schofield, F. H., and Hall, J. A., *Int. Crit. Tables*, **2**, 315-316 (1927); based on Ingersoll, L. R., and Zobel, O. J., "An Introduction to the Mathematical Theory of Heat Conduction," 1913; Neumann, F.[351] Straneo, P.[346]
- VD Van Dusen, M. S., *Int. Crit. Tables*, **5**, 216-217 (1929); based on Lees, C. H.[348]; Abels, H., *Repert. f. Meteor.* (*Wild's, St. Petersburg*), **16**, No. 1 (1893); Andrews, T., *Proc. Roy. Soc.* (*London*), **40**, 544-549 (1886); Forbes, G.[346]; Mitchell, A. C., *Proc. Roy. Soc. Edinburgh*, **13**, 592-596 (1886); Straneo, P., *Nuovo Cim.* (4), **7**, 333-340 (1898) ← *Atti Accad. Lincei* (5), **6**₂: 262-269, 299-306 (1897).

[b] S. Arzybyschew and I. Parfianowitsch[352] find 23.0; P. G. Tait,[353] 21; and SH gives 22 mw/cm·°C.

---

[348] Lees, C. H., *Phil. Trans.*, **204**, 433-466 (1905) → *Proc. Roy. Soc.* (*London*), **74**, 337-338 (1905).

[349] Jakob, M., and Erk, S., *Wiss. Abh. d. P. T. R.*, **13**, 395-409 (1929) = *Z. ges. Kälte-Ind.*, **36**, 229-234 (1929) → *Z. techn. Physik*, **10**, 623-624 (1929).

[350] Pfaff, F., *Sitzb. physik.-med. Soc. Erlangen*, **6**, 155-157 (1874).

[351] Neumann, F., *Phil. Mag.* (4), **25**, 63-65 (1863).

[352] Arzybyschew, S., and Parfianowitsch, I., *Z. Physik*, **56**, 441-445 (1929).

[353] Tait, P. G., *Proc. Roy. Soc. Edinburgh*, **13**, 592-596 (1886).

Snow.

The thermal conductivity (k) and diffusivity (D) of snow vary greatly with the density, and the values obtained by different observers exhibit much discordance (see Table 209).

Table 209.—Thermal Conductivity and Diffusivity of Snow

The thermal conductivity (k) and diffusivity ($D = k/\rho c$), both depend upon the density of the snow; ρ = density, c = specific heat. From a consideration of the data then available, M. S. van Dusen[358] concluded that between 0 and -30 °C, $k = 0.21 + 4.2\rho + 21.6\rho^3$ milliwatt/(cm·°C), $1000D = 2.0 + 0.1/\rho + 10.3\rho^2$ cm²/sec, where ρ gm/cm³ = density of the snow (not of the individual crystals). More recently, J. Devaux[359] has concluded that $k = 0.29(1 + 100\rho^2)$ milliwatt/(cm·°C). Values computed by means of these equations are here entered under VD or D, as may be appropriate.

Unit of $k = 1$ milliwatt/(cm·°C) = 239 10^{-6} g-cal/(cm·sec·°C); of $D = 1$ cm²·sec; of $\rho = 1$ g/cm³. Temp. = 0 °C

Ref.[a]→ ρ	VD	D k	SH	Ref.[a]	Ref.[a]→ ρ	VD	1000D	SH	Ref.[a]
0.11	0.70	0.65	1.07	J	0.125	3.0[b]			
0.125	0.78[b]	0.75			0.19	2.9	2.50		A
0.24	1.52	1.98	1.67	OAY	0.33	3.4	4.60		A
0.25	1.60	2.1	1.88	OAY	0.40	3.9			
0.27	1.8	2.4	1.34	OAY	0.50	4.8			
0.45	4.1	6.2	0.49	J	[c]		4.1		IK
0.50	5.0	7.6	1.3	IZ					

[a] References.
- A — Abels, H., *Rep. Meteor.* (Wild, St. Petersburg), **16**, No. 1 (1893).
- D — Devaux, J.[359]
- IK — Ingersoll, L. R., and Koepp, O. A., *Phys. Rev.* (2), **24**, 92-93 (1924).
- IZ — Ingersoll, L. R., and Zobel, O. J., "An Introduction to the Mathematical Theory of Heat Conduction," 1913.
- J — Jansson, M., *Öfvers. af K. Svenska Vet. Akad., Förh.* (Stockholm), **58**, 207-222 (1901).
- OAY — Okada, T., Abe, K., and Yamada, J., *Proc. Tokyo Math.-Phys. Soc.* (2), **4**, 385-389 (1908).
- SH — Values tabulated by Schofield, F. H., and Hall, J. A., *Int. Crit. Tables*, **2**, 313-315 (1927) and ascribed to the reference here given in the adjacent column.
- VD — Van Dusen, M. S.[358]; equation based on observations of A, IK, J, OAY, Hjelthström, S. A., *Öfvers. af K. Svenska Vet. Akad. Förh.* (Stockholm), **46**, 669-676 (1889), and Neumann, F., *Ann. de chim. et phys.* (3), **66**, 183-187 (1862).

[b] For $\rho = 0.125$, M. Kuroda[360] gives $k = 2.1_6$, $1000D = 8.6$.
[c] Snow densely packed.

From his work on the *névé* on the Mt. Blanc glaciers at elevations of 4.2 to 4.4 km, J. Vallot[354] found that the diurnal variation in temperature did not extend below one meter, at which depth the temperature rises to 0° C only under exceptional conditions; the annual variation does not extend below 6.5 m. He observed the following temperatures, t_1 in the year 1900, t_2 in 1911:

Depth	1	2	3	4	8	10	12	15	meters
t_1	-6.3	-9.1	-11.9						°C
t_2	-0.1	-0.8	-7.3	-12.0	-12.6	-13.2	-12.9	-12.8	°C

[354] Vallot, J., *Compt. rend.*, **156**, 1575-1578 (1913).

M. Kuroda [355] observed that the lowest temperature in a snow-blanket 9 meters thick occurred at a depth of about 4.5 meters, and was $-0.9\,°C$, the temperature of the surface being $+0.2\,°C$, and of the ground $+0.4\,°C$. The density varied from 0.35 g/cm³ at 50 cm to 0.65 at 7 meters.

The distribution of temperature in the snow-blanket and in the underlying soil, near Sodoakylä, Finland, within the Arctic Circle, has been studied over a period of 24 months by J. Keränen.[356]

The Oxford University's Arctic expedition, of 1935-6, to North-East Land found that at a depth of 70 ft. (21 m) in the ice cap the temperature was fairly constant at $0.0\,°C$; and at a somewhat greater depth an unfrozen lake was found.[357]

71. Refractivity of Ice

Over a hundred years ago, Sir David Brewster [361] observed that crystals of ice are optically uniaxial and positive, the index of refraction (ϵ) of the extraordinary ray exceeding that (ω) of the ordinary. A. Bertin [362] has stated that the interference fringes seen when ice is suitably observed in a polarizing microscope are among the most beautiful exhibited by any uniaxial crystal.

H. E. Merwin [363] has concluded from the observations of A. Ehringhaus [364] that between $-3\,°C$ and $-65\,°C$, and 405 mμ and 706 mμ, $\epsilon_\lambda - \epsilon_D = 1.01(\omega_\lambda - \omega_D)$ and $d\omega/dt = -3.8 \times 10^{-5}$ per $°C$, the subscript denoting the wave-length to which the index applies, D denoting the D-lines of sodium at $\lambda = 589.3$ mμ. Hence, $d\epsilon/dt = 3.84 \times 10^{-5}$ per $°C$. From these relations and other data given in Merwin's compilation, the data in Table 210 have been derived. The form of the factors F_ω and F_ϵ was empirically determined for this compilation.

Taking for the density of ice at $-3\,°C$ the value 0.9164 g/cm³ (Tables 200 and 202), and for the indices of refraction for the D-lines the values given in Table 210, the Lorenz "refraction constant" $\left(\dfrac{n^2-1}{\rho(n^2+2)}\right)$ is 0.2097 cm³ per gram for the ordinary, and 0.2105 for the extraordinary ray. The corresponding quantity for water at $20\,°C$ ($n = 1.33300$, $\rho = 0.9982$ g/cm³) is 0.2061, and for water-vapor at 1 atm and $110\,°C$ ($n = 1 + 313.30\,\rho \times 10^{-6}$ (Table 58), $\rho = 0.0005804$ g/cm³) is 0.2088 cm³/g.

Early and rough determinations of ω and ϵ were made by E. Reusch,[365]

---

[355] Kuroda, M., *Sci. Papers Inst. Phys. and Chem. Res., Tokyo*, **12**, 69-81 (1929).
[356] Keränen, J., *Annales. Acad. Sci. Fennicae (A)*, **13**, No. 7 (1920).
[357] Glen, A. R., *Nature*, **139**, 10-12 (1937).
[358] Van Dusen, M. S., *Int. Crit. Tables*, **5**, 216 (1929).
[359] Devaux, J., *Ann. de phys. (10)*, **20**, 5-67 (1933).
[360] Kuroda, M., *Sci. Papers Inst. Phys. and Chem. Res., Tokyo*, **12**, 149-159 (1930).
[361] Brewster, Sir David, *Phil. Trans.*, **1814**, 187-218 (1814); **1818**, 199-273 (1818); *Phil. Mag. (3)*, **4**, 245-246 (1834).
[362] Bertin, A., *Ann. de chim. et phys. (3)*, **69**, 87-96 (1863).
[363] Merwin, H. E., *Int. Crit. Tables*, **7**, 17 (1930).
[364] Ehringhaus, A., *Neues Jahrb. Mineral., Geol.*, Beilage Bd. **41**, 342-419 (1917).
[365] Reusch, E., *Ann. d. Physik (Pogg.)*, **121**, 573-578 (1864).

72. ICE: REFLECTION

and fairly precise ones covering most of the visible spectrum, by C. Pulfrich.[366]

Table 210.—Indices of Refraction of Ice

Adapted from compilation of H. E. Merwin.[363] For long waves, see Tables 218 and 219.

ω, ϵ = index of refraction of the ordinary and the extraordinary ray, respectively. Subscript indicates either the wave-length to which the index refers, D indicating the wave-length of the D-lines of sodium (= 5893A), or the Centigrade temperature. Between -3 and $-65\,°C$ and in the range $\lambda = 4046A$ to $7065A$ the following relations hold good: $\omega_t = \omega_{-3} - (11.4 + 3.8t) \times 10^{-5}$; $\epsilon_t = \epsilon_{-3} - (11.5_2 + 3.84t) \times 10^{-5}$; $\epsilon_\lambda - \epsilon_D = 1.01(\omega_\lambda - \omega_D)$; approximately, $\omega_\lambda = \omega_D + (\lambda_D - \lambda)F_\omega$, $\epsilon_\lambda = \epsilon_D + (\lambda_D - \lambda)F_\epsilon$, where $10^6 F_\omega = 2.07 + 10^{(\lambda_D - \lambda)/4000}$ per angstrom, $10^6 F_\epsilon = 2.09 + 1.01 \times 10^{(\lambda_D - \lambda)/4000}$ per angstrom, all λ's being expressed in angstrom units. The order of approximation may be seen by comparing the tabulated values of $10^4(\lambda_D - \lambda)F_\omega$ with those of $10^4(\omega_\lambda - \omega_D)$ derived from the values of ω; interpolation is facilitated by the use of the values of the F's.

Unit of $\lambda = 1A = 10^{-8}$ cm. Temp. $= -3\,°C$. Index with reference to a vacuum

λ	ω	ϵ	$10^4(\omega_\lambda - \omega_D)$	$10^4(\lambda_D - \lambda)F_\omega$	$10^6 F_\omega$	$10^6 F_\epsilon$
4046 Hg	1.3183	1.3198	93	91.6	4.96	5.02
4358 Hg	1.3159	1.3174	69	69.0	4.49	4.53
4861 H	1.3129	1.3143	39	40.1	3.88	3.92
4916 Hg	1.3126+	1.3140+	36+	37.3	3.82	3.86
5461 Hg	1.3104	1.3118	14	14.5	3.35	3.38
5780 Hg	1.3093+	1.3107	3+	3.5	3.14	3.17
5893 Na	1.3090	1.3104	0	0	3.07	3.10
6234 Hg	1.3079	1.3093	-11	-9.9	2.89	2.92
6563 H	1.3070+	1.3084+	$-19+$	-18.4	2.75	2.78
6908 Hg	1.3063	1.3077	-27	-26.7	2.63	2.65
7065 He	1.3060	1.3074	-30	-30.2	2.58	2.60

72. Reflectivity of Ice and of Snow

Ice.

Throughout the visible spectrum the transparency of ice is so great that the reflectivity (R) can be computed satisfactorily from the index of refraction by means of the formulas given in Section 38. The observed reflectivity passes through a pronounced maximum at $\lambda = 3.2\,\mu$ and again near $\lambda = 13.0\,\mu$. M. Weingeroff [367] suggests that in the latter region there are "residual" rays.

Defining the reflectivity as $R = I_r/I_i$, I_i and I_r being the intensities of the incident and of the specularly reflected radiation, respectively,

---

[366] Pulfrich, C., *Idem (Wied.)*, **34**, 326-340 (1888).
[367] Weingeroff, M., *Z. Physik*, **70**, 104-108 (1931).

E. P. T. Tyndall [368] has given the following values, based upon the observations of G. Bode,[369] angle of incidence about 15°.

λ	1.0	1.5	2.0	2.4	2.6	2.8	3.0	3.2	3.4	3.5	4.0 μ
$100 R$	1.72	1.62	1.62	1.13	0.73	0.70	1.60	5.10	3.90	2.81	1.75

M. Weingeroff [367] has observed the following, the angle of incidence being about 12°:

λ	6.0	7.5–9.0	10.0	10.5	11.0	11.2	11.5	12.0	12.5	13.0	13.8	14.3	15.0–16.0 μ
$100R$	0.8	0.5	0.4	0.5	1.0	1.5	2.0	2.5	3.0	3.5	3.0	2.5	2.0

It has been reported that the intensity of the reflection of x-rays from the (100) plane of ice is reduced about 2.5 per cent by an electric field of 1300 volts/cm parallel to that plane.[370]

Snow.

Freshly fallen, powdery snow on mountains is mat and closely obeys Lambert's law.[359]

Using filtered radiation from a quartz-enclosed mercury arc, and working at a vertical angle of 40° between the arc and the receiving instrument, E. O. Hulburt [371] found the effective (diffuse) reflecting power (R_e) of freshly fallen snow to have the following relative values, that for the region $\lambda = 0.4$ to $0.8\,\mu$ being arbitrarily taken as 100:

λ	0.3 to 0.4	0.4 to 0.8	0.8 to 2.6	2.6 to 7	Beyond 7 μ
R_e	88	100	38	45	65

The albedo of a plane surface is defined as $A = F_r/F_i$, where F_r = total luminous flux reflected by the surface when uniformly illuminated by white light, the total luminous flux incident on the surface being F_i. From the observations of P. G. Nutting, L. A. Jones, and F. A. Elliott,[372] E. P. T. Tyndall [373] concluded that $A = 0.93$ for snow, A for $MgCO_3$ being assumed to be 0.98; J. Devaux [359] gives $A = 0.95$.

These values are much higher than those reported by others. From a long series of observations made near Leningrad, N. N. Kalitin [374] found the maximum value $A = 0.87$ for dazzling, fresh, soft snow fallen the evening before. He quotes the following previously reported values for such maximum:* C. Dorno, 0.89; Abbot and Aldrich, 0.70; A. Ångström, 0.81. The packing of the snow with age decreases the albedo; so does melting of the surface. He found the apparent albedo to lie above 0.45 so long as the ground was completely covered with snow that was not more than

---

* Values ranging from 0.70 to 0.89 have been reported by H. H. Kimball and I. F. Hand.[374a]

[368] Tyndall, E. P. T., *Int. Crit. Tables*, **5**, 256-263 (258) (1929).
[369] Bode, G., *Ann. d. Physik (4)*, **30**, 326-336 (1909).
[370] Német, A., *Helv. Phys. Acta*, **8**, 97-116 (1935).
[371] Hulburt, E. O., *J. Opt. Soc. Amer.*, **17**, 23-25 (1928).
[372] Nutting, P. G., Jones, L. A., and Elliott, F. A., *Trans. Illum. Eng. Soc. (N. Y.)*, **9**, 593-597 (1914).
[373] Tyndall, E. P. T., *Int. Crit. Tables*, **5**, 262 (1929).
[374] Kalitin, N. N., *Monthly Weather Rev.*, **58**, 59-61 (1930).
[374a] Kimball, H. H., Hand, I. F., *Monthly Weather Rev.*, **58**, 280-281 (1930).

slightly soiled. The following are typical of the values he obtained [375]: Loose surface, 0.80; freshly fallen snow, 0.83; dense surface, 0.86; thawed and grainy surface, 0.40; uneven surface, 0.75.

73. Luminescence of Ice

By the luminescence of a substance is meant its emission of light under the existing conditions, and in particular from its interior, as distinguished from reflection by its surface. Several types of luminescence are described in Section 39. For the internal brilliance of a blanket of snow, see Section 75. For reports of the crystalloluminescence of ice, see end of Section 97.

Fluorescence of Ice.

Under this head, phosphorescence, triboluminescence, etc, are included. While exposed to the filtered radiation from radium, the filter being Pt 2 mm thick, ice fluoresces, but less brightly than does water at 20° and under the same conditions. In both cases the luminescence is very weak.[376]

Rayleigh Scattering by Ice.

The vivid blue color of large masses of pure ice has been ascribed by C. V. Raman to the scattering of light by the molecules of the ice, or rather by the slight variations in the concentration of the molecules (cf. Section 39). The purer the ice, the deeper the blue. Slight traces of impurities alter the color very perceptibly.[377]

Raman Scattering by Ice.

For an account of the general characteristics of the Raman effect, see Section 39, and K. W. F. Kohlrausch.[378] Each Raman band for ice is much narrower than the corresponding one for water and corresponds to a slightly smaller value of $\delta\nu$, the difference between the wave-number of the Raman band and that of its exciter.[379] Early observations by I. R. Rao [380] indicated that the intensity of the unresolved prominent band for ice depends upon the frequency of the exciter. He reported as follows, λ_{Hg} being the wave-length of the exciter, and I the intensity of the corresponding Raman-band: $\lambda_{Hg} = 3650A$, $I = 15$; $\lambda_{Hg} = 4047A$, $I = 10$; $\lambda_{Hg} = 5060A$, $I = 5$. No later observation on such variation has been found.

Effect of temperature.—At the temperature of liquid air ($ca. -190\,°C$) the Raman spectrum of ice consists of one intense and fairly sharp line at $\delta\nu = 3090$ cm^{-1} ($\lambda_R = 3.24\,\mu$) and a faint companion at $\delta\nu = 3135$ cm^{-1} ($\lambda_R = 3.19\,\mu$); whereas at temperatures near 0 °C it consists of diffuse lines or bands at $\delta\nu = 3196$ cm^{-1} ($\lambda_R = 3.13\,\mu$) and 3321 cm^{-1} ($\lambda_R =$

---

[375] Kalitin, N. N., *Gerlands Beitrag. z. Geophysik*, **34**, (Köppen Bd. 3), 354-366 (1931).
[376] Mallet, L., *Compt. rend.*, **183**, 274-275 (1926).
[377] Barnes, H. T., "Ice Engineering," pp. 8, 9, 1928.
[378] Kohlrausch, K. W. F., "Der Smekal-Raman Effekt," 1931.
[379] Ganesan, A. S., and Venkateswaran, S., *Indian J. Phys.*, **4**, 195-280 (1929).
[380] Rao, I. R., *Idem*, **3**, 123-129 (1928).

3.01 μ.[381] For interpretations of the change, see also I. R. Rao,[382] who disagrees with Sutherland.

Table 211.—The Raman Spectrum of Ice

$\delta\nu$ is the difference between the wave-number $(1/\lambda)$ of the Raman line and that of the exciting radiation; $\lambda_R = 1/\delta\nu$. Each number or check mark in the columns of relative intensity is placed on the line with the value of $\delta\nu$ that corresponds to the maximum of the line or band as reported by the indicated author. The absence of such number or mark indicates that the author did not report a maximum at that value of $\delta\nu$.

Unit of $\delta\nu = 1$ cm^{-1}; of $\lambda_R = 1\,\mu = 10^{-4}$cm

$\delta\nu$	λ_R	Rao 1928	GaV 1929	Ras 1932	Rao 1934	CBL 1937	Hib 1937
	Ref.[a]→			Relative intensity [b]			
53.5	187			1d			
205	48.8						4
210	47.6					√	
212.1	47.15			5d			
601	16.6						3
2225	4.494						2
3136	3.189						10
3150	3.175					√c	
3156	3.168					√	
3190	3.135	5					
3193	3.132		√				
3196	3.129				55		
3200	3.125	10					
3270	3.058	15					
3300	3.030					√c	
3321	3.011			↑	40		
3330	3.003						8
3390	2.950		√			√c	
3420	2.924			↓			
3549	2.818		√				
5393	1.854		√				

Rao [382] has given the following values for the relative intensity of the Raman scattered light throughout the range $\delta\nu = 2877$ to 3768 cm^{-1}.

$\delta\nu$	2877	3019	3122	3196	3252	3321
I	0	10	36	55	37	40
$\delta\nu$	3321	3389	3466	3538	3636	3768
I	40	39	30	12	5	0

[a] References:
- CBL Cross, P. C., Burnham, J., and Leighton, P. A., *J. Am. Chem. Soc.*, **59**, 1134-1147 (1937).
- GaV Ganesan, A. S., and Venkateswaran, S.[379]
- Hib Hibben, J. H., *J. Chem'l Phys.*, **5**, 166-172 (1937).
- Rao Rao, I. R., 1928,[380] 1934.[382]
- Ras Rasetti, F., *Nuovo. Cim. (N. S.)*, **9**, 72-75 (1932).

[b] The numerical values of the relative intensities have no significance except with reference to others appearing in the same column. When a reference contains no numerical estimate of the relative intensities of the lines or bands recorded, the positions of those lines or bands are indicated by a check mark (√); *d* indicates that the band was recorded as being diffuse; *c* indicates the value of $\delta\nu$ corresponding to a fundamental frequency of a band. The long line in Ras column indicates that Rasetti reported a continuous band extending from $\delta\nu = 3300$ to $\delta\nu = 3420$ cm^{-1}.

74. Diffraction of X-rays by Ice

The diffraction of x-rays by ice has been studied primarily for the purpose of ascertaining the intimate crystalline structure of ice, and most of the reports of such work contain no explicit statement of the values of the individual periodicities observed, or of their relative intensities. The information obtained regarding the crystalline structure is given in Section 60, on the molecular data for ice. Values of the observed periodicities and their relative intensities are given in Table 212.

Table 212.—Diffraction of X-rays by Ice

$d = (\lambda/2)\cdot\sin(\phi/2)$, λ = wave-length of the incident x-rays, ϕ = angle of diffraction at which the intensity of the diffracted radiation passes through a maximum; d characterizes some kind of periodicity in the structure of the crystal. I = relative intensities of the several maxima; w = weak, s = strong, m = medium strong, v = very; *e.g.*, vvs = very, very strong. Unit of d = 1A.

StJ[a] d	Den[a] d	I	−50 to −80° d	I	−85 °C d	BO[a] I	−90 to 105° d	I	−115 to −175° d	I
4.15										
	3.92	10	3.90	w	3.87	w	3.90	w		
	3.67	100	3.63	vvs	3.69	vvs	3.66	s	3.7	m
3.46										
3.30	3.44	20	3.40	m	3.42	m				
	2.68	15	2.64	m	2.66	m				
2.56										
2.34										
	2.26	10	2.26	w	2.28	vw	2.25	s		
	2.065	50	2.05	vs	2.05	vs			2.1	w
1.94	1.92	10	1.90	w	1.92	w	1.91	m		
			1.71	w						
	1.516	15	1.51	w	1.52	w				
			1.45	m						
	1.368	20	1.35	m	1.37	m				
1.30	1.30	2.5								
1.26	1.25	2.5	1.26	vw						
			1.21	vw						
	1.167	5	1.17	vw	1.17	vw				
0.74										

[a] References:
BO Burton, E. F., and Oliver, W. F., *Proc. Roy. Soc. (London) (A)*, **153**, 166-172 (1936).
Den Dennison, D. M., *Phys. Rev. (2)*, **17**, 20-22 (1921).
StJ St. John, A., *Proc. Nat. Acad. Sci.*, **4**, 193-197 (1918).

75. Absorption and Transmission of Radiation by Ice and by Snow

(There seem to be no such data for x-rays, γ-rays, cosmic radiations, or corpuscular radiation. For $\lambda > 4\,\mu$, see Section 80; for scattering by ice, see Section 73.)

[381] Sutherland, G. B. B. M., *Proc. Roy. Soc. (London) (A)*, **141**, 535-549 (1933).
[382] Rao, I. R., *Idem*, **145**, 489-508 (1934).

Table 213.—Monochromatic Absorptivity of Ice

Trans. = per cent transmitted. The absorptivity (k) is defined by the equation $I = I_0 e^{-kx}$, where ($I_0 - I$) is the decrease in the intensity caused by transmission through x cm of ice in the interior of the block.

Unit of $\lambda = 1\,\mu = 10^{-4}$ cm; of $x = 1$ cm; of $k = 1$ cm^{-1}

I. Plane-polarized radiation transmitted perpendicular to the optic axis; $x = 0.5$ cm; λ = wave-length at which the absorptivity passes through a maximum. P.[a]

Ordinary ray [b]			Extraordinary ray [b]		
λ	Trans.	k	λ	Trans.	k
0.79	93	0.145	0.81	55	1.20
0.89	93.5	0.137	0.92	54	1.23
1.02	93	0.145	1.06	46	1.55
1.26	57	1.12	1.29	27	2.62

II. Unpolarized radiation transmitted probably parallel to optic axis.

λ	1	1.50	2.5	4.5	5.2
x	0.1	0.1	0.1	(?)	(?)
Trans.	7[c]	5	0	min	max
k	26.7	23.0	∞		
Ref[a]	B	P	B	B	B

III. K[a] Crystallographic direction is not stated. The reported transmissions by the two blocks are not consistent. If they are combined on the assumption that the blocks differ solely in the value of x, then one finds for the reflectivity (R) a negative value (-5%) and for k the values here given. If R is assumed to be negligible, and only the data for the longer block are used, one obtains the values k_{107}.

λ	0.332	0.346	0.366	0.392	0.416	0.438	0.446
Trans., $x = 10$	97	96	99	99	98	99	98
Trans., $x = 107$	46	46	51	52	54	52	55
$10^4 k$	77	76	68	66	62	65	59
$10^4 k_{107}$	73	73	63	61	58	61	56

IV. CE[a] Crystallographic direction is not stated. Values of nk are reported for the residual rays from the salts indicated. Thickness of ice was $x = 0.0034$ cm. Temp. = $-10\,°C$.

Salt	NaCl	KCl	KBr	TlCl	TlBr	TlI
λ	52	63	83	100	117	152
Trans.	11	13	32	72	85	84
$\lambda k/4\pi$	0.27	0.30	0.22	0.08	0.03	0.03
k	650	600	330	126	30	25

[a] References:
- B Bode, G.[309]
- CE Cartwright, C. H., and Errera, J., *Proc. Roy. Soc. (London) (A)*, **154**, 138-157 (1936)—*Acta Physicochim. URSS*, **3**, 649-684 (1935) → Cartwright, C. H., *Nature*, **136**, 181 (L) (1935).
- K Kalitin, N. N., *Compt. rend. Acad. Sci. URSS (N. S.)*, **9** = 1935₄, 145-146 (1935).
- P Plyler, E. K.[354]

[b] Certain inconsistencies in this and his other paper [386] reporting the same data have resulted in several quotations in which the data for the ordinary ray have been assigned to the extraordinary, and conversely. The assignment here given is that in the author's Table I, which he has informed me is correct.

[c] The absorption is given as 93 per cent, which leads to $k = 26.7$ cm^{-1}, a value far exceeding what one would infer from Plyler's data.

For a given layer of material, the transmitted fraction of the incident radiation depends upon the amount of the radiation that is scattered by the layer, as well as upon the amount that is truly absorbed, that is converted into another form of energy. But this distinction has not been observed in the reporting of experimental data for the absorption of ice, the entire reduction in intensity being described as absorption. The error so produced is probably very small when the ice is clear, except for the shorter wave-lengths, for which measurements of the absorption seem to be lacking. For snow, the scattering is of prime importance; the true absorption is that of the individual ice-crystals, of the ice itself.

Ice.

In the visible spectrum, the absorptivity of ice is certainly small. In the infrared, beyond $\lambda = 1\,\mu$, it is great and entirely analogous to that of water (Section 43), at least as far as $\lambda = 6\,\mu$. A plate of ice 1 mm thick absorbs practically all radiation for which $\lambda \gtrsim 3\,\mu$, and a frozen soap-film cuts off nearly all radiation for which $\lambda \gtrsim 6\,\mu$.[369] For λ greater than about $4\,\mu$, the absorptivity of ice is great.[259, 383]

The extraordinary ray is more strongly absorbed than the ordinary, and the corresponding wave-lengths at which the maxima of the absorption occur are greater in the former than in the latter.[384]

The intercrystallic material, which in some cases was found to be less than 0.0008 cm thick, "has a much higher absorption of infrared light than the ice itself."[385] Plyler concluded that this extra absorption was not due to dissolved salts.

Table 214.—Transmissivity of Ice for Black-Body Radiation
Adapted from data given by S. L. Brown.[387]

Thickness of ice = 3 mm.[a] Transmission = τ per cent; τ_c is defined by an empirical equation ($\tau_c = -18.5 + 0.033t$) constructed for the present compilation. Temperature of the source of radiation is t °C.

t	τ	τ_c	t	τ	τ_c
660	3.4	3.3	865	9.9	10.0
720	5.2	5.3	910	11.7	11.5
790	7.6	7.6	925	11.9	12.0
			960	14.2	13.2

[a] H. Hess[388] has stated that Melloni found that a plate of ice 2.6 mm thick transmitted only 6 per cent of the total radiation incident on it from a Locatelli lamp.

Snow.

In a study of the penetration of radiation into snow and glaciers the

[383] Ångström, A., *Ark. f. Math. Astr., och Fysik*, **13**, No. 21 (1919).
[384] Schaefer, C., and Matossi, F., "Das ultrarote Spektrum," 1930; Plyler, E. K., *J. Opt. Soc. Amer.*, **9**, 545-555 (1924).
[385] Plyler, E. K., *J. Elisha Mitchell Soc.*, **41**, 18 (1925).
[386] Plyler, E. K., *J. Elisha Mitchell Soc.*, **41**, 39-40 (1925).
[387] Brown, S. L., *Phys. Rev.* (2), **21**, 103-106 (1923).
[388] Hess, H., "Die Gletscher," 1904.

following terms are useful: Factor of entrapment (E) = unity minus the albedo (Section 72) = fractional excess of radiation incident upon the surface over that returned by the snow or glacier. Internal illumination (I_i) = sum of the flux of radiation each way through a given unit surface at the place considered; it may be expected to vary with the aspect of the surface. Factor of attenuation (T) = ratio of internal illumination to the illumination (I_s) of the surface of the snow or glacier; $T = I_i/I_s$.

Clean, freshly fallen snow has a perfectly diffusing surface, the light proceeding from it when illuminated being distributed in accordance with Lambert's law,* and entirely independent of the direction of the incident light. In the infrared, especially for $\lambda > 4\ \mu$, it radiates sensibly as an ideal radiator; in the visible spectrum its emissivity is small. Its factor of entrapment (E) is about 0.05 for light, and 0.3 for total energy of sunlight; and its internal brilliance is very nearly independent of the line of sight. The attenuation is logarithmic: $T = T_0 10^{-kx}$, T_0 is very nearly unity, and k is about 0.1 cm^{-1}; x = depth below the surface.

Table 215.—Transmission of Radiation by Snow [375]

The following data refer to direct and diffuse solar radiation; Angstrom vacuum pyranometers were used. The original paper should be consulted. τ_a = percentage of incident radiation that reached the depth d; τ_c = the corresponding percentage for the radiation that actually enters the surface. τ_c measures the true transmissivity, τ_a, the apparent transmissivity.

Unit of $d = 1$ cm; of $\tau = 1$ per cent

Snow→	Dry		Wet
d	τ_a	τ_c	τ_c
2.5	16.0		
5	7.5		8.0
10	2.3	18.5	2.4
15	1.3	5.5	1.1
20	1.0	3.2	
25	0.8	2.2	
40	0.4	1.2	
60	0.2	0.6	

When the sky is clear, the surface temperature of snow not exposed to direct sunlight is, on account of radiation, always below that of the neighboring air, especially at high altitudes; the difference is almost as great as at night, even when the shadow is only a meter square. At night the surface temperature may be 5, 10, or even 15 °C below the temperature of the air; the lower the humidity, the greater the difference. Near midday, the surface in sunlight may melt, although the air temperature is -10 °C. Wind reduces the difference between the temperatures of the air and the surface, and heavy cloud or fog almost obliterates it. When the sky is clear, the diurnal range in the surface temperature may amount to 20 or

---

* Lambert's law: $i = i_0 \cos\theta$, i_0 = intensity of radiation emitted normal to the surface, i = intensity of that emitted at an angle θ to the normal.

30 °C, even when the air temperature remains constant. The amplitude of the diurnal range in temperature decreases exponentially with the depth, and at 30 cm is of the order of 1 °C. The preceding information about snow is from J. Devaux.[359] A prolonged study, extending over two years, of the temperatures at various depths in the snow blanket and in the underlying ground, at Sodankylä, Finland, within the Arctic Circle, has been made and published, with numerous citations, by J. Keränen.[356]

In connection with his study of the cooling of snow during the arctic night of 1916, A. Ångström[383] set up the equation $H = c(dt/dx)$ in which H = total amount of heat, per unit surface and per unit time, received by the snow from the air by conduction and convection, and (dt/dx) = vertical gradient of the air temperature. He called c the convectivity, and found that $c = 0.005$ g-cal per (°C/cm) for the average wind velocity, and that $(dt/dx) = 1/12$ °C/cm when the sky was clear.

Glaciers and Névés.

The emissivity of clean glaciers and *névés* is the same as that of ice. A *névé* and a surface of old, large-grained snow, are each an almost perfectly diffusing surface when clean. For a clean *névé* the factor of entrapment is about 0.4 for light, and 0.5 for the total energy of sunlight; for a clean glacier the corresponding values are about 0.4 and 0.6. The internal brilliance of a glacier is notably greater if the line of sight is toward the surface than for the contrary direction, but the internal illumination (I_i) is almost independent of the aspect of the surface. For a glacier, as for snow, the attenuation is logarithmic: $T = T_0 10^{-kx}$; k is a little greater for red than for green, and its value varies with the structure of the glacier, recorded values varying from 0.008 to 0.032 cm^{-1}. T_0 varies greatly with the nature of the surface, ranging from 0.2 to 0.8; it measures the attenuation produced by the surface layer.[359]

J. Vallot[388a] has found that in the *névé* of the Mt. Blanc glaciers, elevation about 4.3 km, the diurnal variation in temperature does not extend below one meter, nor the annual below 6.5 meters.

76. Emissivity of Ice and of Snow

For radiation greater than 1 μ in wave-length, the absorptivity of ice, like that of water, is great (Section 75), and the reflectivity is small (Section 72); whence one may conclude that ice and snow will radiate nearly as an ideal (black body) radiator, that their emissivities will not be much less than unity (see A. Ångström[383]). The very low value published by K. Siegl[389] is surely incorrect.

E. Schmidt[390] has reported the following values for the emissivity (ϵ) of ice in terms of that of the ideal (black body) radiator, taken as unity:

[388a] Vallot, J., *Compt. rend.*, 156, 1575-1578 (1913).
[389] Siegl, K., *Sitz.-b. Akad. Wiss., Wien (Abt. IIa)*, 116, 1203-1230 (1907).
[390] Schmidt, E., *Forsch. Gebiete Ingenieurw.*, 5, 1-5 (1934).

Wet ice at 0 °C, $\epsilon = 0.966 \pm 0.003$; transparent ice at -9.6 °C frozen to brass, $\epsilon = 0.965 \pm 0.003$, being the same whether the thickness of the ice was 0.4 mm or 0.8 mm; white frost at -9.6 °C and 0.1 to 0.2 mm thick, $\epsilon = 0.985 \pm 0.03$, being the same whether the frost was deposited on brass or on ice.

77. Photoelectric Emission by Ice

When the illumination is that produced by the radiation from an electric spark between aluminum terminals, and filtered by a thin plate of fluorite and not more than 3 mm of air at atmospheric pressure, the photo-electric emission of electrons by ice is 280 times as great as that by water, and 0.70 times that by CuO.[391] Its variation with the filtration is shown in Table 216.

Table 216.—Relative Photoelectric Sensitivity of Ice

Adapted from A. L. Hughes [392] based on the observations of W. Obolensky.[391]

Filtered radiation from an Al-spark; sensitivity $= S$. For the fluorite filtered radiation, the S of CuO is taken as 143. $\lambda_{min} =$ shortest wavelength contained in the filtered beam.

Unit of $\lambda = 1 m\mu = 10A = 10^{-7}$ cm

Filter	λ_{min}	S
Fluorite [a] (CaF$_2$)	125	100
Quartz (SiO$_2$)	145	40
Quartz and air	177	50
Calcite (CaCO$_3$)	220	0.02
Glass	330	0

[a] With not more than 3 mm of air at a pressure of 1 atm.

78. Absorption Spectrum of Ice

In the region $\lambda = 6000A$ to 6μ, the absorption spectrum of ice is analogous to that of water, but the several wave-lengths at which the absorption passes through a maximum are each somewhat greater than the corresponding one for the liquid.[393]

When the optical structure of the specimen is uniform and the path of the radiation is perpendicular to the optic axis, the absorption has a maximum at each of the following wave-lengths (see E. K. Plyler [384]):

Ordinary * ray,	$\lambda = 0.79$	0.89	1.02	1.26 μ
Extraordinary * ray,	$\lambda = 0.81$	0.92	1.06	1.29 μ

*Certain inconsistencies in this and Plyler's other paper [386] reporting the same data have led to an interchange of the terms "ordinary" and "extraordinary" in several quotations of these data. The assignment here given is that in the author's Table I, which he has informed me is correct.

[391] Obolensky, W., *Ann. d. Physik (4)*, **39**, 961-975 (1912).
[392] Hughes, A. L., *Int. Crit. Tables*, **6**, 68 (1929).
[393] McLennan, J. C., Ruedy, R., and Burton, A. C., *Proc. Roy. Soc. (London) (A)*, **120**, 296-302 (1928); Bode, G., *Ann. d. Physik (4)*, **30**, 326-336 (1909).

In other cases, in which the radiation passed parallel to the optic axis, maxima were observed at $\lambda = 1.50$ (Bode [393] and Plyler), 1.95 and 4.5 μ (Bode).

Using a compound plate composed of portions of two crystals, Plyler observed maxima at $\lambda = 0.77$, 0.85, and 0.99 μ, the path of the radiation being perpendicular to the interface of the crystals.

The ultraviolet absorption by ice has been studied by E. J. Cassell,[394] who found a continuous absorption with a long wave-length limit near 1670A.

The band near 3 μ has been studied by G. Bosschieter and J. Errera,[395] who found for ice only a single band with its maximum at $\lambda = 3.08$ μ and two inflections, one near 2.98 and the other near 3.17 μ. They ascribe the maximum (3.08 μ) to a tridymite structure in which O is surrounded by 4 H's, two being nearer the O than are the other two.

A band near $\lambda = 62$ μ has been reported for ice at $-10\,°C$ by C. H. Cartwright,[396] who thought that its origin is to be sought in the crystalline structure of the ice.

79. Optical Rotation by Ice

When plane-polarized light is passed through ice-VI, the plane of polarization is rotated.[397]

80. Dielectric Properties of Ice

The dielectric properties of ice to be considered here are its dielectric constant (ϵ'), its absorption index (κ) expressed in terms either of the equivalent conductivity ($k_e = \epsilon''\omega/4\pi = n^2\kappa\nu$) or of the phase defect ($\phi = \tan^{-1}(\epsilon''/\epsilon')$), and its dielectric strength. Symbols have been defined, dielectric theories discussed, and formulas derived in Section 49. For the electrical conductivity of ice see Section 81.

Since ice is crystalline, it is to be expected that its dielectric properties will vary with the direction that the applied field makes with the axes of the several crystals. No information bearing upon this subject has been found. All the observations seem to have rested on the tacit assumption that the axes of the individual crystals in the specimens studied had a completely random distribution. It seems improbable that the randomness was complete in any case. Differences between the results of the various observers may rest in part upon differences in the mean orientation of the axes of the crystals with reference to the field.

For theories of the structure of ice as related to the dielectric constant

---

[394] Cassel, E. J., *Proc. Roy. Soc. (London) (A)*, **153**, 534-541 (1935).
[395] Bosschieter, G., and Errera, J., *Compt. rend.*, **205**, 560-562 (1937); superseding *Idem*, **204**, 1719-1721 (1937). See also Errera, J., *Jour. de chim. phys.*, **34**, 618-626 (1937).
[396] Cartwright, C. H., *Nature*, **136**, 181 (L) (1935).
[397] Poulter, T. C., *Phys. Rev. (2)*, **37**, 112 (A) (1931).

see R. H. Fowler,[398] F. C. Frank,[399] W. F. Giauque and J. W. Stout,[400] M. L. Huggins,[401] C. P. Smyth,[402] A. Német.[403]

Dielectric Constant of Ice.

O. Blüh [404] has discussed the accord between the various theories and the observed values of ϵ for a number of substances, the data for water, ice, and steam being considered in detail; a bibliography of 172 titles is given. W. Ziegler [405] also has reviewed the subject, giving a bibliography of 159 entries, and J. Errera [406] has given an exposition of theory and in the last paper cited a summary of his work on the dielectric polarization of solids.

G. Oplatka [407] found that ice frozen from water that was not extremely well freed from gas and kept gas-free during the freezing contained large space charges, whereas pure gas-free ice contained none. Under suitable conditions the presence of a space charge may increase the effective dielectric constant 30-fold. He believed that none of the ice used by his predecessors in their study of its dielectric constant was gas-free.

C. P. Smyth and C. S. Hitchcock [408] have reported that for ice frozen from a $0.0002 M$ solution of KCl (1 KCl to 278 000 H_2O = 1 g KCl to 67 100 g water) ϵ' is greater than that of pure ice, and the ϵ' vs. t graphs show hysteresis at the lower frequencies, the ϵ' for increasing temperatures being less than that for decreasing. The specific conductivity (k) of the solution was $10^5 k = 2.2$ (ohm·cm)$^{-1}$.

E. J. Murphy [409] has found "no indication of an abrupt disappearance of the polarization responsible for the high dielectric constant of ice at any temperature above $-139\,°C$." The main effect of lowering the temperature "appears to be an exponential increase of the relaxation time of the polarized condition of the dielectric."

In addition to the data given in the following tables and graphs, a few measurements at an unstated temperature have been reported in insufficient detail by H. Brommels.[410]

The most extended series of measurements of ϵ' at various frequencies for ice at various temperatures are those by J. Errera,[411] C. P. Smyth and C. S. Hitchcock,[408] H. Wintsch,[412] and E. J. Murphy.[409] The last alone gives values for temperatures below $-70\,°C$, but his observations are dis-

---

[398] Fowler, R. H., *Proc. Roy. Soc. (London) (A)*, **149**, 1-28 (1935).
[399] Frank, F. C., *Trans. Faraday Soc.*, **32**, 1634-1647 (1936).
[400] Giauque, W. F., and Stout, J. W., *J. Am. Chem. Soc.*, **58**, 1144-1150 (1936).
[401] Huggins, M. L., *J. Phys'l Chem.*, **40**, 723-731 (1936).
[402] Smyth, C. P., *Chem'l Rev.*, **19**, 329-361 (1936).
[403] Német, A., *Helv. Phys. Acta*, **8**, 97-116 (1935).
[404] Blüh, O., *Physik. Z.*, **27**, 226-227 (1926).
[405] Ziegler, W., *Physik. Z.*, **35**, 476-503 (1934).
[406] Errera, J., *Compt. rend.*, **179**, 155-158 (1924); *Bull. Sci. Acad. Roy. Belg. (5)*, **12**, 327-329 (1926); *Physik. Z. Sowj.*, **3**, 443-468 (1933).
[407] Oplatka, G., *Helv. Phys. Acta*, **6**, 198-209 (1933).
[408] Smyth, C. P., and Hitchcock, C. S., *J. Am. Chem. Soc.*, **54**, 4631-4647 (1932).
[409] Murphy, E. J., *Trans. Electroch. Soc. (Amer.)*, **65**, 133-142 (1934).
[410] Brommels, H., *Comment. Phys.-Math. Soc. Fennica (Helsingfors)*, **1**, No. 19 (1922).
[411] Errera, J., *Jour. de Phys. (6)*, **5**, 304-311 (1924).
[412] Wintsch, H., *Helv. Phys. Acta*, **5**, 126-144 (1932).

80. ICE: DIELECTRIC PROPERTIES

cordant with those of the others in at least two particulars: (1) his values at -7.1 to $-45.8\,°C$ at low frequencies are much higher, ϵ' being over 95

Table 217.—Drude-Debye Constants for the Dielectric Constant of Ice

$$\epsilon = \epsilon' - i\epsilon'', \quad \epsilon' = \epsilon_0 + (\epsilon_1 - \epsilon_0)/(1 + a^2\nu^2); \nu_s = 1/a, \lambda_s = ca, \tau = \frac{a}{2\pi} \times$$

$(\epsilon_0 + 2)/(\epsilon_1 + 2)$, λ_s = transition wave-length. See Section 49, p. 356. It has been found empirically that $a = \alpha e^{-\beta t}$. Each of the three extended series of observations now available lead to a different set of values for ϵ_0, ϵ_1, α, and β, the most consistent being those by SH.[408]

Unit of a and of $\tau = 1$ sec; of $\nu_s = 1$ cycle/sec; of $\lambda_s = 1$ km. Temp. $= t\,°C$

t	$10^4 a$	$10^8 a^2$	ν_s	λ_s	$10^6 \tau$

I. Smyth and Hitchcock (1932).[408] $\epsilon_0 = 3.0$, $\epsilon_1 = 74.6$, $\alpha = 116.0$ microsec; $\beta = 0.1015\,(°C)^{-1}$.

t	$10^4 a$	$10^8 a^2$	ν_s	λ_s	$10^6 \tau$
-0	1.160	1.346	8620	34.8	1.205
-2	1.421	2.019	7037	42.6	1.48
-5	1.927	3.713	5189	57.8	2.00
-10	2.202	4.849	4541	66.1	2.29
-15	5.217	27.22	1917	156.5	5.42
-20	8.833	78.02	1126	265	9.17
-25	14.66	214.6	682	440	15.23
-30	24.37	593.9	410	731	25.3
-40	67.23	4520	149	2017	69.9
-50	185.6	34450	53.9	5560	193
-60	512.1	262200	19.52	15360	532
-70	1413	1997000	7.08	42360	1467

II. Wintsch (1932).[412] $\epsilon_0 = 7.5$, $\epsilon_1 = 73.0$, $\alpha = 141.2$ microsec., $\beta = 0.0906\,(°C)^{-1}$.

t	$10^4 a$	$10^8 a^2$	ν_s	λ_s	$10^6 \tau$
-0	1.412	1.994	7082	42.3	2.84
-5	2.221	4.933	4502	66.6	4.48
-10	3.492	12.194	2864	104.7	7.04
-20	8.64	74.65	1157	259	17.40
-30	21.20	449.4	472	636	42.7
-40	52.95	2803.7	188.8	1587	106.6
-50	130.3	16980	76.7	3907	262.4

III. Errera (1924).[406] $\epsilon_0 = 3.0$, $\epsilon_1 = 77.2$, $\alpha = 182$ microsec., $\beta = 0.090\,(°C)^{-1}$.

t	$10^4 a$	$10^8 a^2$	ν_s	λ_s	$10^6 \tau$
-0	1.82	3.312	5491	54.6	1.83
-2	2.178	4.744	4591	65.2	2.19
-5	2.855	8.151	3503	86.5	2.87
-10	4.477	20.04	2248	134.2	4.50
-20	11.01	121.2	908	330	11.07
-22	13.20	174.1	758	396	13.3
-25	17.27	298.2	579	518	17.4
-30	27.08	733.4	369	812	27.2
-37	50.96	2597	196	1528	51.2
-40	66.61	4437	150	1998	66.9
-50	164.0	26900	61.0	4914	165

IV. Murphy (1934)[409] gives $10^5 \tau_M = 1.85 e^{-0.106 t}$ where $\tau_M = \tau(\epsilon_1 + 2)/(\epsilon_0 + 2) = a/2\pi$. Whence $10^6 a = 116 e^{-0.106 t}$. See remarks in text.

Table 218.—Dielectric Constant of Ice: Observed and Computed

(See Table 219 and Figures 10 and 11 for other observed values.)

Observed (Obs) values are from C. P. Smyth and C. S. Hitchcock [408]; computed (Calc) values are those defined by $\epsilon' = 3.0 + 71.6(1 + a^2\nu^2)^{-1}$, ν = frequency, $10^4 a = 1.160 e^{-0.1015t}$ sec. Temp. = t °C; Dif = Obs. − Calc. The ice was formed of "conductivity water" for which $10^6 k = 2(\text{ohm-cm})^{-1}$.

Unit of $a = 1$ sec; of $\nu = 1$ kilocycle/sec; of $\epsilon' = 1$ cgse unit. Temp. = t °C.

$t \to$	−1	−3	−5	−10	−20	−30	−40	−50	−60	−70
$10^8 a^2 \to$	1.6512	2.4743	3.7133	4.8488	78.022	593.9	4520	3447	262246	1996600
ν	Obs Calc Dif	Obs Calc Dif	Obs Calc Dif	Obs Calc Dif	Obs Calc Dif	Obs Calc Dif	Obs Calc Dif	Obs Calc Dif	Obs Calc Dif	Obs Calc Dif
0.3	73.7 74.5 −0.8	74.2 74.4 −0.2	73.6 74.4 −0.8	74.8 74.3 +0.5	71.9 70.0 +1.9	46.2 49.7 −3.5	16.3 17.1 −0.8	6.17 5.24 +0.93	4.35 3.30 +1.05	4.00 3.04 +0.96
0.5	73.6 74.3 −0.7	73.5 74.2 −0.7	73.8 74.0 −0.2	73.6 73.8 −0.2	65.9 62.9 +3.0	31.3 31.8 −0.5	8.78 8.82 −0.04	4.65 3.82 +0.83	3.92 3.11 +0.81	3.59 3.01 +0.58
1	72.5 73.4 −0.9	72.4 72.9 −0.5	72.5 72.0 +0.5	69.4 71.2 −1.8	45.2 43.2 +2.0	14.6 13.3 +1.3	5.37 4.55 +0.82	3.82 3.21 +0.61	3.50 3.03 +0.47	3.33 3.00 +0.33
5	51.6 53.7 −2.1	46.3 47.2 −0.9	40.2 40.1 +0.1	24.4[a] 35.3 −10.9	7.60 6.49 +1.11	4.06 3.48 +0.58	3.32 3.06 +0.26	3.21 3.01 +0.20	3.12 3.00 +0.12	2.99 3.00 −0.01
20	12.3 12.4 −0.1	10.4 9.6 +0.8	8.34 7.52 +0.82	5.46 6.51 −1.05	3.57 3.23 +0.34	3.15 3.03 +0.12	3.06 3.00 +0.06	3.04 3.00 +0.04	3.00 3.00 0	2.97 3.00 −0.03
60	4.3 4.18 +0.1	4.2 3.78 +0.4	3.71 3.53 +0.18	3.33 3.41 −0.08	3.12 3.03 +0.09	3.04 3.00 +0.04	3.03 3.00 +0.03	3.02 3.00 +0.02	3.00 3.00 0	2.98 3.00 −0.02

[a] Though so printed, this value seems to be wrong.

80. ICE: DIELECTRIC PROPERTIES

Table 219.—Dielectric Constant of Ice
(See also Figures 10 and 11.)

At $-5\,°C$ $\epsilon' = -0.08 + 0.34 \log_{10}\nu$ cgse units, if $10^7 < \nu < 10^9$ cycles/sec.[413] E. J. Murphy[409] has reported the following high values, as read from his graphs, unit of ν being 1 cycle/sec: $-90\,°C$, ballistic method, $\epsilon' = 150$; $-45.6\,°C$, $\nu = 15$, $\epsilon' = 87$; $-7.1\,°C$, $\nu = 300$, $\epsilon' = 95$; for other values, consult his paper.

Unit of $\lambda = 1$ km; of $\nu = 1$ kilocycle/sec; of $\epsilon' = 1$ cgse. Temp. $= t\,°C$

I. Smyth and Hitchcock, 1932. See Table 218.

II. Wintsch,[a] 1932, read from his graphs. Water was thrice distilled, collected in quartz, and boiled just before freezing.

$t \rightarrow$ ν	-5	-6	-10	-20 ϵ'	-30	-40	-50
0.05	85.3		85.2	82.0	74.0		
0.65	74.0		71.2	58.5	37.0	21.8	17.0
1.00		69.7	65.5	44.5	23.5	15.1	12.4
1.13	69.2		64.5	41.8	21.8	14.2	11.2
1.60	65.5		57.0	30.0	16.0	11.0	9.3
2.00		59.5	51.4	23.6	12.9	9.6	8.5
3.00		48.9	38.7	16.4	10.0	8.0	
3.50	46.5		33.3	13.8	8.7	7.5	7.5
4.00		40.2	29.6	12.4	8.9	7.3	
5.00		33.2	24.0	10.2	8.0	6.8	
5.10	34.8		23.0	10.0	7.0	6.3	6.2
6.00		28.0	19.4	9.0	7.1		
7.00		24.1	16.4				
8.00		21.0	14.5				

III. Errera,[a] 1924, certain typographical errors corrected.

$t \rightarrow$ λ	ν	-2	-5	-22^b ϵ'	-37^c	-47.5
680	0.441	77.3	76	43.5		7.4
465	0.645	76	74	34.3	10.15	4.15
430	0.698			31.5	8.3	3.68
294	1.020	73.4	72.6	19.4	4.6	3.22
196	1.53	69	65.8	13	4.15	
79[d]	3.80	47.2	39.8	5.06		
54	5.56	30.6	25.4	4.6	3.2	2.76
38.5	7.80	23.2	16.6	3.4	3	
28.5	10.5	15.2	11.6	3.1		
18.5	16.2	7.8	5.6			
8	37.5	4.6	4.4	2.3		2.3
1.1	273		3.86	2.2		

IV. Granier,[a] 1924, $t = -12\,°C$: from water having $10^6 k = 1.54$ (ohm·cm)$^{-1}$.

λ	ν	ϵ'	λ	ν	ϵ'
70000	0.0043	153	17.6	17	3.8
6000	0.050	100	5.9	51	2.35
940	0.320	86	1.15	260	2.05
194	1.55	56	0.045	6700	2.05
56	5.40	12			

[413] Curtis, H. L., and Defandorf, F. M., *Int. Crit. Tables*, **6**, 78 (1929); from Gutton, C., *Compt. rend.*, **130**, 1119-1121 (1900).

V. Adapted from *International Critical Tables; CD.*[a]

t	ε'	t	ε'	t	ε'
$\nu = 0.050$; Ths[a]		−70	41.5	$\nu = 0.320$; DF'[a]	
−2	94	−80	31.5	−7	51
−10	95.2	−90	20.2	−47	3.6
−18	96.5	−100	14.5	$\nu =$ audio; A[a]	
−182	3.	−110	8.6	−80	3.8[f]
$\nu = 0.120$[e]	DF, FD[a]	−120	6.1	$\nu = 10000$; Thg[a]	
−20	59.5	−130	4.7	−2	3.4
−30	59.0	−140	3.5	−5	2.8[g]
−40	58.5	−150	2.7	$\nu = 100000$; BK[a]	
−50	56.0	−165	2.43	−190	1.8[h]
−60	49.5	−185	2.43		

[a] References:
    A   Abegg, R., *Ann. d. Physik (Wied.)*, **62**, 249-258 (1897).
    BK Behn, U., and Kiebitz, F., "Boltzmann Festschrift," p. 610-617, 1904.
    CD Curtis, H. L., and Defandorf, F. M.[413]
    DF Dewar, J., and Fleming, J. A., *Proc. Roy. Soc. (London)*, **61**, 2-18 (1897).
    DF' Dewar, J., and Fleming, J. A., *Idem*, **62**, 250-266 (1898).
    Er   Errera, J.[411]
    FD Fleming, J. A., and Dewar, J., *Proc. Roy. Soc. (London)*, **61**, 316-330 (1897).
    Gr   Granier, J., *Compt. rend.*, **179**, 1313-1318 (1924).
    Thg Thwing, C. B., *Z. Physik. Chem.*, **14**, 286-300 (1894).
    Ths Thomas, P., *Phys. Rev.*, **31**, 278-290 (1910).
    W   Wintsch, H.[412]

[b] All of these −22 °C values are out of line with the others (see Fig. 9); it seems probable that the temperature is misprinted. If it was actually somewhere between −25 and −27 °C the values would about fit.

[c] This temperature was printed −27 °C, but −37 °C makes these values fit with the others; probably a misprint.

[d] This λ was printed as 97, but 79 is required to bring the values into line with the others. The ν's have been computed by the compiler, and the 3.80 corresponds to the 79.

[e] The condenser was charged and discharged 120 times a second by means of a vibrating contact-maker controlled by a tuning fork.

[f] At 5 megacycles/sec this same value (3.8) was found throughout the range 0 to −24 °C.[414]

[g] At −5 °C and 10 megacycles/sec C. Gutton[413] found $\epsilon' = 2.3$.

[h] At −4.5 °C and 83 megacycles/sec B. de Lenaizon and J. Granier[415] found $\epsilon' = 2.17$.

---

at −7.1 °C and nearly 90 at −45.8 °C, and his ballistic values at −90 °C are about 150; (2) his graph shows that the logarithm of Debye's τ (see page 355 is linear in the reciprocal of the absolute temperature, whereas the other sets of data just mentioned indicate that it is linear in the temperature. True, Murphy gives for τ an expression that requires the logarithm to be linear in the temperature, but that expression is a mere approximation, compromising with the graph and limited by him to temperatures above −46 °C. It should be noticed that his τ is so defined as to be $(\epsilon_1 + 2)/(\epsilon_0 + 2)$ times as great as Debye's.

From the data by each of the other three, the present compiler has determined graphically and by cut-and-try methods the three constants (ϵ_0, $\epsilon_1 - \epsilon_0$, and a) occurring in the Drude-Debye isothermal relation $\epsilon' = \epsilon_0 + (\epsilon_1 - \epsilon_0)/(1 + a^2\nu^2)$ (see eq. 9, Section 49), and has found in each

---

[414] Abegg, R., *Ann. d. Physik (Wied.)*, **65**, 229-236 (1898).
[415] de Lenaizon, B., and Granier, J., *Compt. rend.*, **180**, 198-199 (1925).

80. ICE: DIELECTRIC PROPERTIES

case that the logarithm of a is linear in t; a is proportional to Debye's τ. The values of all three sets of constants and of the α and β in $\log a = \log \alpha - \beta t$ are given in Table 217. It will be noticed that the three β's

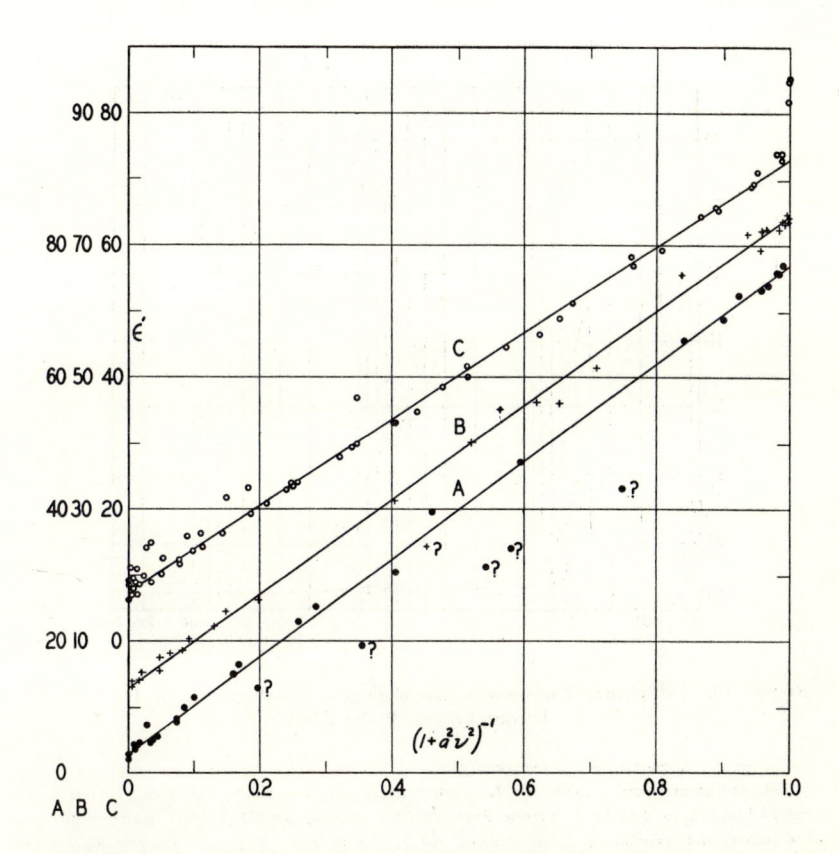

FIGURE 9. Dielectric Constant of Ice: Variation of ϵ' with $(1 + a^2 \nu^2)^{-1}$.

The observed values of ϵ' given by Errera (A), Smyth and Hitchcock (B), and Wintsch (C), and contained in Tables 218 and 219 are plotted against the reciprocal of $(1 + a^2\nu^2)$, the value of a being in each case determined from the constants given in Table 217. The origin of the scale of ϵ' is shifted from curve to curve, each scale being appropriately marked. The 5 questioned values of Errera's all refer to the temperature published as -22 °C; if the actual temperature was somewhere between -25 and -27 °C these points would lie near the line.

differ but little, but the α's differ greatly, suggesting that the three samples of ice differed significantly in some manner. It seems possible that some single value of β might be used in all three cases, and the α's be adjusted so as to obtain a satisfactory agreement with the observations, but time for testing this is not now available. It will also be noticed that, as in the case of water, the values that must be used for ϵ_0 and ϵ_1 do not agree respec-

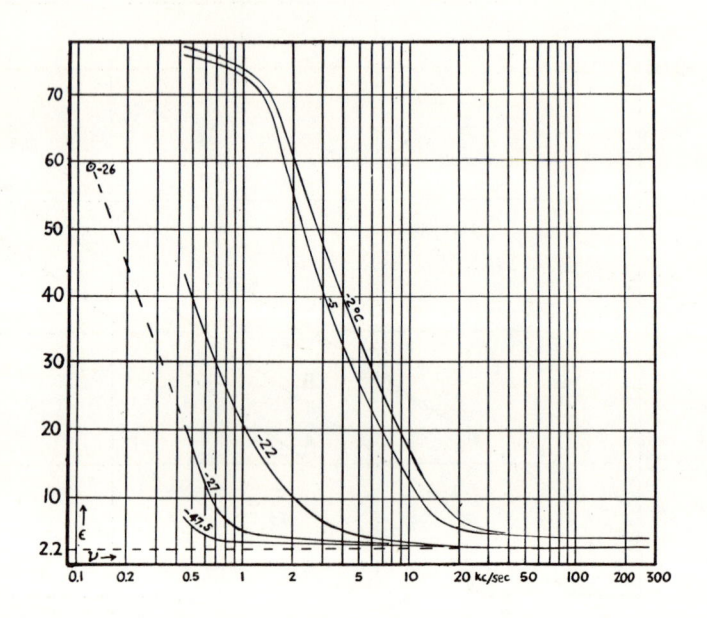

FIGURE 10. Isothermal Variation of the Dielectric Constant (ϵ') of Ice with the Frequency (ν) of the Field.

Unit of $\epsilon' = 1$ cgse; of $\nu = 1$ kilocycle/sec.

[Adapted from compilation by H. L. Curtis and F. M. Defandorf, *Int. Crit. Tables*, **6**, 78 (1929)—based upon data by J. Errera, *Jour. de Phys. (6)*, **5**, 304-311 (1924)—with the addition of an observation (circle) by J. A. Fleming and J. Dewar, *Proc. Roy. Soc. London*, **61**, 316-330 (1897).]

tively with the square of the optical index and with the static value of ϵ'. After the values of the constants had been obtained, each value of ϵ' of each set was plotted against the reciprocal of the corresponding value of $(1 + a^2\nu^2)$, using for a the value defined by the derived values of α and β. These graphs are shown in Figure 9. It will be noticed that all the values of ϵ' given by Smyth and Hitchcock, whatever the temperature and fre-

80. ICE: DIELECTRIC PROPERTIES

quency may be, lie quite satisfactorily along a right line, with a single exception. Those read from Wintsch's graphs do likewise except at the extremities, but the spread is greater; and Errera's values, corrected for two obvious misprints (see notes to Table 219) fit well, excepting the set

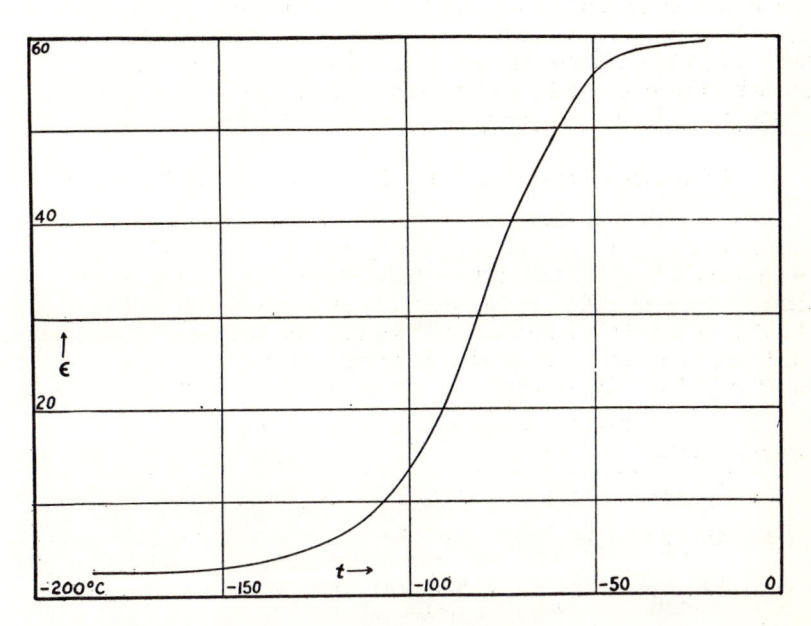

FIGURE 11. Thermal Variation of the Dielectric Constant (ϵ') of Ice for 120 Charges and Discharges per Second.

[From compilation by H. L. Curtis and F. M. Defandorf, *Int. Crit. Tables*, **6**, 79 (1929), based upon data by J. Dewar and J. A. Fleming, *Proc. Roy. Soc. London*, **61**, 2-18 (1897), and by J. A. Fleming and J. Dewar, *Idem*, **61**, 316-330 (1937). Cf. Table 219, Section V.]

for $-22\,°C$, which are consistently lower. If the temperature given in his table as $-22\,°C$ was actually somewhere between -25 and $-27\,°C$, these points also would lie close to the line. It seems probable that this is another misprint. In Table 218 each of the values of ϵ' given by Smyth and Hitchcock is compared with the corresponding one as computed from the constants obtained from their complete set of values.

Dielectric Absorption of Ice.

(For definitions of terms and symbols see Section 49.) The value of the dielectric absorption of ice is commonly indicated by means of either the phase defect $\phi = \tan^{-1}(\epsilon''/\epsilon')$ or the apparent conductivity $k_a = \epsilon''\nu/2$ cgse units $= (c^2\epsilon''\nu/2)10^{-9}(\text{ohm·cm})^{-1}$, ϵ'' being expressed in cgse units, and c = velocity of light.

The values of ϵ'' and of ϵ' can be computed from the constants given in Table 217, and from them ϕ and k_a may be obtained. These calculations have been carried through for the observations of Smyth and Hitchcock and are given in Tables 220 and 222. Whether the experimentally determined values of ϕ and k_a are entirely free from the effects of such true conductivity as the ice may have had, is not entirely clear.

Table 220.—Phase Defect for Ice: Observed and Computed

$\phi = \tan^{-1}(\epsilon''/\epsilon')$. Calculated (Calc) values are those defined by the formulas $\epsilon' = \epsilon_0 + (\epsilon_1 - \epsilon_0)/(1 + a^2\nu^2)$ and $\epsilon'' = (\epsilon_1 - \epsilon_0)a\nu/(1 + a^2\nu^2)$ with $\epsilon_0 = 3.0$, $\epsilon_1 = 74.6$; $10^4 a = 1.160 e^{-0.1015 t}$ sec. Temp. $= t$ °C. See text. Observed (Obs) values are those derived from the values given by C. P. Smyth and C. S. Hitchcock [408] for the equivalent conductivity and the dielectric constant. The specific electrical conductivity (k) of the water used was $10^6 k = 2(\text{ohm·cm})^{-1}$.

Unit of $\nu = 1$ cycle/sec; of $\phi = 1°$ of arc. Temp. $= t$ °C

$\nu/1000 \rightarrow$	0.3	0.5	1	5	20	60
t	Obs Calc Dif	Obs Calc Dif	Obs Calc Dif	Obs Calc Dif	Obs Calc Dif	Obs Calc Dif
−1	4.0 2.3 +1.7	4.8 3.5 +1.3	8.0 7.0 +1.0	31.2 31.6 −0.4	61.8 62.8 −1.0	64.4 65.5 −1.1
−3	2.4 2.6 −0.2	4.8 4.3 +0.5	9.2 8.6 +0.6	36.8 36.3 +0.5	63.7 65.0 −1.3	63.7 63.3 +0.4
−5	3.3 3.2 +0.1	5.6 5.3 +0.3	11.4 10.5 +0.9	41.4 41.7 −0.3	64.8 66.6 −1.8	62.4 60.1 +2.3
−10	4.6 3.6 +1.0	11.1 6.0 +5.1	18.0 11.9 +6.1	53.0 45.2 +7.8	64.4 67.2 −2.8	54.5 57.6 −3.1
−20	14.5 14.2 +0.3	25.2 22.8 +2.4	40.4 39.4 +1.0	65.1 67.1 −2.0	54.2 51.4 +2.8	32.8 24.0 +8.8
−30	38.9 34.5 +4.4	53.6 47.8 +5.8	61.2 62.1 −0.9	58.5 59.2 −0.7	30.3 25.9 +6.4	18.2 9.3 +8.9
−40	60.5 59.0 +1.5	66.1 65.8 +0.3	62.7 66.4 −3.7	36.8 34.7 +2.1	14.7 10.0 +4.7	14.3 3.4 +10.9
−50	63.9 67.1 −3.2	58.4 63.4 −5.0	48.5 50.2 −1.7	18.0 14.4 +3.6	7.3 3.7 +3.6	7.7 1.2 +6.5
−60	49.8 54.6 −4.8	38.3 41.9 −3.6	29.5 24.8 +4.7	10.4 5.3 +5.1	4.1 1.3 +2.8	6.8 0.4 +6.4
−70	38.0 29.1 +8.9	27.5 18.6 +8.9	20.4 9.6 +10.8	7.6 1.9 +5.7	1.9 0.5 +1.4	6.4 0.2 +6.2

81. ICE: ELECTRICAL CONDUCTION

Table 221.—Phase Defect for Ice

For definitions of terms and symbols see Section 49; $\tan \phi = \epsilon''/\epsilon'$. The electrical conductivity (k) of the water used by Wintsch was not stated; it was triply distilled and collected in quartz. For that used by Granier $10^6 k = 1.5 \text{(ohm·cm)}^{-1}$.

Unit of $\nu = 1$ kilocycle/sec; of $\phi = 1°$ of arc. Temp. $= t \,°C$

I. Smyth and Hitchcock, 1932, see Table 220.

II. Wintsch,[412] read from his graphs.

$t \rightarrow$ ν	−5	−6	−10	−20 ϕ	−30	−40	−50
0.05	1.2		1.7	3.4	8.5		
0.5		10.4	13.0	24.5	40.0	48.0	51.7
0.650	12.2		15.0	29.0	43.5	51.3	53.5
1.0		16.4	20.8	37.7	51.0	54.3	54.3
1.13	16.4		22.0	40.0	52.2	54.0	53.0
1.60	20.2		28.5	52.5	55.1	53.0	51.0
2.0		26.0	33.1	51.5	55.8	52.5	49.0
3.0		34.6	42.0	56.4	53.3	47.8	43.2
3.5	36.0		46.0	57.5	53.5	45.8	40.5
4.0		41.0	47.8	58.1	52.7	44.4	39.8
5.0		45.7	52.1	59.0	51.2	41.5	38.0
5.1	44.5		52.7	59.1	51.0	41.0	37.5
6.0		49.6	55.6	59.5			
7.0		52.5	58.0				
8.0		54.0	58.8				

III. Granier,[416] 1924; $t = -12\,°C$.

ν	0.050	0.320	1.550	5.40	17.0	51.0	260	6700	64000
ϕ	28	32	50	68	68	54	18	0.7	0.5[a]

[a] This value (0.5) is from Granier.[417]

Dielectric Strength of Ice.

P. Thomas [417a] has found that in a uniform alternating field applied at the rate of about 600 volts per second (frequency = 1000 cycles/sec) ice broke down when the field reached the value of 11,000 volts/cm; $\epsilon' = 86.4$ cgse, conductivity $= 1.4 \times 10^{-8} \text{(ohm·cm)}^{-1}$; temperature is not stated.

81. ELECTRICAL CONDUCTIVITY OF ICE

In any discussion of the electrical conductivity of ice it is quite essential to recognize the several distinct types of effect that contribute to the observed effective conductivity.

In most cases, the value assigned to the (effective) conductivity is that derived from the resistance R which must be placed in parallel with a pure capacitance C in order to obtain an exact equivalent of the actual ice-condenser under the existing conditions. If the ice-condenser were merely a leaky condenser—if the ice were composed of an ideal, nonconducting,

---

[416] Granier, J., *Compt. rend.*, **179**, 1313-1318 (1924).
[417] Granier, J., *Bull. Soc. Fr. des Elec.* (4), **3**, 333-482 (1923).
[417a] Thomas, P., *J. Franklin Inst.*, **176**, 283-301 (1913).

Table 222.—Apparent Electric Conductivity of Ice: Observed and Computed

Observed (Obs) values are those reported by C. P. Smyth and C. S. Hitchcock [408]; calculated ones (Calc) are those defined by the formulas $\epsilon'' = (\epsilon_1 - \epsilon_0)a\nu/(1 + a^2\nu^2) = (\epsilon_1 - \epsilon')/a\nu$ and $k_a = \nu\epsilon''/2$ cgse $= (\nu\epsilon''/18) \times 10^{-11}(\text{ohm}\cdot\text{cm})^{-1}$ in which $\epsilon_0 = 3.0$, $\epsilon_1 = 74.6$, $10^4 a = 1.160 e^{-0.1015 t}$ sec. Temp. $= t$ °C, and $\epsilon' = \epsilon_0 + (\epsilon_1 - \epsilon_0)/(1 + a^2\nu^2)$. See text of Section 49. The electrical conductivity (k) of the water was $10^6 k = 2(\text{ohm}\cdot\text{cm})^{-1}$.

Unit of $k_a = 1$ (ohm·cm)$^{-1}$; of $\nu = 1$ cycle/sec. Temp. = t °C

$10^8 k_a$

$t \rightarrow$		-1			-3			-5			-10			-20			-30			-40			-50			-60			-70	
$\nu/1000$	Obs	Calc	Dif	Obs	Calc	Dif	Obs	Calc	Dif	Obs	Calc	Dif	Obs	Calc	Dif	Obs	Calc	Dif	Obs	Calc	Dif	Obs	Calc	Dif	Obs	Calc	Dif	Obs	Calc	Dif
0.3	0.086	0.046	+0.040	0.052	0.056	−0.004	0.07	0.069	0.00	0.10	0.078	+0.02	0.31	0.296	+0.01	0.62	0.570	+0.05	0.48	0.474	+0.01	0.21	0.207	0.0	0.086	0.077	+0.009	0.052	0.028	+0.024
0.5	0.17	0.127	+0.04	0.17	0.155	+0.01	0.20	0.190	+0.01	0.40	0.216	+0.18	0.86	0.735	+0.12	1.18	0.974	+0.21	0.55	0.544	+0.01	0.21	0.212	0.0	0.086	0.078	+0.008	0.052	0.028	+0.024
1	0.57	0.503	+0.07	0.65	0.611	+0.04	0.81	0.739	+0.07	1.25	0.834	+0.42	2.14	1.97	+0.17	1.48	1.40	+0.08	0.58	0.579	0.0	0.24	0.214	+0.03	0.11	0.078	+0.03	0.069	0.028	+0.041
5	8.85	9.05	−0.20	9.60	9.65	−0.05	9.86	9.93	−0.07	9.0	9.88	−0.88	4.55	4.28	+0.27	1.84	1.62	+0.22	0.69	0.59	+0.10	0.29	0.214	+0.08	0.16	0.078	+0.08	0.11	0.028	+0.08
20	25.6	26.9	−1.3	23.3	22.9	+0.4	19.7	19.4	+0.3	12.7	17.2	−4.5	5.5	4.49	0.0	2.06	1.63	+0.43	0.89	0.59	+0.30	0.43	0.214	+0.22	0.24	0.078	+0.16	0.11	0.028	+0.08
60	29.9	30.5	−0.6	27.5	25.0	+2.5	23.7	20.5	+3.2	15.6	18.0	+7.6	6.7	4.50	+2.2	3.33	1.63	+1.70	2.58	0.59	+1.99	1.37	0.214	+1.16	1.2	0.078	+1.1	1.1	0.028	+1.1

For additional data and remarks see Section 81.

81. ICE: ELECTRICAL CONDUCTION

electrically perfectly elastic dielectric interpenetrated along the lines of electric force by threads of an ideal conductor—then R would be the combined resistance of those threads, and the conductivity computed from it would be the actual conductivity of the ice, which may be designated by k.

But the problem is not so simple. Unless the ice is exceedingly pure and gas-free, which last has probably not been the case in any of the work, it will acquire a space-charge under the action of the field.[407] With constant fields this will act as a kind of polarization, adding itself to the polarization of the electrodes. With alternating fields, it may constitute a quite significant part of the actual current. The energy dissipated by the alternating concentration of this charge now nearer one electrode and now nearer the other will contribute to R, giving rise to a new term (k_c) in the effective conductivity. It seems reasonable to expect that both k and k_c will steadily decrease as the temperature falls, and that for a given temperature, k_c will increase with the frequency (ν), but k will not.

Furthermore, the molecules of H_2O are polar. Hence, they will tend to place their electrical axes parallel to the impressed field.* The dissipation involved in such reorientation will also contribute to R, and hence to the effective conductivity.[418, pp. 89-108] Denote this component by k_p. As the temperature is decreased, it is to be expected that both the resistance to the rotation involved in such reorienation will be increased and the extent of the rotation will be decreased. The former will increase, and the latter will decrease, the dissipation. Whence, one should expect the dissipation, and consequently k_p, to pass, in general, through a maximum, and then to decrease to zero, the frequency being constant and not too great. If the frequency is very high, the molecules may not have time to rotate through an appreciable angle, and k_p will be zero. (See also p. 504.)

Thus it is evident that the effective conductivity (k_e) of ice is, in general, made up of at least three terms, $k_e = k + k_c + k_p$. The static conductivity is k, suitable correction being applied for such polarization as may exist; at intermediate frequencies, k_p may be the dominant term; and at high frequencies, k_c may be supreme. The data at present available do not suffice for a complete separation of these three terms, but they are consistent with the ideas just expressed (see Table 223). For example, Johnstone's static values for ice ($10^8 k = 2.80$ (ohm·cm)$^{-1}$ at 0 °C and 0.026 at -19 °C, electrolytic polarization eliminated) are not only much lower than the conductivity ($10^8 k = 71$ at 17.9 °C) of the water from which the ice was frozen, but are also lower than that ($10^8 k = 4$) of the purest water obtained by Kohlrausch. On the other hand, Smyth and Hitchcock found at 60 kc/sec and -1 °C the great value $10^8 k_e = 29.9$, over 7 times that of the purest water, and actually 15 per cent of that ($10^8 k = 200$) of the water from which it was frozen; the value decreased continuously with the tem-

---

* P. Debye[418] has shown that even if the maximum dielectric constant (80) were due entirely to such orientation only 1 molecule in 5 million need follow the field.
[418] Debye, P., "Polar Molecules," p. 106, New York, Reinhold Publishing Corp., 1929.

perature. It seems that here k_e is the dominant term. In the range $\nu = 0.3$ to 1 kc/sec, the same authors found that k_e passes through a maximum, the value at the maximum decreasing with ν. Thus, at $\nu = 1$ kc/sec, $10^8 k_e = 0.57$ at -1 °C, 2.14 (max.) at -20 °C, and 0.069 at -70 °C, all lower than that for the purest water. Here k_p is dominant. Why the values at -1 °C and $\nu = 0.3$ to 1 kc/sec are so much lower than the static one found by Johnstone at 0 °C is not clear.

For a complete interpretation of the variation of k_e with the tempera-

Table 223.—Electrical Conductivity of Ice
(See also Table 224.)

The text should be consulted. The conductivity of the water from which the ice was formed is indicated in each case. It seems probable that the k_e of SH and of G is essentially k_p.

Unit of $k = 10^{-8}$ (ohm·cm)$^{-1}$; of $\nu = 1$ kilocycle/sec. Temp. $= t$ °C

	Static: $k_{water} = 71$[a] at 17.9 °C (ICT)[b]			
t	0	-4	-10	-19
k_e	2.8	0.23	0.11	0.026

$\nu \rightarrow$	0.3	0.5	1	5	20	60
t			$k_{water} = 200$ (SH)[b] k_e			
-1	0.086	0.17	0.57	8.85	25.6	29.9
-3	0.052	0.17	0.65	9.62	23.3	27.5
-5	0.07	0.20	0.81	9.86	19.7	23.7
-10	0.10	0.40	1.25	9.0	12.4	15.6
-20	0.31	0.86	2.14	4.55	5.5	6.7
-30	0.62	1.18	1.48	1.84	2.06	3.33
-40	0.48	0.55	0.58	0.69	0.89	2.58
-50	0.21	0.21	0.24	0.29	0.43	1.37
-60	0.086	0.086	0.11	0.16	0.24	1.2
-70	0.052	0.052	0.069	0.11	0.11	1.1

$t = -12$ °C				$k_{water} = 154$ (G)[b]					
ν	0	0.0043	0.050	0.320	1.55	5.40	17.0	51.0	260	6700
k_e[c]	0.004	0.02	0.13	1.09	6.0	9.1	9.4	9.4	9.8	9.4

[a] See J. H. L. Johnstone.[419]
[b] References:
    G   Granier, J.[416]
    ICT Bjerrum, N.[422] Based on observations of J. H. L. Johnstone.[419]
    SH Smyth, C. P., and Hitchcock, C. S.[408]
[c] Computed by the compiler from Granier's data.

ture, consideration must be given to the progressive melting discussed by Buchanan (p. 469).

J. H. L. Johnstone [419] has stated that the effects of polarization are great, are not entirely electrolytic, and are difficult to eliminate. The observations of G. Oplatka [407] indicate much the same. Johnstone used potential leads, and measured their potentials electrostatically. The only measurements at exceedingly low temperatures seem to be those by Dewar

[419] Johnstone, J. H. L., *Proc. Trans. Nova Scotian Inst. Sci.*, **13**, 126-144 (1912).
[420] Fleming, J. A., and Dewar, J., *Proc. Roy. Soc. (London)*, **61**, 316-330 (1897).

81. ICE: ELECTRICAL CONDUCTION

and Fleming (Table 224). The method employed is not described, but it appears to have been a static one, the resistance being derived by means of Ohm's law from the steady current and the applied voltage. There is no

Table 224.—Thermal Variation of the Electrical Resistance of Ice

The several sets of values of R are not comparable.

It seems probable that the values attributed to Wintsch (read from his graphs) represent mainly the component (k_p) arising from dielectric absorption. Those attributed to Dewar and Fleming seem to have been inferred from the constant impressed voltage and the observed current; the values are merely approximate, and the two sets, having been obtained with different vessels, are not comparable; the first set refers to ice from ordinary distilled water, the second to that from especially pure water; t_p is the temperature inferred from their platinum thermometer, and is lower than the actual temperature on the centigrade scale. Frequency is ν.

Unit of $R = 1$ ohm; of $\nu = 1$ kilocycle/sec. Temp. $= t$ °C

I. H. Wintsch.[412]

$t \rightarrow$ ν	−5	−6	−10	−20	−30	−40	−50
				$R/1000$			
0.05	57500		42400	19000	7750		
0.65			342	202	184	233	278
1.00		198	172	118	139	202	249
1.13	184		144	108	133	194	240
1.60	112		86	81	116	180	224
2.00		76	65	70	111	172	216
3.00		42	42	59	102	159	198
3.50	36		35	56	99	152	192
4.00		31	32	53	96	148	182
5.00		24	28	49	92	139	168
5.10	23		28	49	92	138	178
6.00		20	26	48			

II. J. Dewar and J. A. Fleming.[423] Fleming and Dewar.[420]

Distilled				Pure			
t_p	$R/10^6$	t_p	$R/10^6$	t_p	$R/10^6$	t_p	$R/10^6$
−70.7	43.4	−93.2	282	−10.2	1	−92.0	1200
−75.0	42.8	−95.2	353	−19.2	3	−138.2	2000
−82.3	46.3	−98.8	470	−26.1	15	−152.1	2500
−84.4	53.4	−108.4	706	−27.6	40	−206	25000
−86.3	66.5	−126.0	1130	−33.2	250		
−88.2	91.4	−135.0	1570	−42.1	260		
−88.8	118.0	−172.0	5670	−47.0	410		
−91.9	209	−200.0	26200	−68.2	1200		

way in which the actual conductivity can be inferred from the resistances they tabulated. They remark: "Above a certain temperature there is a relatively rapid increase in the conductivity of the ice, as it rises in temperature."[420] A single series of observations at an unstated frequency has

[421] Brommels, H., *Comment. Phys.-Math. Soc. Sci. Fennica (Helsingfors)*, **1**, No. 19 (1922).
[422] Bjerrum, N., *Int. Crit. Tables*, **6**, 152 (1929).
[423] Dewar, J., and Fleming, J. A., *Proc. Roy. Soc. (London)*, **61**, 2-18 (1897).

been published in insufficient detail by H. Brommels,[421] the values of $10^8 k_e$ increasing from 38.9 at $-1.2\,°C$ to 47.3 at $-17.3\,°C$. H. Wintsch[412] used triply distilled water condensed and collected in quartz, but it was probably not completely gas-free when frozen; his data for k_e were given by graphs only.

82. Miscellaneous Electrical Data for Ice

Pyro-electric Effect.

J. Smithson[424] has observed that hail frequently consists of two hexagonal pyramids joined base to base. *"One of the pyramids is truncated,* which leads to the idea that ice becomes electrified on a variation of its temperature, like tourmaline, silicate of zinc, etc."* This is the only mention of the probability of ice being pyroelectric that has come to the compiler's attention, though the observation of J. M. Adams,[425] that the ice-crystal is asymmetric with respect to its basal plane, indicates the same thing. Such asymmetry indicates the existence of piezo-electric properties also.

83. Magnetic Susceptibility of Ice

Like water, ice is diamagnetic. G. Foex[426] has found that at the moment of freezing, the numerical value of the specific susceptibility (χ) decreases by 2.4 per cent. In their compilation, K. Honda, T. Ishiwara, T. Soné, and M. Yamada[427] give $\chi = -0.699$ micro-cgsm for the entire range 0 to $-120\,°C$, based on the observations of T. Ishiwara.[428] Taking the density of ice as $0.9168\ g/cm^3$ at $0\,°C$, this gives for the volume susceptibility (κ) at $0\,°C$ the value $\kappa = -0.641$ micro-cgsm.

More recently B. Cabrera and H. Fahlenbrach[429] have reported observations indicating that $\chi = -0.7019(1 + 0.000667t)$ micro-cgsm, and that the change on freezing is 2.2 per cent. This temperature coefficient, nearly 6 times that for water, is entirely incompatible with Ishiwara's observations, which extended to $-120\,°C$, whereas Cabrera and Fahlenbrach did not go below $-60\,°C$.

[424] Smithson, J., *Ann. Philos. (N. S.)*, **5**, 340 (1823).
[425] Adams, J. M., *Proc. Roy. Soc. (London) (A)*, **128**, 588-591 (1930) → *Phys. Rev. (2)*, **36**, 788 (A) (1930).
[426] Foex, G., See Piccard, A., *Arch. sci. phys. et nat. (4)*, **35**, 209-231, 340-359, 458-482 (1913).
[427] Honda, K., Ishiwara, T., Soné, T., and Yamada, M., *Int. Crit. Tables*, **6**, 354-366 (356) (1929).
[428] Ishiwara, T., *Science Rep. Tôhoku Univ., Sendai (1)*, **3**, 303-319 (1914).
[429] Cabrera, B., and Fahlenbrach, H., *Am. Soc. Esp. Fis. y Quim.*, **31**, 401-411 (1933).

III. Multiple-phase Systems

84. Surface-tension of Water

The number of articles treating of surface-tension and its measurement is very great, but in many cases, most unfortunately, the author of the article is not sufficiently acquainted with the mathematical derivation of the formula employed in obtaining the value of the surface-tension from the observed quantities to be able to appreciate its true significance. As a consequence, the experimental conditions realized by him frequently fail to accord with those demanded by the formula used, and the value of his discussion of the work, whether of himself or of another, is seriously impaired. Furthermore, and as a result of his failure to check its derivation, he occasionally uses a formula that is actually wrong, one involving a misprint or an algebraic error; and, in some cases, he merely guesses at the value of certain small corrections.

As a consequence, any mere assemblage of the various values published for the surface-tension of a given substance—such an assemblage as is commonly given in tables of constants—is of no assistance in enabling one to form an idea either of the most probable value of the surface-tension of that substance, or of the variability of its apparent surface-tension under good laboratory conditions, or of the possible dependence of its apparent surface-tension upon the method employed in measuring it.

Before such information can be obtained, each determination must be studied individually and in every detail, including the derivation of the formulas and their applicability to the experimental conditions actually realized. This involves great labor. In general, every determination based upon observations and computations that have been published without sufficient detail to enable one to make such a critical study should be summarily discarded as valueless; so should those for which the experimental conditions depart from those demanded by the formulas, unless the numerical value of the effect of such departure can be satisfactorily determined. I know of no publication of such a study of the existing data for surface-tension. Comparisons of selected groups of observations, of course, exist; and personal estimates of the most probable value of the surface tension of each of certain substances have been published from time to time. These estimates are presumably based upon some such detailed study as that just mentioned, but in some cases it is obvious that the study fell far short of what should be desired.

For values of the volumes of water menisci, see Tables 286 and 287.

Factors Possibly Affecting Surface-tension.

That the surface-tension of a liquid varies with the temperature and to a less extent with the nature of the overlying gas, and that its apparent value is greatly affected by even slight contamination of the surface, are well known facts. The first two are considered elsewhere, Tables 225 and 226, and text (p. 524). Of the third, nothing more need be said, as the data to be presented supposedly refer to uncontaminated gas-liquid surfaces.

But from time to time questions arise regarding the possible dependence of the observed surface-tension of a liquid upon other factors. They have to do with (1) the method used, (2) the material of the tube (in the method of capillary rise), (3) the effect of proximity to a solid wall, (4) the age of the (uncontaminated) surface, (5) the effect of illumination, (6) of electrification, (7) of a magnetic field, (8) of prolonged contact with catalysts. These, especially in their relation to water, will be considered here.

(1). In comparing the results obtained by different methods, no effect arising either from the use of erroneous formulas or from a failure to secure the conditions demanded by the formulas used in the computations need be considered, for such effects result from mere blunders, and the data involved should be discarded unless the effects of the blunder can be eliminated. This greatly reduces the amount of data to be compared. Those left give no certain evidence of any difference that can unquestionably be attributed to a difference in the method, but a more careful comparison of the several available methods is much to be desired. It is entirely possible that the results obtained by dynamic methods may differ from those by static methods, and that the experimental details involved in some methods introduce unanticipated effects.

(2). In 1894, P. Volkmann[1] found that the height to which water rises in a glass tube is independent of the nature of the tube; and recently, E. K. Carver and F. Hovorka[2] have found, contrary to the announced results of S. L. Bigelow and F. W. Hunter,[3] that the same is true for tubes of glass, zinc, copper, and silver.

(3) It has long been recognized that the density, and even the structure, of a liquid in the immediate neighborhood of a solid may differ from that at a great distance from all solids. Any such difference would probably result in the tension of the gas-liquid surface near a solid wall being different from that elsewhere. In that case the form of the surface will differ from that corresponding to a surface of uniform tension, and, consequently, the value of the surface-tension as computed from the observations will differ from that for the surface far from solid walls, since all such computations are based on the assumption that the form of the surface is that corresponding to a surface of uniform tension. Such an effect

---

[1] Volkmann, P., *Ann. d. Physik (Wied.)*, **53**, 633-663 (1894).
[2] Carver, E. K., and Hovorka, F., *J. Am. Chem. Soc.*, **47**, 1325-1328 (1925).
[3] Bigelow, S. L., and Hunter, F. W., *J. Phys'l Chem.*, **15**, 367-380 (1911).

might greatly exceed the dependence of the tension upon the nature of the solid. In the case of capillary rise, it would become of ever-increasing importance as the diameter of the tube is reduced.

W. A. Patrick and N. F. Ebermann [4] have published observations which they interpret as indicating that the pressure of the vapor in equilibrium with the concave liquid-gas surface in a tube of very small diameter (a few microns) is less than that computed for a surface of the same curvature and the tension characteristic of a flat surface of the liquid far from solid walls (for formula, see p. 568). This suggests that the tension is greater for the smaller surface, presumably on account of the presence of the solid walls. But the actual significance of their observations is not entirely clear, and the interpretation of them is correspondingly difficult. In at least some cases, the interpretations offered may overlook essential factors, as pointed out by D. J. Woodland and E. Mack, Jr.[5] That the observations are not to be accounted for by the mere curvature of the surface is indicated by the earlier observation of N. Gudris and L. Kalikowa [6] that the partial pressure of the vapor in equilibrium with air-suspended water droplets 0.1 to 1.0 μ in diameter is equal to that computed by Kelvin's (Thomson's) formula (p. 568). Furthermore K. W. v. Nägeli [7] had found that the pressure required to drive water from tubes 3 to 9 μ in diameter is about the same as would be inferred from observations on much larger tubes. He used freshly drawn tubes, and stated that older tubes always gave off air, forming minute bubbles in the water, which impeded the flow.

The subject is intimately connected with the least thickness of an adsorbed layer that has the same vapor pressure as does the liquid in bulk, and with the least thickness of a layer of water that exhibits true viscosity and has a viscosity that is the same as that of the liquid in bulk, although it is not identical with either of these. L. J. Briggs [8] studied the adsorbed layer of water on quartz when in equilibrium with atmospheres of various relative humidities at 30 °C. He found that at 99 per cent humidity the amount of that adsorbed layer that could be removed by heating to 110 °C corresponded to a thickness of 0.027 μ. Similarly, I. R. McHaffie and S. Lenher [9] observed that the vapor-pressure of adsorbed water films varied with the thickness unless that exceeded several hundreds of molecules if on glass, and several tens of molecules if on platinum (300 molecules = 0.09 μ, approximately). And S. H. Bastow and F. P. Bowden [10, 11] have observed that even at 0.1 °C a film of water only 0.2 μ thick, flowing

---

[4] Patrick, W. A., and Ebermann, N. F., *Idem*, **29**, 220-228 (1925). Cf. Shereshefsky, J. L., *J. Am. Chem. Soc.*, **50**, 2966-2980, 2980-2985 (1928); Latham, G. H., *Idem*, **50**, 2987-2997 (1928).
[5] Woodland, D. J., and Mack, E., Jr., *J. Am. Chem. Soc.*, **55**, 3149-3161 (1933).
[6] Gudris, N., and Kalikowa, L., *Z. Physik*, **25**, 121-132 (1924).
[7] v. Nägeli, K. W., *Sitz.-ber. Bayer Akad. Wiss. München*, 1866₁, 353-376 (358) (1866).
[8] Briggs, L. J., *J. Phys'l Chem.*, **9**, 617-640 (1905).
[9] McHaffie, I. R., and Lenher, S., *J. Chem. Soc. (London)*, **127**, 1559-1572 (1925).
[10] Bastow, S. H., and Bowden, F. P., *Proc. Roy. Soc. (London) (A)*, **151**, 220-233 (1935); Cf. *Idem*, **134**, 404-413 (1932).
[11] Bowden, F. P., and Bastow, S. H., *Nature*, **135**, 828 (L) (1935).

Table 225.—Surface-tension of Water
(For thermal variations, see also Table 226.)

All values at temperatures not exceeding 100 °C refer to an air-water surface at atmospheric pressure, the water being, presumably, saturated with air, and the air with water-vapor. Those above 100 °C refer to the surface separating pure water from its pure vapor. For the effect of a change in the gas, see Table 229.

As primary standard for the ICT values, Young and Harkins accepted for the tension of a water-air surface at 20 °C the value 72.75 ± 0.05 dyne/cm, which they derived from the determinations of T. W. Richards and L. B. Coombs [29] which were corrected by T. W. Richards and E. K. Carver [30] to yield 72.72; of W. D. Harkins and F. E. Brown [31] giving 72.80; of T. W. Richards and E. K. Carver [30] giving 72.73; and of T. F. Young and P. L. K. Gross (unpublished) giving 72.80, all determined by the rise in capillary tubes.[32]

At the critical point the surface-tension does not become zero when the meniscus vanishes.[33]

In contrast to the other values tabulated below, the TB ones show an anomaly near 13 °C.

For sea-water containing s grams of salts per kg, $\gamma_s = \gamma_o + 0.0221s$ dyne/cm, where γ_s and γ_o are the air-liquid surface-tensions of the sea-water and of pure water, respectively, both at the same temperature.[34] If $s = 35$, the average amount of salts, then $\gamma_s - \gamma_o = 0.77$ dyne/cm.

γ = surface-tension; $a^2 = 2\gamma/(\rho - \sigma)g$; g = acceleration of gravity (here taken as 980.665 cm/sec^2); ρ = density of liquid; σ = density of the adjacent gas (here taken, the gas being air, as $0.001200(293.1)/(273.1 + t) = 0.3517/(273.1 + t)$ g/cm^3, the temperature being t °C); $\Delta \equiv \gamma - \gamma_c$ where $\gamma_c = 75.64 - 0.13910t - 0.0003000t^2$ dyne/cm is an empirical formula which represents the ICT values fairly well; τ is the tolerance assigned by Young and Harkins.

Unit of γ, τ, and $\Delta = 1$ dyne/cm = 1 gram·sec^{-2} = 0.10197 mg*/mm; of $a^2 = 1$ cm^2 = 100 mm^2. Temp. = t °C

Ref[a]		ICT			Moser		Warren		TB
t	γ	τ	a^2	γ_c	γ	100Δ	γ	100Δ	100Δ
−8	76.9₀	0.3	0.1574	76.73₄					
−5	76.4₂	0.2	0.1562	76.32₈					
0	75.6₄	0.1	0.1544₈	75.64₀	75.62[b]	−2	75.94	+30	+36
+5	74.9₂	0.1	0.1529₉	74.93₆	74.86	−8	75.19	+25	−44
10	74.22	0.05	0.1516₀	74.21₉	74.12	−10	74.43	+21	−72
11	74.07	0.05	0.1513₁	74.07₄	73.96	−11			
12	73.93	0.05	0.1510₃	73.92₈	73.80	−13			
13	73.78	0.05	0.1507₅	73.78₁	73.65	−13			
14	73.64	0.05	0.1504₈	73.63₄	73.50	−13			
15	73.49[c]	0.05	0.1501₉	73.48₆	73.35	−14	73.65	+16	−37
16	73.34	0.05	0.1499₁	73.33₇	73.20	−14			
17	73.19	0.05	0.1496₃	73.18₈	73.04	−15			
18	73.05	0.05	0.1493₇	73.03₉	72.89	−15			
19	72.90	0.05	0.1490₉	72.88₉	72.73	−16			
20	72.75	0.05	0.1488₁	72.73₈	72.58	−16	72.86	+12	+13

84. SURFACE TENSION

Table 225.—(Continued)

Ref.[a]		ICT			Moser		Warren		TB
t	γ	τ	α^2	γc	γ	100Δ	γ	100Δ	100Δ
21	72.59	0.05	0.1485₂	72.58₇	72.43	−16			
22	72.44	0.05	0.1482₄	72.43₅	72.27	−16			
23	72.28	0.05	0.1479₅	72.28₂	72.10	−18			
24	72.13	0.05	0.1476₈	72.12₉	71.96	−17			
25	71.97	0.05	0.1473₈	71.97₄	71.81	−16	72.09	+12	+3
26	71.82	0.05	0.1471₁	71.82₀	71.65	−17			
27	71.66	0.05	0.1468₃	71.66₅	71.50	−16			
28	71.50	0.05	0.1465₄	71.51₀	71.34	−17			
29	71.35	0.05	0.1462₇	71.35₄	71.19	−16			
30	71.18	0.05	0.1459₇	71.19₇	71.03	−17	71.33	+13	+5
35	70.38	0.05	0.1445₆	70.40₇	70.23	−18	70.54	+13	−4
40	69.56	0.05	0.1431₃	69.59₆	69.42	−18	69.73	+13	−10
45	68.74	0.05	0.1417₃	68.77₂	68.59	−18	68.83	+6	
50	67.91	0.05	0.1403₂	67.93₅	67.75	−18	68.02	+9	
55	67.05	0.05	0.1388₇	67.08₂	66.84	−24	67.14	+6	
60	66.19	0.05	0.1374₁	66.21₄	66.04	−17	66.24	+3	
70	64.4₂	0.1	0.1344₉	64.43₃	64.28	−15	64.51	+8	
80	62.6₁	0.1	0.1315	62.59₂	62.50	−9	62.69	+10	
90	60.7₅	0.2	0.1284	60.69₁	60.68	−1	60.80	+11	
100	58.8₅	0.2	0.1253	58.73₀	58.80	+7			
110	56.8₉[d]	0.2		56.70₉					
120	54.8₉[d]	0.2		54.62₈					
130	52.8₄[d]	0.3		52.48₇					

[a] References:

ICT   From compilation of T. F. Young and W. Harkins,[35] based upon the accepted value at 20 °C (see head of table) and the observations of C. Brunner,[36] W. J. Humphreys and J. F. Mohler,[37] J. L. R. Morgan and C. E. Davis,[38] J. L. R. Morgan and A. McD. McAfee,[39] W. Ramsay and J. Shields,[40] T. W. Richards, C. L. Speyers, and E. K. Carver,[41] H. Sentis,[42] S. Sugden,[43] and P. Volkmann.[44]

Moser   Moser, H., *Ann. d. Physik (4)*, **82**, 993-1013 (1927).
TB   Timmermans, J., and Bodson, H., *Compt. rend.*, **204**, 1804-1807 (1937); values were read from their graph.
Warren   Warren, E. L., *Phil. Mag. (7)*, **4**, 358-386 (1927).

[b] For water at 0 °C, G. Schwenker[45] has found $\gamma = 75.59_7$ with an estimated uncertainty not exceeding 0.044 per cent.

[c] By a method involving an impact of two jets, W. N. Bond[46] found for a water-air surface that was renewed about 80 times a second, $\gamma = 73.83 \pm 0.13$ dyne/cm at 15 °C.

[d] These values are for water in contact with its own pure vapor.

---

[12] Bulkley, R., *Bur. Stand. J. Res.*, **6**, 89-112 (RP264) (1931).
[13] Bohr, N., *Phil. Trans. (A)*, **209**, 281-317 (1909).
[14] Hiss, R., *Diss.*, Heidelberg, 1913.
[15] Schmidt, F., and Steyer, H., *Ann. d. Physik. (4)*, **79**, 442-464 (1926).
[16] Kleinmann, E., *Idem*, **80**, 245-260 (1926).
[17] Young, T. F., and Harkins, W. D., *Int. Crit. Tables*, **4**, 446-475 (474) (1928).
[18] Lenard, P., *Sitz. Heidelberger Akad. Wiss. (A)*, **5**, No. 28, pp. 16-23 (1914).
[19] Seitz, E. O., *Ann. d. Physik (5)*, **1**, 1099-1108 (1929).
[20] Buchwald, E., and König, H., *Idem*, **23**, 557-569 (1935).
[21] Grumbach, A., and Schlivitch, S., *Compt. rend.*, **181**, 241-243 (1925).
[22] Auer, H., *Z. Physik*, **66**, 224-228 (1930).
[23] Johner, W., *Helv. Phys. Acta*, **4**, 238-280 (1931) = *Diss.*, Bern, 1930.
[24] Piccard, A., *Arch. Sci. phys. et nat. (4)*, **35**, 209-231, 340-359, 458-482 (1913).
[25] Liebknecht, O., and Wills, A. P., *Ann. d. Physik (4)*, **1**, 178-188 (1900).
[26] Quincke, G., *Idem (Pogg.)*, **160**, 560-588 (586) (1877); *Idem (Wied.)*, **24**, 347-416 (376-377) (1885).
[27] Brunner, C., *Idem (Pogg.)*, **79**, 141-144 (1850) reporting observations of J. R. A. Mousson.
[28] Baker, H. B., *J. Chem. Soc. (London)*, **1927**, 949-958 (1927).
[29] Richards, T. W., and Coombs, L. B., *J. Am. Chem. Soc.*, **37**, 1656-1676 (1915).
[30] Richards, T. W., and Carver, E. K., *Idem*, **43**, 827-847 (1921).

Table 226.—Thermal Variation of the Surface-tension of Water

For the value of the surface-tension at each of a selected number of temperatures, see Table 225. For the values of the molecular surface energy at various temperatures, see Table 227.

In this table are given for the air-water surface the values of: I. The temperature derivative $(d\gamma_e/dt)$ as defined by those empirical equations of the form $\gamma_e = \gamma_0(1 - at - bt^2) = \gamma_0(1 - \delta)$ that have been found to represent, respectively, the corresponding sets of data given in Table 225 (the TB data are not here included, having appeared after the completion of this table). II. The several values of $\delta(=(\gamma_0 - \gamma_e)/\gamma_0)$ corresponding to the same three equations and to others of the same form that may be found in other compilations. III. The several values of δ corresponding to certain other types of interpolation equations.

Unit of $\gamma = 1$ dyne/cm; of $d\gamma/dt = 1$ dyne·cm⁻¹ per 1 °C. Temp. $= t$ °C

I. Temperature derivative $d\gamma_e/dt$. $\gamma_e = \gamma_0(1 - at - bt^2)$. Dr. Domke [47] concluded that the data available in 1902 indicated that at 20 °C $d\gamma/dt = -0.151$.

Ref.[a]→	ICT	M	Wa	Ref.[a]→	ICT	M	Wa
$1000a$→	1.83₉	1.95₀	1.91₁	$1000a$→	1.83₉	1.95₀	1.91₁
$10^6 b$→	3.97	2.62	3.33	$10^6 b$→	3.97	2.62	3.33
γ_0→	75.64	75.62	75.91	γ_0→	75.64	75.62	75.91
t	\-\-\-	$-d\gamma_e/dt$	\-\-\-	t	\-\-\-	$-d\gamma_e/dt$	\-\-\-
−5	0.136	(0.146)	(0.143)	45	0.166	0.166	0.168
0	0.139	0.148	0.145	50	0.169	0.168	0.170
+5	0.142	0.150	0.148	55	0.172	0.170	0.173
10	0.145	0.152	0.150	60	0.175	0.172	0.176
15	0.148	0.154	0.153	70	0.181	0.176	0.180
20	0.151	0.156	0.155	80	0.187	0.180	0.186
25	0.154	0.158	0.158	90	0.193	0.184	0.191
30	0.157	0.160	0.160	100	0.199	0.188	0.196
35	0.160	0.162	0.163	110	0.205	(0.192)	(0.201)
40	0.163	0.164	0.165	120	0.211	(0.196)	(0.206)

II. Values of $\delta \equiv (\gamma_0 - \gamma_e)/\gamma_0$; $\gamma_e = \gamma_0(1 - at - bt^2)$.

Ref.[a]→	ICT	M	Wa	F	S	RSC
$1000a$→	1.83₉	1.95₀	1.91₁	1.90₂	2.02₆	2.08₉
$10^6 b$→	3.97	2.62	3.33	2.50	0	3.29
γ_0→	75.64	75.62	75.91			75.89
t	\-\-\-	\-\-\-	100δ	\-\-\-	\-\-\-	\-\-\-
−10	−1.80	(−1.9₃)	(−1.9)	(−1.9)	−2.0	−2.1
−5	−0.9₁	(−0.9₇)	(−1.0)	(−0.9)	−1.0	−1.0
0	0	0	0	0	0	0
+5	+0.9₃	+0.9₀	+1.0	+1.0	+1.0	+1.0
10	1.8₈	1.9₀	1.9	1.9	+2.0	2.1
20	3.8₄	4.0₂	4.0	3.9	4.1	4.3
25	4.8₅	5.0₆	5.0	4.9	5.1	5.4
30	5.8₈	6.1₂	6.0	5.9	6.1	6.6
40	8.0₀	8.2₆	8.2	(8.0)	8.1	8.9
60	12.4₇	12.6₉	12.7	(12.3)	12.2	13.7
80	17.2₅	17.3₆	17.4	(16.8)	16.2	18.8
100	22.3₆	22.3₁	(22.4)	(21.5)	20.3	24.2
120	27.8₁	(27.2₀)	(27.7)	(26.4)	24.3	29.8

84. SURFACE TENSION

Table 226.—(Continued)

III. Richards, Speyer, and Carver [41] have proposed also the formula $\gamma_e = K(1 - T/T_c)^\alpha$, with $\alpha = 1.2$; T and T_c are, respectively, the absolute temperatures corresponding to γ_e and to the critical point of water, K is a constant. That value of α is obviously incorrect (cf. columns 3 and 4); if they recorded the cotangent instead of the tangent of the slope of the logarithmic graph, then α should be $1/1.2 = 0.83$; that value leads to the values in column 5. They represent the RSC data fairly well. The ICT data cannot be satisfactorily represented by an equation of that form, the logarithmic graph being curved; they are, however, fairly well represented by the formula $\gamma_e = K(1 + ct + dt^2)(1 - T/T_c)^{0.849}$. For convenience we may write $(1 + ct + dt^2) \equiv 1 + \epsilon = f(t)$. Then $\frac{\gamma_0 - \gamma_e}{\gamma_0}$ $(\equiv \delta) = 1 - f(t)(1 - t/t_c)^{0.849}$ where t and t_c are the centigrade temperatures corresponding to T and T_c, respectively; $t_c = 374.0\,°\text{C}$; $T_c = 647.1\,°\text{K}$.

Weinstein (We) has proposed the formula $\gamma_e = 73.49(1 - 0.001458t)\rho$ dynes/cm; whence we find $\delta(\equiv(\gamma_0 - \gamma_e)/\gamma_0) = 1 - (1 - 0.001458t)\rho/\rho_0$.

In the following tabulation, experimental data taken from the preceding section of this table are given in columns 2 and 3, and in the other columns are values computed by the formulas just given. It is obvious that only column 6 accords satisfactorily with 2; the values in 4 and 8 are entirely unsatisfactory.

1 Ref.[a]→ α→ t	2 ICT	3 RSC	4 RSC 1.2 ———	5 0.83 100δ	6 0.849 ———	7 100ϵ[b]	8 We 100δ
−10	−1.80	−2.1	−3.2	−2.2	−1.7	−0.55	
−5	−0.91	−1.0	−1.6	−1.1	−0.9	−0.27	−0.7
0	0	0	0	0	0	0	0
+5	+0.93	+1.0	+1.6	+1.1	+0.9	+0.25	+0.7
10	1.88	2.1	3.2	2.2	1.8	0.48	1.5
20	3.84	4.3	6.4	4.5	3.7	0.88	3.1
25	4.85	5.4	8.0	5.6	4.7	1.04	3.9
30	5.88	6.6	9.5	6.7	5.7	1.20	4.8
40	8.00	8.9	12.7	9.0	7.8	1.44	6.6
60	12.47	13.7	18.9	13.6	12.3	1.70	10.3
80	17.25	18.8	25.1	18.2	17.1	1.64	14.1
100	22.36	24.2	31.1	22.8	22.2	1.28	18.1
120	27.81	29.8	37.1	27.6	27.6	0.60	22.1

[a] References:
- F — Forch, C., *Ann. d. Physik (4)*, **17**, 744-762 (1905).
- ICT — See Table 225.
- M — Moser, H., *Ann. d. Physik (4)*, **82**, 993-1013 (1927).
- RSC — Richards, T. W., Speyer, C. L., and Carver, E. K.[41]
- S — Sentis, H., *Jour. de Phys. (3)*, **6**, 183-187 (1897); *Ann. Univ. Grenoble*, **27**, 593-624 (1915).
- Wa — Warren, E. L., *Phil. Mag. (7)*, **4**, 358-386 (1927).
- We — Weinstein, B., *Metron. Beitr. (Norm. Aich. Komm.)*, No. 6 (1889).

[b] From the formula $\gamma_e = K(1 + \epsilon)(1 - T/T_c)^{0.849}$; $\epsilon \equiv 0.000515t - 0.0000038\overline{7}_5 t^2$.

between solid walls, has a true viscosity that is, within experimental error (say 10 per cent), the same as for water in bulk.

Furthermore, in the experiments of R. Bulkley,[12] on the viscous flow of liquids through very fine capillaries, the effect of the walls in modifying the pertinent properties of the liquid was inappreciable, although he could have detected it had it been equivalent to the production of a stationary film only 0.03 μ thick.

All these indicate that the effect of the walls extends no farther than a very small fraction of a micron, and, consequently, will not affect the surface-tension at more distant points.

See also p. 527.

(4). In 1909, N. Bohr [13] concluded that the tension of a clean gas-liquid surface does not change after it is 0.06 sec old. More recently, R. Hiss,[14] F. Schmidt and H. Steyer,[15] and E. Kleinmann [16] have investigated younger surfaces, seeking for evidence that the tension of a newly formed surface exceeds that of an equally clean surface of greater age. All used the same general method, and found an apparent progressive decrease in the tension, equilibrium being reached after a few milliseconds. Whether the effect observed was due to an actual decrease in the tension or was a secondary phenomenon due to an unsatisfactory technique is not entirely clear. T. F. Young and W. D. Harkins [17] seem inclined to ascribe it to the latter. P. Lenard [18] has given reasons for expecting the tension to decrease as the surface ages.

More recently, the subject has been studied by E. O. Seitz [19] and by E. Buchwald and H. König.[20] The latter used a novel method. Each found that the surface-tension decreased as the surface aged, the decrease being approximately exponential.

(5). A. Grumbach and S. Schlivitch [21] report that illuminating a vapor-water surface does not change its tension. The tension of certain other liquid surfaces is changed by illumination.

(6). The few reported observations on the tension of electrified gas-water surfaces give no indication of any effect of electrification, other than those resulting from electrostatic repulsion.

(7). No one has succeeded in showing that a magnetic field has any effect upon the surface tension. H. Auer [22] has recently reported that the application of a horizontal magnetic field of 20,000 gauss to the air-water meniscus in a vertical capillary does not change the surface tension by as

[31] Harkins, W. D., and Brown, F. E., *Idem*, 41, 499-524 (1919).
[32] *Cf.* Lenard, P., v. Dallwitz-Wegener, R., and Zachmann, E., *Ann. d. Physik (4)*, 74, 381-404 (1924).
[33] Winkler, C. A., and Maass, O., *Can. J. Res.*, 9, 65-79 (1933).
[34] Krümmel, O., "Handbuch der Ozceanographie," Vol. 1, pp. 280-281, 1907.
[35] Young, T. F., and Harkins, W., *Int. Crit. Tables*, 4, 446-475 (447) (1928).
[36] Brunner, C., *Ann. d. Physik (Pogg.)*, 70, 481-529 (1847).
[37] Humphreys, W. J., and Mohler, J. F., *Phys. Rev.*, 2, 387-391 (1895).
[38] Morgan, J. L. R., and Davis, C. E., *J. Am. Chem. Soc.*, 38, 555-568 (1916).
[39] Morgan, J. L. R., and McAfee, A. McD., *Idem*, 33, 1275-1290 (1911).
[40] Ramsay, W., and Shields, J., *J. Chem. Soc. (London)*, 63, 1089-1109 (1893).
[41] Richards, T. W., Speyers, C. L., and Carver, E. K., *J. Am. Chem. Soc.*, 46, 1196-1207 (1924).

much as 10 parts in a million. He varied the temperature from 15 to 92 °C. See also, W. Johner,[23] A. Piccard,[24] O. Liebknecht and A. P. Wills,[25] G. Quincke,[26] and C. Brunner.[27]

(8). It has been stated that prolonged contact of water with such catalysts as charcoal, thoria, and platinum so modifies it that its surface-tension is increased.[28]

Molecular Surface Energy.

The molecular surface energy is $\gamma_M = \gamma(M/\rho)^{2/3}$ by definition, γ being the surface-tension of the liquid in contact with its own pure vapor, M the molecular weight of the vapor, and ρ the density of the liquid. It is the mechanical work required to increase the area of the surface by an amount equal to the area of one face of the cube of liquid of mass equal to one mole of the vapor, the temperature being maintained constant by a suitable addition of heat.

The number of molecules contained in such a cube of liquid is the same for every liquid having the same fixed ratio between its molecular weight and the molecular weight of its vapor; and the number of molecules contained in the surface layer of area equal to the face of that cube and of thickness equal to a fixed multiple of the mean distance between adjacent molecules is likewise the same for every such liquid, provided the variation in the mean distance between adjacent molecules as the surface is approached is the same function of the distance from the surface, all distances being expressed in terms of the mean distance between adjacent molecules in the body of the liquid; *i.e.*, in terms of the length of an edge of the cube. In particular, if the molecular weight of each liquid is the same as that of its vapor, and if the mean distance between adjacent molecules does not change as the surface is approached, then each such cube contains one mole of liquid, and each surface layer of area equal to one face of the cube and of thickness equal to a fixed multiple of the mean distance between adjacent molecules (*i.e.*, of thickness proportional to an edge of the cube) will contain the same number of molecules of the liquid; and enlarging the liquid surface by an area equal to the face of the cube will bring the same number of molecules from the interior of the liquid into the surface layer, whatever the liquid may be.

It is for these reasons that M/ρ may be called the molecular volume, $(M/\rho)^{2/3}$ the molecular area, and $(M/\rho)^{1/3}$ the molecular distance. Their molecular significance is independent of the nature of the substance. For the same reasons it seemed probable that fundamental relations would be discovered more readily by studying γ_M rather than γ.

---

[42] Sentis, H., *Ann. Univ. Grenoble*, **27**, 593-624 (1915).
[43] Sugden, S., *J. Chem. Soc. (London)*, **119**, 1483-1492 (1921); **121**, 858-866 (1922).
[44] Volkmann, P., *Ann. d. Physik (Wied.)*, **56**, 457-491 (1895).
[45] Schwenker, G., *Ann. d. Physik (5)*, **11**, 525-557 (1931).
[46] Bond, W. N., *Proc. Phys. Soc. (London)*, **47**, 549-558 (1935).
[47] Domke, *Wiss. Abh. Norm.-Aich. Komm.*, **3**, 1-99 (1902).

It will be noticed that, contrary to the implications of statements to be found in certain texts, neither the conception of molecular surface energy (γ_M) nor the definition of $(M/\rho)^{2/3}$ as the molecular area has anything directly to do with a sphere.

In 1886, R. Eötvös [48] propounded a number of relations based upon supposed analogies and the idea of corresponding states, culminating in the conclusion that $d\gamma_M/dt$ is a constant, independent of the nature of the substance and of the temperature. This relation he tested by means of existing data, and found it to be satisfied in many cases, the value of the derivative being quite close to -2.12 ergs per gram-molecular area and per 1 °C. Whence the negative derivative has been called the Eötvös constant; and is often denoted by k_E; $k_E \equiv -d\gamma_M/dt$. He appears to have made no attempt to test the validity of the intermediate relations upon which this final one was based, and W. Ramsay and J. Shields [49] showed that they did not accord with observations.

Inspired by the work of Eötvös, but starting from a different set of supposed analogies, Ramsay and Shields [49] reached the conclusion that γ_M should be proportional to $(t_c - t)$; that is, to the temperature measured downward from the critical temperature. But in order to secure agreement with experimental data, it was necessary to decrease that temperature by an empirically determined amount (θ), usually about 5 or 6 °C, and then, in order to make $\gamma_M = 0$ when $t = t_c$ they introduced an exponential term, giving finally the relation $\gamma_M = k_E [t_c - t - \theta(1 - 10^{-\lambda(t_c-t)})]$. Except for the exponential term, which is negligible unless $(t_c - t) < 30$ °C, this leads at once to the relation found by Eötvös; it is merely a special case of the integral of his relation.

They confirmed the conclusion of Eötvös, that k_E has essentially the same constant value for each of many substances, the average being 2.12_1. For such substances, they assumed that the molecular weight is the same in the liquid as in the vapor phase.

But for water and certain other substances k_E is much less than 2.12, and varies with the temperature, increasing as the temperature rises.[50] This they ascribed to an association of some of the molecules, causing the effective molecular weight of the liquid to be xM, where x, called by them the constant of association, is greater than unity. Then, for $(t_c - t) > 30$ °C, their relation becomes $x^{2/3}\gamma_M = k [t_c - t - \theta]$, where $k = 2.12_1$.

Whence, $x^{2/3}d\gamma_M/dt + (2/3)x^{-1/3}\gamma_M \dfrac{dx}{dt} = -k$, $k_E (\equiv -d\gamma_M/dt) = kx^{-2/3} + (2/3)(\gamma_M/x)dx/dt$, and $x^{2/3} = k \div [k_E - (2/3)(\gamma_M/x)dx/dt]$. As dx/dt is negative, the two terms in the right-hand member of the second expression conspire to make $k_E < k$. The determination of x, dx/dt, and θ in terms of k, γ_M, and k_E is, in general, difficult, if not impossible. But

---

[48] Eötvös, R., *Ann. d. Physik (Wied.)*, **27**, 448-459 (1886).
[49] Ramsay, W., and Shields, J., *Phil. Trans. (A)*, **184**, 647-674 (1893).
[50] Ramsay, W., and Shields, J., *J. Chem. Soc. (London)*, **63**, 1089-1109 (1893).

84. SURFACE TENSION

Ramsay [51] observed that, at least in certain cases, γ_M can be expressed in the form $\gamma_M = k'(t_c - t - \theta)/[1 + \mu(t_c - t)]$. If it is assumed that this θ is the same as the θ in the relation $\gamma_M x^{2/3} = 2.12(t_c - t - \theta)$, then $x^{2/3} = 2.12 [1 + \mu(t_c - t)]/k'$. This is Ramsay's procedure, and has been used in deriving the values of x_R given in Table 227. For the ICT data

Table 227.—Molecular Surface Energy of Water

See text for references, discussion of the significance of the several quantities, etc.

$\gamma_M \equiv \gamma(M/\rho)^{2/3}$; $k_E \equiv -d\gamma_M/dt$; k_{E_0} = value of k_E tabulated in ICT; $M = 18.0154$, the formula-weight of H_2O; ρ = density of water as given in Tables 93 and 255; γ = tension of the water-air, or of the water-vapor, surface, the latter for the Ramsay data and for the ICT data above 100 °C; $\Delta_{10} \equiv (\gamma_M)_t - (\gamma_M)_{t+10}$; $x_a \equiv (2.12_1/k_E)^{1.5}$; $x_{am} = (21.2_1/\Delta_{10})^{1.5}$; x_R = association constant computed by Ramsay's method (it is of little value, see text). In computing the ICT data, the values called γ_c in Table 225 have been used; in computing the Ramsay data, his observed values of γ were used (they are all much too small); k_E has been computed from the values of γ and ρ, and of their variations with t, by means of the relation given in the text.

Unit of $\gamma_M = 1$ erg per g-molecular area; of $k_E = 1$ erg per g-molecular area, per 1 °C; of $(M/\rho)^{2/3} = 1$ cm²; x_a, x_{am}, and x_R are pure numbers. Temp. = t °C.

t	k_{E_0}	k_E	x_a	γ_M	Δ_{10}	x_{am}	x_R	γ_M	Δ_{10}	x_{am}	x_R	$(M/\rho)^{2/3}$
0		0.980	3.17	519.86	9.72	3.22	1.46	503.16	8.69	3.8	1.53	6.8728
10		0.968	3.24	510.14	9.68	3.25	1.43	494.47	8.73	3.8	1.50	6.8734
20		0.972	3.22	500.46	9.77	3.19	1.40	485.74	9.50	3.3	1.47	6.8803
25	1.0₃	0.977	3.20	495.58								6.8856
30		0.984	3.15	490.69	9.93	3.12	1.37	476.24	9.96	3.1	1.44	6.8921
40		1.004	3.06	480.76	10.15	3.03	1.34	466.28	9.21	3.3	1.41	6.9079
50		1.029	2.96	470.61	10.42	2.90	1.31	457.07	10.39	2.9	1.38	6.9274
60		1.055	2.85	460.19	10.72	2.78	1.28	446.68	10.34	2.9	1.35	6.9501
70	1.0₇	1.085	2.73	449.47	11.05	2.66	1.25	436.34	10.19	3.0	1.32	6.9758
80		1.125	2.59	438.42	11.41	2.53	1.22	426.15	11.61	2.5	1.29	7.0044
90		1.160	2.46	427.01	11.80	2.41	1.19	414.54	11.50	2.5	1.27	7.0357
100	1.1₈	1.200	2.35	415.21	12.36	2.24	1.16	404.04	11.56	2.5	1.24	7.0698
110		1.248	2.22	402.85	12.76	2.14	1.13	392.48	11.87	2.4	1.21	7.1038
120	1.2₇	1.297	2.08	390.09	13.21	2.03	1.10	380.61	11.25	2.6	1.18	7.1409
130		1.344	1.98	376.88			1.09	369.36			1.15	7.1805

for water, $k' = 3.618$ ergs per g-molecular area, per 1 °C, $\theta = 58.2$ °C, and $\mu = 0.00321$ (°C)⁻¹; for the Ramsay data for water, $k' = 2.994$ ergs per g-molecular area, per 1 °C, $\theta = 43.7$ °C, and $\mu = 0.00258$ per 1 °C. In each case $t_c = 374.0$ °C. The constants for the Ramsay data have been completely recalculated, as his calculation was based upon $t_c = 358.1$ °C, and upon densities that differ slightly from those in Table 93 and those derivable from Table 255.

[51] Ramsay, W., *Proc. Roy. Soc. (London)*, **56**, 171-182 (1894) = *Z. physik Chem.*, **15**, 106-116 (1894).

It should, however, be realized that Ramsay's procedure is very arbitrary. There is no reason whatever for believing that the two θ's must have the same value. Rather the contrary; for the empirically determined value of the θ in the expression for γ_M for water is about 10 times as great as the value found for substances for which $k_E = 2.12$. (Ramsay found a value only about 4 times as great, but that was because he used too low a value for the critical temperature.) Furthermore, the value (x_R) so found for the association constant of water is much less than 2, while we now have other reasons for believing that the association constant of water is at least as great as 2. Consequently, little, if any, weight should be attached to the actual, numerical values of x_R. I doubt if they are, in general, worth the time required to compute them. They seem to be of no more value than the much more readily computed quantities $x_a \equiv (2.12_1/k_E)^{1.5}$ and $x_{am} \equiv \{2.12_1\tau/(\gamma_{M,t} - \gamma_{M,t+\tau})\}^{1.5}$, which may be called the coefficients (actual and mean, respectively) of apparent association. If the Eötvös-Ramsay relation applied and x were independent of t, then $x_a = x$; while if x varied with t, x_{am} would be the mean value of x over the range t to $t + \tau$. Also at those high temperatures, if such exist, at which x_a and x_{am} are independent of t, $x_a = x_{am}$ and each is actually the coefficient of association as defined by the Eötvös-Ramsay relation. The value of $k_E \equiv -d\gamma_M/dt$ may be determined either directly from a formula expressing the variation of γ_M with t, or by means of the relation $-d\gamma_M/dt = (M/\rho)^{2/3} \{(2/3)(\gamma/\rho)(d\rho/dt) - d\gamma/dt\}$.

A theoretical discussion of the surface energy of liquids has recently been published by H. Margenau.[52]

Angle of Contact.

The angle of contact (the angle including the liquid) between an air-water, or a vapor-water, surface and a glass surface covered with a film of water is zero.[53]

In general, the contact angle between a gas-liquid surface and any solid covered by a film of the liquid is zero. When the solid is not covered by a film of the liquid, the angle of contact is, in general, variable, the line of contact exhibiting a reluctance to move over the solid. (For a suggested explanation, see *Miscellanea*—p. 526.) In such cases, the characteristic angle of contact is commonly taken as the mean of the greatest and the least equilibrial angle that can be obtained; the former occurs when equilibrium has been attained at the end of an advance of the line of contact toward the uncovered ("dry") portion of the solid; the latter, at the end of an advance in the opposite direction. Angles computed from measurements made on stationary sessile drops are of little value unless they are the means of such greatest and least values.

[52] Margenau, H., *Phys. Rev. (2)*, **38**, 365-371 (1931).
[53] Young, T. F., and Harkins, W. D., *Int. Crit. Tables*, **4**, 434 (1928), on the strength of the observations of: Anderson, A., and Bowen, J. E., *Phil. Mag. (6)*, **31**, 143-148 (1916); Bosanquet, C. H., and Hartley, H., *Idem*, **42**, 456-461 (1921); Richards, T. W., and Carver, E. K., *J. Am. Chem. Soc.*, **43**, 827-847 (1921); Sentis, H., *Jour. de Phys. (3)*, **6**, 183-187 (1897).

It seems that the contact angle is sensitive to changes in the structure of the solid, F. E. Bartell, J. L. Culbertson, and M. A. Miller [54] having found that it has a relatively large value (30 to 80°) for Pyrex and silica when there are great internal strains, but a zero value when the solids have been well annealed. They found also that the value of the angle for brass, as well as for Pyrex, can be changed by compressing the solid. See also Bartell and Miller.[55]

Table 228.—Contact Angle between Air-water Surfaces and Dry Solids

θ is the angle containing the limiting wedge of liquid.

If a solid is wet, *i.e.*, is thoroughly covered with a layer of water, then $\theta = 0$.

Unit of $\theta = 1°$ of arc. Temp. $= t$ °C

Solid	t	θ	Ref.[a]
Gold		68 ± 4	BM
Platinum		63 ± 4	BM
Copper sulfide		0 [b]	DW
Azobenzene	14	77	BH, ICT, M
Apple wax		74	M
Paraffin	14	106.7	BH, ICT
Paraffin		104.6	A
Paraffin		105	TL, KB
Paraffin		110 ± 6	KB
Talc	25	86	BZ
Glyptol resin ("Glyptol 1350")	25	61	BZ
DeKhotinsky cement (hard)	25	106	BZ
Carnauba wax	25	107	BZ
Shellac	25	107	BZ
"Opal wax 20"	25	119	BZ
"Night Blue" 36% of monolayer		35 [c]	VV
Plates coated with oleic acid, see			L, ICT
Plates coated with various (65) substances, see...			N, ICT
Carbon tetrachloride (liquid), see			CA, ICT

[a] References:

A	Ablett, A., *Phil. Mag. (6)*, **46**, 244-256 (1923).
BH	Bosanquet, C. H., and Hartley, H., *Idem*, **42**, 456-461 (1921).
BM	Bartell, F. E., and Miller, M. A., *J. Phys'l Chem.*, **40**, 889-894 (1936).
BZ	Bartell, F. E., and Zuidema, H. H., *J. Am. Chem. Soc.*, **58**, 1449-1454 (1936).
CA	Coghill, W. H., and Anderson, C. O., *U. S. Bur. Mines Tech. Paper* No. **262**, 1923.
DW	DeWitt, C. C., *J. Am. Chem. Soc.*, **57**, 775-776 (L) (1935).
ICT	Young, T. F., and Harkins, W. D., *Int. Crit. Tables*, **4**, 434 (1928).
KB	Kneen, E., and Benton, W. W., *J. Phys'l Chem.*, **41**, 1195-1203 (1937).
L	Langmuir, I., *Trans. Faraday Soc.*, **15**, Pt. 3, 62-74 (1920).
M	Mack, G. L., *J. Phys'l Chem.*, **40**, 159-167 (1936).
N	Nietz, A. H., *Idem*, **32**, 255-269 (1928).
TL	Talmud, D., and Lubman, N. M., *Z. Physik. Chem. (A)*, **148**, 227-232 (1930).
VV	Voet, A., and Van Elteren, J. F., *Rec. Trav. Chim. Pays-Bas*, **56**, 923-926 (1937).

[b] If copper sulfide is ground under water, $\theta = 0$; but if the same ground surface is exposed to air, θ becomes finite.

[c] For a drop of water on a glass plate on which is an adsorbed layer of Night Blue (purest Nachtblau from Dr. Gruebber, Leipzig), $\theta = 35°$ if the adsorbed layer ≧ 36 per cent of a monolayer. On decreasing the coverage below 36 per cent, θ decreases sharply but continuously, becoming 0 (complete wetting of the plate) when the coverage is 32 per cent of a monolayer.

---

[54] Bartell, F. E., Culbertson, J. L., and Miller, M. A., *J. Phys'l Chem.*, **40**, 881-888 (1936).
[55] Bartell, F. E., and Miller, M. A., *J. Phys'l Chem.*, **40**, 889-894, 895-904 (1936).

Effect of Overlying Gas upon the Surface-tension.

The conclusions of B. Tamamushi [56] regarding the effect of the overlying gas upon the tension of the gas-liquid surface appear to be in direct conflict with those of A. Ferguson,[57] but the conflict may be less serious than it appears.

Ferguson used Jaeger's method, based upon the pressure required to blow and detach a bubble, and compared the tension when the gas is air with that when it is CO_2. In each case the pressure of the gas was very slightly greater than 1 atm. He "found that whether the liquids were gas-free or partially or completely saturated with either or both of the gases employed, the result was always the same, *viz.*: the *difference* between the surface-tensions liquid-air and liquid-CO_2 remained the same, but the *absolute* values of the surface tensions increased slightly as the liquid became more and more saturated, finally reaching a steady value." [57, p. 407]

Tamamushi used the method based upon the rise of the liquid in capillary tubes, and studied the effects of several gases upon the surface-tension of each of several liquids. His observations show that the tension of the surface separating a gas-saturated liquid from the gas itself, saturated with the vapor of the liquid, is in every case less than that of the surface separating the pure liquid from its pure vapor. Furthermore, he states that his data, all of which are for pressures not far from 1 atm, satisfy quite closely the empirical relation $(\gamma_v - \gamma_g)/\gamma_v = KC^{1/3} + b$, where γ_v is the surface-tension of the pure liquid in contact with its pure vapor, γ_g is that of the gas-saturated liquid in contact with the moist gas, C is the mass of gas dissolved in 100 units (mass) of the liquid, and K and b are constants depending upon the liquid, but not upon either the nature or the density of the overlying gas. His data indicate that for water $K = 0.0175$, $b = -0.00052$. Actually, the relation is not strictly linear, but the numerical values of K and b decrease as C becomes smaller.

According to Tamamushi's formula, γ_g decreases as C increases, whereas Ferguson states that it increases as the liquid becomes more nearly saturated with the gas. This discrepancy should be investigated with great care. It seems possible that it arises in this manner: Tamamushi did not study the variation of γ_g with the pressure of the gas, which in each case for water lay between 0.73 and 1.0 atm. Hence, his C's are, to a first approximation, proportional to the solubilities of the several gases. In fact, his data for water are more accurately represented by the relation $(\gamma_v - \gamma_g)/\gamma_v = 1.54 S^{1/3}$, where S = mass of gas dissolved in 100 units (mass) of liquid when the pressure of the gas is 1 atm, than they are by the one he gives. This relation does not conflict with Ferguson's observations. It states that, when the pressure does not vary greatly, the relative depression of the tension is proportional to the cube root of the solubility of the gas. Ferguson's statement is to the effect that for a given

---

[56] Tamamushi, B., *Bull. Chem. Soc. Japan*, 1, 173-177 (1926).
[57] Ferguson, A., *Phil. Mag. (6)*, 28, 403-412 (1914).

gas and liquid the depression decreases slightly as the amount of gas dissolved in the liquid increases. These two statements are not contradictory, but supplementary.

J. L. R. Morgan and C. E. Davis,[58] working exclusively with an air-water surface, found that saturating the water with air increased the tension of the surface. This agrees with Ferguson's observations. At 0 °C, the increase in going from an unstated initial condition to saturation is given as 0.16 per cent.[58, p. 557, ftn. 2]

In the absence of numerical data for the variation of γ_g with the pressure of the gas, it is impossible to reduce the available data to the basis of a constant pressure. Both Ferguson's and Richards and Carver's data obviously refer to a pressure that is very nearly 1 atm; Stocker's data seem to refer to a much lower pressure; the pressures given in Table 229 for Tamamushi's data have been computed from the values he gives for C.

Table 229.—Effect of Overlying Gas upon the Surface-tension of Water

Adapted, with additions, from the compilation of T. F. Young and W. D. Harkins.[59]

Remarks in the text should be considered. γ_v = tension of the surface separating the pure liquid from its pure vapor; γ_g = that of the surface separating the liquid saturated with the indicated gas from the gas saturated with the vapor of the liquid. When the gas is air, γ_a is written for γ_g. $\Delta_v \equiv (\gamma_v - \gamma_g)/\gamma_v$, $\Delta_a = (\gamma_a - \gamma_g)/\gamma_a$; that is, $\gamma_g = \gamma_v(1 - \Delta_v) = \gamma_a(1 - \Delta_a)$. p = pressure of the moist gas.

Unit of Δ = 1 per cent; of p = 1 atm. Temp. = t °C

Gas	p	t	Δ_v	Δ_a	Ref.[a]
Air	1	20	0.22		T
Air	1	20	0.027[b]		RC
CO_2	1	15		1.1	F
CO_2	0.8	18	0.83	0.61[c]	T
CO_2	low	20		1.0	S
N_2O	0.9	25.2	0.75[b]		T
H_2S	0.7	15.2	1.19[b]		T
H_2	low	20		0.0	S
Various organic gases					K

[a] References:
F   Ferguson, A.[57]
K   Kőrán, V., *Rec. trav. chim. Pays-Bas*, **44**, 466-475 (1925).
RC Richards, T. W., and Carver, E. K., *J. Am. Chem. Soc.*, **43**, 827-847 (1921).
S   Stocker, H., *Z. physik. Chem.*, **94**, 149-180 (1920).
T   Tamamushi, B.[56]

[b] This RC value of Δ_v for air is based upon their reported values, γ_v = 72.75 and γ_a = 72.73; and the values of Δ_v for N_2O and H_2S are those reported by T for the respective concentrations of 0.145 and 0.334 g of gas per 100 g of water. From these concentrations and the solubilities of the gases (1.17 and 4.52 g per kg of water per atm) the accompanying values of p have been derived. From the same reports (RC and T), Young and Harkins[59] derived the following values of Δ_v for p = 1 atm: RC, air 0.03₅; T, N_2O 0.85; T, H_2S 1.29.

[c] Computed from Δ_v = 0.22 for air, and 0.83 for CO_2.

---

[58] Morgan, J. L. R., and Davis, C. E., *J. Am. Chem. Soc.*, **38**, 555-568 (1916).
[59] Young, T. F., and Harkins, W. D., *Int. Crit. Tables*, **4**, 474 (1928).

526   PROPERTIES OF ORDINARY WATER-SUBSTANCE

Miscellanea.—For volume of the water meniscus, see Tables 286 and 287.

Floating bubbles and drops.—Bubbles and drops of water are frequently observed floating upon an air-water surface. In the case of bubbles, the wall of the bubble must be of such a nature that its tension increases as its thickness decreases, which seems to demand that it be a compound film, that it is not pure water.[60] In the case of drops, a blanket of the surrounding medium (air in the case here considered), separating the drop from the surface on which it floats, seems necessary; see L. D. Mahajan,[61] J. B. Seth, C. Anand, and L. D. Mahajan,[62] M. Katalinić,[63] W. and A. R. Hughes,[64] O. Reynolds,[64a] T. H. Hazlehurst and H. A. Neville.[65] As is well known an object that is not wetted by water may float on its surface, although much denser than water. The surface is depressed by it and the vertically upward component of the surface-tension supports it. Such seems to be the explanation of the floating of mercury droplets reported by N. K. Adam.[66]

Depression under reduced pressure.—Years ago, K. W. v. Nägeli[67] observed that when there is placed under the receiver of an air-pump a vessel of water into which dips a vertical capillary tube, and the receiver is exhausted, then, under suitable conditions, the meniscus in the capillary is depressed. He showed that this depression is in large part due to the excess of the existing vapor pressure over the meniscus above that over the surface in the large vessel, this excess being due to the resistance encountered by the vapor in streaming through the tube. In many cases, this seemed sufficient to account for the observations. But under certain conditions, especially when the pump was worked rapidly, the depression quite significantly exceeded all that could be accounted for by such difference in the vapor pressure. He also observed that the meniscus in the capillary descended at the same rate as that in a second capillary of the same size, but closed below, and filled to the same distance from the top. This, in connection with the well-known tendency for a stationary meniscus to become stuck to the tube, led him to suggest that as the boundaries of a liquid are approached the molecules assume an arrangement that is more orderly and less mobile than that in the interior. The liquid is thus enclosed in a relatively immobile sheath many molecules thick. The adherence of the sides of the sheath to the walls of the capillary anchors the cap

---

[60] Cf. Lord Rayleigh's remarks in the article, "Capillary Action," in "Encyclopedia Britannica," 11th ed., vol. 5, p. 267, 1910.
[61] Mahajan, L. D., *Z. Physik,* **90**, 663-666 (1934); **84**, 676 (1933); **81**, 605-610 (1933); **79**, 389-393 (1932); *Koll. Z.,* **66**, 22-23 (1934); **69**, 16-21 (1934); *Nature,* **126**, 761 (1930); **127**, 20 (erratum) (1931); *Phil. Mag. (7),* **10**, 383-386 (1930).
[62] Seth, J. B., Anand, C., and Mahajan, L. D., *Phil. Mag. (7),* **7**, 247-253 (1929).
[63] Katalinić, M., *Z. Physik,* **38**, 511-512 (1926); *Nature,* **127**, 627-628 (1931).
[64] Hughes, W. and A. R., *Nature,* **129**, 59 (1932).
[64a] Reynolds, O., "Papers on Mechanical and Physical Subjects," vol. **1**, 413-414, Cambridge Univ. Press, 1900 ← *Proc. Manchester Lit. Phil. Soc.,* **21**, 1-2 (1882).
[65] Hazlehurst, T. H., and Neville, H. A., *J. Phys'l Chem.,* **41**, 1205-1214 (1937).
[66] Adam, N. K., *Nature,* **123**, 413 (1929).
[67] v. Nägeli, K. W., *Sitz-Ber. Bayer. Akad. Wiss. München,* **1866 I**, 353-376, 473-492, 597-627 (1866).

forming the meniscus, and as the outer molecules evaporate from the cap a corresponding number is added to its interior face, and thus the cap may be slowly depressed down the tube, perhaps to a point below any that can be accounted for by the vapor pressure alone. The original articles should be studied, and the subject reinvestigated.

Transition layers.—W. D. Harkins and H. M. McLaughlin [68] have published values for the thickness of the hypothetical film of pure water that forms the air-liquid surface of an aqueous solution of NaCl. The values given vary from 4.0 to 2.3A, depending upon the concentration (1A = 10^{-8} cm). But F. Lark-Horovitz and J. E. Ferguson [69] have reported observations that indicate that the surface layer of such a solution is not pure water, but a solution much more dilute than the body of the liquid.

From a study of the reflection of polarized light by a water surface, C. V. Raman and L. A. Ramdas[70] concluded that the transition layer between water in bulk and its overlying vapor is 5.3A. They quote the late Lord Rayleigh as having inferred from similar observations a thickness of 3.0A. Similar observations by J. H. Frazer [71] on the reflection from a glass surface covered with various amounts of adsorbed water indicate that the transition from a glass surface to a water surface is complete when the thickness of the adsorbed water is 3A.

Surface films.—Interesting summaries of the properties of foreign films upon the surface of water have been published by A. Marcelin,[72] and by H. E. Devaux.[73]

Certain data for films of water adsorbed on solids, and estimates of the thickness of the layer of water that may be modified by the action of an adjacent solid, have been considered already (p. 513). For additional observations bearing on the subject, see B. Derjaguin,[74] T. Ihmori,[75] J. M. Macaulay,[76] S. Procopiu,[77] B. H. Wilsdon, D. G. R. Bonnell, and M. E. Nottage.[78]

Relations between the surface-tension and other properties.—Among the various suggested relations between the surface-tension and other properties of a liquid may be mentioned the empirical ones announced by P. Walden,[79] by R. K. Sharma,[80] and by D. Silverman and W. E. Roseveare,[81] and the theoretical one by S. C. Bradford.[82]

[68] Harkins, W. D., and McLaughlin, H. M., *J. Am. Chem. Soc.,* **47**, 2083-2089 (1925).
[69] Lark-Horovitz, F., and Ferguson, J. E., *Phys. Rev. (2),* **42**, 907 (A) (1932).
[70] Raman, C. V., and Ramdas, L. A., *Phil. Mag. (7),* **3**, 220-223 (1927).
[71] Frazer, J. H., *Phys. Rev. (2),* **33**, 97-104 (1929).
[72] Marcelin, A., "Solutions superficielles, fluides à deux dimensions, et stratifications monomoléculaires," 163 pp., 86 Figs., bibliog. of 90 entries. Presses Univ. de France, Paris, 1931.
[73] Devaux, H. E., *Jour. de Phys. (7),* **2**, 237-272 (1931). Bibliog. of 95 entries.
[74] Derjaguin, B., *Nature,* **138**, 330-331 (L) (1936).
[75] Ihmori, T., *Ann. d. Physik (Wied.),* **31**, 1006-1014 (1887).
[76] Macaulay, J. M., *Nature,* **138**, 587 (L) (1936).
[77] Procopiu, S., *Compt. rend.,* **202**, 1371-1373 (1936).
[78] Wilsdon, B. H., Bonnell, D. G. R., and Nottage, M. E., *Trans. Faraday Soc.,* **31**, 1304-1312 (1935); **32**, 570 (1936); *Nature,* **135**, 186-187 (L) (1935).
[79] Walden, P., *Z. physik. Chem.,* **65**, 129-225, 257-288 (1908); **66**, 385-444 (1909); *Z. Elektrochem.,* **14**, 713-724 (1908).
[80] Sharma, R. K., *Chem. Abs.,* **20**, 2267 (1926) ← *Quart. Jour. Indian Chem. Soc.,* **2**, 310-311 (1925).

The most important, however, is that which S. Sugden [83] called the "parachor" and denoted by P. He defines it by means of the formula $P = \gamma^{1/4} M/(D - d)$, where γ = surface-tension of the liquid-vapor surface, M = molecular weight of the vapor, and D and d are the densities of the liquid and vapor, respectively. For many substances, P is almost independent of the temperature, and its value can be obtained by summing certain constants, each characteristic of a chemical element or of a type of structure that enters into the make-up of the molecule. Sugden has stated that within 3 per cent $P = 0.78 V_c$, where V_c is the critical volume of a gram-mole. See also, N. K. Adam.[84]

Movements of bubbles.—Although the phenomena accompanying the motion of bubbles in a liquid depend in part upon the surface tension of the liquid-gas boundary, their discussion is scarcely pertinent to the present compilation. Several papers treating of them have, however, happened to come to the compiler's attention.[85]

Voltaic effects.—Voltaic effects suggesting that the capillary layer or, more exactly, the portion of the liquid that is elevated in a tube, as a result of capillary action, differs voltaically from the liquid in bulk have been reported by E. Torporescu.[86]

Stability of doubly gas-faced liquid films.—A clean, uncoated film of a pure liquid, each face in contact with a gas, is incapable of static equilibrium. A dynamic equilibrium, resulting from localized evaporation and streaming, is however possible, and has been recently considered by H. A. Neville and T. H. Hazlehurst, Jr.[87]

85. Solubility of Selected Gases in Water

Definitions and Symbols.

1. By the *solubility* of substance A in substance B is meant the amount of A that must be added to a unit amount of B in order to produce a solution that will be in equilibrium with an excess of A under the existing conditions. Such a solution is said to be *saturated* with A under those conditions. The definition of solubility is not concerned with the state of A after solution has occurred, although that will, in general, affect the magnitude of the solubility, and may perhaps give rise to special effects.

[81] Silverman, D., and Roseveare, W. E., *J. Am. Chem. Soc.*, **54**, 4460 (1932).
[82] Bradford, S. C., *Phil. Mag.* (6), **48**, 936-947 (1924).
[83] Sugden, S., *J. Chem. Soc. (London)*, **125**, 1177-1189 (1924).
[84] Adam, N. K., "The Physics and Chemistry of Surfaces," Oxford Univ. Press, 1930.
[85] Bryn, T., *Forsch. Gebiete Ingenieurw.*, **4**, 27-30 (1933); Hoefer, K., *Mitt. Forsch.-arb. Gebiete Ingenieurw.*, **138**, 1-47 (1913); Miyagi, O., *Tech. Rep. Tôhoku Imp. Univ. (Sendai)*, **5**, 135-167 (1925); Schriever, W., and Evans, J. F., *Phys. Rev.* (2), **43**, 372 (A) (1933); Bošnjaković, F., *Techn. Mechan. u. Thermod.*, **1**, 358-362 (1930); Meyer, J., *Z. Elektroch.*, **15**, 249-252 (1909); Hattori, S., *Rep. Aeronaut. Res. Inst., Tôkyô Imp. Univ.*, **9**, No. 115, 161-193 (1935); Laby, T. H., and Hercus, E. O., *Proc. Phys. Soc. (London)*, **47**, 1003-1008-1011 (1935); O'Brien, M. P., and Gosline, J. E., *Ind. Eng. Chem.*, **27**, 1436-1440 (1935).
[86] Torporescu, E., *Bull. Math. et Phys., Bucarest*, **6**, 40-41 (1936); *Compt. rend.*, **202**, 1672-1674 (1936).
[87] Neville, H. A., and Hazlehurst, T. H., Jr., *J. Phys'l Chem.*, **41**, 545-551 (1937).
[88] *Int. Crit. Tables*, **3**, 254-255 (1928).
[89] Loomis, A. G., *Int. Crit. Tables*, **3**, 255-261 (1928).

85. SOLUBILITY OF GASES

2. By the *coefficient of absorption* of a gas in a liquid is meant the rate at which the solubility of the gas in it increases with the partial pressure of the gas in the overlying gaseous phase, the temperature remaining

Table 230.—Mean Coefficients of Absorption (0 to P) of Selected Gases by Water

With but few exceptions the values in this table have been derived from the mean molecular coefficients given in Table 232. The two are connected through the relations

$$r = xM_g/(1-x)M_l \text{ and } f = xM_g/(M_l - (M_l - M_g)x)$$

where M_g and M_l are the formula-weights of the gas and of H_2O, respectively, x = mole-fraction of the gas in the solution, r and f = mass of gas dissolved in unit mass of water, and in unit mass of the solution, respectively (see p. 535+). The difference in the units of pressure used in the two tables must be considered. For all the gases in this table, excepting CO_2 and NH_3, x is negligibly small in comparison with unity, and consequently $r = f$.

Unless otherwise indicated, the values here given apply when the partial pressure (P) of the gas is 1 atm; and, excepting NH_3, it is probable that r/P and f/P are independent of P if that does not much exceed 1 atm. But no data are available for the radioactive gases except at very low pressures. For variation of the coefficients with the pressure, see Tables 233 and 234.

Units: Of r/P [f/P] = 1 mg of gas per kg of water [of solution] per atm; of λ = 1 cm³ of gas under existing conditions per cm³ of solution. Temperature = t °C; M_l = 18.0154.

I. **Noble gases.**

An Actinon (actinium emanation). At exceedingly low partial pressures λ = 2.[93]
Rn Radon (radium emanation). See end of Section II.
Tn Thoron (thorium emanation). At exceedingly low partial pressures, λ = 1.[93] See also A. Klaus.[94]

Gas→ M_g→ t	Argon 39.91				Helium 4.00	
				$r/P = f/P$		
0	94.1	103.0		(1.69)	1.725	
5		90.46		1.647		
10	74.8	80.60		1.606	1.768	
15		73.07	66.1	1.571		1.59
20	62.3	67.04	59.9	1.539	1.777	1.57
25		62.18	56.0	1.504		1.55
30	53.5	58.15	53.3	1.476	1.791	1.54
35		54.58				
40	48.5	51.57			1.836	
45		48.84				
50		46.35			1.932	
Ref.₁[a]	W97, W06	E	L	CEB	A[b]	L
Ref.₂[a]	V27	V27		V27	V27	
a_0[c]		−110			−110	
a_1[c]		+1.38			+1.38	
a_2[c]		−9.7			−10.5	

[90] Metschl, J., *J. Phys'l Chem.*, **28**, 417-437 (1924).
[91] Manchot, W., *Z. anorg. allgem. Chem.*, **141**, 38-4 (1924).

Table 230.—(Continued)

Gas→ M_g→ t	Krypton 82.9	— Neon — 20.2		Xenon 130.2
		$r/P = f/P$		
0	370.0	9.8		1406
5				1186
10	295.5	10.8		1005
15			9.7	862
20	231.5	26.1	9.4	741
25			9.1	
30	188.8	49.1	8.9	570
40	160.2	95.2		467
45.45				424
50	141.6[d]	89.7		
Ref.[1a]	A	A[b]	L	A
Ref.[2a]	V27	V25, V27		V27
a_0[c]	−108	−110		−103
a_1[c]	+1.35	+1.38		+1.30
a_2[c]	−9.4	−9.9		−9.0

II. Simple gases.

(For noble gases, except radon, see Section I.)

1 Gas→ M_g→ t	2	3 H_2 Hydrogen 2.0154	4	5 Argon-free N_2[e] 28.016	6	7 "Atmospheric" N_2[f] 28.016
			$r/P = f/P$			
0	1.936	1.931[e]	28.99	28.90	29.50	29.43
1	1.919	1.912			28.81	28.72
2	1.902	1.893			28.14	28.03
3	1.885	1.874			27.49	27.34
4	1.870	1.856			26.89	26.69
5	1.853	1.839	25.85	25.64	26.29	26.06
6	1.838	1.821			25.73	25.47
7	1.823	1.805			25.18	24.87
8	1.808	1.789			24.65	24.31
9	1.793	1.773			24.16	23.78
10	1.780	1.759	23.29	22.87	23.70	23.27
11	1.765	1.745			23.24	22.80
12	1.751	1.732			22.82	22.34
13	1.737	1.718			22.41	21.89
14	1.725	1.706			22.03	21.50
15	1.712	1.694[e]	21.30	20.72	21.66	21.08
16	1.701	1.682			21.30	20.70
17	1.688	1.672			20.96	20.34
18	1.676	1.663			20.62	19.99
19	1.665	1.649			20.32	19.66
20	1.654[g]	1.638[g,e]	1.638[g]	19.01[g]	20.01	19.35
21	1.644	1.626			19.73	19.05
22	1.634	1.616			19.45	18.77
23	1.624	1.604			19.19	18.48
24	1.614	1.592			18.93	18.22
25	1.605[g]	1.582[g,e]	1.582[g]	17.68[g]	18.67	17.98
26	1.597	1.572			18.44	17.72
27		1.562			18.20	17.49
28		1.552			17.98	17.27
29		1.543			17.76	17.05
30		1.535	17.26	16.53	17.56	16.83
35		1.508	15.97	15.52	16.60	15.79

85. SOLUBILITY OF GASES

Table 230.—(Continued)

1 Gas→ t	2 H₂ Hydrogen	3	4 Argon-free N₂[g]	5	6 "Atmospheric" N₂[f]	7
			$r/P = f/P$			
40		1.490	15.53	14.68	15.77	14.92
45		1.475	14.79	14.02	15.01	14.26
50		1.463	14.12	13.55	14.34	13.79
60		1.463				13.01
70		1.47				12.49
80		1.48				12.32
90		1.49				12.30
100		1.50				12.26
Ref₁[a]	Ti	W91a, W92b			Fox	W91b, W92b
Ref₂[a]		W92a				W92a

Gas→ M_g→ t	O₂ Oxygen[h] 32.000			O₃ Ozone[i] 48.000	Rn Radon[j] 222.	
			$r/P = f/P$			
0		69.82	70.29	1373.	5049.	5130.
1		67.94	68.37			4920.
2		66.15[e]	66.55			4710.
3		64.42	64.82			4530.
4		62.78	63.19			4360.
5		61.21	61.64	1220.	4080.	4190.
6		59.68	60.17			4020.
7		58.26	58.76			3870.
8		56.88	57.41			3720.
9		55.57	56.12			3580.
10		54.30	54.87	1075.	3339.	3450.
11		53.11	53.69			3320.
12		51.96	52.55			3180.
13		50.85	51.47			3070.
14		49.81	50.44			2960.
15		48.80	49.43	924.		2860.
16		47.85	48.47			2750.
17		46.93	47.55			2650.
18		46.04	46.67	974.[k]		2560.
19		45.20	45.80			2470.
20	44.36	44.68[e]	44.98	709.	2360.	2380.
21	43.55	43.62	44.20			2300.
22	42.76	42.86	43.45			2240.
23	42.05	42.12	42.72			2160.
24	41.24	41.42	42.02			2100.
25	40.53	40.73[e]	41.35	583.		2040.
26	39.87	40.08	40.70			1980.
27	39.20	39.41	40.07			1920.
28	38.57	38.78	39.47			1850.
29	37.99	38.13	38.89			1830.
30	37.40	37.50	38.33	445.3	1780.	1780.
35	35.05		35.80	323.6		1580.
40	33.18		33.67	222.3	1410.	1420.
45	31.53		31.88	149.1		1300.
50	30.20		30.37	97.1[l]	1180.	1200.
60	28.26			0	1050.	1030.
70	26.77				960.	920.
80	25.84				880.	860.
90	25.42				840.	830.
100	25.31				800.	820.[m]
Ref₁[a]	W91b	W89, W92b	Fox	Mt	StM	Sz
Ref₂[a]	W06	W92a, W88				

Table 230.—*(Continued)*

III. Gaseous compounds.

In all cases the partial pressure (P) of the gas either was very nearly 1 atm or lay in a region, including 1 atm, throughout which r/P and f/P are independent of P. Excepting NH_3, $(\partial r/\partial P)_t$ and, consequently, $(\partial f/\partial P)_t$ and $(\partial x/\partial P)_t$, are constant from a very low pressure to a value of P that is well above 1 atm.

1 Gas→ M_g→ t	2 CO[n] 28.000 r/P	3 CO$_2{}^o$ 44.000 r/P	4	5 CO$_2{}^o$ 44.000	6 f/P	7 $10^{-6}r/P$	8 NH$_3$[p] 17.031 $10^{-6}f/P$
0	44.19	3364.		3352		0.895	0.472
1	43.17	3232.		3221		0.872	0.464
2	42.17	3109.		3100		0.846	0.457
3	41.19	2998.		2989		0.820	0.450
4	40.25	2892.		2884		0.799	0.443
5	39.34	2789.		2781		0.778	0.438
6	38.45	2696.		2689		0.759	0.431
7	37.60	2605.		2599		0.741	0.425
8	36.76	2518.		2511		0.720	0.419
9	35.97	2431.		2425		0.703	0.412
10	35.19	2346.		2342		0.685	0.407
11	34.46	2267.		2262		0.668	0.401
12	33.76	2198.		2192		0.652	0.394
13	33.07	2127.		2123		0.636	0.388
14	32.42	2060.		2056		0.620	0.383
15	31.80	1999.		1994		0.604	0.376
16	31.20	1976.		1935		0.587	0.369
17	30.62	1880.		1877		0.572	0.364
18	30.05	1824.		1821		0.560	0.359
19	29.56	1770.		1768		0.548	0.353
20	20.02[g]	1722.	1648.	1718	1645	0.534	0.348
21	28.55	1672.		1669		0.522	0.342
22	28.22	1623.	1582.	1621	1580	0.510	0.337
23	27.65	1579.		1576		0.500	0.333
24	27.24	1536.	1512.	1534	1509	0.488	0.328
25	26.84[g]	1493.[g]		1492		0.479	0.324
26	26.45	1452.	1439.	1451	1438	0.465	0.320
27	26.08	1419.		1417		0.456	0.313
28	25.73	1384.	1367.	1382	1366	0.448	0.309
29	25.40	1346.		1345			
30	25.08	1317.	1294.	1312	1293		
32			1224.		1222		
34			1151.		1151		
35	23.59	1170.		1168			
40	22.35	1049.		1048			
45	21.32	951.		949			
50	20.42	862.		862			
60	18.91	717.		716			
70	18.40						
80	18.38						
90	18.4						
100	18.4						
Ref.$_1$[a]	W01	Bohr	Ku	Bohr	Ku		
Ref.$_2$[a]	W92a	Bu	Bu	Bu	Bu		

85. SOLUBILITY OF GASES

Table 230.—(Continued)

[a] References: See end of this Section.

[b] The values in this column increase with t, suggesting an error in the observations.

[c] Coefficients in the formula $M_g{}^{-0.5}\log_{10}\lambda = (273a_1 - a_0)/T + 2.3a_1\log_{10}T + a_2$, where T °K is the absolute temperature.[95]

[d] For krypton at 60 °C, $r/P = 132.2$.

[e] This is argon-free atmospheric nitrogen (see next note). The values in columns 4 and 5 have been respectively computed from those of columns 6 and 7.[96] Apparently the only corresponding data available in 1927 for chemically prepared nitrogen were those reported by Braun (1900), Just (1901), and Adeney and Becker (1919); see Coste, who seems to have overlooked Just's work. None since that date have come to the attention of the compiler. Of these, Loomis included only Just's, which are here given in footnote g. Some of the values by the others are here given, subscripts indicating the gas; they do not accord well with those given in the main table.

	Braun					Adeney and Becker			
t	0	15	20	25	t	2.5	3.5	20	25
$(r/P)_H$	2.130	1.854	1.715	1.576	$(r/P)_O$	63.29	....	43.46	40.51
$(r/P)_N$	27.22	22.41	20.31	17.94	$(r/P)_N$	....	27.82	19.90	18.78

[f] Atmospheric nitrogen is the residue of air from which O_2, CO_2, NH_3, and H_2O have been removed.

[g] Just's data (J) yield the following values of r/P, all included in Loomis's compilation (ICT). The nitrogen was prepared chemically, not from the atmosphere.

t	H_2	N_2	CO	CO_2
20	1.67₅	19.90	30.18	
25	1.64₀	18.77	27.59	1488

And Findlay et al. (F) give for CO_2 at 25 °C the value 1480 for the range 270 to 1350 mm-Hg.

[h] The data in column 2 were obtained by measuring the volume of O_2 absorbed by a given volume of water; those in 3, by titrating the O_2 contained in air-saturated water.

[i] The data for ozone refer probably to a partial pressure of about 50 mm-Hg. By an indirect computation, Rothmund (Ro) found 1057 at 0 °C.

[j] The data for radon refer to an exceedingly low partial pressure, of the order of 0.01 μ-Hg. R. W. Boyle[97] has shown that, for Rn dissolved in water, x/P is constant over the range 0.8 to 0.008 μ-Hg. There seems to be no data for the solubility of Rn at higher pressures. S. Meyer (StM) stated that the values in his table are based on the observations of Boyle, H, Ko, Mache, Ra, and Tr. The coefficients of Valentiner's formula (see note c) for Rn are $a_0 = -95$, $a_1 = 1.20$, $a_2 = -8.25$.

[k] This value for O_3 is from (FT).

[l] At 55 °C, $r/P = 56.0$ for O_3.

[m] This value for Rn is for $t = 97$ °C.

[n] For CO, $r/P = f/P$, essentially.

[o] The values in columns 3 and 4, respectively, represent the same data as those in 5 and 6.

[p] The values in columns 7 and 8 represent the same data and apply solely to $P = 1$ atm. The ICT formulas by means of which they were computed ($\log_{10}K = 0.05223A/T + B$, $A = -937_6$ and $B = 4.98_7$ from 0 to 10 °C, and $A = -1074_6$ and $B = 5.23_8$ from 14 to 28 °C) seem to have been derived from the observations of G. Calingaert and F. E. Huggins, Jr.,[98] E. Klarmann,[99] F. M. Raoult,[100] and A. Smits and S. Postma.[101]

---

[92] Findlay, A., and associates, *J. Chem. Soc. (London)*, **97**, 536-561 (1910); **101**, 1459-1468 (1912); **103**, 636-645 (1913); **107**, 282-284 (1915).

[93] v. Hevesy, G., *Physik. Z.*, **12**, 1214-1224 (1911)—*J. Phys'l Chem.*, **16**, 429-450 (1912).

[94] Klaus, A., *Physik. Z.*, **6**, 820-825 (1905).

[95] Valentiner, S., *Z. Physik*, **42**, 253-264 (1927).

[96] See Loomis, *Int. Crit. Tables*, **3**, 256 (1928).

[97] Boyle, R. W., *Phil. Mag. (6)*, **22**, 840-854 (1911).

[98] Calingaert, G., and Huggins, F. E., Jr., *J. Am. Chem. Soc.*, **45**, 915-920 (1923).

[99] Klarmann, E., *Z. anorg. allgem. Chem.*, **132**, 289-300 (1924).

constant. This quantity does not appear in the tabulations except incidentally where it happens to coincide with the mean coefficient. That frequently occurs.

3. By the *mean coefficient of absorption* of a gas in a liquid when the partial pressure of the gas is increased from p_1 to p_2 is meant $(S_2 - S_1)/(p_2 - p_1)$, where S_1 and S_2 are the solubilities at p_1 and p_2, respectively, the temperature being the same in each case. The quantity commonly tabulated and called (unfortunately) the coefficient of absorption is the

Table 231.—Solubility of Air in Water
Adapted from the compilation by A. G. Loomis.[102]

(See also, Table 232, part III. For solubility of air in sea-water, see Table 235.)

By "air" is meant atmospheric air that has been freed from CO_2 and NH_3. r = mass of gas that is contained in a unit mass of water when in equilibrium with air at a pressure of 1 atm and at the indicated temperature.

v_0 = the volume under standard conditions (0 °C and 1 atm) of the gas that is contained in a unit volume of water when in equilibrium with air at a pressure of 1 atm and at the indicated temperature; $v_0 = r\rho_1/\rho_0$.

Σ = total volume of gas (0 °C and 1 atm) = sum of the corresponding values in the next two preceding columns.

The molecular weight of air has been taken as 28.96; the density of O_2 under standard conditions, as 1.4290_4 g per liter, and that of "atmospheric" N_2 (including the inert gases), as 1.2568 g per liter.[103]

Unit of r = 1 mg gas per kg water; of v_0 = 1 ml (0 °C, 760 mm-Hg) of gas per liter of water; pressure = 1 atm. Temp. = t °C

Ref.,[a] → Gas → t	W01 Air[b]	Air[c]	W04 O_2	N_2, A, etc.	Σ	W04 O_2	N_2, A, etc.	Σ	$\left(\dfrac{100\ O_2}{\Sigma}\right)_{v_0}$
			r				v_0		
0	37.27	37.95	14.56	23.87	38.43	10.19	18.99	29.18	34.91
1	36.34	36.99	14.16	23.26	37.42	9.91	18.51	28.42	34.87
2	35.40	36.02	13.78	22.68	36.46	9.64	18.05	27.69	34.82
3	34.53	35.14	13.42	22.12	35.54	9.39	17.60	26.99	34.78
4	33.71	34.28	13.06	21.59	34.65	9.14	17.18	26.32	34.74
5	32.91	33.44	12.73	21.08	33.81	8.91	16.77	25.68	34.69
6	32.12	32.68	12.40	20.59	32.99	8.68	16.38	25.06	34.65
7	31.37	31.93	12.10	20.11	32.21	8.47	16.00	24.47	34.60
8	30.66	31.22	11.80	19.66	31.46	8.26	15.64	23.90	34.56
9	29.97	30.53	11.52	19.23	30.75	8.06	15.30	23.36	34.52
10	29.32	29.88	11.25	18.82	30.07	7.87	14.97	22.84	34.47
11	28.71	29.26	10.99	18.42	29.41	7.69	14.65	22.34	34.43
12	28.11	28.66	10.75	18.05	28.80	7.52	14.35	21.87	34.38
13	27.53	28.10	10.51	17.68	28.19	7.35	14.06	21.41	34.34
14	27.00	27.54	10.28	17.33	27.61	7.19	13.78	20.97	34.30
15	26.49	27.03	10.07	17.00	27.07	7.04	13.51	20.55	34.25
16	25.99	26.53	9.86	16.67	26.53	6.89	13.25	20.14	34.21
17	25.51	26.06	9.66	16.36	26.02	6.75	13.00	19.75	34.17
18	25.07	25.59	9.46	16.07	25.53	6.61	12.77	19.38	34.12
19	24.61	25.16	9.28	15.78	25.06	6.48	12.54	19.02	34.08

[100] Raoult, F. M., *Ann. de chim. et phys.* (5), **1**, 262-274 (1874).
[101] Smits, A., and Postma, S., *Proc. Akad. Wet. Amsterdam*, **17**, 182-191 (1914).
[102] Loomis, A. G., *Int. Crit. Tables*, **3**, 255-261 (257-258) (1928).

85. SOLUBILITY OF GASES

Table 231.—*(Continued)*

Ref.[a]→ Gas→ t	W01 Air[b]	Air[c]	O₂	W04 N₂, A, etc.	Σ	O₂	W04 N₂, A, etc.	Σ	$\left(\dfrac{100\,O_2}{\Sigma}\right)_{v_0}$
			r				*v₀*		
20	24.22	24.74	9.11	15.51	24.62	6.36	12.32	18.68	34.03
21	23.81	24.44	8.92	15.24	24.16	6.23	12.11	18.34	33.99
22	23.42	23.95	8.75	14.99	23.74	6.11	11.90	18.01	33.95
23	23.06	23.58	8.59	14.73	23.32	6.00	11.69	17.69	33.90
24	22.71	23.20	8.44	14.48	22.92	5.89	11.49	17.38	33.86
25	22.34	22.83	8.28	14.24	22.52	5.78	11.30	17.08	33.82
26	22.04	22.51	8.13	14.02	22.15	5.67	11.12	16.79	33.77
27	21.72	22.16	7.98	13.80	21.78	5.56	10.94	16.50	33.73
28	21.43	21.83	7.83	13.56	21.39	5.46	10.75	16.21	33.68
29	21.13	21.50	7.69	13.32	21.02	5.36	10.56	15.92	33.64
30	20.86	21.19	7.55	13.10	20.65	5.26	10.38	15.64	33.60
35	19.55								
40	18.48								
45	17.66								
50	17.00								
60	15.98								
70	15.31								
80	14.96								
90	14.86								
100	14.97								

[a] References: See end of this Section.
[b] Calculated from data for O_2 and N_2 with correction for constant amount of argon.[104]
[c] Calculated from the O_2-content of water saturated with air and of the air expelled from the saturated water by heating.

---

mean coefficient for the range 0 to p; *i.e.*, it is S/p. Throughout the range in which S/p is independent of p, the mean coefficient coincides with the coefficient itself. When S is expressed in terms of the mole-fraction (x) of the gas in solution, we shall call S/p the *mean molecular coefficient of absorption* over the range 0 to p.

4. By the *coefficient of solubility* (λ) of a gas in a liquid is meant the volume of the gas, as measured under the conditions existing in the gas phase, that is contained in unit volume of the saturated solution. This terminology essentially agrees with ordinary practice, but in *International Critical Tables* λ is called the Ostwald absorption coefficient, and the definition is so worded as to make it appear as a partition coefficient.

The solubility of a gas may be expressed in several ways, to each of which corresponds a different set of values for the several coefficients. Thus a complex and somewhat confusing terminology has arisen; this condition is aggravated by a lack of unanimity regarding the actual significance of the terms and symbols commonly employed. For this reason the definitions just given will be adhered to, and in the presentation of the data but one (λ) of the symbols commonly used for denoting the coefficients will be employed. The symbols that will be used are as follows:

[103] See *Int. Crit. Tables*, **3**, 3 (1928).
[104] See Fox, C. J. J., *Trans. Faraday Soc.*, **5**, 68-87 (1909).

$r \equiv m_g/m_l$ = ratio of the mass of the dissolved gas to the mass of the pure liquid in which it is dissolved.

$f \equiv m_g/(m_l + m_g)$ = ratio of the mass of the dissolved gas to the total mass of the solution.

$x \equiv n_g/(n_l + n_g)$ = mole-fraction of the gas in the solution = ratio of the number of gfw of the gas in solution to the sum of that number and the number of gfw of the liquid in which the gas is dissolved. The formula of the gas is to be taken as that pertinent to its pure gaseous state.

$\lambda = V_g/V_s$ = the volume, under the existing conditions of temperature and pressure, of the gas contained in unit volume of the solution.

It will be noticed that r, f, x, and λ are all measures of solubility. When the densities of the solution (ρ_s), of the pure liquid (ρ_l), and of the pure gas (ρ_g), all under the existing conditions of temperature and pressure, are known, then r, f, x, and λ can be readily interconverted, the formula weights (M_g, M_l) of the gas and the liquid being known. If r is negligible as compared with unity, then $r = f = M_g x/M_l$; r is so negligible in every case considered in this section, excepting only CO_2 and NH_3.

p_g and p_v mm-Hg, or P_g and P_v atm, are the partial pressures, in the gas phase, of the gas and of the vapor of the liquid, respectively; $p_g = 760 P_g$, $p_v = 760 P_v$.

$P_t = P_g + P_v$ = total pressure.

V_0 = volume of the dissolved gas at 0 °C and 1 atm, $V_0 \rho_0 = m_g$; if the gas were ideal and the actual temperature of the system were t °C, then $V_g = V_0(273.1 + t)/(273.1)P_g$.

V_l = volume of the pure liquid, at the existing temperature and pressure, contained in the solution, $V_l \rho_l = m_l$.

ρ_0 = density of the pure gas at 0 °C and one normal atm.

The relations connecting the symbols used in the *International Critical*

Table 232.—Mean Molecular Coefficient of Absorption (0 to p) of Selected Gases by Water

Adapted from the compilation by A. G. Loomis.[105]

x/p is the reciprocal of the quantity designated by K in Loomis's compilation; $x = n_g/(n_g + n_l)$, where n_g and n_l denote the number of formula-molecules of gas and of H_2O, respectively, contained in a given amount of the solution that is in equilibrium with the gas under the partial pressure p, the rest of the pressure arising from water-vapor with which the gas is saturated. In general, x/p varies with p; but if p does not much exceed 760 mm-Hg it is probable that x/p is independent of p for all the gases in this table except NH_3. But no data are available for Rn except at very low pressures.

Except as indicated, the values given here apply when $p = 760$ mm-Hg.

Unit of $x/p = 1$ formula-molecule of gas per 10^9 formula-molecules of (gas + H_2O) contained in the solution per mm-Hg. Temperature = t °C.

[105] Loomis, A. G., *Int. Crit. Tables*, **3**, 255-261 (1928).

85. SOLUBILITY OF GASES

Table 232.—(Continued)

I. Noble gases.

Rn, Radon (radium emanation). See end of Section II—Simple gases.

1 Gas→ t	2 Argon	3 Argon	4 Helium[a]	5 x/p	6 Krypton	7 Neon[a]	8 Xenon
0	55.9	61.16	(10)	10.22	105.8	11.5	256
5		53.73	9.76				216
10	44.4	47.87	9.52	10.48	85.5	13.0	183
15		43.40	9.31				157
20	37.0	39.82	9.12	10.53	66.2	14.8	135
25		36.93	8.91				
30	31.8	34.54	8.75	10.65	54.0	17.5	103.8
35		32.42					
40	28.8	30.63		10.88	45.8	22.9	85.0
45		29.01					77.2[b]
50		27.53		11.45	40.5	34.0	
60					37.8		
Ref.1[e]	W97, 06	E	CEB	A	A	A	A
Ref.2[e]		V27		V27	V27	V25, 27	V27

II. Simple gases.

For noble gases, excepting radon, see section I.

1 Gas→ t	2 H₂ Hydrogen	3 H₂ Hydrogen	4 N₂ Argon-free[d] x/p	5 N₂ Argon-free[d]	6 N₂ Atmospheric[e]	7 N₂ Atmospheric[e]
0	22.77	22.72	24.53	24.46	24.96	24.90
1	22.57	22.49			24.37	24.29
2	22.37	22.27			23.81	23.70
3	22.18	22.04			23.26	23.13
4	21.99	21.83			22.75	22.58
5	21.80	21.63	21.87	21.70	22.24	22.05
6	21.62	21.42			21.77	21.55
7	21.44	21.23			21.30	21.04
8	21.26	21.04			20.86	20.57
9	21.09	20.85			20.44	20.12
10	20.93	20.69	19.71	19.35	20.05	19.69
11	20.76	20.53			19.67	19.29
12	20.60	20.37			19.31	18.90
13	20.44	20.21			18.96	18.52
14	20.29	20.07			18.64	18.19
15	20.14	19.93	18.02	17.53	18.32	17.84
16	20.01	19.79			18.02	17.51
17	19.86	19.66			17.74	17.21
18	19.72	19.56			17.45	16.91
19	19.58	19.40			17.19	16.63
20	19.46[f]	19.26[f]	16.66[f]	16.08[f]	16.93	16.37
21	19.34	19.13			16.69	16.12
22	19.22	19.00			16.46	15.88
23	19.10	18.86			16.23	15.64
24	18.99	18.73			16.02	15.42
25	18.88[f]	18.61[f]	15.54[f]	14.96[f]	15.80	15.21
26	18.78	18.49			15.52	14.99
27		18.37			15.40	14.80
28		18.26			15.21	14.61
29		18.15			15.03	14.42
30		18.05	14.61	13.99	14.86	14.24

Table 232.—(Continued)

1 Gas→ t	2 H₂ Hydrogen	3 H₂ Hydrogen	4 N₂ Argon-free[d] x/p	5 N₂ Argon-free[d]	6 N₂ Atmospheric[e]	7 N₂ Atmospheric[e]
35		17.74	13.82	13.13	14.04	13.36
40		17.52	13.14	12.42	13.35	12.62
45		17.35	12.52	11.87	12.70	12.07
50		17.21	11.95	11.46	12.13	11.65
60		17.21				11.00
70		17.3				10.57
80		17.4				10.43
90		17.5				10.4
100		17.7				10.5
Ref.₁[e]	Ti	W91a, W92b			Fox	W91b, W92b
Ref.₂[e]		W92a				W92a

1 Gas→ t	2 O₂ Oxygen[g]	3 O₂ Oxygen[g]	4 O₂ Oxygen[g] x/p	5 O₃ Ozone[h]	6 Rn Radon[i]	7 Rn Radon[i]
0		51.72	52.07	678.0	539.1	548.
1		50.33	50.64			526.
2		48.99	49.30			503.
3		47.72	48.02			484.
4		46.51	46.81			466.
5		45.34	45.66	602.4	435.7	447.
6		44.21	44.57			429.
7		43.16	43.52			413.
8		42.13	42.52			397.
9		41.16	41.57			382.
10		40.22	40.65	530.8	356.5	368.
11		39.34	39.77			354.
12		38.49	38.93			340.
13		37.66	38.13			328.
14		36.86	37.36			316.
15		36.15	36.62	456.4		305.
16		35.45	35.90			294.
17		34.77	35.22			283.
18		34.10	34.57	480.8[j]		273.
19		33.48	33.93			264.
20	32.86	32.88	33.32	350.1	252.	254.
21	32.26	32.31	32.74			246.
22	31.68	31.75	32.18			239.
23	31.10	31.20	31.64			231.
24	30.55	30.68	31.12			224.
25	30.03	30.18	30.63	287.8		218.
26	29.53	29.69	30.15			211.
27	29.04	29.20	29.68			205.
28	28.57	28.73	29.24			200.
29	28.13	28.25	28.81			195.
30	27.70	27.78	28.39	219.9	190.	190.
35	25.96		26.52	160.8		169.
40	24.58		24.94	109.8	151.	152.
45	23.36		23.62	47.96[k]	126.	128.
50	22.37		22.50	73.64		139.
60	20.93			0	112.	110.
70	19.83				102.	98.
80	19.14				94.	92.
90	18.83				90.	89.
100	18.75				86.	88.[l]
Ref.₁[e]	W91b	W89, W92b	Fox	Mt	StM	Sz
Ref.₂[e]	W06	W92a				

85. SOLUBILITY OF GASES

Table 232.—(Continued)

III. Air and gaseous compounds.

In all cases here considered, the partial pressure either was very nearly 1 atm or lay in a region, including 1 atm, throughout which x/p remains essentially constant; excepting NH_3, x/p is independent of p from very low pressures up to a value well above 1 atm.

1 Gas→ t	2 Air[m]	3	4 CO x/p	5 CO_2	6	7 NH_3[n] $10^{-5}x/p$
0	30.51	31.06	37.41	1810		6.40
1	29.75	30.28	36.55	1739		6.31
2	28.98	29.48	35.70	1673		6.21
3	28.28	28.76	34.87	1613		6.11
4	27.59	28.06	34.08	1556		6.02
5	26.94	27.37	33.31	1501		5.94
6	26.30	26.75	32.55	1451		5.86
7	25.68	26.14	31.83	1402		5.78
8	25.09	25.56	31.12	1355		5.69
9	24.53	24.99	30.45	1308		5.61
10	23.99	24.46	29.79	1263		5.53
11	23.49	23.95	29.17	1220		5.45
12	23.00	23.46	28.58	1183		5.37
13	22.53	23.00	28.00	1145		5.29
14	22.10	22.54	27.45	1109		5.21
15	21.68	22.12	26.92	1076		5.13
16	21.27	21.72	26.41	1044		5.04
17	20.88	21.33	25.92	1012		4.96
18	20.52	20.95	25.44	982		4.90
19	20.14	20.59	25.02	953		4.83
20	19.82	20.25	24.57[f]	927	887	4.75
21	19.49	19.92	24.17	900		4.68
22	19.17	19.60	23.89	874	852	4.61
23	18.88	19.30	23.41	850		4.55
24	18.59	18.99	23.06	827	814	4.48
25	18.29	18.69	22.72[f]	804[f]		4.42
26	18.04	18.42	22.39	782	775	4.34
27	17.78	18.14	22.08	764		4.28
28	17.54	17.87	21.78	745	736	4.23
29	17.30	17.60	21.50	725		
30	17.07	17.35	21.23	709	697	
35	16.00		19.97	630		
40	15.13		18.92	565		
45	14.46		18.05	512		
50	13.91		17.29	464		
60	13.08		16.01	386		
70	12.53		15.58			
80	12.24		15.56			
90	12.2		15.6			
100	12.2		15.6			
Ref$_1$[e]	W01	W01	W01	Bohr	Ku	
Ref$_2$[e]			W92a			

[a] In contrast with the others, the values in columns 5 and 7 increase with t, suggesting a serious error in A's data.

[b] This xenon value refers to $t = 45.45$ °C.

[c] References: See end of this section.

[d] This is argon-free atmospheric nitrogen (see Table 230, notes e and f). Loomis states that the data for columns 4 and 5 have been respectively derived from those of 6 and 7.

Table 232.—(Continued)

a Atmospheric nitrogen is the residue of air from which O_2, CO_2, NH_3, and H_2O have been removed.

f Just's data (J) yield the following values of x/p, all included in Loomis's compilation.[105] The nitrogen was prepared chemicaly, not from the atmosphere.

t	H_2	N_2	CO	CO_2
20	19.7₂	16.84	25.55	
25	19.3₀	15.88	23.36	802

And Findlay *et al.* (F) give for CO_2 at 25 °C the value 797, p ranging from 270 to 1350 mm-Hg.

g See Table 230, note *h*.

h Probably for $p = 50$ mm-Hg. By an indirect computation, Rothmund (Ro) found for ozone $x/p = 522.2$ at 0 °C.

i The pressure for radon was of the order of 0.01 μ-Hg (see Table 230, note *j*).

j This value for O_3 is from FT.

k At 55 °C, $x/p = 26.67$ for O_3.

l This value for Rn is for 97 °C.

m The values in column 2 have been computed from the coefficients of absorption of O_2 and of N_2, with due attention to the argon content of the atmosphere (see Fox); those in column 3, from the O_2 content of water saturated with air, and from the O_2 content of the air expelled from the saturated water by boiling.

n These values have been computed by means of the formulas given by Loomis, and apply only to $p = 760$ mm-Hg (see Table 230, note *p*). For solubility of NH_3 at very low pressure, see remarks under SdW and Wijs in Table 233, note *k*.

Table 233.—Effect of Pressure on the Solubility of Gases in Water

Adapted, with additions, from the compilation by A. G. Loomis.[105]

(See also Table 234.)

$\lambda = V_g/V_s =$ the volume of gas contained in unit volume of solution in equilibrium with the overlying gas, both volumes being measured under the existing conditions of p and t.

$r = m_g/m_l =$ ratio of the mass of the dissolved gas to the mass of the pure liquid in which it is dissolved.

$f = m_g/(m_g + m_l) = r/(1 + r) =$ mass of gas per unit mass of solution.

$f_g =$ mass of the pure gas per unit mass of the gas phase.

$m = \rho_g \lambda = \rho_s f =$ mass of gas per unit volume of solution.

$x = n_g/(n_g + n_l) =$ mole fraction of the gas in the solution.

$\delta_1 \equiv (\lambda' - \lambda)/\lambda$, where $\lambda' =$ value of λ that is computed from the corresponding value of r/P in Table 230, and both λ and λ' refer to $P = 1$ atm and to the indicated temperature (t °C).

P atm and p mm-Hg = partial pressure of the gas; $p = 760P$.

$P_t =$ total pressure.

The values of α given by Loomis have been reconverted to λ's by multiplication by $(1 + t/273.1)$; see text, p. 550.

Unit of P and $P_t = 1$ atm; of $p = 1$ mm-Hg; of $r/P = 1$ g of gas per atm per kg of water; of $f/P = 1$ g of gas per atm per g of solution; of $x = 1$ formula-mole per cent; of $\lambda = 1$ cm³ of gas under existing conditions per liter of solution; of $\rho_g \lambda / P = 1$ g of gas per atm per liter of solution; of $\delta = 1$ per cent. Temperature = t °C.

85. SOLUBILITY OF GASES

Table 233.—(Continued)

Gas→	H₂ Hydrogen[a]					O₂ Oxygen[b]	
t→	19.5	23[c]	t→	25	t→	23.0	25.9
p	λ (ICT)[d]		P	1000 r/P	p	λ (ICT)[d]	
760	(17.9₇)	(17.3₆)	25	1.57	760	(29.4₀)	(28.4₆)
900	17.9₆		50	1.56	900	29.3₉	28.4₆
1000	17.9₆	17.3₆	100	1.55	1000	29.3₈	28.4₆
1500	17.9₅	17.3₅	200	1.52	1500	29.3₇	28.4₄
2000	17.9₅	17.3₅	400	1.48	2000	29.3₄	28.3₉
2500	17.9₂	17.3₄	600	1.43₅	2500	29.2₇	28.3₀
3000	17.9₀	17.3₃	800	1.40₀	3000	29.1₅	28.2₀
3500	17.8₆	17.3₁	1000	1.36₆	3500	28.9₉	28.0₈
4000	17.8₀	17.2₇			4000	28.8₁	27.9₅
4500	17.7₀	17.1₉	Formula[e]		4500	28.6₃	27.8₁
5000	17.5₈	17.0₉			5000	28.4₅	27.6₇
5500	17.4₄	16.9₇			5500	28.2₄	27.5₁
6000	17.2₉	16.8₃			6000	28.0₄	27.3₄
6500	17.1₁	16.6₈			6500	27.8₂	27.1₆
7000	16.9₃	16.5₂			7000	27.6₀	26.9₇
7500	16.7₀	16.3₃			7500	27.3₈	26.7₈
8000	16.4₆	16.1₂			8000	27.1₅	26.5₄
Ref[d]	C	C	Ref[d]	WGH₁	Ref[d]	C	C
δ₁	9.6	11.8			δ₁	8.9	8.1

Gas→	N₂ Nitrogen[f]						
t→	19.4	24.9	t→	25	50	75	100
p	λ[g] (ICT)[d]		P	1000r/P (WGH₂)[d,h]			
760	(16.2₀)	(14.9₇)	25	18.4₀	13.6₄	12.7₂	13.3₂
900	16.1₇	14.9₅	50	16.8₆	13.3₂	12.3₆	12.9₀
1000	16.1₅	14.9₄	100	15.81	12.64	11.83	12.33
1500	16.0₈	14.8₇	200	14.16	11.44	11.83	11.40
2000	15.9₉	14.8₂	300	12.76	10.56	10.06	10.61
2500	15.9₂	14.7₃	500	11.10₆	9.30₄	8.96₂	9.50₂
3000	15.8₄	14.6₄	800	9.58₈	8.16₁	7.91₁	8.38₆
3500	15.7₆	14.5₄	1000	8.94	7.65₇	7.42₀	7.82₃
4000	15.6₇	14.4₆	t→	0	25	50	80
4500	15.5₇	14.3₆	P	1000r/P (GK)[d,i]			
5000	15.4₈	14.2₇	100	18.2	13.4	12.5	11.7
5500	15.3₇	14.1₇	125	17.6	14.4	12.4	11.5
6000	15.2₅	14.0₈	200	20.0	17.2	15.6	14.2
6500	15.1₃	13.9₈	300	15.0	20.0	12.5	11.9
7000	15.0₂	13.8₉	t→	100	144	169	
			P	1000r/P (GK)[d,i]			
7500	14.8₉	13.7₉	100	11.9	12.8	13.5	
8000	14.7₅	13.6₉	125	11.7	13.0	15.2	
Ref[d]	C	C	200	14.0	16.8	20.6	
δ₁	3.1	4.2	300	12.1	14.4	16.0	

Gas→	CO		Gas→		CO₂[j]		
t→	17.7	19.0	t→	20	35	35	60
p	λ (ICT, C)[d]		P	ρ₀λ/P (ICT, Sa)[d]			
760	(27.8₈)	(27.1₉)	25	1.29₀			
900	27.8₇	27.1₉	30	1.20₀			
1000	27.8₆	27.1₇	35	1.13₆			
1500	27.8₄	27.1₃	40	1.08₇	0.66₀	0.74₇	0.4₂
2000	27.8₂	27.0₉	45	1.05₁	0.66₄	0.75₁	0.4₀
2500	27.8₀	27.0₄	50	1.01₉	0.66₈	0.75₃	0.39₅
3000	27.7₇	26.9₉	53	1.00₀			
3500	27.7₄	26.9₄	55		0.66₈	0.75₃	0.39₁
4000	27.7₀	26.8₉	60		0.67₂	0.75₇	0.38₉
4500	27.6₅	26.8₃	65		0.67₆	0.75₇	0.38₉

Table 233.—*(Continued)*

Gas→	CO		Gas→	CO₂?			
t→	17.7	19.0	t→	20	35	35	60
p	λ (ICT, C)^d		P	$\rho_g\lambda$/P (ICT, Sa)^d			
5000	27.6₁	26.7₈	70			0.75₉	0.38₉
5500	27.5₆	26.7₂	80				0.39₅
6000	27.5₀	26.6₄	90				0.40₇
6500	27.4₄	26.5₅	100				0.41₅
7000	27.3₇	26.4₃	110				0.42₉
7500	27.2₆	26.2₉					
8000	27.1₅	26.1₅					
δ₁	−7.8	−7.2					

Gas→	CO₂			Gas→	NH₃^k		
t→	60	100	100	t→	0	10^l	20^l
P	$\rho_g\lambda$/P (ICT, Sa)^d			p	r/P(ICT)^d		
40	0.4₂			300		104₉	746.₂
45	0.4₁			500		822	626.₈
50	0.41₁			700		714.₇	551.₉
55	0.40₉			900	84₂	644.₈	499.₂
60	0.40₉	0.214		1000	83₁		
65	0.41₄	0.208		1100	82₄	601.₇	462.₆
70	0.40₉	0.204		1200	81₆		
80	0.40₉	0.199₀	0.15₀	1300	81₂	571.₉	433.₉
90	0.41₇	0.199₃	0.16₇	1400	80₈		
100	0.43₁	0.199₀	0.17₇	1500	80₅	550.₁	412.₉
			0.18₄				
110	0.44₅	0.199₃	0.18₉	1600	80₂		
120		0.199₃	0.19₀	1700	79₅		
130		0.199₃	0.18₇	1800	78₀		
140		0.199₁	0.18₀	Ref.^{d}	(NP)	(Foote)	(Foote)
150		0.199₃					
160		0.196₈					

Gas→				NH₃^k			
t→	20	30^l	40	t→	0	20	40
p	r/P (ICT)^d			p	Density; unit=1 g/liter		
300		537.₅		800		878	899
500		468.₁		1000	839	868	890
700	540	421.₁		1200	828	858	881
800	517		312	1400	817	851	874
900	496	384.₇		1600	807	845	868
1000	478		293	1800	798	840	864
1100	462	358.₄		2000		836	860
1200	448		274.₁	2200		831	856
1300	436	337.₅		2400		828	852
1400	424		256.₁	2600		824	848
1500	413	319.₉		2800		822	844
1600	403		241.₃	3000		819	841
1700	392			3200		816	837
1800	382		229.₃	3400			833
1900	375			Ref.^{d}	ICT (NP)		
2000	365		219.₃	t→	25^k	t→	25^k
2100	356.₁			p	r/P	p	r/P
2200	347.₈		211.₁	3.76	1051	28.0	972
2300	339.₉			6.5	1017	31.75	987
2400	330.₀		203.₉	9.25	1018	40.6	968
2500	326.₆			9.45	1007	48.5	937
2600	320.₄		197.₆	13.5	985	120.1	804
2700	314.₄			17.19	985	173.9	769
2800	309.₃		191.₅	23.7	983		
2900	304.₅			Ref.^{d}	Br, Hou	Ref.^{d}	Br, Hou

85. SOLUBILITY OF GASES

Table 233.—(Continued)

Gas → NH$_3$k

t→	20	30^l	40
p		r/P (ICT)^d	
3000	300.$_2$		185.$_4$
3100	295.$_9$		
3200	292.$_1$		180.$_1$
3400			175.$_3$
3600			170.$_4$
Ref.^{d}	NP	Foote	NP

	−25		−30		−35		−45		−55		−65		−75	
x	p	f/P	p	f/P	p	f/P	p	f/P	p	f/P	p	f/P	p	f/P
25.6	31	6.0	23	7.0	16.5	11.3	9	21						
32.7	61.5	3.9	45	5.3	31.5	7.6	16	15	8	30	4	60		
35.9	79.5	3.3$_1$	59	4.5	43.5	6.0$_5$	21.5	12.2	11	24				
39.7	110	2.6$_5$	83	3.5$_2$	61	4.7$_8$	30.5	9.6	14	21	8.5	34		
51.5	271.5	1.40	207	1.8$_4$	155	2.4$_5$	83	4.5$_9$	42	9.0	19.5	19.5	13	29
54.7	335.5	1.21	257	1.58	194	2.09	106	3.8$_2$	54	7.5	25	16	17	24
62.7			399	1.17	305.5	1.52	172.5	2.7$_0$	92	5.1	45.5	10.2	31	15.1
66.3	597	0.83	469	1.05	360	1.38	205	2.41	110	4.2	55	9.0	38	12.9
69.5	668	0.78	525	0.98	407	1.27	233	2.23	127	4.0$_5$	65	7.9	45.5	11.4
74.6	774	0.72	610	0.91	472	1.18	274	2.04	149	3.7$_5$	77	7.3	53	10.5
77.8			650	0.90	505	1.16	295	1.98	162	3.62	84	7.0	58	10.1
84.1			736	0.86	573	1.11	334	1.90	185	3.42	96	6.6	67	9.5
88.0			774	0.86	602	1.10	355	1.87	196	3.39	102	6.6	72	9.2
100.0					699.5	1.09	409.5	1.86	226	3.36	117.5	6.4$_7$	82.5	9.2

Ref.^{d} Smits and Postma (1914), republ. by Postma (1920).

I. L. Clifford and E. Hunter (1933). The values for f/P_t have been derived from the immediately preceding values of f and P_t. t_0 and t_{100} = temperatures at which $100f = 0$ and 100, respectively.

t→	60	80	90	100	110	120	130	140	150
100f					P$_t$				
0	0.197	0.467	0.692	1.000	1.414	1.96	2.67	3.57	4.70
5	0.439	0.90	1.30	1.82	2.41	3.22	4.27	5.55	7.14
10	0.717	1.48	2.00	2.75	3.58	4.70	6.15	7.83	10.02
15	1.084	2.19	2.91	3.88	5.03	6.55	8.42	10.70	13.58
20	1.559	3.05	4.12	5.31	6.91	8.83	11.33	14.37	18.6
25		4.14	5.55	7.18	9.20	11.72	15.04	19.20	
30		5.55	7.12	9.48	11.98	15.46	19.8		

t→	60	80	90	100	110	120	130	140	150
100f					1000 f/P$_t$				
5	114.0	55.6	38.5	27.5	20.7	15.53	11.70	9.01	7.00
10	139.5	67.6	50.0	36.4	27.9	21.30	16.26	12.76	9.98
15	138.5	68.5	51.6	38.7	29.82	22.90	17.81	14.01	11.05
20	128.4	65.6	48.5	37.7	28.94	22.66	17.65	13.91	10.75
25		60.4	45.0	34.82	27.17	21.31	16.62	13.02	
30		54.1	42.1	31.64	25.04	19.41	15.15		

P$_t$→	0.02	0.2	0.5	1	2	4	6	8	10
t					100f				
170								0.3	3.4
160								3.1	6.6
150							2.6	6.4	10.1
140						1.0	5.9	10.2	14.0
130						4.2	9.6	14.1	18.0
120					0.15	7.7	13.5	18.3	22.0
110					2.9	11.5	17.6	22.4	26.3
100				0	6.2	15.5	22.0	26.7	30.8
90				2.9	10.0	19.8	26.6	31.4	35.6
80			0.3	6.1	14.0	24.4	31.4	36.4	40.6
70				10.0	18.5	29.2	36.6	41.9	46.0
60			6.2	14.4	23.4	34.2	41.9	47.5	52.0
50		2.7		19.0	28.5	39.7	47.5	54.1	60.0

PROPERTIES OF ORDINARY WATER-SUBSTANCE

Table 233.—(Continued)

Gas → P_t → t	0.01	0.2	0.5	1	NH$_3$^{k} 2 — 100f	4	6	8	10
40		5.5	15.0	24.0	33.7	45.5	54.0	62.0	70.2
30		10.0		29.1	39.3	52.5	62.0	73.5	87.0
20		15.0	26.0	34.6	45.6	60.2	74.6	94.6	
10	1.7	20.0		40.6	52.6	71.5			
0	5.0	25.3	37.2	47.3	61.4	94.7			
−10	9.5	31.8	43.1	55.3	75.0				
−20	15.0		51.2	65.4					
−30	21.0		60.0	86.0					
−40	27.5		75.3						
−50	34.0								
−60	41.5								
−70	51.0								
−79.8	62.8								
−82.9	75.5								
t_0 →	17.7	60.4	81.7	100	120.6	144.1	159.3	171	180.5
t_{100} →		−61.0	−46.3	−33.2	−18.5	−1.5	+9.7	18.5	25.3

t → 100f	60	80	90	100	110 — 1000 f_g	120	130	140	150
5	564	478	485	456	418	393	377	347	313
10	758	699	675	645	615	600	582	538	315^m
15	841	812	771	753	737	727	696	672	664
20	913	870	845	828	814	792	768	757	755
25		908	891	879	862	832	820		
30		933	924	913	890				

t → 100f	60	80	90	100	110 — 1000 f_g/P_t	120	130	140	150
5	1285	531	373	250.5	173.5	122.0	88.3	62.6	43.8
10	1057	472	338	234.5	171.8	127.6	94.6	68.7	31.4^n
15	776	371	265	194.0	146.5	111.0	82.6	62.8	48.9
20	586	285	205	156.0	117.8	89.7	67.6	52.7	40.6
25		219	160.5	122.5	93.7	71.0	54.5		
30		168	129.8	96.3	74.3				

P_t → t	0.2	0.5	1	2 — 1000 f_g	4	6	8	10
170							23	205
160							207	363
150						195	389	515
140					98	400	555	648
130					330	571	680	742
120				15	522	699	773	811
110				290	663	784	840	870
100				506	760	848	895	917
90			318	675	836	900	933	951
80		45	526	803	900	942	962	974
70			700	892	948	970	982	988
60		636	825	948	973			
50	380		902	975				
40	655	875	945	984				
30	800		970					
20	905	975	985					
10	955		992					
0	980							
−10	990							
t_0 →	60.4	81.7	100	120.6	144.1	159.3	171	180.5
t_{100} →	−61.0	−46.3	−33.2	−18.5	−1.5	+9.7	18.5	25.3

85. SOLUBILITY OF GASES

Table 233.—(Continued)

[a] See also Table 234.

[b] F. N. Speller (Sp) has given certain graphical representations of data obtained by others for the solubility of O_2 in water (ICT). Additional data have been recently reported by FTHP, dubiety = ± 5 per cent.

[c] Drucker and Moles (DM) found $\lambda = 19.6_2$ cm³/l for H_2 at 25 °C and total pressures of 560 to 730 mg-Hg (ICT). Additional data have been recently reported by FTHP, dubiety = ± 5 per cent, and by V. V. Ipatieff, S. J. Droujina-Artemovitsch, and V. F. Fikomiroff [106] for the ranges 20 to 140 atm and 0.5 to 45 °C.

[d] References: See end of this section. Sets of data taken from Loomis's compilation (ICT) are each accompanied by an additional reference to its original source.

[e] WGH₁ state that between 50 and 1000 atm their data for H_2 may be represented within their estimated error by the formula $r/\rho_0 = 24.4 + 17.2P - 0.00196P^2$ g of H_2 per kg of water, ρ_0 being the density of H_2 at 0 °C and 1 atm; but that at lower pressures, the formula requires impossibly high values of r/ρ_0.

[f] See also Table 234 and Coste's discussion of available data. DM found for N_2 at 25 °C $\lambda = 15.6_2$ cm³/l over the range 270 to 830 mm-Hg (ICT). Additional data have been recently reported by FTHP, dubiety = ± 5 per cent.

[g] Nitrogen prepared chemically; not derived from air.

[h] Purified commercial nitrogen; 99.9 per cent N_2, the rest being argon with traces of CO_2. When P is constant, r has a minimum near 70 °C.

[i] Commercial nitrogen purified by passage over Cu at 450 °C and over soda-lime.

[j] Additional data have been recently reported by FTHP, dubiety = ± 5 per cent.

[k] Scheffer and de Wijs (SdW) have stated that $p = 0.758m (1 + 0.00270m)$ mm-Hg if $t = 25$ °C and $m \gtrless 2.7$ g of NH_3 per liter of solution; $m = \rho_g\lambda$. This formula demands that r/P shall increase to a finite limit (1003) as p is indefinitely decreased. That limit is 5 per cent lower than the value found by Br for $p = 3.76$ mm-Hg (see elsewhere in this section of the table). The ICT gives Wijs as the source of the information here credited to SdW; the compiler has not seen Wijs's dissertation. For suggested explanations of the variation of r/P and of f/P with P, see G. Calingaert and F. E. Huggins, Jr.,[107] and E. Klarmann.[108]

[l] Numbers in this column are values of r/P_t not of r/P.

[m] This should probably be 515, agreeing with the corresponding value under $P_t = 10$ in the last portion of this table.

[n] This should probably be 51.4; see note m regarding the value from which this was derived.

Tables [88] with those just defined are as follows, the names used in those Tables to designate each being given after its equation:

$\alpha = V_0/V_lP_g = (\rho_l/\rho_0)(r/P_g)$, Bunsen absorption coefficient,

$\beta = \alpha P_g/P_t = (\rho_l/\rho_0)(r/P_t)$,

$\gamma = \alpha/\rho_l = r/\rho_0 P_g$, Kuenen absorption coefficient,

$\delta = 100\alpha\rho_0 = 100r\rho_l/P_g$, Raoult absorption coefficient,

$\lambda = \lambda = V_g/V_s = r\rho_s/\rho_g(1 + r)$, Ostwald absorption coefficient,

$K = p_g/x = p_g + (M_g/M_l)(\rho_l/\rho_0\alpha) = p_g(1 + M_g/M_lr)$, Henry's law constant. The unit of pressure is 1 atm for α, β, γ, δ, and 1 mm-Hg for K.

Miscellanea.

Attention must be called to a very common, but fundamentally erroneous, interpretation of the coefficients of absorption. Let $p_g = p$ mm-Hg. It is very frequently stated that $\alpha = 760V_0/V_lp$ is the volume, under

---

[106] Ipatieff, V. V., Droujina-Artemovitsch, S. J., and Fikomiroff, V. F., *Jour. de Phys.* (7), **3**, 512D (A) (1932) ← *J. chim. gén. Russe* (Russian), **1**, 594-597 (1931).

[107] Calingaert, G., and Huggins, F. E., Jr., *J. Am. Chem. Soc.*, **45**, 915-920 (1923).

[108] Klarmann, E., *Z. anorg. allgem. chem.*, **132**, 289-300 (1924).

Table 234.—Effect of Pressure on the Solubility of A, H$_2$, He, and N$_2$ in Water

Additional data for H$_2$ and N$_2$ are in Table 233.

$r = m_g/m_l$ = ratio of the mass of the dissolved gas to the mass of pure water in which it is dissolved.

P = partial pressure of the gas; P_t = total pressure.

Unit of P and P_t = 1 atm; of r = 1 mg of gas per kg of water. Temp. = t °C

Argon[a]

$t \rightarrow$ P	0.2 r/P	$P_t \rightarrow$ t	100 r/P_t	$P_t \rightarrow$ t	200 r/P_t	$P_t \rightarrow$ t	300 r/P_t
1	91.8	65.0	12.27	50	11.29	50	10.72
25	92.0	80.0	12.22	80	10.93	70	10.11
50	89.9	125.0	14.98	100	11.41	105	10.83
75	83.3	180.0	20.56	150	13.58	135	13.04
100	77.6	210.0	22.72	200	20.55	165	16.28
125	72.2[b]	240.0	25.35	240	27.38	230	25.27

Hydrogen[a]

$P \rightarrow$ t	25	50	75	100 r/P	150	200	300
0	1.928	1.920	1.919	1.914	1.898	1.882	1.839
10	1.751	1.742	1.741	1.736	1.721	1.706	1.672
20	1.617	1.608	1.607	1.604	1.587	1.572	1.545
30	1.533	1.524	1.523	1.518	1.503	1.488	1.467
40	1.486	1.477	1.476	1.472	1.457	1.443	1.422
50	1.462	1.454	1.452	1.449	1.435	1.422	1.407
60	1.457	1.455	1.451	1.447	1.435	1.424	1.406
70	1.472	1.469	1.467	1.463	1.451	1.442	1.422
80	1.511	1.507	1.503	1.498	1.489	1.477	1.458
90	1.576	1.568	1.556	1.552	1.544	1.529	1.511
100	1.659	1.639	1.624	1.622	1.606	1.593	1.564

$P \rightarrow$ t	400	500	600	700 r/P	800	900	1000
0	1.800	1.768	1.742	1.717	1.687	1.653	1.618
10	1.640	1.614	1.589	1.568	1.544	1.520	1.494
20	1.520	1.497	1.476	1.459	1.439	1.420	1.401
30	1.445	1.424	1.407	1.389	1.373	1.356	1.342
40	1.403	1.385	1.368	1.351	1.336	1.321	1.310
50	1.386	1.368	1.351	1.334	1.318	1.305	1.295
60	1.387	1.371	1.351	1.336	1.320	1.307	1.295
70	1.404	1.387	1.368	1.352	1.336	1.322	1.308
80	1.436	1.417	1.397	1.381	1.367	1.352	1.336
90	1.483	1.461	1.448	1.424	1.411	1.393	1.375
100	1.537	1.515	1.497	1.478	1.458	1.437	1.418

Helium[a]

$P \rightarrow$ t	25	50	75	100 r/P	150	200	300
0	1.650	1.669	1.660	1.649	1.631	1.612	1.573
25	1.539	1.547	1.526	1.516	1.511	1.506	1.475
50	1.589	1.587	1.582	1.576	1.548	1.548	1.518
75	1.744	1.746	1.739	1.731	1.717	1.702	1.669

$P \rightarrow$ t	400	500	600	700 r/P	800	900	1000
0	1.533	1.498	1.462	1.426	1.390	1.356	1.325
25	1.446	1.419	1.393	1.367	1.342	1.318	1.296
50	1.498	1.469	1.441	1.418	1.394	1.370	1.345
75	1.636	1.603	1.570	1.540	1.514	1.491	1.473

Table 234.—(Continued)

Hydrogen and nitrogen,[a] mixture at 0 °C and 1 atm contains 76.42 vol. % of H_2. They conclude that within a few per cent each gas is dissolved as though the other were absent. v_0 cm³ = volume (0 °C and 1 atm) of the mixed gases that is dissolved per gram of water at 25 °C; P atm = partial pressure of the hydrogen and nitrogen mixture in the gas phase; Δ = probable error of v_0.

P	50	100	200	400	600	800	1000
v_0	0.8349	1.643	3.209	6.068	8.809	11.327	13.724
Δ_0	0.0005	0.001	0.003	0.002	0.004	0.006	0.014
1000 v_0/P	16.70	16.43	16.04	15.17	14.68	14.16	13.72

[a] References:

    Argon: B. Sisskind and I. Kasarnowsky.[109]
    Hydrogen: R. Wiebe and V. L. Gaddy[110]; the effect of the H_2 upon the vapor pressure of the water was considered and allowed for by the observers.
    Nitrogen: A. W. Saddington and N. W. Krase.[111] They believe the numerical values reported by J. Basset and M. Dodé [112] to be unreliable on account of an oiling of the surface; they have not been included in this compilation.
    Helium: R. Wiebe and V. L. Gaddy.[112a]
    Hydrogen and nitrogen mixture: R. Wiebe and V. L. Gaddy.[113]

[b] For this value, $t = 3$ °C.

---

standard conditions, of the gas that "is dissolved in one volume of the solvent when the partial pressure of the gas is 760 mm," and similarly for the other coefficients. That statement is fundamentally erroneous. The observations tell us only what happens when $p_g = p$ mm-Hg. Multiplying by 760 does not increase our information, but merely changes the unit in terms of which the data are expressed. Before multiplication by 760, α is the mean coefficient per mm-Hg over the range 0 to p mm-Hg; after the multiplication it is the mean coefficient per atmosphere over the range 0 to p mm-Hg, and nothing more. To infer from the observations that the latter value of α is the volume actually dissolved when $p_g = 760$ mm-Hg involves an assumption—either that p_g itself is 760 mm-Hg, or that α is independent of p throughout the range from p mm-Hg to 760 mm-Hg. Although in many cases the latter is undoubtedly true within the limits of experimental error, it is not always true, and the tacit incorporation of such assumptions into the definitions, either of the coefficients or of the symbols, is most undesirable. It introduces such confusion as may be found in *International Critical Tables*, vol. 3, p. 256, top of column 2, where the definition of α (p. 254) requires one to infer that at a fixed temperature the amount of hydrogen dissolved when its partial pressure is 760 mm-Hg varies all the way from 16.0 to 14.7, depending upon the conditions of observation. Of course what the data actually signify is that the mean coefficient per atmosphere is 16.0 over the range 0 to 1100 mm-Hg, and 14.7 over the range 0 to 8200. In this case it is obvious that the tacit

---

[109] Sisskind, B., and Kasarnowsky, I., *Z. anorg. allgem. Chem.*, 200, 279-286 (1931).
[110] Wiebe, R., and Gaddy, V. L., *J. Am. Chem. Soc.*, 56, 76-79 (1934).
[111] Saddington, A. W., and Krase, N. W., *Idem*, 56, 353-361 (1934).
[112] Basset, J., and Dodé, M., *Compt. rend.*, 203, 775-777 (1936).
[112a] Wiebe, R., and Gaddy, V. L., *J. Am. Chem. Soc.*, 57, 847-851 (1935).
[113] Wiebe, R., and Gaddy, V. L., *Idem*, 57, 1487-1488 (1935).

Table 235.—Solubility of Atmospheric Gases in Sea-water

In the first three sections of this table are given the amounts of the indicated gases contained in a unit volume of sea-water at the indicated temperature and salinity when it is in equilibrium with air at the partial (dry) pressure of 1 atm and composed, by volume, of 79.09 per cent atmospheric nitrogen, 20.90 per cent O_2, and 0.01 per cent CO_2, the atmospheric nitrogen containing 1.185 per cent by volume of argon. They are numerically equal to the solubilities per atmosphere of partial air pressure.

In the last sections are given data for computing the "combined" CO_2 contained in sea-water in equilibrium with the air just specified. That depends upon the alkalinity (y) of the water as well as upon the partial pressure of CO_2 in the air, but is independent of the salinity. Throughout the range found in sea-water, the amount of the combined CO_2 is directly proportional to y, the factor of proportionality (b) varying with the temperature and with the amount of CO_2 contained in the air.

The values in this table have been taken directly from J. J. C. Fox.[114] They agree with the tables, of different form, given by O. Krümmel [115] and based on Fox's work. Krümmel states that N_2 is essentially in equilibrium throughout the depth of the sea, but O_2 is in excess near the surface, and in defect in the depths.

Fox gives the following equations, included in the compilation by D. F. Smith [116]: For N_2, $v_0 = 18.639 - 0.4304t + 0.007453t^2 - 0.0000549t^3 - Cl(0.2172 - 0.00718t + 0.0000952t^2)$ cm³ per liter; for O_2,[a] $v_0 = 10.291 - 0.2809t + 0.006009t^2 - 0.0000632t^3 - Cl(0.1161 - 0.003922t + 0.0000631t^2)$ cm³ per liter. In each case the air is assumed to have the composition already stated, and a partial pressure = 1 atm.

J. H. Coste [117] has discussed the solution of atmospheric nitrogen in sea-water.

$v_0 \equiv (v_t p/760)(273.1/T)$, where v_t = volume, at T °K and p mm-Hg, of the gas that is contained in unit volume of sea-water when in equilibrium with air in which the partial pressure of the gas is p mm-Hg. It is the volume that v_t would occupy at 0 °C and 1 atm if the gas were ideal.

Cl = number of grams of Cl per kg of sea-water; the amount of total salts (S) per kg of sea-water is $S = 1.812$ Cl.

x_0 = volume at 0 °C and 760 mm-Hg of the total amount of CO_2 contained in unit volume of sea-water; $x_0 = x_f + x_c$ where x_f is the corresponding volume of "free" CO_2, and x_c is that of "combined" CO_2; $x_f = pv_0$ where p atm is the partial pressure of the CO_2 in the air, and the appropriate value of v_0 is obtained from the third section of this table;

---

[a] The sign of the t^3 term is incorrectly printed as + both in Fox's paper and in Smith's compilation. Fox's formula for b, not given here, also contains some typographical error.

[114] Fox, J. J. C., *Trans. Faraday Soc.*, 5, 68-67 (1909) ← *Conseil perm. int. pour l'explor. de la mer*, Publ. de Circons. No. 41, 1907; No. 44, 1909.
[115] Krümmel, O., "Handbuch der Ozeanog.," Vol. 1, J. Engelhorn, Stuttgart, 1907.
[116] Smith, D. F., *Int. Crit. Tables*, 3, 271-283 (272) (1928).
[117] Coste, J. H., *J. Phys'l Chem.*, 31, 81-87 (1927).

85. SOLUBILITY OF GASES

Table 235.—(Continued)

$x_c = by$, y being the alkalinity, and the values of b being given in the last sections of this table.

y = alkalinity = number of mg OH ions per liter of sea-water completely deprived of CO_2.

Unit of $v_0 = 1$ cm³ (0 °C, 1 atm) per liter of sea-water; of Cl = 1 g Cl per kg of sea-water

Nitrogen

$t \to$ Cl	0	4	8	12	16	20	24	28
0	18.64	17.02	15.63	14.45	13.45	12.59	11.86	11.25
4	17.77	16.27	14.98	13.88	12.94	12.15	11.46	10.89
8	16.90	15.51	14.32	13.30	12.44	11.70	11.07	10.52
12	16.03	14.75	13.66	12.72	11.93	11.25	10.67	10.16
16	15.18	14.00	13.00	12.15	11.43	10.81	10.27	9.81
20	14.31	13.24	12.34	11.57	10.92	10.36	9.87	9.44

Oxygen

$t \to$ Cl	0	4	8	12	16	20	24	28
0	10.29	9.26	8.40	7.68	7.08	6.57	6.14	5.75
4	9.83	8.85	8.04	7.36	6.80	6.33	5.91	5.53
8	9.36	8.45	7.69	7.04	6.52	6.07	5.67	5.31
12	8.90	8.04	7.33	6.74	6.24	5.82	5.44	5.08
16	8.43	7.64	6.97	6.43	5.96	5.56	5.20	4.86
20	7.97	7.23	6.62	6.11	5.69	5.31	4.95	4.62

"Free" CO_2, per 0.0001 atm partial pressure

$t \to$ Cl	0	4	8	12	16	20	24	28
0	0.1713	0.1473	0.1283	0.1117	0.0987	0.0877	0.0780	0.0700
2	0.1690	53	63	03	73	67	73	·0.0690
4	0.1667	23	43	0.1089	59	57	66	80
6	0.1644	03	23	75	45	47	59	70
8	0.1621	0.1383	03	61	31	37	52	60
10	0.1598	63	0.1183	47	17	27	45	50
12	0.1575	43	63	33	03	17	38	40
14	0.1552	23	43	19	0.0889	07	31	30
16	0.1529	03	23	05	75	0.0797	24	20
18	0.1506	0.1283	03	0.0991	61	87	17	10
20	0.1483	0.1263	0.1083	0.0977	0.0847	0.0777	0.0710	0.0600

"Combined" CO_2

For combined CO_2, $x_c = by$ cm³ (0 °C, 760 mm-Hg) per liter of sea-water, the unit of y being that stated at head of table; p is partial pressure of CO_2 in the overlying air, unit = 0.0001 atm.

$t \to$ p	0	2	4	6	8	10	12	14
1.5	1.1673	1.1562	1.1451	1.1340	1.1229	1.1118	1.1007	1.0896
2.0	2021	920	818	717	615	513	412	1.1310
2.2	2140	1.2042	1.1943	845	746	647	549	450
2.4	2249	153	1.2051	1.1961	865	769	673	577
2.6	348	259	160	1.2066	1.1973	879	785	691
2.8	437	346	254	162	1.2070	1.1978	887	795
3.0	518	428	338	248	158	1.2068	1.1978	888
3.2	592	504	415	327	238	150	1.2062	1.1973
3.4	658	571	484	397	310	223	136	1.2049
3.6	718	633	547	461	376	290	205	119
3.8	773	689	603	519	434	350	265	180
4.0	822	739	655	572	488	405	322	238

Table 235.—*(Continued)*
"Combined" CO_2

$t \rightarrow$ p	0	2	4	6	8	10	12	14
				—— b ——				
4.5	1.2928	848	767	687	606	525	445	364
5.0	1.3019	1.2942	864	787	709	632	555	477
5.5	105	1.3032	1.2958	885	812	739	666	592
6.0	197	130	1.3062	1.2995	1.2928	1.2861	794	726
6.5	1.3306	1.3246	1.3187	1.3127	1.3067	1.3008	1.2948	1.2889

$t \rightarrow$ p	16	18	20	22	24	26	28
				—— b ——			
1.5	1.0785	1.0674	1.0563	1.0452	1.0341	1.0230	1.0119
2.0	1.1209	1.1107	1.1005	1.0904	802	701	599
2.2	352	253	154	1.1056	1.0957	1.0859	760
2.4	481	385	289	193	1.1097	1.1001	1.0905
2.6	597	504	410	316	222	128	1.1035
2.8	703	611	519	428	336	244	152
3.0	798	708	618	528	438	348	258
3.2	885	796	708	620	531	443	354
3.4	1.1962	875	788	701	614	527	440
3.6	1.2033	1.1948	862	777	691	605	520
3.8	096	1.2011	927	842	757	673	588
4.0	155	071	1.1988	1.1905	821	738	654
4.5	284	203	1.2122	1.2042	1.1961	1.1881	800
5.0	400	322	245	168	1.2090	1.2013	1.1935
5.5	519	446	373	300	226	153	1.2080
6.0	659	592	525	458	390	323	256
6.5	1.2829	1.2769	1.2710	1.2650	1.2591	1.2531	1.2471

assumptions written into the definition do not apply to the case in hand. Such confusion is not peculiar to the *International Critical Tables,* but is to be found throughout the literature treating of the absorption of gases, and in most compilations of such data.

From the relations connecting α and λ with r it is obvious that $\alpha = (\rho_l \rho_g / \rho_s \rho_0) \{(1 + r)\lambda / P_g\}$. If the amount of gas dissolved is small, then the density (ρ_s) of the solution is essentially the same as that (ρ_l) of the pure liquid, and $(1 + r)$ is essentially unity. Under such conditions $\alpha = \rho_g \lambda / \rho_0 P_g$, essentially. Furthermore, if the gas were ideal, then $\rho_g / \rho_0 P_g$ would equal $273.1/(273.1 + t)$ and we could write $\alpha = (273.1)\lambda / (273.1 + t)$, which is the relation used by A. G. Loomis [89] in reducing to α the data which were initially expressed in terms of λ. Actually the gases are not ideal, and in certain cases their departure from that condition must be considered.

For that reason, data so converted by him have, for this compilation, been reconverted to λ by multiplication by $(1 + t/273.1)$, his interpolations being accepted without question. The data have been extrapolated, when necessary, to $P_g = 1$ atm, so as to obtain comparisons with the corresponding values computed from those of r/P_g obtained by others. In some cases marked differences exist.

The supersaturation of liquids with gases has been studied by

85. SOLUBILITY OF GASES

J. Metschl,[90] certain errors that frequently occur in the measuring of the solubility of gases have been discussed by W. Manchot,[91] and the effect of colloids and of fine suspensions upon the solubility of gases has been investigated by A. Findlay and associates.[92]

For rates of solution and of aeration of quiescent columns of water, see Section 86.

References.

In this list are given those references which have been indicated at some place in the section solely by the name of the author or by a symbol. The names and symbols are listed in alphabetical order.

In certain tables there are two kinds of references, designated as Ref₁ and Ref₂, respectively. The first includes those from which the values were obtained; the second, those that should also be considered by one especially interested. In ICT, these Ref₂ references followed the instruction "cf. also."

Symbol	Author and Reference
A	v. Antropoff, A., *Z. Elektroch.*, **25**, 269-297 (1919).
AB	Adeney, W. E., and Becker, H. G., *Proc. Roy. Duvlin Soc. (N. S.)*, **15**, 385-404, 609-628 (1919) = *Phil. Mag. (6)*, **38**, 317-339 (1919); Idem, **39**, 385-404 (1920).
Bohr	Bohr, C., *Ann. d. Phys. (Wied.)*, **68**, 500-525 (1899).
Boyle	Boyle, R. W., *Phil. Mag. (6)*, **22**, 840-854 (1911).
Br	Breitenbach, W. E. (See Hou).
Braun	Braun, L., *Z. physik. Chem.*, **33**, 721-739 (1900).
Bu	Buch, K., *Soc. Sci. fennica. Com. physico-math.*, **2**, No. 16 (1924).
C	Cassuto, L., *Physik. Z.*, **5**, 233-236 (1904).
CEB	Cady, H. P., Elsey, H. M., and Berger, E. V., *J. Am. Chem. Soc.*, **44**, 1456-1461 (1922).
Coste	Coste, J. H., *J. Phys'l Chem.*, **31**, 81-87 (1927) (discussion only).
DM	Drucker, K., and Moles, E., *Z. physik. Chem.*, **75**, 405-436 (1910).
E	Estreicher, T., Idem, **31**, 176-187 (1899).
F	Findlay, A., and Creighton, H. J. M., *J. Chem. Soc. (London)*, **97**, 536-561 (1910). Findlay, A., and Howell, O. R., Idem, **107**, 282-284 (1915). Findlay, A., and Shen, B., Idem, **101**, 1459-1468 (1912). Findlay, A., and Williams, T., Idem. **103**, 636-645 (1913).
Foote	Foote, H. W., *J. Am. Chem. Soc.*, **43**, 1031-1038 (1921).
Fox	Fox, C. J. J., *Trans. Faraday Soc.*, **5**, 68-87 (1909).
FT	Fischer, F., and Tropsch, H., *Ber. deuts. chem. Ges.*, **50**, 765-767 (1917).
FTHP	Frolich, K., Tauch, E. J., Hogan, J. J., and Peer, A. A., *Ind. Eng. Chem.*, **23**, 548-550 (1931).
GK	Goodman, J. B., and Krase, N. W., *Ind. Eng. Chem.*, **23**, 401-404 (1931).
H	Hofmann, R., *Physik. Z.*, **6**, 337-340, 695 (1905).
Hou	Hougen, O. A., *Chem. Met. Eng.*, **32**, 704-705 (1925) includes previously unpublished data by W. E. Breitenbach).
ICT	Loomis, A. G., *Int. Crit. Tables*, **3**, 255-261 (1928).
J	Just, G., *Z. physik. Chem.*, **37**, 342-367 (1901).
Ko	Kofler, M., *Sitz. Akad. Wiss. Wien (Abt. 2a)*, **121**, 2169-2180 (1912).
Ku	Kunerth, W., *Phys. Rev. (2)*, **19**, 512-524 (1922).
L	Lannung, A., *J. Am. Chem. Soc.*, **52**, 68-80 (1930).
Mache	Mache, H., *Sitz. Akad. Wiss. Wien (Abt. 2a)*, **113**, 1329-1352 (1904).
Mt	Mailfert (l'abbé), *Compt. rend.*, **119**, 951-953 (1894).
NP	Neuhausen, B. S., and Patrick, W. A., *J. Phys'l Chem.*, **25**, 693-720 (1921).
P	Postma, S., *Rec. des trav. chim. Pays-Bas*, **39**, 515-536 (1920).
Ra	Ramstedt, E., *Le Radium*, **8**, 253-256 (1911).
Raoult	Raoult, F. M., *Ann. de chim. et phys. (5)*, **1**, 262-274 (1874).
Ro	Rothmund, V., "Nernst Festschrift," pp. 391-394, 1912.
Sa	Sander, W., *Z. physik. Chem.*, **78**, 513-549 (1912).
SdW	Scheffer, F. E. C., and de Wijs, H. J., *Rec. des trav. chim. Pays-Bas*, **44**, 655-662 (1925).
SP	Smits, A., and Postma, S., *Proc. Akad. Wet. Amsterdam*, **17**, 182-191 (1914).
Sp	Speller, F. N., *J. Franklin Inst.*, **193**, 515-542 (1922).
StM	Meyer, S., *Sitz. Akad. Wiss. Wien (Abt 2a)*, **122**, 1281-1294 (1913).
Sz	Szeparowicz, M., Idem, **129**, 437-454 (1920).
Ti	Timofejew, W., *Z. physik. Chem.*, **6**, 141-152 (1890).
Tr	v. Traubenberg, H. R., *Physik. Z.*, **5**, 130-134 (1904).
V25	Valentiner, S., "Preuss. Bergakad. Clausthal. Festschrift," p. 414, 1925.
V27	Valentiner, S., *Z. Physik*, **42**, 253-264 (1927).
W88	Winkler, L. W., *Ber. deuts. chem. Ges.*, **21**, 2843-2854 (1888).

W89	Winkler, L. W., *Idem,* **22,** 1764-1774 (1889).
W91a	Winkler, L. W., *Idem,* **24,** 89-101 (1891).
W91b	Winkler, L. W., *Idem,* **24,** 3602-3610 (1891).
W92a	Winkler, L. W., *Z. physik. Chem.,* **9,** 171-175 (1892).
W92b	Winkler, L. W., *Math. Naturw. Ber. aus Ungarn,* **9,** 195-215 (1892).
W97	Winkler, L. W., *Kisérleti Chemia,* **1,** 854 (1897).
W01	Winkler, L. W., *Ber. deuts. chem. Ges.* **34,** 1408-1422 (1901).
W04	Winkler, L. W., G. Lunge's "Chem.-Techn. Untersuchungmethoden," 5 ed., vol. 1, pp. 768-836 (822), 1904. Also, Vol. 1, p. 573, 1921 ed.
W06	Winkler, L. W., *Z. physik. Chem.,* **55,** 344-354 (1906).
WGH₁	Wiebe, R., Gaddy, V. L., and Heins, C., Jr., *Ind. Eng. Chem.,* **24,** 823-825 (1932).
WGH₂	Wiebe, R., Gaddy, V. L., and Heins, C., Jr., *J. Am. Chem. Soc.,* **55,** 947-953 (1933). Extending and superseding *Ind. Eng. Chem.,* **24,** 927 (1932).
Wijs	Wijs, H. J., *Thesis,* Delft, 1923.

86. Rate of Solution of Gases in Water

When a quiescent column of water, whether fresh or salt, is exposed to a gas, there is, in general, a vertical streaming which mixes the upper layers with the lower ones. This occurs even when the gas is initially saturated with water-vapor and the entire system is maintained at a uniform and contant temperature, and is much more pronounced when there is evaporation, with its attendant cooling and enhanced concentration of the surface layer. See W. E. Adeney and H. G. Becker,[118] W. E. Adeney, A. G. G. Leonard, and A. Richardson,[119] H. G. Becker and E. F. Pearson,[120] and W. E. Adeney.[121]

The streaming is more uniform and rapid in a salt solution than in pure water, is more rapid above 10 °C than below, and varies with the concentration of the salt. For solutions of NaCl the streaming is most rapid when the concentration is about 1 per cent. It extends to a depth of at least 10 feet.[119]

Data on the progressive aeration of quiescent and initially air-free water are given in Table 236.

When a liquid is kept thoroughly mixed while gas is entering it through a fixed surface, the net rate at which the gas enters the liquid is $dm/d\tau = (\alpha P - \beta c)A$, where dm is the amount entering through an area A in the time $d\tau$, P = partial pressure of the gas, c is the concentration of the gas in the liquid, and α and β are two coefficients; α may be called the entrance, and β the exit, coefficient. If the volume of the liquid is V and the total amount of gas in it is m, then $c = m/V$, and the equation may be written either $dm/d\tau = (\alpha P - \beta m/V)A$ or $dc/d\tau = (\alpha P - \beta c)(A/V)$. If the subscripts 0 and ∞ indicate that the values correspond to $\tau = 0$ and to $\tau = \infty$, respectively, *i.e.*, the initial and the saturation values, then $c - c_0 = (c_\infty - c_0)(1 - e^{-\beta A\tau/V})$; calling $c - c_0 \equiv \Delta c$, this becomes $\Delta c/c_\infty = [(c_\infty - c_0)/c_\infty](1 - e^{-\beta A\tau/V})$. The equation still holds good if all the c's are replaced by m's. Also, the initial rate of solution when $c_0 = 0$ is $(dc/d\tau)_0 = \alpha PA/V$, $(dm/d\tau)_0 = \alpha PA$; at saturation, $(dc/d\tau) = 0$, $c = c_\infty$,

---

[118] Adeney, W. E., and Becker, H. G., *Sci. Proc. Roy. Dublin Soc. (N. S.),* **16,** 143-152 (1920) = *Phil. Mag. (6),* **42,** 87-96 (1921).
[119] Adeney, W. E., Leonard, A. G. G., and Richardson, A., *Sci. Proc. Roy. Dublin Soc. (N. S.),* **17,** 19-28 (1922) = *Phil. Mag. (6),* **45,** 835-845 (1923).
[120] Becker, H. G., and Pearson, E. F., *Sci. Proc. Roy. Dublin Soc. (N. S.),* **17,** 197-200 (1923).
[121] Adeney, W. E., *Idem,* **18,** 211-217 (1926) = *Phil. Mag. (7),* **2,** 1140-1148 (1926).

86. RATE OF SOLUTION OF GASES

Table 236.—Aeration of Quiescent Water [126]

The water was quiescent and initially air-free. v_0 = volume (at 0 °C and 760 mm-Hg) of air per liter of water at a point 12 cm below the free surface; v_s = value of v_0 at saturation, when $\tau = \infty$; τ = duration of the exposure; temperature = 13.5 °C.

The formula $(v_s - v)/v_s = e^{-k\tau}$ will not fit the observations unless k is a function of τ, decreasing at an ever decreasing rate as τ increases.

With reference to the solution by water of CO_2 from the atmosphere, J. Johnston [127] has stated that "even in an unstirred liquid contained in an open beaker the process is substantially complete in about ten minutes."

Unit of τ = 1 hr; of v_0 = 1 ml/liter; of v_0/v_s = 1%

τ	v_0	v_0/v_s	τ	v_0	v_0/v_s	τ	v_0	v_0/v_s
2	5.0	21.4	20	11.6	49.6	70	16.0	68.4
4	6.3	26.9	25	12.4	53.0	80	16.4	70.1
6	7.5	32.0	30	13.2	56.4	90	16.9	72.2
8	8.2	35.0	40	13.9	59.4	100	17.4	74.3
10	8.9	38.0	50	14.7	62.8	∞	23.4	100.0
15	10.3	44.0	60	15.4	65.8			

Table 237.—Entrance Coefficient of Gases into Water

The entrance coefficient is the α occurring in the expression $dm/d\tau = (\alpha P - \beta c)A$; whence $\alpha = \beta m_\infty/PV$; m_∞/PV is a mean coefficient of absorption between 0 and P, write $s \equiv m_\infty/PV$. The following values of βs have been computed from the values of β in Table 238 and those of s ($\equiv \rho_l f/P$) in Tables 230 and 231, ρ_l being the density of the solution; for our present purposes, ρ_l may be taken as 1 g/cm³, and f/P as r/P. It will be noticed that α and βs are nearly independent of the temperature.

Unit of α and βs = 10^{-3}mg/min·cm²·atm. Temp. = t °C

Gas→	O_2	N_2	CO_2	Air	Gas→		$CO_2{}^a$	
t			βs		t	βs		α
0	24.3	9.9	256	12.5	0.1	256		245
5	23.7	10.1	266	12.7	9.9	266		249
10	24.2	10.3	268	12.8	22	259		33[b]
20	24.6	10.8	262	12.9	34.6	245		217
30	24.6	11.2	253	13.2	43.1	234		219
40	24.6	11.7	242	13.6				

[a] From C. Bohr [122] excepting at 22 °C.
[b] From H. G. Becker [124] who gives, among others, the following approximate values for the initial rates of solution at 22 °C. It is believed that the volumes were measured at 22 °C and under the pressures (P) at which the absorption occurred, and it is on this assumption that the values of P here tabulated were computed from his two sets of rates. The value of α has been taken as (rate/P). It will be noticed that his value for CO_2 is entirely inconsistent with those of Bohr.

Gas	H_2	CO_2	Cl_2	H_2S	SO_2	HCl	
Rate	0.0043	0.0168	0.0984	0.0950	2.00	54.3	cc/min·cm²
Rate	0.000023	0.000198	0.00187	0.000866	0.0343	5.3	g/hr·cm²
P	0.991	0.0994	0.0986	0.0984	0.0967	0.992	atm
α	0.00038	0.033	0.316	0.146	5.90	89.	mg/min·cm²·atm

[122] Bohr, C., *Ann. d. Physik (Wied.)*, **68**, 500-525 (1899).
[123] Adeney, W. E., and Becker, H. G., *Sci. Proc. Roy. Dublin Soc. (N. S.)*, **15**, 385-404, 609-628 (1918-1919) = *Phil. Mag.* (6), **38**, 317-337 (1919); **39**, 385-404 (1920).

Table 238.—Exit Coefficient of Gases from Water

The exit coefficient is the β occurring in the expression $c - c_0 = (c_\infty - c_0)(1 - e^{-\beta A \tau/V})$ (see text). $\beta = \beta'(T - T') = \beta'(t - t')$; the tabulated values of β have been computed from the tabulated values of β' and T', which were published in the articles cited; it has been assumed that $T = 273 + t$. The value of β for air from sea-water is about 5 per cent less than that for air from distilled water.[123, first paper]

Unit of $\beta = 1$ cm/min. Temp. $= t$ °C

Gas →	O_2	N_2	CO_2	Air
$10^4 \beta'$	96	103	38.1	99
T'	237	240	253	239
t'	−36	−33	−20	−34
t			β	
0	0.346	0.340	0.076	0.336
5	0.384	0.391	0.095	0.386
10	0.442	0.443	0.114	0.436
20	0.538	0.546	0.152	0.534
30	0.634	0.649	0.190	0.634
40	0.730	0.752	0.229	0.733
Ref.[a]	AB	AB	B	AB

A. Guyer and B. Tobler [128] have reported the following values. Temperature of the bulk of the liquid was 20 °C.

Gas	β
Acetylene	4.32
CO_2	3.84
H_2S	3.76
SO_2	0.93
NH_3	0.09

[a] References:
AB Adeney, W. E., and Becker, H. G.[123, second paper]
B   Bohr, C.[122]

Table 239.—Absorption of Oxygen by a Thin Film of Water [121]

It is assumed that the water is initially free of O_2. If $c =$ concentration at the time τ, $c_\infty = c$ when $\tau = \infty$, $A =$ area of the water surface, $V =$ volume of the film, and $\beta = 0.0096(t + 36)$ cm/min,[a] then $c/c_\infty = 1 - e^{-\beta A \tau/V}$. The values of τ tabulated below correspond to $V/A = 0.05$ cm; hence if the absorption occurs on one side only and there is no loss from the other side (a thin film of water on a non-porous solid upon which the gas is not adsorbed), the film is 0.05 cm thick, but if absorption occurs on both sides, the thickness is 0.10 cm. The formula assumes that there is no evaporation, that the film does not mix with any other liquid, that β and c_∞ have the same values for the film as for an extended volume of thoroughly mixed water, and that the concentration is the same throughout the entire volume of the film. When appropriate values are given to c_∞, the tabulated values apply to atmospheric oxygen as well as to pure oxygen, and to sea-water as well as to pure water.

87. DIFFUSION OF GASES

Table 239.—(Continued)

Unit of $\tau = 1$ sec; of $c/c_\infty = 1\%$; temp. $= t$ °C; $V/A = 0.05$ cm

$c/c_\infty \!\downarrow \;\; t\!\rightarrow$	0	5	10	15	20	25	30
10	1.0	0.8	0.8	0.7	0.7	0.6	0.5
20	2.1	1.8	1.6	1.4	1.3	1.3	1.2
30	3.2	2.7	2.3	2.0	1.8	1.7	1.6
40	4.5	4.0	3.6	3.3	2.9	2.7	2.4
50	6.0	5.2	4.6	4.2	3.9	3.6	3.3
60	8.0	7.2	6.4	5.7	5.0	4.6	4.3
70	10.5	9.3	8.2	7.3	6.7	6.1	5.8
80	14.0	12.5	11.0	10.0	9.0	8.3	7.7
90	19.8	17.5	15.6	14.2	13.0	11.9	10.8
99	39.6	35.0	31.2	28.4	26.0	23.8	21.6

^a In the paper here cited this unit is incorrectly given as "per sec"; cf. with tables and graphs in AB reference of Table 238.

---

and $\alpha = \beta c_\infty/P = \beta m_\infty/PV$; c_∞/P and m_∞/PV are mean coefficients of absorption between 0 and P. See C. Bohr,[122] W. E. Adeney and H. G. Becker,[123] H. G. Becker,[124] and W. E. Adeney.[121]

For values of α and β, see Tables 237 and 238.

According to L. A. Klufschareff,[125] a superficial layer of machine oil does not affect the rate of solution of CO_2 into water, but colloid particles, whether acid or neutral, increase both the rate and the amount of the solution of CO_2.

87. Diffusion of Gases in Water

The diffusivity, or coefficient of diffusion, of a solute in a liquid is the quantity D in the equation $(dq/d\tau)_x = -D(dc/dx)_\tau dydz$, where dq is the amount of the solute which in the time $d\tau$ passes in the positive direction of x through the area $dydz$ at x, the concentration at x increasing in the direction of x at the rate $(dc/dx)_\tau$; unless the contrary is stated, both q and c are specified in terms of the same unit amount of solute, and the unit of volume involved in c is the cube of the unit of length appearing in dx, dy, and dz. . Whence, $(dc/d\tau)_x = D(d^2c/dx^2)_\tau$ if D is independent of x; in general, D depends upon c, and hence, upon x. In that case $(dc/d\tau)_x = \{d[Ddc/dx]/dx\}_\tau = D(d^2c/dx^2)_\tau + (dc/dx)_\tau^2(dD/dc)$. The temperature is supposed to be uniform and constant.

---

[124] Becker, H. G., *Ind. Eng. Chem.*, **16,** 1220-1224 (1924).

[125] Klufschareff, L. A., *Jour. de Phys.* (7), **3,** 513D (1932) ← *Jour. chim. appl. Russe*, **4,** 425-428 (1931).

[126] Chappuis, P., *Trav. et mém. bur. int. poids et mes.*, **14,** (B) 1-163, and D.1-D.63 (1910).

[127] Johnston, J., *J. Am. Chem. Soc.*, **38,** 947-975 (1916).

[128] Guyer, A., and Tobler, B., *Helv. Chim. Acta*, **17,** 550-555 (1934). See also, *Idem*, **17,** 257-271 (1934).

Table 240.—Diffusivities of Selected Gases in Water
Adapted from the compilation of H. R. Bruins.[129]

The data for NH_3 represent a mean value of the diffusivity (D) over an ill-defined range of concentration, the diffusion having been from a solution of initial concentration c into water initially NH_3-free. The others refer to exceedingly dilute solutions.

Unit of $D = 10^{-5}$ cm²/sec; of $c = 1$ gfw per liter. Temp. $= t$ °C

Gas	t	D	Ref.₁[a]	Ref.₂[a]	Gas	t	c	D	Ref.₁[a]	Ref.₂[a]
H₂	10	4.₃	Hu 8	Ba	NH₃	5	0.₇	1.2₄	Sch	Ex
	16	4.₇	Hu 7	Ex		5	3.₅	1.2₄	Sch	Ha
	21	5.₂	Hu 7	Ha		8	1.0	1.3₀	Ar	Mu
N₂	22	2.0₂	Hu 7	Ex		8	Sat.[b]	1.0₈	Hu 8	Vo
Rn	18	1.1₄[c]	Ro	Wa		10	Sat.[b]	1.1₄	Hu 8	
CO₂	10	1.4₆	Hu 7, 8, St	Ex		12	1.0	1.6₄	Ar	
	15	1.6₀	Hu 7, 8, St	Ha		15	1.0	1.7₇	Ab	
	18	1.7₁[d]	Ca			15	Sat.[b]	1.2₆	Hu 8	
	20	1.7₇	Hu 7, 8, St	Wr						

H₂ W. W. Ipatieff, Jr., W. P. Teodorovitsch, and S. I. Druschina-Antemovitsch[130] have reported that, within the limits 35 to 100 atm, D for H₂ is independent of the pressure and has the following values:

t	15	25	35	45	°C
$10^5 D$	2.49	3.37	4.22	5.69	cm²/sec.

[a] References: Under Ref.₁ are the references from which the associated data were obtained; under Ref.₂ are supplementary references pertinent to the gas but not necessarily to the particular values of t and c given in this table.

Ab	Abegg, R., *Z. physik. Chem.*, **11**, 248-264 (1893).
Ar	Arrhenius, S., *Idem*, **10**, 51-95 (1892).
Ba	Barus, C., *Carnegie Inst. of Washington, Publ. No.* **186**, 1-88 (1913).
Ca	Carlson, T., *J. Am. Chem. Soc.*, **33**, 1027-1032 (1911); *Medd. f. K. Vetensk. Nobel-inst.*, **2**, No. 6 (1913).
Ex	Exner, F., *Ann. d. Physik (Pogg.)*, **155**, 321-336, 443-464 (1875).
Ha	Hagenbach, A., *Idem (Wied.)*, **65**, 673-706 (1898).
Hu7	Hüfner, G., *Idem*, **60**, 134-168 (1897).
Hu8	Hüfner, G., *Z. physik. Chem.*, **27**, 227-249 (1898).
Mu	Müller, J., *Ann. d. Physik*, **43**, 554-567 (1891).
Ro	Róna, E., *Z. physik. Chem.*, **92**, 213-218 (1917).
Sch	Scheffer, J. D. R., *Idem*, **2**, 390-404 (1888).
St	Stefan, J., *Sitzb. Akad. d. Wiss. Wien (Abt. II)*, **77**, 371-409 (1878).
Vo	Voigtländer, F., *Z. physik. Chem.*, **3**, 316-335 (1889).
Wa	Wallstabe, F., *Physik. Z.*, **4**, 721-722 (1903).
Wr	v. Wroblewski, S., *Ann. d. Physik. (Wied.)*, **2**, 481-513 (1877); **4**, 268-277 (1878); **7**, 11-23 (1879); **8**, 29-52 (1879).

[b] Diffusion from water saturated with NH₃ at a pressure of 1 atm.

[c] $1.1_4 \pm 0.07$.   [d] $1.7_1 \pm 0.03$.

---

88. Pressure-Volume-Temperature Associations for Saturated Water and Steam

In this section are assembled those pressure-volume-temperature associations that are characteristic of water in equilibrium with its own pure vapor, and of water-vapor in equilibrium with pure water. See also Table 260.

When such equilibrium exists, each phase is said to be saturated with

[129] Bruins, H. R., *Int. Crit. Tables*, **5**, 63-76 (1929).

[130] Ipatieff, W. W., Jr., Teodorovitsch, W. P., and Druschina-Antemovitsch, S. I., *Z. anorg. allgem. Chem.*, **216**, 66-74 (1933).

reference to the other, and its state is more briefly described by the single word "saturated," as in "saturated water," "saturated steam." *

Critical Data.

The critical data of a substance are those that are characteristic of it at its critical point. And the critical point in the usual p-v diagram is defined by the associated temperature and pressure that marks the vertex of the curve that bounds the area in which both phases (liquid and vapor) may coexist and each may be distinctly segregated. There $(\partial p/\partial v)_t = 0$ and $(\partial v/\partial t)_p = \infty$. At that point the distinction between the two phases vanishes. Consequently, the meniscus separating the phases vanishes, and so does the latent heat of transition. Each of these phenomena has been used as a criterion for determining when the critical point has been attained; the first, the more frequently.

But things are not quite so simple as one might infer. For example, E. Schröer [132] has stated that the densities of the two phases do not become equal as the system passes through the critical point, although the meniscus vanishes; though mixed by stirring, segregation occurs when the stirring is stopped. Only after the temperature has been raised to about 0.5 °C above the critical point does the system become homogeneous. Similar observations have been reported in some detail by O. Maass and A. L. Geddes,[133] who state that a difference in densities of the upper and the lower layers persisted after 6 hours of continuous stirring. This difference, at temperatures above the critical, decreases with rise in temperature, ultimately vanishing. Also, the difference is destroyed if the system is expanded while slightly above the critical temperature; and also if the upper portion of the system is cooled below the temperature of the lower one. Once the difference of density is destroyed, the system remains homogeneous however the temperature and pressure may be varied, provided that the temperature does not fall below the critical one. Such differences in

---

* Over 50 years ago, W. Ramsay and S. Young [181] proposed that the "curve representing the relations between volumes and temperatures at the vapor-pressures corresponding to the temperatures" be called the "orthobaric" curve. This has led to an occasional use of the adjective orthobaric to modify the name of a property for the purpose of indicating that the values considered are solely those characteristic of a given liquid when subjected to the equilibrium pressure of its own pure vapor; as in "orthobaric density." But the word does not appear in any of the general English dictionaries; and its significance, as defined by Ramsay and Young, is unknown to many who are interested in the data for saturated liquids, and is not at all self-evident. There seems to be little reason for using it, especially as the same purpose can be served by qualifying the name of the liquid or its class by the adjective "saturated," in accordance with a custom long established for vapors. However, should an adjective applicable to the property, instead of the substance, be needed, then either autopiestic or, less appropriately, autobaric would be preferable to orthobaric, as the first component—"auto"—directly suggests that the pressure in mind arises from, and is characteristic of, the substance itself, while "ortho" does not.

[181] Ramsay, W., and Young, S., *Phil. Trans.*, **177**, 123-156 (135-136) (1886).
[132] Schröer, E., *Z. physik. Chem.*, **129**, 79-110 (1927).
[133] Maass, O., and Geddes, A. L., *Phil. Trans. (A)*, **236**, 303-332 (1937).

density had previously been observed by G. Teichner[134]; see also J. Traube,[135] who developed a theory based on the idea that the liquid molecule differed from the gas one in size but not in mass, and that at the critical point the two types of molecules were miscible in all proportions, becoming identical only at a higher temperature. Maass and Geddes refer to other papers bearing on the subject.

Furthermore, C. A. Winkler and O. Maass[136] have reported observations indicating that the surface energy does not become zero when the meniscus vanishes.

And H. L. Callendar[137] has reported that differences in density can be detected as far as 6 °C above the critical temperature (374 °C), that the latent heat of vaporization does not vanish until that higher temperature is reached, and that small amounts of air in the water cause serious difficulties. These conclusions were not entirely borne out by the work of A. Egerton and G. S. Callendar[138]; and the careful work of W. Koch,[139] specially

Table 241.—Critical Constants of Water

The values preferred at present by the National Bureau of Standards are these:

$t_c = 374.15$ °C, $p_c = 225.65$ kg*/cm² = 218.39 atm, $v_c^* = 3.1$ cm³/g.[142]

The several sets of values reported since 1900, excepting H. L. Callendar's (see text), are given below. The assigning of two values to a given reference indicates that the authors state no more than that the true value lies between those limits.

1 atm = 1.03323 kg*/cm² = 1.01325 bars

Unit→ Year	1 °C t_c	1 atm	1 kg*/cm²	1 cm³/g v_c^*	Ref[a]
1904	374	...	...	...	TT
1910	374.07; 374.62	218	225	...	HB
1927	374.20 ± 0.20	...	...	...	S
1928	374.0	217.72	224.95	2.5	ICT
1931	374.11	218.53	225.79	3.086	KS
1932	374; 374.5	...	...	...	EC
1932	374	219	226	2.8; 3.0	NB
1934	374.11+	218.167	225.416	...	SKG
1937	374.23	218.26	225.51	3.066	E
1937	374.15	218.39	225.65	3.1	OSG

For the viscosity at the critical point, see p. 66.

E. H. Riesenfeld and T. L. Chang[143] have given for ordinary water

[134] Teichner, G., *Ann. d. Physik (4)*, **13**, 595-610, 611-619 (1904) = *Diss.*, Wien, 1903.
[135] Traube, J., *Z. anorg. Chem.*, **37**, 225-242 (1903).
[136] Winkler, C. A., and Maass, O., *Can. J. Res.*, **9**, 65-79 (1933).
[137] Callendar, H. L., *Proc. Roy. Soc. (London) (A)*, **120**, 460-472 (1928); *Engineering (London)*, **126**, 594-595, 625-627, 671-673 (1928).
[138] Egerton, A., and Callendar, G. S., *Phil. Trans. (A)*, **231**, 147-205 (1932).
[139] Koch, W., *Forsch. Gebiete Ingenieurw.*, **3**, 189-192 (1932).
[140] Jakob, M., *Physik. Z.*, **36**, 413-414 (1935).
[141] Harand, J., *Monatsh. Chem.*, **65**, 153-184 (1934).

88. SATURATED WATER AND STEAM: P-V-T

Table 241.—*(Continued)*

the following values in the neighborhood of t_c for the density (ρ) of saturated vapor and of saturated liquid. The mean of the two densities corresponding to the same t is almost linear in t, and by extrapolation leads to $\rho_c = 0.329$ g/cm^3, or $v_c^* = 3.04$ cm^3/g. As tabulated, the values proceed in regular order from vapor saturated at 364.0 °C to liquid saturated at 365.4 °C. Unit of $\rho = 1$ g/cm^3; of $v^* = 1$ cm^3/g; temp. $= t$ °C.

t	ρ	v^*	t	ρ	v^*
364.0	0.157	6.37	373.9	0.392	2.55
373.1	0.244	4.10	373.7	0.398	2.51
374.0	0.271	3.69	373.4	0.402	2.49
374.0	0.288	3.47	372.9	0.409	2.44
374.0	0.317	3.15	373.4	0.410	2.44
374.3	0.322	3.10	371.8	0.440	2.27
374.2	0.329	3.04	367.2	0.478	2.09
374.2	0.343	2.92	366.5	0.481	2.08
374.2	0.344	2.91	365.4	0.490	2.04
374.2	0.354	2.82	365.4	0.491	2.04

[a] References.

- EC Egerton, A., and Callendar, G. S.[138]
- E   Eck, H., *Tätigkeit d. Phys. Techn. Reichs.*, p. 32, 1936, = *Physik. Z.*, **38**, 256 (1937).
- HB Holborn, L., and Baumann, A., *Ann. d. Phys.* (4), **31**, 945-970 (1910).
- ICT *Int. Crit. Tables*, **3**, 248 (1928). Compilation by A. F. O. Germann and S. F. Pickering.
- KS Keyes, F. G., and Smith, L. B., *Mech. Eng.*, **53**, 132-135 (1931).
- NB v. Nieuwenburg, C. J., and Blumendal, H. B., *Rec. Trav. chim. Pays-Bas*, **51**, 707-714 (1932).
- OSG Osborne, N. S., Stimson, H. F., and Ginnings, D. C., *J. Res. Nat. Bur. Stand.*, **18**, 389-448 (RP983) (1937).
- S   Schröer, E., *Z. physik. Chem.*, **129**, 79-110 (1927).
- SKG Smith, L. B., Keyes, F. G., and Gerry, H. T., *Proc. Am. Acad. Arts Sci.*, **69**, 137-168 (1934).
- TT Traube, J., and Teichner, G., *Ann. d. Phys.* (4), **13**, 620-621 (1904).

---

undertaken for their investigation, has given no confirmation of them, not even of the serious effects caused by a small admixture of air. One would be inclined to think that Callendar's conclusions are erroneous were it not for the other observations that have been mentioned, but they, and especially those of Maass and Geddes, suggest that the failure in confirmation may have resulted from some difference in the procedures followed.

See also M. Jakob [140] and J. Harand [141]; the latter observed the region of the meniscus with a "schlieren" microscope and found the fluid to be nonhomogeneous immediately after the meniscus had vanished.

Saturated Vapor.

(a) *Vapor-pressure.—Definition.*—The pressure of a vapor that is in equilibrium with its liquid depends not only upon the temperature, but also upon the shape of the liquid surface and upon the presence or absence of a foreign gas, the effect of such gas varying with its partial pressure. That particular pressure of the vapor, unmixed with a foreign gas, that is in equilibrium with a flat surface of its own pure liquid is properly called the

---

[142] Osborne, N. S., Stimson, H. F., and Ginnings, D. C., *J. Res. Nat. Bur. Stand.*, **18**, 389-448 (RP983) (1937).

[143] Riesenfeld, E. H., and Chang, T. L., *Z. phys. Chem. (B)*, **30**, 61-68 (1935).

560   PROPERTIES OF ORDINARY WATER-SUBSTANCE

vapor-pressure of the liquid, and is presumably the quantity that is given in tables of vapor-pressures. The equilibrial pressure under other conditions, though quite frequently called by the same name is preferably designated in a different manner.

Past History, Effect of.—Menzies and his associates have concluded from their own observations that, when the temperature of water is changed, any accompanying change in the internal state of the water is completely attained practically instantly.[144]

Furthermore, as certain observations by T. C. Barnes and his associates [145] have indicated that water from freshly melted ice is more conducive to the growth of certain biological organisms than is either ordinary water or water from freshly condensed steam, and this has been interpreted as arising from the presence of an excess of trihydrol [146] in the freshly melted ice, A. W. C. Menzies [147] has compared the vapor-pressure of water from freshly melted ice with that from freshly condensed steam, finding no difference. He cannot accept the trihydrol explanation (see also A. W. C. Menzies [148]).

Catalysts, Effect of.—Although H. B. Baker [149] has reported a progressive change in the vapor-pressure of water in contact with such catalysts as charcoal, thoria, and platinum black, Menzies and his associates [150] seem to have shown that Baker's conclusions cannot be accepted.

Gas, Effect of.—The effect of an inert gas upon the equilibrial pressure of the vapor of a liquid in contact with it is twofold: (1) the direct

Table 242.—Vapor-pressure of Water (atm): −5 to +374 °C [156]

(See also International Skeleton Steam Tables, 1934, (Table 260) and Tables 243, 244, and 245.)

These values, computed by means of formula (1) are believed to be the most accurate available for $t \leqq 100\ °C$, for which range they were chosen for the International Skeleton Steam Tables of 1934 (see Table 260); for lower temperatures the slightly different values given in the latter tables are to be preferred. Their differences from those here given are shown in Table 246. All values refer to a flat surface of pure water in contact with its own pure vapor. For the effect of inert gases, see text (p. 560) and Tables 243 and 245.

In the Osborne-Meyers paper the pressures are expressed in kg*/cm² as well as in atm, and are given for each 1 °C; also in lb*/in² for each 1 °F.

[144] West, W. A., and Menzies, A. C. W., *J. Phys'l Chem.*, **33**, 1893-1896 (1929).
[145] Barnes, T. C., *Proc. Nat. Acad. Sci.*, **18**, 136-137 (1932); Lloyd, F. E., and Barnes, T. C., *Idem*, **18**, 422-427 (1932); Barnes, T. C., and Jahn, T. L., *Idem*, **19**, 638-640 (1933); Barnes, T. C., and Larson, E. J., *J. Am. Chem. Soc.*, **55**, 50-59 (1933); Barnes, T. C., and Jahn, T. L., *Quart. Rev. Biol.*, **9**, 292-341 (1934); Barnes, T. C., *Science*, **79**, 455-457 (1934); **81**, 200-201 (1935).
[146] Barnes, H. T., and T. C., *Nature*, **129**, 691 (1932).
[147] Menzies, A. W. C., *Proc. Nat. Acad. Sci.*, **18**, 567-568 (1932).
[148] Menzies, A. W. C., *Science (N. S.)*, **80**, 72-73 (1934).
[149] Baker, H. B., *J. Chem. Soc. (London)*, **1927**, 949-958 (1927).
[150] West, W. A., and Menzies, A. W. C., *J. Phys'l Chem.*, **33**, 1893-1896 (1929); Wright, S. L., and Menzies, A. W. C., *J. Am. Chem. Soc.*, **52**, 4699-4708 (1930); Menzies, A. W. C.[148]

88. SATURATED WATER AND STEAM: P-V-T

Unit of $P = 1$ atm $= 1.01325$ bars $= 1.03323$ kg*/cm². Temp. $= (t_1 + t_2)$ °C, int. centigrade scale

$t_1 \downarrow \quad t_2 \rightarrow$	0	1	2	3	4	5	6	7	8	9
−10	0.0060273	0.006479	0.006960	0.007473	0.008019	0.004162	0.004487	0.004835	0.005207	0.005604
−0	.012102	.012936	.013821	.014759	.015752	.008600	.009218	.009875	.010574	.011315
+0	.023042	.024508	.026056	.027688	.029409	.016804	.017917	.019094	.020338	.021653
10	.041831	.044293	.046881	.049599	.052452	.031222	.033133	.035144	.037261	.039489
20	.072748	.076718	.080873	.085222	.089770	.055446	.058588	.061883	.065337	.068956
30	.121698	.12787	.13431	.14102	.14802	.094526	.099497	.10469	.11012	.11578
40	.196560	.20584	.21549	.22553	.23596	.15531	.16291	.17082	.17906	.18764
50	.307520	.32107	.33512	.34969	.36479	.24679	.25805	.26973	.28186	.29446
60	.467396	.48665	.50657	.52717	.54846	.38043	.39664	.41342	.43080	.44878
70						.57047	.59322	.61672	.64099	.66605
80	.691923	.71863	.74619	.77463	.80396	.83421	.86540	.89755	.93068	.96482
90	1.00000	1.0362	1.0735	1.1120	1.1515	1.1922	1.2341	1.2772	1.3215	1.3670
100	1.41389	1.4621	1.5116	1.5624	1.6147	1.6684	1.7236	1.7802	1.8384	1.8981
110	1.95038	2.0223	2.0868	2.1530	2.2209	2.2905	2.3619	2.4351	2.5101	2.5870
120	2.66583	2.7466	2.8293	2.9140	3.0008	3.0896	3.1806	3.2737	3.3690	3.4665
130										
140	3.56630	3.6684	3.7728	3.8797	3.9889	4.1006	4.2148	4.3316	4.4509	4.5728
150	4.69746	4.8248	4.9549	5.0877	5.2234	5.3620	5.5035	5.6480	5.7955	5.9460
160	6.09964	6.2564	6.4164	6.5796	6.7461	6.9159	7.0891	7.2658	7.4459	7.6295
170	7.81669	8.0075	8.2020	8.4002	8.6022	8.8080	9.0176	9.2312	9.4487	9.6703
180	9.89596	10.126	10.360	10.598	10.840	11.087	11.338	11.594	11.854	12.119
190	12.3881	12.662	12.941	13.224	13.513	13.806	14.104	14.407	14.715	15.029
200	15.3472	15.671	16.000	16.334	16.674	17.020	17.370	17.727	18.089	18.457
210	18.8304	19.210	19.595	19.986	20.384	20.787	21.197	21.613	22.035	22.463
220	22.8978	23.339	23.787	24.241	24.702	25.170	25.645	26.126	26.615	27.110
230	27.6130	28.123	28.640	29.164	29.695	30.234	30.780	31.334	31.895	32.465
240	33.0424	33.627	34.220	34.821	35.430	36.047	36.672	37.305	37.947	38.597
250	39.2557	39.923	40.599	41.284	41.977	42.679	43.390	44.111	44.840	45.579
260	46.3264	47.083	47.850	48.626	49.412	50.207	51.012	51.827	52.652	53.486
270	54.3313	55.187	56.052	56.927	57.813	58.710	59.616	60.534	61.463	62.402
280	63.3521	64.313	65.286	66.269	67.264	68.270	69.288	70.317	71.358	72.411
290	73.4752	74.552	75.640	76.741	77.854	78.978	80.116	81.266	82.429	83.605
300	84.7931	85.994	87.209	88.436	89.677	90.931	92.198	93.480	94.774	96.083
310	97.4057	98.742	100.09	101.46	102.84	104.23	105.64	107.06	108.50	109.95
320	111.422	112.91	114.40	115.92	117.45	118.99	120.56	122.13	123.73	125.34
330	126.963	128.61	130.27	131.94	133.64	135.35	137.08	138.82	140.59	142.37
340	144.168	145.99	147.82	149.68	151.55	153.45	155.36	157.29	159.24	161.21
350	163.205	165.22	167.25	169.30	171.38	173.48	175.59	177.74	179.90	182.08
360	184.294	186.53	188.784	191.07	193.372	195.70	198.064	200.45	202.861	205.30
370	207.772	210.272	212.804	215.370	217.978					

Table 243.—Vapor-pressure of Water (mm-Hg): 0 to −16 °C

From compilation by E. W. Washburn.[163]
(See also Tables 242 and 244.)

The following values have been computed by E. W. Washburn [164] from the values derived by himself for the vapor pressure of ice at the corresponding temperatures (p. 598 and Table 264), the relation between the two being taken as that indicated by the formula

$$\log_{10}\left(\frac{p_w}{p_i}\right) = -\frac{1.1489t}{273.1 + t} - 1.330(10^{-5}t^2) + 9.084(10^{-8}t^3)$$

which is based upon the following values:
Specific heat of water $= (1.0092 - 0.001080t + 0.000036t^2)$ cal$_{15}$ per g.°C
Specific heat of ice [a] $= (0.5052 + 0.001861t)$ cal$_{15}$ per g.°C
Latent heat of fusion of 18.015 g of ice at 0°C $= 1435.5$ cal$_{15}$ ($= 333.48$ j)
Gas-constant, $R = 1.9869$ cal$_{15}$ per g-mole·°C $= 8.315_2$ j/g-mole·°C

The values so obtained for the vapor pressure exceed those given by L. Holborn, K. Scheel, and F. Henning,[165] and based upon work done at the Physikalisch-Technischen Reichsanstalt, by amounts ranging from 0 at 0 °C to 0.008 mm-Hg at −15.9 °C.

When the vapor is mixed with such an amount of atmospheric air that the total pressure is one atm, the vapor pressure exceeds that here given by a small amount, Δp, defined by the equation $\Delta p/p = 0.024/(273 + t)$.* The numbers in this table refer to pure water in contact with its pure vapor. At the end of each line is given, under D, the corresponding average increase in pressure per 0.1 °C decrease in temperature.

Unit of $P = 1$ mm-Hg; of $D = 0.001$ mm-Hg. Temp. $= (t_1 + t_2)$ °C

$t_1 \downarrow \; t_2 \rightarrow$	−0.0	−0.1	−0.2	−0.3	−0.4	−0.5	−0.6	−0.7	−0.8	−0.9	D
−0	4.579	4.546	4.513	4.480	4.448	4.416	4.385	4.353	4.320	4.289	−32.1
−1	4.258	4.227	4.196	4.165	4.135	4.105	4.075	4.045	4.016	3.986	−30.2
−2	3.956	3.927	3.898	3.871	3.841	3.813	3.785	3.757	3.730	3.702	−28.3
−3	3.673	3.647	3.620	3.593	3.567	3.540	3.514	3.487	3.461	3.436	−26.3
−4	3.410	3.384	3.359	3.334	3.309	3.284	3.259	3.235	3.211	3.187	−24.7
−5	3.163	3.139	3.115	3.092	3.069	3.046	3.022	3.000	2.976	2.955	−23.2
−6	2.931	2.909	2.887	2.866	2.843	2.822	2.800	2.778	2.757	2.736	−21.6
−7	2.715	2.695	2.674	2.654	2.633	2.613	2.593	2.572	2.553	2.533	−20.1
−8	2.514	2.495	2.475	2.456	2.437	2.418	2.399	2.380	2.362	2.343	−18.8
−9	2.326	2.307	2.289	2.271	2.254	2.236	2.219	2.201	2.184	2.167	−17.7
−10	2.149	2.134	2.116	2.099	2.084	2.067	2.050	2.034	2.018	2.001	−16.2
−11	1.987	1.971	1.955	1.939	1.924	1.909	1.893	1.878	1.863	1.848	−15.3
−12	1.834	1.819	1.804	1.790	1.776	1.761	1.748	1.734	1.720	1.705	−14.3
−13	1.691	1.678	1.665	1.651	1.637	1.624	1.611	1.599	1.585	1.572	−13.1
−14	1.560	1.547	1.534	1.522	1.511	1.497	1.485	1.472	1.460	1.449	−12.4
−15	1.436	1.425	1.414	1.402	1.390	1.379	1.368	1.356	1.345	1.334	−11.3

[a] In the *Weather Review* the number 0.5052 was incorrectly printed as 0.5952.
* This coefficient (0.024) differs from that (0.020) given by Washburn [163, 164]; the latter is incorrect.

---

[151] Cf. W. Thomson (Lord Kelvin), *Phil. Mag.* (4), **42**, 448-452 (1871).
[152] v. Helmholtz, R., *Ann. d. Phys. (Wied.)*, **27**, 508-543 (1886).
[153] Wilson, C. T. R., *Phil. Trans. (A)*, **189**, 265-307 (1897).

88. SATURATED WATER AND STEAM: P-V-T

effect of the pressure that the gas exerts, and (2) the effect of the gas that is dissolved in the liquid. The first is an increase, and the second a decrease, in the pressure of the vapor.

If p_g = partial pressure of the inert gas at the surface of the liquid, p_0 = vapor-pressure of the liquid in the absence of the inert gas, $\Delta p =$

Table 244.—Vapor-pressure of Water (millibars): 0 to −50 °C

Adapted from L. P. Harrison.[166] (See also Table 243.)
Computed by means of the formulas (p. 598 and Table 243) given by E. W. Washburn,[164] and for each 0.5 °C.

Unit of P = 1 millibar = 1000 dyne/cm^2 = 0.75 mm-Hg. Temp. = $(t_1 + t_2)$ °C

$t_2 \rightarrow$ t_1	0	−1	−2	−3	−4	−5	−6	−7	−8	−9
0	6.105	5.677	5.275	4.898	4.546	4.217	3.909	3.620	3.352	3.100
−10	2.866	2.648	2.444	2.255	2.079	1.915	1.763	1.622	1.490	1.370
−20	1.257	1.153	1.056	0.967	0.885	0.809	0.739	0.674	0.615	0.560
−30	0.510	0.464	0.421	0.383	0.347	0.314	0.285	0.257	0.233	0.210
−40	0.189	0.170	0.153	0.138	0.124	0.111	0.0994	0.0890	0.0795	0.0710
−50	0.0633									

Table 245.—Vapor-pressure of Water (mm-Hg): 0 to 102 °C

(Values in Table 242 are to be preferred.)

From compilation by E. W. Washburn.[167] Cf. also Table 242.

These values are based upon observations made at the Physikalisch-Technischen Reichsanstalt, which were reduced to the temperature scale [a] of the Reichsanstalt in 1919, and interpolated, as below, by L. Holborn, K. Scheel, and F. Henning.[165] Reproduction is by permission of Vieweg & Sohn.

When the vapor is mixed with such an amount of atmospheric air that the total pressure is 1 atm, the vapor pressure exceeds that here given by the amount Δp, defined as follows: $\Delta p/p = 0.000775 - 3.13(10^{-6}t)$ if $0 \leq t \leq 40$ °C; $\Delta p/p = 0.000652 - 8.75(10^{-7}p)$ if $50 \leq t < 90$ °C, p being expressed in mm-Hg. The numbers in this table refer to pure water in contact with its pure vapor. At the end of each line is given, under D, the corresponding average increase in the pressure on going from one tabulated value to the next; and under δ is given the amount by which the entry in the zero column exceeds the corresponding value in Table 242, cf. Table 246.

At the triple point corresponding to equilibrium between water, water-vapor, and ice-I, $t = 0.0099$ °C, $p = 4.5867$ mm-Hg.[168]

---

[a] Defined by the melting point of ice (0°), the boiling point of water (100°), and the boiling point of sulphur (taken as 444.55°), all under the pressure of one normal atmosphere; interpolation between these points being made by means of a resistance-thermometer of platinum.[165, p. 7]

[153a] Gudris, N., and Kulikowa, L., *Z. Physik*, **25**, 121-132 (1924).
[154] Hulett, G. A., *Z. physik. Chem.*, **42**, 353-368 (1902-1903).
[155] McHaffie, I. R., and Lenher, S., *J. Chem. Soc. (London)*, **127**, 1559-1572 (1925).

PROPERTIES OF ORDINARY WATER-SUBSTANCE

Unit of $P = 1$ mm-Hg; of D and $\delta = 1$ in last tabulated digit of P. Temp. $= (t_1 + t_2)$ °C

$t_2 \rightarrow$ t_1	0.0	0.1	0.2	0.3	0.4	0.5	0.6	0.7	0.8	0.9	D	δ
0	4.579	4.613	4.647	4.681	4.715	4.750	4.785	4.820	4.855	4.890	34.7	−1
1	4.926	4.962	4.998	5.034	5.070	5.107	5.144	5.181	5.219	5.256	36.8	
2	5.294	5.332	5.370	5.408	5.447	5.486	5.525	5.565	5.605	5.645	39.1	
3	5.685	5.725	5.766	5.807	5.848	5.889	5.931	5.973	6.015	6.058	41.6	
4	6.101	6.144	6.187	6.230	6.274	6.318	6.363	6.408	6.453	6.498	44.2	
5	6.543	6.589	6.635	6.681	6.728	6.775	6.822	6.869	6.917	6.965	47.0	+7
6	7.013	7.062	7.111	7.160	7.209	7.259	7.309	7.360	7.411	7.462	50.0	
7	7.513	7.565	7.617	7.669	7.722	7.775	7.828	7.882	7.936	7.990	53.2	
8	8.045	8.100	8.155	8.211	8.267	8.323	8.380	8.437	8.494	8.551	56.4	
9	8.609	8.668	8.727	8.786	8.845	8.905	8.965	9.025	9.086	9.147	60.0	
10	9.209	9.271	9.333	9.395	9.458	9.521	9.585	9.649	9.714	9.779	63.5	+11
11	9.844	9.910	9.976	10.042	10.109	10.176	10.244	10.312	10.380	10.449	67.4	
12	10.518	10.588	10.658	10.728	10.799	10.870	10.941	11.013	11.085	11.158	71.3	
13	11.231	11.305	11.379	11.453	11.528	11.604	11.680	11.756	11.833	11.910	75.6	
14	11.987	12.065	12.144	12.223	12.302	12.382	12.462	12.543	12.624	12.706	80.1	
15	12.788	12.870	12.953	13.037	13.121	13.205	13.290	13.375	13.461	13.547	84.6	+17
16	13.634	13.721	13.809	13.898	13.987	14.076	14.166	14.256	14.347	14.438	89.6	
17	14.530	14.622	14.715	14.809	14.903	14.997	15.092	15.188	15.284	15.380	94.7	
18	15.477	15.575	15.673	15.772	15.871	15.971	16.071	16.171	16.272	16.374	100.0	
19	16.477	16.581	16.685	16.789	16.894	16.999	17.105	17.212	17.319	17.427	105.8	
20	17.535	17.644	17.753	17.863	17.974	18.085	18.197	18.309	18.422	18.536	111.5	+23
21	18.650	18.765	18.880	18.996	19.113	19.231	19.349	19.468	19.587	19.707	117.7	
22	19.827	19.948	20.070	20.193	20.316	20.440	20.565	20.690	20.815	20.941	124.1	
23	21.068	21.196	21.324	21.453	21.583	21.714	21.845	21.977	22.110	22.243	130.9	
24	22.377	22.512	22.648	22.785	22.922	23.060	23.198	23.337	23.476	23.616	137.9	
25	23.756	23.897	24.039	24.182	24.326	24.471	24.617	24.764	24.912	25.060	145.3	+27
26	25.209	25.359	25.509	25.660	25.812	25.964	26.117	26.271	26.426	26.582	153.0	
27	26.739	26.897	27.055	27.214	27.374	27.535	27.696	27.858	28.021	28.185	161.0	
28	28.349	28.514	28.680	28.847	29.015	29.184	29.354	29.525	29.697	29.870	169.4	
29	30.043	30.217	30.392	30.568	30.745	30.923	31.102	31.281	31.461	31.642	178.1	
30	31.824	32.007	32.191	32.376	32.561	32.747	32.934	33.122	33.312	33.503	187.1	+32
31	33.695	33.888	34.082	34.276	34.471	34.667	34.864	35.062	35.261	35.462	196.8	
32	35.663	35.865	36.068	36.272	36.477	36.683	36.891	37.099	37.308	37.518	206.6	
33	37.729	37.942	38.155	38.369	38.584	38.801	39.018	39.237	39.457	39.677	216.9	
34	39.898	40.121	40.344	40.569	40.796	41.023	41.251	41.480	41.710	41.942	227.7	

88. SATURATED WATER AND STEAM: P-V-T

t	0	1	2	3	4	5	6	7	8	9	D	
35	42.175	42.409	42.644	42.880	43.117	43.355	43.595	43.836	44.078	44.320	238.8	+35
36	44.563	44.808	45.054	45.301	45.549	45.799	46.050	46.302	46.556	46.811	250.4	
37	47.067	47.324	47.582	47.841	48.102	48.364	48.627	48.891	49.157	49.424	262.5	
38	49.692	49.961	50.231	50.502	50.774	51.048	51.323	51.600	51.879	52.160	275.0	
39	52.442	52.725	53.009	53.294	53.580	53.867	54.156	54.446	54.737	55.030	288.2	
40	55.324	55.61	55.91	56.21	56.51	56.81	57.11	57.41	57.72	58.03	30.2	+36
41	58.34	58.65	58.96	59.27	59.58	59.90	60.22	60.54	60.86	61.18	31.6	
42	61.50	61.82	62.14	62.47	62.80	63.13	63.46	63.79	64.12	64.46	33.0	
43	64.80	65.14	65.48	65.82	66.16	66.51	66.86	67.21	67.56	67.91	34.6	
44	68.26	68.61	68.97	69.33	69.69	70.05	70.41	70.77	71.14	71.51	36.2	
45	71.88	72.25	72.62	72.99	73.36	73.74	74.12	74.50	74.88	75.26	37.7	+4
46	75.65	76.04	76.43	76.82	77.21	77.60	78.00	78.40	78.80	79.20	39.5	
47	79.60	80.00	80.41	80.82	81.23	81.64	82.05	82.46	82.87	83.29	41.1	
48	83.71	84.13	84.56	84.99	85.42	85.85	86.28	86.71	87.14	87.58	43.1	
49	88.02	88.46	88.90	89.34	89.79	90.24	90.69	91.14	91.59	92.05	44.6	

$t \backslash P$	0	1	2	3	4	5	6	7	8	9	D	
50	92.51	97.20	102.09	107.20	112.51	118.04	123.80	129.82	136.08	142.60	568.7	+2
60	149.38	156.43	163.77	171.38	179.31	187.54	196.09	204.96	214.17	223.73	843.2	−1
70	233.7	243.9	254.6	265.7	277.2	289.1	301.4	314.1	327.3	341.0	121.4	0
80	355.1	369.7	384.8	400.6	416.8	433.6	450.9	468.7	487.1	506.1	170.7	−1

$t \backslash P$	0.0	0.1	0.2	0.3	0.4	0.5	0.6	0.7	0.8	0.9	D	
90	525.76	527.76	529.77	531.78	533.80	535.82	537.86	539.90	541.95	544.00	202.9	−10
91	546.05	548.11	550.18	552.26	554.35	556.44	558.53	560.64	562.75	564.87	209.4	
92	566.99	569.12	571.26	573.40	575.55	577.71	579.87	582.04	584.22	586.41	216.1	
93	588.60	590.80	593.00	595.21	597.43	599.66	601.89	604.13	606.38	608.64	223.0	
94	610.90	613.17	615.44	617.72	620.01	622.31	624.61	626.92	629.24	631.57	230.0	
95	633.90	636.24	638.59	640.94	643.30	645.67	648.05	650.43	652.82	655.22	237.2	−10
96	657.62	660.03	662.45	664.88	667.31	669.75	672.20	674.66	677.12	679.59	244.5	
97	682.07	684.55	687.04	689.54	692.05	694.57	697.10	699.63	702.17	704.71	252.0	
98	707.27	709.83	712.40	714.98	717.56	720.15	722.75	725.36	727.98	730.61	259.7	
99	733.24	735.88	738.53	741.18	743.85	746.52	749.20	751.89	754.58	757.29	267.6	
100	760.00	762.72	765.45	768.19	770.93	773.68	776.44	779.22	782.00	784.78	275.7	0
101	787.57	790.37	793.18	796.00	798.82	801.66	804.50	807.35	810.21	813.08	283.9	+6

amount by which the equilibrial pressure of the vapor has been increased by the presence of the inert gas, ρ = density of the liquid, T °K = absolute temperature, R = the gas-constant expressed in appropriate units, M = molecular weight of the vapor, and n_g and n_l = number of molecules of gas and of liquid, respectively, that make up the liquid phase, then if Δp is small, it is quite closely given by the formula $\Delta p/p_0 = M p_g/\rho RT - n_g/(n_g + n_l)$. A more accurate formula is $\log_e[(n_g + n_l)p'/n_l p_0] = M p_g/\rho' RT$, where $p' = p_0 + \Delta p$, and ρ' is the density of the liquid at a pressure midway between p_0 and $(p' + p_g)$. It must be remembered that p_g is the actual partial pressure at the surface of the liquid, and that n_g is the actual number of molecules dissolved in the liquid; if the liquid is boiling, these values are quite different from what they would be under other conditions. Under suitable boiling conditions they approach, and may become, zero.

Apparently no direct determination of the equilibrial pressure of the

Table 246.—Comparison of Smoothed Values for the Vapor-pressure of Water (mm-Hg, atm)

In this table several sets of smoothed values for the vapor-pressure (P) of water are compared with those preferred ones (P_{OM}) defined by formula (1) of Osborne and Meyers, and given in Table 242. Some of these sets represent values computed by means of smoothing equations that have been published; all have been taken from published tables of adjusted values. None of them are directly observed values. Graphical comparisons of the observed values have been published by N. S. Osborne,[169] and by N. S. Osborne and C. H. Meyers.[156]

$P = P_{OM} + \delta$, where P_{OM} is the value defined by the Osborne-Meyers formula. When P has been published with less than 5 significant figures (the number here given for P_{OM}) or in such units that the last specified digit does not suffice to determine the fifth digit when P is expressed in the unit here used, then the surplus or undetermined digits in δ are depressed; when the significant figures in δ are all depressed, they are preceded by a 0 in the normal position. For example, the KS value for 160 °C is $6.0996 + 0_4 = 6.100$ atm.

Unit of $p = 1$ mm-Hg; of $P = 1$ atm $= 1.01325$ bars $= 1.03323$ kg*/cm²; of $\delta = 1$ unit in last place in p_{OM} or P_{OM}

References[a]→ t	p_{OM}	ICT₁	IST[b] —δ—	SKG₂	References[a]→ t	p_{OM}	ICT₁	IST[b] —δ—	SKG₂	SM
0	4.5807	−1₇	+0₄	+8₆	50	92.490	+2₀	+2₀	+13	−22₀
5	6.5360	+7₀		+11₀	55	118.04	0		0	−17
10	9.1975	+11₅	+6₅	+14₆	60	149.39	−1	+2	−1	−26
15	12.771	+17		+18	65	187.56	−2		−2	−37
20	17.512	+23	+16	+22	70	233.72	−0₂	0	−3	−28
25	23.729	+27		+24	75	289.13	−0₃	0	−3	−35
30	31.792	+32	+24	+25	80	355.22	−1₂	0	−4	−32
35	42.139	+36		+26	85	433.56	+0₄		−5	−2
40	55.288	+36	+29	+23	90	525.86	−10	0	−4	+8
45	71.840	+4₀		+20	95	634.00	−10		−3	

[156] Osborne, N. S., and Meyers, C. H., *J. Res. Nat. Bur. Stand.*, 13, 1-20 (RP691) (1934)→ *Mech. Eng.*, 56, 207-209 (1934).

88. SATURATED WATER AND STEAM: P-V-T

Table 246.—(Continued)

References[a] → t	P_{OM}	OSFG	SKG$_1$	KS	ICT$_2$ — δ —	JF	WT	EC	C
100	1.0000	0	0		0		0		
110	1.4139	−1	0		0		0		
120	1.9594	−1	+1		0		+1		
130	2.6658	0	+4		+2		+4		
140	3.5663	+1	+5		+0$_7$		+0$_7$		
150	4.6975	+2	+8		+0$_5$		+0$_5$		
160	6.0996	+4	+11	+0$_4$	+0$_4$		+0$_4$		
170	7.8167	+4	+13	+2$_3$	+0$_3$		+1$_3$	+79	
180	9.8960	+2	+14	−3$_0$	−1$_0$	+0$_{50}$	0$_0$	+126	
190	12.388	0	+2	−10	−2	0$_0$	−1	+12	
200	15.347	0	+2	−12	−6	−0$_7$	−4	+8	+13$_5$
210	18.830	0	+1	+12	−7	−0$_6$	−4	+5	+17$_8$
220	22.898	−1	0	−8	−9	−0$_0$	−5	+5	+22$_3$
230	27.613	−1	0	−23	−10	−1$_0$	−5	+8	+26$_7$
240	33.042	0	−1	−35	−15	−1$_0$	−9	+6	+31$_1$
250	39.256	0	−2	−25	−22	−2$_0$	−15	−4	+35$_4$
260	46.326	+2	−3	−15	−26	−2$_4$	−17	−1	+38$_5$
270	54.331	+3	−4	−20	−40	−3$_5$	−28	+14	+39$_3$
280	63.352	+1	−6	−41	−57	−5$_5$	−43	+14	+37$_3$
290	73.475	−3	−9	−47	−5$_5$	−5$_5$	−3$_5$	−4	+32$_6$
300	84.793	−5	−14	+4	−1$_3$	−0$_1$	+0$_7$	−11	+17$_3$
310	97.406	−4	−21	+30	−0$_6$	+0$_7$	+2$_4$	+8	+8$_7$
320	111.42	0	−3	+5	+1		+4	+2	−1$_7$
330	126.96	0	−3	+10	+3		+7	−1	−4$_6$
340	144.17	0	−4	+16	+3		+7	−2	−7$_9$
350	163.20	0	−3	+16	−4		+1	+2	−9$_6$
360	184.29	0	−2	+15	−22		−16	+1	−9$_5$
370	207.77	+1	−3	+7	−28		−21	+7	+1$_5$
371	210.27	+1		+9	−29		−21	+9	
372	212.80	+1	−5	+12	+29		−21	+8	
373	215.37	0		+15	+28		−20	+8	
374	217.98	−2	−10	+20	+26		−18	+7	+13$_8$

[a] References:

C   Callendar, H. L., *Proc. Inst. Mech. Eng.*, **1929**, 507-527 (1929).
EC   Egerton, A., and Callendar, G. S., *Phil. Trans. (A)*, **231**, 147-205 (A698) (1932).
ICT$_1$ From compilation by Washburn, E. W.[107] = WT; see Table 245.
ICT$_2$ From the compilation by Keyes, F. G., *Int. Crit. Tables*, **3**, 233 (1928), and derived by him from WT by the application of corrections required to make the temperature scale correspond to that on which the boiling point of sulphur under a pressure of 1 atm is 444.60 °C instead of 444.55 °C, which is assumed in WT.
IST   Int. Skeleton Steam Tables, 1934, see Table 260.
JF   Jakob, M., and Fritz, W., *Techn. Mech. u. Thermodyn.*, **1**, 173-183, 236-240 (1930); *Forsch. Gebiete Ingenieurw.*, **4**, 295-299 (1933).
KS   Keyes, F. G., and Smith, L. B., *Mech. Eng.*, **52**, 124-127 (1930).
OSFG   Osborne, N. S., Stimson, H. F., Fiock, E. F., and Ginnings, D. C., *Bur. Stand. J. Res.*, **10**, 155-188 (RP523) (1933).
SKG   Smith, L. B., Keyes, F. G., and Gerry, H. T., *Proc. Amer. Acad. Arts. Sci.*, **69**, 137-168 (1934); they publish two smoothing equations—(1) for $t > 100$ °C and (2) for $t < 100$ °C, as here indicated by subscripts—and extended tables of values of p and of dp/dt.
SM   Smith, A., and Menzies, A. W. C., *Ann. d. Phys. (4)*, **33**, 971-978 (1910).
WT   Holborn, L., Scheel, K., and Henning, F.[165]

[b] Above $t = 100$ °C the IST values are identical with P_{OM}.

---

vapor of water in contact with an inert gas is available, but insofar as it is allowable to regard water-vapor as an ideal gas the increase Δp may be computed by means of the formula $\Delta p/p_0 = (\rho' - \rho_0)/\rho_0$, where p_0 is the value of the vapor-pressure of water when no inert gas is present, ρ_0 is the corresponding density of the pure saturated vapor, and ρ' is the observed density of water-vapor saturated in the presence of the inert gas; all corresponding to the same temperature. These values of $\Delta p/p_0$ may be read directly from the values of ρ'/ρ_0 given in Table 252. It will be noticed that the several sets of comparable values there given are discordant both with one another and with the values computed by means of the preceding formula.

Curvature of Surface, Effect of.—If p = equilibrial pressure of the vapor in contact with a spherical surface of the liquid (radius = r), p_0 = that of the vapor in contact with a flat surface, γ = surface-tension of the liquid, and the other symbols have the significance already stated, then $(p - p_0)/p_0 = \pm [\gamma \rho_0/(\rho - \rho_0)p_0](1/r_1 + 1/r_2) = \pm 2\gamma/\rho RTr$ approximately, if $(p - p_0)$ and $(\rho - \rho_0)$ are small, and $r_1 = r_2 = r$, r_1 and r_2 being the principal radii of curvature of the surface; the plus sign is to be taken if the concavity of the surface is toward the liquid.[151] (If r is so small that $p - p_0$ is not small in comparison with p_0, then the more exact expression $\log_e(p/p_0) = (\gamma/RT\rho)(1/r_1 + 1/r_2)$ must be used; see R. v. Helmholtz [152] or C. T. R. Wilson.[153]) Obviously, the unit of mass appearing in R must be the same as that in ρ. For water at 20 °C (T = 293.1 °K), γ = 72.75 dynes/cm, ρ = 0.9984 g/cm^3, R = 8.315 × 10^7 erg/g-mole·°C = 4.615 × 10^6 erg/g·°C, whence $r(p - p_0)/p_0 = -0.001077\,\mu$ if the concavity is turned away from the liquid (1 μ = 0.001 mm). Whence the following:

r	1000	100	10	5	1	0.5 μ
$(p_0 - p)/p_0$	0.0001	0.0011	0.0108	0.0215	0.108	0.215%

Certain observations by W. A. Patrick and his associates indicate that over concave menisci in very small tubes (r = a few microns) the value of $(p_0 - p)$ exceeds that demanded by this formula. They are inconsistent with the observations of N. Gudris and L. Kulikowa,[153a] and seem to be

Table 247.—Thermal Rate of Variation in the Vapor-pressure of Water (atm) [156]

These values, computed by means of formula (2), are believed to be the most accurate available; they supersede the slightly different ones previously published by Osborne, Stimson, Fiock, and Ginnings.[170] In Table 248 certain of these values are expressed in mm-Hg, and other sets of values are compared with them.

The authors give values for each °C in kg*/cm^2 as well as in atm, and for each °F in lb*/in^2. And Osborne, Stimson, and Ginnings [171] give the values of $T(dP/dT)$ in Int. joules/cm^3 for these same values of dP/dT (steps not greater than 5 °C; range, 100 to 374.15 °C).

[157] Adamson, A., *Mem. Proc. Manchester Lit. Phil. Soc.*, **76**, 1-9 (1931).
[158] Batschinski, A., *Nature*, **119**, 198 (1927).
[159] Fischer, V., *Z. Physik*, **43**, 131-151 (1927).
[160] Hofbauer, P. H., *Atti pontif. Acc. sci. nuovi Lincei*, **84**, 353-363, 581-586 (1931).
[161] Kiréeff, V., *Jour. de Phys.* (7), **3**, 150D (A) (1932) ← *Jour. chim. phys. Russe*, **1**, 241-248 (1930).

88. SATURATED WATER AND STEAM: P-V-T

Unit of $dP/dt = 1$ milliatm/°C. 1 milliatm = 1.0332 g*/cm² = 1.0136 millibar = 0.760000 mm-Hg.
Temp. = $(t_1 + t_2)$ °C (Int. scale)

$t_2 \rightarrow$ t_1	0	1	2	3	4	5	6	7	8	9
					dP/dt					
−10	0.4373	0.4662	0.4968	0.5292	0.5633	0.3146	0.3364	0.3595	0.3840	0.4099
0	.8103	.8593	.9109	.9651	1.0220	.5993	.6372	.6772	.7193	.7636
+10	1.4207	1.5063	1.5891	1.6758	1.7665	1.0818	1.1445	1.2102	1.2792	1.3514
20	2.4015	2.5240	2.6516	2.7846	2.9230	1.8612	1.9602	2.0636	2.1715	2.2841
30	3.8796	4.0618	4.2509	4.4472	4.6510	3.0672	3.2172	3.3733	3.5356	3.7043
40						4.8624	5.082	5.309	5.544	5.788
50	6.041	6.303	6.574	6.854	7.144	7.444	7.754	8.074	8.405	8.747
60	9.100	9.465	9.841	10.229	10.630	11.043	11.468	11.907	12.359	12.825
70	13.305	13.799	14.307	14.830	15.368	15.922	16.492	17.077	17.679	18.297
80	18.932	19.585	20.255	20.943	21.649	22.374	23.118	23.881	24.663	25.465
90	26.288	27.131	27.995	28.880	29.786	30.715	31.666	32.639	33.636	34.655
100	35.699	36.766	37.858	38.974	40.115	41.282	42.474	43.693	44.938	46.210
110	47.509	48.836	50.19	51.57	52.99	54.43	55.90	57.40	58.93	60.49
120	62.08	63.70	65.35	67.04	68.76	70.51	72.29	74.11	75.96	77.84
130	79.76	81.72	83.71	85.74	87.80	89.89	92.03	94.20	96.41	98.66
140	100.94	103.26	105.62	108.03	110.74	112.95	115.47	118.03	120.64	123.28
150	125.96	128.69	131.46	134.28	137.13	140.04	142.98	145.97	149.00	152.08
160	155.20	158.37	161.58	164.85	168.15	171.51	174.91	178.36	181.86	185.40
170	189.00	192.64	196.34	200.08	203.87	207.72	211.61	215.56	219.55	223.60
180	227.70	231.85	236.06	240.32	244.63	248.99	253.41	257.89	262.42	267.00
190	271.64	276.33	281.08	285.89	290.75	295.67	300.65	305.68	310.77	315.92
200	321.13	326.40	331.73	337.11	342.56	348.06	353.63	359.25	364.94	370.69
210	376.50	382.37	388.30	394.30	400.36	406.48	412.67	418.92	425.23	431.61
220	438.05	444.56	451.13	457.77	464.47	471.24	478.07	484.98	491.95	498.98
230	506.1	513.3	520.5	527.8	535.2	542.6	550.2	557.8	565.4	573.2
240	581.0	588.8	596.8	604.8	612.9	621.0	629.3	637.6	646.0	654.4
250	662.9	671.6	680.2	689.0	697.8	706.7	715.7	724.8	733.9	743.2
260	752.5	761.8	771.3	780.8	790.5	800.2	809.9	819.8	829.8	839.8
270	849.9	860.1	870.4	880.7	891.2	901.7	912.3	923.0	933.8	944.7
280	955.7	966.8	977.9	989.2	1000.5	1011.9	1023.4	1035.0	1046.8	1058.6
290	1070.4	1082.4	1094.5	1106.7	1119.0	1131.4	1143.9	1156.4	1169.1	1181.9
300	1194.8	1207.8	1220.9	1234.1	1247.4	1260.8	1274.3	1288.0	1301.7	1315.6
310	1329.5	1343.6	1357.8	1372.1	1386.6	1401.1	1415.8	1430.6	1445.5	1460.6
320	1475.7	1491.0	1506.4	1522.0	1537.7	1553.5	1569.5	1585.6	1601.9	1618.3
330	1634.8	1651.5	1668.3	1685.3	1702.5	1719.8	1737.3	1755.0	1772.8	1790.8
340	1809.0	1827.3	1845.9	1864.6	1883.6	1902.8	1922.2	1941.8	1961.6	1981.7
350	2002.0	2022.5	2043.4	2064.5	2085.9	2107.6	2129.6	2151.9	2174.6	2197.7
360	2221.2	2245.0	2269.3	2294.1	2319.4	2345.2	2371.6	2398.7	2426.6	2455.3
370	2485.0	2515.9	2548.6	2584.1	2630.7					

inconsistent with the results obtained in another field by R. Buckley (see discussion on pp. 513, 518).

Tension, Effect of.—G. A. Hulett [154] has reported observations which he

Table 248.—Comparison of Thermal Rates of Variation in the Vapor-pressure of Water (mm-Hg)

$dp/dt = (dp/dt)_{OM} + \delta$ where $(dp/dt)_{OM}$ is the corresponding value from Table 247.

1 mm-Hg = 1.3158 milliatm = 1.3564 g*/cm² = 1.3332 millibars.

Unit of $(dp/dt) = 1$ mm-Hg/°C; of $\delta = 1$ unit in last place in $(dp/dt)_{OM}$

t	Ref[a]→ $(dp/dt)_{OM}$	SKG	WT	JF δ	J	OSFG	t	Ref[a]→ $(dp/dt)_{OM}$	JF	J δ	OSFG
0	0.3323	+6	+12				200	244.0	0	−2	0
10	.6158	+7	+20				210	286.1	0	0	0
20	1.080	+5	+6				220	332.9	0		0
30	1.825	0	+1	−1			230	384.6	−1		0
40	2.948	0	−1	−7			240	441.6	−6		0
50	4.591	−1	−4	−5			250	503.8	−2		+1
60	6.916	−2	−8	−10			260	571.9	−8		+1
70	10.11	0	−1	−3			270	645.9	−9		+1
80	14.38	−1	0	0			280	726.3	−10		−3
90	19.98	0	−1	+2			290	813.5	+5		−2
100	27.13	+1	+1	+3	−1		300	908.0	+40		−1
110	36.11	+1	+1	0	−1		310	1010.4	−5		+2
120	47.18	+1	+1	0	0		320	1121.5			+2
130	60.62	+1	0	+2	+1		330	1242.4			+1
140	76.71	0	−3	−7	+1		340	1374.8			−3
150	95.73	−2	−10	+3	+1		350	1521.5			−3
160	118.0		−3	0	0		360	1688.1			+6
170	143.6		−5	+1	0		365	1782.4			+11
180	173.0		−8	0	−2	0	370	1888.6			−3
190	206.4			−1	−1	0	374	1999.3			−181

H. Moser.[172]

t	$(dp/dt)_{OM}$	δ	t	$(dp/dt)_{OM}$	δ	t	$(dp/dt)_{OM}$	δ
96.5	24.434	+55	98	25.563	−45	99.5	26.732	−38
97	24.806	+13	98.5	25.948	−54	100	27.131	−9
97.5	25.182	−20	99	26.338	−53	100.5	27.534	+38

[a] References:
J   Jakob, M., *Forsch.-Arb. Gebiete Ingenieurw.*, **310**, 9-19 (1928).
JF   Jakob, M., and Fritz, W., *Techn. Mech. Thermodynam.*, **1**, 173-183, 236-240 (1930).
OSFG   Osborne, N. S., Stimson, H. F., Fiock, E. F., and Ginnings, D. C., *Bur. Stand. J. Res.*, **10**, 155-188 (RP523) (1933).
SKG   Smith, L. B., Keyes, F. G., and Gerry, H. T., *Proc. Amer. Acad. Arts Sci.*, **69**, 137-168 (1934).
WT   Holborn, L., Scheel, K., and Henning, F., "Wärmetabellen," 1919.

Table 249.—Temperature of Saturated Water-vapor: 0.0075 to 225 kg*/cm²

Adapted from a table by N. S. Osborne and C. H. Meyers [156] and on the same basis as the data in Table 242. The authors give a similar table in terms of lb*/in² and °F, but do not give the Δ's. (See also Table 253.)

[162] Smith, L. B., Keyes, F. G., and Gerry, H. T., *Proc. Amer. Acad. Arts Sci.*, **69**, 137-168 (1934); **70**, 319-364 (1935).

88. SATURATED WATER AND STEAM: P-V-T

Table 249.—*(Continued)*

Values of t for intermediate values of P may be computed by means of the well-known formula:

$$f(x+h) = f(x) + k\left\{\Delta_1(x) - \frac{1-k}{2}\Delta_2(x) + \frac{(1-k)(2-k)}{3!}\Delta_3(x) - \frac{(1-k)(2-k)(3-k)}{4!}\Delta_4(x) + \cdots\right\}$$

where $k = h/s$, s being the length of the successive (equal) steps in the table, and Δ_n being the nth difference (p. 656). The first and second differences are tabulated on the same lines as the P's to which they correspond. They will be sufficient except where Δ_2 is varying rapidly; in which case, higher differences should be computed and used.

For example, taking one of the worst cases, find the temperature at which the saturation pressure is 0.0077213 kg*/cm². The nearest tabular value below the assigned pressure is $P = 0.0075$, corresponding to $t = 2.590$ °C, $1000\Delta_1 = 4117$, $1000\Delta_2 = -828$, $1000\Delta_3 = +292$, $1000\Delta_4 = -135$, $1000\Delta_5 = +77$; $h = 0.0002213$, $s = 0.0025$, hence $k = 0.0884$. Putting these in the formula, one obtains the value

$$t = 2.590 + \frac{0.0884}{1000}\{4117 + 377 + 85 + 29 + 13 + \cdots\} = 2.998_5 \text{ °C}$$

Whereas it is seen from Table 242 that at 3 °C $P = 0.007473$ atm = 0.0077213 kg*/cm². Hence, the computed value differs from the true by only 0.001_5 °C. Had all differences beyond the second been neglected, the value 2.987 °C would have been found.

Similarly for $P = 0.0185124$ kg*/cm² one obtains $t = 15.997$ °C if only the first and second differences are used; 16.000 °C if the third is included. From Table 242 it is seen that at 16 °C, $P = 0.017917$ atm = 0.0185124 kg*/cm². Here the computed value is in error by less than 0.001 °C when only 3 differences are used.

Unit of $P = 1$ kg*/cm² $= 0.967841$ atm $= 735.559$ mm-Hg $= 0.90665$ bar. Temp. $= t$ °C, Int. scale

P	t	$1000\,\Delta_1$	$-1000\,\Delta_2$	P	t	$1000\,\Delta_1$	$-1000\,\Delta_2$
0.0075	2.590	4117	828	0.065	37.302	1370	83
.0100	6.707	3289	536	.070	38.672	1287	72
.0125	9.996	2753	379	.075	39.959	1215	63
.0150	12.749	2374	280	.080	41.174	1152	58
.0175	15.123	2094	220	.085	42.326	1094	51
.0200	17.217	1874	174	.090	43.420	1043	46
.0225	19.091	1700	142	.095	44.463	997	
.0250	20.791	1558	119	.100	45.460	1871	140
.0275	22.349	1439		.11	47.331	1731	120
.030	23.788	2587	304	.12	49.062	1611	103
.035	26.375	2283	236	.13	50.673	1508	90
.040	28.658	2047	190	.14	52.181	1418	78
.045	30.705	1857	154	.15	53.599	1340	70
.050	32.562	1703	129	.16	54.939	1270	63
.055	34.265	1574	111	.17	56.209	1207	55
.060	35.839	1463	93	.18	57.416	1152	51

Table 249.—*(Continued)*

P	t	1000 Δ₁	−1000 Δ₂	P	t	1000 Δ₁	−1000 Δ₂
0.19	58.568	1101	46	4.3	145.538	839	15
.20	59.669	1055	41	4.4	146.377	824	15
.21	60.724	1014	39	4.5	147.201	809	14
.22	61.738	975	36	4.6	148.010	795	13
.23	62.713	939	32	4.7	148.805	782	13
.24	63.652	907	31	4.8	149.587	769	13
.25	64.559	876		4.9	150.356	756	
.26	65.435	1671	99	5.0	151.112	1477	45
.28	67.106	1572	86	5.2	152.589	1432	41
.30	68.678	1486	77	5.4	154.021	1391	40
.32	70.164	1409	68	5.6	155.412	1351	36
.34	71.573	679	61	5.8	156.763	1315	35
.35	72.252	662		6.0	158.078	1280	33
.36	72.914	1280	56	6.2	159.358	1247	30
.38	74.194	1224		6.4	160.605	1217	29
.40	75.418	2849	256	6.6	161.822	1188	27
.45	78.267	2593	208	6.8	163.010	1161	27
.50	80.860	2385	176	7.0	164.171	1134	24
.55	83.245	2209	149	7.2	165.305	1110	24
.60	85.454	2060	128	7.4	166.415	1086	22
.65	87.514	1932	112	7.6	167.501	1064	22
.70	89.446	1820	98	7.8	168.565	1042	20
.75	91.266	1722	88	8.0	169.607	1022	20
.80	92.988	1634	78	8.2	170.629	1002	18
.85	94.622	1556	71	8.4	171.631	984	18
.90	96.178	1485	61	8.6	172.615	966	18
.95	97.663	1424		8.8	173.581	948	16
1.0	99.087	2677	194	9.0	174.529	932	16
1.1	101.764	2483	166	9.2	175.461	916	15
1.2	104.247	2317	142	9.4	176.377	901	15
1.3	106.564	2175	125	9.6	177.278	886	14
1.4	108.739	2050	109	9.8	178.164	872	
1.5	110.789	1941	98	10.0	179.036	2122	79
1.6	112.730	1843	86	10.5	181.158	2043	72
1.7	114.573	1757	78	11.0	183.201	1971	65
1.8	116.330	1679	71	11.5	185.172	1906	62
1.9	118.009	1608	65	12.0	187.078	1844	57
2.0	119.617	1543	58	12.5	188.922	1787	53
2.1	121.160	1485	55	13.0	190.709	1734	49
2.2	122.645	1430	49	13.5	192.443	1685	47
2.3	124.075	1381	46	14.0	194.128	1638	44
2.4	125.456	1335	44	14.5	195.766	1594	40
2.5	126.791	1291	38	15.0	197.360	1554	39
2.6	128.082	1253	38	15.5	198.914	1515	37
2.7	129.335	1215	35	16.0	200.429	1478	34
2.8	130.550	1180	32	16.5	201.907	1444	32
2.9	131.730	1148	31	17.0	203.351	1412	32
3.0	132.878	1117	29	17.5	204.763	1380	29
3.1	133.995	1088	27	18.0	206.143	1351	28
3.2	135.083	1061	25	18.5	207.494	1323	27
3.3	136.144	1036	24	19.0	208.817	1296	26
3.4	137.180	1012	24	19.5	210.113	1270	24
3.5	138.192	988	22	20.0	211.383	1246	23
3.6	139.180	966	21	20.5	212.629	1223	22
3.7	140.146	945	19	21.0	213.852	1201	22
3.8	141.091	926	20	21.5	215.053	1179	21
3.9	142.017	906	17	22.0	216.232	1158	19
4.0	142.923	889	17	22.5	217.390	1139	19
4.1	143.812	872	18	23.0	218.529	1120	19
4.2	144.684	854	15	23.5	219.649	1101	17

88. SATURATED WATER AND STEAM: P-V-T

Table 249.—(Continued)

P	t	1000 Δ₁	−1000 Δ₂	P	t	1000 Δ₁	−1000 Δ₂
24.0	220.750	1084	17	78	291.864	881	8
24.5	221.834	1067	17	79	292.745	873	8
25.0	222.901	1050	15	80	293.618	865	9
25.5	223.951	1035	15	81	294.483	856	8
26.0	224.986	1020	15	82	295.339	848	7
26.5	226.006	1005	15	83	296.187	841	8
27.0	227.011	990	13	84	297.028	833	8
27.5	228.001	977	13	85	297.861	825	7
28.0	228.978	964	14	86	298.686	818	7
28.5	229.942	950	12	87	299.504	811	7
29.0	230.892	938	12	88	300.315	804	7
29.5	231.830	926		89	301.119	797	7
30.0	232.756	1817	45	90	301.916	790	7
31	234.573	1772	42	91	302.706	783	6
32	236.345	1730	39	92	303.489	777	7
33	238.075	1691	38	93	304.266	770	5
34	239.766	1653	37	94	305.036	765	7
35	241.419	1616	33	95	305.801	758	6
36	243.035	1583	32	96	306.559	752	6
37	244.618	1551	32	97	307.311	746	6
38	246.169	1519	29	98	308.057	740	6
39	247.688	1490	28	99	308.797	734	
40	249.178	1462	28	100	309.531	1452	22
41	250.640	1434	25	102	310.983	1430	21
42	252.074	1409	25	104	312.413	1409	20
43	253.483	1384	24	106	313.822	1389	20
44	254.867	1360	23	108	315.211	1369	20
45	256.227	1337	22	110	316.580	1349	18
46	257.564	1315	21	112	317.929	1331	18
47	258.879	1294	21	114	319.260	1313	18
48	260.173	1273	19	116	320.573	1295	17
49	261.446	1254	20	118	321.868	1278	16
50	262.700	1234	17	120	323.146[a]	1262	17
51	263.934	1217	19	122	324.408	1245	15
52	265.151	1198	16	124	325.653	1230	15
53	266.349	1182	18	126	326.883	1215	15
54	267.531	1164	15	128	328.098	1200	15
55	268.695	1149	16	130	329.298[a]	1185	14
56	269.844	1133	14	132	330.483	1171	13
57	270.977	1119	16	134	331.654	1158	14
58	272.096	1103	14	136	332.812	1144	13
59	273.199	1089	13	138	333.956	1131	13
60	274.288	1076	14	140	335.087[a]	1118	12
61	275.364	1062	13	142	336.205	1106	13
62	276.426	1049	12	144	337.311	1093	11
63	277.475	1037	14	146	338.404	1082	12
64	278.512	1023	11	148	339.486	1070	12
65	279.535	1012	10	150	340.556[a]	1058	11
66	280.547	1002	13	152	341.614	1047	11
67	281.549	989	11	154	342.661	1036	10
68	282.538	978	11	156	343.697	1026	11
69	283.516	967	10	158	344.723	1015	11
70	284.483	957	10	160	345.738[a]	1004	9
71	285.440	947	11	162	346.742	995	11
72	286.387	936	9	164	347.737	984	9
73	287.323	927	10	166	348.721	975	10
74	288.250	917	9	168	349.696	965	9
75	289.167	908	9	170	350.661[a]	956	10
76	290.075	899	9	172	351.617	946	8
77	290.974	890	9	174	352.563	938	10

Table 249.—(Continued)

P	t	1000 Δ₁	−1000 Δ₂	P	t	1000 Δ₁	−1000 Δ₂
176	353.501	928	9	202	364.914	823	8
178	354.429	919	8	204	365.737	815	8
180	355.348ᵃ	911	9	206	366.552	807	7
182	356.259	902	8	208	367.359	800	8
184	357.161	894	8	210	368.159	792	6
186	358.055	886	9	212	368.951	786	9
188	358.941	877	8	214	369.737	777	6
190	359.818ᵃ	869	8	216	370.514	771	8
192	360.687	861	8	218	371.285	763	8
194	361.548	853	7	220	372.048	755	8
196	362.401	846	9	222	372.803	747	
198	363.247	837	7	224	373.550	370	
200	364.084ᵃ	830	7	225	373.920		

ᵃ These values agree with those published by W. Koch [173] in 1934, but differ from those he published earlier [174] by amounts ranging from −0.04 to +0.10 °C.

---

regarded as indicating that the subjecting of water to a hydrostatic tension decreases its vapor-pressure. An effect of that kind is to be expected, but it seems probable that his observations are complicated by the phenomena considered in the preceding paragraphs.

Solute, Effect of.—The partial pressure of water-vapor in equilibrium with an aqueous solution is less than that in equilibrium with pure water at the same temperature (see p. 582). For dilute solutions the reduction in vapor-pressure is approximately proportional to the molecular concentration of the solution.

Adsorbed Water, Effect of.—I. R. McHaffie and S. Lenher [535] have measured the amount of water-vapor adsorbed on glass and on platinum for each of a series of associated temperatures and pressures, and have concluded that the equilibrium pressure is less than the vapor pressure of water unless the adsorbed layer is several hundred molecules thick, if on the glass, and several tens of molecules thick, if on platinum.

Formulas.—Numerous formulas connecting the temperature and the vapor pressure of water have been proposed. The most satisfactory is the one (1) proposed by N. S. Osborne and C. H. Meyers [156] and shown by them to fit the available observations from −5 °C to the critical point.

$$\log_{10} P = A + \frac{B}{T} + \frac{Cx}{T}(10^{Dx^2} - 1) + E(10^{Fy^{5/4}}) \qquad (1)$$

the unit of the saturation pressure (P) being 1 int. atm, the corresponding temperature being t °C (int. scale) and the other quantities having these values: $T = t + 273.16$, $x = T^2 − K$, $y = 374.11 − t$, $A = +5.4266514$, $B = −2005.1$, $C = +1.3869(10)^{-4}$, $D = +1.1965(10)^{-11}$, $K = +$

[163] Washburn, E. W., *Int. Crit. Tables*, **3**, 210-212 (1928).
[164] Washburn, E. W., *Monthly Weather Rev.*, **52**, 488-490 (1924).
[165] Holborn, L., Scheel, K., and Henning, F., "Wärmetabellen," Braunschweig, Vieweg & Sohn, 1919.
[166] Harrison, L. P., *Monthly Weather Rev.*. **62**. 247-248 (1934).

88. SATURATED WATER AND STEAM: P-V-T

293700, $E = -0.0044$, $F = -0.0057148$. From (1), (2) is obtained.

$$\frac{dP}{dT} = P\left[-\frac{B + Cx(10^{Dx^2} - 1)}{T^2} + 2C\{10^{Dx^2}(1 + 2Dx^2 \log_e 10) - 1\} - \frac{5}{4}EFy^{1/4}\,10^{Fy^{5/4}} \log_e 10\right]\log_e 10 \qquad (2)$$

Other recent formulas are those proposed by A. Adamson,[157] A. Batschinski,[158] V. Fischer,[159] P. H. Hofbauer,[160] V. Kiréeff,[161] and L. B. Smith, F. G. Keyes, and H. T. Gerry.[162]

Certain other formulas constructed for use in interpolating between the directly observed precise values are given in the following tables.

Density and Specific Volume of Saturated Water-vapor.

The pressure, and consequently the density, of the vapor that is in equilibrium with water at a given temperature depends upon the form of the water surface and upon the presence or absence of a foreign gas (p. 559+).

Except where the contrary is stated, the following data refer to pure water-vapor in equilibrium with a flat surface of pure water.

C. H. Meyers [175] has found that the specific volume (v^*) of a saturated vapor may be calculated, from the vapor-pressure (p) and an approximate value of the specific volume (v') of the liquid, by means of formula (3), p_c being the critical pressure, and A an empirical constant characteristic of the liquid. It fails if $p/p_c > 0.25$. For water, $A = 0.651$.

$$\log_{10}(1 - pv^*/RT)\cdot(1 - pv'/RT) = A \log_{10}(p/2.718p_c) \qquad (3)$$

Other interpolation formulas of various types may be found in the sources mentioned in Table 250.

Table 250.—Specific Volume of Saturated Water-vapor
(See also Table 241.)

The published values of the specific volume are here represented each by its defect (δ); *i.e.*, by the amount by which it falls short of v_c^*, defined by the formula $v_c^* = 4.555(273.1 + t)/P_{\text{sat}}$, the value of P_{sat} being obtained from Table 242 (except as noted) and 4.555 cm³atm/g·°K being the value accepted by the *International Critical Tables* for the gas-constant (R) of a gas of molecular weight 18.0154 (H_2O). That is, v_c^* is essentially the ideal specific volume for H_2O, the actual specific volume

[167] Washburn, E. W., *Int. Crit. Tables*, **3**, 211-212 (1928).
[168] Prytz, K., *Jour. de Physique (2)*, **3**, 353-364 (1893); *Math.-fys. Medd. Danske Videnskab. Selskab*, **11**, No. 2 (1931).
[169] Osborne, N. S., *Mech. Eng.*, **55**, 116-117 (1933).
[170] Osborne, Stimson, Fiock and Ginnings, *Bur. Stand. Res.*, **10**, 155-158 (RP523) (1933).
[171] Osborne, Stimson, and Ginnings, *J. Res. Nat. Bur. Stand.*, **18**, 389-448 (RP983) (1937).
[172] Moser, H., *Ann. d. Phys. (5)*, **14**, 790-808 (1932).
[173] Koch, W., *Forsch. Gebiete Ingenieurw.*, **5**, 257-259 (1934).
[174] Koch, W., *Idem*, **3**, 1-10 (1932).
[175] Meyers, C. H., *Bur. Stand. J. Res.*, **11**, 691-701 (RP616) (1933).

Table 250.—(Continued)

being $v^* = v_o^* - \delta$; e.g., at 300 °C the O value for the specific volume is $v^* = 30.786 - 9.142 = 21.644$ cm³/g.

Interpolation is facilitated by making use of the fact that the variation of δ with the temperature is in general small.

With the possible exception of HM, E, and C, each of these sets of data represents the result of an attempt to obtain the most accurate values that can be deduced from all available data. That is, there is a mass of experimental data that is common to them all.

The O values, communicated by Dr. Osborne in May, 1938, are to be preferred.

Unit of v_o^* and $\delta = 1$ cm³/g. Temp. = t °C (Int. scale)

t	Source[a]→ v_o^*	O	KSG	JF	HM	E δ	IST	M	WT	C	ICT
0	206389	103	89	79	289		79	68	1389		
10	106554	139	144	114 •			144	74	654		
20	57941	109	117	101			117	68	281		
30	33005	78	83	75			83	61	125		
40	19605	60	62	60			62	54	65		
50	12093	48	48	48	56		48	46	43		
60	7719.1	41	40.8	42			40.8	41	33		
70	5082.0	36	35.7	37			35.7	37	32		
80	3441.2	32.6	32.0	33			32.0	32.8	31		
90	2390.3	29.2	28.8	30			28.8	29.4	29		
100	1699.5	26.4	26.3	27.4	27.3		26.3	26.7	29		22
110	1234.2	24.0	24.1	25.1			24.1	24.3			26
120	913.85	22.0	22.20	23.2			22.20	22.29			24.0
130	688.76	20.5	20.55	21.5			20.55	20.54			21.7
140	527.63	19.0	19.10	19.9			19.10	19.04			19.5
150	410.26	17.8	17.80	18.6	17.68		17.80	17.72			17.9
160	323.43	16.6	16.67	17.3			16.67	16.58			16.3
170	258.20	15.57	15.65	16.2			15.65	15.56			15.1
180	208.56	14.7	14.76	15.3			14.76	14.68			13.7
190	170.28	13.92	13.96	14.4			13.96	13.89			
200	140.42	13.22	13.24	13.56	13.07		13.24	13.19	14.0		
210	116.86	12.59	12.62	12.84			12.62	12.56			
220	98.091	12.02	12.021	12.21	11.89		12.021	11.99			
230	82.991	11.51	11.508	11.64			11.508	11.49			
240	70.732	11.05	11.048	11.14			11.048	11.03			
250	60.697	10.64	10.636	10.69	10.572		10.636	10.63		10.41	
260	52.416	10.27	10.267	10.29			10.267	10.26		9.90	
270	45.532	9.93	9.939	9.93	9.924		9.939	9.92		9.47	
280	39.768	9.63	9.646	9.62			9.646	9.626		9.12	
290	34.908	9.377	9.386	9.35			9.386	9.358		8.80	
300	30.786	9.142	9.161	9.12	9.192		9.161	9.120		8.55	
310	27.268	8.952	8.968	8.913	8.994		8.968			8.36	
320	24.246	8.794	8.808	8.751	8.811		8.808			8.23	
330	21.637	8.669	8.685	8.627	8.640		8.685			8.17	
340	19.371	8.591	8.607	8.556	8.483		8.607			8.19	
350	17.391	8.58	8.589	8.554		8.598	8.589			8.32	
360	15.648	8.71	8.664	8.672		8.688	8.685			8.63	
365	14.852	8.85		8.817		8.844					
366	14.698	8.90									
367	14.546	8.96									
368	14.395	9.02									
369	14.246	9.09									
370	14.099	9.17	9.006	9.11		9.145	9.102			9.25	
371	13.953	9.27					9.192				
372	13.808	9.41	9.200	9.33		9.364	9.310				
373	13.665	9.61					9.483				
374	13.522	10.05	9.829			9.892	9.874			9.73	11.0[b]
374.11	13.506		10.308								
374.15	13.500	10.4									
374.23	13.510[c]					10.444					
375	13.27[d]									9.77	
377	12.92[d]									9.59	
380	12.27[d]									9.44	
380.5	11.98[d]									9.40	

88. SATURATED WATER AND STEAM: P-V-T

Table 250.—(Continued)

a Sources:
- C   Callendar, H. L., *Proc. Inst. Mech. Eng.*, **1929**, 507-527 (1929).
- E   Eck, H., *Ber. Tätigkeit Phys.-Techn. Reichs. im 1936*, p. 32 → *Physik. Z.*, **38**, 256 (1937).
- HM   Havlíček, J., and Miškovský, L., *Helv. Phys. Acta*, **9**, 161-207 (1936).
- ICT   *Int. Crit. Tables*, **3**, 234 (1928). Compilation by Keyes, F. G., based on work of Knoblauch, O., Linde, R., and Klebe, H., *Forsch. Gebiete Ingenieurw.*, **21**, 33-55 (1903), which agrees closely with WT, but with consideration of data by Battelli, A., *Ann. chim. et phys.* (6), **26**, 394-425 (1892); (7), **3**, 408-431 (1894), Dieterici, C., *Ann. d. Phys. (Wied.)*, **38**, 1-26 (1889), and Perot, A., *Ann. chim. et phys.* (6), **13**, 145-190 (1888).
- IST   International Skeleton Steam Table, 1934, see Table 260.
- JF   Jakob, M., and Fritz, W., *Physik. Z.*, **36**, 651-659 (1935). Superseding Jakob, M., *Forsch. Geb. Ing.*, **310**, 9-19 (1928), and Jakob, M., and Fritz, W., *Z. Ver. deuts. Ing.*, **73**, 629-636 (1929); *Techn. Mech. Thermodynam.*, **1**, 173-183, 236-240 (1930); *Forsch. Gebiete Ingenieurw.*, **4**, 295-299 (1933).
- KSG   Keyes, F. G., Smith, L. B., and Gerry, H. T., *Proc. Amer. Acad. Arts Sci.*, **70**, 319-364 (1935). Superseding Smith, L. B., and Keyes, F. G., *Mech. Eng.*, **54**, 123-124 (1932).
- M   Meyers, C. H., *Private communication*; values computed by means of formula (3).
- O   Osborne, N. S., *Private communication*, May, 1938. These values supersede all others previously published by Osborne and his associates; in particular they extend to lower temperatures the table by Osborne, Stimson, and Ginnings, *J. Res. Nat. Bur. Stand.*, **18**, 389-448 (RP983) (1937) and agree with that table for $t > 110$ °C.
- WT   Holborn, L., Scheel, K., and Henning, F., "Wärmetabellen," 1919.

b From compilation by A. F. O. Germann and S. F. Pickering.[176]
c The critical pressure published by Eck (225.5_1 kg*/cm² = 218.26 atm) was used in the computing of this value.
d Callendar's value for P_{sat} was used in the computing of this value.

Table 251.—Density of Saturated Water-vapor

To obtain the density of saturated water-vapor, take the reciprocal of the specific volume as given in Table 250, using the particular value of δ that seems most appropriate. The values given in this table were so obtained from the O series, and will serve to indicate the order of magnitude to be expected.

Unit of $\rho = 1$ g per cm³. Temp. = t °C

t	ρ	t	ρ	t	ρ
0	0.0000048476	80	0.00029338	250	0.019976
10	0.0000093972	100	0.0005977	300	0.04620
20	0.0000172912	120	0.0011213	350	0.1135
30	0.000030370	140	0.001966	360	0.1441
40	0.000051164	160	0.003259	370	0.203
50	0.00008302	180	0.005158	374	0.267
60	0.00013024	200	0.007862	374.15	0.32

Table 252.—Density of Water-vapor Saturated in the Presence of a Foreign Inert Gas

In the first section of the table, under the title "Ideal case," are given values of ρ'/ρ_0 computed by means of the formula $\log_e(\rho'/\rho_0) = P'/RT$. They are the values that would be obtained were water-vapor an ideal gas, water incompressible and of unit density, and the inert ideal gas insoluble. Departures from these ideal conditions will reduce (see p. 560+) the value of ρ'/ρ_0, but the effect will be small in all cases covered by this table. In no case will ρ' be less than ρ_0. These values may serve as a norm against

[170] Germann, A. F. O., and Pickering, S. F., *Int. Crit. Tables*, **3**, 248 (1928).

Table 252.—(Continued)

which to compare the experimental data given in Section II. Marked discrepancies exist.

Pertinent data published by I. R. McHaffie [177] exhibit strange and unexplained variations with the pressure; they are not included in this table.

For the moisture content of the gas phase in equilibrium with aqueous solutions of NH_3, see Table 233.

ρ' = mass of water-vapor per liter of actual gas phase; ρ_0 = value of ρ' when no gas except water-vapor is present (its value for any temperature may be found from Table 250); P' = partial pressure of the inert gas; P (or p) = total pressure (gas + vapor) corresponding to ρ'.

Unit of P and P' = 1 atm = 1.03323 kg*/cm²; of p = 1 kg*/cm²; of ρ' and ρ_0 = 1 mg water-vapor per liter of actual gas phase. Temp. = t °C

I. Ideal case (see heading of table).

$t\rightarrow$ P'	25	37.5	50	100	150	200	230
				ρ'/ρ_0			
10	1.007	1.007	1.007	1.006	1.005	1.005	1.004
50	1.038	1.036	1.036	1.030	1.026	1.023	1.022
100	1.076	1.073	1.070	1.061	1.053	1.048	1.045
150	1.117	1.112	1.107	1.092	1.081	1.072	1.068
200	1.159	1.152	1.146	1.125	1.109	1.097	1.091
300	1.247	1.236	1.226	1.193	1.168	1.149	1.140
400	1.343	1.327	1.312	1.265	1.231	1.204	1.191
500	1.445	1.424	1.405	1.342	1.296	1.261	1.244
600	1.556	1.528	1.503	1.423	1.365	1.321	1.299
700	1.675	1.640	1.609	1.510	1.438	1.384	1.357
800	1.803	1.760	1.722	1.601	1.515	1.450	1.418
900	1.940	1.889	1.843	1.698	1.595	1.518	1.481
1000	2.089	2.028	1.973	1.801	1.680	1.590	1.547

II. Experimental values.

Gas→ $t\rightarrow$ P	H_2 50	N_2 50	Barlett[a,b] 50	$3H_2 + N_2$ 37.5	25	W.G.[a] Gas→ $t\rightarrow$ P	H_2 100
			ρ'/ρ_0				ρ'/ρ_0
100	1.076	1.284	1.130	1.135	1.164	25	1.018
200	1.153	1.551	1.245	1.255	1.337	50	1.04
300	1.206	1.700	1.315	1.345	1.475	100	1.09
400	1.242	1.793	1.368	1.422	1.585	200	1.19
500	1.276	1.876	1.425	1.481	1.662	400	1.40
600	1.308	1.947	1.470	1.530	1.728	600	1.66
700	1.341	1.989	1.506	1.570	1.789	800	1.97
800	1.365	2.021	1.540	1.598	1.831	1000	2.35
900	1.399	2.044	1.567	1.616	1.875		
1000	1.431	2.053	1.582	1.643	1.909		

F. Pollitzer and E. Strebel.[a]

Gas→ $t\rightarrow$ p	H_2[c] 49.9	70.1	Air[c] 49.9	70.1	$t\rightarrow$ p	49.9 ρ'/ρ_0	CO_2 $t\rightarrow$ p	70.1 ρ'/ρ_0
		ρ'/ρ_0						
10		1.026	1.011	1.024	40.0	1.747	30.5	1.421
50	0.996	1.041	1.121	1.121	55.5	2.228	39	1.580
100	1.115	1.060	1.258	1.243	59.0	2.597	52.4	1.929
150	1.252	1.154	1.395	1.364	87.0	4.404		

[177] McHaffie, I. R., *Phil. Mag. (7)*, **3**, 497-510 (1927).

Table 252.—(Continued)

A. W. Saddington and N. W. Krase.[a]

Gas→				N$_2$				
P→	100			200			300	
t	ρ'	ρ'/ρ$_0$	t	ρ'	ρ'/ρ$_0$	t	ρ'	ρ'/ρ$_0$
50	123.6	1.489	50	146.3	1.763	50	197.5	2.038
80	381.7	1.302	85	511.6	1.447	75	460.7	1.904
100	647.4	1.083	150	2960.	1.162	100	912.6	1.527
150	2710.	1.064	190	7550.	1.180	115	1450.	1.503
190	5880.	0.919	225	14840.	1.163	145	3240.	1.445
230	13300.	0.951				165	5320.	1.449
						230	16400.	1.172

[a] References:
  Bartlett, E. P., *J. Am. Chem. Soc.*, **49**, 65-78 (1927); Pollitzer, F., and Strebel, E., *Z. physik. Chem.*, **110**, 768-785 (1924); Saddington, A. W., and Krase, N. W., *J. Am. Chem. Soc.*, **56**, 353-361 (1934).
  W.G. Wiebe, R., and Gaddy, V. L., *Idem*, **56**, 76-79 (1934).
[b] Data discussed by J. J. van Laar.[178]
[c] Interpolation by the compiler.

Saturated Liquid

Boiling Point.

The boiling point of a liquid is the temperature at which the vapor-pressure of the liquid is equal to the pressure to which the liquid is subjected. It is commonly said to be the temperature of the liquid when it is steadily boiling in such a manner that it is thoroughly intermixed with bubbles of vapor. If the rate of heating is low, if the temperature is measured at a point far from the surface through which the heat is supplied, and if the mixing is ideal and the thermometer is suitably screened from radiation, then the temperature of the boiling liquid approaches closely to the boiling point as defined in the first sentence, and may coincide with it. But under other conditions the temperature of the boiling liquid lies above the true boiling point.[179] The temperature indicated by a suitably screened thermometer immersed in the vapor above the boiling liquid is that at which the vapor at the existing pressure is in equilibrium with the condensed vapor; it is not the same as the boiling point unless the composition of the condensed vapor is the same as that of the boiling liquid.[180]

Unless—owing to the presence of nuclei, of submerged solids that are poorly wetted by the liquid and have uneven surfaces penetrated by cavities, or to some other cause—there are in the interior of the liquid free spaces in which the vapor can collect, the liquid can be raised considerably above the boiling point before boiling begins, especially if protected from shocks of all kinds.* F. Donny[182] heated air-free water under a pressure of only a few centimeters of mercury to 135 °C without its boiling. At that tem-

---

* Since this was written, an important paper by J. Aitken,[181] expressing these same general views, has come to the compiler's attention.

[178] van Laar, J. J., *Z. physik. Chem. (A)*, **145**, 207-219 (1929).
[179] Jakob, M., and Fritz, W., *Forsch. Gebiete Ingenieurw.*, **2**, 435-447 (1931).
[180] Cf. Schreber, K., *Z. techn. Physik*, **9**, 277-285 (1928).
[181] Aitken, J., *Trans. Roy. Scottish Soc. Arts*, **9**, 240-287 (1875).
[182] Donny, F., *Ann. de chim. et phys. (3)*, **16**, 167-190 (1846) = *Ann. d. Phys. (Pogg.)*, **67**, 562-584 (1846) ← *Bruxelles Mém. Couron.*, **17**, (1843-1844).

Table 253.—Boiling Points of Water [156]
(Osborne and Meyers)

These values, computed by means of formula (1), are believed to be the best available. Others are compared with these in Table 254.

Each of these values (the temperature at which the saturated vapor exerts the indicated pressure) is the boiling point of water when subjected to the corresponding pressure p exerted by an overlying inert gas when the conditions are such that the boiling has removed from the water all of the previously dissolved gas, and has blanketed the surface of the water with a layer of its own pure vapor. Under other conditions the temperature is somewhat less, lying between this value and that defined by the formula given in the head-matter of Table 245.

Examples: The boiling point when $p = 590$ mm-Hg is 93.058 °C; when $p = 614$ mm-Hg it is 94.1319 °C.

Unit of p $(= p_1 + p_2) = 1$ mm-Hg = 1.33322 millibars. Temp. = $(t_1 + t_2)$ °C, Int. scale

$p_2 \rightarrow$		0	1	2	3	4	5	6	7	8	9
p_1	t_1						1000 t_2				
500	88	678	730	782	834	886	938	990	1042	1093	1144
510	89	196	247	298	350	401	452	502	553	604	655
520	89	705	756	806	856	907	957	1007	1057	1107	1157
530	90	206	256	306	355	405	454	503	553	602	651
540	90	700	749	798	846	895	944	992	1041	1089	1138
550	91	186	234	282	330	378	426	474	521	569	617
560	91	664	712	759	806	854	901	948	995	1042	1089
570	92	136	182	229	276	322	369	415	462	508	554
580	92	600	646	692	738	784	830	876	922	967	1013
590	93	58	104	149	195	240	285	330	375	420	465
600	93	510.0	554.8	599.6	644.3	688.9	733.5	778.0	822.4	866.8	911.2
610	93	955.4	999.6	1043.8	1087.9	1131.9	1175.9	1219.8	1263.6	1307.4	1351.1
620	94	394.8	438.4	482.0	525.5	568.9	612.3	655.6	698.9	742.1	785.2
630	94	828.3	871.3	914.3	957.2	1000.1	1042.9	1085.7	1128.4	1171.0	1213.6
640	95	256.3	298.7	341.1	383.4	425.7	468.0	510.2	552.3	594.4	636.5
650	95	678.5	720.4	762.3	804.1	845.9	887.6	929.3	970.9	1012.5	1053.9
660	96	95.4	136.8	178.2	219.5	260.7	301.9	343.1	384.2	425.2	466.2
670	96	507.2	548.0	588.9	629.7	670.4	711.1	751.7	792.3	832.9	873.4
680	96	913.8	954.2	994.6	1034.9	1075.1	1115.3	1155.5	1195.6	1235.6	1275.6
690	97	315.6	355.5	395.4	435.2	474.9	514.6	554.3	593.9	633.5	673.0
700	97	712.5	751.9	791.3	830.7	870.0	909.2	948.4	987.6	1026.7	1065.7
710	98	104.8	143.7	182.7	221.6	260.4	299.2	337.9	376.6	415.3	453.9
720	98	492.5	531.0	569.5	607.9	646.3	684.6	722.9	761.2	799.4	837.6
730	98	875.7	913.8	951.9	989.9	1027.8	1065.7	1103.6	1141.4	1179.2	1217.0
740	99	254.7	292.4	330.0	367.5	405.1	442.6	480.0	517.4	554.8	592.1
750	99	629.4	666.7	703.9	741.0	778.1	815.2	852.3	889.3	926.2	963.1
760	100	0	36.8	73.6	110.4	147.1	183.8	220.4	257.0	293.6	330.1
770	100	366.6	403.0	439.4	475.8	512.1	548.4	584.6	620.8	657.0	693.2
780	100	729.3	765.3	801.3	837.3	873.3	909.2	945.0	980.8	1016.6	1052.4
790	101	88.1	123.8	159.4	195.0	230.6	266.1	301.6	337.1	372.5	407.9

Table 254.—Comparison of Values for the Boiling Points of Water

The preferred value is t_{OM} taken from Table 253; each of the other values may be obtained by adding to t_{OM} the corresponding value of $\delta/1000$. For example, the value given by C(V) for the boiling point when $p = 660$ mm-Hg is $96.095 - 0.013 = 96.082$ °C.

C. S. Cragoe [185a] has recommended that, in the absence of more accurate data, the excess, $\Delta t = t' - t$, of the boiling point (t') under pressure p above that (t) under the pressure (A) of one normal atm be computed by

[183] Henrick, F. B., Gilbert, C. S., and Wismer, K. L., J. Phys'l Chem., 28, 1297-1304 (1924).

88. SATURATED WATER AND STEAM: P-V-T

Table 254.—*(Continued)*

means of the formula $\phi \Delta t = (273.1 + t') \log_{10}(p/A)$ in which ϕ for a given substance varies but slowly with t', and for any one of the 8 groups, into which he finds all substances for which data are available may be assorted, ϕ varies linearly with the temperature of the normal boiling point. For water at 100 °C, $\phi = 5.79$; it is with this value that the data in column C have been computed.

Other formulas proposed for water may be put in the form $t - 100 = a\pi + b\pi^2 + c\pi^3 + d\pi^4$ where $\pi = (p - 760)/1000$, the pressure being p mm-Hg. Certain of these are included in this table, the values of the coefficients and of the limiting pressures being given at the foot of the appropriate column. Numerous other formulas of the same type have been proposed; some of the more recent ones may be found in papers by L. B. Smith, F. G. Keyes, and H. T. Gerry,[185b] J. A. Beattie and B. E. Blaisdell,[185c] and W. Świętosławski and E. R. Smith.[186]

Unit of $p = 1$ mm-Hg $= 1.33322$ millibars; of $\delta = 0.001$ °C; of $t_{OM} = 1$ °C (Int. scale)

Reference[a]		Mo	C(V)	ZB	WT	Mu	HH	B	C
p	t_{OM}				— δ —				
550	91.186		−34		+5		+3		−19
560	91.664		−32		+5		+1		−23
580	92.600		−29		+6		+1		−10
600	93.510		−25		+5		0		−10
620	94.395		−21		+5		0		−9
640	95.256		−17		+4		0		−2
660	96.095		−13		+4	(+5)	+1		−1
680	96.914	−4	−10	−5	+2	+3	+3	+10	−1
690	97.316	−5	−9	−6	+2	+2	+1	+9	0
700	97.712	−3	−6	−5	+3	+3	+2	+9	0
710	98.105	−3	−6	−5	+2	+3	0	+7	0
720	98.492	−2	−4	−4	+2	+3	−1	+6	0
730	98.876	−2	−4	−4	+1	+2	−1	+4	0
740	99.255	−1	−3	−3	+1	+2	−1	+3	0
750	99.629	0	−1	−1	+1	+2	+1	+2	0
760	100.000	0	0	0	0	0	0	0	0
770	100.367	0	+1	+1	−1	−2	0	−2	0
780	100.729	0	+3	+2	−1	−4	+2	−3	−1
790	101.088	(−2)	+4	+3	−1	(−8)	+1	−5	−1
a		36.87	36.970	36.971		36.7		36.697	
b		−22.00	−19.795	−30.263		−23.0		−20.459	
c		0	22.83	6.695		0		16.39	
d		0	0	0		0		−14.3	
p_{min}		690	560	683		680		680	
p_{max}		780	820	832		780		800	

[a] References:
  B   Broch, O. J., *Trav. et Mém. Bur. Int. Poids et Mes.*, **1**, A43-A48 (1881).
  C   Cragoe, C. S.[185a]
  C(V)   Volet, C., *Trav. et Mém. Bur. Int. Poids et Mes.*, **18**, (1930). (Formula was fitted to P. Chappuis' observations, of which some had not been published before.)
  HH   Holborn, L., and Henning, F., *Ann. d. Phys.* (4), **26**, 833-883 (1908).

[184] Hoyt, C. S., and Fink, C. L., *J. Phys'l Chem.*, **41**, 453-456 (1937).
[185] Rosanoff, M. A., and Dunphy, R. A., *J. Am. Chem. Soc.*, **36**, 1411-1418 (1914).
[185a] Cragoe, C. S., *Inst. Crit. Tables*, **3**, 246-247 (1928).

Table 254.—*(Continued)*

Mo	Moser, H., *Idem (5)*, **14**, 790-808 (1932).
Mu	Formula given by Mueller, E. F., in *Int. Crit. Tables*, **1**, 53 (1926) and recommended by the International Bureau of Weights and Measures in 1927 in connection with the International Temperature Scale. See: *C. R. des Septième Conf. Gén. Poids et Mes.*, **1927**, 56-58, 94-99 (1928). Burgess, G. K., *Bur. Stand J. Res.*, **1**, 635-640 (RP22) (1928).
WT	Derived from Table 245, from "Wärmetabellen."
ZB	Zmaczynski, A., and Bonhoure, A., *Compt. rend.*, **189**, 1069-1070 (1929) ← *Jour. de Phys. (7)*, **1**, 285-291 (1930).

perature the vapor-pressure of water is over 3 atm. The initiation of boiling under such conditions results in an explosion of the entire volume of water. F. B. Kenrick, C. S. Gilbert, and K. L. Wismer [183] report that they have heated water in a capillary tube and at atmospheric pressure to 270 °C.

Likewise, in the absence of condensation nuclei, a vapor can be cooled, without condensation, below the temperature corresponding to equilibrium between it and its liquid (see p. 633).

Effect of a Solute.—The boiling point of a solution exceeds that of the pure solvent by an amount Δ, and we may write $\Delta = nE$, where n is the number of effective gram-molecules contributed by the solute per kilogram of the solvent. E is characteristic of the solvent but independent of the nature of the solute if that is not significantly volatile at the boiling point; it varies with the concentration, and its limiting value (E_0) as n approaches zero is frequently called the ebullioscopic constant of the solvent. If the solute is neither associated nor dissociated, and does not affect the molecular aggregation of the solvent, then $n = 1000m/FW \equiv N$, where m is the mass of the solute dissolved in the mass W of the solvent, and F is the molecular weight of the solute. If each molecule of the solute is dissociated into two parts, and if the solvent is not affected, then $n = 2N$; and similarly in other cases. The value of E_0 is $RM_wT^2/1000L$, where R is the gas-constant (= 8.315 joules/g-mole·°K), M_w = molecular weight of the vapor of the solvent, T °K = absolute temperature of the boiling point of the solution, and L = latent heat of vaporization of the pure solvent at T °K. (R and L must refer to the same unit of mass.) For very dilute aqueous solutions, $T = 373.1$ °K, $L = 2256.6$ j/g, and E_0 for water is $8.315M_w \times (373.1)^2/1000(2256.6M_w) = 0.513$ °C per (effective g-mole of solute per kg of water). Experimental data agree with this value as well as one should expect.

(C. S. Hoyt and C. L. Fink [184] state that the best values now in use for the ebullioscopic constants are those published by M. A. Rosanoff and R. A. Dunphy.[185] Those authors, using $v^* = 1.651$ liters/gram for the specific volume of water-vapor saturated at 100 °C, obtained $E_0 = 0.518$ for water. But the best value of v^* now available is 1.673 (see Table 250); had they used that, retaining the same values as before for the other con-

---

[185b] Smith, L. B., Keyes, F. G., and Gerry, H. T., *Proc. Amer. Acad. Arts Sci.*, **69**, 137-168 (1934).
[185c] Beattie, J. A., and Blaisdell, B. E., *Idem*, **71**, 361-374 (1937).
[186] Świętosławski, W., and Smith, E. R., *J. Res. Nat. Bur. Stand.*, **20**, 549-553 (RP1088) (1938).

88. SATURATED WATER AND STEAM: P-V-T

stants, they would have found $E_0 = 0.511$. Hoyt and Fink, using $R = 1.976 \times 4.185 = 8.270$ joules/g-mole·°K, obtained $E_0 = 0.510$ for water. Had they used the value ($R = 8.315$) adopted by the *International Critical Tables*, and used in this compilation, retaining the other constants unchanged, they would have found $E_0 = 0.513$.)

Density and Specific Volume of Saturated Water.

The best published extended series of values of the specific volume of saturated water is that contained in the International Skeleton Steam

Table 255.—Specific Volume of Saturated Water
(See also Table 241.)

The preferred values (v^*) are given in full, adjacent to the t-column; for other values merely the excess (δ) of each above the preferred value is given, the unit of δ being that of the last place in the tabulated value of v^*. For example, the ICT value for 120 °C is $1.0603 - 0.0011 = 1.0592$; and Eck's value at the critical point is $3.1 - 0.034 = 3.066$. The difference between the significance of the decimal points (in the δ values and in the v^* values) should be remembered.

The pressure is always that of the saturated vapor.

Unit of $v^* = 1$ cm³/g; of $\delta = 1$ unit in the last tabulated place in v^*; of $p = 1$ kg*/cm². Temp. $= t$ °C (Int. scale)

I. Temperature given.

Source[a] →	IST	SK		IST = SK	ICT		OSG	Eck	IST
t	v^*	δ	t	v^*	δ	t	v^*	δ	δ
0	1.00021	+2	100	1.0435	−1	330	1.562		−0.1
10	1.00035	+21	110	1.0515	−5	340	1.640		+0.8
20	1.00184	+20	120	1.0603	−11	350	1.741	+4	+5.8
30	1.00442	+11	130	1.0697	−17	355	1.808	+2	
40	1.00789	+2	140	1.0798	−22	360	1.894	0	+12.6
50	1.0121	0	150	1.0906	−25	365	2.016	−2	
60	1.0171	−1	160	1.1021	−24	366	2.048		
70	1.0228	−2	170	1.1144	−20	367	2.083		
80	1.0290	−1	180	1.1275	−13	368	2.124		
90	1.0359	−1	190	1.1415	−5	369	2.170		
100	1.0435	−1	200	1.1565	+4	370	2.225	−14	+6
			210	1.1726	+13	371	2.293		+4
			220	1.1900	+19	372	2.38	−3.3	+0.1
			230	1.2087	+13	373	2.51		−0.8
			240	1.2291	+1₀	374	2.80	−15.0	−1
			250	1.2512	+0₈	374.15	3.1	−0.34[b]	
			260	1.2755	−1₅				
			270	1.3023	−5₃				
			280	1.3321					
			290	1.3655					
			300	1.4036					
			310	1.4475					
			320	1.4992					

II. Pressure given. S[a]

p	50	100	150	200	225
v^*	1.286	1.451	1.649	1.990	3.06₀

Sources:

Eck   Eck, H., *Ber. Tätigkeit Phys.-Techn. Reichs.* in 1936, p. 32 → *Physik. Z.*, **38**, 256 (1937).
ICT   *Int. Crit. Tables*, **3**, 234 (1928). Compiled by Keyes, F. G., and based on the data of Hirn, G. A., *Ann. Chim. et Phys.* (4), **10**, 32-92 (1867), Ramsay, W., and Young, S., *Phil. Trans.* (A), **183**, 107-130 (1892), and Waterston, J. J., *Phil. Mag.* (4), **26**, 116-134 (1863).
IST   International Skeleton Steam Table, 1934 (see Table 260). The values below 100 °C have been derived from the Chappuis and the Thiesen values (see text); those for $t = 100$ to 360 °C are the same as the SK values.
OSG   Osborne, N. S., Stimson, H. F., and Ginnings, D. C., *J. Res. Nat. Bur. Stand.*, **18**, 389-448 (RP983) (1937); confirmed by Osborne in May, 1938.
S   Schlegel, E., *Z. techn. Phys.*, **14**, 105-107 (1933).
SK   Smith, L. B., and Keyes, F. G.[188] This paper supersedes that of Keyes and Smith,[190] and of Smith and Keyes, *Mech. Eng.*, **55**, 114-116 (1933).

[b] This is for Eck's value at the critical point, which he places at 374.23 °C. In their compilation [189] A. F. O. Germann and S. F. Pickering give 2.5 for the critical volume; and by extrapolation Keyes and Smith [190] found 3.086 cm³/g.

Table of 1934,[187] given in Table 260. For temperatures below 100 °C they have been derived from the determinations at one atm by Chappuis and by Thiesen (see Table 93) by applying a suitable correction for the isothermal compression from saturation to a pressure of one atm, and by changing the unit of volume from 1 ml to 1 cm³. From 100 to 360 °C they are identical with those of the extended table published by L. B. Smith and F. G. Keyes,[188] except for a rounding off to a smaller number of digits. The Smith and Keyes table was computed by means of an empirical formula (4) fitted to their observations above 100 °C and taking into consideration known values for lower temperatures.

$$v^* = [v_c + a(t_c - t)^{1/3} + b(t_c - t) + c(t_c - t)^4]/[1 + d(t_c - t)^{1/3} + e(t_c - t)] \text{ cm}^3/\text{g} \qquad (4)$$

where $v_c = 3.197500$, $t_c = 374.11$ °C, $a = -0.3151548$, $b = -0.001203374$, $c = +7.48908/10^{13}$, $d = +0.1342489$, $e = -0.003946263$.

For temperatures above 330 °C the slightly different values given by Osborne, Stimson, and Ginnings (1937, OSG of Table 255) are to be preferred.

For the density take the reciprocal of the specific volume.

89. Thermal Energies of Saturated Water and Saturated Steam

In this section are given values of the enthalpy (heat content) and of the entropy of both saturated water and saturated steam, and the specific heat of steam continuously saturated. The quantity frequently called the specific heat at saturation really refers to the unsaturated state, being merely the limit approached by the specific heat of the unsaturated phase as the temperature approaches that of saturation. Its values are not given here,

---

[187] See *Mech. Eng.*, **57**, 710-713 (1935).
[188] Smith, L. B., and Keyes, F. G., *Proc. Amer. Acad. Arts Sci.*, **69**, 285-314 (1934) → *Mech. Eng.*, **56**, 92-94 (1934).
[189] Germann, A. F. O., and Pickering, S. F., *Int. Crit. Tables*, **3**, 248 (1928).
[190] Keyes, F. G., and Smith, L. B., *Mech. Eng.*, **53**, 132-135 (1931).

89. SATURATION: THERMAL ENERGY

but will be found in the corresponding sections treating of the unsaturated phases.

For a recent review of the experimental data, see E. F. Fiock.[191]

Table 256.—Enthalpy of Saturated Water and of Saturated Steam

The OSG values to to be preferred. They are given directly, the others by the amounts (ΔH) by which they exceed the corresponding OSG values. Example: The value given by KSG for the saturated vapor at 0 °C is 2500.00 + 1.86 = 2501.86.

H is the excess of the enthalpy ($E + pv$) above its value for saturated water at 0 °C. Subscripts l and v indicate that the value refers to the liquid and to the vapor, respectively.

Unit of H and ΔH = 1 Int. joule/g.   Temp. = t °C   (Int. scale)
Values adopted for International Skeleton Steam Tables, 1934, are given in Table 260.

Ref.[a] →	OSG H_l	O	K	TS	Sch	C	OSG H_v	O	KSG	HM	K	C
t				ΔH_l					ΔH_v			
0	0.000	0	0.00	0.00	0.0	0.00	2500.00	0	+1.86	+0.3		
10	42.028	0			+0.1		2518.50	0	+1.61			
20	83.833	0	−0.03	−0.16	+0.1		2536.81	0	+1.43			
30	125.675	0			0.0		2554.97	0	+1.30			
40	167.454	0	−0.09	−0.01	−0.4		2572.99	0	+1.18			
50	209.247	0	−0.16		−0.2		2590.83	0	+1.03	−0.9		
60	251.072	0	−0.20	+0.09	−0.3		2608.44	0	+0.83			
70	292.943	0			−0.2		2625.77	0	+0.62			
80	334.877	0	−0.29	0.00	−0.3		2642.75	0	+0.58			
90	376.893	0			−0.4		2658.31	0	+0.18			
100	418.76	0.25	−0.16	−0.16	0.0		2675.42	−0.14	−0.17	−1.9		
110	460.99	0.26			0.0		2690.85	+0.05	−0.20			
120	503.34	0.31	−0.2	+0.24	0.0		2705.56	+0.22	−0.23			
130	545.94	0.29			0.6		2719.62	+0.28	−0.32			
140	588.73	0.29	−0.2	+0.45	0.3		2732.95	+0.29	−0.45			
150	631.76	0.30	−0.1		0.1		2745.49	+0.30	−0.65	−0.7		
160	675.07	0.30	+0.1	+0.76	−0.1		2757.18	+0.31	−0.94			
170	718.69	0.31			−0.1		2767.87	+0.31	−1.28			
180	762.68	0.32	0.0	+0.76	−0.6		2777.29	+0.31	−1.48			
190	807.07	0.32			−0.6		2785.35	+0.32	−1.54			
200	851.91	0.33	+0.8	+0.14	−1.4	+0.5	2791.99	+0.33	−1.53	+0.9		10.6
210	897.27	0.33			−2.0	0.8	2797.11	+0.33	−1.46			12.7
220	943.21	0.33	+0.8	−0.32	−2.7	1.3	2800.62	+0.32	−1.40	+2.3		14.2
230	989.80	0.32			−3.6	1.6	2802.38	+0.32	−1.25			16.6
240	1037.13	0.32	+0.2	−0.77	−6.1	2.0	2802.25	+0.33	−1.14			17.6
250	1085.30	0.32	+0.1		−8	2.4	2800.07	+0.32	−1.07	+3.3		19.3
260	1134.44	0.32	0.0	−0.88	−10	2.7	2795.61	+0.33	−1.03			21.7
270	1184.69	0.33	−0.5		−15	3.5	2788.61	+0.32	−1.04	+2.2		23.7
280	1236.25	0.32	−1.4	+2.75	−18	4.2	2778.72	+0.33	−1.09			25.2
290	1289.35	0.33			−23	4.8	2765.52	+0.32	−1.19			27.5
300	1344.29	0.32	−1.8	+4.41	−29	5.5	2748.38	+0.32	−1.33	−1.1	−4.0	29.1
310	1401.50	0.32	−2.1		−36	6.5	2726.42	+0.32	−1.36	−1.3		31.0
320	1461.56	0.32	−1.9		−36	7.5	2698.81	+0.32	−1.49	−0.9		32.7
330	1525.36	0.32	−2.1		−38	8.6	2664.65	+0.32	−2.14	+1.0		33.7
340	1594.23	0.31	−2.2		−34	9.8	2621.23	+0.32	−2.79	+6.8		33.6
350	1670.56	0.32	−4.6		−32	11.3	2563.42	+0.31	−0.84[b]		−5.3	34.9
360	1760.86	0.32	−1.9		−30	18.1	2480.28	+0.33	+10.99[b]		−2.2	38.1
365	1816.88	0.32					2420.12	+0.33				
366	1829.72	0.32					2405.42	+0.31				
367	1843.38	0.32					2389.42	+0.32				
368	1858.08	0.32					2371.84	+0.31				
368	1873.99	0.32					2352.23	+0.33				
370	1891.69	0.32	−0.9		−20	28.8	2329.99	+0.32			−0.5	67.8
371	1911.88	0.31					2303.98	+0.33				
372	1936.11	0.32	−1.3				2272.15	+0.32			−1.6	
373	1968.21	0.33					2229.19	+0.32				
374	2031.07	0.33			+60	−13.5[c]	2145.67	+0.34				174.6[c]
374.15	2083.27	0.32					2083.27	0.32				

[191] Fiock, E. F., Bur. Stand. J. Res., 5, 481-505 (RP210) (1930) → Mech. Eng., 52, 231-242 (FSP-52-30) (1930).

586   PROPERTIES OF ORDINARY WATER-SUBSTANCE

Table 256.—(Continued)

a References and remarks:

C   Callendar, H. L., *Proc. Inst. Mech. Eng.*, **1929**, 507-527 (1929).
HM   Havlíček, J., and Miškovský, L., *Helv. Phys. Acta*, **9**, 161-207 (1936).
K   Koch, W., *Forsch. Gebiete Ingenieurw.*, **5**, 138-145 (1934). *Z. Ver. deuts. Ing.*, **78**, 1160 (1934).
KSG   Keyes, F. G., Smith, L. B., and Gerry, H. T., *Proc. Amer. Acad. Arts Sci.*, **70**, 319-364 (1935). (Included in the 1935 volume, but actually published in 1936.) The authors accepted $H_v = 2675.35$ Int. joules/g at 100 °C and the expression $c_{p\to 0} = 1.47198 + 7.5566(10^{-4})T + 47.8365/T$ based on Gordon's computations and attributed to Keenan, and from these and their own volumetric measurements they derived the values of H_v here given.
O   Osborne, N. S., *Private communication*, 1938. (These values take precedence over the corresponding OSG ones.)
OSG   Osborne, N. S., Stimson, H. F., and Ginnings, D. C. Values for $t < 100$ have been privately communicated by Osborne, 1938, and are subject to slight revision as the work progresses; the others are from a longer table in *J. Res. Nat. Bur. Stand.*, **18**, 389-448 (RP983) (1937). These supersede similar values published by Osborne, Stimson, and Fiock, *Bur. Stand. J. Res.*, **5**, 411-480 (RP209) (1930) → *Mech. Eng.*, **52**, 191-220 (FSP-52-28) (1930) and by Osborne, Stimson, and Ginnings, *Mech. Eng.*, **56**, 94-95 (1934); *Idem*, **57**, 162-163 (1935).
Sch   Schüle, M., *Z. Ver. deuts. Ing.*, **55**, 1506-1512, 1561-1567 (1911).
TS   Trautz, M., and Steyer, H., *Forsch. Gebiete Ingenieurw.*, **2**, 45-52 (1931) → Steyer, H., *Z. Ver. deuts. Ing.*, **75**, 601 (1931).

b These values were obtained by KSG by extrapolation.

c Callendar, holding that the true critical temperature is 380.5 °C, extends his observations to higher temperatures: 275 °C, $H_v = 2296.9$, $H_l = 2051.5$; 377 °C, $H_v = 2242.0$, $H_l = 2055.8$; 380 °C, $H_v = 2077.5$; $H_l = 2029.3$; and 380.5 °C, $H_v = H_l = 1968$.

Table 257.—Specific Heat of Saturated Water-vapor [192]

c_s = amount of heat per g that must be added in order to increase the temperature by 1 °C while the condition of saturation is maintained by suitably adjusting the volume.

t	58	93	148 °C
c_s	−5.9	−5.0	−0.33 (joule/g) per 1 °C.

Table 258.—Entropy of Saturated Water and of Saturated Steam

(Values adopted for the International Steam Tables, 1934, are given in Table 260.)

The OSG values are to be preferred. They are given directly, the others by the amounts (ΔS) by which they exceed the corresponding OSG values. Example: For the saturated vapor at 0 °C M gives the equivalent of $S_v = 9.132 - 0.0037 = 9.1283$.

The values under C and M illustrate the magnitude of the errors in the tables current ten years ago.

S is the excess of the entropy above its value for saturated water at 0 ° C. The subscripts l and v indicate that the corresponding values refer to the liquid and to the vapor, respectively.

[192] Leduc, A., *Int. Crit. Tables*, **5**, 83 (1929).

89. SATURATION: THERMAL ENERGY

Table 258.—(Continued)

Unit of S and of $\Delta S = 1$ Int. joule/g. °C.   Temp. $= t$ °C (Int. scale)

Ref.[a]→ t	OSG S_l	Liquid K	C $10^4\Delta S_l$	M	OSG S_v	Vapor C	M $10^4\Delta S_v$
0	0	0		0	9.132		−37
10	0.1511			+1	8.884		−69
20	0.2962	+2		+3	8.656		−84
30	0.4363			4	8.446[e]		−82
40	0.5719	−1		5	8.253		−78
50	0.7032			11	8.074		−59
60	0.8305	0		11	7.909		−34
70	0.9543			17	7.756		−15
80	1.0746	0		15	7.613		+12
90	1.1918			7	7.480		+40
100	1.3059	−24[b]		9	7.3536		+85
110	1.4174			12	7.2371		+107
120	1.5265	+6		14	7.1278		+128
130	1.6332			19	7.0248		+149
140	1.7378	+7		20	6.9276		+162
150	1.8404			24	6.8355		+166
160	1.9412	+7		25	6.7480		+170
170	2.0403			27	6.6643		+161
180	2.1380	+11		25	6.5836		+156
190	2.2342			26	6.5055		+141
200	2.3293	+15	+23	18	6.4295	+245	+118
210	2.4233		+25	11	6.3554	270	+84
220	2.5164	+15	+36	+2	6.2827	297	+45
230	2.6088		+38	−14	6.2112	326	−10
240	2.7007	+1	+48	−33	6.1404	351	−77
250	2.7922		+46	−60	6.0699	382	−159
260	2.8835	−6	+62	−93	5.9992	415	−252
270	2.9749		+78	−137	5.9279	441	−351
280	3.0668	−18	+88	−193	5.8553	473	−463
290	3.1594		+100	−260	5.7807	498	−575
300	3.2534	−29	+123	−334	5.7032	524	−696
310	3.3495	−36	+121	−399	5.6214	542	−799
320	3.4482	−39	+139	−448	5.5341	639	−869
330	3.5504	−36	+151	−498	5.4392	581	−912
340	3.6583	−35	+190	−539	5.3332	586	−882
350	3.7763	−68	+245	−605	5.2091	591	−797
360	3.9150		+306	−568	5.0512	634	−658
365	4.0002		+375		4.9455	728	
366	4.0196				4.9203		
367	4.0401				4.8930		
368	4.0620				4.8634		
369	4.0858				4.8305		
370	4.1121		+471	−123	4.7936	1062	−506
371	4.1422				4.7509		
372	4.1787				4.6695		
373	4.2282				4.6321		
374	4.3308		−247	+902	4.5079	2663	−869

[a] References and remarks:
- C   Callendar, H. L., *Proc. Inst. Mech. Eng.*, **1929**, 507-527 (1929).
- K   Koch, W., *Forsch. Gebiete Ingenieurw.*, **5**, 138-145 (1934); *Z. Ver. deuts. Ing.*, **78**, 1160 (1934).
- M   Mollier, R., "Neue Tabellen u.s.w.," 5 ed., 1927.
- OSG Osborne, N. S., Stimson, H. F., and Ginnings, D. C. The values for $t < 100$ are from Osborne, Stimson, and Fiock, *Bur. Stand. J. Res.*, **5**, 411-480 (RP209) (1930)→ *Mech. Eng.*, **52**, 191-220 (FSP-52-28) (1930); the others are from a revised and

Table 258.—*(Continued)*

more extended table by Osborne, Stimson, and Ginnings, *J. Res. Nat. Bur. Stand.,* **18,** 389-448 (RP983) (1937) which takes precedence over the former.

b Koch's published value (0.3114 Int. steam cal/g) for the saturated liquid at 100 °C seems to involve a misprint; if it were 0.3119, then $10^4 \Delta S_l$ would be -3 joules/g·°C, which is not very discordant with the adjacent values.

c From fundamental constants and spectroscopic data, A. R. Gordon and Colin Barnes [193] computed the excess of the entropy of saturated steam at 30 °C above that at absolute zero to be 54.39 cal/g-mole(°C) = 12.642 j/g(°C).

90. Steam-tables and Diagrams

Although many steam-tables and diagrams, each giving several types of steam-engineering data pertaining to one or more of the systems—water, steam, water and steam—have been published, there was in 1929 no generally accepted consistent set of data from which complete steam-tables could be constructed. Tables based on different sources differed disconcertingly, although it is probable that any one of several was good enough for most technical purposes, provided its data were not mixed with those from another source. Any mixing had to be done with circumspection.

As a result of this condition, an international steam-table conference was held in London in 1929, another in Berlin in 1931, and a third in this country in 1934. A skeleton steam-table was adopted at the first, and was revised and somewhat extended at the second; and a more extended one was adopted at the third. The last, which supersedes all others, is here given as Table 260. A review of the better work on steam prior to the London conference has been published by H. N. Davis and J. H. Keenan.[194]

In Table 259 are listed some of the more recent steam tables and diagrams and of the reports of extended work pertaining thereto. Extracts from much of the latter may be found distributed through the several sections devoted to the properties concerned.

The calorimetric measurements in this field have been recently reviewed by E. F. Fiock [195]; and many pertinent data based on measurements made at the Physikalisch-Technischen Reichsanstalt are given by L. Holborn, K. Scheel, and F. Henning in their "Wärmetabellen" (1919).

Numerous equations of state have been proposed for water-vapor, but none has been generally accepted as entirely satisfactory. Some will be found in the references given elsewhere in this section, and some are discussed in the following papers, listed in chronological order: Linde, R., *Forsch. Gebiete Ingenieurw.,* **21,** 57-92 (1905), Eichelberg, *Idem,* **220,** 1-31 (1920), Tumlirz, O., *Sitz. Akad. Wiss. Wien (Abt. 2a),* **130,** 93-133 (1921), Strauven, M., *Rev. univers. des mines (6),* **16,** 289-301, 363-376 (1923), Callendar, H. L., *World Power,* **1,** 274-280, 325-328 (1924), Nesselmann, K., *Z. physik. Chem.,* **108,** 309-340 (1924), Hausen, H., *Forsch. Gebiete Ingenieurw.,* **2,** 319-326 (1931).

[193] Gordon, A. R., and Barnes, Colin, *J. Phys'l Chem.,* **36,** 1143-1151 (1932).
[194] Davis, H. N., and Keenan, J. H., *Proc. World Eng. Cong., Tokyo, 1929,* **4,** 239-264 (1931).
[195] Fiock, E. F., *Bur. Stand. J. Res.,* **5,** 481-505 (RP210) (1930) = *Mech. Eng.,* **52,** 231-242 (FSP-52-30) (1930).

90. STEAM TABLES

Table 259.—A List of Some Recent Steam Tables and Diagrams and of Reports of Extended Work Pertaining Thereto

1927: Mollier, R., "Neue Tabellen und Diagramme für Wasserdampf," 5 ed.: Berlin, Springer.

1928: Callendar, H. L., "Steam tables and equations, extended by direct experiment to 4000 lb*/sq in. and 400 °C," *Proc. Roy. Soc. London (A)*, **120**, 460-472 (1928).
Jakob, M., "Die Verdampfungswärme des Wassers und das spezifische Volumen von Sattdampf für Temperaturen bis 210 °C," *Forsch.-Arb. Gebiete Ingenieurw.*, Heft 310, 9-19 (1928) = *Wiss. Abh. d. Phys.-Techn. Reichs.*, **12**, 435-446 (1929).

1929: Callendar, H. L., "Extended steam tables," *Proc. Inst. Mech. Eng.*, **1929**, 507-527 (1929).
Int. Steam-Table Conference, "Skeleton steam table," *Engineering (London)*, **128**, 751-752 (1929) = *Z. Ver. deuts. Ing.*, **73**, 1856-1858 (1929) = *Mech. Eng.*, **52**, 120-122 (1930).
Jakob, M., and Fritz, W., "Die Verdampfungswärme des Wassers und das spezifische Volumen von Sattdampf zwischen 210 und 250 °C," *Z. Ver. deuts. Ing.*, **73**, 629-636 (1929) = *Wiss. Abh. d. Phys.-Techn. Reichs.*, **13**, 93-111 (1930).
Keenan, J. H., "A revised Mollier chart for steam, extended to the critical point," *Mech. Eng.*, **51**, 109-115 (1929).

1930: Callendar, H. L., "Extended H-Φ diagram for saturated and superheated steam," London, E. Arnold & Co.
Fiock, E. F., "A review of calorimetric measurements on thermal properties of saturated water and steam," *Bur. Stand. J. Res.*, **5**, 481-505 (RP210) (1930).
Jakob, M., and Fritz, W., "Die Verdampfungswärme des Wassers und das spezifische Volumen von Sattdampf im Bereich bis 310 °C (100.7 at)": *Techn. Mechan. Thermodynam.*, **1**, 173-183, 236-240 (1930).
Keenan, J. H., "Steam Tables and Mollier Diagrams," Am. Soc. Mech. Eng., New York.
Osborne, N. S., Stimson, H. F., and Fiock, E. F., "A calorimetric determination of thermal properties of saturated water and steam from 0 °C to 270 °C," *Bur. Stand. J. Res.*, **5**, 411-480 (RP209) (1930).
Speyerer, H., and Sauer, G., "Vollständige Zahlentafel und Diagramme für das spezifische Volumen des Wasserdampfes bei Drücken zwischen 1 und 270 at," VDI-Verlag, Berlin.

1931: Fritz, W., "Ergebnisse der kalorimetrischen Messungen des amerikanischen Bureau of Standards on Wasser und Sattdampf zwischen 0 und 270 °C (56 at)," *Forsch. Gebiete Ingenieurw.*, **2**, 41 (1931).
Int. Steam-Table Conference (2nd), "Rahmentafeln für Wasserdampf nebst Erläuterungen," *Z. Ver. deuts. Ing.*, **75**, 187-188 (1931) = *Engineering (London)*, **131**, 296-297, 393 (1931) = *Mech. Eng.*, **53**, 287-290 (1931).
Jakob, M., "Callendars letztes Dampfdiagramm," *Forsch. Gebiete Ingenieurw.*, **2**, 192 (1931).
——, "Steam research in Europe and America," *Engineering (London)*, **132**, 143-146, 518-521, 550-551, 651-653, 684-686, 707-709, 744-746, 800-804 (1931).
Keenan, J. H., "Thermal properties of compressed liquid water," *Mech. Eng.*, **53**, 127-131 (1931).
——, "Abridged edition of Steam Tables and Mollier Diagram," Am. Soc. Mech. Eng., New York.
Moss, H., "The revised Callendar steam tables," London, E. Arnold & Co.

1932: Keenan, J. H., "A steam chart for second-law analysis," *Mech. Eng.*, **54**, 195-204 (1932).
Knoblauch, O., Raisch, E., Hausen, H., and Koch, W., "Tabellen und Diagramme für Wasserdampf, berechnet aus der spezifischen Wärme" (2 ed), München, R. Oldenbourg.

Table 259.—(Continued)

Sugawara, S., "New formulae and tables for steam," *Mem. Col. Eng. Kyoto Imp. Univ.*, **7**, 17-48 (1932).

Willis, P. A., Hawkins, G. A., and Potter, A. A., "A comparison of recent steam tables," *Power*, **75**, 841-843 (1932) → *Mech. Eng.*, **54**, 581 (1932).

1933: Jakob, M., and Fritz, W., "Die Verdampfungswärme des Wassers und das spezifische Volumen von Sattdampf in Bereich von 100 bis 150 at," *Forsch. Gebiete Ingenieurw.*, **4**, 295-299 (1933).

Osborne, N. S., Stimson, H. F., Fiock, E. F., and Ginnings, D. C., "The pressure of saturated water vapor in the range 100 to 374 °C," *Bur. Stand. J. Res.*, **10**, 155-188 (RP523) (1933).

Pflaum, W., and Schulz, W., "i,s-Diagramm für Wasserdampf bis 2800 °C mit Berücksichtigung der Dissoziation," *Forsch. Gebiete Ingenieurw.*, **4**, 116-118 (1933).

Schlegel, E., "Ein i,s-Diagramm für Wasser bis zu 400 at Druck in den Grenzen 0-370 °C," *Z. techn. Physik*, **14**, 105-107 (1933).

1934: Green, A. M., Jr., "Early U. S. steam tables. A historical summary of tabulations published in this country prior to 1921," *Mech. Eng.*, **56**, 715-717, 764 (1934).

Justi, E., "Spezifische Wärme technischer Gase und Dämpfe bei höheren Temperaturen," *Forsch. Gebiete Ingenieurw.*, **5**, 130-137 (1934).

Koch, W., "Die spezifische Wärme des Wassers von 0 bis 350 ° C und vom jeweiligen Sättigungsdruck bis 260 kg*/cm²," *Idem*, **5**, 138-145 (1934).

——, "Wärmeinhalt von Wasser und Wasserdampf," *Idem*, **5**, 257-259 (1934) → *Z. Ver. deuts. Ing.*, **78**, 1160 (1934).

Osborne, N. S., and Meyers, C. H., "A formula and tables for the pressure of saturated water vapor in the range 0 to 374 °C," *J. Res. Nat. Bur. Stand.*, **13**, 1-20 (RP691) (1934).

Smith, L. B., and Keyes, F. G., "The volumes of unit mass of liquid water and their correlation as a function of pressure and temperature," *Proc. Am. Acad. Arts Sci.*, **69**, 285-312 (1934).

Smith, L. B., Keyes, F. G., and Gerry, H. T., "The vapor pressure of water," *Idem*, **69**, 137-168 (1934).

1935: Bridgman, P. W., "The pressure-volume-temperature relations of the liquid and the phase diagram of heavy water," *J. Chem'l Phys.*, **3**, 597-605 (1935). (Contains data for ordinary water also.)

Int. Steam-Table Conference (3d), "The Third International Conference on Steam Tables," *Mech. Eng.*, **57**, 710-713 (1935) = *Engineering (London)*, **140**, 372-373, 393 (1935) = *Z. Ver. deuts. Ing.*, **79**, 1359-1362 (1935).

Jakob, M., and Fritz, W., "Die Verdampfungswärme des Wassers und das spezifische Volumen von Sattdampf bis zu 202 kg*/cm² (365 °C)," *Physik. Z.*, **36**, 651-659 (1935).

Justi, E., and Lüder, H., "Spezifische Wärme, Entropie und Dissoziation technischer Gase und Dämpfe," *Forsch. Gebiete Ingenieurw.*, **6**, 209-216 (1935).

Keyes, F. G., Smith, L. B., and Gerry, H. T., "The specific volume of steam in the saturated and superheated condition together with derived values of the enthalpy, entropy, heat capacity and Joule-Thomson coefficients," *Proc. Am. Acad. Arts Sci.*, **70**, 319-364 (1935). (Included in the 1935 volume, but actually issued in 1936).

1936: Havlíček, J., and Miškovský, L., "Versuche der Masaryk-Akadamie der Arbeit in Prag über die physikalischen Eigenschaften des Wassers und des Wasserdampfes," *Helv. Phys. Acta*, **9**, 161-207 (1936).

Keenan, J. H., and Keyes, F. G., "Thermodynamic Properties of Steam," New York, John Wiley & Sons.

Kirschbaum, E., "Wärme und Stoffaustausch im Mollierschen i,x-Bild," *Forsch. Gebiete Ingenieurw.*, **7**, 109-113 (1936).

1937: Beattie, J. A., and Blaisdell, B. E., "An experimental study of the absolute temperature scale. III. The reproducibility of the steam point. The

Table 259.—(Continued)

effect of pressure on the steam point," *Proc. Am. Acad. Arts Sci.*, **71**, 361-374 (1937).

Bridgman, P. W., "The phase diagram of water to 45,000 kg*/cm²," *J. Chem'l Phys.*, **5**, 964-966 (1937).

Eck, H., "Zustandsgrössen des Wassers im kritischen Gebiete," *Ber. Tätigkeit Phys.-Techn. Reichs. im 1936 = Physik. Z.*, **38**, 256 (1937).

Heck, R. C. H., "The Keenan and Keyes steam tables," *Mech. Eng.*, **59**, 97-100 (1937).

Osborne, N. S., Stimson, H. F., and Ginnings, D. C., "Calorimetric determination of the thermodynamic properties of saturated water in both the liquid and gaseous states from 100 to 374 °C," *J. Res. Nat. Bur. Stand.*, **18**, 389-448 (RP983) (1937).

Tilton, L. W., and Taylor, J. K., "Accurate representation of the refractivity and density of distilled water as a function of temperature," *Idem.*, **18**, 205-214 (RP971) (1937).

Table 260.—International Skeleton Steam-Tables, 1934

The following values were adopted by the Third International Steam-Tables Conference held in the United States of America [a] in 1934,[196] and involve the following units and conversion factors: Inch, pound, atmosphere, and bar, all as already defined in Table 1; the international steam-table calorie, designated as "IT-cal" defined by the relation: 1000 IT-cal = 1/860 international kilowatthour; 1 international watt = 1.0003 (absolute) watt; 1 British thermal unit = 251.996 IT-cal; temperatures are expressed in terms of the international (centigrade) temperature scale, the zero being regarded as equivalent to 273.16 °K. These lead to the following equivalents: 1 kg*/cm² = 14.2233 lb*/in² = 0.967841 atm; 1 cm³ = 3.53146₇ 10⁻⁵ft³; 1 m³/kg = 16.0185 ft³/lb; 1 IT-cal = 4.18605 Int. j = 21.447 10⁻³lb*·ft³/in² = 41.3255 atm·cm³ = 4.1873 joules.

The enthalpy or total heat is defined as "the heat content in excess of that contained by the liquid at zero degree centigrade and saturation pressure."[197]

P = pressure, v^* = specific volume, H = enthalpy (total heat); subscripts l and s indicate that the data refer to the liquid and to the vapor, respectively; τ = tolerance, expressed in units of the last place of the tabulated value; Tol. is the actual tolerance, for the entire column of values above it, unless another tolerance is indicated.

Observed values t_{crit} = 374.11 °C (M.I.T.), 374.2 ± 0.1 °C (P.T.R.).

Unit of P = 1 kg*/cm²; of v^* = 1 cm³/g; of H = 1 IT-cal/g; of τ = 1 unit in last place.
Temp. = t °C

I. Saturated phases: Liquid (l), vapor (s).

t	P	τ	$v_l{}^*$	τ	$v_s{}^*$	τ	H_l	τ	H_s	τ
0	0.006228	6	1.00021	5	206310	210	0	0	597.3	7
10	0.012513	10	1.00035	10	106410	110	10.04	1	601.6	7
20	0.023829	20	1.00184	10	57824	58	20.03	2	605.9	6
30	0.043254	30	1.00442	10	32922	33	30.00	2	610.2	5
40	0.075204	38	1.00789	10	19543	19	39.98	2	614.5	5

[196] *Mech. Eng.*, **57**, 710-713 (1935).
[197] London Conf.; *Mech. Eng.*, **52**, 120-122 (1930).

Table 260.—(Continued)

P	τ	v_l*	τ	v_s*	τ	H_l	τ	H_s	τ
50	0.12578	6	1.0121	2	12045	12	49.95	3	618.9	5
60	0.20312	10	1.0171	2	7678.3	77	59.94	3	623.1	5
70	0.31775	16	1.0228	2	5046.3	50	69.93	3	627.3	5
80	0.48292	24	1.0290	2	3409.2	34	79.95	4	631.4	5
90	0.71491	36	1.0359	2	2361.5	24	89.98	5	635.3	5
100	1.03323	0	1.0435	2	1673.2	17	100.04	5	639.1	5
110	1.4609	10	1.0515	4	1210.1	12	110.12	6	642.7	5
120	2.0245	13	1.0603	4	891.65	89	120.25	6	646.2	5
130	2.7544	16	1.0697	4	668.21	67	130.42	7	649.6	5
140	3.6848	21	1.0798	4	508.53	51	140.64	7	652.7	6
150	4.8535	32	1.0906	4	392.46	39	150.92	8	655.7	7
160	6.3023	42	1.1021	4	306.76	31	161.26	8	658.5	8
170	8.0764	53	1.1144	4	242.55	24	171.68	9	661.0	8
180	10.225	7	1.1275	4	193.80	19	182.18	9	663.3	9
190	12.800	8	1.1415	4	156.32	16	192.78	10	665.2	9
200	15.857	8	1.1565	4	127.18	13	203.49	10	666.8	9
210	19.456	8	1.1726	4	104.24	10	214.32	11	668.0	9
220	23.659	9	1.1900	4	86.070	86	225.29	11	669.0	9
230	28.531	10	1.2087	4	71.483	71	236.41	12	669.4	9
240	34.140	12	1.2291	4	59.684	60	247.72	12	669.4	9
250	40.560	13	1.2512	4	50.061	50	259.23	13	668.9	9
260	47.866	15	1.2755	4	42.149	42	270.97	18	667.8	9
270	56.137	17	1.3023	4	35.593	36	282.98	19	666.0	9
280	65.457	20	1.3321	4	30.122	30	295.30	20	663.6	9
290	75.917	22	1.3655	5	25.522	30	307.99	20	660.4	9
300	87.611	24	1.4036	5	21.625	35	320.98	30	656.1	10
310	100.64	3	1.4475	5	18.300	35	334.63	40	650.8	12
320	115.12	3	1.4992	5	15.438	35	349.00	50	644.2	14
330	131.18	4	1.5619	5	12.952	35	364.23	60	636.0	16
340	148.96	4	1.6408	5	10.764	35	380.69	70	625.6	18
350	168.63	4	1.7468	6	8.802	35	398.9	8	611.9	20
360	190.42	5	1.9066	40	6.963	40	420.8	8	592.9	20
370	214.68	5	2.231	21	4.997	100	452.3	15	559.3	30
371	217.26	10	2.297	26	4.761	100	457.2	15	553.8	35
372	219.88	11	2.381	34	4.498	110	462.9	22	547.1	40
373	222.53	11	2.502	53	4.182	120	471.0	35	538.9	45
374	225.22	11	2.79	15	3.648	120	488.0	50	523.3	50

II. Compressed water.

t→ P	0	50	100	150	200	250	300	350
				v_l*				
1	1.00016	1.01210						
5	0.9999	1.0119	1.0432	1.0906				
10	0.9997	1.0117	1.0431	1.0902				
25	0.9989	1.0110	1.0422	1.0893	1.1556			
50	0.9977	1.0099	1.0409	1.0877	1.1532	1.2495		
75	0.9965	1.0088	1.0397	1.0861	1.1508	1.2452		
100	0.9952	1.0077	1.0385	1.0845	1.1485	1.2410	1.3979	
125	0.9940	1.0067	1.0372	1.0829	1.1462	1.2369	1.3877	
150	0.9929	1.0056	1.0360	1.0814	1.1439	1.2330	1.3782	
200	0.9905	1.0035	1.0337	1.0784	1.1395	1.2255	1.3612	1.671
250	0.9882	1.0015	1.0314	1.0755	1.1353	1.2184	1.3462	1.604
300	0.9859	0.9995	1.0291	1.0726	1.1312	1.2117	1.3327	1.557
350	0.9837	0.9975	1.0269	1.0698	1.1272	1.2054	1.3207	1.521
400	0.9814	0.9956	1.0247	1.0670	1.1234	1.1994	1.3097	
Tol.→	0.0002[b]	0.0002	0.0002	0.0002	0.0003	0.0004	0.0007	0.002

90. STEAM TABLES

Table 26C.—*(Continued)*

$t \rightarrow$ P	0	50	100	150	200 H_l	250	300	350
1	0.023	49.97						
5	0.120	50.05	100.11	150.92				
10	0.240	50.15	100.20	151.00				
25	0.599	50.45	100.46	151.21	203.6			
50	1.20	50.96	100.90	151.58	203.8	259.2		
75	1.79	51.46	101.34	151.95	204.1	259.2		
100	2.39	51.96	101.78	152.32	204.3	259.2	320.5	
125	2.98	52.46	102.22	152.69	204.6	259.3	319.9	
150	3.57	52.96	102.65	153.06	204.8	259.3	319.3	
200	4.74	53.96	103.57	153.82	205.2	259.4	318.4	393.1
250	5.90	54.96	104.46	154.57	205.8	259.5	317.6	387.6
300	7.05	55.96	105.35	155.33	206.2	259.7	317.0	384.0
Tol.→	0.01[c]	0.03	0.05	0.08	0.1[d]	0.1[d]	0.3	0.8

III. Superheated (*i.e.*, dilated) steam.

$t \rightarrow$ P	100 v_s^*	τ	150 v_s^*	τ	200 v_s^*	τ	250 v_s^*	τ	300 v_s^*	τ
1	1730	1	1975	2	2216	2	2454	2	2691	3
5					433.8	4	484.1	5	533.2	5
10					210.4	2	237.6	2	263.3	3
25							89.0	1	101.1	1
50									46.41	7
75									27.48	5

$t \rightarrow$ P	350 v_s^*	τ	400 v_s^*	τ	450 v_s^*	τ	500 v_s^*	τ	550 v_s^*	τ
1	2928	3	3164	3	3400	3	3636	4	3872	4
5	581.6	6	629.6	6	677.4	7	725.0	7	772.5	8
10	288.2	3	312.7	3	337.0	3	361.1	4	385.1	4
25	112.1	1	122.6	1	132.7	1	142.7	1	152.6	2
50	53.12	8	59.05	9	64.60	9	69.92	10	75.10	12
75	33.22	7	37.78	8	41.83	8	45.62	9	49.25	10
100	23.03	5	27.05	5	30.41	6	33.45	7	36.32	7
125	16.66	3	20.53	4	23.52	5	26.14	5	28.55	6
150	11.98	2	16.10	3	18.90	4	21.25	4	23.36	5
200			10.31	3	13.05	3	15.11	3	16.87	3
250			6.366	13	9.46	2	11.39	2	12.96	3
300			3.02	1	6.98	2	8.90	2	10.35	2

$t \rightarrow$ P	100 H_s	τ	150 H_s	τ	200 H_s	τ	250 H_s	τ	300 H_s	τ
1	639.2	5	663.2	5	686.5	6	710.1	6	734.0	12
5					681.9	10	706.7	10	731.5	12
10					675.1	10	702.1	11	728.0	12
25							687.8	11	718.0	12
50									698.4	12
75									672.6	12

$t \rightarrow$ P	350 H_s	τ	400 H_s	τ	450 H_s	τ	500 H_s	τ	550 H_s	τ
1	758.0	12	782.4	12	807.2	12	832.3	12	857.8	20
5	756.1	12	780.8	12	805.9	12	831.3	12	856.9	20
10	753.5	12	778.9	12	804.5	12	830.1	12	855.9	20
25	746.3	12	773.3	12	800.0	12	826.5	12	852.6	20
50	732.9	12	763.1	12	791.6	12	819.9	12	847.3	20
75	717.6	12	752.1	12	783.2	12	813.1	12	841.8	20
100	699.5	12	740.0	13	774.5	13	806.0	13	836.1	20
125	676.7	12	726.9	13	765.2	13	799.1	13	830.3	20
150	646.8	15	712.1	13	755.3	13	791.8	13	824.4	20

Table 260.—(Continued)

t → P	350 H_s	τ	400 H_s	τ	450 H_s	τ	500 H_s	τ	550 H_s	τ
200			676.5	20	733.4	20	776.0	20	812.0	25
250			622.5	25	707.5	25	758.8	25	798.9	30
300			524.5	30	677.5	25	739.7	25		

[a] The first conference was held in London, *Engineering (London)*, **128**, 751-752 (1929), *Mech. Eng.*, **52**, 120-122 (1930), *Z. Ver. Deuts. Ing.*, **73**, 1856-1858 (1929); the second, in Berlin, *Z. Ver Deuts. Ing.*, **75**, 187-188 (1931), *Mech. Eng.*, **53**, 287-290 (1931), *Engineering (London)*, **131**, 296-297, 393 (1931).
[b] For $t = 0$ and $P = 1$, Tol. = 0.00005.
[c] For $t = 0$ and $P = 1, 5, 10,$ and 25, Tol. = 0.005.
[d] For $P = 250$ and $t = 200$ and 250, Tol. = 0.2; for $P = 300$ and $t = 200$ and 250, Tol. = 0.3.

91. Fugacity and Activity of Water

The value of the fugacity of water, as that term is used by G. N. Lewis and M. Randall,[198] is $f = P_s e^{-(g-w)}$ where P_s is the pressure of the vapor when saturated at the temperature (T °K) considered, and g and w are the values of two definite isothermal integrals, one (g) determined solely by the vapor (gas), and the other (w) solely by the liquid (water); $RTg = \int_{P_0}^{P_s} (\delta/P) dP$ and $RTw = \int_{P_s}^{P'} v_w^* dP$, where $\delta = RT - Pv^*$, v being the specific volume of the vapor at T °K and under the pressure P, P_0 is any such low pressure that it is reasonable to assume that δ is negligible if $P \leqq P_0$, v_w^* is the specific volume of water at the same temperature and under the pressure P, P' is the pressure at which the fugacity is desired, and R is the gas-constant. Obviously, the units of v^*, v_w^*, and P must be so chosen that the unit of each of the products Pv^* and $v_w^* dP$, and also of δ, shall be the same as that of the RT with which it is associated.

The expression for w may conveniently be broken into two parts, one involving integration from P_s to A, the other from A to P', where A denotes the pressure of 1 atm. These integrals we shall denote by w' and w'', respectively;

$$RTw' = \int_{P_s}^{A} v_w dP, \quad RTw'' = \int_{A}^{P'} v_w^* dP, \quad w = w' + w''.$$

If $P' = A$, $w'' = 0$ and the fugacity at 1 atm is $f_1 = P_s e^{-(g-w')}$; if $P' = P_s$, $w' = -w''$ and the fugacity at saturation is $f_s = P_s e^{-g}$. Also $f = P_s e^{-(g-w'-w'')} = f_1 e^{w''}$; Lewis and Randall call the quantity $e^{w''}$ the activity and denote it by a; thus getting the relation $f = af_1$.

Values for f and for a of water at 25, 37.5, and 50 °C, and at various pressures between 1 and 1000 atm have been published by M. Randall and B. Sosnick.[199] Those values are quoted by the former in his compilation,[200]

---

[198] Lewis, G. N., and Randall, M., See their "Thermodynamics and the Free Energy of Chemical Substances," 1923. Also, Tunnell, G., *J. Phys'l Chem.*, **35**, 2885-2913 (1931).
[199] Randall, M., and Sosnick, B., *J. Am. Chem. Soc.*, **50**, 967-980 (975) (1928).
[200] Randall, M., *Int. Crit. Tables*, **7**, 232 (1930).

but are affected by computational errors and an assumption. These affect two of the 5 significant digits given in the values of f, and should be carefully considered.

Their procedure was to compute f_1 and a, and to derive $f(=af_1)$ from them. Any error or uncertainty in f_1 will recur, in full force, in each other value of f for that temperature; on that, will be superposed any error in the corresponding value of a.

As there seem to be no suitable experimental data for the density of dilated water-vapor at these temperatures, the value of g can be obtained only by the use of an assumed equation of state. They used Berthelot's equation, taking $\delta = \dfrac{9}{128}\left(\dfrac{RPT_c}{P_c}\right)\left\{6\left(\dfrac{T_c}{T}\right)^2 - 1\right\}$, and took for the critical constants $P_c = 217.5$ atm, $T_c = 647.1\,°K$, essentially those given in the *International Critical Tables* and in this compilation. As extrapolation always introduces an uncertainty, it is imperative that we consider carefully the probable suitability of this equation for our present purposes.

Probably the best test is to examine how well this equation will reproduce, at such temperatures, the known density of the saturated vapor. If the reproduction is close, it is probable that the equation is fairly satisfactory; in the contrary case, and especially if the difference between the two values increases rapidly as the temperature is reduced, the equation must be condemned as unsatisfactory, and the results of computations based upon it must be regarded as of doubtful value. The quantities of interest in such a comparison are given in Table 261, where also are given those computed by means of a new equation of state, derived from observations at and above 100 °C, and taking into consideration the specific heat as well as the density. That equation [201] yields

$$\delta = \left(\dfrac{54.8}{v^*} - \dfrac{42.6}{v^{*2}}\right)T - \dfrac{81.8(10)^3}{v^*} + \dfrac{38.7(10)^6}{v^*T} + \dfrac{32.5(10)^{28}}{v^{*2}T^9}\ \text{cm}^3\text{atm/g}$$

It will be noticed (Table 261) that at the higher temperatures (50 to 200 °C) the values derived from Naumann's equation do not differ greatly from those given by the observations; while those from Berthelot's are only about half as great as the experimental values at 50 °C, and 3/4 as great at 200 °C. At lower temperatures, both sets depart markedly from the experimental values, the Berthelot values at 25 °C being only 1/5 as great as the experimental. The Naumann value always lies between the other two. Whence, we may conclude that neither of these equations of state is suitable for use in computing g for temperatures below 50 °C, and that Berthelot's is quite unsuitable even at 50 °C.

The conservative course is to restrict oneself to the statement that g is probably near, and does not exceed, the value given in the column Obs. For such small quantities we may for our present purposes take $e^{-g} = 1 - g$. Then the expression for the fugacity of water at 1 atm becomes $f_1 =$

[201] Naumann, F., *Z. Physik Chem. (A)*, **159**, 135-144 (1932).

$P_s e^{-(g-w')} = P_s(1-g)e^{w'}$, where $1000g$ does not exceed 3.0, 2.4, and 3.4, respectively, at 25, 37.5, and 50 °C. At those same temperatures, $e^{w'} = 1 + w'$, where $1000w' = 0.72$, 0.67, and 0.60, respectively. Whence at 50 °C, for example, $f_1 = P_s(1.00060 - g) \gtrless P_s(1.00060 - 0.0034) = 0.9972P_s$. Putting into the expression for f_1 the values of P_s (0.031258, 0.063637, and 0.12172 atm) corresponding, respectively, to 25, 37.5, and 50 °C, we find that f_1 is near, and not less than 0.03119, 0.06353, and 0.12138 atm, respectively. It seems certain that f_1 is not greater than the value derived from Berthelot's equation; viz., 0.03126, 0.06361, and 0.12158. Hence, one may say that at 25, 37.5, and 50 °C, the values of f_1 lie within the ranges 0.03119 to 0.03126, 0.06353 to 0.06361, and 0.12138 to 0.12158 atm, respectively, and probably much nearer the lower values.

Table 261.—A Comparison of Some Bases for an Estimate of the Fugacity of Saturated Water

(See text for discussion, symbols, and formulas.)

The fugacity at saturation is $f_s = P_s e^{-g}$, $RTg = \int_{P_0}^{P_s} (\delta/P) dP$, $\delta \equiv RT - Pv^*$, $\Delta \equiv \delta/P$, v^* is the specific volume of the vapor; subscript s indicates that the quantity refers to the saturated vapor. B indicates that the value is based on Berthelot's equation of state, N on Naumann's, Obs. on observed data.

Unit of v_s^* and $\Delta_s = 1$ cm³/g; of $\delta_s = 1$ cm³atm/g; g is dimensionless. Temp. $= t$ °C

t	v_s^*	Δ_s Obs.	N	B	δ_s Obs.	N	B	$10^4 g$ Obs.[a]	N	B
0	205000	1₄₈₉	61	31	8.9	0.37	0.19	71	3	1
10	105900	5₂₂	54	28	6.3	0.65	0.34	51	5	3
20	57660	20₈	46	27	4.8	1.1	0.62	36	8	5
25	43310	13₀	47	26	4.1	1.4	0.8	30	11	6
30	32880	9₁	46	25	3.8	1.0	1.0	28	14	8
37.5	22178	54	40	24	3.4	2.5	1.5	24	18	11
50	12050	41	38	22	5.0	3.1	2.7	34	32	18
100	1671	28	26	16	28.	26.	16.	168	149	96
140	507.7	19.8	20.0	13.1	70.6	71.4	46.7	376	363	248
200	127.1	13.4	14.3	9.8	206.	220.	150.	951	924	730

[a] Here it is assumed that $\Delta = \Delta_s$ at all values of P less than P_s. Observations above 100 °C indicate that Δ is almost independent of P.

It will be noticed that the values here given as derived from Berthelot's equation differ from the published values to which reference has been made. That is because in obtaining those values the wrong sign was inadvertently given to g. That error affects the entire table.

The values for the activity (a) which are given by Randall and Sosnick are said to have been derived from the values of v_w^* found by P. W. Bridgman.[202] But those data, being given only at intervals of 500 kg*/cm² in the pressure, are not entirely satisfactory for use in the determination

[202] Bridgman, P. W., Proc. Am. Acad. Arts Sci., **48**, 307-362 (1913).

of the integral w''. Using a linear interpolation over each of the two intervals (1 to 500, and 500-1000 kg*/cm²), which will give a value somewhat too great, the graph of v^* vs. P being convex toward the axes, the present compiler obtained from those data values of a that, at the higher pressures, were significantly less (18 and 24 units in the last place at 1000 atm and at 25 and 50 °C, respectively) than those given in the table cited. This indicates an error in those tabulated values. He did not attempt to check the values at 37.5 °C.

Table 262.—Activity and Fugacity of Water at 50 °C

At 50 °C the fugacity of water is $f = P_s(1.00060 - g)a$, where P_s = pressure of the vapor at saturation, a = activity of water at the pressure considered, $1.00060 = e^{w'}$, and g is determined by the extent to which the behavior of the vapor departs from that of an ideal gas. It is probable that for water-vapor at 50 °C, g is close to, and does not exceed, 0.0034. Taking $P_s = 0.12172$ atm, $f = 0.12179(1 - g)a$ atm.

Unit of P and of f = 1 atm; a is dimensionless

P	a	f/(1−g)	P	a	f/(1−g)
1	1	0.12179	600	1.5018	0.18290
100	1.0703	0.13035	700	1.6058	0.19557
200	1.1459	0.13956	800	1.7165	0.20905
300	1.2266	0.14939	900	1.8345	0.22342
400	1.3126	0.15986	1000	1.9601	0.23872
500	1.4042	0.17102			

Consequently, no values from the table cited are tabulated in this compilation, but values of a for water at 50 °C have been derived from Amagat's data (Table 95), which are available in steps of 50 atm, and are given, with the corresponding value of f, in Table 262. They differ by a maximum of 3 units in the last place from those obtained by the compiler from Bridgman's data, linear interpolation being used.

92. Pressure-temperature Associations for Equilibrium Between an Ice and Another Phase

In this section are assembled all those pressure-temperature associations that are characteristic of each of the several types of ice when in such equilibrium as can exist between it and pure water-vapor, pure water, and each of the other types of ice. For convenience, a few values of the specific volume and density of water-vapor in equilibrium with ice are also given. For the density of water and ice, see Sections 32 and 67.

Triple Points.

The triple points of a substance are the temperatures at which three phases of it can coexist in equilibrium. To each such temperature corresponds a unique pressure.

Table 263.—Triple Points of the Water-substance
(Adapted from the compilation [a] by P. W. Bridgman ICT.[203])

The symbols in the first column indicate the three phases that are in equilibrium at the temperature and pressure indicated in the next two columns; G = vapor, L = liquid (water), I, II, III, V, VI, and VII indicate, respectively, six types of ice.

Unit of $P = 1$ atm. Temp. $= t$ °C

Phases			t	P
G	I	L[b]	+0.0099	0.00603
I	L	III	−22.0	2047
I	II	III	−34.7	2100
II	III	V	−24.3	3397
III	L	V	−17.0	3417
V	L	VI	+0.16	6175
VI	L	VII	+81.6	21700 [204]

[a] Based on Bridgman, P. W., *Proc. Am. Acad. Arts Sci.*, **47**, 439-558 (1912); *Z. anorg. Chem.*, **77**, 377-455 (1913); *Z. physik. Chem.*, **89**, 252-253 (1915); with a consideration of Tammann, G., "Kristallisieren and Schmelzen," 1903; *Z. physik. Chem.*, **84**, 257-292 (1913); **88**, 57-62 (1914); "Aggregatzustände," 1922.
[b] See p. 604 ($P = 4.58$ mm-Hg = 0.00603 atm), also head text of Table 245 ($P = 4.5867$ mm-Hg = 0.0060351 atm).

Ice and Water-vapor

The only type of ice that can exist in equilibrium with water-vapor is the usual one, ice-I. At each temperature below 0 °C there is a unique pressure of the vapor at which equilibrium exists between ice-I and the vapor. That pressure is called the vapor-pressure of ice, or the saturation pressure of water-vapor in contact with ice-I. At other pressures one of these phases grows at the expense of the other without the intervention of the liquid phase—ice sublimes. But certain observations by H. T. Barnes and W. S. Vipond [205] suggest that the vapor from ice is for a brief interval in a state somewhat different from its permanent one (see p. 000).

Vapor-pressure of Ice-I.—By integrating the Clausius-Clapeyron equation ($dp/dt = L/(v^* - V^*)T$) and determining the several constants from certain selected data, mentioned later, E. W. Washburn [206] derived formula (1).

$$\log_{10} p = -\frac{2445.5646}{T} + 8.2312 \log_{10} T - 1677.006 (10^{-5}) T + 120514 (10^{-10}) T^2 - 6.757169 \qquad (1)$$

Here p mm-Hg is the vapor-pressure of ice-I at t °C, and $T = 273.1 + t$. The coefficient of the first term of the right-hand member is determined by the value of the latent heat of ice at 0 °C, and was derived from C. Dieterici's value [206a] for the latent heat of vaporization of water, and the value

[203] Bridgman, P. W., *Int. Crit. Tables*, **4**, 11 (1928).
[204] Bridgman, P. W., *Chem'l Phys.*, **5**, 964-966 (1937).
[205] Barnes, H. T., and Vipond, W. S., *Phys. Rev.*, **28**, 453 (1909).
[206] Washburn, E. W., *Monthly Weather Rev.*, **52**, 488-490 (1924).
[206a] Dieterici, C., *Ann. d. Physik (Wied.)*, **37**, 494-508 (1889).
[207] Dickinson, H. C., and Osborne, N. S., *Bull. Bur. Stand.*, **12**, 49-81 (S248) (1915).

found by H. C. Dickinson and N. S. Osborne [207] for the latent heat of fusion of ice, both at 0 °C; that of the second term was derived from Dickinson and Osborne's [207] value for the specific heat of ice-I, and upon an estimated value (0.457 cal$_{20}$/g·°C) for the specific heat of water vapor, both at 0 °C; those of the third and fourth, from two experimental values of p: That found by S. Weber [208] at -100 °C, and that by K. Scheel and

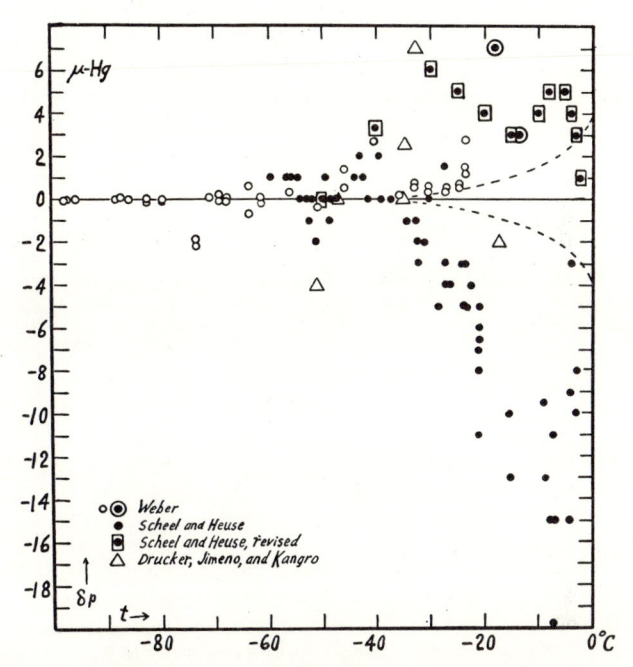

FIGURE 12. Vapor Pressure of Ice-I: Deviations of Observations from the Values Defined by Formula (1).

[Adapted from E. W. Washburn, *Monthly Weather Rev.*, **52**, 488-490 (1924).]
$\delta p = p_{calc} - p_{obs}$; unit of $\delta p = 1$ μ-Hg $= 0.001$ mm-Hg. The dotted curves indicate the change produced in δp by an error of 0.01 °C in the temperature.
Weber = Weber, S., *Com. Phys. Lab. Leiden*, No. **150**, 3-52 (1915).
Scheel and Heuse = Scheel, K., and Heuse, W., *Ann. d. Physik (4)*, **29**, 723-737 (1909).
Scheel and Heuse, revised = the preceding as revised by Holborn, L., Scheel, K., and Henning, F., "Wärmetabellen," 1919.
Drucker, Jiméno, and Kangro = Drucker, C., Jiméno, E., and Kangro, W., *Z. physik. chem.*, **90**, 513-552 (1915).

W. Heuse [209] at -50 °C; and the constant term was so determined as to make p take the well-established value 4.579 mm-Hg at 0 °C.

Washburn was of the opinion that the experimental values in the range -50 °C to near 0 °C are unreliable and should be rejected, the errors arising perhaps from uncertainties in the temperatures. See Fig. 12.

C. F. Marvin [210] has discussed the experimental determinations made prior to 1909.

Table 264.—Vapor-pressure of Ice-I

(Adapted from a table given by E. W. Washburn,[211] the values being those defined by formula (1). See text and Fig. 12. Values below −99 °C have been computed by the compiler.)

p_s = pressure of pure water-vapor that is in equilibrium with pure ice-I at t °C. If the vapor is mixed with atmospheric air, total pressure being P, then the pressure of the vapor when equilibrium exists will be $p = p_s + \Delta p$, where $\Delta p = V^*(P - p)p_s/RT$, V^* being the volume of unit mass of ice at T °K, and R being the gas-constant. For all practical purposes this is equivalent to $100 \, \Delta p/p_s = 24/T$ when $P = 1$ atm.†

Unit of p_s = 1 mm-Hg. Temp. = t °C = $(t_1 + t_2)$ °C

$t_2 \rightarrow$ t_1	−0.0	−0.1	−0.2	−0.3	−0.4	−0.5	−0.6	−0.7	−0.8	−0.9
−0	4.579	4.542	4.504	4.467	4.431	4.395	4.359	4.323	4.287	4.252
−1	4.217	4.182	4.147	4.113	4.079	4.045	4.012	3.979	3.946	3.913
−2	3.880	3.848	3.816	3.785	3.753	3.722	3.691	3.660	3.630	3.599
−3	3.568	3.539	3.509	3.480	3.451	3.422	3.393	3.364	3.336	3.308
−4	3.280	3.252	3.225	3.198	3.171	3.144	3.117	3.091	3.065	3.039
−5	3.013	2.987	2.962	2.937	2.912	2.887	2.862	2.838	2.813	2.790
−6	2.765	2.742	2.718	2.695	2.672	2.649	2.626	2.603	2.581	2.559
−7	2.537	2.515	2.493	2.472	2.450	2.429	2.408	2.387	2.367	2.346
−8	2.326	2.306	2.285	2.266	2.246	2.226	2.207	2.187	2.168	2.149
−9	2.131	2.112	2.093	2.075	2.057	2.039	2.021	2.003	1.985	1.968
−10	1.950	1.934	1.916	1.899	1.883	1.866	1.849	1.833	1.817	1.800
−11	1.785	1.769	1.753	1.737	1.722	1.707	1.691	1.676	1.661	1.646
−12	1.632	1.617	1.602	1.588	1.574	1.559	1.546	1.532	1.518	1.504
−13	1.490	1.477	1.464	1.450	1.437	1.424	1.411	1.399	1.386	1.373
−14	1.361	1.348	1.336	1.324	1.312	1.300	1.288	1.276	1.264	1.253
−15	1.241	1.230	1.219	1.208	1.196	1.186	1.175	1.164	1.153	1.142
−16	1.132	1.121	1.111	1.101	1.091	1.080	1.070	1.060	1.051	1.041
−17	1.031	1.021	1.012	1.002	0.993	0.984	0.975	0.966	0.956	0.947
−18	0.939	0.930	0.921	0.912	0.904	0.895	0.887	0.879	0.870	0.862
−19	0.854	0.846	0.838	0.830	0.822	0.814	0.806	0.799	0.791	0.783

Unit of p_s = 0.001 mm-Hg. Temp. = t °C = $(t_1 + t_2)$ °C

$t_2 \rightarrow$ t_1	−0	−1	−2	−3	−4
−20	776.	705.	640.[a]	580.	526.
−30	285.9	257.5	231.8	208.4	187.3
−40	96.6	86.2	76.8	68.4	60.9
−50	29.5$_5$	26.1	23.0	20.3	17.8
−60	8.0$_8$	7.0$_3$	6.1$_4$	5.3$_4$	4.6$_4$
−70	1.94	1.67	1.43	1.23	1.05
−80	0.40	0.34	0.29	0.24	0.20
−90	0.07$_0$	0.05$_8$	0.04$_8$	0.04$_0$	0.03$_3$
−100	0.009$_9$	0.008$_1$	0.006$_5$	0.005$_3$	0.004$_3$
−110	0.001$_0$[b]	0.0008$_8$	0.0006$_9$	0.0005$_5$	0.0004$_3$

† This coefficient (24) differs from that (20) given by Washburn[211]; the latter is incorrect.

[208] Weber, S., *Comm. Phys. Lab. Leiden*, No. 150, 3-52 (1915).
[209] Scheel, K., and Heuse, W., *Ann. d. Physik (4)*, 29, 723-737 (1909).
[210] Marvin, C. F., *Monthly Weather Rev.*, 37, 3-9 (1909).
[211] Washburn, E. W., *Int. Crit. Tables*, 3, 210-211 (1928) ← *Monthly Weather Rev.*, 52, 488-490 (1924).

92. ICE: P-T PHASE EQUILIBRIUM

Table 264.—*(Continued)*

$t_2 \rightarrow$ t_1	-5	-6	-7 p_s	-8	-9
-20	476.	430.	389.	351.	317.
-30	168.1	150.7	135.1	120.9	108.1
-40	54.1	48.1	42.6	37.8	33.4
-50	15.7^a	13.8	12.1	10.6	9.2$_5$
-60	4.0$_3$	3.4$_9{}^a$	3.0$_2$	2.6$_1$	2.2$_5$
-70	0.90	0.77	0.66	0.56	0.47
-80	0.17	0.14	0.12	0.10	0.08$_4$
-90	0.02$_7$	0.02$_2$	0.01$_8$	0.01$_5$	0.01$_2$
-100	0.003$_4$	0.002$_8$	0.002$_2$	0.001$_8$	0.001$_4$
-110	0.0003$_4$	0.0002$_6$	0.0002$_0$	0.0001$_6$	0.0001$_2$

^{a} C. Dei[213] has obtained experimentally the values $p_s = 0.58 \pm 0.023$ mm-Hg at -22.3, 0.0166 ± 0.0008 at -55, and 0.0037 ± 0.0002 at $-66\,°C$, which essentially agree with those (0.62, 0.0157, 0.0035) given by this table.

^{b} S. Dushman's estimate[212] of p_s at $-110\,°C$ is 0.75×10^{-6} mm-Hg, somewhat smaller than the 1.0×10^{-6} required by Washburn's equation.

Density and Specific Volume of Vapor Saturated with Respect to Ice. At such low vapor-pressures, the vapor may for most practical purposes be regarded as ideal. Its ideal specific volume at $t\,°C$ may be computed

Table 265.—Ideal Specific Volume and Density of Vapor Saturated with Respect to Ice

Computed by means of the formula $v^* = 3461.8(273.1 + t)/p_s$ cm^3/g, where $v^* =$ specific volume, p_s mm-Hg $=$ pressure of vapor in equilibrium with ice-I at $t\,°C$ (see Table 264); $\rho =$ density of the vapor.

Unit of $v^* = 1$ m^3/g; of $\rho = 1$ mg/m^3

t	v^*	ρ	t	$10^{-3}v^*$	100ρ
-0	0.2065	4843	-60	0.1427	701
-5	0.3195	3129	-65	0.290	344
-10	0.5026	1990	-70	0.612	166
-15	0.8036	1244	-75	1.34	75
-20	1.308	765	-80	3.06	33
-25	2.168	461	-85	7.3	14
-30	3.670	272.5	-90	18.0	5.6
-35	6.345	157.6	-95	47	2.1
-40	11.22	89.1	-100	130	0.77
-45	20.36	49.1	-105	385	0.26
-50	37.85	26.42	-110	1330	0.075
-55	72.3	13.8	-115	3900	0.025

by means of the relation $v^* = 4.555(760)(273.1 + t)/p_s = 3461.8 \times (273.1 + t)/p_s$ cm^3/g, the unit of p_s being 1 mm-Hg, and the values of p_s (the vapor-pressure of ice) being taken from Table 264. But the actual specific volume is somewhat smaller than the ideal. A tentative determination by M. Knudsen[214] indicated that the effective molecular weight

---

[212] Dushman, S., *Int. Crit. Tables*, **1**, 92 (1926).
[213] Dei, C., *Atti acc. Lincei (6)*, **12**, 119-124 (1930).
[214] Knudsen, M., *Ann. d. Physik (4)*, **44**, 525-536 (1914).

of water-vapor saturated at −75 °C is 21.1, which corresponds to a specific volume about 15 per cent less than the ideal.

Ice and Water.

Melting-point of Ice

By the melting-point of ice is meant the temperature at which ice is in equilibrum with the adjacent water under the existing pressure. It varies with the pressure and with the purity of the water.

Except as the contrary is indicated, the following data refer to pure ice in contact with pure water, the system being subjected to a uniform hydro-

Table 266.—Absolute Temperature of the Ice-point

The value that was used for computational purposes in the *International Critical Tables*, and that is so used in this compilation except where another is indicated, is this: Ice-point = 0 °C = T_0 °K = 273.1 °K. But it is quite certain that the correct value of T_0 is somewhat greater than that, probably about 273.16 °K. For a recent discussion of the subject, see KT.[a]

Unit of T_0 = 1 °K

Year	T_0	Ref[a]	Year	T_0	Ref[a]
1907	273.174[b]	Buckingham	1930	273.16	HO
1915	273.09	Henning	1931	273.116[d]	Keyes
1919	273.09[c]	HSH	1931	273.16	KK
1921	273.20	HH	1932	273.16[e]	KS
1922	273.135 ± 0.005	Keyes	1933	273.215	Jacyna
1923	273.13 ± 0.01	ST	1934	273.16	KSG
1925	273.15	Roebuck	1935	273.22	Jacyna
1929	273.14 ± 0.01	Schames	1936	273.144	KT
1929	273.16	HO	1936	273.16	Roebuck
1929	272.79[d]	Roebuck			

[a] References:

    Buckingham, E., *Bull. Bur. Stand.*, **3**, 237-293 (S57) (1907).
    Henning, F., "Die Grundlage...der Temperaturmessung," p. 80, Vieweg, 1915.
HH   Henning, F., and Heuse, W., *Z. Physik*, **5**, 285-314 (1921).
HO   Heuse, W., and Otto, J., *Ann. d. Physik* (5), **2**, 1012-1030 (1929); **4**, 778-780 (1930).
HSH Holborn, L., Scheel, K., and Henning, F., "Wärmetabellen," 1919.
    Jacyna, W., *Acta Phys. Polon.*, **2**, 419-424 (1933); **4**, 243-268 (1935) → *Z. Physik*, **97**, 107-112 (1935).
    Keyes, F. G., *J. Am. Soc. Refrig. Eng.*, **8**, 505-508 (1922); *Proc. Am. Acad. Arts Sci.*, **66**, 349-355 (1931).
KK   Kirkwood, J. G., and Keyes, F. G., *Phys. Rev.* (2), **37**, 832-840 (1931).
KS   Keyes, F. G., and Smith, L. B., *Mech. Eng.*, **54**, 125-126 (1932).
KSG Keyes, F. G., Smith, L. B., and Gerry, H. T., *Mech. Eng.*, **56**, 87-92 (1934). Also, Smith, L. B., Keyes, F. G., and Gerry, H. T., *Proc. Am. Acad. Arts Sci.*, **69**, 137-168 (1934).
KT   Keesom, W. H., and Tuyn, W., *Comm. K. Onnes Lab. Leiden*, Suppl. 78, 1-85 (1936) ← *Trav. et Mem. Bur. Int. Poids et Mes.*, **20**, (1936).
    Roebuck, J. R., *Proc. Am. Acad. Arts Sci.*, **60**, 537-596 (1925); **64**, 287-334 (1929); *Phys. Rev.* (2), **50**, 370-375 (1936).
    Schames, L., *Z. Physik*, **57**, 804-807 (1929).
ST   Smith, L. B., and Taylor, R. S., *J. Am. Chem. Soc.*, **45**, 2124-2128 (1923).

[b] From a consideration of his own and preceding determinations, Buckingham concluded that 273.13 "is probably not far from the true value."

[c] This value of T_0 was derived from the value (0.0036618) given in the "Wärmetabellen" (p. 20) for the coefficient of expansion of an ideal gas.

[d] The computation by which Roebuck obtained the value 272.79 has been criticized by Keyes, who by a recomputation from the same data obtained the value 273.116.

[e] In the paper here referred to, Keyes and Smith suggested that T_0 "may not be far from 273.16."

92. ICE: P-T PHASE EQUILIBRIUM

static pressure. For the effect of differential stresses, see Sections 64 (pp. 430-431) and 97 (p. 646). Five types of ice (I, III, V, VI, and VII) must be considered.

Ice-I: Normal Melting-point and Triple-point.—By definition, the melting-point of ice-I (ordinary ice) in contact with water saturated with air

Table 267.—Melting-points of the Ices under Hydrostatic Pressure

Adapted from the compilation by P. W. Bridgman (ICT)[a, 221] with the addition of dP/dt from his paper,[222] (1912) and of such other data as are so indicated. The data for ice-V and ice-VI extend into the supercooled regions. For phase diagram, see Section 93.

Unit of $P = 1$ atm $= 1.01325$ bars $= 1.0332$ kg*/cm². Melting-point $= t$ °C

P	t	dP/dt	P	t	dP/dt
— Ice-I —			*— Ice-VI —*		
0.00603	+0.0099[b]		4640	−15.0	+96.4
1	0.0	−134.3	5110	−10.0	+103.0
590	−5.0	−112.0	5620	−5.0	+110.2
1090	−10.0	−95.4	6160	0.0	+116.2
1540	−15.0	−82.2	6175	+0.16[e]	
1910	−20.0	−71.8	6780	+5.0	+121.8
2047	−22.0[c]		7390	10.0	+128.5
653	−5.53[f]		8040	15.0	+135.5
831	−7.46[f]		8710	20.0[g]	+143.8
1064	−9.75[f]		10250	30.0	+162.0
1104	−10.42[f]		11990	40.0	+182.7
1313	−12.74[f]		13970	50.0	+208.7
1546	−15.66[f]		16150	60.0	+235.1
			18530	70.0	
			21090	80.0	
— Ice-III —			*— Ice-VII[h] —*		
2047	−22.0[c]	+180	21700	81.6[i]	108.4
2430	−20.0	+238	23200	95.3	121.6
2820	−18.5	+310	25200	110.3	134.8
3417	−17.0[d]	+430	27100	124.1	144.5
			29000	137.1	152.7
— Ice-V —			31000	149.5	161.2
3040	−20.0	+119	32900	161.1	170.6
3417	−17.0[d]		34800	172.1	180.5
3680	−15.0	+135	36800	182.5	190.8
4360	−10.0	+152	38700	192.3	202.0
5270	−5.0	+174			
6160	0.0	+201			
6175	+0.16[e]				

[a] Based on P. W. Bridgman[222] with a consideration of G. Tammann.[223]
[b] Triple point (vapor, water, ice-I); see Table 263. The increase (dT) produced in the melting point of ice by a small increase (dP) in the hydrostatic pressure can be computed by Clapeyron's equation: $dT/dP = T(\Delta v^*/L)$, where $T =$ absolute temperature of the melting-point, $\Delta v^* =$ excess of the specific volume of water (the phase obtained by increasing the temperature) over that of ice, and $L =$ latent heat of

[215] Thomson, J., *Trans. Roy. Soc. Edinburgh,* **16**, 575-580 (1849); Thomson, W. (Lord Kelvin), *Phil. Mag. (3),* **37**, 123-127 (1850).
[216] Moser, H., *Ann. d. Physik (5),* **1**, 341-360 (1929).
[217] Michels, A., and Coeterier, F., *Proc. Akad. Wet. Amsterdam,* **30**, 1017-1020 (1927).
[218] Goossens, B. J., *Arch. néerl.* **20**, 447-454 (1886).

Table 267.—*(Continued)*

fusion; all at the temperature T. For ice at 0 °C, $T = 273.1$ °K, $\Delta v^* = 0.0906$ cm^3/g, $L = 333.6$ joules/g; whence, $dT/dP = -0.00742$ °K/bar $= -0.00752$ °K/atm. H. Moser [224] observed -0.00748 °K/atm. There is a further depression of 0.0024 °K/atm due to the solubility of air in water.

   c Triple point: water, ice-I, ice-III.
   d Triple point: water, ice-III, ice-V.
   e Triple point: water, ice-V, ice-VI.
   f Observations by G. Tammann.[225]
   g L. H. Adams [226] found that water and ice-VI are in equilibrium at 25 °C if $P = 9630$ bars $= 9504$ atm.
   h P. W. Bridgman [227]; units changed by the compiler.
   i Triple point: water, ice-VI, ice-VII.

---

at a pressure of one normal atmosphere, but otherwise pure, the entire system being subjected to a uniform hydrostatic pressure of one normal atmosphere, is 0 °C. This is often called the normal melting-point of ice; also, the ice-point.

The effect of the dissolved air is to lower the melting-point by 0.0024 °C (Table 268), and the direct effect of the pressure (1 atm) is to lower it by 0.0075 °C,[215] see Table 267. Hence, if it were possible to reduce the pressure to zero, then the melting-point of ice-I in contact with pure, air-free water would be +0.0099 °C. Actually, the pressure cannot be reduced below that of the vapor-pressure of water at the temperature of equilibrium, which must equal the vapor-pressure of ice at the same temperature. That pressure is equivalent to 4.58 mm-Hg, which corresponds to a negligible (0.00004$_5$ °C) lowering of the melting-point. Consequently, the temperature at which the system (water, vapor, ice-I) is in equilibrium is +0.0099 °C; this is called its triple-point.

H. Moser [216] has reported that the triple-point is experimentally reproducible to within 0.0005 °C, and has recommended that it, rather than the ice-point, be used for fixing the 0 °C point of thermometric instruments. See also A. Michels and F. Coeterier [217] and B. J. Goossens.[218] But J. L. Thomas [219] of this Bureau has found that the ice-point can be "readily reproduced to a few ten thousandths of a degree centigrade," and W. P. White [220] has described procedures by which it can be reproduced and held constant to within 0.0001 °C for a day or more.

Effect of a Solute.—When an aqueous solution, other than a eutectic mixture, begins to freeze, the crystals that form are pure ice (see pp. 401 +

---

   [219] Thomas, J. L., *Bur. Stand. J. Res.*, **12**, 323-327 (RP658) (1934).
   [220] White, W. P., *J. Am. Chem. Soc.*, **56**, 20-24 (1934).
   [221] Bridgman, P. W., *Int. Crit. Tables*, **4**, 11 (1928).
   [222] Bridgman, P. W., *Proc. Am. Acad. Arts Sci.*, **47**, 439-558 (1912) → *Z. anorg. Chem.*, **77**, 377-455 (1913); *Z. physik. Chem.*, **89**, 252-253 (1915).
   [223] Tammann, G., "Aggregatzustände," 1922; "Krystallisieren und Schmelzen," 1903; *Z. physik. Chem.*, **84**, 257-292 (1913); **88**, 57-62 (1914).
   [224] Moser, H., *Ann. d. Physik* (5), **1**, 341-360 (1929).
   [225] Tammann, G., *Z. physik. Chem.*, **72**, 609-631 (1910).
   [226] Adams, L. H., *J. Am. Chem. Soc.*, **53**, 3769-3813 (1931).
   [227] Bridgman, P. W., *J. Chem'l Phys.*, **5**, 964-966 (1937).

and 414). The temperature at which these crystals are in equilibrium with their mother liquor is the melting-point of pure ice when bathed in that liquor, and is lower than the melting-point of pure ice in contact with pure water. It is commonly called the freezing-point of the

Table 268.—Melting-points in Aqueous Solutions of Certain Gases

Only those gases for which the solubilities have been given in this compilation (Tables 230 and 231) are considered. The computed values of Δ have been derived from those solubilities at 0 °C by means of the relation $\Delta = 1.859N$. The experimental data, taken from the compilation by R. E. Hall and M. S. Sherrill,[228] illustrate variations in the values of Δ/N when the assumptions that the solution is dilute and that there is no chemical reaction between the solute and solvent are not fulfilled.

F = formula-weight of the gas, N = number of gfw of gas dissolved in 1 kg of water, Δ °C = depression of the melting-point below what it would have been in the absence of the dissolved gas, the total pressure on the water being the same in both cases, P = partial pressure of the gas when its saturated aqueous solution at 0 °C contains N gfw of gas per kg of water.

Gas	Computed Δ. $P = 1$ atm F	$10^3 N$	$10^3 \Delta$	Gas	Experimental Δ N	Δ/N	Ref[a]
Air	28.96	1.29	2.4[b]	CO₂	0.07	2.12	GF
A	39.91	2.4	4.3	NH₄OH	0.006	2.11	J
He	4.00	0.43	0.8		0.01	2.03	J
Kr	82.9	4.46	8.3		0.02	1.97	J
Ne	20.2	0.48	0.9		0.05	1.96	J
Rn	222.	22.8	42.	NH₃[c]	0.5	1.86	R, P, S
Xe	130.2	10.8	20.1		2.0	1.9	R, P, S
H₂	2.015	0.96	1.8		5.0	2.0	R, P, S
N₂	28.02	1.03	1.9		10.0	2.2	R, P, S
O₂	32.00	2.18	4.0		15.0	2.5	R, P, S
O₃	48.00	28.6	53.		20.0	2.9	R, P, S
CO	28.00	1.58	2.9[d]		25.0	3.3	R, P, S
CO₂	44.00	76.4	142.		30.0[e]	3.7	R, P, S
NH₃	17.03	52600.[f]					

[a] References:
GF Garelli, F., and Falciola, P., *Gaz. chim. ital.*, **34**, II, 1-12 (1904).
J   Jones, H. C., *Z. physik. Chem.*, **12**, 623-656 (1893).
P   Postma, S., *Rec. trav. chim. Pays-Bas*, **39**, 515-536 (1920).
R   Rupert, F. F., *J. Am. Chem. Soc.*, **32**, 748-749 (1910), superseding **31**, 866-868 (1909).
S   Smits, A., and Postma, S., *Proc. Akad. Wet. Amsterdam*, **17**, 182-191 (1914).

[b] If the formula-weight for air is taken as 30.00, the computed value of 1000Δ will be 2.3, as given by H. W. Foote and G. Leopold,[229] and by H. W. Foote.[230]

[c] This set of values is taken from p. 261 of Hall and Sherrill's report[231]; they give another, and very slightly different, set on p. 255.

[d] Apparently the only available experimental value for the depression produced by CO is that by P. Falciola,[232] quoted in Hall and Sherrill's report (p. 255). He did not determine the concentration of the solution, has not reported the partial pressure of

[228] Hall, R. E., and Sherrill, M. S., *Int. Crit. Tables*, **4**, 254-264 (1928).
[229] Foote, H. W., and Leopold, G., *Am. J. Sci. (5)*, **11**, 42-46 (1926).
[230] Foote, H. W., *Int. Crit. Tables*, **4**, 6 (1928).
[231] Hall and Sherrill, *Int. Crit. Tables*, **4**, 254-264 (1928).
[232] Falciola, P., *Gaz. chim. ital.*, **39**, I, 398-405 (1909).

Table 268.—(Continued)

the CO (it was probably about 1 atm), and gives but the single value $\Delta = 0.015\ °C$. As the solubility data indicate that $1000N = 1.58$ when $P = 1$ atm, this value of Δ leads to $\Delta/N = 9.5$, a very surprising value.

[e] Eutectic aqueous solution of NH_4OH; the eutectic point $(-\Delta)$ is $-111\ °C$ (Rupert places it below $-120\ °C$; Postma, at $-100.3\ °C$). According to Postma the composition of the eutectic is 34.5 per cent NH_3 by weight; i.e., $N = 30.9$. At higher concentrations the solution consists of $(NH_4)_2O$ in NH_3OH; and beyond that, of NH_3 in $(NH_4)_2O$ (Rupert).

[f] This is not an aqueous solution of NH_3, but a dilute solution of water in NH_3OH.

Table 269.—Melting-point in Sea-water

The data in terms of the salt-content (s) are from O. Krümmel [233] and are to be preferred to those in terms of σ_0, by O. Pettersson.[234] The latter are approximately represented by $t = 0.017 - \sigma_0/14.15$.

Salt content = s, density = $1 + 10^{-3}\sigma$; σ_0, σ_t = value of σ at $0\ °C$, at $t\ °C$; melting-point (equilibrium between pure ice and sea-water) = $t\ °C$.

Unit of $s = 1$ g salt per kg sea-water; of density = 1 g/ml

s	$-t$	σ_t	s	$-t$	σ_t	s	$-t$	σ_t	s	$-t$	σ_t
1	0.055	0.72	11	0.587	8.80	21	1.129	16.87	31	1.683	24.96
2	0.108	1.52	12	0.640	9.60	22	1.184	17.67	32	1.740	25.76
3	0.161	2.34	13	0.694	10.41	23	1.239	18.49	33	1.797	26.58
4	0.214	3.15	14	0.748	11.22	24	1.294	19.29	34	1.853	27.39
5	0.267	3.96	15	0.802	12.02	25	1.349	20.10	35	1.910	28.21
6	0.320	4.75	16	0.856	12.84	26	1.405	20.91	36	1.967	29.02
7	0.373	5.57	17	0.910	13.64	27	1.460	21.71	37	2.024	29.83
8	0.427	6.38	18	0.965	14.45	28	1.516	22.52	38	2.081	30.65
9	0.480	7.19	19	1.019	15.25	29	1.572	23.34	39	2.138	31.46
10	0.534	8.00	20	1.074	16.07	30	1.627	24.14	40	2.196	32.27
σ_0		4.1			10.4			12.2			14.8			24.4			27.1
$-t$		0.27			0.715			0.85			1.025			1.715			1.895

---

solution, but the solution does not freeze as a unit unless it is saturated with the solute at the temperature at which freezing occurs. Under that condition, it freezes without change in concentration, and the resulting solid is an intimate mixture of solute and ice. A solution of that concentration is called a eutectic solution.

If n is the number of effective gram-molecules contributed by the solute per kilogram of the solvent when the depression of the melting-point is Δ, we may write $\Delta = nE$, where E is characteristic of the solvent but independent of the nature of the solute, E varies with the concentration, and its limiting value (E_0) as n approaches zero is frequently called the cryoscopic constant of the solvent. By the number of "effective" molecules is meant the number of entities that individually participate in the thermal molecular agitation of the solution. If the solute is neither associated nor dissociated, and does not affect the molecular aggregation of the solvent, then $n = 1000 \times m/FW \equiv N$, where m is the mass of the solute dissolved in the mass W of the solvent, and F is the formula-weight of the solute. If each molecule

---

[233] Krümmel, O., "Handb. d. Ozeanographie," Vol. 1, 1907.
[234] Pettersson, O., *Vega*-expedition; see *Beibl. Ann. d. Physik*, **7**, 834-841 (1883).

of the solute is dissociated into two parts, and the solvent is not affected, then $n = 2N$; and similarly in other cases.

The quantity commonly tabulated is not E, but Δ/N, which equals E only when $n = N$ and the state of molecular aggregation of the solvent is not affected by the presence of the solute.

The value of E is $RM_w T^2/1000L$, where R = gas-constant, M_w = molecular weight of the vapor of the solvent, T = absolute temperature $(273.1 + t)$ of the actual melting-point (t °C), and L = latent heat of fusion of the pure solvent at t °C; R and L must refer to the same unit of

Table 270.—Pressure-temperature Associations for Equilibrium Between the Several Pairs of Types of Ice

Adapted from the compilation[a] of P. W. Bridgman (ICT)[221] with addition of dP/dt from another paper.[222] (1912) See also Table 271 and Fig. 12.

P and t are the associated pressure and temperature, and dP/dt is the derivative under the condition that equilibrium is continually maintained.

Unit of P = 1 atm = 1.0332 kg*/cm² = 1.0132 bars. Temp. = t °C

P	t	dP/dt
Ice-I and Ice-II		
1736	−75.0	8.95[b]
1826	−65.0	8.95
1916	−55.0	8.95
2006	−45.0	8.95
2094	−35.0	8.95
2100	−34.7[c]	
Ice-I and Ice-III		
2049	−60.0	+5.2
2091	−50.0	+1.9
2108	−40.0	−0.6
2100	−34.7[c]	
2087	−30.0	−3.1
2047	−22.0[c]	
2035	−20.0	−5.1
Ice-II and Ice-III		
2100	−34.7[c]	
2160	−34.0	104
2450	−31.0	126
2820	−28.0	149
3260	−25.0	183
3397	−24.3[c]	
Ice-II and Ice-V		
3397	−24.3[c]	
3460	−25.0	−65.2
3680	−28.0	−65.2
3880	−31.0	−65.2
4070	−34.0	−65.2

P	t	dP/dt
Ice-III and Ice-V		
3358	−35.0	2.66
3383	−30.0	2.66
3395	−25.0	2.66
3397	−24.3[c]	
3409	−20.0	2.66
3417	−17.0[c]	
Ice-V and Ice-VI		
6162	−20.0	0.77
6166	−15.0	0.77
6169	−10.0	0.77
6172	−5.0	0.77
6176	0.0	0.77
6175	+0.16[c]	
Ice-VI and Ice-VII[d]		
19050	−80.0	34.8
19720	−60.0	31.9
20320	−40.0	28.6
20860	−20.0	24.2
21290	0.0	17.9
21530	+20.0	8.20
21630	40.0	3.38
21670	60.0	0.97
21680	80.0	0
21680	81.6[c]	0

[a] Based on P. W. Bridgman,[222] with a consideration of G. Tammann.[223]
[b] Bridgman (1912) gives dP/dt = 8.35 kg*/cm².°C = 8.08 atm/°C, which does not agree with the published values of P and t. Cf. note g to Table 272.
[c] Triple point, see Table 263.
[d] P. W. Bridgman[227]; units changed by the compiler.

mass. For aqueous solutions so dilute that t may be taken as zero, $L = 333.6$ joules/g (value at 0 °C), $T = 273.1$ °K, $R = 8.315$ joules/g-mole·°K, and the cryoscopic constant is $E_0 = 8.315 M_w (273.1)^2/1000 \times (333.6 M_w) = 1.859$ °K per (g-mole per kg of water). Experimental data agree with this value as closely as should be expected. For more concentrated solutions, E takes other values, depending upon the melting-point.

Ice and Ice.

The known pressure-temperature associations for equilibrium between the members of the several pairs of ices are given in the preceding table.

93. Phase Diagram for Water and the Ices

Although P. W. Bridgman and G. Tammann agree regarding the main features of the phase diagram for water and the several ices, they seem to

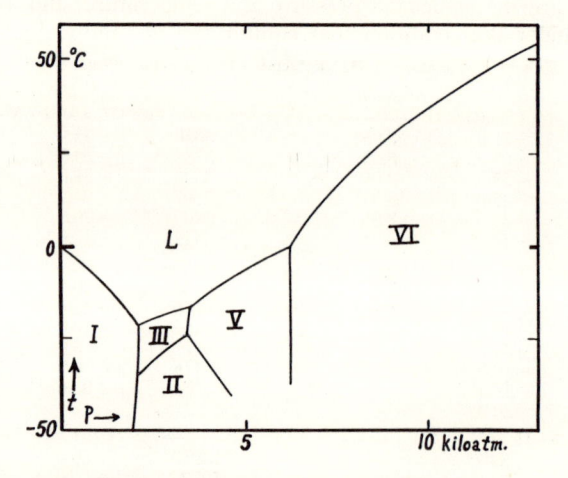

FIGURE 13. Phase Diagram for Water and the Ices (Bridgman).

[From *Int. Crit. Tables*, **4**, 17 (1928).]
  The regions marked, I, II, III, V, and VI, are those in which the ices commonly designated by those symbols exist; L is the corresponding region for water. Bridgman has recently obtained evidence that an unstable form (IV) may arise within the region normally occupied by V, but cannot persist in the presence of V; and he has found a new stable form (VII) with triple-point L-VI-VII at 81.6 °C and 22400 kg*/cm². He has carried the pressure to 45000 kg*/cm² without obtaining evidence of any other form. See text.
  Unit of $P = 1000$ atm $= 1033.2$ kg*/cm² $= 1013.2$ megadyne/cm².

differ irreconcilably regarding the possible existence of the various unstable forms announced by Tammann, and that introduces slight differences in the locations assigned by them for certain of the triple points and of the equilibrium curves in the neighborhood of those points. It seems desirable for some one else to repeat the work of each, even though no evidence confirming the existence of Tammann's Ice-IV and other unstable forms has been presented for many years.

Bridgman's complete diagram of 1912 is reproduced in Fig. 13. Since then he has extended his observations to 45,000 kg*/cm², finding another stable form (VII) at pressures beyond 22,400 kg*/cm² on the liquid-solid line,[235] and an essentially unstable one, which he calls IV, that forms within the region normally occupied by V, but vanishes as soon as V appears.[236]

The corresponding numerical data given elsewhere in this compilation are based primarily on the work of Bridgman, for two reasons: (1) The program adopted for this compilation requires that the data be based as far as possible upon the *International Critical Tables,* and Bridgman's data are given in them. (2) The form in which his data have been presented is more readily adapted to this compilation than is that in which Tammann's data appear.

Those interested in the subject should consult the original articles by G. Tammann,[237] by G. Tammann and E. Schwarzkopf,[238] by G. Tammann and W. Jellinghaus,[239] and by P. W. Bridgman.[222, 235, 236, 239a]

94. Surface Charges on Water and on Ice

At the interface between two substances there exists, in general, an electrical double-layer, the surface of one of the substances being charged positively and that of the other negatively; there is a contact difference of potential between the substances. Under the action of an impressed electrical field, the two components of this layer are urged in opposite directions, giving rise to the phenomena of cataphoresis and electric endosmosis. In many cases the components may be more or less separated by mechanical means also, giving rise to so-called frictional electricity and to the electrification produced by the spraying of a liquid in a gas, by the bubbling of a gas through a liquid, by the shattering of drops or streams of liquid (waterfall electricity), etc. It is with such phenomena that this section is concerned.

M. Faraday [239b] observed that when a stream of air containing minute drops of water flowed with pressure over surfaces of wood, brass, and certain other substances, those surfaces became negatively charged, but they did not become charged if the air was dry, even though the air contained uncondensed steam (¶2130, 2132). On the contrary, if the surface was of ice it became positively charged if the air was not dry (¶2131). These experiments are sometimes quoted as referring to the frictional electricity

---

[235] Bridgman, P. W., *J. Chem'l Phys.,* **5**, 964-966 (1937).

[236] Bridgman, P. W., *Idem,* **3**, 597-605 (1935).

[237] Tammann, G., *Ann. d. Physik (4),* **2**, 1-31 (1900); "Kristallisieren und Schmelzen," 1903; *Z. physik. Chem.,* **72**, 609-631 (1910); **84**, 257-292 (1913); **88**, 57-62 (1914); "Aggregatzustände," 1922.

[238] Tammann, G., and Schwarzkopf, E., *Z. anorg. allgem. Chem.,* **174**, 216-224 (1928).

[239] Tammann, G., and Jellinghaus, W., *Idem,* **174**, 225-230 (1928).

[239a] Bridgman, P. W., *Proc. Am. Acad. Arts Sci.,* **48**, 307-362 (1913); *Z. physik. Chem.,* **86**, 513-524 (1913); *J. Franklin Inst.,* **177**, 315-332 (1914).

[239b] Faraday, M., "Exp. Res. in Elec.," Vol. **2**, pp. 106-126 (18th series, §25), 1843.

generated by the rubbing of water on the solid, but it seems that the drops may have been shattered, which would have introduced complications. But F. Fairbrother and F. Wormwell [240] have reported that a rod of ice merely dipping into water becomes positively charged with respect to the water, whether the water is above or below 0 °C. This indicates that friction between ice and water will result in giving ice a positive charge, the same as Faraday observed in his experiments.

The interpretation of experiments on cataphoresis, electric endosmosis, electrification by bubbling, spraying, and shattering of drops and streams of liquid is difficult, there being many disturbing effects to be considered and eliminated. As a consequence, much of the subject is still in dispute, and it is not practical to give in this compilation more than a few rather general statements. Those interested should refer to the original papers, some of which are listed at the end of this section.

A mere rush of gas over the surface of a liquid gives rise to no charge unless the surface is ruptured. No charge arises from the simple condensation of a vapor nor from simple evaporation nor even from the violent boiling of water, not even though the surface were initially charged. That charge remains on the surface. When the interface between air and distilled water is suddenly ruptured, the net charge acquired by the air is always minus; that by the water is always plus (Gilbert and Shaw, 1925).* And J. J. Thomson (1894)* found that drops of water that have splashed from a wet plate are uncharged in water-vapor, charged positively in air and negatively in H_2; increasing the temperature increases the charge, the gas being air.

The water nuclei observed at the foot of a waterfall are negative if their diameters are under 80A (1A = 10^{-8} cm), positive if between 80 and 150A, and uncharged if the diameters exceed 150A. To explain this, A. Bühl (1928)* has suggested that there are three electrified layers at an air-water surface: A negatively charged layer at the surface, a positively charged one about 80A below the surface, and a second negatively charged one about 150A below the surface. It is not clear how such a distribution can be maintained.

In her study of the cataphoresis of small particles in water, Miss Newton (1930)* observed that the particles (various oils and solids) always moved as if they were negatively charged, and that all of the same size had the same mobility if they were essentially spherical; irregularly shaped particles moved more slowly.

All these types of observation indicate that at an air-water interface the water is positively and the air negatively electrified, but estimates of the amount of the charge differ. It is, however, known that the amount of charge, and even its sign, may be greatly affected by the presence either of solutes or of surface impurities, and that it varies with the nature of the gas.

* For complete reference see the following list of articles.
[240] Fairbrother, F., and Wormwell, F., *J. Chem. Soc. (London)*, **1928**, 1991-1997 (1928).

94. SURFACE CHARGE

The following articles will serve as an introduction to a study of these phenomena:

Alessandrini, E., *Nuovo Cim. (5)*, **4**, 389-402 (1902); Alty, T., *Proc. Roy. Soc. London (A)*, **106**, 315-340 (1924); Becker, A., *Jahrb. d. Radioak.*, **9**, 52-111 (1912) (bibliography of 60 entries); Bühl, A., *Ann. d. Physik (4)*, **84**, 211-244 (1927), **87**, 875-908 (1928), *Kolloid Z.*, **59**, 346-353 (1932) (bibliography of 32 papers that have appeared since Becker's list); Busse, W., *Ann. d. Physik (4)*, **76**, 493-533 (1925); Currie, B. W., and Alty, T., *Proc. Roy. Soc. London (A)*, **122**, 622-633 (1929); Gilbert, H. W., and Shaw, P. E., *Proc. Phys. Soc. London*, **37**, 195-214 (1925) (bibliography of about 100 references); v. Helmholtz, H., *Ann. d. Physik (Wied.)*, **7**, 337-382 (1879); Jurišić, P. J., *Chem. Abs.*, **22**, 1516 (1928) ← *Biochem. Z.*, **189**, 294-301 (1927); Lachs, H., and Biczyk, J., *Roczniki Chemji*, **11**, 362-375 (1931) (Polish, German summary); Lenard, P., *Ann. d. Physik (Wied.)*, **46**, 584-636 (1892), *Sitz. Heidelberger Akad. Wiss. (A)*, **5**, No. 28 (1914); Lignana, M., *Atti d. R. Acc. Sci. Torino*, **65**, 276-281 (1930); McTaggart, H. A., *Phil. Mag. (6)*, **27**, 297-314 (1914); Mooney, M., *Phys. Rev. (2)*, **23**, 396-411 (1924); Newton, D. A., *Phil. Mag. (7)*, **9**, 769-787 (1930); Nolan, P. J., *Idem.* **1**, 417-428 (1926); Nolan, J. J., *Proc. Roy. Irish Acad.*, **37**, 28-39 (1926); Quincke, G., *Ann. d. Physik (Pogg.)*, **113**, 513-598 (1861); Reuss, F. F., *Mém. de la soc. imp. naturalistes à Moscou*, **2**, 327-337 (1809) → Wiedeman's "Die Lehre von der Electricität," 2 ed., Vol. 1, p. 993, 1893; Ruff, O., Niese, G., and Thomas, F., *Ann. d. Physik (4)*, **82**, 631-638 (1927); Simpson, G. C., *Phil. Trans. (A)*, **209**, 379-413 (1909); Thomson, J. J., *Phil. Mag. (5)*, **37**, 341-358 (1894); Zeleny, J., *Phys. Rev. (2)*, **44**, 837-842 (1933).

More recent articles:

Chalmers, J. A., and Pasquill, F., *Phil. Mag. (7)*, **23**, 88-96 (1937); Chapman, S., *Physics*, **5**, 150-152 (1934), *Phys. Rev. (2)*, **45**, 135-136 (A) (1934), **49**, 206 (A) (1936), **51**, 145 (A) (1937), **52**, 184-190 (1937), **53**, 211 (A) (1938); Gilford, C. L. S., *Phil. Mag. (7)*, **19**, 853-878 (1935); Gostkowski, K., *Acta Phys. Polon.*, **3**, 343-345 (1934); Gott, J. P., *Proc. Cambridge Phil. Soc.*, **31**, 85-93 (1935); Kemp, I., *Trans. Faraday Soc.*, **31**, 1347-1357 (1935); Malarski, T., *Acta Phys. Polon.*, **3**, 43-74 (1934); Milhoud, A., *Compt. rend.*, **198**, 1586-1589 (1934), **200**, 1091-1093 (1935); Mukherjee, J. N., Chaudhury, S. G., and Ghosh, B. N., *Kolloid-Beih.*, **43**, 417-463 (1936); Procopiu, S., *Compt. rend.*, **202**, 1371-1373 (1936); Terada, T., and Yamamoto, R., *Proc. Imp. Acad. Japan*, **11**, 214-215 (1935).

IV. Phase Transition

95. ENERGY CHANGES ACCOMPANYING PHASE TRANSITION

Table 271.—External Work and Change in Volume during Phase Transition

Δv^* = increase in specific volume when the transition takes place in the direction indicated by the arrow and at the indicated pressure (P) and temperature (t °C) under which the two phases are in equilibrium.

The data for water → vapor are based on the values given in Tables 242, 250, and 255; the others have been adapted from P. W. Bridgman's compilation [1] and other papers.[2] The data for the triple-points as given in the 1935 paper essentially agree with those in his earlier paper,[3] which are the ones here given.

Unit of P = 1 atm = 1.0132 bars; of Δv^* = 1 cm³/g; of $P\Delta v^*$ = 1 joule/g. Temp. = t °C

t	P	Δv^*	$P\Delta v^*$	t	P	Δv^*	$P\Delta v^*$
\multicolumn{4}{c	}{—— Water → Vapor ——}	\multicolumn{4}{c}{—— Ice-I → Ice-II ——}
0	0.006027	+206290	+126.0	−34.7f	2100	−0.2178	−46.34
100	1	1672.1	169.4	−35	2094	−0.2177	−46.19
200	15.347	126.04	196.0	−45	2006	−0.2170	−44.10
250	39.26	48.81	194.2	−55	1916	−0.2162	−41.97
300	84.79	20.24	173.9	−65	1826	−0.2154	−39.85
350	163.2	7.07	116.9	−75	1736	−0.2146	−37.74
\multicolumn{4}{c	}{—— Water → Ice-I ——}	\multicolumn{4}{c}{—— Ice-I → Ice-III ——}
0	1	+0.0900a	+0.0091	−20	2035	−0.1773	−36.56
− 5	590	0.1016	6.07	−22b	2047	−0.1818	−37.70
−10	1090	0.1122	12.39	−30	2087	−0.1919	−40.58
−15	1540	0.1218	19.01	−34.7f	2100	−0.1963	−41.76
−20	1910	0.1313	25.41	−40	2108	−0.1992	−42.54
−22b	2047	0.1352	28.04	−50	2091	−0.2023	−42.86
\multicolumn{4}{c	}{—— Water → Ice-III ——}	−60	2049	−0.2049	−42.53
−17.0c	3417	−0.0241	− 8.34	\multicolumn{4}{c}{—— Ice-II → Ice-III ——}
−18.5	2820	−0.0301	− 8.60	−24.3g	3397	+0.0145	+ 4.99
−20.0	2430	−0.0371	− 9.13	−25.0	3260	0.0148	4.89
−22.0b	2047	−0.0466	− 9.66	−28.0	2820	0.0164	4.69
\multicolumn{4}{c	}{—— Water → Ice-V ——}	−31.0	2450	0.0179	4.44
0.16d	6175	−0.0527	−32.9	−34.0	2160	0.0206	4.50
0.0	6160	−0.0527	−32.89	−34.7f	2100	0.0215	4.58
− 5.0	5270	−0.0603	−32.20	\multicolumn{4}{c}{—— Ice-II → Ice-V ——}
−10.0	4360	−0.0679	−29.99	−24.3g	3397	−0.0401	−13.80
−15.0	3680	−0.0754	−28.12	−25.0	3460	−0.0401	−14.05
−17.0c	3417	−0.0788	−27.28	−28.0	3680	−0.0401	−14.95
−20.0	3040	−0.0828	−25.50	−31.0	3880	−0.0401	−15.77
\multicolumn{4}{c	}{—— Water → Ice-VI ——}	−34.0	4070	−0.0401	−16.54
−15	4640	−0.0980	−46.1	\multicolumn{4}{c}{—— Ice-III → Ice-V ——}
−10	5110	−0.0960	−49.8	−17.0c	3417	−0.0547	−18.94
− 5	5620	−0.0938	−53.4	−20.0	3409	−0.05469	−18.89
0	6160	−0.0916	−57.1	−24.3g	3397	−0.0546	−18.79
+ 0.16d	6175	−0.0916	−57.3	−25.0	3395	−0.05461	−18.78
5	6780	−0.0884	−60.7	−30.0	3383	−0.05454	−18.68
10	7390	−0.0844	−63.2	−35.0	3358	−0.05446	−18.53
15	8040	−0.0798	−65.0				

[1] Bridgman, P. W., *Int. Crit. Tables*, **4**, 11 (1929).
[2] Bridgman, P. W., *J. Phys'l Chem.*, **3**, 597-605 (1935); **5**, 964-966 (1937).
[3] Bridgman, P. W., *Proc. Am. Acad. Arts Sci.*, **47**, 438-558 (1912).

95. PHASE TRANSITION: ENERGY

Table 271.—(Continued)

P	Δv^*	$P\Delta v^*$	t	P	Δv^*	$P\Delta v^*$
____Water → Ice = VI____				____Ice-V → Ice-VI____		

Water → Ice = VI

	P	Δv^*	$P\Delta v^*$
20	8710	−0.0751	−66.3
30	10250	−0.0663	−68.9
40	11990	−0.0590	−71.7
50	13970	−0.0523	−74.0
60	16150	−0.0477	−78.0

Water → Ice-V

	P	Δv^*	$P\Delta v^*$
52.5	14518	−0.0508	−74.7
57.2	15485	−0.0478	−74.9
66.0	17421	−0.0424	−74.7
73.8	19357	−0.0376	−73.8
80.8	21292	−0.0335	−72.3
81.6[c]	21680	−0.0330	−72.4

Water → Ice-VII

	P	Δv^*	$P\Delta v^*$
81.6[e]	21680	−0.0910	−200.0
95.3	23228	−0.0879	−206.9
110.3	25164	−0.0847	−215.7
124.1	27100	−0.0817	−224.6
137.1	29035	−0.0789	−231.4
149.5	30971	−0.0763	−239.3
161.1	32906	−0.0738	−246.1
172.1	34842	−0.0715	−253.0
182.5	36778	−0.0694	−258.9
192.3	38714	−0.0674	−264.8

Ice-V → Ice-VI

t	P	Δv^*	$P\Delta v^*$
+ 0.16[d]	6175	−0.0389	−24.34
0.0	6176	−0.03886	−24.32
− 5.0	6172	−0.03866	−24.07
−10.0	6169	−0.03847	−23.94
−15.0	6166	−0.03828	−23.81
−20.0	6162	−0.03809	−23.68

Ice-VI → Ice-VII

t	P	Δv^*	$P\Delta v^*$
−80.0	19047		
−60.0	19715		
−40.0	20325		
−20.0	20857		
0.0	21292	−0.0567	−122.6
+20.0	21534	−0.0570	−124.5
40.0	21631	−0.0573	−125.5
60.0	21670	−0.0576	−126.5
80.0	21680	−0.0580	−127.5
81.6[e]	21680	−0.0580	−127.5

[a] See also Table 201.
[b] Triple-point of water, ice-I, and ice-III.
[c] Triple-point of water, ice-III, and ice-V.
[d] Triple-point of water, ice-V, and ice-VI.
[e] Triple-point of water, ice-VI, and ice-VII.
[f] Triple-point of ice-I, ice-II, and ice-III.
[g] Triple-point of ice-II, ice-III, and ice-V.

Latent Heat of Phase Transition.

By the latent heat (L) involved in a given change in phase, is meant the amount of heat absorbed per unit of mass isothermally and reversibly transformed in the indicated direction. The two phases are assumed to be and to remain in mutual equilibrium.

Latent Heat of Vaporization.—For a review of the several determinations of the latent heat of vaporization of water prior to 1930, see E. F. Fiock.[4] Osborne and his associates have redetermined the latent heat from 50 °C almost to the critical point, and have concluded that the best formulation for $t \leqq 100$ °C is that of expression (1),

$$L = 1585.19\left(\frac{374.15 - t}{100}\right)^{0.404} - 36.75304\left(\frac{310 - t}{100}\right)^{1.73} + 17.9218 \times \left(\frac{165 - t}{100}\right)^{2.2} \qquad (1)$$

the unit being 1 Int.joule/g, and temperature being t °C on the international scale.[5] This supersedes previous formulations and tables of L published by these workers, including that of Fiock,[4] Osborne, Stimson, and Fiock,[6] Fiock and Ginnings,[7] and Osborne, Stimson, and Ginnings.[8]

---

[4] Fiock, E. F., *Bur. Stand. J. Res.*, **5**, 481-505 (R:P210) (1930).
[5] Osborne, N. S., Stimson, H. F., and Ginnings, D. C., *J. Res. Nat. Bur. Stand.*, **18**, 389-448 (RP983) (1937).

Their work, however, is continuing at the two extremes of the temperature range. As a result, the extreme values given by (1) are subject to revision; an idea of the size of the changes that may be expected can be obtained from Table 272, containing the most recent revision. Another recent formulation, differing from (1), has been proposed by M. Jakob and W. Fritz [9]; the values defined by it are indicated in the JF column of Table 272.

Of the older data given in Table 272 for comparison, those of H. L. Callendar are unique in that they are based upon a formulation that assumes that identity between liquid and vapor does not occur until the temperature reaches 380.5 °C, 6.5 °C above the critical temperature at which the meniscus vanishes.[10]

A. W. Smith [11] has reported that the value found for the latent heat of vaporization when determined from a slow evaporation from a still surface is about 0.75 per cent greater than when it is determined from the evaporation produced by actual boiling. He thought that the difference was probably real. If it is, then it indicates that the vapor as it leaves the liquid is polymeric, but quickly breaks down into ordinary water-vapor (cf. the next paragraph, *Sublimation*).

Latent Heat of Sublimation.—Previous to the observations of H. T. Barnes and W. S. Vipond [12] it was assumed that the latent heat of sublimation of ice-I at 0 °C was the sum of the latent heat of fusion of ice-I at 0 °C and the latent heat of vaporization of water at the same temperature; and the same is quite generally assumed today. This assumption implies that the vapor in immediate contact with ice is identical with that which is in equilibrium with water, which second assumption need not be true. In fact, Barnes and Vipond reported it false. They stated that when vapor is removed very quickly from ice, the latent heat is only 2540 joules/g, whereas if it is removed slowly the latent heat is 2930 joules/g. The first is only 46 joules/g (=1.6%) greater than the latent heat of vaporization of water; whereas the second exceeds the sum of the latent heats by 102 joules/g (=3.5%). These figures, of which the compiler has found no confirmation, indicate that the vapor as it leaves the ice is polymeric, and only later breaks down into ordinary water vapor, absorbing heat in the process. Barnes thought that it was probable that in ordinary evaporation from ice and snow the change of the vapor from the solid into ordinary vapor takes place just outside the surface, and under ordinary

---

[6] Osborne, N. S., Stimson, H. F., and Fiock, E. F., *Bur. Stand. J. Res.*, **5**, 411-480 (RP209) (1930) = *Trans. Am. Soc. Mech. Eng.*, **52**, 191-220 (FSP-52-28) (1930).

[7] Fiock, E. F., and Ginnings, D. C., *Idem*, **8**, 321-324 (RP416) (1932).

[8] Osborne, N. S., Stimson, H. F., and Ginnings, D. C., *Mech. Eng.*, **56**, 94-95 (1934); **57**, 162-163 (1935).

[9] Jakob, M., and Fritz, W., *Physik Z.*, **36**, 651-659 (1935). Also Jakob, M., *Mech. Eng.*, **58**, 643-660 (1936).

[10] For his reasons, see Callendar, H. L., *Proc. Roy. Soc. London (A)*, **120**, 460-472 (1928); *Proc. Inst. Mechan. Eng.* 1529, 507-527 (1929). See also p. 558+.

[11] Smith, A. W., *J. Opt. Soc. Amer.*, **10**, 711-722 (1925).

[12] Barnes, H. T., and Vipond, W. S., *Phys. Rev.*, **28**, 453 (A) (1909).

circumstances the difference between the two would escape detection in vapor pressure measurements.[13]

Latent Heat of Fusion.—The determinations prior to 1871 of the latent heat of fusion of ice at 0 °C have been reviewed by A. W. Smith,[14] those between 1870 and 1913 by H. C. Dickinson, D. R. Harper, and N. S. Osborne [15]; and recently the entire subject has been again reviewed by A. W. Smith.[11] In the second paper, Smith's table summarizing the early determinations is republished; those values need not detain us. Four values between 1870 and 1913 have to be considered—those obtained by A. W. Smith,[14] by A. D. Bogojawlensky,[16] by U. Behn,[17] and by C. Dieterici.[18] These, after correction of the last two by W. A. Roth,[19] and the reduction of all to the same basis by Dickinson, Harper, and Osborne, lie between 79.59 and 79.69, and their mean is 79.62 cal_{15}/g (= 333.21 joule/g) at 0 °C. In the same paper, Dickinson, Harper, and Osborne publish the results of a very careful determination by themselves, embracing 21 sets of observations and referring to ice from several sources. Their results lie between 79.57 and 79.68, mean = 79.63 cal_{15}/g, which, when corrected for the use of a slightly erroneous value for the specific heat of ice,[20] becomes 79.69 cal_{15}/g (= 333.66 joule/g)* at 0 °C. Additional very careful determinations by Dickinson and Osborne [20] gave 333.63 joule/g at 0 °C. More recently O. Maass and L. J. Waldbauer,[21] and O. Maass and W. H. Barnes [13, p. 30] have obtained the lower values † 79.42 and 79.40 cal/g, about 332.5 joule/g; the particular calorie used is not stated, and the researches appear to have been much less elaborate than those carried out at the Bureau of Standards. The value obtained by that Bureau and published in the *International Critical Tables* is given in Table 272.

As long ago as 1850, C. C. Person [23] pointed out that there is an incipient melting of ice before the temperature has risen to the true melting point, and that to ignore this may seriously affect the value obtained for the latent heat of fusion. J. Y. Buchanan [24] called attention to the same thing, and showed that such melting arises from the presence of impurities

---

* The energy was measured electrically and reduced to cal_{15} on the assumption that 1 cal_{15} = 4.187 joules. This values, which is greater than the one (4.185) now accepted, was used in the reverse conversion.

† These seem to have been derived from the work of W. H. Barnes and O. Maass [22]; see remarks in Section 69.

[13] Barnes, H. T., "Ice Engineering," p. 33, 1928.
[14] Smith, A. W., *Phys. Rev.*, **17**, 193-232 (1903).
[15] Dickinson, H. C., and Harper, D. R., and Osborne, N. S., *Bull. Bur. Stds.*, **10**, 235-266 (S209) (1913).
[16] Bogojawlensky, A. D., *Schrift. Dorpater Naturf. Ges.*, **13**, (1904).
[17] Behn, U., *Ann. d. Physik (4)*, **16**, 653-666 (1905).
[18] Dieterici, C., *Idem*, **16**, 593-620 (1905).
[19] Roth, W. A., *Z. physik. Chem.*, **63**, 441-446 (1908).
[20] See Dickinson, H. C., and Osborne, N. S., *Bull. Bur. Stds.*, **12**, 49-81 (S248) (1915).
[21] Maass, O., and Waldbauer, L. J., *J. Am. Chem. Soc.*, **47**, 1-9 (1925).
[22] Barnes, W. H., and Maass, O., *Can. J. Res.*, **3**, 70-79 (1930).
[23] Person, C. C., *Ann. Chim. et Phys. (3)*, **30**, 73-81 (1850).
[24] Buchanan, J. Y., *Proc. Roy. Inst'n Grt. Brit.*, **19**, 243-276 (1908); *Proc. Roy. Soc. Edinburgh*, **14**, 129-149 (1887) → *Nature*, **35**, 608-611; **36**, 9-12 (1887).

Table 272.—Latent Heat of Change in Phase

The latent heat (L) is the amount of heat absorbed per unit of mass isothermally and reversibly transformed in the direction indicated by the arrow; Δv^* = increase in the specific volume associated with that transformation; P = pressure at which the two phases are in equilibrium at the indicated temperature; $\Delta E = L - P\Delta v^*$ = increase in internal energy, and L/T = increase in entropy, each per unit of mass transformed. The values of $P \cdot \Delta v^*$ have been taken from Table 271.

For the water → vapor data the O values are to be preferred. They are given directly. The values corresponding to any of the other sets of data will be obtained by adding the appropriate value of ΔL to the corresponding O value of L; e.g., the KSG value for 10 °C is 2476.48 − 0.30 = 2476.18; the WT value for 0 °C is 2500.00 − 11 = 2489.

Unit of L, $P \cdot \Delta v^*$, and $\Delta E = 1$ Int. joule/g = 0.2389 cal$_{15}$/g. Temp. = t °C = $(273.1 + t)$ °K

I. Water → Vapor.

Ref[a]→ t	O L	OSG	KSG	Eck	Koch ΔL	WT	JF	C	ICT
0	2500.00		+0.15			−11	− 3.9		− 6
10	2476.48		−0.30			− 7	− 2.5		− 5
20	2452.93		−0.51			− 5	− 1.5		− 5
30	2429.30		−0.42			− 3	− 1.0		− 4
40	2405.54		−0.40			− 2	− 0.3		− 4
50	2381.58		−0.21			0	− 0.2		− 2
60	2357.37		+0.03			0	+ 0.2		0
70	2332.83		+0.22			0	+ 0.1		0
80	2307.87		+0.31			0	− 0.5		+ 1
90	2282.42		+0.33			0	− 1.0		+ 2
100	2256.37	+0.41	+0.27			0	− 1.4		+ 2
110	2229.64	+0.23	+0.16			− 1	− 2.2		0
120	2202.13	+0.07	+0.05			− 2	− 2.7		− 3
130	2173.67	−0.02	+0.01			− 4	− 3.2		− 4
140	2144.22	0	−0.03			− 4	− 3.9		− 6
150	2113.74	0	−0.14			− 4	− 4.3		− 7
160	2082.12	+0.02	−0.34			− 2	− 5.1		− 8
170	2049.17	−0.02	−0.56			+ 1	− 5.6		−10
180	2014.61	+0.01	−0.66			+ 5	− 5.7		−15
190	1978.28	0	−0.65				− 5.4		
200	1940.08	−0.01	−0.56				− 5.3	+10.1	
210	1899.84	0	−0.84				− 4.8	+11.9	
220	1857.41	−0.02	−0.31				− 4.2	+12.8	
230	1812.58	+0.02	−0.21				− 3.4	+14.9	
240	1765.13	0	−0.18				− 2.4	+15.5	
250	1714.77	0	−0.25				− 1.9	+16.9	
260	1661.17	+0.02	−0.43				− 1.0	+19.0	
270	1603.92	+0.01	−0.74				− 0.2	+20.1	
280	1542.47	+0.01	−1.13				+ 0.5	+20.9	
290	1476.17	0	−1.60				+ 1.0	+22.7	
300	1404.09	−0.01	−2.01		−2.2		+ 2.0	+23.6	
310	1324.92	+0.01	−2.12		−2.9		+ 2.9	+24.5	
320	1237.24	−0.01	−2.02		−2.3		+ 4.0	+25.1	
330	1139.29	−0.01	−2.15		−2.0		+ 4.0	+25.0	
340	1027.00	0	−2.01		−1.4		+ 4.0	+23.5	
350	892.85	−0.01	−0.70	+ 2.0	−0.8		+ 3.0	+22.0	
355	812.98			+ 1.2	−0.1				
360	719.43	−0.02	+5.30	+ 2.3	−0.2		+ 2.7	+20.0	
365	603.25	0		+ 2.4	+0.4		+ 3.8		
366	575.69	+0.02							
367	546.04	−0.01							
368	513.79	+0.01							
369	478.24	+0.02							
370	438.30	+0.02		+ 5.8	+0.4		+ 8.8	+39.0	
371	392.10	+0.13							
372	336.04	+0.10		+13.1	−0.3		+13.5		
373	260.98	−0.19							
374	114.61	+0.09		+54.5				+188.1[1]	
374.15	0	0		0					

95. PHASE TRANSITION: ENERGY

Table 272.—(Continued)

II. Ice → Vapor (See text, *Sublimation*.)

III. Water → Ice; Ice → Ice. P. W. Bridgman.[a]

t	L	P·Δv*	ΔE	L/T	t	L	P·Δv*	ΔE	L/T
\multicolumn{5}{c}{Water → Ice-I}	\multicolumn{5}{c}{Ice-I → Ice-II[g]}
0	−333.6[b]	+0.0091	−333.6	−1.22	−34.7[h]	−42.3	−46.3	4.0	−0.177
−5	−308.5	6.1	−314.6	−1.15	−35	−42.48	−46.19	3.71	−0.178
−10	−284.8	12.4	−297.2	−1.08	−45	−40.50	−44.10	3.60	−0.177
−15	−261.8	19.0	−280.8	−1.01₄	−55	−38.59	−41.97	3.38	−0.177
−20	−241.4	25.4	−266.8	−0.95₄	−65	−36.70	−39.85	3.15	−0.177
−22[e]	−234.8	28.0	−262.8	−0.93₄	−75	−34.77	−37.74	2.97	−0.175
\multicolumn{5}{c}{Water → Ice-III}	\multicolumn{5}{c}{Ice-I → Ice-III}
−17.0[d]	−257	−8.3	−249	−1.00₄	−20	23.4	−36.6	60.0	0.092
−18.5	−240	−8.6	−231	−0.94₃	−22[e]	21.8	−37.7	59.5	0.087
−20.0	−226	−9.1	−217	−0.89₃	−30	14.6	−40.6	55.2	0.060
−22.0[e]	−213	−9.7	−203	−0.84₈	−34.7[h]	9.2	−41.8	51.0	0.039
\multicolumn{5}{c}{Water → Ice-V}	−40	2.9	−42.5	45.4	0.012
+0.16[e]	−293.4	−33.0	−260.4	−1.07₄	−50	−8.8	−42.9	34.1	−0.039
0.0	−293	−33	−260	−1.07₃	−60	−23.0	−42.5	19.5	−0.108
−5.0	−285	−32	−253	−1.06₃	\multicolumn{5}{c}{Ice-II → Ice-III}
−10.0	−276	−30	−246	−1.04₉	−24.3[i]	70.7	5.0	65.7	0.284
−15.0	−265	−28	−237	−1.02₃	−25.0	68.2	4.9	63.3	0.275
−17.0[d]	−261	−27	−234	−1.02₈	−28.0	60.7	4.7	56.0	0.248
−20.0	−253	−26	−227	−1.00₀	−31.0	55.2	4.4	50.8	0.228
\multicolumn{5}{c}{Water → Ice-VI}	−34.0	51.9	4.5	47.4	0.221
−15.0	−247	−46	−201	−0.95₇	−34.7[h]	51.5	4.6	46.9	0.216
−10.0	−264	−50	−214	−1.00₄	\multicolumn{5}{c}{Ice-II → Ice-V[k]}
−5.0	−281	−53	−228	−1.04₉	−24.3[i]	67.0	−13.8	80.8	0.269
0	−295	−57	−238	−1.08₀	−25.0	67.0	−14.0	81.0	0.270
+0.16[e]	−294	−57	−237	−1.07₅	−28.0	66.1	−15.0	81.1	0.270
5.0	−303	−61	−242	−1.09₀	−31.0	65.2	−15.8	81.0	0.269
10.0	−311	−63	−248	−1.09₉	−34.0	64.4	−16.5	80.9	0.269
15.0	−316	−65	−251	−1.09₇	\multicolumn{5}{c}{Ice-III → Ice-V}
20.0	−320	−66	−254	−1.09₁	−17.0[d]	−3.7	−18.94	15.1	−0.014₇
30.0	−330	−69	−261	−1.08₉	−20.0	−3.72	−18.89	15.17	−0.014₇
40.0	−342	−72	−270	−1.09₂	−24.3[i]	−3.7	−18.79	15.0	−0.015₂
50.0	−357	−74	−283	−1.10₅	−25.0	−3.64	−18.78	15.14	−0.014₇
60.0	−379	−78	−301	−1.13₈	−30.0	−3.56	−18.68	15.12	−0.014₇
					−35.0	−3.47	−18.53	15.06	−0.014₆
\multicolumn{5}{c}{Water → Ice-VI}	\multicolumn{5}{c}{Ice-V → Ice-VI}
52.5	−333.5	−74.7	−258.8	−1.024	+0.16[e]	−0.8	−24.34	23.5	−0.0031
57.2	−336.5	−74.9	−261.6	−1.019	0.0	−0.83	−24.32	23.49	−0.0030
66.0	−339.1	−74.7	−264.4	−1.009	−5.0	−0.82	−24.07	23.25	−0.0031
73.8	−345.3	−73.8	−271.5	−0.995	−10.0	−0.80	−23.94	23.14	−0.0030
80.8	−352.0	−72.3	−279.7	−0.995	−15.0	−0.77	−23.81	23.04	−0.0030
81.6[f]	−354.5	−72.4	−282.1	−0.999	−20.0	−0.76	−23.68	22.92	−0.0030
\multicolumn{5}{c}{Water → Ice-VII}	\multicolumn{5}{c}{Ice-VI → Ice-VII}
81.6[f]	−354.5	−200.0	−154.5	−0.999	0.0	−23.8	−122.6	+98.8	−0.087
95.3	−398.0	−206.9	−191.1	−1.080	20.0	−13.8	−124.5	+110.7	−0.047
110.3	−444.4	−215.7	−228.7	−1.159	40.0	−6.3	−125.5	+119.2	−0.0048
124.1	−474.6	−224.6	−250.0	−1.195	60.0	−2.1	−126.5	+124.4	−0.0015
137.1	−500.1	−231.4	−268.7	−1.219	80.0	0	−127.5	+127.5	0
149.5	−526.9	−239.3	−287.6	−1.247	81.6[f]	0	−127.5	+127.5	0
161.1	−554.1	−246.1	−308.0	−1.276					
172.1	−582.6	−253.0	−329.6	−1.309					
182.5	−610.2	−258.9	−351.3	−1.339					
192.3	−642.4	−264.8	−377.6	−1.380					

[a] References:

B   Bridgman, P. W.[2, 3] From the 1937 paper come the second set of data for water to ice-VI and all for ice-VII. The triple-point data given in the 1935 paper agree essentially with those in the 1912 one.

C   Callendar, H. L., *Proc. Inst. Mech. Eng.*, 1929, 507-527 (1929).

Eck   Eck, H., *Tätigkeit Phys.-Techn. Reichs. im 1936*, p. 32 = *Physik. Z.*, **38**, 256 (1937).

ICT   Compilation by Smith, A. W., and Bridgeman, O. C., *Int. Crit. Tables*, **5**, 138 (1929).

JF   Jakob, M., and Fritz, W., *Physik. Z.*, **36**, 651-659 (1935). Supersedes similar data by Jakob, M., *Wiss. Abh. Phys.-Techn. Reichs.*, **12**, 435-446 (1928) = *Forsch.-Arb. Gebiete Ingenieurw.*, **310**, 9-19 (1928) and by Jakob and Fritz. *Wiss. Abh. Phys.-Techn. Reichs.*, **13**, 93-111 (1928) = *Z. Ver. deuts. Ing.*, **73**, 629-636 (1929); *Tech. Mech. Thermodynam.*, **1**, 173-183, 236-240 (1930); *Forsch. Gebiete Ingenieurw.*, **4**, 295-299 (1933).

Koch   Koch, W., *Idem*, **5**, 257-259 (1934) → *Z. Ver. deuts. Ing.*, **78**, 1160 (1934).

KSG   Keyes, F. G., Smith, L. B., and Gerry, H. T., *Proc. Am. Acad. Arts Sci.*, **70**, 319-364 (1935).

Table 272.—(Continued)

O   Osborne, N. S., *Private communication*, 1938.
OSG   Osborne, N. S., Stimson, H. F., and Ginnings, D. C.,[5] superseding both *Mech. Eng.*, **57**, 162-163 (1935) and Osborne, Stimson, and Fiock.[6]
WT   Holborn, L., Scheel, K., and Henning, F., "Wärmetabellen," 1919.

[b] This latent heat of fusion of ice-I is from the compilation by R. de Forcrand and L. Gay [28] and is based on observations by H. C. Dickinson and N. S. Osborne.[29]

[c] Triple-point of water, ice-I, and ice-III.

[d] Triple-point of water, ice-III, and ice-V.

[e] Triple-point of water, ice-V, and ice-VI.

[f] Triple-point of water, ice-VI, and ice-VII.

[g] In computing these values of L for ice-I to ice-II, Bridgman used $dP/dt = 8.08$ atm/°C instead of the value (8.95) defined by the values of P in Table 270, presumably in order to make the L's satisfy the triple-point condition. He used the same value for all temperatures. It is obvious that the values given for -34.7 °C are discordant with those given for lower temperatures.

[h] Triple-point of ice-I, ice-II, ice-III.

[i] Triple-point of ice-II, ice-III, and ice-V.

[k] The values of ΔH given in Bridgman's Table XXI are obviously inconsistent with the values of Δv and of the adjusted values of dP/dt given in the same table. The values of L here given have been computed by the compiler from those values of Δv and adjusted dP/dt, and the values of ΔE have been changed accordingly.

[l] Callendar gives the following values for the latent heat of vaporization at higher temperatures: 375 °C, 245.3; 377 °C, 186.3; 380 °C, 48.1; and 380.5 °C, 0 joule/gram.

---

included in the ice. Nevertheless, this source of error has frequently been overlooked. It has, however, been carefully considered in the more elaborate of the recent determinations.

Latent Heat of Ice to Ice.—The values given in Table 272 for the latent heat involved in the transition of one form of ice to another are based almost exclusively upon the work of P. W. Bridgman [25] and have been computed by means of Clapeyron's equation: $L = T \cdot \Delta v^* \cdot dP/dT$, where T °K ($\equiv 273.1 + t$ °C) is the absolute temperature, and Δv^* is the increase in specific volume when the transition takes place in the indicated direction.

Miscellanea.—P. Walden [26] has announced that ML/T has nearly the same value (13.5 cal/°K per g-mole = 56.5 joule/°K per g-mole) for all normal liquids when they freeze; *i.e.*, if the molecular weight (M) of the liquid is 18.015 the increase in entropy during transition is $L/T = 3.14$ joule per gram·°K. Values of L/T are included in Table 272.

From the fact that L/T has, roughly, the same value whether water freezes to ice-I, ice-III, ice-V, or ice-VI, G. Tammann concluded that these four types of ice are isometric, differing only in the distance between adjacent molecules, not in the grouping of atoms in the molecules.[27]

[25] Bridgman, P. W., *Proc. Am. Acad. Arts Sci.*, **47**, 439-558 (1912).
[26] Walden, P., *Z. Elektrochem.*, **14**, 713-724 (1908).
[27] See Tammann, G., "Agregatzustände," p. 143-144, 1922.
[28] de Forcrand, R., and Gay, L., *Int. Crit. Tables*, **5**, 131 (1929).
[29] Dickinson, H. C., and Osborne, N. S., *Bull. Bur. Stand.*, **12**, 49-81 (S248) (1915).

Disposable Energy from Isopiestic Change in Phase.

The maximum amount of external work that can be obtained from an isothermal change in phase under a constant pressure is $W_{TP} + P \cdot \Delta v$, where Δv is the increase in volume. As an amount of work equal to $P \cdot \Delta v$ must be expended against the pressure P, only the amount W_{TP} is disposable for other purposes. Proceeding as in Section 6, and using the same constants as were used there, one obtains formulas (2), (3), and (4),

Table 273.—Disposable Energy from Isopiestic Change in Phase

When the change occurs isothermally and at a constant pressure, the disposable energy is $W_{TP} = w - f(T,P)$; $f(T,P) = +0.008315T \log_e (P/A)$ for each of the changes water → vapor and ice-I → vapor, and $f(T,P) = + \int_A^P (\Delta v)_T dp$ for water → ice-I. The following values have been computed by means of formulas (2), (3), and (4); $w_g = w/18.0154$. For the change water → vapor, w is exactly zero at 100 °C; the finite value (-0.04) defined by formula (2) arises from errors in the constants used in deriving the formula.

Unit of w = 1 kj/gfw-H$_2$O; of w_g = 1 j/g. Temp. = t °C = T °K

Change →		water → vapor		ice-I → vapor		water → ice-I	
t	T	w	w_g	w	w_g	w	w_g
−30	243.1	−15.33	−851	−15.92	−882	+0.587	+32.6
−20	253.1	−14.07	−781	−14.48	−804	+0.413	+22.8
−10	263.1	−12.83	−712	−13.04	−724	+0.213	+11.8
0	273.1	−11.60	−644	−11.60	−644	0	0
+10	283.1	−10.39	−577				
20	293.1	−9.18	−510				
40	313.1	−6.82	−379				
60	333.1	−4.51	−250				
80	353.1	−2.24	−125				
100	373.1	−0.04	−2				

which are indeed merely the differences between the equations there obtained for the corresponding pairs of phases, excepting the last term in (4), which was there ignored as negligible with respect to the large value of W for the formation of the individual phase. As before, A denotes the pressure of 1 atm, temperature = t °C = T °K. Values computed by means of those equations are given in Table 273.

Water to vapor:

$(W_{TP}) = -11.599 + 0.12216t - 2.697(t/1000)^2 + 1.5473(t/1000)^3 - 10.517 \times \{(T/273.1)\log_e(T/273.1) - t/273.1\} - 0.008315T \log_e(P/A)$ kj/gfw   (2)

Ice-I to vapor:

$(W_{TP}) = -11.599 + 0.14417t - 71.622(t/1000)^2 + 1.5473(t/1000)^3 + 9.941 \times \{(T/273.1)\log_e(T/273.1) - t/273.1\} - 0.008315T \log_e(P/A)$ kj/gfw   (3)

Water to ice-I:

$$(W_{TP}) = -0.02201t + 68.925(t/1000)^2 - 20.458\{(T/273.1)\log(T/273.1)$$
$$- t/273.1\} - \int_A^P (\Delta v)_T dp \text{ kj/gfw} \quad (4)$$

96. Vaporization and Condensation

With certain restrictions which will appear, this section may be said to deal exclusively with kinetic phenomena. In it are considered those phenomena that accompany the transition, direct or reverse, between water-vapor and a condensed phase—water or ice—and that cannot be derived directly and solely from observations made under equilibrium conditions. Among the data that are thus excluded are: pressure-volume-temperature associations for (a) dilated water-vapor, Section 14, (b) water and steam at saturation, Section 88, (c) ice and saturated vapor, Section 92; steam-tables and diagrams, Section 90; energy changes accompanying phase transition, Section 95.

Two distinct classes of problems have to be distinguished. One is concerned with the escape of molecules from the denser phase, and with their capture by it. The other has to do with the net transfer of substance from one phase to another under certain specific conditions—with the lack of balance between the escape and the capture of molecules under those adventitious conditions. Problems of the first class are the more fundamental; those of the second are of the greater technical importance, and to them the terms "evaporation" and "condensation" will be restricted, in accordance with common usage.

Escape and Capture of Molecules.

General Relations.—Let α = the ratio of the number of vapor molecules that are caught by the surface of the denser phase (liquid or solid) to the number that strike it in the same time, and m_s = total mass of the vapor molecules that strike unit area of the surface in unit time when the vapor is saturated with reference to the surface. Then the total mass of those vapor molecules that enter the denser phase, per unit area and per unit time, under those conditions will be αm_s, and the corresponding number, m_e, that escape will be the same, $m_e = \alpha m_s$. All three quantities depend upon the temperature of the surface itself.

The quantity α, here called coefficient of capture, is sometimes called the accommodation coefficient, and the failure of the surface to catch all the molecules striking it is sometimes described as a reflection of the molecules.

From a knowledge of the temperature and of the molecular weight and pressure of the saturated vapor, m_s can be readily computed within a small range of error, depending upon the departure of the vapor from the ideal state, and that error can be allowed for whenever other conditions justify the trouble. Hence α can be determined if m_e can be measured. Such

96. VAPORIZATION AND CONDENSATION

is the procedure followed, it being assumed that m_e and α are each independent of the pressure of the overlying vapor, and that the vapor and the surface are at the same temperature. Unless the radius of curvature of the surface is great with reference to the mean free path of the molecules of vapor, m_e and α may be expected to depend upon that curvature. In the rest of this section it will be assumed that the surface is essentially plane.

If n = the number of molecules per unit volume of the saturated vapor and $\bar{v}$ = their mean translational velocity, then the number (n_s) that strike one side of a unit area per unit of time is* $n_s = n\bar{v}/4 = NP_{sat}/(2\pi MRT)^{0.5}$, and $m_s = n_s M/N$ (see Table 12). For water, $10^{-18} n_s = 6.24\, P_{sat}/T^{0.5}$ molecules per cm²sec, and $m_s = 0.1857 P_{sat}/T^{0.5}$ mg per cm²sec, the unit of P_{sat} being 1 dyne/cm², and the temperature being T °K.

The determination of m_e is exceedingly difficult, resting upon extrapolation from observed rates of evaporation that are less than 0.005 and often less than 0.00002 as great as m_e, and as yet, only inferior limits to m_e have been obtained. The observed rate of evaporation is limited by (1) the rate at which the vapor can be removed from the surface of the denser phase, and (2) the rate at which heat can be supplied to the surface (see Table 274). The first is very seriously limited by the presence of a stagnant layer of gas (or vapor) that always clings to the surface, and through which the vapor passes by the slow process of diffusion (see p. 624).

That m_e is finite, that at a given temperature there is a definite limit to the rate of evaporation of a given liquid, has long been recognized.[30]

Coefficient of Capture.—The only available estimates of α for water seem to be those of T. Alty,[31] of T. Alty and C. A. Mackay,[32] and of T. Alty and F. H. Nicoll,[33] of which that of 1935 supersedes all the others. In that, the temperature of the surface is inferred from the observed surface tension, and the value found for α is 0.036 for a surface temperature of 10 °C. The corresponding value of m_e is 4.9 mg/sec·cm², which represents a thermal current of 12.13 watts/cm² through the surface, which in turn could be supplied by conduction through the water only if the temperature gradient were 2100 °C/cm. Obviously, convection must play a most important part in supplying the necessary heat. Their earlier and less accurate work (Alty, 1931; Alty and Nicoll, 1931) indicated that the value

---

* If all the quantities are expressed in cgs units, then the unit of n_s will be 1 molecule/cm²sec; if all except P_{sat} are in such units and the unit of P is q dynes/cm², then the unit of n_s will be q molecules/cm²sec. The quantity RT is of the nature of pressure times specific volume; if the unit of this pressure as well as that of P_{sat} is q dynes/cm², those of mass and volume being 1 g-mole and 1 cm³, respectively, then that of n_s will be $q^{0.5}$ molecules/cm²sec. If the unit of pressure is 1 mm-Hg, $q =$ 1333.22 and $q^{0.5} = 36.513$.

[30] See Mache, H., *Sitzb. Akad. Wiss. Wien (Abt. IIa)*, **119**, 1399-1423 (1910); *Z. Physik*, **107**, 310-321 (1937), and for earlier observations related to this subject, Winklemann, A., *Ann. d. Physik (Wied.)*, **22**, 1-31 (1884); **23**, 203-227 (1884); **26**, 105-134 (1885); **35**, 401-410 (1888); **36**, 93-114 (1889).
[31] Alty, T., *Proc. Roy. Soc. London (A)*, **131**, 554-564 (1931); *Nature*, **130**, 167-168 (1932); *Phil. Mag. (7)*, **15**, 82-103 (1933).
[32] Alty, T., and Mackay, C. A., *Proc. Roy. Soc. London (A)*, **149**, 104-116 (1935).
[33] Alty, T., and Nicoll, F. H., *Can. J. Res.*, **4**, 547-558 (1931).

of α for water decreases as the temperature increases. In that work the temperature of the surface was assumed to be that indicated by a thermojunction.

Alty and Nicoll [33] concluded that $\alpha = 1$ for benzene (C_6H_6) at 30 °C, and R. Marcelin found $\alpha > 0.1$ for ether $[(C_6H_5)_2O]$ and for carbon disulphide (CS_2)[34]; for nitrobenzene ($C_6H_5NO_2$), solid naphthalene ($C_{10}H_8$), and solid iodine, he found that α ranged from 0.035 to 0.25, increasing with the temperature, which was varied from 40 to 60 °C.[35]

Alty has now extended his observations to include the case of vapors striking each its own crystalline phase, and finds that in every case examined, whether the condensed phase is liquid or crystalline, $\alpha = 1$ if the dipole moment of the substance is zero, but is small if the dipole moment is great.[36] Ice was not among the solids studied.

Temperature Adjustment.—It seems that, whereas, only a very small fraction of the vapor molecules striking a water surface enter it, nevertheless "all of them reach temperature equilibrium with the surface before re-evaporating" into the vapor.[37] These authors call the coefficient that measures the approach to such equilbrium the accommodation coefficient.

Change in Association.—In the process of changing from one phase to another there is, in general, a change in the degree of association of the molecules. The observations (p. 000) of A. W. Smith [11] on the vaporization of water, and those of H. T. Barnes and W. S. Vipond [12] on that of ice, indicate that in each case the vapor is polymeric as it leaves the denser phase, but quickly breaks down into ordinary (unassociated) water-vapor. This suggests that the change in association will usually occur in the phase into which the molecules that are being considered are entering, rather than in the one from which they come.

Evaporation.

An extensive, annotated bibliography of evaporation, chronologically arranged and covering the years 1670 to the early portion of 1909, was published by Mrs. Grace J. Livingston [38]; and a theoretical treatment of sublimation has been given by S. Miyamoto.[39]

In the following, the term evaporation will be used to denote the net loss of substance from the surface of the denser phase (water or ice) in a given time, sometimes per unit surface, sometimes for a given total surface, as may appear.

Superheating.*

It is improbable that the free surface of a liquid can be heated above

---

* Note is on p. 623.
[34] Marcelin, R., *J. chim. phys.*, **10**, 680-690 (1912).
[35] Marcelin, R., *Compt. rend.*, **158**, 1674-1676 (1914).
[36] Alty, T., *Proc. Roy. Soc. London (A)*, **161**, 68-79 (1937) → *Nature*, **139**, 374 (L) (1937).
[37] Alty, T., and Mackay, C. A., *Proc. Roy. Soc. London (A)*, **149**, 104-116 (1935).
[38] Livingston, Mrs. Grace J., *Monthly Weather Rev.*, **36**, 181-186, 301-306, 375-381 (1908); **37**, 68-72, 103-109, 157-160, 193-199, 248-252 (1909).
[39] Miyamoto, S., *Trans. Faraday Soc.*, **29**, 794-797 (1933).

96. VAPORIZATION AND CONDENSATION

the temperature at which the pressure of the vapor in equilibrium with it equals the total existing pressure on the surface, but the bulk of the liquid can be readily superheated (p. 579), and will always become superheated before boiling occurs. The increase in vaporization that accompanies the growth of the bubbles draws upon the liquid for heat, which can be supplied only if the temperature of the liquid is higher than that of the surface of the bubble. This raises the question: Does the escaping vapor have its normal temperature—the temperature at which the vapor and a plain surface of the liquid will be in equilibrium when the partial pressure of the vapor is equal to the total pressure of the existing gas phase? It is generally believed that its temperature will be normal, but certain observations by Jakob and Fritz were, at least for a time, taken as indicating that the temperature of the escaping vapor is abnormally high. That interpretation seems to have rested on the assumption that the escaping vapor must have the temperature of the liquid in bulk, whereas its temperature must be that of the surface of the bubble, which is lower. F. Bošnjaković [40] has shown that the observations are entirely consistent with the vapor having its normal temperature, if each bubble may be regarded as surrounded by a layer of stagnant water $17\,\mu$ thick, through which heat passes by conduction only. H. B. Reitlinger [40a] found that, when water is expanded by passing it through a suitable nozzle, it does not suddenly vaporize when its pressure has been reduced to that of the vapor saturated at the temperature of the water, but only when the pressure has been reduced still lower.

Some Factors Affecting Evaporation.

Curvature of Surface.—At a given temperature the equilibrial pressure of the vapor in contact with a concave surface of its liquid is less than that over a flat surface. If the surface is a section of a sphere of radius r, the fractional decrease in the pressure will be $\Delta p/p = 2\gamma M/\rho RTr$ approximately (p. 568), where $\gamma =$ surface tension, $\rho =$ density of the liquid. Consequently, evaporation from such a surface will be slower, and condensation upon it will be greater, than on a flat surface at the same temperature and in contact with vapor of the same density. And for a given common density of vapor there will, when possible, be a transfer of substance, by evaporation, from the flat to the concave surface when they are kept at the same temperature.

Over a convex surface the equilibrial pressure is greater than that over a flat one at the same temperature, and the several effects just mentioned are modified accordingly.

The liquid within capillary spaces is subjected to tension if the con-

---

* Jakob, M., and Fritz, W., *Techn. Mech. Thermod.*, **1**, 173-183, 236-240 (1930); *Forsch. Gebiete Ingenieurw.*, **2**, 435-447 (1931); Jakob, M., *Chem. Apparat.*, **19**, 109-111 (1932); Schreber, K., *Dinglers Polytech. J.*, **345**, 189-191 (1930); **346**, 21-27, 41-46, 61-64 (1931); *Z. techn. Physik*, **14**, 81-85 (1933); see also, Jakob, M., *Mech. Eng.*, **58**, 643-660 (1936); Fritz, W., and Ende, W., *Physik. Z.*, **37**, 391-401 (1936); Fritz, W., and Homann, F., *Idem*, **37**, 873-878 (1936).
[40] Bošnjaković, F., *Techn. Mech. Thermod.*, **1**, 358-362 (1930).
[40a] Reitlinger, H. B., *Compt. rend.*, **198**, 2290-2292 (1934).

cavity of the capillary surface is directed away from the liquid (pressure, if towards), and this may modify both the coefficient of capture and the surface tension. Such effects will probably be inappreciable unless the spaces are very small. If the radius of the space were $1\,\mu$ ($=0.0001$ cm) the tension for water would be of the order of one atmosphere. G. A. Hulett [41] has reported observations which he thought indicated that water subjected to hydrostatic tension evaporated less slowly than water not under such tension; and W. A. Patrick and N. F. Ebermann [42] have published observations which they thought indicated that the pressure of vapor in equilibrium with a very concave surface of its liquid is less than that computed by means of the formula just given, unless to γ is assigned a value in excess of that found for much flatter surfaces (pp. 513 and 568). But the interpretation of each of these sets of observations is difficult. See also, p. 631, M. Polanyi,[43] and A. N. Frumkin.[43a]

Blanketing Layers and Surface Films.—The rate at which vapor can actually leave the surface of a denser phase is very seriously limited by the presence of a stagnant layer of gas or of vapor that always clings to the surface, and through which the escaping vapor must pass by the slow process of diffusion. The thickness of this layer varies with the conditions. For a freely exposed surface in a wind or a stream of gas, it is estimated to be a millimeter or less.[44]

R. Marcelin [45] has conclude dthat, under steady conditions, the surface of a liquid has the temperature at which it will be in equilibrium with the adjacent vapor at its existing partial pressure. That is, a liquid is continually blanketed by a layer of its saturated vapor. Presumably the same is true of a solid. But it must be remembered that whenever there is an evaporative loss from the denser phase the temperature of the surface will be less than that of the bulk of the substance.

Of those who have studied the effect of thin surface films, and especially of monomolecular films, upon the rate of evaporation of a liquid may be mentioned G. Hedestrand,[46] N. K. Adam,[47] E. K. Rideal,[48] and I. and D. B. Langmuir.[49] The work of Rideal and that of the Langmuirs have been discussed by T. Alty.[50] For various reasons, none of this work is suitable for an estimation of the coefficient of capture (p. 621). Hedestrand and Adam found that the presence of a monomolecular film produced very little

---

[41] Hulett, G. A., *Z. physik. Chem.*, **42**, 353-368 (1903).
[42] Patrick, W. A., and Ebermann, N. F., *J. Phys'l Chem.*, **29**, 220-228 (1925). Cf. Shereshefsky, J. L., *Am. Chem. Soc.*, **50**, 2966-2980, 2980-2985 (1928); Latham, G. H., *Idem*, **50**, 2987-2997 (1928).
[43] Polanyi, M., *Physik. Z. Sowj.*, **4**, 144-154 (1933).
[43a] Frumkin, A. N., *Idem*, **4**, 154-155 (1933).
[44] Jeffreys, H., *Phil. Mag. (6)*, **35**, 270-280 (1918); Giblett, M. A., *Proc Roy. Soc. London (A)*, **99**, 472-490 (1921).
[45] Marcelin, R., *J. chim. phys.*, **10**, 680-690 (1912); see also *Compt. rend.*, **154**, 587-589 (1912); **158**, 1419-1421 (1914).
[46] Hedestrand, G., *J. Phys'l Chem.*, **28**, 1245-1252 (1924).
[47] Adam, N. K., *Idem*, **29**, 610-611 (1925).
[48] Rideal, E. K., *Idem*, **29**, 1585-1588 (1925).
[49] Langmuir, I., and D. B., *Idem*, **31**, 1719-1731 (1927).
[50] Alty, T., *Proc. Roy. Soc. London (A)*, **131**, 554-564 (1931).

effect upon the observed rate of evaporation; that being so very much smaller (1/50000) than m_s (see Table 274), no certain conclusion can be drawn from it regarding the effect of the film upon m_e. At a much greater rate of evaporation (0.004m_s), Rideal found that such films produced a marked effect, in some cases reducing the rate by 50 per cent. The Langmuirs discuss these observations, interpreting them in terms of the resistance offered to the passage of the vapor through 3 distinct layers: (1) a layer of stagnant water below the film, (2) the surface film, and (3) the layer of stagnant air above the water. They suggest that the surface film itself offers little, if any, resistance to the passage of the vapor; but from its effect on the surface-tension it prevents the irregular surface streamings that would otherwise exist and that would prevent the formation of a stagnant layer of water at the surface. The effect of the film upon the rate of evaporation is, in their opinion, a secondary one, arising from the attendant formation of a stagnant layer of water, of which they estimate the thickness in one case to be 0.2 mm.

Convection.—That convection currents in the liquid may, and in most cases will, play an important role in the rate of evaporation, even when that is only a small fraction of m_s is obvious from the enormous magnitude of the temperature gradient (Table 274) that is required if the necessary heat is to be supplied solely by conduction through the liquid. See also the latter portion of the preceding paragraph.

Wind.—A wind blowing along the surface tends to sweep away the vapor, and in that way to increase the evaporation. Formulas expressing this effect will be found in Table 275, and illustrations of it in several places in this section. To the references there given, may be added M. Centnerszwer, C. Wekerówna, and Z. Majewska.[51]

Aspect of Surface.—Other things being the same, the evaporation from a vertical surface is nearly twice as rapid as that from a horizontal one.[52]

Electric Charge.—J. R. Sutton[53] has quoted various opinions and observations regarding the effect of electrically charging the vessel from which a liquid is evaporating. By actual observation, he found that the rate of evaporation was the same in all cases, whether the vessel was insulated, earthed, or charged.

Cooling by Evaporation.

The surface from which evaporation is occurring is cooler than it would otherwise be, and that on which condensation is occurring is warmer. Indeed, it is obvious from Table 274 that thermal conduction through water is totally unable to prevent a very great cooling of the surface when evaporation can proceed entirely unhindered. Even under laboratory conditions the cooling is far greater than many would expect. For example, T. Alty[50]

---

[51] Centnerszwer, M., Wekerówna, C., and Majewska, Z., *Bull. Int. Acad. Polonaise (Cracovie) (A)*, **1932**, 369-382 (1932).
[52] Hinchley, J. W., *J. Soc. Chem. Ind.*, **41**, 242T-246T (1922). See also Hilpert, R., *Forschungsheft*, 355 (1932).
[53] Sutton, J. R., *Sci. Proc. Roy. Dublin Soc.*, **11**, 137-178 (1907).

has found by extrapolation that when bodies of water maintained at 60, 40, and 18 °C evaporate into dry air, then the temperatures of their surfaces, as measured by means of a thermojunction, lie below those of the body of the water by 28, 22.5, and 12 °C, respectively. And H. G. Becker [55] has reported that when water is evaporating in free air and from a vessel immersed in a bath kept at 100 °C, then the temperature of the water (not of its surface, but of the water in bulk) will be 70 °C in still air, 60 °C in a moderate draft, and only 54 °C in a strong draft; and that if the temperature of the water is to be kept at 100 °C under those air conditions, then the bath must be heated to 170, 197, and 215 °C, respectively.

Obviously, the increased cooling produced by blowing air over a water surface depends upon the initial humidity of the air, and may be increased by heating the air.[56]

Table 274.—Some Data Pertaining to the Evaporation of Water
See also text (p. 620+) and Table 12.

n_s = number of molecules of saturated water-vapor that strike a flat area of 1 cm² in 1 sec; m_s is the aggregate mass of these molecules; p_{sat} = saturation pressure at temperature t °C; h_s = heat that must be supplied per cm²sec to compensate for the evaporation of m_s of water per cm²sec; dt/dx = temperature gradient that must exist in the water near the surface when the rate of evaporation is m_s and all the heat h_s is supplied by conduction through the water, the thermal conductivity at those gradients being assumed to be the same as under usual conditions. The values of h and of dt/dx have been computed from m_s by means of the data in Tables 272 and 130. If the unit of p_{sat} is 1 mm-Hg, then $n_s = 10^{21}(8.32)p_{sat}/\sqrt{T}$, and $m_s = 0.2476 p_{sat}/\sqrt{T}$ grams, T °K being the absolute temperature. (See p. 621.)

Unit of p_{sat} = 1 mm-Hg; of n_s = 10²² molecules per (cm²sec); of m_s = 1 g per (cm²sec); of h_s = 1 kilowatt per cm²; of dt/dx = 10⁵ °C per cm = 10 °C/μ

t	p_{sat}	n_s	m_s	h_s	dt/dx
0 °C	4.58	0.23	0.069	0.17	0.31
5	6.54	0.33	0.097	0.24	0.43
10	9.21	0.46	0.136	0.33	0.58
15	12.79	0.63	0.187	0.46	0.79
20	17.54	0.85	0.254	0.62	1.06
25	23.76	1.15	0.341	0.83	1.4
30	31.82	1.52	0.453	1.10	1.8
40	55.32	2.60	0.774	1.86	3.0
50	92.5	4.3	1.28	3.0	4.8
60	149.4	6.8	2.02	4.8	7.3
80	355.1	15.7	4.68	10.8	15.7
100	760.0	32.8	9.74	22.0	....

Rate of Evaporation.

As the rate of evaporation depends upon the existing conditions, which

[55] Becker, H. G., *Sci. Proc. Roy. Dublin Soc. (N. S.)*, **17**, 241-248 (1923).
[56] Scott, A. W., *J. Roy. Techn. Coll. Glasgow*, **2**, 620-629 (1932).

may vary widely, formulas of various types have been used to represent it, each assuming the existence of certain specific conditions. Some of these formulas have been assembled in Table 275; comments regarding them will be found in the accompanying notes. In general, the two phases are assumed to have a common temperature, and no attention is paid to the fact that the surface is quite significantly cooler than the rest. The surface temperature, though unknown, is definitely fixed by the conditions of the problem, and thus implicitly enters into the values found for the coefficients in the empirical equations.

The effective partial pressure (p_0) of the vapor in the layer adjacent to the liquid surface (that is, the partial pressure that must be assumed if the removal of the vapor is to be accounted for by pure diffusion) has been studied by H. Mache [57] for the case of water contained in vertical cylinders. As the length (h) of the cylinder above the surface of the water increases, the rate of evaporation decreases, and p_0 approaches the saturation pressure corresponding to the temperature of the system (Table 276.[58]

The rate of evaporation, per unit area, from large areas (lakes, etc.) is about 2/3 of that from small pans, and that of sea-water is about 5 per cent less than that of fresh water.[59]

Phenomena associated with the evaporation of very small drops are considered below (p. 631). For evaporation from snow, see Table 281. To references given elsewhere, may be added M. Allen.[60]

Table 275.—Formulas for the Rate of Evaporation

(Adapted from compilation by A. C. Egerton,[61] with additions.)

Quite recently, F. G. Millar [62] has discussed previously proposed formulas, and has derived a new one which is thought to be sounder and more generally applicable. The original paper should be consulted.

A = area of the evaporating surface, $V \equiv v_1 p_1/P$, where v_1 is the total volume of vapor (as measured at $T\ °K$ and pressure p_1) that leaves the surface in unit time, P = total pressure of the gas phase, $T\ °K$ = temperature of the system, p_s = pressure of the vapor when saturated at $T\ °K$, p_0 = partial pressure of the vapor in the blanket adhering to the surface, p = partial pressure of the vapor in the gas (air) to which the evaporation occurs (in that blowing over the surface, or at the upper and open end of the cylinder containing the liquid), h = distance from the surface of the liquid to the open (upper) end of the vertical cylinder containing it, r = radius of a circular cylinder or of a spherical drop, a and b = principal semi-axes of the transverse section of an elliptical cylinder, w = velocity of gas

---

[57] Mache, H., *Sitz.-b. Akad. Wiss. Wien (Abt. IIa)*, **119**, 1399-1423 (1910).
[58] See also, Trautz, M., and Müller, W., *Ann. d. Physik (5)*, **22**, 333-352 (1935).
[59] Egerton, A. C., *Int. Crit. Tables*, **5**, 54 (1929).
[60] Allen, M., *Proc. Nat. Acad. Sci.*, **10**, 88-92 (1924).
[61] Egerton, A. C., *Int. Crit. Tables*, **5**, 53-55 (1929).
[62] Millar, F. G., *Can. Meteorolog. Memoirs*, **1**, 43-65 (1937).

Table 275.—(Continued)

(air) parallel to the evaporating surface, D = coefficient of interdiffusion of the vapor and gas (air) corresponding to T and P, m = total mass evaporated per unit of time, $m = VPM/RT$ where M = molecular weight of the vapor, and R = the universal gas-constant per g-mole.

$V = C_1 D \log_e \{(P-p)/(P-p_0)\}$. If p_0 is small in comparison with P, this expression for V is essentially equal to $V = C_1 D(p_0 - p)/P$, and $m = C_1 DM(p_0 - p)/RT$. If p_0 is small and $p = 0$ (air perfectly dry), then $V = C_1 D p_0/P$.

Unit of a, b, r, and $h = 1$ cm; of $A = 1$ cm^2; of $D = 1$ cm^2/sec; of $V = 1$ cm^3/sec; of $w = 1$ m/sec; of m as indicated

I. Theoretical formulas. Unit of $m = 1$ g/sec.

Gas turbulent and streaming[a]

Circular area, $h = 0$.
If w is vanishingly small, m is proportional to r; if w is finite, $m = 39.5 \rho_0 (D_e w r^3)^{1/2}$ where ρ_0 = density of the vapor in the blanketing layer, and D_e = effective coefficient of diffusion.

Gas quiescent

Condition[b]	h	C_1	Ref[c]
Cylinder, Elliptical	0	$4\sqrt{ab}$	1
Cylinder, Circular	0	$4r$	1
Cylinder, Circular	h	$4\{(h^2 + r^2)^{1/2} - h\}$	2
Cylinder, Circular	$> 2r$	$\pi r^2/h$	3
Cylinder, Any	$> 2r$	A/h	3
Concentric Spheres		$4\pi r_1 r_2/(r_2 - r_1)$	4

II. Empirical formulas. Unit of $m = 1$ g/hr.
Circular cylinder[d]: $m = 1000 C_2 r^n$, where $C_2 = 5 + 25 e^{-2h}$ and $n = 2.0 - 0.60 e^{-h}$.
Flat rimless surfaces: $m = A k_1 (1 + k_2 w) f$, f being a function of p, p_s, and P.

k_1	k_2	f	Ref[c]
....	....	$p_0(p_s - p)/P^n p_s$	5. Laval
2.62	0.85	$\{(p_s - p)/P\}^{1/2}$	6. Hinchley
2.36	0.44	$(p_s - p)/P$	6. Himus and Hinchley
0.027M	2.24	p_s/P	7. Hine
0.49	2.24	p_s/P	7. Hine; for water
1.27	1.12	$(p_s - p)/P$	8. FitzGerald
....	0.063	......	9. Bigelow
....	$0.293/w^{1/2}$	......	10. Grunsky
....		$(p_s + p_a - 2p)/P$	11. Marvin
1.68	0.764	$(p_s - p)/P$	12. Lurie and M.

[a] Derived by H. Jeffreys [63] for flat circular surfaces not surrounded by an elevated rim. Of the theoretical relations that have been proposed, these probably apply the most closely to practical conditions. The value of D_e depends upon the turbulence and convection; it should be determined experimentally in each case. That of ρ_0 is the mass of vapor per unit volume at the top of the thin layer (1 mm or less in thickness) beyond which, on account of turbulence and convection, the concentration of the vapor decreases very slowly with the elevation. In the open, w may reach 0.4 m/sec, and D_e reach 1000 cm^2/sec; then the formula applies if $10 \lessgtr r \lessgtr 25\,000$ cm. Indoors, w may be 0.04 m/sec and $D_e = 1$ cm^2/sec; then the limits of validity are $1 \lessgtr r \lessgtr 2500$ cm. (In the *International Critical Tables*, **5**, 54 (1929) these limits are incorrectly given as areas 250 m^2 to 10 cm^2 and 25 m^2 to 1 cm^2.) The corresponding expression $m = C_3 \rho_0 \sqrt{D_e w l^3}$, in which C_3 is a form-factor and l is a linear dimension defining the size of the surface, applies to flat surfaces of any shape, provided

[63] Jeffreys, H., *Phil. Mag.* (6), **35**, 270–280 (1918).

Table 275.—(Continued)

that the dimensionless quantity wl/D_v lies between 4 and 10 000, the unit of w here being 1 cm/sec.

[b] Condition of evaporation, whether from the bottom of a vertical cylinder of height h above the surface of the liquid or from a sphere.

[c] References and remarks. References cover both theory and observation.

1. Mache, H.,[57] v. Pallich, J., *Sitzb. Ak. Wiss. Wien (Abt. IIa)*, **106**, 384-410 (1897); Stefan, J., *Ann. d. Physik (Wied.)*, **17**, 550-560 (1882), **41**, 725-747 (1890), *Sitzb. Ak. Wiss. Wien (Abt. IIa)*, **65**, 323-363 (1872); Renner, O., *Ber. deuts. bot. Ges.*, **29**, 125-132 (1911); Winklemann, A., *Ann. d. Physik (Wied.)*, **35**, 401-410 (1888). When $p = 0$, $p_0 = p_{sat}$, and p_{sat}/P is small, this expression reduces to $V = 4rDp_{sat}/P$, which was given by Dalton, J., *Mem. Manchester Lit. Phil. Soc.*, **5**, 535-602 (1802).

2. Brown, H. T., and Escombe, F., *Phil. Trans. (B)*, **193**, 223-291 (1900); Laval, E., *Jour. de Phys. (2)*, **1**, 560-561 (1882) ← *Mém. Soc. Sci. phys. et nat., Bordeaux (2)*, **5**, 107+ (1882); Thomas, N., and Ferguson, A., *Phil. Mag. (6)*, **34**, 308-321 (1917); Vaillant, P., *Compt. rend.*, **150**, 689-691, 1048-1051 (1910); *Jour. de Phys. (5)*, **1**, 877-891 (1911); Winklemann, A., *Ann. d. Physik (Wied.)*, **35**, 401-410 (1888).

3. LeBlanc, M., and Wuppermann, G., *Z. physik. Chem.*, **91**, 143-154 (1916); Marcelin, R.[45] and *Compt. rend.*, **158**, 1674-1676 (1914); Vaillant, P., *Compt. rend.*, **150**, 689-691, 1048-1051 (1910); Winklemann, A.[64]

4. Stefan, J., *Sitzb. Ak. Wiss. Wien (Abt. IIa)*, **65**, 323-363 (1872), *Ann. d. Physik (Wied.)*, **17**, 550-560 (1882), **41**, 725-747 (1890); Langmuir, I., *Phys. Rev. (2)*, **12**, 368-370 (1918); Houghton, H. G., *Physics*, **4**, 419-424 (1933); Fuchs, N., *Physik. Z. Sowj.*, **6**, 224-243 (1934). Here r_1 = radius of the evaporating sphere, and r_2 = radius of a concentric spherical shell at which the partial pressure of the vapor is continuously kept equal to p. The transfer of vapor is by diffusion only, the gas being completely quiescent. When $r_2 \gg r_1$, $C_1 = 4\pi r_1$, approximately, and if τ denotes the time, then $-d(r_1^2)/d\tau = 2DM(p_0 - p)/\rho RT$, a quantity independent of r_1. The observed rates of evaporation of single spheres in large volumes of gas, though of low order of precision, agree with these relations; see Sresnevski, B., *Zhurnal Russ. fiz.-khim. obshchestvo*, **14**, 420-469, 483-509 (1882), **15**, 1-10 (1883); Morse, H. W., *Proc. Am. Acad. Arts Sci.*, **45**, 363-367 (1910); Gudris, N., and Kulikowa, L., *Z. Physik*, **25**, 121-132 (1924); *J. Russ. Phys. Chem. Soc. (Phys.)*, **56**, 167-175 (1924); Whytlaw-Gray, R., and Whitaker, H., *Proc. Leeds Phil. Lit. Soc.*, **1**, 97-103 (1926); Topley, B., and Whytlaw-Gray, R., *Phil. Mag. (7)*, **4**, 873-888 (1927); Houghton, H. G., *Physics*, **4**, 419-424 (1933). See also p. 631.

5. Laval, *Jour. de Phys. (2)*, **1**, 560-561 (1882) ← *Mém. Soc. Sci. Phys. et Nat. Bordeaux (2)*, **5**, 107+ (1882), has stated that n varies with the gas, but not with the temperature; p_0 = pressure of the vapor in contact with the surface.

6. Hinchley, J. W., *J. Soc. Chem. Ind.*, **41**, 242T-246T (1922); Himus, G. W., and Hinchley, J. W., *Idem*, **43**, 840-845 (1924). The purpose of this work was to obtain data and formulas that would be of value to the chemical engineer. Hinchley stated that his formula applies if $w \gtrless 0$, and that it is not in error by more than 10 per cent if $t > 60\,°C$; whereas H. G. Becker[55] has stated that when $w = 0$ and $t > 90\,°C$ then the values defined by this formula are too low. See also, Hill, L., and Hargood-Ash, D., *Proc. Roy. Soc. London (B)*, **90**, 438-447 (1919).

7. Hine, T. B., *Phys. Rev. (2)*, **24**, 79-91 (1924). For a circular surface 30 cm in radius. Water was not considered in setting up the formula containing M; the formula here given for water was not given by Hine, but is obtained from his formula by setting $M = 18.0154$. See also de Heen, P., *Bull. Sci. Acad. Roy. Belg. (3)*, **21**, 11-24 (1891).

8. FitzGerald, D., *Trans. Am. Soc. Civ. Eng.*, **15**, 581-646 (1886). This and the next two expressions refer to evaporation from large outdoor areas of water, ice, or snow. The value of p_s for dry ice and snow is not the same as for water at the same temperature. For observed evaporation from snow, see Table 281. See also Giblett, M. A., *Proc. Roy. Soc. London (A)*, **99**, 472-490 (1921).

9. Bigelow, F. H., *Monthly Weather Rev.*, **36**, 24-39 (1908); it applies to the same conditions as the preceding reference (8). Marvin, C. F., *Idem*, **37**, 57-61 (1909) regarded this value for k_2 as merely a first approximation.

10. Grunsky, C. E., *Monthly Weather Rev.*, **60**, 2-6 (1932). (Discussion by C. F. Marvin on p. 6); it applies to the conditions stated in reference 8. His complete formula is $E = E'(1 + 0.293\sqrt{w})(1 + 0.108H)$, unit of $w = 1$ m/sec, of $H = 1$ km, H = altitude above sea-level; it is intended for general use in computing the evaporation from lakes, reservoirs, etc. (see Table 278.) Note: As k_2w is $0.293\sqrt{w}$, $k_2 = 0.293/w^{0.5}$, as given in the table.

11. Marvin, C. F., *Monthly Weather Rev.*, **37**, 57-61 (1909). The p_a in this formula is the partial pressure of the water-vapor in the air when that is saturated at its existing temperature; p/p_a is the relative humidity.

12. Lurie, M., and Mikhailoff, N., *Ind. Eng. Chem.*, **28**, 345-349 (1936) observed the rate of evaporation from a surface that was flush with the floor of a rectangular wind-tunnel. The total pressure (P) was always that of the atmosphere, and its value was merged with their constant C. The compiler has changed the unit of area and has taken the P out of the C, to make their expression conform to the others in this table.

[d] In general, C_2 and n depend upon h, upon the temperature, pressure, and humidity of the air in which the cylinder is immersed, and upon the wind velocity. The values here given apply to indoor conditions ($w = 0$) with $t = 15$ to $20\,°C$, $P = 749$ to 787 mm-Hg, relative humidity = 56 to 74 per cent. [Thomas, N., and Ferguson, A., *Phil. Mag. (6)*, **34**, 308-321 (1917)].

Table 276.—Effective Partial Pressure of Blanketing Vapor [57]

Water is contained in a vertical tube 2.67 mm in internal diameter, the top of the tube being h mm above the bottom of the water meniscus; the rate of evaporation into a dry gas is observed, and p_0 is the value that must be assigned to the partial pressure of the vapor at the surface of the liquid if the removal of vapor is to be accounted for by pure diffusion up the tube; t °C is the temperature of the system, assumed uniform; p_{sat} is the pressure of water-vapor that is saturated with reference to a flat water surface at temperature t; t_0 °C is the temperature at which the vapor-pressure is p_0. See also: A. Winklemann [64, 64a] and P. Vaillant. [65]

Unit of p_0 and $p_{sat} = 1$ mm-Hg; of $h = 1$ mm. Temp. $= t$ °C

Gas→	— H₂ —		Air	— H₂ —		Air
t→	92.4	65.5	92.4	92.4	65.5	92.4
p_{sat}→	575.6	191.8	575.6	575.6	191.8	575.6
h		$p_{sat} - p_0$			$t - t_0$	
30	48.6	6.2	29.2	2.3	1.0	1.4
60	26.9	3.2	15.5	1.3	0.4	0.7
90	18.6	2.1	10.5	0.9	0.25	0.5
120	14.2	1.6	8	0.6	0.2	0.4
150	11.5	1.3	6.4	0.5	0.15	0.3

Table 277.—Various Observed Rates of Evaporation of Water

The following data are in addition to those which have been summarized by their observers in the formulas given in Table 275; w = wind velocity; m, m_0 = rate of evaporation with and without wind, respectively; e = thickness of water layer removed by evaporation; c = an arbitrary constant.

Unit of $w = 1$ m/sec; of m and $m_0 = 1$ mg/cm²hr = 0.24 mm/day; of $e = 1$ mm/day. Temp. $= t$ °C

	⎯ Becker[a] ⎯		⎯ Hedestrand[a] ⎯		Sutton[a]
w	2.54	5.08	cw	m	South Africa
t	⎯ m/m_0 ⎯		1	10.75	Dry table land
50	2.8	3.8	1.5	14.15	Average for year
80	2.0	2.5	3.0	17.15	151.9 cm/yr
100	1.7	2.2	4.5	18.21	= 4.16 mm/day

Evaporation from the oceans (Wüst[a]).

Zone N	{ 80	70	60	50	40	30	20	10
	{ 70	60	50	40	30	20	10	0
Ocean				e				
Atlantic	0.2	0.3	1.0	1.8	2.5	3.2	3.4	2.5
All Oceans	0.2	0.3	1.0	1.8	2.5	3.0	3.1	2.6
Zone S	{ 0	10	20	30	40	50	60	World
	{ 10	20	30	40	50	60	70	Mean
Ocean				e				
Atlantic	3.3	3.2	2.9	2.3	1.5	0.6	0.2	2.18
All Oceans	2.9	3.1	2.9	2.3	1.5	0.6	0.2	2.24

[a] References:

Becker, H. G.:[55] Hedestrand, G.[46] $t = 20$ °C; Sutton, J. R.:[53] Wüst, G., *Meteor Z.*, **38**, 188-190 (1921) (review) ← *Veröffentl. Inst. Meereskunde, Berlin (N. F.) geogr.-naturw. Reihe*, Heft **6**; 1920, 95 S; see also Kleinschmidt, E., *Meteor. Z.*, **38**, 205-208 (1921).

[64] Winklemann, A., *Ann. d. Physik (Wied.)*, **22**, 1-31 (1884); **23**, 203-227 (1884); **26**, 105-134 (1885); **36**, 93-114 (1889).
[64a] Winklemann, A., *Ann. d. Physik (Wied.)*, **35**, 401-410 (1888).
[65] Vaillant, P., *Jour. de Phys.* (5), **1**, 877-891 (1911).

96. VAPORIZATION AND CONDENSATION

Table 278.—Evaporation from Large Outdoor Areas of Water [66]

$E = E_1(1 + 0.04\sqrt{w})[1 + 3.3(10^{-5})H]$, unit of $w = 1$ mile/day, of $H = 1$ ft.
$= E_1(1 + 0.293\sqrt{w_1})[1 + 10.8(10^{-5})H_1]$, unit of $w_1 = 1$ m/sec, of $H_1 = 1$ m.
$H =$ altitude above sea-level; $t =$ mean monthly temperature; $w =$ mean monthly wind velocity; $E =$ annual mean of the evaporation.
The value of E_1 is that given below for E when $w = 0$.

Unit of $E = 0.001$ inch/day $= 0.0254$ mm/day; of $w = 1$ mile/day $= 0.0186$ m/sec.
Temp. $= t_F$ °F $= t_C$ °C

t_F	t_C	$w \rightarrow$ 0	100	200	300 $E(H=0)$	400	500
20	−6.7	6.5	9.1	10.2	11.0	11.7	12.5
25	−3.9	9.0	12.6	14.1	15.2	16.2	17.0
30	−1.1	11.5	16.1	17.9	19.4	20.7	21.7
35	+1.7	15.0	21	23.5	25.4	20.7[a]	28.4
40	4.4	20.0	28	31.3	33.8	36.0	37.8
45	7.2	26.5	37	41.5	44.8	47.7	50.1
50	10.0	36.0	50	56.5	60.8	64.8	68.0
55	12.8	50.0	70	78.3	84.5	90.0	94.5
60	15.6	70.5	99	110	119	127	132
65	18.3	97	136	152	164	175	184
70	21.1	127	178	199	215	229	240
75	23.9	160	224	250	271	288	302
80	26.7	196	274	307	332	353	371
85	29.5	232	325	364	392	418	438
90	32.2	270	378	423	457	486	510

[a] So printed, but probably should be 27.0.

Small Drops.

See especially N. Fuchs.[67] A drop that is several microns in radius evaporates in still air in accordance with the formula given in Table 275; if r_2 is very great, then $d(r_1^2)/d\tau = -2DM(p_0 - p)/\rho RT$, a quantity independent of the radius, r_1, τ is the time, p_0 is the vapor-pressure at the surface of the drop, and ρ is the density of the liquid. See also preceding text (pp. 623 and 629). That is, the area of the surface of the drop decreases linearly with the time, if p_0 remains constant; p is controllable and is assumed to be constant. But as the radius becomes smaller, of the order of 1 μ (0.001 mm) or less, evaporation becomes slower; drops that in accordance with the preceding formula should vanish in a few seconds, may last for hours.[68]

Gudris and Kulikowa, using drops 1 μ to 0.1 μ in radius, and determining the vapor-pressure at which the size of the drop remained constant, found that the saturation pressure with reference to them is that determined from the radius and the surface-tension in accordance with Kelvin's (W. Thomson's) formula (p. 568). They regarded the reduction in the rate of evaporation as the radius became smaller as an age effect, and attributed it to an absorption of the surrounding gas; it was inappreciable

[66] Grunsky, C. E., *Monthly Weather Rev.*, **60**, 2-6 (1932).
[67] Fuchs, N., *Physik. Z. Sowj.*, **6**, 224-243 (1934).
[68] Gudris, N., and Kulikowa, L., *Z Physik*, **25**, 121-132 (1924); *Chem. Abs.*, **19**, 3186 (1925); *J. Russ. Phys. Chem. Soc. (Phys.)*, **56**, 167-175 (1924); Fuchs, N.[67]

in H_2. For other gases they published curves from which have been read the following values connecting the radius (r) of the drop with the time τ:

Unit of $\tau = 1$ min; of $r = 0.01\mu = 10^{-6}$ cm

────── Air ──────	────── 30 Air + 70 H_2 ──────	────── CO_2 ──────
τ $\quad$ r	τ $\quad$ r $\quad$ τ $\quad$ r	τ $\quad$ r $\quad$ τ $\quad$ r
5 $\quad$ 94	7 $\quad$ 99 $\quad$ 30 $\quad$ 84	5 $\quad$ 89 $\quad$ 20 $\quad$ 86
22 $\quad$ 89	15 $\quad$ 94 $\quad$ 37.5 $\quad$ 79	10 $\quad$ 87.5 $\quad$ 45 $\quad$ 86
37.5 $\quad$ 84	22.5 $\quad$ 89 $\quad$ 45 $\quad$ 74	15 $\quad$ 86.5

On the other hand, D. J. Woodland and E. Mack, Jr.,[69] have reported observations that indicate that $-dm/d\tau = C(r + \delta)$, instead of the Cr required by the linear relation between the area and τ. They think that these observations indicate that the effective removal of the vapor occurs from a surrounding shell of saturated vapor, and not from the surface of the drop itself, the thickness of the shell being δ. Their observations lead to the values $\delta = 0.52\ \mu$ for n-dibutyl phthalate, and $\delta = 1.1\ \mu$ for n-dibutyl tartrate.

C. Barus[70] has stated that, in dust-free air saturated with water-vapor and left undisturbed, the dissipation of very small fog particles "by evaporation is enormously more important than by subsidence"; in his case the two were about equal when the diameter (d) of the particle was $3\ \mu$. "Fog particles precipitated on solutional nuclei (phosphorus) evaporate" to water nuclei which persist without other loss than by subsidence. Those precipitated on nuclei of water-vapor evaporate almost without residue, the persisting nuclei being only 0.4 per cent when $d = 1.6\ \mu$, and 3.6 per cent when $d = 32\ \mu$. "These fog particles vanish into the wet air from which they were precipitated and the experiment may be repeated indefinitely. Relatively more water nuclei persist as the fog particles evaporated are larger."

The evaporation of drops in a stream of air of velocity w m/sec has been studied by T. Namekawa and T. Takahashi[71] who found that $dr/d\tau = -2.53\ [(p_0 - p)/r] \cdot [1 + 2.1w^{1/2}] \cdot 10^{-7}$ if $r < 1$ mm and $w < 2$ m/sec. (See also Y. Takahasi[72] and E. G. Zak.[72])

Condensation.

For those condensation data and phenomena in which condensation may be considered as merely negative vaporization, reference should be made to the preceding pages devoted to vaporization, one exception being Table 281 treating of condensation upon snow.

When water-vapor condenses on a surface that is chilled to $-110\ °C$ or lower, the deposited ice is vitreous.[73] Condensation on extended surfaces at higher temperatures, depending, as it does, upon both the nature

---

[69] Woodland, D. J., and Mack, E., Jr., *J. Am. Chem. Soc.*, **55**, 3149-3161 (1933).
[70] Barus, C., *Am. J. Sci. (4)*, **25**, 409-412 (1908).
[71] Namekawa, T., and Takahashi, T., *Mem. Coll. Sci. Kyoto (A)*, **20**, 139-146 (1937).
[72] Takahasi, Y., *Sci. Abs. (A)*, **40**, 265 (1937) ← *Geophys. Mag., Tokyo*, **10**, 321-330 (1936); Zak, E. G., *Chem. Abs.*, **31**, 3360 (1937) ← *Zhur. Geofiz.*, **6**, 452-465); 466-473 (1936).
[73] Burton, E. F., and Oliver, W. F., *Proc. Roy. Soc. London (A)*, **153**, 166-172 (1936) → *Nature*, **135**, 505-506 (L) (1935).

of the surface and local peculiarities thereof, is not considered in this compilation.

Supersaturation.—It is probable that water-vapor can never become supersaturated (supercooled) in the immediate presence of water or of ice, but when no condensed phase is present it can be considerably supercooled if it contains no dust or other nucleus on which condensation can begin. See the following section: *Nuclear Condensation.* Supercooling is always understood to be that with reference to vapor in equilibrium with a flat surface of water.

C. F. Powell [74] has remarked that Callendar and Nicholson [74a] pointed out that the steam in the cylinder of a steam engine might be supersaturated, and that supersaturated steam might exist in a steam turbine, as has since been pointed out by others. "According to Callendar, the steam passes through the turbine so quickly that thermal equilibrium cannot be maintained. It becomes supersaturated, and no appreciable condensation takes place until the cloud-limit is reached, when nuclei are produced in enormous numbers." (Cloud-limit = supersaturation at which condensation on uncharged nuclei begins. See Table 279.) He comments on erroneous values assumed by Callendar and by H. M. Martin for the supersaturation at the cloud-limit, and from his own observations computes the data in Table 280.

When steam is expanded by passage through a simple convergent-divergent nozzle, the steam becomes supersaturated; condensation does not occur until the steam has reached the condition approximately represented by th 3.5 per cent moisture line on the Mollier enthalpy-entropy diagram, and then drops 6.2A in radius are formed.[75] For the corresponding phenomenon for water, see preceding text (p. 623).

Nuclear Condensation.—In the study of nuclear condensation, the customary procedure is to cool a mixture of gas and vapor by an adiabatic expansion, the amount of expansion that just suffices to produce condensation being determined by trial. The expansion (E) is defined as the ratio of the expanded volume (v_2) to the volume (v_1) before expansion ($E = v_2/v_1$). If $\gamma = c_p/c_v$ = ratio of the principal specific heats of the gas-vapor mixture, and if T_1 °K is the absolute temperature of the mixture before expansion, then the temperature (T_2 °K) after expansion will be given by the relation $T_1/T_2 = (v_2/v_1)^{\gamma-1}$. The conditions are usually such that the vapor in the mixture of volume v_1 and temperature T_1 is saturated with reference to a flat surface of its liquid. The corresponding vapor pressure (p_1) may be found from tables of vapor pressure; the pressure (p_e) of the vapor after expansion is given by the relation $p_e/p_1 = (v_1/v_2)^{\gamma}$; and vapor pressure ($p_2$) corresponding to saturation at T_2 °K may be

---

[74] Powell, C. F., *Proc. Roy. Soc. London (A)*, **119**, 553-577 (1928).

[74a] Callendar, H. L., and Nicolson, J. T., *Min. Proc. Inst. Civ. Eng. (London)*, **131**, 147-206-268 (1897).

[75] Yellott, J. I., *Trans. Am. Soc. Mech. Eng.*, **56**, 411-427-430 (FSP-56-7) (1934); Yellott, J. I., and Holland, C. K., *Idem*, **59**, 171-183 (FSP-59-5) (1937); Jakob, M., *Z. techn. Physik*, **16**, 83-86 (1935).

found from tables. The ratio $S = p_e/p_2$ is called the supersaturation produced by the expansion. The value of the expansion (E) depends only on the volumes v_1 and v_2, but that of S depends also upon the nature of the inert gas and of the vapor.

As E is gradually increased, an initial and four other stages of condensation can be distinguished. Starting with air taken directly from the atmosphere and saturated, one obtains a dense cloud of drops when E is

Table 279.—Condensation of Water-vapor on Nuclei

E = expansion, S = supersaturation, each at the beginning of condensation on nuclei of the nature indicated by the subscript ($-$ = negative ions, $+$ = positive ions, 0 = uncharged nuclei). For exact definition of E and of S, see text. The values of S are unaffected by the nature of the admixed uncondensible gas; those for E refer to vapor mixed with air; t_1 and t_2 °C = temperature before and after expansion, respectively.

I. Adapted from L. B. Loeb (ICT)[88]; t_1 = 18 °C.

E_-	S_-	E_+	S_+	E_0	S_0	References[a]
1.25	4.15	1.31	5.8	1.38	7.9	Wilson
1.29				1.42		Donnan
1.265		1.314		1.366		Przibram
1.270		1.32		1.31	6	Andrén
1.251						Laby

II. C. F. Powell (1928).[76] Values corrected for evaporation from walls, see text.

t_1	E_-	t_2	S_-	t_1	E_0	t_2	S_0
7				7	1.375	−26.4	8.95
18	1.245	− 6.5	3.98	18	1.370	−16.4	7.80
35[b]	1.235	+10.0	3.44	35	1.314	+ 3.2	5.07
50	1.226	+24.7	2.96	50	1.286	+19.1	3.74
77	1.218	+50.5	2.52	77	1.252	+47.0	2.87

III. Volmer and Flood.[a] Values averaged by the observer over ranges of 2 °C or less.

t_1	E_-	t_2	S_0	t_1	E_-	t_2	S_0
29.2	1.266	1.7	4.18	14.7	1.288	−12.2	4.98
18.	1.276	−9.4	4.85				

[a] References:
Andrén, L., *Ann. d. Physik* (4), **52**, 1-71 (1917); Anderson, E. X., and Froemke, J. A., *Z. physik. Chem. (A)*, **142**, 321-350 (1929); Donnan, F. G., *Phil. Mag.* (6), **3**, 305-310 (1902); Laby, T. H., *Phil. Trans. (A)*, **208**, 445-474 (1908); Powell, C. F.;[76] Przibram, K., *Jahrb. Radioak.*, **8**, 285-308 (1911) Bibliography of 135 titles; Volmer, M., and Flood, H., *Z. physik. Chem. (A)*, **170**, 273-285 (1934); Wilson, C. T. R.[78a]

[b] E. X. Anderson and J. A. Froemke (*loc. cit.*) found $E_- = 1.201$ and $S_- = 3.0$ at $t_1 = 25$ °C.

---

only slightly greater than unity. Allowing these to subside, and repeating the process, and so continuing, one presently reaches a stage at which such small expansions produce no condensation. This terminates the initial stage.

Continuing the process with gradually increasing values of E, no condensation (except on the walls) occurs until $E = 1.25$, when a few drops of rain form in the interior of the gas. This is stage 1. A further increase in E merely increases the number of drops—the heaviness of the rain—until $E = 1.38$ (stage 2). Beyond $E = 1.38$ a persistent cloud of small drops is formed (stage 3); the number of drops increases rapidly as E increases beyond 1.38, each drop becoming correspondingly smaller, and presently diffraction colors border the image of a source of light seen through the cloud, and change as the size of the drops decreases (stage 4).

In the initial stage—generally not counted, but regarded as a cleansing process—condensation occurs on dust and other, presumably large, nuclei that are not being continually replaced. Repeated condensations sweep these out of the gas. At $E = 1.25$ condensation begins to occur on negative ions; at higher values, on positive ions; and at $E = 1.38$ on uncharged nuclei which "have been identified with the associated molecules present in water-vapour."[76]

Powell[76] found that in general the expansion actually realized is not simply adiabatic, but is attended by evaporation from the walls of the vessel, which causes the density of the expanded vapor to exceed that corresponding to simple expansion. He corrected his observations for this effect; the values so corrected are given in Table 280. He has concluded that ordinary room temperatures are to be preferred to others as initial temperatures in the investigation of atomic phenomena by means of the cloud method.

Table 280.—State of Water-vapor at the Cloud-limit [76]

Values above 47 °C were obtained by extrapolation based on $S_0 = 1$ at the critical point. Callendar's equations for steam were used. H = heat content, P = vapor pressure, S_0 = supersaturation = P_0/P_{sat}, V = specific volume, ϕ = entropy, the subscripts $_0$ and $_{sat}$ indicate, respectively, that the value is that at which condensation on uncharged nuclei begins (the cloud-limit), and that corresponding to equilibrium with a flat surface of water. As usual, H and ϕ are measured from water at 0 °C.

Unit of $P = 1$ lb*/in² $= 0.0680$ atm $= 68.95$ kdyne/cm²; of $V = 1$ ft³/lb $= 62.429$ cm³/g; of $H = 1$ lb-cal/lb $= 1$ g-cal/g; of $\phi = 1$ lb-cal/lb·°K $= 1$ g-cal/g·°K. Temp. $= t$ °C.

t	P_{sat}	S_0	V_{sat}	V_0	P_0	H_0	ϕ_0
0 °C	0.0892	5.40	3275.9	606.5	0.4811	594.02[a]	1.9899
10	0.1789	4.40	1693.8	584.9	0.7855	598.72[a]	1.9529
20	0.3399	3.70	922.19	249.2	1.254	603.32[a]	1.9174
30	0.6162	3.25	525.8	161.8	1.992	607.85[a]	1.8819
40	1.070	3.02	312.4	103.5	3.214	612.30[a]	1.8441
50	1.789	2.81	192.7	68.59	4.995	616.67[a]	1.8098
60	2.887	2.35	122.9	48.59	7.250	620.98[a]	1.7821
75	5.586	2.22	66.20	29.82	12.31	627.21[a]	1.7448
90	10.161	1.99	37.81	19.00	19.99	633.19[a]	1.7074
100	14.69	1.86	26.79	14.40	26.96	637.02	1.6872
120	28.81	1.66	14.27	8.595	47.11	644.22	1.6450
140	52.48	1.51	8.143	5.392	78.91	650.70	1.6068
160	89.80	1.39	4.923	3.542	122.7	656.85	1.5755
180	145.6	1.29	3.127	2.408	185.6	662.24	1.5423
200	225.2	1.20	2.074	1.728	266.0	667.58	1.5205

[a] For $P_0 < 20$, the values of H_0 fall on the line of 2 per cent wetness.

[76] Powell, C. F., *Proc. Roy. Soc. London (A)*, **119**, 553-577 (1928). See also Anderson, E. X., and Froemke, J. A., *Z. physik. Chem. (A)*, **142**, 321-350 (1929).

Optical methods for determining the size of suspended water droplets have been discussed by J. G. Wilson [77] who has concluded that the radii (r) of the drops responsible for the colors observed by C. T. R. Wilson [78] when water-vapor was condensed by expansion (E) were as follows, the unit being 0.01 $\mu = 10^{-6}$ cm: Brilliant green, E 1.412, r 135; blue-green, E 1.416, r 130; brilliant blue, E 1.418, r 119; purple, E 1.420, r 105; red, E 1.426, r 84; reddish yellow, E 1.430, r 77; orange-white, E 1.436, r 60; whitish, E 1.448, r 48; greenish white, E 1.454, $r \leqq 45$.

These estimates of the sizes of the drops that cause the observed diffraction effects should not be confused with those that C. T. R. Wilson [78] has made of the equivalent sizes of the nuclei upon which the condensations began.

M. Akiyama [79] has reported that 50 per cent of the charged recoil atoms of actinium-A do not act as nuclei of condensation for water-vapor at supersaturations in the neighborhood of those at which the vapor condenses on ordinary positive ions.

P. I. Dee [80] has published a diagram that facilitates the determinations of the quantities required in the interpretation of such adiabatic expansions of air saturated with water-vapor as are here considered.

G. Stüve [81] has concluded that in natural atmospheric condensation, gaseous nuclei give only drops of water, nuclei consisting of soluble salts give drops if condensation begins at temperatures above $-20\,°C$, but stars of snow if below $-20\,°C$, and insoluble hygroscopic nuclei give needles of ice at all temperatures below $0\,°C$.

Much is yet to be learned about the natural condensation of atmospheric moisture. Fogs may occur in air that is unsaturated; they may be absent from air that is saturated with water-vapor.[82] The nature of atmospheric nuclei of condensation has been considered by Bennett,[82] H. Landsberg,[83] J. H. Coste and H. L. Wright,[84] C. Junge,[85] and H. Köhler,[86] to mention only those that have happened to come to my attention. Salt from the ocean is generally believed to be the most abundant of the natural nuclei, but human activities—fires, furnaces, etc.—contribute droplets of nitreous acid and probably some of sulphuric acid.[84] It has been frequently reported that the volumes of the drops of rain, and perhaps of fog also, are simple multiples of a few primary sizes; which suggests that the larger ones are formed by the coalescence of the smaller ones.[86, 87]

---

[77] Wilson, J. G., *Proc. Cambridge Phil. Soc.*, **32**, 493-498 (1936).
[78] Wilson, C. T. R., *Phil. Trans. (A)*, **189**, 265-307 (282) (1897).
[79] Akiyama, M., *Compt. rend.*, **187**, 341-342 (1928).
[80] Dee, P. I., *Proc. Cambridge Phil. Soc.*, **28**, 93-98 (1932).
[81] Stüve, G., *Gerlands Beitr. zu Geophys. (Köppen Bd. 1)*, **32**, 326-335 (1931).
[82] Bennett, M. G., *Sci. Abstr. (A)*, **37**, 259 (1934) ← *J. Roy. Meteor. Soc.*, **60**, 3-14 (1934).
[83] Landsberg, H., *Monthly Weather Rev.*, **62**, 442-445 (1934).
[84] Coste, J. H., and Wright, H. L., *Phil. Mag. (7)*, **20**, 209-234 (1935).
[85] Junge, C., *Gerlands Beitr. zu Geophys.*, **46**, 108-129 (1935).
[86] Köhler, H., *Arkiv. Mat., Astron., Fysik,* **24 A**, No. 9 (1934).
[87] Gold, E., *Nature*, **133**, 102 (L) (1934); Köhler, H., *Trans. Faraday Soc.*, **32**, 1152-1161 (1936); Marki, E., *Meteor. Z.*, **54**, 173-183 (1937).
[88] Loeb, L. B., *Int. Crit. Tab.*, **6**, 117 (1929).

97. FREEZING AND MELTING

Table 281.—Condensation on Snow in the Open [88a]

Observations at high latitudes; data are as applicable to evaporation as to condensation. C = amount of condensation in the time τ, p = partial pressure of water-vapor in the air, p_{sat} = pressure of water-vapor saturated with reference to the surface of the snow. Rolf [88a] finds $C = a + b(p - p_{sat})\tau$; the values of a and b vary with the season (mean temperature?), and if the ground is only partly covered with snow, with the time of day; t is the temperature of the air. The observations were admittedly rough, and in some cases differed by 100 per cent from the values defined by his equation.

Unit of C = 1 mm of water, of τ = 1 hr, of p and p_{sat} = 1 mm-Hg

Winter, ground covered, $t = +0.9$ to -27.5 °C, $a = 0$, $b = 0.0174$
Spring, ground covered, $t = +3.1$ to $+8.7$ °C, $a = -0.0010$, $b = 0.0168$.
Summer, ground partly covered, a and b variable as follows:

Time of day	1000a	1000b	Time of day	1000a	1000b
8:25 to 11:00	+9 ± 2	+29 ± 1	15:15 to 17:00	20 ± 3	24 ± 3
11:15 to 13:00	18 ± 2	28 ± 2	17:15 to 20:00	8 ± 2	25 ± 2
13:15 to 15:00	16 ± 4	29 ± 3	20:15 to 8:10	0.6 ± 1.2	15.7 ± 1.1

Condensation on Metals.—The conclusion of O. Reynolds [89] that "there is no limit to the rate at which pure steam will condense but the power of the surface to carry off the heat" is practically correct, but, as he pointed out, when an uncondensing gas is mixed with the steam, the surface on which the condensation occurs becomes blanketed with a layer of gas that is relatively poor in steam, and through which the steam must pass by diffusion in order to reach the surface. That greatly reduces the rate of condensation. Furthermore, condensation may occur either as isolated drops or as a continuous film of water, depending upon the surface conditions.[90] Phenomena relating to the condensation of steam while flowing through cooled metal tubes, though of great technical importance, scarcely fall within the scope of this compilation.

97. Freezing and Melting

(See also Section 59. For external work and change in volume on freezing or melting, and for latent heat, see Section 95; for melting temperature, Section 92.)

Ice Needles.

As freezing continues after a volume of water has become completely surrounded by ice, or enclosed between rigid walls and an ice-sheet adherent to those walls, the pressure of the water increases, and may rupture the bounding sheet of ice. If the freezing is proceeding very rapidly, as when the water is much supercooled before the freezing begins, the rupture may consist of one or more small perforations. The peripheral portions of the

---

[88a] Rolf, B., *Arkiv. Mat., Astron., Fysik*, **9**, No. 35 (1914).
[89] Reynolds, O., *Proc. Roy. Soc. London (A)*, **21**, 275-283 (1873).
[90] Schmidt, E., Schurig, W., and Sellschop, W., *Tech. Mechan. Thermod.*, **1**, 53-63 (1930).

issuing water will freeze, thus building up a tube of ice, which may grow with surprising speed, sometimes straight but more often curved or abruptly bent, and may attain a length of several centimeters. When the growing ceases, the contents of the tube freezes, converting the whole into a needle of ice. Such growths are not at all uncommon when there has been much supercooling. Their growth is of exactly the type described by H. Erlenmeyer [90a] in explaining the growth of hair-like crystals sometimes observed to form in salt crystals that are creeping over solid surfaces. It seems probable that the "long crystal" mentioned by T. Alty [90b] and the unusual one reported by O. Bally [91] were formed in this way; and the same may have been true of some of those reported by J. Meyer and W. Pfaff.[92] But in the case they mention, of a needle that grew from the wall of an empty bulb connected to one containing water, it seems that the growth must have been by condensation, as they suggest. In view of the low thermal conductivity of ice and of water-vapor, and of the large amount of heat liberated when water-vapor is converted to ice, it would seem that the growth of a needle by such condensation would be extremely slow. Furthermore, the large specific volume of the vapor would restrict the size of a needle, formed by condensation, to a small value, whenever the growth occurred after the water had become completely covered with ice. For example, the volume of the needle that can be produced by the complete freezing of 1 cm^3 of vapor saturated with reference to water at $-15\,°C$ (sp. vol. = 622 000 cm^3/g) cannot exceed that of a cylindrical needle 0.1 mm in diameter by 0.49 mm in length; since, owing to the vapor pressure of ice at $-15\,°C$, only about 2.2 per cent of the total volume of the vapor can so freeze, there would have to be available 45 cm^3 of the vapor saturated with reference to water at $-15\,°C$ if a spicule of that small size (0.1 mm by 0.49 mm) is to be actually formed in that manner.

It is not uncommon for such needles to be spoken of as monocrystals, but the compiler has yet to see experimental evidence justifying such a description.

Supercooling of Water.

[NOTE: Since this section was written, G. Tammann and A. Büchner [93] and J. Meyer and W. Pfaff [92] have studied the subject, and have interpreted their observations in terms of the ideas that they had developed in the course of their studies of other substances. The two groups differ mainly in that Meyer and Pfaff hold that the primordial nuclei upon which the ice is first formed are always solids, foreign to the molten substance, whereas Tammann and Büchner think that they may be formed from the molten substance itself. Each group seems to think that the growth of, or

---

[90a] Erlenmeyer, H., *Helv. Chim. Acta*, **13**, 1006-1008 (1930).
[90b] Alty, T., *Phil. Mag. (7)*, **15**, 82-103 (1933).
[91] Bally, O., *Helv. Chim. Acta*, **18**, 475-476 (1935).
[92] Meyer, J., and Pfaff, W., *Z. anorg. allgem. Chem.*, **224**, 305-314 (1935).
[93] Tammann, G., and Büchner, A., *Z. anorg. allgem. Chem.*, **222**, 371-381 (1935).

97. FREEZING AND MELTING

upon, the nucleus is slow until a certain critical size is attained; then it becomes rapid—visible freezing occurs. If this were true, then the time that elapses between the instant that a specimen of water of given form and size is immersed in a cold bath and the instant that visible freezing begins would be a matter of prime importance; and that time, or its equivalent, is observed and reported by each. None of this necessitates a rewriting of this section.

But more recent observations by the compiler have shown that certain of the impressions conveyed by this section are entirely wrong, although in entire harmony with the reports cited. On account of that harmony it has seemed desirable to let the section stand as written, and to request the reader to bear in mind the following facts, ascertained by the compiler.

When water is protected from the atmosphere, as by sealing it in a glass bulb, it freezes spontaneously at a fixed temperature that is independent both of the rate of cooling and of the time that the bulb has been held at a low temperature. The temperature of this spontaneous freezing varies from specimen to specimen, and for any one specimen may exhibit oscillations and may drift, but in many cases it remains constant within a few tenths of a degree centigrade for months at a time. The volume of each specimen used was about 8 cm^3, and it was contained in a bulb of about twice that size. It is not at all difficult to obtain specimens of this size that can be supercooled to -14 °C, and one has been repeatedly cooled to -21 °C. In no case was any precaution taken to keep the water quiescent; in fact, the supercooled water can be poured with impunity over the entire interior of the bulb. Violent splashing will cause the supercooled water to freeze, but sharp rapping of the exterior of the bulb is without effect. On the other hand, a very gentle wiping of the glass-water interface will cause freezing at a temperature well above that at which the specimen freezes spontaneously. All the observations so far obtained are consistent with the opinion that, within the range of temperature covered (0 to -21 °C), the presence of foreign solids is essential to the spontaneous freezing of the specimen; in this respect they agree with the conclusions of Meyer and Pfaff. They also show conclusively that the time required for a specimen to freeze when subjected suddenly to a given condition of chilling is not a factor of prime importance; it is the temperature, that is reached by some portion of the specimen, that determines the freezing; and that temperature is a characteristic of the specimen, varying from one to another as do the foreign solids that serve as "nuclei." [94]

Under suitable conditions, water can be cooled below its so-called freezing point (the melting point of ice) without becoming solidified. In that state it is said to be supercooled. If a bit of ice, no matter how minute, be touched to supercooled water, freezing on the surface of the ice begins at once, and proceeds rapidly until enough latent heat has been freed to raise the temperature to the so-called freezing point. It then stops unless heat is being abstracted from the mixture.

[94] See Dorsey, N. E., *J. Res. Nat. Bur. Stand.*, **20**, 799-808 (RP1105) (1938).

Such seeding with a suitable crystal seems to be the only means by which the freezing of supercooled water can be initiated at will. And that freezing is simply a growth of ice from the crystal into the water. The various other methods that have been suggested, some of which are mentioned below, are so variable in their action and have been shown to fail so signally in certain instances that one cannot seriously regard any of them as having been shown to be more than secondarily involved in the initiation of such freezing as may have occurred when it was employed. Under some conditions it may have facilitated the initiation, but it can scarcely be regarded as having been primarily involved in the initiation itself.

There are, however, certain conditions that appear to favor the supercooling of water, to reduce the likelihood of freezing being initiated. A. Mousson [95] reported that water can be more readily supercooled when it exists (1) as small drops on surfaces that are not wetted (velvet, finely dusted surfaces, certain leaves, etc.), (2) bubbles [?] in a fog, (3) in narrow capillary tubes, (4) as a thin layer between glass plates that are clamped together (if the plates were not clamped, if the upper one was merely laid upon the under, the water between them froze), than when it exists in bulk; and when protected from mechanical shocks, than when subjected to them. He drew the general conclusion that whatever impedes a rearrangement of the particles of the water facilitates its supercooling. L. Dufour [96] reported that the placing of a layer of oil upon the surface of the water is a very uncertain means for facilitating supercooling, but that drops of water, especially when small, suspended in a liquid of the same density could be markedly supercooled, even when subjected to violent deformations; he occasionally carried it to $-20\,°C$. S. W. Young and R. J. Cross [97] have stated that long-continued heating facilitates the supercooling of a liquid.

In his discussion of the subject, Dufour [96] seems to have failed to distinguish between the removal of that which impedes and the imposition of that which causes. The result of an effective impediment is the same whether or no a cause otherwise efficacious is present, but in the absence of such a cause, the removal of the impediment changes nothing. Many later investigators of supercooling have likewise failed to see the distinction, and much confusion has resulted. In the following paragraphs an attempt is made to give the opinions and the points of view of the several authors cited; the reader should constantly bear in mind the distinction just drawn, and interpret the statements accordingly.

H. T. Barnes [98] has stated that agitation, with the presence of dust or suspended matter and, particularly, of dissolved air, makes supercooling almost impossible. With the avoidance of those conditions he has cooled

---

[95] Mousson, A., *Ann. d. Physik (Pogg.)*, **105**, 161-174 (1858).
[96] Dufour, L., *Arch. des sci. phys. et nat. (N. P.)*, **10**, 346-371 (1861) → *Ann. d. Physik (Pogg.)*, **114**, 530-554 (1861); letter appended to *Arch. des sci. phys. et nat. (N. P.)*, **11**, 22-30 (1861).
[97] Young, S. W., and Cross, R. J., *J. Am. Chem. Soc.*, **33**, 1375-1388 (1911).
[98] Barnes, H. T., "Ice Formation," pp. 95-97, 1906.

water in open flasks to $-6\,°C$. He writes: "Curiously enough, once the freezing point was passed .... it was certain that several degrees below that point would be reached without ice forming, and at $-3°$ or $-4°$ quite violent agitation was required for solidification to take place .... It seems harder to pass the freezing point without ice forming than to continue the cooling beyond this temperature. The degree of instability reaches such a critical state, however, beyond five or six degrees, that extraordinary precautions have to be taken for further cooling."

This relative stability of the slightly supercooled water as compared with the great instability of that cooled to $-6\,°C$ or lower led to the suggestion that a metastable condition, in which crystallization can be initiated only by seeding with ice, existed for a few degrees below zero, and that at lower temperatures a truly labile condition existed. But S. W. Young,[99] S. W. Young and R. J. Cross,[100] and S. W. Young and W. J. von Sicklen [100a] seem to have shown that no such distinction exists. The difference is merely one of degree; the whole supercooled "field is labile and crystallization may be brought about in any portion of it by the production of sufficient mechanical shock." Young and von Sicklen found freezing to attend mechanical shock when the temperature of the water was as high as $-0.02\,°C$.

Observations reported in the three papers just mentioned, and especially those in the first, indicate that the frequency with which the freezing of supercooled water accompanied the friction of solid on solid within the water, varied with the nature of the solid.

Contrary to the observations of H. T. Barnes [98] are those of R. Pictet (mentioned by Oltramare [101]), who found that supercooled water in a stoppered flask can be violently agitated without its freezing, even though the temperature be $-19\,°C$. The flask was half full. The present compiler has observed the same for a few cubic centimeters of supercooled water at $-14\,°C$ in a sealed bulb of about twice the volume of the water.

Observations on the supercooling of drops of water suspended in a liquid of the same density, and of water in capillary tubes have been reported by T. Borovik-Romanova,[102] who gave the following values for the temperature (t) at which freezing began in capillary tubes (presumably of glass) of diameter d, and for the ranges over which the observed values were spread:

d	1.57	0.24	0.15	0.06	mm
t	-6.4	-13.5	-14.6	-18.5	°C
range	1.3	2.5	4	1.6	°C

At a much earlier date, H. C. Sorby [103] had made similar observations, carrying the supercooling in tubes to $-16\,°C$, and observing, partly in

---

[99] Young, S. W., *J. Am. Chem. Soc.*, **33**, 148-162 (1911).
[100] Young, S. W., and Cross, R. J., *Idem*, **33**, 1375-1388 (1911).
[100a] Young, S. W., and von Sicklen, W. J., *Idem*, **35**, 1067-1078 (1913).
[101] Oltramare, G., *Arch. sci. phys. et. nat. Genève (3)*, **1**, 487-501 (1879).
[102] Borovik-Romanova, T., *Chem. Abs.*, **19**, 3186 (1925) ← *J. Russ. Phys. Chem. Soc. (Phys. Part)*, **56**, 14-22 (1924) (Russian).
[103] Sorby, H. C., *Phil. Mag. (4)*, **18**, 105-108 (1859).

company with Tyndall, that even at temperatures near −20 °C the liquid*
occurring in natural cavities in quartz does not freeze, the diameter of the
cavity being about 0.25 mm (0.01 in.). He observed further, that even in
capillary tubes water will freeze at a temperature very near 0 °C if the
water is at any point in contact with ice, and that it melts at 0 °C. And
C. Despretz,[104] using water thermometers with bulbs several lignes in
diameter (1 ligne = 2.25+ mm), had followed the dilatation of water
(unfrozen) to −20 °C. He stated that Blagden (no citation) had cooled
water to −6 °C, and Gay-Lussac (no citation) to −12 °C. G. Oltramare [101] has stated, without citation, that both Pictet and Dufour had cooled
water to −40 °C. This is the greatest supercooling that the compiler has
found mentioned; he has not succeeded in finding a paper by either Pictet
or Dufour reporting this value.

Of the more recent investigations of supercooling, may be mentioned
the following: H. A. Miers and Miss F. Isaac,[106] using water sealed in
tubes "which were vigorously and continuously shaken by hand in a bath
of brine" cooled at the rate of 2 °C per hour, found freezing to occur
between −1.6 and −2 °C, averaging −1.9 °C, at which temperature, they
concluded, "pure water freezes spontaneously, *i.e.*, in the absence of ice
particles." They remark that the index of refraction is a maximum at
about the same temperature.† When the tubes contained loose bits of solids
(glass, garnet, lead) rubbing with friction on the walls as the tubes were
shaken, freezing might occur as high as −0.4 °C. W. H. Martin [107] has
reported that water that had been repeatedly redistilled *in vacuo* and without ebullition could be cooled in a 2-mm tube to −26 °C, whereas ordinary
distilled water froze at −11 °C under the same cooling conditions. G. V.
Lange [108] has cooled water in 0.1 mm capillaries to −18 °C, the cooling
having taken 10 hours.

That very small water droplets suspended in the air may be cooled to
very low temperatures without freezing, is indicated by the well-known
fact "that the most brilliant coronas—those of multiple rings and large
diameter—usually are formed by very high clouds whose temperature often
must be far below freezing." [109]

L. Hawkes [110] has stated that the deposit on the cooling pipes in a room
maintained at −17 to −22 °C "was found to be a mixture of water drops

---

* Sir Humphry Davy [105] had previously found that every such clear, colorless,
liquid inclusion which he had examined consisted of nearly pure water, and with few
exceptions, was under less than atmospheric pressure.

† The maximum value of the index probably lies much nearer 0 °C, see p. 280+.

[104] Despretz, C., *Ann. chim. phys.* (2), **70,** 5-81 (1839) → *Compt. rend.,* **4,** 124-130 (1837) → *Ann. d. Physik (Pogg.),* **41,** 58-71 (1837).

[105] Davy, Sir Humphry, *Phil. Trans.,* **112,** 367-376 (1822) = *Ann. chim. phys.,* **21,** 132-143 (1822) = *Annals Philos. (N. S.),* **5,** 43-49 (1823).

[106] Miers, H. A., and Isaac, Miss F., *Report Brit. Assoc. Adv. Sci.,* **1906,** 522, (1906).

[107] Martin, W. H., *Trans. Roy. Soc. Canada,* III (3), **7,** 219-220 (1913).

[108] Lange, G. V., *Jour. de phys.* (7), **1,** 406D (1929) ← *Bull. de l'Inst. Agronom, Kharkow,* **8-9,** 107-108 (1929).

[109] Humphreys, W. J., "Physics of the Air," 2nd ed., p. 534, 1929.

[110] Hawkes, L., *Nature,* **124,** 225-226 (1929).

and ice—this at $-22\,°C$." The statement seems to imply that the water drops were at $-22\,°C$, although in contact with ice; such surely was not the case. The presence of drops indicates continuing condensation, and had the temperature of the drops been directly observed there is no doubt that it would have been found to lie above $0\,°C$.

Both this and his preceding note [111] are entitled "Super-cooled Water," and refer to the apparently vitreous solid that Beilby obtained when a small drop of water was rapidly chilled to a temperature some 15 degrees below $0\,°C$ (see p. 396). That is not the state of supercooling with which we are here concerned. We are now concerned solely with water in its fluid state.

The freezing of water in such capillary systems as soils, sand, and silica has been discussed by E. A. Fisher.[112]

Superheating of Ice.

Although water can be supercooled, there is no evidence that ice can be superheated—heated above its melting point (pp. 405 and 604). Nevertheless, the temperature of a well-stirred intimate mixture of ice and water will differ slightly from $0\,°C$ if heat is either withdrawn or supplied very rapidly. H. T. Barnes[113] states that "the ice itself shares in the temperature elevation or depression" and explains the departure from $0\,°C$ by the inability of the ice "to freeze or melt rapidly enough to keep up the heat exchange. The velocity of crystallization and of melting is finite, and is the determining factor in the temperature of the two phases when coexisting." But that the temperature of the ice itself should rise above $0\,°C$ under such circumstances seems most improbable, and is, indeed, contrary to the general experience which Barnes[113, p. 90] expresses thus: "It seems to be impossible to superheat a solid with respect to a liquid." It is more likely that the ice does not rise above zero, and that it is protected from the action of the surrounding water at higher temperature by a thin, closely adherent blanket of colder water. In the reverse case, a corresponding blanket of warmer water, heated by the latent heat freed as the water freezes, will protect the ice from excessive chilling, and may keep it at zero even though the temperature of the bulk of the water is slightly lower. It seems probable that in the case of ordinary ice (ice-I) the actual rate of melting, and perhaps that of freezing also, is determined by the thickness and the thermal conductivity of such blankets, rather than by any inherent slowness with which the substance can change from one phase to the other (cf. p. 624).

In this connection, the excitement created in 1880-1882 by T. Carnelley's extravagant claim to have heated ice above $100\,°C$ may be of interest.[114]

[111] Hawkes, L., *Nature*, **123**, 244 (1929).
[112] Fisher, E. A., *J. Phys'l Chem.*, **28**, 360-367 (1924).
[113] Barnes, H. T., "Ice Engineering," p. 2-3, 1928.
[114] See Carnelley, T., *Nature*, **22**, 434-435, 510-511 (1880); **23**, 341-344 (1881); *Chem. News*, **42**, 130, 313 (1880); *Proc. Roy. Soc. London*, **31**, 284-291 (1880-81); *Ber. deut. chem. Ges.*, **13**, 2406-2407 (1880); *J. Chem. Soc.*, **41**, 317-323 (1882). Pettersson, O., *Ber. deut. chem. Ges.*, **13**, 2141-2144 (1880); *Nature*, **24**, 167-169 (1881). Meyer, L., *Ber. deut. chem. Ges.*, **13**, 1831-1833 (1880); **14**, 718-722 (1881). Wüllner, A., *Ann. d. Physik (Wied.)*, **13**, 105-110 (1881). And many other articles in *Nature*, vols. 22, 23, and 24. Certain of the articles are entitled "Hot Ice."

Rate of Freezing and of Melting.

The terms "rate of freezing," "quickness of freezing," "velocity of freezing," and their equivalents are essentially vague and indefinite when they stand alone, and have been used in different senses at different times. They may be used (1) to cover the rate of thickening of an ice sheet upon a pond, or (2) for the rate at which ice forms on an exposed surface of water, or (3) for the time that elapses between the exposure of water to chilling conditions and the initial appearance of ice in the water, or (4) for the rate at which crystallization proceeds along a narrow column of supercooled water; and they have been used in all these ways.

1. The rate of thickening of an ice sheet has been consideerd in Section 59 (p. 407).

2. In an exposition of his trihydrol theory of ice formation, H. T. Barnes [115] refers to certain experiments in which he periodically removed and measured the amount of ice that had formed on the surface of a tank of water since the preceding removal. He found that the rate of freezing, as so measured, decreased from one period to the next, finally becoming zero. If the water were heated to room temperature and then cooled again, ice would form as before. He regarded these observations as indicating that "water may be exhausted of its ice-forming power," "that a nucleus is required for the colloidal ice mass, and after exhausting these nuclei, the formation of ice is rendered difficult," and that time is "required for the restoration of the trihydrol in solution and at the temperature of freezing it is considerably slower than at higher temperatures." Except for a figure and its legend, given by T. C. Barnes and T. L. Jahn,[116] the details of these experiments seem to have remained unpublished. That is most unfortunate. The little that has been published is quite unconvincing and suggests that due attention was not given to important details. For example, the amount of ice that will form in a given time, once freezing is initiated, will be greatly affected by the heat capacity and initial temperature of the tank and its contents, and by the rate at which heat is abstracted from them by the chilling arrangement. But we are given no information that will enable one to form an estimate of these quantities, or of their variations from time to time; and there is no suggestion that they need to be considered. Moreover, the statement regarding the slow restoration of the trihydrol cannot be accepted without much better evidence than has been given us; in fact, it is most improbable.

3. T. C. Barnes and T. L. Jahn [116, 117] reported that, under the same conditions and starting from the same temperature, water from freshly melted ice freezes more quickly than that from freshly condensed steam. And they interpret this as confirming the conclusion of H. T. Barnes [115] that near and below 0 °C the recovery of equilibrium between the several polymers of H_2O is slow. Here again information regarding experimental

---

[115] Barnes, H. T., *Scientific Monthly*, 29, 289-297 (1929).
[116] Barnes, T. C., and Jahn, T. L., *Quart. Rev. Biol.*, 9, 292-341 (1934).
[117] Barnes, T. C., and Jahn, T. L., *Proc. Nat. Acad. Sci.*, 19, 638-640 (1933).

97. FREEZING AND MELTING

details is meager, and satisfactory checks are wanting. There is no indication that any attention has been paid to the great variations that have been found in the extent to which water can be supercooled; variations that force one to suspect, at least, that the extent of possible supercooling truly

Table 282.—Velocity of Crystallization of Supercooled Water

By the velocity of crystallization, we here mean the linear velocity, v, with which freezing, initiated at one end, proceeds along the length of a glass tube filled with supercooled water and continuously immersed in a bath at the same temperature, t, as the water at the instant that freezing began. The internal diameter of the tube is d, the wall thickness is w.

Unit of $v = 1$ mm/sec; of d and $w = 1$ mm. Temp. $= t$ °C

Walton and Judd.[a]

| $d \rightarrow$ | 7 | | 11 | | | 3.5 | |
| $w \rightarrow$ | 2.5 | | 1.5 | | | 3.25 | |
$-t$	v	$-t$	v		$-t$	v
2.00	5.27					
3.61	8.07					
4.67	11.9					
5.86	17.8					
6.18	19.1	6.17	39.2		6.10	24.0
7.10	44.4				6.60	25.2
7.50	51.3	7.65	63.0		7.58	32.2
8.19	69.2					
8.38	85.5				8.58	39.1
9.07	114.0	9.92	91.6			

Tumlirz[a]   Hartmann[a]   TB[a]

| $d \rightarrow$ | 18 | | 2 to 3 | | | 1.2 | |
| $w \rightarrow$ | | | | | | 0.8 | |
$-t$	v	$-t$	v		$-t$	v
0.74	0.37	0.5	2.3			
1.12	1.44	0.8	3.3			
1.40	2.20	1.0	4.0			
1.54	2.76	1.5	6.0			
1.62	2.92	1.9	8.0			
2.00	3.32	1.9	8.5			
2.40	4.49	2.0	9.7			
2.54	5.24	2.0	9.7			
2.67	5.58	3.5	20.0		3.2	11.1[b]
2.71	5.77	3.5	20.3		4.2	16.2[b]
2.90	7.06	5.0	29.2		5.2	23.8[b]
3.20	7.47	5.0	20.2		6.2	30.5
3.49	10.23	7.0	46.2		7.2	41.3
3.64	11.28	7.0	46.7		8.3	52.0
4.14	16.93				9.3	55.0
4.20	18.15				10.3	61.2
4.60	22.07				11.3	70.3
					12.3	84.0
					13.4	96.8

[a] References:
   Hartmann, R., *Z. anorg. allgem. Chem.*, **88**, 128-132 (1914).
   TB   Tammann, G., and Büchner, A., *Idem*, **222**, 12-16 (1935).
   Tumlirz, O., *Sitz. Akad. Wiss Wien (Abt. IIa)*, **103**, 266-276 (1894). As quoted by Walton and Judd.
   Walton, J. H., and Judd, R. C., *J. Phys'l Chem.*, **18**, 722-728 (1914).

[b] For temperatures above -6 °C they used tubes 3 mm in internal diameter; the wall thickness was not found reported.

varies from specimen to specimen, depending upon some inclusion foreign to the water itself, and not merely upon the thermal treatment of the water. See remarks in the note at the beginning of this Section, p. 638.

4. As the rate of melting of ice immersed in water depends upon the rate at which heat is delivered to the ice, so the linear velocity (see Table 282) with which freezing proceeds along the length of a tube filled with supercooled water and immersed in a bath at a constant temperature measures the rate at which heat is removed, rather than a characteristic property of the water-substance. But it is possible that, with the available facilities, heat cannot be removed from ice at a rate that is greater than some fixed amount determined by the characteristics of water. In that case the linear velocity of the freezing will approach a maximum as the rate of abstraction of heat is increased, and that maximum will be determined by some property, or group of properties, of water. The recorded observations give no indication of such a limiting value.

For a discussion of the linear velocity of crystallization, see the references given in Table 282 and the recent papers by R. Kaischew and I. N. Stranski [118] and T. Förster.[119]

Rate of Melting: Effect of Tension.—The effect of tensile stress upon the rate at which ice melts when exposed to air slightly above 0 °C has been studied by O. Fabian.[120] Using cylinders of ice, all of the same size (diameter = 5.4 cm) but differently loaded, and all exposed simultaneously in a room in which the air temperature was 0.9 °C, he found as follows:

Load	0	25	50	kg*
Loss in weight	20	17.5	13	g/hr

Crystalloluminescence.

Statements [121] to the effect that Pontus, in 1833, observed that water luminesces when it freezes—that it exhibits crystalloluminescence—appear to be incorrect.

The announcement of Pontus's observation [122] states that when a glass bulb with a small tubular neck 1 or 2 cm long is completely filled with water, wrapped with cotton soaked with ether, and placed in a receiver which is then exhausted, then a spark, visible in full daylight, jumps from the neck some moments before freezing occurs ("quelques instans avant la congélation une *étincelle bien visible en plein jour* s'échappe du petit tube qui termine l'ampoule"). This is certainly not the description of crystalloluminescence; and that no such brilliant light accompanies either the freezing of water or the formation of frost, however rapid the process, is to be inferred from the total absence of any mention of it in the voluminous records treating of those processes. Any one can readily satisfy himself

---

[118] Kaischew, R., and Stranski, I. N., *Z. Physik. Chem. (A)*, **170**, 295-299 (1934).
[119] Förster, T., *Idem*, **175**, 177-186 (1936).
[120] Fabian, O., *Repert. Exper. Physik (Carl)*, **12**, 397-404 (1876).
[121] Trautz, M., *Z. physik. Chem.*, **53**, 1-111 (1905). Mellor, J. W., "A Comprehensive Treatise on Inorganic and Theoretical Chemistry, vol. **1**, p. 465, London, Longmans & Co., 1922.
[122] Pontus, *J. chim. méd.*, **9**, 429-430 (1833) → *Ann. d. Physik (Pogg.)*, **28**, 637 (1833).

that if there is any crystalloluminescence involved in such freezing it is certainly very weak and can in no sense be confounded with the phenomenon observed by Pontus. His entire description indicates that the spark observed was a secondary phenomenon.

98. Transition of Ice to Ice

(See also, Types of Ice, Section 57.)

The numerical data of various kinds pertaining to the transition of ice to ice having been given elsewhere (external work and change in volume and latent heat, Section 95; transition temperature, Table 270; phase diagram, Section 93), only a few descriptive items remain for this section.

Ice-I can be carried into the domains in which ice-II and ice-III are stable, but ice-II cannot be carried into that of either ice-III or ice-V; ice-III can be carried into the domains of ice-I, ice-II, and ice-V; ice-IV occurs between the domains of ice-III and ice-VI, within that in which ice-V is stable, but it is totally unstable with reference to ice-V, vanishing entirely if ice-V appears (B600)*; ice-V can be carried into the domains of ice-II and ice-III, and into that of ice-VI at temperatures well below that of the triple point, but near that point it cannot be carried the slightest distance into the domain of ice-VI; and ice-VI can be carried far into that of ice-V, and can be kept there for a considerable time without changing into ice-V.

The velocity with which one type of ice changes to another, when carried over into the pressure-temperature domain in which the second is the stable form, ranges from explosive rapidity near the triple point of higher temperature to extreme sluggishness at lower temperatures. For example, when ice-III is formed from ice-I at temperatures above -30 °C, the reaction runs "with explosive velocity," sometimes producing a sharp, audible click (B478); but at -70 °C the change is so slow as to be not appreciable within 4 hours, even though the pressure be several hundreds of kg*/cm² from the equilibrium one (B476). Since the latent heat varies but little with the temperature, it is obvious that something else is of prime importance in the regulation of the speed of transition. What it is, is not known (B535).

The possibility of carrying ice-VI into the domain of ice-V seems to depend upon conditions that one might expect to be quite unessential. For example: At temperatures above -25 °C, but below the melting line, ice-VI could regularly be carried far into the domain of ice-V, but if bits of Jena glass were placed in the water, then ice-V promptly appeared (B503-506). At lower temperatures ice-V could be obtained from either ice-II or ice-III without the presence of glass (B506).

For further details, see the papers on which these statements rest—

* For such references, see end of this section.

P. W. Bridgman.[123] Such references as (B593-506), occurring in the text, refer to the pages of these papers. All but one (B600) refer to the first; that one, obviously, to the second.

99. Miscellaneous Changes Accompanying Phase Transition

Here are assembled those changes accompanying phase transition that do not fit satisfactorily into the preceding sections, and in general, only those changes that have been directly observed. Others will be found recorded in the pertinent sections, or may be derived from data given therein.

Table 283.—Change in Refraction with Change in Phase

From the observations of others, P. Hölemann [124] has computed the following values of the molecular refraction: $R \equiv M(n^2 - 1)/\rho(n^2 + 2)$, where M = molecular weight (18.0154), n = index of refraction, and ρ = density. Subscripts v and l indicate that the value refers to the vapor and to the liquid, respectively.

Unit of λ = 1 mμ = 10^{-7} cm; of R = 1 cm^3/g-mole

λ	R_v	R_l	100 $(R_l - R_v)$
435.8	3.8262	3.7851	4.11
467.8	3.8041	3.7637	4.04
480.0	3.7969	3.7571	3.98
501.6	3.7854	3.7463	3.91
508.6	3.7820	3.7427	3.93
546.1	3.7660	3.7269	3.91
587.6	3.7510	3.7124	3.95
643.9	3.7371	3.6965	4.06
656.3	3.7344	3.6931	4.13
667.8	3.7321	3.6903	4.18

Table 284.—Change in Absorption Spectrum with Change in Phase

The bands in the absorption spectrum of water vapor are more numerous than those in the spectrum of either water or ice, and have a very complicated "fine structure"; those in the spectrum of water and of ice are nearly devoid of "fine structure," and those of water may have sharp edges whereas those of ice do not. The wave-length corresponding to the maximum of absorption in a given band common to all three phases increases as the substance passes from vapor to liquid to solid. (cf. MRB.[a]) This increase in λ on passing from one phase to the next is here called $\Delta\lambda$.

For ice there is a band at $\lambda = 4.75\ \mu$, for water one at 4.7, but there is no corresponding band for water-vapor (E1[a]).

For the vapor the bands at $\lambda = 1.44$ and $2.00\ \mu$ are stronger than for water, but the reverse is true of those at $\lambda = 0.97$ and $1.20\ \mu$ (Co[a]).

---

[123] Bridgman, P. W., *Proc. Am. Acad. Arts Sci.*, **47**, 441-558 (1912); *J. Phys'l Chem.*, **3**, 597-605 (1935); *Idem*, **5**, 964-966 (1937).
[124] Hölemann, P., *Z. physik. Chem. (B)*, **32**, 353-368 (1936).

99. CHANGE ON TRANSITION

Table 284.—(Continued)

Unit of λ and $\Delta\lambda = 1\ \mu = 10^4 A = 10^{-4}$cm

Vapor 100 °C λ	Water λ	Δλ	Ref[a]	Vapor 220 °C	Vapor 120 °C λ	Water 97 °C	Water 20 °C	Ref[a]
1.404	1.475	0.071	Dr	1.414	1.414	1.468	1.475	St
1.885	1.970	0.085	Dr	1.881	1.881	1.945	1.954	St
2.661	2.916	0.225	Pa	2.600	2.620		2.950	St

Water λ (0 °C)	Ice Ord	Ice Ext	Ord Δλ	Ext	Ref[a]
0.745	0.79	0.81	0.045	0.065	Pl
0.845	0.89	0.92	0.045	0.075	Pl
0.98	1.02	1.06	0.04	0.08	Pl
1.215	1.26	1.29	0.045	0.075	Pl

[a] References:
- Co   Collins, J. R., *Phys. Rev. (2)*, **20**, 486-498 (1922).
- Dr   Dreisch, T., *Z. Physik*, **30**, 200-216 (1924).
- El   Ellis, J. W., *Phil. Mag. (7)*, **3**, 618-621 (1927).
- MRB McLennan, J. C., Ruedy, R., and Burton, A. C., *Proc. Roy. Soc. London (A)*, **120**, 296-302 (1928).
- Pa   Paschen, F., quoted by Dr, presumably from *Ann. d. Physik (Wied.)*, **53**, 334-336 (1894).
- Pl   Plyler, E. K., *J. Opt. Soc. Amer.*, **9**, 545-555 (1924).
- St   Stansfeld, B., *Z. Physik*, **74**, 460-465 (1932).

Raman Spectra.

The displacements of the most prominent lines or bands as the phase and temperature are changed are shown in Table 285. I. R. Rao [125] regards these changes as arising from changing proportions of the molecules (H_2O), $(H_2O)_2$, and $(H_2O)_3$ present in the substance. To $(H_2O)_3$ he ascribes $\lambda_R = 3.13\ \mu$ ($\delta\nu = 3195$ cm^{-1}); to $(H_2O)_2$, $\lambda_R = 2.93\ \mu$ ($\delta\nu = 3413$ cm^{-1}); and to (H_2O), $\lambda_R = 2.77\ \mu$ ($\delta\nu = 3610$ cm^{-1}). On the other hand, G. B. B. M. Sutherland [126] has attempted to explain the changes in terms of but two types of molecules—H_2O and $(H_2O)_2$—, ascribing the $\delta\nu = 3200$ cm^{-1} line to $(H_2O)_2$.

In the spectrum of the vapor there is only a single prominent line ($\delta\nu = 3655$ cm^{-1}), which is fine and sharp; in that of water, there are no prominent sharp lines, but there are two prominent bands, one very broad and complex; in that of ice near 0 °C, there are two bands, each much narrower than the water bands, and at about −190 °C there are two fairly sharp lines, one intense.[127]

Water of crystallization gives a Raman spectrum that is very similar to that of ice, but the lines are sharper.[128]

---

[125] Rao, I. R., *Proc. Roy. Soc. London (A)*, **145**, 489-508 (1934).

[126] Sutherland, G. B. B. M., *Idem*, **141**, 535-549 (1933).

[127] Bhagavantam, S., *Indian J. Phys.*, **5**, 49-57 (1930). Cabannes, J., and de Riols, J., *Compt. rend.*, **198**, 30-32 (1934). Daure, P., and Kastler, A., *Idem*, **192**, 1721-1723 (1931). Ganesan, A. S., and Venkateswaran, S., *Indian J. Phys.*, **4**, 195-280 (1929). Rao, I. R., *Idem*, **3**, 123-129 (1928); *Nature*, **125**, 600 (1930); *Proc. Roy. Soc. London (A)*, **145**, 489-508 (1934); *Phil. Mag. (7)*, **17**, 1113-1134 (1934). Kohlrausch, K. W. F., "Der Smekal-Raman Effekt," Springer, Berlin, 1931.

[128] Cabannes, J., and de Riols, J.,[127] Ganesan, A. S., and Venkateswaran, S.,[127] Krishnan, K. S., *Indian J. Phys.*, **4**, 131-138 (1929); Kohlrausch, K. W. F.,[127] and many others.

Table 285.—Change in Raman Spectrum with Change in Phase

Unit of $\delta\nu = 1$ cm^{-1}; of $\lambda_R = 1$ $\mu = 10^{-4}$ cm

Phase		$\delta\nu$					λ_R		
I. I. R. Rao.[125]									
Ice		3196	3321				3.13	3.01	
Water 0 °C			3321	3502				3.01	2.86
Water 98 °C				3466					2.88
Water-vapor					3655[a]				2.74[a]
II. A. S. Ganesan and S. Venkateswaran.[127]									
Ice		3193	3391	3549	5349		3.13	2.95	2.82 1.85
Water	2355	3199	3453	3609	5502	4.25	3.13	2.90	2.77 1.82
III. Several observers.									
Water		3214	3440	3604			3.11	2.91	2.77
Water-vapor					3655				2.74

[a] Measurement by P. Daure and A. Kastler.[127]

Magnetic Susceptibility.

At the request of Piccard, G. Foex measured relatively the specific susceptibility (χ) for the same specimen of water when frozen and when liquid. He found that at the moment of freezing the numerical value of χ decreased by 2.4 per cent of its value for the liquid.[129] T. Ishiwara [130] reported a minute change in the same direction, and of approximately the same magnitude, and more recently, B, Cabrera and H. Fahlenbrach [131] found 2.2 per cent.

[129] Piccard, A., *Arch. Sci. phys. et nat. (4),* **35,** 209-231, 340-359, 458-482 (1913).
[130] Ishiwara, T., *Sci. Rep. Tôhoku Univ., Sendai (1),* **3,** 303-319 (1914).
[131] Cabrera, B., and Fahlenbrach, H., *An. Soc. Esp. Fis. y Quim.,* **31,** 401-411 (1933).

V. Miscellanea

100. Miscellaneous Phenomena and Data

Penetration of Solids by Water.

If either glass or quartz is exposed for 5 or 10 minutes to water (or to certain other liquids) at a pressure of 15 000 atmospheres, or over, and the pressure is then suddenly released, the glass or quartz will be broken, perhaps shattered. This is explained by the gradual penetration of the compressed liquid into the solid, which is unable to withstand the resulting stress when the outer pressure is removed.[1]

Thermal Anomalies of Water.

M. Magat[2] has concluded that the existing data indicate that most of the physical properties of water exhibit thermal anomalies in the neighborhood of 35 to 40 °C.

A. P. Wills and G. F. Boeker[3] and S. Seely[4] report that the thermal variation of the magnetic susceptibility of water is anomalous in the range 35 to 55 °C, the anomaly being marked at each extreme.

G. Tammann[5] reports that the existing data show that each of many properties of water has either a maximum or a minimum value near 50 °C.

J. Timmermans and H. Bodson[6] find that their determinations of the surface tension of water show a clear anomaly at 13 °C.

Impact of Solids upon Water.

The resistance offered by water to the impact upon its surface of solids of various forms has been studied by S. Watanabe.[7]

Table 286.—Volume of the Water Meniscus: Special

(See also Table 287.)

By the volume of the water meniscus is meant the volume (v) of water that lies above the horizontal plane that is tangent to the bottom of the meniscus in a vertical, cylindrical tube of circular cross-section (radius $= r$), the angle of contact being zero. The quantity $l \equiv v/\pi r^2$ may be called the

---

[1] Poulter, T. C., and Wilson, R. O., *Phys. Rev. (2)*, **40**, 877-880 (1932).
[2] Magat, M., *Jour. de Phys. (7)*, **6**, 179-181 (1935); **6**, 64S-65S (1935); *Trans. Faraday Soc.*, **33**, 114-120 (1937). Especially the first.
[3] Wills, A. P., and Boeker, G. F., *Phys. Rev. (2)*, **46**, 907-909 (1934).
[4] Seely, S., *Idem*, **52**, 662 (L) (1937).
[5] Tammann, G., *Z. anorg. allgem. Chem.*, **235**, 49-61 (1937).
[6] Timmermans, J., and Bodson, H., *Compt. rend.*, **204**, 1804-1807 (1937).
[7] Watanabe, S., *Sci. Papers Inst. Phys. Chem. Res. (Tokyo)*, **23**, 118-137, 202-209, 249-255 (1934).

Table 286.—*(Continued)*

equivalent height of the meniscus; the values of l commonly given for water in tables of constants are those (l_B) determined experimentally by Robert Bunsen, and published in his "Gasometrische Methoden," 1877 [8]; they are all too small. The following values of l and v have been computed in part from the table given by F. A. Gould [9] and in part from that published by S. Sugden.[10] Graphical interpolation and smoothing were employed.

In the experimental work of W. Bein [11] the angle of contact was not zero.

$a^2 \equiv 2\gamma/(\rho - \sigma)g = 0.15$ cm² at 15 °C, 0.14 cm² at 50 °C (see Table 225); γ = surface tension. If $2r < 1$ mm, l is essentially equal to $r/3$; if $2r > 5$ cm, then, within less than 0.5 per cent, $v/r = 0.438$ cm² when $a^2 = 0.14$ cm², and 0.470 cm² when $a^2 = 0.15$ cm².

Unit of $a^2 = 1$ mm² $= 0.01$ cm²; of r and of $l = 1$ mm; of $v = 1$ mm³ $= 0.001$ cm³

$a^2 \rightarrow$ $2r$	14 —l—	15	l_B	14 —v—	15
1	0.16	0.16		0.13	0.13
2	0.32	0.32		1.04	1.04
3	0.47	0.47		3.3	3.3
4	0.60	0.61		7.5	7.7
5	0.72	0.73		14.1	14.3
6	0.82	0.83		23.2	23.5
7	0.90	0.92		34.6	35.4
8	0.97	0.99		48.8	49.8
9	1.00	1.00		64	64
10	1.06	1.07		83	84
11	1.10	1.12		104	106
12	1.12	1.15		127	130
13	1.13	1.17		150	155
14	1.14	1.18	1.10	175	181
15	1.12	1.16	1.03	199	206
16	1.13	1.17	0.97	226	234
17	1.12	1.16	0.91	254	264
18	1.12	1.16	0.87	284	295
19	1.11	1.15	0.84	316	326
20	1.10	1.14	0.82	346	358
22	1.08	1.12		409	426
24	1.03	1.09		467	493
26	0.98	1.04		523	552
28	0.94	0.99		575	610
30	0.89	0.94		628	665
32	0.84	0.90		678	722
34	0.80	0.86		729	779
36	0.76	0.82		779	833
38	0.73	0.78		825	885
40	0.69	0.74		870	936
42	0.66	0.71		916	985
44	0.63	0.68		960	1032
46	0.60	0.65		1004	1080
48	0.58	0.62		1048	1127
50	0.56	0.60		1092	1174

[8] Bunsen, Robert, "Gesammelte Abhand.," vol. **2**, p. 364, Leipzig, 1904.
[9] Gould, F. A., *Int. Crit. Tables*, **1**, 73 (1926).
[10] Sugden, S., *J. Chem. Soc. (London)*, **119**, 1483-1492 (1921).
[11] Bein, W., *Z. Inst.-kunde*, **48**, 161-163 (1928).

Table 287.—Volume of the Water Meniscus: General
(Adapted from A. W. Porter.[12] See also Table 286.)

These values for vertical circular cylinders and for cones diverging upwards are based upon the tables of Bashforth and Adams [13] if $r/\beta < 5$, and upon the formula derived by the late Lord Rayleigh [14] for large tubes if $r/\beta \gtrsim 5$. They assume that the contact angle is zero, otherwise they apply to any fluid, an appropriate value being assigned to β. Here, $\beta^2 \equiv \gamma/(\rho - \sigma)g = a^2/2$, a^2 having the same significance as in Table 286; γ = surface tension; 2ϕ = vertex angle of the cone; r = radius of the tube where met by the meniscus; h = the elevation of that intersection above the (horizontal) plane that is tangent to bottom of the meniscus; v = volume of liquid lying above that plane; v_a = value of v as obtained from Table 286.

Unit of β and $a = 1$ mm; of $v = 0.001$ cm³ = 1 mm³

I. Cylinders.

r/β	$2\beta^2(=a^2)\to$ $v/\pi\beta^3$	— 14 — v	v_a	— 15 — v	v_a
0.8853	0.2053	11.9	12	13.2	13
1.8687	1.4092	82.0	81	90.9	89
2.1688	1.9857	115.5	115	128.1	127
2.4074	2.4966	145.3	144	161.1	160
2.9192	3.6847	214.4	211	237.8	234
3.1646	4.3264	251.7	247	279.2	274
5	9.077	528.1	535	585.7	592
6	11.4635	667.0	672	739.7	747
7	13.7109	797.7	803	884.7	894
8	15.8528	922.4	923	1022.9	1028
9	17.928	1043.1	1040	1156.8	1157
10	19.966	1161.7	1156	1288.4	1286

II. Cones.

r/β	— $\phi=30°$ — h/β	$v/\pi\beta^3$	r/β	— $\phi=45°$ — h/β	$v/\pi\beta^3$
0.7916	0.4292	0.0425	0.6658	0.2660	0.0067
1.3507	0.6617	0.2053	1.1780	0.4388	0.0796
1.745	0.7841	0.4272	1.5566	0.5414	0.1826
2.0428	0.8563	0.6570	1.8484	0.6059	0.2847
2.2804	0.9031	0.8773	2.088[a]	0.6492	0.4234
3.0372	1.0025	1.7650	2.4464	0.7030	0.6563
			2.5923	0.7206	0.7644
			2.8376	0.7456	0.9080

[a] This appears once as 2.088 and again as 2.0833.

Table 288.—Radiation from an Ideal Black-body Radiator
(Adapted from F. E. Fowler.[14a])

The energy radiated per unit time by a unit area of an ideal (black-body) radiator at $t\,°C$ is $10^n R$. If there is present another body not at

---

[12] Porter, A. W., *Phil. Mag. (7)*, **14**, 694-700 (1932). The study is continued and extended in Porter, A. W., *Trans. Faraday Soc.*, **29**, 702-707, 1307-1309 (1933); *Phil. Mag. (7)*, **17**, 511-517 (1934).
[13] Bashforth and Adams, "An Attempt to Test the Theory of Capillary Action," Cambridge Univ. Press, 1883.
[14] Rayleigh, Lord, *Proc. Roy. Soc. London (A)*, **92**, 184-195 (1916).
[14a] Fowle, F. E., "Smithsonian Physical Tables," 8th revised edition, p. 313, Table 307, Washington, 1933.

Table 288.—*(Continued)*

absolute zero (-273 °C), then radiation will be received from it, and the net loss of energy by the ideal radiator will be correspondingly less than R. $10^n R = \sigma T^4$, where $\sigma = 5.73 \cdot 10^{-5}$ erg/cm²sec, $T = (273 + t)$ °K, temperature $= t$ °C; 1 kcal $= 1000$ cal $= 4185$ joules $= 4185 \cdot 10^7$ ergs.

Unit→	1 erg/cm²sec		1 cal/cm²sec		1 kcal/cm²hr	
t	R	n	R	n	R	n
0	3.19	5	7.66	−3	2.76	2
100	1.11	6	2.66	−2	9.58	2
200	2.87	6	6.89	−2	2.48	3
300	6.18	6	1.48	−1	5.33	3
400	1.18	7	2.83	−1	1.02	4
500	2.05	7	4.92	−1	1.77	4
600	3.33	7	7.99	−1	2.88	4
700	5.14	7	1.23	0	4.43	4
800	7.60	7	1.82	0	6.55	4
900	1.11	8	2.66	0	9.58	4
1000	1.50	8	3.60	0	1.30	5
1100	2.04	8	4.87	0	1.75	5
1200	2.70	8	6.45	0	2.32	5
1300	3.51	8	8.39	0	3.02	5
1400	4.49	8	1.07	1	3.86	5
1500	5.66	8	1.36	1	4.90	5
1600	7.05	8	1.69	1	6.07	5
1700	8.68	8	2.08	1	7.47	5
1800	1.06	9	2.53	1	9.11	5
1900	1.28	9	3.06	1	1.10	6
2000	1.53	9	3.67	1	1.32	6

Vision under Water.

R. E. Cornish [15] has described a spectacle lens that facilitates vision by an eye immersed in water. Such immersion greatly reduces the refraction of the eye, the index of refraction of the aqueous humour being nearly that of water.

Sea-water.

Properties of sea-water that are analogous to those of pure water will be found in the appropriate sections for water, if given at all in this compilation. Data pertaining to its composition and temperature are given here.

The composition of sea-water varies from place to place, and from time to time, depending upon the evaporation and the inflow of fresher water from streams, ice-bergs, and precipitation. G. Wüst [16] has stated that the salinity (S) of the surface layer of the sea far from shore is given by the formula $S = 35.74 + 0.0126(E\text{-}P)$, where E and P are, respectively, the rate of evaporation and of precipitation (unit of $S = 1$ g salt per kg of sea-water; of E and $P = 1$ mm per day). For values of E, see Table 277. Sea-water averages about 35 g of salts per kg; all chemical elements are

[15] Cornish, R. E., *J. Opt. Soc. Amer.*, **23**, 430 (1933).
[16] Wüst, G., *Meteor. Z.*, **38**, 188-190 (1921) ← *Veröffentl. Inst. Meereskunde* (N. F.), geog.-naturw. Reihe, Heft **6**, (1920).

represented, most of them by very minute amounts. The relative amounts of the more abundant are shown in Table 289.

In the surface layers of the oceans the salinity increases from about 35.1 at the equator to about 36 near latitude 25°, and then decreases, reaching about 30 near 70° N and 33 near 70° S; in general the salinity is somewhat greater in the southern hemisphere than at the corresponding latitude in the northern one. Our knowledge of the distribution of salt throughout the depth of the oceans is very imperfect.[17]

The mean temperature of the surface layers of the oceans is about 27 °C at the equator, 20 °C at latitude 30°, 4.8 °C at 60° N, 0.0 °C at 60° S, and −1.7 °C at 80° N and 80° S; in general the temperature is lower in the southern hemisphere than in the northern. The mean temperature of the oceans from top to bottom is about 4.8 °C at the equator, 3 °C at latitude 45°, −0.6 °C at 75° N, and +0.9 °C at 75° S.[17]

Table 289.—Composition of the Salt of Sea-water [18]

Cl = total mass of chlorine per unit mass of sea-water; m_s and Cl_s = mass of the indicated salt and of the associated chlorine, respectively, per unit mass of sea-water; S = salinity = total m_s. From the values tabulated, it follows that S = 34.4 g/kg, Cl = 18.99 g/kg, S = 1.812Cl, Cl = 0.552S. The actual value of S is subject to variations, but the ratios m_s/S, and S/Cl are essentially constant.

Unit of S, Cl, and m_s = 1 g/kg

Salt	m_s	m_s/S	Cl_s
NaCl	26.9	0.783	16.33
MgCl₂	3.2	0.094	2.38
MgSO₄	2.2	0.064	
CaSO₄	1.4	0.039	
KCl	0.6	0.017	0.28
Rest	0.1	0.003	
Total	34.4	1.000	18.99

Surprises.

Here are listed a few of the things referring to water that seem to the compiler to be thought-provoking. They are merely those he happened to jot down. They, and probably others equally worthy of a place in this list, have been considered elsewhere in connection with related phenomena.

1. The density of water that has stood in contact with carbon or with thoria is abnormal [19] (p. 225).

2. The vapor pressure of water in contact with catalysts is abnormally great. Prolonged heating, followed by a return to the initial temperature, affects that increase, and weeks may be required for it to return to its pristine value.[20] The existence of this effect has been questioned (p. 560).

[17] Krümmel, O., "Handb. d. Ozeanog.," Bd. 1, 1907.
[18] Krümmel, O., and Ruppin, E., *Wiss. Meeresunters. (N. F.)*, 9, (Abt. Kiel), 27-36 (1906).
[19] Peel, Robinson, and Smith, *Nature*, 120, 514-515 (1927).
[20] Baker, H. B., *J. Chem. Soc. (London)*, 1927, 949-958 (1927).

3. Under the same conditions as in the preceding, the **degree of association** of the water, as computed by the method of Ramsay and Shields from the temperature coefficient of the surface tension, is abnormally great, and shows similar variations after prolonged heating.[20] The existence of this effect has been questioned (p. 172).

4. Different samples of nominally identical water may have different **indices of refraction**[21] (p. 279).

5. The **density of ice-I** seems to be subject to well-marked variations (p. 463).

Interpolation.

If the values $f_0, f_1, f_2, f_3 \ldots$ of $f(x)$ corresponding respectively to the values $x_0, x_1, x_2, x_3 \ldots$ are known, the x's progressing by equal steps ($x_1 = x_0 + s, x_2 = x_1 + s, x_3 = x_2 + s, \ldots$), then the value of $f(x_n + h)$, $x_n + h$ lying between x_n and $x_n + s$, may be found by means of formula (1)

$$\begin{aligned}f(x_n + h) &= f_n + k a_n + \frac{k(k-1)}{2!} b_n + \frac{k(k-1)(k-2)}{3!} c_n \\&\quad + \frac{k(k-1)(k-2)(k-3)}{4!} d_n + \ldots \\&= f_n + k \left\{ a_n - \frac{1-k}{2} b_n + \frac{(1-k)(2-k)}{6} c_n \right. \\&\quad \left. - \frac{(1-k)(2-k)(3-k)}{24} d_n + \ldots \right\}\end{aligned} \quad (1)$$

in which the symbol ! (read "factorial") indicates that the continued product of all the integers from one to that appearing before the symbol is to be taken ($4! = 1 \times 2 \times 3 \times 4 = 24$); $k = h/s$, a quantity that is less than unity; and $a_n, b_n, c_n, d_n, \ldots$ are the successive tabular differences (Δ) associated with x_n, as given by the following scheme, in which $a_0 = f_1 - f_0$, $a_1 = f_2 - f_1 \ldots b_0 = a_1 - a_0, b_1 = a_2 - a_1, \ldots c_0 = b_1 - b_0, c_1 = b_2 - b_1, \ldots$, etc. The subtraction must always be made in the direction here indicated, and the proper sign must be given to the difference.

x	$f(x)$	Δ_1	Δ_2	Δ_3	Δ_4	Δ_5	etc.
x_0	f_0	a_0	b_0	c_0	d_0	.	
x_1	f_1	a_1	b_1	c_1	d_1	.	
x_2	f_2	a_2	b_2	c_2	d_2	.	
x_3	f_3	a_3	b_3	c_3	d_3	.	
.	.	.	.	.	.		
.	.	.	.	.	.		
x_n	f_n	a_n	b_n	c_n	d_n	.	
.	.	.	.	.	.		
.	.	.	.	.	.		
.	.	.	.	.	.		

[21] Damien, B. C., *Ann. Sci. École Norm. Sup. (2)*, **10**, 233-304 (272-278) (1881) → *Jour. de Phys. (1)*, **10**, 198-202 (1881). v. d. Willigen, V. S. M., *Arch. Mus. Teyler*, **1**, 74-116, 161-200, 232-238 (1868).

Index

This is a subject index. That is, each page reference following an entry indicates the location of information that bears in some manner on the subject covered by the entry, no matter what may be the actual wording of the text. In many cases the index entry is a more comprehensive term than the pertinent one in the portion of the text to which reference is made.

In order to facilitate the use of the text, the same information has, in many cases, been indexed under several entries, each being in some manner appropriate. Nevertheless, the user will now and again fail to find an entry that he expects. In such cases the information sought may be found under a synonym of that entry, or under some more comprehensive term; if not so found, the section, or sections, in which it should be given may be found by reference to the Table of Contents. If this also fails, the inference is that the compilation does not contain the information sought.

A main entry that is followed by subordinate (indented) ones carries a page reference only if it refers to a main section of the compilation or if there are pertinent items not specifically covered by any of the subordinate entries, and only for such items. Such partial references are enclosed by parentheses.

Symbols used to denote mathematical and physical quantities and units of measure are not indexed individually. Those not explained in the text where used should be sought under the word "symbols."

Explanatory words indicating the nature of the quantity or of the information given, the field concerned, the phase of the substance, the independent variable (p, t, v, etc.) are enclosed in parentheses. Inversion of the natural order of the words is indicated in the usual manner, by capitalizing the first letter of what would naturally be the initial word, and placing a comma before it, thus: Association, Molecular = Molecular association.

The same index entry may cover many pages. If these are consecutive, then only the first page is given, unless there seems to be danger that the user may fail to notice over how many pages the pertinent information extends; in which case, either the extreme pages are indicated or the initial page is followed by a +. If the pages are not all consecutive, then the initial page, modified as just described, is given for each group of consecutive pages.

If a term is defined wherever used, its definition is, in general, not specifically indexed. In other cases, the page on which the definition occurs is indexed and followed by a d; if it is desired to indicate that other pertinent information is to be found on the same page the d is enclosed in parentheses; otherwise it is not. If the same term is defined in each of several sections, then reference to each of these definitions is usually given, thus relieving the user of the necessity of turning to a distant portion of the volume.

The following abbreviations are used: $cf.$ = compare; d = definition is on this page; eff. = effect of, or on; $q.v.$ = which see = seek in the index the term preceding the $q.v.$; see+ = see also; a + placed after a page number means "and following pages"; vapor = water-vapor.

α (dissociation) 26d
α-particles:
— Dissociation 36, 178
— Vapor 56
— Water 178
Absorption:
— Anomalous 352d
— β-rays 178
— Cosmic rays; see γ-rays.
— Dielectric; see Dielectric absorption.
— Electric 352
— Electric waves (water):
— — Anomalous 329[h], 352(d), 360, 362[e], 368
— — Bands 329[h], 360, 363, 368
— — Gamma rays, see γ-rays.
— — Gases in water (see+ Solubility):
— — — Air (t) 534, 539
— — — Coefficients 529d, 534d, 535d, 545d
— — — Mean (t) 529, 536
— — — O_2 (thin films) 554
— — — Pressure (eff., p, t) 540, 546
— — — Rate 552
— — — Sea-water 548
— — Index 351d, 371
— — Neutrons by water 178
— — Optical and thermal radiation:
— — — Air (moist) 127, 130, 131
— — — CO_2 126
— — — Filters using water 334
— — — Glacier 493
— — — Ice (491):
— — — — Intercrystalline 491
— — — — Monochromatic 490
— — — — Transmission (total) 491
— — — Snow (491+, 493):
— — — — Sunlight 492
— — — Solutions 330
— — — Vapor (125+, 136+):
— — — — Air (moist) 127, 130, 131
— — — — Diathermacy 130
— — — — Monochromatic 125+, 128
— — — Water (natural) (334+):
— — — — Daylight 337
— — — — Effective 336
— — — — Monochromatic 335
— — — — Water (pure) (325+):
— — — — — Film (thin) 329
— — — — — Filters 334
— — — — — Glass (eff.) 329
— — — — — Monochromatic 326
— — — — — Pressure (eff.) 332, 346
— — — — — Scattering 347
— — — — — Solar radiation 333
— — — — — Temperature (eff.), 332, 340
— — — — — Transmission (total) 332, 346
— — Sound:
— — — Gases, Moist 70
— — — Vapor 70
— — — Water 196
— Spectrum; see Spectrum.
— Thermal; see Absorption: Optical.
— X-rays; see γ-rays.
Absorptivity; see Absorption.
Acceleration of gravity (normal) 7
Accommodation coefficients (two) 620d, 622d

Acoustics (see+ Sound):
— Ice 460
— Vapor 68
— Water 191
Active gas (glow discharge) 158
Activity (Lewis and Randall):
— Defined 594
— Water 594+, 597
Adhesiveness of ice 428
Adsorbed water:
— Vapor pressure 574
Aeration of water 553
Age: Effect on:
— Evaporation 631
— Photoelectric emission 339
— Surface-tension 518
Agitation of electrons (thermal) 59
Air:
— Composition 548
— Convectivity 493(d)
— Effect:
— — Conduction (electric) 375, 377
— — Density of vapor 578
— — Density of water 198, 251
— — Dielectric constant (water) 358
— — Hydrogen-ion concentration 377
— — Melting point 604
— — Specific heat 257
— — Spectrum (rotation) 148
— — Surface-tension 524
— — Tensile rupture (water) 179, 180
— — Vapor-pressure (ice) 600
— — Vapor-pressure (water) 560+, 562, 563
— — Viscosity 190
— Moist air:
— — Absorption (radiation) 127, 130, 131
— — Absorption (sound) 70
— — Conduction (thermal) 123
— — Diathermacy 130
— — Ratio of specific heats 111
— — Velocity of sound 71
— — Viscosity 66, 67
— — Molecular, weight 534
— — Solubility 534, 539
— — Solubility in sea-water 548
— — Solution (rate) 553
— — Supersaturation of 633+
Air-water surface (electric charge) 609
Albedo (486d):
— Snow 486
— Water 299
Alkalinity (sea-water) 549d
Alpha-particles; see α-particles, supra.
Angle, Brewsterian 296d
Angle of contact 522
Angle of repose 428(d)
Angstrom (unit) 6d
Anisotropy of molecule:
— Oxygen 50
— Vapor 49
— Water 177
Anode glow 157d, 161
Anomalies (thermal, water) 175, 651
Anticipatory effects 167, 282, 398
Antistokes spectrum 303d
Arc (electric) in water 383

INDEX

Architecture (water) 161d, 172
Area (sectional) of molecule 41+
Arrangement of data 1
Association, Molecular:
— Change with phase 622
— Constant (Ramsay and Shields) 520
— Degree 169(d)
— Ices 424
— Vapor 54, 90, 601+
— Water (see+ Linkage 161+):
— — Equilibrium 169, 170, 256, 262, 279, 282, 385+, 560, 644
— — Evidences 166
— — Ice-content 169
Asymmetry (ice crystals) 401, 418
Atmosphere, Normal (value) 6
Atmospheric gases; see Air.
Atmospheric nuclei 419, 636
Atomic beams (reflected by ice) 428
Attachment of electrons (probability) 59
Attenuation of radiation:
— Factor 492d
— Glacier 493
— Névé 493
— Snow 492
— Water 337
Autobaric 557d
Autopiestic 557d
Avogadro's number (N) 9

β-rays:
— Absorption (water) 178
— Luminescence 317
Band spectrum; see Spectrum.
Bar (unit) 6d
Beams, Atomic 428
Bending of ice 434
Bernal-Fowler theory (water, ice) 165+, 174
Berthelot's equation of state 595
Beta-rays; see β-rays, supra.
Biological observations 171
Birefringence:
— Electric 381
— Magnetic 395
— Natural; see Refraction: Ice.
Black-body radiation:
— Intensity (t) 653
— Transmission:
— — Ice 491
— — Moist air 130
— — Water 332
Boiling point (579d):
— Formulas 580+
— Nuclei required 579
— Solute (eff.) 582
— Superheating (water) 579, 622
— Values 580
Boltzmann's constant (k) 8(d)
Bombardment, Electron (vapor) 33+, 57
Break-down voltage:
— Ice 505
— Vapor 154
— Water 384
Brewsterian angle 296d
Brightness in:
— Glacier 493
— Sea 337
— Snow 492
Brittleness of ice 444
Brush discharge in water 383
Bubbles:
— Floating 526
— Movement of 528
Bubbling, Electrification 61, 609
Bunsen, absorption coefficient 545d

Cabannes-Daure effect 302(d)
Callendar's equation of state 55, 78
Callendar's theory of water 164
Calorie (254d):
— Mechanical equivalent 254
— Value used 6, 255

Capacity, Specific inductive; see Dielectric constant.
Capillarity; see Surface-tension.
Capture, Coefficient 620d, 621.
Catalyst, Effect on:
— Surface-tension 519
— Vapor-pressure 560
Cataphoresis 609
Cathode (definitions):
— Drop in potential 157d
— Glow 156d
Cavitation 191
Cavities:
— Freezing in 406, 642
— Liquid in quartz 189, 642
Cell theory of ice 402+
Čerenkov effect 304, 317
Characteristic wave-length 356(d)
Charcoal (eff. on temp. max. density) 276
Charge, Electrical:
— Evaporation (eff.) 625
— Space charge (ice) 496, 507
— Surface 609
— Surface-tension (eff.) 518
Clapeyron's equation 603, 618
Cleavage of ice 403, 415, 424
Cloud-limit 633(d), 635
Coaggregation volume 55d
Coefficient (definitions. For values see name of property):
— Absorption:
— — Apparent (radiation) 325d
— — Fluorescent 323
— — Gases 529d, 534d, 535d, 545d
— — Optical 126d
— — Photoelectric 323d
— — True ..... due to scatter 323d
— Accommodation (two) 620d, 622d
— Capture 620d
— Diffusion of gases in:
— — Gases 72d
— — Liquids 555d
— Entrance 552d
— Exit 552d
— Friction 428d
— Joule-Thomson 119d, 269d
— Scattering 323d
— Solubility 535d
Cohésion diélectrique 154d
Collisions of molecules (number) 38
Collodion filters 319
Colloids, Effect on:
— Rate of solution 555
— Solubility 551
Color of:
— Crater Lake 348
— Ice 398, 487
— Sea 348
— Sea-ice 423
— Water:
— — Change near 0° C 167, 398
— — Scattered light 347
— — Transmitted light 346
— — Turbidity, Effect 335, 347
Column, Depression of capillary 526
Combination of H_2 and O_2:
— Bibliography (brief) 11
— Energy, disposable 19(d):
— — Ice 25
— — Vapor 22
— — Water 24
— Heat of 12
Composition, Stoichiometric (water) 12
Compressed (water) 198d
Compressibility; see Compression.
Compression:
— Adiabatic (water):
— — Compressibility:
— — — Sea-water 252+
— — — Water 246
— — Heat 271
— — Increase in temperature 269+
— Dihydrol 168

PROPERTIES OF ORDINARY WATER-SUBSTANCE

Compression (Cont'd):
— Isothermal:
— — Decrease internal energy (water) 269
— — — Dihydrol 168
— — — Heat (water) 268
— — Ice 471
— — — Trihydrol 168
— — Vapor; see Volume, Specific.
— — Water:
— — — Decrease internal energy 269
— — — — Dilated water 241
— — — — Heat 268
— — — — Natural 252
— — — — Pure 239+
— — — — Unpolymerized 168
— Linear (ice):
— — Descriptive 431
— — Melting, Internal 431
— — Strength 448
— — Yielding, Rate 432
— Trihydrol 168
Compton effect 323d, 325
Concentration:
— Hydrogen-ion 377
— Ionic (symbol defined) 377
Condensation of vapor (620+, 632):
— Atmospheric nuclei 419, 636
— Atmospheric vapor 418+
— Cloud-limit, 633d, 635
— Metals, On 637
— Nuclear 633
— Snow, On 637
— Supercooling 633d
— Supersaturation (633, 634d):
— — Engine and turbine, In 633
Conductance, Specific (see+ Conduction, electric) 374d
Conduction (electric):
— Ice (505+):
— — Conductivity 504, 506, 508
— — Resistance 509
— — Space charge 496, 507
— — Types of conduction 507
— Vapor 154
— Water (natural):
— — Rain-water 380
— — Sea-water 380
— Water (pure) (373):
— — Air (eff.) 375, 377
— — Conductivity 373
— — Conductivity, Equivalent 375
— — Hydrogen-ion concentration 377
— — Ionization (electrolytic) q.v.
— — Ionization exponent 377+
— — Ionization product 377+
— — X-rays (eff.) 373
Conduction (thermal):
— Air, Moist 123
— Ice:
— — Bulk 482
— — Crystal, Single 481
— — Snow 483
— — Relation to other properties 123
— Sea-water 275
— Snow 483
— Theory 274
— Vapor 121
— Water 273
Conductivity (see+ Conduction):
— Apparent 352d, 504, 506
— Dipole 352d
— Equivalent 375d
— Volume 351, 374
Constant:
— Association 520
— Avogadro's (N) 9
— Boltzmann's (k) 8
— Cotton-Mouton 395
— Cryoscopic 606d, 608
— Dissociation 26
— Ebullioscopic 582(d)
— Eötvös 520
— Equilibrium (ionization) 377+

Constant (Cont'd):
— Havelock 382
— Henry's law 545
— Kerr 381
— Lorenz (refraction) 484(d)
— Mobility 60
— Planck's (h) 8
— Ramsay and Shields 520
— Sutherland 40, 63
— Verdet 388
Contact angle 522(d)+
Contact charge (electrical) 609
Contraction (ice) 423, 468+
Convection 552, 621, 625
Convectivity (air over snow) 493(d)
Conversion factors 6
Cooling (evaporation) 625
Copper (permeable to vapor) 74
Corona (electric, in water) 383
Corpuscular radiation and:
— Ice 428
— Vapor 56
— Water 178
Cosmic radiation:
— Absorption by water 322
Cotton-Mouton constant 395(d)
Crater Lake (color) 348
Critical-point data (557+):
— Definition (crit. pt.) 557d
— Dihydrol 168
— Trihydrol 168
— Viscosity 66
— Water-substance (557+):
— — Criteria 557
— — Values 558
Crookes' dark space (156d):
— Electron distribution 159
— Magnetic field (eff.) 158
Cryoscopic constant 606d, 608
Crystallographic structure (ice) 398, 425, 434
Crystalloluminescence 646
Crystals of ice (see+ name of property):
— Asymmetry 401, 418
— Growth 409, 415
— Monocrystals (production) 415
— Negative 406(d)
— Orientation:
— — Optic axis 408, 409+, 415
— — Secondary axes 411
— — Structure 398, 434
Curvature of surface (eff.):
— Evaporation 623
— Vapor-pressure 513, 568, 623, 631
Cybotactic state 162(d), 301(d), 321

Dark space (glow discharge); see Crookes; Faraday.
Daylight; see Sunlight.
Debye:
— Dipole equations 353+, 357
— Dipole theory 352, 357
— Drude-Debye relations 355
Defect:
— Phase; see Phase defect.
— Pressure 88, 91
— Specific volume 81-85
Deformability:
— Ice (429+):
— — Brittleness 444
— — Compression q.v.
— — Extension 432
— — Flexure 434
— — Flowing 422, 430, 437
— — Hardness 450
— — Penetration 436
— — Plasticity; see Viscosity.
— — Poisson's ratio 446
— — Punching 435
— — Recovery (stress) 443
— — Rigidity 446
— — Shearing strength 449
— — Sustaining power (ice sheet) 457

INDEX

Deformability: Ice (Cont'd):
— — Tensile strength 447
— — Viscosity q.v.
— — Young's modulus 444, 447
— Molecule 47, 50, 54, 119, 149
— Snow (458):
— — Density 460, 462, 484
— — Hardness 459, 460
— — Strength 460
— — Tamping (eff.) 459
Degassing (ultrasonic) 196
Dense ice 395+, 397
Density:
— Dihydrol 168
— Glacier 462
— Ice (462):
— — Ice-I at 1 atm. 462-468
— — Ice-I, III, V, VI, VII, at melting points 467
— — Ice-II, III, VI, not at melting points 468
— — Solutions, From 423, 463, 469, 470
— — Variability 462+
— Mercury (normal, cm-Hg) 7
— Névé 462
— Nitrogen, Atmospheric 534
— Oxygen 534
— Snow 460, 462, 484
— Trihydrol 168
— Vapor (see+ Volume, Specific):
— — Gas (eff.) 560+, 577
— — Saturated (ice) 601
— — Saturated (water) 575, 577, 591
— Water (see+ Volume, Specific) (198+):
— — Abnormal values 202+
— — Air (eff.) 198, 251
— — Atmospheric pressure 199
— — C and Th (eff.) 225
— — Maximum 275
— — Melting points (p) 467
— — Saturated 583, 591
— — Sea-water 248, 277
— — Uniformity 198
— — Variability 202, 206, 225
Depolarization (radiation):
— Factor 50d, 300d
— Vapor 134
— Water 305, 306, 308
Depolymerization (dissolved ice, heat of) 168
Depression of capillary column 526
Depression of melting point 605+
Deuterium content natural water 12, 202
Deuterium oxide v
Diamagnetic properties; see Susceptibility.
Diameter (molecular) 39, 41
Diathermacy; see Absorption, Optical and thermal.
Dielectric absorption (352d):
— Ice (504):
— — Apparent conductivity 352d, 504, 506
— — Phase defect 495d, 504+
— — Relaxation time 355d, 496
— Water 363, 368
Dielectric constant (350d):
— Ice (495+):
— — Dispersion 497-503
— — Drude-Debye constants 497
— — Gas (eff.) 496
— — Hysteresis 496
— — Relaxation time 496+
— — Solutions (ice from) 496
— — Space charge 496
— — Saturation 354, 358
— — Sea-water 368
— — Vapor (151):
— — — Data (numerical) 152
— — — Debye's equation 46-47, 151, 353
— — Water (357+):
— — — Absorption bands 329^h, 360, 363, 368
— — — Absorption index 371
— — — Air (eff.) 358
— — — Anomalous dispersion 329^h, 352d, 360, 362^e, 368
— — — Conductivity (eff.) 351

Dielectric constant: Water (Cont'd):
— — Debye's equation 353, 357
— — Debye's theory 352+
— — Dispersion 329^h, 351, 360, 362^e
— — Electric field (eff.) 354, 358
— — Film, Thin 358
— — Frequency, Variation with 359+
— — Glass (eff.) 360
— — Polarizability 366, 368
— — Pressure, Variation with 367
— — Saturation 354, 358
— — Sea-water 368
— — Tables (v)360, (p)368, (t)364, 366
— — Temperature, Variation with 363
— — Transition wave-length 372, 497
— — Variability 358
— Wave-length (characteristic) 356(d)
Dielectric loss, see Dielectric absorption.
Dielectric strength:
— Ice 505
— Vapor 154
— Water 384
Dielectrics (types) 351
Diffraction (x-rays):
— Ice:
— — Periodicities 489
— — Unit cell 425
— Water:
— — Intensities 321
— — Periodicities 320
— — Rings 319
— — Temperature (eff.) 321
Diffusion (gas):
— Coefficient 72d, 555(d)
— Hydrogen through Cu 75
— Vapor:
— — Copper, Through 74
— — Gas, Into 72
— — Rubber, Through 75, 76
— — Solids, Through 73, 76
— Water, Into 555
Diffusion (radiation) (see+ Scattering):
— Glacier and névé 493
— Snow 492
Diffusivity (gas); see Diffusion (gas).
Diffusivity (thermal) (see+ Conduction, thermal):
— Ice 482
— Snow 483
Dihydrol 164d, 168
Dilated (vapor, water) 78d, 198d
Dipole conductivity 352
Dipole moment:
— Dissolved water 175, 176
— Vapor 46, 53
— Water 175
Dipole theory 352
Discharge (electrical):
— Vapor:
— — Glow discharge q.v.
— — Sparking potential 155
— — Strength (electrical) 154
— Water 341, 383
Disintegration of ice 401, 403, 415, 420, 457
Dispersion (see+ Refraction; Dielectric constant):
— Electric waves 329^h, 351(d), 355, 359
— Verdet's constant 389
Disposable energy; see Energy.
Dissociation:
— α-rays, By 36, 178
— Constant 26d
— Fraction=α 26d
— Vapor (see+ Ionization):
— — Bombardment, By 33, 57
— — Glow discharge, In 34, 158+
— — Ionic 33+
— — Photochemical 33
— — Thermal 25, 28, 34
— Water:
— — Arc, In 383
— — Ionic 35
— — Photochemical 35

PROPERTIES OF ORDINARY WATER-SUBSTANCE

Dissociation: Water (Cont'd):
—— Spark, In 341, 384
Distortion (molecule, rotation) 54, 119, 149
Diurnal range; *see* Temperature.
Documentation 1
Double refraction; *see* Birefringence.
Drops:
— Age and evaporation 631+
— Evaporation 623, 627, 631
— Floating 526
— Freezing 640-642
— Persistence 631
— Shells of vapor 632
— Vapor-pressure 513, 568, 623, 631
Drude-Debye constants (ice) 497
Drude-Debye relations 355
Dust-free water (preparation) 318
Dyne (unit) 7d

Ebullioscopic constant 582(d)
Elasticity (ice):
— Limit 447-448, 453, 457
— Poisson's ratio 446
— Recovery from stress 443
— Rigidity 446
— Shear 447
— Young's modulus 444, 447
Electric field (eff. on), *cf.* Electrification:
— Birefringence 381
— Dielectric constant 354, 358
— Reflection 486
— Refraction 125, 283, 381
— Surface-tension 518
— Viscosity 189
Electrical discharge; *see* Discharge.
Electrification by bubbling, etc. 61, 609
Electrification (eff. on):
—‌ Evaporation 625
— Surface-tension 518
Electron:
— Attachment (probability) 59
— Bombardment of vapor 33, 57+
— Configuration 54
— Distribution (Crookes' space) 159
— Emission (photo-electric):
— — Ice 494
— — Water 339
— Energy of agitation 59
— Free path 59
— Ionization 33, 57+, 158
— Luminescence (excited) 317
— Velocity 59
Electron-volt 57(d)
Electro-optics (Kerr effect) 381
Emission, Photo-electric:
— Ice 494
— Water 339
Emission (radiation):
— Glacier 493
— Ice 493
— Névé 493
— Snow 492, 493
— Spectrum (vapor) 149
— Vapor 131
— Water 339
Emissivity; *see* Emission.
Empty water; *see* Dust-free water.
Endosmosis, Electrical 609
Energy, Disposable (20d):
— Formation (19):
— — Formulas (general) 19+
— — Ice 25
— — Vapor 22
— — Water 24
— Ionization 36
— Phase transition 619
— Relation to "free energy" 20
Energy expended in:
— Dissociation (glow discharge) 35, 160, 161
— Exciting radiation 321
— Ionization of vapor 33, 57, 161
Energy, Free (constant pressure) (20d):
— Ice 480

Energy, Free (Cont'd):
— Relation to disposable energy 20
— Vapor 117
— Water 264, 480
Energy, Internal (change in):
— Isothermal compression (water) 269
— Phase transition 616
Energy, Molecular surface (capillarity) 519
Energy of molecules:
— Rotational 33, 44
— Translational 39, 40
Energy, Technical 20d
Energy, Thermal:
— Ice 474
— Vapor 91, 584
— Water 256, 584
Engine cylinder (supersaturation in) 633
Enthalpy (*H*) (8d):
— Cloud limit 635
— Ice 265, 480
— Vapor, Dilated 92, 111, 114, 115
— Vapor, Saturated 585
— Water, Compressed 265, 480
— Water, Saturated 265, 585
Entrance coefficient 552d, 553
Entrapment, Factor 492d
Entrapment of radiation 492, 493
Entropy:
— Cloud limit 635
— Ice 265, 480
— Increase on phase transition 616
— Vapor, Dilated 92, 116, 117
— Vapor, Saturated 116, 586
— Water, Compressed 264, 267, 480
— Water, Saturated 264, 586
Eötvös constant 520(d)
Equation of state (vapor) (78):
— Berthelot's 595
— Callendar's 55, 78
— Keyes, Smith, and Gerry's 78
— Linde's 78
— Naumann's 595
— Wohl's 90
Equilibrium:
— Constant (electrolytic) 377, 379[b]
— Ice and ice 607
— Ice and vapor 598
— Ice and water 467, 602
— Polymers in water 169, 170, 256, 262, 279, 282, 285+, 560, 644
— Vapor and water 556-575, 579-583
Equivalent:
— Conductivity 375(d)
— Electrolytic 375d
— Height of meniscus 651(d)
— Mechanical (of calorie) 254
— Units, Of 6
Erg (unit) 7
Eutectic solution 606d
Evaporation (*see*+ Vaporization) (620, 622(d)+):
— Bibliography 622
— Cooling by 625
— Drops 623, 627, 631
— Effect of:
— — Age 631
— — Aspect of surface 625
— — Charge (electric) 625
— — Convection 621, 625
— — Curvature of surface 623
— — Films on surface 624, 630
— — Layers, Blanketing 623, 624, 630, 632
— — Stress 624
— — Wind 625
— Rate 621, 624, 626, 630, 631
— Sea-water 627, 630
— Superheating (water) 579, 622
—‌ Temperature of surface 624-626
Excitation of radiation by cosmic and γ-rays 304, 321, 325
Excited atoms and molecules (interaction with) 57, 60
Exit coefficient 552d, 554

INDEX

Expansion (adiabatic, decrease in t):
— Vapor 119
— Water 269, 277
Expansion (cloud chamber) (633d):
— Not adiabatic 635
— Stages 634
— Tables 634
Expansion (thermal, cubical):
— Dihydrol 168
— Ice (465, 468):
— — Incipient melting 423
— — Isopiestic 469
— — Solutions, From 469, 470
— Trihydrol 168
— Vapor; see Volume, Specific.
— Water (see+ Compression, Water):
— — Compressed:
— — — Formulas 234, 235
— — — Isopiestic 199-225, 230, 232
— — Dilated 241
— — Saturated 583
Expansion (thermal, linear, ice) 472
Exponent (ionization) 377(d)-379
Extension of ice (tension) 432
Extrusion of ice 440-442

Faraday dark space 156d
Faraday effect; see Verdet constant.
Festigkeit elektrische 154d
Films (see+ Layers):
— Oil (eff., solution of gas) 555
— Stability (liquid) 528
— Surface film (thickness):
— — Evaporation 623, 624, 632
— — Surface tension 512+, 527
— — Viscosity 189
— Water:
— — Absorption of O_2 554
— — Absorption of radiation 329
— — Dielectric constant 358
— — Rigidity 190
Filters of water (radiation) 334
Fine-structure (spectrum, vapor) 137-149
Flexure of ice 434
Floating drops and bubbles 526
Flow of ice 422, 430, 437+
Flowers of ice 405
Fluctuations, Theory 320, 347
Fluorescence:
— Characteristics 300
— Ice 487
— Vapor 134
— Water 304
Fog, Viscosity 66
Force, Range of molecular; see Molecular: Force.
Formation:
— Disposable energy (20d):
— — Formulas (general) 19+
— — Ice 25
— — Vapor 22
— — Water 24
— Frazil ice 407
— Frost 419
— Heat:
— — Equations (general) 12
— — Ice 18
— — Vapor 14
— — Water 17
— Ice q.v.
— Ice-sheet q.v.
— Needle ice 407
— Needles of ice 417, 637
— Sea-ice 423
— Snow 419
— Water (synthesis); see Combination of H_2 and O_2.
Forms of ice and snow 398-424
Fraction dissociated 26(d)+
Frazil ice 407
Free energy; see Energy, Free (constant pressure).

Free path (electrons, molecules) (37-41):
— Effective 37(d), 39, 40
— Electrons 59
— Initial, Total 39
— Sutherland 37(d)+, 39
Free time (molecular) 39
Freezing (see+ Melting) (637):
— Anticipatory effects 167, 282, 398
— Cavities, In 406, 642
— Droplets, Of 640-642
— Pressure exerted by 397, 449
— Rate 644
— Supercooled water, 396, 416, 637+
— Tension (eff.) 646
— Tubes, In 641, 645, 646
Freezing-point; see Melting-point 602-608
Friction, Sliding (ice) 428
Frictional electricity 609
Frost 419
Fugacity of water 594(d)
Fusion; see Melting; Heat, Latent.

γ-rays (also Cosmic rays):
— Absorption (water) 321
— Compton effect (water) 325
— Excitation by 304, 321, 325
— Interaction with matter 321+
— Luminescence (excited) 304, 321
— Scattering of:
— — Coefficient 323d
— — Compton effect 325
— — Distribution (angular) 324
— — Water, By 321
Gamma rays, see γ-rays, supra.
Gas (see+ Air):
— Absorption by water; see Absorption; Solution.
— Activity in glow discharge 158
— Degassing (ultrasonic) 196
— Diffusion in vapor 72
— Effect on:
— — Absorption of sound 70
— — Boiling-point 582
— — Density of vapor 560+, 577
— — Dielectric constant 358, 496
— — Emissivity (vapor) 132, 133
— — Melting-point 604
— — Surface-tension 524
— — Tensile rupture 179, 180
— — Vapor-pressure 560+, 562, 563, 577
— — Viscosity 190
— Melting-point of solutions of 605
— Rate of solution of 552
— Solubility of, q.v.
Gas constant:
— Boltzmann's (molecular, k) 8
— Gram-mole value (R) 9
Gibbs function (20d):
— Ice 265, 480
— Vapor 117
— Water 265, 480
Glacier:
— Attenuation (radiation) 493
— Brilliance in 493
— Density 462
— Emissivity 493
— Entrapment (radiation) 493
— Flow of 421+, 437+
— Grains 420
— Lightning strokes 422
— Temperature (diurnal range) 422, 493
— Transmission of radiation 492
— Viscosity 453, 454, 456
Glass:
— Effect on absorption (radiation, water) 329[h]
— Effect on dielectric constant 360
— Penetration by water 651
Glow discharge:
— Activity of gas 158
— Anode glow (thickness) 161
— Crookes' dark space:
— — Electron distribution 159
— — Magnetic field (eff.) 158

664 PROPERTIES OF ORDINARY WATER-SUBSTANCE

Glow discharge: (Cont'd)
— Current-voltage relation 160
— Definitions 156
— Dissociation 34, 158+
— — Distribution 161
— — Energy 35, 160, 161
— — Nature 159+
— — Rate 158
— — Various data 160
— Electrons, Distribution 159
— Energy expended:
— — Dissociation, In 35, 160, 161
— — Distribution of 161
— Magnetic field (eff.) 158
— Negative column 156d
— Positive column:
— — Spectrum 159
— — Vanishing 157
— Spectrum of positive column 159
g-mole 8d
Gram calorie:
— Defined 254
— Equivalents 6, 254, 255
Gram-formula weight (gfw) 7
Gram-mole (8d):
— Number of molecules in (N) 9
— Number per cm³ 38
Gram-weight (g*) 7(d)
Gravity, Acceleration of (normal; g) 7
Growth of ice:
— Crystal 409, 415
— Sheet 407

h (Planck's) 8
Hail 418
Hardness:
— Ice 450
— Snow 459, 460
Havelock constant 382
Heat:
— Capacity; see Heat, Specific.
— Conductivity; see Conduction (thermal).
— Content; see Enthalpy.
— Latent:
— — Fusion:
— — — Ice 562, 615-618
— — — Trihydrol 168
— — Internal of expansion:
— — — Vapor 119
— — — Water 269
— — Phase transition 613(d)
— — Sublimation 614
— — Vaporization:
— — — Dihydrol 168
— — — Ice-I 614
— — — Trihydrol 168
— — — Water 613, 616
— Mechanical equivalent 254
— Specific:
— — Dihydrol 168
— — H₂ (Randall's choice) 15
— — Ice (474):
— — — Apparent 474, 477
— — — Dissolved ice 168
— — — Molecular 265, 478, 480
— — — True 479
— — — Washburn's choice 562
— — Internal (vapor) 102
— — O₂ (Randall's choice) 15
— — Ratio of:
— — — Vapor 110
— — — Water 262, 264
— — — Relation to other properties 123
— — — Sea-water 263ᵇ, 272
— — — Trihydrol 168
— — Vapor (dilated) (91+):
— — — Formulas 92, 95, 100
— — — Integral ($c_p \to_0$) 95
— — — Internal 102
— — — Molecular (C_p) 99, 117
— — — Molecular (Randall's choice, C_p) 15
— — — Molecular (C_v) 105, 107
— — — Pressure constant 95+, 111

Heat, Specific, Vapor (Cont'd):
— — — Pressure = 0 95
— — — Ratio $\gamma = c_p/c_v$ 110
— — — Volume constant 101, 103+, 119
— — Vapor (saturated) 586
— — Water (264):
— — — Air (eff.) 257
— — — Difference ($c_p - c_v$) 262, 264
— — — Formulas 264
— — — Mean (1 atm) 259
— — — Molecular (C_p) 264, 480
— — — Molecular (Randall's choice) 18
— — — Pressure constant 257, 480
— — — Pressure constant (Washburn's choice) 562
— — — Ratio $\gamma = c_p/c_v$ 262, 264
— — — Saturation, At 260, 263
— — — Sea-water 263ᵇ, 272
— — — Unpolymerized water 168
— — — Volume constant 257, 261, 262
— Total 591d
Heat of:
— Compression (water):
— — Adiabatic 271
— — Isothermal 268
— Depolymerization (dissolved ice) 168
— Formation:
— — Equations (general) 12
— — Ice 18
— — Vapor 14
— — Water 17
— Ionic dissociation 33
— Sublimation 614
— Vaporization 168, 613, 616
Heating by sunlight 333
Heavy water (Deuterium oxide) v
Height of meniscus, Equivalent 651(d)
Henry's law constant 545d
Homogeneous ice (production) 414
Hot ice 643
Hydrogen:
— Bonds unequal in H₂O 358
— Diffusion through Cu 75
— Specific heat (Randall's choice) 15
Hydrogen-ion concentration 377+
Hydrol 164d, 168
Hydrone 173d
Hydronol 173d
Hydroxyl-ion (moments of inertia) 46

Ice (see+ name of property):
— Adhesiveness 428
— Appearance of:
— — Ice-I 398
— — Ice-III 396
— Bending 434
— Brittleness 444
— Cell theory 402
— Cleavage 403, 415, 424
— Colloidal 165, 407
— Color 398, 487
— Contraction 423, 468, 470
— Crystals q.v.
— Deformability q.v.
— Dense ice 395, 397
— Depolymerization (dissolved ice) 168
— Disintegration 401, 403, 415, 420, 457
— Extrusion 440-442
— Flexure 434
— Flowers of ice 405
— Flowing 422, 430, 437
— Formation 398+, 407-412, 414, 424
— Forms of 398+
— Frazil ice 407
— Freezing; see Ice, Formation; Freezing.
— Friction 428
— Frost 419
— Glaciers q.v.
— Hail 418
— Hardness 450
— Homogeneous ice (production) 414
— Hot ice 643
— Icebergs 420

INDEX

Ice (Cont'd):
— Icicles 409, 417
— Intercrystalline material 402, 491
— Internal melting 401, 403+, 431
— Molecular data 424
— Monocrystals (production) 415
— Needle ice 407
— Needles of ice 417, 637
— Negative crystals 406(d)
— Nuclei for formation 638, 644
— Opalescence 415
— Penetration 436
— Piezo-electric 510
— Pressure exerted by (maximum) 397, 449
— Punching 435
— Purity 403, 414
— Pyro-electric 510
— Quincke's theory 401+
— Radon content 414
— Recovery from stress 443
— Recrystallization 412, 438
— Regelation 412, 437
— Sawing of 444
— Sea-ice 423
— Sheet ice 407, 457
— Snow q.v.
— Stretching of 432
— Structure:
— — Bulk 401
— — Crystals 173, 398, 425, 434
— — Molecule 425
— Superheating of 643
— Thrust by 397, 449
— Types of 395
— Vitreous 396
— Welding of 437+
Ice-content of water:
— Amount 169
— Depolymerization (heat of) 168
— Specific heat 168
Ice-I (see+ name of property):
— Appearance 398
— Defined 395
— Two forms 399
Ice-III (see+ name of property), Appearance, 396
Ice⇌ice:
— Equilibrium 607
— Latent heat 616, 618
— Transition 396, 647
Ice-point (602d):
— Absolute temperature 602
— Air (eff.) 604
— Reproducibility 604
Ice-sheet:
— Forming of 407
— Orientation of crystals 408, 409
— Strength 457
— Thickening 408, 409
Ice⇌Vapor 598
Ice⇌Water 467, 602
Icebergs 420
Icicles 409, 417
Illumination:
— Effect on surface-tension 518
— Glaciers (internal) 493
— Internal (defined) 492d
Impact (solids, water) 651
Inclusions, Liquid (quartz) 189, 642
Index of absorption 351d, 371
Index of refraction; see Refraction.
Inductive capacity; see Dielectric constant.
Inertia, Moments of:
— OH-ion 46
— Vapor molecule 44, 144
Intercrystalline material (ice):
— Absorption of radiation 491
— Evidence for 402
International:
— Calorie (unit) 254d, 591d
— Joule (unit) 5
— Steam-table 591
Interpolation 656

Inversion temperature (Joule-Thomson) 270, 277
Ionization:
— Electrolytic (water) (377+):
— — Air (eff.) 377
— — Equilibrium constant 379^b
— — Exponent 377(d)-379
— — Formulas 379^a
— — Product 377d, 378
— Electrons, By (vapor):
— — Energy 33, 57, 161
— — Glow discharge 158
— — Miscellanea 59
— — Number of ions 59
— — Potential 33, 57
— Energy, Disposable 36
— Excited atoms, etc. 57, 60
— Exponent 377(d)-379
— Heat of (water) 35
— Hydrogen-ion concentration 377(d)
— pH 377(d)+
— Potential 33, 57
— Product 377d, 378
Ions:
— Conductivity, Equivalent 375d, 376
— Mobility 60
— Nature (vapor) 58
— Production 57+, 609
Isometrics:
— Vapor 85
— Water 225, 238
Isopiestics (vapor) 79
Italics (use) 3

Joule (unit) 5
Joule-Thomson coefficient (119d, 269d):
— Inversion temperature (water) 270, 277
— Vapor 119
— Water 270
Joule-Thomson effect 256d

Kerr constant 381d
Kerr effect (optical) 381
Ketteler's dispersion formula 280, 291
Keyes, Smith, and Gerry's eq. of state 78
Kilocalorie, International 254d, 591d
Kinetic data (vapor molecules) 40
Kreusler corrected 329
Kuenen absorption coefficient 545d

Lag of:
— Equilibrium (molecular species) 170, 256, 279, 385, 560, 644
— Faraday effect 388
Lambert's law 492
Latent heat; see Heat, Latent.
Layers, Boundary (see+ Films):
— Blanketing water 527, 623, 624, 630, 632
— Transition (water, vapor) 296, 527
— Water and solid 189, 512+
Light, Effect of; see Illumination.
Lightning and glaciers 422
Linde's equation of state 78
Linear expansion (ice) 472
Linkage of molecules (water) 161(d)
Liquid inclusions (quartz) 189, 642
Liquid, Intercrystalline (ice) 402, 491
Lorenz refraction constant 484d
Luminescence:
— Cabannes-Daure effect 302
— Čerenkov effect 304, 317
— Definitions 300
— Depolarization 300
— Ice:
— — Crystalloluminescence 646
— — Fluorescence 487
— — Raman scattering (487):
— — — Intensity 487, 488
— — — Spectrum 488
— — — Temperature (eff.) 487
— — Rayleigh scattering 487
— Mechanical excitation 317

PROPERTIES OF ORDINARY WATER-SUBSTANCE

Luminescence (Cont'd):
— Nitrogen (moist) 134
— Types (characteristics):
— — Crystalloluminescence 646
— — Fluorescence 300
— — Longitudinal scattering 306
— — Phosphorescence 300
— — Raman scattering 302
— — Rayleigh scattering 301, 347
— — Tyndall scattering 301
— — X-ray scattering 321
— Vapor:
— — Fluorescence 134
— — Raman scattering 135
— — Rayleigh scattering 134
— — X-ray scattering 136
— Water:
— — β-ray luminescence 317
— — Depolarization 305, 306, 308
— — Electron luminescence 317
— — Fluorescence 304
— — γ-ray luminescence 304, 321, 325
— — Longitudinal scattering 306
— — Mechanical luminescence 317
— — Phosphorescence; see Fluorescence.
— — Raman scattering q.v.
— — Rayleigh scattering q.v.
— — Tyndall scattering 301
— — X-ray scattering 321, 324

Magnetic field (effect):
— Crookes' dark space 158
— Refraction (birefringence) 395
— Surface-tension 518
— Susceptibility (water) 384
— Viscosity 189
Magnetic properties; see name of property.
Magnetization, Coefficient of; see Susceptibility.
Magneto-optics; see Verdet's constant.
Mass:
— Molecule of H_2O 38
— Vapor striking surface 38
Maxwell's theory of viscosity 452
Mechanical equivalent of calorie 254
Melting (see+ Freezing) (637):
— Internal (compression) 431
— Internal (radiation) 403, 405
— Latent heat; see Heat, Latent.
— Progressive (eff.):
— — Conduction (electric) 508
— — Expansion (thermal) 423, 470
— — Specific heat 474
— — Stress (eff.) 430, 431, 437+, 646
Melting-point (602d):
— Absolute temperature 602
— Air (eff.) 604
— Aqueous solutions 604+
— Depression (solutions) 605+
— Normal 603(d)+
— Pressure, Under 467, 603
— Sea-water 606
— Solute (eff.) 604
— Solution of gas 605
Meniscus of water:
— Equivalent height 651
— Volume 651
Mercury, Normal density (cm-Hg) 7
Micron (unit, μ) 10
Mobility constant 60d
Mobility of ions in:
— Air 60
— Vapor 60
Model of molecule 51, 144
Moist air; see Air, Moist.
Mole; see Gram-mole.
Mole fraction 536d, 540d
Molecular (see+ Molecule):
— Absorption coefficient (gas) 529(d)+
— Anisotropy 46, 177
— Area 519
— Association; see Association.
— Collisions (frequency) 39

Molecular (Cont'd):
— Conductivity, Equivalent 375d
— Data (miscellaneous):
— — Ice 424
— — Vapor 37, 144
— — Water 161
— Distance 519d
— Force (range) 56, 189, 191, 512+
— Gas-constant:
— — Boltzmann's (k) 8
— — For g-mole (R) 9
— — Kinetic data 40
— Scattering (light) 134, 301, 305+, 347, 487
— Scattering (x-rays) 319
— Size 39, 41, 43, 52, 144
— Specific heat; see Heat, Specific.
— Structure of ice-I 425
— Surface energy 519d, 521
— Volume 519d
— Weight:
— — Air 534
— — H_2O (M) 9, 12
— — H_2O (Rossini's choice) 17
— — Vapor at $-75°$ C 602
Molecule:
— Anisotropy:
— — Vapor 49
— — Water 177
— Area (sectional vapor) 41
— Association (see+ Association, Molecular):
— — Ice 424
— — Vapor 54, 90, 601+
— — Water 161, 166-172, 276, 560
— Capture by liquid 620
— Collisions per sec 39
— Deformability 47, 50, 54, 119, 149
— Depolymerization (dissolved ice, heat) 168
— Diameter 39, 41
— Dipole moment q.v.
— Electron configuration 54
— Energy (thermal agitation, vapor):
— — Rotational 33, 44
— — Translational 39, 40
— Escape from water 552+, 554
— Force (range of) 56, 189, 190+, 512+
— Free path q.v.
— Free time 39
— Inertia, Moment of (vapor) 44, 144
— Kinetic data 40
— Mass 38
— Model (vapor) 51, 144
— Moment of dipole; see Dipole moment.
— Moment of inertia:
— — OH-ion 46
— — Vapor 44, 144
— Momentum (angular, vapor) 44
— Number of:
— — cm³ (per) 38
— — Colliding per sec 39
— — g-mole, Per (N) 9
— — Striking surface 38
— Orientation of (water) 162+
— Packing of (water) 162+
— Polarizability q.v.
— Reflection from liquid 620
— Rotation; see Reorientability 353, 357
— Size 39, 41, 43, 52, 144
— Structure; see Molecule, Model.
— Velocity 38, 40
— Vibrations (types) 142
— Weight 38
Moment of dipole; see Dipole moment.
Moment of inertia (molecule); see Molecule, Moment of inertia
Momentum of molecule, Angular (vapor) 44
Monocrystals (production) 415

Nägeli's theory (capillarity) 526
Naumann's equation of state 595
Needle ice 407(d)
Needles of ice 417, 637
Negative column 156d
Negative crystals 406(d)

INDEX

Neutrons (absorption, scattering) 178
Névé 462, 483, 493
Nitrogen (atmospheric):
— Defined 533f
— Density 534
Nitrogen, Moist (luminescence) 134
Normal:
— Acceleration of gravity (g) 7
— Atmosphere (atm) 6
— Density of mercury (cm-Hg) 7
— Melting-point 603
Nuclei:
— Atmospheric (condensation) 419, 636
— Charge on (spraying, splashing) 609
— Condensation on 633
— Effect on boiling 579
— Evaporation of drops on 632
— Ice formation, Of 638, 644
Number:
— Avogadro's (N) 9
— Collisions of molecules 39
— g-moles per cm^3 38
— Molecules per g-mole (N) 9
— Molecules striking surface (gas) 38

Ocean; see Sea.
OH-ion:
— Life 148, 159
— Moments (inertia) 46
— Occurrence 151
— Rotation 33, 142
Oil (eff., solution of gas) 555
Opalescence (ice) 415
Optical rotation:
— Magnetic; see Verdet's constant.
— Natural:
— — Ice 495
— — Water 350
Orientation:
— Ice crystals 408, 409, 411, 415
— Molecules (water) 162
Orthobaric (term discussed) 557
Ostwald absorption coefficient (gas) 535d, 545d
Oxygen (see+ name of property):
— Absorbed by thin film 554
— Anisotropy 50
— Density 534
— Interaction with vapor 70
— Specific heat (Randall) 15

Packing (molecules, water) 162+
Parachor 528(d)
Path, Free; see Free path.
Penetration of:
— Ice 436
— Solar radiation; see Sunlight
— Solids by water 651
Period covered by compilation vi
Periodicities, X-ray:
— Ice 489
— Water 320
Permeability, Magnetic; see Susceptibility.
Permeability of solids:
— Copper (vapor) 74
— Glass (water) 651
— Miscellaneous solids 76
— Quartz (water) 651
— Rubber (vapor) 75, 76
pH 377(d)+
Phase defect (351d):
— Ice 495d, 504, 505
— Water 369, 372
Phase diagram 608
Phase equilibria; see Equilibrium.
Phase transition:
— Change on transition:
— — Association 622
— — Energy (internal) 616
— — Enthalpy; see Heat, Latent.
— — Entropy 616
— — Refraction 648

Phase transition: Change on transition (Cont'd):
— — Spectrum 648, 649
— — Susceptibility (magnetic) 650
— — Volume 612
— — Energy, Disposable 619
— — External work 612
— — Ice to ice 396, 647
— — Latent heat (613d):
— — — Fusion 168, 615, 616
— — — Ice to ice 617, 618
— — — Sublimation 614
— — — Vaporization 168, 613, 616
— — Rate (ice to ice) 647
— — Rate (water to ice) 644
— — Water↔ice 637
— — Water↔vapor 620
Phosphorescence; see Fluorescence.
Photochemical dissociation:
— Vapor 33
— Water 35
Photoelectric effects:
— Emission (electrons):
— — Ice 494
— — Water (339):
— — — Age (eff.) 339
— — — Conductivity (eff.) 340
— — — Intensity 340
— Voltaic effects (water) 339
Piezo-electric effect (ice) 510
Plan of compilation v
Planck's constant (h) 8
Plastic solid 452d
Plasticity of ice-I; see Viscosity.
Poise (unit) 9d
Poisson's ratio (ice) 446
Polarizability (dielectric) 353d, 366, 368
Polarizability of molecule (47d, 49d, 177d):
— O-ion 49
— Vapor 49, 53
— Water 177
Polarization (electric, ice) 508
Polarization (light, reflected) 296
Polarization (light, scattered):
— Vapor 134
— Water 304, 305, 306, 308
Polymerization; see Association.
Polymers (water, equilibrium) 169, 170, 256, 262, 279, 282, 385+, 560, 644
Positive column (156d):
— Spectrum 159
— Vanishing of 157
Potential (electrical):
— Cathode drop 157d
— Contact; see Charge
— Ionization 33, 57
— Sparking (vapor) 155
Potential, Thermodynamic (p constant) 20d
Power, Rotatory; see Rotation, Optical.
Power factor; see Phase defect.
Prefixes for units 4
Pressure:
— Critical 558
— Exerted by:
— — Freezing water 397, 449
— — Vapor (saturated); see Pressure saturated vapor
— — Water (isometric) 225
— Internal 181(d)
Pressure defect (dilated vapor) 88, 91
Pressure saturated vapor:
— Ice 598
— Water (559d):
— — Adsorbed water 574
— — Cloud limit 635
— — Drops 513, 568, 623, 631
— — Effect of:
— — — Air 560+, 562, 563
— — — Catalysts 560
— — — Curvature of surface 513, 568, 623, 631
— — — Gas 560+, 562, 563, 577
— — — History 560
— — — Solute 560-563, 574, 577, 582
— — — Temperature 568, 570

Pressure saturated vapor: Water: Effect of (Cont'd):
— — Tension 570
— — Formulas 562, 574
— — Thermal slope 568, 570
— — Values:
— — — −5 to +374° C (atm) 560
— — — 0 to −16° C (mm-Hg) 562
— — — 0 to −50° C (millibars) 563
— — — 0 to +102° C (mm-Hg) 563
— — — Comparisons (mm-Hg, atm) 566
Pressure water (compressed):
— Compression, Adiabatic:
— — Natural water 254
— — Pure water 246
— Compression, Isothermal:
— — Natural water 252
— — Pure water 239, 241, 242, 245
— Isometrics (thermal) 225, 238
— Specific volume (see+ Volume, Specific):
— — Pure water 203, 207
— — Sea-water 248
P-T associations (ice, phase equilib.) (597):
— Ice to ice 607
— Melting point (602d) of ice in:
— — Pure water 603
— — Sea-water 606
— — Solutions 604+
— Triple points q.v.
— Vapor pressure (ice) 598
P-V-T associations (see+ Volume Specific):
— Ice 462
— Snow 462
— Vapor, Dilated 78, 593
— Vapor, Saturated 556, 591, 601
— Water, Compressed, 198, 592
— Water, Saturated 556, 591
Product, Ionization 377d, 378
Punching of ice 435
Purity of ice 403, 414
Pyro-electric effect (ice) 510

Quartz:
— Liquid inclusions 189, 642
— Penetration by water 651
Quasicrystalline structure (water) 162, 165
Quincke's theory of ice 401

Radian (unit) 9
Radiation:
— Absorption, q.v.
— Black-body 653
— Corpuscular (interaction with matter):
— — Ice 428
— — Vapor 56
— — Water 178
— Cosmic 322, 325
— Emissivity; see Emission (radiation).
— Excited in water; see Luminescence.
— Filters using water 334
— Ideal radiator 653
— Luminescence q.v.
— Reflection q.v.
— Refraction q.v.
— Solar; see Sunlight.
— Transmission; see Absorption.
Radiator, Ideal 653
Radon in ice 414
Rain-water:
— Electric conduction 380
— D_2 content 202
Raman scattering:
— Change with phase 649
— Characteristics 302
— Ice (487):
— — Intensity 487
— — Spectrum 488
— — Temperature (eff.) 487
— Vapor 135
— Water (307):
— — Analysis 312, 313
— — Crystallization (water of) 649

Raman scattering: Water (Cont'd):
— — Intensity 308
— — Polarization 308
— — Solute (eff.) 311
— — Spectrum 314, 316
— — Temperature (eff.) 309
Ramsay's procedure (surface-tension) 521
Ramsay and Shields' relation 520
Randall's choice for specific heat:
— H_2, O_2, and vapor 15
— Water 18
Range:
— α-particles:
— — Vapor 56
— — Water 178
— Molecular force 56, 189, 191, 513
— Temperature (diurnal); see Temperature.
Raoult absorption coefficient (gas) 545d
Rate of:
— Dissociation (glow discharge) 158
— Evaporation 621, 624, 626, 630, 631
— Freezing 644
— Solution (gas) 552
— Transition (ice to ice) 647
— Yielding of ice q.v.
Ratio of specific heats:
— Vapor 110
— Water 262, 264
Rayleigh scattering:
— Characteristics 301, 347
— Ice 487
— Sea-water 305
— Vapor 134
— Water (305):
— — Color 348
— — Intensity 306
— — Polarization 306
Recovery (ice, stress) 443
Recrystallization:
— Described 412
— Migratory 412
— Pressure, Under 438
— Temperature of 412
References (symbols) 2
Reflection:
— Acoustic:
— — Ice 462
— — Ocean bottoms 196
— Atomic beams by ice 428
— Molecules by liquid 620
— Radiation:
— — Albedo:
— — — Snow 486(d)
— — — Water 299(d)
— — Formulas 296
— — Ice 485
— — Snow 486
— — Water (296):
— — — Polarization 296
— — — Reflectivity 298
— — — Sea-water 299, 300
— — — Surface scattering 297
— — — Transition layer (thickness) 296, 527
— Sound; see Reflection: Acoustic.
— Vapor from liquid 620
Reflectivity 296(d), 485(d)
Refraction:
— Change with phase 648
— Dihydrol 168
— Dispersion; see Refraction, Index.
— Dispersion formulas 124, 279, 287, 291
— Double; see Birefringence.
— Electric field (eff.) 125, 283, 381
— Ice (484):
— — Index; see Refraction, Index.
— — Positive uniaxial 484
— — Temperature (eff.) 484, 485
— — Uniaxial 484
— Index (351d):
— — Ice 485
— — Reduction (air to vacuum) 282
— — Vapor 124
— — Water 279-295

INDEX

Refraction (Cont'd):
— Lorenz constant 484d
— Natural waters 295
— Sea-water 295
— Trihydrol 168
— Vapor 123
— Water (279):
— — Birefringence:
— — — Electric 381
— — — Magnetic 395
— — Dispersion formulas 279, 287
— — Electric field (eff.) 283, 381
— — Index; see Refraction, Index.
— — Natural waters 295
— — Pressure (eff.) 294
— — Sea-water 295
— — Temperature (eff.) (280, 288):
— — — Coefficient 281, 290, 293
— — — Formulas 280, 291
— — — Index (t) 279-295
— — Variability 279
Regelation 412, 437
Relaxation time:
— Dielectric 355d, 496
— Rigidity 447
— Viscosity 452
Reorientability (molecule):
— Free 353
— Restricted 357
Repose, Angle of 428d
Resistance (electric); see Conduction (electric).
Resistivity (acoustic, water) 195(d)
Resistivity (electric); see Conduction (electric).
Rigidity:
— Ice 446
— Water 190
Rim of sea 300(d), 349(d)
Röntgen rays; see X-rays.
Rotation (molecules); see Reorientability.
Rotation (optical):
— Magnetic; see Verdet's constant.
— Natural:
— — Ice 495
— — Water 350
Rotation terms (spectrum, vapor) 137-149
Rubber (permeability to vapor) 75, 76
Rupture, Tensile (water) 179

Salinity (249d, 654):
— Sea-ice 423
— Sea-water 380, 654
Saturated (defined):
— Solution 528d
— Vapor 556d
— Water 198d, 556d
Saturation (dielectric) 354, 358
Sawing ice, 444
Scattering (γ-rays, x-rays):
— Coefficient 323d
— Compton effect 323d, 325
— Distribution (angular) 324
— Ice 489
— Vapor (x-rays) 136
— Water 319, 321
Scattering (light):
— Longitudinal 306
— Molecular 134, 301, 305+, 347, 487
— Raman q.v.:
— — Ice 487
— — Vapor 135
— — Water 307
— Rayleigh q.v.:
— — Ice 487
— — Vapor 134
— — Water 305, 347
— Sky 347
— Snow 491
— Surface (water) 297
— Theory of fluctuations 320, 347
— Tyndall 301
Scattering (neutrons, water) 178
Scattering (sound waves) by toluene 197

Scope of compilation vi
Sea:
— Brightness (internal) 337
— Color 348
— Light in 337, 349
— Reflection 299, 300
— Rim of 300, 349
Sea bottom (reflection, sound) 196
Sea-ice (423):
— Color 423
— Formation 423
— Melting (progressive) 423
— Salinity 423
— Volume (maximum) 423, 470
Sea-salt (composition) 655
Sea-water:
— Absorption (radiation) 335
— Alkalinity 549d
— Atmospheric gases in 548
— Brightness, Underwater 337
— Carbon dioxide 548
— Color of sea 348
— Composition:
— — Salt 655
— — Surface layers 655
— — Water 654
— Compressibility 252+
— Conduction:
— — Electrical 380
— — Thermal 275
— Daylight (penetration) 334, 337, 349
— Density 248, 277
— Dielectric constant 368
— Evaporation 627, 630
— Expansion (adiabatic) 277
— Freezing point; see Melting-point.
— Gases, Atmospheric (solubility) 548
— Heat, Specific 263[b], 272
— Light in (spectrum) 337
— Maximum density 277
— Melting-point in 606
— N_2 in 548
— O_2 in 548
— Penetration of daylight 334, 337, 349
— Reflection 299, 300
— Refraction 295
— Resistance (sound) 196
— Salinity 380, 654
— Scattering of light 305
— Secchi disk (visibility) 338
— Solubility (atmospheric gases) 548
— Sound (resistance) 196
— Sound (velocity) 194
— Spectral composition (light in) 337, 349
— Surface-tension 514
— Temperature:
— — Maximum density 277
— — Oceans 655
— Velocity of sound 194
— Viscosity 188
— Volume, Specific 248
Secchi disk (visibility) 338
Shear (ice) 446, 449, 454
Shear (water) 190
Size of molecules 39, 41, 43, 52, 144
Skating 428
Skeleton steam-tables 591
Sky light 347
Slipperiness (ice) 428
Smekal-Raman effect 302
Snow:
— Absorption (radiation) 491
— Albedo 486
— Brilliance, Internal 492, 493
— Condensation on 637
— Conduction (thermal) 483
— Deformability 458
— Density 460, 462, 484
— Diffusion (radiation) 492
— Diffusivity (thermal) 483
— Forms 419
— Formation 419
— Hardness 459, 460

PROPERTIES OF ORDINARY WATER-SUBSTANCE

Snow (Cont'd):
— Radiation 492, 493
— Reflection 486
— Strength 460
— Tamping (eff.) 459
— Temperature (surface) 492
Snow-blanket (variation with depth):
— Density 460, 462, 484
— Hardness 459, 460
— Range in temperature (diurnal) 483, 492, 493
— Temperature 483, 492
Solar radiation; see Sunlight.
Solid wall (eff.):
— Surface-tension 512
— Viscosity 189
Solubility (gases in water; cf. Solution of gases) (528d, 545+):
— Air 534, 539
— Atmospheric gases (sea-water) 548
— Coefficients:
— — Definitions 528
— — Mean 529, 536
— Colloids (eff.) 551
— Pressure (eff.) 540, 546
— Sea-water, Solubility in 548
— Supersaturation 550
— Suspensions (eff.) 551
— Symbols 528
Solute (effect):
— Boiling-point 582:
— Dielectric constant:
— — Ice 496
— — Water 358, 360
— Maximum density (water) 276
— Melting-point 604
— Raman spectrum 311
— Vapor-pressure 560-563, 574, 577, 582
— Volume, Specific (ice) 423, 470
Solution, Eutectic 606d
Solution of gases (see+ Solubility):
— Aeration 553
— Coefficients 528d+, 552d
— Colloids (eff.) 551, 555
— Convection 552
— Entrance coefficient 553
— Exit coefficient 554
— Film, In 554
— Formulas 552
— Melting-point in 604, 605
— O_2 534, 554
— Oil layer (eff.) 555
— Process 552
— Rate 552
— Solubility, q.v.
— Streaming 552
Sound:
— Absorption:
— — Vapor (moist gases) 70
— — Water 196
— Churning of water (eff.) 196
— Reflection:
— — Ice 462
— — Ocean bottoms 196
— Resistivity (water) 195
— Scattering by toluene 197
— Velocity:
— — Air, Moist 71
— — Ice 461
— — Sea-water 194
— — Vapor 68
— — Water (191+):
— — — Frequency (eff.) 195
— — — Maximum (75° C) 191
— — — Natural waters 193
— — — Pure water 192
— — — Temperature (eff.) 191, 192
Space-charge (ice) 496, 507
Spark in water 341, 384
Sparking potential (vapor) 155
Specific heat; see Heat, Specific.
Specific inductive capacity; see Dielectric constant.
Specific resistance; see Conduction (electric).

Specific volume; see Volume, Specific.
Spectrum:
— Absorption:
— — Ice 494
— — Vapor 136
— — Water 341
— — Air (eff.) 148
— Analyses:
— — Vapor 142
— — Water 312+, 343
— Antistokes 303d
— Brush discharge (water) 383
— Change with phase 648, 649
— Corona (water) 383
— Emission (vapor) 149
— Fine structure (vapor) 137-149
— Glow discharge (vapor) 159
— Ice 494
— Interpretation; see Spectrum; Analyses.
— Positive column 157, 159
— Raman (see+ Raman scattering):
— — Change with phase 649
— — Ice 488
— — Vapor 135
— — Water 307, 313, 314, 316, 649
— Rotation terms (vapor) 137-149
— Sea, Light in 337, 349
— Spark in water 341
— Stokes 303d
— Temperature (eff.) 151, 346
— Vapor:
— — Absorption (136+):
— — — Analyses 142, 144
— — — Bands 137, 145, 149
— — — Fine structure 137-149
— — — Lines 137
— — — Molecular constants 144
— — — Pressure (eff.) 142
— — — Rotation terms 146
— — — Transparent regions 141
— — Emission 149
— — Glow discharge 159
— Water:
— — Absorption 341
— — Analysis 343
— — Pressure (eff.) 346
— — Sparks in 341, 384
— — Temperature (eff.) 346
Spraying, Electrification by 61, 609
Sprungwellenlange 356
State, Equation of; see Equation of state.
Steam; see vapor.
Steam-diagrams (references) 588
Steam-tables:
— International skeleton 591
— List of 589
Steradian (unit) 9d
Stoichiometric composition (water) 12
Stokes spectrum 303d
Strength, Dielectric; see Dielectric strength.
Strength, Electrical 154d
Strength of ice:
— Compression, Linear 448
— Shear 449
— Sheet 457
— Tension 447
Strength of snow 460
Strength of water, Tensile 179
Stress, Effect on:
— Evaporation 624
— Melting 430-432, 437+, 646
Stress, Recovery from (ice) 443
Stretching:
— Ice 432
— Water 179, 241
Structure:
— Ice:
— — Bulk 401
— — Crystal 398, 434
— — Molecule 425
— Lattice:
— — Ice 174, 398, 425, 434
— — Water 167, 174, 319

INDEX

Structure (Cont'd):
— Water (161+):
— — Anomalies (thermal) 175, 651
— — Capillary spaces, In 528
— — Persists above critical temperature 175
Sublimation (latent heat) 614
Sunlight, Penetration:
— Glacier 493
— Heating 333
— Snow 492
— Water:
— — Natural 334, 336, 349
— — Pure 333
Supercooled water:
— Freezing 396, 416, 638
— Viscosity 183, 185, 189, 396
Supercooling of:
— Vapor 633
— Water 396, 638
Superheating of:
— Ice 643
— Water 579, 622
Supersaturation:
— Air with vapor 633, 634(d)
— Cloud chamber (stages) 634
— Cloud Limit 635
— Steam 633
— Water with gas 550
Surface:
— Aspect (effect on evaporation) 625
— Charge (electrical) 609
— Cooling (evaporation) 625
— Curvature (effect of):
— — Evaporation 623
— — Vapor-pressure 513, 568, 623, 631
— Energy (molecular) 519
— Films and layers 189, 296, 512, 527, 528, 555, 623, 624, 630, 632
— Scattering (light, water) 297
— Temperature; see Temperature: Surface.
Surface-tension (511+):
— Angle of contact 522
— Bubbles (movement) 528
— Depression of column 526
— Dihydrol 168
— Effect of:
— — Age 518
— — Catalysts 519
— — Electrification 518
— — Gas 524
— — Illumination 518
— — Magnetic field 518
— — Method 512
— — Solid wall 512
— Eötvös constant 520
— Films (surface) 527
— Films, Liquid (stability) 528
— Floating bubbles and drops 526
— Forces (range) 512+
— Formulas (temperature) 516
— Meniscus (height, volume) 651
— Molecular surface energy 519
— Nägeli's theory 526
— Ramsay and Shield's (relation, constant) 520
— Relation to other properties 527
— Sea-water 514
— Temperature (variation with) 514, 516
— Trihydrol 168
— Values 514, 521
— Vapor-pressure (curved surface) 513, 568, 623, 631
— Voltaic effect 528
Surprises 655
Susceptibility (magnetic):
— Change with phase 650
— Dihydrol 168
— Ice 510 ·
— Permeability (relation to) 384
— Specific susceptibility (relation to) 384
— Trihydrol 168
— Water (384+):
— — Field (magnetic, eff.) 384
— — Formulas (temperature) 386

Susceptibility (Magnetic): Water (Cont'd):
— — Thermal history (eff.) 385
— — Values 385, 386
Sustaining power (ice sheet) 457
Sutherland:
— Constant 40, 63
— Diameter 39
— Free path 37, 39
Symbols:
— Compound names 4
— Italics 3
— Prefixes (units) 4
— References, Used in 2
— Solubility of gases 528
— Synthetic units 3
— Table of 6
— Units 3, 6
Synthesis (see+ Combination):
— Energy, Disposable 19+
— Heat of 12+
Synthetic symbols (units) 3
Systems of units 4

Tamping of snow 459
Technical energy 20d
Temperature:
— Adiabatic compression (water) 269+
— Adjustment (vapor and liquid) 622
— Critical 557
— Decrease on adiabatic expansion 119, 269, 277
— Diurnal range; see Temperature: Range.
— Ice and water 643
— Ice-point 602
— Inversion (Joule-Thomson) 270, 277
— Isentropic increase (compression, water) 272
— Maximum density (water) 275
— Maximum volume (ice, solutions) 423, 470
— Range (diurnal):
— — Glacier 422, 493
— — Névé 483, 493
— — Snow 484, 492
— Recrystallization 412
— Saturated vapor 570
— Sea 655
— Surface temperature:
— — Evaporation 624-626
— — Snow 492
— Vapor:
— — Escaping 623
— — Saturated 570
Tension, Surface; see Surface-tension.
Tension on ice:
— Melting 646
— Strength 447
— Yielding 432
— Young's modulus 444, 447
Tension on water:
— Compressibility (dilated water) 241
— Rupture 179
— Vapor pressure 570
Thermal conduction, diffusivity, expansion; see the nouns.
Thermodynamic potential, p constant 20d
Thickness:
— Blanketing layers 623, 624, 632
— Shell of vapor (drops) 632
— Stagnant layers 623, 624
— Transition layers 296, 527
— Wall film 189, 512+
Thrust of ice (maximum) 397, 449
Time, Free; see Free time.
Time, Relaxation; see Relaxation time.
Transition:
— Frequency 356d
— Layers, q.v.
— Phase, see Phase transition.
— Wave-length 356(d)+, 372, 497
Transmission (radiation); see Absorption.
Trihydrol 164d, 168
Trihydrol as ice-forming nuclei 644
Triple points (597d):
— Constants 563, 598, 603, 604

672 PROPERTIES OF ORDINARY WATER-SUBSTANCE

Triple points (Cont'd):
— Reproducibility 604
Tubes, Freezing in 641, 645, 646
Turbine, Supersaturation in 633
Turbidity 334-338, 347
Tyndall on internal melting 405
Tyndall scattering (radiation) 301
Types of ice 395

Ultrasonics (see+ Sound), Degassing by 196
Uniformity of water 198
Union of hydrogen and oxygen 11
Units (3):
— Compound names 4
— Equivalents 6
— Prefixes 4
— Symbols 3, 6
— Synthetic 3
— Systems 4
Unpolymerized water:
— Compressibility 168
— Specific heat 168

Vapor (see+ name of property):
— α-particles in 56
— Equation of state, q.v.
— Escaping (temperature) 623
— Molecular data, Miscellaneous 37, 144
— Reflection from liquid 620
Vapor-pressure; see Pressure.
Vaporization (620+):
— Accommodation (coefficient, two) 620d, 622d
— Association (change in) 622
— Capture (coefficient) 620(d), 621
— Convection 621, 625
— Evaporation, q.v.
— Latent heat 168, 613
— Limiting factors 621
— Steam-tables 588
— Temperature adjustment 622, 623
Variability:
— Density of ice 462
— Samples of water (206):
— — Density 202, 206, 225
— — Dielectric constant 358
— — Refraction 279
Velocity of:
— Molecules 38, 40
— Sound, q.v.
Verdet's constant (water) (388):
— Dispersion 389
— Faraday effect 388d
— Lag 388
— Temperature (eff.) 388, 390
— Values 390
Vibrations in ice:
— Types 460
— Velocity:
— — Flexure 460
— — Longitudinal 461
— — Transverse 461
Vibrations of molecules (types) 142
Virial constant 168
Viscosity:
— Critical point 66
— Dihydrol 168
— Fog 66
— Ice 451, 453d
— — Elastic limit 453, 457
— — Glacier 453, 454, 456
— — Load (eff.) 456
— — Rate of shear (eff.) 454
— — Relaxation time 452d
— — Values 453-457
— — Maxwell's theory 452
— Moist air 67
— Relation to other properties 123, 190
— Sea-water 188
— Stationary film (thickness) 189
— Trihydrol 168
— Vapor (61+):

Viscosity: Vapor (Cont'd):
— — Fog 66
— — Formulas 62
— — Moist air 67
— — Sutherland constant 40, 63
— — Values 64-67
— Water (182+):
— — Effect of:
— — — Electric field 189
— — — Gas, Dissolved 190
— — — Magnetic field 189
— — — Pressure 186
— — — Solid, Adjacent 189
— — Range of force 189
— — Sea-water 188
— — Supercooled water 183, 185, 189, 396
— — Values 66, 183-189
— Weinberg's formula and observations 453
Vision under water 654
Vitreous ice 396
Volume:
— Coaggregation 55
— Critical 558
— Meniscus 651
— Molecular 519d
Volume, Specific (see+ Density):
— Change with phase 612
— Defect of (vapor) 81-85
— Ice:
— — Melting-points, At 467
— — Solutions, From 423, 470
— — Maximum (ice, solutions) 423, 470
— — Minimum (water) 275
— — Sea-water 248
— Vapor:
— — Cloud-limit 635
— — Dilated 79, 593
— — Gas (eff.) 560+, 577
— — Saturated (ice) 601
— — Saturated (water) 575, 577, 591
— Water (see+ Compression):
— — Compressed water 198d
— — Formulas 234
— — Values:
— — — Compressed water 198, 203, 248, 467, 592
— — — Saturated water 583, 591
— — — Sea-water 248
Volume conductivity 374d

Water (see+ name of property):
— α-particles in 178
— Bernal-Fowler theory 165, 174
— Callendar's theory 164
— Deuterium content 12, 202
— Dust-free 318
— Molecular data 161
— Quasicrystalline 162, 165
— Sea-water, q.v.
— Stoichiometric composition 12
— Thermomechanical properties 274
— Uniformity 198
— Variability 202, 206, 225, 279, 358
Waterfall electricity 609
Water-vapor; see name of property; also Vapor ization.
Waves:
— Acoustic; see Sound.
— Electric; see Dielectric constant.
— Ice 460
Weight, Molecular:
— Air 534
— H_2O (M) 9, 12, 17
— Vapor (−75° C) 602
Weinberg's formula (viscosity) 453
Welding of ice 437+
Wohl's equation of state 90
Work (external, phase transition) 612

X-rays:
— Absorption (321+):
— — Coefficient (water) 322

INDEX

X-rays (Cont'd):
— Diffraction:
— — Ice:
— — — Crystal structure 173, 398, 425
— — — Periodicities 489
— — Water 173, 319
— Effect on electric conduction (water) 373
— Reflection (water) 297
— Scattering by vapor 136
— Scattering by water 319, 321

Yielding of ice:
— Compression (linear) 431+
— Flexure 434
— Pressure 436, 437
— Punching 435
— Shear 449, 454
— Tension 432
Young's modulus 444, 447

American Chemical Society
MONOGRAPH SERIES
PUBLISHED
No.
1. The Chemistry of Enzyme Action (Revised Edition)
   By K. George Falk.
2. The Chemical Effects of Alpha Particles and Electrons (Revised Edition)
   By Samuel C. Lind.
3. Organic Compounds of Mercury
   By Frank C. Whitmore. (Out of Print)
4. Industrial Hydrogen
   By Hugh S. Taylor.
5. Zirconium and Its Compounds
   By Francis P. Venable.
6. The Vitamins (Revised Edition)
   By H. C. Sherman and S. L. Smith.
7. The Properties of Electrically Conducting Systems
   By Charles A. Kraus.
8. The Origin of Spectra
   By Paul D. Foote and F. L. Mohler. (Out of Print)
9. Carotinoids and Related Pigments
   By Leroy S. Palmer.
10. The Analysis of Rubber
    By John B. Tuttle.
11. Glue and Gelatin
    By Jerome Alexander. (Out of Print)
12. The Chemistry of Leather Manufacture (Revised Edition)
    By John A. Wilson. Vol. I and Vol. II.
13. Wood Distillation
    By L. F. Hawley. (Out of Print)
14. Valence and the Structure of Atoms and Molecules
    By Gilbert N. Lewis. (Out of Print)
15. Organic Arsenical Compounds
    By George W. Raiziss and Jos. L. Gavron.
16. Colloid Chemistry (Revised Edition)
    By The Svedberg.
17. Solubility (Revised Edition)
    By Joel H. Hildebrand.
18. Coal Carbonization
    By Horace C. Porter. (Revision in preparation)
19. The Structure of Crystals (Second Edition) and Supplement to Second Edition
    By Ralph W. G. Wyckoff.
20. The Recovery of Gasoline from Natural Gas
    By George A. Burrell.
21. The Chemical Aspects of Immunity (Revised Edition)
    By H. Gideon Wells.
22. Molybdenum, Cerium and Related Alloy Steels
    By H. W. Gillett and E. L. Mack.
23. The Animal as a Converter of Matter and Energy
    By H. P. Armsby and C. Robert Moulton.
24. Organic Derivatives of Antimony
    By Walter G. Christiansen.
25. Shale Oil
    By Ralph H. McKee.
26. The Chemistry of Wheat Flour
    By C. H. Bailey.
27. Surface Equilibria of Biological and Organic Colloids
    By P. Lecomte du Noüy.

American Chemical Society Monograph Series

PUBLISHED

No.
28. **The Chemistry of Wood**
    By L. F. Hawley and Louis E. Wise.
29. **Photosynthesis**
    By H. A. Spoehr. (Out of Print)
30. **Casein and Its Industrial Applications (Revised Edition)**
    By Edwin Sutermeister and F. L. Browne.
31. **Equilibria in Saturated Salt Solutions**
    By Walter C. Blasdale.
32. **Statistical Mechanics as Applied to Physics and Chemistry**
    By Richard C. Tolman. (Out of Print)
33. **Titanium**
    By William M. Thornton, Jr.
34. **Phosphoric Acid, Phosphates and Phosphatic Fertilizers**
    By W. H. Waggaman.
35. **Noxious Gases**
    By Yandell Henderson and H. W. Haggard. (Out of Print)
36. **Hydrochloric Acid and Sodium Sulfate**
    By N. A. Laury.
37. **The Properties of Silica**
    By Robert B. Sosman.
38. **The Chemistry of Water and Sewage Treatment**
    By Arthur M. Buswell. (Revision in preparation)
39. **The Mechanism of Homogeneous Organic Reactions**
    By Francis O. Rice.
40. **Protective Metallic Coatings.** By Henry S. Rawdon.
    Replaced by **Protective Coatings for Metals.**
41. **Fundamentals of Dairy Science (Revised Edition)**
    By Associates of Rogers.
42. **The Modern Calorimeter**
    By Walter P. White.
43. **Photochemical Processes**
    By George B. Kistiakowsky.
44. **Glycerol and the Glycols**
    By James W. Lawrie.
45. **Molecular Rearrangements**
    By C. W. Porter.
46. **Soluble Silicates in Industry**
    By James G. Vail.
47. **Thyroxine**
    By E. C. Kendall.
48. **The Biochemistry of the Amino Acids**
    By H. H. Mitchell and T. S. Hamilton. (Revision in preparation)
49. **The Industrial Development of Searles Lake Brines**
    By John E. Teeple.
50. **The Pyrolysis of Carbon Compounds**
    By Charles D. Hurd.
51. **Tin**
    By Charles L. Mantell.
52. **Diatomaceous Earth**
    By Robert Calvert.
53. **Bearing Metals and Bearings**
    By William M. Corse.
54. **Development of Physiological Chemistry in the United States**
    By Russell H. Chittenden.
55. **Dielectric Constants and Molecular Structure**
    By Charles P. Smyth. (Out of Print)

American Chemical Society Monograph Series
PUBLISHED

No.
56. **Nucleic Acids**
    By P. A. Levene and L. W. Bass.
57. **The Kinetics of Homogeneous Gas Reactions**
    By Louis S. Kassel.
58. **Vegetable Fats and Oils**
    By George S. Jamieson.
59. **Fixed Nitrogen**
    By Harry A. Curtis.
60. **The Free Energies of Some Organic Compounds**
    By G. S. Parks and H. M. Huffman.
61. **The Catalytic Oxidation of Organic Compounds in the Vapor Phase**
    By L. F. Marek and Dorothy A. Hahn.
62. **Physiological Effects of Radiant Energy**
    By H. Laurens.
63. **Chemical Refining of Petroleum**
    By Kalichevsky and B. A. Stagner. (Out of Print)
64. **Therapeutic Agents of the Quinoline Group**
    By W. F. Von Oettingen.
65. **Manufacture of Soda**
    By T. P. Hou.
66. **Electrokinetic Phenomena and Their Application to Biology and Medicine**
    By H. A. Abramson.
67. **Arsenical and Argentiferous Copper**
    By J. L. Gregg.
68. **Nitrogen System of Compounds**
    By E. C. Franklin.
69. **Sulfuric Acid Manufacture**
    By Andrew M. Fairlie.
70. **The Chemistry of Natural Products Related to Phenanthrene (Second Edition with Appendix)**
    By L. F. Fieser.
71. **The Corrosion Resistance of Metals and Alloys**
    By Robert J. McKay and Robert Worthington.
72. **Carbon Dioxide**
    By Elton L. Quinn and Charles L. Jones.
73. **The Reactions of Pure Hydrocarbons**
    By Gustav Egloff.
74. **Chemistry and Technology of Rubber**
    By C. C. Davis and J. T. Blake.
75. **Polymerization**
    By R. E. Burk, A. J. Weith, H. E. Thompson and I. Williams.
76. **Modern Methods of Refining Lubricating Oils**
    By V. A. Kalichevsky.
77. **Properties of Glass**
    By George W. Morey.
78. **Physical Constants of Hydrocarbons**
    By Gustav Egloff. Vol. I.
79. **Protective Coatings for Metals**
    By R. M. Burns and A. E. Schuh.
80. **Raman Effect and its Chemical Applications**
    By James H. Hibben.
81. **Properties of Water**
    By Dr. N. E. Dorsey.
82. **Mineral Metabolism**
    By A. T. Shohl.

American Chemical Society Monograph Series

IN PREPARATION

Piezo-Chemistry
By L. H. Adams.

The Chemistry of Coordinate Compounds
By J. C. Bailar.

Water Softening
By A. S. Behrman.

The Biochemistry of the Fats and Related Substances
By W. R. Bloor.

Phenomena at the Temperature of Liquid Helium
By E. F. Burton.

Ions, Dipole Ions and Uncharged Molecules
By Edwin J. Cohn.

Carbohydrate Metabolism
By C. F. Cori and G. T. Cori.

The Refining of Motor Fuels
By G. Egloff.

Organometallic Compounds
By Henry Gilman.

Animal and Vegetable Waxes
By L. W. Greene.

Surface Energy and Colloidal Systems
By W. D. Harkins and T. F. Young.

Raw Materials of Lacquer Manufacture
By J. S. Long.

Chemistry of Natural Dyes
By F. Mayer.

Acetylene
By associates of the late J. A. Nieuwland.

Photochemical Reactions in Gases
By W. A. Noyes, Jr., and P. A. Leighton.

Catalog of Ring Systems
By A. M. Patterson.

Furfural and other Furan Compounds
By F. N. Peters, Jr., and H. J. Brownlee.

Aliphatic Sulfur Compounds
By E. Emmet Reid.

Electrical Precipitation of Suspended Particles from Gases
By W. A. Schmidt and Evald Anderson.

Dipole Moments
By C. P. Smyth.

Aluminum Chloride
By C. A. Thomas.

Precise Electric Thermometry
By W. P. White and E. F. Mueller.

Colloidal Carbon
By W. B. Wiegand.

The Chemistry of Leather Manufacture (Supplement))
By J. A. Wilson.

Measurement of Particle Size and Its Application
By L. T. Work.